Topley and Wilson's

Principles of Bacteriology, Virology and Immunity

Seventh Edition in four volumes

Volume 3

W. W. C. Topley, 1886–1944

Topley and Wilson's

Principles of Bacteriology, Virology and Immunity

Seventh Edition in four volumes

Volume 3

WILLIAMS & WILKINS
Baltimore

General Editors

Sir Graham Wilson
MD, LLD, FRCP, FRCPath, DPH, FRS

Formerly Professor of Bacteriology as Applied to
Hygiene, University of London, and Director of
Public Health Laboratory Service, England and
Wales.

Sir Ashley Miles CBE
MD, FRCP, FRCPath, FRS

Deputy Director, Department of Medical
Microbiology, London Hospital Medical College,
London.
Emeritus Professor of Experimental Pathology,
University of London, and formerly Director of the
Lister Institute of Preventive Medicine, London.

M. T. Parker
MD, FRCPath, Dip Bact

Formerly Director, Cross-Infection Reference
Laboratory, Central Public Health Laboratory,
Colindale, London.

Volume 3

Bacterial diseases

Edited by

G. R. Smith

© **G. S. Wilson, A. A. Miles and M. T. Parker 1984**

First published 1929

Seventh edition in four volumes 1983 and 1984

Library of Congress Cataloging in Publication Data

Topley, W. W. C. (William Whiteman Carlton), 1886–1944.
   Topley and Wilson's Principles of bacteriology, virology, and immunity.
   Bibliography: p.
   Includes index.
   Contents: v. 1. Introduction to microbiology and immunity – v. 2. Systematic bacteriology – v. 3. Bacterial diseases – v. 4. Virology.
   1. Medical microbiology – Collected works. 2. Immunology – Collected works. I. Wilson, Graham S. (Graham Selby), Sir, 1895–   .   II. Miles, A. Ashley (Arnold Ashley), Sir, 1904–   .   III. Parker, M. T. (Marler Thomas) IV. Principles of bacteriology, virology, and immunity. [DNLM: 1. Bacteriology. 2. Immunity. 3. Viruses. QW 4 T675p]
   QR46.T6 1983   1′.01   83–14501
   ISBN 0-683-09066-6

To EM, BRP and to the memory of JW

Printed in Great Britain

Volume Editor

G. R. Smith PhD, MRCVS, DVSM, Dip Bact
Head, Department of Infectious Diseases, Nuffield
Laboratories of Comparative Medicine, Institute
of Zoology, The Zoological Society of London.

Contributors

Joyce D. Coghlan BSc, PhD
Formerly Director, Leptospira Reference
Laboratory of the Public Health Laboratory
Service, Colindale, London.

L. H. Collier MD, DSc, FRCP, FRCPath
Professor of Virology at The London Hospital
Medical College, Honorary Consultant in
Virology in the Tower Hamlets District Health
Authority.

Brian I. Duerden BSc, MD, MRCPath
Professor of Medical Microbiology, University of
Sheffield Medical School and Honorary
Consultant Microbiologist, Children's Hospital,
Sheffield.

Richard J. Gilbert MPharm, PhD, MPS, MRCPath,
Dip Bact
Director, Food Hygiene Laboratory, Central
Public Health Laboratory, Colindale, London.

John M. Grange MD, MSc
Reader in Microbiology, Cardiothoracic Institute,
University of London.

Roger J. Gross MA, MSc, MIBiol
Top Grade Microbiologist, Division of Enteric
Pathogens, Central Public Health Laboratory,
Colindale, London.

B. P. Marmion DSc (Lond), MD (Lond), FRCP,
FRCPath(UK), FRSE
Clinical Professor (Virology), University of
Adelaide, South Australia; Late Professor of
Bacteriology, Edinburgh University.

M. T. Parker MD, FRCPath, Dip Bact
Formerly Director, Cross-Infection Reference
Laboratory, Central Public Health Laboratory,
Colindale, London.

G. L. Ridgway BSc, MD, MRCPath
Consultant Clinical Microbiologist, University
College Hospital, Honorary Senior Lecturer,
University College, London.

Diane Roberts BSc, PhD
Principal Grade Microbiologist, Food Hygiene
Laboratory, Central Public Health Laboratory,
Colindale, London.

G. R. Smith PhD, MRCVS, DVSM, Dip Bact
Head, Department of Infectious Diseases, Nuffield
Laboratories of Comparative Medicine, Institute
of Zoology, The Zoological Society of London.

J. W. G. Smith MD, FRCPath, FFCM, Dip Bact
Director, National Institute for Biological
Standards and Control, London.

A. E. Wilkinson MB, BS, FRCPath
Formerly Director, Venereal Diseases Reference
Laboratory (PHLS), The London Hospital,
London.

Sir Graham Wilson MD, LLD, FRCP, FRCPath,
DPH, FRS
Formerly Professor of Bacteriology as Applied to
Hygiene, University of London, and Director of
Public Health Laboratory Service, England and
Wales.

General Editors' Preface to 7th edition

After the publication of the 6th edition in 1975 we had to decide whether it would be desirable to embark on a further edition and, if so, what form it should take. Except for the single-volume edition of 1936, the book had always appeared in two volumes. We hesitated to alter this arrangement but reflection made us realize that a change would be necessary.

If due attention was to be paid to the increase in knowledge that had occurred during the previous ten years two volumes would no longer be sufficient. Not only had the whole subject of microbiology expanded greatly, but some portions of it had assumed a disciplinary status of their own. Remembering always that our primary concern was with the causation and prevention of microbial disease, we had to select that part of the newer knowledge that was of sufficient relevance to be incorporated in the next edition without substantial enlargement of the book as a whole.

One of the subjects that demanded consideration was virology, which would have to be dealt with more fully than in the 6th edition. Another was immunology. Important as this subject is, much of it is not directly concerned with immunity to infectious disease. Moreover, numerous books, reviews and reports were readily available for the student to consult. What was required by the microbiologist and allied workers was a knowledge of serology, and by the medical and veterinary student a knowledge of the mechanisms by which the body defends itself against attack by bacteria and viruses. We resolved, therefore, to provide a plain straightforward account of these two aspects of immunity similar to but less detailed than that in the 6th edition.

The book we now present consists of four volumes. The first serves as a general introduction to bacteriology including an account of the morphology, physiology, and variability of bacteria, disinfection, antibiotic agents, bacterial genetics and bacteriophages, together with immunity to infections, ecology, the bacteriology of air, water, and milk, and the normal flora of the body. Volume 2 deals entirely with systematic bacteriology, volume 3 with bacterial disease, and volume 4 with virology.

To this last volume we would draw special attention.

It contains 27 chapters describing the viruses in detail and the diseases in man and animals to which they give rise, and is a compendium of information suitable alike for the general reader and the specialist virologist.

The first two editions of this book were written by Topley and Wilson, and the third and fourth by Wilson and Miles. For the next two editions a few outside contributors were brought in to bridge the gap that neither of us could fill. For the present edition we enlisted a total of over fifty contributors. With their help every chapter in the book has been either rewritten or extensively revised. This has led to certain innovations. The author's name is given at the head of each chapter; and each chapter is prefaced by a detailed contents list so as to afford the reader a conspectus of the subject matter. This, in turn, has led to a shortening of the index, which is now used principally to show where subjects not obviously related to any particular chapter may be found. A separate but consequently shorter index is provided for each of the first three volumes and a cumulative index for all four volumes at the end of volume 4. Each volume will be on sale separately. As a result of these changes we shall no longer be able to ensure the uniformity of style and presentation for which we have always striven, or to take responsibility for the truth of every factual statement.

We are fortunate in having Dr Parker, who has been associated with the 5th and 6th editions of the book, as the third general editor of all four parts of this edition and as editor of volume 2. Dr Geoffrey Smith with his extensive knowledge of animal disease has greatly assisted us both as a contributor and as editor of volume 3. Dr Fred Brown, formerly of the Animal Virus Research Institute, Pirbright, has organized the production of volume 4, and Professor Heather Dick the immunity section of volume 1.

Two small technical matters may be mentioned. Firstly, in volume 2 we have retained many of the original photomicrographs and added others at similar magnifications because they portray what the student sees when he looks down an ordinary light microscope in the course of identifying bacteria. Electronmicrographs have been used mainly to illustrate general statements about the structure of the organisms under consideration. Secondly, all temperatures are given in degrees Celsius unless otherwise stated.

Apart from those to whom we have just expressed our thanks, and the authors and revisers of individual chapters, we are grateful to the numerous workers who have generously supplied us with illustrations; to Dr N. S. Galbraith and Mrs Hepner at Colindale for furnishing us with recent epidemiological information; to Dr Dorothy Jones at Leicester for advice on the *Corynebacterium* chapter and Dr Elizabeth Sharpe at Reading for information about *Lactobacillus*; to Dr R. Redfern at Tolworth for his opinion on the value of different rodent baits; to Mr C. J. Webb of the Visual Aids Department of the London School of Hygiene and Tropical Medicine for the reproduction of various photographs and diagrams; and finally to the Library staff at the London School and Miss Betty Whyte, until recently chief librarian of the Central Public Health Laboratory at Colindale, for the continuous and unstinted help they have given us in putting their bibliographical experience at our disposal.

GSW
AAM

Volume Editor's Preface

The infectious diseases—bacterial, mycoplasmal, chlamydial and rickettsial—that affect man and animals form the subject matter of Volume 3. Here is to be found the practical application of the detailed bacteriology in the preceding volume, with particular emphasis being placed on epidemiology, diagnosis, prophylaxis and control. Some sections, such as those on mycoplasmal, chlamydial, bacteroides, pasteurella and campylobacter infections, have been extensively re-written to take account of the rapidly expanding knowledge in human and animal disease, or of a greater veterinary importance than that reflected in earlier editions. The remainder have been carefully scrutinized, partly re-written, and brought up to date. The descriptions of 'new' diseases, such as enteritis in man due to *Yersinia enterocolitica*, *Clostridium difficile* or *Campylobacter*, Legionnaires' disease, infant botulism, and contagious equine metritis, are based on the vast literature that has already accumulated without being so lengthy as to destroy the balance of the book.

Particular attention has been paid to the so-called opportunist infections caused by organisms that were previously regarded as non-pathogenic but are now known to be responsible for much infection and cross-infection in hospitals. Again, an entirely new chapter has been devoted to the infections caused by the somewhat neglected group of anaerobic non-sporing gram-negative bacteria. Advances have been made in recent years in the method of cultivating these organisms, and as a result their isolation in the routine clinical laboratory is already more frequent than it has been in the past. Knowledge of the food-borne diseases has greatly increased of late years, requiring a fuller description of the part played not only by some of the organisms mentioned above, but also by the enterotoxic action of well known members of the food-poisoning group. A much needed chapter on epidemiology has been added to serve as a suitable introduction to the description of the various infectious diseases that follows. As in previous editions, due regard has been paid to the more important historical aspects of infectious disease, and certain sections describing early investigations have been retained because of their special interest or significance in medical and veterinary bacteriology.

For a description of the numerous infectious diseases caused by viruses, the reader is referred to Volume 4.

London
1984   GRS

Contents of volume 3

Bacterial diseases

Contents of volumes 1, 2 and 4

Contents of volume 4
Virology

48

General epidemiology

Graham Wilson

Introductory

Epidemiology is a subject covering a wide field without any natural boundaries. It has been defined in different terms by various workers according to the aspect on which, they consider, most emphasis should be laid (see Evans 1979). For practical purposes it deals with the nature, distribution, causation, mode of transfer, prevention, and control of disease. Though in the past it has been concerned with the study of infectious disease, it is now being increasingly applied to the non-communicable diseases, such as diabetes, coronary occlusion, nutritive disorders, cancer, industrial and road accidents, and even to health (see Morris 1975). Both for information and technique it draws on the biological, social and mathematical sciences (Rojas 1964).

Epidemiology is the counterpart of clinical medicine, taking the population group or herd as the unit of study rather than the individual patient. The clinician observes differences between the patient and healthy persons of similar age and sex: the epidemiologist compares the occurrence of disease in the population group in question with that in other groups differing in various respects—race, climate, diet, occupation, and general environment. The clinician deals with absolute numbers, the epidemiologist with rates—prevalence, incidence, morbidity, mortality and so on—expressed as ratios of cases to the population.

The functions of the epidemiologist in relation to infectious diseases are to determine the rates, and in this way to act as a bureau of information to be drawn upon by students and administrators of community medicine; to investigate outbreaks of disease, their cause, source, reservoirs, mode of transfer, attack and mortality rates; to make *ad hoc* inquiries into the occurrence of the less common diseases; to devise and institute surveillance programmes, usually by field studies in order to ascertain the prevalence of selected diseases and the effect that control measures, if any, are having on them; to consider the means of preventing future outbreaks and of controlling existing outbreaks; and to carry out controlled trials of vaccines or drugs. The information on which rates are determined and calculated often requires analysis by statistical methods; and investigation, surveillance and vaccination programmes den.and the cooperation of the microbiologist.

For descriptive purposes it is convenient to divide epidemiology into two types: investigational and administrative, distinguished by the nature of the problems concerned and the approach to them. The **investigational epidemiologist** makes observations either on naturally occurring phenomena, or on controlled phenomena experimentally produced. Thus, he may observe that anthrax in cattle in Southern Europe and

1

the Middle East is an epidemic disease occurring in the summer months, whereas in Northern Europe it is a sporadic disease occurring in the winter months; or that the virus of foot-and-mouth disease can infect at a distance, whereas that of the clinically almost indistinguishable vesicular disease of swine requires close contact between infected animals; or that the mode of infection with tularaemia or Q fever varies from country to country. In such instances his task is to try and find out the cause of these disparities, explaining them in scientifically acceptable terms. Occasionally, when the natural phenomena are too complex, he may resort to the experimental method to test a particular hypothesis. For example, he may devise methods to decide whether bubonic plague in rats is spread by fleas or by some other means; or whether infection in yellow fever in man is spread by air, or mosquitoes; or babesiosis in cattle by contagion or by ticks. Whether engaged in natural or experimental inquiry the investigational epidemiologist should ideally be located in a microbiological laboratory where he can obtain first-hand information on the presence and distribution of pathogenic organisms in the locality and, when necessary, their phage, sero, or bacteriocine type. The staff of the microbiological laboratory can also be of help in participating in the field inquiries that are so often called for, taking specimens and examining them promptly under the most suitable conditions.

The **administrative epidemiologist**, on the other hand, should be located in a Ministry, Department, or other governmental body that deals primarily with the health of the nation. His task is to collect and analyse all the information he can get, much of it coming from hospital and microbiological laboratories throughout the country, some of it supplied by the police, newspapers, veterinary workers, or other sources. From this information, but chiefly from returns of births and deaths received by a central office, he works out rates of various sorts so as to enable him to determine what

priority to give to preventive or control measures, such as those for combating particular diseases, for cutting down infant or perinatal mortality, for improving the nutritional state of certain classes of the population, for instituting vaccination programmes, and for the welfare of the community as a whole. He is concerned not only with legislative practice at home, but with quarantine diseases on an international scale, supplying regular information to the World Health Organization. Together with his agricultural colleagues he must impose restrictions on the importation of contaminated food or raw materials, and keep an eye on the diseases brought into the country by immigrants or animals.

The investigational epidemiologist is essentially a field worker. He needs to have a good background of clinical experience, since the probable nature of a disease must often be recognized before the microbiologist knows what specimens to take and what tests to carry out on them. He should have spent some time in the study of community medicine and its impact on society, and should have a general knowledge of scientific methods, and, above all, a lively curiosity and imagination.

In contrast, the administrative epidemiologist is an office worker. If not medically qualified, he must be familiar with problems of health and disease. He needs to make use of the simpler statistical methods, to understand the working of the legislative machinery, to have a well balanced mind so as to be able to offer sound practical advice to the Government on the nature and priority of the measures to be taken, bearing in mind the probable behaviour and reaction of the people. Finally, he must keep under review both the infectious and the non-infectious diseases, the latter of which, by virtue of their frequent chronicity and the invalidity they cause, are in the more developed countries with ageing populations assuming increasing importance.

Analysis of observations

In studying the epidemiology of any infectious disease one has to take into account (a) the prime causative agent or parasite and its habitat; (b) the mode of transmission of the agent from its habitat or reservoir to the human or animal subject; and (c) the susceptibility of the subject and the various factors that contribute towards it.

Identification of the causative agent is generally the task of the microbiologist, but identification of its habitat or reservoir requires a combination of microbiologist and epidemiologist. Habitat and reservoir

are not necessarily the same. The anthrax bacillus has its habitat or source in the soil, but man becomes infected indirectly from animals, which constitute for him the reservoir of infection.

The mode of transmission of the infective agent is not always easy to determine. For example, nearly 20 years elapsed between the discovery of the organism of Malta fever and the finding that it was carried by goats' milk. Incidentally, this illustrates the fact that in the prevention or control of a disease knowledge of the way in which it is spread is often of greater value than knowledge of the causative agent itself. The discovery of *Brucella melitensis* had no effect on the incidence of the disease in the military and naval forces in Malta; stoppage of goats' milk in their diet brought it immediately to an end. In tracing the pathway of spread by air, water, milk, food, direct contact, insect vector or other ways the epidemiologist and microbiologist should work hand in hand (see Wilson 1974).

Study of the third component, that is the degree of susceptibility of the subject exposed to the risk of infection and the factors affecting it, is one of the most complex tasks in epidemiology. The cause of a disease is always multifactorial. The parasite cannot cause disease unless it is brought into contact in sufficient numbers at a sufficient rate (velocity of infection) not only with the subject but with the appropriate receptive system of that subject, and unless the subject is susceptible to the particular infection concerned. Numerous predisposing factors influence both the transmission of the agent and the susceptibility of the subject, rendering it no easy task for the epidemiologist and statistician to decide which are the main determining ones. The ideal, of course, is to compare two populations as alike as possible in their age and sex distribution, their customs and mode of life, their diet, their socio-economic level, and every other respect except one, so that the presumption is that any difference noted between them in the incidence of disease is dependent on the one respect in which they differ. Under natural conditions this is usually impossible, but experimentally, as for example in trials of vaccines, a close approximation to it can often be attained. It cannot be emphasized too strongly that, wherever many factors are concerned in the causation of disease, great caution must be exercised before concluding that any single one of them is mainly responsible. In Denmark, for instance, when BCG vaccination of children was extensively used, the death rate from tuberculosis fell from 174 per 100 000 in 1918 to 16 per 100 000 in 1950. In Iceland, where BCG was not used, it fell from 189 per 100 000 in 1918 to 20 per 100 000 in 1950. In the absence of the figures from Iceland the temptation to conclude that BCG was responsible for the enormous decline in the tuberculosis death rate in Denmark would have been very strong (see Sigurdsson and Edwards 1952). Potential fallacies are numerous. The obvious cause is not necessarily the correct one. For example, the close association of paratyphoid fever with the eating of cream buns seemed to leave no doubt that the synthetic cream used was responsible for the infection. Investigation, however, showed that the synthetic cream, as it came from the factories, was almost sterile, but that on the baker's premises it became contaminated with processed egg containing *Salmonella paratyphi*. The real source of the infection was the egg; the synthetic cream acted merely as a passive vehicle. Or again, in diseases such as scarlet fever and diphtheria in which infection usually occurs by the respiratory route, the possibility of infection by some unusual route may be overlooked. In an outbreak of scarlet fever in Copenhagen in 1935, for example, where there were 10 000 cases, the milk-borne nature of the infection was not realized for three weeks.

When an association between a disease and some factor is judged to be real, Austin and Werner (1974) suggest the use of five criteria in helping to decide whether the factor causes the disease, the disease causes the factor, or another factor causes both: (a) consistency of the association; (b) strength of the association; (c) specificity of the association; (d) temporal relation of the factor to the association; (e) coherence of the association with other knowledge.

In the causation of a particular disease two main factors may apparently be playing a part; which of these so-called confounding variables (Austin and Werner 1974) is the more important may be determined by the method of cross-classification. In the causation of cervical cancer, for example, both age and parity are concerned, as is shown in Table 48.1.

Table 48.1 The risk of cervical cancer according to age and parity

No. of pregnancies	Age group in years			
	13–34	35–54	55+	All ages
1–2	2	3	4	3
3–5	5	6	7	6
6+	8	9	10	9
All	5	6	7	6

After Austin and Werner (1974).

In each age category it will be seen that the rates increase with parity; and that in every parity group the rates increase with age. Both factors are operative, but parity has the stronger effect.

When only two divisions of a factor are examined, such as 'with' and 'without', a simple 2×2 or fourfold contingency table may be used. The table has four

Table 48.2 The risk of dying within six years, in men aged 60–64, according to smoking habits

	Non-smokers	Pipe smokers	Total
Dead	117	54	171
Alive	950	348	1298
Total	1067	402	1469
Per cent dead	11.0	13.4	

After Snedecor and Cochran (1967).

cells, and the numbers go inside these and not in the margin (Table 48.2). The difference between non-smokers and pipe smokers is 2.4 per cent. Whether this is attributable to a sampling error or to a real difference in the death-rates of the two groups must then be determined by the usual tests for significance.

Investigation of outbreaks

The distribution between an *outbreak* and an *epidemic* is one of degree rather than of kind. An outbreak tends to be localized to a small area, or confined to a particular group of persons such as schoolchildren, diners at a restaurant or on a train, or participants at a conference coming from a distance. An epidemic, on the other hand, spreads over a wide area, such as a county, a region, the whole country, or as a pandemic over several countries.

In the investigation of an outbreak the aim of the epidemiologist must be to establish as far as possible the nature, distribution, causation, and mode of spread of the disease in question. He will ascertain the number of cases, the size of the population at risk, and, if there are any deaths, the case-fatality rate. He will study all likely causative factors, and will call on the microbiologist in the public health laboratory or in the regional veterinary laboratory to help him in establishing the agent responsible and its mode of transfer. Though there are many hundreds of parasitic agents, the body has only a few methods of response, so that without laboratory investigation it is often impossible to tell which organism is concerned. More-over, for tracing the source of the organism, bacterial finger-printing by serological, bacteriophage, or bacteriocine methods may be essential. Serotyping, for instance, played an important part in the incrimina-tion of spray-dried egg as a cause of food poisoning during the second world war (see Report 1947).

In 1941 and again in 1942 scattered cases of **typhoid fever** were observed in the **Home Counties** of England. The probability that they were infected from the same source was confirmed by finding that the organisms isolated from them all belonged to phage type D4— an uncommon type in the country. This led to the incrimination of milk coming from a wholesaler, one of whose suppliers was a farmer in the West of Eng-land who was found to be a chronic typhoid carrier (Bradley 1943).

Phage-typing likewise played a crucial part in the outbreak of typhoid fever among the nurses in the Orthopaedic Hospital at **Oswestry** in 1948 (Bradley *et al.* 1951). The epidemiological evidence pointed strongly to a milk-borne infection. This was strengthened by finding that the stream which was used for watering the cows and washing the milk churns on one of the farms was contaminated with the faecal discharge of a typhoid carrier. The conclusion that milk was acting as the vehicle of infection appeared on epidemiological grounds to be justified. This conclusion, however, was shown to be erroneous when it was found that the phage type of the typhoid bacillus isolated from the carrier was different from that isolated from the patients. It may be added that this outbreak, whose cause at the time was a mystery, had an extraordinary sequel. Some years later typhoid cases caused by the same Vi-phage type as that infect-ing the Oswestry nurses cropped up in various parts of the country. Investigation showed that they had followed the consumption of canned meat coming from a processing plant in the Argentine. At this plant the chlorination system for disinfecting the water used for cooling the autoclaved cans had been out of action for some time, with the consequence that the raw heavily contaminated water from the river had been used untreated. The strains of typhoid bacillus isolated from the patients (type 34) proved to be of the same phage type as those from the Oswestry nurses, sug-gesting that the nurses had probably been infected not from milk but from canned meat. Examination of the diet sheets confirmed this suspicion, showing that canned meat had in fact been served to the nurses on the particular day on which infection had apparently occurred (Anderson and Hobbs 1973).

Surveillance

The term surveillance has almost as many definitions as that of epidemiology. Literally it means to watch over. In practice, the watch is kept over diseases not over individuals. Its purpose is to keep track of (a) their prevalence; (b) changes in their distribution according to incidence, clinical form, mortality, age, sex, place, season, or other variable; and (c) their response to such preventive and control measures as are being taken. The ultimate aim, of course, is eradi-cation or, where this is impracticable, reduction of their incidence to as low a level as possible. Surveil-lance may be carried out on a regional, national, or even international scale, and is usually a long-term project. Short-term surveillance, however, on a limited scale may be necessary when a disease is introduced into a country that is normally free of it; in such instances watch will be continued till no further cases occur and the disease has been eliminated. This was the practice with smallpox.

Again, short-term surveillance is recommended of children who have been exposed to common infectious diseases, such as measles, rubella, varicella and mumps. Contacts need not be kept away from school so long as there is a close watch over them, and they are isolated as soon as they show symptoms of the disease or reach the infectious stage (see Illingworth 1967; Raška 1983).

Acute diseases

Numerous special uses of surveillance may be mentioned. One of the most important is to be on the look-out for other diseases closely simulating the one that is being studied. Raška (1964), for example, refers to the confusion between pertussis and parapertussis. Though these two diseases resemble each other clinically, they differ in their epidemiology and in their prevention. Pulmonary disease caused by the tuberculoid group of mycobacteria may easily be mistaken for tuberculosis; and vesicular disease of swine for foot-and-mouth disease.

In a country with a well developed public health laboratory service the infectious diseases should be kept under such close supervision that no new disease could be introduced without its being rapidly detected. This, of course, implies the cooperation of district and regional health officers with a good system of notification and intercommunication.

Serological epidemiology

A useful method of surveillance consists in the routine collection of sera from patients suffering from what is or appears to be a new disease. These sera should be stored at $-70°C$. When, perhaps some years later, the disease occurs again, the patient's serum may be compared with the stored serum, and the identity of the new disease with that of the previous disease may be established.

A striking instance of this is afforded by the history of **influenza**. In 1957 a worldwide epidemic occurred caused by a new virus referred to as the Asian influenza virus. How this had arisen was a matter of conjecture. Examination, however, of a large number of sera from persons of different ages revealed the presence in elderly persons who had passed through the 1890–92 epidemic of antibodies to the haemagglutination-inhibiting (HI) antigen of the Asian strain. Apparently this antigen had formed part of the virus responsible for the 1890–92 epidemic and had lain dormant in man or taken refuge in an animal to reappear nearly 70 years later (see Masurel 1969; also Stuart-Harris and Schild 1976).

Another instance is that of **Legionnaires' disease** which occurred in outbreak form at a conference in Philadelphia in 1976. This was thought to be a new disease (see Chapter 73), but the examination of sera taken some years previously showed the presence of antibodies to the causative organism, indicating that the disease had occurred before but had not been recognized.

Stored sera are also of value in detecting the presence of infection when only subclinical cases are occurring.

Surveillance is likewise of value in a disease such as **influenza**, the causative virus of which is subject to varietal changes. By following the distribution, it may be possible to detect the advance of a particular pathogenic variety from one part of the world to another; and, with a knowledge of the previous exposure of the population to this or to a closely similar variety and the consequent immunological response, to forecast an epidemic of the disease and, if time allows, to prepare a suitable vaccine against it.

Apart from actual disease, a new vehicle carrying pathogenic microorganisms may be imported into the country. A notable example of this occurred during the second world war with the introduction from America of spray-dried egg containing various organisms of the *Salmonella* group. Such a mishap was not unexpected, since chickens were known to be the biggest reservoir of salmonella infection, and the temperature to which spray-dried egg was subjected during production was believed to be insufficient to destroy all vegetative organisms. This expectation soon proved to be correct. As the first shipment arrived at Liverpool, samples of the egg were distributed to a number of laboratories and were found to contain salmonellae. In spite of precautionary measures that were taken, cases of food poisoning occurred in the population; and the fact that these were caused not by indigenous strains of the organism but by strains that had not previously been met with in the country left no doubt that the egg was responsible.

Continued surveillance of patients who are convalescing or have recovered from a particular disease may bring to light complications whose connection with the disease might not otherwise have been suspected. Cirrhosis of the liver, for instance, may be a late complication of hepatitis. The occurrence of multiple cases of hepatitis B may be related to blood transfusion 3–5 months previously. Clinical observation alone may reveal the not infrequent occurrence of rheumatic fever after a streptococcal infection; but surveillance of large numbers of children suffering from various forms of streptococcal disease are necessary to ascertain the proportion that suffer from this complication, and to decide whether it is large enough to justify the use of penicillin as a routine prophylactic agent.

The frequency with which drug resistance is acquired in the treatment of various diseases is likewise information that surveillance alone can provide. This necessitates a continuous watch, since different microbes respond in different ways to different drugs, and the frequency with which changes occur varies in time and place. Under certain conditions streptococci, for example, rapidly become resistant to the sulphonamides, and tubercle bacilli to isoniazid. On the other hand, only a small proportion of gonococci become resistant to penicillin, and then only in a small degree; and pneumococci remain permanently susceptible to this antibiotic.

Biocoenosis

One field in which surveillance has played a large part and in which its continued use is required is that which has been exploited most in Soviet Russia under the term **biocoenosis** (bios = life, kainos = common). This is really an ecological study of the interaction in nature of various living creatures—rodents and other mammals, birds, reptiles, and arthropods—from which man may become infected or of which he himself forms a part. In Russia the study of wild animals in so-called *natural foci* was started in 1938 by Pavlovsky, and was linked with plans for developing the virgin territories of Siberia and the Far East. Since that date it has formed a subject of intensive research, more than 300 workers from the Gamaleya and other institutes being engaged in it at any one time. Diseases with natural foci in various parts of the world fall under a number of headings: (a) viral—tick-borne and mosquito-borne encephalitis, haemorrhagic fever, lymphocytic choriomeningitis, rabies; (b) bacterial and chlamydial—plague, tularaemia, brucellosis, psittacosis; (c) rickettsial—Rocky Mountain spotted fever, South-East Asian scrub typhus, rickettsial pox, Q fever, endemic rickettsioses of Siberia and the Far East; (d) spirochaetal—tick-borne relapsing fever, Persian tick-borne typhus; (e) leptospiral—Weil's disease and numerous diseases due to different species of *Leptospira*; (f) parasitic—sleeping sickness, cutaneous and visceral forms of leishmaniasis; (g) helminthic—trichinosis, schistosomiasis, echinococcosis. Such a variety of diseases calls for a like variety of investigators—medical and veterinary microbiologists and epidemiologists, and biologists including parasitologists, ornithologists, entomologists and botanists.

Foremost in the creation of natural foci is the character of the landscape and the local topography, permitting the stable circulation of the infectious agent of a given disease. Thus, observation has shown that leptospiral diseases and Japanese encephalitis occur mainly in meadowland, rickettsial diseases in the Steppes, tick-borne encephalitis in the forests, leishmaniasis and tick-borne spirochaetosis in the deserts and semi-deserts. River valleys are specially favourable habitats of wild animals and birds, providing natural foci for such diseases as haemorrhagic fever, leptospirosis, tularaemia, and tick-borne encephalitis. Other factors influencing the choice of landscape are the climate, the available food and water supplies, and the type and abundance of vegetation.

Within any particular landscape there exist what are referred to as micro-reservoirs of infection where natural foci may persist for many years. Burrows of jackals, foxes, badgers and porcupines are common reservoirs of numerous diseases. Caves harbour flying bats, small lizards, and over 100 species of ectoparasites of wild animals transmitting the causal agents of leishmaniasis, tick-borne exanthematic typhus, sandfly fever, and other diseases.

Human activity in developing virgin lands may lead to the formation of natural foci close to, and sometimes even within, the villages on which buildings are being erected. Man involuntarily attracts populations of wild vertebrates, blood-sucking insects and ticks, which adopt a new mode of life. Many of these foci are established on pasture land near the villages where cattle are feeding, and furnish a reservoir of infection for such zoonoses as Q fever, tularaemia, brucellosis, leptospirosis, and haemorrhagic fever.

This description, at some length, of the subject of biocoenosis, affords an illustration of how wide and how varied a field is covered by surveillance. The purpose of the intensive investigation that the Russians have fostered is, of course, to determine what preventive measures need to be taken for the human population—casual visitors such as trappers, hay makers, and fishermen, for whom protective clothing and repellents suffice, and permanent and semi-permanent residents such as geologists, surveyors, miners and roadmakers for whom vaccination is required. The Russians are by no means the only people who have studied the diseases associated with wild life in virgin territory. Extensive work was carried out previously by investigators of plague, tularaemia, and Rocky Mountain spotted fever in North America, and by parasitologists and virologists in Africa of the diseases caused by arboviruses. (For references to biocoenosis, see Petrishcheva 1966, Pavlovsky (undated), Sinneker 1976.)

Vaccination

The epidemiologist is often called upon to decide whether vaccination is likely to be of value in the prevention of an infectious disease, such as diphtheria, poliomyelitis, measles, rubella, mumps or tetanus, and, if it is carried out, to determine its efficacy. The arguments for and against vaccination may be evenly balanced, as they have been with whooping cough, so that a decision may ultimately rest on economic or even political grounds. Determining the effect of vaccination, however, requires careful surveillance, account being taken of the nature of the vaccine used, the number and size of the doses given, the interval between the doses, the age at which they are administered, the proportion of the population vaccinated, and the length of time that protection, if any, lasts. The results, moreover, in terms of morbidity and mortality, have to be compared with those in a control population resembling in every way that of the vaccination group. As would be expected, different diseases respond differently to vaccination. What is important to ascertain is not only the apparent number of subjects protected, but the proportion of the whole population that it is necessary to vaccinate in order to convey herd immunity, i.e. protection of the unvaccinated as well as the vaccinated group. In a disease such as diphtheria, in order to attain this, something like 70 per cent need to be vaccinated. In measles, a figure of even 98 per cent may be ineffective. This difference depends on the velocity with which infection spreads. In practice complete herd immunity cannot be guaranteed, though poliomyelitis may perhaps be regarded as an exception. The reason is that, as Fox and his colleagues (1971) point out, random mixing does not occur except in small populations. There are always multiple small groups in which intermixing is common within but not between the groups. Hence, no matter how high a proportion of the population is immune or is immunized, there will always be pockets containing sufficient susceptibles among whom contacts are frequent, thus ensuring that the epidemic potential remains high.

Difficulty in deciding the effect of vaccination may arise when another disease, clinically indistinguishable, mimics that against which vaccination is being performed. In such circumstances, notifications may be quite misleading. For example, in a trial of influenza vaccine carried out some years ago, laboratory examination showed that a proportion of the notified cases were caused not by the influenza virus but by a coxsackie virus. To avoid confusion of this sort, microbiological surveillance should always be sought. Another example is that of pertussis and parapertussis, and a further one typhoid and paratyphoid fever.

For studying the dynamics of acute bacterial disease some epidemiologists devise what they refer to as *models* (see Cvjetanović *et al.* 1978). By this term they mean a simplified description of the disease process that can be used for planning vaccination and other control programmes, their relative cost in relation to effectiveness, the nature and scope of further investigations required, education and training schemes and so on. The model is built up on all the relevant information available, such as the incubation period, the duration of sickness, the length of the infectious stage, the morbidity and mortality rates at various ages, the frequency of relapses, the proportion of subclinical cases and of temporary and permanent carriers, the velocity or 'force' of infection, the effect of atmospheric, seasonal, and other environmental conditions and the immunological state of the population. On such a basis it may be possible to hazard a forecast of the occurrence of future epidemics. Useful as such models may be, based as they are on well defined characters, they cannot be regarded as more than hypotheses; nor are they necessarily superior to non-mathematical models in affording a guide to action.

Chronic diseases

So far, we have been dealing with acute disease, but epidemiology is equally adapted to chronic disease, though the methods of study tend to be somewhat different (see Barrett-Connor 1979). The term 'acute' implies a swift onset and short course, whereas 'chronic' means slow progression and long duration. Many chronic diseases are of unknown cause so that the length of the incubation period cannot be assessed. Some appear to result from continued bad personal hygiene, such as excessive drinking and smoking, improper diet, and lack of exercise. In the causation of others there is probably a strong genetic factor. So far as we know, too, chronic diseases are generally nontransmissible. Because many possible factors may play a part in their causation, more advanced methods of mathematical analysis are necessary than with acute diseases; in this task the computer is of great help.

A major tool in the study of chronic disease is the population survey in which special groups, communities or entire populations are examined for a variety of characters and diseases. Surveys of this sort may be either retrospective or prospective. Both forms have been used, for example, in studying the effect of cigarette smoking on the production of lung cancer. In the retrospective survey the number of lung cancer patients giving a history of smoking is compared with that in a similar group of non-smokers. A disadvantage of this form is the impossibility of working out rates of incidence of lung cancers among smokers, since the size of the population from which they were drawn is unknown. This difficulty is avoided in the prospective, or so-called longitudinal, form of survey, since two groups of known size—smokers and non-smokers—are followed up from the start. Prospective studies, such as the Framlingham study (see Dawber *et al.* 1963), which disclosed the relation of high blood pressure and of raised content of plasma cholesterol to the genesis of coronary disease, may be carried on for years and are necessarily expensive, but they do provide information not easily obtainable in any other way.

Prospective studies are not limited to chronic disease. Thus, the New York and Seattle Watch Studies have been used for determining the frequency among infants of respiratory and enteric illnesses due to viruses (see Fox *et al.* 1972). (For a description of epidemiological methods in the study of chronic diseases, see Report 1967.)

Definition of terms

At the risk of being elementary, it may be well to define certain terms used in epidemiology.

Prevalence is not to be confused with *incidence*. *Prevalence* denotes the proportion of persons in a defined population which at any given time are, or have been, affected with a particular disease. *Incidence* on the other hand, is a measure of the frequency with which a particular disease occurs during a stated period of time, usually a year. *Prevalence* represents the accumulated incidence of new cases during previous years from which have been subtracted deaths from all causes (Paul 1958): *incidence* is a longitudinal section that takes account only of *new* cases during a period of time.

Again, the *incubation period*, which is the time between the occurrence of infection and the appearance of manifest symptoms or signs of the disease, differs from the *serial interval*, which is the time elapsing between the clinical onset of successive cases infected in series (Hope-Simpson 1948). Sartwell (1966) prefers to speak of the *generation time*, meaning by this the average interval between the receipt of infection and the time when the patient is most likely to transmit infection. This differs somewhat from the time between successive cases in series, as defined by Hope-Simpson (1948). It is usually shorter than the incubation period, though in syphilis it is longer. The *latent period* is a term used by Fox, Hall and Elveback (1970) to denote the time between the initiation of infection and the first shedding of the agent. This again is shorter than the incubation period. In some of the viral diseases, for instance, the patient may be infectious before any sign of clinical illness has appeared. It may be observed that many of the acute viral contagious diseases behave as if the infectious period is short, even though virus can be detected for some days (Sartwell 1966).

Mortality rate refers to the proportion of deaths in the population within a given time, usually a year; *fatality rate* to the proportion of patients that die of a particular disease. A mortality rate obtained by calculating the proportion of deaths in a given time to the whole population is known as the *crude mortality rate*. For more exact purposes a mortality rate adjusted for age and sex is often used and is called a *standardized mortality rate*. The *morbidity rate* corresponds to the mortality rate, indicating the proportion of cases of a particular disease or group of diseases in the population during a given time. Both rates are expressed as a rule as numbers per 100 000 of the population; and as both are ratios, i.e. cases or deaths/population, both numerator and denominator must be known as accurately as possible.

The term *attack rate* is often used, particularly in outbreaks of food poisoning, to denote the proportion of those exposed to a given risk who become ill.

The term '*cohort*' was originally used by Andvord (1930) and later by Frost (1939) to denote a population characterized by birth within a given period of time and followed up for a certain number of years or throughout life. As an example we may choose the death-rate from tuberculosis in England and Wales during the past 100 years. During 1850-60 the standardized mortality rate was 1438 per 100 000; by 1950 it had dropped to about 60 per 100 000. What was the explanation of this fall? Examination of the figures for groups, or cohorts, born at 10-year intervals during this period shows not only that the mortality rate had fallen greatly but the maximum mortality had shifted from infancy to old age. Cohort analysis makes it clear that during this period the whole behaviour of the disease altered greatly, and that the chance of any one person dying of tuberculosis depended on the date of his birth. It may be added that the term 'cohort' is sometimes used to denote groups not characterized by birth within a given period of time but by some other characteristic (Rose and Barker 1978); but both Springett (1979) and Jacobs (1979) regard this as an unjustifiable extension of the original term.

Mathematical studies   Numerous statisticians have made a mathematical study of the epidemic process. See, for instance, Kermack and McKendrick (1927) on the mathematical theory of epidemics; the problem of endemicity (Kermack and McKendrick 1932); application of the theory of probabilities to a study of *a priori* pathometry (Ross 1915-16, Ross and Hudson 1916-17); and the mathematical theory of infectious diseases (Bailey 1975). [Reference may be made to a series of papers by Galbraith, Forbes and Mayon-White (1980) on the changing patterns of communicable disease in England and Wales.]

References

Anderson, E. S. and Hobbs B. C. (1973) *Israeli J. med. Sci.* **9**, 162.

Andvord, K. F. (1930) *Norsk. Mag. Laegevid.* **91**, 642.

Austin, D. F. and Werner, S. B. (1974) *Epidemiology for the Health Services.* Charles C Thomas, Springfield, Illinois.

Bailey, N. T. J. (1975) *The Mathematical Theory of Infectious Diseases and its Applications.* Charles Griffin, London.

Barrett-Connor, E. (1979) *Amer. J. Epidem.* **109**, 245.

Bradley, W. H. (1943) *Brit. med. J.* **i**, 438.

Bradley, W. H., Evans, W. I. and Taylor, I. (1951) *J. Hyg., Camb.* **49**, 324.

Cvjetanović, B., Grab, B. and Uemura, K. (1978) *Dynamics of Acute Bacterial Diseases.* World Health Organization, Geneva.

Dawber, T. R., Kannel, W. B. and Lyell, L. P. (1963) *Ann. N.Y. Acad. Sci.* **107**, 539.

Evans, A. S. (1979) *Amer. J. Epidem.* **109**, 379.

Fox, J. P., Elveback, L., Scott, W., Gatewood, L. and Ackerman, E. (1971) *Amer. J. Epidem.* **94**, 179.

Fox, J. P., Hall, C. E. annd Cooney, M. R. (1972) *Amer. J. Epidem.* **96**, 270.

Fox, J. P., Hall, C. E. and Elveback, L. R. (1970) *Epidemiology: Men and Disease*. Collier-Macmillan Ltd, London.

Frost, W. H. (1939) *Amer. J. Hyg.* **30,** 91.

Galbraith, N. S., Forbes, P. and Mayon-White, R. T. (1980) *Brit. med. J.* **ii,** 427, 489, 546.

Hope-Simpson, R. E. (1948) *Lancet* **ii,** 755.

Illingworth, R. S. (1967) *Brit. med. J.* **iv,** 41.

Jacobs, A. L. (1979) *Brit. med. J.* **i,** 266.

Kermack, W. O. and McKendrick, A. G. (1927) *Proc. Roy. Soc., A.* **115,** 700; (1932) *Ibid.* **138,** 55.

Masurel, N. (1969) *Lancet* **i,** 907.

Morris, J. N. (1975) *Uses of Epidemiology*. Churchill Livingstone, Edinburgh.

Paul, J. R. (1958) In: *Clinical Epidemiology*, p. 84. University of Chicago Press.

Pavlovsky, Y. N. (undated) *Human Diseases with Natural Foci*. Foreign Languages Publishing House, Moscow.

Petrishcheva, P. A. (1966) *USSR Ministry publ. Hlth, Moscow*, p. 1.

Raška, K. (1964) *J. Hyg. Epidem. Microbiol. Immunol.* **8,** 137; (1983) *Rev. infect Dis.* **5,** 1112.

Report (1947) *Spec. Rep. Ser. med. Res. Coun., Lond.* No. 360; (1967) *World Hlth Org., Technical Report Series*, No. 365.

Rojás, R. A. (1964) *Curso de Epidemiologia*. Universidad de Chile, Santiago.

Rose, G. and Barker, D. J. P. (1978) *Brit. med. J.* **ii,** 1558.

Ross, R. (1915–16) *Proc. Roy. Soc., A.* **92,** 204.

Ross, R. and Hudson, H. B. (1916–17) *Proc. Roy. Soc., A.* **93,** 212, 225.

Sartwell, P. E. (1966) *Amer. J. Epidem.* **83,** 204.

Sigurdsson, S. and Edwards, P. Q. (1952) *Bull. World Hlth Org.* **7,** 153.

Sinneker, H. (1976) *General Epidemiology*. John Wiley, London.

Snedecor, G. W. and Cochran, W. G. (1967) *Statistical Methods*. Iowa State University Press.

Springett, V. H. (1979) *Brit. med. J.* **i,** 126.

Stuart-Harris, C. H. and Schild, G. C. (1976) *Influenza. The Viruses and the Disease*, p. 136. Edward Arnold, London.

Wilson, G. S. (1974) *Brit. vet. J.* **130,** 207.

49

Actinomycosis, actinobacillosis, and related diseases

Graham Wilson

Introductory

The term actinomycosis was originally given by Bollinger in 1877 to a disease in cattle characterized by the formation of a hard wooden swelling of the tongue or a fusiform enlargement of the jaw. In the lesions of this disease he observed numerous opaque slightly yellowish bodies of coarsely granular or mulberry-like appearance, which were found to consist of a fungus. His colleague Harz, a botanist, to whom he referred for an opinion on this organism, decided that it was a true mould and suggested for it the name of ray fungus or *Actinomyces bovis*. A similar mycotic disease in man was described the following year by Israël (1878, 1879); and in 1879 Ponfick suggested that the two diseases were identical. In 1891, Wolff and Israël isolated from two cases in human beings an anaerobic non-pigmented branching organism—now known as *Actinomyces israeli*. In 1902 Lignières and Spitz described a new disease in cattle, distinguished from actinomycosis chiefly by its tendency to invade the lymph glands. From the affected animals they cultivated a very small, non-branching, gram-negative bacillus, to which they gave the name actinobacillus.

The disease they called actinobacillosis. As pointed out in Chapter 22, the causative bacillus, *Actinobacillus lignieresi*, is systematically related far more to the *Pasteurella* than to the *Actinomyces* group; but, as the lesions it produces resemble so closely those of actinomycosis, we think it best to consider it in the present chapter.

As early as 1874 Vandyke Carter drew attention to a disease in India known as Madura foot. Microscopical examination revealed the presence of a fungus, but attempts to cultivate it were unsuccessful till 1894, when Vincent succeeded in isolating an aerobic branching filamentous organism, similar in many respects to *Actinomyces bovis*, which he called *Streptothrix madurae*—now known as *Nocardia madurae*.

From the disease known as 'Farcy of cattle', Nocard in 1888 cultivated an aerobic branching filamentous organism, *Nocardia farcinica*. Similar branching organisms have been encountered in a number of diverse lesions in human beings and other animals.

From this brief description, it will be realized that there are several diseases from which actinomyces-like organisms have been isolated. In one group—much the more numerous—caused by anaerobic organisms, they grow in the form of colonies having a ray structure; such colonies are seen in pus as granules or *Drusen*. In another group, caused by aerobic organisms, they form a felted mycelium without either radial arrangement or, as a rule, granule formation. Though both forms of disease are referred to in clinical practice as actinomycosis, there is something to be said for restricting this term to the form caused by the anaerobic actinomycetes, and using the term nocardiosis for the form caused by the aerobic actinomycetes. The formation of clubs in the tissues is not confined to *Actinomyces*; sometimes it occurs in chronic granulomata of pigs and cattle caused by such organisms as *Staphylococcus aureus* and *Corynebacterium pyogenes*.

Classification of Diseases in Man and Animals caused by Actinomyces, Nocardia, and Actinobacillus

ACTINOMYCOSIS. Due to anaerobic organisms
   Actinomyces israeli in man and *A. bovis* in cattle. True ray fungus disease. Granules or *Drusen* in tissues.
   A. eriksoni in man. No granule formation.
   A. propionicus in man. No granule formation.

A. baudeti in cats and dogs. Granules formed in tissues.

NOCARDIOSIS. Due to aerobic organisms
   (1) Non-acid-fast group. Madura foot—granules found in tissues. Other lesions in man and animals—no granules found in tissues.
   (2) Acid-fast group—chiefly pulmonary and abdominal infections in man. No granules formed in tissues. Farcy of cattle—no granules formed in tissues.

ACTINOBACILLOSIS. Due to bacilli
   (1) *Actinobacillus lignieresi* of Lignières and Spitz. Actinobacillosis in cattle. Granules or *Drusen* formed in tissues.
   (2) *Actinobacillus actinoides* of Theobald Smith. Broncho-pneumonia of calves; pneumonia of rats. No granules formed in tissues.
   (3) *Actinobacillus equuli*. Joint-ill of foals.
   (4) *Streptobacillus moniliformis*. Rat-bite fever (one type) in man, and infective arthritis in mice. No granules formed in tissues.
   (5) *Actinobacillus piliformis*. Tyzzer's disease in mice. No granules formed in tissues.

Actinomycosis in man

Epidemiology

Actinomycosis in man is not a common disease. In England and Wales records of the Public Health Laboratory Service included 368 cases during the 12 years 1957 to 1968—an incidence of 0.665 per million a year. In Scotland 186 cases were recognized during the 13 years 1936 to 1948—an incidence of 3 per million a year (Porter 1953). According to van der Hoeden's (1939) observations in the Netherlands the disease is three times as common in males as in females, and ten times as common in rural districts as in large towns. In the Scottish series of cases males were affected 3–4 times as often as females, but the proportion of cases in country districts and small towns was only 1.8 times as great as in the large towns (Porter, pers. comm.). In the English and Welsh series the number of males affected was twice the number of females, and the incidence, so far from being higher in the country districts, was nearly twice as high in the cities and large towns as in the smaller urban and rural districts. All three series, however, agree in showing that about 80 per cent of cases occur in persons over 20 years of age. In Scotland the attack rate was ten times as high in agricultural workers as in other occupied males. In England and Wales, on the other hand, the highest incidence was in factory workers, housewives and children. More than half the cases in this series were in dental patients.

The cervico-facial region is the commonest situation affected. Table 49.1, compiled from miscellaneous sources, indicates the distribution of the lesions. The fatality of the disease treated with iodides or by surgical measures depends largely on the site. Of 10 patients suffering from cervico-facial infection observed by Colebrook (1921), 9 recovered: on the other hand 12 out of 14 with thoracic and abdominal infections died. In 87 cases recorded by Harbitz and Gröndahl (1911) the lesions proved fatal or remained unhealed in 42. Since the advent of sulphonamide and antibiotic therapy the mortality in England and Wales, which up to 1942 had been rising steadily—probably owing to improved diagnosis—has fallen almost precipitously, the annual number of deaths being reduced from 82 in 1941 to an average of 6 a year during the decade 1957–66.

At one time when actinomycosis was thought to be caused by a saprophytic aerobic organism described by Bostroem in 1891, infection was supposed to follow wounding of the mucosa in cattle by the contaminated awns of grasses or barley spikes and in man by some foreign body that provided a suitable nidus for growth. But now, when Bostroem's organism is regarded as a contaminant, and actinomycosis as a disease caused by an obligatory parasitic anaerobic organism, infection is believed to be of autogenous

Table 49.1   Distribution of actinomycotic lesions in man

Site	Illich-Colbrook series	Porter's series	Lentze's series	PHLS series 1957–68	Total	Percentage
Cervico-facial	250	118	933	232	1533	76.0
Abdominal	99	35	20	66	220	10.9
Thoracic	65	16	45	37	163	8.1
Other situations or doubtful	41	13	13	33	100	5.0
Total	455	182*	1011	368	2016	100.0

* Information on 4/186 cases not available.

origin. Nevertheless, wounds of the oral mucosa in cattle may well predispose to the disease.

Strong evidence in favour of this belief comes from the work of Naeslund (1925, 1926, 1929) who, in a study of the flora of the human mouth, succeeded in isolating both aerobic and anaerobic types of *Actinomyces*. The anaerobic types were indistinguishable from *Actinomyces israeli*, and were more frequently met with than the aerobic types. They were, moreover, isolated from a high proportion of salivary calculi, thus supporting Söderlund's (1921) view that the formation of these calculi must be regarded as a specific result of actinomycotic infection. The same organisms also seemed to play an important part in the development of dental tartar. The work of Lord and Trevett (1936), Emmons (1938), Sullivan and Goldsworthy (1940), and Slack (1942) left little doubt of the occurrence of true *Actinomyces* strains, indistinguishable from the Wolff-Israël type, in the human mouth, notably in diseased tonsils, carious teeth, and pyorrhoeal pus. During the histological examination of chronically inflamed tonsils Grüner (1969) found colonies of *Actinomyces* in 17 out of 103 cases. The constituent organisms, which were gram positive, were in the form of granules having the hyphae arranged like spokes in a wheel. Numerous similar findings are on record. If this organism is accepted as an inhabitant of the human mouth, the factors predisposing to the development of actinomycosis remain to be determined. Possibly suitable necrotic foci, such as septic teeth, have to be present in which the organisms can develop under anaerobic conditions, or other organisms may be necessary to favour their growth (Holm 1950). Trauma often seems to play a part. In Porter's (1953) series 27 of the 121 cervico-facial cases had a history of dental extraction or facial injury, and 28 of the 35 abdominal cases had a history of appendicectomy, gastric perforation, or a blow on the body wall. Harvey, Cantrell and Fisher (1957) likewise record a frequent history of mucosal injury. There are reports of three cases in which the disease followed a human bite (see Robinson 1944).

Bacteriology

In the tissues *Actinomyces* grows in the form of colonies or *Drusen*, which macroscopically have the appearance of small so-called sulphur granules. A section suitably stained reveals the presence of a central mass of partly necrotic material, in which branching filaments may be recognized, and a peripheral zone of swollen bodies looking like clubs and arranged radially. The clubs are surrounded by pus cells, outside of which there are aggregations of mononuclear cells and dense masses of fibrous tissue. The colony often presents a characteristically scalloped edge. The mode of formation of the clubs is still under dispute, some workers maintaining that they represent the swollen extremities of the filaments, others that they are due to a reaction on the part of the tissues. According to Pine and Overman (1963) the club consists of a hypha surrounded by a capsule formed by the organism itself. The granule is essentially a mycelial mass cemented together by a polysaccharide-protein complex and impregnated with calcium phosphate derived from the

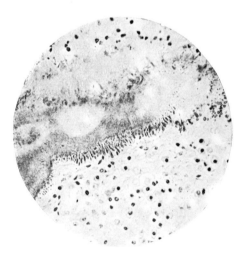

Fig. 49.1   Part of the edge of a colony of *Actinomyces bovis* in the jaw of an ox, showing the peripheral zone of clubs (× 350).

host. This interpretation is borne out by Kubo, Osada, and Konno (1980). The disease spreads by contiguity; the surrounding tissues, no matter of what type, are invaded and destroyed.

The pus discharged from these lesions is not usually abundant; it is yellow, often bloody, thick or thin, and sometimes tough and viscous. When shaken up with water in a test-tube it will be seen to contain small granules, which rapidly sink to the bottom. They vary greatly in size, but are generally about 0.25 to 1 mm in diameter; they are opaque, spherical bodies, greyish-white by reflected and yellowish by transmitted light, and have an oily appearance. Generally they are soft and can be readily crushed, but in cattle they may be hard or even calcified. In pus from old lesions they are darker in colour, and not infrequently brown. Micros-copical examination of these granules reveals the presence of a felted mycelium, together with rod-shaped and coccoid bodies, surrounded by radially disposed clubs. The filaments are irregularly gram positive; the clubs are usually gram negative. By electronmicro-scopy Kubo and his colleagues (1980) find that the sulphur granules of actinomycosis are large, oval, or horseshoe-shaped, and contain a number of gram-positive filaments or short rod-like hyphae beneath the clubs. The clubs themselves are composed of a varying number of fine granules arranged around a central hypha. Clubs are said to be more conspicuous in lesions of cattle than in those of man.

Early attempts to cultivate these organisms led to discrepant results, but the work of Wright (1905) in Boston served to put the bacteriology of the disease on a sound footing. From 13 cases in man and from 2 in cattle he isolated an anaerobic organism of the Wolff-Israël type. Though unable to set up progressive disease by inoculation of laboratory animals, he con-cluded from his own work and from a critical résumé of the literature that the Wolff-Israël organism was the true cause of human actinomycosis. Work by sub-sequent observers, such as Henry (1910), Colebrook (1920), and Bosworth (1923) in Britain, Harbitz and Gröndahl (1911) in Norway, and Magnusson (1928) in Sweden, tended more and more to confirm Wright's conclusions. Colebrook showed that the serum of severely infected human patients might agglutinate this organism; and Magnusson (1928), working with *Actinomyces bovis*, found that, though the disease could not be satisfactorily reproduced in small ani-mals, it was possible to set up typical actinomycotic lesions by the inoculation of pure cultures into cattle. That it has not been cultured from all cases of the disease is probably explained by technical difficulties, especially when the lesion is invaded by secondary organisms, and by the not infrequent sterility of gran-ules from old lesions.

More recent observers have brought strong evi-dence to show that actinomycosis in man is due to a mixed infection with *Actinomyces israeli* and other,

mainly anaerobic, organisms. Of these, one of the commonest is the organism described by Klinger in 1912—now known as *Actinobacillus actinomycetem-comitans*—and subsequently reported on by Cole-brook (1920) and Bayne-Jones (1925). Holm (1950) put forward this synergistic view, and German workers have supported it. In the 283 cases of Heinrich and Pulverer (1959) *Actinob. actinomycetemcomitans* was found in 90. In every one of the 1011 cases reported on by Lentze (1967), with three easily explic-able exceptions, *Actinomyces israeli* was accompanied by other organisms, aerobic or anaerobic. Among the aerobic group staphylococci and α-haemolytic strep-tococci were by far the most frequent. Among the anaerobic group *Actinob. actinomycetemcomitans*, *Fu-sobacterium necrophorum*, *Bacteroides melaninogeni-cus*, other fusobacteria, and *Leptotrichia* all figured prominently. Like Holm (1950), Lentze regards acti-nomycosis as a polybacterial infection. These accom-panying organisms appear to act partly by lowering the redox potential, thus enabling *Actinomyces to grow more freely, and partly by the* secretion of collagen-liquefying enzymes that facilitate invasion of the tissues. Brede (1959) found that hyaluronidase and other depolymerizing enzymes were formed by nearly all the accompanying organisms he studied, but not by a single one of 88 strains of *Actinomyces israeli*. *Streptococcus milleri* (see Chapter 29) is not infre-quently isolated from abscesses of the jaw with or without *Actinomyces israeli*.

Apart from *Actinomyces israeli*, one or two other anaerobic species have from time to time been in-criminated in the causation of human disease. For example, *Actinomyces propionicus* may be respon-sible for lachrymal canaliculitis and the formation of lachrymal concretions (Pine and Hardin 1959, Buchanan and Pine 1962); and *Actinomyces eriksoni* may give rise occasionally to abscesses in the lung or subcutaneous tissue (Georg *et al.* 1965).

Diagnosis, prophylaxis and treatment

Diagnosis is made by microscopical and cultural ex-amination of the affected tissues. The granules should be crushed between slides and stained by Gram's method. If calcified, they should first be treated with hydrochloric acid. The presence of a mycelium of gram-positive filaments surrounded by radially dis-posed gram-negative clubs is characteristic of actino-mycosis. The identity of the organisms in clinical material and in cultures can be determined by the fluorescent antibody technique in which absorbed sera are used (Lambert *et al.* 1967, Blank and Georg 1968). Cultures are made from fresh young granules only; after thorough washing, they are seeded into a number of tubes containing blood broth (Gordon 1920), glucose agar, or serum agar; incubation is carried out under both aerobic and anaerobic conditions, preferably in an atmosphere to which 5–10 per cent CO_2 has been added. Rosebury, Epps and Clark (1944) prefer to streak the unwashed granule or the pus on to four plates in series of Bacto brain heart infusion agar and to incubate in an anaerobic atmosphere containing 5 per cent CO_2. Lentze (1967) strongly recommends the use of Fortner's (1928) method of anaerobic cultivation, i.e. growth in the presence of *Serratia marcescens* in a sealed petri dish; and Heinrich and Korth (1967) describe the use of a semi-defined medium that gives results superior to those on ascitic agar. For the microscopical demonstration of actinomyces in dental plaque, serotype-specific antisera prepared against whole cells and labelled with a fluorescein dye may be used (Marucha *et al.* 1978); and for their cultural isolation a medium containing metronidazole and cadmium sulphate is recommended by Kornman and Loesche (1978). Primary cultures are frequently contaminated; single colonies should therefore be picked off to obtain pure cultures. Examination of the patient's serum for agglutinating, haemagglutinating, or complement-fixing antibodies may be tried, but the result is sometimes negative, and false positive reactions may be met with in other diseases such as tuberculosis and streptococcal infections (Holm and Kwapinski 1959). (For a detailed description of the various serological methods used in the diagnosis of diseases caused by strains of *Actinomyces* and fungi, see Mackenzie *et al.* 1980). In pulmonary actinomycosis in man the sputum is tough, often viscous, and sometimes haemorrhagic; the typical granules can be found on examination (Harbitz and Gröndahl 1911). Most of the abdominal cases commence in the appendix. In the rare cases of actinomycotic meningitis, the primary lesions are generally in the lungs and bronchial glands (Henry 1910).

As we are still ignorant of the factors determining the development of the disease, it is difficult to lay down prophylactic measures against it. It is clear, however, that the discharge from the lesions should be considered dangerous, and every care taken to prevent its coming into contact with man or animals.

Though vaccine therapy was reported on favourably in the past (Colebrook 1921, Negroni 1937), treatment of the disease is now carried out mainly with penicillin, alone or in addition to sulphonamide or streptomycin (Nichols and Herrell 1948). *Actinomyces israeli* is sensitive to all the common antibiotics—penicillin, terramycin, chloramphenicol, aureomycin, and streptomycin—the degree of sensitivity diminishing in this order (Garrod 1952). Harvey, Cantrell and Fisher (1957) recommend surgical removal of the affected tissues combined with large doses of penicillin injected daily for 12–18 months after the operation. That treatment with antibiotics is not always successful may be due partly to the impenetrability of the granules, and partly to the presence of resistant accompanying organisms. *Actinob. actinomycetemcomitans*, for example, though susceptible to the tetracyclines, is resistant to penicillin and erythromycin, and moderately resistant to streptomycin and chloramphenicol; and *Bacteroides melaninogenicus*, though susceptible to penicillin G and the tetracyclines, is resistant to streptomycin and chloramphenicol (Lentze 1967). For this reason the tetracyclines are probably the drug of choice (Heinrich 1960). Treatment must be kept up for at least three months (Weese 1975). King and White (1981) report rapid improvement in pulmonary actinomycosis after a 2-week course of rifampicin.

For a general description of actinomycosis the reader is referred to the monograph by Cope (1938), to the proceedings of the Symposium edited by Heite (1967), and to the book by Bronner and Bronner (1971).

Anaerobic actinomycotic infections of animals

Among domestic animals cattle are much the most frequently affected. The disease occurs rarely in pigs, sheep, goats and dogs. Most carnivora are resistant. We have met with one case in a mouse, in which the liver showed typical actinomycotic lesions.

In *cattle* the characteristic lesion produced by *Actinomyces bovis* is a fusiform swelling of the jaw—often referred to as lumpy jaw. Like actinomycosis in man, this disease manifests the features of spread by contiguity, sinus formation, presence of granules in the pus, absence of involvement of the lymphatic system and failure to metastasize.

The extent to which cattle are affected is difficult to ascertain, largely because the distinction between this disease and actinobacillosis has often been disregarded. Most workers, such as Griffith (1916), Hülphers

(1923), and Magnusson (1928), believe that *A. bovis* attacks almost exclusively the jaw bone, and that the so-called lingual actinomycosis or 'wooden tongue' is really actinobacillosis. Though Albiston and Pullar (1934) state that the soft tissues—tongue, lymph, salivary glands, muscle—may be affected by *A. bovis*, the balance of evidence is against them. Crude slaughterhouse returns do not distinguish between the two diseases, but the morbidity figures given by Jelenevski (see Hutyra and Marek 1926) for various abattoirs between 1896 and 1911 appear to refer mainly to true actinomycosis of the jaw. They are as follows: Berlin 0.3 per cent, Vienna 0.01 per cent, Moscow 3.34 per cent, Kiev 0.67 per cent and Warsaw 0.65 per cent of animals slaughtered. Even in the same country the prevalence of the disease varies from one part to

Table 49.2   Bacteriological findings in 'actinomycosis' of cattle in Sweden and the Netherlands (modified from Magnusson 1928)

Site of lesion	No. of cases examined	No. from which pure cultures obtained	Organisms isolated in pure culture			
			Actinobacillus	*Actinomyces*	*Corynebacterium pyogenes*	*Staph. aureus*
Jaw	145	120	7 (5.8)	85 (70.8)	28 (23.3)	0 (0)
Tongue	96	90	86 (95.6)	0 (0)	0 (0)	4 (4.4)
Lymphatic glands	37	32	32 (100)	0 (0)	0 (0)	0 (0)
Skin and soft parts of head	73	65	63 (96.9)	2 (3.1)	0 (0)	0 (0)
Udder	32	31	0 (0)	0 (0)	0 (0)	31 (100)
Other tissues mainly lungs and viscera	13	13	7 (53.8)	0 (0)	5 (38.5)	1 (7.7)
Total	396	351	195 (55.6)	87 (24.8)	33 (9.4)	36 (10.2)

Figures in parentheses are percentages of cases in which a pure culture was obtained.

another. Thus, in the Netherlands, van der Hoeden (1939) gives the frequency of lesions in slaughtered cattle as ranging from 0.42 per 1000 in Gouda to 12.17 in Amsterdam. More exact figures based on bacteriological examinations were reported by Hülphers (1923) for Stockholm, Gunst (1927) for Amsterdam, and Magnusson (1928) for Malmö (Table 49.2).

From this table it will be seen that rather over half the cases diagnosed in the slaughterhouse as 'actinomycosis' were caused by *Actinobacillus lignieresi*, a quarter by *Actinomyces bovis*, and another quarter by *Corynebacterium pyogenes* and *Staphylococcus aureus*. Magnusson considers it doubtful whether the lesions in the jaw from which *Corynebacterium pyogenes* was isolated were really caused by this organism; it may well have been merely a secondary invader of lesions due primarily to *Actinomyces bovis*. In Great Britain actinomycosis of the jaw in cattle is of low sporadic incidence. Actinomycosis of the udder is caused exclusively by staphylococci. As was pointed out by Magrou (1919), these organisms can give rise to typical actinomycotic granules.

True actinomycosis in *swine* is uncommon. The so-called actinomycotic lesions in the udder of sows are usually caused by *Actinobacillus lignieresi*, staphylococci, or *Corynebacterium pyogenes*. The lesions themselves are surrounded by granulation tissue, and in the pus granules may be found. The disease caused by staphylococci, whether in cattle or in swine, is best referred to as *botriomycosis* (see Chapter 60).

In *cats*, and less often in *dogs*, abscesses in the subcutaneous tissues following bites are sometimes caused by a microaerophilic organism known as *Actinomyces baudeti* (see Chapter 22). Sometimes the serous membranes, the lungs, bones, or central nervous system are affected. In subcutaneous abscesses multiple fistulae discharge pus coloured rose by blood and containing yellowish or reddish-brown granules that are never calcified. Microscopically, the granules consist of a felted mass of filaments at the periphery of which are coronets of basophilic gram-positive rounded or oval bodies differing greatly in appearance from the clubs of *Actinomyces bovis* or *israeli* (Brion 1939, Prévot *et al.* 1951, Brion *et al.* 1952).

Nocardiosis in man and animals

Due to organisms of non-acid-fast group

(1) Madura disease

This disease, which is peculiar to man, was first thoroughly investigated by Vandyke Carter of Bombay in 1874. It is common in India, and is widely endemic in tropical and subtropical countries. The chronic granulomatous lesions are most frequent on the feet, owing to the habit of labourers of walking barefooted. They are characterized by swelling, suppuration, and sinus formation. The pus contains granules, which in some cases are yellowish in colour, in others black. These granules are generally larger than those seen in actinomycosis—up to 1 mm in diameter; they may be aggregated into masses looking not unlike fish-roe (Kanthack 1893). Under the microscope, those from the pale or ochroid variety of the disease consist of a central irregularly staining felted mycelium, showing true branching; sometimes they are surrounded by a ray formation of club-like structures. Vincent (1894) cultivated from them an aerobic filamentous organism, which gave a pink or a rose coloration on potato (see Chapter 22). It proved to be practically non-pathogenic for laboratory animals. He called it *Streptothrix madurae*; its correct name is *Nocardia madurae*. In Mexico 94 per cent of the cases are said to be due to *Nocardia brasiliensis*—a member of the acid-fast group—and only 4 per cent to *N. madurae* (Gonzáles Ochoa 1962). The disease is not contagious. Infection occurs from the soil and spreads by contiguity: lymph nodes are not usually affected.

Wright (1898) found that the granules from the black or melanoid type of the disease consisted of a mycelium of hyphae or fungoid elements, more or less degenerated, embedded in a hyaline brown-coloured refringent substance, which itself formed more or less of a reticulum. The pigment could be dissolved by sodium hypochlorite solution, leaving the mycelium unaltered. Cultures resulted in the development of a mould, consisting of a branching septate mycelium, and forming typical mould-like colonies. Wright concluded that the pale variety of the disease was due to an organism of the *Streptothrix* (*Nocardia*) group, and the black variety to a mould or hyphomycete. This view is supported by the observations of Tribedi and Mukherjee (1939). It is probable that several types of organism are responsible for the disease, and that the pale variety, at least, may be caused either by a nocardia or by a fungus.

(2) Other lesions

Nocardiae of the non-acid-fast group have been described by numerous authors in various lesions, such as of the bladder and prostate by Cohn (1913), in the purulent cornea by Namyslowski (1912), the cerebrospinal fluid of children suffering from meningitis by Gerbasi (1927), and fatal infections of the lungs by Biggart (1934) and Lynch and Holt (1945). Pelletier (see Laveran 1906) saw a case in a young negress of Senegal who had a tumour of the left knee as large as a person's head and riddled with fistulae. The discharge of greyish pus contained a large number of red granules 0.5 mm q in which Laveran found gram-positive coccoid organisms arranged in zoogloeae. From a similar tumour of the thorax Thiroux and Pelletier (1912) isolated a filamentous branching organism now known as *Nocardia pelletieri* (see Chapter 22).

A few cases of visceral disease have been described

in *animals*. For example, Silberschmidt (1899) isolated a strain called *Nocardia caprae*, from three goats; Dean (1900) a strain from a submaxillary abscess in the horse; and Ginsberg and Little (1948) a strain from an abdominal lesion in the dog.

Due to organisms of the acid-fast group

Man

Cases due to this group of nocardiae appear to be much commoner than those due to the non-acid-fast group. Eppinger (1891) described a case of brain abscess in a glass-grinder due to a branching filamentous organism. Old calcified abscesses were found in the bronchial and supraclavicular glands, and pseudotubercles were noticed in the lungs and pleura. The organism grew abundantly on ordinary media under aerobic conditions (see Chapter 22). Injection into rabbits and guinea-pigs proved fatal in 1 to 4 weeks; *post mortem* the lungs, liver, and spleen were studded with small white nodules. He called the organism *Cladothrix asteroides* (*Nocardia asteroides*) and the disease Pseudotuberculosis cladothrichica. Later Mc-Callum (1902) recorded a case of peritonitis in a negro child due to infection with the same organism. Murray and his colleagues (1961) collected records of 179 cases of nocardial infection in the literature, most of which were probably due to acid-fast organisms. The disease is commoner in males than females. It may be acute or chronic, localized or generalized, and may or may not be superimposed on an underlying disease. Not infrequently it is activated by corticosteroid therapy. Thirty per cent of cases affect the respiratory system alone, 30 per cent are disseminated, 14 per cent affect the skin and subcutaneous tissues, 5 per cent the brain alone, and a rather higher percentage brain and lung together (Murray *et al.* 1961). Metastatic abscesses are common. Infection probably occurs by inhalation; the first lesions are in the peribronchial nodes. Clinically the disease resembles an acute or subacute pneumonia, and generally lasts about six months. The sputum is thick, sticky, and purulent. Pathologically the lesions usually consist of multiple, often confluent abscesses. There is extensive necrosis without capsulation and with very little granulation or fibrous tissue. No granules are seen, and there is an absence of burrowing or sinus formation such as occurs in actinomycosis caused by the anaerobic type (Weed *et al.* 1955). These cases are liable to be confused with tuberculosis. Goldsworthy (1937) recommends that the organisms should be sought for in the sputum of all patients with obscure or unusual chronic infections of the lungs, especially when repeated attempts to demonstrate the tubercle bacillus have failed. The organisms may be decolorized in the ordinary Ziehl-Neelsen process, but their filamentous morphology is rendered strikingly evident by Gram's strain. Cultivation of the organism, and the exclusion of tuberculosis by inoculation of the sputum into guinea-pigs, should effect a differential diagnosis. Contaminants can often be eliminated by incubation at 40–50° (Curry 1980).

Most infections are caused by *Nocardia asteroides*. *N. brasiliensis* and *N. caviae* usually give rise to cutaneous infections that take the form of a mycetoma. These are characterized by swelling of the skin and subcutaneous tissues with multiple sinus tracts. Nocardial lesions, like those of actinomycosis, extend by eroding the tissue barriers including bone, brain, and the vascular system. The sinuses that are formed discharge an exudate containing granules composed of microcolonies.

According to Drake and Henrici (1943) infected rabbits and guinea-pigs show the delayed sensitivity type of reaction to a crude extract of a culture of *N. asteroides*. However, this antigen cross-reacts with other antigens, especially those of tuberculosis, and does not seem to have been made use of in human diagnosis.

For treatment penicillin is of little value. Reliance must be placed on the sulphonamides, with minocyline as an alternative in allergic patients (Curry 1980). (For recent reviews, see Rosett and Hodges 1978 and Curry 1980.)

Animals

Numerous cases of infection with this group of organisms have been reported in animals. *N. asteroides* causes various lesions in *dogs*; Ginsberg and Little

(1948) isolated this organism from a lesion of the throat.

Cattle farcy. Farcin du boeuf This disease is characterized by the appearance—generally on the medial surface of the extremities—of firm, painless perivascular nodes, which later suppurate. When incised, they discharge a whitish, odourless mass resembling soft cheese (Report 1913). The regional lymph nodes become converted into firm, painful tumours. It is a chronic disease lasting for a year or more, and marked in the later stages by severe cachexia. From the pus of an affected animal Nocard (1888) isolated an aerobic branching organism, known as *Nocardia farcinica*, but probably the same as that now called *Nocardia asteroides*. Intravenous injection of pure cultures into a cow and a sheep resulted in the development of miliary nodules. The disease is not common in Europe; it is very prevalent in Guadeloupe, and has been recorded in India, and in Kenya (Daubney 1927).

Apart from farcy, other organs may be affected. Evans (1918) recorded an infection of the udder of cows and isolated the organism from 18 out of 21 samples of milk. Pier, Mejia and Willers (1961) found *N. asteroides* to be causing severe and persistent infection of the udder, often accompanied by high fever, rupture of the gland, and rapidly progressive fibrosis. By use of skin-hypersensitivity and complement-fixation tests, in conjunction with cultures from the udder, Pier and Enright (1962) consider it possible to pick out infected animals from a herd. There is no response to therapy, and such animals should be culled.

Actinobacillosis

In 1902 Lignières and Spitz described under the term actinobacillosis a disease that had occurred in epizoötic form amongst cattle in the Argentine during the summer of 1900–01. Once a herd had become infected the disease spread rapidly, attacking 15–25 per cent, sometimes even 50 per cent, of the animals in the course of a few weeks. After a short time the epizoötic died down, but sporadic cases, generally of a mild form, occurred at intervals. The main animals attacked were cattle; sheep were rarely, and horses never, attacked. In cattle all ages were affected, from calves of 1 month to animals several years old.

In about 80 per cent of the cases the lesion was in the subcutaneous tissue of the neck; in about 5 per cent in the tongue. With the exception of the lungs, the internal organs were not often affected. The most striking difference from actinomycosis lay in the frequency with which the regional lymphatic nodes were invaded. In the neck the lesion commences as a rounded subcutaneous nodule of fibrous consistency, adherent to the skin, with a smooth but later protu-

berant nodular surface. During the next few days the lesion grows rapidly, softens in the centre, and becomes converted into a cold abscess. In this state it persists for weeks or months, gradually enlarging. Eventually it ulcerates and discharges a highly viscous, milky-white or slightly green, almost odourless pus, containing little greyish-white opaque granules up to 0.4 mm in diameter. The pus is so viscous that it is rarely discharged completely; it forms crusts on the surface of a fungating ulcer, which persists indefinitely. The abscesses are usually multiple, and frequently occur in different situations simultaneously—such as the tongue, glands, and lungs. Fibrous tissue is formed around the lesions—as in actinomycosis. On examination the granules are seen to be simple or composite. They consist of tufts of radiating clubs varying in size from about $16 \times 8\,\mu m$ to $99 \times 24\,\mu m$. They are stained best by glycerol picrocarmine; in a few seconds the clubs take on a yellow colour, while the pus appears pink. No mycelium is found. If, however, the granule is stained with carbol fuchsin, it is sometimes possible to distinguish small bacilli in the granular detritus that occupies the centre. Both clubs and bacilli are gram negative. A very pretty picture is obtained by staining with Ziehl-Neelsen and decolorizing cautiously with 1 per cent H_2SO_4. The interior of the colony is blue, while the clubs, being weakly acid-fast, are stained red.

Cultures made from the granules revealed the presence of a very small gram-negative, non-motile, coccobacillus, which Lignières and Spitz called the actinobacillus—now called *Actinobacillus lignieresi*. (For description see Chapter 22.) The organism appears to be a frequent inhabitant of the rumen, and may be found on the tongue of normal animals (Phillips 1961, 1964). In Great Britain the disease is less acute, and the lesions are more localized, though the lymph nodes are almost invariably affected.

Reproduction of the disease in animals

Subcutaneous inoculation of a pure culture into cattle leads to the formation of an abscess similar to that occurring in the natural disease, though not all strains are virulent. Fluctuation is apparent about the 10th day. The abscess enlarges, and eventually the skin gives way, allowing a small quantity of pus to exude. Fungating masses spring up around the opening and obstruct drainage. The animals gradually waste. The pus contains typical granules. Feeding has not been successful in transmitting the disease. A local abscess develops after subcutaneous inoculation of pure cultures into pigs and sheep. In horses a large oedematous swelling is formed, which discharges pus after a few days.

The work of Lignières and Spitz has been confirmed by workers in several countries. Nocard (1902) rec-

Fig. 49.2 A colony of *Actinobacillus lignieresi* in the tongue of an ox, showing the long, finger-shaped clubs ($\times 350$).

ognized the disease in France. He pointed out that in actinomycosis of cattle the granules in the pus were yellowish in colour and often calcified, whereas in actinobacillosis they were greyish-white and rarely calcified. Griffith (1916) examined the diseased tongues and lymphatic glands of 44 cattle slaughtered in Britain. Of these no fewer than 40 had lesions of actinobacillosis. From 23 of these he isolated a gram-negative bacillus which had the characters of *Actinobacillus lignieresi*. One of the cultures inoculated subcutaneously into a calf caused a local lesion containing typical granules. The remaining 4 cases, affecting the lower jaw, were due to *Actinomyces bovis*. In an examination of 34 specimens of 'actinomycosis' in cattle slaughtered at Islington, Bosworth (1923) recognized *Actinobacillus lignieresi* 21 times; from 17 of the cases he isolated it in pure culture. The other 13 specimens were examples of actinomycosis. Of 85 445 tongues imported from the Argentine and examined by MacFadden (1913) in London, 4949 or 5.8 per cent showed lesions, most of which were probably due to the actinobacillus.

Diagnosis

The important points are: (1) The location: practically all jaw lesions appear to be due to *Actinomyces bovis*, whereas lesions of the soft parts—tongue, cheek, gum, palatal mucosa, skin, and lymphatic glands—are due to the actinobacillus. (2) The frequent involvement of the lymph glands in actinobacillosis; in actinomycosis the glands are swollen only as the result of secondary infection from open lesions. (3) Granules are rarely found in the glands in actinomycosis, very frequently in actinobacillosis. (4) In actinobacillosis the granules are smaller, lobulated, paler, and rarely calcified: in actinomycosis the granules are larger, oval or horseshoe-shaped, darker, and often calcified. (5) In actinobacillosis the granules consist of a central mass of detritus containing minute gram-negative bacilli surrounded by long, radially disposed clubs: in actinomycosis recently formed granules contain a central gram-positive mycelium, together with gram-positive rods and coccoid forms, surrounded by short, radially disposed clubs; older granules may show a few gram-positive bacillary and granular elements embedded in a gram-negative matrix. (6) In actinobacillosis cultures reveal the presence of a small gram-negative, aerobic cocco-bacillus: in actinomycosis a gram-positive, branching, filamentous organism, often assuming rod forms in culture, not unlike *Corynebacterium diphtheriae*, is isolated under anaerobic conditions. During life agglutinins may be found in the serum of cattle with actinobacillosis in a titre of 1/160 or over, often with a prozone; but in nearly half the cases the titre does not exceed the limit found in normal animals (Phillips 1965).

Actinobacillosis may affect *sheep*. According to

Tunnicliff (1941), Christiansen in 1917 described a pyaemic infection of the heads of sheep due to an organism which he called *Bacterium purifaciens*, but which was probably the actinobacillus. Thomas (1931), in South Africa, was the first to associate the disease in sheep with *Actinobacillus lignieresi*. The disease was relatively mild and occurred in animals fed on prickly pear. Lesions occur primarily in the lips, tongue, and cheeks, but the mandibular, retropharyngeal, mediastinal, and bronchial glands may also be affected. Single or multiple areas of granulation tissue are found on section, with burrowing irregular cavities containing sticky bright yellow pus, sometimes showing, on microscopical examination, ray-like formations and clubs. The disease in sheep has also been recorded by Ravaglia (1934) in Italy and Marsh and Wilkins (1939) in the United States.

Prophylactically, Lignières and Spitz (1902) found that segregation and killing of infected animals was the best procedure. Horse antiserum was employed in the treatment of some animals—though without conspicuous success. Iodine is said to be of much more value in the treatment of actinobacillosis than of actinomycosis.

Actinobacillosis is rare in man. According to Flamm and Wiedermann (1962) seven cases have been reported in the literature, but the identification of the causative organism with *Actinob. lignieresi* has not been entirely satisfactory.

More recently Vandepitte and his colleagues (1977) compiled a table of 19 cases in which *Actinobacillus actinomycetemcomitans* was isolated, mainly from cases of subacute endocarditis, in the absence of *Actinomyces*.

Pneumonia in calves due to *Actinobacillus actinoides*

Smith (1918, 1921) described an epizoötic bronchopneumonia in calves, most common in the 2nd and 3rd months of life, characterized by areas of consolidation, focal necroses, and other lesions. From the lungs of a number of fatal cases in two separate outbreaks he isolated an organism which he called *Bacillus actinoides* (see Chapter 22). In some cases it was present in pure culture, in others it was associated with *Corynebacterium pyogenes*, *Bordetella bronchiseptica*, or other organism. On microscopical examination of the lungs, it was seen in the form of slender bacilli situated at the periphery of necrotic areas, aggregated into masses in the alveoli and alveolar ducts, or in the proliferating epithelium that occupied the smaller bronchioles. On coagulated horse serum, tiny whitish flocculi appeared in the water of condensation; these consisted of radiating filaments terminating in clubs. The organism was non-pathogenic to laboratory animals, but intratracheal injection into calves was sometimes followed by the appearance of

small necrotic foci in the lungs, indistinguishable from those observed in the natural disease. Jones (1922) isolated it from the pneumonic lungs of white rats.

Joint-ill of foals due to *Actinobacillus equuli*

This organism appears to be responsible for joint-ill of foals, sleepy foal disease (Edwards 1931, 1932, Report 1949, Maguire 1958), and for neonatal deaths from septicaemia. Mares usually abort shortly before term, and the foals die within the first four days of life. *Post mortem*, the lesions are those of acute nephritis, arthritis, and pneumonia. *A. equuli* can be isolated in pure culture from the viscera and stomach contents of the foal. The disease is seldom met with in Great Britain, but is common in North America, South Africa, New Zealand, and Japan. It affects individual animals rather than herds. The mares apparently carry the organism in the throat and gut, but healthy carriers are difficult to detect (Baker 1972).

Rat-bite fever

For many years there was great confusion over the aetiology of this disease. Schottmüller (1914), Blake (1916), Tileston (1916), and others described the isolation of a streptothrix—*Streptothrix muris ratti*—from the blood of patients suffering from rat-bite fever; Japanese workers (see Chapter 74), on the other hand, brought evidence to show that the disease was due to a spirochaetal organism known as *Spirochaeta morsus muris* or *Spirillum minus*. There seems to be no question now that both organisms may be responsible, and that rat-bite fever may be of at least two different types. The evidence for this conclusion is furnished by a careful study of the literature.

In 1925 Levaditi, Nicolau and Poincloux in France described an organism, to which they gave the name *Streptobacillus moniliformis*. It was isolated from the blood of a laboratory worker who was suffering from an acute disease characterized by fever, sore throat, papular erythema, and multiple arthritis. The following year Parker and Hudson (1926) gave an account of an epidemic disease occurring in Haverhill, Massachusetts. This disease they referred to as Erythema arthriticum epidemicum or Haverhill fever. An organism was isolated—*Haverhillia multiformis*—which was subsequently proved to be identical with *Streptobacillus moniliformis*. In neither instance was there any suspicion of infection having occurred through rats. In 1933 Strangeways accidentally encountered the same organism in mice that had died after intraperitoneal inoculation with the blood of rats infected with *Trypanosoma equiperdum*. Investigation showed her that *Streptobacillus moniliformis* was an inhabitant of the nasopharynx of the rat, in which it normally led an apparently harmless existence. A very similar organism had already been reported by Tunnicliff (1916) in the lungs of white rats affected with broncho-pneumonia. It would therefore appear that *Streptobacillus moniliformis* is no other than the *Streptothrix muris ratti* of the earlier workers. Both organisms appear to be closely related to the *Actinobacillus* group. Since this organism is pathogenic for man and is a normal parasite of the rat, it is not difficult to understand how it may be responsible for human infection. *Spirillum minus* is also a normal parasite of the rat, and may likewise cause rat-bite fever. We shall restrict ourselves at the moment to the description of the type of disease caused by *Streptobacillus moniliformis*. For the type caused by *Spirillum minus* see Chapter 44.

Rat-bite fever due to *Streptob. moniliformis* is an irregularly relapsing, febrile, septicaemic disease characterized by metastatic arthritis and morbilliform and petechial cutaneous eruptions. The incubation period, as a rule, is under 10 days, and may be as short as 3 days. The disease is often ushered in by a chill and a rapid rise of temperature. The wound usually heals without exacerbation. There is extreme prostration, severe generalized muscular pain and tenderness, headache, weakness, and a widespread morbilliform or petechial eruption. Generalized enlargement of the lymph glands may occur. Non-suppurative arthritis, affecting one joint after another, is a characteristic feature of the disease. There is a leucocytosis with a high proportion of neutrophils. After a few days the temperature falls by crisis, and the disease then assumes a relapsing type, febrile paroxysms occurring at irregular intervals for weeks or months. The case fatality is not known with certainty, but is probably round about 10 per cent. Little is known of the post-mortem appearances, but Blake's case showed an ulcerative endocarditis and a subacute myocarditis. Diagnosis during life is made by cultivation of the causative organism from the blood during a febrile attack. About 20 ml of blood are distributed partly into tubes of 20 per cent serum broth and partly over a number of plates of Loeffler's serum. The cultures may be incubated aerobically or anaerobically, preferably in the presence of 5 per cent CO_2. According to Brown and Nunemaker (1942) agglutinins may be demonstrable in the patient's blood serum in a titre varying between 1/160 and 1/5120. A suitable antigen may be prepared by formolizing a broth culture of a laboratory strain of *Streptob. moniliformis* that forms a homogeneous sediment. Though infection usually follows a rat-bite, the outbreak of Haverhill fever studied by Parker and Hudson (1926) was traced to the consumption of *raw milk*, as was also another outbreak affecting 86 persons described by Place and Sutton (1934). How the milk becomes contaminated it is difficult to say, but the excretion of the organism in the urine of infected mice and perhaps rats affords a possible explanation. Laboratory workers are occasionally infected (Levaditi *et al.* 1925, Holden and MacKay 1964, Gledhill 1967).

The differential diagnosis of rat-bite fever due to *Streptob. moniliformis* from that due to *Spirillum minus* was studied by Allbritten, Sheely, and Jeffers (1940). The actinobacillary type has an incubation period usually under 10 days, the wound does not generally become inflamed when the temperature rises, the febrile paroxysms are less regularly periodic, the rash is of the small macular or petechial type, arthritis is a prominent feature of the disease, leucocytosis is common, and the benefit from arsenic in treatment is doubtful. In the spirillar type, on the other hand, the incubation period is usually longer than 10 days, the local wound becomes inflamed when the temperature rises and may suppurate, the febrile paroxysms are more regular, the rash is of the large macular or papular type, arthritis is rare, leucocytosis is less common, and arsenic is specific in treatment. Reference, however, to the excellent review by Brown and Nunemaker (1942) suggests that, though some symptoms are commoner in one form of the disease than the other, both organisms may give rise to the same clinical picture. For treatment, Roughgarden (1965), who reviewed the literature on antimicrobial therapy, recommends penicillin in a dosage of not less than 400 000 to 600 000 units a day for at least a week; but some patients do well on tetracycline, and others on a combination of streptomycin or tetracycline with penicillin.

Infective arthritis of mice

The natural disease in laboratory mice caused by *Streptobacillus moniliformis* was described by Levaditi, Selbie and Schoen (1932) and Mackie, van Rooyen and Gilroy (1933), and in wild mice by Williams (1941). It may occur in either sporadic or epidemic form, and may be of an acute septicaemic or a more chronic polyarticular type. The mortality is high. Infection probably results either through bites, or through contamination with urine, which may contain the organism. In acute cases the animal appears ill and has a lustreless coat. The eyes may be glued together by a semipurulent conjunctival discharge. Death occurs in a few days, and nothing characteristic is found at necropsy. The subacute or chronic type is characterized by polyarthritis and myocarditis, the former evidenced by swelling of the joints, the latter by cyanosis of the tail and extremities. One of the commonest manifestations is oedematous swelling of the feet and legs, sometimes confined to one or both hind limbs. Ulceration of the feet may occur, but gangrene is rarely observed. Paralysis of the hind limbs, enlargement of lymphatic glands, subcutaneous swellings, and keratitis are other manifestations that are not infrequently encountered. The conjunctivitis may suggest infection with *Erysipelothrix muriseptica*, the swelling and ulceration of the feet ectromelia. Diagnosis is made by culture. The organism can often be isolated from the blood, even in chronic cases, and from the spleen, liver, glands and joints. Serum agar or Loeffler's serum should be used. Experimental reproduction of the disease in normal mice is readily accomplished by inoculation of pure cultures (see Chapter 22).

Tyzzer's disease

For an account of this disease in mice, see Chapter 22.

References

Albiston, H. E. and Pullar, E. M. (1934) *Aust. vet. J.* **10,** 146.
Allbritten, F. F., Sheely, R. F. and Jeffers, W. A. (1940) *J. Amer. med. Ass.* **114,** 2360.
Baker, J. R. (1972) *Vet. Rec.* **90,** 630.
Bayne-Jones, S. (1925) *J. Bact.* **10,** 569.
Biggart, J. H. (1934) *Bull. Johns Hopk. Hosp.* **54,** 165.
Blake, F. C. (1916) *J. exp. Med.* **23,** 39.
Blank, C. H. and Georg, L. K. (1968) *J. Lab. clin. Med.* **71,** 283.
Bollinger. (1877) *Zbl. med. Wiss.* **15,** 481.
Bostroem, E. (1891) *Beitr. path. Anat.* **9,** 1.
Bosworth, T. J. (1923) *J. comp. Path.* **36,** 1.
Brede, H. D. (1959) *Zbl. Bakt.* **174,** 110.
Brion, A. (1939) *Rev. Méd. vét.* **91,** 121.
Brion, A., Goret, P. and Joubert, L. (1952) *vi Congr. int. Patol. comp., Madrid,* **i,** 47.
Bronner, M. and Bronner, M. (1971) *Actinomycosis,* 2nd ed. John Wright & Sons Ltd, Bristol.
Brown, T. M. and Nunemaker, J. C. (1942) *Bull. Johns Hopk. Hosp.* **70,** 201.
Buchanan, B. B. and Pine, L. (1962) *J. gen. Microbiol.* **28,** 305.
Cohn, T. (1913) *Zbl. Bakt.* **70,** 290.
Colebrook, L. (1920) *Brit. J. exp. Path.* **1,** 197; (1921) *Lancet* **i,** 893.
Cope, Z. (1938) *Actinomycosis.* Oxford University Press.
Curry, W. A. (1980) *Arch. intern. Med.* **140,** 818.
Daubney, R. (1927) *J. comp. Path.* **40,** 195.
Dean, G. (1900) *Trans. path. Soc., Lond.* **51,** 26.
Drake, C. H. and Henrici, A. T. (1943) *Amer. Rev. Tuberc.* **48,** 184.
Edwards, P. R. (1931) *Kentucky agric. Exp. Sta. Bull.* No. 320; (1932) *J. infect. Dis.* **51,** 268.
Emmons, C. W. (1938) *Publ. Hlth Rep., Wash.* **53,** 1967.
Eppinger, H. (1891) *Beitr. path. Anat.* **9,** 287.
Evans, A. C. (1918) *J. infect. Dis.* **23,** 373.
Flamm, H. and Wiedermann, G. (1962) *Z. Hyg. InfektKr.* **148,** 368.
Fortner, J. (1928) *Zbl. Bakt.* **108,** 155.
Garrod, L. P. (1952) *Brit. med. J.* **i,** 1263.
Georg, L. K., Robertstad, G. W., Brinkman, S. A. and Hicklin, M. D. (1965) *J. infect. Dis.* **115,** 88.
Gerbasi, M. (1927) *Zbl. Bakt.* **101,** 369.
Ginsberg, A. and Little, A. C. W. (1948) *J. Path. Bact.* **60,** 563.
Gledhill, A. W. (1967) *Lab. Anim.* **1,** 73.
Goldsworthy, N. E. (1937) *J. Path. Bact.* **45,** 17; (1938) *Ibid.* **46,** 207.
González Ochoa, A. (1962) *Lab. Invest.* **11,** 1118.

Gordon, M. H. (1920) *Brit. med. J.* **i,** 435.
Griffith, F. (1916) *J. Hyg., Camb.* **15,** 195.
Grüner, O. P. N. (1969) *Acta path. microbiol. scand.* **76,** 239.
Gunst, J. A. (1927) *Tijdschr. Diergeneesk.* **54,** 497, 552.
Harbitz, F. and Gröndahl, N. B. (1911) *Beitr. path. Anat.* **50,** 193.
Harvey, J. C., Cantrell, J. R. and Fisher, A. M. (1957) *Ann. intern. Med.* **46,** 868.
Heinrich, S. (1960) *Zbl. Bakt.* **177,** 255.
Heinrich, S. and Korth, H. (1967) In: *Krankheiten durch Aktinomyzeten und verwandte Erreger,* p. 16. Ed. by H. J. Heite. Springer-Verlag, Berlin.
Heinrich, S. and Pulverer, G. (1959) *Zbl. Bakt.* **176,** 91.
Heite, H. J. (1967) *Krankheiten durch Aktinomyzeten und verwandte Erreger.* Springer-Verlag, Berlin.
Henry, H. (1910) *J. Path. Bact.* **14,** 164.
Hoeden, J. van der (1939) *Tijdschr. Diergeneesk.* **66,** 53.
Holden, F. A. and MacKay, J. C. (1964) *Canad. med. Ass. J.* **91,** 78.
Holm, P. (1950) *Acta path. microbiol. scand.* **27,** 736.
Holm, P. and Kwapinski, J. B. (1959) *Acta path. microbiol. scand.* **45,** 107.
Hülphers (1923). *See* Magnusson (1928).
Hutyra, F. and Marek, J. (1926) *Special Pathology and Therapeutics of the Diseases of Domestic Animals.* **1.** Chicago and London.
Israël, J. (1878) *Virchows Arch.* **74,** 15; (1879) *Ibid.* **78,** 421.
Jones, F. S. (1922) *J. exp. Med.* **35,** 361.
Kanthack, A. A. (1893) *J. Path. Bact.* **1,** 140.
King, J. W. and White, M. C. (1981) *Arch. intern. Med.* **141,** 1234.
Klinger, R. (1912) *Zbl. Bakt.* **62,** 191.
Kornman, K. S. and Loesche, W. J. (1978) *J. clin. Microbiol.* **7,** 514.
Kubo, M., Osada, M. and Konno, S. (1980) *Nat. Inst. Anim. Hlth Q.* **20,** 53.
Lambert, F. W., Brown, J. M. and Georg, L. K. (1967) *J. Bact.* **94,** 1287.
Laveran, (1906) *C. R. Soc. Biol.* **61,** 340.
Lentze, F. (1967) In: *Krankheiten durch Aktinomyzeten und verwandter Erreger,* p. 1. Ed. by H. J. Heite. Springer-Verlag, Berlin.
Levaditi, C., Nicolau, S. and Poincloux, P. (1925) *C. R. Acad. Sci.* **180,** 1188.
Levaditi, C., Selbie, F. R. and Schoen, R. (1932) *Ann. Inst. Pasteur* **48,** 308.
Lignières, J. and Spitz, G. (1902) *Bull. Soc. Méd. vét.* **20,** 487, 546.
Lord, F. T. and Trevett, L. D. (1936) *J. infect. Dis.* **58,** 115.
Lynch, J. P. and Holt, R. A. (1945) *Ann. intern. Med.* **23,** 91.
McCallum, W. G. (1902) *Zbl. Bakt.* **31,** 529.
MacFadden, A. W. J. (1913) *Rep. med. Offr Hlth Port Lond.* Nov. 13th.
Mackenzie, D. W. R., Philpot, C. M. and Proctor, A. G. J. (1980) *Publ. Hlth Lab. Service, Monograph Series* No. 12.
Mackie, T. J., Rooyen, C. E. van and Gilroy, E. (1933) *Brit. J. exp. Path.* **14,** 132.
Magnusson, H. (1928) *Acta path. microbiol. scand.* **5,** 170.
Magrou, J. (1919) *Ann. Inst. Pasteur* **33,** 344.
Maguire, L. C. (1958) *Vet. Rec.* **70,** 989.
Marsh, H. and Wilkins, H. W. (1939) *J. Amer. vet. med. Ass.* **94,** 363.
Marucha, P. T., Keyes, P. H., Wittenberger, C. L. and London, J. (1978) *Infect. Immun.* **21,** 786.

Murray, J. F., Finegold, S. M., Froman, S. and Will, D. W. (1961) *Amer. Rev. resp. Dis.* **83**, 315.

Naeslund, C. (1925) *Acta path. microbiol. scand.* **2**, 110, 244; (1926) *Ibid.* **3**, 637; (1929) *Ibid.* **6**, 66, 78.

Namyslowski, B. (1912) *Zbl. Bakt.* **62**, 562.

Negroni, P. (1937) *Folia biol.* Nos. 79–82, p. 344.

Nichols, D. R. and Herrell, W. E. (1948) *J. Lab. clin. Med.* **33**, 521.

Nocard, E. (1888) *Ann. Inst. Pasteur* **2**, 293; (1902) *Bull. Soc. Méd. vét.* **20**, 695.

Parker, F. and Hudson, N. P. (1926) *Amer. J. Path.* **2**, 357.

Phillips, J. E. (1961) *J. Path. Bact.* **82**, 205; (1964) *Ibid.* **87**, 442; (1965) *Ibid.* **90**, 557.

Pier, A. C. and Enright, J. B. (1962) *Amer. J. vet. Res.* **23**, 284.

Pier, A. C., Mejia, M. J. and Willers, E. H. (1961) *Amer. J. vet. Res.* **22**, 502.

Pine, L. and Hardin, H. (1959) *J. Bact.* **78**, 164.

Pine, L. and Overman, J. R. (1963) *J. gen. Microbiol.* **32**, 209.

Place, E. H. and Sutton, L. E. (1934) *Arch. intern. Med.* **54**, 659.

Porter, I. A. (1953) *Brit. med. J.* **ii**, 1084.

Prévot, A. R., Goret, P., Joubert, L., Tardieux, P. and Aladame, N. (1951) *Ann. Inst. Pasteur* **81**, 85.

Ravaglia, F. (1934) *Nuova Vet.* **12**, 20.

Report (1913) *Rep. med. Offr. Hlth Port Lond.* Nov. 13th; (1949) 30th *Ann. Rep. Animal Hlth Trust* p. 27.

Robinson, R. A. (1944) *J. Amer. med. Ass.* **124**, 1045.

Rosebury, T., Epps, L. J. and Clarke, A. R. (1944) *J. infect. Dis.* **74**, 131.

Rosett, W. and Hodges, G. R. (1978) *Amer. J. med. Sci.* **276**, 279.

Roughgarden, J. W. (1965) *Arch. intern. Med.* **116**, 39.

Schottmüller, H. (1914) *Derm. Wschr.* **58**, Suppl. p. 77.

Silberschmidt. (1899) *Ann. Inst. Pasteur* **13**, 841.

Slack, J. (1942) *J. Bact.* **43**, 193.

Smith, T. (1918) *J. exp. Med.* **28**, 333; (1921) *Ibid.* **33**, 441.

Söderlund, G. (1921) *Acta chir. scand.* **53**, 189.

Strangeways, W. I. (1933) *J. Path. Bact.* **37**, 45.

Sullivan, H. R. and Goldsworthy, N. E. (1940) *J. Path. Bact.* **51**, 253.

Thiroux, A. and Pelletier, J. (1912) *Bull. Soc. Path. exot.* **5**, 585.

Thomas, A. D. (1931) *17th Rep. Director vet. Serv. Anim. Industr., Onderstepoort* Pt. 1, p. 215.

Tileston, W. (1916) *J. Amer. med. Ass.* **66**, 995.

Tribedi, B. P. and Mukherjee, B. N. (1939) *Brit. J. Surg.* **27**, 256.

Tunnicliff, E. A. (1941) *J. infect. Dis.* **69**, 52.

Tunnicliff, R. (1916) *J. infect. Dis.* **19**, 767.

Vandepitte, J., Geest, H. de and Jousten, P. (1977) *J. clin. Path.* **30**, 842.

Vincent, H. (1894) *Ann. Inst. Pasteur* **8**, 129.

Weed, L. A., Andersen, H. A., Good, C. A. and Baggenstoss, A. H. (1955) *New Engl. J. Med.* **253**, 1137.

Weese, W. C. (1975) *Arch. intern. Med.* **135**, 1562.

Williams, S. (1941) *Med. J. Aust.* **28**, 357.

Wolff, M. and Israël, J. (1891) *Virchows Arch.* **126**, 11.

Wright, J. H. (1898) *J. exp. Med.* **3**, 421; (1905) *J. med. Res.* **13**, 349.

50

Erysipelothrix and listeria infections

Geoffrey Smith and Graham Wilson

Infections caused by *Erysipelothrix rhusiopathiae*

Swine erysipelas

Swine erysipelas is caused by the slender gram-positive bacillus, *Erysipelothrix rhusiopathiae*. Adolescent animals seem to be most frequently affected, but it occurs at all ages except during the first 3 months of life, when it is uncommon. The finer breeds of pigs are particularly susceptible. Four clinical types of the disease are described (Craig 1926). (1) The acute or septicaemic form. After an incubation period of 1 to 5 days, illness commences with prostration and high fever, anorexia, thirst, occasional vomiting, and conjunctivitis. Twenty-four hours later bright or dark red patches appear on the skin over the ears, snout, neck, abdomen, and inner sides of the forelegs. The case fatality is about 80 per cent, death occurring usually in 3 or 4 days. (2) The urticarial form, or 'the diamonds'. This is a mild form accompanied by slight malaise and fever, and characterized by the eruption on the 2nd or 3rd day of well defined, quadrangular or rhombic patches on the skin of the sides, back, and buttocks; the patches are slightly swollen, deep red or violet in colour, sometimes with a pale centre, from 1–5 cm in diameter, and may become covered with dry crusts which are afterwards cast off. Recovery occurs in a few days; death is unusual. (3) The chronic or cardiac form. This may follow either of the previous types, or may arise independently. Warty vegetations develop on one of the heart valves, generally the mitral. Death may occur suddenly, or the animals may live for several weeks with symptoms of cardiac insufficiency. (4) The joint or arthritic form. This may follow the urticarial form, accompany the cardiac form, or arise independently. The joints of the limbs become enlarged, owing to a synovitis; the animals are stunted, prefer the recumbent posture, and when induced to get up, walk in a stiff fashion on their toes with the back arched. This form is not usually fatal, but interferes with growth and fattening.

Morbid anatomy

In animals dying of the septicaemic form the spleen is enlarged, the lymphatic glands are congested, the gastric and intestinal mucosae are reddened, and the lungs are oedematous; there are sometimes small haemorrhages on the serous and mucous membranes and beneath the endocardium. The bacilli are found in small numbers in the blood, spleen, kidneys, and other organs, and in the secretions and excretions. In the cardiac form they are chiefly found in, and may be confined to, the vegetations on the valves. *E. rhusiopathiae* can sometimes be isolated from cases of polyarthritis in which the joints have been affected for a year or more (Hughes 1955, Shuman 1959); in other cases no micro-organisms can be isolated (see Lamont 1979). Erysipelothrix arthritis of pigs is of interest in relation to human rheumatoid arthritis (see Ajmal 1970).

Epidemiology

Swine erysipelas is mainly sporadic in Britain, but in many countries it causes epidemics and serious economic loss. It occurs particularly in valleys and low-lying areas, and during the summer and autumn months.

The natural disease of pigs apparently results from ingestion of infected material, though infection by the skin appears possible. The organisms invade the blood stream and are excreted mainly in the urine and faeces—often before clinical evidence of disease is apparent (Wood 1967). They have been found in the gall bladder of animals that have recovered from the mild type of the disease (Pitt 1908). They may become localized in the tonsils and, when there is an accompanying bacteraemia, they may be persistently shed in the urine and faeces (Wood 1967). Wellmann (1955) believes that subclinical infections leading to immunity are not infrequent and play a large part in the epidemiology of the disease.

The bacilli are very resistant to the changes associated with putrefaction, and may remain alive for months in carcasses of buried animals (Lösener 1896). There is no convincing evidence, however, that they are capable of a saprophytic existence in soil (Doyle 1960, Lamont 1979). In experiments with soil, Wood (1973) found that the organisms died out at a logarithmic rate under all conditions tested; low temperature tended to prolong survival. It is possible that infection occurs by means of food and water contaminated with soil, but the disease is transmitted mainly by carrier pigs (Doyle 1960). *E. rhusiopathiae* has often been isolated from the tonsils and faeces of apparently healthy pigs (Doyle 1960, Lamont 1979); the strains isolated belong to numerous serotypes (Kucsera and Gimesi 1976, Wood and Harrington 1978) and are often pathogenic for mice, pigeons, or pigs (Damme and Devriese 1976, Kucsera and Gimesi 1976, Wood and Harrington 1978). Other species such as sheep, rats, chickens and turkeys can also act as carriers.

An analysis (Lamont 1979) of published information on serotyping (serotypes N and 1–16) allowed several tentative conclusions: serotypes 1 and 2 predominated among strains from diseased pigs, serotype 2 being an especially common cause of arthritis; a wide range of serotypes occurred among

isolates from healthy pigs and fish body surfaces, though serotype 2 predominated. Szent Iványi (1952) and Heuner (1958) found serotype 1 to be more common than serotype 2. There are at least 22 serotypes (Nørrung 1979).

The disease can be experimentally reproduced in swine by inoculation with pure cultures (see Chapter 23).

Diagnosis

The bacteriological diagnosis of swine erysipelas is made by microscopic examination of the blood during life, and of the blood and viscera after death, for the characteristic slender gram-positive bacilli. If the organisms are too few to be seen microscopically, they may be cultivated on agar, in broth, and in gelatin stabs. When contaminated or putrefying material alone is available for examination, the mouse inoculation test should be performed (see Chapter 23), or the material may be plated on 5 per cent blood agar at pH 6.8 containing sodium azide 0.1 per cent and crystal violet 0.001 per cent (Packer 1943). From acute cases smooth colonies alone develop, from chronic cases a mixture of rough and smooth (Redlich 1932). The agglutination test may be of value, particularly in animals showing evidence of joint involvement. The antigen should be prepared from a smooth strain (see Schoening and Creech 1936), and may be stabilized with 0.0015 per cent of thymol in saline (Sikes and Tumlin 1967). A titre of 1 in 40 after the fifth day is indicative of infection.

Other serological methods include a growth-inhibition test (Wellman 1955), agglutination of sheep erythrocytes sensitized with bacterial antigens (Poole and Counter 1973), and a test for neuraminidase-neutralizing antibodies (Müller and Seidler 1975). During life examination of the blood reveals a leucopaenia, an increase in the mononuclear cells, and the presence of many atypical large mononuclear and lymphocytoid cells (Egehoj 1938). An extract, called Rusiopatin L, prepared by alkaline hydrolysis of the bacterial bodies, is said to be of diagnostic value as a skin test in chronic cases (Parnas 1957).

Prophylaxis and treatment

General prophylactic methods are similar to those for anthrax (see Chapter 54, and Doyle 1960).

Pasteur and Thuillier (Pasteur 1882, Pasteur and Thuillier 1883) introduced vaccine treatment. Their first vaccine consisted of an attenuated rabbit passage strain and their second, given 12 days after the first, of a hypervirulent pigeon passage strain. This method caused losses of up to 5 or 10 per cent in the finer breeds of pigs. To overcome this disadvantage, Lorenz (1893, 1894, 1896) used sero-vaccination, in which a dose of antiserum was injected followed after about 4 and 14 days by injections of living virulent bacilli. The objection to this method is that some of the inoculated animals may remain healthy carriers and spread infection to other animals. Live attenuated vaccines (see Gledhill 1959, Shuman and Wood 1970) are in use in Eastern Europe and Japan.

The method now commonly used is based on the work of Traub (1947) in Germany, who confirmed the observation of a Hungarian worker, Birô, that some strains of *E. rhusiopathiae* form a soluble protective substance in broth culture. This can be adsorbed on aluminium hydroxide and combined with the formolized bacteria. The resultant vaccine, when properly prepared (see Eissner 1952), gives very satisfactory results in the field. One dose is said to protect pigs for 4-6 months, two doses for 8-12 months (Flückiger 1952). Experimental evidence suggests that some protection is given against heterologous strains as well as against the homologous strain (Gledhill 1952, Shuman *et al.* 1965).

An injection of 10-30 ml of antiserum was formerly recommended for affected and in-contact pigs. Serum treatment has been replaced by the use of antibiotics such as penicillin.

Infections in sheep and other animals, including birds

In sheep, polyarthritis (Poels 1913, Cornell and Glover 1925, Marsh 1931, Stamp and Watt 1947, Lamont 1979) is the most common manifestation of *E. rhusiopathiae* infection. It occurs in many countries, in lambs aged several days to several months. The serotypes responsible are often those that occur in porcine infections (Marsh 1933a, Lamont 1979). The epidemiology is poorly understood. According to Marsh (1933b) the umbilicus and docking and castration wounds are sometimes the routes of entry. Septicaemia in sheep has been described (Christiansen 1919, Cornell and Glover 1925); ulcerative endocarditis, though known (Stamp and Watt 1947, Chineme *et al.* 1973), is rare. Cutaneous infection (Whitten *et al.* 1948) occasionally affects the feet of sheep that have recently been dipped. Infections in cattle and horses are known. Goudswaard and co-workers (1973) reported endocarditis and arthritis in dogs; the strain of *E. rhusiopathiae* isolated differed from strains found in pigs. Natural infections of mice were originally described by Koch (1880) as *mouse septicaemia*.

Septicaemic infection in many avian species has been reported (Beaudette and Hudson 1936, Jensen and Cotter 1976). Such infection in turkey poults, especially males, is well known; other birds affected include ducks, chickens, pheasants, peacocks, and parrots. The serotypes isolated from domestic poultry include serotypes 1 and 2 (Murase *et al.* 1959b). Jensen and Cotter (1976) described a large outbreak in eared grebes on the Great Salt Lake.

(For reviews of *E. rhusiopathiae* infection in animals see Gledhill 1959, Doyle 1960, Eissner and Ewald 1973, Lamont 1979.)

Erysipeloid and other human infections

Infection of man with the bacillus of swine erysipelas was first described by Rosenbach in 1884 (see Rosenbach 1909). Though, according to Verge (1933), a cutaneous, an intestinal, and a generalized form of the disease may occur, the first type, usually referred to as 'erysipeloid', is undoubtedly the commonest. It is met with in cooks, kitchen workers, butchers, and in those who handle fish, game, or cheese (Klauder 1926, 1932, 1938, King 1946). Epidemics of the disease have been reported. Stefansky and Grünfeld (1930) described an outbreak at Odessa affecting about 200 persons engaged in handling freshwater fish brought by sailing ships from the Dnieper and Bug; and Lawson and Stinnett (1933) in the United States reported 247 cases of erysipeloid occurring among workers employed in sawing and polishing bones in the manufacture of buttons.

Klauder (1932) says that infection can invariably be traced to contact with animals, fish, shell-fish, dead matter of plant or animal origin, or matter derived from animals such as hides, pelts, bones, and manure. The organism usually gains access through injuries of the skin, including the bites of animals, fish, and crustacea. Where the organism comes from, however, is not so clear. Direct transmission to man from swine suffering from erysipelas appears to be uncommon (Bierbaum and Gottron 1929), but it is sometimes met with in veterinary students (Morrill 1939). The organism may be present in the slimy coating of various fish (Schoop 1936), but the fish probably become contaminated after being caught (Murase *et al.* 1959a, Schoop and Stoll 1966, Shewan 1971). The bacilli are very resistant to salting and putrefaction, and survive for a long time outside the body.

However the organism reaches the skin, it gives rise, after an incubation period of 1-4 days, to an erysipeloid lesion characterized by a sharply defined, slightly elevated, purplish-red zone, which extends peripherally at the same time as it fades and desquamates centrally. The rash is accompanied by considerable swelling, itching, burning, and pain. Regional lymphangitis and lymphadenitis may develop. Arthritis of the finger joints is not unusual. Relapses are common, and second attacks may occur. The milder forms last 2-4 weeks, but the disease may continue for months. Death rarely occurs, though Russell and Lamb (1940) have described one fatal case of endocarditis in which the organisms were isolated from the blood six times during life and from the endocardial vegetations at autopsy (see also Klauder *et al.* 1943, McCarty and Bornstein 1960). A case of cranial abscess in a veterinarian was reported by Thjötta (1943).

The bacilli are best cultivated by the inoculation of a piece of excised skin into 1 per cent glucose broth (Barber *et al.* 1946, Sneath *et al.* 1951). Alternatively, tissue fluid may be obtained for culture by injecting saline into the skin at the border of the lesion and re-aspirating it without withdrawal of the needle.

Most cases respond favourably to antibiotics such as penicillin.

Infections caused by *Listeria monocytogenes*

Listeria monocytogenes gives rise to disease in varying forms in man and in a wide range of animals throughout the world. It has been isolated from more than 50 species of wild and domestic animals including birds, fish, crustaceans, and ticks (Murray 1955, Jones and Woodbine 1961, Gray 1962, Gray and Killinger 1966, Seeliger and Finger 1976). The organism is suspected of being responsible at times for widespread disease in wild animals, such as the periodic epidemics that destroy the lemmings in Northern Canada (Gray 1958*a*) and the plague-like disease of gerbilles in South Africa (Pirie 1927). Seeliger and Finger (1976) refer to more than 5000 proved cases of listeriosis in man. The very young are particularly susceptible; old age, neoplastic diseases, and treatment with corticosteroids or cytotoxic substances may act as predisposing factors. In all animals pregnancy seems to predispose to disease, leading often to metritis with death and premature expulsion of the fetus. Monocytosis, which is a characteristic feature in the rabbit, also occurs in rodents and man, but not in other species.

L. monocytogenes has frequently been isolated from environmental sources such as silage (Gray 1960), soil, plants, mouldy fodder, wildlife feeding grounds, sewage effluent, and broiler chicken carcasses (Seeliger *et al.* 1965, Kampelmacher and van Noorle Jansen 1975, Weis and Seeliger 1975, Gitter 1976). It also occurs in the faeces of human beings (Kwantes and Isaac 1971, Ortel 1971), regardless of any association with animals (Kampelmacher *et al.* 1976), and in the intestinal contents of chickens (Dijkstra 1978), pigs, and other species. Seeliger and his colleagues (1965) regard the organism primarily as a ground saprophyte, multiplying in soil and infecting animals and man by means of food contaminated with soil. Listeriosis is no longer thought to be principally a zoonosis; transmission from animals to man probably occurs only rarely. The epidemiology is poorly understood. *L. monocytogenes* can sometimes be isolated from the mesenteric lymph nodes of apparently healthy animals (Höhne *et al.* 1975, Hyslop 1975).

There are at least six serotypes of *L. monocytogenes* and a number of subtypes (Seeliger 1975). According to Seeliger and Finger (1976) serotypes 1/2a and 4b account for 97 per cent of the strains from human and animal disease. These serotypes are sometimes isolated from faecal and environ-mental sources—sources that also provide most of the isolates of serotypes other than 1/2a, 4a and 4b. Environmental sources sometimes yield haemolytic strains; such strains are non-pathogenic for experimental animals (see Chapter 23) and are not known to be associated with disease.

Disease in animals

Rabbits

In 1926 Murray, Webb and Swann at Cambridge described a natural disease of rabbits characterized by a large-mononuclear leucocytosis, and caused by a small gram-positive bacillus to which they gave the non-committal name *Bact. monocytogenes*. This organism is now known as *Listeria monocytogenes* and the disease as *infective mononucleosis*.

Young rabbits, between 1 and 3 months old, are chiefly affected. They are undersized, usually develop a very distended belly, undergo progressive emaciation, and often die suddenly. At post-mortem examination the subcutaneous tissue is found to be oedematous, looking like jelly. Clear serous fluid is abundant in the pericardium, pleura, and peritoneum. As a rule the lungs are oedematous, and contain one or more red infarcts, the result of infected emboli. The liver is congested and sometimes contains small, round, pale-yellow or grey areas of focal necrosis, 1.5 mm or less in diameter. The spleen is usually small, shrunken, pale pink, and rather tough; but in acute cases, and in cases in which there is extensive necrosis of the liver, the spleen is normal in size or slightly enlarged, soft, and of a dark purple colour. The adrenals are sometimes soft and diffluent. The mesenteric lymph nodes are always enlarged and oedematous. Culture of the organism is difficult; it may be successful from the mesenteric nodes; the heart's blood is usually sterile. Triple phosphate crystals are present in the serous exudates.

The disease, which is not common under natural conditions, can be reproduced experimentally, and is often characterized by monocytosis and focal necroses of the liver (Chapter 23).

Cattle

Bovine encephalitis caused by *L. monocytogenes* was

recognized by Jones and Little (1934) in New Jersey. It is one of the most frequently diagnosed forms of listeria infection in animals and, in the United States of America, one of the greatest economic importance. The disease is usually subacute or chronic; recovery is not unusual, but the brain is left damaged and the animals do not thrive. Lesions occur mainly in the white matter. During the acute stage lassitude, anorexia, loss in general condition, fever, drooping of the head to one side, protrusion of the tongue, and discharge of an abundant viscous secretion from the nose and mouth are noticeable. Sometimes there are circling movements, bilateral conjunctivitis, or lesions of the skin. An apparent association between silage feeding and the occurrence of listeriosis was first mentioned by Olafson (1940). Gray (1960) and others have isolated *L. monocytogenes* from silage; high quality silage is probably less likely than poor quality silage to become contaminated (Gouet *et al.* 1977). Clinically the disease may be confused with rabies, Aujeszky's disease, or viral encephalitis.

Pregnant animals may suffer from metritis during the later months. The fetus is expelled, often in a macerated condition, or it may die after premature birth showing necrotic lesions in the viscera (for references see Young and Firehammer 1958). The cows themselves suffer only slightly, and their fertility is unaffected. The organisms not infrequently settle in the udder, cause mastitis of varying degrees of severity, and are excreted in the milk for weeks or months on end (Hyslop and Osborne 1959, Gitter *et al.* 1980). Young animals sometimes suffer from septicaemia associated with miliary necrotic lesions in the liver (see Jones and Woodbine 1961). *L. monocytogenes* has been isolated from cases of keratoconjunctivitis in cattle and sheep (Kummeneje and Mikkelsen 1975, Morgan 1977).

Sheep and goats

Listeria infection in sheep was recognized by Gill (1933, 1937) in New Zealand to be responsible for what was known as the *circling disease*. In the United States Jungherr (1937) met with a similar disease in Connecticut and Biester and Schwarte (1939) in Iowa. In Great Britain Paterson (1940) isolated *L. monocytogenes* from abortion in sheep, and Young (1956) from lambs affected with circling disease in the Shetlands. In young lambs septicaemia is common, but in older animals meningo-encephalitis or encephalomyelitis is seen. The disease runs an acute course usually proving fatal within a week. Clinically, fever, depression, and nutritional disturbances are followed by such symptoms and signs as torticollis, trismus, grinding of the teeth, nasal discharge, inco-ordination, drooping of the head to one side, circling movements, and ultimately pareses and collapse. At post-mortem examination focal infiltrations with mononuclear and polymorphonuclear cells, areas of necrosis, and perivascular cuffing are seen histologically. Goats suffer in much the same way and, like sheep, are liable to abort during pregnancy.

Swine

Listeria infection of swine was reported by Biester and Schwarte (1940) in the United States, and has since been observed in many countries. It usually takes the form of an acute disease characterized by meningoencephalitis. The symptoms are fever, inco-ordination, trembling, and a stilted gait, affecting particularly the forelegs. Some animals suffer from dyspnoea and cough. Abortion may occur during pregnancy. Clinical disease, however, in swine is not common. It may be added that Gray and Killinger (1966) regard septicaemia as the most frequent form of listeria infection in swine, and think that many of the cases of meningo-encephalitis are really due to hog cholera.

Rodents

Apart from rabbits, in which the disease has already been described, *L. monocytogenes* may affect guinea-pigs, mice, merions, ferrets, chinchillas, gerbilles, and various other rodents (see Paterson 1939, Seeliger 1958). The infection is usually generalized and accompanied by focal necroses in the viscera, particularly the liver and adrenals. Monocytosis may be present during life.

Fowls

Tenbroeck (1932) isolated the organism from a sporadic disease of chickens in New Jersey. Seastone (1935), also in the United States, and Paterson (1937) in England recorded similar findings. Kwantes and Isaac (1971) found it on the skin of 20 out of 35 chicken carcasses purchased for domestic use in Swansea; Gitter (1976) isolated it from 10 out of 68. Hens, ducks, geese, turkeys, and other birds may be affected. The disease occurs sporadically or epizoötically and may be acute or chronic. Healthy carriers are not infrequent. There is nothing characteristic about the symptomatology, but a striking feature at post-mortem examination is often the presence of multiple areas of degeneration in the heart muscle, sometimes accompanied by pericardial effusion, focal necroses in the liver and kidneys, and generalized oedema. Occasionally the central nervous system is affected.

Other animals

Encephalitis occasionally occurs in horses, and septicaemia in foals. Infection has also been reported in dogs, cats, lemmings, skunks, racoons, and numerous other animals (see Seeliger 1958). In lemmings no obvious lesions are said to be found after death (Murray 1955).

(For reviews of listeria infection in animals, see Seeliger and Linzenmeier 1953, Murray 1955, Roots and Strauch 1958, Gray 1958*a*, Jones and Woodbine 1961, Vanini and Moro 1964, Gray and Killinger 1966, Report 1966, 1975.)

Diagnosis in animals

Diagnosis of listeria infection during life is usually impossible, and serological findings are difficult to interpret. Even at necropsy the causative organism may be difficult to isolate. When the central nervous system is affected, the medulla should be macerated mechanically, suspended in broth, and plated on to tryptose agar. After 18–24 hours' incubation the plates should be examined with a binocular microscope under oblique illumination for the characteristic bluish-green colonies (see Chapter 23). If the first attempt is unsuccessful, the medullary suspension should be refrigerated for some days or weeks and recultured at intervals (Gray 1957, 1958*a*, 1962). In aborted fetuses the organisms can often be isolated from the stomach, liver, blood, and membranes after mechanical maceration. Cultivation from the necrotic lesions in the liver and spleen is recommended in fowls and rodents. Faeces may be homogenized in saline, the supernatant fluid injected into mice, and cultures made from the liver and spleen (Sandvik and Skogsholm 1962).

Alternatively successive dilutions may be made and plated on to tryptose agar or blood agar. For contaminated material Gray, Stafseth and Thorp (1950) recommend the use of broth containing 0.05 per cent potassium tellurite, followed by plating after 6–24 hours at 37° on tryptose agar. This method, though useful, cannot be relied on, because potassium tellurite inhibits many strains of *L. monocytogenes*. Beerens and Tahon-Castel (1966) advocate the use of a meat extract tryptose medium containing 0.004 per cent nalidixic acid to which horse blood is added when the plates are poured. This medium inhibits nearly all gram-negative bacteria except *Pseudomonas aeruginosa*. Its value is attested by Kampelmacher (1967). A similar effect is obtained by 3.75 per cent potassium thiocyanate (Lehnert 1960, Wood 1969). Kramer and Jones (1969) favour nutrient broth containing nalidixic acid, thallous acetate, and glucose. A mixture of thallous acetate and nalidixic acid is likewise favoured by Khan and his colleagues (1973) and by Leighton (1979); these two reagents in tryptose phosphate broth are said to inhibit the growth of both gram-positive and gram-negative organisms. (For other media see Durst and Berencsi 1975, Mavrothalassitis 1977.) Seeliger and his colleagues (1970), who contribute a very useful article on the technique of isolation, regard storage of the material for 26 weeks at 4° as the best method of enrichment. This should be followed by cultivation on tryptose agar, with or without addition of blood and nalidixic acid. Developing colonies should always be exam-

ined by Henry's method of oblique transmitted illumination (see Chapter 23) and tested serologically. Studies on isolation techniques are summarized by Ralovich (1975). Specific identification is aided by the use of fluorescent antibody staining (Nelson and Shelton 1963, Khan *et al.* 1977).

Disease in man

Since the original description of cases by Burn (1933–4, 1935, 1936) in the United States and by Gibson (1935) in Scotland, the ability of *L. monocytogenes* to cause disease in man has been increasingly recognized. Writing in 1966 Gray and Killinger said that well over 1000 confirmed cases had occurred throughout the world; but Seeliger, Emmerling and Emmerling (1968) had records of 2004 human cases occurring in Germany alone between 1950 and 1966. The organism seems to attack the weaklings, particularly newborn infants and elderly persons suffering from some underlying disease, and women under the stress of pregnancy (see Louria *et al.* 1967). How infection occurs is unknown. *L. monocytogenes* is widely distributed in soil, sludge fertilizer, silage, plants, slaughterhouse waste and stream water. Seeliger and his colleagues (1965) regard it as a ground saprophyte multiplying in soil, and suggest that man becomes infected through soil-contaminated food. This view has gained increasing acceptance. The majority of cases are sporadic. In an outbreak in Canada (Schlech *et al.* 1983) 41 cases were associated with the consumption of contaminated coleslaw. Larsson and co-workers (1978) reported the transfer of infection between infants in hospital. Listeriosis is occasionally transmitted from animals to man. Raw milk is a potential source of infection. Cutaneous infection has been reported in veterinary surgeons who have handled material from aborting cows (see Seeliger and Finger 1976). Infection from human carriers is regarded as improbable (Robin *et al.* 1966), though indirect infection seems possible. Bojsen-Møller (1972) brings evidence to suggest that infection occurs by the mouth, and that the organisms are excreted in the faeces, usually for only a few days. In Copenhagen he estimates that 1 per cent of the normal population and 4–5 per cent of slaughterers are excreters at any one time. Kwantes and Isaac (1971) isolated *L. monocytogenes* from 32 out of 5000 samples of faeces, and Ortel (1971) in Germany from the stools of 6 out of 125 adults and 4 out of 378 premature infants and children. Kampelmacher and his colleagues (1976) isolated *L. monocytogenes* from 8.8 per cent of 1337 samples from 207 pregnant women. Of 22 women from farms, 92 with pets, and 93 with no animal contact, 32, 39 and 34 per cent respectively gave positive results. The disease is protean in its manifestations. It may affect the throat, the lymphatic nodes, the central nervous system, the pregnant uterus, or give rise to a septicaemic disease resembling typhoid fever. Reports are at variance on the frequency with which these different forms occur. Many workers would give precedence to meningitis and encephalitis, but judging by Potel's (1958) experience disease of the pregnant woman or her infant is the commonest. On rare occasions *L. monocytogenes* has been associated with endocarditis and with enterocolitis in infants (Larsson *et al.* 1978). *Listerial endocarditis* presents the usual features of subacute infective endocarditis. It is said to be becoming more frequent. Cardiac abnormalities and prosthetic appliances seem to be predisposing causes.

Saxbe (1972) makes the interesting suggestion that Queen Anne suffered from this disease or, at least, was a carrier of the organism. Despite 17 pregnancies, including 11 abortions, 1 stillbirth, 2 neonatal deaths, and one child who suffered from meningitis, convulsions and obstructive hydrocephalus and died at 11 years of age, she left no surviving children.

The so-called anginose-septic form is sometimes accompanied by conjunctivitis or by swelling of the cervical and submandibular lymph nodes. It may be mistaken for glandular fever, and indeed *L. monocytogenes* has been cultivated from the blood of several cases of this disease (Nyfeldt 1929, 1958, Webb 1943). There is reason, however, to doubt whether it plays any significant part in the clinical picture. Infection of the central nervous system results in meningitis, or, much less often, encephalitis.

Meningitis is most frequent during the first three weeks of life and in persons over 40 years of age; in fact these two groups account for nearly 90 per cent of the cases (Gray and Killinger 1966). In Nieman and Lorber's (1980) series meningitis accounted for 55 per cent of the 186 adult cases reviewed. Most of these occurred in the 6th and 7th decades of life. Clinically, besides a high temperature and stiffness of the neck, listerial meningitis tends to be characterized by ataxia, tremors, seizures, and fluctuating consciousness. Onset is sudden, and death follows within 24 or 48 hours. Post-mortem examination reveals a suppurative leptomeningitis and ependymitis, purulent otitis media, focal necrosis and patchy degeneration of the liver, swollen and congested spleen, focal necrosis of the adrenals, atelectasis, focal pneumonia and bronchiolitis. In the much less common *encephalitic form* there is a sudden onset of hemiparesis or cranial nerve palsy. In infants, 16 per cent of cases of meningitis are said to be due to *Listeria*. Infection presumably occurs in the vaginal canal during birth.

Primary bacteraemia accounts for about a quarter of listerial infections in adult life. Pregnant and alcoholic patients are affected, as well as those suffering from malignant disease and those who are receiving immunosuppressive therapy. Most of the patients are under 50 years of age, and most respond well to treatment. In pregnancy some cases are so mild as to be almost inapparent; in others there are fever and rigors suggestive of pyelonephritis. Usually the child is born normally but when the metritis is severe it may be expelled prematurely or dead. The mother always recovers. The infant, on the other hand, when it is infected, invariably dies of septicaemia—*granulomatosis infantiseptica*. At post-mortem examination miliary granulomatous nodules are to be found in the spleen, liver, adrenals, lungs, the mucosa of the oesophagus, stomach, and small intestine, and in the throat and meninges. Reiss, Potel and Krebs (1951) first isolated *L. monocytogenes* from this disease; several other cases have since been described (Erdmann and Potel 1953, Hood 1961). The organisms are present in the meconium, the bile, and the viscera, and have been isolated from the blood of the fetus (Victoria *et al.* 1967). *L. monocytogenes* has been held responsible for some cases of repeated abortion. In these women it is said to have been isolated from the cervical secretion (Rappaport *et al.* 1960), though Rabau and David (1963) were unable to find it in any cases of abortion. By the immunofluorescence technique, however, it was found in 3 per cent of vaginal secretions from healthy pregnant women in New York State (Hays and Miller 1966).

Diagnosis in man

During life diagnosis is best made by culture of the organism from the blood, bone marrow, or cerebrospinal fluid. In meningeal infections there is a polymorphonuclear leucocytosis, and blood cultures are positive in about 75 per cent of cases. The protein in the spinal fluid is raised; the sugar content is only slightly lowered; polymorphonuclear cells abound, and are followed by a predominance of lymphocytes. A mononuclear leucocytosis is uncommon except in the glandular form of the disease. In cases occurring during pregnancy, the organism may be isolated from the blood and urine of the mother before delivery, from the placenta and lochia, and from the meconium of the infant. Material taken *post mortem* from the liver, spleen, lymph nodes, brain, or other organs should be examined by the methods already described on p. 28.

Serological examination is often helpful, but the findings have to be interpreted with great care and can seldom be conclusive. Partial antigens of *L. monocytogenes* are shared with staphylococci, *Streptococcus faecalis*, and some corynebacteria such as *C. pyogenes* (Seeliger 1953, 1955, 1958). H and O agglutinins should be titrated separately, and no significance should be attached to a titre under 1 in 320. A rising or falling titre is suggestive. The complement-fixation test should be used to confirm the agglutination results. In both tests absorption of low-titre sera with staphylococci adds to the significance of the results (Seeliger 1962). According to Schierz and Burger (1966) the specific antibody to *L. monocytogenes* is a protein. This can be adsorbed on to the tanned red cells of the sheep and demonstrated by haemagglutination. Castañeda's antigen-fixation test is recommended as a rapid screening procedure (Castañeda 1950, Seeliger *et al.* 1969). More recently, Delvallez and his colleagues (1979) have extracted and purified a surface soluble antigen that appears to be specific to the genus *Listeria*. It is not shared by any other bacterial species. The use of this antigen and of an antiserum prepared against it may well prove to be of value in the diagnosis of listerial infections.

Treatment

Penicillin and ampicillin given in large doses are the drugs of choice (Seeliger and Matheis 1969, Nieman and Lorber 1980). Next to them come the tetracyclines and erythromycin. The tetracyclines must be given with care, since they may induce teeth and bone injuries in infants and children.

(Reviews of human listeria infection will be found in Kaplan 1945, Gray 1958a, b, Potel 1958, Roots and Strauch 1958, Seeliger 1958, Trüb *et al.* 1963, Kampelmacher and van Noorle Jansen 1964, 1980, Vanini and Moro 1964, Gray and Killinger 1966, Robin *et al.* 1966, Louria *et al.* 1967, Bojsen-Møller 1972, Seeliger and Finger 1976, Robertson 1977, Nieman and Lorber 1980, and Report 1980; and of the disease in man and animals in Report 1966, 1975.)

References

Ajmal, M. (1970) *Vet. Bull.* **40**, 1.

Barber, M., Nellen, M. and Zoob, M. (1946) *Lancet* **i**, 125.

Beaudette, F. R. and Hudson, C. B. (1936) *J. Amer. vet. med. Ass.* **88**, 475.

Beerens, H. and Tahon-Castel, M. M. (1966) *Ann. Inst. Pasteur* **111**, 90.

Bierbaum, K. and Gottron, H. (1929) *Derm. Z.* **57**, 5.

Biester, H. E. and Schwarte, L. H. (1939) *J. infect. Dis.* **64**, 135; (1940) *J. Amer. vet. med. Ass.* **96**, 339.

Bojsen-Møller, J. (1972) *Acta path. microbiol scand., B*, Suppl. No. 229.

Burn, C. G. (1933–34) *Proc. Soc. exp., Biol., N.Y.* **31**, 1095; (1935) *J. Bact.* **30**, 573; (1936) *Amer. J. Path.* **12**, 341.

Castañeda, M. R. (1950) *Proc. Soc. exp. Biol., N.Y.* **73**, 46.

Chineme, C. N., Slaughter, L. J. and Highley, S. W. (1973) *J. Amer. vet. med. Ass.* **162**, 278.

Christiansen, M. (1919) *Maanedsskr. Dyrlaeg.* **31**, 242.

Cornell, R. L. and Glover, R. E. (1925) *Vet. Rec.* **5**, 833.

Craig, J. F. (1926) *Annu. Congr. nat. vet. Med. Ass. G.B.I.*, p. 163.

Damme, L. R. van and Devriese, L. A. (1976) *Zbl. VetMed.* **B23**, 74.

Delvallez, M., Carlier, Y., Bout, D., Capron A. and Martin, G. R. (1979) *Infect. Immun.* **25**, 971.

Dijkstra, R. G. (1978) *Tijdschr. Diergeneesk.* **103**, 229.

Doyle, T. M. (1960) *Vet. Rev. Annot.* **6**, 95.

Durst, J. and Berencsi, G. (1975) *Zbl. Bakt. I. Abt. Orig. A*, **232**, 410.

Egehoj, T. (1938) *Vet. Rec.* **50**, 60.

Eissner, G. (1952) *Mh. prakt. Tierheik.* **4**, 401.

Eissner, G. and Ewald, F. W. (1973) *Rotlauf.* VEB Gustav Fischer Verlag, Jena.

Erdmann, G. and Potel, J. (1953) *Z. Kinderheilk.* **73**, 113.

Flückiger, G. (1952) *Tierärztl. Umsch.* **7**, 84.

Gibson, H. J. (1935) *J. Path. Bact.* **41**, 239.

Gill, D. A. (1933) *Vet. J.* **89**, 258; (1937) *Aust. vet. J.* **13**, 46.

Gitter, M. (1976) *Vet. Rec.* **99**, 336.

Gitter, M., Bradley, R. and Blampied, P. H. (1980) *Vet. Rec.* **107**, 390.

Gledhill, A. W. (1952) *J. gen. Microbiol.* **7**, 179; (1959) *Infectious Diseases of Animals, Diseases due to Bacteria*, Vol. 2, p. 651. Ed. by A. W. Stableforth and I. A. Galloway, Butterworth Scientific Publications, London.

Goudswaard, J., Hartman, E. G., Janmaat, A. and Huisman, G. H. (1973) *Tijdschr. Diergeneesk.* **98**, 416.

Gouet, P., Girardeau, J. P. and Riou, Y. (1977) *Anim. Feed Sci. Technol.* **2**, 297.

Gray, M. L. (1957) *Zbl. Bakt.* **169**, 373; (1958a) Listeriosen. *Zbl. VetMed.*, Beiheft 1, p. 90. Paul Parey, Berlin; (1958b) *Agric. exp. Sta. Montana St. Coll.* Paper No. 423, Journal Series; (1960) *Science* **132**, 1767; (1962) *Ann N. Y. Acad. Sci.* **98**, 686.

Gray, M. L. and Killinger, A. H. (1966) *Bact. Rev.* **30**, 309.

Gray, M. L., Stafseth, H. J. and Thorp, F. (1950) *J. Bact.* **59**, 443.

Hays, R. C. and Miller, J. K. (1966) See *Report* 1966, p. 373.

Heuner, F. (1958) *Arch. exp. VetMed.* **12**, 40.

Höhne, K., Loose, B. and Seeliger, H. P. R. (1975) See *Report* 1975, p. 127.

Hood, M. (1961) *Pediatrics, Springfield* **27**, 390.

Hughes, D. L. (1955) *Brit. vet. J.* **111**, 183.

Hyslop, N. St G. (1975) See *Report* 1975, p. 94.

Hyslop, N. St G. and Osborne, A. D. (1959) *Vet. Rec.* **71**, 1082.

Jensen, W. I. and Cotter, S. E. (1976) *J. Wildl. Dis.* **12**, 583.

Jones, F. S. and Little, R. B. (1934) *Arch. Path.* **18**, 580.

Jones, S. M. and Woodbine, M. (1961) *Vet. Rev. Annot.* **7**, 39.

Jungherr, E. (1937) *J. Amer. vet. med. Ass.* **91**, 73.

Kampelmacher, E. H. (1967) *Lancet* **1**, 165.

Kampelmacher, E. H., Maas, D. E. and Noorle Jansen, L. M. van (1976) *Zbl. Bakt. I. Abt. Orig. A*, **234**, 238.

Kampelmacher, E. H. and Noorle Jansen, L. M. van (1964) *Ber. Rijks-Inst. Volgezondh Utrecht*, p. 148; (1975) See *Report* 1975, p. 66; (1980) *Zbl. Bakt. I. Abt. Orig. A*, **246**, 211.

Kaplan, M. M. (1945) *New Engl. J. Med.* **232**, 755.

Khan, M. A., Seaman, A. and Woodbine, M. (1973) *Zbl. Bakt. I. Abt. Orig. A*, **224**, 362; (1977) *Ibid.* **239**, 62.

King, P. F. (1946) *Lancet* **ii**, 196.

Klauder, J. V. (1926) *J. Amer. med. Ass.* **86**, 536; (1932) *J. industr. Hyg.* **14**, 222; (1938) *J. Amer. med. Ass.* **111**, 1345.

Klauder, J. V., Kramer, D. W. and Nicholas, L. (1943) *J. Amer. med. Ass.* **122**, 938.

Koch, R. (1880) *Investigations into the Etiology of Traumatic Infective Diseases.* New Sydenham Society, London.

Kramer, P. A. and Jones, D. (1969) *J. appl. Bact.* **32**, 381.

Kucsera, G. and Gimesi, A. (1976) *Magy. Állatorv. Lap.* **31**, 253.

Kummeneje, K. and Mikkelsen, T. (1975) *Nord. VetMed.* **27**, 144.

Kwantes, W. and Isaac, M. (1971) *Brit. med. J.* **iv**, 296.

Lamont, M. H. (1979) *Vet. Bull.* **49**, 735.

Larsson, S., Cederberg, A., Ivarsson, S., Svanberg, L. and Cronberg, S. (1978) *Brit. med. J.* **ii**, 473.

Lawson, G. B. and Stinnett, M. S. (1933) *Sth. med. J.* **26**, 1068.

Lehnert, C. (1960) *Zbl. Bakt.* **180,** 350.

Leighton, I. (1979) *Med. Lab. Sci.* **36,** 283.

Lorenz (1893) *Zbl. Bakt.* **13,** 357; (1894) *Ibid.* **15,** 278; (1896) *Ibid.* **20,** 792.

Lösener, W. (1896) *Arb. ReichsgesundhAmt.* **12,** 448.

Louria, D. B. *et al.* (1967) *Ann. intern. Med.* **67,** 261.

McCarty, D. and Bornstein, S. (1960) *Amer. J. clin. Path* **33,** 39.

Marsh, H. (1931) *J. Amer. vet. med. Ass.* **78,** 57; (1933a) *Ibid.* **82,** 584; (1933b) *Ibid.* **82,** 753.

Mavrothalassitis, P. (1977) *J. appl. Bact.* **43,** 47.

Morgan, J. H. (1977) *Vet. Rec.* **100,** 113.

Morrill, C. C. (1939) *J. infect. Dis.* **65,** 322.

Müller, H. E. and Seidler, D. (1975) *Zbl. Bakt.* I. Abt. Orig. A, **230,** 51.

Murase, N., Suzuki, K., Isayama, Y. and Murata, M. (1959a) *Jap. J. vet. Sci.* **21,** 215.

Murase, N., Suzuki, K. and Nakahara, T. (1959b) *Jap. J. vet. Sci.* **21,** 177.

Murray, E. G. D. (1955) *Canad. med. Ass. J.* **72,** 99.

Murray, E. G. D., Webb, R. A. and Swann, M. B. R. (1926) *J. Path. Bact.* **29,** 407.

Nelson, J. D. and Shelton, S. (1963) *J. Lab. clin. Med.* **62,** 935.

Nieman, R. E. and Lorber, B. (1980) *Rev. infect. Dis.* **2,** 207.

Nørrung, V. (1979) *Nord. VetMed.* **31,** 462.

Nyfeldt, A. (1929) *C. R. Soc. Biol.* **101,** 590; (1958) Listeriosen. *Zbl. VetMed.,* Beiheft 1, p. 86. Paul Parey, Berlin.

Olafson, P. (1940) *Cornell Vet.* **30,** 141.

Ortel, S. (1971) *Zbl. Bakt.* **217,** 41.

Packer, R. A. (1943) *J. Bact.* **46,** 343.

Parnas, J. (1957) *Z. ImmunForsch.* **114,** 186.

Pasteur, L. (1882) *C. R. Acad. Sci.* **95,** 1120.

Pasteur and Thuillier (1883) *C. R. Acad. Sci.* **97,** 1163.

Paterson, J. S. (1937) *Vet. Rec.* **49,** 1533; (1939) *Ibid.* **51,** 873; (1940) *Vet. J.* **96,** 327.

Pirie, J. H. H. (1927) *Publ. S. Afr. Inst. med. Res.* **3,** 163.

Pitt, W. (1908) *Zbl. Bakt.* **46,** 100.

Poels, J. (1913) *Folia microbiol., Delft* **2,** 1.

Poole, G. M. and Counter, F. T. (1973) *Appl. Microbiol.* **26,** 211.

Potel, J. (1958) Listeriosen. *Zbl. VetMed.,* Beiheft 1, p. 70. Paul Parey, Berlin.

Rabau, E. and David, A. (1963) *Lancet* **i,** 228.

Ralovich, B. (1975) See *Report* 1975, p. 286.

Rappaport, F., Rabinovitz, M., Toaff, R. and Krochik, N. (1960) *Lancet* **i,** 1273.

Redlich, E. (1932) *Z. InfektKr. Haustiere* **42,** 300.

Reiss, H. J., Potel, J. and Krebs, A. (1951) *Z. inn. Med.* **6,** 451.

Report (1966) *Proc. 3rd int. Sympos. Listeriosis, Bilthoven;* (1975) *Proc. 6th int. Sympos. Listeriosis, Nottingham Univ.* Leicester University Press; (1980) *Publ. Hlth Lab. Serv. commun. Dis. Rep.* No. 30, p. 4.

Robertson, M. H. (1977) *Post Grad. med. J.* **53,** 618.

Robin, L. A., Magard, H., Alves de Almeida E. and Chaudeur, E. (1966) *Rev. Hyg. Méd. soc.* **14,** 231.

Roots, E. and Strauch, D. (1958) Listeriosen. *Zbl. VetMed.,* Beiheft 1. Paul Parey, Berlin.

Rosenbach, F. J. (1909) *Z. Hyg. InfektKr.* **63,** 343.

Russell, W. O. and Lamb, M. E. (1940) *J. Amer. med. Ass.* **114,** 1045.

Sandvik, O. and Skogsholm, A. (1962) *Acta path. microbiol. scand.* **54,** 126.

Saxbe, W. B. (1972) *Pediatrics, Springfield* **49,** 97.

Schierz, G. and Burger, A. (1966) *Z. med. Mikrobiol.* **152,** 300.

Schlech, W. F. and 10 others (1983) *New Engl. J. Med.* **308,** 203.

Schoening, H. W. and Creech, G. T. (1936) *J. Amer. vet. med. Ass.* **88,** 310.

Schoop, G. (1936) *Dtsch. tierärztl. Wschr.* **44,** 371.

Schoop, G. and Stoll, L. (1966) *Z. med. Mikrobiol.* **152,** 188.

Seastone, C. V. (1935) *J. exp. Med.* **62,** 203.

Seeliger, H. P. R. (1953) *Z. ImmunForsch.* **110,** 252; (1955) *Z. Hyg. InfektKr.* **141,** 15; (1958) Listeriose. *Beit. Hyg. Epidem.,* Heft. 8, 2nd ed.; (1962) *Derm. trop.* **1,** No. 1; (1975) *Acta microbiol. hung.* **22,** 179.

Seeliger, H. P. R., Emmerling, P. and Emmerling, H. (1968) *Dtsch. med. Wschr.* **93,** 2037.

Seeliger, H. P. R. and Finger, H. (1976) In: *Infectious Diseases of the Fetus and Newborn Infant,* p. 333. Ed. by J. S. Remington and J. O. Klein. W. B. Saunders Co., Philadelphia, London, Toronto.

Seeliger, H. P. R., Finger, H. and Klütsch, J. (1969) *Zbl. Bakt.* **211,** 215.

Seeliger, H. and Linzenmeier, G. (1953) *Z. Hyg. InfektKr.* **136,** 335.

Seeliger, H. and Matheis, H. (1969) *Dtsch. med. Wschr.* **94,** 853.

Seeliger, H. P. R., Sander, F. and Bockemühl, J. (1970) *Z. med. Mikrobiol.* **155,** 352.

Seeliger, H. P. R., Winkhaus-Schindl, I., Andries, L. and Viebahn, A. (1965) *Path. et Microbiol., Basel* **28,** 590.

Shewan, J. M. (1971) *J. appl. Bact.* **34,** 299.

Shuman, R. D. (1959) *Lab. Invest.* **8,** 1416.

Shuman, R. D. and Wood, R. L. (1970) *Diseases of Swine,* 3rd edn, p. 508. Ed. by H. W. Dunne. State University Press, Iowa.

Shuman, R. D., Wood, R. L. and Cheville, N. F. (1965) *Cornell Vet.* **55,** 387.

Sikes, D. and Tumlin, T. J. (1967) *Amer. J. vet. Res.* **28,** 1177.

Sneath, P. H. A., Abbott, J. D. and Cunliffe, A. C. (1951) *Brit. med. J.* **ii,** 1063.

Stamp, J. T. and Watt, J. A. A. (1947) *Vet. Rec.* **59,** 30.

Stefansky, W. K. and Grünfeld, A. A. (1930) *Zbl. Bakt.* **117,** 376.

Szent Iványi, T. (1952) *Acta vet. hung.* **2,** 109.

Tenbroeck, C. (1932) *See* Seastone (1935).

Thjötta, T. (1943) *Acta path. microbiol. scand.* **20,** 597.

Traub, E. (1947) *Mh. VetMed.* **2,** 165.

Trüb, C. L. P., Boese, W. and Posch, J. (1963) *Arch. Hyg. Berl.* **147,** 495.

Vanini, G. C. and Moro, S. (1964) *G. Batt. Virol.* **57,** 87, 116, 129, 288, 298.

Verge, J. (1933) *Rev. gén. Méd. vét.* **42,** 65.

Victoria, R., Vanzzini, Q. F. B. V. and Miravete, A. P. (1967) *Rev. Inst. Salud. publ., Mexico* **27,** 129.

Webb, R. A. (1943) *Lancet* **ii,** 4.

Weis, J. and Seeliger, H. P. R. (1975) *Appl. Microbiol.* **30,** 29.

Wellmann, G. (1955) *Zbl. Bakt.* **162,** 265.

Whitten, L. K., Harbour, H. E. and Allen, W. S. (1948) *Aust. vet. J.* **24,** 157.

Wood, D. (1969) In: Shapton and Gould's *Isolation Methods for Microbiologists,* p. 63. Academic Press, London.

Wood, R. L. (1967) *Amer. J. vet. Res.* **28,** 925; (1973) *Cornell Vet.* **63,** 390.

Wood, R. L. and Harrington, R. (1978) *Amer. J. vet. Res.* **39,** 1833.

Young, S. (1956) *Vet. Rec.* **68,** 459.

Young, S. and Firehammer, B. D. (1958) *J. Amer. vet. med. Ass.* **132,** 434.

51

Tuberculosis

John M. Grange

Introductory

Tuberculosis is a disease of great antiquity and has almost certainly caused more suffering and death than any other bacterial infection. Even in 1980, despite the availability of highly effective chemotherapy, many millions of human beings became infected. In the past tuberculosis has been referred to as the 'great white scourge' and, by John Bunyan, as 'the captain of all of these men of death'. The clinical features of both pulmonary and spinal tuberculosis were well described by Hippocrates in about 400 B.C. (see Major 1945); accounts of the disease appeared in ancient Hindu texts (Petersen 1919) and it afflicted neolithic man (Bartels 1907).

The transmissible nature of tuberculosis was clearly established by Jean-Antoine Villemin, a French military doctor. In 1868 Villemin published the results of a series of studies in which he convincingly demonstrated that tuberculosis could be produced in rabbits by inoculating them with tuberculous material from man or cattle. The disease could be passaged from animal to animal and differences in virulence were observed between human and bovine material. In addition Villemin established that scrofula (tuberculous cervical adenitis) and pulmonary tuberculosis were different manifestations of the same disease.

Villemin's prediction that the causative agent of tuberculosis would be isolated was realized in 1882 when Robert Koch succeeded in culturing it on inspissated serum. By a large series of inoculations with pure cultures of the bacillus, several generations removed from the primary one, Koch transmitted the disease to many animals of different species. Koch's classical study established without doubt that the bacillus he isolated was the cause of tuberculosis. To this day its demonstration affords the sole infallible criterion for diagnosing tuberculosis in all of its diverse forms.

In addition to culturing the causative organism, Koch succeeded in staining it by treatment with an alkaline solution of methylene blue for 24 hours. Subsequently Ehrlich improved the technique by using a hot solution of the arylmethane dye fuchsin and it is this technique, slightly modified by Ziehl and Neelsen whose names it bears, that is still used today (for details see Chapter 24).

After Koch's discovery, acid-fast bacilli were isolated from cases of tuberculosis-like disease in various animals and were named after the host from which they were isolated. Five main types of 'tubercle bacilli' were recognized—human, bovine, vole, avian and 'cold-blooded'. The first three are properly regarded as types of the species *Mycobacterium tuberculosis*. The expression 'avian type' is synonymous with *Mycobacterium avium*. The 'cold-blooded' tubercle bacilli comprise two species of rapidly growing mycobacteria, namely *Mycobacterium fortuitum* (synonym *M. ranae*, the frog tubercle bacillus) and *Mycobacterium chelonei* (the turtle tubercle bacillus). Other acid-fast bacilli have been isolated from various environmental sources (see Chapter 24).

Tuberculosis is caused by *M. tuberculosis*. Infections resembling tuberculosis are occasionally caused by species of mycobacteria that are primarily environmental saprophytes. These species have collectively been termed atypical, anonymous, MOTT (Mycobacteria Other than Typical Tubercle), environmental, nyrocine, or tuberculoid bacilli.

The principal differences between *M. tuberculosis* and other culturable mycobacteria lie in their relation to disease in man. Being an obligate pathogen, *M. tuberculosis* is always transmitted from host to host although direct contact is not essential. In contrast human beings infected with the other mycobacteria rarely transmit the infection to others.

The human type of *M. tuberculosis* is transmitted almost exclusively in the sputum and cough spray of open cases of pulmonary tuberculosis and gains access to the body by inhalation of infective droplets. Consequently the initial lesion is usually in the lung, from which organisms may be disseminated to other organs by the lymphatic or blood stream. The bovine type, on the other hand, is often spread to man by the ingestion of infected raw milk or cream. The primary lesions are usually in the tonsil or the intestinal wall, leading to infection of the cervical and mesenteric lymph nodes respectively and of more distant organs. Cases of pulmonary tuberculosis due to the inhalation of the bovine type of bacilli expectorated by infectious cattle occasionally occur (Sigurdsson 1945) and this is the regular means of transmission of disease among the animals themselves (see p. 53). As recently as 1944, the bovine type was found to be responsible in Great Britain for 30 to 35 per cent of cases of non-pulmonary tuberculosis in persons under 15 years, and between 20 and 25 per cent of cases at all ages (Wilson *et al.* 1952). The disease has since been virtually eliminated by the eradication of infected cattle and the introduction of heat treatment (pasteurization) of milk.

The epidemiology of tuberculosis

Epidemiological studies on tuberculosis are concerned with the natural transmission of the disease in a community and the effect of control measures.

The prevalence of tuberculosis in a community

Before effective treatment became available the annual death-rate from tuberculosis was a good indicator of active disease (Yelton 1946); and post-mortem studies revealed the prevalence of latent tuberculosis in various age groups (Burkhardt 1906). Since the advent of effective chemotherapy, attention has turned to diagnosis by tuberculin testing, radiography and bacteriological studies. These methods are far from ideal.

The annual tuberculosis infection rate may be calculated from the results of tuberculin testing (see Styblo 1980). A simple method for estimating it in low-prevalence areas was described by Bleiker (1974). It is the best indicator for evaluating the extent of tuberculosis in the community and is, moreover, independent of the impact of control measures (Styblo 1978).

The dissemination of disease depends on the duration of the infectious stage. Studies carried out in the pre-chemotherapy era indicated that about 65 per cent of patients with open tuberculosis died within 4 years of diagnosis; and the average survival time of those not cured was only 14 months (Lindhardt 1939, Springett 1971). Even in the absence of chemotherapy many patients with open tuberculosis reverted to a quiescent state and appeared to have been cured. Studies by Springett (1971) and the National Tuberculosis Institute, Bangalore (Report 1974c) showed the proportion of such 'cured' cases to be 26 and 33 per cent respectively.

Natural trends in the incidence of tuberculosis

In the absence of specific tuberculosis control programmes, the incidence of the disease in a community may be affected by many factors, including the density of population, the extent of overcrowding and the general standard of living and health care. The natural trend must be considered when evaluating the impact of BCG vaccination, chemotherapy, and other specific anti-tuberculosis measures. In most European countries the decline in tuberculosis during the first half of this century was about 5 per cent annually. The introduction of mass BCG programmes had only a small effect, but the introduction of case-finding and effective treatment of infectious patients accelerated it by 7 to 8 per cent annually (Styblo and Meijer 1978). The situation in the developing countries is much more variable, but in most of them the annual risk of infection has remained constant or declined very slowly (Bleiker and Styblo 1978).

Transmission of infection

It is well established that patients with sputum which is positive on direct smear examination are the principal sources of infection (Grzybowski and Allen 1964, Rouillon *et al.* 1976). Smear-negative patients, whether culture-positive or not, are of very low infectivity. The risk depends greatly on the closeness of contact as well as the infectiousness of the source case. A study by van Geuns and his colleagues (1975) in Holland clearly demonstrates that the risk of infection is much higher among household contacts than among social or occupational contacts (Table 51.1).

On the other hand, the spread of the disease in the community may occur by minimal contact with a highly infectious person. Rao and his colleagues (1980) described how 187 of 3764 children aged between 8 and 11 became infected after contact with a diseased swimming-pool attendant. These authors stressed the importance of vigorous contact-tracing in the control of tuberculosis.

The incidence of tuberculosis in a community is dependent on the number of persons infected by each source case. Open cases of tuberculosis in America were estimated to infect 2 or 3 persons (Johnston and Wilbrik 1974) and in Holland between 7 and 10 persons (van Geuns *et al.* 1975, Styblo 1978). The infectious rate of untreated smear-positive cases was between 5 and 10 persons annually (Rouillon *et al.* 1976). One of the aims of control programmes is therefore to detect these smear-positive cases. These, however, often progress rapidly and regular radiographic ex-

Table 51.1   The risk of infection according to degree of contact with a source case

Contact	Risk of infection (per cent) in patients who were	
	smear +ve	smear −ve
Household contact	20.2	1.1
Near relative or friend	3.7	0.2
Contacts at work	0.3	−

aminations succeed in detecting only between 13 and 24 per cent (Toman 1976). Most infectious cases are not diagnosed until after symptoms have developed. The premature collapse of clinical interest and awareness in low-prevalence areas leads to serious delays in diagnosis (Toman 1976, Grange 1979).

Epidemiological significance of chemotherapy

Modern curative chemotherapy rapidly renders patients non-infectious even though they may continue to excrete culturable bacilli in their sputum (Gunnels *et al.* 1974, Riley and Moodie 1974). The exact interval between the commencement of chemotherapy and a loss of infectivity is difficult to estimate owing to the low infectivity rate and the impossibility of establishing the exact date of infection in a household contact. For practical purposes, however, the patient may be regarded as being non-infectious as soon as chemotherapy is started (see Rouillon *et al.* 1976).

The impact of chemotherapy on the incidence of infection depends on the effectiveness of the therapeutic regimen, the compliance of the patients, and the percentage of patients in whom a diagnosis is made and treatment begun (Bleiker 1974). Only curative chemotherapy reduces the incidence of chronic sputum-positive cases in the community; regimens in which the drugs or supervision are inadequate have no effective impact. In addition the average duration of infectivity before the commencement of treatment is of great relevance.

The exact extent of the contribution made by chemotherapy in reducing the incidence of tuberculosis is difficult to dissociate from the effects of improved standards of living, improved medical care, and BCG vaccination. It is, however, noteworthy that the annual decline in the incidence of infection rose from 5 per cent to 12 per cent in developing countries when chemotherapy was introduced, whether or not BCG was also used (Rouillon *et al.* 1976).

Endogenous and exogenous infection in post-primary tuberculosis

The widely held view that all cases of post-primary tuberculosis are due to the reactivation of a dormant primary focus of infection is based on the unjustified assumption that contact with *M. tuberculosis* induces a life-long absolute immunity to further infection. Certainly BCG vaccination, although not strictly analogous with infection by a virulent organism, induces an immunity that is neither absolute nor permanent. The incidence of tuberculosis in tuberculin-positive persons, i.e. those who have experienced previous contact with *M. tuberculosis*, is higher among those exposed to sources of infection than among those not so exposed, thus indicating that reinfection may occur

(Plunkett *et al.* 1940, Israel and DeLien 1942, Heimbeck 1949, Meyer 1949). Moreover, drug-resistant strains are sometimes isolated from patients who acquired their primary infection before the introduction of antituberculous agents (Thomas *et al.* 1954), and bacteria of different phage types are sometimes isolated during successive attacks of tuberculosis in the same patient (Raleigh and Wichelhausen 1973).

Epidemiological evidence suggests that much post-primary tuberculosis is due to endogenous reactivation (Springett 1951, Wilkins 1956, Stead 1967). The conclusions of the International Union Against Tuberculosis (1974) go so far as to state that exogenous reinfection is very rare. On the other hand, Tinker (1959) and Gryzbowski and Allen (1964) both estimated that 30 per cent of their cases of post-primary tuberculosis resulted from fresh infection. Canetti (1968) pointed out that in countries in which the prevalence of tuberculosis was low most cases of post-primary tuberculosis must be of endogenous origin, whereas in regions where the incidence of disease was high much reinfection was of exogenous origin (see also Canetti *et al.* 1972). In such areas most primary infections occur in children; adult cases are almost all post-primary infections. When the prevalence of disease in such an area is rapidly reduced by means of an antituberculosis campaign the incidence of adult tuberculosis, surprisingly, drops (Styblo 1978: appendix III).

Factors affecting the incidence of, and mortality from, tuberculosis

Workers in the mining, quarrying, sand blasting, tool grinding and earthenware industries are at an increased risk from tuberculosis because of exposure to metallic and stone dust (Gye and Kettle 1922; Kettle 1924, 1930; Hart and Aslett 1942). The higher incidence of tuberculosis among printers and factory workers (Stewart and Hughes 1951, Logan and Benjamin 1957) than among fishermen and agricultural workers may be ascribed to the increased risks of cross-infection. Occupational exposure to tuberculous patients is an obvious hazard, but in infected hospital staff the disease is usually diagnosed and treated in its early stages (Theodore *et al.* 1956). Pathology laboratory staff are at greater risk of infection than the normal population (Reid 1957) but modern safety measures and adequate training should minimize this. Post-mortem attendants are at an even higher risk (Meade 1948).

Tuberculosis is commoner among unskilled workers than among professional and executive workers. Tinker (1959) found that the lower economic classes in London were more prone to infection than the higher. Many of the patients lived alone in lodging houses, were poorly nourished, and smoked and drank excessively. Among the older patients there were many more men than women. This was said to be because men tended to congregate in public houses; women were less gregarious than men and cared for themselves better. Factors favouring infection in men were held to be risks of transmission at work (McDonald 1952) and the damaging effect of smoking (Lowe 1956). It is not known whether there is any innate difference between the sexes in their susceptibility to tuberculosis. Although Lurie and his colleagues (1949*a,b,c*) found that large doses of oestrogen retarded the spread of artificially induced tuberculosis in rabbits, whereas chorionic gonadotrophin had the opposite effect, it is doubtful whether hormonal activity modifies the disease in man.

A low prevalence of tuberculosis in a community is likely to result in decreased medical awareness of the disease. In a survey in Britain (Report 1971) unsatisfactory medical attention was thought to have contributed to 149 of 263 deaths from tuberculosis.

Clinical features and nomenclature of tuberculosis in man

Human tuberculosis is a disease of numerous different manifestations, but there is an underlying pattern, which is here briefly depicted. Infection occurring in a person never previously exposed to *M. tuberculosis* is termed *primary tuberculosis*. Vaccination by BCG may be regarded as an artificially induced primary infection. The primary infection consists of a lesion at the site of entry of the organism, which is the lung, tonsil, intestine or, rarely, the skin.

This lesion is termed the *primary focus* or Ghon focus (Ghon 1916) and together with the infection of the regional lymph nodes is referred to as the *primary complex* (Ranke 1928). In some cases, particularly in infection of the tonsil and in infants, the primary focus is extremely small and the enlarged lymph nodes are the only observable manifestations. Lymph node involvement is, as a rule, of greater magnitude and frequency in children than in adults.

Haematogenous spread of the bacilli causes more widespread disease such as miliary tuberculosis, or infections of the central nervous system, genito-urinary tract, or bones and joints (see Davies 1971); it is

particularly liable to cause fatal disease in children under 3 years of age.

Post-primary tuberculosis develops in the tuberculin-positive subject either as a result of re-activation of organisms in a 'healed' primary lesion or of exogenous reinfection (see p. 35). It was at one time referred to as the 'adult type'. This terminology is a relic from the time when tuberculosis was so prevalent in the developed countries that most persons had acquired a primary tuberculous infection during childhood or adolescence. Most adult patients showed the features of post-primary infection. This no longer holds true where the prevalence of tuberculosis is low and many patients are not infected until late in life. The widespread use of BCG vaccination has also complicated the picture, as this procedure induces a primary lesion.

Spread of infection to lymph nodes and more distant organs is less frequent in post-primary tuberculosis than in primary infections. Large progressive cavitating lesions frequently develop in the apical lobes of the lungs and their walls contain numerous bacilli. Some of these bacilli reach the lower parts of the lung via the air passages and lead to the formation of secondary caseating lesions. Although haematogenous and lymphatic spread is uncommon in post-primary disease, tuberculous ulcers caused by infected sputum may develop in the larynx and mouth; and swallowed sputum may lead to similar ulcers in any part of the alimentary tract (for further details see Pagel *et al.* 1964).

Post-primary tuberculosis usually occurs five or more years after the primary infection (Styblo 1978) and affects children as well as adults (Davies 1971).

The term 'open tuberculosis' is applied to those cases in which bacilli are detectable in the sputum. This is often due to a cavity being in direct contact with a part of the bronchial tree so that bacilli gain access to the sputum. Open cases are less commonly encountered in children and adolescents than in older patients.

Although tuberculosis is extremely variable in its clinical manifestations in man there is, nevertheless, a basic sequence of events referred to by Wallgren (1948) as 'the timetable of tuberculosis'. This pattern, which

Table 51.2   The 'time-table of tuberculosis'

Time after infection	Clinical features
3–8 weeks	Conversion to tuberculin sensitivity
3 months	Miliary disease Tuberculous meningitis
3–6 months	Pleural effusions Segmental pulmonary lesions
1 year	First appearance of post-primary pulmonary lesions, renal lesions or bone and joint disease

was also discussed by Davies (1971), was established by observations on children in the pre-chemotherapeutic era. The sequence of events is shown in Table 51.2. Lung lesions with cavitation rarely occur during the first year after infection, except in puberty when the primary disease may progress directly to a post-primary type of infection. Tuberculosis of bones and joints usually becomes clinically evident between one and three years after infection.

For a general review of the clinical features of human tuberculosis see Davies (1971) and Grange (1980). A number of useful reviews of tuberculosis in various parts of the body have been published: central nervous system (Kocen 1977, Parsons 1979), genito-urinary tract (O'Boyle and Gow 1976, Gow 1979), bones and joints (Murley 1971, Kemp 1976), skin (Kounis and Constantinidis 1979), lymph nodes (Iles and Emerson 1974, Newcombe 1979) and female genital tract (Sutherland 1979).

The pathogenesis and immunology of tuberculosis

The clinical manifestation and eventual outcome of tuberculosis—as of all infectious processes—depend on the virulence of the pathogen and the nature of the host's immune response.

Virulence of *Mycobacterium tuberculosis*

Two different mechanisms play a part: the synthesis of toxic substances, and the ability of the bacilli to survive within the phagocytic cells. The expression of virulence is affected by the occurrence of a specific immune response and by the non-specific innate resistance of the host to infection. Thus in the rabbit the bovine type of *M. tuberculosis* is much more virulent than the human type; the Asian type is of low virulence in the guinea-pig; and the virulence of the vole type is expressed only in voles, shrews, and wood mice. Strains of the human or bovine type isolated from skin lesions are often attenuated for animals (Report 1911) although passage through susceptible hosts causes some of them to revert to full virulence (Griffith 1957). Likewise bovine strains of *M. tuberculosis* isolated from horses are often of reduced virulence for rabbits and calves (Konyha and Kreier 1971).

Toxic substances synthesized by *M. tuberculosis*

Several toxic substances have been isolated from tubercle bacilli; especially important are cord factor (Bloch 1950) and sulpholipids (Middlebrook *et al.* 1959). The structure and toxic properties of these compounds are discussed by Goren (1972), Goren and Brennan (1979), Kato and Goren (1974) and Goren and co-workers (1974*a, b*). It is doubtful whether these lipids are directly associated with virulence (Grange *et al.* 1978).

Intracellular survival

Virulence of mycobacteria is almost certainly linked to the ability of the bacteria to survive within macrophages. This is partly due to their ability to prevent fusion of the phagosomes and lysosomes (Lowrie *et*

al. 1979). There is also an association between virulence and resistance to hydrogen peroxide, which is one of the macrophage's defence mechanisms (Mitchison *et al.* 1963, Nair *et al.* 1964, Grange *et al.* 1977, 1978, Jackett *et al.* 1978). Strains highly resistant to isoniazid are attenuated and sensitive to hydrogen peroxide owing to a loss of peroxidase activity (Gayathri Devi *et al.* 1975).

Mode of infection in man

Disease caused by types of *M. tuberculosis* other than the bovine type almost always begins in the lung as a result of inhalation of bacilli. The bovine type often enters the body by penetrating the tonsil or intestinal mucosa.

Mycobacterium tuberculosis may gain access to the body by direct inoculation, particularly when contaminated needles or knives are handled. These infections are extremely rare nowadays, but in the past they were seen in butchers, laundresses and post-mortem attendants. At one time skin tuberculosis was an occupational hazard of anatomists—*prosectors' wart*. Outbreaks of tuberculosis due to the use of contaminated syringes and needles during the course of immunization against diphtheria, pertussis or typhoid fever have been reported (see Report 1939, Tamura *et al.* 1955, Wilson 1967). Tuberculous abscesses have been observed after the injection of penicillin, and it was suggested that such abscesses were the result of carriage of bacilli to the injection site by macrophages in an already infected patient—an example of the so-called *fixation abscess* (Blacklock and Williams 1957). It is probable, however, that many of them were due to mycobacteria contaminating the fluid used to dissolve the penicillin. Thus several post-injection abscesses have been shown to be caused by the rapidly growing species *M. fortuitum* and *M. chelonei* (see Wolinsky 1979). In addition, abscesses due to *M. chelonei* have been caused by injections of contaminated histamine solution (Inman *et al.* 1969) and combined vaccine (Borghans and Stanford 1973).

Most early workers assumed that pulmonary tuberculosis was due to inhaled bacilli (Cornet 1889, Flügge 1899). Others, notably von Behring (1904) and Calmette (1923), believed that the disease in adults was due to dissemination from a primary lesion in the intestine acquired during infancy or childhood. This view was largely based on animal experiments (Calmette and Guérin 1905, 1906*a,b*) in which pulmonary lesions developed after intra-oesophageal inoculation of bacilli. The observations of White and Minett (1941), however, brought evidence to show that such lesions resulted from aspiration of bacilli into the trachea during inoculation.

The conclusion that pulmonary tuberculosis was due to inhalation of bacilli gained strong support from the thorough studies of Ghon (1916) in Vienna and of Opie (1917*a,b*, 1924*a,b*, 1925) in America on the anatomical distribution of lesions in pulmonary tuberculosis in children. Comparisons of post-mortem findings with radiological findings showed that primary lesions occurred in the substance of the lungs, and that the regional lymph nodes were affected only secondarily. Had the bacilli reached the lungs from the intestine via the lymphatics, the lymph nodes would have been invaded before the substance of the lungs. Occasionally cases of pulmonary disease caused by bovine strains follow the inhalation of cough spray from tuberculous cattle (Sigurdsson 1945). Spread of infection from patient to patient was postulated (Collins *et al.* 1981).

Mechanism of infection

Cornet (1889) and Chaussé (1914, 1916) concluded that the chief source of infection was dust arising from dried sputum or dust liberated by brushing or shaking clothes contaminated by droplets expelled during coughing. Flügge (1899) and members of his school (Moeller 1899, Heymann 1901, Findel 1907, Ziesche 1907, Hollmann 1924), on the other hand, maintained that the bacilli were spread in the form of minute *moist* droplets expelled from tuberculous patients during speaking, coughing and sneezing. Wells (1934) noted that the liquid content of small droplets rapidly evaporated leaving desiccated 'droplet nuclei' which were small enough to remain suspended in the air for a considerable period of time (see Chapter 9). The relative importance of dust particles, moist droplets or droplet nuclei in causing pulmonary infection in man is uncertain, but the most dangerous particles are probably droplet nuclei containing only a few bacilli (Riley and O'Grady 1961). Larger particles do not reach the lung parenchyma.

The risk of infection is related to the number of bacilli in the air. Riley and colleagues (1962) showed that, on average, tuberculosis wards contained one infective particle in every 15 000 to 20 000 cubic feet of air. A few patients were particularly infectious, and

one patient with laryngeal tuberculosis contributed one infectious particle to every 200 cubic feet of air in a ward. The risk of infection is particularly high when people are crowded together in conditions of poor ventilation, for example in a submarine (Houk *et al.* 1968).

Innate immunity to tuberculosis

Very little is known concerning the mechanism of the naturally occurring resistance of various mammalian species, including man, to the different types of *M. tuberculosis*. Numerous attempts have been made to relate resistance against tuberculosis to age, sex, ethnic origin, and genetic constitution. Such studies are difficult to interpret in view of the part played by factors such as personal hygiene, diet, occupational exposure to infection, overcrowding, and environmental exposure to agents liable to cause lung damage. The role of natural immunization by 'environmental' mycobacteria is discussed below.

Ethnic factors   Certain races appear to be more susceptible than others to infection. Cummins (1908, 1912) showed that in the Egyptian Army from 1902–1906 the incidence of tuberculosis in the Sudanese was double that in the Egyptians themselves; that towards the end of the nineteenth century the incidence in the Indian Army was five times higher among the Gurkhas than among the Mahrattas; and that in the West Indies the disease was between 4 and 10 times more common among native soldiers than among British troops. The incidence of clinically evident tuberculosis in nurses working in the same hospital was found to be two-and-a-half times higher among those of Irish and Welsh origin than among those of English origin (Report 1948).

Heredity   Tuberculosis often runs in families but to what extent this is due to genetic factors, exposure to a source case within the household, or to pure chance is difficult to determine.

Diehl and von Verschuer (1933) in Germany studied 106 twin pairs in which one or both suffered, or had suffered, from tuberculosis. In 26 (70 per cent) of 37 monozygotic (identical) pairs the tuberculous history was the same in each twin, and in 11 (30 per cent) it was different. The corresponding figures for 69 dizygotic (non-identical) pairs were 17 (25 per cent) and 52 (75 per cent) respectively. A re-examination of the twins 20 years later yielded essentially the same results (von Verschuer 1955) and similar findings were reported in Switzerland (Vehlinger and Künsch 1938). Lange (1935) suggested that Diehl and Verschuer's findings could be explained by the fact that identical twins—more than non-identical twins—tend to be subjected to the same environmental conditions in early life. Against this, however, is the increasing similarity of the response of identical twins to tuberculosis as age advances.

Kallman and Reisner (1943) in the City and State of New York studied the occurrence of tuberculosis, in patients over 15 years of age, among 616 twin partners, 930 full siblings (both parents in common), 74 half siblings (one parent in common), 688 parents and 226 marriage partners of twin patients. Of the 308 twin pairs 78 were identical and 230 were non-identical. The chances of the disease developing in relatives of infected patients were: marriage partners 7.1 per cent; half siblings 11.9 per cent; parents 16.9 per cent; full siblings 25.5 per cent. The chances of tuberculosis developing therefore increased in proportion to the degree of genetic relationship. The differences could not be explained in terms of variations in exposure to source cases. Simonds (1957, 1963) studied 202 twin pairs, and found that both twins had the disease in 29.6 per cent of the identical pairs and 12.8 per cent of the non-identical pairs. The significantly higher incidence of tuberculosis in identical twin contacts than in non-identical was difficult to explain in terms other than genetic ones.

Differences in susceptibility to tuberculosis have also been observed between different breeds of cattle. Carmichael (1938) found that the incidence of tuberculosis among the Hamitic cattle in Ankole was much higher than among the Zebu cattle in Uganda even though the former variety appeared to be better adapted to the environment. Of the Hamitic cattle, 80 per cent reacted to tuberculin, as against less that 1 per cent of the Zebu cattle, which were also found to be almost totally resistant to experimental inoculation. Lurie (1941) observed that six inbred colonies of rabbits differed widely in their susceptibility to bovine strains of *M. tuberculosis* whether administered as an aerosol or inoculated parenterally. This variation was due to differences in the bactericidal power of phagocytic cells (Lurie *et al.* 1952).

Early events after infection

Organisms soon pass from the site of infection to the regional lymph nodes via the lymphatic vessels and to more distant organs via the blood stream (see Davies 1971). Experimentally, bacilli were found to reach the spleen of guinea-pigs 48 to 96 hours after the injection of a small intracutaneous dose but within 1 hour after a large dose. Bacilli were found in the blood within a few hours and histological changes in the regional lymph nodes were demonstrable 64 hours after inoculation (Soltys and Jennings 1950). Mycobacteria are initially engulfed by polymorphonuclear leucocytes (PMN) but rapidly released. In contrast to their behaviour in macrophages, the bacilli in PMN do not inhibit the fusion of lysosomes with phagosomes; nevertheless PMN are unable to destroy them (Smith *et al.* 1979). Within 24 hours the bacteria are phagocytosed by macrophages which, by that time, are the predominant cells in the lesion (Adams 1975, Closs and Haugan 1975). The alveolar macrophage appears to play only a small defensive role against *M. tuberculosis*. Even in sensitized subjects it is not the alveolar macrophages but the recruited bloodborne macrophages that destroy the organisms (Lefford 1980).

Granulomas

The characteristic feature of the early lesion is the epithelioid granuloma or tubercle. Granulomas are a feature of many chronic infections including brucellosis, syphilis, coccidioidomycosis, leishmaniasis, and mycobacterial diseases. They are also formed in some diseases of unknown aetiology, notably Crohn's disease and sarcoidosis. The precise definition of a granuloma is not easy; Emori and Tanaka (1978) define the early granuloma as 'a chronic, compact aggregation of activated macrophages around a granulomagenic agent'. The granulomas induced by mycobacteria are of the 'high turnover' type (Spector 1971) and depend on a continuous supply of fresh monocytes, although cell division within the granuloma probably also occurs. Resistance to infection, but not necessarily specific immunity, is strongly associated with granuloma formation (Emori and Tanaka 1978, Turk 1979).

In an initial infection the granulomas are almost certainly induced directly by mycobacterial components. Wax D was found to be granulomagenic in guinea-pigs and rabbits (White *et al.* 1958) but Moore and Myrvik (1974) failed with this substance in rabbits. Other possible granulomagenic agents, including phosphatides, cord factor and an unidentified glycolipid have been described, but many such studies have been criticized on the grounds that the components were often contaminated. More recently, however, Emori and Tanaka (1978) have found that synthetic muramyl dipeptide, which is identical with part of the peptidoglycan of the mycobacterial cell wall, has granulomagenic properties in guinea-pigs and rats; the use of a synthetic molecule ruled out the possibility of contamination by other cell-wall components. (For further details of granuloma formation, and the cells concerned, see Spector 1969, Adams 1974, Galindo *et al.* 1974, Warfel 1978, van der Rhee *et al.* 1979.)

The role of the macrophage

The macrophage is one of the principal cells in the immune system. It plays a vital role in the induction of the immune response as well as being a principal effector cell on account of its phagocytic properties. The importance of the antigen-specific activation of macrophages in tuberculosis and *Listeria monocytogenes* infection in mice was demonstrated by Mackaness (1967). (See also Grange 1980.)

The development of tuberculin positivity

After primary infection there is a period of bacillary multiplication before clinical disease develops. The first manifestation is a conversion to tuberculin positivity, which may be accompanied by fever, less often by erythema nodosum, and rarely by phlyctenular conjunctivitis (Wallgren 1948). The period preceding tuberculin conversion probably varies according to the route of infection. After intradermal BCG vaccination it ranges from 3 to 12 weeks. At Lübeck, in Germany, where virulent *M. tuberculosis* was accidentally given orally in place of BCG, 72 children became ill; 4 showed symptoms in 2–3 weeks, 54 in 4–8 weeks, 11 in 9–11 weeks and 3 in 12–17 weeks (Report 1935*b*). The commonest time, therefore, was 4–8 weeks. The same conclusion was reached by Wallgren (1948).

The length of time required for the development of tuberculin positivity or overt disease after respiratory infection is more uncertain. Observations on students briefly exposed to a highly infectious patient (Oeding 1946, Bergqvist 1948) showed that the development of tuberculin sensitivity appeared to take a minimum of 3 weeks, and that overt disease was rare before 6 weeks.

Cell-mediated immunity and delayed hypersensitivity

The relation between cell-mediated immunity (CMI) and delayed hypersensitivity (DH), exemplified by the classical tuberculin reaction, has been the subject of much controversy. It has been claimed on the one hand that CMI and DH are different manifestations of the same immune process, and on the other that the two phenomena are dissociable. It has also been claimed that DH may antagonize the CMI response. The opposing viewpoints in this debate were reviewed in three editorials in the American Review of Respiratory Disease (Lefford 1975, Salvin and Neta 1975, Youmans 1975). The main issue was whether hypersensitivity was responsible for increased resistance to *M. tuberculosis*.

Koch (1891*a*) observed that an intradermal injection of virulent *M. tuberculosis* led to a progressive systemic infection in the guinea-pig. A further intradermal injection of bacilli during the infection led to a local necrotic reaction with destruction and shedding of the organisms. The elicitation of the so-called Koch phenomenon had no effect on the progress of the systemic disease. Nevertheless, it became generally accepted that DH was of benefit to the host—a view that is still prevalent. Römer (1908) observed that tuberculin hypersensitivity in animals was associated with increased resistance to reinfection, provided that the hypersensitivity reaction was not too intense. Other early workers (Austrian 1913, Krause and Willis 1919) showed that the induction of tuberculin hypersensitivity in guinea-pigs reduced their resistance to subsequent challenge with *M. tuberculosis*. Wilson and his colleagues (1940) also observed a lack of correlation between CMI and DH in guinea-pigs. Although a moderate degree of DH was associated with protection, animals with a high degree were less protected than those with none.

These findings were supported by the clinical observations of Turner (1953) who found that quiescent tuberculosis was more likely to be reactivated in patients with either a negative or a strongly positive tuberculin reaction than in those showing a moderate response.

The opposing viewpoint was expressed by Mackaness (1967), however, who regarded DH and CMI as different manifestations of the same basic process, because both were induced by injection of mycobacteria, both were stimulated by a second challenge, and both were transferable by lymphocytes. It has been maintained that mycobacterial fractions differ in respect of the response they elicit. Thus Raffel (1950) induced DH with tuberculoprotein administered with Wax D, without increasing the immunity; whereas Youmans and Youmans (1969) found that injection of mycobacterial RNA and ribosomes produced immunity but not DH. Although both CMI and DH are transferable by lymphocytes and not by serum, it is now appreciated that the lymphocyte population is heterogeneous. The T lymphocytes are divisible into several functional groups by their possession of surface differentiation antigens, and antisera to such antigens have been used to dissociate *in vitro* the various forms of cellular hypersensitivity in the guinea-pig (Godfrey and Koch 1980).

Rook and Stanford (1979) suggested that the controversy surrounding the protective role of DH has arisen because this term is applied indiscriminately to more than one phenomenon. They demonstrated that mycobacterial infections lead to the development of three types of 'delayed' skin-test reaction: the poorly understood reaction of Jones and Mote (1934); an early developing non-necrotic reaction corresponding to the macrophage activation described by Mackaness (1967); and a late developing necrotic reaction corresponding to the Koch phenomenon. In mice the more pathogenic mycobacteria either suppress or fail to elicit the non-necrotic reaction which, being associated with macrophage activation, is appropriate for the control of systemic disease. The necrotic reaction, on the other hand, appears to be more beneficial when the infection is restricted to the skin. It is relevant that of many strains of mice, the C57BL strain is the most resistant to subcutaneous challenge with *M. tuberculosis* but the most susceptible to intravenous challenge.

The immune spectrum in tuberculosis

The outcome of infection by *M. tuberculosis* varies enormously from patient to patient. At one extreme the infection is subclinical and merely causes a conversion to tuberculin positivity; at the other extreme a progressive and potentially fatal disseminated infection occurs. Lenzini and his colleagues (1977) attempted to classify cases of tuberculosis according to an immunological spectrum similar to that occurring in leprosy. Thus reactive cases, showing a good cell-mediated immune response, comparatively few bacteria and a good response to treatment, and unreactive cases, resembling lepromatous leprosy in showing numerous organisms in the apparent absence of a cell-mediated immune response, were described. Bhatnagar and his colleagues (1977) also described a spectrum of immune responses in tuberculosis with self-limiting subclinical cases at one end and miliary tuberculosis at the other. Proudfoot (1971), however, showed that miliary disease was accompanied by an immune response leading to the formation of millet-seed-like granulomas; in cryptic disseminated disease, necrotic foci teem with organisms, but there is little or no cellular reaction. The latter disease, which is often rapidly fatal, is sometimes referred to as 'diffuse', in contrast to 'nodular' tuberculosis, or as the *Yersin* type of disease on account of its histological similarity to *Yersinia pseudotuberculosis* infection in the guinea-pig (Skinsnes 1968).

There is no doubt that immunologically reactive self-limiting cases of tuberculosis frequently occur, but it is unlikely that there is a form of chronic unreactive tuberculosis analogous to lepromatous leprosy—the tubercle bacillus is far too toxic. Patients with cryptic disseminated tuberculosis and negative skin tests are extremely ill, and usually die. In contrast tuberculin-negative patients with pulmonary tuberculosis do not on the whole differ clinically from tuberculin-positive patients (Kent and Schwartz 1967, McMurray and Echeverri 1978, Kardjito and Grange 1980). Anergy with respect to skin-testing responses does not therefore necessarily indicate a failure of effective immune response.

There appears to be no direct parallel between the spectra of immune response in tuberculosis and leprosy (Rook and Stanford 1979, Kardjito and Grange 1980). In tuberculosis, activation of the macrophages containing the bacilli is advantageous, whereas in leprosy a cytotoxic response, releasing bacilli from cells which cannot be activated, is desirable.

Anergic reactions in tuberculosis

von Pirquet (1906) used the term *allergy* to imply an altered immune response to an antigen, and the term *anergy* to mean a failure of the immune system to respond to the antigen. Jadassohn (1914) suggested that anergy might not always be due to a failure of the immune system to respond but might be a positive suppressive reaction. In 1908 Pickert and Löwenstein postulated that the lack of cutaneous sensitivity in tuberculin-negative healthy persons was due to circulating 'antikutins'. It is now well appreciated that suppressive reactions are vital for the overall control of both the humoral and cell-mediated immune responses.

Anergy in mycobacterial diseases is due to the presence of 'blocking factors', the generation of suppressor T lymphocytes, or a non-specific or specific trapping of antigen-sensitive lymphocytes in lymph nodes (Rook 1976). Mycobacterial cell-wall components, including Wax D, cause macrophages to liberate soluble mediators which modify T lymphocyte function. These mediators include dialysable inhibiting factor (DIF) and interleukin 1 (lymphocyte activating factor, LAF). Whereas DIF is directly inhibitory, interleukin 1 is inhibitory when produced in large amounts (Rook and Stewart-Tull 1976). Nakamura and Tokunaga (1980) found that some strains of mice produced T suppressor cells which inhibited delayed-type skin reactions after infection with BCG. The suppression was much more evident when the animals were infected intravenously than when they were infected intradermally. As discussed on p. 40 there may be an advantage in suppressing DH, as opposed to CMI, in sys-temic mycobacterial infections but not in dermal infections. In anergic cases of tuberculosis the blastogenic response of peripheral blood lymphocytes to mitogens or purified protein derivative (PPD) is reduced (McMurray and Echeverri 1978). This is due to a non-specific tendency for lymphocytes to be retained in the lymph nodes and a superimposed trapping of specific antigen-sensitive T lymphocytes. Thus although antigen-sensitive lymphocytes are not detectable in the peripheral blood in anergic cases, they are demonstrable in the lymph nodes (Rook *et al.* 1976). This trapping, the mechanism and function of which is poorly understood, explains the temporary anergy seen in some cases of tuberculosis.

Skin-testing studies (Shield *et al.* 1977) show that sensitization by mycobacteria varies from country to country, and that there is a threshold of sensitization above which anergy occurs. Thus in Burma, where sensitization by slow-growing mycobacteria is common, 13 per cent of patients with tuberculosis are shown to be anergic by testing with PPD and sensitins prepared from a wide range of mycobacterial species.

In addition to the mechanisms discussed above, the CMI and DH reactions in tuberculosis are affected by malignant disease, especially Hodgkin's disease, sarcoidosis, autoimmune disease, anti-inflammatory or immunosuppressive therapy, acute viral infections, and malnutrition (see Nauta 1979, Caplin 1980).

The effect of migration on immunity to tuberculosis

Persons from undeveloped rural areas are very susceptible to tuberculosis if they move to a more developed area (Cummins 1908, Westernhöffer 1911, Borrel 1920, Bushnell 1920). In many such patients the disease is widespread, with lymph node infection and various other extrapulmonary manifestations, in contrast to the slowly progressive pulmonary disease that usually occurs in adults. A similar phenomenon has been observed more recently among Asian immigrants in Great Britain. The incidence of tuberculosis among immigrants from Pakistan and India in 1975 was 77 and 63 times higher respectively than that in the non-immigrant population, and was highest among those who had been resident in Great Britain for less than 5 years (Report 1975c). Many of these patients suffered from extrapulmonary or non-pulmonary disease. In the London Borough of Brent during 1975 there were 101 cases of such disease in a total of 290 patients, of whom 211 were immigrants from Africa or Asia (Grange *et al.* 1977). Many cases of disease in Asian adult patients have massive or widespread infection of the lymph nodes but a minute or undetectable primary focus (Report 1980b). A similar high incidence of lymphadenitis among Asian immigrants (Indians, Chinese, Philippinos) has been reported in Canada (Enarson *et al.* 1979) and among Turkish 'guest workers' in West Germany (Lukas 1977). Although tuberculosis in Asian immigrants is often due to strains of low virulence for guinea-pigs, no relation between the type of organism and the nature of the disease in man has been demonstrated (Grange *et al.* 1976, 1977). Cole (1981) has demonstrated immunological differences between newly arrived immigrants from Asia on the one hand and longer established immigrant and non-immigrant persons on the other. The peripheral blood lymphocytes from the former group reacted less than lymph node lymphocytes in in-vitro assays of DH, whereas no such difference was demonstrable in the latter group. These findings suggested that lymphocytes reacting to *M. tuberculosis* antigens were trapped in the lymph nodes. It was not, for ethical reasons, possible to obtain lymph nodes from healthy newly arrived immigrants, so it was uncertain whether this phenomenon was a cause or a result of the unusual nature of the infection that occurs in such persons.

Diagnosis

Tuberculosis can be definitely diagnosed only by isolating the causative organism, *M. tuberculosis*, in pure culture. Often, however, clinical and radiological investigations supported by direct microscopy and skin testing must suffice.

Microscopical examination

Although microscopy alone does not usually distinguish between *M. tuberculosis* and other mycobacteria, it is nonetheless a rapid and simple means of detecting those cases of pulmonary disease that are a source of infection (see p. 34). To be detected microscopically, there must be between 5000 and 10 000 bacilli in 1 ml of sputum. (Cruikshank 1952, Yeager *et al.* 1967, Hobby *et al.* 1973, Rouillon *et al.* 1976.)

The techniques used for staining mycobacteria are based on the resistance of the organisms to decolorization by acids after staining by an arylmethane dye. The most widely used staining technique is that bearing the names of Ziehl and Neelsen which is described in detail in Chapter 24.

Fluorescence microscopy was introduced by Hagemann (1937*a,b*) who originally used berberine sulphate as the dye but later (1938) recommended auramine. Truant and his colleagues (1962) used two arylmethane dyes, auramine O and rhodamine B, together. This combined staining method detected acid-fast bacilli in 358 out of 3000 samples of sputum as against only 274 by the Ziehl-Neelsen method (Somlo *et al.* 1969).

In view of the insensitivity of microscopical techniques many authors have attempted to concentrate the bacteria in the specimens. The techniques described include chemical concentration (Grysez and Bernard 1920, Faisca 1921, Douglas and Meanwell 1925, Mayer 1926), chemical flocculation (Hanks *et al.* 1938), and flotation (Andrus and Mac-Mahon 1924, Pottenger 1931, Edwards *et al.* 1936, Smith 1938, Hanks and Feldman 1940). Centrifugation is, however, the most widely used concentration method and is applicable to sputum, urine, cerebrospinal fluids and aspirates from the serous cavities. Sputum should be homogenized and liquefied by the use of *N*-acetyl-L-cysteine (Kubica *et al.* 1963) or some other mucolytic agent.

Microscopical examination is subject to certain inherent errors. It is important to avoid contamination of specimens, collecting pots, reagents, slides, and other equipment by saprophytic mycobacteria or other acid-fast material such as bacterial or fungal spores. Tap water is especially likely to be contaminated by mycobacteria (Wilson 1933, Engel *et al.* 1980). Full practical details for establishing a microscopy service are given by Vestal (1975) and by the International Union Against Tuberculosis (1977).

Cultural methods

The media used are described in Chapter 24. Löwenstein-Jensen medium is popular for isolating human strains of *M. tuberculosis* and most other mycobacteria, and Stonebrink's (1958) pyruvic acid egg medium for the isolation of bovine strains. Uncontaminated specimens such as cerebrospinal and pleural fluids and biopsy material may, after centrifugation or homogenization where appropriate, be used to inoculate the media.

Most specimens require pre-treatment with an agent which selectively kills organisms other than mycobacteria, such as antiformin (a mixture of sodium hydroxide and sodium hypochlorite; Uhlenhuth and Xylander 1908), sodium hydroxide (Petroff 1915), sulphuric or hydrochloric acid (Löwenstein 1924), oxalic acid (Corper and Uyei 1930), Jungmann's iron-sulphuric acid-peroxide reagent (Jungmann and Gruschka 1938, Nassau 1942) and quaternary ammonium detergents (Wagener and Reuss 1953, Krasnow and Wayne 1969). Other agents include alkaline lauryl sulphate (Tacquet *et al.* 1967, Šula 1968) trisodium phosphate (Corper and Stoner 1946) and cetylpyridinium bromide (Mokhtari 1980). For the preferential isolation of slowly growing mycobacteria, Stanford and Paul (1973) used a biphasic technique in which contaminants were destroyed by sodium hypochlorite, and rapidly growing mycobacteria by a short period of treatment with trisodium phosphate. None of these methods is ideal, and there is no universally accepted method. Acids are preferable to alkalis for the decontamination of urine as *Pseudomonas* species are partly resistant to alkali. Homogenization of sputum is improved by pancreatin or the mucolytic agent *N*-acetyl-L-cysteine (Kubica *et al.* 1963). The time required for decontamination ranges from 15 minutes with 4 per cent NaOH to 24 hours with trisodium phosphate. After decontamination the agent is neutralized by adjusting the pH or by diluting the specimen in phosphate buffer, and the organisms are concentrated by centrifugation. After treatment with sodium triphosphate, the organisms may be precipitated with calcium chloride and the sediment redissolved in a solution of sodium citrate (Darzins and Pukite 1964, Meidl and Harlacher 1969).

Two simplified culture techniques have been introduced. Sputum decontaminated by means of 4 per cent NaOH may be inoculated directly on to acidified egg media (Ogawa 1949, Marks 1959). In the 'sputum swab' method of Nassau (1958) swabs soaked in sputum are treated successively with oxalic acid and sodium citrate solutions before the media are inoculated. This method is useful under difficult conditions in developing countries (see p. 247 of Pagel *et al.* 1964).

As an adjunct or alternative to decontamination procedures antibacterial agents that do not affect mycobacteria may be incorporated in the media (Corper and Cohn 1946, Brotherston *et al.* 1961, Mitchison *et al.* 1972).

Inoculated culture media are usually incubated at 35–37° for 8 weeks, with weekly inspection for growth. Most strains of *M. tuberculosis* appear within 4 weeks, but they may not be visible for 8 weeks or more if they originated from patients treated with anti-tuberculous agents (Kennedy *et al.* 1958). Other mycobacterial species often take longer to appear than *M. tuberculosis*. This applies not only to very slow growing species such as *Mycobacterium xenopi* but also, paradoxically, to rapidly growing strains.

Cultural techniques may detect as few as 10–100 organisms per ml of sputum (Yeager *et al.* 1967). Colonies appearing on the culture media are shown to be mycobacteria by means of Ziehl-Neelsen staining but must then be identified by the techniques described in Chapter 24. Strains of *M. tuberculosis* are clearly identifiable by their lack of pigment, failure to grow at 25° and sensitivity to *p*-nitrobenzoic acid (Yates and Collins 1979).

Specimens for culture should always be collected directly into sterile containers and care must be taken to avoid contamination of the outside of the vessel. When the specimen is to be sent to the laboratory by post, the postal regulations must be strictly observed.

Ideally sputum should be submitted as three early-morning specimens although a single specimen taken at any time is only slightly less likely to give a positive result on cultural examination (Andrews and Radakrishna 1959). When sputum is not expectorated, the fasting stomach contents may be examined (Armand-Delille and Vibert 1927) or the larynx may be swabbed (Grass 1931). Neither technique is as sensitive as culture of even scanty sputum samples. Bronchial lavage is said to be superior to both gastric lavage and laryngeal swabbing (Lees *et al.* 1955) but is unpleasant for the patient and carries an element of risk. Tytler (1945) held that culture of faeces, when carried out carefully, was slightly less sensitive than culture of the stomach contents. In contrast Saenz and Costil (1936) found *M. tuberculosis* in the faeces in only a small proportion of cases of advanced pulmonary disease or of intestinal tuberculosis, and rarely in non-advanced disease. In our opinion the culture of faeces for the diagnosis of tuberculosis is a waste of time and resources.

Excretion of *M. tuberculosis* in the urine in cases of renal tuberculosis is intermittent; for this reason a 3-day pool of early morning specimens is recommended. Bacilli may be concentrated by centrifugation or, alternatively, by membrane filtration (St Hill and Gow 1966). Pleural fluid may be prevented from clotting by the addition of two drops of 20 per cent sodium citrate solution for every 10 ml of fluid; it is then centrifuged to concentrate the organisms. Biopsy specimens are homogenized by grinding with sterile phosphate-buffered saline and sand in a Griffith tube.

M. tuberculosis is isolated from blood in only a very small proportion of cases of severe progressive tuberculosis; consequently, there is no point in culturing blood in this disease. Claims that mycobacteria could be isolated from the blood in tuberculosis and a range of other diseases were almost certainly based on gross errors of technique. (See Wilson 1933, and Report 1935*a*.)

Animal inoculation tests

The susceptibility of the guinea-pig to infection by *M. tuberculosis* is extremely high, and for many years guinea-pig inoculation had an important place in diagnosis. There are, however, several disadvantages in its use. The animals are costly to purchase and maintain, some die from intercurrent disease, and the method is time consuming. Disease takes from 4 to 6 weeks to develop and it is then necessary to identify the causative agent by cultural techniques. The excretion of *M. tuberculosis* in the urine and faeces of infected animals is hazardous to health and the cost of providing facilities for keeping such animals is very high.

The value of guinea-pig inoculation has been questioned by several investigators (Marks 1972). Holm and Lester (1947) found that the isolation of human strains on culture media failed in 8.6 per cent of cases whereas guinea-pig inoculation failed in 22.2 per cent; when the bovine type was present the failure rates were 16.6 and 12.2 per cent respectively. Isolation techniques for bovine strains have since been greatly improved by the use of pyruvate-containing media (Stonebrink 1957). Furthermore, up to 10 per cent of strains of *M. tuberculosis* isolated in Great Britain are attenuated for the guinea-pig (Grange 1979). Marks (1972) has also shown that guinea-pig inoculation is no more sensitive than modern cultural techniques for the isolation of *M. tuberculosis*. For this reason, together with the disadvantages listed above, the method is now virtually obsolete. Adequate in-vitro methods are also available for the identification of *M. tuberculosis* and for distinguishing BCG from virulent strains (Yates *et al.* 1978).

Serological diagnosis

Although the literature is extensive and almost every form of immunoassay has been applied to this disease, no test has proved satisfactory owing to the unacceptably high incidence of misleading results (Coates 1980). (For a review see Grange 1983.)

The tuberculin test

Tuberculin: preparation and standardization

Tuberculin may be defined as a substance, or mixture of substances, derived from *M. tuberculosis*, capable of eliciting a delayed-type hypersensitivity reaction in sensitized persons or animals. The term is often applied to any mycobacterial skin-testing reagent, but some workers prefer to use the more general terms *sensitins* or *elicitins*. The original tuberculin was prepared by Robert Koch (1891*b*) from 5 per cent glycerol broth in which *M. tuberculosis* had been grown for 6–8 weeks. After the removal of bacilli by filtration through a clay or kieselguhr candle, the filtrates were concentrated by evaporation in a boiling waterbath to a tenth of their original volume. The resulting clear brown syrupy fluid, termed Old Tuberculin (OT), contained 40–50 per cent glycerol, which acted as a preservative. Subsequently synthetic media replaced glycerol broth and the resultant material was known as Heat-Concentrated Synthetic Medium (HCSM) Tuberculin.

Long and Seibert (1926) demonstrated that the activity of tuberculin resided in its protein content. Consequently attempts were made to extract the active component by protein-precipitating agents including ammonium sulphate and trichloracetic acid (Seibert 1934, 1941*a,b*, 1944, Seibert *et al.* 1934, 1938, Seibert and Glenn 1941). These studies led to the production of a *Purified Protein Derivative* (PPD) consisting of proteins with estimated molecular weights of about 10 500. More recently a major protein constituent of tuberculin has been found to have a molecular weight of 9700 and to contain 89 amino acids (Kuwabara 1975). Despite its name, PPD is not pure protein but contains small amounts of nucleic acid and carbohydrate (Patterson 1959).

Nevertheless PPD is a purer preparation of protein than OT. The difference in purity is of significance only when high concentrations of reagents—over 100 tuberculin units (TU)—are used. OT then causes some non-specific reactions (Paterson and Leech 1954).

Batches of PPD or OT are standardized by comparing the diameters of the reactions induced in the skin of sensitized guinea-pigs with those elicited by international standard preparations (Römer 1909, Okell and Parish 1927, Parish 1938), or by chemical estimation of their protein contents (Lesslie 1962). The qualitative variation between batches is avoidable by the preparation of very large batches such as the widely used RT23 preparation (Magnusson and Bentzon 1958).

Tuberculin, whether PPD or OT, is heat stable but is susceptible to inactivation by light (Report 1955). The protein is adsorbed to glass; but this is preventible by including the detergent Tween 80 (polyoxyethylene sorbitan mono-oleate) in the diluting fluid (Landi *et al.* 1966, 1970). Isotonic phosphate-buffered saline, pH 7.38, containing Tween 80 is recommended as a diluent by the World Health Organization (Report 1962).

Other skin-testing reagents

These fall into two classes: tuberculins prepared from mycobacteria other than *M. tuberculosis*; and reagents with either more specificity or a greater ability to discriminate between active tuberculosis and past sensitization.

Reagents analogous to PPD have been made from several other species of mycobacteria. These include *M. avium* (McCarter *et al.* 1938) and its *intracellulare* variant (Palmer and Edwards 1962), *Mycobacterium scrofulaceum* Gause strain (Grzybowski and Brown 1967), and *M. xenopi* (Beck 1972).

These reagents have been used to investigate the specificity of weak tuberculin reactions. A reaction to such a PPD greater than that to human PPD has been considered to indicate a non-tuberculous sensitization (Palmer and Edwards 1968). Certain non-mycobacterial antigens, notably of the fungus *Histoplasma capsulatum*, may also induce a cross-reaction to tuberculin. In areas where this fungus is present a histoplasmin test should be carried out in addition to the tuberculin test (Edwards *et al.* 1958, Duboczy 1968).

Attempts have been made to develop more specific reagents by fractionation (Glenchur *et al.* 1965, Wilhelm and Römer 1977) but no such preparation has replaced the standard reagents (see review by Daniel and Janicki 1978).

Both Old Tuberculin and PPD contain denatured antigens. This denaturation is due in part to autolysis of proteins occurring during culture but in greater part to the heating (Seibert 1928). Additional denaturation almost certainly occurs during the precipitation of protein so that one eminent immunologist frequently refers to PPD as a 'rotted, boiled and pickled antigen'.

The problem has been overcome by the production of the so-called 'endotuberculins' from disrupted living bacilli instead of from filtrates. These reagents were originally sterilized by treatment with ether or by heating to 75° (Larson *et al.* 1961), but subsequently bacteria-excluding filters were used (Larson *et al.* 1966). This method was used by Stanford and his colleagues (1975) to produce a range of *New*

Tuberculins, (not to be confused with Koch's New Tuberculin).

The first of these was Burulin, a skin-test reagent for *Mycobacterium ulcerans* infection. As this mycobacterium could not be grown in broth, it was grown on solid Löwenstein-Jensen medium, washed thoroughly in various solvents to remove all traces of medium, and subjected to ultrasonic disintegration. Fragments of bacteria were removed by centrifugation; the supernatant fluid was sterilized by membrane filtration and standardized by a spectroscopic estimation of its protein content. Autolysis, heat denaturation and protein precipitation were thus avoided. In field studies the reagent was highly specific for patients in the reactive stage of *M. ulcerans* infection and elicited no cross-reaction in patients with leprosy or tuberculosis.

Methods for eliciting the tuberculin reaction

Experimentally, sensitivity to tuberculin may be tested in several ways: by parenteral administration to elicit a general reaction, by instillation into the conjunctival sac to elicit an ocular reaction or, in mice, by injection into the footpad with subsequent measurement of increases in foot thickness. This last method is very sensitive and has permitted a detailed analysis of the response to tuberculin to be made (Rook 1978, Rook and Stanford 1979). In man, some form of skin test is almost always used.

The original skin test of von Pirquet (1907) consisted in making a small scratch in the skin with a sharp instrument through a drop of tuberculin. Mantoux (1910) injected a known amount of tuberculin intracutaneously, and some later workers (Griffiths *et al.* 1965, Dull *et al.* 1968) used a needleless jet-injection apparatus. Tuberculin may be applied to the skin, after light scarification with sandpaper, either in the form of a jelly (Deane 1946) or dried on a filter paper —the *patch test* of Vollmer and Goldberger (1937, 1939*a,b*). Several workers have introduced tuberculin into the skin by needle puncture, with either a single wide-bore needle (Trambusti 1928), a spring-loaded multiple-puncture apparatus (Heaf 1951), or a multiple-tine apparatus bearing dried tuberculin (Rosenthal 1958, 1966).

The two tests in regular use today are the Mantoux and Heaf tests, the former on account of its accuracy and the latter on account of its simplicity. The tine test has also gained some popularity since the introduction of commercially available tine disks, which are particularly labour saving for testing single patients or small groups.

There are, however, differences in opinion as to its reliability. Lunn and Johnson (1978) found that the Mantoux test gave positive results in 50 per cent of 307 subjects, with induration of at least 10 mm in diameter occurring in 35 per cent. The tine test gave positive results in a mere 4 per cent of the same group, a further 16 per cent being doubtfully positive. In contrast Sinclair and Johnston (1979) found the tine test to be only marginally less sensitive than the Mantoux

test. Other workers found that the tine test detected between 88 and 97 per cent of Mantoux reactors (Rosenthal 1961, Capobres 1962, Badger *et al.* 1963). It is essential to allow the tines to remain in the skin long enough for the PPD to dissolve in the tissue fluids. Caplin (1980) considered the test was adequate for screening purposes but that a negative result should be confirmed by a Mantoux test.

The Mantoux test is the most satisfactory of all tests for epidemiological studies, as an exact amount of material can be introduced into the skin and the degree of reactivity can be accurately assessed by measuring the diameter of induration. The testers must be trained to carry out intradermal injections correctly. Use of the Heaf test requires very little training; the instrument propels six needles into the skin to a depth of 2 mm and the punctures are made through a drop of PPD (2 mg per ml) placed on the skin. Positive results are recorded as follows: grade 1 (four or more papules at the puncture sites); grade 2 (confluence of the papules into a ring); grade 3 (a single large plaque); and grade 4 (a plaque with vesicle formation or central necrosis). The test is read after 72 hours but may be read satisfactorily even after 7 days (Report 1959*b*). Although lacking the precision of the Mantoux test, which can be used to confirm doubtful reactions, the Heaf test is satisfactory for routine use (Stewart *et al.* 1958).

The dosage of tuberculin, whether OT or PPD, may be expressed either as a dilution of the pure product; by weight, taking 0.1 ml of a 1/100 dilution of OT to correspond to 1 mg; or in unit form, 1 unit (TU) being equivalent to 0.1 ml of a 1/10 000 dilution of the International Standard OT, or to 0.000 02 mg of PPD.

The usual dose used in the Mantoux test is 5 TU of OT or PPD, and an induration of at least 5 mm in diameter after 48–72 hours is regarded as positive. Negative reactors may be retested with 100 TU, but positive reactions thus obtained may be non-specific (see p. 44).

The size of the tuberculin reaction is dependent upon factors which influence the persistence of injected antigen at the test site (see Pepys 1979). (For a detailed account of tuberculin tests see Caplin 1980.)

Reactions elicited by mycobacterial sensitins

The skin-test reaction to tuberculin is frequently cited as the classical example of the type IV, or delayed-type, hypersensitivity of Coombs and Gell (1975). It is now apparent that mycobacterial sensitins elicit several responses mediated by quite distinct immunological mechanisms (Pepys 1979, Rook and Stanford 1979, Kardjito and Grange 1980). The nature and significance of three of these reactions, namely the Jones-Mote, 'Listeria-type' and 'Koch type' reactions are discussed on p. 48.

In our opinion the tuberculin test should be critically re-evaluated with respect to both the antigen and the significance of the responses. In particular, there is a need to distinguish between the responses that are related to protection and those that are not.

Epidemiological use of the tuberculin test

The value of the tuberculin test for assessing the prevalence of tuberculous infection was first realized by von Pirquet (1909) in Vienna. Examining 1134 clinically non-tuberculous children, mainly from poor families among which tuberculosis was common, he found that the proportion of positive reactors rose from 0 per cent in the first two years of life to 34 per cent at the age of 5, and 91 per cent at the age of 14. In other European cities 97 per cent of adults reacted positively (Mantoux 1910, Calmette *et al.* 1911). These findings agreed well with post-mortem studies carried out at that time on patients dying from various causes. Thus tuberculous lesions, though uncommon in infants, were found in over 90 per cent of adults (Naegeli 1900, Burkhardt 1906); hence the adage '*Jedermann hat am Ende ein Bisschen Tuberculose*'. (Eventually everyone gets some tuberculosis.) As the incidence of tuberculosis fell throughout the more prosperous countries, so the number of positive reactors declined; and the test indicated in which regions and circumstances there was a high risk of infection. Thus children from wealthy rural areas became tuberculin-positive at a slower rate than those from urban areas. The presence of an open case of tuberculosis in the household significantly increased the conversion rate (Dow and Lloyd 1931). In time doubts arose as to the specificity of the tuberculin reaction; and it was suspected that various non-pathogenic mycobacteria could induce sensitivity to tuberculin. Thus McCarter and co-workers (1938) found that many students who reacted to human PPD also reacted, though less strongly, to PPD made from *M. avium* or *Mycobacterium smegmatis*. By carefully measuring the size of reactions, Edwards and Palmer (1953) concluded that most reactions of less than 10 mm in diameter were non-specific. Some small reactions have, however, been reported in elderly persons with a history of tuberculosis (Caplin *et al.* 1958).

Despite these disadvantages, tuberculin-testing is regarded as the best way to assess the annual infection rate of tuberculosis in communities in which BCG is not used (Styblo 1978).

In recent years there has been a renewal of interest in the use of sensitins to study the effect of contact with various mycobacteria in the environment of man and animals. Abrahams (1970) had suggested that such contact might modify the host's immune reactions to subsequent challenges with mycobacterial antigens.

Diagnostic use of the tuberculin test

The tuberculin test is not a reliable diagnostic tool in tuberculosis, as it fails to distinguish active from healed tuberculosis. A positive test is a strong indicator of disease in non-vaccinated children in developed countries, especially when a previous test was negative. In some countries such as the USA, where tuberculosis is uncommon and BCG is not widely used, a conversion to tuberculin-reactivity is of some diagnostic value in adults. Comstock and Woolpert (1978) considered that the tuberculin test was able to detect infection by *M. tuberculosis* in the USA provided that the test was performed and read correctly, and that small doses of PPD were used to avoid the production of sensitization by the skin-test antigen itself. Grzybowsky (1979), commenting on this statement, emphasized the need for further evaluation of the tuberculin test in order to define its practical use more closely. Comstock and Woolpert (1979), in what must be the shortest communication in the annals of science, concurred with this view.

The tuberculin test and BCG vaccination

The lack of a relation between tuberculin positivity and protective immunity invalidates the usefulness of the tuberculin test in the assessment of the success, or otherwise, of vaccination. Indeed the practice adopted by some physicians of repeating the vaccination until a negative tuberculin reaction becomes positive may even elicit a harmful immune response. For similar reasons it is inadvisable to vaccinate persons who are already tuberculin-positive as a result of subclinical infection. Low-grade reactions are usually due to contact with mycobacteria other than *M. tuberculosis* and in Great Britain it is recommended that health service employees showing such reactions (less than 5 mm of induration on Mantoux testing or grade 1 on Heaf testing) should be vaccinated (Report 1978*e*).

The prophylaxis of tuberculosis

Protection from tuberculosis may be afforded by public health measures, by vaccination, or by chemoprophylaxis.

Prevention of infection

Before the introduction of effective chemotherapy, tuberculosis was preventible only by general public health measures (see Griffith and Denaro 1956). Segregation of infected patients was at one time practised (Stallybrass 1949), but the introduction of effective chemotherapy has altered the approach to prevention (see p. 35). The risk of infection from cattle has been almost eliminated in most of the developed countries y a combination of pasteurization of milk and the building up of tuberculin-tested herds (see p. 53). At the present time the major health hazard is the undiagnosed smear-positive case of tuberculosis.

Vaccination

Villemin (1868) predicted the isolation of the specific tuberculosis 'virus' and the production of a vaccine. After the isolation of the tubercle bacillus by Koch in 1882 numerous attempts were made in animals and man to vaccinate against tuberculosis (see review by Rosenthal 1957). None of the early vaccines proved satisfactory, and the general opinion grew that dead tubercle bacilli had little or no protective action. This conclusion has been challenged by numerous workers, particularly Weiss and Wells (1960), who maintained that under appropriate conditions dead vaccines were as effective as living BCG vaccine (see review by Weiss 1959). The same opinion was expressed by Dubos (1964) for tubercle bacilli killed by ethylene oxide. Youmans and Youmans (1969), however, from their experiments on mice, were convinced that there was a qualitative difference in immunogenicity between living and dead cells, owing to the presence in the living cells of a potent labile substance contained in the ribosomes.

A great deal of work has been carried out to determine the chemical nature of the immunogenic fraction in the tubercle bacillus. The results were reviewed by Crowle (1958), who concluded that such immunizing action as there was appeared to be contained principally in the phosphatide fraction present in a methanolic extract of autoclaved acetone-washed air-dried bacilli—the methylic antigen of Nègre and Boquet (1925). A similar immunizing effect could be obtained by a preparation of cell walls coated with a mineral oil (Brehmer et al. 1968). Youmans and Youmans (1969) attributed immunogenicity to a labile RNAase-sensitive substance found in the mycobacterial ribosomes obtained by ultracentrifugation. Smith and co-workers (1968), however, who reviewed the literature of the previous ten years, found no agreement among different workers as to which fraction was the most immunogenic. The results depended on the method used to measure immunity, the way in which the fractions were extracted, the amount of bacillary residue in the fractions, the use of adjuvants, intercurrent infection, non-specific resistance, and the type of statistical analysis. They recommended that in future trials a standard reference vaccine should always be used alongside the test preparation, and that the cri-

teria of protective potency should be agreed upon. The unsatisfactory nature of the existing methods was borne out by a comparative trial in which nine laboratories tested the potency of five vaccines, each using its own method. The results varied so greatly that they might equally well have been obtained by simple randomization (Wiegeshaus et al. 1971). Other species of mycobacteria have also been used as vaccines. Friedmann (1903) advocated the use of the 'Turtle Tubercle Bacillus' (*M. chelonei*); its use continued sporadically for many years (see, for example, Fowler 1930). Avian tubercle bacilli were used for a short time in an attempt to vaccinate cattle against tuberculosis (McFadyean et al. 1913).

Living virulent tubercle bacilli were used on a small scale by Webb and Williams (1911; see also Baldwin and Gardner 1921), but this practice was soon given up as being too dangerous. The use of virulent bacilli for therapeutic purposes was advocated by von Aichbergen (1937), who rubbed them into the scarified skin in order to produce lupoid lesions. This procedure was based on the observation that patients suffering from lupus vulgaris seldom contracted serious lung disease.

Attempts to produce a living attenuated vaccine began with the studies of Dixon (1889). Subsequently von Behring and his associates produced an attenuated strain of human *M. tuberculosis* for use as a vaccine in cattle. This vaccine was widely used until it was found that vaccinated animals excreted living human-tubercle bacilli in their milk (Griffith 1913).

BCG vaccine

BCG (Bacille Calmette-Guérin) was produced by Calmette and Guérin in Lille as a result of a chance observation that a bovine strain of *M. tuberculosis* lost its virulence when subcultured repeatedly on a medium containing ox bile; the organism was transferred to fresh medium every three weeks from 1908 to 1921. The vaccine was first administered orally to human infants by Weill-Hallé in 1921 (see Irvine 1957). There was much initial opposition to it, and it received a severe setback when, in 1930, 73 children died at Lübeck because virulent tubercle bacilli were accidentally administered instead of BCG (Report 1935b). Calmette died in 1933 'saddened by the ingratitude of men' (Guérin 1957).

The vaccine, which requires careful standardization (see Levy et al. 1968) and must be protected from sunlight (Edwards et al. 1953), may be dispensed moist or freeze-dried. Although originally administered orally, it is now usually given by intracutaneous injection, but dermal inoculation by multiple puncture (Birkhaug 1944, Rosenthal et al. 1961a,b), by jet injection (see Power 1967, Dam et al. 1970), or by scarification (Nègre and Bretey 1947) is preferred by some workers.

Table 51.3   The protective effect of BCG vaccine as observed in eight major studies

Group studied	Period of study	Age range	Protection (per cent)
North American Indian	1935–38	0–20 years	80
Chicago, USA	1937–48	3 months	75
Georgia, USA	1947	6–17 years	0
Illinois, USA	1947–48	Young adults	0
Puerto Rico	1949–51	1–18 years	31
Georgia, USA	1950	5 years	14
Great Britain	1950–52	14–15 years	78
South Indian	1950–55	All ages	31

An evaluation of the protective role of BCG vaccination is not easy. In particular it is very difficult to provide an identical control group. Before 1970, eight major controlled field trials of BCG were carried out. They are summarized by Hitze (1980) and in Report (1972). The results, shown in Table 51.3, varied greatly, the estimated degree of protection ranging from 0 to 80 per cent.

In Great Britain, the Medical Research Council Tuberculosis Vaccines Clinical Trials Committee organized an extensive study that began in 1950. Almost 55 000 children aged 14 to $15\frac{1}{2}$ years in three English cities were vaccinated with BCG or the vole tubercle bacillus (see below). During the following five years these two vaccines gave 78 per cent and 81 per cent protection respectively. After 10–15 years protection waned to 59 per cent and 73 per cent respectively. All types of tuberculosis were prevented, but there was no evidence that vaccinated persons who developed the disease were less severely affected than non-vaccinated persons.

By contrast, a recent study in India of 260 000 subjects did not show any evidence of a protective effect of BCG over a seven-and-a-half-year period (Report 1980a). Several reasons have been advanced to explain the variations between these two major trials, including differences in the potency of the vaccine, genetic and nutritional factors, and the effect of prior contact with mycobacteria in the environment. Rook and Stanford (1979) have shown that there are two types of cell-mediated immune response to mycobacteria, both of which produce positive skin tests. These are termed the Listeria-type and Koch-type reactions; the former is protective, but the latter is less so and may even be antagonistic. Some mycobacterial species induce the Listeria type and others the Koch type; both types are enhanced by subsequent BCG vaccination. It has therefore been postulated that the protective efficacy of BCG in a community is predetermined by the mycobacterial type in the environment (Stanford et al. 1981). On the other hand, Clemens and his colleagues (1983) analysed eight of the major controlled trials and found that those carried out under the best conditions showed the best protection, whereas those carried out under poorly controlled

conditions showed the poorest protection. Bias or statistical incompetence may therefore have contributed to the poor results.

Although BCG reduces the incidence of clinical disease, including tuberculous meningitis, in children after a primary infection, there is little or no evidence that it affords any protection against post-primary tuberculosis in adults (Dahlström 1953). The degree of immunity cannot be assessed by the degree of tuberculin sensitivity (Wijsmuller 1966); nor should it be assumed—though it often is—that immunity wanes at the same rate as tuberculin sensitivity, and that when sensitivity is lost revaccination is required (Hart et al. 1967).

In countries where tuberculosis is still prevalent most workers would agree that BCG vaccination should be offered to infants (Report 1979b) and to other specially exposed groups such as medical students and nurses. The value of vaccinating the whole population is more questionable. Where the prevalence of tuberculosis is very low, as in some parts of the United States, vaccination has little to offer. Indeed it nullifies the value of the tuberculin test, which, as the prevalence of tuberculosis declines, becomes more and more useful as a diagnostic and epidemiological tool. Official opinion in the United States is against vaccination (Report 1959a), reliance being placed on general anti-tuberculosis measures, including surveillance, early diagnosis and chemoprophylaxis. Both in the United States and in Great Britain the danger of suffering from clinical tuberculosis is greatest among the tuberculin-positive reactors, especially those with a high degree of reactivity. By the judicious use of the tuberculin test and of radiography those most at risk can be detected and the disease diagnosed at an early stage.

Although BCG may, in some areas, protect against tuberculosis, it does little to reduce the incidence of *infectious* cases of the disease. Most of the cases of primary tuberculosis in children and young adults, which are preventible by BCG vaccination, are smear-negative and therefore non-infectious. The post-primary infectious cases in older persons are not significantly prevented by BCG campaigns. It has been calculated that the effect of BCG in preventing infectious cases is only 0.3 to 2 per cent annually (Styblo and Meijer 1976, 1978).

The decision as to whether a mass BCG campaign is desirable in a given region depends on financial and organizational considerations as well as epidemiological ones. (For a mathematical model for decision making on this subject see Rouillon and Waaler 1976.) Safety is one factor to be considered. Complications following vaccination include local abscesses and regional suppurative lymphadenitis, particularly in infants, localized or generalized disseminated lesions and keloid scars (Lotte *et al.* 1980). In most parts of Euope BCG-induced osteitis is rare (0.58 cases per million vaccinated children), but the incidence is much higher in Sweden and Finland (34.8 and 45.8 per million, respectively (Wasz-Hockert and Lotte 1980). Fatal disseminated BCG infections have been reported (see Šićević 1972) but are extremely rare and are probably restricted to immunodeficient persons.

Vaccination with the vole bacillus

The favourable results of studies by Wells and Brooke (1940) and Birkhaug (1944) on guinea-pigs, and of Griffith and Dalling (1940) and Young and Paterson (1949) on calves, indicated that a living vaccine made with the vole bacillus could give rise to an immunity equal to or greater than that given by BCG. Field trials on human beings have confirmed the truth of this conclusion. In Czechoslovakia Šula (1958) used a live attenuated strain of vole bacillus; in this trial tuberculosis developed in three out of 32 772 persons—mainly infants and children—vaccinated with it, and in 13 out of 76 631 vaccinated with BCG. In Britain, vole vaccine compared very favourably with BCG (see above, and Report 1972). Some batches of vole vaccine caused a small number of cases of lupus 1–2 years after vaccination (Frew *et al.* 1955). These batches, although protective, induced a high degree of tuberculin hypersensitivity. Another batch was equally protective but induced only a low degree of tuberculin hypersensitivity and did not cause lupus (Report 1971, 1972). The Medical Research Council Tuberculosis Vaccines Clinical Trials Committee concluded that it was worth considering the reintroduction of the vole vaccine (Report 1972).

Chemotherapy and chemoprophylaxis of tuberculosis

Chemotherapy has revolutionized the treatment of tuberculosis. Operative intervention, including artificial pneumothorax, resections and thoracoplasty belong to a bygone age. Likewise the traditional sanatorium regimens of rest, fresh air and good food are obsolete. Patients, even those with advanced disease, who live in poor conditions and engage in hard manual labour respond to chemotherapy as well as those treated in hospital, and bed rest is unnecessary (Report 1959a, Fox 1964, 1968, 1972, 1977).

The anti-tuberculous agents include rifampicin, isoniazid, pyrazinamide, streptomycin and other aminoglycosides (viomycin, kanamycin, capreomycin), ethambutol, ethionamide, thiacetazone, para-aminosalicylic acid (PAS) and cycloserine. Of these the first four are bactericidal, whereas the others are bacteristatic. Mitchison and Dickinson (1978) and Grosset (1978a) have shown that rifampicin and pyrazinamide are particularly effective in sterilizing tuberculous lesions. The ability of a drug to sterilize is not directly related to its bactericidal activity. Isoniazid is very effective in causing an initial drop in the viable count of bacilli in the sputum; rifampicin is more effective at eliminating residual organisms, presumably because it can kill dormant organisms in which only short bursts of metabolic activity occur. Rifampicin, alone among the antituberculous agents, retains its bactericidal activity when the multiplication of the bacilli is very slow, provided that the pH is neutral (Grosset 1978b). Pyrazinamide acts only against intracellular organisms at a low pH, and streptomycin destroys only extracellular bacilli in an alkaline environment; thus neither can be regarded as a 'complete' drug.

There is general agreement that the treatment of tuberculosis requires the use of at least two drugs. Resistance to antituberculous agents occurs at a low but constant rate, so that the use of a single drug is likely to lead to the selection of organisms resistant to it. Moreover, as just described, some agents act on only part of the mycobacterial population.

For many years the standard treatment for tuberculosis consisted of streptomycin, isoniazid and PAS. All three drugs were given daily for 3 months, and the latter two for a total of 2 years. In regions where the administration of streptomycin was too costly or impracticable an oral regimen of isoniazid and thiacetazone was used. Since the introduction of rifampicin, much effort has been put into the design of short courses of treatment. A standard regimen used in Great Britain consists in an initial 2-month period of daily rifampicin, isoniazid and ethambutol followed by the latter two drugs for a total of 9 months (British Thoracic and Tuberculosis Association 1976). A similar course was designed in France, but streptomycin was substituted for ethambutol (Brouet and Rousell 1977). These regimens require a total of 270 doses, a considerable reduction in the 3000 doses used previously, when drugs were administered four times a day in the erroneous belief that constantly high concentrations in the serum were essential. More recently even shorter regimens have been designed (Report 1974*a*, 1975*a,b*, 1976*a,b*, 1977, 1978*a,b,c,d*, Fox and Mitchison 1975, Kreis 1976, Fox 1977, 1978). The aim of these investigations has been to devise the shortest possible regimen that will not lead to bacteriological relapses. These regimens have been briefly reviewed by Fox (1978). Most are based on intensive treatment with the four bactericidal drugs rifampicin, isoniazid, pyrazinamide and streptomycin for 2 months, followed by treatment with rifampicin and isoniazid for a further 4 months. By administering the drugs three times each week during the intensive phase, and twice weekly thereafter, the number of doses is reduced to 64 with little loss of efficacy. The advantage of such short-course regimens are the lower cost of the drugs, the lower incidence of side effects, the lesser likelihood that patients who abscond will relapse, and the easier supervision of the patient.

Albert and his colleagues (1976) and Rouillon and her associates (1976) consider that, although it is necessary to supervise the patient *during* modern courses of chemotherapy, very little is gained by extensive follow-up examinations. Furthermore, most cases of relapse are detected by the reappearance of symptoms and not by routine clinical, radiological and bacteriological examinations. A further benefit of modern chemotherapeutic regimens is that the bactericidal drugs rapidly render the patient non-infectious.

Treatment with rifampicin-containing regimens causes a 40-fold diminution of the numbers of bacilli in the sputum within two weeks (Brueggeman *et al.* 1974); and most patients become smear-negative within the same period (Hobby *et al.* 1973). Patients may continue to be culture-positive for much longer (Rouillon *et al.* 1976); but the reduction in infectivity may not be due solely to killing of the bacilli. Loudon and Spohn (1969), for instance, found that the frequency of coughs was 35-fold less after two weeks of therapy. It has been postulated that the concentration of isoniazid in droplet nuclei after evaporation may render them non-infectious (Loudon *et al.* 1969).

The chemotherapeutic regimens described above are suitable for all forms of tuberculosis. Renal tuberculosis has been successfully treated with a 6-month regimen of rifampicin, pyrazinamide and isoniazid (O'Boyle and Gow 1976) and, although not proved by clinical trials, tuberculosis of the central nervous system appears to be susceptible to treatment by these three drugs (Report 1976*b*). Bone and joint tuberculosis also respond well to 6-month regimens. A few failures occur in cases in which the spine is affected; for these a longer period of treatment is recommended (Report 1979*a*).

Some workers supplement antituberculous chemotherapy with steroids, the rationale being that, by reducing the host's resistance, the bacilli multiply more rapidly and thus become more susceptible to the bactericidal action of the drugs. Such therapy, however, does not modify the effect of modern short-course treatment (Fox 1978). Steroids are life-saving in some cases of tuberculosis of the central nervous system, especially when cerebral oedema occurs (Gordon and Parsons 1972); and they reduce the occurrence of effusions in tuberculous pericarditis (Williams and Hetzel 1978).

Chemoprophylaxis

The term chemoprophylaxis should be used only to describe the use of anti-tuberculous agents to prevent non-infected persons from becoming infected. In practice, the term is often applied to the administration of drugs to persons who are infected but show no clinical signs. Such persons fall into two groups: first, contacts, usually children, who have become tuberculin-positive and, second, older persons with radiological evidence of quiescent disease.

True chemoprophylaxis has a very limited place in the prevention of tuberculosis. The best protection in uninfected infants is BCG vaccination and treatment of the source case who, as described above, will rapidly become non-infectious. It is no longer considered necessary to use an isoniazid-resistant mutant of BCG when vaccination and isoniazid therapy are used together (Vandiviere *et al.* 1973).

Children with clinical symptoms should receive a curative drug regimen as described above. Children with symptomless primary lesions run a serious risk of severe disease, including miliary and central nervous system tuberculosis, especially if they are under 3 years of age. Some authorities recommend isoniazid treatment of unvaccinated symptomless children who are known to have become tuberculin-positive (see Ferebee, 1970); this affords about 80 per cent protection (Horvat *et al.* 1979).

The American Thoracic Society (1967) recommended that every unvaccinated adult who is found to respond to 5 TU of PPD with a reaction of more than 10 mm in diameter should receive insoniazid daily for one year. In particular, it was considered that positive reactors with radiological evidence of quiescent disease, with a history of untreated or inadequately treated disease, with other diseases especially diabetes and Hodgkin's disease, and with known exposure to a source case, should be treated. Duboczy (1968) doubted the wisdom of such widespread use of isoniazid; and Krebs and Riska (1976) drew attention to the hepatotoxic effect of isoniazid. As the risk from isoniazid-induced hepatitis increases with age, the American Thoracic Society (1974) set the upper age limit for preventive therapy at 35 years unless additional risk factors existed (see also Comstock and Edwards 1975). The risk of overt disease developing in patients with fibrotic pulmonary lesions varies from country to country but on average is 2.9 cases for every 1000 person-years of risk (Krebs *et al.* 1979).

Isoniazid reduces this risk but there is still no definite policy as to whether such patients should be treated or merely kept under observation in case signs of reactivation occur.

Sensitivity testing

The genetic basis of acquisition of resistance to anti-tuberculous agents is described in Chapter 24, and reference has been made above to the need to administer more than one drug to prevent the selection of resistant strains. Sensitivity testing is used to determine the resistance of strains isolated before the commencement of treatment (primary resistance); or to discover whether resistance arises during treatment (secondary resistance). The two techniques for sensitivity testing in general use are the *Resistance ratio* (Marks 1976) and the *Proportion* (Vestal 1975) methods. In the first method Löwenstein-Jensen medium incorporating doubling dilutions of drugs is inoculated with organisms; the end point is expressed as a ratio of the drug concentration inhibiting the test strain to the concentration inhibiting a known sensitive strain, or to the average concentration inhibiting a number of sensitive strains. In the second method the proportion of bacteria in the culture capable of growing in the presence of a given concentration of

drug is estimated by performing colony counts of sub-cultures on drug-containing and drug-free media. The strain is regarded as resistant when more than 1 per cent of the bacteria grow in the presence of the drug. Both these methods are expensive and time-consuming and require rigorous quality controls if reliable results are to be obtained.

In designing chemotherapeutic regimens, sensitivity testing is essential to determine whether therapeutic failures are due to primary or secondary resistance. The value of sensitivity testing in routine practice is, on the other hand, more questionable. Studies in Hong Kong (Report 1974b) indicated that it contributed practically nothing to the outcome of treatment with streptomycin, isoniazid and PAS. It would, therefore, make no difference to the effect of modern short-course chemotherapy, the intensive phase of which would be completed before the sensitivities were known. Grosset (1978c) has argued that, even in technically advanced countries, the diagnosis and management of tuberculosis is often far from ideal and that manpower and financial resources should be used more profitably than in establishing sensitivity-testing services. In areas where the level of primary resistance to one of the first-line drugs is 5 per cent or less, it is unlikely that sensitivity testing benefits more than 1 in 150 patients (Fox 1977).

Infections in man with other culturable mycobacteria

Many species of mycobacteria, although normally existing as environmental saprophytes, occasionally cause disease in man and animals. Such infections are divisible into those resembling tuberculosis, and two with characteristic features which have led to their being given specific names—*Buruli ulcer* and *swimming-pool granuloma.*

Buruli ulcer

This is, after tuberculosis and leprosy, the most important specific mycobacterial disease in man. The causative organism, *Mycobacterium ulcerans*, is described in Chapter 24. Epidemiological evidence (Barker 1973) strongly indicates that this organism is derived from the environment, although several attempts to isolate it from sources other than infected human tissue have failed. The disease occurs in many tropical areas and in the Bairnsdale district of Australia (Mac-Callum *et al.* 1948). It is prevalent in Uganda (Dodge 1964, Report 1971), where it occurs particularly in swampy areas with a surface water pH of 6.1–6.9 (Stanford and Paul 1973). The thorns of a tall prickly grass, *Echinochloa pyrimidalis*, may introduce the organisms into the skin (Barker 1973).

In its early stage the lesion is a firm painless subcu-

taneous nodule fixed to the skin. It may cause itching; and in Zaire this stage of the disease is known as 'Mputa Matadi'—the 'itching stone'. Histologically the organisms are found in the subcutaneous connective tissue surrounding the fat lobules. As the disease progresses, the fat and the overlying skin become necrotic and deeply undermined ulcers are formed. Any part of the body may be affected but most commonly the limbs. Metastatic bone lesions occur in a minority of patients, usually children. Occasionally massive oedema of the affected limb occurs (Report 1970). In some cases early non-ulcerating lesions resolve; in others the disease progresses for several months or years; eventual healing is usual, but frequently with gross scarring and deformity and crippling due to contractures.

This disease displays some interesting immunological features. During the progressive phase numerous organisms are present in the lesions and there is a minimal cellular infiltration. During this phase the patient does not react to challenge with burulin, a specific skin-testing reagent prepared from *M. ulcerans* (Stanford *et al.* 1975). Thus there is an analogy between this stage of the infection and lepromatous leprosy or anergic tuberculosis. When the healing phase begins, the bacteria become fewer, cellular infiltration and granuloma formation ensue, and the patient reacts to burulin. At this stage the disease resembles tuberculoid leprosy. The reason for this change from anergy to reactivity is unknown.

M. ulcerans is sensitive to rifampicin *in vitro* and in experimentally infected mice (Stanford and Phillips 1972), but clinical trials in man have proved inconclusive. The usual treatment is excision, and either primary closure or skin grafting.

Mycobacterium marinum infection

This uncommon disease is also known as *swimming-pool granuloma* and *fish-tank granuloma*. Most outbreaks have occurred among users of swimming-pools (Linell and Norden 1954, Philpott *et al.* 1963), but cases have also been caused through abrasions acquired while working with fish tanks (Swift and Cohen 1962, Adams *et al.* 1970, Maonsson *et al.* 1970, Heineman *et al.* 1972). The lesions are single or multiple and appear as wart-like excrescences usually on the knees, elbows, or hands (see Adams *et al.* 1970). The disease is usually superficial and self-limiting. More serious lesions such as tenosynovitis and osteomyelitis (Williams and Riordan 1973) are extremely rare.

The value of chemotherapy in *M. marinum* infection is difficult to evaluate in view of the natural tendency of lesions to heal. Antibacterial agents which appear to be clinically useful include tetracycline (Kim 1974), minocycline (Loria 1976), trimethoprim with sulphamethoxazole (Kelly 1976), and rifampicin with ethambutol (Wolinsky *et al.* 1972).

Non-specific infections

In addition to the mycobacteria that cause the diseases referred to above, several species have been isolated from pulmonary disease, lymphadenitis, skin abscesses, soft-tissue abscesses, bone-and-joint lesions, bursitis, tenosynovitis, meningitis, genito-urinary disease, corneal ulcers and disseminated lesions (see review by Wolinsky 1979).

Of the rapid-growing mycobacteria the only common pathogens are *M. chelonei* and *M. fortuitum* (formerly the turtle and frog tubercle bacilli respectively). Most of the pathogenic strains of *M. fortuitum* are of a particular serotype and are limited in their range of saccharolytic activity (Grange and Stanford 1974); the other serotypes and biotypes are commoner in the environment. The literature gives the false impression that infections due to *M. fortuitum* are commoner than those due to *M. chelonei*, but this is because of the frequent failure to identify the latter organism correctly (Grange and Stanford 1974). Other rapidly growing mycobacteria rarely if ever infect man or animals.

Most of the slowly growing mycobacteria, on the other hand, are capable of causing disease. By far the most important species is *M. avium* and its *intracellulare* variant, which is synonymous with the so-called 'Battey Bacillus'. Less frequently encountered mycobacterial causes of non-specific disease are *M. scrofulaceum*, *M. kansasi*, *M. xenopi*, *M. szulgai*, and *M. malmoense*.

The incidence of disease due to these opportunist pathogens is determined by their distribution in the environment, and by the natural or iatrogenic opportunities for the establishment of infection. Their worldwide distribution varies considerably (see Chapter 24) and the incidence of disease varies likewise. Wolinsky (1979) has summarized reports listing the relative incidence of infection due to *M. kansasi* and *M. avium* in different parts of the world. The incidence varies greatly both between and within countries. In contrast to *M. tuberculosis*, these mycobacteria spread from person to person very rarely. Although there are a few anecdotal accounts of infection within families (Beck *et al.* 1963, Onstad 1969, Lincoln and Gilbert 1972), these infections could equally well have been due to common exposure to an environmental source as to spread from person to person.

Predisposing causes of infection by the mycobacteria described above include chronic pulmonary disease, autoaggressive diseases, immunosuppression, and exposure to mycobacteria by trauma or surgical treatment. Occupational dust exposure is an important predisposing factor in mycobacterial lung disease (Schaefer *et al.* 1969, Report 1975*b*, Nel *et al.* 1977). Malignancies, particularly those of the lymphoreticular system, also increase the susceptibility to mycobacterial infections (Ortbals and Marr 1978). For a fuller account of the diseases caused by the so-called 'atypical' (nyrocine; see Grange and Collins 1983) mycobacteria, see Report 1981. Numerous papers of interest on various aspects of tuberculosis will be found in Report (1982.)

Tuberculosis in animals

Frequency and bacteriology

Tuberculosis is widespread throughout the animal kingdom. It is common among certain species of domesticated animals, and, though usually uncommon in wild animals living under feral conditions, it occurs not infrequently in them when they are brought into captivity. Cattle, pigs and fowls are the most frequently affected of stock animals (see Myers and Steele 1969).

Bacteriologically, disease is due to *M. tuberculosis* (human, bovine, and vole types) and *M. avium*. Other species are rarely encountered in birds and mammals but occur in cold-blooded animals.

Tuberculosis in cattle

Tuberculosis was formerly one of the most serious diseases of cattle in Great Britain, and is still a major problem in many parts of the world. In 1946 it was estimated that 17–18 per cent of all cattle and 30–35 per cent of all cows in Great Britain reacted to the tuberculin test (Ritchie 1945-6). About 40 per cent of animals slaughtered in public abattoirs showed macroscopic lesions, and about 0.5 per cent of all dairy cows excreted *M. tuberculosis* in their milk (Report 1932, Savage 1933). The incidence increases with age (see Gofton 1937). The disease is usually progressive; hence all tuberculin-positive reactors are potentially dangerous as a source of infection to other animals (see Francis 1958).

Infection occurs mainly by the aerogenous route and gives rise to a gradually spreading pulmonary lesion with extension of the infection to the pleurae and intrathoracic lymph nodes. Other viscera, including the liver, are sometimes affected. Lesions occur in the udder in advanced disease, but cattle may excrete bacilli in their milk in the absence of histologically demonstrable disease in the udder (Gaiger and Davies 1933, Daigeler 1953).

M. avium occasionally infects cattle but, owing to the virtual eradication of infection with the bovine type of *M. tuberculosis*, its relative importance in Britain has increased greatly (Lesslie and Birn 1970). The lesions are usually localized, but 'open' cases, particularly of mastitis, occur (see Boughton 1969). Infection by the human type of *M. tuberculosis* is rare and never causes progressive disease.

Diagnosis of tuberculosis in cattle

The tubercle bacilli may be demonstrated directly in the animal's secretions or excretions by microscopy, culture, and animal inoculation, or an indirect diagnosis may be arrived at by tuberculin testing.

Demonstration of tubercle bacilli in milk   Tubercle bacilli are sometimes demonstrable in sputum collected from the walls of the byre (Riddoch 1903), but it is more usual to examine milk.

Milk is obtained separately from each quarter of the udder for microscopical examination and processed according to Maitland (1937; see also Francis 1958, p. 68).

The milk is centrifuged at 2500 rpm for 3 minutes; the deposit is spread on glass slides and air-dried for 30 minutes. The slides are heat-fixed, defatted with a mixture of equal parts of alcohol and ether for 15 minutes, and then stained by the Ziehl-Neelsen method. Tubercle bacilli in milk are usually associated with clumps of endothelial cells (Torrance 1922, 1927). It is therefore only necessary to examine the film under low magnification for these clumps, and to reserve the oil-immersion lens for films in which such clumps are present. Microscopy is a sensitive technique which is positive in up to 90 per cent of cases detected by guinea-pig inoculation (Maitland 1950).

Cultivation of the bacilli is a method not often used; it lacks both speed and sensitivity. Care must be taken to avoid contamination of the milk by mycobacteria present in the surrounding environment. Over a quarter of all samples examined by Albiston (1930) and by Chapman and his colleagues (1965) were contaminated by such mycobacteria.

Guinea-pig inoculation is a sensitive way of detecting bovine tubercle bacilli in milk. Fresh milk should be used, as the tubercle bacilli may be killed by substances released by other bacteria such as streptococci (Mattick and Hirsh 1946). Alternatively 2 per cent borax may be used as a preservative (Stevens and Soltys 1952).

The tuberculin test in the diagnosis of bovine tuberculosis   This test is widely used in cattle and is performed by (a) an intradermal injection into the skin of the neck (Great Britain) or the caudal fold (USA) and measurement of the local reaction 3–5 days later, or (b) a subcutaneous injection followed by a series of temperature readings during the following 24 hours, or (c) instillation of tuberculin into the conjunctival sac with observation of conjunctival swelling and discharge 3–10 hours later. Of these three tests, the intradermal is much the most widely used.

As in man, non-specific reactions occur in animals owing to contact with mycobacteria in their environment (Minett 1932, Penso *et al.* 1952, Kazda 1967). Of the species giving rise to cross-reactions, *M. avium* is the most important; hence in Great Britain it is routine practice to use the *single intradermal comparative test*, in which 0.1-ml doses of mammalian and avian PPD are injected into separate sites on the neck (see Green 1946). Animals whose reactions are due to sensitization by *M. avium* react more strongly to the avian than to the mammalian PPD. The interpretation of the test and the factors influencing it are discussed by Ritchie (1964). As in man, negative tuberculin tests are occasionally encountered in animals with advanced disease.

Control of tuberculosis in cattle

Early attempts to eliminate this disease relied on the detection of infected animals by clinical and bacteriological examination. This method was widely practised in Germany from 1900 to 1913 at the instigation of von Ostertag but was found to be ineffective (Ostertag 1927). Such methods fail because diseased animals infect many others before they are detected (Brook 1937).

Effective control methods are based on the detection of disease by tuberculin testing. In the original 'Bang method', initiated in Denmark by Bang in 1894 (see Bang 1928, 1929, 1930, 1932), positive reactors were segregated but for reasons of economy were still used for milk production unless clinically obvious disease developed. When the incidence of positive reactors had declined sufficiently, a compulsory total eradication scheme was introduced in which tuberculin reactors were destroyed and their owners adequately compensated (Nielsen 1947).

The general method now is to test every animal in the herd with tuberculin, to segregate, remove, or slaughter the positive reactors, and to build up the herd exclusively from negatively reacting animals. The tuberculin test must be repeated at 3- or 6-monthly intervals to detect any fresh reactors in the clean herd. Once a sufficiently high proportion of herds in a given area has been freed or almost freed from infection, the area can be scheduled as an accredited or attested area, and the remaining herds are then compulsorily brought into the eradication scheme. This method, when accompanied by complete removal or slaughter of the reacting animals, is often highly successful in eliminating the disease. Clean herds must, however, be tuberculin-tested at intervals to see that infection is not re-introduced. In the United States of America the proportion of animals reacting to tuberculin was lowered between 1917 and 1940 from an average initial rate of about 4 per cent to one of about 0.5 per cent (Myers 1940), to 0.18 per cent by 1943, and to 0.11 per cent by 1954, but rose again to 0.23 per cent by 1959 (Johnson *et al.* 1961). Scandinavia was likewise successful in the use of this method. In Great Britain the Attested Herds scheme came into operation in 1935, was largely in abeyance during the war years, was restarted in 1944, and was then prosecuted with vigour. By October 1960 the whole country had been covered. The incidence of infection, as judged by the tuberculin test, had fallen by then to 0.16 per cent. The fall continued till it levelled out in 1971 at 0.038 per cent. The cost of elimination over the period of 25 years was £130 million.

Total eradication of bovine tuberculosis is impeded by the existence of reservoirs of the disease in wild life. Strains of the bovine type have been isolated from deer, opossums and feral pigs in New Zealand, and from badgers in Britain (Muirhead *et al.* 1974, McDiarmid 1975, Zuckerman 1980, Little *et al.* 1982, Wilesmith *et al.* 1982).

Diseased animals with false-negative tuberculin reactions serve as foci from which whole herds may be infected and, in low prevalence areas, a lack of clinical awareness may delay the diagnosis (Stenius 1949).

Tuberculosis in pigs

Before the elimination of tuberculosis from cattle in Britain about two-thirds of mycobacterial disease in pigs was due to the bovine type of *M. tuberculosis* (Cotchin 1943, Thornton 1949); subsequently *M. avium* became and continued to be predominant (Lesslie *et al.* 1968, Boughton 1969). Disease due to *M. tuberculosis* may be generalized, but infection by *M. avium*, except by some serotypes, usually causes no more than local enlargement of lymph nodes of the head and neck. Microscopical lesions may, however, be found in the lungs and kidneys in 12 per cent of infected animals (Lesslie *et al.* 1968). Strains of *M. avium* from pigs are often difficult to culture especially when present in small numbers, but the number of isolations is increased by the use of mycobactin-containing media (Matthews 1969, Matthews *et al.* 1978). It is therefore necessary to use such media when examining porcine heart-valve xenografts for mycobacteria (Matthews and McDiarmid 1978).

Tuberculosis in other mammals

Bovine strains of *M. tuberculosis* infect horses (Griffith 1937, Konyha and Kreier 1971), sheep (Francis 1958), goats (Soliman *et al.* 1953), antelopes (Robinson 1953, Dillman 1976, Clancy 1977) buffaloes (Guilbride *et al.* 1963), chamois (Bouvier *et al.* 1951), kudus (DeKock 1938), camels, deer and bison (Blood and Henderson 1974), and badgers (Muirhead *et al.* 1974). Infections by both human and bovine types occur in monkeys (McDiarmid 1975) and in cats and dogs (Snider 1971).

The vole type of *M. tuberculosis* has been isolated from the field vole, bank vole, shrew, and wood mouse (McDiarmid 1975). Other small mammals including mice, rats, and squirrels appear to be relatively resistant to mycobacterial infection (see Boughton 1969).

M. avium occasionally causes infection in horses (Lesslie and Davies 1958), deer (McDiarmid 1975) and monkeys (Urbain 1949, Pattyn *et al.* 1967, Smith *et al.* 1973). An organism named *Mycobacterium simiae* was isolated from rhesus monkeys (*Macaca mulatta*) by Karassova and her colleagues (1965); the taxonomic status of this organism is uncertain (Käppler 1968, Weiszfeiler *et al.* 1968).

Tuberculosis in birds

M. avium is an important pathogen of both wild and domesticated birds. Between 1913 and 1933 avian tuberculosis was the commonest post-mortem finding in domestic fowls (Matheson and Wilson 1934). Subsequently, improved standards of husbandry, and the practice of killing birds at a much earlier age, caused a dramatic reduction in the incidence of the disease, so that the proportion of birds showing lesions at post-mortem examination dropped from 14 per cent in 1933 to 0.3 per cent in 1962 (Blaxland 1962). The disease is still common in free-range chickens and occurs, although not so frequently, among turkeys, ducks, geese and waterfowl (Boughton 1969). In the Wildfowl Reserve at Slimbridge, England, *M. avium* serotype 1 was isolated from 48 out of 53 members of the Anatidae (ducks, geese and swans) (Schaefer *et al.* 1973).

Disease due to *M. avium* has been reported in many different species of wild birds including crows (Mitchell and Duthie 1950), sparrows (Plum 1942, Matejka and Kubin 1967) starlings (Bickford *et al.* 1966) and pheasants (McDiarmid 1975). *M. avium* probably causes about 1 per cent of deaths in wild birds; but the prevalence of infection, especially among the gregarious species, may be much higher (Boughton 1969).

Mycobacterium avium causes a generalized disease in birds affecting the liver, spleen, intestines, lungs, bones, joints, peritoneum, kidneys and ovaries (Gallagher 1926). The affected organs contain nodular lesions with a macroscopical appearance of secondary tumour deposits. Histologically there are diffuse, spreading, non-encapsulated foci of epithelioid cells and histiocytes (see Francis 1958). The disease can be diagnosed by the intradermal tuberculin test with avian tuberculin but, as in mammals, non-specific reactions occur (Kazda 1967). The slaughter policy is the best method of dealing with extensive disease in a closed fowl flock. The older birds should be destroyed; fowls seldom become infectious under 18 months of age (Myers and Steele 1969). An eradication plan based on the results of tuberculin testing may be used in collections of valuable birds.

Tuberculosis in *wood pigeons* is of particular interest, as the strains of *M. avium* infecting these birds are very difficult to isolate in culture (Christiansen *et al.* 1946). These strains, like those causing Johne's disease, require mycobactin for growth (McDiarmid 1962, Soltys and Wise 1967). Some isolates resemble typical *M. avium* strains in causing extensive disease in chickens, but also cause Johne's disease when administered to calves (Matthews and McDiarmid 1979).

Not all tuberculosis in birds is due to *M. avium*, but the high body temperature, 42–43°, prevents many species from multiplying (Boughton 1969). Infection due to the human type of *M. tuberculosis* occurs in parrots and, much less often, in canaries but, because of the inability of this organism to grow at temperatures in excess of 39°, it causes only superficial lesions (Feldman 1959). *M. xenopi* is a species that grows well at the body temperature of birds. Although large intravenous doses of *M. xenopi* cause disease and death in the fowl (Pattyn 1966), there are no definite reports of naturally occurring infections in birds. Marks (1964) observed that most human infections due to *M. xenopi* occurred in estuarine or coastal areas, and postulated that sea birds might be the source of infection.

Kazda (1966) isolated an organism, superficially resembling *M. marinum*, from diseased organs of fowls and named it *Mycobacterium brunense*. Immunodiffusion analysis has shown this strain to be a variant of *M. avium* (Stanford and Grange 1974). Other unusual isolates which are also probably variants of *M. avium* are reviewed by Boughton (1969).

Tuberculosis in cold-blooded animals

The rapidly growing species *M. fortuitum* and *M. chelonei* have been termed the 'cold-blooded tubercle bacilli' as they were originally isolated from the frog (Küster 1905) and turtle (Friedmann 1903) respectively. Strains of *M. fortuitum* have been isolated from a terrapin (Wilson 1925) and from gias (Darzins 1950), but this species is normally a soil saprophyte (Grange and Stanford 1974). The original strain of *M. xenopi* was isolated from a tumour-like lesion on the leg of a toad (*Xenopus laevis*) (Schwabacher 1959), but this was also an exceptional occurrence.

Aronson (1926) isolated *M. marinum* from some salt-water fish; *M. fortuitum* has also occasionally been isolated from fish (see Grange and Stanford 1974). In general, mycobacterial infection of cold-blooded creatures is uncommon, but occasional epidemics appear to occur (Darzins 1950).

References

Abrahams, E. W. (1970) *Tubercle, Lond.* **51**, 316.
Adams, D. O. (1974) *Amer. J. Path.* **76**, 17; (1975) *Ibid.* **80**, 101.
Adams, R. M., Remington, J. S., Steinberg, J. and Seibert, J. S. (1970) *J. Amer. med. Ass.* **211**, 457.
Aichbergen, H. K. von (1937) *Beitr. Klin. Tuberk.* **89**, 708.
Albert, R. K., Iseman, M., Sbarbaro, J. A., Stage, A. and Pierson, D. (1976) *Amer. Rev. resp. Dis.* **114**, 1051.
Albiston, H. E. (1930) *Aust. vet. J.* **6**, 123.
American Thoracic Society (1967) *Amer. Rev. resp. Dis.* **96**, 558; (1974) *Ibid.* **110**, 271.
Andrews, R. H. and Radakrishna, S. (1959) *Tubercle, Lond.* **40**, 155.
Andrus, P. M. and MacMahon, H. E. (1924) *Amer. Rev. Tuberc.* **9**, 99.
Armand-Delille, P. F. and Vibert, J. (1927) *Pr. méd.* **35**, 402.
Aronson, J. D. (1926) *J. infect. Dis.* **39**, 315.
Austrian, C. R. (1913) *Bull. Johns Hopk. Hosp.* **24**, 11, 141.
Badger, T. L., Breitweiser, E. R. and Muench, H. (1963) *Amer. Rev. resp. Dis.* **87**, 338.
Baldwin, E. R. and Gardner, L. U. (1921) *Amer. Rev. Tuberc.* **5**, 429.
Bang, B. (1928) *J. Amer. vet. med. Ass.* **72**, 20; (1929) *Tierärztl. Rdsch.* **35**, 745, 766; (1930) *Vet. Rec.* **10**, 557; (1932) *Z. Tuberk.* **64**, 87.
Barker, D. J. P. (1973) *Trans. R. Soc. trop. Med. Hyg.* **67**, 43.
Bartels, P. (1907) *Arch. Anthrop.* **6**, 243.
Beck, A. (1972) *Tubercle, Lond.* **53**, 27.
Beck, A., Keeping, J. A. and Zorab, P. A. (1963) *Tubercle, Lond.* **44**, 378.
Behring, E. von (1904) *Dtsch. med. Wschr.* **30**, 193.
Bergqvist, S. (1948) *Bull. Hyg., Lond.* **23**, 92.
Bhatnagar, R., Malaviya, A. N., Narayanan, S., Rajgopalan,

P., Kumar, R. and Bharadwaj, O. P. (1977) *Amer. Rev. resp. Dis.* **115**, 207.

Bickford, A. A., Ellis, G. H. and Moses, H. E. (1966) *J. Amer. vet. med. Ass.* **149**, 312.

Birkhaug, K. (1944) *Acta med. scand.* **117**, 274.

Blacklock, J. W. S. and Williams, J. R. B. (1957) *J. Path. Bact.* **74**, 119.

Blaxland, J. D. (1962) *Vet. Bull.* **32**, 13.

Bleiker, M. A. (1974) *Bull. Un. int. Tuberc.* **49**, 128.

Bleiker, M. A. and Styblo, K. (1978) *Bull. Un. int. Tuberc.* **53**, 295.

Bloch, H. (1950) *J. exp. Med.* **91**, 197.

Blood, D. C. and Henderson, J. A. (1974) *Veterinary Medicine*, 4th edn, p. 394. Baillière, Tindell and Cassell, London.

Borghans, J. G. A. and Stanford, J. L. (1973) *Amer. Rev. resp. Dis.* **107**, 1.

Borrel, A. (1920) *Ann. Inst. Pasteur* **34**, 105.

Boughton, E. (1969) *Vet. Bull.* **39**, 457.

Bouvier, G., Bergisser, H. and Schweizer, R. (1951) *Schweiz. Arch. Tierheil.* **93**, 689.

Brehmer, W., Anaeker, R. L. and Ribi, E. (1968) *J. Bact.* **95**, 2000.

British Thoracic and Tuberculosis Association (1976) *Lancet* **ii**, 1102.

Brook, G. B. (1937) *Vet. Rec.* **49**, 279.

Brotherston, J. G., Gilmour, N. J. L. and Samuel, J. McA. (1961) *J. comp. Path.* **71**, 286.

Brouet, G. and Rousell, G. (1977) *Rev. franç. Mal. resp.* Suppl. 1, **6**, 5.

Brueggemann, M. W., Moore, C. A. and Loudon, R. G. (1974) *Amer. Rev. resp. Dis.* **109**, 698.

Burkhardt, A. (1906) *Z. Hyg. InfektKr.* **53**, 139.

Bushnell, G. E. (1920) *A Study in the Epidemiology of Tuberculosis.* John Bates, Son and Danielsson, Ltd, London.

Calmette, A. (1923) *Tubercle Bacillus Infection and Tuberculosis in Man and Animals* (Translation). Williams and Wilkins Co., Baltimore.

Calmette, A., Grysez, V. and Letulle, R. (1911) *Pr. méd.* **19**, 651.

Calmette, A. and Guérin, C. (1905) *Ann. Inst. Pasteur* **19**, 601; (1906a) *Ibid.* **20**, 353; (1906b) *Ibid.* **20**, 609.

Canetti, G. (1968) *Ann. N. Y. Acad. Sci.* **154**, 13.

Canetti, G., Sutherland, I. and Svandova, E. (1972) *Bull. Un. int. Tuberc.* **47**, 116.

Caplin, M. (1980) *The Tuberculin Test in Clinical Practice. An Illustrated Guide.* Baillière Tindall, London.

Caplin, M., Silver, C. P. and Wheeler, W. F. (1958) *Tubercle, Lond.* **39**, 84.

Capobres, D. B. (1962) *J. Amer. med. Ass.* **180**, 1130.

Carmichael, J. (1938) *J. comp. Path.* **50**, 383.

Chapman, J. S., Bernard, J. S. and Speight, M. (1965) *Amer. Rev. resp. Dis.* **91**, 351.

Chaussé, P. (1914) *Ann. Inst. Pasteur* **28**, 771; (1916) *Ibid.* **30**, 613.

Christiansen, M., Ottosen, H. E. and Plum, N. (1946) *Skand. Vet. Tidskr.* **36**, 352.

Clancy, J. K. (1977) *Tubercle, Lond.* **58**, 151.

Clemens, J. D., Chuong, J. J. H. and Feinstein, A. R. (1983) *J. Amer. med. Ass.* **249**, 2362.

Closs, O. and Haugan, O. A. (1975) *Acta path. microbiol. scand.* **83**, 51.

Coates, A. R. M. (1980) *Eur. J. resp. Dis.* **61**, 307.

Cole, P. J. (1981) *Pers. Commun.*

Collins, C. H., Yates, M. and Grange, J. M. (1981) *Tubercle, Lond.* **62**, 113.

Comstock, G. W. and Edwards, P. Q. (1975) *Amer. Rev. resp. Dis.* **111**, 573.

Comstock, G. W. and Woolpert, S. F. (1978) *Amer. Rev. resp. Dis.* **118**, 215; (1979) *Ibid.* **119**, 323.

Coombs, R. R. A. and Gell, P. G. H. (1975) In: *Clinical Aspects of Immunology*, 2nd edn, p. 575. Ed. by P. G. H. Gell and R. R. A. Coombs. Blackwell, Oxford.

Cornet, G. (1889) *Z. Hyg. InfectKr.* **5**, 191.

Corper, H. J. and Cohn, M. L. (1946) *Amer. J. clin. Path.* **16**, 621.

Corper, H. J. and Stoner, R. E. (1946) *J. Lab. clin. Med.* **21**, 1364.

Corper, H. J. and Uyei, N. (1930) *J. Lab. clin. Med.* **15**, 348.

Cotchin, E. (1943) *J. comp. Path.* **53**, 310.

Crowle, A. J. (1958) *Bact. Rev.* **22**, 183.

Cruikshank, D. B. (1952) In: *Modern Practice of Tuberculosis*, p. 53. Ed. by T. H. Sellors and J. L. Livingstone. Butterworth, London.

Cummins, S. L. (1908) *Brit. J. Tuberc.* **2**, 35; (1912) *Trans. R. Soc. trop. Med. Hyg.* **5**, 245.

Dahlström, G. (1953) *Acta tuberc. scand.* Suppl. No. 32.

Daigeler, A. (1953) *Tierärztl. Umsch.* **8**, 38.

Dam, H. G. and 10 others (1970) *Bull. World Hlth Org.* **43**, 707.

Daniel, T. M. and Janicki, B. W. (1978) *Microb. Rev.* **42**, 84.

Darzins, E. (1950) *Arch. Inst. bras. invest. Tuberc.* **9**, 29.

Darzins, E. and Pukite, A. (1964) *Zbl. Bakt. I. Abt. Orig.* **193**, 510.

Davies, P. D. B. (1971) *Brit. J. Hosp. Med.* **5**, 749.

Deane, E. H. W. (1946) *Lancet* **i**, 162.

DeKock, G. (1938) *S. Afr. med. J.* **12**, 725.

Diehl, K. and Verschuer, O. F. von (1933) *Zwillingstuberkulose, Zwillingsforschung und erbliche Tuberkulosedisposition.* Gustav Fisher, Jena.

Dillman, J. S. S. (1976) Compiled Final Report, Dept Veterinary Services, Zambia **1**, 129; *Ibid.* **2**, 159.

Dixon, S. E. (1889) *Med. News, Phila.* **55**, 435.

Dodge, O. G. (1964) *J. Path. Bact.* **88**, 167.

Douglas, S. R. and Meanwell, L. J. (1925) *Brit. J. exp. Path.* **6**, 203.

Dow, D. J. and Lloyd, W. E. (1931) *Brit. med. J.* **ii**, 183.

Duboczy, B. O. (1968) *Amer. Rev. resp. Dis.* **98**, 319.

Dubos, R. (1964) *Amer. Rev. resp. Dis.* **90**, 505.

Dull, H. B., Herring, L. L., Calafiore, D., Berg, G. and Kaiser, R. L. (1968) *Amer. Rev. resp. Dis.* **97**, 38.

Edwards, L. B. and Palmer, C. E. (1953) *Lancet* **i**, 53.

Edwards, L. B., Palmer, C. E. and Magnus, K. (1953) *Monogr. Ser. World Hlth Org.* No. 12.

Edwards, P., Lynn, A. and Cutbill, L. J. (1936) *Tubercle, Lond.* **17**, 391.

Edwards, P. Q., Peeples, W. J. and Berger, A. G. (1958) *Pediatrics, Springfield.* **21**, 389.

Emori, K. and Tanaka, A. (1978) *Infect. Immun.* **19**, 613.

Enarson, D., Ashley, M. J. and Grzybowsky, S. (1979) *Amer. Rev. resp. Dis.* **119**, 11.

Engel, H. W. B., Berwald, L. G. and Havelaar, A. H. (1980) *Tubercle, Lond.* **61**, 21.

Faisca, J. B. R. (1921) *C. R. Soc. Biol., Paris* **84**, 1002.

Feldman, W. H. (1959) In: *Diseases of Poultry*, 4th edn, p. 287. Ed. by H. E. Biester and L. H. Schwarte. Iowa State Univ. Press.

Ferebee, S. H. (1970) *Advanc. Tuberc. Res.* **17**, 28.

Findel, H. (1907) *Z. Hyg. InfektKr.* **57**, 104.

Flügge, C. (1899) *Z. Hyg. InfektKr.* **30**, 107.

Fowler, W. C. (1930) *Tubercle, Lond.* **12**, 12.

Fox, W. (1964) *Brit. med. J.* **i**, 135; (1968) *Amer. Rev. resp. Dis.* **97**, 767; (1972) *Bull. Un. int. Tuberc.* **47**, 49; (1977) *Ibid.* **52**, 25; (1978) *Ibid.* **53**, 268.

Fox, W. and Mitchison, D. A. (1975) *Amer. Rev. resp. Dis.* **111**, 325.

Francis, J. (1958) *Tuberculosis in Animals and Man*, Cassel, London.

Frew, H. W. O., Davidson, J. R. and Reid, J. T. W. (1955) *Brit. med. J.* **i**, 133.

Friedmann, F. F. (1903) *Zbl. Bakt.* **34**, 647, 793.

Gaiger, S. H. and Davies, G. O. (1933) *Vet. Rec.* **13**, 900.

Galindo, B., Lazins, J. and Castillo, R. (1974) *Infect. Immun.* **9**, 212.

Gallager, B. W. (1926) *Fmr's Bull. U. S. Dep. Agric.* No. 1200.

Gayathri Devi, B., Shaila, M. S., Ramakrishnan, T. and Gopinathan, K. P. (1975) *Biochem. J.* **149**, 187.

Geuns, H. A. van, Meijer, A. and Styblo, K. (1975) *Bull. Un. int. Tuberc.* **50**, 107.

Ghon, A. (1916) *The Primary Lung Focus of Tuberculosis in Children*, translated by B. D. King. Churchill, London.

Glenchur, H., Fossieck, B. E. C. J. and Silverman, M. (1965) *Amer. Rev. resp. Dis.* **92**, 741.

Godfrey, H. P. and Koch, C. (1980) *Immunology* **40**, 247.

Gofton, A. (1937) *Edinb. med. J.* **44**, 333.

Gordon, A. and Parsons, M. (1972) *Brit. J. Hosp. Med.* **7**, 651.

Goren, M. B. (1972) *Bact. Rev.* **36**, 33.

Goren, M. B. and Brennan, P. J. (1979) In: *Tuberculosis*, p. 63. Ed. by G. P. Youmans. Saunders, Philadelphia.

Goren, M. B., Brokl, O. and Schaefer, W. B. (1974a) *Infect. Immun.* **9**, 142; (1974b) *Ibid.* **9**, 150.

Gow, J. G. (1979) *Brit. J. Hosp. Med.* **22**, 556.

Grange, J. M. (1979) *Brit. J. Hosp. Med.* **22**, 540; (1980) *Mycobacterial Diseases*. Edward Arnold, London; (1983) *Advanc. Tuberc. Res.* **22**, 1.

Grange, J. M., Aber, V. R., Allen, B. W., Mitchison, D. A. and Goren, M. B. (1978) *J. gen. Microbiol.* **108**, 1.

Grange, J. M., Aber, V. R., Allen, B. W., Mitchison, D. A., Mikhail, J. R., McSwiggan, D. A. and Collins, C. H. (1977) *Tubercle, Lond.* **58**, 207.

Grange, J. M. and Collins, C. H. (1983) *Tuberele, Lond.* **64**, 14.

Grange, J. M., Collins, C. H. and McSwiggan, D. (1976) *Tubercle, Lond.* **57**, 59.

Grange, J. M. and Stanford, J. L. (1974) *Int. J. syst. Bact.* **24**, 320.

Grass, H. (1931) *Münch. med. Wschr.* **78**, 1171.

Green, H. H. (1946) *Vet. J.* **102**, 267.

Griffith, A. S. (1913) *J. Path. Bact.* **17**, 323; (1937) *J. comp. Path.* **50**, 159; (1957) *J. Hyg., Camb.* **40**, 673.

Griffith, A. S. and Dalling, T. (1940) *J. Hyg., Camb.* **40**, 673.

Griffith, L. J. and Denaro, S. A. (1956) *Amer. Rev. Tuberc.* **74**, 462.

Griffiths, M. I., Davitt, M. C., Brindle, T. W. and Holme, T. (1965) *Brit. med. J.* **ii**, 399.

Grosset, J. (1978a) *Bull. Un. int. Tuberc.* **53**, 5; (1978b) *Ibid.* **53**, 265; (1978c) *Ibid.* **53**, 200.

Grysez, V. and Bernard, A. (1920) *C. R. Soc. Biol., Paris* **83**, 1506.

Grzybowski, S. (1979) *Amer. Rev. resp. Dis.* **119**, 321.

Grzybowski, S. and Allen, E. A. (1964) *Amer. Rev. resp. Dis.* **90**, 707.

Grzybowski, S. and Brown, M. T. (1967) *Tubercle, Lond.* **48**, 95.

Guérin, C. (1957) In: *BCG Vaccination Against Tuberculosis*. Ed. by S. R. Rosenthal. Churchill, Edinburgh and London.

Guilbride, P. D. L., Rollinson, D. H. L., McNulty, E. G., Alley, J. G. and Wells, E. A. (1963) *J. comp. Path.* **73**, 337.

Gunnels, J. J., Bates, J. H. and Swindoll, H. (1974) *Amer. Rev. resp. Dis.* **109**, 323.

Gye, W. E. and Kettle, E. H. (1922) *Brit. J. exp. Path.* **3**, 241.

Hagemann, P. (1937a) *Dtsch. med. Wschr.* **63**, 514; (1937b) *Zbl. Bakt.* **140**, Beiheft 184; (1938) *Münch. med. Wschr.* **85**, 1066.

Hanks, J. H., Clark, H. F. and Feldman, H. (1938) *J. Lab. clin. Med.* **23**, 736.

Hanks, J. H. and Feldman, H. A. (1940) *J. Lab. clin. Med.* **25**, 886.

Hart, P. D'A. and Aslett, E. A. (1942) *Spec. Rep. Ser. med. Res. Coun., Lond.* No. 243.

Hart, P. D'A., Sutherland, I. and Thomas, J. (1967) *Tubercle, Lond.* **48**, 201.

Heaf, H. (1951) *Lancet* **ii**, 151.

Heimbeck, J. (1949) *Acta tuberc. scand.* Suppl. 21, p. 36.

Heineman, H. S., Spitzer, S. and Pianphongsant, T. (1972) *Arch. intern. Med.* **130**, 121.

Heymann, B. (1901) *Z. Hyg. InfektKr.* **38**, 21.

Hitze, K. (1980) *Bull. Un. int. Tuberc.* **55**, 12.

Hobby, G. L., Iseman, M. D., Holman, A. P. and Jones, J. M. (1973) *Antimicrob. Agents Chemother.* **4**, 94.

Hollmann, R. (1924) *Z. Tuberk.* **41**, 127.

Holm, J. and Lester, V. (1947) *Publ. Hlth Rep., Wash.* **62**, 847.

Horvat, A., Nikolic, D. and Djordjevic-Grba, O. (1979) *Bull. Un. int. Tuberc.* **54**, 69.

Houk, V. N., Rent, D. C., Baker, J. H., Sorenson, K. and Hanzel, G. D. (1968) *Arch. environm. Hlth* **16**, 26.

Iles, P. B. and Emerson, P. A. (1974) *Brit. med. J.* **i**, 143.

Inman, P. M., Beck, A., Brown, A. E. and Stanford, J. L. (1969) *Arch. Derm.* **100**, 141.

International Union Against Tuberculosis (1974) *Bull. Un. int. Tuberc.* **49**, 317; (1977) *Technical Guide for Collection, Storage and Transport of Sputum Specimens and Examination for Tuberculosis by Direct Microscopy*, 2nd edn, I.U.A.T., Paris.

Irvine, N. (1957) *BCG and Vole Vaccination*, 2nd edn. Waterlow, England.

Israel, H. L. and DeLien, H. (1942) *Amer. J. publ. Hlth* **32**, 1146.

Jackett, P. S., Aber, V. R. and Lowrie, D. B. (1978) *J. gen. Microbiol.* **104**, 37.

Jadassohn, J. (1914) *Arch. Derm. Syph., N. Y.* **119**, 10.

Johnson, H. W., Baisden, L. A. and Frank, A. H. (1961) *J. Amer. vet. med. Ass.* **138**, 239.

Johnston, R. F. and Wilbrik, H. K. (1974) *Amer. Rev. resp. Dis.* **109**, 636.

Jones, T. D. and Mote, J. R. (1934) *New Engl. J. Med.* **210**, 120.

Jungmann, K. and Gruschka, T. (1938) *Klin. Wschr.* **17**, 239.

Kallman, F. J. and Reisner, D. (1943) *Amer. Rev. Tuberc.* **47**, 549.

Käppler, W. (1968) *Z. Tuberk.* **129**, 321.

Karassova, V., Weissfeiler, J. and Krasznay, E. (1965) *Acta microbiol. hung.* **12**, 275.

Kardjito, T., Donosepoetro, M. and Grange, J. M. (1981) *Tubercle, Lond.* **62**, 31.

Kardjito, T. and Grange, J. M. (1980) *Tubercle, Lond.* **61**, 231.

Kato, M. and Goren, M. B. (1974) *Infect. Immun.* **10**, 733.

Kazda, J. (1966) *Zbl. Bakt. I. Abt. Orig.* **199**, 529; (1967) *Ibid.* **203**, 92, 190, 199.

Kelly, R. (1976) *Med. J. Aust.* **ii**, 681.

Kemp, H. (1976) *Brit. J. Hosp. Med.* **15**, 39.

Kennedy, H. E., Vandiviere, H. M. and Willis, H. S. (1958) *Amer. Rev. Tuberc.* **77**, 802.

Kent, D. C. and Schwartz, R. (1967) *Amer. Rev. resp. Dis.* **95**, 411.

Kettle, E. H. (1924) *Brit. J. exp. Path.* **5**, 158; (1930) *Proc. R. Soc.* **24**, 79.

Kim, R. (1974) *Arch. Derm.* **110**, 299.

Kocen, R. (1977) *Brit. J. Hosp. Med.* **18**, 436.

Koch, R. (1882) *Berl. klin. Wschr.* **19**, 221; (1891*a*) *Dtsch. med. Wschr.* **17**, 101; (1891*b*) *Ibid.* **17**, 1189.

Konyha, L. D. and Kreier, J. P. (1971) *Amer. Rev. resp. Dis.* **103**, 91.

Kounis, N. G. and Constantinidis, K. (1979) *Practitioner* **222**, 390.

Krasnow, I. and Wayne, L. G. (1969) *Appl. Microbiol.* **18**, 915.

Krause, A. K. and Willis, H. S. (1919) *Amer. Rev. Tuberc.* **3**, 153.

Krebs, A., Farer, L. S., Snider, W. E. and Thompson, N. J. (1979) *Bull. Un. int. Tuberc.* **54**, 65.

Krebs, A. and Riska, N. (1976) *Bull. Un. int. Tuberc.* **51**, 193.

Kreis, B. (1976) *Bull. Un. int. Tuberc.* **51**, 71.

Kubica, G. P., Dye, W. E., Cohn, M. L. and Middlebrook, G. (1963) *Amer. Rev. resp. Dis.* **87**, 775.

Küster, E. (1905) *Münch. med. Wschr.* **52**, 57.

Kuwabara, S. (1975) *J. biol. Chem.* **250**, 2556, 2563.

Landi, S., Held, H. R., Hauschild, A. H. W. and Hilsheimer, R. (1966) *Bull. World Hlth Org.* **35**, 593.

Landi, S., Held, H. R. and Tseng, M. C. (1970) *Bull. World Hlth Org.* **43**, 91.

Lange, B. (1935) *Z. Tuberk.* **72**, 241.

Larson, C. L., Baker, M. B., Baker, R. and Ribi, E. (1966) *Amer. Rev. resp. Dis.* **94**, 257.

Larson, C. L., Ribi, E., Wicht, W. C. and List, R. (1961) *Amer. Rev. resp. Dis.* **83**, 184.

Lees, A. W., Miller, T. J. R. and Roberts, G. B. S. (1955) *Lancet* **ii**, 800.

Lefford, M. J. (1975) *Amer. Rev. resp. Dis.* **111**, 243; (1980) *Infect. Immun.* **28**, 508.

Lenzini, L., Rottoli, P. and Rottoli, L. (1977) *Clin. exp. Immunol.* **27**, 230.

Lesslie, I. W. (1962) *Rep. Hlth Tuberc. Conf., 1962*, p. 168. Chest and Heart Ass., Lond.

Lesslie, I. W. and Birn, K. J. (1970) *Tubercle, Lond.* **51**, 446.

Lesslie, I. W., Birn, K. J., Stuart, P., O'Neill, P. A. F. and Smith, J. (1968) *Vet. Rec.* **83**, 647.

Lesslie, I. W. and Davies, D. R . T. (1958) *Vet. Rec.* **70**, 82.

Levy, F. M., Mande, R., Conge, G., Fillastre, C. and Orssaud, E. (1968) *Advanc. Tuberc. Res.* **16**, 63.

Lincoln, E. M. and Gilbert, L. A. (1972) *Amer. Rev. resp. Dis.* **105**, 683.

Lindhardt, M. (1939) *Acta tuberc. scand.* Suppl. 3, p. 1.

Linell, F. and Norden, A. (1954) *Acta tuberc. scand.* **33**, Suppl., 1.

Little, T. W. A., Swan, C., Thompson, H. V. and Wilesmith, J. W. (1982) *J. Hyg., Camb.* **89**, 211, 225.

Logan, W. P. D. and Benjamin, B. (1957) *Studies on Medical and Population Subjects. Gen. Reg. Off.*, No. 10, H.M.S.O., Lond.

Long, E. R. and Seibert, F. B. (1926) *Amer. Rev. Tuberc.* **30**, 733.

Loria, P. R. (1976) *Arch. Derm.* **112**, 517.

Lotte, A., Wasz-Hockert, O., Poisson, N., Dumitrescu, N. and Verron, M. (1980) *Bull. Un. int. Tuberc.* **55**, 55.

Loudon, R. G., Bumbarner, L. R. and Coffman, G. K. (1969) *Amer. Rev. resp. Dis.* **100**, 172.

Loudon, R. G. and Spohn, S. K. (1969) *Amer. Rev. resp. Dis.* **99**, 109.

Lowe, C. R. (1956) *Brit. med. J.* **ii**, 1081.

Löwenstein, E. (1924) *Wien. klin. Wschr.* **37**, 231.

Lowrie, D. B., Aber, V. R. and Jackett, P. S. (1979) *J. gen. Microbiol.* **110**, 431.

Lukas, W. (1977) *S. Afr. med. J.* **51**, 786 (Abstract).

Lunn, J. A. and Johnson, A. J. (1978) *Brit. med. J.* **i**, 1451.

Lurie, M. B. (1941) *Amer. Rev. Tuberc.* **44**, Suppl. 1, 68, 80, 112.

Lurie, M. B., Abramson, S. and Allison, M. J. (1949*a*) *Amer. Rev. Tuberc.* **59**, 168.

Lurie, M. B., Abramson, S. and Heppleston, A. G. (1952) *J. exp. Med.* **95**, 119.

Lurie, M. B., Abramson, S., Heppleston, A. G. and Allison, M. J. (1949*b*) *Amer. Rev. Tuberc.* **59**, 198.

Lurie, M. B., Harris, T. N., Abramson, S. and Allison, M. J. (1949*c*) *Amer. Rev. Tuberc.* **59**, 186.

MacCallum, P., Tolhurst, J. C., Buckle, G. and Sissons, H. A. (1948) *J. Path. Bact.* **60**, 93.

McCarter, J., Getz, H. R. and Stiehm, R. H. (1938) *Amer. J. med. Sci.* **195**, 479.

McDiarmid, A. (1962) *Diseases in Free Living Wild Animals*, F. A. O. Agriculture Studies no. 57, Rome, p. 10; (1975) *Advanc. vet. Sci.* **19**, 97.

McDonald, J. C. (1952) *Brit. J. soc. Med.* **6**, 259.

McFadyean, J., Sheather, A. L., Edwards, J. T. and Minett, F. C. (1913) *J. comp. Path.* **26**, 327.

Mackaness, G. B. (1967) *Amer. Rev. resp. Dis.* **97**, 337.

McMurray, D. N. and Echeverri, A. (1978) *Amer. Rev. resp. Dis.* **118**, 827.

Magnusson, M. and Bentzon, M. W. (1958) *Bull. World Hlth Org.* **19**, 829.

Maitland, M. L. C. (1937) *Lancet* **i**, 1297; (1950) *J. Hyg., Camb.* **48**, 397.

Major, R. H. (1945) *'Classic Descriptions of Disease'*, 3rd edn. Chas C. Thomas, Springfield, Ill.

Mantoux, D. (1910) *Pr. méd.* **18**, 10.

Maonsson, T., Brehmer-Anderson, E., Wittbeck, B. and Grubb, R. (1970) *Acta derm-venereol. Stockh.* **50**, 119.

Marks, J. (1959) *Mon. Bull. Minist. Hlth Lab. Serv.* **18**, 81; (1964) *Proc. R. Soc. Med.* **57**, 479; (1972) *Tubercle, Lond.* **53**, 31; (1976) *Ibid.* **57**, 207.

Matejka, M. and Kubin, M. (1967) *Vet. Med., Praha* **12**, 491.

Matheson, D. C. and Wilson, J. E. (1934) *Proc. 5th World Poult. Congr., Rome* **3**, 3.

Matthews, P. R. J. (1969) *Res. vet. Sci.* **10**, 104.

Matthews, P. R. J. and McDiarmid, A. (1978) *Lancet* **ii**, 884; (1979) *Vet. Rec.* **104**, 286.

Matthews, P. R. J., McDiarmid, A., Collins, P. and Brown, A. (1978) *J. med. Microbiol.* **11**, 53.

Mattick, A. T. R. and Hirsch, A. (1946) *Lancet* **i**, 417.

Mayer, G. (1926) *Zbl. Bakt., I. Abt. Orig.* **100**, 10.

Meade, G. M. (1948) *Amer. Rev. Tuberc.* **58**, 675.

Meidl, F. and Harlacher, C. (1969) *Arch. Hyg., Berl.* **153**, 143.

Meyer, S. N. (1949) *Acta tuberc. scand.* Suppl. 18.

Middlebrook, G., Coleman, C. and Schaefer, W. B. (1959) *Proc. nat. Acad. Sci., Wash.*, **45**, 1801.

Minett, F. C. (1932) *J. comp. Path.* **45**, 317.

Mitchell, C. A. and Duthie, R. C. (1950) *Canad. J. comp. Med.* **14**, 109.

Mitchison, D. A., Allen, B. W., Carrol, L., Dickinson, J. M. and Aber, V. R. (1972) *J. med. Microbiol.* **5**, 165.

Mitchison, D. A. and Dickinson, J. M. (1978) *Bull. Un. int. Tuberc.* **53**, 254.

Mitchison, D. A., Selkon, J. B. and Lloyd, J. (1963) *J. Path. Bact.* **86**, 377.

Moeller, A. (1899) *Z. Hyg. InfectKr.* **32**, 205.

Mokhtari, Z. (1980) *Bull. Un. int. Tuberc.* **55**, 51.

Moore, V. L. and Myrvik, Q. N. (1974) *Infect. Immun.* **10**, 21.

Muirhead, R. H., Gallagher, J. and Birn, K. J. (1974) *Vet. Rec.* **95**, 552.

Murley, A. H. G. (1971) *Brit. J. Hosp. Med.* **5**, 763.

Myers, J. A. (1940) *Man's Greatest Victory over Tuberculosis.* Chas C. Thomas, Springfield, Ill.

Myers, J. A. and Steele, J. H. (1969) *Bovine Tuberculosis Control in Man and Animals.* Warren H. Green, Inc., St Louis, Mo.

Naegeli, O. (1900) *Virchows Arch.* **160**, 426.

Nair, N. C., Mackay-Scollay, E. M., Ramachendron, K., Selkon, J. B., Tripathy, S. P., Mitchison, D. A. and Dickinson, J. M. (1964) *Tubercle, Lond.* **45**, 345.

Nakamura, R. M. and Tokunaga, T. (1980) *Infect. Immun.* **28**, 331.

Nassau, E. (1942) *J. Path. Bact.* **54**, 443; (1958) *Tubercle, Lond.* **39**, 18.

Nauta, E. H. (1979) In: *Immunological Aspects of Infectious Diseases.* Ed. G. Dick. MTP Press Ltd, England.

Nègre, L. and Boquet, A. (1925) *Ann. Inst. Pasteur* **39**, 101.

Nègre, L. and Bretey, J. (1947) *Vaccination par le BCG par Scarifications Cutanées.* Masson et Cie, Paris.

Nel, E. E., Linton, W. S., Merwe, W. van der, Berson, S. D. and Kleeberg, H. H. (1977) *S. Afr. med. J.* **51**, 779.

Newcombe, J. (1979) *Brit. J. Hosp. Med.* **22**, 553.

Nielsen, F. W. (1947) *Vet. J.* **103**, 252.

O'Boyle, P. and Gow, J. G. (1976) *Brit. med. J.* **i**, 141.

Oeding, P. (1946) *Bull. Hyg., Lond.* **21**, 426.

Ogawa, T. (1949) *Kekkaku* **24**, 45.

Okell, C. C. and Parish, H. J. (1927) *Brit. J. exp. Path.* **8**, 170.

Onstad, G. C. (1969) *Amer. Rev. resp. Dis.* **99**, 426.

Opie, E. L. (1917a) *J. exp. Med.* **25**, 855; (1917b) *Ibid.* **26**, 263; (1924a) *Amer. Rev. Tuberc.* **10**, 249; (1924b) *Ibid.* **10**, 265; (1925) *Widespread Tuberculous Infection of Healthy Individuals and its Significance.* W. B. Saunders, Philadelphia.

Ortbals, D. W. and Marr, J. J. (1978) *Amer. Rev. resp. Dis.* **117**, 39.

Ostertag, R. von (1927) *Berl. tierärztl. Wschr.* **43**, 373.

Pagel, W., Simmonds, F. A. H., MacDonald, N. and Nassau, E. (1964) *Pulmonary Tuberculosis.* Oxford University Press.

Palmer, C. E. and Edwards, L. B. (1962) *Bull. Un. int. Tuberc.* **32**, 373; (1968) *J. Amer. med. Ass.* **205**, 167.

Parish, H. J. (1938) *Tubercle, Lond.* **19**, 337.

Parsons, M. (1979) *Tuberculous Meningitis.* Oxford University Press.

Paterson, A. B. and Leech, F. B. (1954) *Amer. Rev. Tuberc.* **69**, 806.

Patterson, D. S. P. (1959) *Tubercle, Lond.* **40**, 71.

Pattyn, S. R. (1966) *Zbl. Bakt., I. Abt. Orig.* **201**, 426.

Pattyn, S. R., Boveroulle, M. T., Mortelmans, J. and Vercruysse, J. (1967) *Acta zool. path., Antwerp* **43**, 125.

Penso, G., Castelnuovo, G., Gaudiano, A., Princivalle, M., Vella, L. and Zampieri, A. (1952) *R. C. Ist sup. Sanità* **15**, 491.

Pepys, J. (1979) In: *Immunological Aspects of Infectious Diseases.* Ed. G. Dick. MTP Press Ltd, England.

Petersen, W. F. (1919) *Amer. Rev. Tuberc.* **3**, 500.

Petroff, S. A. (1915) *J. exp. Med.* **21**, 38.

Philpott, J. A., Woodburne, A. R., Philpott, O. S., Schaefer, W. B. and Mollohan, C. S. (1963) *Arch. Derm.* **88**, 158.

Pickert, C. and Löwenstein, R. (1908) *Dtsch. med. Wschr.* **2**, 2262.

Pirquet, C. von (1906) *Muench. med. Wschr.* **54**, 1457; (1907) *Berl. klin. Wschr.* **48**, 644, 699; (1909) *J. Amer. med. Ass.* **52**, 675.

Plum, N. (1942) *Skand. Vet. Tidskr.* **32**, 465.

Plunkett, R. E., Weber, G. W., Siegal, W. and Donk, R. R. (1940) *Amer. J. publ. Hlth* **30**, 229.

Pottenger, J. E. (1931) *Amer. Rev. Tuberc.* **24**, 583.

Power, J. G. P. (1967) *J. R. Army med. Cps* **113**, 197.

Proudfoot, A. T. (1971) *Brit. J. Hosp. Med.* **5**, 773.

Raffel, S. (1950) *Experientia* **6**, 410.

Raleigh, J. W. and Wichelhausen, R. (1973) *Amer. Rev. resp. Dis.* **108**, 639.

Ranke, K. E. (1928) *Ausgewählt Schriften z. Tuberkulosepathologie.* Ed. M. and W. Pagel. Springer, Berlin.

Rao, V. R., Joanes, R. F., Kilbane, P. and Galbraith, N. S. (1980) *Brit. med. J.* **281**, 187.

Reid, D. D. (1957) *Brit. med. J.* **ii**, 10, 35.

Report (1911) *Final Rep. roy. Comm. Tuberc.* H.M.S.O. Lond.; (1932) *A Survey of Tuberculosis of Bovine Origin in Great Britain.* People's League of Health, Lond.; (1935a) *Quart. Bull. Hlth Org. L.o.N.* **4**, 818; (1935b) *Arb. Reichsgesundh Amt.* **69**, 1; (1939) *Brit. med. J.* **i**, 480; (1948) *Tuberculosis in Young Adults.* H. K. Lewis, Lond.; (1955) *Bull. World Hlth Org.* **12**, 179; (1959a) *Bull. World Hlth Org.* **21**, 51; (1959b) *Tubercle, Lond.* **40**, 317; (1962) *Bull. Un. int. Tuberc.* **32**, 89; (1970) *Brit. med. J.* **ii**, 390; (1971) *Trans. R. Soc. trop. Med. Hyg.* **65**, 763; (1972) *Bull. World Hlth Org.* **46**, 371; (1974a) *Lancet* **ii**, 237; (1974b) *Tubercle, Lond.* **55**, 169; (1974c) *Bull World Hlth Org.* **51**, 473; (1975a) *Brit. med. J.* **iii**, 698; (1975b) *Amer. Rev. resp. Dis.* **115**, 727; (1975c) *Tubercle, Lond.* **56**, 295; (1976a) *Amer. Rev. resp. Dis.* **114**, 471; (1976b) *Lancet* **i**, 787; (1977) *Amer. Rev. resp. Dis.* **116**, 3; (1978a) *Ibid.* **118**, 39; (1978b) *Ibid.* **118**, 219; (1978c) *Lancet* **ii**, 334; (1978d) *Bull. Un. int. Tuberc.* **53**, 29; (1978e) *Control of tuberculosis in NHS employees.* D.H.S.S. Health Circular HC (78) 3; (1979a) *Bull. Un. int. Tuberc.* **54**, 19; (1979b) *Bull. World Hlth Org.* **57**, 819; (1980a) *Bull. Un. int. Tuberc.* **55**, 14; (1980b) *Brit. med. J.* **281**, 895; (1981) *Rev. infect. Dis.* **3**, 813; (1982) *Amer. Rev. resp. Dis.* **125**, 1–132.

Rhee, H. J. van der, Burgh-de-Winter, C. P. M. van der and Daems, W. T. (1979) *Cell Tissue Res.* **197**, 355.

Riddoch, J. (1903) *J. comp. Path.* **16**, 357.

Riley, L. R. and Moodie, A. S. (1974) *Amer. Rev. resp. Dis.* **110**, 810.

Riley, R. L., Mills, C. C., O'Grady, F., Sultan, L. U., Wittstadt, F. and Shivpuri, D. N. (1962) *Amer. Rev. resp. Dis.* **85**, 551.

Riley, R. L. and O'Grady, F. (1961) *Airborne Infection: Transmission and Control.* MacMillan Co., New York.

Ritchie, J. N. (1945–6) *Proc. R. Soc. Med.* **39**, 216; (1964) *Conquest, Lond.* **52**, 3.

Robinson, F. M. (1953) *J. S. Afr. vet. med. Ass.* **24**, 97.

Römer, P. H. (1908) *Beitr. Klin. Tuberk.* **11**, 79; (1909) *Ibid.* **13**, 1.

Rook, G. A. W. (1976) *Proc. R. Soc. Med.* **69**, 442; (1978) *Nature, Lond.* **271**, 64.

Rook, G. A. W., Carswell, J. W. and Stanford, J. L. (1976) *Clin. exp. Immunol.* **26**, 129.

Rook, G. A. W. and Stanford, J. L. (1979) *Parasite Immunol.* **1**, 111.

Rook, G. A. W. and Stewart-Tull, D. E. S. (1976) *Immunology* **31**, 389.

Rosenthal, S. R. (1957) *BCG Vaccination Against Tuberculosis.* Churchill, London; (1958) *Amer. Rev. Tuberc.* **77**, 778; (1961) *J. Amer. med. Ass.* **177**, 452; (1966) *G. P. Kansas Cy* **34**, 117.

Rosenthal, S. R., Loewinsohn, E., Graham, M. L., Liveright, D., Thorne, M. G. and Johnson, V. (1961*a*) *Paediatrics, Springfield* **28**, 622; (1961*b*) *Amer. Rev. resp. Dis.* **84**, 690.

Rouillon, A., Perdrizet, S. and Parrot, R. (1976) *Tubercle, Lond.* **57**, 275.

Rouillon, A. and Waaler, H. (1976) *Advanc. Tuberc. Res.* **19**, 64.

Saenz, A. and Costil, L. (1936) *Diagnostic Bacteriologique de la Tuberculose.* Masson et Cie, Paris.

Salvin, S. B. and Neta, R. (1975) *Amer. Rev. resp. Dis.* **111**, 373.

Savage, W. G. (1933) *Brit. med. J.* **ii**, 905.

Schaefer, W. B., Beer, J. V., Wood, N. A., Boughton, E., Jenkins, P. A. and Marks, J. (1973) *J. Hyg., Camb.* **71**, 549.

Schaefer, W. B., Birn, K. J., Jenkins, P. A. and Marks, J. (1969) *Brit. med. J.* **ii**, 412.

Schwabacher, H. (1959) *J. Hyg., Camb.* **57**, 57.

Seibert, F. B. (1928) *Amer. Rev. Tuberc.* **17**, 394; (1934) *Ibid.* **30**, 713; (1941*a*) *Ibid.* **44**, 1; (1941*b*) *Bact. Rev.* **5**, 69; (1944) *Chem. Rev.* **34**, 107.

Seibert, F. B., Aronson, J. C., Reichel, J., Clark, L. T. and Long, E. R. (1934) *Amer. Rev. Tuberc.* **30**, Suppl., 713.

Seibert, F. B. and Glenn, J. T. (1941) *Amer. Rev. Tuberc.* **44**, 9.

Seibert, F. B., Pederson, K. O. and Tiselius, A. (1938) *Amer. Rev. Tuberc.* **38**, 399.

Shield, M. J., Stanford, J. L., Paul, R. C. and Carswell, J. W. (1977) *J. Hyg., Camb.* **78**, 331.

Šićević, S. (1972) *Acta paediat., Stockh.* **61**, 178.

Sigurdsson, S. (1945) *Studies on the Risk of Infection with Bovine Tuberculosis to the Rural Population.* Oxford University Press.

Simonds, B. (1957) *Eugen. Rev.* **49**, 1; (1963) *Tuberculosis in Twins.* Pitman Medical, London.

Sinclair, D. J. M. and Johnston, R. N. (1979) *Brit. med. J.* **i**, 1325.

Skinsnes, O. K. (1968) *Ann. N. Y. Acad. Sci.* **154**, 19.

Smith, C. C., Barr, R. M. and Alexander, J. (1979) *J. gen. Microbiol.* **112**, 185.

Smith, C. R. (1938) *Amer. Rev. Tuberc.* **37**, 525.

Smith, D. W., Grover, A. A. and Wiegeshaus, E. H. (1968) *Advanc. Tuberc. Res.* **16**, 191.

Smith, E. K., Hunt, R. D., Garcia, F. G., Fraser, C. E. O., Merkal, R. S. and Karlson, A. G. (1973) *Amer. Rev. resp. Dis.* **107**, 469.

Snider, W. R. (1971) *Amer. Rev. resp. Dis.* **104**, 877.

Soliman, K. N., Rollinson, D. H. L., Barron, N. S. and Spratling, F. R. (1953) *Vet. Rec.* **65**, 421.

Soltys, M. A. and Jennings, A. R. (1950) *Amer. Rev. Tuberc.* **61**, 299.

Soltys, M. A. and Wise, D. R. (1967) *J. Path. Bact.* **93**, 351.

Somlo, A. M., Black, T. C. and Somlo, L. I. (1969) *Amer. J. clin. Path.* **51**, 519.

Spector, W. G. (1969) *Int. Rev. exp. Path.* **8**, 1; (1971) *Ann. Sclavo* **13**, 788.

Springett, V. H. (1951) *Brit. med. J.* **ii**, 144; (1971) *Tubercle, Lond.* **52**, 73.

Stallybrass, C. O. (1949) *Brit. med. J.* **i**, 207.

Stanford, J. L. and Grange, J. M. (1974) *Tubercle, Lond.* **55**, 143.

Stanford, J. L. and Paul, R. C. (1973) *Ann. Soc. belg. Méd. trop.* **53**, 389.

Stanford, J. L. and Phillips, I. (1972) *J. med. Microbiol.* **5**, 39.

Stanford, J. L., Revill, W. D. L., Gunthorpe, W. J. and Grange, J. M. (1975) *J. Hyg., Camb.* **74**, 7.

Stanford, J. L., Shield, M. J. and Rook, G. A. W. (1981) *Tubercle, Lond.* **62**, 55.

Stead, W. W. (1967) *Amer. Rev. resp. Dis.* **95**, 729.

Stenius, R. (1949) *Nord. Med.* **42**, 1219.

Stevens, A. J. and Soltys, M. A. (1952) *Vet. Rec.* **64**, 139.

Stewart, A. and Hughes, J. P. W. (1951) *Brit. med. J.* **i**, 899.

Stewart, C. J., Carpenter, R. G. and McCauley, A. (1958) *Tubercle, Lond.* **39**, 143.

St Hill, C. A. and Gow, J. G. (1966) *Brit. J. Urol.* **38**, 163.

Stonebrink, B. (1957) *Proc. Tuberc. Res. Coun.* **44**, 67; (1958) *Acta tuberc. scand.* **35**, 67.

Styblo, K. (1978) *Bull. Un. int. Tuberc.* **53**, 141; (1980) *Advanc. Tuberc. Res.* **20**, 1.

Styblo, K. and Meijer, J. (1976) *Tubercle, Lond.* **57**, 17; (1978) *Bull. Un. int. Tuberc.* **53**, 283.

Šula, L. (1958) *Tubercle, Lond.* **39**, 10; (1968) *Bull. World Hlth Org.* **39**, 647.

Sutherland, A. M. (1979) *Brit. J. Hosp. Med.* **22**, 569.

Swift, S. and Cohen, H. (1962) *N. Engl. J. Med.* **267**, 1244.

Tacquet, A., Tison, F., Devulder, B. and Roos, P. (1967) *Bull. Un. int. Tuberc.* **39**, 21.

Tamura, M., Ogawa, G., Sagawa, I. and Amano, S. (1955) *Amer. Rev. Tuberc.* **71**, 465.

Theodore, A., Berger, A. G. and Palmer, C. E. (1956) *J. chron. Dis.* **4**, 111.

Thomas, O. F., Borthwick, W. M., Horne, N. W. and Crofton, J. W. (1954) *Lancet* **i**, 1308.

Thornton, H. (1949) *Textbook of Meat Inspection.* Baillière, Tindall & Cox, London.

Tinker, C. M. (1959) *J. Hyg., Camb.* **57**, 367.

Toman, K. (1976) *WHO Chronicle* **30**, 51.

Torrance, H. L. (1922) *Vet. Rec.* **34**, 289; (1927) *Ibid.* **39**, 875.

Trambusti, B. (1928) *Boll. Ist. sieroter. Milano* **7,** 371.

Truant, J. P., Brett, W. A. and Thomas, W. (1962) *Bull. Henry Ford Hosp.* **10,** 287.

Turk, J. L. (1979) *J. R. Soc. Med.* **72,** 243.

Turner, H. M. (1953) *Tubercle, Lond.* **34,** 155.

Tytler, W. H. (1945) *Tuberculosis: A symposium of Current Thought and Practice in Great Britain,* p. 32. U.N.R.R.A., London.

Uhlenhuth, P. and Xylander, J. (1908) *Berl. klin. Wschr.* **14,** 1346.

Urbain, A. (1949) *Bull. Acad. vét. Fr.* **22,** 349.

Vandiviere, H. M., Dworski, M., Melvin, I. G., Watson, K. A. and Begley, J. (1973) *Amer. Rev. resp. Dis.* **108,** 301.

Vehlinger, E. and Künsch, M. (1938) *Beitr. Klin. Tuberk.* **92,** 275.

Verschuer, O. F. von (1955) *Dtsch. med. Wschr.* **80,** 1635.

Vestal, A. (1975) *Health, Education and Welfare Publication* no. (CDC) 79–8230, USA.

Villemin, J. A. (1868) *Études Experimentales et Cliniques sur la Tuberculose.* Baillière et Fils, Paris.

Vollmer, H. and Goldberger, E. W. (1937) *Amer. J. Dis. Child.* **54,** 1019; (1939*a*) *Ibid.* **57,** 1272; (1939*b*) *Ibid.* **58,** 527.

Wagener, K. and Reuss, U. (1953) *Zbl. Bakt., I. Abt. Orig.* **160,** 90.

Wallgren, A. (1948) *Tubercle, Lond.* **29,** 245.

Warfel, A. H. (1978) *Exp. molec. Path.* **28,** 163.

Wasz-Hockert, O. and Lotte, A. (1980) *Bull. Un. int. Tuberc.* **55,** 65.

Webb, G. B. and Williams, W. W. (1911) *J. Amer. med. Ass.* **57,** 1431.

Weiss, D. W. (1959) *Amer. Rev. resp. Dis.* **80,** 340, 495, 676.

Weiss, D. W. and Wells, A. Q. (1960) *Amer. Rev. resp. Dis.* **81,** 518.

Weiszfeiler, J., Jokay, I., Karczag, E., Almassy, K. and Somos, P. (1968) *Acta microbiol. hung.* **15,** 69.

Wells, A. Q. and Brooke, W. S. (1940) *Brit. J. exp. Path.* **21,** 104.

Wells, W. F. (1934) *Amer. J. Hyg.* **30,** 611.

Westernhöffer, E. (1911) *Berl. klin. Wschr.* **48,** 1063.

White, E. G. and Minett, F. C. (1941) *Brit. J. Tuberc.* **35,** 69.

White, R. G., Bernstock, A., Johns, R. G. S. and Lederer, E. (1958) *Immunology* **1,** 54.

Wiegeshaus, E. H., Harding, G., McMurray, D., Grover, A. A. and Smith, D. W. (1971) *Bull. World Hlth Org.* **45,** 543.

Wijsmuller, G. (1966) *Bull. World Hlth Org.* **35,** 459.

Wilesmith, J. W., Little, T. W. A., Thompson, H. V. and Swan, C. (1982) *J. Hyg., Camb.* **89,** 195.

Wilhelm, G. and Römer, G. (1977) *Zbl. Bakt., I. Abt. Orig.* **239,** 379.

Wilkins, E. G. (1956) *Brit. med. J.* **i,** 883.

Williams, C. S. and Riordan, D. C. (1973) *J. Bone Jt Surg.* **55,** 1042.

Williams, I. P. and Hetzel, M. R. (1978) *Thorax* **33,** 816.

Wilson, G. S. (1925) *J. Path. Bact.* **28,** 69; (1933) *Spec. Rep. Ser. med. Res. Coun., Lond.,* No. 102; (1967) *The Hazards of Immunization.* Athlone Press, Lond.

Wilson, G. S., Blacklock, J. W. S. and Reilly, L. V. (1952) *Non-pulmonary Tuberculosis of Bovine Origin in Great Britain and Northern Ireland.* Nat. Ass. Prevent. Tuberculosis, Lond.

Wilson, G. S., Schwabacher, H. and Maier, I. (1940) *J. Path. Bact.* **50,** 89.

Wolinsky, E. (1979) *Amer. Rev. resp. Dis.* **119,** 107.

Wolinsky, E., Gomez, F. and Zimpfer, F. (1972) *Amer. Rev. resp. Dis.* **105,** 964.

Yates, M. D. and Collins, C. H. (1979) *Ann. Microbiol. (Inst. Pasteur)* **130B,** 13.

Yates, M. D., Collins, C. H. and Grange, J. M. (1978) *Tubercle, Lond.* **59,** 143.

Yeager, H., Lacy, J., Smith, L. S. and Lemaistre, C. A. (1967) *Amer. Rev. resp. Dis.* **95,** 998.

Yelton, S. E. (1946) *Publ. Hlth Rep., Wash.* **61,** 1144.

Youmans, G. P. (1975) *Amer. Rev. resp. Dis.* **111,** 109.

Youmans, G. P. and Youmans, A. S. (1969) *J. Bact.* **97,** 134.

Young, J. A. and Paterson, J. S. (1949) *J. Hyg., Camb.* **47,** 39.

Ziesche, H. (1907) *Z. Hyg. InfektKr.* **57,** 50.

Zuckerman, S. (1980) *Badgers, Tuberculosis and Cattle.* HMSO, Lond.

52

Leprosy, rat leprosy, sarcoidosis, and Johne's disease
Graham Wilson and Geoffrey Smith

Leprosy

History and distribution

Leprosy may be defined as 'a chronic mycobacterial disease, infectious in some cases, primarily affecting the peripheral nervous system and secondarily involving skin and certain other tissues' (Jopling 1978). The disease is of great antiquity. Where or how it arose is not known. Rogers and Muir (1925) think that it appeared first in the northern belt of Central Africa; and Fiennes (1978) suggests that it originated from contact with the water buffalo, which suffers from skin lesions packed with acid-fast bacilli. The disease was described in Asia Minor about 345 B.C.; thence it spread gradually westwards, till in the Middle Ages practically every country in Europe was affected. During the fourteenth and fifteenth centuries leprosy in Europe declined rapidly; about the same time, however, it was carried to the Western Hemisphere.

In the present century the disease has appeared in Oceania, where it has spread widely. The highest incidence is in tropical countries, particularly (1) Equatorial Africa; (2) the long belt stretching from India, Assam and Burma, through the Malay States and Indo-China to the islands of the Pacific; (3) the north-eastern part of South America; and (4) the Northern Territory of Australia. In spite of the increased attention given to the control and treatment of the disease, it is doubtful whether the world incidence has been much affected. Though it has declined in Norway, the Philippines, and Hawaii, it has apparently increased in parts of Asia and Africa and in Brazil. The number of cases in the world was estimated by Bechelli and Domínguez (1966) to be between 10 and 11 million, of which 3.8 million were in Africa and 6.5 million in Asia. Owing to the antagonism between leprosy and tuberculosis, leprosy is said to disappear in countries in which tuberculosis is prevalent and to be capable of extension only in countries in which tuberculosis is at a minimum (Chaussinand 1959).

Leprosy is caused by an acid-fast bacillus, *Mycobacterium leprae*, discovered by Hansen in 1868 (see Chapter 24).

Suggestive evidence has been brought to show that the rat leprosy bacillus may at times cause disease in man (Marchoux 1922, 1933, Adler 1937, Burnet 1938*a*, *b*), but this still awaits conclusive proof.

Clinical forms and nomenclature

The nomenclature of leprosy is rather confusing. In the past the disease was classified according to the clinical form into (*a*) nodular, and (*b*) neural or maculo-anaesthetic. Clinical manifestations, however, vary, and a more fundamental classification is based on the immunological reaction of the patient. This follows one or other of two main patterns—that of high resistance and that of low resistance. In patients with a high resistance the disease affects the nerves and the overlying skin. It is benign, chronic and sometimes self-healing; and the lepromin reaction is strongly positive. This is now known as the *tuberculoid* form, and corresponds roughly to the former neural or maculo-anaesthetic form. In patients with low resistance the disease is severe, systemic, and progressive, affecting not only the nerves, skin, and lymph nodes, but also the eyes, nose, mouth and larynx. The lepromin reaction is negative. This is now known as the *lepromatous* form, and corresponds roughly to the former nodular form.

The two new terms are also associated with the histopathological picture, from which in fact they are primarily derived. The tuberculoid form is characterized by the presence of few bacilli and a strong cellular response, the lepromatous form by many bacilli and little tissue response (Binford 1969). In the tuberculoid form there are raised erythematous cutaneous lesions, or neuritic lesions characterized by hypertrophy and sclerosis; histologically the lesions resemble those of tuberculosis, and acid-fast bacilli are comparatively scanty. In the lepromatous form the skin lesions consist of granulomatous tissue containing the characteristic vacuolated lepra cells of Virchow together with large numbers of acid-fast bacilli. In both forms, except in diffuse lepromatous infiltration of the skin, the brunt of the infection is borne by the sensory cutaneous nerves, along the fine fibres of which the bacilli spread (Khanolkar 1959, Weddell *et al.* 1959).

This classification is, of course, not exclusive. An intermediate form—often known as the *dimorphous* form—occurs, partaking of characters of each of the two main forms and usually accompanied by a variable but never strong lepromin reaction. This form may pass into either of the other two. Judging the degree of resistance by the lepromin test, by the therapeutic response in bacteriologically positive cases, and by the stability of the infection if it is at one of the poles, Ridley and Jopling (1966) suggest a classification into five groups. (*See also* Rabello 1978, Dharmendra 1980.)

Leprosy is an extremely chronic disease and may be of lifelong duration. After it has lasted for a variable length of time it may retrogress as the result of a naturally acquired immunity. On the other hand, after remaining chronic for years, it may take on an acute course. The mortality from leprosy itself is negligible; most patients die from complications of one sort or another or from intercurrent disease. Though the lesions in leprosy are mainly confined to the skin and underlying nerves, the blood stream may be invaded, with resulting foci in the liver, spleen, adrenals, testicles or bone marrow, and excretion of the organisms in the milk (Pedley 1968). There is reason to believe that much of the tissue damage is due to toxic sub-

stances liberated from the big reservoir of dead bacilli (Rees 1964, Browne 1967).

During the course of the disease acute exacerbations, called *lepra reactions*, may occur manifested by a change in the clinical condition and in the appearance of the lesions. The terminology of these reactions is still unsettled (see Ramu and Dharmendra 1980). Jopling (1978) recognizes two types: type I, seen mainly in border-line leprosy, is due to a change in cell-mediated immunity, for either better or worse, and is characterized by erythema and swelling of the skin lesions, and pain and tenderness of the swollen nerves; type 2, occurring almost exclusively in lepromatous leprosy, particularly during treatment, is due to an antigen-antibody reaction, and is characterized by the occurrence of crops of brightly coloured erythematous nodules resembling those in erythema nodosum, and usually by systemic disturbances.

Epidemiology

The disease occurs where opportunities for close skin contact are common. It is particularly frequent therefore among the lower economic classes who live under unhygienic overcrowded conditions. Cochrane (1959*a*) recognizes a *village pattern* in which the disease is widespread throughout the village, and a *house pattern* in which the incidence in the village as a whole is low but in which a number of houses contain many cases. Males are more often affected than females. In endemic areas infection seems to occur mainly in childhood, though, owing to the long incubation period, the disease is often not diagnosed till early adult life. Adults appear to be far less susceptible than children. This view is supported by the infrequency of conjugal leprosy, and the failure of attempts to infect volunteers experimentally. Both Cochrane (1959*b*) and Badger (1959) regard the disease as fairly highly contagious. Cochrane considers that most persons who come into close contact with open cases of leprosy become infected, but not all of them show clinical manifestations of the disease. That is, the infection rate is high, but the clinical attack rate is low. Moreover, the clinical lesions may disappear and the patient undergo natural cure. In the Philippines it was estimated that about 78 per cent of children lost all signs of active disease before they reached adult life. Unfortunately the absence of a diagnostic skin test precludes more exact information on the frequency and age of infection. Aycock and Gordon (1947) regard hereditary disposition as more important in determining the contraction of the disease than duration of exposure. The incubation period is usually between two and four years, but is said to extend occasionally even up to 40 years.

Judged by the time of appearance of lesions in contact cases in India it ranges from 1½ to 4½ years, being shorter in females than in males (Prasad and Ali 1967). Jochmann

(1913) quotes Gairdner's experience of two European children in Trinidad who were vaccinated with lymph from pustules on an apparently healthy child coming from a leprous family and who showed signs of leprosy 2 to 3 years later. In two American sailors who were tattooed in Melbourne at the same time by the same operator lesions appeared in the tattooed site likewise after 2½ years (Porritt and Olsen 1947).

The way in which leprosy is spread is uncertain. The restriction of the lesions to the skin suggests that the bacilli gain entrance through cuts, abrasions and fissures—a view that is supported by the case of the two sailors just recorded. Chaussinand (1955) regards infected nasal mucus as the principal vehicle of infection in tropical and sub-tropical countries. In malignant cases many thousand bacilli may be seen in films made from the nose. Respiratory infection, therefore, by inhalation of droplets or droplet nuclei may well be of more importance than infection by the skin. Patients suffering from the lepromatous form constitute the main source of infection; most tuberculoid cases are non-infectious. Blood-sucking flies and other insects are suspected of carrying infection, particularly since Narayanan and his colleagues (1972) found leprosy bacilli in mosquitoes captured in the dwellings of lepromatous patients. The experimental transmission of infection to mice and armadillos is described in Chapter 24.

Diagnosis

The best method of diagnosis is by *skin biopsy*. The skin over the affected site is cleansed with alcohol, a fold of skin is picked up with forceps exerting sufficient pressure to stop the circulation, and a piece the size of a grain of rice is cut off with a bistouri or curved scissors. Blood and tissue debris are removed from the cut surface, and the pulp is spread on a slide and stained with Ziehl-Neelsen. Chaussinand (1955), who used this method, found acid-fast bacilli in the cutaneous lesions of 43 per cent of tuberculoid and 100 per cent of lepromatous cases. Alternative methods of examining the skin are preferred by some workers (see Cochrane 1959d). If necessary, several sites should be examined, including the lobe of the ear, the skin of the forehead, chin, cheeks, nose and buttocks as well as the skin over suspicious lesions. Smears are made from the edge rather than from the centre of the lesions. The pieces of excised skin may also be fixed and embedded in paraffin, and sections cut and stained for acid-fast bacilli in the usual way. When the bacilli cannot be found in the skin, they should be sought for in the *nasal mucosa*. The nostrils are cleared of mucus with cotton swabs. With the aid of a torch and a speculum the mucosa over the inferior turbinates on both sides is scraped with a blunt scalpel or small round curette (Jopling 1978). Film preparations are then made (Fig. 52.1). Odd acid-fast bacilli may be found in the nose, but in the lepromatous form the

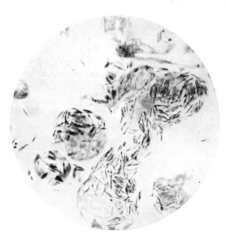

Fig. 52.1   Leprosy bacilli in the tissues. The bacilli are seen in dense clumps, mostly intracellular (× 1000).

organisms are present in large numbers, usually arranged in globular clusters. Since leprosy bacilli are less acid-fast than tubercle bacilli, decolorization of films stained with Ziehl-Neelsen should be carried out gently, as, for instance, with a mixture of 1 per cent HCl and 70 per cent alcohol applied for 3–5 seconds, or longer if necessary. In Chaussinand's series acid-fast bacilli were found in the nasal mucosa of 8.7 per cent of tuberculoid and 72.7 per cent of lepromatous cases. When there is any doubt whether organisms seen in films or sections are genuine leprosy bacilli, a suspension of the ground-up material should be injected into guinea-pigs. If they are leprosy bacilli, there will be no result; if, on the other hand, they are tubercle bacilli, the animals will suffer from the usual form of inoculation tuberculosis. A diagnosis of leprosy is seldom justified unless one of the two cardinal signs is present: (*a*) clinical signs of nerve involvement, (*b*) the demonstration of leprosy bacilli in the skin or nasal mucosa (Cochrane 1959c).

According to Rees and Waters (1963), live leprosy bacilli stain uniformly and are electron-dense: dead bacilli stain irregularly, showing fragmentation, beading, or granularity, and are only partly electron-dense. In lepromatous lesions over 50 per cent of the bacilli are dead, and in tuberculoid lesions an even higher proportion.

No serological method has yet proved satisfactory in the diagnosis of leprosy. The complement-fixation test is non-specific (Row 1925–26). The Wassermann reaction may be positive in the complete absence of syphilis.

Mitsuda reaction

This reaction, alternatively called the *lepromin reaction*, was described by Mitsuda in Japan in 1919 (see also Mitsuda 1923). The test is carried out by the intracutaneous injection of 0.1 ml of a boiled extract of leprous tissues—so-called *integral lepromin*; or more frequently now with purified bacilli from the tissues of artificially infected armadillos. A positive reaction takes two forms—an early and a late one. The *early reaction* is characterized by an acute localized area of inflammation with congestion and oedema appearing usually in 24 to 48 hours and tending to disappear in 3–4 days; the *late reaction* by local infiltration of the skin starting at about 7 days and proceeding to the formation in 3 to 4 weeks of a nodule that may in some cases undergo central necrosis and ulceration and take several weeks to heal. The early or Fernández (1940) reaction corresponds to the tuberculin test: that is to say it is of the delayed allergic type; clinically it is considered to be of no significance. The late nodular infiltration or Mitsuda reaction proper, which is regarded clinically as the criterion of a positive test, corresponds to the Kveim reaction in sarcoidosis (see p. 68). Hart and Rees (1967) refer to it as the manifestation of a 'delayed delayed' type of hypersensitivity to antigens of biological origin. In both the early and the late reactions mycobacterial and skin antigens appear to take part. Dharmendra (1942), who prepared a purified form of lepromin containing a higher proportion of leprosy bacilli—the so-called *bacillary lepromin*—showed that the early reaction was a response mainly to bacillary protein. Protein preparations do not sensitize the patient; but tissue preparations which give a strong late reaction do tend to sensitize, so that a patient negative to a first test may be positive to later tests (Report 1960*a*). (For the preparation of a standardized form of bacillary lepromin see Hanks *et al.* 1970, Draper 1980.) Both the Fernández and the Mitsuda reactions are generally negative in the lepromatous and strongly positive in the tuberculoid form of leprosy. The lepromin reaction is often given by healthy persons living in non-endemic countries, and is therefore of no diagnostic value. Cummins and Williams (1934), for example, at Cardiff found that 24 of 25 tuberculin-positive healthy men reacted positively to the test. Prognostically, the Mitsuda reaction is regarded as indicative of resistance to infection. Tuberculin-positive subjects react to the test more often than tuberculin-negative; but leprosy does not appear to sensitize to tuberculin, even though the tubercle and the leprosy bacilli share common antigens. Leprosy patients who are infected with the tubercle bacillus react to tuberculin; in the lepromatous form their sensitivity may be depressed (see Wade 1950). It is a remarkable fact that lepromatous patients, though failing to give a positive skin reaction to the leprosy bacillus, are able to do so in response to antigens of other mycobacteria, showing that this failure cannot be due to simple anergy (Rees 1964).

Histologically the positive Mitsuda reaction consists of an organized granuloma or of a tuberculoid formation in process of organization. Many epithelioid and some giant cells

are seen, but bacilli are few and granular. In a negative reaction bacilli are present in great numbers, along with histiocytes and vacuolated macrophages (Report 1962, Beiguelman 1967).

Immunologically, normal human antibody responses are unimpaired in leprosy. Cell-mediated immunity, however, is depressed, least of all in the tuberculoid and most in the lepromatous form (*see* Turk 1969). This can be shown not only by the Mitsuda test, but also by the lymphocyte transformation test carried out with the leprosy bacillus as antigen. In the lepromatous form no transformation occurs, whereas in the tuberculoid form a strong reaction is obtained. Histological examination of the lymph nodes of lepromatous patients shows a depletion of lymphocytes in the paracortical areas and their replacement by macrophages. In the tuberculoid form, on the other hand, there is abundant proliferation of lymphocytes in these areas. The fact that persons who have long been in contact with lepromatous patients give a poorer response than normal persons suggests that they have been subclinically infected (Godal *et al.* 1972, Menzel *et al.* 1980). The usual failure of such contacts to suffer from leprosy is presumably due to the development of immunity to the bacillus. (For further references to the Mitsuda reaction see Chaussinand 1955, 1959, Hale *et al.* 1955, Cochrane 1959*b*, Hanks 1961, Jopling 1978, Dharmendra 1981.)

Prophylaxis and treatment

The control of leprosy should follow lines similar to those used in tuberculosis and other communicable diseases. A policy of rigid segregation leads to concealment of the disease and defeats its own end. Early diagnosis and treatment, coupled with a policy of partial segregation of the patients, their direction into occupations where they will not be a danger to the community, the transference of their children to the homes of healthy parents, 3-monthly surveillance of domestic contacts, improvement of housing conditions, and education of the public are most likely to be successful. In Norway where leprosy patients could be ordered by law to live in precautionary isolation away from their families and immediate surroundings— with exceptions for married couples who wanted to live together—the disease was practically eliminated within 80 years (Vogelsang 1957). Of the 10 million or so lepers in the world only 18 per cent were estimated to be under treatment (Bechelli and Domínguez 1966).

Numerous vaccines have been tried in the treatment of leprosy, such as tuberculin, acid-fast bacilli from leprous lesions, the organisms in lepra nodules themselves, autolysed tubercle bacilli, and nastin prepared by the ethereal extraction of an acid-fast streptothrix—all without avail. A few patients have improved for a time, presumably as the result of a non-specific inflammatory or pyrexial reaction, but no lasting benefit has been substantiated. Nor is there any reason to believe that therapeutic vaccination is likely to be successful in leprosy when it has been such a failure in other diseases. As has already been pointed out, a positive Mitsuda reaction is associated with an increased resistance to leprosy. Since it is often possible to convert a negative into a weak positive reaction by the injection of BCG, prophylactic vaccination is recommended by some leprologists. The justification for this is a little doubtful. Cochrane (1959*b*), for example, considers that the type of disease probably determines the nature of the allergic response rather than the reverse. On the other hand, if it is true that tuberculosis confers some protection against leprosy—though on this the evidence is conflicting (see Lowe and McNulty 1953, Hale *et al.* 1955, Chaussinand 1959, Brown *et al.* 1966)—BCG vaccination may render a person who is exposed to leprosy less liable to contract the disease if he becomes infected. At present reports of its value are conflicting. In Uganda the protection rate was 80 per cent (Stanley *et al.* 1981); in Burma no difference was noticed between vaccinated and unvaccinated children (Bechelli *et al.* 1974). Shield and co-workers (1980) attribute this failure to the inhibitory action of *Mycobacterium scrofulaceum*, which is common in Burma but absent from Uganda. Though the protective effect of BCG vaccination cannot be predicted, it is probably wise for the present to use it for child contacts, especially infants born into leprous families. Time will show whether a bacillary vaccine made from the infected tissues of armadillos will prove more effective.

Chaulmoogra oil, which was used for many years in treatment, has been displaced by the sulphones. The drug most widely used is 4,4-diaminodiphenyl sulphone (Dapsone or DDS); it is given by the mouth and excreted in the urine. Like the sulphonamides it appears to act by interfering with growth of the bacilli. Its activity can be reversed by *p*-aminobenzoic acid. It is only slightly toxic, giving rise occasionally to anae-mia, dermatitis, hepatitis, and psychosis, and the dosage has to be worked up gradually (see Browne 1968). Jopling (1978) recommends a daily dose of 50–100 mg by mouth. Smaller doses are effective but probably encourage the development of dapsone resistance. Treatment of tuberculoid and indeterminate cases should be continued for 3 years; of borderline cases for 9–12 years; and of lepromatous cases for life, though on half dosage (Browne 1968). In patients who are sensitive to dapsone, or are infected with dapsone-resistant organisms, another drug, such as thiacetazone, one of the long-acting sulphonamides, or rifampicin, which is very effective but expensive, may be given. Patients should be followed up regularly, and skin smears made every 6–12 months. Combined drug therapy is often advisable, so as to lower the danger of dapsone resistance. Under treatment with DDS lepromatous patients show rapid clinical improvement (Lowe 1950), but seldom become bacteriologically negative in under two years; more often five years or longer are required (Palencia *et al.* 1954, Bushby 1959). The success of treatment can be followed bacteriologically by estimating the number and proportion of live and dead bacilli (Ridley 1967). As has already been pointed out, dead bacilli play a significant part in the causation of lesions in leprosy, and it is therefore not surprising that bacteriologically successful treatment may at times aggravate the clinical manifestations of the disease.

It may be added that chemoprophylaxis for children who are contacts of lepromatous or other bacteriologically positive cases of leprosy is on trial; the preliminary results have revealed only a partial protection (Report 1970, 1971).

Rat leprosy

Rat leprosy is a disease that was first described by Stefansky (1903) in 1901 at Odessa. Rats were being slaughtered in large numbers, consequent on the outbreak of human plague, and rat leprosy was found in 4–5 per cent of them. The incidence reported by other observers varies considerably.

In the United States McCoy (1913) found it in 186 out of about 200 000 rats caught in San Francisco, Cal., an incidence of 0.093 per cent; and Wherry (1908) found it in 20 out of 9631 rats caught in Oakland, Cal., an incidence of 0.21 per cent. In Japan, Ota and Asami (1932) found it in 0.7 per cent of *Rattus norvegicus* and in 0.1 per cent of *Rattus rattus*; and Yamamoto, Sato and Sato (1936) found it in 1.24 per cent of 2573 rats caught in Tokyo. According to Marchoux (1933), about 5 per cent of the sewer rats in Paris harboured rat leprosy bacilli in the lymph nodes, but in only about 0.6 per cent of the rats were generalized lesions present. In Batavia 9.3 per cent of about 10 000 rats were said by Lampe and de Moor (1936) to have lesions of the lymph nodes, but not more than 0.23 per cent of the animals had manifest skin leprosy.

The disease exists in two forms—the glandular type and the musculo-cutaneous type—but there appears to be no sharp division between them. In the *glandular type* one or more of the groups of subcutaneous lymphatic nodes—inguinal, axillary and cervical—is enlarged, hard and whitish. On section the nodes are uniform and hard; there are no nodules or necrotic areas visible macroscopically. Microscopically, the capsule and trabeculae are thickened; the sinuses are filled with dense aggregations of irregularly polygonal cells, which have a large nucleus and much cytoplasm—probably macrophages; the cytoplasm is packed with acid-fast bacilli, so that the contours of the cell-body are often invisible. A few giant cells, with several peripheral nuclei, containing numerous bacilli in their cytoplasm are also visible. Some organisms may be found free.

In the *musculo-cutaneous type* the rat is emaciated; the skin presents one or more irregularly round or oval areas of alopecia, commonest on the head; occasionally ulcers are seen, about 0.5 cm in diameter, covered with a mealy-looking discharge containing acid-fast bacilli. At the site of these areas of alopecia the skin is atrophic. The subcutaneous tissues are devoid of fat, and show the presence of a greyish-white, granular material, which Stefansky regarded as altered muscle tissue; it contains numerous acid-fast bacilli. Sometimes nodular masses are found in the muscles, covered with particles of stretched atrophic skin. Stefansky believed that the primary lesion began in the subcutaneous muscle fibres, and sometimes spread to the skeletal muscles. Histologically, numerous acid-fast bacilli are found in the corium and subcutaneous tissue; some are free, but most are situated in the cells of the granulation tissue, and in large cells rich in cytoplasm, which resemble the lepra cells of Virchow, but contain no vacuoles. Acid-fast bacilli invade the muscle fibres, collect around the nuclei, and lead to destruction of the tissue. Lesions of the internal organs are rare, but McCoy (1913) states that nephritis is very common; the kidneys are enlarged, yellowish-brown, friable, and often cystic, but no acid-fast bacilli are found in them. In an examination of 186 leprous rats, McCoy (1913) noted alopecia in 47.5 per cent, cutaneous ulcers in 22 per cent, diffuse subcutaneous infiltration in 97.9 per cent, enlarged lymphatic nodes in 87 per cent, and nephritis in 53.8 per cent.

Inside the cells the bacilli are said to be distributed not in cigar-like bundles as are human leprosy bacilli, but at random like pins in a packet; and instead of displacing the nucleus they arrange themselves round it (Marchoux 1933).

Reproduction of the disease in animals

Dean (1905) was successful in transmitting the disease to rats by inoculation of suspensions of infected skin and lymph nodes, and in passing it from rat to rat. The disease ran a slow course, lasting from 6 months to a year. *Post mortem*, after subcutaneous inoculation, there was sometimes a local lesion containing semi-caseous material and acid-fast bacilli; in these cases the regional lymphatic nodes were sometimes invaded by bacilli. After intraperitoneal inoculation the lesions were more extensive; the epiploon was infiltrated with grey masses of opaque material, which consisted of collections of cells containing bacilli. The liver was not usually affected, but in some animals there were small, pale, sharply circumscribed areas, composed chiefly of acid-fast bacilli. The capsule of the spleen was often thickened and contained bacilli, but in only one case was the organ itself involved. Experimentally the disease can also be reproduced by scarification of the skin (Marchoux 1933, Marchoux and Chorine 1938), by intranasal instillation (Wayson and Masunaga 1935), and by conjuctival infection (Marchoux *et al.* 1935), but which of these routes is mainly responsible for natural infection is not known. Growth of the organisms in the tissues is very slow, and no multiplication is evident for 6–8 weeks (Elek and Hilson 1956). Experimentally infected rats

may excrete the bacilli in the faeces and urine (Fielding 1945).

Rat leprosy can be transmitted readily to *mice* by intraperitoneal inoculation, and by the corneal route (Goulding *et al.* 1953, Robson and Smith 1959). The *Syrian hamster* is likewise susceptible (Balfour-Jones 1937). Monkeys and rabbits can sometimes be infected by intracerebral inoculation, but not guinea-pigs (Sellards and Pinkerton 1938).

Leprosy in other animals

Cat leprosy Lawrence and Wickham (1963) in Australia described a disease in cats characterized by granulomatous lesions of the skin and subcutaneous tissues, often associated with ulceration and with enlargement of the lymph nodes. The granulomata contained numerous epithelioid cells packed with acid-fast bacilli and giant cells of the Langhans type. Typical lesions could be reproduced in rats by subcutaneous injection with material from lymph nodes. The cause appears therefore to be the rat leprosy bacillus. The disease in cats has also been reported in England (Wilkinson 1964).

Lepra bubalorum is an infectious disease of the buffalo seen in Indonesia. According to Lobel (1936), who studied the disease, it was first noted by Kok and Roesli in 1926. It runs a protracted course and is characterized by multiple cutaneous nodules containing acid-fast bacilli resembling human leprosy bacilli in their appearance and grouping into globular formations. There are no systemic disturbances, and the skin lesions undergo slow resorption. The bacilli have not been cultured, nor has the disease been reproduced experimentally, even in the buffalo itself.

Lepra bovina is the name given to a leprosy-like disease seen in Indonesian cattle, though only two cases have so far been recognized. Besides the cutaneous nodules, seen in the leprosy of buffaloes, lesions occur in some of the internal organs (Kraneveld and Roza 1954).

Machicao and La Placa (1954) described a leprosy-like disease of **frogs** (*Pleurodema cinerea* and *P. marmoratus*) in Bolivia. Of 663 frogs examined at La Paz 129 (19.5 per cent) were found to be affected. The visceral lesions, which were distributed mainly in the liver and digestive tract, consisted of firm, yellowish-white, translucent nodules, round, ovoid or irregular in shape, ranging from 0.3–3 mm in diameter, and often projecting from the surface. Subcutaneous lesions up to 4 mm in diameter were also present; the skin over them was devoid of pigment and frequently ulcerated. Histologically, the granulomata were made up of histiocytes without giant cells, without caseation, and without a surrounding ring of lymphocytes. Acid-fast bacilli, straight, 3–5 μm long, with parallel sides and rounded ends, were found in dense clumps within the cells resembling the globi of leprotic lesions. The organism could not be cultivated. The disease is transmissible to frogs of the same but not of different species.

Sarcoidosis

This disease is defined in histo-pathological terms as a 'disease characterized by the presence in all of several affected organs and tissues of non-caseating

epithelioid-cell granulomas, proceeding either to resolution or to conversion into hyaline connective tissue' (Mitchell *et al.* 1977). It is of world-wide distribution (see Present and Siltzbach 1967). Its causation is still a mystery. Case-to-case transmission is unknown, and the main reason for believing it to be infective is the report of Mitchell and Rees (1969) that it can be transmitted to mice by injection into the footpad. The disease may take the form of bilateral hilar adenopathy, usually undergoing spontaneous retrogression, or of sarcoids in the skin, lymph nodes, liver, eyes, bones, and occasionally the nervous system. Histologically, the picture is one of a granuloma characterized by follicular aggregations of epithelioid cells, the presence of giant cells, the absence of caseation, and the occurrence of hyalinization during healing. At one time it was thought to be a modified form of tuberculosis (see Scadding 1960, 1967), but evidence against this view is cumulative and impressive (Buck 1961, Siltzbach 1969). Among other things the prevalence of the disease remains constant in countries in which the tuberculosis death-rate is falling; the disease cannot be transmitted to guinea-pigs; tubercle bacilli can be demonstrated in only a minority of cases; the tuberculin reaction is weak or negative; the Kveim reaction is usually positive; BCG vaccination affords no protection (Sutherland *et al.* 1965); and drugs active against tuberculosis are useless in sarcoidosis. The relative failure to respond to tuberculin is associated with the disease and is not of constitutional origin; before the development of clinical manifestations patients respond to tuberculin in the usual way (Suth-

erland *et al.* 1965). This disposes of the view that sarcoidosis is a modified form of tuberculosis occurring in a patient of lowered reactivity. The anergy is thought to be due to a tuberculin-neutralizing factor in the blood (Wells and Wylie 1949, Daddi and Gialdroni-Grassi 1966).

Diagnosis is made partly by x-ray examination of the chest, partly by biopsy (see Lillington and Jamplis 1963), and partly by the skin test described by Kveim (1941) and intensively studied by Siltzbach (1961, 1966). It consists in the intradermal injection of a suspension of sarcoid tissue, followed 3–4 weeks later by a biopsy at the site of the reaction to see whether the histological picture is characteristic of the disease. Provided the suspension is made from the spleen and not from lymph nodes and is properly standardized, the reaction is positive in about three-quarters of patients with active disease, but in a lower proportion during the stage of late fibrosis (James 1966). Apart from Crohn's disease (Mitchell *et al.* 1969), the test is specific for sarcoidosis. False positive reactions are few. An in-vitro method of performing the test is described by Hardt and Wanstrup (1969). (For a summary of papers given at the 5th international conference on sarcoidosis at Prague in 1969 see Report 1969, and for a bibliography of the disease, 1878–1963, see Mandel *et al.* 1964. For a biographical history of the eponymy of sarcoidosis, see Scadding 1981. For oncogenic viruses in sarcoid see Fatemi-Nainie *et al.* 1982.)

Johne's disease

Johne's disease (paratuberculosis) is a specific enteritis affecting cattle, sheep, goats and deer, and occasionally other species such as yaks, camels and buffaloes. The causative agent is an acid-fast organism known as Johne's bacillus (*Mycobacterium johnei*). The disease was first described by Johne and Frothingham in 1895 at Dresden, who regarded it as a peculiar form of tuberculosis, possibly due to the avian type of bacillus.

Johne's disease occurs in many parts of the world, mainly in temperate zones but also in the tropics. It affects particularly young cows, but no age is exempt. Since, however, the incubation period is lengthy—up to about 18 months—it is uncommon in cows less than 2 years old. Certain breeds of cattle, such as the Jersey, are particularly susceptible. Sheep are less often attacked, though according to McEwen (1939) the disease may exact an annual toll of 5 per cent of the animals. Very occasional cases have been reported in horses (see Rankin 1956). In Great Britain infection is widespread in cattle; Taylor (1951*a*) estimated that 25–30 per cent were latently infected. As a rule it remains latent, but in about 10 per cent of animals it is progressive and proves fatal. Johne's disease is economically one of the most important of the bovine diseases.

Natural infection probably occurs through the ingestion of fodder that has been soiled with faeces of infected animals. It has been shown experimentally that Johne's bacillus may remain alive for several months in infected faeces and ditch water exposed to ordinary atmospheric conditions (Lovell, Levi and Francis 1944). Infection may occur in fetal life when the placenta of the dam is diseased (Doyle 1958).

In cattle the disease runs a chronic course, subject often to long periods of intermission, and characterized by failure to thrive, a falling off in the milk yield, diarrhoea of varying severity, and progressive emaciation; there is no fever. Recovery is rare.

Pathologically, the main lesions are situated in the last 50 feet of the small intestine, and in the neighbourhood of the ileocaecal junction, but sometimes the whole gut is affected. The intestine is thick and rigid, looking not unlike a hose pipe; the mucosa is greatly thickened and is thrown into regular corrugations, resembling the convolutions of the cerebrum. The mucosa is smooth and pale or pink in colour, and covered with slimy material; on the surface of the corrugations it is dotted with red spots or haemorrhages; between the folds it often has a warty appear-

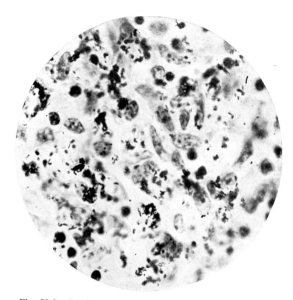

Fig. 52.2 Section of intestine of cow with Johne's disease. The bacilli are seen in dense clumps, mostly intracellular (× 1000).

ance. Occasionally small nodules are seen, due to enlargement of solitary lymphoid follicles. There is no ulceration of the mucosa, and the peritoneal surface appears normal. The mesenteric lymph nodes are usually enlarged, oedematous and pigmented. Histologically, the mucosa, and to a less extent the submucosa, are infiltrated with lymphoid and epithelioid cells, which are responsible for the thickening. Near the surface of the gut the mucosa is absolutely structureless; all traces of nuclei and cell outlines have disappeared. Giant cells are rare; there is no caseation, ulceration, fibrosis, or calcification. Short acid-fast bacilli, often in enormous numbers, are found in the mucosa lying between the glands and in the lymphoid tissue of the solitary follicles; they may invade the submucosa, and very occasionally the underlying muscle layer. The bacilli are arranged in dense clumps, and may be intra- or extra-cellular. Johne and Frothingham were of the opinion that they were primarily intracellular in position, but that owing to the subsequent disintegration of the cell, they were set free in the tissue. The cellular reaction around the bacilli is diffuse, not localized as in tuberculosis. Bacilli can generally be found microscopically in the mesenteric glands. By cultural methods they can be shown to be widespread throughout the body, being distributed in the liver, spleen, lung, and many of the lymph nodes in the cervical and abdominal regions (Taylor 1951a). The reproductive tract may be infected, and the organism may be excreted in the semen of bulls (Larsen and Kopecky 1970). Biochemical aspects of the disease have been reviewed by Patterson and Allen (1972).

In *sheep* the clinical and pathological manifestations

are more varied than in cattle, corresponding to some extent to the three different types of infecting organism (see Taylor 1951b, Stamp and Watt 1954, and Chapter 24). The disease is commonest in animals between 3 and 5 years of age, and is characterized by loss of condition without necessarily accompanying diarrhoea. Thickening of the intestinal mucous membrane is often only slight in animals infected with the non-pigmented strains; in those infected with pigmented strains the mucosa of the small intestine is bright yellow in colour and the intestinal wall frequently thickened. Calcification of the lesions sometimes occurs in sheep and goats. Infection with pigmented strains has not been reported in goats.

The disease can be reproduced in both cattle and sheep by feeding or by parenteral inoculation with pure cultures of Johne's bacillus (see Chapter 24). After infection some animals clear themselves; some become carriers and excrete the organism; and some carriers eventually become clinical cases. Dissemination of the organisms throughout the body is said to occur (Gilmour 1976).

The *bacteriology* of the disease is described in Chapter 24.

Diagnosis and prophylaxis

Differential diagnosis has to be made from strongylosis, coccidiosis, and particularly tuberculous enteritis. The bacilli may be demonstrated microscopically in the faeces of 25–30 per cent of clinical cases, and occasionally in milk (Smith 1960). In advanced cases they may be found in a piece of pinched-off rectal mucosa. Fluorescence microscopy may help in recognition of the organisms. Positive results are of value mainly for confirming the clinical diagnosis (Doyle 1956). In microscopically negative cases the cultural method may be tried. Cameron (1956) advises treatment of the faeces with 5 per cent oxalic acid or 20 per cent antiformin for 30 minutes at 37°, followed by 0.01 per cent malachite green for 10 minutes and inoculation on to a medium containing 50 per cent egg yolk with phlei extract added (see Chapter 24). Treatment with oxalic acid followed by inoculation on to a serum agar medium containing antibiotics is said to give better results (Report 1960b). At post-mortem examination the bacilli are best demonstrated in the ileocaecal mucosa and associated lymph nodes, either microscopically or culturally by use of the oxalic acid method (Taylor 1950). The *complement-fixation test* was reported on favourably by Sigurdsson (1945, 1946, 1947) in sheep and by Hole (1958) in cattle. A positive reaction is obtained in a high proportion of clinically affected animals and is therefore useful for confirmatory purposes. Only about 25 per cent of carriers react positively, and the test cannot therefore be used for the diagnosis of latent infection. A considerable drawback is that a positive result is sometimes obtained in apparently healthy animals from non-infected herds (Chandler 1955, Rankin 1961). Some such false-positive reactions are due to infection with *Corynebacterium renale* (Gilmour and Goudswaard 1972). Fluorescent-antibody and complement-fixation techniques were shown by Goudswaard and his colleagues (1976) to be more sensitive than tests based on haemagglutination, haemagglutination-lysis, and immunodiffusion. The tuberculin test is negative with bovine, but sometimes positive with avian tuberculin (see Bang 1909, 1914). A positive reaction may likewise follow the intradermal injection of *johnin*—a tuberculin-like substance prepared from glycerine broth cultures of Johne's bacillus (see M'Fadyean *et al.* 1912, Twort and Ingram 1913, Dunkin 1933). Owing to the low and fluctuating degree of allergy, however, in Johne's disease, the test is unreliable; it may be positive in the absence of lesions demonstrable on post-mortem examination, and it may be negative in the presence of gross lesions (see McEwen 1939, Report 1941). Hole (1958) sums up the position by saying that a considerable proportion of animals in which infection cannot be demonstrated react to both johnin and avian tuberculin, and that many animals found to be infected do not react to either.

The disease is usually fatal, but numerous instances of apparent recovery or of long intermissions in its course are known. In some apparently recovered animals post-mortem examination reveals the presence of intense cellular reactions in the mesenteric glands, which seem to result in the disappearance of the bacilli. Infection in such animals probably remains latent and never gives rise to clinical symptoms. There is no reliable method of detecting symptomless carriers.

Since the disease appears to be spread mainly by the contamination of water, foodstuffs and pasture land with the faeces of infected animals, preventive measures will include the destruction of diseased animals, suitable disposal of their excreta, disinfection of byres, the ploughing-up of infected pasture, and the segregation of young stock till their first lactation. If this is impracticable, pasture land should be regarded as potentially dangerous for at least a year, and should not be used for the grazing of sheep or cattle. These measures may suffice to eradicate the disease (see Edwards 1947), but cannot be relied on completely. Since in pregnant animals the fetus may become infected, calves from cows suffering from the disease should not be reared. In areas of the world, such as Colorado, where the disease occurs in wild animals—Rocky Mountain big-horn sheep and goats—eradication may prove difficult (Thoen and Muscoplat 1979).

According to Vallée, Rinjard and Vallée (1934), subcutaneous vaccination of non-infected animals with 5-10 mg of living virulent cultures of Johne's bacillus affords a high degree of protection against the disease. Unfortunately the vaccine sensitizes animals to tuberculin, and may interfere with a tuberculosis eradication scheme. Using living non-virulent bacilli suspended in equal parts of olive oil and liquid paraffin to which pumice powder was added as an irritant, Doyle (1964) found during an experience of 20 years that vaccination of calves 3-4 weeks old, combined with the usual hygienic precautionary measures, afforded a high degree of protection against natural infection. Revaccination is probably desirable from time to time, but in sheep, according to Gilmour and Angus (1973), may precipitate a massive cellular response leading to extensive macroscopic lesions. In Iceland a heat-killed phenolized vaccine suspended in mineral oil is said to have reduced the average annual specific mortality rate in lambs by about 85 per cent (Sigurdsson 1960). Wilesmith (1982) found that in herds in which the calves were vaccinated for 4 years the disease was eliminated.

No treatment for the disease itself has so far afforded any promise. Reviews of the disease are given by Doyle (1956), Sigurdsson (1956) and Hole (1958).

References

Adler, S. (1937) *Lancet* **ii**, 714.
Aycock, W. L. and Gordon, J. E. (1947) *Amer. J. med. Sci.* **214**, 329.
Badger, L. F. (1959) In: *Leprosy in Theory and Practice*, p. 51. Ed. by R. G. Cochrane. John Wright & Sons Ltd, Bristol.
Balfour-Jones, S. E. B. (1937) *J. Path. Bact.* **45**, 739.
Bang, O. (1909) *Zbl. Bakt.* **51**, 450; (1914) *Proc. 10th int. vet. Conf., Lond.* **3**, 157.
Bechelli, L. M. and Domínguez, V. M. (1966) *Bull. World Hlth Org.* **34**, 811.
Bechelli, L. M. *et al.* (1974) *Bull. World Hlth Org.* **51**, 93.
Beiguelman, B. (1967) *Bull. World Hlth Org.* **37**, 461.
Binford, C. H. (1969) *Amer. J. clin. Path.* **51**, 681.
Brown, J. A. K., Stone, M. M. and Sutherland, I. (1966) *Brit. med. J.* **i**, 7.
Browne, S. G. (1967) *Trans. R. Soc. trop. Med. Hyg.* **61**, 265; (1968) *Brit. med. J.* **iii**, 725.
Buck, A. A. (1961) *Amer. J. Hyg.* **74**, 137, 152, 174, 189.
Burnet, É. (1938a) *C. R. Acad. Sci.* **207**, 690; (1938b) *Arch. Inst. Pasteur Tunis* **27**, 327.
Burnet, É. and Jadfard, H. (1939) *Bull. Acad. Méd., Paris* **122**, 383.
Bushby, S. R. M. (1959) In: *Leprosy in Theory and Practice*, p. 188. Ed. by R. G. Cochrane. John Wright & Sons Ltd, Bristol.
Cameron, J. (1956) *J. Path. Bact.* **71**, 223.
Chandler, R. L. (1955) *N.Z. vet. J.* **3**, 145.
Chaussinand, R. (1948) *Acta trop., Basle* **5**, 160; (1955) *La Lèpre*, 2nd ed., Expansion Scientifique Française; (1959) *Ann. Inst. Pasteur* **97**, 125.
Cochrane, R. G. (1959a) *Leprosy in Theory and Practice*, p. 1. John Wright & Sons Ltd, Bristol; (1959b) *Ibid.* p. 114; (1959c) *Ibid.* p. 120; (1959d) *Ibid.* p. 368.
Cummins, S. L. and Williams, E. M. (1934) *Brit. med. J.* **i**, 702.

Daddi, G. and Gialdroni-Grassi, G. (1966) *Rev. Tuberc., Paris* **30**, 1033.

Dean, G. (1905) *J. Hyg., Camb.* **5**, 99.

Dharmendra (1942) *Leprosy in India* **14**, 122; (Ed.) (1980) *Leprosy*, vol. 1, p. 319. Kothari Medical Publishing House, Bombay; (1981) *Ibid.* vol. 2.

Doyle, T. M. (1956) *Vet. Rec.* **68**, 869; (1958) *Ibid.* **70**, 238; (1964) *Ibid.* **76**, 73.

Draper, P. (1980) See Latapí *et al.*, p. 97.

Dunkin, G. W. (1933) *J. comp. Path.* **46**, 159.

Edwards, S. J. (1947) *Vet. Rec.* **59**, 211.

Elek, S. D. and Hilson, G. R. F. (1956) *J. Path. Bact.* **72**, 427.

Fatemi-Nainie, S., Anderson, L. W. and Cheevers, W. P. (1982) *Virology* **120**, 490.

Fernández, J. M. M. (1940) *Int. J. Leprosy* **8**, 1.

Fielding, J. W. (1945) *Med. J. Aust.* **i**, 473.

Fiennes, R. N. T-W- (1978) *Zoonoses and the Origins and Ecology of Human Disease*, pp. 26–27. Academic Press, London.

Gilmour, N. J. L. (1976) *Vet. Rec.* **99**, 433.

Gilmour, N. J. L. and Angus, K. W. (1973) *J. comp. Path.* **83**, 437.

Gilmour, N. J. L. and Gardiner, A. C. (1969) *J. comp. Path.* **79**, 71.

Gilmour, N. J. L. and Goudswaard, J. (1972) *J. comp. Path.* **82**, 333.

Godal, T., Lofgren, M. and Negassi, K. (1972) *Int. J. Leprosy* **40**, 243.

Goudswaard, J., Gilmour, N. J. L., Dijkstra, R. G. and Beek, J. J. van (1976) *Vet. Rec.* **98**, 461.

Goulding, R., Robson, J. M. and Rees, R. J. W. (1953) *Lancet* **i**, 423.

Hale, J. H., Molesworth, B. D., Grove-White, R. J., Sambamurthi, C. M. and Russell, D. A. (1955) *Int. J. Leprosy* **23**, 139.

Hanks, J. H. (1961) *Symposium on Research in Leprosy*, p. 36, Leonard Wood Memorial, Baltimore.

Hanks, J. H., Abe, M., Nakayama, T., Tuma, M., Bechelli, L. M. and Domínguez, V. M. (1970) *Bull. World Hlth Org.* **42**, 703.

Hardt, F. and Wanstrup, J. (1969) *Acta path. microbiol. scand.* **76**, 493.

Hart, P. d'A. and Rees, R. J. W. (1967) *Brit. med. Bull.* **23**, 80.

Hole, N. H. (1958) *Advanc. vet. Sci.*, **4**, 341. Academic Press, Inc., New York.

James, D. G. (1966) *Lancet* **ii**, 633.

Jochmann, G. (1913) *Pocken und Vaccinationslehre*. Alfred Hölder, Wien.

Johne and Frothingham (1895) *Dtsch.-Z. Thiermed.* **21**, 438.

Jopling, W. H. (1978) *Handbook of Leprosy*, 2nd edn. Heinemann Medical Books Ltd, London.

Khanolkar, V. R. (1959) In: *Leprosy in Theory and Practice*, p. 78. Ed. by R. G. Cochrane, John Wright & Sons Ltd, Bristol.

Kraneveld, F. C. and Roza, M. (1954) *Docum. neer. indones. Morb. trop.* **6**, 303.

Kveim, A. (1941) *Nord. Med.* **9**, 169.

Lampe, P. H. J. and Moor, C. E. de (1936) *Geneesk. Tijdschr. Ned.-Ind.* **76**, 1619.

Larsen, A. B. and Kopecky, K. E. (1970) *Amer. J. vet. Res.* **31**, 255.

Latapí, F., Saúl, A., Rodríguez, O., Malacara, M. and Browne, S. G. (Eds) (1980) Leprosy. *Proc. XI int. Leprosy Congr. 1978.* Excerpta Medica, Amsterdam.

Lawrence, W. E. and Wickham, N. (1963) *Aust., vet. J.* **39**, 390.

Lillington, G. A. and Jamplis, R. W. (1963) *Ann. intern. Med.* **59**, 101.

Lobel, L. W. M. (1936) *Int. J. Leprosy* **4**, 79.

Löfgren, S. and Lundbäck, H. (1950) *Acta med. scand.* **138**, 71.

Lovell, R., Levi, M. and Francis, J. (1944) *J. comp. Path.* **54**, 120.

Lowe, J. (1950) *Lancet* **i**, 145.

Lowe, J. and McNulty, F. (1953) *Leprosy Rev.* **24**, 61.

McCoy, G. W. (1913) *Publ. Hlth Bull., Wash.* No. 61, p. 27.

McEwen, A. D. (1939) *J. comp. Path.* **52**, 69.

M'Fadyean, J., Sheather, A. L. and Edwards, J. T. (1912) *J. comp. Path.* **25**, 217.

Machicao, N. and La Placa, E. (1954) *Lab. Invest.* **3**, 219.

Mandel, W., Thomas, J. H., Carman, C. T. and McGovern, J. P. (1964) Bibliography on Sarcoidosis 1878–1963. *Publ. Hlth Serv. Publication* No. 1213. US Dept Hlth, Educ. & Welfare.

Marchoux, E. (1922) *Bull. Acad. Méd., Paris* **87**, 545; (1933) *Rev. franc. Derm. Vénérol.* **9**, 323.

Marchoux, E. and Chorine, V. (1938) *Ann. Inst. Pasteur* **61**, 296.

Marchoux, E., Chorine, V. and Koheclin, D. (1935) *Ann. Inst. Pasteur* **55**, 632.

Menzel, S., Bjune, G. and Kronvall, G. (1980). *See* Latapí *et al.*, p. 152.

Mitchell, D. N., Cannon, P., Dyer, N. H., Hinson, K. F. W. and Willoughby, J. M. T. (1969) *Lancet* **ii**, 571.

Mitchell, D. N. and Rees, R. J. W. (1969) *Lancet* **ii**, 81.

Mitchell, D. N., Scadding, J. G., Heard, B. E. and Hinson, K. F. W. (1977) *J. clin. Path* **30**, 395.

Mitsuda, K. (1919) *Jap. J. Derm. Urol.* **19**, 697. Translated and reprinted in *Int. J. Leprosy* (1953, **21**, 347); (1923) *IIme Conférence internationale de la Lèpre*, p. 219. Strasbourg.

Narayanan, E. *et al.* (1972) *Lepr. Rev.* **43**, 188, 194.

Nethercott, S. E. and Strawbridge, W. G. (1956) *Lancet* **ii**, 1132.

Ota, M. and Asami, S. (1932) *C. R. Soc. Biol.* **111**, 287.

Palencia, L., Montaño, R. and Varela, G. (1954) *Rev. Inst. Salub. Enferm. trop.* **14**, 133.

Patterson, D. S. P. and Allen, W. M. (1972) *Proc. R. Soc. Med.* **65**, 998.

Pedley, J. C. (1968) *Lepr. Rev.* **39**, 201.

Porritt, R. J. and Olsen, R. E. (1947) *Amer. J. Path.* **23**, 805.

Prasad, K. V. N. and Ali, P. M. (1967) *Indian J. med. Res.* **55**, 29.

Present, D. H. and Siltzbach, L. E. (1967) *Amer. Rev. resp. Dis.* **95**, 285.

Rabello, F. E. (1978) See Latapí *et al.*, p. 63.

Ramu, G. and Dharmendra (1980) In: *Leprosy*, vol. 1, p. 108. Ed. by Dharmendra. Kothari Medical Publishing House, Bombay.

Rankin, J. D. (1956) *J. Path. Bact.* **72**, 689; (1961) *Res. vet. Sci.* **2**, 89.

Rees, R. J. W. (1964) *Progr. Allergy* **8**, 224.

Rees, R. J. W. and Waters, M. F. R. (1963) *Pathogenesis of Leprosy*, p. 39. Ciba Foundation Study Group, No. 15.

Report (1941) *J. Hyg., Camb.* **41**, 297; (1960a) *World Hlth Org., Tech. Rep. Ser.* No. 189; (1960b) *Animal Health Services in Great Britain 1958*, p. 37. H.M.S.O., London; (1962) *Trop. Dis. Bull.* **61**, 495; (1969) *Amer. Rev. resp. Dis.* **100**, 888; (1970) *World Hlth Org., Tech. Rep. Ser.* No. 459; (1971) *Chron. World Hlth Org.* **25**, 178.

Ridley, D. S. (1967) *Trans. R. Soc. trop. Med. Hyg.* **61,** 596.

Ridley, D. S. and Jopling, W. H. (1966) *Int. J. Leprosy* **34,** 255.

Robson, J. M. and Smith J. T. (1959) *Brit. J. exp. Path.* **40,** 33.

Rogers, L. and Muir, E. (1925) *Leprosy.* Bristol.

Row, R. (1925–26) *Trans. R. Soc. trop. Med. Hyg.* **19,** 407.

Scadding, J. G. (1960) *Brit. med. J.* **ii,** 1617; (1967) *Sarcoidosis.* Eyre & Spottiswoode, London; (1981) *J. R. Soc. Med.* **74,** 147.

Sellards, A. W. and Pinkerton, H. (1938) *Amer. J. Path.* **14,** 421.

Shield, M. J., Stanford, J. L. and Rook, G. A. W. (1980) *See* Latapí *et al.,* p. 8.

Sigurdsson, B. (1945) *J. Immunol.* **51,** 279; (1946) *Ibid.* **53,** 127; (1947) *Ibid.* **55,** 131; (1956) *Bact. Rev.* **20,** 1; (1960) *Amer. J. vet. Res.* **21,** 54.

Siltzbach, L. E. (1961) *J. Amer. med. Ass.* **178,** 476; (1966) *Lancet* **ii,** 695; (1969) *Practitioner* **202,** 613.

Smith, H. W. (1960) *J. Path. Bact.* **80,** 440.

Stamp, J. T. and Watt, J. A. (1954) *J. comp. Path* **64,** 26.

Stanley, S. J., Howland, C., Stone, M. M. and Sutherland, I. (1981) *J. Hyg., Camb.* **87,** 233.

Stefansky, W. K. (1903) *Zbl. Bakt.* **33,** 481.

Sutherland, I., Mitchell, D. N. and Hart, P. d'A. (1965) *Brit. med. J.* **ii,** 497.

Taylor, A. W. (1950) *J. Path. Bact.* **62,** 647; (1951*a*) *Vet. Rec.* **63,** 776; (1951*b*) *J. Path. Bact.* **63,** 333.

Thoen, C. O. and Muscoplat, C. C. (1979) *J. Amer. vet. med. Ass.* **174,** 838.

Turk, J. L. (1969) *Bull. World Hlth Org.* **41,** 779.

Twort, F. W. and Ingram, G. L. Y. (1913) *A Monograph on Johne's Disease.* London.

Vallée, H., Rinjard, P. and Vallée, M. (1934) *C. R. Acad. Sci.* **199,** 1074.

Vogelsang, T. M. (1957) *Int. J. Leprosy* **25,** 345.

Wade, H. W. (1950) *Int. J. Leprosy* **18,** 373.

Wayson, N. E. and Masunaga, E. (1935) *Publ. Hlth Rep., Wash.* **50,** 1576.

Weddell, G., Jamison, D. and Palmer, E. (1959) In: *Leprosy in Theory and Practice,* p. 96. Ed. by R. G. Cochrane. John Wright & Sons Ltd, Bristol.

Wells, A. Q. and Wylie, J. A. H. (1949) *Lancet* **i,** 439.

Wherry, W. B. (1908) *J. infect. Dis.* **5,** 507.

Wilesmith, J. W. (1982) *Brit. vet J.* **138,** 321.

Wilkinson, G. T. (1964) *Vet. Rec.* **76,** 833.

Withers, F. W. (1959) *Vet. Rec.* **71,** 1150.

Yamamoto, K., Sato, M. and Sato, Y. (1936) *Jap. J. Derm. Urol.* **40,** 28.

53

Diphtheria and other diseases due to corynebacteria

Graham Wilson and Geoffrey Smith

Introductory

The modern conception of diphtheria as a clinical syndrome may be regarded as dating from the observations of Bretonneau in 1826. Its conception as a disease with a single bacterial cause dates from the publication of Loeffler's classical paper in 1884. Klebs had described the bacillus in the diphtheritic membrane in the previous year, but he had brought forward no convincing evidence of its aetiological role. Loeffler demonstrated the bacillus, which we now know as *Corynebacterium diphtheriae*, in the throats of 13 of 22 diphtheria patients, and isolated it from 6 of them in pure culture. With these strains he inocu-lated a variety of laboratory animals. Rats and mice proved completely resistant. Guinea-pigs died in 2 to 5 days after subcutaneous inoculation, showing at necropsy a localized area of haemorrhagic oedema, a pleural effusion that was often blood-stained, areas of pulmonary consolidation, and congestion of the adrenals. False membranes were sometimes produced by intratracheal inoculation of rabbits and pigeons, similar to those produced by Trendelenburg in 1869 and Oertel in 1871 with material direct from human patients. Though the lesions caused by the diphtheria bacillus were highly characteristic, attempts to reproduce a paralytic disease in experimental animals, including fowls, were a failure. One highly significant point was noted; in most of those animals which died the bacilli could be recovered only from the tissues in the neighbourhood of the site of infection. Cultures taken from the lesions in the internal organs or tissues usually remained sterile. Loeffler concluded that death was due not to the dissemination of bacilli throughout the body, but to the effect of some substance elaborated in the local lesion.

The next important step in our knowledge of diphtheritic infection was the discovery by Roux and Yersin in 1888 that sterile filtrates from cultures of the diphtheria bacillus would kill guinea-pigs with lesions identical with those which follow the injection of the living organism. This demonstration of the presence of a potent extracellular toxin afforded an explanation of Loeffler's observation that death was not associated with any spread of bacteria beyond the local lesion. In rabbits inoculated with such filtered cultures Roux and Yersin noted the occurrence of late paralysis.

The lesions produced in experimental animals have been described (Chapter 25, p. 96). It will be recalled that diphtheritic toxaemia in the guinea-pig is associated with fatty degeneration of the heart muscle and diaphragm and with congestion and haemorrhage in the adrenal glands. In man there is reason to believe that the direct action of the toxin on the heart muscle is one of the chief causes of death (Bolton 1905, Dudgeon 1906, Jaffé 1920, Andrewes *et al.* 1923); adrenal changes are much less evident, though histological examination may reveal the presence of congestion with small extravasations of blood and degenerative changes in the cortical and medullary cells (Andrewes *et al.* 1923).

We have already referred to Loeffler's observation that the diphtheria bacillus can be recovered from the local experimental lesion but not usually from the internal organs. It has generally been held that very little invasion of the body tissues occurs apart from the site of the local lesion. The organism has rarely been isolated from the blood stream during life; only three positive results appear in the records of 313 cases of diphtheria studied by Leede, Roedelius and Reiche in Hamburg (see Andrewes *et al.* 1923). According to Wildführ (1949), however, this is because cultures have been made too late. Wildführ himself, who took blood from 202 contacts of diphtheria patients, isolated virulent diphtheria bacilli from 31 of them up to three days before the appearance of the local membrane; in three of them bacilli were present in the blood at the time of appearance of the membrane and in only one of them after it had appeared. The patients were suffering from constitutional symptoms—fever, malaise, headache and nausea—and in most of them a slight swelling and redness of the tonsils could be detected. At autopsy, individual observers claim to have demonstrated the organism in various internal organs, but

the only viscus from which it can commonly be cultivated is the lung—and this organ is more likely to be infected by direct inhalation than from the blood stream (see Report 1923).

The evidence, as a whole, clearly suggests that natural diphtheria in man, like experimental diphtheria in the guinea-pig, rabbit, or pigeon, is essentially a toxaemia in which individual resistance or susceptibility to infection, and recovery or death, are largely determined by the presence or rapid production of antitoxin. Tissue invasion, when it occurs, appears to be confined to the prodromal stage of the disease.

Bacteriology

Apart from the two variants, *Corynebacterium belfanti* and *C. ulcerans*, all cases of diphtheria are caused by *C. diphtheriae*. As has been pointed out in Chapter 25, three types of this organism are recognized.

The *gravis* and *intermedius* types tend to be associated with a high case-fatality rate; infections with the *mitis* type, on the other hand, are milder.

The differences recorded are very striking. For example, the incidence of complications varies both qualitatively and quantitatively. Table 53.1, taken from Cooper and his colleagues (1936), makes this clear.

Table 53.1 Complications and deaths associated with different types of diphtheria bacilli

Type	Total cases	Haemor-rhagic (per cent)	Paralytic (per cent)	Laryngeal (per cent)
Gravis	2313	3.5	17.0	2.3
Intermedius	1993	3.7	9.9	1.3
Mitis	1488	0.4	4.5	7.5

It would appear that paralytic complications are commonest in *gravis* infections, and obstructive lesions in the air passages in *mitis* infections, and that haemorrhagic complications are far more frequent in infections due to the *gravis* or *intermedius* than to the *mitis* type. So far as albuminuria is concerned, Anderson and his colleagues (1933a) record a 52.6 per cent incidence among their *gravis* cases, 52.1 per cent among their *intermedius* cases, and 28.4 per cent among their *mitis* cases.

There are also differences in the case-fatality rate associated with the different types. McLeod (1943), who collected information on about 25 000 cases from different parts of the world, gives the following figures:

8.1 per cent in 11 492 *gravis* cases.
7.2 ,, ,, ,,   6807 *intermedius* cases.
2.6 ,, ,, ,,   6858 *mitis* cases.

The sharp difference between the severity of the *mitis* cases and those due to infection with the *gravis* and *intermedius* types emerges clearly from the available records. It is much less clear that there is any consistent difference between the severity of *gravis* and *intermedius* infections. There is at least a suggestion (Robinson and Marshall 1934, 1935) that whichever of these two types assumes temporary dominance during a particular epidemic prevalence gives rise to the higher case-fatality rate.

All these figures, it should be noted, refer to cases treated with antitoxin; under these conditions, the mortality associated with *mitis* infections is usually low. According to Cooper and his colleagues (1936), all three types may be met with in phases of exalted and of diminished virulence, the *gravis* at its lowest corresponding to the *mitis* at its highest, and the *intermedius* at its highest surpassing the *gravis* in its lower or moderate phases of virulence.

Wide variations occur in the frequency of the different types at different times and in different places, and sometimes in the same place from year to year (for references see 5th edition p. 1670). There is evidence to show that the *gravis* and to a less extent the *intermedius* types are able to spread more rapidly than the *mitis* type in naturally immune or artificially immunized populations. Why this should be so is problematical.

It has been shown that the production of toxin is related to the iron content of the medium in which the organisms are growing (Pappenheimer and Johnson 1936, Mueller 1941; see also Groman and Judge 1979). The observations of Zinnemann (1943, 1946), however, indicate that the greater severity of *gravis* and *intermedius* infections is due not to this cause, but to the greater amount of toxin produced by the *gravis* and *intermedius* types in the human body. The existence of qualitative differences in the toxins of the three types, as postulated by O'Meara and his colleagues (1947), has not been confirmed.

Mention has already been made (Chapter 25, p. 97) of the conversion in the laboratory of avirulent into virulent strains under the action of the bacteriophage. How often such a change occurs under natural conditions, it is impossible to say.

Disappearance of diphtheria bacilli from the throat

During the course of the disease the diphtheria bacillus can usually be isolated from the affected site. In non-fatal cases the organism is gradually eliminated from the body. The rate at which this occurs has been studied by a number of workers.

Hartley and Martin (1919–20) observed the rate of disappearance of diphtheria bacilli from the throat of 457 young adults suffering from faucial diphtheria. The method used was to take a swab on the day of admission, and every 7th day thereafter, till three successive negative results were obtained. The time at which any patient ceased to carry *C. diphtheriae* was calculated as being half-way between the last positive and the first of three successive negative swabs. The figures were complete for the first 50 days, after which some of the patients were no longer under observation. The results were plotted in the ordinary way, with either the numbers still carrying, or the logarithms of those numbers, as ordinates, and time as abscissae. It was found that the latter curve after the first five days from admission, gave a close approximation to a straight line. The general form of the equation of such a curve is

$$\frac{\log n_1 - \log n_2}{t_2 - t_1} = K,$$

where n_1 and n_2 are the numbers carrying at times t_1 and t_2. If we take time as measured from the fifth day after admission, this may be expressed in the form

$$\log n = \log n' - Kt,$$

where n is the number carrying on any given day, n' is the number carrying on the 5th day, and t is the time interval in days measured from the fifth. The equation to the curve which gave the best fit was calculated by Greenwood as

$$\log n = 2 \cdot 6002 - 0 \cdot 0218 \, t.$$

The observed and calculated number of carriers for the first 50 days were as given in Table 53.2.

Table 53.2 Disappearance of the diphtheria bacillus from the throat (Hartley and Martin 1919–20)

Days after admission	Number carrying (observed)	Number carrying (calculated)
5	392	—
10	302	310
15	232	242
20	194	189
25	156	147
30	118	115
35	92	89
40	70	70
45	52	54
50	41	42

The agreement is obviously very close, and such a relation between lapse of time and the disappearance of bacilli suggests that this disappearance is due to the operation of a large number of small causes acting independently, in other words to chance. It would seem that a person who has carried *C. diphtheriae* for 6 weeks or more is, on the average, just as likely to become free during the next few days as another who has carried for only a week. There are, however, as Hartley and Martin point out, a few persons with obviously unhealthy tonsils who carry persistently, but cease to carry after tonsillectomy.

It may be noted that the constant 0.0218 in the formula quoted above indicates that about 5 per cent of those carrying on any one day have ceased to carry on the day following. In order to determine whether this rate of clearing held true in general, or applied only to this particular sample, figures collected from various published reports were analysed in the same way. It was found that the general form of the curve was very similar in each case, *i.e.* the logarithmic relation between carrier rate and time held good, but the value of the constant varied very widely, such figures as 0.018, 0.093, and 0.202 being obtained. Some adjustment is necessary because, in certain of these series, the test of clearance was a single negative swab, but the value of the constant in Hartley and Martin's series would only be altered to 0.032 on this basis, so that the rate of clearing under different conditions varies considerably—a fact which seems to merit further investigation.

A larger and equally illuminating series of figures recording the apparent rate of disappearance of diphtheria bacilli in convalescents at the North-Eastern Hospital was published by Thomson, Mann and Marriner (1928–29), who took nose as well as throat swabs for examination (Table 53.3). Cases in which the bacilli persisted for over 12 weeks from the beginning of the illness were regarded as carriers. It was found that infection disappeared more quickly from the throat than from the nose, and that approximately 90 per cent of those who became carriers were infected in the nose or nose and throat as opposed to 10 per cent of those infected only in the throat. Disappearance of the organisms occurred quicker in older than in younger children.

Table 53.3 Proportion of convalescents becoming negative each week (Thomson *et al.* 1928–9)

Time (weeks)	Throat		Nose, or nose and throat	
	No.	%	No.	%
Within 2nd	861	49.9	298	24.0
3rd	255	14.8	159	12.8
4th	254	14.7	184	14.8
5th	138	8.0	158	12.7
6th	86	5.0	136	11.0
7th	46	2.7	97	7.8
8th	41	2.4	52	4.2
9th	18	1.0	51	4.1
10th	15	0.9	45	3.6
11th	5	0.3	32	2.6
12th	7	0.4	28	2.3
Total	1726	100.1	1240	99.9

In a similar investigation at Liverpool, Wright (1941*a*) observed that clearance was most rapid in patients infected with the *intermedius* type and slowest in those infected with the *gravis* type. Three consecutive negative swabs were occasionally followed by a positive, showing that they did not ensure freedom of the patient from infection on discharge. Persistence of diphtheria bacilli, especially in the nose, was found to be favoured by an associated infection with haemolytic streptococci (Boissard and Fry 1941, 1942). The more rapid clearance of the *intermedius* type may perhaps explain why epidemic spread of this type is much less frequent than that caused by the *gravis* type (McLeod 1950).

Epidemiology

The behaviour of diphtheria under natural conditions has been studied by bacteriologists and epidemiologists using as their chief tools the Schick test and the throat swab. The picture presented to us in this way is far more complete than that which we possess for any other disease. This is why we selected it as an illustrative example in our general discussion of herd infection and herd immunity in Chapter 18. There is no need here to do more than amplify that account by the addition of particulars which were not relevant to the more general discussion.

Diphtheria occurs throughout the year, but both the morbidity and the mortality rates are highest during the winter months. Before the introduction of mass immunization it had an *age incidence* which, though varying somewhat in different places, was always highest in the pre-school child. The disease was uncommon during the first year of life, and became increasingly frequent until the maximum incidence was reached some time between the 2nd and the 5th year. A gradual decrease occurred among the age group 5–10, and a more rapid fall between 10 and 15 years, after which the probability of contracting diphtheria became small (Frost 1928). In countries, however, in which a high proportion of the pre-school population is artificially immunized against the disease, the incidence has undergone a shift to the higher age groups; and though children are still predominantly attacked, as many as 25 per cent of cases may occur in adults. In early life boys and girls are more or less equally affected, but later—particularly above the age of 15 years—the incidence is considerably higher in females (Report 1948*a*). The *case-fatality rate* varies from one outbreak to another, but in untreated persons it tends to be round about 7 per cent. During childhood it varies inversely with the age of the patient, being highest in infancy and lowest in adolescence; this may be associated to some extent with the higher rate in early life of complications, especially those affecting the circulatory system (Report 1948*a*).

Clinical diphtheria is less common in the tropics than in temperate climates. This may be because of the higher proportion of persons in the tropics with antitoxic immunity than in corresponding age groups in a country such as the United States; or as Frost (1928) suggests, to inherent racial or environmental differences between the two groups.

Mode of spread

The relation of diphtheria to population density—so far as it is possible to measure that elusive variable—is not altogether clear. Over considerable periods of its history the disease has shown a preference for rural as opposed to urban conditions. On the other hand, a high ratio of urban population seems to favour a continuously high level of diphtheria (see McLeod 1950). It is, however, typically a disease of households, schools, and institutions where children are herded together at susceptible ages; a particular school or institution may constitute an endemic centre of the disease for months or years. Closeness and duration of contact seem to play a major part in determining the spread of the disease. There is good evidence that contact of the type which is exemplified by children sleeping in a single dormitory is much more dangerous than casual or momentary contact during working hours (Dudley 1923). Even so, the disease is often curiously capricious in its striking power, and the attack rate among those exposed to risk varies widely from one outbreak to another. Under conditions in which most cases of clinical diphtheria are removed to an isolation hospital spread of the disease appears to be due chiefly to carriers. According to Wildführ (1949) diphtheria bacilli are often detectable in the throat during the incubation period, and it is possible that these precocious carriers are as dangerous as convalescent carriers, or as carriers that remain well throughout.

The possibility of spread by *fomites* has to be considered. From the case of the child's box of bricks, reported by Abel in 1893, there have been instances in which it seemed possible to incriminate a particular article as a disseminator of infection. Very occasionally, as with the common supply of penholders recorded by Dudley (1923), such a mode of transmission may have been partly operative. In general, however, the balance of evidence would appear to be heavily against the view that fomites play any significant role in the spread of the disease; and there is nothing to suggest that stringency of disinfection would make any appreciable impression on its incidence. An exception must, however, be made for *dust*, which observations by Wright and his colleagues at Liverpool (Crosbie and Wright 1941, Wright, Shone and Tucker 1941) showed to be a possible source of danger in hospitals and other institutions. Crosbie and Wright were able to isolate diphtheria bacilli in large numbers from floor dust in the neighbourhood of diphtheria patients, and to show that they might remain fully virulent for as long as five weeks. They were of the opinion that floor dust distributed by sweeping and blanket dust liberated during the making of beds may play an even more important part in cross-infection in diphtheria wards than the direct transmission of droplets from patient to patient. Measures to reduce aerial infection by dust, such as the application of heavy oil to the floors (van den Ende, Lush and Edward 1940, Thomas 1941) and to the blankets (van den Ende and Thomas 1941), hold out some promise of success, though so far no simple method for destroying diphtheria bacilli in the dust itself has yet been devised.

It was at one time believed that certain *domestic animals* were concerned in the spread of diphtheria, but most of the evidence on which this belief was based has been shown to be erroneous.

Klein's (1889) statements on diphtheria in the *cat* were incorrect; Savage (1919–20), in fact, showed that this animal is very resistant to infection, though one or two suggestive cases are on record (Simmons 1920). In *hens* confusion was at one time caused by fowl diphtheria, which bears a superficial resemblance to the human disease, but which is now known to be due to infection with a filtrable virus (see Chapter 87). Litterer (1925) recorded a small outbreak of disease in fowls closely simulating human diphtheria in which he isolated virulent diphtheria bacilli; but it seems improbable that these birds, even if they do occasionally become infected from human sources, play any serious part in the epidemiology of the disease. Virulent diphtheria bacilli have likewise been isolated from the *horse*, and from the *monkey* (Dold and Weigmann 1934, Ramon and Erber 1934), but here again such occurrences may be regarded as pathological curiosities.

The *cow* is in a different category, since the general consumption of milk renders any infection of this animal peculiarly dangerous. In spite of earlier records to the contrary, it is now reasonably certain that the cow does not naturally suffer from any form of infection with *C. diphtheriae*, and that the only lesion of epidemiological importance in this animal is the occasional infection of pre-existing superficial ulcers on the teats from the hands of a milker who is suffering from the disease or is carrying virulent bacilli. The existence of these lesions was recorded by Dean and Todd (1902), Ashby (1906), and Henry (1920), and the authors have likewise met with them on one occasion; the virulence of the bacilli was fully established. Such infected wounds provide a persistent focus from which the milk may be contaminated. Milk infected in this way is likely to contain a larger number of diphtheria bacilli than when contaminated directly from a human case or carrier.

It may be added that the incrimination of the milk rests, in almost all instances, on epidemiological evidence. The isolation of *C. diphtheriae* from milk, even when a particular supply is under grave suspicion, has in the past been very uncommon; nevertheless a few successful results have been reported (Bowhill 1899, Eyre 1899, Marshall 1907, McSweeney and Morgan 1928). With the introduction of tellurite medium the technical difficulties of isolation are greatly reduced, as is shown by the reports of Fry (1941) and Goldie and Maddock (1943). There remains, however, the difficulty that the actual milk which conveyed the infection is seldom, if ever, available for examination at the time when attention has been drawn to it, so that unless the contamination continues for some days an infected sample is unlikely to be forthcoming. It must be emphasized that the identification of a diphtheria-like bacillus isolated from milk requires the most stringent tests. No organism derived from such a source should be reported as *C. diphtheriae* until its character has been established by a properly controlled virulence test.

The *secular incidence* of diphtheria is subject to considerable fluctuation, rising in a given area at one time to a maximum and then falling over a period of years to a much lower level. In the first half of the nineteenth century diphtheria was endemic in Europe and the United States. Between 1850 and 1860 there sprang, apparently from a focus in France, a pandemic with a high case-fatality rate. After 1885 a consistent decline began, though the incidence of the disease still remained high. It is perhaps only natural to ascribe the decreased incidence that has been widely observed of recent years to the effect of prophylactic inoculation, but there is little doubt that other factors have been at work as well, such as the natural immunization accompanying latent and subclinical infections. The typing of diphtheria bacilli has not been carried out long enough to provide us with all the information we need: but the results, so far as they go, suggest that in a partly vaccinated population the occurrence of an epidemic wave may be associated with the replacement of a *mitis* or *intermedius* type of organism by a *gravis* (Bøe and Vogelsang 1948, Ruys and Noordam 1949, Hartley *et al.* 1950, McLeod 1950). In a fairly well immunized population in which diphtheria is gradually dying out, the reverse phenomenon may be observed, namely, the replacement of a predominantly *gravis* type by a *mitis* (Report 1949); this supports the conclusion reached on other grounds that the *mitis* type is better able than the more toxic organisms to establish a commensal relationship with the host (see Leete 1945). (For a general study of the epidemiology of diphtheria, see W. T. Russell 1943.)

The diagnosis of infection in diphtheria

The laboratory diagnosis of infection with diphtheria depends on the demonstration of the causative organism. Though we are concerned here more with general principles than with technical details, it will be convenient to describe the current practice in most British laboratories. In the past chief reliance was placed on examination of a direct smear, and on the microscopic appearance of the growth on a Loeffler slope culture. Since the differentiation by McLeod and his colleagues (Anderson *et al.* 1931, 1933*a, b*) of three different colonial types of diphtheria bacilli on tellurite blood agar, and since the agreement reached by numerous

workers on the advantage of using a selective medium, the study of the colonies developing on tellurite plates has assumed increasing importance in the laboratory diagnosis of diphtheria. Not only does the isolation of diphtheria bacilli in the form of single colonies carry greater assurance than their demonstration morphologically in a film containing numerous other organisms, but the recognition of their type—*gravis, intermedius,* or *mitis*—proves often of epidemiological and sometimes of clinical interest. It should be borne in mind that occasional cases of diphtheria are caused by *C. ulcerans.* This organism closely resembles the *gravis* type; it is distinguished from it by the characters listed elsewhere (Chapter 25, p. 100).

We may now outline the general procedure to be followed and the interpretation of the results in looking for the diphtheria bacillus in suspected cases of the disease, and in convalescents, contacts, and suspected carriers.

Suspected cases of clinical diphtheria

By examination of a *smear* made directly from the swab and stained preferably by the wet-film technique, it is sometimes possible to make a presumptive diagnosis of diphtheria. Though this method is advocated by a few workers, the majority of bacteriologists do not regard it with favour. It is restricted in its usefulness to throat swabs from children, and even in these the proportion of positive results in true cases of diphtheria is often fairly low. Though it should not do so, a negative result serves as an excuse to many practitioners to withhold serum for treatment. A preliminary negative report on a direct smear has frequently to be followed by a positive report on culture, thus confusing the practitioner and leading him to doubt the accuracy of the laboratory's findings. Moreover, the time taken in preparing and examining direct smears may often be spent more profitably in other ways. The value of the direct film, however, may be increased by using the fluorescent antibody technique (Whitaker *et al.* 1961).

In the routine *cultivation* of swabs from suspected cases of diphtheria it is advisable to inoculate each swab on to a slope of Loeffler's serum, a plate of tellurite blood agar, and, for the detection of haemolytic streptococci, on to an ordinary blood agar plate. All three cultures should be examined 12–15 hours later. Positive results obtained on both Loeffler and tellurite, or on tellurite alone, can be reported safely at once. Characteristic findings on Loeffler in the presence of a negative tellurite plate usually justify a presumptive positive report. When, however, the results are ambiguous or negative, it is better to send an interim doubtful report, and await the result of subculture or of further incubation of the tellurite plate before a final negative report is made. In cases of this sort the finding of Vincent's spirilla and fusiform bacilli in a direct film or of haemolytic streptococci on a blood agar plate is often of value in suggesting the true nature of the infection, though it should be remembered that Vincent's organisms are often present in large numbers in the normal mouth.

This is not the place to discuss the relative merits of the various tellurite media that have been proposed. In different laboratories preference is expressed for the particular type of tellurite blood agar medium described by McLeod and his co-workers (Anderson *et al.* 1931), by Clauberg (1929, 1931, 1933, 1935, 1936, 1939), by Horgan and Marshall (1932), by Neill (1937), by Glass (1937), by Hoyle (1941), or by some other worker; or for the particular type of tellurite serum agar described by Kerrin and Gaze (1937), by Handley (1949) or by Whitley and Damon (1949*a, b*). German practice favours a combination of fluid enrichment and solid selective media (Kröger 1955, Sachse 1955). So far, no medium can be regarded as perfect. We have given our reasons (Chapter 25, p. 94) for preferring a simple medium such as Hoyle's. Whatever medium is used, however, there will always be a small proportion of colonies that cannot be recognized within 24 hours; a final negative report should therefore not be issued for 2 days after receipt of the swab. Doubtful colonies may have to be tested in sugar media, and in this regard it should be borne in mind that occasional *mitis* strains ferment sucrose. In some countries, for example Brazil, sucrose-fermenting virulent strains of diphtheria bacilli are quite common, rendering it necessary to test them on dextrin and glycerol (Christovão 1957). Confirmation of suspicious colonies may be accelerated by use of the urea-hydrolysis test (see Chapter 25, p. 95), the results of which can sometimes be read within 30 minutes; a positive result practically excludes the diphtheria bacillus, though unfortunately a negative one does not necessarily exclude a diphtheroid. *C. ulcerans*, whose colonies closely resemble those of the *gravis* type of diphtheria bacillus, can be rapidly distinguished by its ability to hydrolyse urea and its failure to reduce nitrates (Cook and Jebb 1952).

The combined use of a Loeffler slope and a tellurite blood-agar culture enables a high proportion of all cases of diphtheria to be diagnosed bacteriologically within 18–24 hours. Early investigators, such as Weigmann and Degn (1932), Kairies (1933), Lewis (1933), Kemkes (1934), and Hasenbach (1935), found tellurite medium to give a slightly higher proportion of true positive results than Loeffler's serum. The experience of numerous workers has since confirmed their results, and has shown that in *cases* of diphtheria the causative bacillus may be demonstrated by tellurite in about 5–10 per cent more swabs than by Loeffler. The routine use of a Loeffler slope culture has, however, numerous advantages. It shows whether the swab was suitable for examination; failure of growth suggests that the patient's throat had been treated recently with an an-

tiseptic, or that the swab had been improperly taken or had been too long in transmission to the laboratory, or that the bacilli were being inhibited by the presence of staphylococci (see Chapter 30, pp. 231–2). Development of diphtheria bacilli is more rapid on Loeffler than on tellurite, and a tentative diagnosis can therefore be made earlier—often within 8 hours (see Daniels *et al.* 1951).

An even more rapid method originally devised by Folger in 1902 and recommended by Solé (1934) and many other workers, consists in using a cotton-wool swab impregnated with sterile horse or ox serum. The swab is dipped in the serum; the excess is squeezed out against the side of the tube; and the serum is then coagulated by gentle heating over a flame or in the oven at 80° for 15 minutes. The diphtheritic membrane is touched with the side of the swab, which is then incubated for 2–4 hours at 37°. A smear is made and stained with Albert's, Neisser's or Ponder's stain. Quite often a morphological diagnosis can be made at this stage. After preliminary incubation the swab can be streaked on to a Loeffler slope and a tellurite plate and the growth examined the following day (see Brahdy *et al.* 1935).

Occasionally strains of diphtheria bacilli are met with that are unusually sensitive to potassium tellurite, and may not develop on primary culture in the presence of this substance; to avoid missing these, a Loeffler serum culture is a valuable safeguard. On the other hand, the presence of apparently typical diphtheria bacilli on Loeffler in the complete absence of diphtheria colonies on tellurite must always be regarded with suspicion, and a final positive report should never be sent till diphtheria bacilli have been isolated in pure culture by plating the Loeffler slope growth on to chocolate blood agar or tellurite blood agar.

The advantage of a tellurite plate lies in its selective action in eliminating most of the organisms that grow on other media and providing discrete colonies, rather than the usual confluent growth of a Loeffler slope. The concentration of tellurite must be such that it is not too inhibitory on the one hand or insufficiently selective on the other. Usually 0.04 per cent of potassium tellurite is about right. By careful examination of the colonial form under a binocular dissecting microscope and of the morphology of the organisms in a wet-stained film, or even by nigrosin (Fleming 1941), many workers of experience can assign correctly the majority of cultures to the *gravis, intermedius,* or *mitis* types; but it is always wise to confirm this preliminary identification by means of fermentation and, if necessary, virulence tests. A Gram stain is often useful, helping, as it does, to distinguish between the weakly staining diphtheria bacillus and the strongly staining diphtheroid bacilli.

Shone, Tucker, Glass and Wright (1939), who made a very careful study at Liverpool of the comparative merits of Loeffler and tellurite in the diagnosis of diphtheria cases, found that of 1501 strains of diphtheria bacilli 92.4 per cent were recognized on both a Loeffler slope and a tellurite blood agar plate, 2.3 per cent on Loeffler only, and 5.3 per cent on tellurite only. The failures of the Loeffler medium seem to be due partly to the scantiness of the diphtheria bacilli in some cases and the consequent difficulty of finding them among other bacteria, and partly to the atypical appearance of certain strains, especially those of the *gravis* type. The chief disadvantage of the tellurite medium lay in the slow growth of some strains of the *mitis* type. Cooper and his colleagues (1940), who made a similar series of observations in McLeod's laboratory at Leeds, found that a tellurite blood agar plate yielded at least 10 per cent more positive results than a Loeffler slope, and led to less likelihood of missing severe cases. On Loeffler the very short form of *intermedius* type of diphtheria bacillus was found to be difficult to distinguish from unusually granular Hofmann strains. The results on Loeffler's medium at Liverpool were more favourable than those obtained in many other laboratories, probably because Wright and his colleagues had spent two years' intensive work checking the accuracy of their microscopical diagnosis. In the hands of less skilled observers the proportion of true positive results on Loeffler would have been lower, and a number of false positives would almost certainly have been recorded (see Cruickshank 1943). Specific and type identification of the colonies can be made by slide agglutination with polyvalent and with *gravis, intermedius,* and *mitis* type sera. Controls should always be put up with saline and with normal rabbit serum (Calalb *et al.* 1967). In the study of circumscribed outbreaks bacteriophage typing may be of value (Toshach *et al.* 1977).

Convalescents, contacts, and carriers

In swabs from convalescents, contacts, and carriers diphtheria bacilli are generally fewer than in those from cases. The use of tellurite medium in their detection is not only an advantage; it is almost imperative. Numerous workers have found that approximately twice as many positive results are obtained on this medium as on Loeffler's serum. The development of the colonies is, however, often slow, and a negative report should rarely be made till the plates have been incubated for 48 hours. In contacts diphtheria bacilli may sometimes be isolated during the incubation period (Wildführ 1949).

Virulence tests

At one time the only way to ascertain whether a strain of diphtheria bacillus was virulent, and therefore potentially infective, was by inoculating it into a guinea-pig. Though this is still true if complete assurance is to be obtained, the number of virulence tests to be carried out may be greatly reduced by noting the type of diphtheria bacillus to which the strain belongs. Experience now in many laboratories has shown that

nearly all strains of the *gravis* and *intermedius* type are virulent but that some strains of the *mitis* type are avirulent. Thus Parish (1936) reports 96.9 per cent, 99.5 per cent, and 99.7 per cent of virulent cultures among large samples of *gravis* strains examined by different workers, and 94.1 per cent, 99.1 per cent, and 99.9 per cent for *intermedius* strains, as against 80.0 per cent, 86.6 per cent, and 87.5 per cent for strains of *mitis* type. Stallybrass (1936), quoting Wright's findings in Liverpool, reports one avirulent culture among 145 *gravis* strains, four avirulent cultures among 116 *intermedius* strains, but no fewer than 100 avirulent cultures among 231 *mitis* strains. The proportion of avirulent strains varies also with the origin of the culture. Thus strains of all three types are invariably virulent when isolated from cases of acute diphtheria. Avirulent strains are encountered among convalescents, contacts, and carriers, particularly in those with nasal and other extra-faucial infections. Strains of the *gravis* and *intermedius* type seldom need to be tested for virulence unless derived from long-standing carriers, or from special areas where non-virulent strains are known to be common (see McLeod and Robinson 1948, Edward and Allison 1951). Strains of the *mitis* type from contacts, carriers, and persistently positive convalescent cases will generally require testing. The strains, however, that most require testing for virulence are atypical strains, especially those found in carriers (Cooper *et al.* 1940).

Three methods of testing for virulence are in use—the subcutaneous, the intracutaneous, and the plate method. The technique of the first two tests varies with different workers. In the subcutaneous test it is usual to inoculate between an eighth and the whole of an 18-hour Loeffler slope culture in a $15\cdot2 \times 1\cdot6$ cm test-tube into a guinea-pig. The importance of washing off the growth with a suitable suspending medium such as broth rather than with saline, which may kill the organisms, has been stressed by Holt and Wright (1940). These observers likewise recommend the chest as the preferential site of inoculation, since oedema is said to be more noticeable here than in the thigh. To make certain that the strain under test forms the toxin specific to the diphtheria bacillus it is wise to inoculate two guinea-pigs simultaneously, one of which is protected by means of a suitable dose of antitoxin, such as 500 units given intraperitoneally on the previous evening.

The intradermal, or Römer (1909), method has the advantage of allowing several strains to be tested on the same animal. A Loeffler slope culture is washed off with 1 to 4 ml of broth, depending on the amount of growth, and inoculated intradermally in a quantity of 0.2 ml into the shaven side of an albino guinea-pig. A known virulent and a known avirulent strain should always be inoculated as controls. To protect the animal from death during the development of the characteristic skin reactions, 20 to 40 units of refined anti-diphtheria serum should be injected simultaneously into the peritoneal cavity. A guinea-pig protected by the intraperitoneal injection of 500 units of serum on the previous evening should likewise be inoculated intradermally with the same strains as those used for the test animals. This precaution is desirable, especially with atypical strains and with cultures of doubtful purity, so as to prevent confusion being caused by necrotic lesions due to other species of *Corynebacterium* or to staphylococci.

The plate method is an in-vitro method introduced by Elek (1948, 1949). A wide strip of filter paper impregnated with diphtheria antitoxin containing 1000 units per ml is implanted in an agar plate while the medium is still fluid. When the agar has set, the organism to be tested is streaked at right angles to the strip. If toxin is produced, it precipitates the antitoxin in the form of double arrow-headed lines, the exact configuration and thickness of which depend on the occurrence of optimal proportions of toxin and antitoxin in different parts of the plate. This test has the advantage of rapidity, since with most strains the results can be read within 24 hours. Good agreement was reported between the in-vitro and in-vivo methods (King *et al.* 1949, Ouchterlony 1949). The numerous precautions that should be taken to avoid error are described by Maniar and Fox (1968) and Bickham and Jones (1972).

According to Christovão and Cotillo (1966), toxin circulating in the blood of a suspected case of diphtheria may be demonstrated by a haemagglutination test in which antitoxin is adsorbed on to human group O red cells.

The diagnosis of susceptibility or immunity to diphtheria: Schick test

The test that first enabled us to distinguish between susceptible and resistant members of the human population was introduced by Schick (Michiels and Schick 1913*a*, *b*, Schick 1913). It is performed as follows.

The test toxin is stored in a relatively concentrated solution, and when required for use is diluted so that one Schick dose is contained in 0.2 ml. Part of this diluted toxin is heated to 70° for 5 minutes, to serve as a control. Into the flexor surface of one arm 0.2 ml of the unheated toxin is injected intradermally, care being taken that the injection is made into the substance of the dermis in such a way as to raise a definite bulla. Into the flexor surface of the opposite arm is injected 0.2 ml of the heated toxin. The reaction to the toxin develops at a varying rate. In some persons a positive reaction is evident after 24 hours, but in others it is not fully developed for a week. From the observations of Downie and his colleagues (Report 1942), the best times to make readings would appear to be the 4th and the 7th days; if only one reading can be made, the 7th day should be selected. The following types of reaction may be observed to the toxin and the control:

(*a*) The negative reaction—no reaction of any kind in either arm.

(*b*) The positive reaction—no reaction in the control arm.
 In the test arm a circumscribed red flush appears generally after 24 to 36 hours, reaching its maximum development on the 4th to 7th day. At this time there is a circular area, 1–2 cm in diameter, slightly raised above the general surface of the skin. This slowly fades during the following few days or more, leaving an area of brownish pigmentation, with a desquamating epidermis.

(*c*) The pseudo-reaction (negative).—This reaction, which is of the characteristic allergic type, develops equally in both arms during the first 24 hours. It is less sharply circumscribed than the true positive reaction, and fades much more rapidly. By the 4th day it has usually disappeared, leaving some slight degree of reddish or brownish discoloration.

(*d*) The combined reaction (pseudo + positive).
 The control arm shows the succession of changes referred to under the pseudo-reaction. The reaction in the test arm is almost indistinguishable from that in the control arm during the first 24 hours. After this time the reaction in the test arm continues to develop, while that in the control arm commences to fade. By the 4th day the difference between the two arms is usually quite distinctive.

The positive and combined reactions are generally regarded as indicative of susceptibility. More precisely they show that antitoxin is either absent from the circulating blood or present in only a very small quantity. At one time it was taught that a positive reaction showed that there was less, and a negative reaction that there was more, than 1/30 unit of antitoxin per ml, but further experience has rendered it clear that either reaction may occur over a fairly wide range of antitoxin values (see Jensen 1931, Fraser and Halpern 1935, Leach and Pöch 1935, Parish and Wright 1938, Thelander 1940, Downie *et al.* 1941, Craig 1962). A positive reaction may be registered with as much as 1/30 unit per ml and a negative reaction with less than 1/1000 unit. In general, however, a positive reaction is uncommon with 1/100 unit per ml or more, and rare with 1/30 unit or more; a negative reaction seldom occurs with less than 1/500 unit (see also Vahlquist and Högstedt 1949). In Brazil, Cotillo and his colleagues (1966) found agreement between the Schick and the antitoxin tests in about 90 per cent of vaccinated and about 80 per cent of unvaccinated subjects. In explanation of the discordances they suggested that the two tests might be measuring a different kind of immunity.

The Schick test, within its limits, is a quantitative one; and the test dose must therefore be defined, and the test toxin standardized (see Chapter 18).

As a possible alternative to the Schick test, the amount of circulating antitoxin in a person's blood may be determined by the Römer technique. This is clearly a more accurate method, and entails a single withdrawal of blood compared with the intradermal injections and their observation during the following few days. The necessary animal inoculations might be supposed to prevent the adoption of this method as a routine. Jensen (1933*b*), however, who used rabbits instead of guinea-pigs for the intracutaneous injections, thus increasing the number of tests that could be carried out on a single animal, considered it the method of choice for both convenience and accuracy. A second alternative for the titration of circulating antitoxin is a haemagglutination test in which diphtheria toxin is adsorbed on to tanned red cells of human group O (Guedes *et al.* 1966).

The evidence now available leaves no reasonable doubt that a person who reacts negatively to the Schick test, or who has been shown to possess an equivalent amount of circulating antitoxin, is comparatively immune to all ordinary risks of infection. It must be made clear, however, that the degree of immunity possessed by schick-negative reactors is far from absolute, and that, in fact, quite large numbers of cases of diphtheria have been reported in immunized persons many of whom were probably schick-negative. Such cases are usually due to infection with the *gravis* type of diphtheria bacillus (Cooper *et al.*

1936, McLeod 1943, Hartley *et al.* 1950) and seldom prove fatal. Partly for this reason it is undesirable to give certificates of immunity to individual children based on the results of negative Schick tests after immunization.

We may note here the single instance of a direct experiment on man which has, so far as we are aware, been placed on record. Guthrie, Marshall and Moss (1921) collected eight volunteers and determined their reaction to the Schick test; four reacted positively, two negatively, and two gave pseudo-reactions which were read as negative. The throat of each of these eight volunteers was swabbed with a virulent culture of *C. diphtheriae.* The organism established itself on the tonsils of the four positive reactors, and of three of the four negative reactors. The four positive reactors developed clinical diphtheria; the three negative reactors became temporary carriers without showing any sign of the disease.

In discussing the nature and origin of the normal antibodies in Chapter 11, we have already described the observations of Zingher and of others on the results obtained in the application of the Schick test to large samples of the normal population, and have considered the opposing theories of the origin of normal antitoxin as the result of latent infection, or as a phase of normal biochemical ontogeny, in the absence of any specific stimulus; and we have expressed our personal view that it is impossible to account for the observed distribution of natural antitoxic immunity among the population in general, and among certain isolated groups in particular, without assuming that natural immunization plays a highly important role.

Prophylaxis in diphtheria

Our present methods of controlling diphtheria as a herd disease depend on the isolation of patients, the segregation of recognized carriers, and prophylactic immunization. The last method comprises both active immunization by means of an antigenic preparation now usually referred to as diphtheria prophylactic, and passive immunization conferred by the injection of antitoxic serum. Passive immunization, because of its transitory nature, has a very restricted application and will be dealt with later. For all ordinary field purposes active immunization is employed. An adequate understanding of the principles that govern the application of this method to the control of the disease is so important to the medical practitioner and the medical officer of health that the relevant findings must be set out in some detail. We shall begin by describing the various forms of diphtheria prophylactic and their standardization, proceed to a consideration of their application under varying conditions in

practice, and then discuss some of the results that have followed their use in mass immunization.

The various forms of diphtheria prophylactic

As early as 1892 von Behring and Wernicke showed that susceptible animals might be safely immunized by inoculating them with increasing doses of living culture after a protective dose of antitoxic serum. Six years later Nikanaroff reported successful results from the inoculation of a few doses of toxin neutralized with antitoxin; and Dreyer, in 1900, proposed the use of toxin-antitoxin mixtures during the early stages of antitoxin production in horses. For various reasons there was a long lag before similar methods were applied to man; and it was not till 1913, when von Behring reported the successful use of a toxin-antitoxin mixture for the immunization of children, that inoculation against diphtheria entered the field of prophylactic medicine. From then onwards the use of such mixtures for the active immunization of children developed very rapidly, particularly in the United States under the influence of Park and Zingher (Park 1913, 1918, 1922, Park and Zingher 1915, 1916, Zingher 1921, 1922). Toxin-antitoxin mixtures were found to be not without danger (see Park and Schroder 1932, Wilson 1967), and after the discovery by Glenny and Südmersen (1921), Glenny and Hopkins (1923), and Ramon (1923, 1926, 1928) that diphtheria toxin could be so altered by treatment with formalin as to deprive it of its toxicity without destroying its antigenic power, formol toxoid, or anatoxin as it is called in French-speaking countries, gradually began to replace the older mixtures. The further demonstration by Glenny (1930) that formol toxoid could be improved by precipitation with alum has now placed at our disposal an agent of high antigenic potency which, when properly prepared and used, can be relied upon to give a fairly high degree of protection against diphtheria. (For review of prophylactic agents, see Hartley 1949.)

Toxin-antitoxin mixture In preparations of this class sufficient antitoxin is added to neutralize the toxic effects of the toxin without completely destroying its antigenic activity. The toxin, that is to say, is under-neutralized. Though very satisfactory results have been obtained by the use of such mixtures, a few accidents have shown that the union of toxin and antitoxin is unstable and is liable to be disturbed by unsuitable methods of preparation or of storage. In one instance, in which five deaths and 40 serious reactions followed the use of this agent (Forbes 1927), it was found that, in forgetfulness of the Danysz phenomenon, the toxin had been added in fractions to obtain the correct toxin-antitoxin ratio. In another instance a particular phial of toxin-antitoxin mixture that had been frozen was found to have caused severe constitutional reactions, whereas other phials of the same batch that had not been frozen had been used without any untoward results. Further investigations revealed that the freezing of such mixtures frequently resulted in a considerable rise in toxicity. Pope (1927) has shown that this phenomenon is dependent, in part at least, on a separation of the phenol during freezing; this leads to a local concentration greatly in excess of the 0.5 per cent or so added for preservative purposes, and to a consequent destruction of antitoxin, which is more sensitive than toxin to phenolization. Observations recorded by Robinson and White (1928) indicate that, even in the absence of added phenol, dissociation of toxin-antitoxin mixtures may occur on freezing, the degree of dissociation depending apparently on the concentration of the mixture. A dose of toxin equivalent to 12 MLD for a guinea-pig may be fatal to a child (Schmidt 1942). (For a fuller description see Wilson 1967.)

Formol toxoid. FT This is prepared by treatment of culture filtrates with formalin in a concentration

depending on the amount of amino-nitrogen present. The pH of the toxin is adjusted to 7.4–7.6, and the formolized mixture is incubated for 3–4 weeks at 37–38° until the product, as judged by the intracutaneous inoculation of guinea-pigs and rabbits, is devoid of all residual toxicity (Barr, Pope, Glenny and Linggood 1941). The chemical changes concerned in the conversion of toxin into toxoid are still imperfectly understood, but reference may be made to the observations on this subject of Kissin and Bronstein (1928, 1930), Hewitt (1930), Bunney (1931), and Schmidt (1933*a, b, c*). Partial toxicity is said to be regained in some batches if they are stored at a high temperature, such as 37°. Residual formalin in a concentration of 0.006 per cent inhibits reversion, but one below 0.002 per cent does not (Akama *et al.* 1971).

Toxoid-antitoxin mixture This was introduced to avoid the use of toxin. It was employed successfully for many years, but was displaced by the more effective APT.

Toxoid-antitoxin floccules. TAF This is a suspension of the precipitate of floccules that are formed when toxoid and antitoxin are mixed in appropriate neutralizing amounts. The elimination of much of the non-specific serum protein and bacillary products present in toxoid-antitoxin mixtures results in a product that combines effectiveness with a minimum of unpleasant local effects (see Glenny and Pope, 1927, Swyer 1931, Dudley 1932, Underwood 1934, and others), but it has the disadvantage of containing serum.

Alum-precipitated toxoid. APT The use of alum-precipitated toxoid for active immunization in man followed the demonstration by Glenny (1930) that this reagent was extremely effective in inducing immunity in experimental animals (see also Glenny and Barr 1931, Glenny *et al.* 1931). Its high potency appears to be due to delayed absorption of the precipitated toxoid from the site of inoculation, with a consequent prolongation of the antigenic stimulus. It consists essentially of a suspension of the washed precipitate produced by the addition of alum to formol toxoid. Considerable care has to be exercised over its preparation, and exposure of the toxoid to acid or to phenol, such as is inherent in the method described by Watson, Taggart and Shaw (1941), must be avoided (Glenny *et al.* 1926, Hartley 1935*a*, Barr *et al.* 1941). Fortunately Glenny and his colleagues (Barr, Pope, Glenny and Linggood 1941) have provided a detailed description of how APT should be made, and the resulting product has been proved by field trials to be both effective and reliable (see also Pope and Linggood 1939, and Seal and Johnson 1941 on purification of toxoid and APT).

Purified toxoid, aluminium phosphate precipitated. PTAP This product arose out of an attempt by Holt (1947, 1948, 1950) to prepare an accurately reproducible prophylactic agent of high potency and stability. To purify and concentrate the toxoid he used precipitation with magnesium hydroxide and cadmium chloride followed by fractionation with ammonium sulphate. The purified product was adsorbed under optimal conditions on to aluminium phosphate, which, unlike the aluminium hydroxide employed in the preparation of APT, precipitated only the protein. Preliminary field trials suggested that it was rather more potent than APT in stimulating immunity (Holt and Bousfield 1949), but it caused much the same degree of reaction (Report 1954*b*). In laboratory tests it was found by Greenberg (1955) to be the most potent of all diphtheria prophylactics.

Phosphate toxoid. PT   This is a modified form of PTAP introduced by Tasman and van Ramshorst (1952). It likewise is very potent.

Toxoid-antitoxin floccules, aluminium phosphate precipitated. ADF   This preparation consists of toxoid-antitoxin floccules dissolved in N/20 NaOH and adsorbed on to aluminium phosphate (Mason 1950, 1951). It is a highly potent antigen, causes fewer and less severe reactions in adults than do FT or APT, but has the disadvantage of containing serum.

Active immunization

To produce a high degree of immunity against diphtheria by prophylactic inoculation, account must be taken of the nature of the prophylactic, the dosage, and the interval between the doses. Other factors, such as the latent immunizability of the population (Dudley, May and O'Flynn 1934), and possibly the volume of the fluid in which the prophylactic agent is injected (Hartley 1935*b*), may influence the result, but need not be considered here.

Of chief importance is the **antigenic quality of the prophylactic.** As was pointed out in the last section, methods of standardization are still imperfect, and unless a high standard is set, batches of prophylactic may be issued that are unable to stimulate a satisfactory immune response. Even an increase in dosage as much as fivefold may fail to atone for a qualitatively poor antigen (Fulton *et al.* 1942).

Several careful studies have been made on the efficacy of different types of prophylactic agent in immunization against diphtheria (for references see 5th edition p. 1684). The findings may be summarized by saying that two properly spaced doses of APT or PTAP result in an average schick-conversion rate of about 98 per cent, whereas with other prophylactics, excluding PT and ADF, this high rate may be closely approached, but is seldom equalled.

The choice of prophylactic must be determined not only by its immunizing quality but also by the degree of local or constitutional *reaction* to which it gives rise on injection. In general, toxoids precipitated with alum or aluminium phosphate are more irritant than plain formol toxoid or toxoid-antitoxin mixtures. Reactions are commoner in persons giving a negative, a

pseudo, or a combined response to the Schick test than in those responding to the toxin alone, and are seen more often in older children and adults than in infants or young children. It is wise, therefore, when immunizing subjects over 10 years old to use either one of the less irritant prophylactics, such as TAF, or, if an alum-precipitated toxoid is selected, to give it in small dosage. To judge the subject's sensitivity, the Schick test may be carried out beforehand. The response to this test, however, does not afford an infallible guide to the probable reaction to diphtheria prophylactic. For this reason the *Moloney* test (Moloney and Fraser 1927) was introduced. An injection of 0.2 ml of a 1 in 20 dilution of formol toxoid is made intradermally. Any person giving more than a minimum local reaction should be immunized cautiously. This test likewise is by no means infallible and in practice is seldom used (see Boyd 1946).

The irritant effect of diphtheria prophylactic may be either of specific or of non-specific origin. Adult subjects tested with a purified, but by no means pure, toxoid fraction and with a fraction containing non-toxic diphtheria proteins reacted sometimes to one, sometimes to the other, sometimes to both, and sometimes to neither; reactions were much more common in schick-negative than in schick-positive subjects. Reactions to either are evidence of latent immunity, since a single small injection of diphtheria prophylactic is followed by brisk antitoxin production (Lawrence and Pappenheimer 1948, Pappenheimer and Lawrence 1948). Alum-precipitated toxoids cause a higher degree of sensitivity than do plain toxoids (Ben-Efraim and Long 1957).

On the whole it is wise in older subjects to carry out a preliminary Schick test. Schick-positive subjects should be given a small first dose, followed, if there is little or no reaction, by one or two further doses at a suitable interval. Schick-negative subjects and those responding to the heated toxoid may be regarded as potentially immune and need not, as a rule, be injected; if, however, like doctors, nurses or school teachers, they may be exposed to an unusually high risk of infection, they should be given a single small dose, preferably of TAF or purified formol toxoid. It may be noted that even the small amount of toxin contained in a schick-test dose sensitizes the tissues and results in a more rapid response to a prophylactic antigen given later (Bousfield and Holt 1962).

Dosage varies with the nature of the prophylactic. Formol toxoid, toxoid-antitoxin mixture, and toxoid-antitoxin floccules are best administered in three doses—FT in doses of 0.2–0.5 ml, 0.5 ml and 1.0 ml, and the other two in doses of 0.5 ml, 1.0 ml and 1.0 ml. Alum-precipitated toxoid and PTAP can be safely relied upon to produce a corresponding, or even higher, degree of immunity in two doses of 0.2 ml and 0.5 ml. Though a single dose of either of these products may give quite a high schick-conversion rate, it is undesirable to rely on one injection alone (see Report 1940). Glenny (see Barr and Glenny 1945) advocates a large initial dose followed by smaller second or third doses in order to lay the foundation of a good immunity. Though this method may be followed in the immunization of infants and young children, it is not to be recommended for older age groups on account of the severe tissue reactions that may follow the injection of a large dose of toxoid into latently immune subjects. For reinforcing immunity in subjects who have been primarily immunized or who, on the evidence of the Schick test, have some degree of basal immunity, a single dose of any of the usual prophylactics is sufficient. The size of the dose will depend on which prophylactic is chosen and on the age and probable reactivity of the subject (see above). So long as some degree of basal immunity is present, even a slight stimulus, such as that conveyed by the minute amount of toxin in a schick-test dose, may, as already stated, suffice to bring about a conversion from the schick-positive to the schick-negative state (Fraser 1931, Barr and Parish 1950). When a rapid production of antitoxin is required—as in persons recently exposed to infection—formol toxoid is the prophylactic of choice; otherwise APT or PTAP in a dose of 0.1–0.5 ml or TAF in a dose of 0.2–0.5 ml should be given.

The **interval between the doses** has been shown by numerous workers to play a decisive part in determining the degree of antibody response (for references see Lewis 1941). In a schick-positive subject who has had no previous experience of the diphtheria bacillus, the first dose of prophylactic has the effect merely of sensitizing the antibody-forming apparatus. The second dose, provided time has been allowed for adequate sensitization to take place, is followed rapidly by the appearance of antitoxin in the circulating blood and the conversion of the schick-positive to the schick-negative state. Sensitization is a slow process and till it is complete the full effect of the second dose of antigen will not be experienced; once, however, it is established, it is relatively immaterial how long the second dose is delayed.

With FT and TAF the customary interval between the doses is 2–3 weeks, with APT and PTAP 4–6 weeks, but longer intervals—up to 3 months—are permissible and even beneficial if the child is not likely to be in contact with diphtheria in the meantime. Judged by experiments with PTAP in rabbits, 8–12 weeks appears to be preferable to 4 weeks (Khabas *et al.* 1959).

The **age** at which immunization is performed should depend primarily on the age distribution of diphtheria in the particular country concerned. Where diphtheria is prevalent and is common in the first year of life, immunization should be begun about the 6th month. Before this time the antibody response is to some extent interfered with by maternal antitoxin in the infant's blood, though good results may be obtained in individual subjects at this age, or even earlier, if three widely spaced doses of APT are given (Barr *et al.* 1950). In countries such as Great Britain and Denmark, in which diphtheria has for all practical purposes ceased to occur during the first year of life, it is wiser to defer immunization till the infant is a year old in order that its resistance may be greatest during the 2nd, 3rd and 4th years, when the risk of contracting the disease is highest. If the infant is immunized at 3–6 months of age, it will need a reinforcing injection at 18 months, and again on entering school, instead of only one at 5 years of age (see Report 1953). In practice the age at which immunization is begun is often determined by administrative and other reasons. In order to cut down the number of injections, diphtheria prophylactic is frequently combined with pertussis vaccine and, because pertussis is usually most fatal during the first year of life, the mixed preparation is given early, starting before the infant is 3 months old (see Bell 1948, Sant' Agnese 1950). Three injections of the combined vaccine are required, and a reinforcing injection at 12–18 months is imperative (Barr *et al.* 1955). Diphtheria toxoid may be combined with tetanus toxoid, with or without pertussis vaccine (see Chapter 64), for routine immunization (North and Patterson 1953, Ramon 1955, Levine *et al.* 1961). For a brief discussion of the advantages and disadvantages of combined vaccines, see Chapter 67, and for schedules of immunization, see the Ministry of Health's Report (1965, 1968).

The **duration of immunity** is dependent not only upon the type of prophylactic, but upon the subsequent exposure of the inoculated subject to infection with the diphtheria bacillus. In a community in which there is much diphtheria or a high carrier rate the immunizing mechanism is likely to receive more or less frequent stimuli, which will have the effect of reinforcing the immunity conferred by the prophylactic. On the other hand, in communities with little or no experience of diphtheritic infection, the antitoxin content of inoculated subjects may be expected to fall progressively (see Jensen 1933*a*, *b*) and the Schick reaction to revert to positive. It is not surprising, therefore, that the results recorded by different workers are not in complete accord. Some observers, for example, like Park (1922), working in densely populated cities, have had few Schick relapses over a period of many years, whereas others, working in areas where there has been little diphtheria infection, have had high relapse rates. Thus Fraser and Brandon (1936), in Canada, found that of children who had received three doses of formol toxoid five years previously, and who had not been exposed to infection since, 34 per cent had reverted to the schick-positive state. Similar results are recorded by Schwartz and Janney (1938) and Volk and Bunney (1942) in the United States, and by Sigurjónsson (1939) in Iceland (see also Parish and

Okell 1928, Cooke 1936, Jones 1936–37, Christie 1940, Nevius and McGrath 1940, Duke and Stott 1943, Sellers *et al*. 1945). Generally speaking, we may say that in children properly immunized with two doses of APT and not exposed to diphtheritic infection the schick-relapse rate should not exceed 5 per cent per year. Since immunity cannot be relied upon to last indefinitely, it is wise to re-inoculate children every 4 or 5 years. For this purpose a single dose of prophylactic is sufficient, since there is abundant evidence to show that a single re-inoculation of persons, even of those who have relapsed to the schick-positive state, is followed by a rapid rise in the antitoxin content of the blood (see Parish and Okell 1928, Parish and Wright 1938, Fraser 1940, Volk and Bunney 1942, Wishart *et al*. 1944). Alternatively, oral immunization may be recommended. The administration of two lozenges, a week apart, each containing 1500 Lf of purified diphtheria toxoid and 1500 units of hyaluronidase, is said to give a satisfactory antibody response (Cockburn *et al*. 1961, Report 1963*b*). In practice, a single lozenge containing 1100–3000 Lf doses given to children 10–12 years old who had been primarily immunized in infancy failed to yield a satisfactory response; about 15 per cent of the children still had less than 0.01 unit of antitoxin in their blood serum 2–3 weeks later (Galbraith *et al*. 1966).

Scheibel and his colleagues (1962) doubt the need for routine reinforcing doses. In an examination of 235 adult males who had received three doses of toxoid adsorbed on to aluminium hydroxide 12–16 years previously, they found that all but two contained more than 0.01 unit of antitoxin per ml of serum. (See also Scheibel *et al*. 1966.)

It may be noted that many patients who have recovered from diphtheria remain schick-positive (see Warin 1940). Since second attacks of the disease are not uncommon, such persons may be regarded as susceptible, and inoculated or re-inoculated with the rest. Why an attack of diphtheria is not always followed by immunity is still doubtful. That it is not due mainly to suppression of the antibody-forming mechanism by the administration of large quantities of antitoxin therapeutically seems clear from Madsen's (1939) observation that 13 out of 26 untreated patients failed to form detectable antitoxin as the result of their attack.

The desirability of a **preliminary Schick test** to find out whether inoculation is required, and of a subsequent Schick test to find out whether immunization has been successful, is open to question. Space does not allow a full discussion of the subject, and we must therefore restrict ourselves to a summary of the conclusions we have reached from our own observations and from those of other workers.

In a mass immunization campaign a preliminary Schick test, entailing as it does two inoculations and an extra visit for each child, seems to us to be justified on only two conditions, firstly that a high proportion

of the children are schick-negative, and secondly that inoculation is restricted to schick-positive reactors. In areas in which diphtheria has not been prevalent and in which inoculation has not been widely practised, a high proportion of the children are generally found to be schick-positive. Under such conditions schick testing is wasteful of both time and material. It is all the more so if, as many experienced workers recommend, the schick-negative reactors are inoculated at the same time, usually with a single dose, to reinforce their failing immunity. For practical purposes, therefore, we see no advantage in carrying out a preliminary Schick test on a pre-school or school population. The susceptible members require inoculation anyhow, and the immune members will come to no harm through being given two doses of prophylactic but will, on the contrary, have their immunity strengthened.

In adults, the position is different. Mass immunization is seldom called for, so that the extra time required for a preliminary Schick test is a less important consideration. A fairly high proportion will probably be found to be immune; and the argument for inoculating adults who are schick-negative is not so strong as for children who are still at an age when susceptibility is relatively great. An even more important factor, however, is that the inoculation of adult schick-negative or pseudo-reactors with full doses of prophylactic, especially APT, is likely to be followed by a severe local and even constitutional reaction. As already mentioned, it is wise, therefore, before inoculating adults, to carry out a preliminary Schick test, and to restrict inoculation to schick-positive reactors, unless the subject is, like a doctor, nurse, or school teacher, likely to be exposed to an unusually high risk of infection.

By itself a **post-inoculation Schick test** is of value in learning whether the child has been satisfactorily immunized, though it affords no guarantee against the development of diphtheria. It is carried out best 2–4 months after the last inoculation, when the antitoxin content of the blood is at its height. The test may be usefully applied to persons who are exposed to special risk of infection, but is seldom justifiable in mass immunization campaigns. There is, however, a strong case to be made out for testing a sample of the inoculated population in order to gain assurance that a high proportion, such as 98 per cent, have responded satisfactorily. So many variables enter into the development of schick-immunity that this should never be taken for granted.

To compare the antigenic potency of two different prophylactics or of different modes of administration of the same prophylactic, it is desirable to estimate the **schick-conversion rate.** This necessitates preliminary as well as post-inoculation schick-testing, and is calculated on the proportion of schick-positive subjects in each group that are rendered schick-negative. Unless a preliminary test is made, the proportion of original schick-negative reactors in each group is unknown, and a difference in their distribution will affect the result of the comparison.

Complications directly attributable to inoculation against diphtheria are uncommon. McCloskey (1950) in Australia brought evidence to suggest that diphtheria toxoid, particularly when combined with pertussis vaccine, might activate a latent poliomyelitis infection. This was confirmed by the special investigation carried out by the Medical Research Council in England and Wales, which showed that the risk was greatest with combined and with alum-precipitated vaccines and least with plain toxoid and TAF (Report 1956). It is wise therefore to suspend diphtheria immunization during an epidemic of poliomyelitis, and to avoid the use of combined vaccines in children who have not already been immunized with poliomyelitis vaccine. Aseptic meningitis has been described, but it must be a rare sequel (Stillerman 1948). Sterile abscesses—or more correctly cysts—occur in a small proportion of persons injected with alum-precipitated products, but they generally disappear spontaneously. They are regarded as an expression of sensitivity to the protein of the diphtheria bacillus, occurring in persons who are immune to the disease (Volk *et al.* 1954). (For a fuller description of the accidents and complications that have followed immunization against diphtheria, see Wilson 1967.)

The protective effect of active immunization

Though a high schick-negative or schick-conversion rate affords a strong presumption that immunization has been satisfactory, the ultimate court of reference in determining the efficacy of prophylactic inoculation must be the behaviour of immunized persons when exposed to the risk of natural infection.

The numerous observations that have been made in Europe and America leave no doubt that active immunization, properly performed, is highly effective in protecting the individual against diphtheria, and in eliminating the disease from closed communities in which all the susceptible inmates are immunized. The evidence for this is set out in detail on pages 1821–23 of the 6th edition, and there is no reason to repeat it here. Protection is not complete, as is shown by the occurrence of diphtheria cases, and even occasional deaths, in immunized persons (Glover and Wright 1942). (See also Gibbons 1945, Fanning 1947, Report 1957.) In such instances it has generally been found that infection has been caused by organisms of the *gravis* or *intermedius* types, and that the case-fatality rate has been considerably less than in non-immunized persons (see Dudley *et al.* 1934, Parish and Wright 1935, Bynoe and Helmer 1945, Hartley *et al.* 1950).

The results of mass immunization

We may now turn to the question whether prophylac-

tic inoculation, controlled when necessary by schick-testing, will reduce significantly or eliminate altogether the incidence of diphtheria in the community at large, when applied on the widest possible scale. It must be realized that the emphasis is on the final limiting clause. There is no reason to doubt that, if all susceptibles in the community could be immunized, the reduction in the incidence of diphtheria would be of the same order as that attained in closed or semi-closed communities. But we are dealing with practical politics, and what we want to know is the probable effect of an administratively possible policy. We are here faced with a problem of great statistical difficulty. It will not do to cite a few instances in which the inauguration of mass immunization of school children has been associated with a decline in the general diphtheria morbidity rate, because we know quite well that most infective diseases are subject to secular trends in frequency, and a falling rate for any one of them might well have no significant association with any particular preventive measure adopted at the time the fall occurred.

Woods (1928), for instance, pointed out that the fall in the diphtheria morbidity rate which followed the introduction of mass immunization in New York in 1918 could not be accepted as certainly significant. The rate was already falling before the immunization began; and if a straight line is fitted to the trend of mortality for the years prior to 1917, and is then extrapolated beyond 1918, this extended line approximates very closely to that which would describe the actual course of events during the following decade (see also Madsen and Madsen 1956).

Godfrey (1932), in an admirable review of the available evidence, discussed the possible reasons for such divergent results as those recorded in New York City and Philadelphia, where immunization was followed by a striking decrease in morbidity, and in Newark, Detroit and Buffalo, where similar measures failed. Tracing in detail the course of events in eleven different districts in New York State he arrived at the conclusion that the crucial point was to be found in the immunization of the child of pre-school age (the age group 1–5). The earlier attempts at mass immunization were, for obvious administrative reasons, confined mainly to children of school age (the age groups 5–9 and 9–14). Godfrey could find no evidence that the immunization of 50 per cent or over of the children in these age groups produced a definite fall in the incidence of diphtheria in the community as a whole. When, however, the immunization of 30 per cent or more of the age group 1–5 was superimposed on that of the later age groups there was an immediate and striking decline in the general prevalence of the disease. Godfrey found only two exceptions to this rule in the eleven areas from which his data were obtained. (See also Lee 1931, Deadman and Elliott 1933.)

Logan (1950) approached the problem in a rather

different way. Taking the figures for measles, whooping cough, scarlet fever and diphtheria in England and Wales, he drew curves for each disease based on the quinquennial death-rates per million under 15 years of age from 1866 to 1940, and then extended the curves annually to 1949, comparing the observed with the expected rates. These two rates corresponded closely in whooping cough. In measles the observed death-rate fell more rapidly than the expected, in scarlet fever more rapidly still, and in diphtheria a little more rapidly even than in scarlet fever. The curves for these last two diseases were fairly similar, but Logan found that, whereas the notifications of scarlet fever fell by only 11 per cent between 1938–40 and 1949, those of diphtheria fell by 91 per cent. The fall in scarlet-fever mortality, therefore, was almost entirely the result of reduced fatality. In diphtheria, on the other hand, there was a great reduction in both incidence and mortality. Logan concluded that, after the introduction of widespread immunization against diphtheria late in 1940, the diphtheria mortality fell to an extent quite different from that for whooping cough and measles and in a manner quite different from that for scarlet fever. On the basis of the 1866–1940 trend, 1427 children might have been expected to die from diphtheria in 1949: in fact, only 67 died.

Figure 53.1 shows that the death-rate from diphtheria in England and Wales fell from 300 per million in 1901 to 0.73 per million in 1951. The greater part of

this enormous decrease occurred after the introduction of mass immunization at the end of 1940 (Logan 1952). The fall was progressive and in 1959 not a single death from diphtheria was recorded for the whole country. The experience of many other countries has been similar (see Ramon 1957, Tasman and Lansberg 1957).

It must be emphasized that neither in the individual nor in the herd does protective inoculation confer absolute immunity to the disease. Several outbreaks have occurred in immunized populations which leave no doubt that the protective effect of inoculation has been somewhat exaggerated in the past. In the outbreak of diphtheria in Halifax, Nova Scotia, for instance, 21 per cent of the cases occurred in persons who had given a primary schick-negative reaction or who had had three doses of toxoid within the previous 3 years (Gibbons 1945). In the Detroit outbreak of 1956 over 75 per cent of the kindergarten children, and in one school 94 per cent, had a history of immunization in early infancy or during the pre-school years (Report 1957). Again, Fanning (1947) records an explosive outbreak of diphtheria in a private day school for girls of whom 94 per cent had a history of protective inoculation; 15 of the 18 cases that occurred were in immunized children. The intensive studies of Hartley and his colleagues on the outbreak on Tyneside and of Tulloch and his colleagues on the outbreak in Dundee (Hartley *et al.* 1950) showed that diphtheria often attacked fully inoculated persons, though it very seldom proved fatal. Once the antibody-producing mechanism is sensitized, infection with the diphtheria bacillus acts as a powerful stimulus leading to a rapid rise in the antitoxin content of the blood. Even though the antitoxin content may be too low to protect against infection, it rises so quickly that it almost invariably protects against death. This is, in fact, the main virtue of diphtheria inoculation for the individual.

Diphtheria immunization confers protection not only on the individual but also on the herd. Once the general level of immunity is raised above a critical level—the level in any given community depending on many different factors—the chain of infection is broken and the disease tends to die out. To achieve this end it would appear that in a fairly densely populated urban community at least 70 per cent of school and pre-school children must be immunized and kept immune by reinforcing injections. Even this may prove insufficient if other factors are unfavourable, such as, for example, invasion with a particularly virulent strain of *gravis* type.

Objections have been raised to diphtheria immunization on the ground that it may lead to an increase in the proportion of healthy carriers and so perpetuate or even aggravate the disease. The falseness of this argument is proved by experience, both in the United Kingdom and America. Thus, before the second world war the carrier rate in London elementary school children varied between about 2.5 and 5.0 per cent. In September, 1942, however, about 2 years after the beginning of the Government's immunization campaign, the carrier rate of virulent diphtheria bacilli was

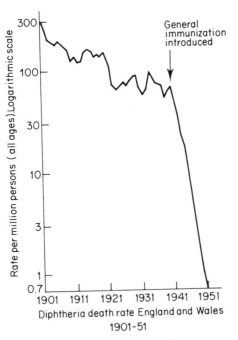

Fig. 53.1 Showing the precipitous fall in the death-rate from diphtheria in England and Wales since the introduction of general immunization. (From Logan 1952.)

only 0.38 per cent, and in February, 1943, it had fallen to 0.19 per cent (Report 1944). A survey of 2100 school children in 1950–51 in Leeds, where immunization had been consistently practised for the previous 10 years, failed to reveal a single carrier of virulent diphtheria bacilli (Bradshaw *et al.* 1952). These figures lend no support to the suggestion that immunization does more harm than good to the herd by increasing the number of healthy carriers. They suggest rather that in practice efforts should be directed not to stopping immunization in order to lower the carrier rate, but to increasing it to a point at which cases of the disease cease to occur (see Report 1945). Figures in other countries point to the same conclusion (see Calomfiresco *et al.* 1964).

In summary, we have in the modern method of anti-diphtheria immunization a measure that is based on well attested immunological principles, makes use of a reagent of proved potency, is supported by the results of animal experiments, is effective in raising the resistance of the individual and in controlling the incidence of the disease in closed or semi-closed communities, affords a moderate protection against an attack of the disease and a high protection against death, and when carried out on a sufficiently extensive scale is capable of leading to almost complete disappearance of diphtheria among the population at large. (For standardization of diphtheria prophylactics see Chapter 19 and p. 94.)

Prophylactic passive immunization

To meet particular and transitory emergencies, such as the accidental admission to a children's ward of a child suffering from diphtheria, the subcutaneous or intramuscular administration of 500–1000 units of antitoxin to each of the other children thus exposed to risk has been found an effective method of stopping the spread of the disease. Provided the children have not been exposed to the infected child for more than 24 hours, a Schick test may be carried out, a preliminary reading taken the following day, and inoculation with serum restricted to those giving a positive reaction. Schick testing, however, seems to us to be of very doubtful value, unless a high proportion of the children are believed to be immune. Many of the susceptibles will not be read as schick-positive within 24 hours, and if a longer time than this is allowed to elapse before serum is injected, there is a risk of diphtheria developing. On the whole, it is more satisfactory to give serum to every child. So long as refined serum is used, trouble from serum reactions will be slight.

Combined active and passive immunization, and the control of institutional epidemics

Combined active and passive immunization consists in giving simultaneous injections into opposite arms

of serum and of diphtheria prophylactic, followed later by a second dose of prophylactic. Its object is to stimulate an active immunity under cover of a passive immunity. The doses recommended are 500 units of refined serum and 0.3–0.5 ml of APT, followed in 2–4 weeks' time by a second dose of 0.5 ml of APT. This method is of particular value when a case of diphtheria occurs in a group of susceptible children in a school, hospital, or other institution, when the extent of the distribution of the infecting agent is unknown, or when further introduction of infection is to be expected. In these circumstances passive immunization alone, which wears off in 2–3 weeks, may be followed by the appearance of diphtheria cases due to infection from healthy carriers. The combination of active and passive immunization results not only in a transitory protection of the exposed children, but in the development of an active immunity that can generally be relied upon to prevent further cases of diphtheria occurring, even in the presence of continued infection. Since, however, there is an intermediate period of relative susceptibility between the wearing off of passive and the development of active immunity, it is wise to swab the entire population and segregate all virulent carriers till a fortnight after the second dose of APT. By this time sufficient active immunity will generally have developed to render almost negligible the risk of further cases occurring. This method proved very successful in the hands of Fulton, Taylor, Wells and Wilson (1941) in the control of school and institutional epidemics. To the paper by these authors and to two papers by Downie and his colleagues (1941, 1948) reference may be made for the rationale and application of combined active and passive immunization.

The advantage of this method in conjunction with swabbing of the entire population and segregation of virulent carriers is that school closure is unnecessary, and that, apart from the carriers who are segregated for 6 weeks, every child can continue its normal education. Little is to be gained by preliminary schick testing of the population, as was recommended by Okell, Eagleton, and O'Brien (1924) in the days before combined active and passive immunization was shown to be practicable, and valuable time may be lost. Moreover, in countries where diphtheria is still prevalent, the danger of the disease developing in schick-negative reactors who are exposed to special risk of infection is by no means remote. On the whole, we think that there is far more to be gained by giving combined active and passive immunization to all children than by immunizing only the schick-positive reactors. The time and labour expended are also less. An exception may be made for adults, who should in any case be immunized with TAF rather than with APT.

In a population that is already well immunized the occurrence of a small group of cases points to a local source of infection, such as an undiagnosed case or a carrier. All close contacts of the patients should be swabbed, and the procedure repeated ten days later, because a small proportion of carriers may be missed on the first occasion. Infected persons should be isolated and treated. Contacts who have not had an injection of diphtheria toxoid for two years should be given a reinforcing dose, and contacts who have never been immunized should receive active and passive immunization. The recommended procedure is described in more detail by Taylor, Tomlinson and Davies (1962).

The treatment of diphtheria with antitoxin

The nature of antitoxic immunity and the general principles that determine its practical application have been discussed in the chapters on immunity. We are here concerned with the methods that are employed in the particular case of diphtheria in man, and the available information on their efficacy.

Antitoxic serum is prepared commercially by the injection of horses with increasing doses of toxoid. In practice, only animals containing natural diphtheria antitoxin in their blood are chosen, since they respond so much better than others to immunization. Injections are made intramuscularly two or three times a week till a dose of half a litre or more can be given. When the content of antitoxin has risen high enough, three bleedings of 8 litres each are taken over a period of eight days. The horse is rested for a week or two and then given a short course of five injections, after which it is bled again. The process may be repeated four or five times. The serum is separated from the clot, filtered, and preserved with an antiseptic such as 0.3 per cent tricresol (Dean 1908). Serum obtained in this way may be expected to contain about 1000 units per ml of antitoxin.

Various workers have attempted to isolate the antitoxin in a state of chemical purity, but without complete success. It is contained in the pseudoglobulin fraction, which is precipitated by half saturation with ammonium sulphate. A greater degree of purification is obtained by the use of proteolytic enzymes to digest much of the unwanted protein (Pope 1938, 1939). The resulting *refined diphtheria antitoxin* is absorbed more rapidly and eliminated more slowly than the ammonium sulphate product (Glenny and Llewellyn-Jones 1938). It is also far less prone to cause serum sickness. Experiments, however, show that it does not pass the placental barrier of the pregnant guinea-pig and reach the fetus (Hartley 1948). Pregnant women, therefore, who contract diphtheria shortly before delivery should preferably be given serum concentrated by the ammonium sulphate method.

Our object in treatment will obviously be to introduce as much antitoxin as we can, at the earliest possible moment, since we must act on the assumption that the beneficial result of our interference will prob-

ably be limited to the neutralization of the toxin that has not yet been firmly anchored to the cells which are susceptible to its action. In support of this view we may cite such experimental results as those recorded by Glenny and Hopkins (1923). For the same reason we should give therapeutic injections of antitoxin intravenously, intraperitoneally or intramuscularly, never subcutaneously, since the relatively slow absorption from the subcutaneous tissues will defeat our main objective. Tasman and his colleagues (1958) plead for the routine use of the intravenous method, pointing out that it ensures an immediate high concentration of antitoxin in the patient's blood and the rapid appearance of antitoxin in the saliva. If the patient is shown to be sensitive, as judged by the reaction to the intracutaneous injection of 0.2 ml of a 1 in 10 dilution of serum, the serum should be given intramuscularly, or alternatively antiserum prepared in the sheep or goat may be tried (van Triet 1959).

It is sometimes taught that the administration of antitoxin is without effect after the 5th or 6th day of disease, a dictum based mainly on the figures recorded by Faber (1904), which show a progressive increase in case fatality from 7.1 per cent in patients receiving antitoxin on the first day of disease, to 17.0 per cent, 21.3 per cent and 19.9 per cent among those treated on the 6th, 7th and later days respectively. It has, however, been pointed out that such hospital statistics are of little value, because there is an important element of selection of which they take no account. Cases admitted to hospital on the 5th day or after will, *ipso facto*, include a large number of severe infections, since the delay in admission will commonly be due either to delayed diagnosis, often resulting from failure to obtain a medical opinion, or to an unwillingness to resort to hospital. In either event those patients who are well on the way to recovery by the 5th or 6th day will never be admitted, whereas those who are seriously ill will be transferred to hospital, either because a late diagnosis has been arrived at, or because the serious condition of the patient has alarmed his friends.

There is no justification for withholding antitoxin because a case is in an advanced stage, though every effort should be made to ensure that it is given early in the disease.

A system of dosage in common use is that advocated by Park (1921) (see Table 53.4); but there has been a tendency to increase the amount of antitoxin given in the severer type of case as the result of the experience of recent years, and this tendency seems likely to continue. Thus Bie (1922), in an admirable review of the evidence in regard to the value of antitoxin treatment in diphtheria, advocates the administration of very large total doses (92 000–170 000 units). Harries (1945) likewise recommends a large single dose, but considers that it should rarely be necessary to give more than 50 000 units, provided part or the whole of the dose is

Table 53.4   Units of antitoxin to be administered to cases of varying grades of severity (after Park)

Age or weight of patient	Mild cases	Early moderate	Late moderate and early severe	Severe and malignant
Under 2 years (4.5–13.5 kg)	2000–3000	3000–5000	5000–10 000	7500–10 000
2–15 years (13.5–40.5 kg)	3000–4000	4000–10 000	10 000–15 000	10 000–20 000
Adults (40.5 kg or more)	3000–5000	5000–10 000	10 000–20 000	20 000–50 000
Route of administration	Intramuscular	Intravenous	Intravenous	Intravenous

given intravenously. Except in very severe cases or cases seen late, 10 000–20 000 units given intravenously should usually suffice.

It might be expected that after some 90 years it would be easy to produce irrefutable statistical evidence of the beneficial effects of the antitoxin treatment of diphtheria in man. The evidence does, in truth, seem decisive to most of us; that it is still possible to bring forward contrary arguments that are not altogether specious is due to the fact that once a strong presumptive case has been made out in favour of a particular therapeutic measure it is not justifiable to continue the period of trial at a risk of human lives.

Of the classical studies from the early days of antitoxin treatment, those recorded by Fibiger (1898) are the most satisfactory, and even they have been criticized on statistical grounds by Rasch (see Madsen and Madsen 1956). For a period of one year all cases of diphtheria admitted to hospital were divided into two groups by separating those admitted on alternate days. All cases admitted on one day were given antitoxin, all those admitted on the next day were treated without antitoxin, and so on throughout the period of trial. The results were as follows: of 239 cases treated with antitoxin 8 died, a fatality of 3.5 per cent; of 245 cases treated without antitoxin 30 died, a fatality of 12.25 per cent. Applying the formula given in Chapter 19, the difference between the case fatality in the two groups is 8.75 per cent and the standard error of this difference is 2.445 per cent, so that the odds against the difference being due to random sampling are about 2000 to 1.

Other results recorded during the early and middle 'nineties—comparisons between the case-fatality rates in hospitals in which antitoxin was given and others in which it was not, or between the experience in one hospital before and after the introduction of antitoxin—all pointed in the same direction; and the period of trial ended with the adoption of antitoxin as a routine method of treatment about 1895.

Though there is every reason to believe that serum treatment is of value for the individual patient, and that it has been to some extent instrumental in lowering the case-fatality rate, it would be wrong to ascribe to it any substantial part in the decline in mortality

that occurred in many European countries about the time of and in the few years following its introduction. Most of this was probably due to secular changes in the severity of the disease (Madsen and Madsen 1956), to improvement in diagnosis which increased the proportion of mild cases notified, and to better treatment particularly of laryngeal cases. The subsequent rise in case-fatality rate that was witnessed, for example in Berlin (Deicher and Agulnik 1927), in spite of intensive serum treatment, indicated that the disease was capable of reasserting its killing power. Suspicion has been cast on the modern refined serum, which is thought to be less potent in combating toxaemia than the crude serum used earlier in the century.

It has been calculated that the greatest quantity of diphtheria toxin present in the entire blood volume of a patient suffering from a severe attack of the disease is only ten times the fatal dose for a guinea-pig (Madsen and Madsen 1956). Even admitting that man is ten times as susceptible as the guinea-pig (Tasman and Lansberg 1957), this quantity should be completely neutralized by a few thousand antitoxin units intravenously. The fact that serum treatment is not uniformly successful suggests that toxin is early fixed to susceptible cells of the body and cannot readily be neutralized by circulating antitoxin.

Attempts have been made to prepare an antibacterial serum to supplement the action of an antitoxin, but the observations of Maitland and his colleagues (1952) failed to reveal any beneficial effect of its use on the case-fatality rate.

The standardization of diphtheria antitoxin

We have already referred to the standardization of diphtheria antitoxin, in our discussion of the toxin-antitoxin reaction (Chapter 19). We noted that the instability of a toxic filtrate renders it quite unsuitable as a standard of reference, and that, for this reason, among others, Ehrlich's original definition of the unit of antitoxin as *the smallest amount of antitoxin that will neutralize* 100 *MLD of toxin*, the guinea-pig being the test animal, had soon to be abandoned. It has now been replaced by a unit defined in terms of a standard antitoxin. Such a standard antitoxic serum, when dried and preserved *in vacuo* in the presence of phosphorus pentoxide, maintains its potency over long periods of time. It has, moreover, become a general principle in biological standardization that a reagent shall, whenever possible, be assayed by comparing its potency with that of a standard preparation of the same reagent, to which some unit value has been assigned by international agreement (see Chapter 19). In the case of diphtheria, Ehrlich's original antitoxin was adopted as the first international standard preparation, and subsequent standards have been characterized in terms of the first (see Report 1923), so that the unit of activity remains constant, though the actual weight of the standard preparation that has unit activity may change with each preparation. The unit of diphtheria antitoxin is therefore contained in that amount of an antitoxic serum that has the same total combining capacity, for toxin and toxoid, as one unit of the standard antitoxin.

The methods employed in measuring the potency of antitoxic sera, and the knowledge that has been acquired in their development and use, have played so large a part in the evolution of our present methods of controlling the disease that, at the cost of some repetition, it is desirable to summarize them here. Three such methods are available. Two of them depend on in-vivo tests, the third on an in-vitro titration.

The first is Ehrlich's classical method of injecting mixtures of toxin and antitoxin subcutaneously into guinea-pigs (see Prausnitz 1929). It is convenient to employ the end-point of toxaemic death rather than the end-point of complete neutralization—the L+ rather than the Lo dose of toxin. The first procedure is, then, to determine the smallest amount of a suitable toxic filtrate that, when mixed with one unit of standard antitoxin and injected into a 250 g guinea-pig, will, on the average, kill the animal by the 4th day. In practice batches of guinea-pigs are injected with each mixture prepared and the mixture producing a 50 per cent mortality, or thereabouts, is regarded as containing the L+ dose of toxin. The procedure is then reversed. The amount of toxin is held constant at the L+ dose, and the amount of the serum to be tested is varied, again an adequate number of guinea-pigs being used for each dose in the range within which the L+ mixture is expected to fall. The amount of serum in the mixture that gives a 50 per cent mortality within 4 days contains one unit of antitoxin. The number of guinea-pigs that must be inoculated with any such mixture in order to give an estimate of the unit value of an antitoxic serum within a specified margin of error depends on the factors considered in Chapter 19. In practice a rough preliminary titration is made, and the exact value of the serum is then assessed by the use of larger numbers of guinea-pigs.

The second method depends upon the observation of Römer (1909) that the intradermal injection into a guinea-pig of 1/250–1/500 MLD of toxin is followed by a localized swelling and erythema. With slightly larger doses, this erythematous reaction is followed by definite necrosis. The neutralization of toxin by antitoxin prevents this reaction. The great advantage of this method is that it allows the toxicity of several different mixtures to be tested on a single guinea-pig, and thus not only economizes in animals but eliminates much of the difficulty due to differences in susceptibility between one guinea-pig and another. The successive steps in the process of standardization do not differ in any essential from those followed in the Ehrlich method. Varying amounts of a suitable toxin are first mixed with one unit of the standard antitoxin and 0.2 ml of each mixture is injected into the depilated skin of a guinea-pig. The amount of toxin in the mixture that gives a minimal skin reaction is noted. This amount of toxin has been defined by Glenny and Allen (1921) as the Lr dose. A series of mixtures is now prepared in which the Lr dose of toxin is mixed with varying amounts of the serum under test, and 0.2 ml of each of these mixtures is injected intradermally into another guinea-pig. The amount of serum in that mixture that gives a minimal skin reaction contains one unit of antitoxin. In practice it is customary to use fractions of a unit and corresponding fractions of the Lr dose of toxin, in order to avoid the administration of lethal doses of toxin in the mixtures containing toxin in excess.

The third method depends on the observation of Ramon (1922) that a satisfactory measure of the combining power of an antitoxic serum can be obtained by mixing falling amounts of the serum with a constant amount of toxin and noting the ratio of one reagent to the other in the tube that first shows flocculation. This is the method of optimal proportions (see Volume 1). Glenny and Okell (1924) (see also Glenny and Wallace 1925) have suggested that the amount of toxin corresponding to one unit of antitoxin in the mixture that shows optimal flocculation when tested by the Ramon method should be called the Lf dose. The procedure with this in-vitro test is essentially the same as with either of the in-vivo methods. The Lf dose of a suitable toxic filtrate is determined by titration against the standard antitoxin. The amount of the serum under test that gives optimal flocculation with the Lf dose of toxin is determined by a second titration. This amount of serum contains one unit of antitoxin.

We may set out the units or named doses employed in testing diphtheria toxin and antitoxin as follows (see Glenny 1925):

(1) Toxin is measured by its direct toxic effect in the guinea-pig as

      MLD (death on the 4th day)

      MRD (minimal skin reaction).

There is no fixed equivalent amount of antitoxin.

(2) Combining power with *one unit of antitoxin* is measured in the guinea-pig as

L + dose (mixture causes death on the 4th day)

Lo (mixture causes minimal oedema)

Lr (mixture causes minimal skin reaction).

(3) Combining power with one unit of antitoxin is measured *in vitro* as

Lf (mixture gives most rapid flocculation in Ramon test).

The comparison of large numbers of different antitoxic sera—natural or concentrated—by these different methods has brought to light facts of considerable theoretical and practical importance. The L + dose of toxin is always appreciably larger than the Lo dose. The Lr dose of toxin always approximates closely to the Lo dose, as would be expected, since a very small excess of unneutralized toxin will elicit the Römer reaction. The Lf dose is, in general, slightly less than the Lr dose. Glenny (1925) notes the following relation between the various doses of an average toxic filtrate: L + dose = 0.21 ml, Lo dose = 0.18 ml, Lr dose = 0.175 ml, Lf dose = 0.155 ml. The ratio of the Lf dose to any of the in-vivo doses is not constant for all toxic filtrates. Antitoxin gives flocculation with both toxin and toxoid, whereas the determination of the L +, Lo and Lr doses depends on the presence of unneutralized toxin, and is hence affected by differences in the proportion of toxin to toxoid. Moreover, as was shown by Glenny, Pope and Waddington (1925), not only does the Lf/Lr ratio vary from one toxic filtrate to another, when these are tested against the same serum, but the Lf/Lr ratio of a single toxic filtrate varies when it is tested against different antitoxic sera. It follows that, when sera are compared with one another by in-vivo and in-vitro methods, their apparent relative potency may vary according to the method of comparison employed; and, with such sera, the ratio $\dfrac{\text{in-vitro value}}{\text{in-vivo value}}$ will vary from one serum to another. Glenny and his colleagues note that they have obtained in-vitro/in-vivo ratios varying from 0.4 to 2.0 with different antitoxic sera, and that, in general, if the Ehrlich value is considerably higher than the Ramon value, *i.e.* if the in-vitro/in-vivo ratio is low, the serum will be found to give rapid flocculation; whereas, if the Ramon value is higher than the Ehrlich value, the serum will be found to give very slow flocculation and the toxin-antitoxin complex will show considerable dissociation on simple dilution. In their later papers, Glenny and his colleagues use the inverse ratio—in-vivo/in-vitro—so that a serum with a ratio greater than unity has a greater protective action than its flocculation value would lead one to expect.

The inverse ratio is a measure of avidity, a quality of antitoxin which we discussed in Chapter 10. There we noted that avidity can also be measured by the dilution ratio, which, like the serum ratio, was introduced by Glenny and his colleagues. This ratio is the amount of antitoxin required to neutralize the Lr dose

of toxin in a total volume of 2 ml, divided by the amount necessary to neutralize this dose in a volume of 200 ml, as determined by the guinea-pig intracutaneous test. These levels of testing correspond to the Lr/10 and the Lr/1000, and sera for which the amount required to neutralize is greater at the more dilute level (*i.e.* those with a low dilution ratio) are classed as non-avid because their combination with toxin is not firm, and dissociation occurs at high dilutions. The dilution ratio varies not only with the antitoxins tested, but with the toxin used, being lower with culture filtrates that are the more toxic, in terms of the number of MRD per Lr dose (Barr 1949). There is a rough correlation between the two measures of avidity—the in-vivo/in-vitro ratio, and the dilution ratio.

The avidity of antitoxic sera appears to be determined, at least in part, by the type of serum protein with which the antitoxin is associated. Thus (Barr and Glenny 1931a, b, Glenny and Barr 1932, Glenny *et al.* 1932) when successive globulin fractions are precipitated with increasing amounts of ammonium sulphate, the earlier fractions show a higher in-vivo/in-vitro ratio, a higher dilution ratio and a greater curative power in rabbits than the original serum, while the avidity as judged by these tests decreases with each successive fraction precipitated.

Kekwick and his colleagues (Kekwick and Record 1940, Kekwick *et al.* 1941) found that, of the three electrophoretically separable serum globulins in antitoxic horse sera, the α component had no detectable antitoxic activity. The antitoxic β and γ components, which together account for the total activity of the serum, differed from each other in several respects, notably in their flocculation time, the composition of their floccules, and their in-vivo/in-vitro ratio. The γ component flocculates more rapidly in the presence of toxin, and has therefore a higher in-vivo/in-vitro ratio; the floccules formed contain twice as much antitoxin nitrogen as those formed by the β globulin. In the horse's serum the γ component appears earlier during the course of immunization than the β, which however continues to increase as further injections of toxoid are given. The γ antitoxin precipitates at lower salt concentrations than the β antitoxin, and tends therefore, during the ordinary concentration process, to be salted out with the euglobulin fraction and discarded, so that the bulk of the antitoxin remaining in the pseudoglobulin fraction is of the β type. The proteolytic method used in the preparation of refined serum also favours the retention of the β at the expense of the γ antitoxin. It is clear that the proportion of the β and γ components in different batches of therapeutic serum is subject to variation. The suggestion is that the avidity of the serum is determined to some extent by the proportion of γ globulin, and that batches of sera high in this respect are likely to be of greater therapeutic efficacy than others containing mainly β globulin. Such an explanation is not inconsistent with

McSweeney's (1941) observations on the successful results obtained with a particular serum in the treatment of cases of hypertoxic diphtheria (see also Trevan 1941).

The heterogeneity of the antibody proteins in serum and of the combining powers in toxic filtrates raises a number of difficulties in the characterization of antitoxic sera. As we saw in Chapter 19, for a valid comparison of potency of two antitoxins, the dosage-response curves must be substantially parallel—a circumstance we are most likely to find when the two antitoxins are similar in respect of avidity and so forth. Therapeutic antitoxins of high unitage are usually the end-result of prolonged immunization, and are usually avid; and may be compared directly with the highly avid international standard antitoxin. Non-avid antitoxins, even of high unitage, and relatively non-avid antitoxic sera from man and animals that have been exposed to infection, or have undergone an ordinary course of immunization, often have dosage-response curves of a different slope from that of the standard. In these circumstances, the estimate of the relative potency of the standard and test antitoxin will vary with the response level chosen for the assay. Thus, Jerne (1951), working with the neutralization test in the rabbit's skin, records a twenty-fold variation. This difficulty may be overcome by using standards of avidity similar to those of the test antitoxins, or perhaps by changing the in-vivo neutralizing test so that differences in avidity do not affect the slope of the dose-response curve (Miles 1954). But even when we have a reliable measure of unitage we are left with the question as to whether unitage is a sufficient index of clinical efficacy in antitoxins intended for prophylaxis or therapy. It happens that high-potency antitoxins are usually avid, and can be relied on in this respect. There are, however, no agreed measures of avidity by which therapeutic antitoxins can be specified (see, e.g., British Pharmacopoeia 1953). Jerne (1951) suggests the determination of a single association constant for each antitoxin, to be designated an 'avidity constant,' and describes a method for its determination; but the practicability of this proposal remains to be established.

The standardization of diphtheria prophylactics

The flocculation test of Ramon (1922) provides an obvious method of determining the amount of toxoid in any prophylactic, and Glenny, Pope and Waddington (1925) found that the immunizing value of a modified toxin was closely related to its Lf value. It has, indeed, become a common practice to state the dose of toxoid contained in a prophylactic in terms of Lf units, when the actual dose of toxoid is stated at all, which is not often the case. But it would obviously be preferable, were it possible, to measure the actual immunizing potency of all preparations by direct comparison with some arbitrarily selected standard reagent.

The technical difficulties that arise in such a comparison are however very great. For the reasons discussed in Chapter 19, a comparison carried out on a small number of animals is quite valueless; and it is doubtful whether the precise standardization of a reagent used for inducing active immunity is as yet within the realm of practical politics.

The relevant Regulations (1952) in force in Great Britain, under the Therapeutic Substances Act, may be summarized as follows.

Toxicity   Five ml of formol toxoid, TAM or TAF, and 1 ml of APT or PTAP, are injected subcutaneously or intraperitoneally into each of no fewer than 5 healthy guinea-pigs, each weighing from 250 to 350 g. The injection must not cause the death of any of the guinea-pigs from specific intoxication within 20 days.

Immunizing potency   The test for minimum immunizing potency of formol toxoid consists in the injection into each of 10 healthy guinea-pigs of 0.05 ml of FT in 1 ml saline, on two occasions at an interval of not more than 4 weeks. Not later than 3 weeks after the second injection each guinea-pig receives an intracutaneous injection of 0.2 ml of a dilution of diphtheria toxin containing 5 schick-test doses. If, 48 hours later, more than 2 of the 10 guinea-pigs exhibit a positive Schick reaction, the preparation is insufficiently potent. TAM and TAF are tested similarly, except that TAM-inoculated animals are tested with 1 schick-test dose; and in the TAF test, 9 animals are used, and tested with 1 and 2 schick-test doses. In the last case, no more than 1/3rd of the animals must react to 1 schick-test dose, or alternatively, no more than 2/3rd to 2 schick-test doses. For the purposes of the tests, a schick-test dose contains 0.001 Lf and free toxin in an amount such that one-twentieth (0.00005 Lf) or less given intracutaneously into guinea-pigs will, after 48 hours, regularly produce a local reaction of the schick-positive type.

APT and PTAP are diluted in saline so that the equivalent of 1 Lf is present in 1 ml. Each of 10 guinea-pigs receives 1 ml on two occasions at an interval of not more than 4 weeks. Not later than 3 weeks after the second injection, the antitoxin content of the serum of each animal is determined. If the geometric mean of the antitoxin contents is less than 2 units, the preparation is insufficiently potent (see also British Pharmacopoeia 1953, p. 200).

These tests, it will be noted, prescribe limits of toxicity and antigenic efficacy, not standardization in terms of units. Their reliability depends in part upon the maintenance of guinea-pigs of an average susceptibility to toxin and to immunizing agents. Since we cannot know how far laboratory guinea-pigs vary in these respects, the requirements outlined are devised so that the antigenic content of prophylactics that pass the tests is in substantial excess of that necessary for good immunization. The alternative, as we saw in Chapter 19, is to express potency in terms of standards. Although for many years Prigge (1939) in Germany used a standard preparation of diphtheria prophylactic, and measured potency in 'Schutzeinheiten' (protective units), it was only in 1951 (Report 1952) that an international standard for formol toxoid was established (see also Report 1954a).

Comparative assay in animals of diphtheria prophylactics with a standard preparation (see, e.g., Greenberg *et al.* 1945–46, Greenberg and Roblin 1948) presents certain difficulties. The same amount of formol toxoid has a widely different immunizing potency when given alone and with an adjuvant such as alum. Moreover, the slope of the dosage-response curve of a plain and an adsorbed toxoid may differ, so that direct comparisons of two kinds of prophylactic are invalid (see Chapter 19). Clearly, the animal test does not measure the immunizing potency of the toxoid contained in a prophylactic, but of the prophylactic as compounded for use. At least two standards are required, one for plain and one for adsorbed toxoid, and others may be needed if newer forms of prophylactic are devised. Nevertheless, the measurement of immunizing potency of diphtheria prophylactics in units would have many advantages over the current methods, and we may look forward to its establishment as a routine.

The standardization of diphtheria prophylactics on laboratory animals is not an altogether satisfactory method for determining the antigenic potency for human beings (see Hartley 1934). In many ways, tests of potency are so artificial that the translation of potency into protective efficacy is an extremely haphazard process (see Maaløe and Jerne 1952, Miles and Perry 1953). It is therefore desirable, in cases of doubt, to control each batch by measurements on children. For this purpose the schick-conversion rate is suitable (see, e.g., Fulton *et al.* 1942). (For an historical account of immunization against diphtheria, see Dolman 1973, Spink 1978.)

Chemotherapy in treatment of cases

Most strains of diphtheria bacilli are sensitive to penicillin, the *mitis* type being the most and the *gravis* type the least. Treatment of cases with penicillin appears to hasten the disappearance of the organism from the throat. In one small series of 65 acute cases of faucial diphtheria treated with penicillin for 3–6 days in addition to antitoxin, 49 became free from the infecting organism within four days of the end of treatment (Report 1948b). This rate of disappearance is more rapid than that reported by previous workers in cases treated by antitoxin alone.

The treatment of carriers

Thomson, Mann, and Marriner (1928–29) found that a high proportion of pure throat carriers became negative after tonsillectomy. Valuable as this method is, it is limited in its application to throat carriers, who constitute only about 10 per cent of the total. Systemic treatment with penicillin of persistent throat carriers is rather disappointing, but may be worth trying before tonsillectomy is resorted to (see Report 1948b). Erythromycin, given orally in a dose of 200–300 mg every 6 hours for 10 days, was reported on favourably by Wood and O'Gorman (1957). Romagnoli (1955) was successful in clearing a high proportion of nose and throat carriers by spraying the pharynx with a 1/2000–1/3300 suspension of tyrothricin, but it is not clear whether the organisms disappeared more quickly under treatment than they would have done naturally. For the treatment of nasal carriers the inhalation of sulphonamide snuff was recommended by Delafield, Straker and Topley (1941). In our personal experience a mixture of 33 per cent sulphathiazole and 67 per cent magnesium carbonate inhaled six times a day cures the majority of nasal carriers within a fortnight. When an associated haemolytic streptococcal infection is present it may be advisable to use pure sulphanilamide snuff, as recommended by Boissard and Fry (1942).

Other diseases caused by corynebacteria

Besides causing infections of the upper respiratory tract, the diphtheria bacillus may sometimes be found as a secondary invader in skin wounds and in lesions of the vagina. During the 2nd world war Liebow and his colleagues (1946) saw numerous cases among soldiers evacuated from the Solomon Islands and the Philippines of deep punched-out ulcers containing toxigenic diphtheria bacilli. Among the natives they found a big reservoir of cutaneous diphtheria, which led to early immunization. Gunatillake and Taylor (1968) made similar observations in Ceylon. Diphtheroid bacilli of various types have frequently been isolated from the sites of pathological lesions in man. They have been cultivated from the blood of typhus cases, from the cerebrospinal fluid of cases of general paralysis, from the conjunctiva in various forms of subacute or chronic conjunctivitis, from the external auditory meatus in cases of ear disease, from the pustules of acne vulgaris, from other lesions of the skin, from the urine in cases of subacute or chronic urethritis, and from lymphatic nodes in a variety of diseases, particularly lymphadenoma. In a few instances the evidence suggests that the association is a causative one (see Kaplan and Weinstein 1969). We have already mentioned cases of diphtheria due to *C. ulcerans*. The acne bacillus appears to be partly or wholly responsible for *acne vulgaris* (Smith and Waterworth 1961); and a diphtheroid bacillus, distinguished by the fluorescence of its colonies under a Wood's ultraviolet lamp seems to be aetiologically related to the scaly skin lesions that occur particularly between the toes in the disease known as *erythrasma* (Sarkany *et al.* 1962, Somerville 1970). *Corynebacterium haemolyticum* may cause a mild disease of the upper respiratory tract, or cutaneous and other extrapharyngeal lesions (Maclean *et al.* 1946, Wickremesinghe 1981). The lesions caused by the group of anaerobic diphtheroid bacilli have already been mentioned in Chapter 25; so

also has the diphtheria-like disease due to *C. ulcerans.* At least two cases of lymphadenitis caused by *Corynebacterium ovis* are on record (Battey *et al.* 1968, Hamilton *et al.* 1968). Apart from these there is no satisfactory evidence to suggest that any other corynebacterium plays a significant role as a pathogenic parasite of man, though various species figure prominently in his normal bacterial flora (Harris and Wade 1915, Andrewes *et al.* 1923).

Diphtheroid infections in animals

There are a few diseases which occur naturally in animals other than man and are caused by infection with corynebacteria.

Ulcerative lymphangitis of horses, and pseudotuberculosis of sheep

Both these chronic diseases are the result of infection with *C. ovis* or, as it is sometimes called, *C. pseudotuberculosis* or the Preisz-Nocard bacillus. The characters of this organism and the lesions that it produces in experimental animals have been described in Chapter 25. Unlike *C. diphtheriae* this organism is pyogenic and invasive, as well as toxigenic. The exotoxin produced by it differs from the exotoxin of *C. diphtheriae* both in the character of the lesions produced in experimental animals and in its antigenic relationships. The pathogenesis of the natural disease in horses and sheep seems to be determined mainly by the invasive and pyogenic activities of the causative organism. The part played by the toxin is at present doubtful. *Ulcerative lymphangitis* occurs in the limbs of horses, particularly under conditions of poor hygiene. The legs swell, ulcers appear, and the lymphatic ducts form painful swollen cords. The lesions may spread to the trunk, and abscesses may occur in the internal organs.

Pseudotuberculosis of sheep

This disease—better known perhaps as caseous lymphadenitis—is economically an important disease, particularly in Australia (see Bull and Dickinson 1931, 1933, 1935, Dickinson and Bull 1931, Discussion 1934). It is a chronic disease characterized by caseous abscesses, mainly in the superficial lymph glands. Lesions sometimes occur in the deep lymph glands, lungs, and other organs. The wool becomes dry and lifeless. The abscesses range in size from a millimetre or so in diameter up to 10 centimetres; they are rather gritty and contain greenish pus. Infection takes place through wounds of the skin such as occur in shearing, docking, and castration; contaminated soil is probably often responsible. Focal skin lesions occur and may

progress to lymphadenitis. The presence of the disease is frequently not discovered until after slaughter. For the serological diagnosis of infection caused by *C. ovis*, Zaki (1968) recommends an antihaemolysin inhibition test.

C. ovis occasionally produces lymphangitis and lymphadenitis in cattle; infection is also known in goats and deer. Rare human cases have been described, mainly in persons in contact with live or slaughtered sheep (Henderson 1979).

Infections caused by *C. pyogenes*

This organism often inhabits the mucous membranes of healthy cattle, sheep, goats, and pigs, in which it may cause, by itself or with other agents, suppurative processes including abscess formation, pneumonia, arthritis, endometritis, and mastitis. It is probably the most important and widespread member of the *Corynebacterium* group found in association with animal disease.

Summer mastitis of cattle most frequently affects dry cows and heifers at grass, in warm weather. The usual acute form results in the rapid destruction of udder tissue, with permanent loss of the affected quarter. A purulent exudate, often with a foul smell, may be discharged through the skin of the mammary gland. *C. pyogenes* can be seen as a small diphtheroid bacillus in stained smears of the exudate; other organisms such as *Peptococcus* spp., *Bacteroides* spp., *Fusobacterium necrophorum*, microaerophilic cocci, and *Streptococcus dysgalactiae* are often present also (Lovell, 1959, Sørensen 1978). The disease is sometimes fatal. A weak toxin is formed, which often gives rise to antitoxin production in infected animals; this may be measured by the power of the serum to inhibit the haemolysis of rabbit cells by toxin (Lovell 1939). The toxin is converted into toxoid by treatment with formalin. An alum-precipitated toxoid induces antitoxin formation in animals inoculated intramuscularly, but field trials have given little hope that the administration of toxoid can prevent the disease (Lovell *et al.* 1950). Circulating antitoxin may, however, lower the death rate. The incidence of summer mastitis can be reduced by infusing the udders of dry cows with an appropriate antibiotic.

Infections caused by *C. equi*

Magnusson (1923, 1938) described a pyaemic disease of foals in Sweden caused by a pigment-forming organism, which he called *C. equi*, and which has recently been called *Rhodococcus equi* by Goodfellow and Alderson (1977). The disease, which has since been reported in many parts of the world, is as a rule characterized by a suppurative bronchopneumonia, with greyish-red pus in the bronchi, and an intense purulent inflammation of the pulmonary lymph nodes. In some cases intestinal ulceration occurs, accompanied by large abscesses in the mesenteric lymph nodes. The disease is often fatal. It usually affects foals aged 2 to 6 months, on premises where many births take place. It is not readily transferred from one place to another by movement of animals. The organism can persist for long periods in nature, and infection has for many years been suspected to be associated with soil. Use of a selective medium containing nalidixic acid, novobiocin, cyclohexamide, and potassium tellurite showed that *C. equi* was widespread in the faeces of horses, cattle, and other animals (Woolcock *et al.* 1980).

C. equi has often been isolated from the submaxillary and cervical lymph nodes of normal pigs, and from lymph nodes containing caseous lesions; these, however, are probably not caused by *C. equi*. (Karlson *et al.* 1940, Cotchin 1943, Thal and Rutqvist 1959).

Rare cases of *C. equi* infection have been recognized in cattle, buffaloes, sheep, goats, cats, and reptiles; the associated diseases include lymphadenitis, abscesses in internal organs, pneumonia, and uterine infection (see Barton and Hughes 1980). Several cases of human pulmonary infection have been observed in immunologically defective patients.

Urinary tract infections

C. renale causes bovine cystitis and pyelonephritis, almost always in the female. Both kidneys are often affected and may be greatly enlarged. The renal pelvis and calyces contain blood-tinged purulent exudate and necrotic debris. Necrosis with suppuration occurs in the renal papillae, and the medullary lesions may extend into the cortex. The walls of the bladder and ureters are thickened, and the bladder mucosa may be ulcerated. Large diphtheroid bacilli can be demonstrated in stained smears of sediment from the turbid blood-stained urine of affected animals. Weitz (1947) sometimes found *C. renale* in the vagina of healthy cows. At least three serological type exist (Yanagawa *et al.* 1967, Yanagawa and Honda 1978). Type I includes the American reference strain of *C. renale*. Type II may occur in the urine and vagina of normal cows

and occasionally produces disease. Type III is found in many parts of the world. It is not carried by normal cows but is often present in the prepuce of normal bulls (Hiramune *et al.* 1975), and may cause severe disease (see also Chapter 25, p. 102). Rojas and Biberstein (1974) believe that diphtheroids associated with ovine posthitis in Australia and California are *C. renale*.

Porcine cystitis and pyelonephritis caused by an anaerobic diphtheroid-like bacillus, *C. suis*, was described by Soltys and Spratling (1957) (see also Soltys 1961) in pregnant sows. Of 38 infected pigs, studied by Narucka and Westendorp (1973), 16 were non-pregnant. Boars are seldom affected but not infrequently carry the organism in the prepuce (Jones *et al.* 1982). The organism is non-pathogenic to laboratory animals.

Pseudotuberculosis of mice

This is a natural disease of mice caused by *C. kutscheri* (*C. murium*). So far as it is known, no other animal species is susceptible. Latent infection is said to occur in many strains of mice; it can be evoked into activity by a single injection of 10 mg of cortisone (Fauve *et al.* 1964). Because the natural and the experimental host are the same, the lesions of the disease have been described in Chapter 25 (p. 104).

References

Abel, R. (1893) *Zbl. Bakt.* **14,** 756.

Akama, K., Ito, A., Yamamoto, A. and Sadahiro, S. (1971) *Jap. J. med. Sci. Biol.* **24,** 181.

Anderson, J. S., Cooper, K. E., Happold, F. C. and McLeod, J. W. (1933*a*) *J. Path. Bact.* **36,** 169; (1933*b*) *Lancet* **i,** 293.

Anderson, J. S., Happold, F. C., McLeod, J. W. and Thomson, J. G. (1931) *J. Path. Bact.* **34,** 667.

Andrewes, F. W. *et al.* (1923) Monograph on Diphtheria. *Med. Res. Coun. Lond.* p. 275.

Ashby, A. (1906) *Publ. Hlth, Lond.* **19,** 145.

Barr, M. (1949) *J. Path. Bact.* **61,** 85.

Barr, M. and Glenny, A. T. (1931*a*) *J. Path. Bact.* **34,** 539; (1931*b*) *Brit. J. exp. Path.* **12,** 337; (1945) *J. Hyg., Camb.* **44,** 135.

Barr, M., Glenny, A. T. and Butler, N. R. (1955) *Brit. med. J.* **ii,** 635.

Barr, M., Glenny, A. T. and Randall, K. J. (1950) *Lancet* **i,** 6.

Barr, M. and Parish, H. J. (1950) *Mon. Bull. Minist. Hlth Lab. Serv.* **9,** 97.

Barr, M., Pope, C. G., Glenny, A. T. and Linggood, F. V. (1941) *Lancet* **ii,** 301.

Barton, M. D. and Hughes, K. L. (1980) *Vet. Bull.* **50,** 65.

Battey, Y. M., Tonge, J. L., Horsfall, W. R. and McDonald, I. R. (1968) *Med. J. Aust.* **ii,** 540.

Behring, E. von (1913) *Dtsch. med. Wschr.* **39,** 873.

Behring and Wernicke (1892) *Z. Hyg. InfektKr.* **12,** 10.

Bell, J. A. (1948) *J. Amer. med. Ass.* **137,** 1009.

Ben-Efraim, S. and Long, D. A. (1957) *Lancet* **ii,** 1033.

Bickham, S. T. and Jones, W. L. (1972) *Amer. J. clin. Path.* **57,** 244.

Bie, V. (1922) *Acta med. scand.* **56**, 537.

Bøe, J. and Vogelsang, T. M. (1948) *Nord. Med.* **40**, 2253.

Boissard, J. M. and Fry, R. M. (1941) *Publ. Hlth*, **54**, 105; (1942) *Lancet* **i**, 610.

Bolton, C. (1905) *Lancet* **i**, 278.

Bousfield, G. and Holt, L. B. (1962) *Mon. Bull. Minist. Hlth Lab. Serv.* **21**, 31.

Bowhill, T. (1899) *Vet. Rec.* **11**, 586.

Boyd, J. S. K. (1946) *Lancet*, **ii**, 195.

Bradshaw, D. B., Dixon, C. W., Mawson, F. M., Turner, G. H. and Zinnemann, K. S. (1952) *Lancet*, **i**, 558.

Brahdy, M. B., Lenarsky, M., Smith, L. W. and Gaffney, C. A. (1935) *J. Amer. med. Ass.* **104**, 1881.

Bretonneau, P. (1826) *Des Inflamm. Spéc. du Tissu Muqueux et en Partic. de la Diph. etc.* Crevot, Paris.

British Pharmacopoeia (1953) Pharmaceutical Press, London.

Bull, L. B. and Dickinson, C. G. (1931) *Aust. J. exp. biol. med. Sci.* **8**, 45; (1933) *Aust. vet J.* **9**, 82; (1935) *Ibid.* **11**, 126.

Bunney, W. E. (1931) *J. Immunol.* **20**, 47.

Bynoe, E. T. and Helmer, D. E. (1945) *Canad. J. publ. Hlth* **36**, 135.

Calalb, G., Stănică, E., Storian, C. and Meitert, E. (1967) *Roman. med. Rev.* **11**, 79.

Calomfiresco, A. *et al.* (1964) *Archs roum. Path. exp. Microbiol.* **23**, 1053.

Christie, A. (1940) *J. Pediat.* **17**, 502.

Christovão, D. de A. (1957) *Arch. Faculd. Hig. Saúd. públ. Univ. São Paulo* **11**, 97, 115.

Christovão, D. de A. and Cotillo, L. G. (1966) *Arq. Hig. Saúde públ.* **20**, 223.

Clauberg, K. W. (1929) *Zbl. Bakt.* **114**, 539; (1931) *Ibid.* **120**, 324; (1933) *Ibid.* **128**, 153; (1935) *Ibid.* **134**, 271; (1936) *Ibid.* **135**, 529; (1939) *Klin. Wschr.* **18**, 1490.

Cockburn, W. C., Bradstreet, C. M. P., Bailey, M. E. and Ungar, J. (1961) *Brit. med. J.* **ii**, 1754.

Cook, G. T. and Jebb, W. H. H. (1952) *J. clin. Path.* **5**, 161.

Cooke, J. V. (1936) *J. Pediat.* **9**, 641.

Cooper, K. E., Happold, F. C., Johnstone, K. I., McLeod, J. W., Woodcock, H. E. de C. and Zinnemann, K. S. (1940) *Lancet* **i**, 865.

Cooper, K. E., Happold, F. C., McLeod, J. W. and Woodcock, H. E. de C. (1936) *Proc. R. Soc. Med.* **29**, 1029.

Cotchin, E. (1943) *J. comp. Path.* **53**, 298.

Cotillo, L. G., Iaria, S. T., Schmid, A. W. and Wilson, D. (1966) *Arq. Hig. Saúde públ, São Paulo*, **20**, 215.

Craig, J. P. (1962) *Amer. J. publ. Hlth* **52**, 1444.

Crosbie, W. E. and Wright, H. D. (1941) *Lancet* **i**, 656.

Cruickshank, R. (1943) *Publ. Hlth, Lond.* **57**, 17.

Daniels, J. B., Johnson, M. P. and MacCready, R. A. (1951) *Canad. J. publ. Hlth* **42**, 185.

Deadman, W. J. and Elliott, F. J. (1933) *Canad. publ. Hlth J.* **24**, 137.

Dean, G. (1908) See *Bacteriology of Diph.* p. 513. Nuttall & Graham-Smith, Cambridge.

Dean, G. and Todd, C. (1902) *J. Hyg., Camb.* **2**, 194.

Deicher, H. and Agulnik, F. (1927) *Dtsch. med. Wschr.* **53**, 825.

Delafield, M. E., Straker, E. and Topley, W. W. C. (1941) *Brit. med. J.* **i**, 145.

Dickinson, C. G. and Bull, L. B. (1931) *Aust. vet. J.* **8**, 83.

Discussion. (1934) *Proc. R. Soc. Med.* **27**, 1335.

Dold, H. and Weigmann, F. (1934) *Z. Hyg. InfektKr.* **116**, 154.

Dolman, C. E. (1973) *Canad. J. publ. Hlth*, **64**, 317.

Downie, A. W., Glenny, A. T., Parish, H. J., Smith, W. and Wilson, G. S. (1941) *Brit. med. J.* **ii**, 717.

Downie, A. W., Glenny, A. T., Parish, H. J., Spooner, E. T. C., Vollum, R. L. and Wilson, G. S. (1948) *J. Hyg., Camb.* **46**, 34.

Dreyer, G. (1900) *Exper. Undersogelser over Difterigiftens Toxoner.* Copenhagen.

Dudgeon, L. S. (1906) *Brain* **29**, 227.

Dudley, S. F. (1923) *Spec. Rep. Ser. med. Res. Coun., Lond.* No. 75; (1926) *Ibid.* No. 111; (1932) *Quart. J. Med.* **1**, 213.

Dudley, S. F., May, P. M. and O'Flynn, J. A. (1934) *Spec. Rep. Ser. med. Res. Coun., Lond.* No. 195.

Duke, H. L. and Stott, W. B. (1943) *Brit. med. J.* **ii**, 710.

Edward, D. G. ff. and Allison, V. D. (1951) *J. Hyg., Camb.* **49**, 205.

Elek, S. D. (1948) *Brit. med. J.* **i**, 493; (1949) *J. clin. Path.* **2**, 250.

Ende, M. van den, Lush, D. and Edward, D. G. ff. (1940) *Lancet* **ii**, 133.

Ende, M. van den and Thomas, J. C. (1941) *Lancet* **ii**, 755.

Eyre, J. W. (1899) *Brit. med. J.* **ii**, 586.

Faber, E. E. (1904) *Jb. Kinderheilk.* **59**, 620.

Fanning, J. (1947) *Brit. med. J.* **i**, 371.

Fauve, R. M., Pierce-Chase, C. H. and Dubos, R. (1964) *J. exp. Med.* **120**, 283.

Fibiger, J. (1898) *Hospitalstidende*, 4. Ser. **6**, 309, 337.

Fleming, A. (1941) *J. Path. Bact.* **53**, 293.

Forbes, J. G. (1927) *Spec. Rep. Ser. med. Res. Coun., Lond.* No. 115.

Fraser, D. T. (1931) *Trans. roy. Soc. Can.* **25**, Sect. v, 193; (1940) *Proc. 3rd int. Congr. Microbiol., New York* 1939, p. 803.

Fraser, D. T. and Brandon, K. F. (1936) *Canad. publ. Hlth J.* **27**, 597.

Fraser, D. T. and Halpern, K. C. (1935) *Canad. publ. Hlth J.* **26**, 469; (1937) *J. Immunol.* **33**, 323.

Frost, W. H. (1928) *J. prev. Med., Baltimore* **2**, 325.

Fry, R. M. (1941) *Mon. Bull. Emerg. publ. Hlth Lab. Serv.* Feb., p. 7.

Fulton, F., Taylor, J., Moore, B., Wells, A. Q. and Wilson, G. S. (1942) *Brit. med. J.* **i**, 345, 349.

Fulton, F., Taylor, J., Wells, A. Q. and Wilson, G. S. (1941) *Brit. med. J.* **ii**, 759.

Galbraith, N. S., Bradstreet, C. M. P. and Bailey, E. M. (1966) *Mon. Bull. Minist. Hlth Lab. Serv.* **25**, 110.

Gibbons, R. J. (1945) *Canad. J. publ. Hlth* **36**, 341.

Glass, V. (1937) *J. Path. Bact.* **44**, 235.

Glenny, A. T. (1925) *J. Hyg., Camb.* **24**, 301.

Glenny, A. T. (1930) *Brit. med. J.* **ii**, 244.

Glenny, A. T. and Allen, K. (1921) *J. Path. Bact.* **24**, 61.

Glenny, A. T. and Barr, M. (1931) *J. Path. Bact.* **34**, 131; (1932) *Ibid.* **35**, 91.

Glenny, A. T., Barr, M. and Stevens, M. F. (1932) *J. Path. Bact.* **35**, 495.

Glenny, A. T., Buttle, G. A. H. and Stevens, M. F. (1931) *J. Path. Bact.* **34**, 267.

Glenny, A. T. and Hopkins, B. E. (1923) *Brit. J. exp. Path.* **4**, 283.

Glenny, A. T. and Llewellyn-Jones, M. (1938) *J. Path. Bact.* **47**, 405.

Glenny, A. T. and Okell, C. C. (1924) *J. Path. Bact.* **27**, 187.

Glenny, A. T. and Pope, C. G. (1927) *J. Path. Bact.* **30**, 587.

Glenny, A. T., Pope, C. G. and Waddington, H. (1925) *J. Path. Bact.* **28**, 279.

Glenny, A. T., Pope, C. G., Waddington, H. and Wallace, U. (1926) *J. Path. Bact.* **29**, 31.

Glenny, A. T. and Südmersen, H. J. (1921) *J. Hyg., Camb.* **20**, 176.

Glenny, A. T. and Wallace, U. (1925) *J. Path. Bact.* **28**, 317.

Glover, B. T. J. and Wright, H. D. (1942) *Lancet* **ii**, 133.

Godfrey, E. S. (1932) *Amer. J. publ. Hlth* **22**, 237.

Goldie, W. and Maddock, E. C. G. (1943) *Lancet* **i**, 285.

Goodfellow, M. and Alderson, G. (1977) *J. gen. Microbiol.* **100**, 99.

Greenberg, L. (1955) *Bull. World Hlth Org.* **13**, 367.

Greenberg, L. *et al.* (1945–46) *Bull. Hlth Org. L.o.N.* **12**, 365.

Greenberg, L. and Roblin, M. (1948) *J. Immunol.* **59**, 221.

Groman, N. and Judge, K. (1979) *Infect. Immun.* **26**, 1065.

Guedes, J. da S. *et al.* (1966) *Arq. Hig. Saúde públ.* **20**, 107, 215.

Gunatillake, P. D. P. and Taylor, G. (1968) *J. Hyg., Camb.* **66**, 83.

Guthrie, G. C., Marshall, B. C. and Moss, W. L. (1921) *Johns Hopk. Hosp. Bull.* **31**, 388.

Hamilton, N. T., Perceval, A., Aarons, B. J. and Goodyear, J. E. (1968) *Med. J. Aust.* **ii**, 356.

Handley, W. R. C. (1949) *J. Hyg., Camb.* **47**, 102.

Harries, E. H. R. (1945) *Lancet* **i**, 98.

Harris, W. H. and Wade, H. W. (1915) *J. exp. Med.* **21**, 493.

Hartley, P. (1934) *Wissenschaftl. Woche, Frankfurt a. M.* **3**, 81; (1935a) *Brit. J. exp. Path.* **16**, 460; (1935b) *Ibid.* **16**, 468; (1948) *Mon. Bull. Minist. Hlth Lab. Serv.* **7**, 45; (1949) *J. Pharm. Pharmacol.* **1**, 425.

Hartley, P. *et al.* (1950) *Spec. Rep. Ser. med. Res. Coun., Lond.* No. 272.

Hartley, P. and Martin, C. J. (1919–20) *Proc. R. Soc. Med.* **13** (Sect. Epidem.), 277.

Hasenbach, I. (1935) *Zbl. Bakt.* **134**, 137.

Henderson, A. (1979) *J. med. Microbiol.* **12**, 147.

Henry, J. E. (1920) *J. Amer. med. Ass.* **75**, 1715.

Hewitt, L. F. (1930) *Biochem. J.* **24**, 983.

Hiramune, T., Narita, M., Tomonari, I., Murase, N. and Yanagawa, R. (1975) *Nat. Inst. Anim. Hlth Quart.* **15**, 116.

Holt, H. D. and Wright, H. D. (1940) *J. Path. Bact.* **51**, 287.

Holt, L. B. (1947) *Lancet* **i**, 282; (1948) *Brit. J. exp. Path.* **29**, 335; (1950) *Developments in Diphtheria Prophylaxis.* Wm Heinemann, London.

Holt, L. B. and Bousfield, G. (1949) *Brit. med. J.* **i**, 695.

Horgan, F. S. and Marshall, A. (1932) *J. Hyg., Camb.* **32**, 544.

Hoyle, L. (1941) *Lancet* **i**, 175.

Jaffé (1920) *Arb. Inst. exp. Ther. (Frankfurt a. M.)* **11**, 5.

Jensen, C. (1931) *C. R. Soc. Biol.* **108**, 539, 543, 552, 577, 579; (1933a) *Acta path. microbiol. scand.* **10**, 137; (1933b) *Ibid.* Suppl., No. 14; (1937) *Proc. R. Soc. Med.* **30**, 1117.

Jerne, N. K. (1951) *Acta path. microbiol. scand.* Suppl. No. 87.

Jones, F. G. (1936–37) *J. Lab. clin. Med.* **22**, 576.

Jones, J. E. T., Farries, E. and Smidt, D. (1982) *Dtsch. tierärztl. Wschr.* **89**, 110.

Kairies, A. (1933) *Med. Klin.* **29**, 709.

Kaplan, K. and Weinstein, L. (1969) *Ann. intern. Med.* **70**, 919.

Karlson, A. G., Moses, H. E. and Feldman, W. H. (1940) *J. infect. Dis.* **67**, 243.

Kekwick, R. A. *et al.* (1941) *Lancet* **i**, 571.

Kekwick, R. A. and Record, B. R. (1940) *Brit. J. exp. Path.* **22**, 29.

Kemkes, B. (1934) *Dtsch. med. Wschr.* **60**, 1631.

Kerrin, J. C. and Gaze, H. W. (1937) *J. Hyg., Camb.* **37**, 280.

Khabas, I. M., Ter-Osipova, M. Z. and Kats, I. Z. (1959) *J. Microbiol. Epidem. Immunobiol.* **30**, 94.

King, E. O., Frobisher, M. and Parsons, E. I. (1949) *Amer. J. publ. Hlth* **39**, 1314.

Kissin, D. and Bronstein, L. (1928) *Z. ImmunForsch.* **56**, 11; (1930) *Ibid.* **66**, 210.

Klebs, E. (1883) *Verh. Kongr. inn. Med.* 139.

Klein, E. E. (1889) *Rep. loc. Govt Bd publ. Hlth* **19**, 143.

Kröger, E. (1955) *Zbl. Bakt.* **162**, 79.

Lawrence, H. S. and Pappenheimer, A. M. (1948) *Amer. J. Hyg.* **47**, 226.

Leach, C. N. and Pöch, G. (1935) *J. Immunol.* **29**, 367.

Lee, W. W. (1931) *J. prev. Med., Baltimore* **5**, 211.

Leete, H. M. (1945) *J. Hyg., Camb.* **44**, 184.

Levine, L., Ipsen, J. and McComb, J. A. (1961) *Amer. J. Hyg.* **73**, 20.

Lewis, E. S. (1933) *J. Lab. clin. Med.* **18**, 413.

Lewis, J. T. (1941) *The Principles and Practice of Diphtheria Immunization,* Oxford University Press, London.

Liebow, A. A., MacLean, P. D., Bumstead, J. H. and Welt, L. G. (1946) *Archs intern. Med.* **78**, 255.

Litterer, W. (1925) *Sth. med. J.* **18**, 577.

Loeffler, F. (1884) *Mitt. ReichsgesundhAmt.* **2**, 421.

Logan, W. P. D. (1950) *Med. Offr* **84**, 217; (1952) *Mon. Bull. Minist. Hlth Lab. Serv.* **11**, 50.

Lovell, R. (1939) *J. Path. Bact.* **49**, 329; (1959) In: *Infectious Diseases of Animals: Diseases due to Bacteria,* Vol. 1, p. 239. Ed. by A. W. Stableforth and I. A. Galloway. Butterworths Scientific Publications, London.

Lovell, R., Foggie, A. and Pearson, J. K. L. (1950) *J. comp. Path.* **60**, 225.

Maaløe, O. and Jerne, N. K. (1952) *Annu. Rev. Microbiol.* **6**, 349.

McCloskey, B. P. (1950) *Lancet* **i**, 659.

Maclean, P. D., Liebow, A. A. and Rosenberg, A. A. (1946) *J. infect Dis.* **79**, 69.

McLeod, J. W. (1943) *Bact. Rev.* **7**, 1; (1950) *J. Path. Bact.* **62**, 137.

McLeod, J. W. and Robinson, D. T. (1948) *Lancet* **i**, 97.

McSweeney, C. J. (1941) *Lancet* **i**, 208.

McSweeney, C. J. and Morgan, W. P. (1928) *Lancet* **ii**, 1201.

Madsen, E. (1939) *Acta path. microbiol. scand.* **16**, 113.

Madsen, T. and Madsen, S. (1956) *Dan. med. Bull.* **3**, 112.

Magnusson, H. (1923) *Arch. wiss. prakt. Tierheilk.* **50**, 22; (1938) *Vet. Rec.* **50**, 1459.

Maitland, H. B., Marshall, F. N., Petrie, G. F. and Robinson, D. T. (1952) *J. Hyg., Camb.* **50**, 97.

Maniar, A. C. and Fox, J. G. (1968) *Canad. J. publ. Hlth* **59**, 297.

Marshall, W. E. (1907) *J. Hyg., Camb.* **7**, 32.

Mason, J. H. (1950) *J. Hyg., Camb.* **48**, 418; (1951) *Lancet* **i**, 504.

Michiels, J. and Schick, B. (1913a) *Z. Kinderheilk.* **5**, 255; (1913b) *Ibid.* **5**, 349.

Miles, A. A. (1954) *Fed. Proc.* **13**, 799.

Miles, A. A. and Perry, W. L. M. (1953) *Bull. World Hlth Org.* **9**, 1.

Moloney, P. J. and Fraser, C. J. (1927) *Amer. J. Publ. Hlth* **17**, 1027.

Mueller, J. H. (1941) *J. Immunol.* **42**, 343, 353.

Narucka, U. and Westendorp, J. F. (1973) *Neth. J. vet. Sci.* **5**, 116.

Neill, G. A. W. (1937) *J. Hyg., Camb.* **37**, 552.

Nevius, W. B. and McGrath, A. C. (1940) *Amer. J. Dis. Child.* **59**, 1266.

Nikanaroff, P. J. (1898) *Arch. Ser. Biol., St. Petersb.* **6**, 57.

North, E. A. and Patterson, R. W. (1953) *Med. J. Aust.* i, 800.

Oertel (1871) *Dtsch. Arch. klin. Med.* **8**, 242.

Okell, C. G., Eagleton, A. J. and O'Brien, R. A. (1924) *Lancet* i, 800.

O'Meara, R. A. Q., Baker, R. S. W. and Balch, H. H. (1947) *Lancet* i, 212.

Ouchterlony, Ö. (1949) *Lancet* i, 346.

Pappenheimer, A. M. and Johnson, S. J. (1936) *Brit. J. exp. Path.* **17**, 335.

Pappenheimer, A. M. and Lawrence, H. S. (1948) *Amer. J. Hyg.* **47**, 233, 241.

Parish, H. J. (1936) *Proc. R. Soc. Med.* **29**, 481.

Parish, H. J. and Okell, C. C. (1928) *Lancet* ii, 322.

Parish, H. J. and Wright, J. (1935) *Lancet* i, 600; (1938) *Ibid.* i, 882.

Park, W. H. (1913) *Amer. J. Obstet. Gynaec.* **68**, 1213; (1918) *N.Y. med. J.* **108**, 221; (1921) *J. Amer. med. Ass.* **76**, 109; (1922) *Ibid.* **79**, 1584.

Park, W. H. and Schroder, M. C. (1932) *Amer. J. publ. Hlth* **22**, 7.

Park, W. H. and Zingher, A. (1915) *J. Amer. med. Ass.* **65**, 2216; (1916) *Amer. J. publ. Hlth* **6**, 431.

Pope, C. G. (1927) *J. Path. Bact.* **30**, 301; (1938) *Brit. J. exp. Path.* **19**, 245; (1939) *Ibid.* **20**, 201.

Pope, C. G. and Linggood, F. V. (1939) *Brit. J. exp. Path.* **20**, 297.

Prausnitz, C. (1929) *Mem. on Standardization of Therap. Sera, etc. League of Nations Hlth Org., Geneva.*

Prigge, R. (1939) *Ergebn. Hyg. Bakt.* **22**, 1.

Ramon, G. (1922) *C. R. Soc. Biol.* **86**, 661, 711, 813; (1923) *Ibid.* **89**, 2; (1926) *Arch. Inst. Pasteur Algér.* **4**, 61; (1928) *Ann. Inst. Pasteur* **42**, 959; (1955) *Mono. Inst. nat. Hyg.* No. 6, p. 107; (1957) *Quarante Années de Recherches et de Travaux.* Off. int. Epizoot., Paris.

Ramon, G. and Erber, B. (1934) *C. R. Soc. Biol.* **116**, 726.

Regulations (1952) *Statutory Instruments No. 1937. The Therapeutic Substances Regulations.* H.M.S.O., London.

Report (1923) *Biological Standardization Comm. League of Nations*; (1940) *Amer. J. publ. Hlth* **30**, Suppl. to No. 3, p. 47; (1942) *Mon. Bull. Emerg. publ. Hlth Lab. Serv.* April, p. 1; (1944) *Med. Offr* **71**, 175; (1945) *Mon. Bull. Minist. Hlth, Lond.* **4**, 85; (1948a) *Diphtheria.* Dept Hlth Scotland, H.M.S.O. Edin.; (1948b) *Lancet* ii, 517; (1949) *Mon. Bull. Minist. Hlth, Lond.* **8**, 116; (1952) *World Hlth Org. tech. Rep. Ser.* No. 56, p. 4; (1953) *Ibid.* No. 61; (1954a) *Ibid.* No. 86, p. 8; (1954b) *Mon. Bull. Minist. Hlth Lab. Serv.* **13**, 127; (1956) *Lancet* ii, 1223; (1957) *Amer. J. publ. Hlth* **47**, 751; (1963a) *Active Immunization against Infectious Disease,* Minist. Hlth, London; (1963b) *J. Hyg., Camb.,* **61**, 425; (1965) *Active Immunization against Infective Disease.* Minist. Hlth, London; (1968) *Ibid.*

Robinson, D. T. and Marshall, F. N. (1934) *J. Path. Bact.* **38**, 73; (1935) *Lancet* ii, 441.

Robinson, E. S. and White, B. (1928) *J. Immunol.* **15**, 381.

Rojas, J. A. B. and Biberstein, E. L. (1974) *J. comp. Path.* **84**, 301.

Romagnoli, A. (1955) *Giorn. Malatt. infett. parassit.* **7**, 228.

Römer, P. H. (1909) *Z. ImmunForsch.* **3**, 208.

Roux, E. and Yersin, A. (1888) *Ann. Inst. Pasteur* **2**, 629.

Russell, A. (1943) *Proc. R. Soc. Med.,* **36**, 503.

Russell, W. T. (1943) *Spec. Rep. Ser. med. Res. Coun., Lond.* No. 247.

Ruys, A. C. and Noordam, A. L. (1949) *Amer. J. publ. Hlth* **39**, 185.

Sachse, H. (1955) *Zbl. Bakt.* **162**, 96.

Sant' Agnese, P. A. di (1950) *Amer. J. publ. Hlth* **40**, 674.

Sarkany, I., Taplin, D. and Blank, H. (1962) *Arch. Derm.* **85**, 578.

Savage, W. G. (1919–20) *J. Hyg., Camb.* **18**, 448.

Scheibel, I., Bentzon, M. W., Christensen, P. E. and Biering, A. (1966) *Acta path. microbiol. scand.* **67**, 380.

Scheibel, I., Bentzon, M. W., Tulinius, S. and Bojlén, K. (1962) *Acta path. microbiol. scand.* **55**, 483.

Schick, B. (1913) *Münch. med. Wschr.* **60**, 2608.

Schmidt, H. (1942) *Behringwerk-Mitth.* Hft No. 21.

Schmidt, S. (1933a) *Z. ImmunForsch.* **78**, 27; (1933b) *Ibid.* **78**, 323; (1933c) *Ibid.* **78**, 339.

Schwartz, A. B. and Janney, F. R. (1938) *J. Amer. med. Ass.* **110**, 1743.

Seal, S. C. and Johnson, S. J. (1941) *J. infect. Dis.* **69**, 102.

Sellers, A. H., Baillie, J. H., Cruikshank, J. M. and McKibbon, J. C. (1945) *Canad. J. publ. Hlth* **36**, 390.

Shone, H. R., Tucker, J. R., Glass, V. and Wright, H. D. (1939) *J. Path. Bact.* **48**, 129.

Sigurjónsson, J. (1939) *Z. Hyg. InfektKr.* **122**, 189.

Simmons, J. S. (1920) *Amer. J. med. Sci.* **160**, 589.

Smith, M. A. and Waterworth, P. M. (1961) *Brit. J. Derm.* **73**, 152.

Sørensen, G. H. (1978) *Nord. VetMed.* **30**, 199.

Solé, A. (1934) *Wien. klin. Wschr.* **47**, 713.

Soltys, M. A. (1961) *J. Path. Bact.* **81**, 441.

Soltys, M. A. and Spratling, F. R. (1957) *Vet. Rec.* **69**, 500.

Somerville, D. A. (1970) *J. med. Microbiol.* **3**, 37.

Spink, W. W. (1978) *Infectious Diseases,* p. 168, Dawson & Son Ltd, Folkestone.

Stallybrass, C. O. (1936) *Proc. R. Soc. Med.* **29**, 487.

Stillerman, H. B. (1948) *Amer. J. Dis. Child.* **76**, 33.

Swyer, R. (1931) *Lancet* i, 632.

Tasman, A. and Lansberg, H. P. (1957) *Bull. World Hlth Org.* **16**, 939.

Tasman, A., Minkenhof, J. E., Vink, H. H., Brandwijk, A. C. and Smith, L. (1958) *Leeuwenhoek ned. Tijdschr.* **24**, 161.

Tasman, A. and Ramshorst, J. D. van (1952) *Geneesk. Gids* **30**, 469.

Taylor, I., Tomlinson, A. J. H. and Davies, J. R. (1962) *Roy. Soc. Hlth J.* **82**, 158.

Thal, E. and Rutqvist, L. (1959) *Nord. VetMed.* **11**, 298.

Thelander, H. E. (1940) *Amer. J. Dis. Child.* **59**, 342.

Thomas, J. C. (1941) *Lancet* ii, 123.

Thomson, F. H., Mann, E. and Marriner, H. (1928–29) *Metropol. Asyl. Bd Ann. Rep.* p. 304.

Toshach, S., Valentine, A. and Sigurdson, S. (1977) *J. infect. Dis.* **136**, 655.

Trendelenburg. (1869) *Arch. klin. Chir.* **10**, 720.

Trevan, J. W. (1941) *Lancet* i, 329.

Triet, A. J. van (1959) *Brit. J. exp. Path.* **40**, 559.

Underwood, E. A. (1934) *Lancet* i, 678.

Vahlquist, B. and Högstedt, C. (1949) *J. Immunol.* **62**, 277.

Volk, V. K. and Bunney, W. E. (1939) *Amer. J. publ. Hlth* **29**, 197; (1942) *Ibid.* **32**, 700.

Volk, V. K., Top, F. H. and Bunney, W. E. (1954) *Amer. J. publ. Hlth* **44,** 1314.

Warin, J. F. (1940) *Brit. med. J.* **i,** 655.

Watson, A. F., Taggart, R. A. and Shaw, G. E. (1941) *J. Path. Bact.* **53,** 63.

Weigmann, F. and Degn, J. (1932) *Zbl. Bakt.* **125,** 374.

Weitz, B. (1947) *J. comp. Path.* **57,** 191.

Whitaker, J. A., Nelson, J. D. and Fink, C. W. (1961) *Pediatrics, Springfield* **27,** 214.

Whitley, O. R. and Damon, S. R. (1949a) *Publ. Hlth Rep., Wash.* **64,** 201; (1949b) *Ibid.* **64,** 457.

Wickremesinghe, R. S. B. (1981) *J. Hyg., Camb.* **87,** 271.

Wildführ, G. (1949) *Zbl. Bakt.* **154,** 14.

Wilson, G. S. (1967) *The Hazards of Immunization.* Athlone Press, London.

Wishart, F. O., Waters, G. G. and Horner, C. M. (1944) *Canad. J. publ. Hlth* **35,** 276.

Wood, N. and O'Gorman, G. (1957) *Antibiot. Med. clin. Ther.* **2,** No. 7, p. 3.

Woods, H. M. (1928) *J. Hyg., Camb.* **28,** 147.

Woolcock, J. B., Mutimer, M. D. and Farmer, A-M. T. (1980) *Res. vet. Sci.* **28,** 87.

Wright, H. D. (1941a) *J. Path. Bact.* **52,** 129; (1941b) *Ibid.* **52,** 283.

Wright, H. D., Shone, H. R. and Tucker, J. R. (1941) *J. Path. Bact.* **52,** 111.

Yanagawa, R., Basri, H. and Otsuki, K. (1967) *Jap. J. vet. Res.* **15,** 111.

Yanagawa, R. and Honda, E. (1978) *Int. J. syst. Bact.* **28,** 217.

Zaki, M. M. (1968) *Res. vet. Sci.* **9,** 489.

Zingher, A. (1921) *Arch. Pediat.* **38,** 336; (1922) *J. Amer. med. Ass.* **78,** 1945.

Zinnemann, K. (1943) *J. Path. Bact.* **55,** 275; (1946) *Ibid.* **58,** 43.

54

Anthrax

Graham Wilson and Geoffrey Smith

Introductory

Anthrax is a disease that has been known from antiquity. In earlier days, however, it was not clearly separated from other affections closely simulating it. Maret (1752) and Fournier (1769) defined the clinical type of malignant pustule in man, and Chabert (1780) gave a clear description of anthrax in animals. In 1823 Barthelémy showed that it was transmissible by inoculation. Rayer (1850) described small, non-motile, filiform bodies in the blood of sheep dead of the disease, and confirmed its transmissibility by inoculation. (For references see Hutyra and Marek 1922.) In a series of papers, Davaine (1863a, b, c, 1864) showed that anthrax could be transmitted to sheep, horses, cattle, guinea-pigs, and mice, by the subcutaneous inoculation of infected but not of normal blood; that in such animals the bacilli did not appear in the blood till 4 or 5 hours before death; that in the blood they increased rapidly in numbers, and became filamentous; and that after death they disappeared as soon as putrefaction commenced. He showed, moreover, that the blood of an infected animal, taken before its invasion with the bacilli, was non-infective, but that after invasion it was capable of conveying the disease; that animals fed on infected viscera frequently became infected, but that animals fed on the putrefying organs of non-infected animals did not do so; and that after death from anthrax the spleen, liver, kidneys, lungs, blood, and, to a less extent, other organs contained the bacilli in large numbers. In the same year, Tiegel and Klebs (see Koch 1881) showed that anthrax blood, if filtered through a clay candle, was deprived of its infectivity; the filtrate was innocuous to animals, but the deposit on the filter remained active. These observations showed, as conclusively as could be expected in the absence of cultivation, that anthrax was caused by a living organism that multiplied in the body, invaded the bloodstream, and produced death by septicaemia. To this organism Davaine gave the name of *Bactéridie*. Subsequently Davaine and Raimbert (1864) found the same organism in a malignant pustule in man, thus demonstrating the aetiological identity of the disease in man and animals.

The final proof of the causative role of *Bacillus anthracis* was produced by Koch (1877), who, in a classic masterpiece which brought him suddenly into fame, gave a full account of the organism, described its formation of resistant spores, its cultivation *in vitro*, the reproduction of the disease by injection of pure cultures, and the recovery of the organism from the animals at necropsy.

The subsequent earlier history of anthrax is largely connected with attempts at immunization, first made by Greenfield (1880*a,b*), then by Toussaint, and later with more success by Pasteur, by Sclavo, and by Sobernheim (see section on Vaccination).

(For an excellent detailed review of anthrax, see Sobernheim 1913, and for a shorter review Lincoln *et al.* 1964).

Epidemiology

Anthrax in animals

Anthrax is primarily a disease of animals, from which man is secondarily infected. In order of susceptibility the herbivora rank foremost, notably cattle, sheep, goats, buffaloes, horses, camels, reindeer, and elephants. Of moderate susceptibility are the pig, dog, cat, rat, and man. Birds, with the exception of ostriches, are very resistant. The disease is widespread throughout the world, affecting most countries to some extent or other. Roughly one-third of all outbreaks occur in Europe, and two-thirds in Asia. The United States and Africa suffer much less, and Canada and Australasia hardly at all. There are large enzootic areas where infection is always present, such as China, Ethiopia, Iran, the Congo, Mexico, Colombia, and Bolivia, and numerous other parts of the world where it is so common as to be almost enzootic. Low-lying swampy districts with warm loose moist soils having a pH higher than 6.0 (van Ness 1971), such as are common in the great deltas of the Brahmaputra and the Yangtze Kiang, and to a less extent in the plains of the Danube and the mouth of the Rhône, constitute the most favourable terrain for propagation of the disease. In Europe cattle are mainly affected, in Asia sheep and goats. Though information is far from complete, about 10 000 outbreaks, varying in size, have been recorded annually throughout the world (Kauker 1965). Anthrax occurs in wild animals and is common in certain game parks in southern Africa.

After the Anthrax Order of 1910 came into force, the number of outbreaks in Great Britain ranged for a long time between about 400 and 700 a year (Report 1914–37). More recently it has tended to fall. Thus, in the ten years 1960–69 the average annual number was only 326. In very few outbreaks were more than one or two animals affected. In 1937, for example, there were 743 outbreaks affecting 820 cattle, 30 pigs, 13 horses, 1 sheep, and 15 other animals. The highest figure since 1910 was reached in 1956, when 1245 outbreaks were recorded with 1330 deaths. This is very different from the type of outbreak that occurs in enzootic areas and in other warm countries where the disease is common, such as Southern Europe, Asia Minor, and India. Here large numbers of animals are attacked and whole herds may be decimated. As explained below, this difference in behaviour is due to the different mode of infection.

As a rule, anthrax in animals takes the form of a septicaemia, varying from a sudden apoplectic attack with death occurring in a few minutes after the appearance of the first symptoms, to a less acute type, manifested by fever, and frequently by intestinal disturbances, terminating fatally. In goats and sheep the apoplectic type of disease is usual. In horses and cattle, especially those infected by biting flies, circumscribed cutaneous swellings or carbuncles may sometimes appear, not unlike the malignant pustule of man; in swine and dogs, localization is common on the mucosae, particularly those of the pharynx and larynx. These forms with local manifestations are rarely so fatal as the general septicaemic disease. In animals that recover, mastitis may occur but is probably rare. Weidlich (1934) reported one case in a cow in which apparently normal milk containing anthrax bacilli was secreted for some months. The incubation period of anthrax is usually about 1–5 days.

Mode of infection and spread Anthrax is rarely spread directly from animal to animal. Infection generally occurs by the alimentary tract from ingestion of infected food. During the last stages of the disease in animals, the bacilli are excreted in the urine, faeces, and saliva. At the time of death and for some time afterwards, bloody infected fluid exudes from the openings of the body and soils the neighbouring ground. The bacilli, which in the blood are invariably in the vegetative form, after being voided from the body, soon produce spores under the influence of a suitable temperature and exposure to the atmosphere. These spores are extremely resistant to inimical agencies and may remain alive on the surface of the ground for as long as 12 years (Pasteur 1881*a*); other workers have found spores lasting 35 years or more in dry soils. Cattle and sheep feeding on ground contaminated from an infected carcass, or, as in the case of 'osteophagic' animals (cf. Chapter 72) in phosphorus-deficient areas (Clark 1938), on the carcasses or bones themselves, are liable to be infected. Contamination may be transmitted from one region to another by parts of infected carcasses carried by birds and carnivora (Cameron 1945). Earthworms were considered by Koch (1881) to play little part in transmission.

The view that the infection of animals occurs chiefly by feeding on contaminated pasture land affords a possible explanation of the greater prevalence of anthrax in low-lying, marshy areas, and by the banks of streams, where grasses and decaying vegetable materials are abundant on which the bacillus may grow; of the rise in incidence as soon as the weather becomes warm; and of the absence of an aestival prevalence of the disease in cold countries. It appears, however, that these environmental conditions are favourable, not so much for the proliferation of the contaminating anthrax bacilli, as for transformation into spores of bacilli from infected animals.

Vegetative forms do not survive long in competition with other bacteria—either soil bacteria (Minett and

Dhanda 1941) or the putrefactive bacteria that proliferate in the infected carcass (Minett 1950). In soil, survival is favoured by a neutral or alkaline reaction, a moisture content of at least 60 per cent, and an air temperature between 21° and 37°, which promotes sporulation (Davies 1960, Kauker 1965). In temperate climates, such as that of Great Britain, any spores contaminating the soil probably survive the winter but are destroyed by soil organisms when they germinate in the summer. In infected blood from open carcasses sporulation is fairly rapid at 32°, but at 15–21° it is so slow that the bacilli are destroyed by the growth of saprophytic contaminants (Minett 1950). This probably means that infected blood shed by animals dying during the colder months of the year is soon rendered innocuous.

In the more northerly parts of Europe the incidence of anthrax is highest in the winter months, suggesting a mode of infection other than that from soil. It is true that ground may be contaminated by effluents from tannery and other industrial works, but most outbreaks are attributable to imported foodstuffs, which are used chiefly in the winter months (M'Fadyean 1903b, Brennan 1953). Thus, in Great Britain cattle, which are given artificial foods, suffer more than sheep, which are not. Outbreaks, moreover, are common in previously uninfected farms (Stockman 1911); and in most outbreaks there is a history of artificial feeding. (See also Jackson 1930.) The association of anthrax with artificial foods was evident in Germany, where there were 7181 cases in 1914, and only 743 in 1919, a change attributed to the cessation during the war of importation of artificial foodstuffs (Poppe 1922). A similar association was observed during the 1939–45 war (see Mohr 1960). In Great Britain there were 699 outbreaks in 1939; the number fell to 95 by 1946. With re-establishment of imports of foodstuffs and fertilizers of animal origin from countries where anthrax is endemic, the annual incidence rose gradually to 407 in 1951 and to 1245 in 1956. *Bac. anthracis* is demonstrable in imported bone, meat, blood, fish and maize meal, barley and oil-cake. Bone meal may be heavily contaminated (Davies and Harvey 1953, 1972, Report 1959a). Outbreaks in zoos are often associated with the feeding of contaminated meat (Lyon 1973).

The ingested organisms appear to cause infection via some part of the pharyngeal or alimentary mucosa, especially that of the intestines. The factors determining the precise portal of entry in a given type of the disease are largely unknown, though clearly trauma of some kind may be important. Pasteur (1880) postulated infection of sheep via the mouth and pharyngeal mucosa, and observed greatly increased artificial infection rates when the pharynx was eroded by adding prickly material, such as thistle leaves and barley spikes, to contaminated food.

Another method of infection that may occur,

though less commonly, is by flies. *Stomoxys calcitrans* and mosquitoes, for example, have been shown experimentally to convey anthrax (Poppe 1922). The transmission is mechanical; there is no development of the bacilli in the insect's body (Morris 1918, Sen and Minett 1944).

Summarizing, we may say that in warm countries animals are infected chiefly by contaminated carrion and pasturage, less often by biting flies, and in cold countries by artificial foodstuffs; and that the most frequent mode of infection appears to be through the alimentary tract.

Excluding very mild and latent cases, the case-fatality rate in animals ranges between 75 and 100 per cent.

Anthrax in man

Anthrax in man may be divided into (1) the non-industrial type, affecting shepherds, farmers, veterinary surgeons, knackers, butchers, pathologists, and others coming into close contact with infected animals; and (2) the industrial type, arising from the manipulation of wool, animal hair and bristles, hides, skins and bones, or occurring in other industries such as those dealing with harness, furniture, cutlery, boots, manure, rag-sorting, horn, or grain porterage. The epidemiology of these two types is so different that they must be treated separately.

The first type, which takes the form of *malignant pustule*, is due to contamination of the skin with material from infected animals. It is particularly common, therefore, in countries having a high incidence of anthrax among animals, and is relatively uncommon in countries such as Great Britain, where the incidence among animals is low. Moreover, there is a close correlation between the seasonal incidence of anthrax in animals and in human beings. This was well seen in southern Italy, where the number of cases in man rose almost simultaneously during the summer months with the increased prevalence among animals (C.H.W. Page 1909), and in Bulgaria (Koschucharoff 1938).

The second type of anthrax, which may take the form of malignant pustule or pulmonary disease, is dependent on infection acquired during the industrial treatment of animal products, and shows no special seasonal incidence. The number of cases of external anthrax notified under the Factories Act 1895 in England and Wales in 1901–04 was 208, of which 44 (21.2 per cent) were fatal. There were also 82 non-industrial fatalities. An incidence of this order was maintained until 1919, when, under the Anthrax Prevention Act of 1919, the import of certain classes of potentially contaminated material was prohibited. From 1920 to 1944 the industrial incidence was from half to one-third of the previous figures, with a case fatality of about 12 per cent. There was a further drop in incidence from 1945 to 1958, chiefly in cases attributable to horse hair, hides and bristles. The number of industrial cases reported during this 14-year period was 284 with 9 (3.2 per cent) deaths. A similar decline in deaths from non-industrial anthrax occurred. In both classes, the decline was attributable to the introduction of antibiotic therapy (Report 1959*b*). The total number of cases in England and Wales during the 25 years 1945–69 was 369 and the deaths 17. In the years 1976–80 there were only 14 cases with not a single death. Anthrax is now a rare disease. In the world as a whole about 9000 cases are thought to occur annually (see Kauker and Zettl 1968).

At one time anthrax was not uncommon in the woollen industries of the West Riding of Yorkshire and of Worcestershire, where the dangerous classes of raw wool from Asia Minor and Persia were used, but it has now been largely brought under control. In the horsehair industry those using contaminated hair from China, Russia, and Siberia were chiefly affected. Among the dock porters anthrax was seen mainly in the ports of London and Liverpool. In general the sources of contaminated material are Asia, many African territories, and to a lesser extent South America. Materials from countries with effective anthrax control measures, like Australia, New Zealand and South Africa, are not implicated. (For incidence of human anthrax in Germany, Austria and the USA, see Report 1934–35, Wolff and Heimann 1951, Steele and Helvig 1953, Kendall 1959, Mohr 1960, Kauker and Zettl 1968.)

In all but the woollen industry, the type of anthrax contracted is external, *i.e.* the malignant pustule. As infection occurs from contamination of the skin with the infected animal products, it is natural to expect that the uncovered parts of the body will suffer most severely. Legge (1934) found the following percentages in the location of the pustule in 937 cases: head and face, 44.6; neck, 31.2; upper extremity, 20.4; lower extremity and trunk, 1.9 each. The site of the lesion varies with the nature of the industry. Hide porters are frequently infected on the back of the neck, which is more open than other parts to excoriation. In butchers and others who handle carcasses, the arm is often affected. The face and neck are prone to attack in those using infected shaving brushes (Report 1921, Vincent 1922), and so on. In the woollen industries, the infection is internal, and occurs during the processes of sorting and combing. The worker inhales a large quantity of dust containing anthrax spores, and is hence liable to the pulmonary, and, less commonly, the intestinal, type of anthrax—*woolsorters' disease*. At the beginning of the century, internal anthrax was not infrequent. In Great Britain, of a total of 503 industrial cases recorded during 1900–09, 25 were internal anthrax, and of these 24 proved fatal. In 1937–57 there was a total of 235, of which only 3 were fatal (Report 1959*b*)—an improvement attributable to legislation demanding the sterilization of imported wools, hairs and bristles before their industrial processing.

A close imitation of woolsorters' disease can be produced in monkeys by exposure to artificially generated aerosols of anthrax spores (Gleiser *et al.* 1963, Gochenour *et al.* 1963), or to the natural spore-laden atmosphere of a goat-hair processing plant (Brachman *et al.* 1966). Post-mortem examination shows the presence of haemorrhages in the lungs, haemorrhagic mediastinitis and lymphadenitis, with occasionally haemorrhagic meningitis.

The various products that are liable to give rise to anthrax in man come from regions where the disease is common in animals; their danger is in proportion to the chance of their being contaminated. Hides and wool are frequently contaminated with blood; horse-hair and hog bristles are contaminated from various sources.

Owing to their greater exposure, men suffer from anthrax more than women, but the disease is more fatal in women. The case fatality of cutaneous anthrax depends largely upon the site of infection and the day on which treatment is started. According to Legge (1934) it ranges from about 3 per cent on the forehead to about 24 per cent on the neck. Eurich (1930) estimated the usual case fatality at 5 to 15 per cent. Respiratory and alimentary anthrax are almost invariably fatal. The advent of penicillin, however, has greatly improved the prognosis of the cutaneous form. The incubation period of anthrax is usually short. In the respiratory form it is probably not much more than 24 hours, and even in malignant pustule a lesion is generally evident within 2 days.

Apart from the methods already mentioned, anthrax may be conveyed to man by the bites of insects. Contact infection is uncommon, though it is thought to be responsible for cutaneous anthrax during the dry season in the Gambia (Heyworth *et al.* 1975). Occasionally infection may occur by the food, and give rise to the intestinal form. The bacilli are destroyed by the gastric juice, but the spores escape and multiply in the intestine. Evidence based on tests in laboratory animals suggests that to infect man in this way, large numbers of spores must be present in the food. The intestinal form is not uncommon in Indonesia after the eating of raw or partly cooked meat that has been left at a high temperature enabling the bacilli to multiply (Lincoln *et al.* 1964, Kauker 1965). Milk of animals dying or just dead of the disease may contain the bacilli (M'Fadyean 1909); infection by this method is reported, *e.g.* in Africa, but not in Great Britain where the atmospheric temperature is generally too low to permit sporulation of the bacilli.

It will be seen, therefore, that man is infected with anthrax only as a result of his dealings with animals or with animal products. Methods of prevention must be founded on an understanding of this fact. The infectivity of the anthrax bacillus for man appears to be very low; and it is a common experience that, whenever a case occurs in an industrial establishment, anthrax bacilli can be found, often in large numbers, widely distributed in the environment. The organisms may be inhaled by the workers, without producing obvious disease. In the United States, Carr and Rew (1957) isolated the bacillus from the nose or nasopharynx of 14 of 101 workers in a mill processing goat-hair; and Dahlgren and his colleagues (1960) estimated that in a similar mill, where inhalation anthrax was most frequent, non-immunized persons could inhale some 1300 spores in an 8-hour period without ill effect. That subclinical infection may occur in these circumstances is evident from the demonstration, in 11 of 72 workers in a goat-hair processing mill where anthrax was endemic, of antibody to the protective antigen of *Bac. anthracis* (Norman *et al.* 1960). The risk of infection by inhalation may extend to the neighbourhood of industrial premises (see Brachman *et al.* 1961).

Apart from malignant pustule, and the less common respiratory and intestinal forms of the disease in man, meningitis is occasionally caused by the anthrax bacillus.

Bacteriology of anthrax in man and animals

Anthrax is a disease that, when fatal, invariably terminates in septicaemia. Whatever form the disease takes—the malignant pustule, respiratory, or intestinal—it is characterized by a primary local proliferation of the bacilli. In the external form, the primary lesion is the pustule itself, with extension to the lymph nodes draining the part affected. In the pulmonary and intestinal forms, the distribution of the bacilli in the early stages of natural and experimental disease indicates that the primary lesion occurs not in the lung tissue or the intestinal mucosa but in the mediastinal or mesenteric lymph nodes (Barnes 1947, Ross 1957, Trnka *et al.* 1958, Albrink *et al.* 1960), from which the blood stream is continuously infected (Widdicombe *et al.* 1956). (See also Chapter 41.) The fate of the animal rests on the result of the local attack; if it is resisted by the phagocytes and other defences of the body, recovery rapidly occurs; if the bacilli prove too virulent, they invade the blood stream, and multiply abundantly. Invasion occurs late—generally not more than a few hours before death—and is accompanied by severe toxic manifestations. In the guinea-pig, as we relate in Chapter 41, death has been attributed to secondary shock induced by the bacillary toxin complex, which has a characteristic effect on the lower nephron; and in the rabbit, to a terminal asphyxia, as indicated by a very low oxygen content of the blood. Albrink and Goodlow (1959) found no evidence of specific intoxication like that in the guinea-pig in a number of laboratory animals, including the chimpanzee.

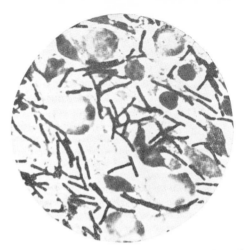

Fig. 54.1   *Bacillus anthracis.* Smear from spleen of experimentally infected guinea-pig, showing the bacilli in large numbers (× 1000).

It is characteristic of anthrax that the bacilli remain confined almost entirely to the blood vessels; they are found in maximum numbers in the capillaries of the liver, lung, kidney, spleen, intestine, and stomach, and in smaller numbers in those of the brain, skin and muscle. Their distribution varies, however, with the animal attacked. In the blood of mice or rabbits, for example, there are few bacilli to be found; in guinea-pigs there may be more bacilli than red blood corpuscles. In the larger animals they are usually plentiful, but pigs and horses may die before the organisms have proliferated sufficiently to be detected microscopically (Stockman 1911). Gibbons and Hussaini (1974) described a case in which a cow aborted 12 days after the start of treatment with penicillin; *Bac. anthracis* was isolated from the 120-day fetus.

In man, the malignant pustule starts as a small area of inflammation. It increases in size at a varying rate. Coagulation necrosis of the centre occurs, leading to the formation of a brown, purplish or black eschar surrounded by an intermediate zone of vesicles filled with clear yellow or sanious fluid, and an outer zone of widespread oedema and induration. There is no true pus and very little pain (Hodgson 1941). The bacilli are most abundant immediately below the central necrotic area. Invasion of the bloodstream may occur, but the number of organisms is seldom as great as in the larger animals (Eurich 1930). Nikiforov (1960) considers that an early transient bacteraemia is a consistent feature of cutaneous infections in man.

Diagnosis

In animals suspected of having died of anthrax, a post-mortem examination should not be made; otherwise blood will be spilt on the neighbouring ground and will provide a source of infection for other beasts. In Great Britain such post-mortem examinations are now illegal. It is sufficient in cattle and sheep to cut off an ear and send it to the laboratory; or alternatively a swab should be soaked in the blood, and several blood films should be prepared for examination. In pigs and horses, however, as death may occur in the early stages of septicaemia when very few organisms are present in the blood, it is advisable to excise a superficial lymphatic gland, and to make smears of any oedematous fluid that may be present. Malignant pustule can often be diagnosed by bacteriological methods early in the disease, but anthrax of the respiratory and alimentary systems can rarely be diagnosed till late.

It is easy to isolate spore-bearing bacilli from natural materials. The task of distinguishing *Bac. anthracis* from the wide variety of saprophytes in the *Bacillus* group is discussed in Chapter 41. Specimens should be obtained as soon as possible after death, since the organisms are destroyed when putrefaction sets in.

Fig. 54.2 *Bacillus anthracis.* Smear from a malignant pustule in man, showing the bacilli in small numbers (× 1000).

Microscopical examination Microscopical examination of fresh material is generally sufficient to enable a provisional diagnosis to be made.

The blood or fluid should be spread on a slide and stained without fixation by Gram's method. A rather thicker film should be fixed imperfectly by three passages through the flame and, after being allowed to cool, stained so as to reveal capsules—by oxidized methylene blue (M'Fadyean 1903a, c 1904) or by Giemsa. The capsules stain reddish-purple with a ragged edge, and the bacilli a deep blue. The bacilli are arranged singly or in pairs; usually they are about 4–

6 μm long, but at times filamentous forms may be seen. Of the saprophytic members of the *Bacillus* group that may be present, only *Bac. megaterium* is likely to be confused morphologically with *Bac. anthracis*. *Clostridium perfringens*, which may invade the blood stream under certain conditions, may also simulate the anthrax bacillus. In fluid from a malignant pustule the bacilli may be scanty; but they are generally present in scrapings from the base of the vesicle. Not too much weight should be given to negative results.

Cultural examination Where possible, fresh material should be used. The blood or tissue juice may be taken on a swab, a sterile thread, a fragment of earthenware, or a piece of gypsum that has been soaked in broth and subsequently sterilized (Strassburg method). The drying of the material prevents the destruction of the bacilli by the bactericidal power of the serum (Eurich 1933), facilitates their rapid sporing, and prevents the growth of other organisms. When wool or hair is to be examined, it should be soaked in a weak solution of KOH, or in a detergent for about 4 hours and then squeezed or teased out. The fluid extract should be heated to 70° for 10 minutes, or according to Jones (1942) at 65° for 5 minutes, before cultivation. Cultures are made on media designed to provide some preliminary distinction of the anthrax bacillus from anthracoid organisms (see Chapter 41). The bacillus grows well on serum media containing bicarbonate, which promotes the development of mucoid colonies characteristic of *Bac. anthracis* (see Leise *et al.* 1958); egg yolk agar, which helps to distinguish the feebly lecithinase-positive anthrax bacillus from the strongly positive *Bac. cereus*, is likewise suitable. Two selective media, one containing haematin and lysozyme (Pearce and Powell 1951), the other propamidine and polymyxin B (Morris 1955), are recommended. On nutrient agar containing 0.05 unit of penicillin per ml, Jensen and Kleemeyer (1953) observed the growth of anthrax bacilli, but not of anthrax-like bacilli, in the form of a 'string of pearls'.

Suspicious colonies may be tested for motility by seeding into one end of a Craigie tube; the anthrax bacillus grows without any evidence of spread by motility.

For enrichment of specimens containing few bacilli, Thomson (1955) advocates incubating the swab in 2 ml fresh defibrinated ox blood; after 6–17 hours at 37° positive cultures contain typical capsulated bacilli. Further examination includes testing for phage susceptibility (see McCloy 1951 and Chapter 41) and pathogenicity.

Pathogenicity tests

When possible, these should be carried out with the pure cultures that have been isolated. The best animals to employ are the guinea-pig and the mouse. When the original material is used, particularly if this is old, putrid, or contaminated, or when wool, hair, or hides are being examined, the animal may die of infection with some organism other than the anthrax bacillus; to avoid this outcome, about 0.5 ml of a 24 hours' broth culture is injected subcutaneously into the thigh. Death generally occurs within 2 to 3 days; *post mortem*, there is a haemorrhagic gelatinous exudate at the site of inoculation, and a large, congested, dark-red spleen (see Chapter 41); microscopically, the bacilli are found in the blood and in smears made from the viscera (Fig. 54.1). Cultures should be put up for confirmation.

It is well to inoculate three or four mice by scratching the skin with a needle dipped in the fluid. As the anthrax bacillus can get through the abraded skin more easily than other contaminating organisms, death of one or more of the mice will probably occur from a pure anthracaemia. This method, first described by Koch, Gaffky and Loeffler (1884), is a good example of the method of purification of cultures by animal inoculation (see Chapter 20). Alternatively, the inoculum is injected intramuscularly into two or three guinea-pigs, which have been passively immunized 24 hours beforehand with 1000 units each of *Cl. perfringens*, *Cl. novyi*, *Cl. septicum* antitoxins, and 500 units of tetanus antitoxin. Of *Bacillus* spp. other than *Bac. anthracis*, *Bac. cereus* is the most likely to be infective, but only in relatively large doses, and without producing the characteristic post-mortem appearances of anthrax.

Wool, hair, or other industrial material is soaked in detergent or alkali as described above; 50 ml or more of the supernatant fluid are centrifuged, and the deposit is inoculated intramuscularly into two or three guinea-pigs passively immunized against tetanus and gas gangrene. The animal test is more delicate than the plate method. Using it, Glynn and Lewis (1912), to whom reference should be made for fuller particulars, isolated anthrax spores from 21.3 per cent of 141 samples of industrial material—mainly hides, wool, hair and bones—supposed to have produced anthrax. Extracts for inoculation of mice made by a commercial detergent at 37°, and heated for 10 minutes at 70°, were positive in 34 of 76 specimens; as against only two positives obtained by plating saline extracts (Biegeleisen *et al*. 1962). The use of fluorescein-labelled antiglobulin for staining young cultures on blood agar or impression smears of tissues from infected mice may appreciably shorten the time to diagnosis (Biegeleisen *et al*. 1962).

Serological diagnosis

As might be expected in a disease with so short an incubation period and duration, there is no specific antibody response of the host that can be used as an index of present infection. For retrospective diagnosis, however, in patients who have recovered under the influence or not of antibiotic therapy the demonstration of toxin-neutralizing antibodies in the serum may be of help (Darlow and Pride 1969). Ascoli (1911) devised a test for the presence of the soluble anthrax antigens that were extractable from the infected tissues with saline at 100°. Such material, layered on the surface of a strong anthrax serum, produces a precipitate at the interface of the two liquids. A normal serum is also tested, and the antiserum reaction controlled with an extract of known *Bac. anthracis*. The antigens concerned are probably both the polypeptide capsular material and the somatic polysaccharides (see Tomcsik and Bodon 1934, Ivánovics 1939). Since the capsular material is antigenically similar to that of many other *Bacillus* spp. and the polysaccharide similar to that, for example, in *Bac. cereus* (Chapter 41), the test is clearly non-specific in any rigorous sense. But the unlikelihood of any member of the *Bacillus* group other than *Bac. anthracis* proliferating in an animal so as to flood the tissues with bacillary antigens gives this 'thermoprecipitin' test of Ascoli's a certain value as an ancillary means of diagnosis, especially in those cases where the bacillus cannot be isolated.

Immunity in anthrax

The investigation of the natural resistance of certain animals to anthrax played a prominent part in the controversy between the cellular and humoral schools of immunity in the early days of bacteriology. Without attempting to adduce all the evidence that was brought forward during this controversy, we shall merely draw attention to some of the salient points. The humoral factor of interest was the anthracidal substance demonstrable in the serum of certain animals. The potency of rat serum in this respect led von Behring to attribute the naturally high resistance of the rat to its action. Metchnikoff and his colleagues, on the other hand, concluded that this undoubted in-vitro activity of rat serum was not operative *in vivo*. Kritschewski and Messik (1930) observed that spores would germinate in the tissues of the rat, their further proliferation being restrained by phagocytes. In the moderately resistant adult dog Metchnikoff demonstrated a close association between the resistance to experimental anthrax and the phagocytic powers of the leucocytes. It is noteworthy in this connection that germ-free rats are far more susceptible to *Bac. anthracis* than conventionally reared animals, the acquisition of whose natural, 'non-specific' immunity to *Bac. anthracis* appears to depend on experience of the normal microbial environment, acting perhaps as a stimulant to the maturation of the RE system (Taylor *et al*. 1961). The phagocytes appear likewise to be implicated in the natural resistance of fowls to anthrax. Pasteur made fowls susceptible by partial immersion in cold water (Pasteur *et al*. 1878), and attributed the effect to the decrease in body temperature from the normal 41–42°, to one at which the bacilli would grow *in vivo*. *In vitro*, however, the bacilli grow in blood even at 43°; and the cooling *in vivo* appears to promote infection by depressing phagocytosis (Wagner 1890). Again, in birds infected by spores, bacilli resulting from spore-germination are rapidly killed by phagocytes (Weyl 1892). The relevance of the serum anthracidins is also diminished by the fact that they occur in species like the rabbit and guinea-pig, which are, respectively, moderately and highly susceptible to *Bac. anthracis*.

Preisz's (1909) observation that extracts of capsulated anthrax bacilli could inhibit serum anthracidins suggests that in infection by virulent capsulated forms the capsular material might protect the organisms in this way. Capsular material, however, is an important determinant in phagocytosis, since capsulated bacilli, unlike non-capsulated forms, are phagocytosis-resistant (Bail and Weil 1911). Bail and Weil observed that normal serum containing anthracidin potentiated the bactericidal activity of leucocytes and that the combined action of the two was not affected by capsular material. Extracts of infected tissue, however, would inhibit the anthracidal action of mixtures of serum and leucocytes from anthrax-susceptible, but not from naturally resistant animals. This inhibiting effect in susceptible animals Bail attributed to bacterial 'aggressins', substances enhancing infectivity. Subsequent research has thrown some light on these phenomena. The leucocytes appear to owe their anthracidal effect to certain basic polypeptides extractable from them (Chapter 41). The capsule of the bacillus, composed largely of non-toxic polyglutamic acid (Chapter 41), is associated with virulence in that, although capsulated avirulent forms occur (see, *e.g.*, Sterne 1937), non-capsulated mutants are avirulent (Bail 1915), in part, presumably, because of their susceptibility to phagocytes. The predominant role in natural immunity attributed by Metchnikoff to leucocytes may therefore be due to their acting as a source of anthracidal substances, and the aggressive action of *Bac. anthracis* to capsular substance. The virulence of the bacillus, however, appears ultimately to depend on the toxin complex synthesized by the bacillus *in vivo*, described by Smith and his colleagues (Chapter 41); and its power to proliferate on the degree to which the toxin antagonizes the leucocytic defence mechanisms—perhaps by interfering with the release of anthracidal substance.

Specific immunity

The toxins and the protective antigens of *Bac. anthracis* are discussed in Chapter 41. It is clear that neither the polysaccharide nor the polyglutamic acid capsular material is immunogenic. Capsulated avirulent organisms do not induce immunity (Sterne 1937, 1946); and non-capsulated variants will do so

(Staub 1949). The occurrence of immunogenic material in sterilized oedema fluid and extracts of infected tissue, first described by Bail (1904), has been confirmed by many workers. This active material was shown by Smith, Keppie and their colleagues (Chapter 41) to be part of the toxin complex produced by the bacillus *in vivo*; the isolated toxin complex, moreover, is immunogenic after toxoiding.

Non-toxic immunizing antigen is also produced in culture under special conditions, not only from capsulated but from non-capsulated variants, and the serum of hyperimmunized rabbits confers passive protection (Gladstone 1946, 1948). Subsequent observations suggest that protective antibody acts by neutralizing the anti-phagocytic effect of the toxin, and perhaps by interfering with capsule formation by the bacillus.

The mechanism of specific immunity is still obscure. In the pathogenesis of the disease resistance to toxin and to bacterial proliferation appear to be distinct (Klein *et al.* 1963*a*). In immunized guinea-pigs that succumb to challenge with a highly virulent strain of anthrax bacillus little or no toxin is demonstrable in the blood, the number of organisms in the blood and spleen varies greatly, and there is no constant relation between the pre-challenge titre of protective antibody and survival (Ward *et al.* 1965).

The immunogenicity of protective antigen can be assayed with some accuracy in guinea-pigs, rats or mice, by using time-to-death as the index of protection (DeArmon *et al.* 1961, McGann *et al.* 1961); and the potency of protective antiserum by estimating the toxicity of toxin-antiserum mixtures in the rabbit skin (Belton and Henderson 1956). Alum-precipitated antigen protects guinea-pigs, mice and rabbits. Monkeys are protected for periods of a year or more; and the human subject, in which the material induces few untoward reactions, responds with the production of circulating antitoxin, especially after reinforcing doses (Wright *et al.* 1954, Tresselt and Boor 1955, Darlow *et al.* 1956).

Prevention

In any case of suspected anthrax in an animal, the carcass should not be opened until samples of blood or, in the case of pigs and horses, exudates or superficial lymph nodes have been examined and anthrax excluded. Anthrax carcasses should be burnt thoroughly, or, when this is not possible, buried in chloride of lime at a depth of 1.8 m (6 feet) in places where spread of contamination by drainage is unlikely. Disinfection of the probably contaminated surroundings must be made with disinfectants—caustic soda, hypochlorites or cresols—applied for several hours, in concentrations that make due allowance for the great reduction in disinfectant power by the organic matter in the environment. Other animals should be kept under observation, and isolated if they appear to be ill or become febrile. In milch herds, milk from animals that become febrile should be withheld from distribution (M'Fadyean 1909). Contacts may be vaccinated and animals suspected of incubating the disease treated with antibiotics and antiserum. In a severe outbreak, contacts may be given antiserum, followed by vaccination.

Where the disease is epidemic, prophylaxis is best assured by vaccination. In countries in which the cases are sporadic, a careful watch should be kept over imported animal feeding stuffs and fertilizers; bone meal should preferably be sterilized by autoclaving (Steele and Helvig 1953). Drainage of moist pasture land should be encouraged.

The prevention of industrial anthrax in man is directed partly to disinfection of the imported animal products, wool, hides, horsehair, and so on, and partly to diminishing the risk of contact with dangerous material in the factories.

For the various methods of disinfection, and for factory legislation, the reader is referred to textbooks of hygiene. Briefly, however, the disinfection of bales and fleeces may be accomplished by preliminary treatment with warm alkali and soap, followed by exposure to a 2 per cent solution of formaldehyde and drying in hot air (Legge 1934; see also Wolff and Heimann 1951). For the disinfection of hides Robertson (1932) recommended the use of H_2S; anthrax spores are destroyed in 7–16 days. It is more usual, however, in Great Britain to treat them with 1 in 5000 $HgCl_2$ and 1 per cent formic acid for 24 hours, or with 2 per cent HCl and 10 per cent NaCl for 48 hours, and then store them for 14 days. Hair and bristle should be autoclaved or exposed to warm formaldehyde vapour. Contaminated overalls can be disinfected by immersion for 18–24 hours in a 4 per cent solution of formalin to which a cationic or ampholytic detergent is added in 1 per cent concentration (Darlow 1970). (For information on other methods and disinfectants, see Hailer and Heicken 1948, 1950.)

Vaccination

In 1878 Burdon-Sanderson and Duguid (see Wilson 1979) at the Brown Animal Sanatory Institution in London found that the inoculation of bovine animals with blood from infected rodents caused a severe but not fatal disease, after which they were partly resistant to reinoculation. The following year Greenfield (1880a) confirmed these findings, and then went on to modify the virulence of the anthrax bacillus by serial cultivation in aqueous humour contained in a glass tube sealed at both ends and incubated at 35°. Under these conditions the culture became less and less virulent till finally it was completely non-virulent for mice. Even the 8th generation when inoculated into cattle protected them against any serious illness caused by the injection of virulent anthrax bacilli (Greenfield 1880b). In 1880 Toussaint in France described how he had lowered the virulence of *Bac. anthracis* in defibrinated blood by heating for 10 minutes at 55°, and how this material had rendered sheep refractory to reinoculation with virulent blood. In 1881, Pasteur (1881b, c) introduced a method of vaccination, founded on the same principle as that which had been successful in diminishing the virulence of the organism of chicken cholera. He found that when a strain of *Bac. anthracis* was cultivated in broth at 42–43° not only did it lose its power of forming spores, but it gradually decreased in virulence, till after 2 to 3 months it was no longer pathogenic, even in the most susceptible animals. Pasteur prepared two vaccines; the first or *premier vaccin* was a subculture in broth from a strain that had been kept at 42–43° for 15 to 20 days; its virulence was such that it could kill mice and young guinea-pigs but was unable to kill adult guinea-pigs or rabbits. The second or *deuxième vaccin* was a subculture after 10 to 12 days; it was more virulent, being able to kill mice, adult guinea-pigs and a certain proportion of injected rabbits (see Mikesell *et al.* 1983).

Arrangements were made with the President of the Agricultural Society of Melun for a trial of this new method under field conditions. Accordingly on 5 May 1881, on a farm at Pouilly-le-Fort, 24 sheep, 1 goat and 6 cows received their first vaccination, and on 17 May they were vaccinated a second time with the more virulent but still attenuated strain. On 31 May, Pasteur and his co-workers, Roux and Chamberland, inoculated each of the vaccinated animals with a virulent culture of anthrax; at the same time a series of control animals consisting of 24 sheep, 1 goat and 4 cows were similarly inoculated. Two days later the vaccinated sheep and the goat were in perfect condition; the control sheep and the goat were all dead. The 4 unvaccinated cows were suffering from severe oedema and fever; the 6 vaccinated cows had neither fever nor oedema. The following day one of the vaccinated sheep died, but at necropsy it was found to be carrying a fetus that had been dead for about a fortnight.

It should be pointed out that Greenfield's observations antedated not only Pasteur's experiments but also those of Toussaint; and that Greenfield's method of attenuating the virulence of the anthrax bacillus by serial cultivation was published even before Pasteur (1880) described a similar method for attenuating the virulence of the fowl cholera bacillus. Without depreciating in any way the value of Pasteur's contribution to the development of an anthrax vaccine, it is only fair to attribute priority in this respect to Greenfield (Wilson 1979, Tigertt 1980). The success of the experiments at Pouilly-le-Fort led to the use of **live vaccines** on a large scale in cattle and sheep in Europe, in South America, and in other countries. Chamberland (1894), from statistics collected in 1882 to 1893, estimated that before vaccination the mortality in sheep was 10, and in cattle 5 per cent; after the introduction of the vaccine, the corresponding mortalities were 0.94 and 0.34 per cent. In Germany, from 1901 to 1908, the total mortality of vaccinated cattle was 0.03 per cent, of horses 0.05 per cent, and of sheep 0.06 per cent (Klimmer 1911). However, it appeared later that anthrax in France was nearly as prevalent as it was before the vaccination era (C.H.W. Page 1909).

There is, of course, no way of determining from results of this kind, whether the animals selected for record before and after vaccination together constituted a homogeneous population, either in their natural susceptibility to the disease, in environmental conditions, or in the degree of risk of infection to which they were subjected; so that no firm conclusion about a change in immunity can be made. Nevertheless, the association subsequently demonstrated in widely disparate localities and conditions between vaccination and decrease in the prevalence of anthrax—sometimes to the point of eradication of the disease—constitutes a substantial confirmation of the value of vaccination.

Attenuated vaccines, however, have been little used in later work, chiefly owing to the variations in virulence of the attenuated cultures. According to Pasteur, the virulence of cultures at 42–43° decreased gradually, so that at a given moment, a vaccine of the correct virulence could be selected. This view has not been established. Moreover, in cultures maintained at 42–43° a variety of bacillary variants appear, some of which are non-capsulated and avirulent, some capsulated and fully virulent, and some of varying virulence that are either spore-bearing or incapable of sporulation (Preisz 1911). Vaccines of attenuated cultures have been replaced by the more stable and more consistently reproducible spore vaccines. A few strains of *Bac. anthracis*, like the Carbazoo and the Boshoff strains, appear to have become stabilized at a suitable degree of attenuation. The Carbazoo strain was used by Mazzucchi (1935) for a spore vaccine suspended in a 2 per cent solution of saponin. The saponin induces a necrotic reaction which appears to limit the spread of the injected organism. Smaller concentrations are effectively adjuvant in their action with attenuated spore vaccines, inducing only a local oedema that promotes the local growth of the organism (Sterne 1948).

These vaccines yielded very promising results in the field, and are now being used extensively in different parts of the world. Sterne (1939) in South Africa introduced a vaccine made of avirulent uncapsulated sporing bacilli in 50 per cent glycerol saline. Its use in 2½ million cattle and several thousand horses and sheep gave as good protection as the saponin vaccine, and had the advantage of being safer and producing less reaction. In Great Britain a combination of the two methods is used, Sterne's strain being suspended in saponin. Judged by experiments on guinea-pigs, a

higher degree of immunity is conferred by vaccination with a non-proteolytic variant of the Sterne strain produced by ultraviolet radiation (Fubra 1966). In the USSR a live vaccine prepared from the ST 1 strain is widely used (see Lincoln *et al.* 1964).

Cienkowsky's method consists in the use of two vaccines of spore-bearing bacilli attenuated by heat and fixed in virulence by repeated passage through gophers (*Zieselmaus*). It has been used chiefly in Russia, and also in Japan (Poppe 1922).

Sobernheim's combined method This is a method of combined active and passive immunization. Sobernheim (1902, 1904, 1913) prepared an antiserum by injection of cattle or sheep with cultures of increasing virulence. For his vaccine, he used a culture slightly attenuated by growth at 42.5°. In practice, simultaneous injections are made, beneath the skin, of 10–16 ml of antiserum and 0.5–1.0 ml of vaccine. This method was extensively adopted in Germany and South America, where it appears to have given good results. It has the merit of requiring injection on only one occasion instead of two; of conferring a passive immunity which protects the animal while an active immunity is developing; and of being attended by very little danger. It can be employed after an epidemic has broken out, without fear of rendering the animals more susceptible to infection.

None of the methods in use confers an immunity for more than 9 months or a year; hence vaccination must be repeated annually. The evidence of the efficacy of vaccination is, as we noted above, circumstantial in that extensive field trials have not been made in which the incidence of the disease in vaccinated and unvaccinated cattle of a single herd are compared; though several spore vaccines have been shown to protect cattle against experimental anthrax (for references see Klein *et al.* 1962). The most striking effect in the field associated with vaccination is observed in districts where the disease is rife. Thus in South Africa, annual mass immunization of millions of cattle kept in the most primitive conditions was followed by disappearance of the disease; after 3–5 years' vaccination, the cattle of a district could be left unvaccinated with little recurrence of the disease (see Sterne *et al.* 1945). Vaccination with live material in countries where the disease is sporadic is sometimes deprecated on the grounds that the infected animals might initiate outbreaks; but there are no records of an attenuated strain reverting to full virulence. Tanner and his colleagues (1978) failed to isolate the anthrax bacillus from the blood or milk of cattle vaccinated with the Sterne strain.

As regards **non-living vaccines,** the protective antigen described by Gladstone (1948) has been produced (see Chapter 41) in amounts and form suitable for limited tests in man and domestic animals (Boor and Tresselt 1955, Schlingman *et al.* 1956, Jackson *et al.* 1957, Puziss and Wright 1963). In cattle alum-precipitated material induces almost as long lasting an immunity as a spore vaccine, and gives rise to none of the accidental deaths that occasionally occur with spore vaccines. In guinea-pigs a course of protective antigen followed 10 days later by a single injection of a spore vaccine induced a very high degree of immunity (Klein *et al*. 1962).

In man vaccines made with protective antigen may prove to be of value for those exposed to the risk of industrial anthrax (see, *e.g.*, Norman *et al*. 1960, Stanley and Smith 1963). There is no method of determining the degree of immunity produced, but a serum titre of 1/64 or above is thought to constitute a significant measure of resistance (Klein *et al*. 1963*b*).

Serum treatment

Several workers have used sera, prepared from hyperimmunized cattle or sheep, for protective and for therapeutic purposes, but speaking generally the serum treatment of animals has never been employed on a large scale.

In man, on the other hand, serum treatment has been extensively used. Sclavo (1895) first immunized sheep, but later (1896, 1898, 1901) he found that the ass gave the most satisfactory results. In 1903 Sclavo collected 164 cases of anthrax treated with serum in Italy; there were 10 deaths or a fatality rate of 6.09 per cent compared with a fatality rate of 24.1 per cent for the whole of Italy (see C.H.W. Page 1909). In Great Britain Eurich (1933) had a fatality rate of only 5 per cent in 200 cases of cutaneous anthrax treated without excision of the pustule. He combined serum treatment—80 ml and 60 ml intravenously on the 1st and 2nd day respectively—with salvarsan. Hodgson (1941, 1944) reported on a series of 107 cases of cutaneous anthrax treated without surgical excision of the lesion. Of 52 that were given serum only, 6 died; of 55 that were given serum plus neoarsphenamine, none died.

Chemotherapy

Since the advent of the sulphonamides and the antibiotics, treatment by serum and arsenical drugs, and the dubious practice of excising the lesion, have fallen largely into disuse. Good results have been obtained with the sulphonamides, of which sulphadiazine is probably the best.

The anthrax bacillus is unusual in being susceptible to penicillin, aureomycin, terramycin, streptomycin, and chloramphenicol (Garrod 1952). Penicillin is the most active and is the drug of choice. It is now used extensively, and in the treatment of cutaneous anthrax it yields very satisfactory results. Occasionally, however, the patient suffers from a febrile reaction while under treatment, and has to be given either serum or another antibiotic. It is to be noted that, though penicillin prevents systemic invasion and toxaemia, it has little effect on the course of the local lesion, which passes through its usual cycle more or less unchanged. Neither serum nor antibiotics appear to have much effect in established cases of pulmonary, intestinal or meningeal anthrax in man.

(For a fuller description of anthrax, see Whitford 1979.)

References

Albrink, W. S., Brooks, S. M., Biron, R. E. and Kopel, M. O. (1960) *Amer. J. Path.* **36,** 457.
Albrink, W. S. and Goodlow, R. J. (1959) *Amer. J. Path.* **35,** 1055.
Ascoli, A. (1911) *Zbl. Bakt.* **58,** 63.
Bail, O. (1904) *Zbl. Bakt.* **36,** 266; (1915) *Ibid.* **75,** 159; **76,** 38, 330.
Bail, O. and Weil, E. (1911) *Arch. Hyg., Berl.* **73,** 218.
Barnes, J. M. (1947) *Brit. J. exp. Path.* **28,** 385.
Belton, F. C. and Henderson, D. W. (1956) *Brit. J. exp. Path.* **37,** 156.
Biegeleisen, J. Z., Cherry, W. B., Skaliy, P. and Moody, M. D. (1962) *Amer. J. Hyg.* **75,** 230.
Boor, A. K. and Tresselt, H. B. (1955) *Amer. J. vet. Res.* **16,** 425.
Brachman, P. S., Kaufmann, A. F. and Dalldorf, F. G. (1966) *Bact. Rev.* **30,** 646.
Brachman, P. S., Pagano, J. S. and Albrink, W. S. (1961) *New Engl. J. Med.* **265,** 203.
Brennan, A. D. J. (1953) *Vet. Rec.* **65,** 255.
Cameron, H. S. (1945) *Cal. agric. Extension Serv. Circular* No. 130.
Carr, E. A. and Rew, R. R. (1957) *J. infect. Dis.* **100,** 169.
Chamberland, C. (1894) *Ann. Inst. Pasteur* **8,** 161.
Chauveau, A. (1880) *C. R. Acad. Sci.* **91,** 648.
Clark, R. (1938) *J. S. Afr. vet. med. Ass.* **9,** 5.
Dahlgren, C. M., Lee, M. B., Decker, H. M., Freed, S. W., Phillips, C. R. and Brachman, P. S. (1960) *Amer. J. Hyg.* **72,** 24.
Darlow, H. M. (1970) *Proc. R. Soc. Med.* **63,** 1019.
Darlow, H. M., Belton, F. C. and Henderson, D. W. (1956) *Lancet* ii, 476.
Darlow, H. M. and Pride, N. B. (1969) *Lancet* ii, 430.
Davaine, C. (1863*a*) *C. R. Acad. Sci.* **57,** 220; (1863*b*) *Ibid.* **57,** 351; (1863*c*) *Ibid.* **57,** 386; (1864) *Ibid.* **59,** 393.
Davaine, C. and Raimbert (1864) *C. R. Acad. Sci.* **59,** 429.
Davies, D. G. (1960) *J. Hyg., Camb.* **58,** 177.
Davies, D. G. and Harvey, R. W. S. (1953) *Lancet* ii, 880; (1972) *J. Hyg., Camb.* **70,** 455.
DeArmon, I. A., Klein, F., Lincoln, R. E., Mahlandt, B. G. and Fernelius, A. L. (1961) *J. Immunol.* **87,** 233.
Eurich, F. W. (1930) *A System of Bacteriology.* Med. Res. Coun., London **5,** 439; (1933) *Brit. med. J.* ii, 50.
Fubra, E. S. (1966) *J. Bact.* **91,** 930.
Garrod, L. P. (1952) *Antibiot. Chemother.* **2,** 689.
Gibbons, D. F. and Hussaini, S. N. (1974) *Nature, Lond.* **252,** 612.
Gladstone, G. P. (1946) *Brit. J. exp. Path.* **27,** 394; (1948) *Ibid.* **29,** 379.
Gleiser, C. A., Berdjis, C. C., Hartman, H. A. and Gochenour, W. S. (1963) *Brit. J. exp. Path.* **44,** 416.
Glynn, E. E. and Lewis, F. C. (1912) *J. Hyg., Camb.* **12,** 227.
Gochenour, W. S., Sawyer, W. D., Henderson, J. E., Gleiser, C. A., Kuehne, R. W. and Tigertt, W. D. (1963) *J. Hyg., Camb.* **61,** 317.

Greenfield, W. S. (1880a) *J. roy. agricult. Soc.*, 2nd series **16**, 273; (1880b) *Proc. R. Soc., Lond.* **30**, 557.

Hailer, E. and Heicken, K. (1948) *Z. Hyg. InfektKr.* **128**, 87, 109; (1950) *Ibid.* **131**, 219.

Heyworth, B., Ropp, M. E., Voos, U. G., Meinel, H. I. and Darlow, H. M. (1975) *Brit. med. J.* **iv**, 79.

Hodgson, A. E. (1941) *Lancet* **i**, 811; (1944) *Ibid.* **ii**, 161.

Hutyra, F. and Marek, J. (1922) *Special Pathology and Therapeutics of the Diseases of Domestic Animals*, 2nd Amer. edn **1**, 1. Chicago and London.

Ivánovics, G. (1939) *Arch. wiss. prakt. Tierheilk.* **74**, 75.

Jackson, F. C., Wright, G. G. and Armstrong, J. H. (1957) *Amer. J. vet. Res.* **18**, 771.

Jackson, R. (1930) *J. comp. Path.* **43**, 95.

Jensen, J. and Kleemeyer, H. (1953) *Zbl. Bakt.* **159**, 494.

Jones, E. R. (1942) *J. Path. Bact.* **54**, 307.

Kauker, E. (1965) *Sitzungsberichte der Heidelberger Akademie der Wissenschaften.* Springer Verlag.

Kauker, E. and Zettl, K. (1968) *Berl. Münch. tierärztl. Wschr.* **76**, 194.

Kendall, C. E. (1959) *J. occupational Med.* **1**, 174.

Klein, F., DeArmon, I. A., Lincoln, R. E., Mahlandt, B. G. and Fernelius, A. L. (1962) *J. Immunol.* **88**, 15.

Klein, F., Haines, B. W., Mahlandt, B. G., DeArmon, I. A. and Lincoln, R. E. (1963a) *J. Bact.* **85**, 1032.

Klein, F., Haines, B. W., Mahlandt, B. G. and Lincoln, R. E. (1963b) *J. Immunol.* **91**, 431.

Klimmer (1911) *Handbuch der Serumtherapie und Serumdiagnostik in der Veterinärmedizin.* Leipzig. (Quoted from Hutyra and Marek 1922.)

Koch, R. (1877) *Cohns Beitr. Biol. Pflanz.* **2**, 277; (1881) *Mitt. ReichsgesundhAmt.* **1**, 49.

Koch, R., Gaffky and Loeffler, F. (1884) *Mitt. ReichsgesundhAmt.* **2**, 147.

Koschucharoff, P. (1938) *Z. ImmunForsch.* **92**, 53.

Kritschewski, I. L. and Messik, R. E. (1930) *Z. ImmunForsch.* **65**, 420.

Legge, T. M. (1934) *Industrial Maladies*, pp. 36–44. Humphrey Milford, London.

Leise, J. M., Friedlander, H., Jaeger, R. and Phillips, R. L. (1958) *Phys. Defence Div., Fort Detrick, Md*, p. 73.

Lincoln, R. E., Walker, J. S., Klein, F. and Haines, B. W. (1964) *Advanc. vet. Sci.* **9**, 327.

Lyon, D. G. (1973) *Vet. Rec.* **92**, 334.

McCloy, E. W. (1951) *J. Hyg., Camb.* **49**, 114.

M'Fadyean, J. (1903a) *J. comp. Path.* **16**, 35; (1903b) *Ibid.* **16**, 346; (1903c) *Ibid.* **16**, 360; (1904) *Ibid.* **17**, 58; (1909) *Ibid.* **22**, 148.

McGann, V. G., Stearman, R. L. and Wright, G. G. (1961) *J. Immunol.* **86**, 458.

Mazzucchi, M. (1935) *12th int. vet. Congr.* **2**, 138.

Mikesell, P. *et al.* (1983) *Amer. Soc. Microbiol. News* **49**, 320.

Minett, F. C. (1950) *J. comp. Path.* **60**, 161; (1952) *Bull. Off. int. Épiz.* **37**, 238.

Minett, F. C. and Dhanda, M. R. (1941) *Indian J. vet. Sci.* **11**, 308.

Mohr, W. (1960) *Landarzt.* **36**, 9.

Morris, E. J. (1955) *J. gen. Microbiol.* **13**, 456.

Morris, H. (1918) *J. comp. Path.* **31**, 134.

Ness, G. B. van (1971) *Science* **172**, 1303.

Nikiforov, V. N. (1960) *J. Microbiol. Epidemiol. Immunobiol.* **31**, 1532.

Norman, P. S., Ray, J. G., Brachman, P. S., Plotkin, S. A. and Pagano, J. S. (1960) *Amer. J. Hyg.* **72**, 32.

Page, C. H. W. (1909) *J. Hyg., Camb.* **9**, 279, 357.

Pasteur, L. (1880) *C. R. Acad. Sci.* **91**, 209; (1881a) *Ibid.* **92**, 209; (1881b) *Ibid.* **92**, 429; (1881c) *Ibid.* **92**, 666.

Pasteur, Joubert and Chamberland. (1878) *Bull. Acad. Méd.*, 2nd Ser. **7**, 432.

Pearce, T. W. and Powell, E. O. (1951) *J. gen. Microbiol.* **5**, 387.

Poppe, K. (1922) *Ergebn. Hyg. Bakt.* **5**, 597.

Preisz, H. (1909) *Zbl. Bakt.* **49**, 341; (1911) *Ibid.* **58**, 510.

Puziss, M. and Wright, G. G. (1963) *J. Bact.* **85**, 230.

Report (1914–37) *Rep. Proc. Dis. Animals Acts, Min. Agric. Fish.*; (1921) *Min. Hlth, Circ.* No. 172; (1934–35) *Amer. publ. Hlth Ass. Year Book, Suppl. Amer. J. publ. Hlth*, 1935 **25**, No. 2; (1959a) *Mon. Bull. Min. Hlth P.H.L.S.* **18**, 16; (1959b) *Report of the Committee of Inquiry on Anthrax.* H.M.S.O., London.

Roberstpon, M. E. (1932) *J. Hyg., Camb.* **32**, 367.

Ross, J. M. (1957) *J. Path. Bact.* **73**, 485.

Schlingman, A. S., Devlin, H. B., Wright, G. G., Maine, R. J. and Manning, M. C. (1956) *Amer. J. vet. Res.* **17**, 256.

Sclavo, A. (1895) *Zbl. Bakt.* **18**, 744; (1896) *Riv. Igiene Sanit. publ.* **7**, 705, 745; (1898) *Ibid.* **9**, 200, 814; (1901) *Ibid.* **12**, 212, 247; (1903) *Ibid.* **14**, 519.

Sen, S. K. and Minett, F. C. (1944) *Indian J. vet. Sci.* **14**, 149.

Sobernheim, G. (1902) *Berl. klin. Wschr.* **39**, 516; (1904) *Dtsch. med. Wschr.* **30**, 948, 988; (1913) In: *Kolle and Wassermann's Handbuch der pathogenen Mikroorganismen*, 2 te Auflage, Bd 3, p. 583.

Stanley, J. L. and Smith, H. (1963) *J. gen. Microbiol.* **31**, 329.

Staub, A-M. (1949) *Ann. Inst. Pasteur* **76**, 331.

Steele, J. H. and Helvig, R. J. (1953) *Publ. Hlth Rep., Wash.* **68**, 616.

Sterne, M. (1937) *Onderstepoort J. vet. Sci.* **8**, 271; *Ibid.* **9**, 49; (1939) *Ibid.* **13**, 313; (1946) *Ibid.* **21**, 41; (1948) *Ibid.* **23**, 157, 165.

Sterne, M., Nicol, J. and Lambrechts, M. C. (1945) *J. S. Afr. vet. med. Ass.* **16**, 12.

Stockman, S. (1911) *J. comp. Path.* **24**, 97.

Tanner, W. B. *et al.* (1978) *J. Amer. vet. med. Ass.* **173**, 1465.

Taylor, M. J., Rooney, J. R. and Blundell, G. P. (1961) *Amer. J. Path.* **38**, 625.

Thomson, P. D. (1955) *J. comp. Path.* **65**, 1.

Tigertt, W. D. (1980) *J. Hyg., Camb.* **85**, 415.

Tomcsik, J. and Bodon, G. (1934) *Z. ImmunForsch.* **83**, 426.

Toussaint. (1880) *C. R. Acad. Sci.* **91**, 135.

Tresselt, H. B. and Boor, A. K. (1955) *J. infect. Dis.* **97**, 207.

Trnka, Z. *et al.* (1958) *Schweiz. Z. allg. Path.* **21**, 1083.

Vincent, C. (1922) *J. infect. Dis.* **31**, 499.

Wagner, K. E. (1890) *Ann. Inst. Pasteur* **4**, 570.

Ward, M. K., McGann, V. G., Hogge, A. L., Huff, M. L., Kanode, R. G. and Roberts, E. O. (1965) *J. infect. Dis.* **115**, 59.

Weidlich, N. (1934) *Wien. tierärztl. Mschr.* **21**, 289.

Weyl, T. (1892) *Z. Hyg. InfektKr.* **11**, 381.

Whitford, H. W. (1979) In: *CRC Handbook Series in Zoonoses*, Vol. II, p. 31. Ed. by J. H. Steele. C.R.C. Press, Inc., Boca Raton, Florida.

Widdicombe, J. G., Hughes, R. and May, A. J. (1956) *Brit. J. exp. Path.* **37**, 343.

Wilson, G. S. (1979) *J. Hyg., Camb.* **82**, 346; **83**, 173.

Wolff, A. H. and Heimann, H. (1951) *Amer. J. Hyg.* **53**, 80.

Wright, G. G., Green, T. W. and Kanode, R. G. (1954) *J. Immunol.* **73**, 387.

55

Plague and other yersinial diseases, pasteurella infections, and tularaemia

Geoffrey Smith and Graham Wilson

Yersinial infections

Y. pestis infection: Plague

Plague has been one of the greatest scourges of the human race. From time to time it has swept over the world in relentless waves, exacting a toll of life probably unequalled by that of any other epidemic disease. In the 5th and 6th chapters of the first Book of Samuel there is an account of a pestilence among the Philistines that some have interpreted as bubonic plague (MacArthur 1952) and others as bacillary dysentery (Shrewsbury 1964). Since the outbreak lasted for at least seven weeks, it is possible that the high mortality in the early stages was due to plague and in the later stages to bacillary dysentery. The description in the 5th chapter of the 1st Book of Kings (Vulgate version) mentions that the illness was characterized by anal discomfort for which the sufferers made themselves seats of skins. (The Hebrew word translated as 'emerods' probably meant swellings, not specifically haemorrhoids.)

Before the Christian era numerous epidemics are on record; during the 1500 years following the birth of Christ there are records of 109 epidemics, including the great plague of Justinian's reign, and the Black Death of the fourteenth century, in which one-third of the population in England and on the Continent died (Ziegler 1969).

Between 1500 and 1720 there were 45 pandemics of plague (Editorial 1925a). During the eighteenth and nineteenth centuries it was comparatively quiescent, though there were big outbreaks at Marseilles in 1720, at Moscow in 1770, and in the Balkan peninsula in 1808 and 1815, and in Egypt between 1783 and 1844. At the close of the last century it sprang once more into activity. Starting in Hong Kong in 1894, it invaded India, Japan, Asiatic Turkey, and European Russia in 1896; the following year it reached Madagascar and Mauritius. In 1899 Arabia, Persia, the Straits Settlements, Austria, Portugal, British South Africa, Egypt, the French Ivory Coast, Portuguese Africa, the Argentine, Brazil, Paraguay, and the Hawaiian Islands were affected. There was a small outbreak at Glasgow in 1900, and in the same year the disease appeared in Sydney in Australia and at San Francisco, California. Undoubtedly the brunt of the disease was borne by India, where in the 20 years 1898–1918 more than 10 million deaths were recorded (White 1918–19). (For a fuller description of plague, see Spink 1978, and for a history of plague in the British Isles, see Shrewsbury 1970 and Morris 1971)

In 1959 less than 300 cases of plague were registered in the whole world, and the disease seemed to be dying out. Part of this spectacular decrease was attributed to natural causes, and part to intensive measures taken against rodents and insects. But during the war in Vietnam the Americans found plague to be endemic in the country, though often unrecognized and unreported. In each of the years 1965, 1966, and 1967 about 5000 cases came to notice (Cavanaugh *et al.* 1968); and the conditions in the country were such as to afford little hope of eliminating the disease (Marshall *et al.* 1967).

One of the most disturbing features in the world situation is the widespread infection of rodents in the Americas, South Africa, and Kenya, constituting what is known as rural, sylvatic, or wild rodent plague in distinction to bubonic plague, which is spread by rats. According to Pollitzer (1954, 1960) nearly 200 species and subspecies of rodents are known to be concerned in the spread of *Yersinia pestis*. For a general review of the history of plague and of its present prevalence the reader is referred to papers and monographs by Hoekenga (1947), Kaul (1949), Link (1954), Pollitzer (1954, 1960), Report (1959, 1960, 1965, 1970), and Knapp (1969).

The causative organism was isolated independently and almost simultaneously by Kitasato and Yersin in 1894 in Hong-Kong (see Chapter 38). Numerous workers, particularly Ogata (1897), Simond (1898), and Gauthier and Raybaud (1902, 1903) were responsible for showing that the disease was primarily one of rodents, and that it was spread to man by the agency of infected fleas. This conception was criticized by several workers (Nuttall 1898, Galli-Valerio 1900, 1903), but was ultimately proved to be correct by the English Plague Commission (Report 1906).

Epidemiology

Bubonic plague

The epidemiology of plague is extremely complex. Infection depends for its maintenance on a great variety of rodents and insects which differ from country to country and from time to time. Ecological studies point to a multiplicity of factors concerned in the fluctuating balance that exists between rodents of greater and lesser degrees of susceptibility to the plague bacillus and in the degree of risk to which man is exposed. Geographical, meteorological, and climatic factors, the conditions of civilization, such as the type of building, the amount of overcrowding, the form of sewerage, the activity of shipping and other forms of transport, the mode of clothing, and the degree of sanitation—these and other factors all have an indirect influence on the qualitative and quantitative distribution of the rodents and insects that act as potential reservoirs and carriers of *Y. pestis*. It would be quite impossible within the compass of this chapter to describe the epidemiology of the disease in each of the countries where it exists. For this, reference must be made to the papers and monographs that have just been quoted. We propose instead to give a few representative but necessarily simplified descriptions, sufficient, we hope, to afford a broad outline of the picture as a whole. So much of the fundamental work on plague was carried out in Bombay that we shall pay particular attention to this, not because it represents the present situation but because it established the importance of the flea in the transmission of infection and because it must always remain as one of the classical examples of experimental epidemiological technique.

Where infection is endemic in sylvatic rodents, it may remain smouldering for a long time before circumstances favour its epidemic spread. During this time human beings in country districts may occasionally contract the disease through handling an infected animal, but generally speaking the disease does not become epidemic in man unless infection spreads to what is sometimes termed a liaison rodent, namely a species of rat that comes into contact with man. An epidemic starting in this way runs a characteristic course. As Daniel Defoe noted in his *Journal of the Plague Year in London, 1665*, there is a pre-epidemic phase in which a few cases occur separated by considerable intervals of time. This is succeeded by the epidemic phase in which a rapid succession of cases occurs, constituting the major portion of the outbreak. Finally there is a decline resembling the commencement, but not so long drawn out. Infection does not occur by direct contact with the sick; it spreads by contiguity from place to place, so that an increasing number of small foci are established. Often the infection seems to be localized to buildings, particularly grain stores, warehouses, or shops; in these, many cases of plague are met with in persons who have no immediate relation to each other. Plague has a predilection for dirty, insanitary dwellings, and more particularly for their ground floors; it attacks the lowest class of the population living under the worst conditions.

Infection may also occur *per saltum*. A focus of plague suddenly appears several miles—sometimes hundreds of miles—from the primary focus, and serves in its turn as the centre from which infection may spread to the surrounding localities. The contiguous mode of spread is dependent on the gradual dissemination of plague amongst the rats; *per saltum* infection results either from the transportation of rats or rat fleas by railways, ships, or other means of communication, or by human beings infested with rat fleas. It is probably owing to importation of plague from cities that villages are infected.

Though bubonic plague has occurred in nearly every part of the globe, it is confined chiefly to the warmer latitudes (Robertson 1923). Extreme heat and dryness of the atmoshere are inimical to its spread; thus in India it occurs during the colder months of the year, when the mean temperature is between 10° and 29.4°, and the air has a high relative humidity, or, as it is sometimes expressed, a low saturation deficiency.

The incubation period of bubonic plague is usually 2–5 days, and the case fatality of the untreated disease 60–90 per cent.

In Bombay the English Plague Commission (Report 1907, p. 724) determined that before the annual outbreak of human plague there was an epizoötic among the rats. First the rats belonging to the species *Rattus norvegicus* (*Mus decumanus*) were affected. After an interval of about 10 days an epizoötic appeared in the rats of the species *Rattus rattus* (*Mus rattus*); and after a further interval of 10 days the human epidemic broke out. From this they inferred that *R. rattus* was infected from *R. norvegicus*, and subsequently conveyed the disease to man. It was further observed that plague persisted in *R. norvegicus* during the off-season—in Bombay from June to December—and flared up at the onset of the colder weather (Fig. 55.1).

The transference of plague from rat to rat and from rat to man occurs almost exclusively by fleas. Chief among these are *Xenopsylla cheopis* and *Nosopsyllus fasciatus*—the rat fleas. In the absence of their specific hosts, both of these types of flea will bite human beings. Towards the end of an attack the bacilli are present in the blood in enormous numbers, and are readily imbibed by the fleas that infest the rat. When the animal dies, the fleas leave the corpse and wait for a suitable opportunity to attach themselves to a fresh host. Meanwhile the bacilli multiply in the proventriculus, often to such an extent as to block it completely, and prevent access of food to the stomach. A flea in this condition is hungry, and when it succeeds in finding a new host, attacks it with vigour. The act of sucking, however, only distends the already contaminated oesophagus, and on the cessation of the pumping act some of the blood is forced back into the

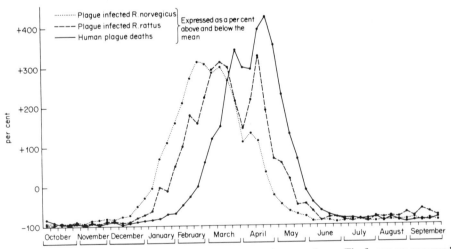

Fig. 55.1   Chart illustrating the sequence of epidemic plague in rats and man. The figures are expressed as percentages above and below the mean. (After the English Plague Commission in India.)

wound (Bacot and Martin, see Report 1914). Sometimes a temporary passage is cleared through the mass of obstructing bacilli; this fails, however, to restore the lost valvular function to the proventriculus; it merely leaves a passage through which the blood can flow out of the stomach as freely as it enters. Hence after a full meal, blood extends from the posterior portion of the stomach to the anterior chamber of the pharyngeal pump. Such a flea is probably more dangerous than one whose proventriculus is completely blocked, since the contents of the stomach can be regurgitated into the wound with greater freedom (Report 1915*b*, Burroughs 1947).

The length of time that a flea can remain *infected* depends on several factors, chief of which are temperature and humidity. In India fleas were found to harbour plague bacilli up to 47 days (Report 1915*a*), in Madagascar for several weeks (Girard 1936), and in the United States for as long as 130 days (Eskey and Haas 1940). Survival of the bacilli in the flea is favoured by a low temperature—about 10°—and a nearly saturated atmosphere; adverse factors are a temperature over 26.7°, or an even lower temperature with a dry atmosphere (Liston 1924). But the time that fleas can remain *infective* is much shorter; most of them die in a day or two, and the limit of infectivity is probably about 7–14 days.

The bacilli in the flea's stomach may be attacked by the phagocytes in the rat's blood which has been ingested; when the temperature rises the phagocytes become more active. This may explain why a flea clears itself more quickly at a high than at a low temperature, and also why, if it is fed on healthy blood after being infected, it clears itself more quickly than when it is starved. The clearance is said to be still more rapid when the flea is fed on the blood of immunized animals (Report 1908, p. 260). According to Cavanaugh and Randall (1959), bacilli in the flea's stomach that are sensitive to phagocytosis may be transformed in the monocytes of normal animals into organisms resistant to phagocytosis by polymorphonuclear leucocytes, and thus prove virulent to the animal host.

Plague in rats

In India the two species mentioned above are chiefly responsible for the spread of plague. *Rattus norvegicus* is the large grey rat; it lives in sewers, stables, and garbage. *Rattus rattus* is the black rat; it is smaller and less fierce than the grey rat, and, living in houses or their immediate neighbourhood, it comes into closer contact with the human population. The natural mode of infection of rats appears to be almost entirely by fleas. It is possible that animals acquire the disease by devouring their dead companions; but considering that it is not easy to infect rats by feeding them on plague material, and that the post-mortem appearances in rats dying naturally of plague differ from those of rats experimentally infected by feeding, it is doubtful whether this is more than a rare occurrence.

During the height of the epizoötic the lesions are those of acute plague, but during and subsequent to the decline, a number of healthy rats are encountered with atypical lesions. One of the commonest of these is a large abscess in the spleen or liver, containing plague bacilli. Swellengrebel and Hoesen (1915) in Java, and Bordas, Dubief and Tanon (1922) in France found a number of rats with atypical lesions containing plague bacilli of lowered virulence; on passage through one or two guinea-pigs they became fully virulent again. Williams and Kemmerer (1923) in New Orleans confirmed the existence of *latent plague* in rats. It has also been found among other rodents in Russia and the United States. Whether these animals act as chronic carriers of the disease and are responsible for keeping alive the infection from one epizoötic to the next is still an open question. Most workers are disposed to attribute more weight in this respect to the survival of plague bacilli in the flea than in its host (Girard 1936, Eskey and Haas 1940).

Experimental transmission of plague in rats   It has been abundantly demonstrated by different workers that plague does not spread from rat to rat by contact in the absence of fleas. On the other hand infection spreads readily in the presence of fleas, even though the animals have no immediate contact with each other. Two experiments of the English Plague Commission will illustrate this.

EXPERIMENT 1. (Report 1910.) Bombay rats were inoculated with plague and mixed with normal rats in godowns. In one set of godowns fleas were excluded, but in the other set they were permitted to multiply. In the flea-free godowns none of the normal rats, which were in continuous contact with the infected rats, contracted plague, whereas in the others infection spread rapidly (Table 55.1).

Table 55.1   Effect of fleas on the mortality of rats in experimental epidemics

	Flea-free godowns	Flea godowns
Inoculated rats put in	84	70
„ „ died of plague	73 or 86·9%	63 or 90%
Normal rats put in	125	125
„ „ died of plague	0	57 or 45·6%
Plague rats eaten	3	1
Fleas present	None	About 24 per rat

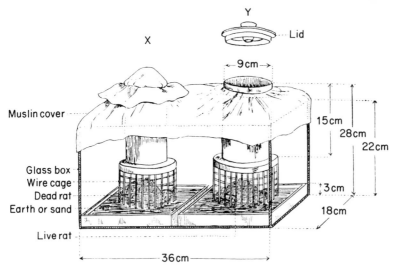

Fig. 55.2 The wire mesh employed was 3 mm; in the figure it is shown much larger, for the sake of clearness.

The rats dying of plague in the flea godowns nearly all showed cervical buboes, indicating that the site of infection had been on the head or neck; this is the situation on which the fleas chiefly congregate.

EXPERIMENT 2. (Report 1906, p. 421.) This was a repetition of an experiment performed by Gauthier and Raybaud (1902, 1903). Two wire-mesh cages were employed, each standing in a shallow tray filled with dry earth or sand. The cages were enclosed in a single glass box, the roof of which was covered with fine muslin impervious to fleas (see Fig. 55.2). Rats were introduced through the lids of the cages, and the lids themselves were covered with fine muslin. Into Cage X was placed a rat, inoculated with plague, together with 10 to 20 fleas. As soon as the rat was dead, a healthy rat was placed in Cage Y. The carcass of the first rat was allowed to remain in X for 8 to 12 hours after the animal's death. During that time fleas transferred themselves to the new rat through the wire mesh of the cages. Immediate contact between the two rats was absolutely prevented. This experiment was repeated several times, and in 69 per cent of instances in which healthy, presumably non-immune, rats were used, the rat in Cage Y contracted plague.

Experiments were also carried out on guinea-pigs (Report 1906, p. 421). A series of 6 godowns was specially constructed by the Commission. The walls were of brick and mortar, nine inches thick; the floors were of concrete on top of a high plinth. Each godown measured internally seven foot by six foot. Leading into the interior were double doors, lined with wire netting, between which was an inspection chamber. The essential difference in the structure of the godowns was in the roofs. Nos. 1 and 2 were roofed with country tiles, placed in four layers on the top of wooden laths.

On the inside of this roof there was a wire netting on a wooden framework in Godown 2, and two layers of wire netting, 10 inches apart, in No. 1, so that while rats could build their nests in the tiles of the roof, they were completely shut off from the interior of the godowns. Mangalore tiles were used for roofing Godowns 3 and 4; these do not afford so good a shelter for rats as country tiles; a single layer of wire netting separated the tiles from the interior. Godowns 5 and 6 were roofed with a single layer of corrugated iron fastened down with cement to the tops of the walls, so that no rats could nest at all. Godowns 1 and 3 had a small roof light; the rest were in darkness (Fig. 55.3).

By allowing three guinea-pigs to run about in each godown for 6 days, and making flea counts every day, it was ascertained that the number of fleas varied in each instance with the accessibility of the roof to rats. Thus in No. 1, 54 fleas; in No. 2, 228; in No. 3, 40; and in No. 4, 70 fleas were caught in 6 days. In Nos. 5 and

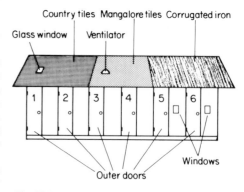

Fig. 55.3

6 fleas were few or absent. Incidentally this experiment brings out the importance of darkness in favouring infestation by fleas.

In these godowns a series of experimental epidemics was initiated among guinea-pigs, generally by introducing a number of animals experimentally infected with plague and adding normal animals to them. For the detailed results, reference must be made to the Commission's Report (1906); the conclusions that were drawn from them were as follows:

(1) Close contact of plague-infected with healthy animals, if fleas were excluded, did not lead to the production of an epizoötic. Since the godowns were never cleaned out, close contact meant contact with faeces, urine, and discharges of infected animals, and the eating of contaminated food.

(2) Close contact of the young with the infected animals failed to infect the former; even the suckling of young animals by infected mothers proved ineffectual.

(3) When fleas were present, an epizoötic broke out, the rate of progress being in direct proportion to the number of fleas.

(4) Infection can take place without any contact with contaminated soil. Thus, in one experiment, guinea-pigs placed in wire cages 2 inches above the ground contracted plague. But when the cages were suspended 2 feet above the ground, none of them became ill, indicating that aerial infection does not occur.

Experiments were made in the plague houses of Bombay, in which guinea-pigs were allowed to run about to act as traps for fleas; the usual number of fleas—mostly rat fleas—caught in this way was about

20 per room. In 29 per cent of instances guinea-pigs left in these rooms for from 18 to 40 hours contracted plague. The situation of the bubo was usually in the cervical region—explicable by the habit that fleas have of collecting round the under-surface of the neck.

In other experiments guinea-pigs were placed in cages, of the pattern shown in Fig. 55.4. One of the cages was surrounded by a 6-inch boundary of tangle-foot; the other by a layer of sand. When left in plague-infected houses, 24 per cent of the animals in the unprotected cages contracted plague whereas not one of the animals in the protected cages did so.

Mode of spread of plague in man

The conclusions drawn from these animal experiments received support from observations in the field. During the early years of the century plague was very prevalent in Bombay. It affected particularly the lower classes; the Parsees and Europeans suffered only slightly. The bacilli gained entrance to the body usually through the skin of the leg, as manifested by the occasional presence of a small vesicle (Simond 1898) resulting from the bite of a flea, and by the development of an inguinal bubo. During the hot season rats, *R. norvegicus* and *R. rattus*, bred freely, producing a large susceptible population. As the cold weather arrived, fleas increased in numbers, and plague broke out in *R. norvegicus*. This was followed by plague in *R. rattus*, and this in turn by plague in man. At the onset of the hot weather in May or June the flea population decreased rapidly, and bubonic plague in rats and man came to an end for the year. The conclusions reached by the English Commission (Re-

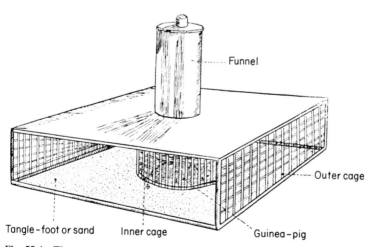

Fig. 55.4 The outer large-mesh wire netting, which has been omitted from one side of the outer cage, was to prevent the tangle-foot from being destroyed, and to prevent rats from coming into close contact with the guinea-pig. The tangle-foot prevented the access of fleas to the inner cage; the sand did not.

port 1908, p. 266) on the seasonal prevalence of plague in Bombay and the Punjab were that the rise of the rat epizoötic, and therefore of the human epidemic, depended on a suitable mean temperature, between 10° and 29.4°, and on a sufficient number of susceptible rats, and rat fleas.

Of more recent years *R. norvegicus* and *R. rattus* have decreased in Bombay and Calcutta, while the number of *Bandicota bengalensis* has increased. This rodent is less heavily infested with *X. cheopis* than are brown and black rats, and correspondingly less dangerous for man (Seal 1960).

Much the same conditions prevailed in the Punjab and the United Provinces as in the Bombay Presidency, but in Madras plague was far less frequent. This was mainly because the predominating flea was *X. astia*, which, as Liston (1924) and Hirst (1926, 1927) showed, is less effective in carrying plague from rat to rat than *X. cheopis*.

The epidemiology of plague varies in different countries, such as Egypt, Java (Williams *et al.* 1980) and Brazil, according to the temperature, rainfall, type of rodent, and species of flea (see 6th ed., p. 2128). The conditions determining its spread are very stringent. The links in the rat–flea–man chain must be delicately adjusted; a slight fault in one link is sufficient to impair the efficiency of the whole chain, or even to break it altogether. Several times during the present century plague has been introduced into England, chiefly at the ports, yet it has uniformly failed to take hold.

In East Suffolk, it is true, it was endemic in rats, ferrets, rabbits, and hares, having been imported probably in grain ships. Between 1906 and 1918 no fewer than 22 human cases occurred, but after that it died out (see Bulstrode 1910–11, Eastwood and Griffith 1914, Macalister and Brooks 1914, van Zwanenberg 1970). (For a history of bubonic plague in the British Isles see Shrewsbury 1970, together with the criticism of his work by Morris 1971.)

Pneumonic plague

When plague is well under way in rats, it may be kept going by aerial dissemination resulting in the pneumonic form of the disease.

The epidemiology of pneumonic plague is different from that of the bubonic form, one of the chief differences being that the former is highly contagious, whereas the latter is not.

Provided the conditions are favourable, primary pneumonic plague may spread with great rapidity. In 1910–11 the outbreak in Manchuria and North China levied a toll of 60 000 lives; in 1917–18 there were 16 000 deaths in South Mongolia and China, and in 1920–21 there were 9000 deaths in Transbaikalia (Wu Lien Teh 1922). The epidemic is independent of rodents; it spreads directly by droplet infection from one person to another. The closer the contact, the more

likely is infection to spread; this accounts for the prevalence of pneumonic plague in the winter, when overcrowding in insanitary dwellings is so common. Strong and his colleagues (1912), who studied the Manchurian epidemic of 1910–11, were impressed with the part played by atmospheric humidity. At a time when the temperature was often below freezing, the air was practically saturated with moisture. Under these conditions the infected droplets persisted considerably longer in the air than they would have done in a dry atmosphere. In India, where the temperature is higher and the saturation deficiency greater, primary pneumonic plague seldom becomes epidemic. This form of the disease is almost invariably fatal (Jettmar 1923). The incubation period is 1–5 days.

In some countries, such as Manchuria, Transbaikalia, and the Kirghiz Steppes, bubonic plague occurs during the warm weather, and pneumonic plague during the winter. The outbreaks of pneumonic plague probably arise from cases of bubonic plague complicated by secondary pneumonia.

Sylvatic or wild plague

Like many other diseases—the leptospiroses, tick-borne relapsing fever, Rocky Mountain spotted fever, tularaemia, cutaneous leishmaniasis, the viral encephalitides—plague persists and flourishes in various endemic foci of the world. It does this by circulating among groups of mammals and blood-sucking insects which together form what the Russians term a *biocenosis* (bios = life, koinos = common) (see Chapter 48). Into such a community man sometimes strays, either as a hunter, prospector, surveyor, mining engineer, or tourist, and contracts the disease; hence the occurrence of sporadic cases in widely separated parts of the country and at particular seasons of the year. The study of these endemic foci of zoönotic disease has been pursued particularly by Pavlovsky (1963) and his school (see Petrischeva 1966) in the USSR, but their existence in other parts of the world has been attested by a wide variety of workers. The following examples will illustrate something of the diversity of host-parasite relationships that are met with in different terrains; for a fuller account see Pollitzer (1954).

Ground squirrels and other American rodents Plague first appeared at San Francisco in 1900. By 1904 it had spread from the rats to the ground squirrels—*Citellus beecheyi*—and had begun to pass eastward from the coast (McCoy 1910). In 1934 it was found that plague had crossed not only the central valleys of California, but also the Sierra Nevada mountains into the Great Basin section of Oregon. A new species of ground squirrel, *Citellus oregonus*, was shown to be infected. As a result of special field surveys it was learnt that by 1940 no fewer than ten states had been invaded. Though plague had not yet spread east of the 100th meridian, the country to the west of this is one of the largest areas of sylvatic plague in the world. Between 1908 and 1966 the disease from this source caused 115 cases and 65 deaths (Caten and Kartman 1968). The rodents forming the main primary reservoirs of infection are ground squirrels, wood rats, and white-tailed prairie dogs; chipmunks, mantled ground squirrels, marmots, voles and mouse-like hamsters (*Peromyscus maniculatus*) provide a more limited reservoir (Kalabukhov 1965). Numerous species of fleas, lice and ticks are concerned in the transfer of infection among these rodents (Eskey and Haas 1940, Report 1965). In *Canada*, plague infection is confined to ground squirrels and has so far been found only in Alberta and Saskatchewan (Humphreys and Campbell 1947). In *South America* five countries—Bolivia, Brazil, Ecuador, Peru, and Venezuela—are infected. The commonest species of rodent carriers are the guinea-pig (*Microcavia* spp.), the tree rat (*Graomys griseo-flavus*), the viscacha (*Lagostomus maximus*), and the hare (*Lepus europeus*). (For a history of plague in the Americas, see Report 1965.)

Gerbilles and multimammate mice In South Africa plague was introduced in 1900–02 by rats in forage vessels coming from South American ports. It is kept going chiefly by gerbilles—the common gerbille (*Tatera brantsi* and *Tatera schinzi*) and the Kalihari or namaqua gerbille (*Desmodillus auricularis*). Large epizoötics occur periodically among these animals, killing off 80–90 per cent, together with a smaller proportion of the associated small carnivora—mongoose and suricate. The rodents are not completely exterminated by the disease; the survivors breed freely, so that in time another large susceptible population grows up which is likewise decimated by plague. During the 25 years 1919–43 more than 900 outbreaks of human plague were reported, mostly on farms. Human plague follows the spread of plague in wild rodents, but irregularly and often at a considerable time interval. Man is infected chiefly from house rats (*Rattus rattus*) and from the semi-domestic multimammate mice (*Mastomys coucha*), which serve as intermediaries for the wild rodents (see Mitchell 1921, Editorial 1925*b*, Kauntze 1926, Davis 1948). Gerbilles (*Tatera indica*) are also responsible in the Uttar Pradesh (the former United Provinces of India) for maintaining endemic plague in the rural areas (Baltazard and Bahmanyar 1960).

Spermophiles and field mice Endemic plague in the Steppes of South-West Russia has been traced to spermophiles—*Spermophilus musicus* and *Sp. rufescens*—and to field mice. In winter when the spermophiles are hibernating, cases of human plague appear to be due to infection from wandering field mice, which take refuge in the thatched roofs of the houses (Zabolotny 1923). Many cases in winter, however, are of the pneumonic type (Radcliffe 1879–80). In certain provinces, such as Saratow, Astrakhan, and the Don

province, marmots are also infected (Tschurilina 1930). The suslik or little sisel (*Citellus pygmaeus*) is responsible for endemic plague in the lowlands to the north of the Caspian Sea (Fenyuk 1960).

Marmots Numerous outbreaks of plague have occurred in Transbaikalia, Mongolia, and Manchuria. The origin of these has been traced to small rodents. These are usually referred to as tarbagans (*Arctomys siberica*) but more accurately as bobac marmots (*Arctomys bobac*), though it seems possible that both species of rodent are concerned. The bobac marmot is a small rodent living in deep burrows. From September to April it hibernates, and the breeding season is in July and August. During April, May, June and September its skin is largely sought after by Russian and Chinese hunters for the manufacture of imitation sable. The marmot harbours fleas—*Ceratophyllus silantievi* Wagner—which will bite man when starved (Wu Lien Teh 1913). In its burrows there are certain passages used for the deposition of excrement, and the corpses of animals that have died during hibernation, as well as skin, bones, and other refuse. These passages are in absolute darkness, and are cool, moist, and of equable temperature; it is probable that they afford suitable conditions for conservation of the plague bacillus during the winter (Jettmar 1923). During the summer, sporadic cases of bubonic plague occur amongst the marmot hunters. In the winter pneumonic plague occurs in epidemic form.

Merions In Kurdistan, South Turkey, North Iraq, and North Syria down to the Euphrates infection is maintained by merions Baltazard and Seydian 1960 Baltazard *et al.* 1960). Two of these—*Meriones persicus* and *M. libycus*—are resistant and two—*M. vinogradovi* and *M. tristrami*—are very susceptible to plague. There are no rats. Human infection is picked up by a worker in the field, conveyed back to the village, and spread from man to man by the human flea, *Pulex irritans*. As this is a poor vector, the epidemic is soon over. The merion (*Meriones meridianus* Pall) is also responsible for infection in the Volga-Ural sands (Kalabukhov 1965). (For further information on plague in animals, see Poland and Barnes 1979).

Geographical distribution of varieties of plague bacillus

In Chapter 38 we have already described the biochemical varieties of *Y. pestis*. These varieties, though of no clinical or diagnostic importance, have an epidemiological interest, in that they tend to be distributed according to a geographical pattern (Devignat 1953). The common rats—*R. norvegicus* and *R. rattus*—carry the *orientalis* variety, as do also many field rodents. Table 55.2 shows in a simplified form the geographical distribution of the different varieties of plague bacillus and some of the commoner rodents that they infect.

Table 55.2 Geographical distribution of field rodents carrying plague and the varieties of *Y. pestis* concerned (partly after Pollitzer 1960)

Rodents	Variety of *Y. pestis*	Locality
Bandicoot (*Bandicota*)	*orientalis*	India, Sri Lanka
Cavy (*Cavia*)	*orientalis*	Argentine, Brazil, Peru
Gerbille (*Tatera*)	mixed	S. Africa, Uttar Pradesh
Ground squirrel (*Citellus*)	*orientalis*	USA
„ „ (*Spermophilus*)	*antigua*	S.E. Russia and Central Asia
Hamster (*Cricetus*)	*antigua*	S.E. Russia
Hare (*Lepus*)	*orientalis*	Argentine
Jack rabbit (*Lepus*)	*orientalis*	California
Marmot (*Arctomys*)	*antigua*	Manchuria, Mongolia, Transbaikalia, Central Asia
Merion (*Meriones*)	*mediavalis*	Kurdistan, Turkey, Iraq
Multimammate mouse (*Mastomys*)	mixed	S. Africa, Kenya, Congo
Tree rat (*Graomys*)	*orientalis*	Argentine
Viscacha (*Lagostomus*)	*orientalis*	Argentine
Wood rat (*Neotoma*)	*orientalis*	California, Utah, Nevada, Arizona
Swamp rat (*Otomys*)	mixed	East, Central, and S. Africa
Grass or Nile rat (*Arvicanthis*)	mixed	East Africa and Belgian Congo
Vole (*Microtus*)	*orientalis*	USA

Diagnosis of plague in rodents

Rats caught in traps—preferably of the cage pattern—should be put into suitable air-tight containers and killed with ether, chloroform, chloropicrine, or ordinary petrol before being examined. The fleas, which are anaesthetized by this process, should be shaken or combed off the animals and collected in vials containing normal saline. When rats cannot be examined locally, the spleen, liver, and any enlarged lymph nodes should be removed and tranferred to Broquet's (1908) fluid (20 per cent glycerolated water) or to the medium recommended by Cary and Blair (1964), before being dispatched to the nearest laboratory. Rodents concerned with sylvatic plague must be obtained by the most appropriate means—collection of corpses, trapping, shooting, or removal from their burrows by the use of long rods, flooding, or digging. Fleas from infested premises may be trapped on decoy animals, caught on tangle-foot, or swept up in the floor dust.

During an epizoötic the diagnosis of plague in rats can be made almost as well macroscopically as by microscopical examination. In *Rattus rattus* and *Rattus norvegicus* the presence of a bubo is the most useful single indication of plague. In the early stage the gland is enlarged, congested, and shows haemorrhagic points on section. When fully developed, it contains an area of grey necrosis confined to the medulla, or occupying the whole gland. The bubo is hard, and can be moved about under the skin with ease. Microscopically, plague bacilli are found in about 99 per cent of buboes; in over half the buboes involution

forms are present. The bacilli are best demonstrated by staining with carbol-thionin; with this dye they are faintly coloured except at the poles, which stain deeply.

The liver frequently shows a patchy distribution of alternate red and yellow areas—best described as mottling. In many instances there are small, grey or whitish areas of necrosis, giving the organ a stippled appearance as if it had been dusted with grey pepper; as a rule they do not project above the surface. The spleen is sometimes enlarged, and is firm and moulded over the stomach; sometimes it contains granules or actual nodules, which may be discrete or confluent. Occasionally a wedge-shaped portion of the spleen is converted into a cheese-like mass. Pleural effusion is a very characteristic feature; it is clear, abundant, and straw-coloured, less often blood-stained. The five most important signs of plague are: (1) bubo; (2) subcutaneous and general congestion; (3) granular liver; (4) congested spleen; (5) pleural effusion (Report 1907). Of these 1, 3, and 5 may persist in rats that have undergone putrefaction, and are then of great diagnostic value.

Microscopically, bacilli are most abundant in the bubo, then in the spleen, and least numerous in the heart's blood (Fig. 55.5). Films should be fixed in alcohol rather than by heat, and attention paid to the polar-stained barrel forms in fresh material, and to the characteristic rings and discs in putrefying material (Kister 1930). The fluorescent antibody technique may be of use in identifying plague bacilli in impression films of the tissues (Moody and Winter 1959). Cultures should be made from these three situations, and single

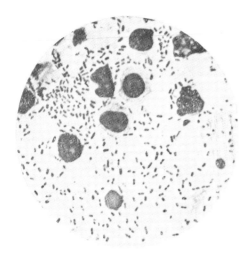

Fig. 55.5 *Yersinia pestis*. Film from inguinal bubo of infected guinea-pig, showing the bipolar-stained ovoid bacilli (× 1000).

colonies picked off and carefully studied. In putrefying animals the leg may be cut above the knee, and the bone marrow aspirated with a syringe, suspended in a little Ringer solution, and plated on to selective media such as sodium sulphite gentian violet agar or copper sulphate agar (see Report 1956), or the medium containing potassium tellurite, novobiocin, erythromycin, and actidione recommended by Morris (1958).

Inoculation of guinea-pigs or mice should be made subcutaneously with a suspension of the bubo or spleen. If the animal has undergone putrefaction, and other organisms are likely to interfere with the inoculation test, pieces of the tissue or of bone marrow should be rubbed on to the shaven abdomen of a guinea-pig. *Y. pestis* is able to penetrate the apparently unbroken skin, whereas very few other organisms can do so. The Commission found that this test failed in only 2 per cent of fresh and 10 per cent of putrid rats.

In putrid animals the *thermoprecipitin test* is sometimes of value (Kraus 1897, Piras 1913, Warner 1914, Philip and Hirst 1917). One part of the finely divided organ is mixed with 5 to 10 parts of distilled water, boiled for 5 minutes, filtered repeatedly through paper or asbestos wool under pressure till clear, and then run on to the top of a specific serum in a small tube. In a positive reaction a precipitate appears in 5 minutes at 37°, increasing to a maximum after 2 hours. Alternatively the organisms may be destroyed by treatment overnight with ether and the water-soluble antigens extracted with saline. In a plague antiserum group precipitins for *Yersinia pseudotuberculosis* and *Francisella tularensis* should be absorbed out previously (Larson *et al.* 1951).

Several diseases may be confused with plague, particularly pasteurellosis and pseudotuberculosis. Buboes are not infrequently caused by infection with strepto-

cocci, staphylococci, coliform, or other pyogenic organisms; and according to Macalister and Brooks (1914) the lesions of plague in rats may be simulated by infection with *Trypanosoma lewisi*. The general appearances of plague in other rodents resemble those in rats. In ground squirrels the distinction between plague and tularaemia must be made by cultivation.

In chronic or resolving plague the chief lesions found are buboes, and abnormalities in the spleen — necrotic areas, abscesses, perisplenitis, scars bisecting or trisecting the organ, and adhesions to the neighbouring tissues.

In survey work the animals should be examined in batches of ten, the glands and spleens being pooled, ground up, and injected into guinea-pigs. Should the plague bacilli present be of low virulence, the animals may not die, and passage through further animals may be necessary. In the search for plague infection in wild rodents it is said to be more satisfactory to collect fleas, or other insects from the animals than to examine the animals themselves (Eskey and Haas 1940). The fleas are pooled in batches of 25 or 50, washed, ground up, suspended in quarter-strength Ringer solution, and injected subcutaneously into white mice or guinea-pigs. However, during the plague season both animals and fleas should be examined, so as to determine the intensity of the epizoötic. (For a more detailed description of the diagnosis of plague see Report 1956.)

In regions where plague infection is widespread, the passive haemagglutination test on rodent sera affords much the most rapid and inexpensive method of detecting plague foci (Report 1970).

Diagnosis of plague in man

During life

A small vesicle is sometimes present on the leg in the early stages of the disease corresponding to the flea-bite. Microscopical examination of the fluid will show the presence of plague bacilli. The bubo, which is generally in the inguinal region, should be aspirated, and the fluid examined microscopically, by culture, and by animal inoculation (Fig. 55.5).

Plague bacilli are frequently present in the patient's blood, where they may be demonstrated microscopically or, preferably, by culture. In the early stages of the disease 34 to 60 per cent of blood cultures have been reported positive by different observers; but Bonebakker (1936), making use of Kirschner's (1934) observation that pure bile constitutes an excellent medium for the growth of plague bacilli in blood, was able to demonstrate their presence in 85 per cent. The degree of bacteraemia is said to fluctuate considerably during the course of the disease (Report 1908, p. 221).

According to Cavanaugh and his colleagues (1968), virulent plague bacilli are sometimes present in the throat of bubonic patients and of symptomless contacts. They are said to be found in the urine of 19–30 per cent of patients, but not in the faeces (Ogata 1897, Report 1908, p. 221). They may be isolated rapidly by making use of selective media, followed by microscopical observation of phage lysis of the developing colonies (Chadwick 1963). The differences between the plague bacillus and the pseudotubercle bacillus, with which it may be confused, are described in Chapter 38.

After death

Microscopical, cultural, and animal inoculation tests should be relied on for diagnosis, the best material being the bubo or the spleen. If the cadaver is comparatively fresh, the tissue suspension should be injected subcutaneously or intraperitoneally into a guinea-pig; the latter method is more reliable when the organisms are few. According to Girard (1952) microscopical examination of material aspirated from the buboes reveals 88 per cent of positive cases; guinea-pig inoculation is required for the remainder. In putrid cadavers the organ should be rubbed on to the shaven abdomen of a guinea-pig or on to its nasal mucosa (Zlatogoroff 1904, Girard 1934), or inoculated into a white rat by pricking the root of the tail with a needle. The thermoprecipitin test may likewise be used.

An alternative method that takes longer than this or than the fluorescent antibody method of Moody and Winter (1959), but that should be useful for confirmatory purposes and for epidemiological surveys, is described by Albizo and Surgalla (1968). Blood agar plates, with or without selective agents and containing antibody to fraction 1, are inoculated from the tissues, incubated for 66 hours, exposed for 1 minute to chloroform vapour to release fraction 1 from the bacilli, and examined for precipitin rings around the colonies 24 hours later.

In frozen cadavers *Y. pestis* can be cultivated from buboes up to 102 days (Zlatogoroff 1904). When a post-mortem examination cannot be made, a finger may be cut off—digitotomy—and the bone marrow examined, or material may be aspirated with a syringe from a bubo or from the lung, liver, spleen, or sternal marrow.

In doubtful, mild or chronic cases, when the bubo is small, hard, and difficult to aspirate, it may be excised and injected into a guinea-pig (Uriarte 1925). For transport of material to the laboratory, the Cary-Blair (1964) medium is recommended. In *pneumonic plague* the sputum is usually thin and stained bright red with blood—not purulent, viscous, and rusty as in pneumococcal pneumonia. It contains large numbers of plague bacilli, demonstrable first by culture and later microscopically.

Of the various *serological means* available for diag-

nosis, Chen and Meyer (1966) strongly favour the complement-fixation test or the more delicate haem-agglutination test, in which tanned sheep red cells on to which fraction 1 of the plague bacillus has been adsorbed are used as antigen. The test may be positive when cultivation tests are negative, and is particularly valuable in the convalescent stage for retrospective diagnosis (Legters *et al.* 1969). (For a more detailed description of the diagnosis of plague see Report 1956, Goldenberg *et al.* 1970, Bahmanyar and Cavanaugh 1976.)

Immunity

After plague has been epidemic among rats for a time it decreases in incidence and ultimately dies out. This process is associated with an increase in the immunity of rats. Rats from a plague-infected area are more resistant to experimental inoculation than those from non-infected areas (Sokhey and Chitre 1937); and the young of resistant animals are said to enjoy a high degree of natural immunity (Liston 1924).

How far resistance is due to an increased genetic immunity in the strict sense, resulting from the operation of natural selection, and how far it is due to the acquirement of active immunity by the young animals as the result of latent infection, occurring perhaps at a time when antibody derived from the placental circulation or the colostrum is still present in the animal's tissues, it is difficult to say, but readers who are interested in this fascinating problem will do well to consult the review of the subject by Hill (1934). On the whole, the evidence points more to the importance of genetic immunity. For example, in Bombay city, over 2000 rats were caught each night and examined at the Haffkine Institute. During two years not a single plague-infected rat was found (Sokhey and Chitre 1937). Again, in the region of Paris, where plague was introduced by infected rats in 1917, the resulting epizoötic gradually declined after about 1920, so that by 1936 not a single rat out of 3525 examined was found to be infected (Joltrain 1936). In view of these findings it is difficult to ascribe the increasing immunity of the rats to the persistence of a benign epizoötic. It seems more probable that a strain of rat is evolved having a greater degree of natural resistance to plague infection than that of the original susceptible animals.

Prophylaxis

Measures against rats should be taken concertedly on an area basis; random killing is useless. All buildings housing food, such as granaries, stores and shops, should be rendered rat-proof. That this can be done successfully was demonstrated in Java, where between 1914 and 1939 over 1 500 000 houses were remodelled (van Loghem 1939). This policy should be combined with (*a*) steps to remove all refuse and other material available for food, and (*b*) a campaign to destroy rats by means of chemical poisons. Some of these, for example, sodium fluoroacetate, phosphorus, arsenious oxide, red squill, and strychnine, though effective, cause great suffering to the animals and their use should be discouraged or forbidden. Warfarin, an anticoagulant that causes death by internal haemorrhage, alphanaphthylurea, which produces pulmonary oedema, and zinc phosphide are much less objectionable (Report 1946). Rats may, however, become resistant to warfarin (Bentley 1967). Norbormide, which is non-toxic to other animals, is a possible substitute when zinc phosphide would be dangerous, but is said to be less effective (Rennison *et al.* 1968). As alternatives, coumatetralyl, diphacinome or bromadiolone, which are usually effective against warfarin-resistant rats, may be used (Redfern and Gill 1980). Any anticoagulant should be alternated with an acute poison, such as zinc phosphide, to retard the possible development of anticoagulant resistance (Davis 1970, Brooks *et al.* 1980, Andrews and Belknap 1983).

The second link in the chain—the flea—is best dealt with by strict personal hygiene and domestic cleanliness aided, when necessary, by the powerful insecticide DDT. In the absence of dirt and litter the fleas have nowhere to breed and, even when they have, the judicious application of DDT (dichlorodiphenyltrichloroethane) may rapidly lower the flea infestation of rats by 80 or 90 per cent. Macchiavello (1946), who used DDT in conjunction with sodium fluoroacetate in combating a localized outbreak of plague in Peru, found that 10 per cent DDT was toxic to the rats, which regularly licked their fur, and that sodium fluoroacetate was toxic to rat fleas, which died by secondary poisoning from ingesting the blood of the poisoned rats. Owing to its extensive use for destroying mosquitoes, fleas have become resistant to DDT in many countries where anti-malaria measures have been conducted on a big scale. In its place diazinon dust, carbaryl, or malathion may be found effective (Cavanaugh *et al.* 1968, Bahmanyar and Cavanaugh 1976). Destruction of fleas should always precede or accompany the campaign against rodents (Report 1970).

The possibility that other insects—bugs, flies, ticks—may share in spreading the disease has been carefully considered (Nuttall 1897, 1898, Report 1915c). There seems little doubt that lice and ticks play a part in the transmission of sylvatic plague (Hampton 1940), but their role in the spread of plague among rats would appear to be negligible.

Man forms the third link. Four measures are usually advocated: (1) Notification of cases or, in countries where this is difficult, registration of deaths. The diagnosis can be confirmed *post mortem* by splenic puncture. (2) Isolation of cases, which, except in pneumonic plague, is of little value, since man plays but a small part in the spread of the disease. (3) Evacuation, which

is sound in theory, since it removes the susceptible population from the infected localities, but is very difficult in practice—at any rate on a large scale. It was, however, applied successfully by Yersin (1899) in Annam, where the contacts were transferred to an adjacent island and the houses burnt down. (4) Vaccination, the chief value of which is, or should be, to protect an exposed population while effective preventive measures, such as the remodelling of houses to exclude rats, are carried out. Should plague break out, the houses should immediately be sprayed inside with 5 per cent DDT to destroy all insects.

As Meyer (1950) points out, the introduction of DDT and its derivatives, of potent rat poisons, and of sulphadiazine and streptomycin have completely changed our outlook on plague.

Control of the *pneumonic* form of the disease may present difficulties, but here again the administration of sulphathiazole to immediate contacts can be relied upon to confer a high degree of protection. Patients should be isolated; doctors and nurses should wear masks (Strong *et al.* 1912); and overcrowding should be prevented in the general population.

The control of *sylvatic plague* presents a formidable problem. In most countries it is usually considered sufficient to prevent wild rodents encroaching on inhabited areas where they may infect the local population of rats. In Soviet Russia, however, serious attempts have been made to exterminate them by poisonous baits combined with fumigation and 'disinsectization' of their burrows. In the area to the north of the Caspian Sea over 100 million acres were treated repeatedly over a period of years to destroy the little sisels. The operation was an enormous one but, judged by the cessation of human epidemics in the area, it was successful (Fenyuk 1960). A similar, though smaller, operation was conducted against the marmots in Transbaikalia.

Readers desiring detailed information on the control of rodents and vectors are referred to excellent accounts by Pollitzer (1954, 1960) and Report (1967).

Vaccines

Animals were successfully vaccinated against plague by several of the early workers (Yersin *et al.* 1895, Report 1899, Report 1910, p. 536). Different species appear to vary in their response to a given vaccine. According to Schütze (1939) a high content of envelope antigen is required in the vaccine if rats and guinea-pigs are to be protected, but in the mouse this antigen is of less importance. These observations may explain some of the discrepancies between the proponents of dead and of living vaccines. Sokhey (Report 1936, 1937), for example, using mice, apparently obtained very good results in the laboratory with a killed vaccine made from a broth culture grown at 27°, which is known to contain very little envelope antigen; whereas

Otten (1936, 1938) using guinea-pigs and rats, and Pirie and Grasset (1938) using rats, found a dead vaccine to be greatly inferior in its protective power to a living vaccine made from a strain of modified virulence. By subjecting cultures or infected guinea-pig tissues to ultrasonic treatment and washing, Keppie, Cocking and Smith (1958) extracted the so-called fraction 1 or virulence antigen from the bacillary envelope. Heated at 60° for 1 hour, the suspension was non-toxic and immunized both mice and guinea-pigs. Crumpton and Davies (1957) found that living smooth strains gave 20 to 100 fold better protection in mice than killed smooth organisms, and ascribed this to the presence of antigen 4 (see Chapter 38). Williams and Cavanaugh (1979) tested the protective value for rats of a vaccine made from the specific envelope or capsular fraction, F1, of the plague bacillus. They compared it with two other vaccines—one a live vaccine prepared from the avirulent EV1 strain, the other a killed US vaccine. All three vaccines afforded similar protection, the degree of which increased with the titre of F1 antibody as measured by the passive haemagglutination test. These varying results on animals tend to confirm the opinion expressed in his review by Burrows (1963) that the factor or factors required for making a good vaccine are still incompletely understood.

It is difficult to know how far the findings in animals are applicable to human beings. In previous editions of this book (see 5th ed., pp. 1975–76) we described the field trials carried out in India with Haffkine's killed vaccine (see Report 1901, Taylor 1933) and in Indonesia with Otten's live vaccine (Otten 1936, 1941). Though favourable results were claimed, subsequent experience has thrown doubt on the merits of either of these two vaccines. Neither live nor dead vaccines seem to confer more than a partial and transitory immunity in man; and, though perhaps justifiable for use in the protection of individual persons and small groups at risk, they are liable to give rise to reactions that render them unsuitable for mass vaccination. (See also Report 1970, Meyer 1970.)

Serum treatment   Antisera have been prepared by numerous workers (see Haffkine 1905) and credited with the ability to lower the death-rate in cases of bubonic plague (Yersin 1897, 1899, Girard 1941, Wagle *et al.* 1941); but no adequate proof of their value has so far been forthcoming, and their use is no longer recommended (Report 1970).

Chemotherapy   Bubonic plague may be successfully treated by a number of different drugs, such as the sulphonamides, streptomycin, tetracycline, and chloramphenicol, but not penicillin. The drug of choice is tetracycline, given in large doses (4–6 g daily) during the first 48 hours (Report 1970). The recommended dose of sulphadiazine is 12 g daily for four to seven days; of streptomycin 0·5 g every four hours intramuscularly for two days, then 0·5 g every six

hours till clinical improvement occurs; and of chloramphenicol 50–75 mg per kilo body weight by the mouth daily up to a total of 20–25 grams. Owing to early recognition of cases and the institution of antibiotic therapy, the case-fatality rate is now said to be only 1–5 per cent as opposed to 60–90 per cent in untreated cases (Marshall *et al.* 1967).

Yersinia pseudotuberculosis infections

These infections, often referred to as 'pseudotuberculosis', should not be confused with pseudotuberculosis (caseous lymphadenitis) of sheep, a disease caused by *Corynebacterium ovis* (see Chapter 25); neither should they be confused with pseudotuberculosis of mice, a disease caused by *Corynebacterium murium* (see Chapter 25).

Animals

In 1883 Malassez and Vignal described a disease in a guinea-pig that had been used for experimental inoculation, which was characterized by the presence of nodules histologically similar to those of tuberculosis, but containing zoogloeal masses of coccoid bacilli. To this disease they gave the name zoogloeal tuberculosis. A similar organism was found by Charrin and Roger (1888), Dor (1888), Nocard (1889), and Zagari (1890) in guinea-pigs dying either naturally or after inoculation with pathological material. Preisz (1894), who made a comparative study of these organisms, concluded that they were identical with a bacillus described by Pfeiffer in 1890 under the name *Bacillus pseudotuberculosis rodentium*, now known as *Yersinia pseudotuberculosis*. Pfeiffer gave a good description of the lesions found in the guinea-pig and the rabbit, and suggested the name 'pseudotuberculosis' for the disease. As Weidenmüller (1968) points out, however, the disease is by no means confined to rodents. In much of Europe, for example, the lagomorphs—particularly hares—are predominantly affected (Report 1968). In many countries, such as the USSR, the disease is endemic in field mice. Rabbits, guinea-pigs, and to a less extent, rats, suffer from infection; and, among birds, pigeons, ducks and turkeys. In domestic animals, such as the cat and the dog, infection is much less common, though it does occur (Pallaske and Meyne 1932); and among farm animals pigs are not infrequently attacked. Sporadic cases of infection, however, occur in a wide variety of mammals and birds (see Mair 1968). The organisms are excreted in the faeces, and infection occurs mainly by the alimentary tract. Domestic animals are infected by contaminated food. Paterson and Cook (1963), for example, found that their breeding stock of guinea-pigs was being infected by green vegetables on which the neighbouring wood pigeons had fed. The pigeons were grossly affected with the disease. Sporadic cases of pseudotuberculosis occur not infrequently in zoological collections, especially in the winter. Tsubokura and co-workers (1976) and others in various parts of the world found *Y. pseudotuberculosis* in the caecal contents and tonsils of a small proportion of apparently healthy pigs; isolations were most frequent in the winter and the strains of the organism often belonged to type 3.

Post mortem in guinea-pigs, there are rounded greyish-white spherical nodules up to 2 or 3 mm in diameter, distributed chiefly in the spleen, liver, and lungs. The nodules are well defined, often protrude from the surface, and on section appear creamy or caseous. Enlarged caseous mesenteric glands are generally present. Microscopically the lesions contain large numbers of short, coccoid or ovoid, bipolar-stained gram-negative bacilli, which can be readily cultivated on agar (see Chapter 38). The disease can be reproduced experimentally in large numbers of different animals and birds (see Pallaske 1933). On injection into guinea-pigs, rabbits, rats, or mice the bacilli prove fatal in 1 to 3 weeks, depending on the dose used and the site of injection. At necropsy after intramuscular inoculation there is a caseous local lesion, enlargement of the regional lymphatic glands, and nodules in the spleen, liver, and lungs. Microscopically, the bacilli are numerous in the local lesion and glands.

According to Cook (1952) they can be demonstrated most readily by making use of their slightly acid-fast property. Smears are stained with a 1/10 dilution of 1 per cent carbol-fuchsin for 30 minutes, washed, decolorized with 0·5 per cent acetic acid for 30 seconds, washed, and counterstained for 10 to 15 seconds with Loeffler's methylene blue. The cytoplasm of cells and cellular debris are stained blue or purplish blue; the cell nuclei and the pseudotubercle bacilli are stained red. Alternatively van der Schaaf (1968) recommends the use of Hansen and Köster's (1936) stain with alkaline methylene blue and safranin; pseudotubercle bacilli show up sky blue against a red background. For isolation of the organisms from the faeces samples are best stored at 4° for a week or more to enable multiplication to occur, and then plated on to Morris's (1958) pasteurella medium, to which mycostatin and crystal violet are added to prevent the growth of moulds and of aerobic spore-bearers (Paterson and Cook 1963).

Man

In man *Y. pseudotuberculosis* may give rise to an acute generalized disease of a septicaemic character, or to acute inflammation of the mesenteric lymph nodes. Secondary immunological complications such as erythema nodosum and in rare instances arthritis may follow (Mair 1977).

The generalized disease is not common. In 1938 Topping and his colleagues recorded the fifth authen-

tic case in medical literature, and Hässig and others (1949) reviewed 17 cases and described two of their own. Adult males are chiefly affected. The characteristic picture is that of a febrile disease with swelling of the liver and spleen. Jaundice increasing in intensity towards the end is often present. Three-quarters of the cases are fatal. Death occurs in 2–3 weeks. At postmortem examination the liver is studded with nodular necrotic foci or abscesses, 1–20 mm in diameter; the spleen shows a diffuse lymphoid hyperplasia. The organisms can be cultivated from the liver nodules after death and sometimes from the blood during life. Bradley and Skinner (1974) suggested that disordered iron metabolism acted as a predisposing factor.

In contrast to the generalized form of the disease, the appendicular form (right iliac fossa syndrome), characterized by acute mesenteric lymphadenitis (Knapp 1954, 1956, 1958, 1959), is by no means infrequent. Several hundred cases have been reported (Report 1968). Children and young adults, especially males, are the main sufferers. Many patients are operated upon for acute appendicitis, but the appendix is found to be normal. In this form spontaneous recovery is usual. The glands are swollen and inflamed. Histologically, they show focal, and later a diffuse, proliferation of reticulocytes, and a leucocytic infiltration which leads to circumscribed abscesses (Schmidt 1959). Daniëls (1973) rejected an earlier view that the lymphadenitis had a haematogenous origin; he considered that lymphatic spread occurred from a transitory lesion in the terminal part of the ileum. The organism is seldom isolated from the faeces. Infection occurs either by contact with an infected animal, as in sportsmen, or from contaminated food (Mollaret 1968). Enteritis caused by *Y. pseudotuberculosis* is rare.

Diagnosis of the appendicular form is made by isolation of the organism from an excised gland, or by the demonstration of agglutinins in the patient's blood.

The agglutination test is carried out with living suspensions of serotypes 1–6, the tubes being read after 4 and 24 hours' incubation at 52°. Under these conditions the H agglutinin titre may range from 1/80 to 1/12 800 in serum taken at the time of the operation. Sera containing antibodies to types 2 and 4 must be absorbed with salmonellae belonging to groups B and D respectively in order to eliminate co-agglutinins. About 90 per cent of cases are due to type 1.

Since 1959, large-scale outbreaks of a disease now known to be epidemic pseudotuberculosis have been recognized in the eastern part of Russia. Somov and Martinevski (1973) estimated that some 5000 cases had occurred. Rodents are thought to be the main source of infection. The manifestations of the disease include fever, a scarlatiniform eruption, polyarthritis, arthralgia, and symptoms associated with lesions of the gastrointestinal tract and liver.

Yersinia enterocolitica infections of man and animals

Frederiksen (1964) proposed the name *Yersinia enterocolitica* for a group of organisms of which the majority had been isolated since 1960 from pseudotuberculosis-like lesions in man, and from animals especially chinchillas, hares, and pigs. They resembled organisms described earlier by Schleifstein and Coleman (1939) in America and Hässig and his colleagues (1949) in Switzerland. By 1970, the pathogenic importance of *Y. enterocolitica* for man had become established in Europe (Knapp and Thal 1963, Winblad *et al.* 1966, Niléhn *et al.* 1968, de Wulf *et al.* 1969, Mollaret 1971). Thereafter the literature contained many reports of isolation of the organism from human patients, animals and the environment in various parts of the world (see Report 1973, 1979, 1980, Morris and Feeley 1976, Bottone 1977). Wauters (1973*a*) commented on the heterogeneous nature of the strains included in the species; five biotypes had been established by the methods of Niléhn (1969) and 34 O-serotypes by an extension of the scheme of Winblad (1967). Many environmental strains were rhamnose-positive and untypable (Morris and Feeley 1976). Brenner and his colleagues (1980) divided strains into the four species *Y. enterocolitica, intermedia, frederikseni,* and *kristenseni* (see Chapter 38).

Diseases and epidemiology

In man, about two-thirds of all cases of *Y. enterocolitica* infection take the form of enteritis (Bottone 1977, Mair 1977), mainly in children under 7 years of age. The disease is characterized by fever, diarrhoea, abdominal pain, and sometimes vomiting, and may be of two or more weeks' duration. It occurs mainly in the autumn and winter.

The appendicular form of disease is less common. It differs from that produced by *Y. pseudotuberculosis* in that it affects children and adults of both sexes and is not infrequently accompanied by diarrhoea; the lesions of terminal ileitis are often more severe (Mollaret 1972), and large pyroninophilic cells are sometimes present in the affected mesenteric lymph nodes (Ahlqvist *et al.* 1971).

Y. enterocolitica septicaemia, a disease associated with high mortality, is comparatively rare. It occurs mainly in elderly patients. Predisposing factors include alcoholism, hepatic cirrhosis, haemodialysis, and treatment with immunosuppressive drugs. Visceral abscesses may occur as a result of septicaemia.

Of the secondary immunological complications of *Y. enterocolitica* infection, erythema nodosum and non-suppurative polyarthritis are not uncommon (Ahvonen 1973, Bottone 1977), especially in adults or in older patients (Winblad 1973). Infection may occasionally produce Reiter's syndrome, subcutaneous abscesses, meningitis, panophthalmitis, and an erysipelas-like illness (Mair 1977). Several reports have suggested an association between the presence of antibody to *Y. enterocolitica* serotype 3 and the occurrence of thyroid disease (Reynolds *et al.* 1978). Bottone (1978) found that atypical strains of *Y. enterocolitica* produced comparatively minor infections of wounds, the conjunctiva, and the alimentary and urinary tracts in patients whose normal resistance had been weakened.

Y. enterocolitica has been found in the faeces of healthy human carriers—usually in members of a family in which a clinical case has occurred (Vandepitte *et al.* 1973).

Organisms said to be *Y. enterocolitica* but differing biochemically and serologically from those associated with the main types of human infection have been isolated from a variety of sources. These include small mammals such as hares and chinchillas, water, milk, meat, and shell-fish (Hanna *et al.* 1976, Highsmith *et al.* 1977, Schiemann 1978, Kapperud 1981). In Europe, Japan, South Africa and Canada, serotype 3 (biotype 4) is the most common cause of *Y. enterocolitica* infection in man. Serotype 9 (biotype 2) is well known in Europe. Weaver and Jordan (1973) found that serotypes 3 and 9 did not predominate in the USA, but that there appeared to be a high prevalence of serotype 8; a variety of other serotypes occurred.

Numerous reports (Ahvonen *et al.* 1973, Esseveld and Goudzwaard 1973, Toma and Deidrick 1975, Bockemühl and Roth 1978, Pedersen and Winblad 1979) have shown that serotypes 3 and 9 sometimes occur in the intestinal contents and tonsils of pigs, especially in autumn and winter. This has led to the suspicion that pigs constitute a reservoir of infection for man. Circumstantial evidence for transmission from pig to man has been produced in certain outbreaks (Rabson and Koornhof 1973), but there is no reason to believe that such transmission occurs commonly. Human serotypes including serotypes 3 and 9 have also been found in the intestinal contents and mesenteric lymph nodes of dogs (Kaneko *et al.* 1977, Pedersen and Winblad 1979), and in lesions of the internal organs of non-human primates (Poelma *et al.* 1977). Mollaret (1972), Schiemann and Toma (1978) and others mentioned the isolation of *Y. enterocolitica* from numerous animal species including guinea-pigs, rats, squirrels, cats, horses, cattle, goats, deer, poultry and wild birds; the serotypes were occasionally those associated with human disease.

Several recorded outbreaks of human infection were thought to have been caused by contaminated food (Askawa *et al.* 1973, Zen-Yoji *et al.* 1973, Olšovský *et al.* 1975). Black and his colleagues (1978) showed convincingly that an outbreak in American schoolchildren was caused by the consumption of chocolate-flavoured milk contaminated with serotype 8. In this outbreak transmission of disease from affected children to their household contacts did not occur; in certain other outbreaks transmission from person to person has been suspected. Vandepitte and co-workers (1973) found that of 639 patients with _Y. enterocolitica_ infection no less than 8.6 per cent were also infected with salmonellae; this seemed to indicate a common source for the two infections.

Diagnosis

Diagnosis may be made by cultural or serological methods.

Isolation of the organism may be attempted from lymph nodes, blood, faeces, or foods. _Y. enterocolitica_ is tolerant to high concentrations of bile salts, and media such as SS agar or MacConkey's agar are therefore useful; the incubation temperature should be 22–29°. Soltész and co-workers (1980) describe a selective medium containing casein hydrolysate, peptone, sodium oxalate, and bile salts. After incubation for 24 or 48 hours the colonies can be recognized by means of a stereomicroscope and oblique transillumination (Wauters 1973_b_, Bottone 1977). Preliminary enrichment in addition to direct plating is essential for heavily contaminated material such as faeces or foodstuffs. A suitable method consists in mixing the sample with buffered saline and holding the mixture at 4–7° for 28 days; subcultures on SS or MacConkey's agar are made at weekly intervals. Isolates are identified as _Y. enterocolitica_ by a small series of tests (Morris and Feeley 1976), and may also be subjected to biotyping by the systems of Niléhn (1969) and Wauters (1970). Serotyping is usually confined to identification of the somatic (O) types, of which there are 34 (Winblad 1967, Wauters _et al._ 1971, 1972). The phage type (Nicolle _et al._ 1973, Niléhn 1973) is partly related to the origin of the strains and to their serological and biochemical characters. Serotype 3 strains can be subdivided by phage-typing.

Serological diagnosis of human infections is practised most frequently in areas where certain serotypes overwhelmingly predominate, for example in Europe where serotypes 3 and 9 account for the majority of cases. Antibodies may not be detectable at the onset of illness, and paired serum samples taken at intervals of 10 days or more should therefore be examined. In agglutination tests made with O-antigen preparations, a titre of at least 1/160 is considered positive. Because of cross-reactions with antibodies against other gram-negative bacteria, OH-antigen preparations are unreliable (Lysy and Knapp 1973). Serological cross-reactions between _Y. enterocolitica_ serotype 9 and _Brucella_ spp. are well known to occur (Ahvonen _et al._ 1969, Hurvell and Lindberg 1973).

(For reviews see Bottone 1981, Swaminathan _et al._ 1982.)

Pasteurella infections

Pasteurella multocida infections of animals

Infections with members of the genus _Pasteurella_ are of great importance in animals but comparatively uncommon in man. The term 'pasteurellosis' (Lignières 1901), although still often used, is somewhat outmoded and misleading; it fails to distinguish between diseases caused by different pasteurella species, or between the various forms of infection that occur.

The observations of Bollinger and of Kitt in 1878 on an epidemic disease affecting wild hogs, deer, and later cattle, in the neighbourhood of Munich constitute the first work of importance on septicaemia caused by this organism. Kitt was successful in isolating the causative organism and in transmitting the disease to mice and pigeons. A similar organism was obtained by Pasteur in 1880 from fowl cholera, by Loeffler in 1882 from swine plague, by Poels (1886) from septic pleuropneumonia of calves, and by numerous workers during the next few years from diseases in other animals. Since the beginning of the present century, swine plague, an acute septicaemia, has become rare (Hudson 1959).

The range of susceptible mammalian and avian species is now known to be very wide, and the organism —sometimes though not always in a state of high virulence— has frequently been isolated from the nasopharyngeal region of healthy animals. Biochemical tests are incapable of determining the host of origin (Heddleston 1976), but in certain instances they may give an indication. Serotyping is of epidemiological interest, but is complicated by the existence of several different methods.

Roberts (1947), by means of a mouse-protection test, established types I, II, III and IV of _P. multocida_. Carter (1955, 1961), by means of an indirect haemagglutination test, established four capsular types (A, B, D, and E), of which type B is generally agreed to be the equivalent of Roberts' type I; an opinion on the relationship between all Roberts' types and the capsular types has been given by Prodjoharjono and his colleagues (1974). Studies reviewed by Namioka (1973) established 11 somatic (O) groups based on antigens that consisted of combinations of several factors. It is now usual to describe the serotype of a strain by stating both the somatic and capsular groups, e.g., serotype 5:A. More recently Heddleston and his colleagues (1972) described a gel-diffusion precipitation method that is of epidemiological value in serotyping strains from poultry. A single serotype in one typing system often represents more than one serotype in another (Brogden and Packer 1979). (See also Chapter 38.)

Certain serotypes tend to be associated with particular animal species and forms of disease, and even with geographical regions. Capsular types A and D are worldwide in distribution.

P. multocida infections in animals are often precipitated by stress factors and the diseases produced are commonly septicaemic or respiratory, although other forms occur. In subacute respiratory disease concurrent infections with other organisms may exist. A strain of *P. multocida* that is of high virulence for one animal species by experimental inoculation may or may not be of high virulence for another naturally susceptible species.

Cattle and other ruminants

In many animal species *P. multocida* infection may occur in the form of septicaemia associated with widespread haemorrhages, but the term *haemorrhagic septicaemia* is used specifically to denote such infection in cattle and buffaloes.

Haemorrhagic septicaemia of cattle and buffaloes This disease (Carter and Bain 1960, Bain 1963, Carter 1967) is of economic importance in many parts of Asia, where it is usually caused by strains belonging to the type I of Roberts (1947), i.e., by capsular type-B strains, many of which are of somatic group 6. In experimental infections such strains have been found highly virulent for buffaloes, mice and rabbits, rather less for cattle, much less for pigs, of variable virulence for sheep and goats, and completely avirulent for chickens. Virulent type-I strains have been found in the nasopharynx of healthy buffaloes. Roberts' type-IV strains are occasionally responsible for Asian outbreaks of septicaemia in cattle and buffaloes, but there is a suspicion that such infections may be secondary to pneumonia. In Central Africa haemorrhagic septicaemia is usually caused by strains of capsular type E. The disease is also known in Southern Europe and the USSR. Cultures still exist of a capsular type-B strain isolated from bison in the Yellowstone National Park by Gouchenour (1924), but type B is thought to occur no longer in North America.

Haemorrhagic septicaemia occurs mainly in the rainy season and often appears to be associated with fatigue. Nasopharyngeal carriers provide a source of infection, and the portal of entry is probably the tonsillar region; animals can be infected experimentally by means of a nasal spray. Once clinical infection is established, death is almost certain. On bacteriological examination large numbers of pasteurellae are found in the saliva, milk, faeces, and urine. Blood smears contain large numbers of coccobacilli that exhibit bipolar staining, and in the terminal stages the disease bears a resemblance to endotoxic shock. Many animals die within 24 hr of the onset of symptoms, but those that live longer may show a variable degree of

pneumonia at necropsy. Sometimes oedematous swellings occur in the region of the throat, neck and brisket due to infiltration of the subcutaneous tissues with a straw-coloured exudate. The main post-mortem features are widespread petechial haemorrhages and swollen and congested lymph nodes. There may also be exudates in the serous cavities, oedema in the laryngeal region, and acute gastroenteritis with blood in the intestinal contents. The lungs are often congested, and if death has not supervened too rapidly there may be some pneumonia with thickened interlobular septa and fibrinous pleurisy.

More than 10 per cent of unvaccinated cattle and buffaloes in India and South-East Asia are said to be immune, and serum from such animals or from artificially immunized animals is capable of affording passive protection to mice against experimental infection. Vaccination is often routinely used, especially before the beginning of the rainy season, in areas where the disease is known to occur. Vaccines consist of formalin-inactivated strains from haemorrhagic septicaemia, usually blended with an oil emulsion or alum-precipitated. Protection is thought to last no more than 6 to 8 months (Report 1977). Nagy and Penn (1976) reported the successful immunization of cattle against haemorrhagic septicaemia by means of capsular antigens of well established purity mixed with aluminium hydroxide.

Respiratory and other infections Respiratory infections with *P. multocida* occur in young cattle, especially those kept under intensive systems of husbandry or subjected to stress by fatigue, excitement or fear, irregular feeding or watering, and overcrowding. Such factors are often present in a particularly damaging combination when animals are transported; and when the distances are long the stress may be very severe. This was well illustrated by an account (Carter 1956) of the practice of transferring beef cattle aged 6–24 months from the range in Western Canada to farms in Eastern Canada; this often gave rise to the disease known in North America as *shipping fever*, i.e. the occurrence of pulmonary infections during or shortly after a journey. Infections of this kind may be single or mixed and are caused by a variety of organisms that include *P. multocida* (capsular type A and occasionally D, and untypable strains), *Pasteurella haemolytica*, corynebacteria, mycoplasmas, chlamydias, and viruses. When *P. multocida* is concerned the pneumonia is often accompanied by thickening of interlobular septa, and by fibrinous pleurisy; occasionally septicaemia may occur. The strains of *P. multocida* responsible are in general less virulent for cattle than are the strains that cause haemorrhagic septicaemia of cattle and buffaloes.

P. multocida has been reported as a cause of bovine mastitis by Lovell (1939) and others, and has been isolated from the synovial fluid of calves suffering from polyarthritis. It has also been isolated from calves with meningoencephalitis as well as from the brain of animals of other species. In sheep the importance of *P. multocida* is greatly overshadowed by that of *P. haemolytica*; the organism is nevertheless sometimes responsible for pneumonia in sheep and goats.

Pigs

Pigs are susceptible to respiratory infection with *P. multocida*, notably with capsular types A and D, but such infection is frequently secondary to infection with other organisms such as mycoplasmas and viruses. The experimental pathogenicity of *P. multocida*, either alone or with other organisms, has been studied in gnotobiotic pigs (I. M. Smith *et al.* 1973) and specific pathogen-free pigs (Kielstein *et al.* 1977). It is suspected that *P. multocida*, particularly toxigenic strains (Rutter and Luther 1983), sometimes contributes to the pathogenesis of atrophic rhinitis of swine (see Chapter 67). Dirks and his colleagues (1973) found that the proportion of pigs from which they could isolate *P. multocida* was higher in herds with atrophic rhinitis than in healthy herds; they also reported the development of snout deformities and turbinate atrophy in a proportion of pigs that had received *P. multocida* intranasally. From experiments in gnotobiotic pigs Rutter and Rojas (1982) concluded that severe atrophic rhinitis might result from combined infections with certain strains of *P. multocida* and *Bordetella bronchiseptica*.

Dogs and cats

J. E. Smith (1955) isolated *P. multocida* from 10 per cent of nasal swabs of 111 dogs killed in accidents or at the owners' request, and from 54 per cent of the tonsils. The organism is likewise common in the mouth and throat of healthy cats. Carter (1977), in proposing five biotypes of *P. multocida* (mucoid, haemorrhagic septicaemia, porcine, canine, and feline), cited studies by J. E. Smith (1958) and others which showed that strains from dogs and cats often differed in certain respects from strains that produced disease in cattle, sheep, pigs, and poultry. In addition, canine and feline strains often differ from each other; for example, most canine strains are of low virulence for mice whereas most feline strains are of much higher virulence. From a study of strains from cats and dogs, Müller and Krasemann (1974) suggested the existence of a relation between virulence and neuraminidase production. Canine and feline strains are often non-capsulated and untypable on the basis of capsular antigens. They do not play an important pathogenic role in their natural hosts except possibly in infections resulting from bites.

Rabbits and rodents

De Kruif (1921, 1922*a, b*, 1923), working with rabbits suffering from *snuffles*—a chronic respiratory infection caused by *P. multocida*—observed the organism in the nose of a large proportion of normal animals. Webster (1924*a, b*) found that before an outbreak there was a rise in the normal carrier rate. The susceptibility of individual rabbits varied greatly, some being highly resistant, others succumbing readily. *P. multocida* infection is endemic in many rabbit colonies and occasionally produces, or is associated with, a number of diseases other than snuffles, such as septicaemia, mastitis, otitis media, encephalitis, and pyometra.

Wright (1936) observed an outbreak of fibrinous serositis in guinea-pigs, and Pirie (1929) and Mitchell (1930) described the so-called *de Aar disease* that affected the veld rodents of South Africa.

An organism well known to infect small laboratory rodents is frequently referred to as *Pasteurella pneumotropica* (Jawetz 1950), although not all agree that it is sufficiently different from *P. multocida* to warrant the rank of a separate species. This organism often inhabits the nasopharynx of animals other than those of specific pathogen-free status. It is a potential respiratory pathogen, but experimental infections are—for reasons that are not well understood—often difficult to produce. It has been isolated from the conjunctival mucosa and from the uterus of symptomless laboratory rodents; it has also been associated with keratoconjunctivitis and abortion. It sometimes produces abscesses and has been reported as causing chronic necrotizing mastitis. Latent infection sometimes occurs in wild rodents, as well as in horses, calves, dogs, cats, man, and other species (Shepherd *et al.* 1982).

Birds

Fowl cholera is a disease of domestic poultry and wildfowl, and outbreaks can cause heavy losses on large commercial chicken and turkey farms. The strains of *P. multocida* responsible are often of exceptionally high virulence for the natural hosts and for laboratory mice by experimental inoculation. Calves, pigs and sheep may be much more resistant (Collins 1977). These strains belong mainly to capsular type A; some belong to type D and others are untypable. Serotypes 8:A and 5:A are of particular importance (Carter 1972, Heddleston *et al.* 1972); these serotypes are thought (Carter 1972) to correspond to Heddleston's types 3 and 1 respectively.

The acute epizootic form of fowl cholera is characterized by an incubation period of about 12 hours, listlessness, and the general symptoms of an acute illness. There may be discharges from the beak and nostrils, and diarrhoea with the passage of bright yellow or green faecal material, often blood-stained. Death follows in a few hours to 3 days. The main lesions are a serofibrinous pericarditis and an acute haemorrhagic enteritis, most noticeable in the duodenum. The disease may also be subacute or chronic, taking the form of nasal catarrh or sinusitis, swelling and oedema of the wattles, or lameness due to arthritis.

Collins (1977), in reviewing immunization against fowl cholera, concluded that acquired resistance was humorally mediated, but a study by Baba and his colleagues (1978) indicated that cell-mediated immunity also played a part. Formolized suspensions of *P. multocida*, with and without adjuvant, have been used as vaccines, and strains of low virulence administered in the drinking water are said to give protection. Such protection is not always serotype specific.

Pasteurella haemolytica infections of animals

The name *Pasteurella haemolytica* was introduced by Newsom and Cross (1932) to designate an organism first described by F. S. Jones (1921), whose isolates were obtained from bovine pneumonic lungs in the USA. The same organism was isolated by Dungal (1931) from sheep pneumonia in Iceland. In Britain similar organisms were obtained from the pneumonic lungs of calves (Tweed and Edington 1930) and sheep (Montgomerie *et al.* 1938), and from the nasal cavity of normal sheep and cattle (Bosworth and Lovell 1944). Reports by others including Hartley and Boyes (1955), Stamp and his colleagues (1955) and G. R. Smith (1960a, 1961a) suggested that septicaemic infections occurred in lambs, either within a few weeks of birth or after five or more months of life.

It is now known that *P. haemolytica* is associated in many parts of the world with pneumonia in sheep and cattle, and with septicaemia in lambs. The evidence indicates that the organism is an important cause of disease. Much remains to be learnt, however, about the factors that influence its pathogenicity and those that affect the susceptibility of animals.

G. R. Smith (1961a) described two types of *P. haemolytica* that differed in a number of respects including their colonial morphology, survival in broth cultures, reactions in certain biochemical tests, antibiotic resistance, and pathogenic effects. These types, designated A and T, are now commonly referred to as biotypes to avoid confusion with the serotypes (Biberstein *et al.* 1960), of which 15 can be distinguished by means of an indirect haemagglutination test. Serotypes 3, 4, 10, and 15 belong to biotype T, and the remainder to biotype A. Certain strains cannot be typed serologically (Aarsleff *et al.* 1970), but their carbohydrate fermentation reactions resemble those of biotype-A strains (Gilmour 1978).

In considering the diseases caused by *P. haemolytica* it should be remembered that certain taxonomic points are still debatable.

Bosworth and Lovell (1944) preferred not to use the name *P. haemolytica*, referring instead to 'haemolytic coccobacilli'. Mráz (1969) considered that the resemblance of *P. haemolytica* to *Actinobacillus lignieresi* was closer than that to *P. multo-*

cida; one of only three essential characters that distinguished *P. haemolytica* from *A. lignieresi* was the production of a second zone of haemolysis on nutrient agar containing blood from lambs not more than a few weeks old (G. R. Smith 1962). Although this unusual property is possessed by both the A and T biotypes, the work of J. E. Smith and Thal (1965) and Biberstein and Francis (1968) suggested that the biotypes might eventually prove sufficiently different from each other to justify the establishment of two separate species.

Healthy sheep and cattle

Nasal carriage of *P. haemolytica*, reported by Bosworth and Lovell (1944), has been confirmed by many others. In a 12-month study of healthy flocks of sheep in the Edinburgh area, Biberstein and his colleagues (1970) found that many nasal strains were untypable, but of the remainder serotype 2 (biotype A) was the most common. The prevalence of various serotypes varied widely from month to month and from flock to flock, and the highest carriage rates occurred in late autumn and late spring. In an abattoir study of apparently healthy sheep, Gilmour and his colleagues (1974) isolated *P. haemolytica* from 95 per cent of tonsils and 64 per cent of nasopharyngeal swabs. Of the tonsillar strains 65 per cent belonged to biotype T; of the nasopharyngeal strains those belonging to biotype T (6 per cent) were greatly outnumbered by biotype-A strains.

Pneumonia in sheep and cattle

In the acute form of the disease known as *enzootic pneumonia* of sheep *P. haemolytica* is invariably present in pneumonic lung tissue in large numbers. Biotype A is almost always responsible and, of all the serotypes, serotype 2 occurs with particular frequency. In the few instances in which biotype T is found, either alone or together with biotype A, its presence is probably often the result of septicaemia.

Enzootic pneumonia affects sheep of all ages. Outbreaks, which occur sporadically and unpredictably, are generally considered to be associated with various forms of stress such as fatigue or inclement weather. The predisposing factors are, however, poorly understood. Mortality rarely exceeds 10 per cent, but sheep that recover clinically may remain in poor condition. Post-mortem examination of sheep that die from acute infection shows that the anterior lobes of the lungs are most commonly affected. The diseased tissue is consolidated and, macroscopically, often resembles liver. The interlobular septa are very obvious and opalescent. The pleura may be thickened, and straw-coloured exudate, often containing fibrin, is present in the pleural cavity. Pericarditis and haemorrhages on the heart and serous surfaces are common findings. It would seem that death results not so much from destruction of lung tissue as from the accumulation of bacteria and their toxic products. The abnormal lung tissue contains necrotic areas and alveoli packed with elongated cells. The alveoli surrounding the main lesion are filled with large macrophages (Stamp and Nisbet 1963).

Except in sheep within a few weeks of birth, the distribution of organisms is essentially thoracic, but rare exceptions occur in which massive lung infection is accompanied by invasion of the liver, spleen, and other sites (G. R. Smith 1961a).

Many workers have tried, as a rule unsuccessfully, to reproduce enzootic pneumonia experimentally by administering *P. haemolytica* cultures or homogenates of diseased lung tissue from naturally infected sheep. Montgomerie and his colleagues (1938), and numerous others since, suggested that a virus or other non-bacterial organism might be concerned in the aetiology. In recent years the agents mentioned have included parainfluenza 3 virus, adenoviruses, and mycoplasmas. Experiments with conventionally reared lambs (Davies *et al.* 1977) and specific pathogen-free lambs (Sharp *et al.* 1978) indicated that exposure to parainfluenza 3 virus several days before exposure to *P. haemolytica* resulted in severe pneumonia, whereas each agent alone produced only a mild infection. Possibly this synergistic effect also plays a part in natural infections, but it seems unlikely to be the only means by which *P. haemolytica* is enabled to produce enzootic pneumonia. G. R. Smith (1964) exactly reproduced fatal enzootic pneumonia in 2-year old sheep by the intrabronchial administration of pure biotype-A cultures, but huge doses of the order of 10^{10}–10^{11} viable bacteria were necessary. However, as will be seen later, the experimental reproduction of biotype-T septicaemia in lambs—a disease that must certainly be regarded as a pure infection with a single agent—also requires the administration of huge doses. The importance of *P. haemolytica* biotype A should not be underestimated, despite the difficulty in reproducing enzootic pneumonia with pure cultures. The striking pathogenic potential of biotype A for the ovine species is apparent from the experiments of G. R. Smith (1960a, 1961b) in which lambs aged 3 weeks were inoculated intraperitoneally; organisms in moderate and very small doses multiplied profusely in animals of this age group and produced death within 36 hr.

In chronic forms of sheep pneumonia *P. haemolytica* may be found in only small numbers in the lung, sometimes together with other organisms. G. E. Jones and his colleagues (1978) considered that a combination of *P. haemolytica* and mycoplasmas was commonly concerned in the aetiology of an atypical pneumonia of sheep described earlier by Stamp and Nisbet (1963). It has been shown experimentally that this atypical pneumonia can seriously affect liveweight gains in lambs (G. E. Jones *et al.* 1982).

Some homologous immunity is produced by vaccinating lambs with sodium salicylate extracts of certain biotype-A serotypes, but satisfactory protection against serotype 2 necessitates the use of killed whole organisms (Gilmour *et al.* 1983).

P. haemolytica is well known to be associated with pneumonia in calves and is considered to play a prominent part in shipping fever in North America (Carter 1967). The many reports of attempts to produce bovine pneumonia experimentally (Jericho and Langford 1978) suggest that calves and sheep present similar problems.

Septicaemia in lambs

Outbreaks fall into two categories that differ in respect of the age of the lambs affected and the biotype of the infecting organism.

Septicaemia that is almost always caused by biotype A occurs in lambs aged between one day and three months. Hartley and Boyes (1955) in New Zealand found that most affected lambs were 1–3 days old, whereas G. R. Smith (1960a) in Scotland found that most were 3–4 weeks old. At necropsy some lambs show serofibrinous peritonitis, others fibrinous pleurisy with small areas of pneumonia. Peritoneal and thoracic lesions sometimes occur in the same animal. Other abnormalities include haemorrhages on the heart and serous surfaces, and multiple small foci of liver necrosis. G. R. Smith (1960a, 1961b) showed that lambs aged 3 weeks were highly susceptible to intraperitoneal or intravenous inoculation with biotype A strains derived from septicaemia or from enzootic pneumonia, but young adult sheep were highly resistant.

Septicaemia caused by biotype T (G. R. Smith 1961a) is an economically important disease of lambs and young adult sheep aged between 5 and 12 months (Stamp *et al.* 1955). Gilmour (1978) stressed that in Scotland it was most common in September, October and November at the time when sheep were being moved from high- to low-ground pastures for the winter, or were being folded on rape or turnips. The first deaths often occur within a few days of the movement of the flock, and outbreaks can be halted almost immediately by changing the grazing. The period between the onset of symptoms and death is usually only a few hours. At necropsy the lungs show congestion and oedema, but pneumonic consolidation is absent except in animals that have died after an unusually protracted illness. Other macroscopic abnormalities include heavily infected necrotic lesions of the tonsil, oesophagus and adjacent tissues (Dyson *et al.* 1981), widespread haemorrhages, enlarged lymph nodes, fibrinous pericarditis, haemorrhagic inflammation of the abomasal mucosa, and small necrotic foci in the liver. On histological examination, bacterial emboli are found in the lung, liver, and spleen. Stamp and his colleagues (1955) showed that the pathological features of the disease could be reproduced in lambs aged 7–9 months by the intravenous injection of strains of *P. haemolytica* freshly isolated from natural cases, but the doses required exceeded 10⁹ organisms. G. R. Smith (1960*b*) found that the lethal dose of heat-killed broth culture of a biotype-T strain given intravenously was only a low multiple of the lethal dose of living culture. Lambs that died from natural infection differed from those that died as a result of the intravenous injection of large doses of living organisms in that their lungs, liver and spleen contained greater numbers of viable bacteria.

Other infections in animals

P. haemolytica is associated with pneumonia and septicaemia in goats (Gourlay and Barber 1960), and Florent and Godbille (1950) described two cases of septicaemia in newborn calves. The organism is a cause of mastitis in ewes (Jubb and Kennedy 1963) and occasionally of meningitis, abscesses and arthritis in lambs. Organisms said to be *P. haemolytica* have been isolated from the respiratory tract and internal organs of chickens, and from the pneumonic lungs of pigs.

Pasteurella infections in man

P. multocida occasionally infects man. Hubbert and Rosen (1970) had records of over 200 cases. It has been isolated from lesions in nearly all parts of the body. Infections range from trivial sepsis to severe wound abscesses accompanied by sloughing and septicaemia.

In the respiratory tract the organism is found in cases of bronchitis, bronchiectasis, pneumonia, empyema, otitis media, sinusitis and mastoiditis; and it may be present in the nose of healthy carriers (see J. E. Smith 1959). Itoh and co-workers (1980) reported positive sputum or blood cultures in seven debilitated patients in a hospital for chronic diseases. Most often the organism appears to be a secondary invader, but it is primarily responsible for lesions caused by cat and dog bites. Francis and his colleagues (1975) presented an analysis of 48 cases associated with wounds inflicted by dogs or cats. Man apparently can become infected from the cough spray of cattle, sheep, pigs, poultry, and cats (J. E. Smith 1959, Carter 1962), but according to Hubbert and Rosen (1970) in about 25 per cent of cases a history of animal contact is lacking. Meningitis caused by *P. multocida* usually appears 1–2 months after accidental or surgical trauma (Controni and Jones 1967); it may also occur in infants. There have been occasional reports of brain abscesses, peritonitis and arthritis. *P. pneumotropica* has been isolated on rare occasions from septicaemia, upper respiratory-tract infection, and lesions caused by dog and cat bites (Rogers *et al.* 1973). (For a review of human infections see Henderson 1963, Hubbert and Rosen 1970.)

Henriksen and Jyssum (1960) first drew attention to the occasional presence of *Pasteurella ureae* in the respiratory tract, and D. M. Jones (1962) recorded the isolation of 18 strains from the sputum of patients suffering mainly from chronic bronchitis or bronchiectasis. Like *P. multocida* it probably plays the part of a secondary invader in such cases. In rare instances it has been isolated from the cerebrospinal fluid of patients with meningoencephalitis.

P. haemolytica appears to be non-pathogenic for man.

Francisella infections: Tularaemia

SYNONYMS: Deer-fly fever, Pahvant Valley fever, Ohara's disease.

In 1911 McCoy described a plague-like disease among the ground-squirrels (*Citellus beecheyi* Richardson) of California. The following year McCoy and Chapin (1912) isolated the causative organism—a very tiny gram-negative bacillus—from naturally infected animals, and reproduced the disease with pure cultures in guinea-pigs. They named it *Bacterium tularense*—from Tulare, the county in which the disease was first observed. It is now called *Francisella tularensis*. The contribution made by Edward Francis has been referred to by Jellison (1972). In 1921 the disease came to be known as tularaemia. It occurs in the northern hemisphere, but not in South America, Africa, or Australia. It is widespread among rodents and lagomorphs in North America, particularly in the western States; cotton-tail rabbits, ground squirrels, hares and jack-rabbits are the animals chiefly infected. In Europe hares and voles constitute the main reservoir (Ringertz and Dahlstrand 1968). Burroughs and his colleagues (1945) give a list of 48 mammals and birds in which natural infection is known to occur. In the United States occasional epizoötics occur in sheep (Jellison and Kohls 1955) and ovine infections are also known in Russia. Other domestic animals are less often affected (Hopla 1974). The lesions found in animals dead of the disease are similar to those of plague; there is a bubo, generally in the cervical, axillary, or inguinal region, containing dry, yellowish, caseous material; the spleen is greatly enlarged, very dark in colour, and contains yellowish-white, discrete, caseous granules up to 1 mm in diameter, projecting slightly above the surface; there are numerous granules in the liver; the lungs are rarely affected; the organisms are present in enormous numbers in the spleen, in small numbers in the liver, bubo, and heart's blood (McCoy and Chapin 1912). Experimentally the disease can be reproduced in ground-squirrels, gophers, guinea-pigs, rabbits, mice, and monkeys: rats are more resistant; cats, dogs, and pigeons appear to be immune. Feeding, nasal instillation, cutaneous, subcutaneous, intraperitoneal, and conjunctival injection are all successful (see Chapter 31). The disease is spread by blood-sucking insects, especially ticks, which inject saliva into the wound (McCoy and Chapin 1912, Francis 1921, Francis and Lake 1922, Parker *et al.* 1924, 1929). According to Parker and Spencer (1926) the organism may be transmitted from infected female ticks to their progeny, though Hopla (1974) denies this. If it is true, it must be one of the few examples known of the hereditary transmission of a bacterium by insects.

The first case of the disease in man was reported in 1914 by Wherry and Lamb in the United States. From 1924 to 1937 inclusive, 8022 cases of infection with 358 deaths were recognized (Olson 1938), though there is reason to believe that numerous other cases occurred that were not diagnosed. The disease has been met with in Japan, where it is known as Ohara's disease (see Ohara 1930), and in most countries of Europe (see Jusatz 1961). Cases are mainly sporadic, but outbreaks have been reported (Tigertt 1962, Popek *et al.* 1969, Dahlstrand *et al.* 1971, Berglund 1973). In the United States a case-fatality rate of about 5 per cent has been reduced by antibiotic therapy to about 1 per cent. It is generally agreed that the case-fatality rate in Russia is lower than that in the USA (Tigertt 1962).

At least two types of *F. tularensis* exist, and in the Nearctic region both occur. They differ in respect of virulence and certain biochemical characters (see Miller 1974). The virulent and less virulent types were designated A and B respectively by Jellison and his colleagues (1961), and *F. tularensis nearctica* and *F. tularensis holarctica* respectively by Olsufjev (1970).

Mode of infection

Infection of man is usually due to contact with infected hares, rabbits, and rodents such as water voles, musk rats, and lemmings. It occurs directly through contamination of the hands with the tissues; or indirectly through tick bites, or by the inhalation either of their dried faeces or of infected dust from hay, straw, or sugar-beet, or by contact with contaminated water. In an analysis of 536 cases in the United States, Assal and his colleagues (1967) traced the origin of infection in 25.9 per cent to ticks, in 13.4 per cent to rabbits, and in 2.6 per cent to other sources, but in over half the cases, i.e. 58.1 per cent, the source could not be identified. Jellison and Kohls (1955) report numerous cases in shepherds, occurring particularly at lambing and shearing time.

In Norway and Sweden infection has been traced to hares (Thjøtta 1931b, Olin 1938), in Germany and South-East Europe to field mice (see Jusatz 1961), and in Soviet Russia to water rats (Sarchi 1930) and to susliks (Tumansky and Kolesnikova 1935). The *lemming fever* of Norway, which follows the consumption of drinking water polluted by the bodies and excreta of the lemming, is thought to be related to tularaemia (Thjøtta 1931a). Several outbreaks in the Moscow area were traced to the drinking of well water polluted by infected rodents (Tsareva 1959). Case-to-case infection is said to be uncommon. In rare instances tularaemia has resulted from a cat bite. The incidence of the disease is mainly in those classes of the population that are brought into contact with infected animals, such as butchers and trappers. Laboratory workers are often attacked (Parker and Spencer 1926). Indeed there is probably no other organism that is so dangerous to work with in the laboratory. Unless scrupulous care is taken to avoid infection of the skin, nose, and conjunctiva, from fluid cultures and inoculated animals, the disease will almost certainly be contracted.

Infection may remain latent, as in water voles and susliks. The organism gains access to the body either through the skin of the fingers, the conjunctiva, or by inhalation of contaminated dust. It is known that in guinea-pigs the organism can pass through the intact skin (Lake and Francis 1922), and there is reason to believe that it may do so in human beings, though infection is probably aided by the presence of small wounds and abrasions. The incubation period of the disease is generally stated to be about 3 days. In the usual ulcero-glandular type of the disease, after an acute onset with headache, rigors, pains and fever, a papule appears at the site of infection, generally on the back of the finger. The papule breaks down and leaves a ragged ulcer. The epitrochlear and axillary glands swell and become painful; they often break down and discharge purulent material. Fever is common for the first 2 or 3 weeks. Convalescence is very slow. When the organism gains entrance through the conjunctiva, ulcers may form on the inner surface of the eyelids, followed by swelling and tenderness of the preauricular and cervical glands. Inhalation leads either to the typhoid form, in which there are no localizing symptoms, or to the pulmonary form, which may be primary or follow the ulcero-glandular disease; the distinction between the primary and secondary forms is often far from clear (Avery and Barnett 1967). In fatal cases pneumonia is common. Death, when it occurs, is commonest during the third week. In those patients who have been examined at autopsy, multiple abscesses in the regional lymphatic glands, the spleen, liver and lungs have often been found (see Francis and Callender 1927, Foulger *et al.* 1932). In laboratory workers the typhoidal form of disease is more common than the ulcero-glandular form.

Diagnosis, prevention, and treatment

Diagnosis in animals is best made by macroscopic, microscopic, cultural, and pathogenicity tests. In the differential diagnosis from plague, account should be taken of the absence of pus at the site of infection, the greater variability in size of the granules in the spleen, the rarity with which the lungs are affected, and the failure of the organism to develop on ordinary media. The presence of specific antigen in the tissues may be demonstrated by a thermoprecipitin test (Larson 1951), and the presence of organisms by immunofluorescence (Karlsson and Söderlind 1973). In man, discharge from the local lesions or glands should be cultivated on special media (see Chapter 38) and injected into guinea-pigs (Francis 1923). The organism may often be demonstrated in the sputum by intraperitoneal inoculation of mice (Larson 1951). The agglutination test is of special value (Francis and Evans 1926). It becomes positive, as a rule, during the 2nd week of the disease, and reaches its maximum between the 4th and 8th weeks. The titre varies from about 1/160 to 1/5000. Sera agglutinating *F. tularensis* at 1/320 or higher often act on *Brucella abortus* as well. In some cases the two organisms may be agglutinated to the same degree. Agglutinins to *F. tularensis* persist in low titre for years after the attack (Ransmeier and Ewing 1941). Massey and Mangiafico (1974) described a microagglutination technique. Curiously enough the organism does not often appear to be present in the blood (Lake and Francis 1922). Intradermal diagnostic tests have been described (Foshay, 1936), and a haemagglutination test has been recommended (Charkes 1959).

Prevention of the disease lies in avoiding contact with potentially infected animals or other infective material. Special care is needed to avoid tick bites. Immunity after an attack is lasting. Cellular mechanisms of immunity, similar to those that protect against other facultatively intracellular bacteria, are thought to play a more important role than humoral mechanisms (Kostiala *et al.* 1975). Prophylactic vaccines (Kadull *et al.* 1950, Hornick and Eigelsbach 1966) are of some value. Non-living vaccines are much less effective than living vaccines (Hambleton *et al.* 1974).

Burke (1977) made an analysis of cases of tularaemia in laboratory workers from 1950 to 1969. During this period, a phenol-killed vaccine was used for the first 10 years and the so-called 'live vaccine strain' (LVS) thereafter. The use of LVS was associated with a striking fall in the incidence of typhoidal tularaemia; the incidence of ulcero-glandular tularaemia was unchanged, but the disease was less severe than it had been in the workers who received killed vaccine.

Both for prophylaxis and therapy streptomycin, which kills the organism, is the drug of choice. The tetracyclines, which are merely bacteristatic, are generally effective provided they are given in sufficient dosage, e.g., 2 g a day, for 14 days (Sawyer *et al.* 1966). (For a detailed account of the occurrence, epidemiology, and diagnosis of the disease in animals, see Olsufyev (1963) and Hopla 1974.)

References

Åarsleff, B., Biberstein, E. L., Shreeve, B. J. and Thompson, D. A. (1970) *J. comp. Path.* **80**, 493.

Ahlqvist, J., Ahvonen, P., Räsänen, J. A. and Wallgren, G. R. (1971) *Acta path. microbiol. scand.* Sect. A, **79**, 109.

Ahvonen, P. (1973) *See* Report 1973, p. 133.

Ahvonen, P., Jansson, E. and Aho, K. (1969) *Acta path. microbiol. scand.* **75**, 291.

Ahvonen, P., Thal, E. and Vasenius, H. (1973) *See* Report 1973, p. 135.

Albizo, J. M. and Surgalla, M. J. (1968) *Appl. Microbiol.* **16**, 1114.

Andrews, R. V. and Belknap, R. W. (1983) *J. Hyg., Camb.* **91**, 359.

Asakawa, Y., Akahane, S., Kagata, N., Noguchi, M., Sakazaki, R. and Tamura, K. (1973) *J. Hyg., Camb.* **71**, 715.

Assal, N., Blenden, D. C. and Price, E. R. (1967) *Publ. Hlth Rep., Wash.* **82**, 627.

Avery, F. W. and Barnett, T. B. (1967) *Am. Rev. resp. Dis.* **95**, 584.

Baba, T., Ando, T. and Nukina, M. (1978) *J. med. Microbiol.* **11**, 281.

Bahmanyar, M. and Cavanaugh, D. C. (1976) *Plague Manual.* World Hlth Org., Geneva.

Bain, R. V. S. (1963) *Haemorrhagic Ssepticaemia*, FAO Agricultural Studies No. 62, Rome.

Baltazard, M. and Bahmanyar, M. (1960) *Bull. World Hlth Org.* **23**, 169.

Baltazard, M., Bahmanyar, M., Mostachfi, P., Eftekhari, M. and Mofidi, C. (1960) *Bull. World Hlth Org.* **23**, 141.

Baltazard, M. and Seydian, B. (1960) *Bull. World Hlth. Org.* **23**, 157.

Bentley, E. W. (1967) *WHO Chron.* **21**, 363.

Berglund, A. (1973) *See* Report 1973, p. 232.

Biberstein, E. L. and Francis, C. K. (1968) *J. med. Microbiol.* **1**, 105.

Biberstein, E. L., Gills, M. and Knight, H. (1960) *Cornell Vet.* **50**, 283.

Biberstein, E. L., Shreeve, B. J. and Thompson, D. A. (1970) *J. comp. Path.* **80**, 499.

Black, R. E. *et al.* (1978) *New Engl. J. Med.* **298**, 76.

Bockemühl, J. and Roth, J. (1978) *Zbl. Bakt. I. Abt. Orig. A* **240**, 86.

Bonebakker, A. (1936) *Geneesk. Tijdschr. Ned. Ind.* **76**, 1890.

Bordas, Dubief and Tanon (1922) *Pr. méd.* **30**, Part II, p. 831.

Bosworth, T. J. and Lovell, R. (1944) *J. comp. Path.* **54**, 168.

Bottone, E. J. (1977) *CRC Crit. Rev. Microbiol.* **5**, 211; (1978) *J. clin. Microbiol.* **7**, 562; [Ed.] (1981) *Yersinia enterocolitica.* CRC Press, Boca Raton, Florida.

Bradley, J. M. and Skinner, J. I. (1974) *J. med. Microbiol.* **7**, 383.

Brenner, D. J., Ursing, J., Bercovier, H., Steigerwalt, A. G., Fanning, G. R., Alonso, J. M. and Mollaret, H. H. (1980) *Curr. Microbiol.* **4**, 195.

Brogden, K. A. and Packer, R. A. (1979) *Amer. J. vet. Res.* **40**, 1332.

Brooks. J. E., Htun, P. T. and Naing, H. (1980) *J. Hyg., Camb.* **84,** 127.

Broquet, C. (1908) *Bull. Soc. Path. exot.* **1,** 547.

Bulstrode, H. T. (1910–11) *40th ann. Rep. loc. Govt Bd, med. Offr's Suppl.*, p. 36.

Burke, D. S. (1977) *J. infect. Dis.* **135,** 55.

Burroughs, A. L. (1947) *J. Hyg., Camb.* **45,** 371.

Burroughs, A. L., Holdenried, R., Longanecker, D. S. and Meyer, K. F. (1945) *J. infect. Dis.* **76,** 115.

Burrows, T. W. (1963) *Ergebn. Hyg. Bakt.* **37,** 59.

Carter, G. R. (1955) *Amer. J. vet. Res.* **16,** 481; (1956) *Canad. J. comp. Med.* **20,** 289; (1961) *Vet. Rec.* **73,** 1052; (1962) *Canad. J. publ. Hlth* **53,** 158; (1967) *Advances in Veterinary Science*, Vol. 11, p. 321. Academic Press, New York; (1972) *Avian Dis.* **16,** 1109; (1977) *Proc. 19th annu. Meeting Amer. Ass. vet. Lab. Diagnosticians, Miami Beach, Florida, 1976*, p. 189.

Carter, G. R. and Bain, R. V. S. (1960) *Vet. Rev. Annot.* **6,** 105.

Cary, S. G. and Blair, E. B. (1964) *J. Bact.* **88,** 96.

Caten, J. L. and Kartman, L. (1968) *J. Amer. med. Ass.* **205,** 333.

Cavanaugh, D. C. *et al.* (1968) *Amer. J. publ. Hlth* **58,** 742.

Cavanaugh, D. C. and Randall, R. (1959) *J. Immunol.* **83,** 348.

Chadwick, P. (1963) *Canad. J. Microbiol.* **9,** 829.

Charkes, N. D. (1959) *J. Immunol.* **83,** 213.

Charrin and Roger, G. H. (1888) *C. R. Soc. Biol.* **5,** 272.

Chen, T. H. and Meyer, K. F. (1966) *Bull. World Hlth Org.* **34,** 911.

Collins, F. M. (1977) *Cornell Vet.* **67,** 103.

Controni, G. and Jones, R. S. (1967) *Amer. J. med. Technol.* **33,** 379.

Cook, R. (1952) *J. Path. Bact.* **64,** 228.

Crumpton, M. J. and Davies, D. A. L. (1957) *Nature, Lond.* **180,** 863.

Dahlstrand, S., Ringertz, O. and Zetterberg, B. (1971) *Scand. J. infect. Dis.* **3,** 7.

Daniëls, J. J. H. M. (1973) *See* Report 1973, p. 210.

Davies, D. H., Dungworth, D. L., Humphreys, S. and Johnson, A. J. (1977) *N.Z. vet. J.* **25,** 263.

Davis, D. H. S. (1948) *Ann. trop. Med. Parasit.* **42,** 207.

Davis, R. A. (1970) *Control of Rats and Mice*, Minist. Agricult. Fish. and Food, Bull. 181. H.M.S.O., London.

Devignat, R. (1953) *Schweiz. Z. allg. Path.* **16,** 509.

Dirks, C., Schöss, P. and Schimmelpfennig, H. (1973) *Dtsch. tierärztl. Wschr.* **80,** 342.

Dor, L. (1888) *C. R. Soc. Biol.* **5,** 449.

Dungal, N. (1931) *J. comp. Path.* **44,** 126.

Dyson, D. A., Gilmour, N. J. L. and Angus, K. W. (1981) *J. med. Microbiol.* **14,** 89.

Eastwood, A. and Griffith, F. (1914) *J. Hyg., Camb.* **14,** 285.

Editorial (1925*a*) *Publ. Hlth Rep., Wash.* **40,** 51; (1925*b*) *Brit. med. J.* **i,** 851.

Eskey, C. R. and Haas, V. H. (1940) *Publ. Hlth Bull., Wash.* No. 254.

Esseveld, H. and Goudzwaard, C. (1973) *See* Report 1973, p. 99.

Fenyuk, B. K. (1960) *Bull. World Hlth Org.* **23,** 263.

Florent, A. and Godbille, M. (1950) *Ann. Méd. vét.* **94,** 337.

Foshay, L. (1936) *J. infect. Dis.* **59,** 330.

Foulger, M., Glazer, A. M. and Foshay, L. (1932) *J. Amer. med. Ass.* **98,** 951.

Francis, D. P., Holmes, M. A. and Brandon, G. (1975) *J. Amer. med. Ass.* **233,** 42.

Francis, E. (1921) *Publ. Hlth Rep., Wash.* **36,** 1731; (1923) *Ibid*, **38,** 1391.

Francis, E. and Callender, G. R. (1927) *Arch. Path. Lab. Med.* **3,** 577.

Francis, E. and Evans, A. C. (1926) *Publ. Hlth Rep., Wash.* **41,** 1273.

Francis, E. and Lake, G. C. (1922) *Publ. Hlth Rep., Wash.* **37,** 83.

Frederiksen, W. (1964) *Proc. 14th scand. Congr. Path. Microbiol., Oslo*, p. 103.

Galli-Valerio, B. (1900) *Zbl. Bakt.* **27,** 1; (1903) *Ibid. Ref.* **33,** 753.

Gauthier, J. C. and Raybaud, A. (1902) *C. R. Soc. Biol.* **54,** 1497; (1903) *Rev. Hyg.* **25,** 426.

Gilmour, N. J. L. (1978) *Vet. Rec.* **102,** 100.

Gilmour, N. J. L., Martin, W. B., Sharp, J. M., Thompson, D. A. and Wells, P. W. (1983) *Res. vet. Sci.* **35,** 80.

Gilmour, N. J. L., Thompson, D. A. and Fraser, J. (1974) *Res. vet. Sci.* **17,** 413.

Girard, G. (1934) *C. R. Soc. Biol.* **117,** 601; (1936) *Ann. Méd. Pharm. colon.* **34,** 235; (1941) *Bull. Soc. Path. exot.* **34,** 37; (1952) *Bull. World Hlth Org.* **2,** 109.

Goldenberg, M. J., Hudson, B. W. and Kartman, L. (1970). In: *Diagnostic Procedures for Bacterial, Mycotic and Parasitic Infections*, 5th edn. Part ii, pp. 422, 439. Ed. by H. L. Bodily, E. L. Updyke and J. O. Mason. Amer. Publ. Hlth Ass., New York.

Gouchenour, W. S. (1924) *J. Amer. vet. med. Ass.* **65,** 433.

Gourlay, R. N. and Barber, L. (1960) *J. comp. Path.* **70,** 211.

Haffkine, W. M. (1905) *Sci. Mem. med. sanit. Dep. India* New Ser., No. 20.

Hambleton, P., Evans, C. G. T., Hood, A. M. and Strange, R. E. (1974) *Brit. J. exp. Path.* **55,** 363.

Hampton, B. C. (1940) *Publ. Hlth Rep., Wash.* **55,** 1143.

Hanna, M. O., Zink, D. L., Carpenter, Z. L. and Vanderzant, C. (1976) *J. Food Sci.* **41,** 1254.

Hansen, K. and Köster, H. (1936) *Dtsch. tierärztl. Wschr.* **44,** 739.

Hartley, W. J. and Boyes, B. W. (1955) *Proc. N. Z. Soc. Anim. Prod.* **15,** 120.

Hässig, A., Karrer, J. and Pusterla, F. (1949) *Schweiz. med. Wschr.* **79,** 971.

Heddleston, K. L. (1976) *Amer. J. vet. Res.* **37,** 745.

Heddleston, K. L., Gallagher, J. E. and Rebers, P. A. (1972) *Avian Dis.* **16,** 925.

Henderson, A. (1963) *Antonie van Leeuwenhoek* **29,** 359.

Henriksen, S. D. and Jyssum, K. (1960) *Acta path. microbiol. scand.* **50,** 443.

Highsmith, A.K., Feeley, J. C. and Morris, G. K. (1977) *Hlth Lab. Sci.* **14,** 253.

Hill, A. B. (1934) *Spec. Rep. Ser. med. Res. Coun., Lond.* No. 196.

Hirst, P. (1926) *Ceylon J. med. Sci.* Sect. D, **1,** 155; (1927) *Ibid.* **1,** 273.

Hoekenga, M. T. (1947) *J. trop. Med. Hyg.* **50,** 190.

Hopla, C. E. (1974) *Advanc. vet. Sci.* **18,** 25.

Hornick, R. B. and Eigelsbach, H. T. (1966) *Bact. Rev.* **30,** 532.

Hubbert, W. T. and Rosen, M. N. (1970) *Amer. J. publ. Hlth* **60,** 1103, 1109.

Hudson, J. R. (1959) *In Infectious Diseases of Animals, Diseases due to Bacteria*, Vol. 2, p. 413. Ed. by A. W. Stableforth and I. A. Galloway. Butterworth Scientific Publications, London.

Humphreys, F. A. and Campbell, A. G. (1947) *Canad. J. publ. Hlth* **38**, 124.

Hurvell, B. and Lindberg, A. A. (1973) *See* Report 1973, p. 159.

Itoh, M., Tierno, P. M., Milstoc, M. and Berger, A. R. (1980) *Amer. J. publ. Hlth* **70**, 1170.

Jawetz, E. (1950) *J. infect. Dis.* 86, 172.

Jellison, W. L. (1972) *Bull. Hist. Med.* **46**, 477.

Jellison, W. L. and Kohls, G. M. (1955) *U.S. Publ. Hlth Monogr.* No. 28.

Jellison, W. L., Owen, C. R., Bell, J. F. and Kohls, G. M. (1961) *Wildl. Dis.* No. 17, 1.

Jericho, K. W. F. and Langford, E. V. (1978) *Canad. J. comp. Med.* **42**, 269.

Jettmar, H. M. (1923) *Z. Hyg. InfektKr.* **97**, 322.

Joltrain, E. (1936) *Bull. Acad. Méd., Paris* **116**, 601.

Jones, D. M. (1962) *J. Path. Bact.* **83**, 143.

Jones, F. S. (1921) *J. exp. Med.* **34**, 561.

Jones, G. E., Field, A. C., Gilmour, J. S., Rae, A. G., Nettleton, P. F. and McLaughlan, M. (1982) *Vet. Rec.* **110**, 168.

Jones, G. E., Gilmour, J. S. and Rae, A. (1978) *J. comp. Path.* **88**, 85.

Jubb, K. V. F. and Kennedy, P. C. (1963) *Pathology of Domestic Animals*, Vol. 1, p. 457, Academic Press, New York and London.

Jusatz, H. J. (1961) *Z. Hyg. InfektKr.* **148**, 69.

Kadull, P. J., Reames, H. R., Corieli, L. L. and Foshay, L. (1950) *J. Immunol.* **65**, 425.

Kalabukhov, N. I. (1965) *J. Hyg. Epidem., Praha* **9**, 147.

Kaneko, K., Hamada, S. and Kato, E. (1977) *Jap. J. vet. Sci.* **39**, 407.

Kapperud, G. (1981). *Acta path. microbiol. scand.* Sect. B, **89**, 29.

Karlsson, K.-A. and Söderlind, O. (1973) *See* Report 1973, p. 224.

Kaul, P. M. (1949) *WHO epidem. vital Statist. Rep.* **2**, 1942.

Kauntze, W. H. (1926) *Bull. Hyg., Lond.* **1**, 66.

Keppie, J., Cocking, E. C. and Smith, H. (1958) *Lancet* **i**, 246.

Kielstein, P., Martin, J. and Janetschke, P. (1977) *Arch. exp. VetMed.* **31**, 609.

Kirschner, L. (1934) *Geneesk. Tijdschr. Ned.-Ind.* **74**, 1141.

Kister (1930) *Zbl. Bakt.* **117**, 433.

Kitasato, S. (1894) *Lancet* **ii**, 428.

Knapp, W. (1954) *Zbl. Bakt.* **161**, 422; (1956) *Z. Hyg. InfektKr.* **143**, 261; (1958) *New Engl. J. Med.* **259**, 776; (1959) *Ergebn. Mikrobiol. Immun Forsch.* **32**, 196.; (1969) *Münch. med. Wschr.* **111**, 2633.

Knapp, W. and Thal, E. (1963) *Zbl. Bakt.* **190**, 472.

Kostiala, A. A. I., McGregor, D. D. and Logie, P. S. (1975) *Immunology* **28**, 855.

Kraus, R. (1897) *Wien. klin. Wschr.* **10**, 736.

Kruif, P. H. de (1921) *J. exp. Med.* **33**, 773; (1922*a*) *Ibid.* **35**, 561; (1922*b*) *Ibid.* **36**, 309; (1923) *Ibid.* **37**, 647.

Lake, G. C. and Francis, E. (1922) *Publ. Hlth Rep., Wash.* **37**, 392.

Larson, C. L. (1951) *J. Immunol.* **66**, 249.

Larson, C. L., Philip, C. B., Wicht, W. C. and Hughes, L. E. (1951) *J. Immunol.* **67**, 289.

Legters, L. J., Coltingham, A. J. and Hunter, D. G. (1969) *Bull. World Hlth Org.* **41**, 859.

Lignières, M. J. (1901) *Ann. Inst. Pasteur* **15**, 734.

Link, V. B. (1954) *Publ. Hlth Mono.* No. 26. Wash.

Liston, W. G. (1924) *Brit. med. J.* **i**, 900, 950, 997.

Loghem, J. J. van (1939) *Bull. kolon. Inst. Amsterdam* **2**, 131.

Lovell, R. (1939) *Vet. Rec.* **51**, 747.

Lysy, J. and Knapp, W. (1973) *See* Report 1973, p. 42.

Macalister, G. H. and Brooks, B. St J. (1914) *J. Hyg., Camb.* **14**, 316.

MacArthur, W. P. (1952) *Trans. R. Soc. trop. Med. Hyg.* **46**, 209.

Macchiavello, A. (1941) *Contribuciones al Estudio de la Peste bubonica en el Nordeste del Brazil.* Pan American Sanitary Bureau, Washington, D. C.; (1946) *Amer. J. publ. Hlth* **36**, 842.

McCoy, G. W. (1910) *J. Hyg., Camb.* **10**, 589; (1911) *Publ. Hlth Bull., Wash.* No. 43, p. 53.

McCoy, G. W. and Chapin, C. W. (1912) *J. infect. Dis.* **10**, 61.

Mair, N. S. (1968) *Int. Symp. Pseudotuberculosis, Paris 1967*, p. 121. S. Karger, Basel; (1977) In: *Topics in Gastroenterology*, No. 5, p. 325. Blackwell, Oxford.

Malassez, L. C. and Vignal, W. (1883) *Arch. Phys. norm. path.*, 3rd Ser. **2**, 369.

Marshall, J. D., Joy, R. J. T., Ai, N. V., Quy, D. V., Stockard, J. L. and Gibson, F. L. (1967) *Amer. J. Epidem.* **86**, 603.

Massey, E. D. and Mangiafico, J. A. (1974) *Appl. Microbiol.* **27**, 25.

Meyer, K. F. (1950) *J. Immunol.* **64**, 139; (1970) *Bull. World Hlth Org.* **42**, 653.

Miller, L. G. (1974) *Canad. J. Microbiol.* **20**, 1585.

Mitchell, J. A. (1921) *J. Hyg., Camb.* **20**, 377; (1930) *J. Hyg., Camb.* **29**, 394.

Mollaret, H. H. (1968) *Int. Symp. Pseudotuberculosis, Paris 1967*, p. 45. S. Karger, Basel; (1971) *Path. et Biol., Paris* **19**, 189; (1972) *Ann. Biol. clin.* **30**, 1.

Montgomerie, R. F., Bosworth, T. J. and Glover, R. E. (1938) *J. comp. Path.* **51**, 87.

Moody, M. D. and Winter, C. C. (1959) *J. infect. Dis.* **104**, 288.

Morris, C. (1971) The Plague in Britain. *Historical J.* **14**, 205.

Morris, E. J. (1958) *J. gen. Microbiol.* **19**, 305.

Morris, G. K. and Feeley, J. C. (1976) *Bull. World Hlth Org.* **54**, 79.

Mráz, O. (1969) *Zbl. Bakt. I. Abt. Orig.* **209**, 349.

Müller, H. E. and Krasemann, C. (1974) *Zbl. Bakt. I. Abt. Orig. A*, **229**, 391.

Nagy, L. K. and Penn, C. W. (1976) *Res. vet. Sci.* **20**, 249.

Namioka, S. (1973) *See* Report 1973, p. 177.

Newsom, I. E. and Cross, F. (1932) *J. Amer. vet. med. Ass.* **80**, 711.

Nicolle, P., Mollaret, H. H. and Brault, J. (1973) *See* Report 1973, p. 54.

Niléhn, B. (1969) *Acta path. microbiol. scand.*, suppl. 206, 1; (1973) *See* Report 1973, p. 59

Niléhn, B., Sjöström, B., Damgaard, K. and Kindmark, C. (1968) *Acta path. microbiol. scand.* **74**, 101.

Nocard (1889) *C. R. Soc. Biol.* **1**, 608.

Nuttall, G. H. F. (1897) *Zbl. Bakt.* **22**, 87; (1898) *Ibid.* **23**, 625.

Ogata, M. (1897) *Zbl. Bakt.* **21**, 769.

Ohara, H. (1930) *Zbl. Bakt.* **117**, 440.

Olin, G. (1938) *Bull. Off. int. Hyg. publ.* **30**, 2804.

Olson, B. J. (1938) *Bull. Off. int. Hyg. publ.* **30**, 2808.

Olšovský, Z., Olšáková, V., Chobot, S. and Sviridov, V. (1975) *J. Hyg. Epidem., Praha* **19**, 22.

Olsufjev, N. G. (1970) *J. Hyg. Epidem., Praha* **14,** 67.

Olsufyev, N. (1963) In: *Human Diseases with Natural Foci,* Ed. by Y. N. Pavlovsky, p. 219. Foreign Languages Publishing House, Moscow.

Otten, L. (1936) *Indian J. med. Res.* **24,** 73; (1938) *Meded. Dienst Volksgezondh. Ned. Ind.* **27,** 111; (1941) *Ibid.* **30,** 61.

Pallaske, G. (1933) *Z. InfektKr. Haustiere* **44,** 43.

Pallaske, G. and Meyn, A. (1932) *Dtsch. tierärztl. Wschr.* **40,** 577.

Parker, R. R., Brooks, C. S. and Hadleigh, M. (1929) *Publ. Hlth Rep., Wash.* **44,** 1299.

Parker, R. R. and Spencer, R. R. (1926) *Publ. Hlth Rep., Wash.* **41,** 1341, 1403.

Parker, R. R., Spencer, R. R. and Francis, E. (1924) *Publ. Hlth Rep., Wash.* **39,** 1057.

Paterson, J. S. and Cook, R. (1963) *J. Path. Bact.* **85,** 241.

Pavlovsky, Y. N. (1963) *Human Diseases with Natural Foci.* Foreign Languages Publishing House, Moscow.

Pedersen, K. B. and Winblad, S. (1979) *Acta path. microbiol. scand.* Sect B, **87,** 137.

Petrischeva, P. A. (1966) *Control of Diseases with a Natural Reservoir.* USSR Minist. publ. Hlth, Moscow.

Pfeiffer, A. (1890) *Zbl. Bakt.* **7,** 219.

Philip, W. M. and Hirst, L. F. (1917) *J. Hyg., Camb.* **15,** 527.

Piras, L. (1913) *Zbl. Bakt.* **71,** 69.

Pirie, J. H. H. (1929) *Publ. S. Afr. Inst. med. Res.* **4,** 218.

Pirie, J. H. H. and Grasset, E. (1938) *S. Afr. med. J.* **12,** 294.

Poelma, F. G., Borst, G. H. A. and Zwart, P. (1977) *Acta zoo. path. antverp.* No. 69, 3.

Poels, J. (1886) *Fortschr. Med.* **4,** 388.

Poland, J. D. and Barnes, A. M. (1979) In: *CRC Handbook Series in Zoonoses* Vol. 1, p. 515, ed. by J. H. Steele. CRC Press Inc., Boca Raton, Florida.

Pollitzer, R. (1954) Plague, *World Hlth Org., Monograph Ser.,* No. 22; (1960) *Bull. World Hlth Org.* **23,** 313.

Popek, K., Kopecná, E., Bieronská, N., Černý, Z., Janíček, B. and Kožušník, Z. (1969) *Zbl. Bakt.* **210,** 502.

Preisz, H. (1894) *Ann. Inst. Pasteur* **8,** 231.

Prodjoharjono, S., Carter, G. R. and Conner, G. H. (1974) *Amer. J. vet. Res.* **35,** 111.

Rabson, A. R. and Koornhof, H. J. (1973) *See* Report 1973, p. 102.

Radcliffe, J. N. (1879–80) *9th Rep. loc. Govt Bd, Med. Offr's Suppl.,* p. 1.

Ransmeier, J. C. and Ewing, C. L. (1941) *J. infect. Dis.* **69,** 193.

Redfern, R. and Gill, J. E. (1980) *J. Hyg., Camb.* **84,** 263.

Rennison, B. D., Hammond, L. E. and Jones, G. L. (1968) *J. Hyg., Camb.* **66,** 147.

Report. (1899) *German Plague Comm., Arb. ReichsgesundhAmt.* **16,** 1; (1901) *Indian Plague Comm.* **5,** 181, H.M.S.O.; (1906) *English Plague Comm., J. Hyg. Camb.* **6,** 421; (1907) *Ibid* **7,** 324; (1908) *Ibid.* **8,** 162; (1910) *Ibid.* **10,** 315; (1914) *Ibid.* **13,** Suppl., p. 423; (1915*a*) *Ibid.* **14,** Suppl., p. 770; (1915*b*) *Ibid.* **14,** Suppl., p. 774; (1915*c*) *Ibid.* **14,** Suppl., p. 777; (1936) *Rep. Haffkine Inst., Bombay,* Govt Central Press; (1937) *Ibid*; (1946) *Infestation Control: Rats and Mice.* Minist. Food, Lond., HMSO.; (1956) *Bull. World Hlth Org.* **14,** 457; (1959) *World Hlth Org., Tec. Rep. Ser.,* No. 165; (1960) *Bull. World Hlth Org.* **23,** No. 2–3, pp. 135–418; (1965) *Pan. Amer. sanit. Bur., Wash., Sci. Publ.* No. 115; (1967) Meet the Rat. *World Hlth,* April; (1968) *Int. Symp. Pseudotuberculosis, Paris,* 1967. S. Karger, Basel; (1970) *WHO Techn. Rep. Ser.,* No. 447; (1973)

Contributions to Microbiology and Immunology, Vol. 2. *Yersinia, Pasteurella and Francisella.* S. Karger, Basel; (1977) *Bull. Off. int. Epizoot.* **87,** 607; (1979) *Contributions to Microbiology and Immunology,* Vol. 5. *Yersinia enterocolitica: Biology, Epidemiology, and Pathology.* S. Karger, Basel; (1980) *Bull. World Hlth Org.* **58,** 519.

Reynolds, M. T., Keane, C. T., Tomkin, G. H., Roberts, J. C., Lenehan, T. J. and Mair, N. S. (1978) *Brit. med. J.* **ii,** 400.

Ringertz, O. and Dahlstrand, S. (1968) *Acta path. microbiol. scand.* **72,** 464.

Roberts, R. S. (1947) *J. comp. Path.* **57,** 261.

Robertson, H. McG. (1923) *Publ. Hlth Rep., Wash.* **38,** 1519.

Rogers, B. T., Anderson, J. C., Paomer, C. A. and Henderson, W. G. (1973) *J. clin. Path.* **26,** 396.

Rutter, J. M. and Luther, P. D. (1983) *Vet. Rec.* **113,** 304.

Rutter, J. M. and Rojas, X. (1982) *Vet. Rec.* **110,** 531.

Sarchi, G. J. (1930) *Zbl. Bakt.* **117,** 367.

Sawyer, W. D., Dangerfield, H. G., Hogge, A. L. and Crozier, D. (1966) *Bact. Rev.* **30,** 542.

Schaaf, A. van der (1968) *Int. Symp. Pseudotuberculosis, Paris* 1967, p. 193. S. Karger, Basel.

Schiemann, D. A. (1978) *Canad. J. Microbiol.* **24,** 1048.

Schiemann, D. A. and Toma, S. (1978) *Appl. environ. Microbiol.* **35,** 54.

Schleifstein, J. and Coleman, M. B. (1939) *N.Y. St. J. Med.* **39,** 1749.

Schmidt, J. (1959) *Arch. Hyg.* **143,** 262.

Schütze, H. (1939) *Brit. J. exp. Path.* **20,** 235.

Seal, S. C. (1960) *Bull. World Hlth Org.* **23,** 283, 293.

Sharp, J. M., Gilmour, N. J. L., Thompson, D. A. and Rushton, B. (1978) *J. comp. Path.* **88,** 237.

Shepherd, A. J., Leman, P. A. and Barnett, R. J. (1982) *J. Hyg., Camb.* **89,** 79.

Shrewsbury, J. F. D. (1964) *The Plague of the Philistines and other Medical-Historical Essays.* Victor Gollancz Ltd, London; (1970) *A History of Bubonic Plague in the British Isles.* Cambridge University Press, Cambridge.

Simond, P.-L. (1898) *Ann. Inst. Pasteur* **12,** 625.

Smith, G. R. (1960*a*) *J. comp. Path.* **70,** 326; (1960*b*) *Ibid.* 429; (1961*a*) *J. Path. Bact.* **81,** 431; (1961*b*) *J. comp. Path.* **71,** 94; (1962) *J. Path. Bact.* **83,** 501; (1964) *J. comp. Path.* **74,** 241.

Smith, I. M., Hodges, R. T., Betts, A. O. and Hayward, A. H. S. (1973) *J. comp. Path.* **83,** 307.

Smith, J. E. (1955) *J. comp. Path.* **65,** 239; (1958) *Ibid.* **68,** 315; (1959) *Ibid.* **69,** 231.

Smith, J. E. and Thal, E. (1965) *Acta path. microbiol. scand.* **64,** 213.

Sokhey, S. S. and Chitre, G. D. (1937) *Bull. Off. int. Hyg. publ.* **29,** 2093.

Soltész, L. V., Schalén, C. and Mårdh, P.-A. (1980) *Acta path. microbiol. scand., B* **88,** 11.

Somov, G. P. and Martinevsky, I. L. (1973) *See* Report 1973, p. 214.

Spink, W. W. (1978) *Infectious Diseases.* Dawson & Son Ltd, Folkestone.

Stamp, J. T. and Nisbet, D. I. (1963) *J. comp. Path.* **73,** 319.

Stamp, J. T., Watt, J. A. A. and Thomlinson, J. R. (1955) *J. comp. Path.* **65,** 183.

Strong, R. P., Teague, O., Barber, M. A. and Crowell, B. C. (1912) *Philipp. J. Sci., B* **7,** 129–270.

Swaminathan, B., Harmon, M. C. and Mehlman, I. J. (1982) *J. appl. Bact.* **52,** 151.

Swellengrebel, N. H. and Hoesen, H. W. (1915) *Z. Hyg. InfektKr.* **79,** 436.

Taylor, J. (1933) *Indian J. med. Res., Memo.* No. 27.

Thjøtta, T. (1931a) *Bull. Hyg., Lond.* **5,** 490; (1931b) *J. infect, Dis.* **49,** 99.

Tigertt, W. D. (1962) *Bact. Rev.* **26,** 354.

Toma, S. and Deidrick (1975) *J. clin. microbiol.* **2,** 478.

Topping, N. H., Watts, C. E. and Lillie, R. D. (1938) *Publ. Hlth Rep., Wash.* **53,** 1340.

Tsareva, M. I. (1959) *J. Microbiol. Epidem. Immunobiol.* **30,** 34.

Tschurilina, A. A. (1930) *Z. Hyg. InfektKr.* **111,** 198.

Tsubokura, M. *et al.* (1976) *Jap. J. vet. Sci.* **38,** 549.

Tumansky, V. and Kolesnikova, Z. (1935) *Rev. Microbiol., Saratov* **14,** 269.

Tweed, W. and Edington, J. W. (1930) *J. comp. Path.* **43,** 234.

Uriarte, L. (1925) *C. R. Soc. Biol.* **92,** 901.

Vandepitte, J., Wauters, G. and Isebaert, A. (1973) *See* Report 1973, p. 111.

Wagle, P. M., Sokhey, S. S., Dikshit, B. B. and Ganapathy, K. (1941) *Indian med. Gaz.* **76,** 29.

Warner, C. E. (1914) *J. Hyg., Camb.* **14,** 360.

Wauters, G. (1970) Thesis. University of Louvain, Belgium; (1973a) *See* Report 1973, p. 38; (1973b) *See* Report 1973, p. 68.

Wauters, G., Le Minor, L. and Chalon, A. M. (1971) *Ann. Inst. Pasteur* **120,** 631.

Wauters, G., Le Minor, L., Chalon, A. M. and Lassen, J. (1972) *Ann. Inst. Pasteur* **122,** 951.

Weaver, R. E. and Jordan, J. G. (1973) *See* Report 1973, p. 120.

Webster, L. T. (1924a) *J. exp. Med.* **39,** 837, 843, 857; (1924b) *Ibid.* **40,** 109, 117.

Weidenmüller, H. (1968) *Int. Symp. Pseudotuberculosis, Paris* 1967, p. 63. S. Karger, Basel.

Wherry, W. B. and Lamb, B. H. (1914) *J. infect. Dis.* **15,** 331.

White, F. N. (1918–19) *Indian J. med. Res.* **6,** 190.

Williams, C. L. and Kemmerer, T. W. (1923) *Publ. Hlth Rep., Wash.* **38,** 1873.

Williams, J. E. and Cavanaugh, D. C. (1979) *Bull. World Hlth Org.* **57,** 309.

Williams, J. E., Hudson, B. W., Turner, R. W., Saroso, J. S. and Cavanaugh, D. C. (1980) *Bull. World Hlth Org.* **58,** 459.

Winblad, S. (1967) *Acta path microbiol. scand.* suppl. 187, 115; (1973) *See* Report 1973, p. 129.

Winblad, S., Niléhn, B. and Jonsson, M. (1966) *Acta path. microbiol. scand.* **67,** 537.

Wright, J. (1936) *J. Path. Bact.* **42,** 209.

Wu Lien Teh (1913) *J. Hyg., Camb.* **13,** 237; (1922) *Ibid.* **21,** 62.

Wulf, C. J. M. A. de, Esseveld, H. and Goudzwaard, C. (1969) *Ned. Tijdschr. Geneesk.* **113,** 665.

Yersin, A. (1894) *Ann. Inst. Pasteur* **8,** 662; (1897) *Ibid.* **11,** 81; (1899) *Ibid.* **13,** 251.

Yersin, Calmette and Borrel (1895) *Ann. Inst. Pasteur* **9,** 589.

Zabolotny, D. (1923) *Ann. Inst. Pasteur* **37,** 618.

Zagari (1890) *Zbl. Bakt.* **8,** 208.

Zen-Yoji, H., Maruyama, T., Sakai, S., Kimura, S., Mizuno, T. and Momose, T. (1973) *Jap. J. Microbiol.* **17,** 220.

Ziegler, P. (1969) *The Black Death.* Collins, London.

Zlatogoroff, S. J. (1904) *Zbl. Bakt.* **36,** 559.

Zwanenberg, D. van (1970) *Med. Hist.* **14,** 63.

56

Brucella infections of man and animals, campylobacter abortion, and contagious equine metritis

Graham Wilson and Geoffrey Smith

Brucella infections

Undulant fever

Ever since the time of Hippocrates a low type of fever characterized by fairly regular remissions or intermissions has been recognized along the Mediterranean littoral. Several names have been applied to it, of which Malta fever and Mediterranean fever are the two best known. In 1887 Bruce reported the discovery of a small micro-organism, which he called *Micrococcus melitensis*, in the spleen of patients dying of the disease. Its injection into monkeys gave rise to a re-mittent fever, sometimes terminating fatally, and the organism was recovered in pure culture from the liver and spleen. Nearly twenty years later Zammit (1905) of the Malta Board of Health showed that the organism also infected goats. Further work rendered it evident that the goat was in fact the natural host of *Brucella melitensis*, and that infection was carried to man by the consumption of raw milk.

In 1918 Evans drew attention to the similarity be-

141

tween the organism isolated from Malta fever and that described by Bang (1897) as the cause of a widespread disease in cattle known as contagious abortion. Her observations were soon confirmed, and in honour of Sir David Bruce the generic name *Brucella* was proposed for the two organisms. The close similarity between *Br. melitensis* and *Br. abortus* suggested the possibility that this latter organism might be pathogenic for human beings. It was not long before Bevan (1921–22) brought evidence to prove the correctness of this surmise. He drew attention to the occurrence in Southern Rhodesia of cases of undulant fever affecting persons who had apparently had no direct or indirect contact with goats. In the areas concerned epizoötic abortion of cattle was prevalent, and it appeared probable that infection was being transferred to man by raw cows' milk. Numerous investigators turned their attention to this subject, and before long cases of undulant fever due to *Br. abortus* had been reported from practically every country in the world.

A third causal agent of this disease is *Brucella suis*— an organism originally described by Traum (1914) in the United States. Its natural habitat is the pig. Man becomes infected chiefly by contact with diseased carcasses, and undulant fever due to this organism has therefore a very definite occupational incidence. As pointed out in Chapter 40, three other species of *Brucella* are now recognized, namely *Br. canis*, *Br. ovis*, and *Br. neotomae*. These findings have necessitated a reconsideration of the nomenclature of the disease. Brucella infections are known to be widespread over the globe, so that the term Malta fever is no longer generally applicable. It is being gradually replaced by the term undulant fever, or even more by the term brucellosis, both of which are free from any geographical connotation.

For the sake of clarity we propose to give a separate description of the epidemiology of undulant fever in man due to the three type organisms. We shall deal first with infections caused by *Br. melitensis*, including in our description the general symptomatology and bacteriology of the disease.

Undulant fever due to *Br. melitensis*

Synonyms: Malta fever, Mediterranean fever, Neapolitan fever, Country fever of Constantinople, New fever of Crete, Rock fever of Gibraltar. French: *Mélitococcie.*

Quoting from the classical monograph of Hughes (1897): 'Clinically, the fever has a peculiarly irregular temperature curve, consisting of intermittent waves or undulations of pyrexia, of a distinctly remittent character. These pyrexial waves or undulations last, as a rule, from 1 to 3 weeks, with an apyrexial interval, or period of temporary abatement of pyrexial intensity between, lasting for 2 or more days. . . . This pyrexia is usually accompanied by obstinate constipation, pro-

gressive anaemia, and debility. It is often complicated with, and followed by, neuralgic symptoms referred to the peripheral or central nervous system; arthritic effusions; painful inflammatory conditions of certain fibrous structures, of a localized nature; or swelling of the testicles.' Numerous clinical types are recognized, such as the malignant, the undulatory, the intermittent, and the irregular type. In addition, there are subclinical infections, which can be diagnosed only by bacteriological methods, and infections that may remain latent for weeks or months before flaring up or retrogressing. For a description of these the reader is referred to Marston (1863), who first separated Malta fever from other fevers, to Hughes (1897), whose clinical study has never been improved upon, and more recently to Rainsford (1935) and Lisbonne and Janbon (1935). Several complications may occur, beyond those already mentioned, including osteomyelitis, which usually take the form of localized bone abscesses, bronchitis, and acute inflammation of one or other of the visceral organs (see Michel-Béchet *et al.* 1939).

The commonest symptoms are probably asthenia, fever, muscular and articular pains, nocturnal sweats—often drenching—anorexia, constipation, nervous irritability, and chills or rigors (Taylor and Hazemann 1932).

The incubation period is variable. It may be as short as a week or as long as several months. Most commonly it is 10–30 days. The duration of the illness varies from a few days to over a year, 3 months being usual. Relapses, however, may occur from time to time over a period of many years. Alice Evans, for example, who contracted the disease in 1922, had recurrent attacks in 1923, 1928, 1931, and 1943, accompanied by ill health and periods of complete incapacity alternating with periods of recovery between the attacks (O'Hern 1977).

Bacteriology

Bacteriologically, the work of Bruce (1887, 1888, 1893), of Hughes (1893, 1897), and of the Mediterranean Fever Commission (Report 1905–07) showed that the disease was a septicaemia. Working in Malta, Bruce (1893) was able on two occasions to grow the organism from the juice obtained by splenic puncture, and subsequent observers showed that, in the early stages of the disease, *Br. melitensis* could frequently be isolated from the peripheral blood. The post-mortem appearances agreed with this interpretation; there was an increase of pericardial fluid, congestion and enlargement of the liver, and great hypertrophy of the spleen. The presence of small areas of congestion in the mucosa and submucosa of the intestine, and the enlargement of the mesenteric glands, the interior of which often consisted of semi-purulent material, were suggestive of an alimentary mode of infection. Cultures taken *post mortem* from the spleen, liver, and lymphatic glands were positive in 100 per cent of cases, from the kidney in 85 per cent, and from the heart blood, pericardial fluid and bone marrow in about 50 per cent (Kennedy 1906). *Br. melitensis* was never found in the saliva or in the sweat of patients, but it was shown to be quite commonly excreted in the urine. Though excretion is often irregular, probably at least 75 per cent of patients suffering from Mediterranean fever pass the specific agent at some time or other in their urine (Kennedy 1905). Excretion may last for weeks and extend onwards into convalescence. The organism is found in the gall-bladder *post mortem* and is probably present in the faeces during life. Eyre (1908) cultivated it from the colon of a case coming to autopsy.

Epidemiology in Malta

The epidemiology of undulant fever due to *Br. melitensis* varies in different countries. Since it has been studied most fully in Malta, we may review briefly the findings of the Mediterranean Fever Commission (Report 1905–07) in this island. The maximum incidence rate in Malta was found to occur in July and August. Generally speaking, it varied directly with the atmospheric temperature and inversely with the rainfall. The most susceptible age was between 11 and 30 years, but the disease occurred at any time of life. The upper classes of the civilian population and the officers of the services suffered more heavily than the lower classes and the non-commissioned ranks respectively. During the 33 years 1859–91 the incidence of the disease varied between 269.5 per 1000 in 1859 and 91.2 per 1000 in 1888. In spite of this comparatively high rate of morbidity, the case-fatality rate was low—about 2 per cent. But the duration of the fever and the prolonged disability following its actual termination combined to render it a potent cause of invalidity.

At the time of the Commission's appointment the mode of infection was still unknown. Several ways were suggested, but little evidence was produced in favour of any particular one. Johnstone (1905), who conducted an epidemiological inquiry into the disease, reported that neither food nor drink appeared to have any clear connection with the spread of the fever; nor did dust infection or direct personal contagion play a part. Similarly Davies (1906) concluded that neither water, milk, nor any other article of food seemed to be responsible in any way; he favoured a theory of contagion, either direct or through the agency of mosquitoes.

It was left to Zammit of the Malta Board of Health, and a member of the Commission on Mediterranean Fever, to indicate the true path of infection. In his first communication (1905) he reported that 5 out of 6 goats which he had examined gave a positive agglutination reaction to *Br. melitensis*. From the blood of two

of these animals, he succeeded in cultivating the actual organism itself. Horrocks (1905) rapidly confirmed this, and further demonstrated the frequent presence of *Br. melitensis* in the milk and the urine of apparently healthy goats. Kennedy (1905) who examined the blood serum of 161 goats from 8 different herds, demonstrated the presence of specific agglutinins in no fewer than 84 or 52.2 per cent of specimens. This high incidence of infection existed not only in the public herds, but in animals that were kept privately under special care.

As it was evident that *goats' milk* was probably the main source of infection, it was desirable to ascertain on a large scale the proportion of infected animals. For this purpose, Zammit's test was found to be of great value. It consists in observing the presence of agglutinins, not in the blood serum, but in the milk of the goats. The proportion of infected goats judged by this test is much lower than that given by the serum agglutination test, since only 60–85 per cent of the animals showing blood agglutinins also contain agglutinins in their milk. Zammit (1906) examined 710 samples of milk, and obtained 133 positive reactions. From every milk giving a strong reaction he was able to isolate *Br. melitensis* in culture. Horrocks and Kennedy (1906), who made numerous investigations both by the serum and by the milk agglutination test, came to the conclusion that 41 per cent of the goats in Malta were infected, and that 10 per cent of those that supplied milk were excreting *Br. melitensis*. It was found difficult to recognize the infected animals on clinical grounds; they did not suffer from a true fever, and though some showed loss of weight, a thinning of the coat, and a short hacking cough, others remained perfectly well and displayed no sign of illness. The amount of milk secreted was likewise no criterion, for the infected animals frequently yielded as much as or more than the non-infected ones.

As most of the milk consumed in the island was goats' milk—cows were few owing to insufficient pasturage—it was easy to see, from these investigations, how the disease might be transmitted to human beings. Steps were taken by both the Navy and the Army, in 1906, to stop the supply of goats' milk to the troops. The result was most striking. Within a year the disease had been practically eradicated, only a few easily explained cases occurring (Table 56.1).

Among the civilian population such radical steps could not be taken and the disease continued to flourish. The incidence, which in 1926 was 26.2 per 10 000 inhabitants, rose to 74.4 in 1934. After a fall during the war years, when the consumption of goats' milk greatly decreased, it rose in 1946 to the peak figure of 81.6. Since that date, owing to the effect of pasteurization of goats' milk and of the increasing proportion of cows' milk consumed, the incidence fell progressively till in 1964 it was only 1.7 per 10 000 (Agius 1965).

Table 56.1   Undulant fever in Malta taken from Eyre 1908, 1912.

	Civil		Navy		Army	
	Cases	Deaths	Cases	Deaths	Cases	Deaths
1901	642	54	252	3	253	9
1902	624	45	354	2	155	6
1903	589	48	339	6	404	9
1904	573	59	333	8	320	12
1905	663	88	270	7	643	16
1906	822	117	145	4	163	2
1907	714	78	12	0	9	1
1908	502	?	6	—	5	?
1909	456	?	10	—	1	?
1910	318	?	3	—	1	?

It should be added that, as in laboratory workers, disease may result from contamination of the skin or mucosae with infective material. Occasional cases also occur suggestive of insect-borne or sexually transmitted infection. According to Tovar (1947) ticks, bugs and fleas can be readily infected with all three species of *Brucella*, and ticks and bugs may continue to excrete the organisms in the faeces for at least three months.

The history of Mediterranean fever is instructive. In the first place it affords an example of a disease whose incidence was not immediately affected by the discovery of the causative agent. In order to prevent an infectious disease, knowledge of the reservoir of infection and of the means by which it is transferred to the animal body is often of greater importance than that of the exact identity of the parasite. Yellow fever affords a case in point. Once infection had been proved to be carried by mosquitoes, eradication of the disease became possible, even before the causative virus had been recognized. It must of course be admitted that proof of the way in which infection spreads may not become possible till the identity of the parasite is known; but apart from immunization of the host, measures for preventing infectious disease are as a rule most profitably directed to control or eradication of the reservoir of infection or to interference with the means by which infection is transmitted to the susceptible host.

The second point of interest is that, in an epidemiological inquiry, it is harder to assess the importance of a given factor when it operates more or less uniformly throughout a population than when it is distributed in such a way as to affect different sections unevenly. Both the epidemiologists who investigated the disease in Malta (Johnstone 1905, Davies 1906) reported that there was no evidence to suggest that infection was conveyed by food. Both of them overlooked the fact that a disease spread by an article of almost universal consumption, such as goats' milk, would show little preference for any one class of the population, if the contamination of the milk were widespread and frequent rather than local and occa-

sional. The clue to the problem was furnished by the bacteriologist. It was this second discovery, that a large proportion of goats contained agglutinins in their blood for *Br. melitensis*, which complemented the first discovery, that of the organism itself, and pointed out the way by which the spread of the disease might be checked.

(For an excellent account of the epidemiology of Malta fever in the Royal Navy, see Dudley 1931).

Epidemiology in other countries

The next most important bacteriological and epidemiological study of the disease was made by Taylor and his colleagues in the south of *France*, working at the Undulant Fever Centre established at Montpellier by the Rockefeller Foundation (see Taylor and Hazemann 1932, Taylor, Lisbonne, and Roman 1932, Taylor, Lisbonne, and Vidal 1935, Taylor *et al.* 1938). The disease was found to be widespread in the southeast of France, where sheep and goats are common and cattle are relatively few. The incidence in males was 2–3 times that in females, and was highest in the 15–45 age group. The occupational incidence was very striking, the majority of the cases occurring in the agricultural community, especially among those engaged in rearing sheep and goats. The seasonal prevalence was at its maximum in the spring, corresponding to the time of lambing and abortion. Though the consumption of goats' milk, and of fresh cheese made from goats' or ewes' milk, undoubtedly played some part in infection (see Carrieu and Lafenêtre 1932, Veloppé and Jaubert 1935), the evidence on the whole suggested that infection most frequently resulted from direct contact with infected animals and manure. In a population, however, living under rather primitive conditions, and exposed to many different sources of infection, it was difficult to ascertain precisely the most important route by which the organisms gained access to the body. The interesting observation was made (Taylor, Vidal, and Roman 1934) that cows in close contact with sheep and goats might become infected with *Br. melitensis*, excrete this organism for months in the milk, and give rise to undulant fever in human beings consuming it in the raw condition.

The disease affects several other countries, including Italy, Spain, Greece, Transcaucasia, Algeria, Tunis Palestine, Arabia, India, South Africa, China, and various parts of South America. Like Malta, Corsica and Sicily are seriously affected. In Britain, owing to the absence of infection in sheep and goats, indigenous disease caused by *Br. melitensis* is unknown. Occasional cases are seen in persons infected abroad; and one small outbreak after the eating of Italian pecorino cheese made from raw sheep's milk is recorded by Galbraith and his colleagues (1969). In the United States there is a certain amount of undulant fever due to *Br. melitensis* in the goat-raising areas, but most of it is due to *Br. abortus* and *Br. suis*.

Undulant fever due to *Br. abortus*

In its general symptomatology and bacteriology this disease resembles that caused by *Br. melitensis*. Though, as will be pointed out later, the ability of *Br. abortus* to give rise to undulant fever is less than that of *Br. melitensis*, once the typical disease has developed, it seems to pursue its course more or less independently of the type of the causative organism. The incubation period is very variable, ranging from a week or so to 6 months or more (see Hardy, Frant, and Kroll 1938). The usual duration of the disease is 13 weeks, and the case fatality about 2 per cent. Besides typical cases, however, a considerable number of so-called subclinical infections occur, characterized by only slight pyrexia and by a variety of different symptoms. Sometimes there is an influenza-like illness lasting for not more than a week, while at others a low intermittent fever may persist for months or even years (see Dalrymple-Champneys 1960). Acute abdominal symptoms are not uncommon, and in several patients an appendicectomy or cholecystectomy has been performed before the real nature of the disease was recognized (Simpson 1930). The disease may become chronic. Spink (1956) in Minnesota reported on 68 untreated patients, mostly infected with *Br. abortus*, who suffered from continued ill health after the acute phase had passed. Thirty of these patients still had symptoms after a year. Localizing manifestations, such as spondylitis, osteitis, cholecystitis, and radiculoneuritis, were common, and many patients passed into a neurasthenic state.

Most cases are due to infection with *Br. abortus* biotype 1, but occasional cases caused by biotype 5, the so-called British melitensis type, are on record (Barrow and Peel 1965, Robertson 1967). Among veterinarians infection with the attenuated vaccinal strain S19 is not uncommon (see Pivnick *et al.* 1966).

Infection during pregnancy is sometimes, but not often, followed by abortion. The organism may be isolated from the placenta and uterus of the mother and from the stomach contents of the fetus (see Kristensen and Holm 1929, Spengler 1929, Alessandrini and Pacelli 1932, Habs 1933, Dalrymple-Champneys 1960).

Incidence

Though the disease is widespread throughout the world, its incidence is difficult to assess. As in tuberculosis, the reported cases may bear very little relation to the real prevalence of infection. Fifty years ago the nearest approach we could make was by noting the proportion of sera from patients with undiagnosed pyrexia that agglutinated *Br. abortus* to a significant titre, and calculating the percentage they formed of the total number of Widal sera examined in the country. Figures obtained in this way from different European countries about 1930 were reproduced in the 6th edition (Table 74.2, p. 2177). They showed that the incidence per million of the population was much the same in Great Britain, Germany, Austria, Holland, and Sweden, but higher in Denmark and Switzerland. Since 1930 contagious abortion of cattle has been practically or completely eradicated in these countries, so that the incidence is now very low indeed.

In the United States undulant fever is known to be very prevalent, but some of the cases are due to infection with *Br. suis* and a few with *Br. melitensis*. In England and Wales cases of undulant fever are reported to the Communicable Diseases Research Centre at Colindale from Public Health and Hospital laboratories throughout the country. During the five years 1976-80 the average number of cases per annun was 47. This is necessarily an underestimate, but as it is impossible to know how many cases were undiagnosed the real number cannot be known. There is reason, however, to believe that it would probably be not greatly above the reported number. Most cases of undulant fever are sporadic, but occasional small outbreaks occur (Elkington *et al.* 1940, Steele and Hastings 1948, Anderson 1950, Henderson 1964).

Latent infections

Besides definite clinical disease, *Br. abortus* may give rise to latent infection. Examination of Wassermann sera in Britain and in Germany showed that in 1930 about 1.5 per cent agglutinated *Br. abortus* to a titre of 1/40 or over. The titre is generally less than 1/160, and high titres, such as those yielded by the majority of sera from patients with undulant fever, are uncommon. A few of these sera are derived from patients actually suffering from the disease. The majority, however, come from persons with no clinical evidence of undulant fever. The probability is that these agglutinins are due to latent infections with *Br. abortus*. This interpretation is supported by two facts. The first is that agglutinins, even in a titre of 1/20, are rarely present in the sera of persons who do not drink raw milk or cream, and who are not exposed to contact with infected animals. The second is that agglutinins are present in quite a high proportion of persons who drink quantities of infected raw milk or cream, or whose duties bring them into close contact with infected animals (Table 56.2). (For references to figures used, see 4th ed., p. 1905.) The figures in the table include the United States and the Argentine, in both of which there is some infection with *Br. suis* and *Br. melitensis*, but it will be seen that in all countries about 10-20 per cent of persons in these special occupational classes possessed agglutinins to *Brucella*. Some of these persons were suffering from undulant fever at the time of examination, or had recently recovered from it, but most of them had no symptoms indicative of active infection. An exception must be made for

Table 56.2   Proportion of sera agglutinating *Br. abortus* to 1/40 or over in persons exposed to infection about the years 1930–35 (figures compiled from several papers in different countries)

Occupational class	Country	No. of sera	No. positive	Percentage positive
Milk and farm employees	Germany	220	31	14.1
	Hungary	63	10	15.9
	New Zealand	104	17	16.4
	Argentine	136	16	11.8
Slaughterers	Great Britain	206	27	13.1
	Hungary	93	21	22.6
	USA	452	62	13.7
	Argentine	1776	191	10.8
Veterinarians	Great Britain	63	13	20.6
	France	28	7	25.0
	Denmark	94	22	23.4
	USA	715	92	12.9
	Argentine	110	29	26.4

veterinary surgeons, several of whom suffered from allergy, which manifested itself in the form of a rash on the arm developing after 'cleansing', i.e. evacuation of the cow's uterus (see Huddleson and Johnson 1930, Haxthausen and Thomsen 1931, Wilson 1932, Thomsen 1937*b*). There is little doubt that the figures given in Table 56.2 underestimate the true number, because in latent infections the direct agglutination test is often negative. By the use of other serological methods it can be shown that in Britain and Eire something like 60–90 per cent of veterinary surgeons show evidence of present or past infection (Kerr *et al.* 1966, Meenan 1967). The figures in Table 56.2 are quoted mainly to show the effect of occupation on the risk of infection with *Br. abortus.* As already pointed out, the eradication of contagious abortion has rendered them no longer applicable.

Infection with *Br. abortus* resembles in many respects that with the tubercle bacillus. Both organisms have a fairly high degree of infectivity enabling them to establish themselves, at least temporarily, in the tissues. Neither of them, however, has a high pathogenicity, so that infection usually remains latent or retrogresses. Only in a small proportion of infected persons are the conditions favourable for the production of clinically detectable disease. Latent infections probably lead to latent immunization, thus explaining why undulant fever does not occur more often than it apparently does in veterinarians. There is also evidence to suggest that persons who are exposed to heavy infection for the first time, or after a long interval of freedom from exposure, are more likely to suffer from the disease than those who are exposed to mild infection more or less continuously. Garrod (1937) drew attention to the susceptibility of *Br. abortus* to the bactericidal action of the gastric juice, and suggested that this might be one reason why undulant fever is not commoner than it is. The possibility that

strains of *Br. abortus* vary naturally in virulence must be considered, but so far there is no evidence to support it (see Birch and Gilman 1935, Olin 1935).

Mode of infection

The mode of infection varies in different countries and in different parts of the same country. The main source is obviously cattle, and man becomes infected by consumption of raw milk or cream, or by contact with infected animals, either alive or dead. Generally speaking, the town population is exposed to infection from milk, whereas the country population is exposed to both milk-borne and direct contact infection.

Numerous records in different countries show that a considerable proportion of samples of raw milk coming from infected herds contain living *Br. abortus* bacilli. In Great Britain about 5 per cent of herd samples were infected (Report 1961). Experience has shown that *consumption of raw infected milk* is attended by a risk of contracting the disease, whereas consumption of the same milk after pasteurization or other effective heat treatment is harmless. In London, where 99 per cent of the milk is heat-treated, and in the large cities of the United States where most of the milk is pasteurized, undulant fever has long been practically unknown. The cases that occurred were mostly among those who drank Tuberculin Tested or other types of raw milk, either at home or on a visit to the country.

Several studies made in institutions and small communities supplied with infected raw milk have shown that a high proportion of the population develop serum agglutinins, that a small proportion suffer from subclinical and clinical forms of undulant fever, and that in most of the persons exposed the infection remains completely latent. Removal of the infected animals, or pasteurization of the milk supply, is followed by cessation of clinical illness and by a gradual disappearance of agglutinins from the sera of latently infected persons. (For references see 4th ed., p. 1906.)

It may be pointed out here that alimentary infection is almost always due to the consumption of raw milk or cream. *Butter* and *cheese*, except when made from untreated milk and consumed within a week or two, are probably inoffensive, since the organisms die out very rapidly as the result of lactic fermentation (Mazé and Césari 1931, Caronna 1936, Taylor *et al.* 1938). Moreover, a large proportion of manufactured butter is made from pasteurized cream in which the organisms are destroyed (see J. Smith, 1934*a*, Pullinger 1935). One curious laboratory outbreak, possibly associated with defective plumbing, is on record (Huddleson and Munger 1940).

Contact infections are mainly occupational, occurring in towns among slaughterers and meat packers, and in the country among the farming population. Since the latter class, in particular, is usually exposed to infection from raw milk, the parts played by the two sources of infection are difficult to determine. In Britain, Dalrymple-Champneys (1960) ascribed about 80 per cent of cases to the consumption of raw milk or cream. Other observers pay rather more attention to contact infection, but the frequency of the two methods must vary with the nature of the population concerned. Contact infection is mainly from cattle, but horses and even dogs may at times constitute the source of infection (Menzani 1932).

The age and sex incidence is a little peculiar. In both town and country, the brunt of the disease is borne by adult males. During the six years 1956-61, of the 494 cases reported to the Public Health Laboratory Service in England and Wales, 69 per cent were in males. The incidence was highest in the 10-50 year age group; 71 per cent of the 494 cases fell into this group. (See also Poole 1975.) Occupationally, farm and dairy workers and veterinary surgeons and a miscellany of other occupations contributed to the male incidence; among females, housewives and school children were mainly affected.

Laboratory infections, though less common than with *Br. melitensis*, are by no means infrequent, and probably occur through contamination of the abraded skin or the conjunctiva with infective material (see Meyer 1933, Meyer and Eddie 1941).

(For general epidemiological studies and reviews of the disease, see Spengler 1929, Dalrymple-Champneys 1929, 1960, Henricsson 1932, Olin 1935, Report 1950, Löffler, Moroni and Frei 1955, Olitzki 1970.)

Undulant fever due to *Br. suis*

Probably the first diagnosed case of this disease was that described by Keefer (1924) in 1922 in the United States, though it was some time before the identity of the causative organism was realized. Clinically and bacteriologically the disease is similar to undulant fever caused by the *Br. melitensis* and *Br. abortus* types. Like *Br. melitensis*, *Br. suis* can be demonstrated more frequently in the bloodstream than *Br. abortus*—possibly because of its greater ease of cultivation.

Epidemiologically, this disease is less extensive than the types already described. It is essentially an occupational disease, occurring among slaughterers and packers who handle the infected carcasses of pigs. The studies of Hardy and his colleagues (1931, 1932), of Huddleson, Johnson, and Hamann (1933*a*), and of Heathman (1934) in the middle western states of North America, have shown that a small amount of overt and a large amount of latent infection occurs in these workers. The organism apparently gains access through the abraded skin, but in one outbreak in an Iowa hog-slaughtering plant air-borne infection was suspected (Hendricks *et al.* 1962). Adult males are almost solely affected (Hasseltine 1930, Hendricks 1955). Occasionally *Br. suis* infects cattle, and milk-borne outbreaks of undulant fever, without any special occupational incidence, are likely to occur (see Beattie and Rice 1934, Horning 1935, Borts *et al.* 1943). Nicoletti, Quinn and Minor (1967) ascribe one human case to infection from a dog; and biotype 4, which infects wild reindeer, is responsible for some cases of undulant fever among Eskimos in Alaska (Huntley *et al.* 1963, Meyer 1964). Brucella infection of pigs is far less common than contagious abortion of cattle. In consequence undulant fever due to *Br. suis* has hitherto been confined mainly to the middle-west of North America, to Brazil, and to the Argentine (Molinelli *et al.* 1948), though it is met with in several other countries. One case has been reported from Eire (Williams *et al.* 1957).

Diagnosis of undulant fever

The occurrence of subclinical and atypical infections renders the clinical diagnosis of undulant fever peculiarly difficult, and all observers are agreed on the fact that large numbers of cases are missed. The final diagnosis can be made only by bacteriological examination. In practice it is well to regard every case of pyrexia of undiagnosed origin as possibly due to brucella infection, and to bear this disease in mind during the investigation of acute and chronic inflammatory illness in which the diagnosis remains doubtful. Help is sometimes afforded by a blood count, which usually shows a secondary anaemia, a mild leucopaenia, and a relative lymphocytosis.

Blood culture

In the febrile stages of the disease, an attempt should be made to isolate the causative organism from the bloodstream.

The patient's blood should be withdrawn while the

temperature is elevated, preferably during the rise of a pyrexial wave. It should be distributed in 3–5 ml quantities into two flasks containing 20–100 ml of glucose broth, serum broth, Bacto tryptose broth, or liver extract broth of pH 7.5. One of the flasks should be incubated aerobically, the other in an atmosphere in which 5–10 per cent of the air has been replaced by CO_2. Addition to the medium of erythritol, 2–8 μmol per ml, may stimulate the growth of abortus and melitensis, though not of suis, strains (Keppie *et al.* 1967). The growth of CO_2-dependent dye-sensitive strains is improved by the addition of 0.1 per cent Tween 40, of 0.5 per cent bovine serum, and of killed brucella organisms (Huddleson 1956, 1964). When contamination of the blood is suspected, 1 in 800 000 ethyl violet with or without an antibiotic mixture of polymyxin B 4 units/ml, bacitracin 10 units/ml, and actidione 0.001 mg/ml may be added to the medium (Morgan 1960, Painter *et al.* 1966, Alton and Jones 1967, Robertson 1967). Subcultures should be made every 3–5 days on to a solid medium such as glucose agar, trypticase soy agar, or tryptose agar—all enriched with 7.5 per cent bovine serum—and these should be incubated in the same atmosphere as the parent culture. Growth is often slow, and no culture should be discarded in less than 4–8 weeks. When fluid blood is not available, blood clot may be ground up and cultured in the usual way (West and Borman 1945). Castañeda's method of culturing blood in a bottle containing both solid and liquid medium saves the trouble of subculturing and has much to commend it (Alton and Jones 1967). (For other methods, see Huddleson 1957, Reusse and Schindler 1957.) Two ml of the patient's blood may be inoculated intraperitoneally into guinea-pigs (Poston 1938).

According to Abramson and his colleagues (1969), who made blood cultures from 476 hospital patients infected with *Br. melitensis*, positive results were obtained most frequently during the acute stage of the disease, when the patient was febrile, at a time of day when the temperature was raised, and in patients whose serum titre was 1/1400 or over.

Though blood culture is positive in a high proportion of *Br. melitensis* and *Br. suis* infections, not more than 10–20 per cent of *Br. abortus* infections prove positive by this method (but see Hosty and Johnson 1953). In *Br. melitensis* infections the organism can often be isolated from the urine, provided repeated samples are examined (Kennedy 1905). Occasionally *Br. abortus* can be demonstrated in the bile (Leavell and Amoss 1931), stools (Smith 1932, Goldstein *et al.* 1936) and tonsils (Carpenter and Boak 1932, Poelma and Pickens 1932).

Agglutination test

The most generally useful method is the agglutination test, first introduced by Wright and Smith in 1897.

Final dilutions of the serum should be put up ranging from 1 in 20 to 1 in 5120 against suspensions of strains corresponding to the prevalent type of infection in the country. Only strains that are absolutely smooth, as judged by the thermoagglutination and acriflavine tests, should be chosen. The organisms should be grown on liver extract or glycerol agar for 2–3 days, washed off with saline, and heated to 55° for 1 hour. The suspension should be standardized to match an opacity tube containing 1000 million *Escherichia coli* per ml and preserved with 0.5 per cent phenol. Suspensions prepared in this way will generally remain satisfactory for 3 months, but the moment signs of commencing stringiness appear they should be discarded. The test should include both an abortus and a melitensis suspension. Tubes should be incubated in a waterbath for 24 hours at 37° or for 18 hours at 50°. For surveys, when large numbers of sera have to be examined, the slide or plate method of agglutination may be used (Huddleson 1939, Alton and Jones 1967). According to Mittal and Tizard (1979*a*) the so-called *growth-agglutination* and *growth-inhibition tests* performed with growing cultures are more sensitive than the conventional tube-agglutination test.

The interpretation of the agglutination test is not always easy, and demands a knowledge of the clinical history and the occupation of the patient. Very much the same considerations affect the interpretation of this test as of that used in the diagnosis of enteric fever, and the reader may well study Chapter 68 in which the relevant factors are discussed in some detail. The propositions we lay down must not be regarded as more than suggestions, applying particularly to Great Britain. Each of them will necessarily be subject to exceptions. (1) A titre of 1/80 or less, in the absence of clinical symptoms, is indicative either of a latent brucella infection, or of a past infection—not necessarily attended by definite disease. (2) A titre of 1/80 or over, in the absence of clinical symptoms or of a recent pyrexial attack, is suggestive of frequent infections, usually occurring in persons drinking large quantities of infected raw milk or exposed to contact with infected animals or carcasses. (3) A titre of 1/80 or over in the presence of pyrexia and other symptoms of disease, occurring in a person whose occupation or habits do not expose him to special risk, is very suggestive of active infection with a member of the *Brucella* group. In persons belonging to the occupational classes referred to, in whom a latent infection is not uncommon, a titre of 1/80 is too low to be of diagnostic significance. On the other hand, a titre of 1/1000 or over is rarely met with except as the result of an active infection, and may usually be regarded as evidence of undulant fever. (4) A titre of 1/20–1/80, in the presence of clinical undulant fever, may likewise be considered as practically diagnostic of this disease. (5) The complete absence of agglutinins from a patient's serum does not exclude the diagnosis of brucella infection. Cases are on record in which a positive blood culture has been obtained in the presence of a negative agglutination reaction (see Gilbert and Dacey 1932, Heathman 1934, Taylor *et al* 1938). (6) A rise in titre in the presence of clinical illness is strongly suggestive of active infection; and a fall in titre after successful antibiotic treatment is of value in retrospective diagnosis. As a rule agglutinins are present in a suggestive titre by the end of the 2nd week, and in frank cases of undulant fever they generally rise to a titre of 1/640 or more. After the attack is over they tend to fall fairly rapidly, and may sink to a low level within 3 months. In chronic cases agglutinins may fluctuate in titre according to the activity of the infection, and may sometimes be no longer demonstrable.

Attention must be called to the not infrequent occurrence of a prozone, sometimes extending to even 1/640 in high-titre sera. For this reason a wide range of dilutions should always be put up (see Priestley 1931, Hirsch 1935).

The explanation of this phenomenon, as well as the failure of serum from some undulant fever patients to agglutinate *Brucella* at all, has been ascribed by various workers to the presence of non-agglutinating or so-called *agglutinin-blocking antibodies* (Griffiths 1947, Schuhardt *et al.* 1951, Wilson and Merrifield 1951). Anderson and his colleagues (Anderson *et al.* 1964, Reddin *et al.* 1965) showed that, in response to experimental infection in calves, two main immunoglobulins—IgG and IgM—appeared in the serum. The IgM or macroglobulins, which were sensitive to 2-mercaptoethanol and to heat (65° for 15 min) appeared first and were chiefly responsible for brucella agglutination. Later came the IgG microglobulins, which were not sensitive to 2-mercaptoethanol or to heat, and which played little part in agglutination but were closely related to the ability of the serum to fix complement. Further observations by Heremans, Vaerman and Vaerman (1963), Wilkinson (1966) and others showed that incomplete antibodies of the IgG and IgA type acted as blocking antibodies, preventing the IgM antibodies from causing agglutination. The attachment of the blocking antibody to the organisms can be demonstrated by the technique described by Coombs, Mourant and Race (1945), in which the organisms after exposure to the patient's serum are centrifuged, washed, resuspended in saline, and treated with the serum of a rabbit immunized against human globulin (Löffler *et al.* 1955; for references see Farnetani 1967). In the acute case of undulant fever the IgM antibodies are said by some workers to be usually in excess of the IgG antibodies, so that, though the blocking phenomenon is evident in low dilutions of serum, agglutination occurs readily in the higher dilutions. In chronic cases, however, the IgM antibody may largely have disappeared and no agglutination may be evident by the direct test. But the existence of IgG antibodies may be revealed either by the Coombs

test or by the complement-fixation test (Kerr *et al.* 1967, 1968, Schassan 1968, Allan *et al.* 1976).

The results of these tests must, however, be interpreted with caution. Williams (1970), for example, sometimes observed strongly positive reactions in veterinarians who were perfectly well, and negative reactions in patients who were clearly suffering from infection. The position is, in fact, far from clear. Serre, Dana and Roux (1970) ascribe the blocking phenomenon to simple excess of antibody. The effect can be removed by dilution, or by suspending the antigen–antibody mixture in 5 per cent saline, or according to Cruickshank (1956) by centrifugation. In their experience real blocking antibody, which they believe to be IgA, is uncommon and is met with mainly in the chronic form of the disease. Diaz and Levieux (1972), on the other hand, believe that the blocking phenomenon results from an excess of non-agglutinating IgG1 which inhibits the activity of the normal agglutinating IgG2 antibody. Moreover, in contrast to the statement made above, experience, particularly of the Russian workers, has shown that IgG antibodies prevail in the acute disease and IgM in the chronic. Lindberg and his colleagues (1982), however, maintain that by use of the lipopolysaccharide antigen of *Br. abortus* a safe diagnosis of acute and chronic brucellosis may be made: the acute disease is marked by a high IgM and a rising IgG titre, and the chronic by a raised IgG titre alone. (See Chapter 11, p. 325.)

Several observers have stated that the agglutinin titre to *Brucella* may be high in patients suffering from enteric fever, tuberculosis, and other febrile diseases. The evidence for this statement is not very satisfactory. Apart from the anamnestic reaction, which is not likely to account for more than a slight raising of the titre (see Amoia 1933), the most common cause of non-specific agglutination is the use of a suspension made from strains that are not absolutely smooth. Such strains are often agglutinated by normal serum, and apparently even more often by the sera of febrile patients. Their use in the past has been very common and has given rise to much confusion in the literature. Smooth strains of *Br. abortus* may, however, be agglutinated to high titre by the serum of patients infected with *Yersinia enterocolitica* type 9 (Ahvonen *et al.* 1969). The reason for this is that *Brucella* and *Yersinia* have O antigens in common. *Yersinia*, however, is motile and gives rise to H antibodies. A rapid H agglutination test, therefore, distinguishes between the two organisms without interference from O antigens (Mittal and Tizard 1979*b*; see also Marx *et al.* 1975 and Lindberg *et al.* 1982). Agglutinins may also appear in the serum of persons injected with cholera vaccine (Eisele *et al.* 1947, Feeley 1969). With these exceptions it is generally permissible to conclude that an elevated titre to a smooth suspension is indicative of infection—latent, active or past—with an organism of the *Brucella* group.

Other tests for antibodies have been described, such as a *precipitin test* (Schlesmann 1932), a *flocculation test* (Hunter and Colbert 1956), and the indirect fluorescent antibody test, which agrees closely with these latter two tests (Edwards *et al.* 1970). Henderson and his colleagues (1976) found that immunofluorescence tests for IgG, IgM and IgA had no advantage over agglutination and complement-fixation tests in the diagnosis of brucella infection in veterinarians; nor had the 2-mercaptoethanol and anti-human globulin (Coombs) tests, which could well be dispensed with. Farrell, Robertson and Hinchliffe (1975) go so far as to say that the diagnosis of chronic brucella infection must rest on the interpretation of the clinical symptoms, since the serological picture thought to be indicative of the chronic disease may be found in persons after an acute infection with no clinical evidence of activity. Lindberg and his colleagues (1982), however, maintain that a safe diagnosis may be made by use of the lipopolysaccharide antigen of *Br. abortus*: the acute stage is marked by a high IgM and a rising IgG titre; the chronic stage by a raised IgG titre alone.

In Castañeda's (1961) *surface fixation test*—sometimes referred to as an *antigen fixation test*—a loopful of the serum to be tested is placed on a slip of absorbent paper over a dried spot of stained brucella antigen. Loopfuls of known positive and negative sera are placed on adjoining spots and the strip is dipped into saline. The saline rises in the strip over the negative serum spot carrying with it a comet-like trail of antigen. At the site of the positive serum the antigen remains fixed. With the test serum varying degrees of fixation are seen.

Haemagglutinins appear early in the disease (Diaz *et al.* 1967), and the *passive haemagglutination test* on the patient's serum seems to be more sensitive than either the agglutination or the complement-fixation test. Magee (1980) found that the *enzyme-labelled immunoabsorbent assay (ELISA) test* was antibody-class specific for IgG, IgM and IgA; and as sensitive as *radioimmunoassay*; the results agreed closely with those of the complement-fixation and Coombs tests. It recognized four groups of sera: (1) those from acute cases having a very high content of IgG and a moderate one of IgM and IgA, (2) those from patients with chronic disease having a high content of IgG but not of IgM or IgA; (3) those with a small content of IgG representing residual antibody; and (4) those with little or no brucella-specific antibody.

The type of infecting organism may often be determined by use of the quantitative absorption test described in Chapter 40, carried out on the patient's serum (see Habs and Sievert 1935).

It may be added that the drinking of raw milk containing antibodies in the absence of live bacilli is not followed in man by the appearance of agglutinins in the blood (see Boak and Carpenter 1929, Peterson 1935); nor is the ingestion of dead brucellae, except perhaps in large doses, likely to give rise to agglutinins in normal persons (Braude *et al.* 1949, McCullough 1949).

Brucellin test   In 1922 Burnet drew attention to an allergic skin test, performed by the intradermal inoculation into the arm of 0.05–0.1 ml of a filtrate of a 20-day-old broth culture of *Br. melitensis* (*melitin*) or *Br. abortus* (*abortin*). A positive reaction is characterized by the appearance in 6 hours of a slightly raised, sometimes tender, oedematous plaque, 2–6 cm in diameter, distinguished in colour from the surrounding skin. Some workers insist, for a positive reaction, on a minimum diameter of 0.5 cm induration and oedema at the end of 48 hours. A central nodule may develop and persist for several days. Pseudo-reactions generally appear rapidly and disappear within 24 hours. Numerous extracts of organisms have been introduced for this test such as *brucellin* (Olin 1935), *brucellergin* (Huddleson 1939), purified brucella protein (Morales-Otero and González 1949), and more recently the *MPB antigen* of Castañeda standardized in human beings (see Alton and Jones 1967), and the *melitin* of Renoux (1965) standardized in guinea-pigs; but as no standard preparation or standard dose has yet been agreed upon, the diagnostic value of the test is still under discussion. Generally speaking the reaction resembles the tuberculin reaction, in denoting the occurrence of infection without giving any clear indication of its activity. In undulant fever it usually becomes positive slowly and increases in intensity during convalescence; it may then remain positive for months or years. Great discretion has to be exercised in interpreting the test. In some countries 20–30 per cent of the normal population may react positively to it. It may be weak or absent in the presence of high-titre antibodies, or conversely it may be strong in the absence of antibodies. It is not a test to be relied upon in the diagnosis of active infection (Bradstreet *et al.* 1970). As in tuberculosis, the occurrence of primary infection may be signalized by an inflammatory response at the site of a previous skin injection—the so-called *delayed reaction* (see Howe *et al.* 1947).

For references to the opsono-cytophagic test of Huddleson, Johnson and Hamann (1933*b*), and the tropin reaction of van der Hoeden (1940) see 4th ed. p. 1911 and 5th ed., p. 2053.

For *post-mortem diagnosis*, cultures should be made from the spleen, liver, kidney and mesenteric nodes, and incubated aerobically and in 10 per cent CO_2. Alternatively, the ground-up organs may be suspended in broth and injected intramuscularly into guinea-pigs.

(For a detailed description of the performance of the various diagnostic tests, see Robertson *et al.* 1980.)

Prophylaxis and treatment

The prevention of the disease in human beings consists ideally in the complete abolition of contact, direct or indirect, with infected animals or their products. Where infection is chiefly milk-borne, this may be comparatively easy. Either total avoidance of milk, or its consumption only after pasteurization, boiling or sterilization, has been found in practice to be highly effective. Where, however, infection occurs mainly by contact, prevention becomes much more difficult.

Vaccination, for use in specially exposed persons, is still in the experimental stage. It is known that the immunogenic constituents of *Br. abortus* are located in the cell wall (Markenson *et al.* 1962, Smith *et al.* 1962*a*), and in the ribosomes (Corbel 1976), but so far a satisfactory vaccine has not been prepared. The vaccine, *Br. melitensis* Rev. 1, which is used for goats and sheep, is too toxic for human use (Pappagianis *et al.* 1966), and has, like *Br. abortus* S19 (see Pivnick *et al.* 1966), given rise to infection in man (Alton and Elberg 1967). (See the review of live and dead vaccines by Roux 1972.)

Treatment of the disease itself is unsatisfactory, and in spite of all measures the fever or various other manifestations may continue for months or even years. The sulphonamides alone are of little value (Wilson and Maier 1939, 1940); but administered in the form of sulphadiazine by the mouth along with streptomycin injected intramuscularly they are more effective (Spink *et al.* 1948, 1949*a*). Tetracyclines appear to be the most potent; combined with streptomycin they are the treatment of choice. Treatment should be thorough, and continued, if possible, for some time after the disappearance of symptoms; but several observers have warned against the danger of too heavy or too prolonged dosage (see Spink 1956, Report 1958, Trever *et al.* 1958).

Shock treatment with brucellin, or serum treatment, is no longer advocated (see 5th ed., p. 2053).

(For general review of brucella infection of man, see Olitzki 1970, Poole 1975; and of man and animals Symposium 1976, Ray 1979, Report 1980*b*; and for diagnosis, treatment and prevention of brucellosis, see Report 1981*b*.)

Brucella infection of sheep and goats

Reference has already been made, under the section dealing with undulant fever due to *Br. melitensis*, to infections of sheep and goats with this organism, and to the diagnosis of the disease by examination of the blood serum and milk serum for agglutinins. The disease exists particularly in countries along the Mediterranean littoral, in South Africa, in parts, notably Arizona, of the United States, and in some areas of the USSR (Karsten 1943). When first introduced into a herd, particularly during pregnancy, infection is frequently followed by a number of abortions, but unless fresh animals are added, it dies down, and subsequent abortions are uncommon. The infection, however, may become chronic, and persist for months or years.

Diagnosis presents difficulties. Even in culturally positive animals the agglutination test may be negative, and in uninfected animals it may occasionally be positive (Unel *et al.* 1969). In vaccinated animals it is useless (Alton 1967). The complement-fixation test is rather more sensitive, and is of value in vaccinated animals, in which complement-fixing antibodies disappear within six months of vaccination. The most specific test is the antiglobulin test, but even this is positive in only about 70 per cent of infected animals (Unel *et al.* 1969). For diagnosis in the individual animal Corbel and Lander (1982) recommend a combination of the serum agglutination, Coombs antiglobulin, and complement-fixation tests with *Br. ovis*. The intradermal melitin test (Huddleson and Johnson 1933, Taylor *et al.* 1935), carried out by injecting mel-

itin (see Alton and Jones 1967) into one of the skin folds between the base of the tail and the margin of the anus and examining for oedema 24 and 48 hours later, indicates that the animal has been infected at some time or other but gives no clue to the present activity of infection. At post-mortem examination the organisms can be isolated most readily from the lymphatic glands—particularly the supramammary and parotid—and the udder.

During life the organisms are often excreted in enormous numbers in the milk and urine; after abortion the vaginal discharge may remain highly infective for some days. Infection probably occurs most often through contamination of the mucosa of the alimentary and respiratory tracts, and of the eye, less often by direct contact through the skin (Report 1958). In Syria dogs are thought to play a big part in its dissemination (Hirchert *et al.* 1967). It seems to be more persistent in goats than in sheep, but in both it may die out spontaneously. Young animals are less susceptible than the sexually mature, and males than females. Cases of natural infection of sheep with *Br. abortus* do occur, but are uncommon (see Stoenner 1951, Allsup 1969).

In New Zealand and Australia an organism referred to as *Br. ovis* (see Chapter 40) gives rise to a disease characterized by epididymo-orchitis in rams, abortion in ewes, and neonatal mortality in lambs (Buddle 1956, Myers 1973).

Experimentally, the disease can be reproduced in

both sheep and goats by feeding and by cutaneous, subcutaneous, intravenous, and intraperitoneal inoculation with virulent cultures of *Br. melitensis*. The disease so produced is often afebrile and, in spite of a septicaemia, symptoms may be lacking. Many of the animals, however, are unthrifty; their coats become thin; their weight diminishes; and a short hacking cough develops. Mastitis and arthritis are not uncommon; male animals may suffer from orchitis, and females, when infected during pregnancy, may abort. Death rarely occurs. Bacteriologically, the organisms are found in the blood for a variable time after infection. They then disappear, first from the peripheral blood and most of the viscera, then from the spleen and kidneys, then from the superficial lymph glands, and last of all from the mammary glands. The milk does not usually become infective for 2–3 months after inoculation or feeding. Agglutinins, sometimes reaching a high titre, are demonstrable in the blood serum, usually within a fortnight or so of infection (see Horrocks and Kennedy 1906, Zammit 1906, Eyre *et al.* 1907).

Experimentally, disease can also be produced in sheep and goats by *Br. abortus* (see Bang 1897, Reinhardt and Gauss 1915, van der Hoeden 1933).

Control of the disease is beset with economic and social difficulties. The spread of infection should be prevented as far as possible by elimination of all animals excreting the organism in the milk or vaginal mucus, by provision of separate boxes at the time of lambing or abortion, by abolition of communal pastures, by restricting the movement of animals from infected herds, and by control over the sale of animals for breeding. Measures of this sort can be only partly successful because, as already pointed out, no means are available for detecting all infected animals.

Vaccination must be used in addition. Various killed vaccines have been tried, with indifferent results, but have now been largely replaced by a vaccine made with the living attenuated Rev. 1 strain of *Br. melitensis* (see Chapter 40). Injected into kids or lambs 10–12 weeks old in a dose of 10^9 cells, it gives rise to transitory production of antibodies and usually to protection of some years' duration (Alton and Jones 1967, Entessar *et al.* 1967). The vaccine is suitable for goats and sheep between 3 and 8 months of age. A smaller dose can be given to pregnant animals without risk of abortion and without apparent diminution in protective power. The smaller dose gives rise to only a transient agglutination response and an insignificant complement-fixation reaction (Alton *et al.* 1972, Jones and Berman 1976, Elberg 1980).

The ideal policy is one of eradication (see p. 158). All animals should be subjected to serological and skin tests, positive reactors should be slaughtered, and breeding should be carried out only from apparently uninfected animals. Herd tests should be made at 3-monthly intervals till no further reactors appear.

It may be noted that abortion of sheep and goats may also be caused by two subspecies of *Camp. fetus* (p. 163), by the chlamydia of enzoötic abortion (Chapter 76), and by two members of the *Salmonella* group—*Salm. abortus ovis* and *Salm. dublin* (Chapter 37). (For review of ovine abortion, see Meyer 1980.)

Contagious abortion in cattle

Avortement épizoötique (French). *Seuchenhaftes Verwerfen* (German).

Contagious abortion in cattle is an infectious disease, due in most cases to *Br. abortus* of Bang, and characterized by inflammatory changes of the uterine mucosa and fetal membranes, resulting as a rule in the premature expulsion of the fetus. Though regarded as contagious by Hutrel d'Arboval (1826) and Youatt (1834) (see Hutyra and Marek 1922), its infective nature was not shown definitely till 1878, when Lehnert transmitted the disease by intravaginal inoculation of pregnant cows with the vaginal discharge and placental tissue of aborting animals. In 1897 Bang, working with Stribolt, demonstrated microscopically a small gram-negative bacillus in the uterine exudate of a cow with impending abortion, and succeeded in isolating it in pure culture. The intravaginal injection of this organism into two pregnant cows gave rise to abortion, and from the uterine exudate the organism was recovered in each case. Bang's work was confirmed by Preisz (1903) and Nowak (1908) on the Continent, by M'Fadyean and Stockman (1909) in this country, and by MacNeal and Kerr (1910) in America.

Other causes of infectious abortion are *Camp. fetus* (p. 161), enteroviruses, *Salm. dublin*, and fungi, such as *Aspergillus fumigatus*.

Epidemiology and bacteriology

Though in the past the disease was widespread throughout Europe, North America, and most other countries in the world where there was a large population of cattle, it is being gradually brought under control. Norway, Sweden, Finland, Denmark, Switzerland, Holland, Czechoslovakia, and Yugoslavia have eradicated it completely, and the United States, Canada, and Australia are progressing rapidly to the same end. Figures of prevalence, therefore, are subject to such change as to render their reproduction misleading. An official survey undertaken in England and Wales in 1960–61 indicated that 25–30 per cent of all dairy herds were infected (Report 1964*a*). The subsequent introduction of an eradication scheme led to a progressive fall in the incidence of the disease and its eventual elimination. England and Wales were declared free from brucellosis on 1 November 1981. Scotland had already been free since 1 January 1980.

Large herds are more often infected than small. On account of the heavy loss from abortions, the frequent subsequent sterility of the animals, and the diminution in the milk yield, it is economically one of the most important diseases affecting cattle (Minett and Martin 1936, Spink *et al.* 1949*b*). The fact, moreover, that infection may be transmitted to man renders the disease of public health interest.

The special predilection of *Br. abortus* for the reproductive system was shown to be due in part to the high concentration of erythritol in the placenta of the cow and the seminal vesicles of the bull (Smith *et al.* 1962*b*, Keppie *et al.* 1965). This substance has a strong stimulating effect on the growth of the organisms. Such a demonstration is of particular interest, since it affords one of the few examples in which we have any explanation for the cause of tissue localization.

Though clearly it is only pregnant animals that can display the typical symptom of abortion, infection may be conveyed by natural channels to cattle of any age and either sex. It does not, however, become established in calves before the first oestrous cycle. The commonest time for abortion to occur is from the 5th to the 8th month of pregnancy; 35.4 per cent of cases occur in the 7th month alone (Wall 1911). Judged from experimental work, the incubation period is variable; according to Mohler and Traum (1911) it may last from 1 to 33 weeks.

Introduced into a fresh herd the disease may spread rapidly, assuming epidemic proportions. Provided that no new animals are imported, it loses its initial severity, passing into an endemic state in which, if no preventive measures are taken, it remains for some years. A pregnant animal becoming infected for the first time generally aborts at an early stage; at its next pregnancy abortion occurs either not at all, or not till a later period; and though abortion may be repeated on a third occasion it is much commoner for the calf to be delivered at full term. Bang (1897) reports that of 83 cows only 30 aborted in 2 successive years, and only 6 three times in succession. From this it may be gathered that an immunity is acquired which is usually sufficient to protect the animal against further attacks of the disease.

Infection, however, generally persists for a long time, often for life. The organisms tend to leave the uterus and settle down in the udder and adjacent lymphatic glands (see Doyle 1935) but invade the uterus again at the time of the next pregnancy. Such animals, whether they abort a second time or not, are liable to shed large numbers of bacilli in the uterine discharge at the time of parturition. When unrecognized as excreters, they afford a potent means of spreading infection in the herd. The later infection occurs during pregnancy, the more likely is the animal to abort (Thomsen 1937*a*). In the *udder Br. abortus* gives rise to small inflammatory foci situated in the alveoli, the interalveolar connective tissue, and along the lactiferous ducts (Runnels and Huddleson 1925, Ridala 1936); the lesions are too small to be detectable by clinical examination. Calves infected during the first few weeks of life rapidly clear themselves; but when they are infected experimentally during the second six months they may not succeed in getting rid of all organisms (Nagy and Hignett 1967). Nevertheless, under natural conditions infection acquired by calves and by heifers before sexual maturity tends to remain latent or to die out; it spares the reproductive organs, is limited to the reticuloendothelial system (see Lübke 1935*a*, *b*), and is followed by a substantial degree of immunity.

Excretion in the milk

From the udder the organisms are excreted in the *milk*. This important fact was first demonstrated by Schroeder and Cotton in 1911. Karsten (1932) found that the proportion of aborting cows in which the milk was subsequently shown to be infected varied on different farms from 24 to 70 per cent, and that in infected herds 19–33 per cent of cows going to full term excreted the organisms in their milk.

Numerous observers have shown that excretion is associated with the presence of agglutinins in the blood serum (Sheather 1923). The higher the serum titre, the more likely are organisms to be present in the milk. Pröscholdt (1932), for example, found that 53 per cent of cows with a serum titre of 1/100 or over were milk excreters. The same author observed that, of 208 cows whose milk was infected, 145 had aborted and 63 had calved normally. Lerche (1931) gave very similar figures. Infection of the milk may be as heavy after normal calving as after abortion (Stockmayer 1936).

The organisms are usually demonstrable in the milk within a week, though in some animals as long as five months may elapse before excretion begins (Wall 1930). Many cows cease excreting after a few weeks, but others remain infected for a year or two, or even permanently. Excretion is irregular and often intermittent.

The number of organisms present fluctuates from day to day. It is often highest at the beginning of lactation and gradually falls, but sometimes it is more abundant and more consistent during the latter part (Morgan and McDiarmid 1960). Except in the first week or two of lactation, when there may be as many as 200 000 per ml, it is uncommon for the milk on withdrawal to contain more than 30 000 per ml (Bang and Bendixen 1932a, Stockmayer 1936); the usual number is 10–2000 per ml (Karsten 1932). Disease is often confined to one or two quarters of the udder. The hind-quarters are more often affected than the fore, and the right hind-quarter seems to be affected more often than the left (Bang and Bendixen 1931, Doyle 1935).

Though the organism commonly responsible for the disease is *Br. abortus*, numerous observations have shown that cattle in close contact with infected goats or sheep may become infected with *Br. melitensis* (see Taylor, Vidal, and Roman 1934, Benussi 1935). The infection seems to be localized mainly in the udder, and abortion is uncommon. Similarly in large hog-raising districts, cattle may become infected with *Br. suis* and excrete this organism in the milk (see Beattie and Rice, 1934, Molinelli and Ithurrat 1934), but, as with *Br. melitensis*, the disease produced tends to be more self-limiting than that caused by *Br. abortus*.

Mode of infection and experimental reproduction of the disease

There has been a long dispute over the commonest mode of infection. Cattle are so easy to infect experimentally, not only by direct inoculation into the tissues, but also by the natural passages, that it is very difficult to decide on the route usually followed in the natural spread of disease. The three main portals are:

(1) THE MOUTH. Bang (1906) and others showed that it was possible to reproduce the disease in pregnant heifers and cows by feeding them with infective material, and therefore suggested that infection probably occurred through contaminated fodder. It is known that the uterine exudate which is voided at the time of the abortion is extremely infective, that the bacilli may remain alive in it for some time—for 7 months if kept in the ice-chest (Bang 1897); that the aborted fetus or fetal membranes deposited on grass and left at atmospheric temperature shielded from direct sunlight may remain infected for as long as 6 months (Bosworth 1934–35); and that unless strict isolation of the animals is practised, there is ample opportunity for contamination of the pasture and other food material. A further source of infection is the milk of the aborting cows, which often contains *Br. abortus* in large numbers. Rats have been accused of spreading infection (Karkadinovsky 1936), but there is little evidence to support this (see Bosworth 1940).

(2) THE VAGINA. In his original experiments Bang (1897) demonstrated the possibility of infecting pregnant cows by intravaginal inoculation of pure cultures. This has been confirmed frequently (Report 1909, M'Fadyean *et al.* 1913, Seddon 1919). It seems, therefore, not improbable that contamination of the vagina with infected soil or litter may be responsible for

naturally acquired infection. A second method by which vaginal infection may occur is sexual intercourse. Bulls may be infected by injection of pure cultures under the prepuce (M'Fadyean *et al.* 1913). Injected intravenously, the organisms may lodge in the testicles and be excreted in the seminal fluid (Seddon 1919). Infection of bulls is not uncommon under natural conditions, and lesions may occur in the testis, epididymis, or vesiculae seminales (see Lambert *et al.* 1963). Bendixen and Blom (1947), who drew attention to the danger of transferring infection to the cow by artificial insemination, isolated *Br. abortus* from the semen of 28 out of 394 bulls. There is also epidemiological evidence to incriminate the bull as a means of spreading the disease (Bang 1897).

(3) THE SKIN AND CONJUNCTIVA. Attention has been drawn to these routes of infection by Bang (1931), Bang and Bendixen (1932b), and Cotton, Buck and Smith (1933). Experimentally it has been found possible to infect animals fairly readily with *Br. abortus* by means of an infected compress left for a variable length of time in contact with the abraded, or even intact, skin. Infection may also follow simple conjunctival instillation of a *Br. abortus* suspension; indeed this is one of the most convenient ways of producing experimental infection.

Which of these three routes is the most common in practice, we have at present no means of telling, though opinion favours the alimentary route. Investigation is required to show what their frequency distribution is under natural conditions.

Diagnosis

In the yellow or dark-brown mucoid exudate found between the uterine mucosa and the chorion, large numbers of *Br. abortus* may be demonstrated microscopically, aggregated into clumps within the polymorphonuclear cells. The organism may likewise be seen in smears made from the gastro-intestinal contents of the fetus. Provided the material sent for examination is fresh and uncontaminated, a provisional diagnosis may often be made microscopically. The use of a differential method of staining, such as that of Hansen or Köster (see Chapter 40), is advisable. According to Smith and his colleagues (1962b), 60–85 per cent of the total number of abortus bacilli are found in the fetal cotyledons. As already pointed out, this localization appears to be due to the presence of erythritol in the placenta.

Cultivation The organism is best grown by seeding glycerol, or liver, agar slopes with the uterine exudate, or with the stomach contents of the fetus, and incubating them in an atmosphere of air containing 5–10 per cent of added CO_2 (Smith 1926, Huddleson *et al.* 1927). (See Chapter 40.) When uncontaminated, the organisms will form characteristic lenticular colonies in the course of 2 or 3 days (Fig. 40.3). Their identity should be confirmed by agglutination. The bacilli are sometimes found in the blood and in the liver of the fetus, but less abundantly than in the stomach. When the material is not pure, it is advisable to use one or more of the selective media containing dyes and antibiotics described on pp. 156–7, and resort at the same time to guinea-pig inoculation. The material is ground up in Ringer's solution, and 0·5 ml injected intramuscularly into two or three guinea-pigs. The animals are killed after 4–8 weeks, the blood is examined for agglutinins, and cultures are made from the sublumbar glands and the spleen (see 'Milk' below). At post-mortem examination of slaughtered cattle the organisms are most likely to be found in the udder, and in the supramammary, iliac, and pharyngeal glands. They may be sought for by direct culture and by guinea-pig inoculation (see Doyle 1935). Any obvious lesions, such as hygroma of the knee (v. d. Hoeden 1932a), should also be examined.

Agglutination reaction

In practice this is the most widely used method of diagnosis.

The test may be carried out by the tube or the plate method. Reference should be made to Alton and Jones (1967) who have described the technique of these tests, together with the preparation and standardization of the antigens. Final dilutions of 1 in 10 to 1 in 2560 should be put up against suspensions of both *Br. abortus* and *Br. melitensis*, and the results expressed, if so desired, in terms of international units per ml as determined by comparison with the International Standard Serum (Report 1964b, Davidson *et al.* 1969, Alton 1971).

In the 2nd International Standard Serum 1 unit is contained in 0.095 52 mg. Each ampoule contains 1000 IU, so that when the contents are dissolved in 1 ml of distilled water the strength of the solution is 1000 IU/ml. A dilution of 1 in 25 therefore contains 40 IU/ml, of 1 in 50 20 IU/ml, of 1 in 100 10 IU/ml. If in a given laboratory using its own antigen the titre of this serum is found to be only 1/500, and the titre of the animal's serum under test to be 1/250, then the number of international units in the animal's serum is taken to be $\dfrac{1000 \times 250}{500} = 500$. Since the agglutination and complement-fixation tests measure different biological activities, separate unitages are assigned to these two tests (see Davidson and Hebert 1978).

Zone phenomena due to the presence of blocking antibodies (see p. 149) are frequently encountered in the tube test. They may be of the pro-zone or the intermediate zone variety (see Priestley 1931), and may sometimes be avoided by the use of 5 per cent saline for dilution of the serum. When the readings remain doubtful the Coombs anti-globulin test should be employed (see p. 149). Non-specific agglutinins are said

to be largely destroyed by heating the serum to 65° for 10 minutes (Amerault *et al.* 1961).

The interpretation of this test is often difficult, and depends on a number of factors such as the age of the animal, previous vaccination or abortion, the presence of known infection in the herd, the stage of pregnancy, and so on.

Calves, even those born of infected mothers, give a negative serum reaction at birth. If, however, they are allowed to suck infected dams within the first 24 hours of life, agglutinins become demonstrable in their serum within about 2 hours. This is due to the absorption of colostrum, in which the concentration of immunoglobulins is 6–8 times that in serum (Cunningham 1977). Calves passively immunized in this way lose their agglutinins very rapidly (McAlpine and Rettger 1925). Even calves that are actively infected and acquire agglutinins some time after birth appear to lose them within 6 months, unless reinfection occurs (Thorp and Graham 1933).

In an animal that is infected during pregnancy the titre generally rises before abortion to 1/200–1/1000 or over. During the following 6 months it tends to fall. If the cow becomes a chronic carrier, the titre usually remains fairly high. About 80 per cent of cows with a persistent titre of 1/200 or over are found to be excreting *Br. abortus* in the milk. A titre of 1/1000 or over is almost diagnostic of udder infection. When animals are followed for any considerable length of time, the agglutinin titre of positive reactors will often be found to decline. This is particularly true of low reactors. Damon (1932), for example, who observed a herd with a mean complement of 225–250 animals over a period of 4 years, found that 27.7 per cent of animals reacting at 1 in 25, 17 per cent of those reacting at 1 in 50, and 5 per cent of those reacting at 1 in 100 or over lost their agglutinins permanently (see also Hadley and Welsh 1931). Huddleson and Smith (1931), however, found that, of a total of 247 animals reacting at 1 in 25 or over, and followed up for a period of 1–8 years, only 4 became permanently negative. Animals that abort usually have a considerably higher titre than non-aborting animals. A few animals may fail to show any agglutinins at the time of abortion, and for a few days afterwards; this is ascribed by Kerr (1968) to the inability of antigens in the fetus to pass the uterine barrier. Most of these animals become positive within 2–3 weeks, but there is some evidence that others may fail completely to produce agglutinins. The vaginal mucus test, however, is said by Kerr to be positive at the time of parturition. Occasional animals are said to excrete *Br. abortus* in the milk and yet fail to show agglutinins in the blood serum or whey (Karsten 1932).

In unvaccinated animals or in animals vaccinated as adults an agglutinin content of 50 IU/ml should be regarded as suspicious, and one of 100 units or over as definite evidence of infection. In animals 30 months old or more vaccinated as calves with the living attenuated S19 strain of *Br. abortus* the corresponding figures are twice these, namely 100 IU/ml as suspicious and 200 IU/ml as definite (Report 1964*b*). In doubtful cases the complement-fixation test is often of value in deciding whether or not an animal is infected (Morgan 1968). When doubt still persists, monthly tests should be made on all suspicious animals so as to detect as soon as possible those that become definitely positive. It must always be remembered that, though a positive result is a practically certain indication of infection, a doubtful or negative one cannot be held to exclude infection (see also Report 1964*b*).

Miller and his colleagues (1976) describe a test for distinguishing brucella agglutinins formed during natural infection from those resulting from vaccination with the S19 strain. *Radioimmunoassay* may be used in the diagnosis of bovine brucellosis. It measures the amount of specific antibody of the IgG1 and IgG2 subclasses, but is insensitive to IgM—a characteristic that may render it more suitable than the seroagglutination and complement-fixation tests for distinguishing infected from vaccinated animals. It is not subject to prozone or ambiguous reactions, both of which interfere with the interpretation of the complement-fixation test (Chappel *et al.* 1976, 1978). For diagnosis of early brucella infection with *Br. abortus*, Wilson, Thornley and Coombs (1977) recommend a solid-phase assay with radioactively labelled antibody. In the diagnosis of caprine brucellosis in Kenya Waghela, Wandera and Wagner (1980) found the Rose Bengal test to be more sensitive and the agar gel immuno-diffusion test to be the more specific when compared with the complement-fixation test.

Infection in *bulls* is by no means always easy to detect. It is advisable to carry out agglutination tests on the blood serum and the semen plasma, and to make cultures from the semen itself (Bendixen and Blom 1947). At post-mortem examination, material from the testicles and seminal vesicles should be inoculated both on to culture media and into guinea-pigs. In the experience of Cedro and his colleagues (1967) 25 per cent of bulls from which brucellae were isolated *post mortem* gave a negative agglutination reaction in the blood serum, but this figure is higher than the experience of other workers would suggest.

For rapid screening purposes the *Rose Bengal plate* or *card agglutination test* may be used (Nicoletti 1967, Morgan *et al.* 1969). It has been widely adopted and, provided positive sera are subjected to a complement-fixation test, it is a most valuable addition to the serological armamentarium (Morgan 1970, Report 1971, Corbel 1972). (For a review of the various serological methods available for the diagnosis of brucella infection in animals, see Alton 1980, Chappel 1980.)

The abortin reaction Abortin is a dark syrupy liquid prepared in much the same way as Old Tuberculin. It was introduced by M'Fadyean and Stockman (1909) as a diagnostic test based on the hypersensitiveness of infected animals. For use it is suitably diluted and injected subcutaneously. In a positive reaction the temperature rises to 40–41°. As a method of diagnosis the use of abortin did not find favour; the authors themselves after extensive trials were disappointed with it (Stockman 1914).

Vaginal mucus agglutination test This test, which was introduced by Kerr (1955), consists in obtaining a sample of vaginal mucus, diluting it 1 in 5 with saline, shaking, leaving for 30 minutes, and titrating the supernatant fluid in the same way as serum. The diluted mucus may be heated in a water-bath at 56° for one hour to destroy contaminants before being mixed with the agglutinating suspension and incubated overnight at 37°. A titre of 1/10 is regarded as diagnostic of infection with a virulent organism. In fluids with high titres a prozone is common. The test may give a negative result at the time of abortion, but soon becomes positive. The highest titres—up to 1/1280—are found from 10 days onwards. The agglutinins persist for several weeks or months. One of the great advantages of this test is that it is unaffected by vaccination with strain S19 (Kerr, Pearson and Rankin 1958). The mucus may be obtained by means of a *vaginal tampon* (see Roberts and Philip 1960), but a pipette is said to give fewer false positive results.

In diagnosis no one test can be relied on completely; when possible a combination of tests should be performed (McGaughey and Hanna 1973). (For comments on the value of the various tests, see Nicoletti 1976.)

Demonstration of *Br. abortus* in milk

Examination for the presence of *Br. abortus* in milk may be conducted by cultural or by animal inoculation methods. It must be remembered that excretion of *Br. abortus* in the milk is often intermittent and that not too much attention should be paid to a single negative result.

Cultivation Cultural methods are most satisfactory with milk samples from individual cows collected under more or less aseptic conditions. Separate quarter samples should be taken, and either kept refrigerated, or preserved with 0.5–1.0 per cent boric acid, till the time of examination. The whole milk, the gravity cream, or the deposit from high-speed centrifugation should be streaked on to a number of plates of liver extract agar, or 2 per cent glycerol agar containing 10–15 per cent of ox serum. Contaminating organisms may be partly suppressed by the addition to the medium of gentian violet and malachite green. The exact concentrations of these dyes vary according to the manufacturer; usually a final concentration of in 100 000 to 1 in 200 000 is satisfactory. The plates should be incubated in 5–10 per cent CO_2, and examined at intervals for 14 days. More selective media, for use with herd samples, can be made by combining antibiotics with dyes. Kuzdas and Morse (1953), who introduced this combination, used a mixture of polymyxin B, actidione, bacitracin, circulin, and crystal violet incorporated in Albimi brucella agar. Serum dextrose agar may be used as a base, or blood agar. Ryan (1967) claimed very good results with a blood agar medium containing bacitracin, polymyxin B sulphate, vancomycin, nalidixic acid, cetrimide, actidione, and mycostatin. Strains, such as *Br. abortus* biotype 2, that are unusually sensitive to dyes and antibiotics are best cultivated on a plain liver agar medium containing 0.1 per cent Tween 40 (Huddleson 1964); but according to Farrell and Robertson (1972) Farrell's medium, which gives excellent results, is not inhibitory to this biotype. According to Brodie and Sinton (1975), enrichment in a fluid medium for five days followed by five days on a solid medium gives better results than direct cultivation on a solid medium.

Animal inoculation If the animal method is chosen, it is advisable to inoculate a mixture, composed of 2 ml of gravity cream and the deposit from 100 ml of milk after high-speed centrifugation, intramuscularly into the hind leg of each of two guinea-pigs. Alternatively 4 ml of whole milk may be injected, 2 ml into each thigh. The animals should be killed about 6 weeks later. At post-mortem examination the femoral and sublumbar glands will be enlarged and pale; the spleen may be enlarged, its surface slightly irregular, and a few small greyish-yellow necrotic foci may be present. The liver may show two or three tiny necrotic foci. The macroscopic lesions are often inconspicuous, and must on no account be relied upon for diagnosis. The blood serum should be tested for agglutinins; a titre of 1/25 or over is highly suggestive of infection. Cultures should be made from the sublumbar glands and spleen, and all suspicious organisms identified by agglutination and other methods. In not all animals containing serum agglutinins is it possible to isolate the organisms from the tissues. On the other hand, it is uncommon to isolate them in the absence of a positive agglutination reaction. Individual guinea-pigs vary in their susceptibility to *Br. abortus*, and it is common to obtain positive results in one animal and negative in the other (see Smith 1932, Plate 1934). When tubercle bacilli are present in the milk simultaneously with *Br. abortus*, the guinea-pig will suffer a double infection. The diagnosis of tuberculosis can be made on the basis of the macroscopic lesions and the demonstration of acid-fast bacilli in the organs. The diagnosis of *Br. abortus* infection can be made on the basis of the agglutination reaction and the cultivation of the organisms from the tissues. Pullinger (1936), however, pointed out that the tubercle bacillus interferes with the development of *Br. abortus* in the animal body, so that the demonstration of this organism may often be unsuccessful in milk coming from cows with tuberculous mastitis.

Though the animal inoculation method is more sensitive than the cultural, it is not uncommon to obtain a positive result by culture when the animal test is negative. This may be due partly to chance distribution of organisms in the inoculum, partly to variation in the susceptibility of different animals, and partly to the fact that some biotypes of *Br. abortus* are comparatively avirulent for the guinea-pig (Farrell and Robertson 1967).

Serological examination of milk Indirect evidence of infection of the milk may be obtained by examination of the whey for agglutinins or complement-fixing antibodies. The whey may be obtained by adding 1 drop of rennet to 5 ml of skim milk, incubating at 58° for 30 minutes, breaking up the clot, and centrifuging. Alternatively 5 ml of carbon tetrachloride may be added to 10 ml of milk, together with a small quantity of rennet. The milk is corked, and shaken for several minutes till the fat is extracted. It is then incubated for 1 hour, and centrifuged (Hall and Learmonth 1933). The advantage of 58° for incubation is that it destroys any complement present. The **whey agglutination test** is carried out in the usual way. Control tests must be put up, since non-specific reactions may occur. A whey agglutinin titre of 1/80 or over in unvaccinated animals affords a high probability that the udder is infected and that *Br. abortus* is being excreted in the milk. The higher the titre, the more probable this is. According to Farrell and Robertson (1968), the **whey complement-fixation test** is a more reliable indicator of excretion of *Br. abortus* in the milk than the whey agglutination test. Of milk samples from individual cows 56 per cent fixing complement at 1 in 10 were culturally positive, 72 per cent at 1 in 40, and 88 per cent at 1 in 640 or higher.

A more rapid method of diagnosis is afforded by the **brucella milk ring test**. This was introduced by Fleischauer (1937) in Germany, and reported on favourably by Danish and Swedish workers (for references see Roepke *et al.* 1950). It is carried out on herd milk, and, though not as sensitive as the serum agglutination test, it is particularly valuable for picking out herds that are probably infected with *Br. abortus*. A drop of a dense suspension of *Br. abortus* stained blue with haematoxylin or red with tetrazolium is added to 1 ml of whole milk in a tube. The tube is gently inverted several times to ensure thorough mixing and incubated for one hour at 37°. The results are read as shown at the top of page 158 (Alton and Jones 1967).

On an average *Br. abortus* can be demonstrated in about 30 per cent of milks giving a positive ring test; the stronger the reaction, the more likely is the milk to be infected. When performed on milk from cans each containing the product of about 10 cows, it is of help in indicating to the public health bacteriologist which animals are most likely to be infected. Milk from individual cows can then be tested before culture or guinea-pig inoculation (see Report 1956). The result

Cream ring	Milk column	Rating
Deeply coloured	White	$+ + + +$
Definitely coloured	Slightly coloured	$+ + +$
Definitely coloured	Moderately coloured	$+ +$
Cream ring and milk column about the same colour		$+$
White or slightly coloured	Definitely coloured	$-$

of the test is disturbed by vaccination during pregnancy but not to any extent by vaccination during calfhood (McDiarmid *et al.* 1958).

Prophylaxis and treatment

At the time of abortion the uterine discharge is highly infectious; the usual measures must therefore be taken for isolation of the animal, and subsequent disinfection of its stall. The fetus and membranes should be burned, or buried in lime. Irrigation of the cow's vagina should be practised with an antiseptic solution for a week or till all discharge has ceased, and not till some considerable time has elapsed since the complete cessation of the discharge should the cow be taken to the bull. If a bull has served an animal that has recently aborted, it is well to irrigate its penis and preputial sheath with an antiseptic solution, in order to prevent active infection of the bull, or passive carriage of the organism to another cow. Calves born of uninfected animals should be given colostrum within 8 hours of birth; otherwise the antibodies, whose titre in colostrum is 6-8 times higher than in serum, will be digested by the gastric juice (Cunningham 1977). Calves born of infected cows are better brought up on milk from another animal. Though there is little or no evidence to show that infection is spread by milking machines (Morgan 1968), it is a wise precaution to milk infected animals last.

Eradication of contagious abortion

The ideal policy is the complete eradication of infection. Various methods are recommended for doing this. The general principle consists in the detection of infected herds by the Ring test, followed by blood serum, whey agglutination, and Ring tests on individual animals, the slaughter, sale, or segregation of positive reactors, or preferably the slaughter of all animals in an infected herd (Wilesmith 1978), and the building-up of a clean, non-infected herd. The degree of success attending this policy varies with a number of factors. With a small or medium-sized herd, particularly when self-contained and protected against infection from water, manure, and other animals, with provision of calving boxes, and re-testing of non-reactors every 2 months so as to eliminate all animals as soon as they become positive, it is often possible to eradicate the infection entirely and to maintain a healthy herd for several years. On the other hand, with very large herds, particularly when not self-contained,

with flying herds, with farms on which adequate accommodation for segregation of infected, and quarantine of newly imported animals is impossible, and with imperfect control over infection from other sources, the results are often disappointing. Generally speaking, if the conditions are favourable, and if the disease is not at the height of its activity, this policy should be adopted, since its success is followed by the improved general health of the herd, better and more regular breeding, and an increased milk supply. A serious problem is presented by the infected non-reactor, which may abort, or excrete *Br. abortus* in the milk, and so contaminate the other animals in the herd. After the dairy herds have been cleaned up comes the more difficult task of dealing with the beef and suckling herds. (For description of various eradication schemes see Kerr 1968; and for a history of the Brucellosis Eradication Scheme in the United States, see Brown 1977, and in the United Kingdom see Morgan 1977, Editorial 1977.) As already stated, England and Wales were declared free from brucellosis from 1 November 1981. Scotland was free from 1 January 1980. (Report 1981*a*.)

Vaccination

Though vaccination of non-pregnant heifers was proposed by Bang in 1906 and by M'Fadyean and Stockman in 1909 (see 2nd edition, p. 1355), it was not till Buck (1930) introduced a strain of modified virulence, *Br. abortus* S19, that vaccination came into its own. The vaccine, whose mode of preparation is described by Alton and Jones (1967), is given in a single dose to calves 6 months old, non-pregnant heifers, and cows not more than 4 months pregnant. Under field conditions about a 90 per cent degree of protection against abortion may be expected lasting up to at least the fifth pregnancy (McDiarmid 1957). Revaccination is unnecessary. Adult cows should not normally be vaccinated; but adult vaccination is justified when infection has been recently introduced into a non-vaccinated herd so as to prevent an 'abortion storm' (McDiarmid 1960).

One disadvantage of this vaccine is that it stimulates the production of agglutinins and so interferes with the serological diagnosis of natural infection. When, however, its use is limited to young non-pregnant animals, it will be found that the agglutinins usually disappear within 6-18 months.

Partly to avoid interference with diagnosis, partly because S19 may be excreted in the milk and give rise to human infection, and partly for other reasons, attention has been paid to killed vaccines. McEwen's rough strain 45/20 of *Br. abortus* (McEwen and Priestley 1938), killed by heat, suspended in a water-in-oil emulsion, and injected in two doses at 6-weeks' interval, gives almost as good protection as S19, with a much lower production of agglutinins (Cunningham and O'Reilly 1968, McDiarmid 1968). Complement-fixing antibodies, however, are said to appear and persist for several months (Morgan and McDiarmid 1968). The results of using this vaccine have been conflicting (Alton 1978). After trial for some time it has now been discontinued officially in the United Kingdom (Report 1978). (For a review of vaccination, see Alton 1978, Jones and Berman 1980.)

Vaccination should always be combined with the institution of hygienic and other measures designed to decrease the total amount of infection in the herd. No vaccine can be relied on in the presence of heavy infection; and no vaccine, alive or dead, can alter the course of the disease in an already infected animal.

Though vaccination cannot eradicate infection, it can pave the way towards eradication. By raising the resistance of the individual animal it limits the multiplication of virulent organisms that do gain access to the tissues. Thus the degree of infection of the uterus and udder is diminished, so that abortion and milk excretion are much less frequent than in uninoculated animals. This, in its turn, lessens the total amount of infective material thrown off, and the danger of infection both of other animals in the herd and of human beings consuming the milk.

Brucella infection of swine

This disease is of much more limited distribution than contagious abortion of cattle. It first attracted serious attention in the large hog-raising districts of the middle western States of North America. According to the observations of Boak and Carpenter (1930), Feldman and Olson (1934), and McNutt (1935) about 2 per cent of animals destined for the abattoirs were infected in this area. The disease has been recognized in several countries, but with a few exceptions it appears to be mainly of sporadic occurrence.

Bacteriologically, all outbreaks appear to have been due to *Br. suis*, and all of them, with the exception of the Danish epizoötic, to the American type (see Chapter 40), though complete information has not always been available on this point. The disease is essentially an infection of the reticulo-endothelial system, with the production of inflammatory disturbances in the glands, joints, and reproductive organs. Bacteraemia is common during the first 2–3 months of infection (Hutchings 1950). Contrary to contagious abortion of cattle in which the females are chiefly affected, the disease in swine attacks mainly the *boars*. Abortion is less frequent than in cattle but may occur early in gestation and be overlooked. Abscess formation is met with in the submaxillary, cervical, inguinal, and popliteal lymph glands. The joints of the hind legs, particularly the knee and hock, are frequently affected; they are swollen, painful, frequently contain fluid, and sometimes ankylose. Epididymo-orchitis is very common. In Makkawejski's series, no fewer than 269 out of 786 boars showed orchitis, usually bilateral. The affected testicles are sometimes of enormous size, reaching 2.95 kg in weight, and purulent foci, which may undergo calcification, are not infrequent. Multiple foci may also be present in the vesiculae

seminales. *Br. suis* may, however, often be demonstrated in the tissues in the absence of macroscopic pathological changes.

In *sows* a disease for which Thomsen (1934) suggests the name '*miliary brucellosis of the uterus*' is frequently met with. The interior of the uterus is studded with small yellowish-white nodules located in the deeper layer of the mucosa and projecting slightly above the surface.

The disease can be readily reproduced by feeding and by conjunctival inoculation. According to Stockmayer (1933), pigs are susceptible to *Br. abortus* as well as to *Br. suis*, though no outbreaks due to *Br. abortus* have yet been recorded in the field. Thomsen believes that natural infection is spread to a considerable extent by copulation, though this of course will account only for transmission of the disease to sexually mature animals. Infection of the conjunctiva by urine, and alimentary infection probably play a part. Sucking pigs are readily infected. It is possible that infection occurs through the milk of the mother, though invasion of the udder is uncommon. In Denmark infection has been introduced into some herds by hares (Bendtsen 1959).

In the field the tube or plate agglutination test affords the best method of **diagnosis**. For screening purposes the card test (see p. 156) is often useful. A titre of 1/100 is regarded as definitely positive, 1/50 as strongly suggestive, and 1/25 as doubtful. Occasionally animals fail to produce agglutinins up to a diagnostic standard. In the abattoir, the blood serum should be examined, and a search made by culture and guinea-pig inoculation for *Br. suis* in the tissues. According to Johnson and Huddleson (1931) and Johnson, Huddleson and Hamann (1933), the organisms are commonest in the spleen, gastric and supra-mammary lymph glands, and the liver. Numerous other organs may be infected, but less frequently.

The disease is of undoubted economic importance, owing particularly to impotence in the boars and abortion or sterility in the sows. But it is probably of even more importance in its effect on public health. Large numbers of cases of undulant fever have been ascribed to it, and it affords a very real hazard for workers in abattoirs and packing houses. So far as reports go, it would seem to be fairly amenable to control by the eradication policy. In Denmark, by the institution of such a policy, including slaughter of infected animals in lightly infected herds, and slaughter of all animals in heavily infected herds, it proved possible to stamp out the disease entirely within a very short time. Johnson, Huddleson and Hamann (1933) in the United States were successful with a less radical policy, consisting of blood-testing once a month, followed by segregation and ultimate elimination of the positive reactors. The seasonal breeding of sows makes control of the disease considerably easier than in cattle. On account of the frequency of non-specific agglutinins in swine sera, eradication is better carried out on the basis of the herd than of the individual animal (Busch and Parker 1972). Attempts at prophylactic vaccination have hitherto been a failure. (For further information on this disease the reader is referred to the excellent monograph of Thomsen 1934; see also Report 1971.)

Brucella infection of horses, dogs, and other animals

Horses

After the work of Rinjard and Hilger (1928) in France, brucella infection of horses was recognized in many other European countries (for references see 5th ed., p. 2066). The incidence in males and females was observed to be much the same, but older horses were more often infected than younger. Since agglutinins are said to be present in normal horses up to a titre of 1/50 or 1/100, estimates of the incidence of infection based on agglutinins alone should be accepted with caution (see Deem 1937, Krüger 1937). The disease may be contracted from other horses or from cattle. Though the infection frequently remains latent, in some animals it is accompanied by the appearance of suppurative lesions, particularly on the head and neck (*poll-evil, fistulous withers*), and less often of bone abscesses, arthritis and tenosynovitis. Their appearance may be preceded by fever and weakness. No history of local trauma is usually obtainable. The inflammation is extremely painful, and lasts for a few days to several months. The abscess may disappear by absorption, but more often it breaks down at multiple points on the surface, and discharges a yellowish vis-cous fluid containing fibrinous clots. Bacteriologically *Br. abortus* can usually be isolated, or sometimes in the United States *Br. suis*, though secondary infection of the fistulae readily occurs. Horses can be infected experimentally with *Br. abortus* by feeding or parenteral inoculation (v. d. Hoeden 1932b, d, Schneller 1934); but according to Fitch, Bishop and Boyd (1932) the local manifestations of the disease cannot be reproduced except by direct inoculation into the neck ligament. The opinion is expressed by van der Hoeden (1932b, d) that poll-evil is essentially an allergic manifestation. It was found, for example, that, if dead abortus bacilli were injected subcutaneously into an infected animal, a local abscess formed from which living organisms could be cultivated. Bacteriological diagnosis of infection is made by the serum-agglutinin test; a titre of 1/200 may be regarded as positive, and 1/100 as suspicious. An allergic test may also be used, the ophthalmic (v. d. Hoeden 1932b, d) or the intradermal (Rossi and Saunié 1934) route being chosen. Horses may convey infection to cattle (White and Swett 1935), and in any eradication scheme the two animals should be kept separate.

Dogs

Most dogs have a fairly high resistance to *Brucella*; infection, when it occurs, tends to remain latent; and cases of manifest disease are mainly sporadic. Abortion may be caused by either *Br. abortus*, *Br. melitensis*, or *Br. suis*. Such was the general picture up to 1963 when an epizoötic form of the disease was observed in the United States. In 1966 the causative organism was isolated and later recognized by Carmichael and Bruner (1968) to be a new species—*Brucella canis* (see Chapter 40). The disease was confined almost entirely to beagles, and was characterized by abortion, still-birth, whelping failures, and vaginal discharge in bitches, and epididymo-orchitis, prostatitis, testicular atrophy, and sterility in dogs. In both sexes the spleen and lymph nodes were enlarged, and there was an absence of fever (Morisset and Spink 1969). Bacteraemia may persist for several months, and the organism is often excreted in the urine of both males and females and in the vaginal discharge. Natural transmission occurs by aborted fetal and placental tissue, by infective vaginal discharges, and by males to females during breeding. House dogs are occasionally infected; but Weber and Schliesser (1978), who examined 1000 sera, found agglutinins to *Br. canis* in only two. (See also Carmichael and George 1976.)

Experimentally, animals can be infected by many different routes. The organism localizes in lymphoid tissue, in which it may remain for several months. Diagnosis of the natural disease is made by culture of the blood and of genito-urinary excretions, and by the demonstration of agglutinins to *Br. canis* in the serum. Treatment is by tetracycline, to which the organism is very susceptible. Recovery leaves behind it immunity. A few human cases of infection with *Br. canis* are known to have occurred (Morisset and Spink 1969, Swenson *et al.* 1972). (For a review of canine brucellosis, see Carmichael *et al.* 1980.)

Other animals   Many other animals, including wild animals, are subject to infection with one or more species of *Brucella*. Apart from **reindeer**, which are subject to infection with *Br. suis* biotype 4 (see Chapter 40), infection tends to assume a latent form.

Little is known about natural infection in **cats**, though it is said that experimental inoculation with *Br. abortus* or *Br. melitensis* may give rise to a severe disease with lesions in the joints and internal organs (Makkawejsky and Karkadinowskaja 1932). Hirchert and his colleagues (1967) isolated *Br. melitensis* from the uterus of an aborting cat.

Grey rats appear to be sometimes infected with *Br. abortus* on farms on which contagious abortion is common (Karkadinovsky 1936), but tissue lesions seldom occur. After experimental feeding, they may excrete the organism for a short time in the faeces and urine (Bosworth 1937, Sandholm 1938). The disease in **desert wood rats** caused by *Br. neotomae* is referred to in Chapter 40.

Infection of **hares** with the Danish type of *Br. suis—*biotype 4—was recognized in Germany by Witte (1941). Pathologically the disease is characterized by nodules of varying size containing purulent necrotic tissue and affecting mainly the testes, mammary glands, and spleen, less often the lungs, liver, and uterus. *Br. suis* can be readily isolated from the lesions, and agglutinins are demonstrable in the blood serum (Bendsten *et al.* 1956).

Szyfres and Tomé (1966) reported infection of **grey foxes** in the Argentine with *Br. abortus* biotype 1. Agglutinins were demonstrated in the blood serum at titres between 1/25 and 1/800.

Fowls are comparatively resistant to brucella infection. Though Emmel and Huddleson (1929, 1930) thought that disease was very common in the United States, experience has not shown this to be true. Infected birds suffer few or no symptoms, and agglutinins, which to be diagnostic should be 1/200 or over (Beller and Stockmayer 1933), disappear early. Egg production may suffer slightly (see Anczykowski 1957, Angus *et al.* 1971).

How infection spreads among wild animals in which the organisms seem to be discharged by none of the natural channels is discussed by Thorpe, Sidwell and Lundgren (1967). For further information on brucella infection in animals the reader is referred to Stableforth (1959), Morgan (1970), Olitzki (1970), Report (1971).

(For further information on brucellosis, see Ray 1979, Report 1980*a,b*, Blobel and Sliesser 1982.)

Campylobacter infections

An account of the various systems of classification and nomenclature applied to campylobacters is given in Chapter 27. Skerman and his colleagues (1980) recommend a system of nomenclature based on that of Véron and Chatelain (1973). Smibert (1974, 1978), whose classification will be used here, recognizes the following species: *C. fetus*, containing three subspecies, namely *fetus*, *intestinalis*, and *jejuni*; *C. sputorum*, also containing three subspecies, namely *sputorum*, *bubulus*, and *mucosalis*; and *C. fecalis*.

Cattle

M'Fadyean and Stockman (Report 1913) described abortion of sheep and cattle in England due to a spirillar organism that was called *Vibrio fetus* by Theobald Smith (1918) but is now known as *Campylobacter fetus*. It later became clear that in cattle the economic importance of the abortions was overshadowed by that of the infertility and early embryonic deaths that also occurred (Plastridge *et al.* 1947, Terpstra and Eisma 1951, J. R. Lawson and MacKinnon 1952).

Campylobacter infection of the bovine reproductive tract occurs in many parts of the world. Venereally transmitted infection is the more common form; it is produced by *C. fetus* subsp. *fetus*, an organism that affects cattle only. An orally transmitted form, occurring sporadically, is produced by *C. fetus* subsp. *intestinalis*, an organism that also affects sheep and human beings. Experiments in cattle, sheep, and small laboratory animals have shown that after oral administration subsp. *intestinalis*, unlike subsp. *fetus*, persists in the intestine and gall bladder and produces bacteraemia (Bryner *et al.* 1964, 1971).

C. fetus subsp. *fetus* occurs on the mucous membrane of the prepuce and glans penis of infected bulls, and sometimes in the urethra, without producing visible signs. Transmission occurs during coitus. In the past, before the danger was fully appreciated, widespread dissemination resulted from the use of artificial insemination. In the female, endometritis, cervico-vaginitis, salpingitis, and inflammatory oedema of the fetal membranes occur. Infection often greatly increases the number of matings required to produce conception; oestrous cycles become irregular and prolonged, and early embryonic deaths take place.

The period of infertility usually lasts for three to five months. Abortion may occur at any stage of pregnancy but is most common between the fifth and sixth months; as much as half of the chorioallantois may sometimes show inflammatory lesions. *C. fetus* subsp. *intestinalis* is probably transmitted by the ingestion of contaminated food and water; it is thought not to occur in the intestinal tract of normal animals (Smibert 1978).

Diagnosis

Diagnosis is best made by culture (see Chapter 27). Material for examination may include fresh preputial washings or smegma, the stomach contents and spleen of aborted fetuses, placenta, allantoic fluid, and cleanly taken vaginal mucus, especially that collected at the time of oestrus. Microscopical examination of stained smears is often helpful. Liess (1958) recommended a sodium thioglycollate medium containing 0.1 per cent agar, 10 per cent blood, and 0.002 per cent brilliant green. Cohn (1962) favoured an oxygen gradient plate containing an uninoculated bottom layer of 1.5 per cent trypticase yeast-extract agar (TYA), a middle layer of 0.8 per cent TYA inoculated with the material under examination, and an upper slanted layer of uninoculated 1.5 per cent TYA. An effective enrichment and transport medium (Winter and Caveney 1978) consisted of bovine serum, inspissated and then broken up, containing 5-fluorouracil, polymyxin B sulphate, brilliant green, nalidixic acid, and cyclohexamide; the air in each rubber-capped bottle was replaced by an atmosphere of O_2 2.5 per cent, CO_2 10 per cent, and N_2 87.5 per cent. Bracewell and Lander (1979) recommended a modification of the medium of Skirrow (1977), consisting of agar containing lysed horse blood, vancomycin, trimethoprim, and polymyxin B sulphate; plates were incubated in an atmosphere containing O_2 5–7 per cent. Serum dextrose antibiotic agar, blood agar, and various other media have also been used. The ability of campylobacters to pass through a membrane filter with an average pore diameter of 0.65 μm may be made use of in examining heavily contaminated material (see Smibert 1978). Unlike *C. fetus* subsp. *jejuni*, the two subspecies *fetus* and *intestinalis* grow at 25° but not at 42–43°. A fluorescent antibody technique is widely used to demonstrate the infecting organism in preputial and other material (Mellick *et al.* 1965, Philpott 1966); but the most reliable method of detecting infection in the bull is by mating with virgin heifers. A useful diagnostic measure is the demonstration of agglutinins in vaginal mucus (see J. R. Lawson 1959). Such agglutinins may take several months to appear, and persist for 2–14 months; they tend to disappear from the mucus at the time of oestrus. The agglutinins in vaginal mucus consist of immunoglobulins of the IgG, IgM, and IgA classes (van Aert *et al.* 1977).

Prophylaxis

Outbreaks are dealt with by enforcing a period of sexual rest lasting a number of months, followed by artificial insemination. Fertility is often regained several months before elimination of the organism from the cervix and vagina. Infected and in-contact bulls may be treated with topical applications of streptomycin and penicillin, and then tested, preferably by mating with virgin heifers.

In female cattle, vaccines containing whole cells or bacterial extracts and mineral oil or other adjuvant are said to have both a protective and a curative effect; and gamma globulin from hyperimmune serum is also both protective and curative (Clark *et al.* 1976, Berg and Firehammer 1978, Schurig *et al.* 1978). Antigenic variation may occur during the course of infection. Vaccination of bulls results in the appearance of incomplete antibody and immunoglobulins of the IgG, IgM, and IgA classes in preputial secretion, and assists in preventing and eliminating infection (Clark *et al.* 1968, van Aert *et al.* 1976, Bracewell and Lander 1979). Bryner and his colleagues (1979) demonstrated considerable variation in the protection given by a number of commercial vaccines to pregnant guinea-pigs against challenge with *C. fetus* subsp. *intestinalis*. Living vaccines have been used (see Smibert 1978).

Other organisms

C. fetus subsp. *jejuni*, which will grow at 42–43°, has been isolated from the faeces of calves and numerous other species, often in the absence of disease. It produces diarrhoea in calves infected experimentally with large doses (Al-Mashat and Taylor 1980) and is probably the same organism as that isolated by Jones and Little (1931) and Jones and co-workers (1931) from calves with diarrhoea; it also represents the *related vibrios* of King (1962), responsible for enteritis in man. Bryner and his colleagues (1972) and other workers have isolated pathogenic campylobacters from the gall bladder of normal cattle and sheep. An organism isolated from the genital tract of cattle (Plastridge *et al.* 1964) resembled the non-pathogenic *C. fecalis* found in sheep faeces by Firehammer (1965). A similar organism from a bovine abomasal lesion produced enteritis in experimentally infected calves (Al-Mashat and Taylor 1981). *C. sputorum* subsp. *bubulus* is a non-pathogenic catalase-negative organism found in the preputial sac and semen of normal bulls, and in the vagina of normal cattle and sheep (Florent 1953, Dozsa 1965).

Sheep

Campylobacter abortion in sheep is transmitted orally, the ram playing no significant role (Firehammer *et al.* 1956, Jensen *et al.* 1957). The disease, which has been recorded in most of the major sheep breeding countries, does not produce infertility. The causative organisms are *C. fetus* subsp. *intestinalis* and subsp. *jejuni*. According to Williams and his colleagues (1976), in the USA subsp. *jejuni* is isolated more frequently than either of the serological types (Berg *et al.* 1971) that belong to subsp. *intestinalis*.

Experimentally, oral dosing of pregnant sheep produces intestinal infection and bacteraemia for a week or more, and sheep in contact with faecal spreaders may become infected (Miller *et al.* 1959, Bryans and Shephard 1961). The bacteraemia leads to placentitis and expulsion of the fetus in animals that are two or more months pregnant (Dennis 1961, Jensen *et al.* 1961). In the natural disease abortion usually occurs in the later stages of pregnancy, often about six weeks or less before the flock is due to lamb.

Fetuses, placentae, and uterine fluids of aborted sheep are heavily contaminated, and the disease spreads rapidly in a flock. An *abortion storm* often follows a few abortions early in the breeding season. The abortion rate varies from 5 to 90 per cent but is usually between 10 and 20 per cent. Wild birds such as pigeons, starlings, sparrows, and blackbirds sometimes carry *C. fetus* subsp. *jejuni* and may help in spreading the disease (Smibert 1969, 1978). Campylobacters were isolated by Meinershagen and coworkers (1965) from the faeces of magpies 213 days after oral infection with ovine strains. *C. fetus* subsp. *jejuni* has also been found in the intestine of a variety of normal mammals, including lambs (Smibert 1965). In aborting sheep the lesions—placentitis, and sometimes grey necrotic foci in the fetal liver—are not pathognomonic. Diagnosis is based mainly on demonstrating the organism, and many of the methods used for cattle are applicable.

Ewes that abort become immune and will breed successfully in the next season. The potential usefulness of vaccination in sheep is limited by our present incomplete knowledge of serotypes, and by the sporadic nature of the disease.

Two non-pathogenic campylobacters are sometimes found in normal sheep. The catalase-negative *C. sputorum* subsp. *bubulus* occurs in the genital tract, and the catalase-positive *C. fecalis* in the faeces (Firehammer 1965).

Pigs

C. fetus subsp. *jejuni* may be found in the intestine of young healthy pigs (Smibert 1978). It is often present, sometimes in large numbers, in pigs with swine dysentery (Davis 1961, Lussier 1962, Söderlind 1965), a disease thought to be due primarily to *Treponema hyodysenteriae* (see Chapter 44); the campylobacters isolated from swine dysentery by Doyle (1944, 1948) were unusual in failing to reduce nitrate. Neill and coworkers (1978, 1979) isolated organisms thought to be

campylobacters from porcine and bovine fetuses; in some respects these organisms resembled *C. fetus* subsp. *fetus* but in others they differed, e.g. they became aerobic on subculture.

C. sputorum subsp. *mucosalis* is associated with a group of related porcine enteropathies—porcine intestinal adenomatosis (PIA), necrotic enteritis (NE), regional ileitis (RI), and proliferative haemorrhagic enteropathy (PHE)—of which PIA is thought to be the primary lesion (G. H. K. Lawson and Rowland 1974, Rowland and Lawson 1974, 1975).

PIA is a transient proliferation of the intestinal mucosa of weaned pigs (Biester and Schwarte 1931). The small and large intestines may be affected and the usual clinical manifestation is a reduction in growth rate. Immunofluorescence and electronmicroscopy show bacteria in the apical cytoplasm of affected epithelial cells, and very large numbers of *C. sputorum* subsp. *mucosalis* can be isolated from the intestinal tissue. This organism does not occur in the intestine of normal pigs but has been isolated from the mouth. The disease has been reproduced by the oral administration of mixtures containing laboratory cultures and diseased tissue homogenate (Roberts *et al.* 1977). Pigs with NE, RI, and PHE—like those with PIA—also harbour *C. sputorum* subsp. *mucosalis* in the intestine. Although many intracellular organisms may be seen in PHE, few are viable (G. H. K. Lawson *et al.* 1979); large amounts of IgA are present in the diseased tissue in both PIA and PHE.

Gebhart and co-workers (1983) isolated an apparently new *Campylobacter* sp. from swine proliferative ileitis.

Man, birds, and other animals

C. fetus subsp. *intestinalis* and subsp. *jejuni* produce systemic infections and enteritis in man (see Chapter 71).

Vibrio-like organisms, often apparently identical with *C. fetus* subsp. *jejuni*, have been isolated from normal and diseased poultry, dressed poultry carcasses, and wild birds (see Smibert 1978). Luechtefeld and her colleagues (1980) found subsp. *jejuni* in the caecal contents of 35 per cent of 445 migratory waterfowl. Skirrow and Benjamin (1980) showed that seagulls and other animals sometimes carried strains that, although thermophilic, differed from subsp. *jejuni* in being resistant to nalidixic acid. Dogs, cats, horses, simian primates, zoo animals, and others have been found to harbour subsp. *jejuni* in the intestine. Some studies, particularly of dogs, suggest that the presence of this organism in the intestine is no more common in animals with diarrhoea than in those without; other studies indicate the opposite (Hosie *et al.* 1979, Fleming 1980, Prescott and Bruin-Mosch 1981). There is some evidence that enteric infection with subsp. *jejuni* may be transmitted to man from animals, particularly puppies and poultry (Skirrow 1977, 1981, Rettig 1979, Blaser *et al.* 1980, Report 1980*b*); but the evidence must remain largely circumstantial until adequate methods for the typing of strains become available. Prescott and Barker (1980) produced mild colitis in gnotobiotic puppies with a human isolate. Fernie and Park (1977) isolated from bank voles organisms that resembled *C. fetus* subsp. *fetus* except that they did not grow at 25°; and from laboratory rats organisms resembling subsp. *jejuni* except that they did not grow in media containing 1.0 per cent glycine.

(For reviews of campylobacter infections in animals see J. R. Lawson 1959, Smibert 1978.)

Contagious equine metritis

This disease, first reported in 1977 on stud farms in the Newmarket area (Platt *et al.* 1977), is also known to occur in Ireland, France, West Germany, the USA, and Australia. The contagious equine metritis organism (CEMO), a micro-aerophilic gram-negative coccobacillus, is sometimes referred to as *Haemophilus equigenitalis* (Taylor *et al.* 1978); this name, however, has not gained universal acceptance. Infection is spread venereally, the stallion acting as a symptomless carrier. When adequate hygienic precautions are not taken mechanical spread can occur, through handling of the genitalia. There is sometimes a mucopurulent vaginal discharge within 48 hours of mating. The essential pathological changes are endometritis with areas of necrosis, and cervicitis. Infection results in a striking reduction in conception rate, and the clitoral fossa may harbour infection long after clinical recovery. Females and males are readily infected experimentally by introducing organisms into the cervix and urethral fossa respectively (Platt *et al.* 1978). Mares can be treated by intra-uterine infusions of benzyl penicillin or ampicillin; and stallions by thorough cleansing of the genitalia with chlorhexidine followed by topical application of nitrofurazone (Powell 1978).

Diagnosis may be confirmed by culturing the causative organism from swabs brought to the laboratory in Stuart's transport medium (Stuart *et al.* 1954). A suitable medium such as chocolate blood agar is seeded and incubated at 37° in a humidified atmosphere containing a supplement of 5–10 per cent CO_2. If necessary, contamination can be controlled by the addition of streptomycin and a fungicide to the medium (Platt *et al.* 1978). Very small colonies are usually visible after 48 hours' incubation. The gram-negative coccobacilli are oxidase-, catalase-, and phosphatase-positive but otherwise very unreactive (Taylor *et al.* 1978). Serological tests (agglutination, antiglobulin, complement-fixation, and passive haemagglutination) show some promise as diagnostic tools (see Powell *et al.* 1978).

Taylor and his colleagues (1979) found agglutinins to the organism of contagious equine metritis in 37.6 per cent of 223 human patients with non-gonococcal urethritis but advised caution in interpreting this finding.

For reviews of the disease see Powell (1981) and Brewer (1983).

References

Abramson, L. A., Shanina, R. Y., Igonina, A. F. and Drankin, D. I. (1969) *Zh. Mikrobiol. Epidem. Immunobiol.* **46**, 18.

Aert, A. van, DeKeyser, P., Brone, E., Bouters, R. and Vandeplassche, M. (1976) *Brit. vet. J.* **132**, 615.

Aert, A. van, DeKeyser, P., Florent, A. F., Bouters, R., Vandeplassche, M. and Brone, E. (1977) *Brit. vet. J.* **133**, 88.

Agius, E. (1965) *Arch. Inst. Pasteur Tunis* **42**, 31.

Ahvonen, P., Jansson, E. and Aho, K. (1969) *Acta path. microbiol. scand.* **75**, 291.

Alessandrini, A. and Pacelli, M. (1932) *Un Pericolo Sociale: Le Brucellosi.* Ann. Igiene, Rome.

Allan, G. S., Chappel, R. J., Williamson, P. and McNaught, D. J. (1976) *J. Hyg., Camb.* **76**, 287.

Allsup, T. N. (1969) *Vet. Rec.* **84**, 104.

Al-Mashat, R. R. and Taylor, D. J. (1980) *Vet. Rec.* **107**, 459; (1981) *Ibid.* **109**, 97.

Alton, G. G. (1967) *J. comp. Path.* **77**, 327; (1971) *Res. vet. Sci.* **12**, 330; (1978) *Aust. vet. J.* **54**, 551; (1980) *WHO/BRUC/80/355.*

Alton, G. G. and Elberg, S. S. (1967) *Vet. Bull.* **37**, 793.

Alton, G. G. and Jones, L. M. (1967) *World Hlth Org. Monogr. Ser.* No. 55.

Alton, G. G., Jones, L. M., Garcia-Carrillo, C. and Trench, A. (1972) *Amer. J. vet. Res.* **33**, 1747.

Amerault, T. E., Manthei, C. A., Goode, E. R. and Lambert, G. (1961) *Amer. J. vet. Res.* **22**, 564.

Amoia, R. (1933) *Soc. int. Microbiol., Boll. Sez. Ital.* **5**, 171.

Anczykowski, F. (1957) *Vet. Bull.* **27**, 512.

Anderson, F. M. (1950) *See* Report 1950, p. 220.

Anderson, R. K., Jenness, R., Brumfield, H. P. and Gough, P. (1964) *Science* **143**, 1334.

Angus, R. D., Brown, G. M. and Gue, C. S. (1971) *Amer. J. vet. Res.* **32**, 1609.

Bang, B. (1897) *Z. Thiermed.* **1**, 241; (1906) *J. comp. Path.* **19**, 191.

Bang, O. (1931) *2me Congr. int. Path. comp.* ii, 269.

Bang, O. and Bendixen, H. C. (1931) *Int. Milchwirtsch. Kongr. Proc.*, p. 117; (1932*a*) *Z. InfektKr. Haustiere* **42**, 81; (1932*b*) *Medlemsbl. danske Dyrlaegeforen.* Aug. 15, No. 1.

Barrow, G. I. and Peel, M. (1965) *Mon. Bull. Minist. Hlth Lab. Serv.* **24**, 21.

Beattie, C. P. and Rice, R. M. (1934) *J. Amer. med. Ass.* **102**, 1670.

Beller, K. and Stockmayer, W. (1933) *Zbl. Bakt.* **127**, 456.

Bendixen, H. C. and Blom, E. (1947) *Vet. J.* **103**, 337.

Bendtsen, H. (1959) *Nord. VetMed.* **11**, 391.

Bendtsen, H., Christiansen, M. and Thomsen, A. (1956) *Nord. VetMed.* **8**, 1.

Benussi, L. (1935) *Ann. Igiene (Sper.)* **45**, 28.

Berg, R. L. and Firehammer, B. D. (1978) *J. Amer. vet. med. Ass.* **173**, 467.

Berg, R. L., Jutila, J. W. and Firehammer, B. D. (1971) *Amer. J. vet. Res.* **32**, 11.

Bevan, L. E. W. (1921–22) *Trans. R. Soc. trop. Med. Hyg.* **15,** 215.

Biester, H. E. and Schwarte, L. H. (1931) *Amer. J. Path.* **7,** 175.

Birch, R. R. and Gilman, H. L. (1935) *J. infect. Dis.* **56,** 78.

Blaser, M. J., LaForce, F. M., Wilson, N. A. and Wang, W.-L. L. (1980) *J. infect. Dis.* **141,** 665.

Blobel, H. and Schliesser, T. (Eds) (1982) *Handbuch der Bakteriellen Infectionen bei Tieren.* Gustav Fischer, Jena.

Boak, R. A. and Carpenter, C. M. (1929) *J. Immunol.* **17,** 65; (1930) *J. infect. Dis.* **46,** 425.

Borts, I. H., Harris, D. M., Joynt, M. F., Jennings, J. R. and Jordan, C. F. (1943) *J. Amer. med. Ass.* **121,** 319.

Bosworth, T. J. (1934–35) *4th Rep. Director Inst. Animal Path., Camb.* p. 65; (1937) *J. comp. Path.* **50,** 345; (1940) *Ibid.* **53,** 42.

Bracewell, C. D. and Lander, K. P. (1979) *St. vet. J.* **34,** 111.

Bradstreet, C. M. P., Tannahill, A. J., Pollock, T. M. and Mogford, H. E. (1970) *Lancet* **ii,** 653.

Braude, A. I., Gold, D. and Anderson, D. (1949) *J. Lab. clin. Med.* **34,** 744.

Brewer, R. A. (1983) *Vet. Bull.* **53,** 881.

Brodie, J. and Sinton, G. P. (1975) *J. Hyg., Camb.* **74,** 359.

Brown, G. M. (1977) *Ann. Sclavo* **19,** 20.

Bruce, D. (1887) *Practitioner* **39,** 161; (1888) *Ibid.* **40,** 241; (1893) *Ann. Inst. Pasteur* **7,** 289.

Bryans, J. T. and Shephard, B. P. (1961) *Cornell Vet.* **51,** 376.

Bryner, J. H., Estes, P. C., Foley, J. W. and O'Berry, P. A. (1971) *Amer. J. vet. Res.* **32,** 465.

Bryner, J. H., Foley, J. W. and Thompson, K. (1979) *Amer. J. vet. Res.* **40,** 433.

Bryner, J. H., O'Berry, P. A., Estes, P. C. and Foley, J. W. (1972) *Amer. J. vet. Res.* **33,** 1439.

Bryner, J. H., O'Berry, P. A. and Frank, A. H. (1964) *Amer. J. vet. Res.* **25,** 1048.

Buck, J. M. (1930) *J. agric. Res.* **41,** 667.

Buddle, M. B. (1956) *Dept Agric., N.Z., Anim. Hlth Div. Publ. No. 217.*

Burnet, E. (1922) *Arch. Inst. Pasteur Afrique nord.* **2,** 187.

Busch, L. A. and Parker, R. L. (1972) *J. infect. Dis.* **125,** 289.

Carmichael, L. E. and Bruner, D. W. (1968) *Cornell Vet.* **58,** 579.

Carmichael, L. E., Flores-Castro and Zoha, S. (1980) *WHO/ZOON/80/135.*

Carmichael, L. E. and George, L. W. (1976) *See* Regamey *et al.* (1976) p. 237.

Caronna, C. (1936) *Azione vet.* **5,** 72.

Carpenter, C. M. and Boak, R. A. (1932) *J. Amer. med. Ass.* **99,** 296.

Carrieu, M. and Lafenêtre, M. (1932) *Le Lait* **12,** 779.

Castañeda, M. R. (1961) *Bull. World Hlth Org.* **24,** 73.

Cedro, V. C. F. *et al.* (1967) *Rev. Invest. agropecuarias* **4,** No. 1.

Chappel, R. J. (1980) *WHO/BRUC/80/353.*

Chappel, R. J., McNaught, D. J., Bourke, J. A. and Allan, G. S. (1978) *J. Hyg., Camb.* **80,** 373.

Chappel, R. J., Williamson, P., McNaught, D. J., Dalling, M. J., and Allan, G. S. (1976) *J. Hyg., Camb.* **77,** 369.

Clark, B. L., Dufty, J. H. and Monsbourgh, M. J. (1968) *Aust. vet. J.* **44,** 530.

Clark, B. L., Dufty, J. H., Monsbourgh, M. J. and Parsonson, I. M. (1976) *Aust. vet. J.* **52,** 362.

Cohn, E. G. (1962) *Anais Microbiol.* **10,** 147.

Coombs, R. R. A., Mourant, A. E. and Race, R. R. (1945) *Brit. J. exp. Path.* **26,** 255.

Corbel, M. J. (1972) *J. Hyg., Camb.* **70,** 779; (1976) *J. Hyg., Camb.* **76,** 65.

Corbel, M. J. and Lander, K. P. (1982) *WHO/BRUC/82/373.*

Cotton, W. E. Buck, J. M. and Smith, H. E. (1933) *J. Amer. vet. med. Ass.* **83,** 91.

Cruickshank, J. C. (1956) *J. Hyg., Camb.* **54,** 562.

Cunningham, B. (1977) *Vet. Rec.* **100,** 522.

Cunningham, B. and O'Reilly, D. J. (1968) *Vet. Rec.* **82,** 678.

Dalrymple-Champneys, W. (1929) *Rep. publ. Hlth med. Subj., Min. Hlth, London,* No. 56; (1960) *Brucella Infection and Undulant Fever in Man.* Oxford University Press, London.

Damon, S. R. (1932) *Amer. J. Hyg.* **16,** 798.

Davidson, I. and Hebert, C. N. (1978) *Bull. World Hlth Org.* **56,** 123.

Davidson, I., Hebert, C. N. and Morgan, W. J. B. (1969) *Bull. World Hlth Org.* **40,** 129.

Davies (1906) *Rep. Comm. medit. Fev.* Part 4.

Davis, J. W. (1961) *J. Amer. vet. med. Ass.* **138,** 471.

Deem, A. W. (1937) *J. infect. Dis.* **61,** 21.

Dennis, S. M. (1961) *Vet. Rev. Annot.* **7,** 69.

Diaz, M. and Levieux, D. (1972) *C. R. Acad. Sci., Paris* **274,** 1593.

Diaz, R., Jones, L. M. and Wilson, J. B. (1967) *J. Bact.* **93,** 1262.

Doyle, L. P. (1944) *Amer. J. vet. Res.* **5,** 3; (1948) *Ibid.* **9,** 50.

Doyle, T. M. (1935) *J. comp. Path.* **48,** 192.

Dozsa, L. (1965) *J. Amer. vet. med. Ass.* **147,** 620.

Dudley, S. F. (1931) *Lancet* **i,** 683.

Editorial (1977) *Vet. Rec.* **101,** 43.

Edwards, J. M. B., Tannahill, A. J. and Bradstreet, C. M. P. (1970) *J. clin. Path.* **23,** 161.

Eisele, C. W., McCullough, N. B., Beal, G. A. and Rottschaefer, W. (1947) *J. Amer. med. Ass.* **135,** 983.

Elberg, S. S. (1980) *WHO/ZOON/80/143.*

Elkington, G. W., Wilson, G. S., Taylor, J. and Fulton, F. (1940) *Brit. med. J.* **i,** 477.

Emmel, M. W. and Huddleson, I. F. (1929) *J. Amer. vet. med. Ass.* **75,** 578; (1930) *Ibid.* **76,** 449.

Entessar, F., Ardalan, A., Ebadi, A. and Jones, L. M. (1967) *J. comp. Path.* **77,** 367.

Evans, A. C. (1918) *J. infect. Dis.* **22,** 580.

Eyre, J. W. H. (1908) *Lancet* **i,** 1677, 1747; (1912) *Ibid.* **i,** 88.

Eyre, J. W. H., McNaught, J. G., Kennedy, J. C. and Zammit, T. (1907) *Rep. Comm. medit. Fev.* Part 6, p. 3.

Farnetani, N. (1967) *Quad. Sclavo Diag.* **3,** 160.

Farrell, I. D. and Robertson, L. (1967) *Mon. Bull. Minist. Hlth Lab. Serv.* **26,** 52; (1968) *J. Hyg., Camb.* **66,** 19; (1972) *J. appl. Bact.* **35,** 625.

Farrell, I. D., Robertson, L. and Hinchliffe, P. M. (1975) *J. Hyg., Camb.* **74,** 23.

Feeley, J. C. (1969) *J. Bact.* **99,** 645.

Feldman, W. H. and Olson, C. (1934) *J. infect. Dis.* **54,** 45.

Fernie, D. S. and Park, R. W. A. (1977) *J. med. Microbiol.* **10,** 325.

Firehammer, B. D. (1965) *Cornell Vet.* **55,** 482.

Firehammer, B. D., Marsh, H. and Tunnicliff, E. A. (1956) *Amer. J. vet. Res.* **17,** 573.

Fitch, C. P., Bishop, L. M. and Boyd, W. L. (1932) *J. Amer. vet. med. Ass.* **80,** 69.

Fleischauer, G. (1937) *Berl. tierärztl. Wschr.* **53,** 527.

Fleming, M. P. (1980) *Vet. Rec.* **107,** 202.

Florent, A. (1953) *C. R. Soc. Biol., Paris* **147,** 2066.

Galbraith, N. S., Ross, M. S., Mowbray, R. R. de and Payne, D. J. H. (1969) *Brit. med. J.* **i,** 612.

Garrod, L. P. (1937) *J. Path. Bact.* **45**, 473.

Gebhart, C. J., Ward, G. E., Chang, K. and Kurtz, H. J. (1983) *Amer. J. vet. Res.* **44**, 361.

Gilbert, R. and Dacey, H. G. (1932) *J. Lab. clin. Med.* **17**, 345.

Goldstein, J. D., Fox, W. W. and Carpenter, C. M. (1936) *Amer. J. med. Sci.* **191**, 712.

Griffitts, J. J. (1947) *Publ. Hlth Rep., Wash.* **62**, 865.

Habs, H. (1933) *Zbl. ges. Hyg.* **28**, 481.

Habs, H. and Sievert, L. (1935) *Dtsch. med. Wschr.* **61**, 1398.

Hadley, F. B. and Welsh, W. E. (1931) *Cornell Vet.* **21**, 27.

Hall, I. C. and Learmonth, R. (1933) *J. infect. Dis.* **52**, 27.

Hardy, A. V., Frant, S. and Kroll, M. M. (1938) *Publ. Hlth Rep., Wash.* **53**, 796.

Hardy, A. V., Jordan, C. F. and Borts, I. H. (1932) *Publ. Hlth Rep., Wash.* **47**, 187;

Hardy, A. V., Jordan, C. F., Borts, I. H. and Hardy, G. C. (1931) *Nat. Inst. Hlth, Wash.* Bull. No. 158.

Hasseltine, H. E. (1930) *Publ. Hlth Rep., Wash.* **45**, 1660.

Haxthausen, H. and Thomson, A. (1931) *Arch. Derm. Syph., Berl.* **163**, 477.

Heathman, L. S. (1934) *J. infect. Dis.* **55**, 243.

Henderson, R. J. (1964) *Mon. Bull. Minist. Hlth Lab. Serv.* **23**, 34.

Henderson, R. J., Hill, D. M., Vickers, A. A., Edwards, J. M. and Tillett, H. (1976) *J. clin. Path.* **29**, 35.

Hendricks, S. L. (1955) *Amer. J. publ. Hlth* **45**, 1282.

Hendricks, S. L., Borts, I. H., Heeren, R. H., Hausler, W. J. and Held, J. R. (1962) *Amer. J. publ. Hlth* **52**, 1166.

Henricsson, E. (1932) *Epizootischer Abortus und Undulantfieber.* Isaac Marcus Boktryckeri-Aktiebolag, Stockholm.

Heremans, J. F., Vaerman, J.-P. and Vaerman, C. (1963) *J. Immunol.* **91**, 11.

Hirchert, R., Lange, W. and Leonhardt, H.-G. (1967) *Berl. Münch. tierärztl. Wschr.* **80**, 45.

Hirsch, W. (1935) *Arch. Schiffs- u. Tropenhyg.* **39**, 30.

Hoeden, J. van der (1931) *Tijdschr. Diergeneesk.* **58**, 1321; (1932a) *Ibid.* **59**, 385; (1932b) *Ibid.* **59**, 612; (1932c) *Ibid.* **59**, 1383, 1446; (1932d) *Z. InfektKr. Haustiere* **42**, 1; (1933) *J. comp. Path.* **46**, 232; (1940) *Tijdschr. Diergeneesk* **67**, 910, 963.

Horning, B. G. (1935) *J. Amer. med. Ass.* **105**, 1978.

Horrocks, W. H. (1905) *Rep. Comm. medit. Fev.* Part 3, p. 84.

Horrocks, W. H. and Kennedy, J. C. (1906) *Rep. Comm. medit. Fev.* Part 3, p. 37.

Hosie, B. D., Nicolson, T. B. and Henderson, D. B. (1979) *Vet. Rec.* **105**, 80.

Hosty, T. S. and Johnson, M. B. (1953) *Amer. J. publ. Hlth* **43**, 30.

Howe, C., Miller, E. S., Kelly, E. H., Bookwalter, H. L. and Ellingson, H. V. (1947) *New Engl. J. Med.* **236**, 741.

Huddleson, I. F. (1939) *Brucellosis in Man and Animals.* Humphrey Milford, London; (1956) *Bact. Proc.* p. 53; (1957) *Bull. World Hlth Org.* **16**, 929; (1964) *J. Bact.* **88**, 540.

Huddleson, I. F., Hasley, D. E. and Torrey, J. P. (1927) *J. infect. Dis.* **40**, 352.

Huddleson, I. F. and Johnson, H. W. (1930) *J. Amer. med. Ass.* **94**, 1905; (1933) *Amer. J. trop. Med.* **13**, 485.

Huddleson, I. F., Johnson, H. W. and Hamann, E. E. (1933a) *J. Amer. vet. med. Ass.* **83**, 16; (1933b) *Amer. J. publ. Hlth* **23**, 917.

Huddleson, I. F. and Munger, M. (1940) *Amer. J. publ. Hlth* **30**, 944.

Huddleson, I. F. and Smith, L. H. (1931) *J. Amer. vet. med. Ass.* **79**, 63.

Hughes, M. L. (1893) *Ann. Inst. Pasteur* **7**, 628; (1897) *Mediterranean, Malta or Undulant Fever.* Macmillan & Co., London.

Hunter, C. A. and Colbert, B. (1956) *J. Immunol.* **77**, 232.

Huntley, B. E., Philip, R. N. and Maynard, J. E. (1963) *J. infect. Dis.* **112**, 100.

Hutchings, L. M. (1950) *See* Report 1950, p. 188.

Hutyra, F. and Marek, J. (1922) *Special Pathology and Therapeutics of Diseases of Domestic Animals*, 2nd Amer. edn, Vol. 1, p. 780. Chicago and London.

Jensen, R., Miller, V. A., Hammarlund, M. A. and Graham, W. R. (1957) *Amer. J. vet. Res.* **18**, 326.

Jensen, R., Miller, V. A. and Molello, J. A. (1961) *Amer. J. vet. Res.* **22**, 169.

Johnson, H. W. and Huddleson, I. F. (1931) *J. Amer. vet. med. Ass.* **78**, 849.

Johnson, H. W., Huddleson, I. F. and Hamann, E. E. (1933) *J. Amer. vet. med. Ass.* **83**, 727.

Johnstone (1905) *Rep. Comm. medit. Fev.* Part 2.

Jones, F. S. and Little, R. B. (1931) *J. exp. Med.* **53**, 835, 845.

Jones, F. S., Orcutt, M. and Little, R. B. (1931) *J. exp. Med.* **53**, 853.

Jones, L. M. and Berman, D. T. (1976) *See* Regamey *et al.* (1976), p. 328; (1980) *WHO/BRUC/80/365.*

Karkadinovsky, J. A. (1936) *C. R. Soc. Biol.* **121**, 1611.

Karsten (1932) *Dtsch. tierärztl. Wschr.* **40**, 689; (1943) *Berl. Münch. tierärztl. Wschr.* No. 21/22, p. 160.

Keefer, C. S. (1924) *Bull. Johns Hopk. Hosp.* **35**, 6.

Kennedy, J. C. (1905) *Rep. Comm. medit. Fev.* Part 3, pp. 56, 71, 91; (1906) *Ibid.* Part 4, p. 92.

Keppie, J., Williams, A. E., Witt, K. and Smith, H. (1965) *Brit. J. exp. Path.* **46**, 104.

Keppie, J., Witt, K. and Smith, H. (1967) *Res. vet. Sci.* **8**, 294.

Kerr, W. R. (1955) *Brit. vet. J.* **111**, 169; (1968) *Some Diseases of Animals Communicable to Man in Britain*, p. 295. Pergamon Press, Oxford.

Kerr, W. R., Coghlan, J. D., Payne, D. J. H. and Robertson, L. (1966) *Lancet* **ii**, 1181.

Kerr, W. R., Payne, D. J. H., Robertson, L. and Coombs, R. R. A. (1967) *Immunology* **13**, 223.

Kerr, W. R., Pearson, J. K. L. and Rankin, J. E. F. (1958) *Vet. Rec.* **70**, 503.

Kerr, W. R. *et al.* (1968) *J. med. Microbiol.* **1**, 181.

King, E. O. (1962) *Ann. N.Y. Acad. Sci.* **98**, 700.

Kristensen, M. and Holm, P. (1929) *Zbl. Bakt.* **112**, 281.

Krüger (1937) *Dtsch. tierärztl. Wschr.* **45**, 182.

Kuzdas, C. D. and Morse, E. V. (1953) *J. Bact.* **66**, 502.

Lambert, G., Manthei, C. A. and Deyoe, B. L. (1963) *Amer. J. vet. Res.* **24**, 1152.

Lawson, G. H. K. and Rowland, A. C. (1974) *Res. vet. Sci.* **17**, 331.

Lawson, G. H. K., Rowland, A. C., Roberts, L., Fraser, G. and McCartney, E. (1979) *Res. vet. Sci.* **27**, 46.

Lawson, J. R. (1959) In: *Infectious Diseases of Animals: Diseases due to Bacteria*, vol. 2, p. 745. Ed. by A. W. Stableforth and I. A. Galloway. Butterworths Scientific Publications, London.

Lawson, J. R. and MacKinnon, D. J. (1952) *Vet. Rec.* **64**, 763.

Leavell, H. R. and Amoss, H. L. (1931) *Amer. J. med. Sci.* **181**, 96.

Lerche (1931) *Z. InfektKr. Haustiere* **38,** 253.

Liess, B. (1958) *Mh. Tierheilk.* **10,** 267.

Lindberg, A. A., Haeggman, S., Karlson, K., Carlson, H. E. and Mair, N. S. (1982) *J. Hyg., Camb.* **88,** 295.

Lisbonne, M. and Janbon, M. (1935) *Encyclopédie Medico-chirurgicale.*

Löffler, W., Moroni, D. L. and Frei, W. (1955) *Die Brucellose als Anthropo-Zoonose.* Springer-Verlag, Berlin.

Lübke, A. (1935*a*) *Z. InfektKr. Haustiere* **47,** 240; (1935*b*) *Arch. wiss. prakt. Tierheilk* **68,** 233.

Luechtefeld, N. A. W., Blaser, M. J., Reller, L. B. and Wang, W.-L. L. (1980) *J. clin. Microbiol.* **12,** 406.

Lussier, G. (1962) *Canad. vet. J.* **3,** 267.

McAlpine, J. G. and Rettger, L. F. (1925) *J. Immunol.* **10,** 811.

McCullough, N. B. (1949) *Amer. J. publ. Hlth* **39,** 866.

McDiarmid, A. (1957) *Vet. Rec.* **69,** 877; (1960) *Ibid.* **72,** 917; (1968) *Some Diseases of Animals Communicable to Man in Britain,* p. 281. Pergamon Press, Oxford.

McDiarmid, A., Findlay, H. T., Jameson, J. E., Phease, R. N., Walker, J. H. C., Jones, L. M. and Ogonowski, K. (1958) *Brit. vet. J.* **114,** 83.

McEwen, A. D. and Priestley, F. W. (1938) *Vet. Rec.* **50,** 1097.

M'Fadyean, J., Sheather, A. L. and Minett, F. C. (1913) *J. comp. Path.* **26,** 142.

M'Fadyean, J. and Stockman, S. (1909) *Rep. Comm. on Epizootic Abortion* Append. to Part I, London.

McGaughey, W. J. and Hanna, J. (1973) *Vet. Rec.* **93,** 246.

MacNeal, W. J. and Kerr, J. E. (1910) *J. infect. Dis.* **7,** 469.

McNutt, S. H. (1935) *J. Amer. vet. med. Ass.* **83,** 183.

Magee, J. T. (1980) *J. med. Microbiol.* **13,** 167.

Makkawejsky, W. N. and Karkadinowskaja, I. A. (1932) *Dtsch. tierärztl. Wschr.* **40,** 229.

Makkawejski, W. N., Karkadinowskaja, I. A. and Micheew, N. I. (1933) *Dtsch. tierärztl. Wschr.* **41,** 321.

Markenson, J., Sulitzeanu, D. and Olitzki, A. L. (1962) *Brit. J. exp. Path* **43,** 67.

Marston, J. A. (1863) *Army med. Dep. statist. Rep.* **3,** 486.

Marx, A., Sandulache, R., Pop, A. and Cerbu, A. (1975) *Ann. Microbiol., Paris* **126B,** 435.

Mazé, P. and Césari, E. (1931) *C. R. Soc. Biol.* **108,** 630.

Meenan, P. N. (1967) *J. Irish med. Ass.* **60,** 41.

Meinershagen, W. A., Waldhalm, D. G., Frank, F. W. and Scrivner, L. H. (1965) *J. Amer. vet. med. Ass.* **147,** 843.

Mellick, P. W., Winter, A. J. and McEntee, K. (1965) *Cornell Vet.* **55,** 280.

Menzani, R. (1932) *Nuova vet.* **10,** 37.

Meyer, K. F. (1933) *J. Bact.* **29,** 43; (1980) *WHO/ZOON/80/134.*

Meyer, K. F. and Eddie, B. (1941) *J. infect. Dis.* **68,** 24.

Meyer, M. E. (1964) *J. infect. Dis.* **114,** 169.

Michel-Béchet, R., Puig, R. and Charvet, P. (1939) *Localisations Viscérales et Aspects Chirurgicaux des Brucelloses.* Masson & Cie, Paris.

Miller, J. K., Kelly, J. I. and Roerink, J. H. G. (1976) *Vet. Rec.* **98,** 210.

Miller, V. A., Jensen, R. and Gilroy, J. J. (1959) *Amer. J. vet. Res.* **20,** 677.

Minett, F. C. and Martin, W. J. (1936) *J. Dairy Res.* **7,** 122.

Mittal, K. R. and Tizard, I. R. (1979*a*) *J. Hyg., Camb.* **83,** 295; (1979*b*) *Res. vet. Sci.* **26,** 248.

Mohler, J. R. and Traum, J. (1911) *28th Ann. Rep. Bur. Anim. Indust.* p. 147.

Molinelli, E. A. and Ithurrat, E. M. F. (1934) *Semana méd.* **41,** Part 2, 176.

Molinelli, E. A., Ithurrat, E. M. F. and Ithurralde, D. (1948) *Rev. Asoc. méd. argent.* **62,** 111.

Morales-Otero, P. and González, L. M. (1949) *Proc. Soc. exp. Biol., N.Y.* **71,** 387.

Morgan, W. J. B. (1960) *Res. vet. Sci.* **1,** 47; (1968) *Some Diseases of Animals Communicable to Man in Britain,* p. 263. Pergamon Press, Oxford; (1970) *J. Dairy Res.* **37,** 303; (1977) *Ann. Sclavo* **19,** 35.

Morgan, W. J. B. and McDiarmid, A. (1960) *Res. vet. Sci.* **1,** 53; (1968) *Vet. Rec.* **83,** 184.

Morgan, W. J. B., Mackinnon, D. J. and Cullen, G. A. (1969) *Vet. Rec.* **85,** 636.

Morisset, R. and Spink, W. W. (1969) *Lancet* **ii,** 1000.

Myers, D. M. (1973) *Appl. Microbiol.* **26,** 855.

Nagy, L. K. and Hignett, P. G. (1967) *Res. vet. Sci.* **8,** 247.

Neill, S. D., Ellis, W. A. and O'Brien, J. J. (1978) *Res. vet. Sci.* **25,** 368; (1979) *Ibid.* **27,** 180.

Nicoletti, P. (1967) *J. Amer. vet. med. Ass.* **151,** 1778; (1976) *See* Regamey *et al.* (1976) p. 131.

Nicoletti, P. L., Quinn, B. R. and Minor, P. W. (1967) *N.Y. St. J. Med.* **67,** 2886.

Nowak, J. (1908) *Ann. Inst. Pasteur* **22,** 541.

O'Hern, E. M. (1977) *Ann. Sclavo* **19,** 12.

Olin, G. (1935) *Studien über das Undulantfieber in Schweden.* Isaac Marcus Boktryckeri-Aktiebolag, Stockholm.

Olitzki, A. (1970) *Immunological Methods in Brucellosis Research,* Nos. 8 and 9. S. Karger, Basel.

Painter, G. M., Deyoe, B. L. and Lambert, G. (1966) *Canad. J. comp. Med.* **30,** 218.

Pappagianis, D., Elberg, S. S. and Crouch, D. (1966) *Amer. J. Epidem.* **84,** 21.

Peterson, C. E. (1935) *J. Lab. clin. Med.* **20,** 727.

Philpott, M. (1966) *Vet. Rec.* **79,** 811.

Pivnick, H., Worton, H. Smith, D. L. T. and Barnum, D. (1966) *Canad. J. publ. Hlth* **57,** 225.

Plastridge, W. N., Williams, L. F. and Petrie, D. (1947) *Amer. J. vet. Res.* **8,** 178.

Plastridge, W. N., Williams, L. F. and Trowbridge, D. G. (1964) *Amer. J. vet. Res.* **25,** 1295.

Plate, G. (1934) *Dtsch. tierärztl. Wschr.* **42,** 537.

Platt, H., Atherton, J. G. and Simpson, D. J. (1978) *Equine vet. J.* **10,** 153.

Platt, H., Atherton, J. G., Simpson, D. J., Taylor, C. E. D., Rosenthal, R. O., Brown, D. F. J. and Wreghitt, T. G. (1977) *Vet. Rec.* **101,** 20.

Poelma, L. J. and Pickens, E. M. (1932) *J. Bact.* **23,** 112.

Poole, P. M. (1975) *Postgrad. Med. J.* **51,** 433.

Poston, M. A. (1938) *Publ. Hlth Rep., Wash.* **53,** 1.

Powell, D. G. (1978) *Equine vet. J.* **10,** 1; (1981) *Adv. vet. Sci.* **25,** 161.

Powell, D. G., David, J. S. E. and Frank, C. J. (1978) *Vet. Rec.* **103,** 399.

Preisz, H. (1903) *Zbl. Bakt.* **33,** 190.

Prescott, J. F. and Barker, I. K. (1980) *Vet. Rec.* **107,** 314.

Prescott, J. F. and Bruin-Mosch, C. W. (1981) *Amer. J. vet. Res.* **42,** 164.

Priestley, F. W. (1931) *J. Path. Bact.* **34,** 81.

Pröscholdt, O. (1932) *Dtsch. tierärztl, Wschr.* **40,** 673.

Pullinger, E. J. (1935) *Lancet* **i,** 1342; (1936) *J. Hyg., Camb.* **36,** 456.

Rainsford, S. G. (1935) *J. R. nav. med. Serv.* **21,** 81.

Ray, W. C. (1979) *CRC Handbook Series in Zoonoses*, Vol. 1, p. 99.

Reddin, J. L., Anderson, R. K., Jenness, R. and Spink, W. W. (1965) *New Engl. J. Med.* **272**, 1263.

Regamey, R. H., Hulse, E. C. and Valette, L. (Eds.) (1976) *49th Int. Symp. on Brucellosis.* S. Karger, Basel.

Reinhardt and Gauss (1915) *Zbl. Bakt.* **28**, 172.

Renoux, G. (1965) In: Regamey, Henneson and Ungar's *Progress in Immunobiological Standardization*, Vol. 1, p. 176. S. Karger, Basel.

Report (1905–07) *Mediterranean Fever Commission*, Harrison & Sons, London; (1909) *Rep. Dep. Comm. Epizootic Abortion*, Part I, *Bd Agric. Fish.*, H.M.S.O., London; (1913) *Rep. Dep. Comm. Epizootic Abortion*, Part III, *Bd Agric. Fish.*, H.M.S.O., London; (1950) Brucellosis. A Symposium. *Amer. Ass. Advanc. Sci.*, Washington, D.C.; (1956) *Mon. Bull. Minist. Hlth Lab. Serv.* **15**, 85; (1958) *World Hlth Org. Tec. Rep. Ser.* No. 148; (1961) *Mon. Bull. Minist. Hlth Lab. Serv.* **20**, 33; (1964a) Animal Disease Surveys, No. 4. *Minist. Agric. Fish. Food*, H.M.S.O. London; (1964b) Brucellosis. *World Hlth Org. techn. Rep. Ser.* No. 289; (1971) *World Hlth Org. techn. Rep. Ser.* No. 464; (1978) *Vet. Rec.*, **102**, 294; (1980a) *WHO/BRUC/80/ several*; (1980b) *Bull. World Hlth Org.* **58**, 519; (1981a) *Vet. Rec.* **109**, 394; (1981b) A Guide to the Diagnosis, treatment and Prevention of brucellosis. *WHO/VPH 81/ 31.*

Rettig, P. J. (1979) *J. Pediat.* **94**, 855.

Reusse, U. and Schindler, R. (1957) *Z. Hyg. InfektKr.* **143**, 578.

Ridala, V. (1936) *Enquiries into the Pathogenic Changes Produced by Br. abortus in the Udder and Certain Organs of the Cow. Vet. Milk Hyg. Inst.*, Univ. Tartu, Esthonia.

Rinjard, P. and Hilger, A. (1928) *Bull. Acad. vét. France* **81**, 272.

Roberts, L., Rowland, A. C. and Lawson, G. H. K. (1977) *Vet. Rec.* **100**, 12.

Roberts, R. M. and Philip, J. R. (1960) *Res. vet. Sci.* **1**, 328.

Robertson, L. (1967) *J. clin. Path.* **20**, 199.

Robertson, L., Farrell, I. D., Hinchliffe, P. M. and Quaife, R. A. (1980) *Publ. Hlth Lab. Service, Monograph Series*, No. 14. H.M.S.O., London.

Roepke, M. H., Paterson, K. G., Driver, F. C., Clausen, L. B., Olson, L. and Wentworth, J. E. (1950) *Amer. J. vet. Res.* **11**, 199.

Rossi, P. and Saunié, L. (1934) *C. R. Soc. Biol.* **115**, 134, 137.

Roux, J. (1972) *Bull. Inst. Pasteur* **70**, 145

Rowland, A. C. and Lawson, G. H. K. (1974) *Res. vet. Sci.* **17**, 323; (1975) *Vet. Rec.* **97**, 178.

Runnels, R. A. and Huddleson, I. F. (1925) *Cornell Vet.* **15**, 376.

Ryan, W. J. (1967) *Mon. Bull. Minist. Hlth Lab. Serv.* **26**, 33.

Sandholm, A. (1938) *Z. InfektKr. Haustiere* **53**, 201.

Schassan, H. H. (1968) *Z. ImmunForsch.* **134**, 424.

Schlesmann, C. (1932) *Klin. Wschr.* **11**, 1711.

Schneller (1934) *Tierärztl. Rdsch.* **40**, 762.

Schroeder, E. C. and Cotton, W. E. (1911) *28th Rep. Bur. Anim. Indust.* p. 139.

Schuhardt, V. T., Woodfin, H. W. and Knolle, K. C. (1951) *J. Bact.* **61**, 299.

Schurig, G. G., Duncan, J. R. and Winter, A. J. (1978) *J. infect. Dis.* **138**, 463.

Seddon, H. R. (1919) *J. comp. Path.* **32**, 1.

Serre, A., Dana, M. and Roux, J. (1970) *Path. et Biol.*, *Paris* **18**, 367.

Sheather, A. L. (1923) *J. comp. Path.* **36**, 255.

Simpson, W. M. (1930) *Ann. intern. Med.* **4**, 238.

Skerman, V. B. D., McGowan, V. and Sneath, P. H. A. (1980) *Int. J. syst. Bact.* **30**, 225.

Skirrow, M. B. (1977) *Brit. med. J.* **2**, 9; (1981) *Vet. Res. Commun.* **5**, 13.

Skirrow, M. B. and Benjamin, J. (1980) *J. Hyg.*, *Camb.* **85**, 427.

Smibert, R. M. (1965) *Amer. J. vet. Res.* **26**, 320; (1969) *Ibid.* **30**, 1437; (1974) In: *Bergey's Manual of Determinative Bacteriology*, 8th edn, p. 207. Ed. by R. E. Buchanan and N. E. Gibbons. Williams and Wilkins, Baltimore; (1978) *Annu. Rev. Microbiol.* **32**, 673.

Smith, H., Keppie, J., Pearce, J. H. and Witt, K. (1962a) *Brit. J. exp. Path.* **43**, 538.

Smith, H., Williams, A. E. Pearce, J. H., Keppie, J., Harris-Smith, P. W., Fitz-George, R. B. and Witt, K. (1962b) *Nature, Lond.* **193**, 47.

Smith, J. (1932) *J. Hyg.*, *Camb.* **32**, 354; (1934a) *Ibid.* **34**, 242; (1934b) *J. comp. Path.* **47**, 125.

Smith, T. (1918) *J. exp. Med.* **28**, 701; (1926) *Ibid.* **43**, 317.

Söderlind, O. (1965) *Vet. Rec.* **77**, 193.

Spengler, G. (1929) *Die Bangsche Krankheit beim Menschen.* Urban und Schwarzenberg, Berlin.

Spink, W. W. (1950) See Report 1950, p. 136; (1956) *The Nature of Brucellosis.* University Minnesota Press, Minneapolis.

Spink, W. W., Hall, W. H., Shaffer, J. M. and Braude, A. I. (1948) *J. Amer. med. Ass.* **136**, 382; (1949a) *Ibid.* **139**, 352.

Spink, W. W., Hutchings, L. M., Mingle, C. K., Larson, C. L., Boyd, W. L., Jordan, C. F. and Evans, A. C. (1949b) *J. Amer. med. Ass.* **141**, 326.

Stableforth, A. W. (1959) Stableforth and Galloway's *Infectious Diseases of Animals*, Vol. 1, p. 53. Butterworth, London.

Steele, J. H. and Hastings, J. W. (1948) *Publ. Hlth Rep.*, *Wash.* **63**, 144.

Stockman, S. (1914) *J. comp. Path.* **27**, 237.

Stockmayer, W. (1933) *Berl. Münch. tierärztl. Wschr.* **49**, 741; (1936) *Z. InfektKr. Haustiere* **49**, 46.

Stoenner, H. G. (1951) *J. Amer. vet. med. Ass.* **118**, 101.

Stuart, R. D., Toshach, S. R. and Patsula, T. M. (1954) *Canad. J. publ. Hlth* **45**, 73.

Swenson, R. M., Carmichael, L. E. and Cundy, K. R. (1972) *Ann. intern. Med.* **76**, 435.

Symposium (1976) *Develop. biol. Standardzn* **31**. S. Karger, Basel.

Szyfres, B. and Tomé, J. G. (1966) *Bull. World Hlth Org.* **34**, 919.

Taylor, C. E. D., Rosenthal,, R. O., Brown, D. F. J., Lapage, S. P., Hill, L. R. and Legros, R. M. (1978) *Equine vet. J.* **10**, 136.

Taylor, C. E. D., Rosenthal, R. O. and Taylor-Robinson, D. (1979) *Lancet* **i**, 700.

Taylor, R. M. and Hazemann, R. H. (1932) *Rev. Hyg.* **54**, 481.

Taylor, R. M., Lisbonne, M. and Roman, G. (1932) *Ann. Inst. Pasteur* **49**, 284.

Taylor, R. M., Lisbonne, M. and Vidal, L. F. (1935) *Mouvement San.* **12**, 51.

Taylor, R. M., Lisbonne, M., Vidal, L. F. and Hazemann, R. H. (1938) *Bull. Hlth Ord. L.o.N.* **7**, 503.

Taylor, R. M., Vidal, L. F. and Roman, G. (1934) *C. R. Soc. Biol.* **116,** 132.

Terpstra, J. I. and Eisma, W. A. (1951) *Tijdschr. Diergeneesk.* **76,** 433.

Thomsen, A. (1934) *Brucella Infection in Swine.* Levin and Munksgaard, Copenhagen; (1937*a*) *J. comp. Path.* **50,** 1; (1937*b*) *Skand. VetTidskr.* **27,** 728.

Thorp, F. and Graham, R. (1933) *J. Amer. vet. med. Ass.* **82,** 871.

Thorpe, B. D., Sidwell, R. W. and Lundgren, D. L. (1967) *Amer. J. trop. Med. Hyg.* **16,** 665.

Tovar, R. M. (1947) *Amer. J. vet. Res.* **8,** 138.

Traum, J. E. (1914) *Rep. Chief. Bur. Anim. Industry.* p. 30.

Trever, R. W., Cluff, L. E., Peeler, R. N. and Bennett, I. L. (1958) *Bull. Johns Hopk. Hosp.* **102,** 46.

Unel, S., Williams, C. F. and Stableforth, A. W. (1969) *J. comp. Path.* **79,** 155.

Veloppé and Jaubert (1935) *Rev. gén. Méd. vét.* **44,** 513.

Véron, M. and Chatelain, R. (1973) *Int. J. syst. Bact.* **23,** 122.

Waghela, S., Wandera, J. G. and Wagner, G. G. (1980) *Res. vet. Sci.* **28,** 168.

Wall, S. (1911) *Z. InfektKr. Haustiere* **10,** 23, 132; (1930) *Proc. 11th int. vet. Congr., London.*

Weber, A. and Schliesser, T. (1978) *Berl. Münch. tierärztl. Wschr.* **91,** 28.

West, D. E. and Borman, E. K. (1945) *J. infect. Dis.* **77,** 187.

White, G. C. and Swett, P. P. (1935) *J. Amer. vet. med. Ass.* **87,** 146.

Wilesmith, J. W. (1978) *Vet. Rec.* **103,** 149.

Wilkinson, P. C. (1966) *J. Immunol.* **96,** 457.

Williams, C. E., Renshaw, H. W., Meinershagen, W. A., Everson, D. O., Chamberlain, R. K., Hall, R. F. and Waldhalm, D. G. (1976) *Amer. J. vet. Res.* **37,** 409.

Williams, E. (1970) *Lancet* **i,** 604.

Williams, T. P., Entwistle, D. M., Masters, P. L. and Woods, A. C. (1957) *Lancet* **ii,** 1203.

Wilson, D. V., Thornley, M. J. and Coombs, R. R. A. (1979) *J. med. Microbiol.* **10,** 281.

Wilson, G. S. (1932) *Vet. Rec.* **44,** 1240.

Wilson, G. S. and Maier, I. (1939) *Brit. med. J.* **i,** 8; (1940) *Ibid.* **i,** 47.

Wilson, M. M. and Merrifield, E. V. O. (1951) *Lancet* **ii,** 913.

Winter, A. J. and Caveney, N. T. (1978) *J. Amer. vet. med. Ass.* **173,** 472.

Witte, J. (1941) *Münch. tierärztl. Wschr.* p. 128.

Wright, A. E. and Smith, F. (1897) *Lancet* **i,** 656.

Zammit, T. (1905) *Rep. Comm. medit. Fev.* Part III, p. 83; (1906) *Ibid.* Part IV, p. 96.

57

Pyogenic infections, generalized and local

M.T. Parker

Introductory

The separation of bacterial infections, or of the bacteria that cause them, into pyogenic and non-pyogenic groups is largely arbitrary, and depends on the observation of the relative frequency of frankly suppurative lesions among all the reactions that together characterize the association between a particular bacterial parasite and a particular animal host. It is quite easy to select some bacterial species, such as *Staphylococcus aureus* or *Streptococcus pyogenes*, as frankly pyogenic, and some others, such as *Mycobacterium leprae*, *Corynebacterium diphtheriae*, or *Clostridium tetani*, as non-pyogenic, in the usual sense; but there are very many pathogenic bacteria that cannot be assigned definitely to one class or the other. Thus, the tubercle bacillus, which usually gives rise to a granulomatous lesion in solid organs, may cause the formation of pus in the urinary bladder, and the typhoid bacillus, which is not ordinarily a pyogenic organism, may cause suppurative cholecystitis or osteitis.

Most of the bacteria responsible for septic lesions are members of the body flora that under normal circumstances are prevented from invading the tissues by the body's natural defences. For infection to occur, predisposing factors that result in the bypassing or inhibition of these defence mechanisms are often necessary. In this chapter we shall note several instances of this; more will be given in Chapter 58 when we discuss hospital-acquired infections, most of which are made possible only by host predisposition.

The bacteria that are common causes of pyogenic infection include several to which we devote separate chapters: streptococci (Chapter 59), staphylococci and gram-positive anaerobic cocci (Chapter 60), several of the gram-negative non-sporing aerobic (Chapter 61) and anaerobic bacilli (Chapter 62). However, nearly all members of the body flora—bacteria, fungi, and even some protozoa—as well as some bacteria that normally live in the natural environment, may in certain circumstances behave similarly. In this chapter we shall describe or refer to the main septic infections that occur in the various organs of the body, paying particular attention to those not dealt with elsewhere in the book, but first we shall discuss generalized infections with pyogenic organisms. We do not have space to detail the treatment of individual infections with antimicrobial substances; for this reference should be made to Chapter 5 and to books such as that of Garrod and his colleagues (1981).

Septicaemia

Bacteraemia, in the sense of the mere presence of bacteria in the blood stream, has been described in several earlier chapters. Septicaemia is here used as a designation for a clinical syndrome associated with a severe bacteraemic infection.

It is sometimes asserted that in septicaemia organisms are multiplying in the blood, whereas in bacteraemia they are derived from infected tissues. Such a distinction clearly cannot be drawn from the results of blood culture; it rests indeed on dubious clinical inference. The large and rapid increase in the number of circulating bacteria that sometimes takes place just before death suggests multiplication in the blood stream; but with this exception, the septicaemic phase of any bacterial infection is probably the expression of a rapid and continuous invasion of the blood stream from the tissues. The fact that bacteraemia often ceases when a single and accessible purulent focus is drained surgically testifies to the efficacy of the clearing mechanism of the blood stream.

The term **pyaemia** is used to designate a type of general infection, as, for example, that seen in the more severe forms of staphylococcal disease, in which metastatic abscesses occur in such situations as the kidney, the brain, or the myocardium. In appendicitis or other suppurative lesions in the abdominal cavity, multiple abscesses may occur in the liver, in direct relation to the portal vein—a condition known as 'portal pyaemia'. In either case the dissemination appears to depend on the occurrence of a suppurative phlebitis at the site of the primary lesion. This extends directly as a progressive thrombosis; and particles of the infected thrombus, becoming detached, are carried by the blood stream until they reach some situation in which the anatomy of the blood vessels determines their impaction, where they give rise to secondary abscesses.

Causative organisms

During the past 40 years there have been considerable changes in the relative importance of particular organisms as causes of septicaemia in man. Before 1940, the pneumococcus, the haemolytic streptococci (mainly *Str. pyogenes*) and *Staph. aureus* were the organisms most often isolated from blood cultures, and pneumococci were the most common causes of fatal septicaemia. Soon after the introduction of effective antibacterial drugs, not only did the mortality rate for pneumococcal and streptococcal septicaemia diminish considerably but their relative frequency was reduced, presumably because many local suppurative infections were treated successfully. However, the introduction of antibiotics had little immediate effect on the frequency of staphylococcal septicaemia and caused only a short-lived reduction in its fatality.

These and subsequent trends are illustrated by the

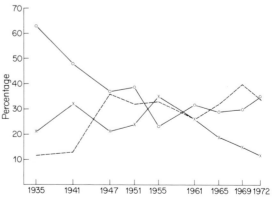

Fig. 57.1   Percentage of streptococci (0——0), *Staph. aureus* (x——x, and gram-negative aerobic bacilli (- - - -) in bacteraemia, 1935–72 (data from McGowan *et al.* 1975).

Table 57.1 Bacteraemia in patients in the Boston City Hospital, 1935-72 (data from McGowan *et al.* 1975)

	1935	1941	1947	1951	1955	1961	1965	1969	1972
Number of cases /1000 admissions	7	10	12	12	16	24	33	26	28
Percentage of cases due to:									
pneumococci	32	31	21	21	12	14	13	15	15
haemolytic streptococci	17	4	3	2	1	5	4	4	6
'viridans' streptococci	14	12	11	9	7	7	6	7	9
enterococci	0	1	2	7	3	6	6	4	5
all streptococci	63	48	37	39	23	32	29	30	35
Staph. aureus	21	32	21	24	35	26	19	15	12
Esch. coli	9	10	15	9	11	10	14	14	14
Klebsiella–Enterobacter	0	0	10	8	9	8	8	14	13
Proteus	2	2	9	12	11	5	5	8	4
Ps. aeruginosa	<1	1	2	3	2	3	5	4	3
all gram–negative bacilli listed above	12	13	36	32	33	26	32	40	34
any other organisms*	4	7	7	4	5	15	17	11	13

* Includes *Serratia* and *Acinetobacter*.

reports of Finland and his colleagues (Finland *et al.* 1959, McGowan *et al.* 1975) from the Boston City Hospital (Fig. 57.1 and Table 57.1). It should be noted that septicaemia was recognized with increasing frequency between 1935 and 1960; whether it had become more common it is impossible to say. At all times, over half of the septicaemic infections were present when the patient was admitted to hospital, but the average age of the patients tended to rise in successive surveys, particularly among those whose infections had developed in hospital.

In retrospect, the Boston figures give a rather different picture of events from that apparent to contemporary observers. During the 1950s, *Staph. aureus* was attracting their interest almost exclusively, but staphylococcal septicaemia showed only a modest increase in relative frequency at that time. After 1960, its importance decreased progressively, and by the end of the decade it had fallen well below the pre-1940 level. On the other hand, the gram-negative aerobic bacilli emerged as an important cause of septicaemia as early as 1947, but this event attracted little attention for over 10 years. Before 1947, *Escherichia coli* was the only gram-negative aerobe of significance as a cause of septicaemia. After this date, members of the *Klebsiella-Enterobacter* and *Proteus* groups together equalled or exceeded *Esch. coli* in relative frequency. From 1960 onwards, *Klebsiella-Enterobacter* and to a lesser extent *Pseudomonas aeruginosa* infections tended to replace *Proteus* infections. Another notable change after 1960 was a considerable increase in the frequency of isolation of organisms other than those listed in the table, notably members of the *Acinetobacter* and *Serratia* groups. Thus, a second significant increase in the total proportion of infections due to gram-negative aerobes dates from around 1960.

Recent surveys of bacteraemia in Britain are reported by Watt and Akubadejo (1967), Williams and

co-workers (1976) and Abeysundere and co-workers (1979). (For enterococcal septicaemia see Schlaes *et al.* 1981).

Effects of septicaemia due to gram-negative aerobes

Septicaemia due to gram-negative bacilli is often accompanied by profound circulatory collapse; after an initial period of fever, sometimes with rigors, there is a sudden or gradual onset of shock, with peripheral cyanosis and hypotension (Weisbren 1951, Bryant *et al.* 1971). These arise directly or indirectly from the action of bacterial endotoxin (Chapter 33), which damages tissue cells and at the same time sets in train a number of deleterious host responses. Injury to the vascular endothelium leads to loss of fluid from the blood and to the release of vasodilatory substances. The body responds to this by intense vasoconstriction and 'cold shock' ensues. These effects are compounded by activation of the alternative complement system and further release of vasoactive substances. Circulation through the small blood vessels is also disturbed by disseminated intravascular coagulation and platelet aggregation. Endotoxin may also cause focal damage to the kidneys, with acute renal failure, and to the lungs, causing alveolar collapse ('shock lung'). (See Corrin 1980, Wardle 1980.)

Endotoxic shock may develop in septicaemia caused by à variety of coliform bacteria, *Proteus* spp., *Ps. aeruginosa*, *Bacteroides* spp. and *Neisseria meningitidis*; the nature of the infecting strain appears to have little effect on the severity of the disease. About one-third of patients with bacteraemic infections are said to exhibit some signs of shock, and over 60 per cent of those with severe hypotensive disease die. The outcome is most unfavourable in elderly patients with severe underlying disease; neutropaenic patients are particularly vulnerable. The most common local pre-disposing conditions are urinary-tract infection, particularly when instrumentation is performed, infection of the peritoneal cavity, burns and infusion sites, and abortion; but in immunosuppressed patients there may be no detectable local infection. Occasionally endotoxic shock follows the infusion of contaminated fluid (Chapter 58).

Endotoxin can be detected in the blood of patients by mixing it with an extract of the amoebocytes of the horseshoe crab (*Limulus*), when coagulation occurs (Levin and Bang, 1964, Levin *et al.* 1970, Reinhold and Fine 1971). Unless endotoxic shock is recognized at an early stage and the bacteraemia is promptly eliminated by appropriate antibiotic treatment the prognosis is very grave. Physiological resuscitation may be supplemented by measures designed to limit the body's responses to endotoxin, such as the administration of corticosteroids or non-steroid prostaglandin inhibitors.

Antiserum prepared by injecting suspensions of boiled enterobacterial cells protects experimental animals against endotoxic shock (Tate *et al.* 1966). This effect is greatest against the homologous strain, but there is significant heterologous protection against the lethal effects of other gram-negative organisms (McCabe *et al.* 1973) presumably by antibody against the common core glycolipid. Vaccines of mutant strains lacking type-specific polysaccharide side-chains appear to give some wide-spectrum protection in both passive and active immunization experiments (Braude *et al.* 1981). Their action appears to be anti-endotoxic rather than antibacterial (Young *et al.* 1975). Antisera have been prepared in human volunteers by vaccination with boiled suspensions of a mutant strain of *Esch. coli* and are under investigation as a means of protecting susceptible patients from endotoxic shock (Ziegler *et al.* 1982). (The use of active immunization to protect burned patients from *Ps. aeruginosa* septicaemia is described in Chapter 61.)

Brief mention must be made of an uncommon but highly fatal septicaemic disease caused by a halophilic vibrio (see Chapter 27) that affects persons with pre-existing hepatic disease, often within 24 hr of eating raw oysters (Blake *et al.* 1979).

Anaerobic septicaemia

Comparable information is not available about anaerobes as causes of septicaemia, but several of them are of importance, particularly members of the Bacteroidaceae (Chapter 62) and the anaerobic cocci (Chapter 60). As well as causing gas gangrene, clostridia may give rise to a purely septicaemic disease; in addition to *Cl. perfringens*, these include *Cl. septicum* and some non-histotoxic clostridia (Alpern and Dowell 1969, 1971). It should be remembered that, in addition to strict anaerobes, the blood may yield organisms that need 5–10 per cent of CO_2 for growth.

Blood culture

We discussed the technique of collecting blood for culture in Chapter 20. Opinions differ about optimal media for the isolation of delicate organisms, including anaerobes, from the blood (Ellner and Stoessel 1966, Rosner 1970, 1972, 1974, 1975, Crawford *et al.* 1974, Forgan-Smith and Darrell 1974, Hall *et al.* 1974*a, b*, Shanson and Barnicoat 1975, Collee *et al.* 1977). When an antimicrobial agent may be present, it should be neutralized or its effects should be diluted out (see Stokes 1974).

The interpretation of results may present difficulties, particularly when the organism isolated is one that is often found on the skin or in the air. Many unfamiliar causes of bacteraemia have undoubtedly been classified as contaminants on first isolation. Clinical considerations may make it undesirable to withhold treatment until confirmation is obtained by repeating the isolation. The presence of two or more organisms in the blood in a single episode of infection is not uncommon (Hochstein *et al.* 1965) and presents formidable difficulties in interpretation. Isolation of the same organism from more than one bottle of medium seeded from the same sample of blood may be a helpful indication of the validity of the result. ('Pseudobacteraemia' due to the contamination of blood cultures from extraneous sources is discussed in Chapter 58.)

Bacterial endocarditis

In bacterial endocarditis, the causative organism multiplies in a vegetation that has formed on a heart valve or a valve prosthesis. Traditionally the disease has been divided into an *acute* form, which develops in the course of a clinically recognizable generalized infection and affects previously undamaged heart valves, and a *subacute* form, which appears without warning and often insidiously, and tends to affect persons who give evidence of previous valvular damage. The acute disease is an occasional complication of infection with a wide variety of organisms, but the subacute disease is caused mainly by non-haemolytic or 'viridans' streptococci and to a lesser extent by coagulase-negative staphylococci. This distinction is, however, far from clear cut; infection of damaged valves may sometimes run an acute course, and this is the rule in infections of artificial valve prostheses; and an increasing proportion of sufferers from the subacute disease do not give a history of previous heart damage.

Causative organisms

Streptococci are by far the most common cause of endocarditis in man. Almost any streptococcus may be responsible, but certain species predominate (Table 57.2). These are α-haemolytic or non-haemolytic streptococci that very seldom give rise to other suppurative infections and cause endocarditis of the subacute type: (1) the 'viridans' streptococci of the mouth,

Table 57.2 Percentage frequency of various streptococci* among 539 isolates from the blood in endocarditis (Britain, 1972–76, M. T. Parker, unpublished)

Streptococcus	Percentage	
pyogenes	0.7	
haemolytic, groups C and G	1.8	
group B	2.2	
pneumoniae	0.6	
milleri	6.7	
enterococci	5.3	
sanguis	17.4	
mitior	18.8	49.4
mitior-like, dextran +	5.8	
other 'viridans'	7.4	
mutans	13.0	
bovis, biotype I	12.8	15.8
bovis, biotype II	3.0	
unclassified	4.6	

*For classification and nomenclature, see Parker and Ball (1976).

Str. sanguis, Str. mitior and strains intermediate in character between these two species; (2) *Str. mutans*, another mouth streptococcus; and (3) *Str. bovis*, particularly the dextran-forming biotype I, a bowel organism.

Streptococci that give rise to suppurative lesions elsewhere in the body are in general less common causes of endocarditis. This may develop in the course of such an infection, and is then often of the acute form. *Str. pyogenes*, haemolytic streptococci of groups C and G, and pneumococci are rare causes of acute endocarditis. Group-B streptococci may be responsible for subacute endocarditis, or acute endocarditis associated with septic infection of the female genital tract or elsewhere. Enterococci (usually *Str. faecalis*) generally cause acute endocarditis in patients with a primary infection of the urinary or female genital tract (Beattie 1954, Toh and Ball 1960, Mandell *et al.* 1970). *Str. milleri* causes subacute endocarditis, but this rarely if ever develops in patients with the visceral abscesses that characterize infection with this streptococcus (Chapter 59).

Staphylococci are the second most common cause of endocarditis, and account for about one-quarter of the cases (see, for example, Moulsdale *et al.* 1980). Most of these are *Staph. aureus* infections associated with acute septicaemia. Coagulase-negative staphylococci are responsible for only about 6 per cent of cases of subacute endocarditis on the natural heart valve, but cause over 20 per cent of cases that occur after the insertion of an artificial heart valve (Karchmer and Swartz 1977). Micrococci (Marples and Richardson 1980) and aerococci (Colman 1967) are uncommon causes of endocarditis.

The list of other organisms that may sometimes cause endocarditis is long and constantly being added to: it includes gonococci, meningococci and other neisseriae; haemophilic bacilli (Horder 1908–09, Schottmüller 1910, Keith and Lyon 1963), including *Haemophilis aphrophilus* (Khairat 1940, Page and King 1966); almost any of the enterobacteria (Car-

ruthers 1977), including salmonellae, and *Ps. aeruginosa*; *Cardiobacterium hominis* (Slotnick and Dougherty 1964); actinomycetes and actinobacilli (Blevins and MacNeal 1946, Page and King 1966); brucellae; *Erysipelothrix* (Lawes *et al.* 1952, Schiffman and Black 1956); members of the *Bacillus* group; various unclassified aerobic coryneform bacilli (van Scoy *et al.* 1977); clostridia and many other anaerobic and microaerophilic bacteria including *Bacteroides* and *Fusobacterium, Campylobacter* (Kilo *et al.* 1965, Loeb *et al.* 1966), *Propionibacterium acnes* and anaerobic cocci (Felner and Dowell 1970). Non-bacterial agents include yeasts, moulds, and even basidiomycetes (Speller and McIver 1971), as well as rickettsiae (Marmion *et al.* 1961, Palmer and Young 1982) and chlamydiae (Levison *et al.* 1971, Ward and Ward 1974).

Bacteriological diagnosis

Organisms from the vegetation are discharged fairly regularly into the blood stream, though the numbers present may at times be small, particularly in the subacute disease. Repeated blood culture is seldom necessary (Werner *et al.* 1967), but as large a volume as is practicable should be cultured under optimal conditions, part of it anaerobically (see p. 173). Incubation should be continued for at least 2 weeks, though growth usually appears within 48 hr.

In a small percentage of cases, no causative organism can be isolated despite strenuous efforts. In addition to instances in which a rickettsia or a chlamydia is the causative agent, these may include infections with a nutritionally exacting organism (for example, see McCarthy and Bottone 1974) or a cell-wall-deficient variant (Mattman and Mattman 1965) that will not grow on the media employed. Others may be cases of non-bacterial thrombotic endocarditis (Hayward 1973).

Endocarditis in animals Various organisms cause endocarditis in domestic animals. In pigs, *Ery. rhusiopathiae* is the most common cause; occasional infections with streptococci of groups D, L and G have been described. Endocarditis in lambs is caused by *Str. faecalis* and in cattle by various group D streptococci. (See Chapter 59 for references to streptococcal endocarditis in animals and p. 175 for endocarditis in opossums.)

Pathogenesis

The formation of the vegetation is still not understood. It is believed to begin by the deposition of platelet thrombi and the subsequent accumulation of fibrin (Angrist and Oka 1963) on areas of the endocardium damaged by previous disease or abnormal pressure effects. These non-bacterial vegetations are subsequently invaded by bacteria from the blood stream, which persist and multiply in the depths of the vegetation where they are protected from the action of phagocytes.

In the rabbit, insertion of a catheter into the heart through the carotid artery results in damage to the aortic valve with the formation of sterile vegetations composed of platelets and fibrin. The subsequent intravenous injection of a dose of

10^6-10^7 staphylococci or 'viridans' streptococci results in a typical bacterial endocarditis (Perlman and Freedman 1971, Durack and Beeson 1972, Durack *et al.* 1973; see also Sande 1976). If the deposition is prevented by the administration of an anticoagulant, the valves nevertheless become infected when 'viridans' streptococci are subsequently given (Hook and Sande 1974); but with *Staph. epidermidis*, anticoagulant treatment significantly inhibits the development of endocarditis (Thörig *et al.* 1977). Captive opossums (*Didelphis virginiana*) frequently suffer from spontaneous endocarditis; after a single intravenous injection of *ca* 10^7 'viridans' streptococci without previous artificial damage to a heart valve, some 50 per cent of opossums develop endocarditis (Rowlands *et al.* 1970, Vakilzadeh *et al.* 1970).

In many cases of endocarditis, disseminated lesions occur elsewhere in the body; these include haemorrhagic spots in the skin, particularly under the nails, and minute focal lesions in the kidneys and brain. In many cases there is also a diffuse glomerulonephritis which may lead to renal failure. These lesions are sterile, and appear to result from the deposition of immune complexes in small blood vessels (Keslin *et al.* 1973, Boulton-Jones *et al.* 1974). High levels of antibody to the infecting bacterium can usually be detected in the blood in bacterial endocarditis, and immune complexes have also been demonstrated (Bayer *et al.* 1976, Mohammed *et al.* 1977).

Predisposing factors

Until about 20 years ago, the subacute form of endocarditis predominated, and the disease affected mainly adolescents and young adults, with a slight preponderance of females; nearly all cases were caused by 'viridans' streptococci, and the chief predisposing factors were rheumatic and congenital heart disease (Cates and Christie 1951). Since 1960, acute cases appear to have become slightly more common, endocarditis is now predominantly a disease of middle-aged and elderly males, and organisms other than 'viridans' streptococci have been isolated with increasing frequency (Kaye *et al.* 1962, Uwaydah and Weinberg 1965, Hughes and Gauld 1966, Finland and Barnes 1970, Cherubin and Neu 1971).

In a later study of streptococcal endocarditis, Parker and Ball (1976) showed the following: some 80 per cent of patients were aged over 35 years and 50 per cent over 55 years; males exceeded females in a ratio of 2:1; in the small number of patients aged less than 35 years, 'viridans' streptococci were responsible for over 70 per cent of infections; in the age-group 35–55 years they caused a rather smaller percentage of infections, the deficiency being made up by an increased percentage of *Str. mutans* infections; in patients aged over 55 years, 'viridans' streptococci caused less than half the infections, and *Str. bovis* nearly one quarter. *Str. milleri* endocarditis, too, was more frequent in the older age-groups. A considerable minority of endocarditis cases in the elderly are caused by faecal streptococci, staphylococci, enterobacteria and *Candida*.

The factors responsible for these changes appear to have been (1) a reduction in the number of infections in young patients with rheumatic heart disease, and an increase in the number of cases both of (2) subacute disease in elderly patients without a history of heart disease in early life, and of (3) acute disease in patients who had developed acute septicaemia in association with serious underlying disease or its treatment. It has been suggested (Hughes and Gauld 1966) that the main predisposing factor in the subacute disease in elderly persons is atherosclerotic damage to heart valves.

Two other predisposing factors of increasing importance deserve mention. Surgical operations on the heart or great vessels in which a prosthesis is inserted are often followed by endocarditis, which may be caused by a variety of bacteria or fungi (Karchmer and Swartz 1977). In many parts of the world, the intravenous self-administration of narcotics has become an important cause of endocarditis in young persons (Dowling *et al.* 1952, Simberkoff 1977). Unlike other forms of endocarditis, this mainly affects the tricuspid valves. The causative organisms show striking geographical variations; the predominant strains have been variously reported as *Staph. aureus*. *Str. faecalis*, α-haemolytic streptococci, *Pseudomonas*, *Serratia* and *Candida*.

Route of infection

There is little doubt that the infecting organisms reach the heart valve via the blood stream. Okell and Elliott (1935) found streptococci in the blood of 12 of 110 patients suffering from moderate or extensive gum disease, but in none of 68 persons with healthy gums. After tooth extraction, the incidence of positive blood cultures rose, but the bacteraemia observed was transitory, the blood becoming sterile, often within ten minutes of being infected. Transient bacteraemia of this kind occurred in 72 of 100 patients with gum disease, and in 12 of 38 patients without obvious gum disease. These findings have been amply confirmed. Transient bacteraemia follows even light manipulation of the teeth. It also occurs after operative manipulation at other sites—the cervix uteri (Schottmüller 1911), the urethra (Barrington and Wright 1930), joints, tonsils, the prostate (Richards 1932), and the lower bowel (Le Frock *et al.* 1973). However, endocarditis can develop in the absence of any such interference. We have drawn attention to the importance of *Str. bovis*—an organism found only in the bowel—as a cause of endocarditis in the elderly. Many patients with *Str. bovis* endocarditis have subsequently been shown to have an undiagnosed colonic lesion (Klein *et al.* 1977, Murray and Roberts 1978).

Prophylaxis and treatment

Every effort should be made to reduce the frequency of transitory bacteraemia in patients who are predisposed to endocarditis by rheumatic heart disease

(Doyle *et al.* 1967), congenital heart disease, previous endocarditis or a valve prosthesis. Therefore, antibiotic prophylaxis should be given to these patients whenever they undergo dental manipulation, tonsillectomy, or operations on the urogenital tract or the bowel. This should be given initially by the parenteral route, should not be begun more than one hour before the operation, and should be of short duration. Treatment with penicillin (Garrod and Waterworth 1962; see also Sprunt 1977) leads within 2 days to a predominance of penicillin-resistant α-haemolytic streptococci in the throat and a consequent risk that endocarditis might develop despite prophylaxis. It should be remembered that patients already on continuous prophylaxis with penicillin for the prevention of rheumatic fever (Chapter 59)—especially when this is given orally—may have a penicillin-resistant throat flora. For further information about the prophylaxis of endocarditis, and for choice of antibiotics for this purpose, see Kaplan (1977), Lowy and Steigbigel (1978) and Report (1982).

Bacterial endocarditis is almost invariably fatal unless a prolonged course of treatment with an appropriate antimicrobial agent or combination of agents is given (see Gray 1975, Wilson *et al.* 1980, Garrod *et al.* 1981). The choice of drug, and its dosage, must be determined by testing its bactericidal action on the infecting organism. It is essential to maintain the concentration of the agent in the blood at a bactericidal level during treatment.

The causative streptococci, except the enterococci, are usually highly sensitive to benzylpenicillin. A 4-week course of this drug in high dosage (> 10⁷ units per day systemically) can usually be relied upon to eliminate infection with sensitive strains (Karchmer *et al.* 1979), but many therapists give an aminoglycoside as well. Most streptococci are moderately resistant to aminoglycosides *in vitro*, but penicillin-aminoglycoside combinations are highly bactericidal (Chapter 29) and are effective in the treatment of endocarditis induced experimentally with penicillin-sensitive strains (Durack *et al.* 1974). Two weeks' treatment with procaine penicillin G and streptomycin intramuscularly eliminated infection in all of 91 patients (Wilson *et al.* 1981).

Enterococcal endocarditis is more difficult to treat. Benzylpenicillin has a poor bactericidal action on enterococci, ampicillin is somewhat better, but amoxycillin is the most

effective of the penicillins (Russell and Sutherland 1975). However, it is necessary to give an aminoglycoside as well.

Penicillins and aminoglycosides act synergically against many enterococci. However, an increasing proportion of strains are highly resistant to aminoglycosides, and some of these are not killed by a combination of penicillin and gentamicin (Moellering *et al.* 1971, 1980). Thus the causative enterococcus may have to be tested for killing by various penicillin-aminoglycoside mixtures.

After adequate treatment, the patient is often cured in the sense that all the organisms are destroyed, but second and subsequent attacks due to a different strain occur in a proportion of cases (Levinson *et al.* 1970). Whether the removal of the teeth reduces the chance of recurrence is uncertain (Hobson and Juel-Jensen 1956, Croxson *et al.* 1971). After bacteriological 'cure', residual mechanical damage to the valves may lead to progressive and ultimately fatal heart failure unless surgical repair or valve replacement is possible. The latter is in some instances performed during the active stage of the disease, if unmanageable heart failure is present or the infection cannot be eradicated by antimicrobial treatment.

More information about endocarditis will be found in various papers in books edited by Kaye (1976) and Kaplan and Taranta (1977).

Infected prostheses in chronic hydrocephalus   Valves used to drain the cerebrospinal fluid into the cardiac atrium or elsewhere often become infected. Chronic fever and splenomegaly develop, with the appearance of a fibrinous vegetation on the distal end of the prosthesis, in some 10 per cent of the patients, usually within 2 months of the operation but occasionally very much later. Even when the end of the appliance is in contact with a heart valve, endocarditis seldom occurs. If the infection is allowed to continue the patient usually dies, often with a chronic glomerulonephritis like that seen in endocarditis. Long-continued treatment with appropriate antibiotics usually fails, but if the prosthesis is removed the infection is easily eradicated.

The causative organism is *Staph. epidermidis* in about 60 per cent and *Staph. aureus* in about 25 per cent of cases (Chapter 60), but a variety of other organisms are occasionally responsible, including *Propionibacterium acnes* (Beeler *et al.* 1976). Gram-negative bacilli are rare causes of this infection unless drainage is into a ureter. (For further information, see Schoenbaum *et al.* 1975.)

Systemic infections of newborn infants

Neonatal septicaemia, which is often associated with meningitis, may be caused by a variety of bacteria, among which *Esch. coli* (Chapter 61) and group-B streptococci (Chapter 59) predominate. Robbins and his colleagues (1974) studied 133 cases of neonatal meningitis, of which 38 per cent were due to *Esch. coli*,

31 per cent to group-B streptococci, and 5 per cent to *Listeria*. Neonatal infections that develop in the first few days of life are usually acquired from the mother. 'Early' neonatal infections with group-B streptococci and *Listeria* are predominantly septicaemic. Those with group-B streptococci are usually acquired from

the birth canal of a healthy mother at the time of delivery, but those with *Listeria* frequently result from intra-uterine infection from a diseased mother. Later neonatal infections by both of these organisms are predominantly meningitic. Salmonellae, group-A streptococci and pneumococci are uncommon causes of maternally acquired systemic disease.

Infections that develop after the first few days of life are usually acquired from sources other than the mother (see Chapter 58 for hospital-acquired neonatal infections). Systemic neonatal infections often occur in premature babies, in whom the normal mechanisms of host defence are not yet fully developed. Recent success in keeping very small babies alive has greatly increased the frequency with which systemic infections are seen; unfortunately some of the infections are attributable to the measures used to ensure the survival of the baby (Chapter 58).

For accounts of systemic infections of the newborn, see Groover and co-workers (1961), Johnston and Sell (1964), Berman and Banker (1966), McCracken and Shinefield (1966) and Overall (1970).

Septic infection of wounds

All wounds are subject to infection, though in general the larger the wound and the more damaged the wounded tissue the greater the risk. Here we shall discuss only the infection of accidental wounds; the consequences of intentional wounding by the surgeon are considered in Chapter 58.

The bacteriology of accidental wounds

The accidental wound, whether inflicted in civil life or in war, is usually contaminated by bacteria present near the wound at the time of infliction, including those from the body surface and clothes of the injured person, and those from the soil, dust and other debris that fall on or are driven into the wound. It will be convenient to make an arbitrary distinction between the bacteria we find in a wound within a few hours of injury which, with a few exceptions, we can regard as *contaminants*, and those found later. The latter, after multiplication has taken place, are said to have *colonized* the wound; the term *infected* is best reserved for colonized wounds in which there are signs of local inflammation.

Wound contamination

If we except wounds grossly soiled with materials which, like road-dust, earth, saliva or faeces, contain large numbers of bacteria, the freshly inflicted wound contains few viable organisms. But, though the numbers are small, a variety of different species is usually present; and few wounds are completely sterile. In general, micrococci and staphylococci other than *Staph. aureus* are most commonly found. Other frequently isolated species are spore-bearing bacilli (both aerobic and anaerobic), enterococci, α-haemolytic and non-haemolytic streptococci, diphtheroids and gram-negative intestinal bacilli (see Pulaski *et al.* 1941, Spooner 1941, Hare and Willits 1942, de Waal 1943, Altemeier and Gibbs 1944). Of the more pathogenic bacteria, *Staph. aureus* is present in up to 20 per cent of wounds, and *Str. pyogenes* in only a small proportion (see Williams and Miles 1945). *Cl. perfringens*, though not uncommon, behaves in most wounds as a saprophyte, and dies out fairly rapidly (see Chapter 63). So, in fact, do most of the contaminating species mentioned above, including *Staph. aureus* and *Str. pyogenes*, especially if dead tissue and foreign matter are removed from the wound.

Wounds in which there is damage to the lower bowel, the oropharyngeal region or the vagina are subject to heavier initial contamination, predominantly with gram-negative non-sporing anaerobes, various streptococci, anaerobic cocci, and, in the case of bowel wounds, enterobacteria and other aerobic gram-negative rods. Animal and human bites are usually heavily contaminated with the flora of the 'donor's' mouth.

Wound colonization and infection

It must be emphasized that potentially pathogenic bacteria may multiply in wounds without any obvious clinical manifestation of disease except perhaps a prolongation of the healing time. Williams and Miles (1949), for example, record that 35.3 per cent of 745 small industrial wounds were latently infected in this way, mainly with *Staph. aureus*. Wounds that show clinical evidence of inflammation may yield more than one potential pathogen, and it may sometimes be difficult to decide which of the organisms present are contributing to this.

When a wound involving only the skin and underlying soft tissues and bone becomes septic, the organism most commonly found is *Staph. aureus* (see, for example, Levaditi *et al.* 1939, Miles *et al.* 1940, Meleney 1943, Pulvertaft 1943). Next in frequency come *Str. pyogenes*, enterobacteria, *Ps. aeruginosa*, and enterococci; the precise frequency varies with the type and age of the wounds studied (see Altemeier 1942, Rustigian and Cipriani 1947). Coagulase-negative staphylococci are often present, and in a small number of wounds are in pure culture (Wilson and Stuart 1965).

Gas gangrene (Chapter 63) is an important complication of large lacerated wounds, particularly when these have been heavily contaminated with soil or faeces. Other clostridia—tetanus (Chapter 64) and botulism organisms (Merson and Dowell 1973)—may multiply in wounds and form their respective toxins; this may occur in quite small wounds, often with little or no signs of local inflammation. Diphtheria bacilli also form toxin in minor skin lesions; in non-immune subjects diphtheritic paralysis may follow.

When a carrier site other than the skin has been penetrated or damaged, the infected wound often has a more complex flora in which non-sporing anaerobic gram-negative bacilli and anaerobic cocci are prominent, along with a variety of streptococci. In addition, most bowel wounds are colonized with enterobacteria. It is thus difficult to determine which components are of significance, but there is now reason to believe that the gram-negative anaerobes play a major part in inflammation in wounds in these regions (see Chapter 62), either acting alone or in synergy with various aerobes. In bite wounds, the anaerobic flora is probably an important cause of the initial acute inflammation; in addition, dog and cat bites are quite often infected with *Pasteurella multocida* (Allott *et al.* 1944, Mair 1966), and this may result in delayed healing and necrosis in deeper tissues.

We have mentioned (Chapter 27) that certain halophilic vibrios cause septic infections. A number of these are wound infections in previously healthy persons in whom a minor injury becomes contaminated with sea-water (Blake *et al.* 1979). These infections are often of some severity, though considerably less than that of the septicaemic disease that may affect predisposed persons when the same organisms are ingested (p. 173).

The frequency of infection in older wounds is far higher than we should expect from the initial contamination rates. Many of the added infections are acquired in hospital. They will be discussed in Chapter 58.

Prevention of wound infection The most important measures for the prevention of infection in wounds sustained outside hospital are prompt cleansing, the surgical removal of dead tissue and foreign matter, and the judicious use of prophylactic chemotherapy if admission to hospital is expected to be delayed.

Other local septic infections

Suppurative lesions of the skin and subcutaneous tissues

In man, these are caused mainly by *Str. pyogenes* (Chapter 59) and *Staph. aureus* (Chapter 60). (See also Chapter 61 for the effects of pseudomonads on the moist skin, and Chapter 62 for the role of bacteroidaceae in chronic ulcers and in infections of glands associated with the skin.)

Acne vulgaris is a chronic disease of the pilosebaceous follicles in adolescent human subjects; no similar disease in other animal species has yet been reported. The primary lesion is non-bacterial; for reasons that are not clear, the epithelial cells lining the follicles become hyperkeratotic and fail to separate normally. The mouth of the follicle is thus blocked, so that sebum and hairs cannot escape. The papules, pustules and nodules of the disease that develops appear to be an inflammatory response to irritation by fatty acids formed from the sebum by bacterial action. The inflamed lesions contain large numbers of propionibacteria, among which *Prop. acnes* predominates, and of coagulase-negative staphylococci. The most widely held view is that *Prop. acnes* is the most important lipolytic agent in acne. (For a general account of the role of bacteria in acne, see Holland *et al.* 1981.)

Mastitis

In man

Puerperal mastitis, and mastitis of newborn babies of either sex, affect the gland proper and are almost invariably caused by *Staph. aureus* (Chapter 60). Non-puerperal mastitis in adults affects the subareolar mammary ducts and usually yields non-sporing anaerobes among which members of the Bacteroidaceae predominate (Chapter 62).

In dairy animals

In Britain, mastitis in cattle is the main concern; in countries in which goats or sheep are a major source of milk similar problems arise.

Causative organisms Before 1950, about two-thirds of cases of bovine mastitis in Britain were caused by streptococci and about one-half by *Str. agalactiae*; *Staph. aureus* and *Corynebacterium pyogenes* were responsible for most of the remainder. After the practice of treating mastitis with antibiotics became general, there was a reduction in the number of cases in which *Str. agalactiae* could be found, and an increase of those yielding no pathogen; the number associated with other streptococci and *C. pyogenes* showed little change, and there was a considerable increase in the frequency of staphylococcal mastitis, which by 1959 accounted for a quarter of the total (Stableforth 1959). Coliform bacilli, rarely seen in mastitis before 1950, later became a common cause of the disease, and infections with *Ps. aeruginosa* and *Klebsiella aerogenes* subsequently appeared in increasing numbers. The two commonest organisms are now *Esch. coli* and *Str. uberis*; together they cause 60–70 per cent of cases. Other organisms occasionally responsible for mastitis include *Mycoplasma bovis*, *Fusobacterium necrophorum*, mycobacteria other than *M. tuberculosis*, pneumococci, and *Listeria monocytogenes* (Gitter *et al.* 1980).

Mastitis due to *Str. agalactiae* is highly transmissible in herds of lactating cattle, but is a greater cause of loss of milk production than of serious disease. It often affects more than one quarter of the udder and tends to become chronic. Mass treatment of herds by the intramammary instillation of penicillin, followed by decontamination of cow-houses, may eliminate infections, but attempts to eradicate the disease from countries or even districts by this means have not been successful. *Str. uberis*, which is now one of the most important aetiological agents of mastitis in Britain, causes a mild disease, but this may become chronic. It tends to affect cattle confined in buildings and yards and is difficult to eradicate because the streptococci live in the bovine gut and can multiply in bedding. *Str. dysgalactiae* may cause a more acute mastitis and also sores on the teats; it is sometimes present, along with *C. pyogenes* in 'summer' mastitis (see below, and Chapter 53). It is, however, one of the less common causes of streptococcal mastitis. The organism has been isolated from the vagina and tonsils of cows in the absence of mastitis (Francis 1941, Daleel and Frost 1967). According to Higgs and his colleagues (1980), strains of *Str. dysgalactiae* differ considerably in their ability to cause mastitis when infused into the cow's udder. *Str. zooepidemicus* is an occasional cause of severe mastitis, sometimes with arthritis. Pneumococcal mastitis has become rare in recent years, but may be clinically severe and is highly transmissible; it is said to be associated particularly with hand milking, which suggests a human source of infection. A few infections with streptococci of groups D, G and L have been described. (See Chapter 59 for rare instances of group-A streptococcal infection of the udder, and Wilson and Salt (1978) for a succinct account of streptococcal mastitis.)

With a few exceptions, streptococci cause mastitis of the lactating udder, but *C. pyogenes* affects the 'dry' udder, usually in the summer or autumn just before calving (see Chapter 53). The disease often becomes chronic, with abscess formation and the development of sinuses. Acute staphylococcal mastitis usually occurs a few days after calving. It is sometimes accompanied by gangrene, and may prove fatal. More often the disease is chronic, and is associated with small areas of induration in one or more quarters. In the milder forms there are only slight changes in the udder, but the milk usually contains an excess of leucocytes. Occasionally a chronic granulomatous lesion—often erroneously described as actinomycosis but correctly called botriomycosis—is seen. In cattle, *Staph. aureus* is a normal inhabitant of the skin, and is present in greatest numbers on the surface of the udder (Davidson 1961). (For further information about staphylococcal mastitis see H. W. Smith 1959.)

Bacteriological diagnosis In its more severe forms mastitis can be detected by clinical means alone, but the chronic types of infection usually call for laboratory assistance. Individual quarter samples of milk are preferred for examination, the fore-milk being the most suitable. The laboratory test chosen may be one designed to detect a change in the milk indicative of clinical infection, or one in which the causative organism is identified. In the first group we may include the microscopic cell count, estimation of the amount of deposit after centrifugation, measurement of the chloride content or pH, the Whiteside and the California mastitis tests (see Stableforth 1959) and the Hotis test (see Throop *et al.* 1957). The cell count may be misleading. *Esch. coli*, for example, may cause an intense neutrophil response in the udder, with the excretion of few pus cells till the acute stage of the disease is over.

For the cultural examination of milk from individual cows, direct inoculation on to a solid medium is recommended. Enrichment methods give no indication of the relative numbers of different organisms present, and are liable to give false positive results from cross-contamination. Plain blood agar may be used when it is desired to detect all mastitis organisms. Blood agar containing crystal violet and aesculin (Edwards 1933), or a somewhat similar medium containing a thallium salt (Wilson and Slavin 1950), are good selective media for streptococci. The CAMP test (see Chapter 29) on aesculin blood agar may be used for the identification of *Str. agalactiae*. For herd tests, composite milk samples may be incubated in the presence of sodium azide and bromcresol purple as a preliminary means of detecting milks containing *Str. agalactiae* (Edwards 1938). Antibiotic-containing selective media such as those used for the isolation of streptococci from human material (Chapter 29) would also be suitable.

Epidemiology and prevention Bovine mastitis is of great economic importance but contributes less than might be expected to human disease. The staphylococci and streptococci of the bovine udder rarely cause septic infection in man. There is little evidence that *Str. agalactiae* is transmitted to man in milk (but for rare examples of milk-borne tonsillitis and scarlet fever due to *Str. pyogenes* see Chapter 59). However, bovine *Staph. aureus* strains may form enterotoxin and have given rise to milk-borne staphylococcal food poisoning in man.

The spread of mastitis in dairy herds is mainly attributable to cross-infection during milking. Failure to prevent this by hygienic practice in the dairy encouraged the lavish use of antibiotics for treatment and mass prophylaxis. As in the human population in hospitals, this has led to a situation in which antibiotic-resistant *Staph. aureus* and various gram-negative aerobes are now the predominant causes of endemic infection in herds. It has also increased the risk of infection with 'free-living' gram-negative aerobes, such as *Ps. aeruginosa*, that can multiply in moist situations in and around the dairy.

Many attempts have been made to prevent staphylococcal mastitis by specific immunological methods. There is general agreement that toxoid alone is ineffective, but French workers claim success with a toxoid-vaccine mixture containing aluminium hydroxide (see Richou and Thieulin 1955). Several British and American workers have used similar methods; some, but not all of them, found that vaccinated animals suffered from acute infections less often than did controls (Spencer *et al.* 1956, Pearson 1959, Slanetz *et al.* 1959, Derbyshire and Edwards 1963, Derbyshire and Smith 1969). Vaccination does not prevent the development of chronic mastitis or result in the elimination of established infections.

For further information on mastitis in domestic animals, see the review by Nai (1957) and the relevant chapters in Stableforth and Galloway's (1959) text-book; the recent literature of the subject is summarized, from 1972 onwards, in the annual publications of the Commonwealth Bureau of Dairy Science and Technology (see Report 1972*b*).

Conjunctivitis

One of the commonest causes of conjunctivitis in man is the Koch-Weeks bacillus or the closely related *Haemophilis influenzae* (Chapter 29). Infections are widespread and may occur in epidemics. Pneumococcal conjunctivitis is less common, often more severe, and mainly confined to children; infections are generally sporadic, but institutional outbreaks have been described. Subacute, or angular, conjunctivitis is due to the Morax-Axenfeld bacillus (Chapter 28). Gonococcal conjunctivitis—ophthalmia neonatorum—is described in Chapter 66; nowadays it is considerably less common in babies than chlamydial infection (see

below). *Staph. aureus* is frequently isolated from 'sticky eye' in the newborn, but how often it is the causative organism is uncertain. *Ps. aeruginosa* occasionally causes severe iridocyclitis or panophthalmitis after minor trauma to the conjunctiva; the organism may be introduced into the eye in contaminated eye drops or fluid used to clean contact lenses.

In addition to trachoma, *Chlamydia trachomatis* causes 'inclusion conjunctivitis'. This is a common infection of the newborn, usually acquired from the mother at birth; in adults it occurs in association with genital infection but may be contracted in swimming-baths (see Chapter 76). Adenovirus infection is a frequent cause of conjunctivitis in children and adults, often occurring in epidemics in association with pharyngitis; epidemics related to attendance at ophthalmic out-patient departments or factory surgeries may result from failure to disinfect instruments. (For epidemic keratoconjunctivitis ('shipyard-eye') see Chapter 97.)

Gigliotti and co-workers (1981) in the USA studied the aetiology of conjunctivitis in patients aged from 1 month to 18 years. A causative organism was identified in 72 per cent of the patients. The commonest agents were haemophili (42 per cent), adenoviruses (20 per cent) and pneumococci (12 per cent). Haemophili predominated in the winter and adenoviruses in the summer.

Suppurative lesions of bones and joints

Acute osteomyelitis occurring independently of a compound fracture is usually caused by *Staph. aureus*. It is a disease of children and young adults, but a few cases in very young children are caused by *Haem. influenzae*. Staphylococcal osteomyelitis is nowadays seldom preceded by a local lesion elsewhere but often occurs at the site of a recent blow (see Glover *et al.* 1982). It is generally believed that the causative organism establishes itself in a focus of local damage in the bone after having reached it via the blood stream. Osteomyelitis is said to have decreased considerably in frequency in the last 40 years or so; in some tropical areas, however, it continues to be very common (see, for example, Jellis 1981). (See also Waldvogel *et al.* 1970, Waldvogel and Vasey 1980.)

Acute suppurative periostitis is rare. When it occurs, it is usually a streptococcal infection. A subacute, or chronic, periostitis occasionally follows typhoid or paratyphoid fever, and is caused by the bacillus responsible for the primary infection. Such abscesses

occur most commonly over the ribs, or the cranium. Osteomyelitis or periostitis due to other salmonellae, particularly *S. dublin*, is not uncommon. The association of osteomyelitis with the sickle-cell trait is mentioned in Chapter 68.

Metastatic purulent arthritis may occur as a sequel to infection with any pyogenic bacterium. *Staph. aureus* is the organism most frequently isolated in older children, but infections with *Haem. influenzae* and streptococci occur between the ages of 6 months and 4 years (Almquist 1970). Occasional infections with pneumococci and meningococci are seen. Kellgren and others (1958) drew attention to the association of suppurative arthritis with rheumatoid arthritis. By no means all of these infections occur in patients who are receiving steroids, or who have had intra-articular injections. In animals other bacteria are often responsible for septic arthritis, as, for instance, *C. pyogenes* and *Ery. rhusiopathiae*.

Suppurative lesions of the central nervous system

Meningitis is considered in Chapter 65. Here we will discuss briefly the aetiology of abscesses in the brain or associated with the meninges. Their bacterial cause varies according to the situation of the lesion and its underlying pathology.

Str. milleri (Chapter 59) is specifically associated with abscesses of the frontal lobe and the overlying meninges, and can be isolated from some 80 per cent of these; no other organism is present with a similar frequency. The development of these abscesses is often preceded by a frontal sinusitis or other infection of the upper respiratory tract (de Louvois *et al.* 1977), but occasionally the infection can be traced to a dental root-abscess (Ingham *et al.* 1978). Abscesses of the temporal lobe, the cerebellum and the associated meninges are usually consequences of chronic otitis media. The flora is often mixed, but members of the Bacteroidaceae are almost constantly present (Ingham *et al.* 1977*a*) and are looked upon as the usual causative organisms (Chapter 62). Abscesses after 'open' fracture of the skull or resulting from the bloodborne spread of organisms from the lungs or elsewhere have a variable flora that may include staphylococci, various streptococci, enterobacteria or *Ps. aeruginosa*, but rather rarely anaerobes. Spinal extradural abscesses are usually caused by *Staph. aureus*. (For methods of examining pus from brain abscesses, see de Louvois 1980.)

Dental caries

The aetiology of this disease is not yet fully understood. The main factors responsible appear to be the nature of the dietary and the action of micro-organisms. As long ago as 1835 Robertson suggested that acids formed from retained food were the primary cause of decalcification and solution of the enamel, and in 1889 Miller attributed their formation to bacterial fermentation of starch and sugar.

The cariogenic potential of dietary carbohydrate, particularly of refined sugar, especially when this is taken between meals, is now generally accepted. The view that it acts only by leading to the formation of lactic acid, though widely held, is not unanimous. The effect of acid is limited by the alkalinity of the saliva, as well as by the structural integrity and proper calcification of the enamel, which in turn depend on dietary factors (Report 1936). It has been suggested that caries may begin by lysis of the phosphoprotein matrix of the enamel, possibly by bacterial phosphatase (King 1944, Little 1962, Kreitzman *et al.* 1969).

Bacterial aetiology

The lactobacilli have some claim to be considered as causative agents in caries by virtue of their ability to produce acid from carbohydrate. They are more abundant in the mouth of persons suffering from dental caries than of normal persons (Kligler 1915, Bowen 1967, Shovlin and Gillis 1969). However, recent attention has been directed to bacteria, most of them streptococci, that are able to form a tough and adhesive type of dextran from the glucose moiety of sucrose while producing acid from the fructose moiety (Chapter 29). The dextran tends to bind together the food debris, epithelial cells, mucus and bacteria to form dental plaque, while acid formed close to the tooth surface leads to destruction of the dentine. Organisms with this ability include *Str. mutans*, *Str. sanguis* and *Actinomyces viscosus*.

The suggestion that *Str. mutans* is the main cause of dental caries was first made by Killian Clarke in 1924. Strong support for this has since come from work on experimental caries in laboratory animals. Organisms found to cause caries in rats or hamsters kept on a diet high in sucrose (Orland *et al.* 1955, Fitzgerald and Keyes 1960, Krasse 1966, Gibbons and Banghart 1968) were nearly all subsequently identified as *Str. mutans*. *Str. sanguis* and most other streptococci have considerably less ability to cause caries in germ-free rats (Drucker and Green 1979). *Str. mutans*, when grown in the presence of dextran or sucrose, undergoes agglutination (Gibbons and Fitzgerald 1969), and this has been thought to aid adherence of the organism to the teeth. However, cariogenicity appears not to be related to this character (Tanzer *et al.* 1979) but is closely associated with the ability to produce the enzyme responsible for the synthesis of a particular dextran, a water-insoluble polymer rich in $1 \rightarrow 3$ bonds (Freedman *et al.* 1979; see also Chapter 29).

Although the formation of the appropriate dextran is an essential prerequisite, there is evidence that a cellular component of *Str. mutans* may play an important role in caries production. Caries develops in monkeys fed a high-sucrose diet, but this can be prevented by injections of a suitable *Str. mutans* vaccine. Vaccines of washed whole or disintegrated cells, but not extracellular preparations rich in dextran-forming enzymes, have a protective action (Bowen *et al.* 1975). The development of immunity is associated with the appearance of serum antibody of the IgG but not of the IgA or IgM class (Lehner *et al*. 1978, 1979; see also Caldwell and Lehner 1982). Treatment of cellular vaccines with proteolytic enzymes removes their immunogenicity. Colman and Cohen (1979) showed that effective immunization was associated with the production of antibody against a particular cell-wall protein. This has since been purified and found to prevent caries in monkeys (Lehner *et al*. 1980).

The relevance of these findings to human dental caries is not yet certain. In man, *Str. mutans* is found in plaque more often in the presence of established caries than in caries-free subjects (Loesche *et al*. 1975), but serial studies that show a relation between the presence of the organism and the initiation of the process have yet to be reported. As Bowden and his colleagues (1979) point out, caries can be conveniently divided into two processes: (a) initiation of the lesion and (b) extension of the lesion. It may well be that different organisms are responsible for each of these.

Prevention

Numerous measures are proposed for the prevention of caries (see Report 1972*a*). Among the most effective is ensuring the presence of fluorine, in a concentration of about 1 part per million, in the drinking water. It is

supposed to act partly by inhibiting glycolytic processes and thus diminishing acid formation on the plaque, and partly by increasing the resistance of the enamel to attack by acid (Goose 1965).

Regular tooth-brushing may fail to remove plaque from the more inaccessible parts. Attempts have therefore been made to inhibit plaque formation by the use of antibacterial agents; a chlorhexidine mouthwash has given promising results (Davies 1974). According to Slee and Tanzer (1979), chlorhexidine and certain other guanides, some cationic detergents and hydroxyquinoline derivatives, and inorganic iodine, kill *Str. mutans* more readily than *Str. sanguis* in artificial plaque. We cannot say whether it will in future become practicable to vaccinate against dental caries.

(For a general review of dental caries and the role of *Str. mutans*, see Hamada and Slade 1980.)

Periodontal disease It is now recognized that dental plaque in or near the gingival sulcus plays an important part in the initiation of gingivitis and periodontal disease. How it does this, and the part played by plaque bacteria, are not clear (see Lehner 1978). The role of gram-negative anaerobes in established periodontal disease is discussed in Chapter 62. For the flora of the mouth and teeth, see Chapter 8, and of the gingival sulcus, a paper by Sanyal and Russell (1978).

Suppurative lesions of the respiratory tract

Infections of the lungs and bronchi are discussed in Chapter 67. *Pleural empyema* usually arises from direct extension from a pulmonary lesion; common pathogens include pneumococci, *Staph. aureus*, gram-negative aerobic bacilli and a variety of non-sporing anaerobes (Bartlett *et al*. 1974, Finland and Barnes 1978). Purulent lesions of the *paranasal sinuses* are usually caused by organisms from the nasopharynx, especially pneumococci and *Haem. influenzae* (Lystad *et al*. 1964) and non-sporing anaerobes, the last often causing foetid sinusitis (Urdal and Berdal 1949, Frederick and Braude, 1974, Brook 1981). Rhinoscleroma and ozaena are discussed briefly in Chapter 61. Pneumococci and *Haem. influenzae* are also the main causes of *acute otitis media* (Halsted *et al*. 1968). In chronic otitis media with perforation of the ear drum, *Ps. aeruginosa* and *Staph. aureus* predominate in the aerobic flora, and peptococci, peptostreptococci

and non-sporing gram-negative bacilli—in that order of frequency—in the anaerobic flora (Brook 1980). Unsuccessfully treated acute otitis media is one of the main causes of acquired deafness in children (Lowe *et al*. 1963).

Invasive infections of the central nervous system often arise from sepsis in the paranasal sinuses or the middle ear: pneumococci usually cause a generalized meningitis (Chapter 65) and *Str. milleri* and non-sporing anaerobes a brain abscess (p. 181). Acute infection of the nasal sinuses with meningococci is believed by some observers to play a part in the genesis of meningococcal meningitis (Chapter 65).

Middle-ear disease is very common in rats, particularly adults; the main organisms implicated appear to be *Actinobacillus actinoides*, streptococci and diphtheroids (Nelson and Gowen 1930).

Intra-abdominal sepsis

Appendicitis

The question whether any of the bacteria found in the inflamed appendix play a part in the genesis of this

disease remains unanswered. Organisms regularly present in the flora of the lower part of the small intestine, including gram-negative non-sporing anaerobes, clostridia, enterobacteria and various streptococci can usually be isolated. Poole and Wilson (1977) found *Str. milleri* much more often in the inflamed than in the normal appendix. Severe salmonella and shigella infections occasionally precipitate an attack of acute appendicitis (Rubenstein and Johnson 1945, White *et al.* 1961).

The bacteriology of peritoneal infection associated with acute appendicitis is discussed below. (For mesenteric lymphadenitis due to *Yersinia enterocolitica* and *Y. pseudotuberculosis* see Chapter 55.)

Cholecystitis

The frankly suppurative form of cholecystitis is generally due to infection with *Esch. coli*, sometimes associated with streptococci. Acute cholecystitis may also follow infection with typhoid or paratyphoid bacilli. The relation between infection of the gall-bladder and the carrier state in typhoid fever is discussed in Chapter 68. Infections of the gall-bladder provide an interesting example of the importance of local conditions, of a semi-mechanical nature, in originating or perpetuating bacterial infection. Bacteria have frequently been isolated from the interior of gall-stones. It is highly probable that a mild bacterial infection so alters the local condition in the gall-bladder as to favour the formation of gall-stones, and that the cholesterol and bile salts of which these are composed are deposited on a central nidus of desquamated cells, or fibrin, containing bacteria. There are also good reasons for believing that the presence of gall-stones greatly increases the liability to bacterial infection; so that we have a vicious circle, each abnormality tending to promote, or perpetuate, the other (see Rains 1962).

We have elsewhere noted the view that the typhoid bacillus gains access to the gall-bladder from the blood vessels of its wall and not through the bile ducts. Many observers have suggested that this is true in the majority of cases of cholecystitis, and that the primary lesion is a focus of infection beneath the epithelial lining. Rosenow (1916) and Wilkie (1928) concluded that a streptococcus was the cause of this lesion, and that coliform bacilli were of importance only as secondary invaders. Rains and his colleagues (1960) isolated an anaerobic actinomycete from the interior of 30 of 57 gall-stones.

Liver abscess It should be noted that non-sporing gram-negative anaerobes, though not often isolated from acute cholecystitis, are a major cause of liver abscesses (Chapter 62). *Str. milleri* is by far the commonest streptococcus to be found in liver abscesses (Bateman *et al.* 1975, Parker and Ball 1976), from which it is often isolated in pure culture.

Peritonitis

Some cases of peritonitis are primary, for example, certain cases in children due to the pneumococcus. Haemolytic streptococcal peritonitis may form part of a septicaemic infection or, in newborn infants, may arise by direct extension from the umbilicus. Peritonitis due to *Staph. aureus* may follow a neonatal skin infection (Beaven 1958). Most cases, however, are secondary to some primary intra-abdominal lesion, such as appendicitis, perforated gastric or duodenal ulcer, or intestinal obstruction, and mixed cultures of non-sporing anaerobes, clostridia, coliform bacilli, faecal streptococci and *Str. milleri* are usually obtained. The results of prophylaxis and treatment with metronidazole leave little doubt that anaerobes play an important part in causing intra-abdominal sepsis associated with lesions of the gastro-intestinal tract (Chapter 62), but this does not exclude the possibility that other organisms are also concerned in the process. Indeed, there is evidence for synergy between aerobes and anaerobes in this and in some forms of wound infection. Eykyn and Phillips (1978) observed that both aerobes and anaerobes disappeared when sepsis resolved in metronidazole-treated patients. Kelly (1978) studied experimental wound infections with *Esch. coli* and *Bacteroides fragilis* in guinea-pigs, and showed that mixtures of subinfective doses of the two organisms gave rise to purulent lesions. Observations on the bacteria remaining in wounds at the end of abdominal operations in man supported this view (Kelly 1980); when both aerobes and anaerobes were present the sepsis rate was 71 per cent, but when anaerobes only, or aerobes only, were grown the percentages were respectively 22 and 13.

Several factors may be concerned in synergy between aerobes and anaerobes. The growth of aerobes in tissue may lower the Eh and so help to initiate the growth of strict anaerobes. According to Socransky and Gibbons (1965), some aerobes produce metabolites that stimulate the growth of *Bacteroides* strains. Anaerobes may produce antiphagocytic substances that protect aerobes from leucocytes (Ingham *et al.* 1977*b*). Onderdonk and his colleagues (1979) showed that metronidazole protected rats from death after the intraperitoneal injection of *Esch. coli* only when *Bacteroides fragilis* was also administered; their experiments suggested that the anaerobe converted metronidazole into a substance that inhibited the growth of *Esch. coli*.

Pelvic peritonitis associated with septic infection of the female genital tract usually yields a mixed flora rather like that of peritonitis of intestinal origin, except that anaerobic cocci are rather more prominent. Puerperal and post-abortive infections are often accompanied by septicaemia, which may dominate the picture when the causative organisms are streptococci of groups A or B, *Cl. perfringens* or non-sporing anaerobes.

Pelvic infections in women not associated with pregnancy are said to be becoming increasingly common (Robinson *et al.* 1981); they appear in most cases to be a consequence of ascending infection of the genital tract and generally to be associated with a sexually transmitted disease. In addition to gonococcal infection, which has long been recognized as a cause of chronic salpingitis, there is now evidence that infections with genital strains of *Chlam. trachomatis* are often responsible for recurrent episodes of pelvic infection, though their frequency appears to show considerable geographical variation (see Chapter 76, and Taylor-Robinson and Thomas 1980). The role of the other organisms commonly found in the lesions whether or not gonococci or chlamydiae are present—various coliform organisms, non-sporing anaerobes and *Ureaplasma*—has not been established.

Vaginitis

Vaginal discharge, when not attributable to uterine infection, may be caused by *Trichomonas vaginalis* or *Candida* spp., but in a considerable proportion of cases of so-called 'non-specific' vaginitis it is difficult to assign a pathogenic role to the organisms isolated. Vaginal carriage of coliform bacilli and *Staph. aureus* is usually symptomless. (See Chapter 60 for toxic shock syndrome associated with the multiplication of *Staph. aureus* in the vagina.)

Considerable attention has been paid in recent years to the possibility that *Gardnerella vaginalis* (Chapter 43) causes vaginal discharge (Gardner and Dukes 1955, McCormack *et al.* 1977, Balsdon *et al.* 1980). This organism is often found in large numbers in malodorous vaginal discharges, in microscopic preparations of which it can be seen adhering to epithelial cells. Although *Gard. vaginalis* is susceptible only to moderately high concentrations of metronidazole, this compound is quite effective in curing cases of non-specific vaginitis in which the organism is present. However, other abnormalities are usually found in the discharge in non-specific vaginitis. In normal women of reproductive age the vaginal pH is generally below 4.5 and lactobacilli predominate (Chapter 8); in candidal vaginitis, too, the pH is usually low and the bacterial flora consists mainly of lactobacilli and aerobic organisms. In non-candidal vaginitis, including cases in which *Trichomonas* is present, on the other hand, the pH is almost invariably above 4.5 and there is generally a heavy mixed flora of strict anaerobes and *Gard. vaginalis* (Taylor *et al.* 1982; see also Spiegel *et al.* 1980). Among the anaerobes, *Bacteroides* of the melaninogenicus-oralis group and gram-positive cocci predominate. The odour of the discharge in non-specific vaginitis has been attributed to the products of fermentation by the anaerobic flora. Several workers have noted the presence of anaerobic curved, vibrio-like, organisms in the vaginal discharge (Moore 1954, Hjelm *et al.* 1981, and others) but it seems unlikely that the authors were all describing the same organism.

Group A streptococci are quite a common cause of profuse vaginal discharge in children.

Infections of the urinary tract

Bacterial infection of the urinary tract may be a simple cystitis or a pyelonephritis. Infection of the renal pelvis may be responsible for or contribute to progressive loss of renal function and eventual renal failure, but it is probably not the most common cause of this. Acute urinary-tract infection causes frequency of micturition and dysuria, and the causative bacterium is present in large numbers in the urine. In acute pyelonephritis there is usually in addition pain and tenderness in the loin, and often pyrexia and bacteraemia. However, infection, whether of the bladder or the renal pelvis, is often symptomless and revealed only by culture of the urine. An acute symptomatic attack may be an isolated incident or may be recurrent; repeated attacks tend to occur in patients with symptomless bacteriuria in the intervening periods. (For terminology, see Report 1979.)

Pathogenesis

Predisposing factors of two sorts, acting alone or together, are important determinants for infection of the urinary tract: (1) local mechanical factors that interfere with the normal flow of urine, and (2) the artificial introduction of bacteria into the bladder or even the upper parts of the urinary tract in the course of catheterization or surgical operation (Chapter 58).

Obstruction to the flow of urine by renal or vesical calculi, enlarged prostate, urethral stricture or vesical neoplasm all predispose to infection. However, *reflux* of urine into the renal pelvis as a result of malfunction of the vesico-ureteric valve is an even more important determinant. The continual flow of urine, and its periodic voiding by micturition, tend to protect the urinary tract from infection. Both obstruction and reflux increase the volume of residual urine after micturition and so counteract this protective mechanism. Despite the importance of these predisposing factors, many infections develop spontaneously in persons who do not give evidence of obstruction or reflux and have not suffered surgical interference to the bladder. This suggests that bacteria may from time to time reach the bladder by physiological means and can then establish themselves there.

With rare exceptions, which include the renal lesion in the enteric carrier (Chapter 68), the human urinary tract is infected by the ascending route. In experimental animals, many organisms become localized in the kidney when given intravenously in high dosage (see p. 1997 of the 6th Edition of this book), but the lesions produced are rather different from those of pyelonephritis in man. It is of greater relevance that pyelonephritis is produced in female rats by infusing coliform or *Proteus* bacilli into the bladder in quite small doses and at a pressure insufficient to damage the kidney (Vivaldi *et al.* 1959, Andersen and Jackson 1961, Cotran *et al.* 1963, Larrsson *et al.* 1980). Infusion of sterile fluid is sometimes followed by the appearance of infection with organisms previously identified in urethral cultures.

Bacterial causes

Urinary-tract infections developing in uncatheterized persons outside hospital are almost invariably due to bacteria from the normal body flora. Coliform bacilli, nearly all of them *Esch. coli*, predominate strongly at most ages. A particular sort of coagulase-negative staphylococcus—a novobiocin-resistant member of biotype 3 of *Staph. saprophyticus* (see Chapter 30)—causes a considerable minority of urinary-tract infections, both cystitis and pyelonephritis, in previously healthy women aged 16–25 years (Pereira 1962, Mitchell 1968, Mabeck 1969, Maskell 1974); it seldom infects males, or females in other age-groups. In the general population, *Proteus* bacilli are a common cause of urinary-tract infection in male children (Saxena and Bassett 1975); in hospitals, however, they often cause serious chronic infections in patients with urinary obstruction or after instrumentation (Chapter 61).

Not all strains of *Esch. coli* appear to be of equal pathogenicity in the urinary tract. Most strains responsible for infection belong to a limited number of O serogroups and have certain acid-polysaccharide K antigens; those that cause symptomatic infections are said to adhere better to epithelial cells from the urinary tract than do faecal strains (see Chapter 61).

The ability of the novobiocin-resistant staphylococcus to cause urinary-tract infection in healthy women appears to be unique among staphylococci and micrococci. Possible mechanisms for its pathogenicity are mentioned in Chapter 60. Other staphylococci, notably *Staph. aureus* and *Staph. epidermidis*, may cause urinary-tract infection, but almost only when introduced artificially into the bladder.

Epidemiology

Since it was recognized that the quantitative culture of freshly voided urine—collected without the use of a catheter—is a reliable means of detecting urinary-tract infection with gram-negative bacilli, even in the female (Kass 1955), there have been many investigations of the frequency of inapparent infection in populations of supposedly healthy persons. It is commoner in females than in males, and its frequency increases with age; 1 per cent of female school children, 3 per cent of adolescent girls, 6 per cent of women in early pregnancy, and 10 per cent or more of older women are infected (Kass 1960, Kunin *et al.* 1960).

Neonatal infection of the urinary tract occurs in about 1 per cent of babies—mostly in males—and usually resolves spontaneously (Littlewood *et al.* 1969). It is sometimes accompanied by septicaemia; and faecal colonization with the infecting strain may be widespread and persistent. Chronic urinary-tract infection in young children has a poor prognosis and often leads to progressive renal insufficiency (Steel *et al.* 1963). It is now generally recognized that urinary-tract infection that becomes established in early childhood is more likely to be associated with permanent and progressive loss of renal function than is infection acquired subsequently. Early infection generally occurs in children with vesico-ureteric reflux. In the more advanced cases of this, intrarenal reflux develops, with backflow of urine into the calices and collecting tubules (Rolleston *et al.* 1974). This results in renal scarring, failure of the kidney to grow, and tubular dysfunction. The extent to which this loss of renal function is attributable to mechanical and bacterial factors is uncertain, but the association between 'reflux nephropathy' and pyelonephritis in this age-group is very close. In bacteriuric schoolgirls, further damage to the kidneys occurs mainly among those with initially scarred kidneys (Report 1978, 1981). Thus, although patients with bacteriuria are more likely than uninfected patients to suffer recurrent symptomatic attacks, these seldom lead to progressive renal damage unless scarring had developed early in life.

Some 5 per cent of pregnant women have bacteriuria, and at least one-quarter of those affected will suffer symptomatic attacks during the pregnancy; possible reasons for this are discussed by McFadyen (1980). Bacteriuria in pregnancy is associated with an increased risk of premature labour, low birth-weight and perinatal mortality.

There is good evidence that *Esch. coli* reaches the bladder from the vulva by the ascending route (see, for example, Grüneberg 1969, Roberts *et al.* 1975). Organisms from the periurethral region can be 'milked' up the female urethra into the bladder (Bran *et al.* 1972), and there is some evidence associating the presence of bacteria in the bladder urine with recent sexual intercourse (Kelsey *et al.* 1979). Swabs of the periurethral region in females often yield considerable numbers of coliform bacilli (O'Grady *et al.* 1970). Kunin and his colleagues (1980) found that episodes of cystitis associated with bacteriuria were often preceded by periurethral colonization, but this often occurred in women who remained healthy, and prospective studies failed to establish that sexual intercourse was a precipitating factor.

Infections with the novobiocin-resistant staphylococcus tend to occur in the age-group in which women first become sexually active (Maskell 1974). Similar staphylococci are found in small numbers on the skin, but a site of heavy carriage in males or females has not yet been identified. According to Hallett and his colleagues (1976), *Proteus* bacilli are commonly present in the preputial sac of uninfected male children.

Abacterial cystitis

Symptoms indistinguishable from those of acute cystitis often occur in the absence of bacteriological evidence of urinary-tract infection (Gallagher *et al.* 1965; see Sanford 1975). Some workers believe that these symptoms are referable to an infection of the urethra or its glands, but in most cases a potential pathogen cannot be found in the urethral swab. O'Grady and his colleagues (1970) observed that over half of women with abacterial cystitis suffered subsequently from a symptomatic urinary-tract infection and considered that many cases of the disease might indeed be caused by coliform bacilli that had invaded the urethra but on this occasion had not reached the bladder. According to Stamm and his colleagues (1980), most women with abacterial cystitis exhibit pyuria, and many of them yield the usual pathogens—*Esch. coli* or coagulase-negative staphylococci—but in numbers that are lower than those usually accepted as evidence of urinary-tract infection (see below); some appear to have urethral infections with *Chlam. trachomatis*.

Laboratory diagnosis

After comparing the numbers of bacteria in freshly voided urine and in urine obtained by suprapubic aspiration, Kass (1955, 1960) concluded that the presence of 10^8 organisms per litre was presumptive evidence of urinary-tract infection, and that if this number was exceeded in two successive samples a correct diagnosis would be made on 96 per cent of occasions. This may be true for enterobacterial infections, but, according to Hovelius and Mårdh (1977), the number of *Staph. saprophyticus* in the bladder urine may not exceed 10^7 per l in some undoubted infections.

Kass's figure of 10^8 organisms per l for 'significance' was established as a means of distinguishing between true infection and extraneous contamination of the specimen during collection, and is applicable to midstream urine collected with reasonable care. Other workers have found it possible to use lower figures when the specimens were collected and examined under close supervision; Gillespie and his colleagues (1960), for example, found that 3×10^6 organisms or more per l provided reliable evidence of infection in male patients. Very much lower figures can be used when specimens are collected by suprapubic puncture. In practice, false positive results are more often attributable to bacterial multiplication during transit to the laboratory than to initial contamination. This can be avoided by refrigeration or by the use of a bacteristatic additive (see, for example, Lauer *et al.* 1979) when the delay in examining the specimen does not exceed 24 hr or so.

Despite the fact that bacterial and abacterial cystitis are clinically indistinguishable, the assumption that bladder bacteriuria represents 'infection' (Report 1979) is widely accepted, and its detection by semi-quantitative methods is a major activity in hospital laboratories. Many rapid screening methods are available, varying in degree of elaboration from the dip-slide method of Naylor and Guttmann (1967) to the bioluminescence method of Johnson and his colleagues (1976). Indirect screening by testing for glucose consumption or nitrate reduction, though applicable for the detection of significant numbers of enterobacteria, gives false negative results with *Staph. saprophyticus*. For the performance and interpretation of bacteriological tests on urine, see Brumfitt *et al.* (1973) and Meers (1978).

Several tests have been proposed for distinguishing between cystitis and pyelonephritis. In *Esch. coli* infections, serum antibody against the infecting strain is said to be formed only when the renal pelvis is affected. O antibody may be detected by agglutination (Percival *et al.* 1964) or indirect haemagglutination, and this and the enterobacterial common antigen (Chapter 33) by direct immunofluorescence or radioimmunoassay (Sanford *et al.* 1978). Retrospective diagnosis may be made by the immunofluorescent identification of the common antigen in kidney sections (Aoki *et al.* 1969). It has been stated that the presence in the urine of bacteria coated with immunoglobulin distinguishes pyelonephritis from cystitis (Thomas *et al.* 1974), but the specificity of this test has been doubted (Hjelm *et al.* 1979, Mundt and Polk 1979).

Treatment

Many single attacks of uncomplicated urinary-tract infection clear spontaneously, and most respond to quite a short course of an appropriate antibiotic (see Greenwood *et al.* 1980). In the presence of gross abnormality of the urinary tract, however, infection is very difficult to eradicate (Garrod *et al.* 1954). In-vitro tests of the sensitivity of the causative organism are used to determine appropriate dosage of an antimicrobial agent. Opinions are divided as to whether the effective concentration is that attainable in the blood or in the urine; the more widely accepted view is that the latter is adequate if the infection is uncomplicated by involvement of the renal parenchyma (see Naumann 1978). However, conventional in-vitro tests of sensitivity are probably not a very good guide to treatment. Greenwood and O'Grady (1977, 1978) describe an in-vitro model in which a dense bacterial suspension is exposed to changing concentrations of antibiotic, while the culture is continuously diluted and subject to periodic discharge. It is thus possible to mimic the effects of various dosage schedules, rates of urinary excretion, and volumes of residual urine on the bacteria. The results obtained give a somewhat different prediction of the relative efficacy of individual antibiotics from that given by conventional tests, and emphasize the importance of frequent and complete emptying of the bladder, and maintaining an adequate concentration of the antibacterial agent during the night hours.

Briefly, a short course of an appropriate agent should be given in symptomatic attacks. Some 80–90 per cent of infections will resolve after a single dose of an agent to which the organism is susceptible *in vitro* —cotrimoxazole, trimethoprim, amoxycillin or a cephalosporin, as the case may be. Failure to eradicate the infection, as determined by bacteriological examination, or a history of repeated attacks, indicates the necessity for careful investigation of the urinary tract, including intravenous pyelography. In the absence of radiological abnormalities, continuous low-dosage chemoprophylaxis with, for example, cotrimoxazole (O'Grady *et al.* 1969, Pearson *et al.* 1979), continued if necessary for several years, with high fluid intake and training in efficient emptying of the bladder, will generally prevent subsequent attacks and may lead to an eventual disappearance of bacteriuria. Opinions are divided upon whether this procedure should be generally adopted, or whether treatment should be given only when symptomatic episodes occur (Ronald and Harding 1981, Stamm *et al.* 1981). Bacteriuria in pregnancy leads to pyelonephritis with such frequency that its detection and treatment, or at least temporary suppression, seems to be justified, though it is not yet clear whether this prevents deleterious effects on the fetus. The question is discussed by Garrod and his colleagues (1981) to whom reference should also be made for information about the use of individual antimicrobial agents, and about the treatment of infection in the anatomically abnormal urinary tract.

Pyelonephritis in *cattle* (see also Chapter 53) is characterized by a membranous necrotic inflammation, which leads to progressive destruction of the kidney; there is often considerable fibrosis of the pelvis of the kidney, the ureter, and the bladder. The disease is produced by the gram-positive diphtheroid bacillus *C. renale*, alone or sometimes mixed with streptococci, coliform bacilli, *C. pyogenes* or other organisms (Glage 1928, Jones and Little 1930). Though it commonly attacks adult and particularly female cattle, the bacillus may be acquired in early life, and is carried in the vagina (Weitz 1947) or urinary tract. It appears to have a predilection for the medulla of the kidney, but it is not clear whether infection reaches the tissue via the blood stream or as an ascending infection from the urinary tract.

References

Abeysundere, R. L., Bradley, J. M., Chipping, P., Rogers, B. T. and Noone, P. (1979) *J. Infect.* **1**, 127.

Allott, E. N., Cruickshank, R., Cyrlas-Williams, R., Glass, V., Meyer, I. H., Straker, E. A. and Tee, G. (1944) *J. Path. Bact.* **56**, 411.

Almquist, E. E. (1970) *Clin. Orthoped.* no. 68, p. 96.

Alpern, R. J. and Dowell, V. R. (1969) *J. Amer. med. Ass.* **209**, 385; (1971) *Amer. J. clin. Path.* **55**, 717.

Altemeier, W. A. (1942) *Surg. Gynec. Obstet.* **75**, 518.

Altemeier, W. A. and Gibbs, E. W. (1944) *Surg. Gynec. Obstet.* **78**, 164.

Andersen, B. R. and Jackson, G. G. (1961) *J. exp. Med.* **114**, 375.

Angrist, A. A. and Oka, M. (1963) *J. Amer. med. Ass.* **183**, 249.

Aoki, S., Imamura, S., Aoki, M. and McCabe, W. R. (1969) *New Engl. J. Med.* **281**, 1375.

Balsdon, M. J., Taylor, G. E., Pead, L. and Maskell, R. (1980) *Lancet* **i**, 501.

Barrington, F. J. F. and Wright, H. D. (1930) *J. Path. Bact.* **33**, 871.

Bartlett, J. G., Gorbach, S. L., Thadepalli, H. and Finegold, S. M. (1974) *Lancet* **i**, 338.

Bateman, N. T., Eykyn, S. J. and Phillips, I. (1975) *Lancet* **i**, 657.

Bayer, A. S., Theofilopoulos, A. N., Eisenberg, R., Dixon, F. J. and Guze, L. B. (1976) *New Engl. J. Med.* **295**, 1500.

Beattie, J. W. (1954) *Brit. med J.* **ii**, 25.

Beaven, D. W. (1958) *Lancet* **i**, 869.

Beeler, B. A., Crowder, J. G., Smith, J. W. and White, A. (1976) *Amer. J. Med.* **61**, 935.

Berman, P. H. and Banker, B. Q. (1966) *Pediatrics, Springfield* **88**, 6.

Blake, P. A., Merson, M. H., Weaver, R. E., Hollis, D. G. and Heublein, P. C. (1979) *New Engl. J. Med.* **300**, 1.

Blevins, A. and MacNeal, W. J. (1946) *Amer. Heart J.* **31**, 663.

Boulton-Jones, J. M., Sissons, J. G. P., Evans, D. J. and Peters, D. K. (1974) *Brit. med. J.* **ii**, 11.

Bowden, G. H. W., Ellwood, D. C. and Hamilton, I. R. (1979) In: *Advances in Microbial Ecology*, Vol. 3, p. 135. Ed. by M. Alexander. Plenum Press, New York and London.

Bowen, W. H. (1967) *J. Path. Bact.* **94**, 55.

Bowen, W. H., Cohen, B., Cole, M. F. and Colman, G. (1975) *Brit. dent. J.* **139**, 45.

Bran, J. L., Levison, M. E. and Kaye, D. (1972) *New Engl. J. Med.* **286**, 626.

Braude, A. I., Ziegler, E. J., McCutchan, J. A. and Douglas, H. (1981) *Amer. J. Med.* **70**, 463.

Brook, I. (1980) *Amer. J. Dis. Child.* **134**, 564; (1981) *J. Amer. med. Ass.* **246**, 967.

Brumfitt, W., Percival, A. and Williams, J. D. (1973) *Estimation of Bacteria and White Cells in the Urine* (Broadsheet No. 80). Association of Clinical Pathologists, London.

Bryant, R. E., Hood, A. F., Hood, C. E. and Koenig, M. G. (1971) *Arch. intern. Med.* **127**, 120.

Caldwell, J. and Lehner, T. (1982) *J. med. Microbiol.* **15**, 339.

Carruthers, M. M. (1977) *Amer. J. med Sci.* **273**, 203.

Cates, J. E. and Christie, R. V. (1951) *Quart. J. Med.* **20** N.S., 93.

Cherubin, C. E. and Neu, H. C. (1971) *Amer. J. Med.* **51**, 83.

Clarke, J. K. (1924) *Brit. J. exp. Path.* **5**, 141.

Collee, J. G., Duerden, B. I. and Brown, R. (1977) *J. clin. Path.* **30**, 609.

Colman, G. (1967) *J. clin. Path.,* **20**, 294.

Colman, G. and Cohen, B. (1979) In: *Pathogenic Streptococci*, p. 214. Ed. by M. T. Parker. Reedbooks, Chertsey, Surrey.

Corrin, B. (1980) *J. clin. Path.* **33**, 891.

Cotran, R. S., Vivaldi, E., Zangwill, D. P. and Kass, E. H. (1963) *Amer. J. Path.* **43**, 1.

Crawford, J. J., Sconyers, J. R., Moriarty, J. D., King, R. C. and West, J. F. (1974) *Appl. Microbiol.* **27**, 927.

Croxson, M. S., Altmann, M. M. and O'Brien, K. P. (1971) *Lancet* **i**, 1205.

Daleel, E. E. and Frost, A. J. (1967) *Brit. vet. J.* **123**, 232.

Davidson, I. (1961) *Res. vet. Sci.* **2**, 22.

Davies, R. M. (1974) In: *The Normal Microbial Flora of Man,*

p. 99 (*Soc. appl. Bact. Symp.* no. 3). Ed. by F. A. Skinner and J. G. Carr. Academic Press, London and New York.

Derbyshire, J. B. and Edwards, S. J. (1963) *Vet. Rec.* **75**, 1208.

Derbyshire, J. B. and Smith, G. S. (1969) *Res. vet. Sci.* **10**, 559.

Dowling, H. F., Lepper, M., Caldwell, E. R. and Spies, H. W. (1952) *Medicine, Baltimore* **31**, 155.

Doyle, E. F., Spagnuolo, M., Taranta, A., Kuttner, A. G. and Markowitz, M. (1967) *J. Amer. med. Ass.* **201**, 807.

Drucker, D. B. and Green, R. M. (1979) In: *Pathogenic Streptococci*, p. 206. Ed. by M. T. Parker. Reedbooks, Chertsey, Surrey.

Durack, D. T. and Beeson, P. B. (1972) *Brit. J. exp. Path* **53**, 44, 50.

Durack, D. T., Beeson, P. B. and Petersdorf, R. G. (1973) *Brit. J. exp. Path.* **54**, 142.

Durack, D. T., Pelletier, L. L. and Petersdorf, R. G. (1974) *J. clin. Invest.* **53**, 829.

Edwards, S. J. (1933) *J. comp. Path.* **46**, 211; (1938) *Ibid.* **51**, 250.

Ellner, P. D. and Stoessel, C. J. (1966) *J. infect. Dis.* **116**, 238.

Eykyn, S. J. and Phillips, I. (1978) *J. antimicrob. Chemother.* **4**, suppl. C, 75.

Felner, J. M. and Dowell, V. R. (1970) *New Engl. J. Med.* **283**, 1188.

Finland, M. and Barnes, M. W. (1970) *Ann. intern. Med.* **72**, 341; (1978) *J. infect. Dis.* **137**, 274.

Finland, M., Jones, W. F. and Barnes, M. W. (1959) *J. Amer. med. Ass.* **170**, 2188.

Fitzgerald, R. J. and Keyes, P. H. (1960) *J. Amer. dent. Ass.* **61**, 9.

Forgan-Smith, W. R. and Darrell, J. H. (1974) *J. clin. Path.* **27**, 280.

Francis, J. (1941) *Vet. J.* **97**, 243.

Frederick, J. and Braude, A. I. (1974) *New Engl. J. Med.* **290**, 135.

Freedman, M. L., Birkhed, D. and Tanzer, J. M. (1979) In: *Pathogenic Streptococcus*, p. 212. Ed. by M. T. Parker, Reedbooks, Chertsey, Surrey.

Gallagher, D. J. A., Montgomerie, J. Z. and North, J. D. K. (1965) *Brit. med. J.* **i**, 622.

Gardner, H. L. and Dukes, C. D. (1955) *Amer. J. Obstet. Gynec.* **69**, 962.

Garrod, L. P., Lambert, H. P. and O'Grady, F. (1981) *Antibiotic and Chemotherapy*, 5th edit. Churchill Livingstone, Edinburgh and London.

Garrod, L. P., Shooter, R. A. and Curwen, M. P. (1954) *Brit. med. J.* **ii**, 1003.

Garrod, L. P. and Waterworth, P. M. (1962) *Brit. Heart J.* **24**, 39.

Gibbons, R. J. and Banghart, S. (1968) *Arch. oral Biol.* **13**, 297.

Gibbons, R. J. and Fitzgerald, R. J. (1969) *J. Bact.* **98**, 341.

Gigliotti, F., Williams, W. T., Hayden, F. G. and Hendley, J. O. (1981) *J. Pediat.* **98**, 531.

Gillespie, W. A., Linton, K. B., Miller, A. and Slade, N. (1960) *J. clin. Path.* **13**, 187.

Gitter, M., Bradley, R. and Blampied, P. H. (1980) *Vet. Rec.* **107**, 390.

Glage, F. (1928) In: *Kolle and Wassermann's Hdb. path. Mikroorg.*, 3te Abt., 1928–29, **6**, 563.

Glover, S. C., McKendrick, M. W., Padfield, C., Geddes, A. M. and Dwyer, N. St J. P. (1982) *Lancet* **i**, 608.

Goose, D. H. (1965) *Publ. Hlth* **79**, 131.

Gray, I. R. (1975) *Quart. J. Med.* **44** n.s., 449.

Greenwood, D., Kawada, Y. and O'Grady, F. (1980) *Lancet* **i**, 197.

Greenwood, D. and O'Grady, F. (1977) *Brit. med. J.* **ii**, 665; (1978) *J. antimicrob. Chemother.* **4**, 113.

Groover, R. V., Sutherland, J. M. and Landing, B. H. (1961) *New Engl. J. Med.* **264**, 1115.

Grüneberg, R. N. (1969) *Lancet* **ii**, 766.

Hall, M., Warren, E. and Washington, J. A. II (1974a) *Appl. Microbiol.* **27**, 187; (1974b) *Ibid.* **27**, 699.

Hallett, R. J., Pead, L. and Maskell, R. (1976) *Lancet* **ii**, 1107.

Halsted, C., Lepow, M. L., Balassanian, N., Emmerich, J. and Wolinsky, E. (1968) *Amer. J. Dis. Child.* **115**, 542.

Hamada, S. and Slade, H. D. (1980) *Microbiol. Rev.* **44**, 331.

Hare, R. and Willits, R. E. (1942) *Canad. med. Ass. J.* **46**, 23.

Hayward, G. (1973) *Brit. med. J.* **2**, 706, 764.

Higgs, T. M., Neave, F. K. and Bramley, A. J. (1980) *J. med. Microbiol.* **13**, 393.

Hjelm, E., Forsum, U. and Frödin, L. (1979) *J. clin. Path.* **32**, 1206.

Hjelm, E., Hallén, A., Forsum, U. and Wallin, J. (1981) *Lancet* **ii**, 1353.

Hobson, F. G. and Juel-Jensen, B. E. (1956) *Brit. med. J.* **ii**, 1501.

Hochstein, H. D., Kirkham, W. R. and Young, V. M. (1965) *New Engl. J. Med.* **273**, 468.

Holland, K. T., Ingham, E. and Cunliffe, W. J. (1981) *J. appl. Bact.* **51**, 195.

Hook, E. W. III and Sande, M. A. (1974) *Infect. Immun.* **10**, 1433.

Horder, T. J. (1908-09) *Quart. J. Med.* **2**, 289.

Hovelius, B. and Mårdh, P.-A. (1977) *Acta path. microbiol. scand.* **B85**, 427.

Hughes, P. and Gauld, W. R. (1966) *Quart. J. Med.* **35**, 511.

Ingham, H. R., High, A. S., Kalbag, R. M., Sengupta, R. P., Tharagonnet, D. and Selkon, J. B. (1978) *Lancet* **ii**, 497.

Ingham, H. R., Selkon, J. B. and Roxby, C. M. (1977a) *Brit. med. J.* **ii**, 991.

Ingham, H. R., Sisson, P. R., Tharagonnet, D., Selkon, J. B. and Codd, A. A. (1977b) *Lancet* **ii**, 1252.

Jellis, J. E. (1981) In: *The Staphylococci*, p. 225. Ed. by A. Macdonald and G. Smith. The University Press, Aberdeen.

Johnson, H. H., Mitchell, C. J. and Curtis, G. D. W. (1976) *Lancet* **ii**, 400.

Johnston, R. B. and Sell, S. H. (1964) *Pediatrics, Springfield* **34**, 473.

Jones, F. S. and Little, R. B. (1930) *J. exp. Med.* **51**, 909.

Kaplan, E. L. (1977) *Circulation,* **56**, 139A.

Kaplan, E. L. and Taranta, A. V. (Eds.) (1977) *Infective Endocarditis*. The American Heart Association, Dallas, Texas.

Karchmer, A. W., Moellering, R. C. Jr., Maki, D. G. and Swartz, M. N. (1979) *J. Amer. med. Ass.,* **241**, 1801.

Karchmer, A. W. and Swartz, M. N. (1977) In: *Infective Endocarditis*, p. 58. Ed. by E. L. Kaplan and A. V. Taranta. The American Heart Association, Dallas, Texas.

Kass, E. H. (1955) *Trans. Ass. Amer. Phys.,* **69**, 56; (1960) In: *The Biology of Pyelonephritis*, p. 399. Ed. by E. L. Quinn and E. H. Kass. Churchill, London.

Kaye, D. (Ed.) (1976) *Infective Endocarditis*. University Park Press, Baltimore, Md.

Kaye, D., McCormack, R. C. and Hook, E. W. (1962) *Antimicrob. Agents Chemother. 1961*, p. 37.

Keith, T. A. III and Lyon, S. A. (1963) *Amer. J. Med.* **34**, 535.

Kellgren, J. H., Ball, J., Fairbrother, R. W., and Barnes, K. L. (1958) *Brit. med. J.* **i.** 1193.

Kelly, M. J. (1978) *J. med. Microbiol.* **11**, 513; (1980) *Ann. Roy. Coll. Surg. Engl.* **62**, 52.

Kelsey, M. C., Mead, M. G., Grüneberg, R. N. and Oriel, J. D. (1979) *J. med. Microbiol.* **12**, 511.

Keslin, M. H., Messner, R. P. and Williams, R. C. Jr (1973) *Arch. intern. Med.* **132**, 578.

Khairat, O. (1940) *J. Path. Bact.* **50**, 497.

Kilo, C., Hagemann, P. O. and Marzi, J. (1965) *Amer. J. Med.* **38**, 962.

King, J. D. (1944) *Brit. med. Bull.* **2**, 222.

Klein, R. S., Recco, R.A., Catalano, M. T., Edberg, S. C., Casey, J. I. and Steigbigel, N. H. (1977) *New Engl. J. Med.* **297**, 800.

Kligler, I. J. (1915) *J. allied dent. Soc.* **10**, 141, 282, 445.

Krasse, B. (1966) *Arch. oral Biol.* **11**, 429.

Kreitzman, S. N., Irving, S., Navia, J. M. and Hanis, R. S. (1969) *Nature, Lond.* **223**, 520.

Kunin, C. M., Polyak, F. and Postel, E. (1980) *J. Amer. med. Ass.* **243**, 134.

Kunin, C. M., Southall, I. and Paquin, A. J. (1960) *New Engl. J. Med.* **263**, 817.

Larsson, P., Kaijser, B., Baltzer, I. M. and Olling, S. (1980) *J. clin. Path.* **33**, 408.

Lauer, B. A., Reller, L. B. and Mirrett, S. (1979) *J. clin. Microbiol.* **10**, 42.

Lawes, F. A. E., Durie, E. B., Goldsworthy, N. E. and Spies, H. C. (1952) *Med. J. Aust.* **i**, 330.

Le Frock, J. L., Ellis, C. A., Turchik, J. B. and Weinstein, L. (1973) *New Engl. J. Med.* **289**, 467.

Lehner, T. (1978) *Proc. R. Soc. Med.* **71**, 161.

Lehner, T., Russell, M. W. and Caldwell, J. (1980) *Lancet* **i**, 995.

Lehner, T., Russell, M. W., Challacombe, S. J., Scully, C. M. and Hawkes, J. E. (1978) *Lancet* **i**, 693.

Lehner, T., Russell, M. W., Scully, C. M., Challacombe, S. J. and Caldwell, J. (1979) In: *Pathogenic Streptococci*, p. 215. Ed. by M. T. Parker. Reedbooks, Chertsey, Surrey.

Levaditi, C., Gérard-Moissonnier, Bréchot, H. and Tournay, R. (1939) *Bull. Acad. Méd., Paris* **122**, 371.

Levin, J. and Bang, F. B. (1964) *Bull. Johns Hopk. Hosp.* **115**, 265.

Levin, J., Poore, T. E., Zauber, N. P. and Oser, R. S. (1970) *New Engl. J. Med.* **283**, 1313.

Levinson, M. E., Kaye, D., Mandell, G. L. and Hooke, E. W. (1970) *J. Amer. med. Ass.* **211**, 1355.

Levison, D. A., Guthrie, W., Ward, C., Green, D. M. and Robinson, P. G. C. (1971) *Lancet* **ii**, 844.

Little, K. (1962) *Nature, Lond.* **193**, 388.

Littlewood, J. M., Kite, P. and Kite, B. A. (1969) *Arch. Dis. Childh.* **44**, 617.

Loeb, H., Bettag, J. L., Yung, N. K., King, S. and Bronsky, D. (1966) *Amer. Heart J.* **71**, 381.

Loesche, W. J., Rowan, J., Straffon, L. H. and Loos, P. J. (1975) *Infect. Immun.* **11**, 1252.

Louvois, J. de (1980) *J. clin. Path.* **33**, 66.

Louvois, J. de, Gortvai, J. and Hurley, R. (1977) *Brit. med. J.* **ii**. 981.

Lowe, J. F., Bamforth, J. S. and Pracy, R. (1963) *Lancet* **ii**, 1129.

Lowy, F. and Steigbigel, N. H. (1978) *Amer. Heart J.* **96**, 689.

Lystad, A., Berdal, P. and Lund-Iversen, L. (1964) *Acta otolaryng., Stockh.* Suppl. 188, p. 390.

Mabeck, C. E. (1969) *Lancet* **ii**, 1150.

McCabe, W. R., Greely, A., DiGenio, J. and Johns, M. A. (1973) *J. infect. Dis.* **128** (suppl.), S284.

McCarthy, L. R. and Bottone, E. J. (1974) *Amer. J. clin. Path.* **61**, 585.

McCormack, W. M., Hayes, C. H. and Rosner, B. (1977) *J. infect. Dis.* **136**, 740.

McCracken, G. H. and Shinefield, H. R. (1966) *Amer. J. Dis. Child.* **112**, 33.

McFadyen, I. R. (1980) *Proc. R. Soc. Med.* **73**, 227.

McGowan, J. E., Barnes, M. W. and Finland, M. (1975) *J. infect. Dis.* **132**, 316.

Mair, N. S. (1966) In: *Some Diseases of Animals Transmissible to Man in Britain*, p. 47. Ed. by O. Graham-Jones. Pergamon Press, London.

Mandell, G. L., Kaye, D., Levison, M. E. and Hook, E. W. (1970) *Arch. intern. Med.* **125**, 258.

Marmion, B. P. *et al.* (1961) *Brit. med. J.*, **ii**, 1264.

Marples, R. R. and Richardson, J. F. (1980) *J. med. Microbiol.* **13**, 355.

Maskell, R. (1974) *Lancet* **ii**, 1155.

Mattman, L. H. and Mattman, P. E. (1965) *Arch. intern. Med.* **115**, 315.

Meers, P. D. (1978) *The Bacteriological Examination of Urine* (Publ. Hlth Lab. Serv. Monogr. No. 10). HMSO, London.

Meleney, F. L. (1943) *Ann. Surg.* **118**, 171.

Merson, M. H. and Dowell, V. R. Jr (1973) *New Engl. J. Med.* **289**, 1005.

Miles, A. A., Schwabacher, H., Cunliffe, A. C., Ross, J. P., Spooner, E. T. C., Pilcher, R. S. and Wright, J. (1940) *Brit. med. J.* **ii**, 855.

Miller, W. D. (1889) *Die Mikroorganismen der Mundhöhle*. G. Thieme, Leipzig.

Mitchell, R. G. (1968) *J. clin. Path.* **21**, 93.

Moellering, R. C. Jr, Murray, B. E., Schoenbaum, S. C., Adler, J. and Wennersten, C. B. (1980) *J. infect. Dis.* **141**, 81.

Moellering, R. C. Jr, Wennersten, C., Medrek, T. and Weinberg, A. N. (1971) *Antimicrob. Agents Chemother. 1970*, 335.

Mohammed, I., Ansell, B. M., Holborow, E. J. and Bryceston, A. D. M. (1977) *J. clin. Path.* **30**, 308.

Moore, B. (1954) *J Path. Bact.* **67**, 461.

Moulsdale, M. T., Eykyn, S. and Phillips, I. (1980) *Quart. J. Med.* **49**, 315.

Mundt, K. A. and Polk, B. F. (1979) *Lancet* **ii**, 1172.

Murray, H. W. and Roberts, R. B. (1978) *Arch. intern. Med.* **138**, 1097.

Nai, D. D. (1957) *Atti Soc. ital. Sci. vet.* **11**, 33.

Naumann, P. (1978) *J. antimicrob. Chemother.* **4**, 9.

Naylor, G. R. E. and Guttmann, D. (1967) *J. Hyg., Camb.* **65**, 367.

Nelson, J. B. and Gowen, J. W. (1930) *J. infect. Dis.* **46**, 53.

O'Grady, F., Chamberlain, D. A., Stark, J. E. (1969) *Post. Grad. med. J.* **45**, suppl. 61.

O'Grady, F. W., Richards, B., McSherry, M. A., O'Farrell, S. M. and Cattell, W. R. (1970) *Lancet* **ii**, 1208.

Okell, C. C. and Elliott, S. D. (1935) *Lancet* **ii**, 869.

Onderdonk, A. B., Louie, T. J., Tally, F. P. and Bartlett, J. G. (1979) *J. antimicrob. Chemother.* **5**, 201.

Orland, F. J. *et al.* (1955) *J. Amer. dent. Ass.* **50**, 259.

Overall, J. C. Jr (1970) *J. Pediat.* **76**, 499.

Page, M. I. and King, E. O. (1966) *New Engl. J. Med.* **275**, 181.

Palmer, S. R. and Young, S. E. J. (1982) *Lancet* **ii**, 1448.

Parker, M. T. and Ball, L. J. (1976) *J. med. Microbiol.* **9**, 275.

Pearson, J. K. L. (1959) *J. Dairy Res.* **26**, 9.

Pearson, N. J., Towner, K. J., McSherry, A. M., Cattell, W. R. and O'Grady, F. (1979) *Lancet* **ii**, 1205.

Percival, A., Brumfitt, W. and Louvois, J. de (1964) *Lancet* **ii**, 1027.

Pereira, A. T. (1962) *J. clin. Path.* **15**, 252.

Perlman, B. B. and Freedman, L. R. (1971) *Yale J. Biol. Med.* **44**, 206.

Poole, P. M. and Wilson, G. (1977) *J. clin. Path.* **30**, 937.

Pulaski, E. J., Meleney, F. L. and Spaeth, W. L. C. (1941) *Surg. Gynec. Obstet.* **72**, 982.

Pulvertaft, R. J. V. (1943) *Lancet* **ii**, 1.

Rains, A. J. H. (1962) *Brit. med. J.* **ii**, 685.

Rains, A. J. H., Barson, G. J., Crawford, N. and Shrewsbury, J. F. D. (1960) *Lancet* **ii**, 614.

Reinhold, R. B. and Fine, J. (1971) *Proc. Soc. exp. Biol., N.Y.* **137**, 334.

Report (1936) *Spec. Rep. Ser. med. Res. Coun., Lond.* No. 211; (1972*a*) *Tech. Rep. Ser. World Hlth Org.* No. 494; (1972*b*) Mastitis Literature Survey, Commonwealth Bureau of Dairy Science and Technology, Shinfield, Reading (and annually thereafter); (1978) Cardiff–Oxford Bacteriuria Study Group, *Lancet* **i**, 889; (1979) Med. Res. Coun. Bacteriura Committee, *Brit. med. J.* **ii**, 717; (1981) Newcastle Covert Bacteriuria Research Group, *Arch. Dis. Childh.* **56**, 585; (1982) *Lancet* **ii**, 1323.

Richards, J. H. (1932) *J. Amer. med. Ass.* **99**, 1496.

Richou, R. and Thieulin, G. (1955) *Rec. Méd. vét.* **131**, 73.

Robbins, J.B., McCracken, G. H., Gotschlich, E. C., Ørskov, F., Ørskov, I. and Hanson, L. A. (1974) *New Engl. J. Med.* **290**, 1216.

Roberts, A. P., Linton, J. D., Waterman, A. M., Gower, P. E. and Koutsaimanis, K. J. (1975) *J. med. Microbiol.* **8**, 311.

Robertson, W. (1835) *A Practical Treatise on the Human Teeth, Etc.* Quoted from Miller (1889).

Robinson, N., Beral, V. and Ashley, J. S.A. (1981) *J. Epid. commun. Hlth* **35**, 265.

Rolleston, G. L., Maling, T. M. J. and Hodson, C. J. (1974) *Arch. Dis. Childh.* **49**, 531.

Ronald, A. R. and Harding, G. K. M. (1981) *Ann. intern. Med.* **94**, 268.

Rosenow, E. C. (1916) *J. infect. Dis.* **19**, 527.

Rosner, R. (1970) *Appl. Microbiol.* **19**, 281; (1972) *Amer. J. clin. Path.* **57**, 220; (1974) *Appl. Microbiol.* **28**, 245; (1975) *J. clin. Microbiol.* **1**, 129.

Rowlands, D. T. Jr, Vakilzadeh, J., Sherwood, B. F. and Le May, J. C. (1970) *Amer. J. Path.* **58**, 295.

Rubenstein, A. D. and Johnson, B. B. (1945) *Amer. J. med. Sci.* **210**, 517.

Russell, E. J. and Sutherland, R. (1975) *J. med. Microbiol.* **8**, 1.

Rustigian, R. and Cipriani, A. (1947) *J. Amer. med. Ass.* **133**, 224.

Sande, M. A. (1976) In: *Infective Endocarditis*, p. 11. Ed. by D. Kaye. University Park Press, Baltimore.

Sanford, B. A., Thomas, V. L., Forland, M., Carson, S and Shelokov, A. (1978) *J. clin. Microbiol.* **8**, 575.

Sanford, J. P. (1975) *Annu. Rev. Med.* **26**, 485.

Sanyal, B. and Russell, C. (1978) *Appl. envir. Microbiol.* **35**, 670.

Saxena, S. R. and Bassett, D. C. J. (1975) *Arch. Dis. Childh.* **50**, 899.

Schiffman, W. L. and Black, A. (1956) *New Engl. J. Med.* **255**, 1148.

Schlaes, D. M., Levy, J. and Wolinsky, E. (1981) *Arch. intern. Med.* **141**, 578.

Schoenbaum, S. C., Gardner, P. and Shillito, J. (1975) *J. infect. Dis.* **131**, 543.

Schottmüller, H. (1910) *Münch. med. Wschr.* **57**, 617, 697; (1911) *Ibid.* **58**, 557.

Scoy, R. E. van, Cohen, S. N., Geraci, J. E. and Washington, J. A. II (1977) *Proc. Mayo Clin.* **52**, 216.

Shanson, D. C. and Barnicoat, M. (1975) *J. clin. Path.* **28**, 407.

Shovlin, F. E. and Gillis, R. E. (1969) *J. dent. Res.* **48**, 356.

Simberkoff, M. S. (1977) In: *Infective Endocarditis*, p. 46. Ed. by E. Kaplan and A. V. Taranta. The American Heart Association, Dallas, Texas.

Slanetz, L. W., Bartley, C. H. and Allen, F. E. (1959) *J. Amer. vet. med. Ass.* **134**, 155.

Slee, A. M. and Tanzer, J. M. (1979) In: *Pathogenic Streptococci*, p. 203. Ed. by M. T. Parker. Reedbooks, Chertsey, Surrey.

Slotnick, I. J. and Dougherty, M. (1964) *Leeuwenhoek Ned. Tijdschr.* **30**, 261.

Smith, H. W. (1959) In: *Infectious Diseases of Animals* **2**, 557. Ed. by A. W. Stableforth and I. A. Galloway. Butterworth, London.

Socransky, S. S. and Gibbons, R. J. (1965) *J. infect. Dis.* **115**, 247.

Speller, D. C. E. and MacIver, A. G. (1971) *J. med. Microbiol.* **4**, 370.

Spencer, G. R., Stewart, J. H. and Lasmanis, J. (1956) *Amer. J. vet. Res.* **17**, 594.

Spiegel, C. A., Amsel R., Eschenbach, D., Schoenknecht, F. and Holmes, K. K. (1980) *New Engl. J. Med.* **303**, 601.

Spooner, E. T. C. (1941) *Brit. med. J.* **ii**, 477.

Sprunt, K. (1977) In: *Infective Endocarditis*, p. 17. Ed. by E. L. Kaplan and A. V. Taranta. The American Heart Association, Dallas, Texas.

Stableforth, A. W. (1959) In: *Infectious Diseases of Animals*, **2**, 589. Ed. by A. W. Stableforth and I. A. Galloway. Butterworth, London.

Stableforth, A. W. and Galloway, I. A. (Eds.) (1959) *Infectious Diseases of Animals*, Butterworth, London.

Stamm, W. E., McKevitt, M., Counts, G. W., Wagner, K. F., Turck, M. and Holmes, K. K. (1981) *Ann. intern. Med.* **94**, 251.

Stamm, W. E. *et al.* (1980) *New Engl. J. Med.* **303**, 409.

Steel, R. E., Leadbetter, G. W. and Crawford, J. D. (1963) *New Engl. J. Med.* **269**, 883.

Stokes, E. J. (1974) *Blood Culture Technique* (Ass. clin. Path. Broadsheet No. 81). Association of Clinical Pathologists, London.

Tanzer, J. M., Freedman, M. L. and Fitzgerald, R. J. (1979)

In: *Pathogenic Streptococci*, p. 211. Ed. by M. T. Parker. Reedbooks, Chertsey, Surrey.

Tate, W. J., Douglas, H. and Braude, A. I. (1966) *Ann. N.Y. Acad. Sci.* **133**, 746.

Taylor, E., Blackwell, A. L., Barlow, D. and Phillips, I. (1982) *Lancet* **i**, 1376.

Taylor-Robinson, D. and Thomas, B. J. (1980) *J. clin. Path.* **33**, 205.

Thomas, V., Shelokov, A. and Forland, M. (1974) *New Engl. J. Med.* **290**, 588.

Thörig, L., Thompson, J. and Eulderink, F. (1977) *Infect. Immun.* **17**, 504.

Throop, B. T., Swanson, E. W. and Mundt, J. O. (1957) *Amer. J. vet. Res.* **18**, 93.

Toh, C. C. S. and Ball, K. P. (1960) *Brit. med. J.* **ii**, 640.

Urdal, K. and Berdal, P. (1949) *Acta Otolaryngol.* **37**, 20.

Uwaydah, M. M. and Weinberg, A. N. (1965) *New Engl. J. Med.* **273**, 1231.

Vakilzadeh, J., Rowlands, D. T., Sherwood, B. F. and Le May, J. C. (1970) *J. infect. Dis.* **122**, 89.

Vivaldi, E., Cotran, R., Zangwill, D. P. and Kass, E. H. (1959) *Proc. Soc. exp. Biol., N.Y.* **102**, 242.

Waal, H. L. de (1943) *Edinb. med. J.,* **50**, 577.

Waldvogel, F. A., Medoff, G. and Swartz, M. N. (1970) *New Engl. J. Med.* **282**, 198, 260, 346.

Waldvogel, F. A. and Vasey, H. (1980) *New Engl. J. Med.* **303**, 360.

Ward, C. and Ward, A. M. (1974) *Lancet* **ii**, 734.

Wardle, E. N. (1980) *J. clin. Path.* **33**, 888.

Watt, P. J. and Akubadejo, O. A. (1967) *Brit. med. J.* **i**, 210.

Weisbren, B. A. (1951) *Arch. intern. Med.* **88**, 467.

Weitz, B. (1947) *J. comp. Path.* **57**, 191.

Werner, A. S., Cobbs, C. G., Kaye, D. and Hook, E. W. (1967) *J. Amer. med. Ass.* **202**, 199.

White, M. E. E., Lord, M. D. and Rogers, K. B. (1961) *Arch. Dis. Childh.* **36**, 394.

Wilkie, A. L. (1928) *Brit. J. Surg.* **15**, 450.

Williams, G. T., Houang, E. T., Shaw, E. J. and Tabaqchali, S. (1976) *Lancet* **ii**, 1291.

Williams, R. E. O. and Miles, A. A. (1945) *J. Path. Bact.,* **57**, 27; (1949) *Spec. Rep. Ser. med. Res. Coun., Lond.* No. 266.

Wilson, C. D. and Salt, G. F. H. (1978) In: *Streptococci*, p. 143. Ed. by F. A. Skinner and L. B. Quesnel. Academic Press, London.

Wilson, C. D. and Slavin, G. (1950) *J. comp. Path.* **60**, 230.

Wilson, T. S. and Stuart, R. D. (1965) *Canad. med. Ass. J.* **93**, 8.

Wilson, W. R., Nichols, D. R., Thompson, R. L., Giuliani, E. R. and Geraci, J. E. (1980) *Amer. Heart J.* **100**, 689.

Wilson, W. R., Thompson, R. L., Wilkowskie, C. J., Washington, J. A. II, Giuliani, E. R. and Geraci, J. E. (1981) *J. Amer. med. Ass.* **245**, 360.

Young, L. S., Stevens, P. and Ingram, J. (1975) *J. clin. Invest.* **56**, 850.

Ziegler, E. J. *et al.* (1982) *New Engl. J. Med.* **307**, 1225.

58

Hospital-acquired infections

M.T. Parker

Introductory

Most other chapters in this volume are concerned with bacterial diseases caused by groups of organisms that are taxonomically related or cause similar clinical effects. In this chapter we consider all the microbial diseases that occur in man under a single circumstance—after entering hospital. Many of the diseases to which we refer are dealt with in greater detail in other chapters in this volume, or in Vol. 4. Here we shall describe particularly the pattern of hospital-acquired infection, the special circumstances in hospitals that favour the occurrence of infection, the source of the infecting organisms and their usual routes of spread, and the general principles upon which preventive measures should be based.

Definitions

The term 'hospital-acquired infection' is generally taken to include only infections that lead to clinical illness; this is the sense in which we shall use it. The symptomless acquisition of a potential pathogen at a carrier site or in a wound ('colonization') is thus excluded. [It should be noted that we use the word 'sepsis' for an acute inflammatory disease—usually with pus formation—whether or not septicaemia (Chapter 57) is present.] The term 'nosocomial infection', which is widely employed in the USA, is synonymous with hospital-acquired infection. 'Iatrogenic infection' is a rather narrower term referring only to those hospital-acquired infections that are directly attributable to medical or surgical treatment.

Earlier definitions usually required that the infection should be a direct consequence of presence in a hospital: either that the infecting organism had been acquired in the hospital or that the illness could be attributed to a procedure to which the patient had been subjected there. Definitions now generally used include all microbial diseases that develop in hospital patients after the expiry of the supposed incubation period of the disease. The result of this is to include many infections that might well have occurred had the patient not been admitted to hospital, for example, infections in terminally ill patients with organisms from their own body flora, and infections acquired at birth from the mother. This tends to inflate estimates of the frequency of hospital-acquired infection but has the advantage of defining a group of infections that are, at least theoretically, preventible by action taken in the hospital. Such a broad definition cannot, however, be reasonably applied to out-patients or hospital staff, in whom only infections caused by hospital-acquired organisms or associated with specific procedures or accidents are recognizable.

History

Interest in infection acquired in hospital began long before micro-organisms had been implicated as their cause. Clinical observations led some workers to conclude that certain pyogenic diseases were caused by a communicable agent that had been introduced into a susceptible site in the patient. Striking successes in reducing the frequency of puerperal infection and post-operative wound sepsis by measures designed to prevent the spread of the supposed agent provided powerful evidence for its existence. Implicit in the nature of the preventive measures introduced were hypotheses about the source and routes of spread of the infective agent. Thus, from the beginning, attempts to control infection and studies of their epidemiology went hand in hand, to their mutual advantage (see p. 217). (For nineteenth-century references, see Cruickshank 1944, Walter 1948, Williams 1956, Hurley and de Louvois 1980.)

1. The first stage in this history may be said to have begun with the investigations of Semmelweis (1861) into the prevention of puerperal fever. He was the first to show that this was an infectious disease and the first to show how it could be prevented. Though his conclusions were not accepted at the time, they were amply confirmed by microbiological studies twenty years or so later (see Chapter 59). The introduction of 'antiseptic' surgical methods by Joseph Lister (1867), influenced as he was by Pasteur's demonstration that micro-organisms were responsible for fermentation, was empirical

in its application to wound sepsis. Their use, and that of von Bergmann's more convenient 'aseptic' techniques, resulted in such a dramatic reduction in sepsis rates that the scientific study of wound infection waned for over a quarter of a century.

2. In the last years of the nineteenth century interest was transferred to children's hospitals, in which mortality rates from hospital-acquired scarlet fever, diphtheria and measles were very high. The investigators were clinicians (Hutinel 1894, Grancher 1900) whose main concern was to prevent the spread of infection in wards. Their purely epidemiological studies provided a basis for methods of isolation used in infectious-disease hospitals in the ensuing decades (p. 218).

3. Between 1880 and the end of the century, haemolytic streptococci, *Staphylococcus aureus* and *Pseudomonas aeruginosa* were shown to cause septic diseases, but their role in hospital acquired infection was seldom recognized except in epidemics among babies (p. 195). Serious interest in the bacteriology of wound sepsis did not begin until the first world war (1914–18) when the failure of aseptic surgical methods to prevent infection in gunshot wounds became obvious. It was soon shown that tetanus and gas gangrene usually resulted from initial wound contamination. The same conclusion was at first too easily accepted in respect of sepsis in large open wounds. Late in the war, however (see Fleming and Tytler 1923), it became clear that many of these were caused by added infection acquired in the wards, and the progressive colonization of wounds by haemolytic streptococci was demonstrated.

4. In the years between the two world wars there was little interest in wound infection in civilian practice, though

Cruickshank (1935) showed that the spread of haemolytic streptococci in burns units was similar to that observed earlier in gunshot wounds. There were important studies of the spread of haemolytic streptococci in scarlet-fever wards (Allison and Brown 1937) and ear, nose and throat departments (Okell and Elliott 1936).

5. During the second world war (1939–45) studies of infection in war wounds were resumed with vigour. The importance of added infection in wards was now fully accepted, and both haemolytic streptococci and *Staph. aureus* were established as important causes of this (see, for example, Miles *et al.* 1940, McKissock *et al.* 1941, Spooner 1941). Bacteriological studies revealed that a considerable proportion of normal persons were carriers of *Staph. aureus*, that colonized but uninflamed wounds were a rich source of organisms, and that gram-positive pathogens tended to accumulate on surfaces and in dust in rooms occupied by patients.

6. Antibiotics active against pyogenic organisms began to be widely used towards the end of the second world war and raised hopes that wound sepsis would no longer cause serious difficulties. The rapid spread of resistant strains of *Staph. aureus* in hospitals from 1946 onwards led to a considerable expansion of interest in this organism, and to investigations of the spread of resistant strains among patients and staff in hospitals. Some years later (1954–55) clinicians in many countries observed a striking increase in the severity of staphylococcal infections. Although, as we shall see (Chapter 60), this was not directly attributable to the antibiotic resistance of the predominant strain, the attention of hospital doctors was for the next few years focused almost entirely on infections with resistant staphylococci. This situation changed quickly after 1960, partly because effective antibiotics for the treatment of severe infections with the prevalent resistant staphylococci had become available, and partly because other more pressing problems had arisen.

7. Soon after the introduction of antibiotics, another change in the pattern of hospital-acquired infection began, but its significance was not recognized until some years later. Various gram-negative aerobic bacilli appeared as increasingly common causes of serious septic infection (Chapter 57). Before 1945, *Escherichia coli* had been a moderately important but little studied cause of sepsis. By the early 1950s it had been joined by *Proteus* and *Klebsiella* strains and by *Ps. aeruginosa*, which were more often resistant to antibiotics available at the time. Transfer of interest from staphylococci to gram-negative aerobes began early in the 1960s; soon afterwards the prevalent gram-negative strains were seen to be acquiring additional antibiotic resistances, and a number of 'new' gram-negative bacilli were assuming importance in hospitals (see Chapter 61). Bacteraemic infections with gram-negative aerobes became increasingly common, and tended to occur in patients with serious underlying disease (Freid and Vosti 1968). After 1970 there was increasing evidence that these organisms often colonized patients who were receiving antibiotics and so became endemic in hospital wards.

8. Recent advances in medical and surgical techniques have improved the outlook for many patients suffering from severe or potentially fatal diseases. However, the procedures by which the illnesses can be cured or alleviated frequently carry a very high risk of infection, which often threatens life or renders the treatment ineffective. Gross impairment of the body's defence mechanisms may be an inevitable consequence of or even an essential element in the treatment of the patient's underlying disease. The prevention of infection in immunodeficient patients has thus now become a matter of considerable importance.

Epidemiology

Many of the infectious diseases encountered in other types of residential institution—communal nurseries, boarding schools, military camps and so on—may on occasion be introduced into hospitals. Outbreaks of food- or water-borne disease and other forms of enteritis, of virus infections of the respiratory tract, and of the infectious fevers of childhood in the hospital population tend to resemble those among persons of similar age in other institutions, though their consequences for some of the hospital patients may be more serious. Most hospital-acquired infections are, however, of quite a different character; they are consequences of the physical condition of the patient, the treatment to which he or she is subjected in hospital, or a combination of these. In most cases they are caused by micro-organisms that seldom give rise to disease of a similar nature or severity in healthy persons outside hospital. These organisms may be derived from one of three sources: (1) from another person; (2) from the patient him- or herself; or (3) from a non-human source.

Infection from another person is usually described as *cross-infection* by British authors; it must be taken to include instances in which the acquired organism first colonizes the patient and disease develops some time later, for example, after a surgical operation. A considerable proportion of hospital-acquired infections are *auto-* or *self-infections*, caused by organisms present at a carrier site or lying dormant in the tissues when the patient entered hospital and subsequently activated by underlying disease or treatment. Under *infections from non-human sources* we include those caused by organisms not recently derived from a person; but we would exclude those caused by organisms of human origin that have survived for a limited time, but not multiplied, in a dry situation in the hospital. The most important group of infections by organisms not of human origin, so defined, is that caused by gram-negative aerobic bacilli that live an 'independent' existence in moist situations in the hospital, where under suitable circumstances they multiply extensively. Of lesser importance are those caused by highly resistant organisms—generally spore-bearers—from soil or animal faeces, and occasional animal pathogens, that may be introduced into the hospital from outside. (For more information about sources and routes of infection see p. 200.)

Epidemic, endemic and sporadic infection

Infections may appear in hospitals in groups—often described as *epidemic* when cases are closely spaced in time and *endemic* when they occur over an extended period—or they may occur singly (*sporadic* cases). The distinction between these categories may be difficult to make without careful epidemiological investigation.

The earliest bacteriological studies of institutional epidemics were made on groups of babies, in which short-term fluctuations in the incidence of septic diseases were often observed. Thus, epidemics of pemphigus caused by *Staph. aureus* (Almquist 1891) and of fatal bacteraemia due to *Ps. aeruginosa* (Wassermann 1901) were recognized at quite an early date. The first well investigated epidemic of surgical wound infection due to *Staph. aureus* was reported in 1939 by Devenish and Miles, but the recognition of outbreaks of post-operative sepsis continued for a long time to be highly subjective. As late as the 1950s some workers were reporting sepsis rates for supposedly non-epidemic periods that considerably exceeded those given by others for epidemics (see Williams *et al.* 1966).

After 1935, typing methods became available for one after another of the organisms that are common causes of septic infection, making it possible to recognize associated groups of infections caused by a single bacterial strain. It then became apparent that in many of these incidents the organism was widely disseminated among symptomless contacts of the clinical cases. Antibiotic-resistant strains of *Staph. aureus*, for example (Chapter 60), were often transmitted by symptomless carriers over considerable periods of time, causing clinical infections in numbers that varied widely from strain to strain. The source of the individual infection was thus often difficult to identify. Typing also showed that infections from a single human source might be few in number and widely spaced in time. Thus, a distinction between epidemic and endemic incidents based only on the time-course of an outbreak might be misleading.

If the term 'epidemic' is used it would be best to restrict it to outbreaks arising from a single source—a person or a contaminated object or substance—over whatever period of time. The term 'endemic incident' would then be used to describe the more common situations in which a single strain of an organism may be acquired from one of a number of different human sources.

Stamm and his colleagues (1981), summarizing their conclusions from a study of 252 hospital 'epidemics' in the USA, observed a considerable change in the relative importance of various pathogens during the previous 25 years. In 1956–60, 64 per cent of epidemics were caused by *Staph. aureus* and

only 10 per cent were of septic infections caused by gram-negative aerobic bacilli; in 1976–80 the corresponding percentages were respectively 13 and 30. Epidemics of diarrhoea associated with *Salmonella* and *Esch. coli* formed 23 per cent of the whole in the earlier period and only 13 per cent in the later. No hepatitis B epidemics were recorded in 1956–60 but they formed 8 per cent of those recorded in 1976–80. The same authors reported that in the years 1971–80 *Esch. coli*, *Ps. aeruginosa*, klebsiellae, *Proteus* and enterococci were more often responsible for endemic than for epidemic incidents; the reverse was true of *Salmonella*, *Serratia* and *Enterobacter*.

The term 'sporadic' can be applied with certainty only to self-infections in which the causative organism was detected in the flora of the patient at the time of admission to hospital. It can be applied, though with rather less confidence, to single infections caused by well characterized strains in wards or departments in which all infections are monitored bacteriologically. Its use is often extended to infections with organisms that are commonly present at carrier sites in members of the general population and are considered to be non-communicable in hospitals, such as the gram-negative non-sporing anaerobes. This is a convenient general rule, but may have to be modified if subsequent investigation changes our views about the communicability of some of these organisms, for example, highly antibiotic-resistant strains of *Staph. epidermidis* (Chapter 60).

'Pseudo-epidemics' An increase in the frequency of clinical infection does not always indicate the appearance or spread of a single infecting organism. Thus, in some 'epidemics' of minor skin sepsis in newborn babies a variety of different strains of *Staph. aureus* may be present in the lesions; the increased sepsis rate may be attributable to an undetected deterioration in the standards of routine skin care, or to the discontinuation of an effective preventive measure, e.g. the use of hexachlorophane (see Chapter 60). We are aware of one supposed epidemic of gas gangrene attributable to a change in the technique of leg amputation that led to increased tension in the sutured wound.

A significant proportion of incidents reported as epidemics prove on careful investigation to be spurious. According to Weinstein and Stamm (1977), 11 per cent of 181 'epidemics' recorded in the USA were of this sort; the false conclusion could be attributed either to incorrect clinical or epidemiological observation or to errors in the collection or examination of bacteriological specimens. Of 20 such incidents, three were the result of clinical misdiagnosis, six to an erroneous conclusion that infections had been acquired in hospital or that the infection rate had increased, eight to the contamination of specimens during collection or processing in the laboratory, two to misidentification of organisms, and one to an undetected laboratory cause.

'Pseudobacteraemia' Epidemics of this arise when blood cultures from a number of patients are contaminated with a single organism from an extraneous source. Numerous instances have been recorded; they have been attributed to imperfect sterilization of blood-culture medium with the result that spore-bearing organisms survive, or to the use of solutions heavily contaminated with gram-negative aerobic bacilli at some stage in the blood-collection procedure. The contaminated solution may be applied to the skin of the patient or the hands of the operator before venepuncture, or may be in a separate container, e.g. of anticoagulant, into which a portion of blood is discharged before inoculating the blood-culture medium through the same needle. One unfortunate consequence of 'pseudobacteraemia' is to lead to unnecessary antibiotic treatment.

The causative micro-organisms

Very many species of micro-organisms—bacteria, viruses, fungi and protozoa—have been recognized as causes of infection in hospital patients. Their number has increased greatly in recent years (von Graevenitz and Sall 1975, Fraser 1981, Mitchell 1983), and the list given in Table 58.1 is far from exhaustive. However, quite a small number of species of bacteria—*Staph. aureus*, a few species of gram-negative aerobic bacilli and the non-sporing gram-negative anaerobes—together account for the majority of infections.

Certain organisms, to which we shall refer as '*conventional*' pathogens, P in Table 58.1, cause similar diseases in hospital patients and in healthy members of the general population (p. 194). However, most of the organisms that are important causes of hospital-acquired infection very seldom give rise to clinical disease in healthy persons. Many of them invade the tissues in the presence of local predisposing factors that permit the organism to by-pass the natural defences of the body surface. We shall refer to these organisms as '*conditional*' pathogens—C in Table 58.1; examples include many of the gram-negative non-sporing bacilli that cause septic infections, the anaerobes in general, some of the streptococci, and *Staph. epidermidis*. Conditional pathogens in the normal flora of the patient are responsible for many of the sporadic auto-infections. In hospitals, antibiotic-resistant strains of some of the conditional pathogens spread widely among the patients and give rise to epidemics and endemic prevalences of a type rarely seen in other institutions. Certain of the conventional pathogens—designated P(C)—cause severe infections much more often in hospital patients than in others; *Staph. aureus* and some of the pyogenic streptococci are examples of this.

We reserve the term '*opportunist*' pathogen for

Table 58.1   Pathogenicity of micro-organisms for hospital patients

Organism	Pathogenicity*		Organism	Pathogenicity*	
Bacteria			**Viruses**		
Staph. aureus	P	(C)	Hepatitis B	P	
Staph. epidermidis		C	Respiratory viruses	P	
Streptococci: group A	P		Enteric viruses, rotaviruses	P	
:groups B, C and G	P	(C)	Lassa and haemorrhagic fevers	P	
:*Str. milleri*, enterococci,			Rubella	P	
other non-haemolytic		(C)	Measles	P	(O)
:pneumococci	P		Varicella	P	(O)
Anaerobic cocci		C	Vaccinia	P	(O)
			Herpes simplex and EB virus	P	(O)
Histotoxic clostridia		C	Cytomegalovirus	P	(O)
Cl. tetani		C			
Bacteroidaceae		C	**Fungi**		
			Candida	P	(O)
Salmonella	P		*Cryptococcus*	P	(O)
Shigella	P		*Histoplasma*	P	(O)
Enteropathogenic *Esch. coli*	P		*Coccidioides*	P	(O)
Other *Esch. coli*		C	Moulds	?	O
Proteus, Providencia		C			
Klebsiella, Enterobacter, Serratia		C	**Protozoa**		
Ps. aeruginosa, other pseudomonads		C	*Pneumocystis*	?	O
Flavobacterium, Acinetobacter		C	*Toxoplasma*	P	(O)
Haem. influenzae	P		*Acanthamoeba*	P	(O)
Legionella	P				
Coryne. diphtheriae	P				
Other coryneform bacteria,		C			
Propionibacterium					
Listeria	P	(O)			
Myco. tuberculosis	P				
Other mycobacteria		C			
Nocardia	?	O			

*P = 'conventional' pathogen; causes disease in healthy persons.
C = 'conditional' pathogen; causes significant disease only in predisposed persons.
O = 'opportunist' pathogen; causes systemic disease only in immunodeficient persons.
(C) = 'conventional' pathogen, but infections more frequent and more often severe in predisposed persons.
(O) = 'opportunist' pathogen, but occasional systemic infections in apparently normal persons; some are common causes of mild or localized diseases in the general population.

organisms that very rarely cause systemic disease in the apparently healthy but often do so in patients with a severe depression of one or more of the body's general defence mechanisms (below). A few of the opportunists—designated O—appear to cause disease only in immunodeficient persons. Most of them, designated (O), occasionally give rise to systemic disease in non-hospital patients in whom immunodeficiency has not been demonstrated, though its presence is often difficult to exclude; and many of them are common causes of mild or localized infections in the general population, for example, varicella, herpes and several other viruses. However, many of the systemic infections in immunodeficient patients are caused by organisms that are not opportunist pathogens as we have defined these, but are conditional pathogens which, in such patients, are able to invade the tissues without prior damage to skin or mucous membranes, or even conventional pathogens such as pneumococci, *Haemophilus influenzae*, legionellae, salmonellae and *Mycobacterium tuberculosis*.

Determinants for hospital-acquired infection

Whether a patient becomes infected in hospital, and whether clinical consequences ensue, are determined by the interplay of several independent factors. In addition to the presence of potentially pathogenic organisms, these include the susceptibility of the patient to them, the opportunities for them to reach a body site from which they can invade the tissues and, if the patient is receiving an antimicrobial agent, whether the organisms to which he or she is exposed are resistant to it.

Increased susceptibility

A considerable proportion of patients are already prone to infection when they enter hospital, and others are rendered more susceptible to it by procedures or medication that they undergo after admission.

Local Pre-existing wounds or other surface lesions may render patients susceptible to invasion by micro-organisms. Surgical operations and a variety of other traumatic procedures used for diagnosis or treatment provide further opportunities for organisms to enter the tissues. Surface defences may also be by-passed by the accidental injection of organisms into the tissues or even the blood stream, and also by common mishaps that result in their being introduced—often in considerable numbers—into normally sterile 'external' spaces, such as the bladder, the conjunctiva or the lower respiratory tract, in which they easily establish themselves.

General The components of the body's general defences against micro-organisms (Chapter 17) include the activities of phagocytes, humoral immunity and cell-mediated immunity. One or more of these

defence mechanisms are impaired in a variety of diseases, as also by treatment with certain drugs and by irradiation. The complex states of immunodeficiency seen in some patients are thus a product of abnormalities due to underlying disease and those induced by treatment. Systemic disease in immunodeficient patients may be a consequence of invasion by organisms from the outside of the body, often in the absence of a surface lesion, or may arise from the activation of a latent infection. The onset is often insidious, with few physical signs in the earlier stages. In many cases several different micro-organisms can be isolated from the blood or internal organs.

Depression of granulocyte production, or the disordered function of granulocytes, is one of the more common causes of increased susceptibility to infection in hospital patients. It occurs in some forms of leukaemia and in other bone-marrow diseases, and is often reinforced by drug treatment for these conditions. According to Bodey and his colleagues (1966), serious infections begin to appear in leukaemic patients under drug treatment when the granulocyte count falls to 100 per μl and become progressively more common when it decreases further. The increased susceptibility to bacterial infection associated with diabetic acidosis is said to be caused by disordered granulocyte function. Infections in leucopaenic patients are generally bacterial, and are caused mainly by *Staph. aureus*, *Esch. coli* and other gram-negative aerobes, and less often by *Candida* and various moulds. Two organisms from the normal body flora appear to cause disease only in granulocytopaenic patients: the so-called '*JK organisms*' are coryneform skin bacilli that are resistant to nearly all antibiotics except vancomycin and give rise to cellulitis, abscesses and sometimes septicaemia (Pearson *et al.* 1977, Gill *et al.* 1981, Young *et al.* 1981); and the CO_2-requiring gram-negative bacillus, *Capnocytophaga ochracea*, often found in the periodontal space, may cause ulcerative stomatitis and septicaemia (Forlenza *et al.* 1980). Defects in humoral immunity are seen in multiple myeloma, chronic lymphatic leukaemia and sickle-cell disease, and after splenectomy. They are particularly associated with septicaemic infections caused by pneumococci and *Haemophilus influenzae*. Depression of cell-mediated immunity occurs at some stages in Hodgkin's disease, in other malignant conditions of the lymphatic system, and in acute leukaemia in children; it is also induced artificially in the recipients of transplanted organs. It is characteristically associated with generalized infections with *Listeria monocytogenes*, salmonellae and tubercle bacilli, and with varicella virus, cytomegalovirus, *Cryptococcus neoformans* and *Toxoplasma gondii*; it also predisposes to pneumonia caused by *Pneumocystis carinii* and legionallae. There is some evidence from in-vitro tests for a deficiency of cell-mediated immunity in untreated malignant disease of visceral organs. However, the associated infections usually appear to arise as a consequence of obstruction or ulceration and to be caused by members of the normal flora of the region; in the thorax and abdomen these usually include gram-negative anaerobic bacilli.

Forms of treatment that cause serious immunodeficiency include the use of cytotoxic drugs and antimetabolites, of immunosuppressive agents, including corticosteroids, and heavy irradiation. There is also in-vitro evidence to suggest that major surgical operations, severe burns, and even anaesthesia, may cause temporary immunosuppression (Bruce and Wingard 1971, Park *et al.* 1971).

(For general accounts of infection in immunosuppressed patients, see Young 1981 and Cohen 1983, and various chapters in books edited by Allen 1976, Burke and Hildick-Smith 1978, Bodey and Rodriguez 1979, and Rubin and Young 1981.)

Susceptibility in newborn babies   In-vitro evidence suggests that newborn babies are somewhat immuno-deficient, but this is usually not of great significance in the otherwise healthy full-term infant; in the prematurely born, however, the deficiency may be profound and persists until around the date of normal birth. It manifests itself mainly in increased susceptibility to infection by various gram-negative aerobic bacilli, especially *Esch. coli* and *Ps. aeruginosa*, and to *Staph. aureus* and group B streptococci. Several abnormalities in the immunity mechanisms of newborn and particularly of premature infants have been described (see, for example, Medici and Gatti 1978); these include: (1) a poor inflammatory response, attributed to deficiency of Hageman's serum factor XII; (2) low serum levels of various components of complement, leading among other things to reduced opsonization; (3) defective function of granulocytes, notably decreased powers of migration, chemotaxis and intracellular killing; and (4) absence of IgM, which does not cross the placenta, and is said to be important in preventing invasion by *Esch. coli* and similar organisms.

Sources of infection specific to hospitals

Patients are likely to be exposed to potentially pathogenic micro-organisms from sources that are of relatively minor importance in their homes or even in other types of institution.

Human   Hospitals admit infectious patients for treatment, and also generate further sources of infection when other patients acquire a communicable pathogen, whether or not they become ill in consequence of this. Infectious-disease hospitals are designed to segregate patients with certain communicable diseases. Many patients who would formerly have been sent to such hospitals are now nursed at home, but there is a growing tendency to treat some of them, albeit in smaller numbers and for shorter periods, in general hospitals, for example, children with diarrhoea in paediatric wards, in which facilities for physical segregation and 'isolation' nursing are often less readily available. Infectious-disease hospitals have seldom been used consistently to house patients liable to spread pyogenic infections, and provision for their segregation in general hospitals is seldom adequate (p. 218). In recent years opportunities for the spread of infection in hospitals have been increased by the progressive replacement of multi-purpose wards by special units designed for the treatment of patients

with similar complaints, for example, burns wards, urological units, and intensive-care units, in which similar procedures are often applied serially to a number of patients by the same staff team, providing numerous opportunities for contact spread of infection. Members of the staff, though important as vectors in inter-patient spread, are less often of significance as sources of infection unless they are suffering from clinical infection; heavy skin carriers of *Staph. aureus* in operating theatres perhaps provide an exception to this rule.

Non-human   The current importance of moist objects and fluids as sources of certain gram-negative aerobic bacilli is mainly attributable to recent changes in hospital practice, notably (1) the increasing use of 'invasive' techniques for diagnosis and treatment, (2) the introduction of complicated medical devices that are difficult or even impossible to clean and decontaminate, and (3) the central production of fluid medicaments and weak solutions of 'disinfectants' and their storage under conditions that permit subsequent bacterial multiplication (see Chapter 61).

Widespread use of antimicrobial agents

When in hospital the patient is likely to encounter highly antibiotic-resistant strains of a number of bacterial pathogens, including staphylococci (Chapter 60), various gram-negative aerobic bacilli—coliform bacteria, pseudomonads and other non-fermentative organisms (Chapter 61)—and sometimes salmonellae (Chapter 38 and p. 213).

In this chapter we are not directly concerned with the treatment of infections with resistant organisms, but we should note that, though resistance restricts the choice of agents available for treatment, the continuing development of new agents and chemical modification of existing ones have so far kept pace with the appearance of fresh resistances. There have yet been few occasions on which no effective agent was available for the treatment of a large group of hospital-acquired bacterial infections. However, the new agents tend to be very expensive and can be effectively deployed only when good laboratory support is available. This situation can be met, at a price, in rich countries, but is leading to serious difficulties in the 'third' world (Report 1982).

Many hospital patients respond poorly to treatment with an apparently appropriate antibiotic. The view that resistant organisms are inherently more virulent than the corresponding sensitive strain has been repeatedly advanced, especially in respect of *Staph. aureus* (Chapter 60), but is not generally substantiated. The fact that resistant organisms tend to be isolated more often from fatal than from non-fatal infections in hospital can be explained in another way: the patients who succumb usually belong to infection-prone groups, are often in wards in which antibiotic usage is high, and themselves tend to have received much antibiotic treatment.

About one in four of all patients who enter a hospital receive courses of antibiotics during their stay (for references, see Report 1982), but in some wards or departments this proportion is much greater. Profound changes take place in the flora of antibiotic-treated patients; sensitive organisms are often eliminated from the bowel, skin and oropharynx and are replaced by resistant organisms that are usually acquired by cross-infection from other persons in the hospital. Strains of a number of potentially pathogenic organisms, particularly of *Staph. aureus* and various gram-negative aerobic bacilli, that are resistant to widely used antimicrobial agents tend to accumulate in hospitals and may be selected by the administration of any one of the agents to which they are resistant. Thus, heavy and varied exposure of the flora of patients to antimicrobial agents, together with frequent person-to-person transfer of resistant organisms, leads to the endemic persistence of multiple-resistant strains. The plasmid resistance determinants (R factors) in many of the gram-negative organisms, which often code for resistance to several unrelated antimicrobial agents, are readily transferable between bacterial strains, including members of a number of different genera. Thus, a resistance or combination of resistances may appear in a hospital in one organism and later be transferred to quite a different one, which may then establish itself endemically in the population (Lowbury *et al.* 1969, Schaberg *et al.* 1981).

There are some situations, however, in which the appearance of resistant organisms after antimicrobial treatment is not attributable to cross-infection. For example, when penicillin is given, α-haemolytic streptococci resistant only to this antibiotic often become predominant in the throat, but these organisms can be detected there in small numbers before the beginning of treatment (Phillips *et al.* 1976). The appearance of other single resistances may result from the selection of spontaneous mutants, e.g. when *Staph. aureus* infections are treated with novobiocin or fusidic acid. The appearance of resistance in a previously sensitive pathogen, for example, of chloramphenicol resistance in the course of treating a case of typhoid fever with this antibiotic (Datta *et al.* 1981), may result from the transfer to it of an R factor or its transposable resistance determinant from another organism in the patient's flora.

Superinfection   Heavy overgrowth of gram-negative bacteria at surface sites is a common occurrence in hospital patients, and is an important prelude to the development of invasive infection in predisposed persons. In premature babies, and in immunodeficient patients in general, organisms appear to enter the tissues through intact skin or mucous membrane; invasion of the blood stream, usually with symptoms of endotoxic shock (Chapter 57), may lead rapidly to death. Massive colonization of the lower respiratory tract by gram-negative bacilli, and to a lesser extent by *Staph. aureus*, is a common finding in severely ill or moribund patients (Report 1966), most of whom have received antibiotics. In many of these patients a diagnosis of 'pneumonia' is made, but it is very difficult to assess the part played by these organisms even in the patients who show evidence of pulmonary infection at necropsy (but see 'gram-negative bacillary necrotizing pneumonia' on p. 212).

When hospital patients are given antibiotics, gram-negative bacilli with a very wide pattern of resistance frequently become predominant in the aerobic flora of the mouth and pharynx and in the sputum (Louria and Kaminski 1962, Pollack *et al.* 1972) and on the skin (Lowbury 1969). In the faeces, less resistant *Esch. coli* strains tend to be replaced by more resistant ones and by other enterobacteria, notably klebsiellae (Rose and Schreier 1968; Chapter 61). The endemic presence of these organisms is undoubtedly attributable to frequent antibiotic use; and the colonization of the treated patient is facilitated by the elimination of sensitive components of the bacterial flora, but some workers doubt whether this is the only determinant, at least for the overgrowth of gram-negative bacilli in the respiratory tract (Johanson *et al.* 1969, Valenti *et al.* 1978). This appears to occur with particular frequency in bedridden and terminally ill patients, in whom physiological factors, such as inability to clear fluid from the respiratory passages, may be of significance.

Resistant organisms that are not playing a pathogenic role may nevertheless have deleterious effects by forming antibiotic-destroying enzymes that prevent the successful treatment of infections by other organisms present at the same site—an action described by Maddocks and May (1969) as 'indirect pathogenicity'. Thus, the formation of β-lactamase by enterobacteria or pseudomonads may limit the efficacy of ampicillin in eliminating sensitive strains of *Haem. influenzae* from chronic bronchial infections and of benzylpenicillin in controlling epidemics of group-A streptococcal infections of burns.

Diarrhoeal diseases attributable to antibiotic administration are discussed on p. 214.

Infection rates

The total infection rate for a hospital is a figure of little significance; it tends to reflect the sort of patients who are admitted and the activities of the hospital rather than its success in preventing infection. Infection rates for particular types of patient, and information about the sorts of infection they acquire, are, however, of great value in drawing attention to unsolved problems. Comparisons of infection rates in groups of closely matched patients, though they present practical difficulties, are an essential means of assessing the value of individual preventive measures (p. 217).

When we measure an infection rate we count the number of persons in a given group who become ill, not the number who are colonized by a potential pathogen. Infection rates are thus measured at the bedside, or as second best from the patients' clinical records, and not in the laboratory. Strictly defined clinical criteria for the presence of infection are essential (see, for example, Report 1980a, 1981b). Laboratory reports should be used only to determine the probable cause of clinically recognized infections. In studying the infection rate in a particular class of patient, we may decide that the main risk is associated with a certain procedure or event; it is then logical to calculate the rate for this, for example, post-operative, post-partum or neonatal infection rates. If the risk is deemed to be continuous, the rate can be calculated per admission or discharge, or better as per week of stay in hospital.

The main difficulty is to obtain groups of strictly comparable patients for study. For example, patients undergoing surgical operations are a heterogeneous group (p. 207). Attempts to secure groups of patients uniform in respect of age, underlying disease and other factors known to affect the risk of infection on whom to study the effects of precautionary measures may be successful only at the expense of greatly reducing the size of the sample and widening the confidence limits of the figures obtained. This consideration assumes particular importance when the initial infection rate is not very great and the expected reduction of it by the measure under investigation is only moderate (Lidwell 1963). Thus, a statistically significant reduction (P < 0.05) can be detected by comparing two groups each of 60 subjects if a preventive measure halves a sepsis rate of 30 per cent, but groups each of 3600 subjects will be needed if it causes only a 25 per cent reduction in a sepsis rate of 3 per cent. If matching of test and control groups for all predisposing factors is impracticable it may be possible to correct for these by means of multiple-regression analysis (Lidwell 1961, Davidson *et al.* 1971b).

Several attempts have been made to estimate the *incidence* and define the main types of infection in large samples of patients in hospitals in Europe and North America. Despite considerable differences in the methods of enquiry, the results obtained were in general conformity. Among nearly 7000 patients in medical wards in 13 British hospitals, 5.1 per cent suffered one or more episodes of infection during their stay; of these, nearly 20 per cent were septic lesions of skin or wounds, some 40 per cent were infections of the lower respiratory tract, and 24 per cent affected the urinary tract (Report 1965a). In an extensive sample survey of medical and surgical patients in 338 hospitals in the USA (Haley *et al.* 1981), infections developed in 5.2 per cent of all patients (6.6 infections per 100 patients); the expected associations with age, severe underlying disease, immunosuppression and various forms of instrumentation were observed. Daschner (1981a, b), in a survey of nearly 40 000 admissions to a university health centre in Germany, recorded a total infection rate of 4.4 per cent; he emphasized, as have other observers, the great differences between hospital departments, with rates of 1 per cent in the ear, nose and throat department, 6.4–7.4 per cent in surgical wards, and 12.4 per cent in adult and 16.3 per cent in infant intensive-care wards. For all the patients, 26 per cent of infections were of skin and wounds, 15 per cent were pneumonic, 21 per cent septicaemic and 27 per cent were of the urinary tract. In a multi-hospital study extending over 10 years, little variation in the yearly infection rate was observed (Allen *et al.* 1981). (For other studies, see Wenzel *et al.* 1976a, b, and for the incidence of postoperative wound sepsis, p. 207.)

For large-scale studies of the incidence of infection,

frequent clinical examination of the patients, or at least the careful scrutiny of case records, is necessary. This can be very time-consuming and results become available slowly. Many workers therefore favour studies of *prevalence*, in which patients are seen only once and infections present on the day of inspection are recorded ('cross-sectional surveys'; Ayliffe *et al.* 1977). Very large numbers can be accumulated with relatively little effort and the results analysed quickly. The total sepsis rates recorded in this way are considerably higher than in incidence studies, but the pattern of infection revealed, and its relation to well recognized predisposing factors, are similar. Prevalence rates of 9–12 per cent have been recorded in multi-hospital studies in Sweden (Bernander *et al.* 1978), Denmark (Jepsen and Mortensen 1980) and Britain (Report 1981b). In the British study of some 18 000 patients in 77 hospitals in 1980, the total prevalence rate was 9.2 infections per 100 patients surveyed; 31 per cent of the infections were of the skin or wounds, 12 per cent of the lower respiratory tract, and 45 per cent of the urinary tract.

Of the ill-effects of hospital-acquired infection, prolongation of stay in hospital is the most easily measured. This has been variously estimated as 7 days for sepsis in 'clean' operation wounds (Report 1960), 17 days for infections in orthopaedic wounds (Davies and Cottingham 1979) and 14 days for all types of infections (Freeman *et al.* 1979, Haley *et al.* 1980). It is difficult to assess the contribution of infection to the death of patients in hospital (Report 1966) because the course of the patient's underlying disease had infection not supervened can in many cases not be predicted.

Sources and routes of infection

The clinical and bacteriological study of individual outbreaks of infection provides a good deal of anecdotal information about the sources and routes of transmission of infection in hospitals. To this may be added inferences from the distribution of the causative organisms among patients and in their surroundings. The picture that often emerges is of more than one possible source and several feasible routes of spread. Quantitative data about the relative importance of these are scanty and often subject to controversy; the little that we can rely upon is derived from controlled studies of the effect of eliminating sources and blocking routes of transmission on the incidence of clinical infection (p. 217).

Sources

Infection from a human source may be autogenous, or it may be from another person, though usually not directly; more often it is transferred indirectly from donor to recipient via the air or on an object, which may be looked upon as an *intermediate reservoir* of the infecting organism. Non-human sources of infection include contaminated fluids and moist objects in or on which certain gram-negative aerobic bacilli can live and multiply almost indefinitely; these may be looked upon as *independent reservoirs* of infection, i.e. independent of recent contamination from a human source. It should be noted, however, that some organisms with the ability to live independently at moist inanimate sites also frequently colonize antibiotic-treated hospital patients. Thus, for example with klebsiellae (Chapter 61), a complex series of events may occur in which, a strain from an independent reservoir first causes a common-source epidemic, then spreads to other patients by cross-infection, and from one of these to another fluid in which it multiplies.

In Table 58.2 we summarize the sources from which bacterial infections can be acquired, indicating as + the more common and as ∓ the occasional sources.

Entries in the first and third columns indicate respectively the ability of organisms of human origin to cause self-infection and cross-infection; those in the second column are for transmission from mother to baby at or around the time of birth. With self-infections, the unqualified + and ∓ signs are used for organisms that form part of the normal flora of patients when admitted to hospital and the [+] sign for organisms frequently acquired by patients at carrier sites whilst in hospital. The symbol (∓) indicates organisms that may cause self-infection by reactivation of quiescent disease, usually in immunodeficient patients; this type of self-infection also occurs with a number of viral and other non-bacterial 'opportunist' pathogens not shown in the table.

A number of organisms that, on the evidence of frequent self-infection, must be assumed to be pathogenic for hospital patients appear seldom to cause cross-infection. In the case

Table 58.2 Sources of bacteria that cause infections in hospitals

	Human sources			Inanimate sources	Other animals
	Self	Mother to baby	Other persons		
Gram-positive cocci					
Staph. aureus	[+]	∓	+	+(D)	
Staph. epidermidis	+	∓			
Streptococcus: group A	∓	∓	+	+(D)	
: group B	+	+	∓		
: *milleri*, enterococci, other non-haemolytic	+				
: pneumococci	+	∓	∓		
Anaerobic cocci	+				
Anaerobic bacilli					
Histotoxic clostridia	+			∓(D)	
Cl. tetani	?			+(D)	
Bacteroidaceae	+				∓
Gram-negative aerobic bacilli					
Salmonella	(∓)	∓	+	+(W;S)	∓
Shigella		∓	+	+(W;S)	
Enteropathogenic *Esch. coli*		∓	+	+(W;S)	
Other *Esch. coli*	+	∓	∓	+(W;S)	
Proteus, Providencia	[+]	+	+	+(W;S)	
Klebsiella; 'respiratory'	+			+(W;S)	
Other *Klebsiella; Serratia, Enterobacter*	[+]		+	+(W;M)	
Ps. aeruginosa	[+]	∓	+	+(W;M)	
Other pseudomonads	[+]		+	+(W;M)	
Flavobacterium, Acinetobacter	[+]		+	+(W;M)	
Legionella				+(W;M) (?D)	
Gram-positive aerobic bacilli					
Coryne. diphtheriae			+		
Other coryneform bacteria	+			+(D)	
Listeria	(∓)	+	∓		
Myco. tuberculosis	(∓)	∓	+	+(D)	
Other mycobacteria	(∓)				
Nocardia				+(D)	∓

+ = Frequest source; ∓ = occasional source.
Self-infection: [+] indicates that the infecting organism has often been acquired at a carriage site after admission to hospital; and (+) that auto-infection usually results from reactivation of latent disease.
Inanimate sources: (D) = persistence of the organism at 'dry' environmental sites; (W;S) = persistence only at 'wet' sites, usually without significant multiplication; (W;M) = multiplication (and usually very long persistence) at 'wet' sites.

of the Bacteroidaceae, this is usually attributed to their rapid death on exposure to air. With some more robust organisms, such as the clostridia of gas gangrene, *Str. milleri*, and the anaerobic cocci, infections are generally confined to damaged tissue in close proximity to a rich source of organisms; the reason for this is not clear, but perhaps this is the only circumstance in which sufficient organisms to cause infection reach the injured tissue. It is, however, surprising that pneumococci, widely distributed in the human population and as invasive as group A streptococci, so seldom cause surgical wound infection.

Occasionally pathogens of *animal origin* may enter the hospital in contaminated materials: spores of tetanus bacilli (Chapter 64), salmonellae in foods and certain oral medicaments prepared from animal products (p. 213), and mycobacteria (Tyras *et al.* 1978), actinobacilli (Lalonde and Hand 1980), and *Coxiella burneti* (Goffin *et al.* 1981) in porcine heart valves.

Pathogens in the inanimate environment of patients Bacteria (mainly gram positive) that have a considerable resistance to desiccation and are therefore often found at dry environmental sites are designated (D) in Table 58.2; most of them will also survive as well in wet situations unless overgrown by other organisms. The symbol W is used to designate bacteria—mainly gram negative—with lesser resistance to drying but considerable ability to survive under moist conditions. Among these are organisms—(W;S)—that persist well but do not multiply in fluids with a minimal content of organic nutrients, and others—(W;M)—that can multiply in these circumstances. As we shall see (Chapter 61), many of the organisms designated (W;M) also survive, and sometimes multiply, in weak solutions of certain disinfectants; they also grow well at temperatures considerably below that of the body. These differences are of some importance in determining the distribution of the various pathogens in the environment of patients.

Dry environmental sites The greater ability of gram-positive organisms than gram-negative aerobic bacilli to survive on dry surfaces is only relative and is confined to the initial period of exposure. In the first hour, *Staph. aureus* may perhaps halve in numbers while *Esch. coli* decreases by 90–99 per cent; in subsequent hours and days the numbers of both decrease more slowly but at about the same rate (Lidwell and Lowbury 1950, Lowbury and Fox 1953, Pettit and Lowbury 1968). Thus, under normal circumstances airborne bacteria of human origin are almost entirely gram positive, but when very large numbers of gram-negative bacilli are liberated into the air, for example by patients with infected burns, sufficient will survive to be detected by air sampling. Contamination of surfaces by contact may lead to the local deposition of millions of gram-negative bacilli per cm^2, and significant numbers of them may persist for days. The same may be true of extensive areas of floor after treatment

with a heavily contaminated mop or with wash water in which gram-negative organisms have proliferated (Ayliffe *et al.* 1967).

The bacterial flora of surfaces in hospital rooms is constantly being added to by organisms from various sources; these die out at different rates. At most times an equilibrium is reached, with gram-positive cocci and aerobic spore bearers predominating; washing or disinfecting of floors temporarily reduces their numbers, but these return to the original level within a few hours (Ayliffe *et al.* 1966, 1967), mainly by the transfer of organisms on footwear. The equilibrium may be disturbed when a heavy disperser of a particular organism occupies the room; counts of *Staph. aureus* in burns units may by 10–100 times greater than in general wards (Ransjo and Hambraeus 1982). If the organism being dispersed is gram positive it may be detectable in ward dust for weeks after the patient has left the room, but there is little evidence that persons who subsequently occupy the room become infected. There is a strong clinical impression that gram-positive cocci that cause septic infections have been fairly recently derived from an animate source, though clear laboratory evidence of loss of infectivity after long survival in dry situations is lacking. *Staph. aureus* from ward dust often exhibits poor and delayed initial growth and temporarily increased susceptibility to inhibitory chemicals in laboratory media. Group A streptococci in blanket dust rapidly lose infectivity for man by the respiratory route (Chapter 59); whether this is also true for wound infection is not known. The ability to detect a gram-positive pathogen on surfaces and in dust thus may not be of great significance. Most attention should be paid to recent contamination, and to sites from which organisms are easily transferred to a susceptible site in the patient.

The re-suspension of dust from surfaces by dry sweeping, shaking of bedding or the use of incorrectly designed vaccum cleaners (Rogers 1951) greatly increases the total number of bacteria in the air. Attempts to prevent this by good housekeeping methods form part of normal hospital practice, but special dust-suppressive measures in wards appear to have little effect on surgical sepsis rates (Clarke *et al.* 1954). Few organisms are dislodged from floors by air movement, but walking on them results in significant resuspension; however, the calculations of Hambraeus and her colleagues (1978) suggested that resuspended dust contributed at most 15 per cent of *Staph. aureus* in the air of an operating theatre. Bacteria on walls are generally fewer in number than on floors except at places likely to have been contaminated by hands (Ayliffe *et al.* 1967). The organisms are usually firmly adherent to the surface; walls are thus of significance as sources of contamination only when touched.

No evidence has been obtained of a relationship between total bacterial counts at dry environmental sites and the risk of infection. The consensus view is that a good standard of houshold cleanliness of general ward surfaces is all that is required, and that routine bacteriological monitoring of them is not an effective method of controlling cleaning practice. Surfaces liable to be touched by infection-prone patients, or to be a source of inadvertent contamination for the hands of hospital staff, may require special attention. (For techniques for sampling surfaces for bacteria see Favero *et al.* 1968.)

So far we have considered only non-sporing bacterial pathogens of human origin, for which dry surfaces provide only a temporary reservoir. *Clostridial spores* form a more permanent component of their flora. Although usually present only in quite small numbers, *Cl. perfringens* and *Cl. tetani* can be found in the dust in most hospital rooms, but they rarely appear to be responsible for infection. Most of the *Cl. perfringens* probably comes from the faeces or clothing of patients but the contribution from outside air is not easy to estimate. The extreme rarity of surgical tetanus in countries in which the sterility of instruments and other materials used at operation is strictly controlled suggests that infection from intra-hospital sources, including human faeces, is very unlikely. Infection from air entering the operating theatre from outside the building is still a possibility that merits consideration (see p. 207), particularly in countries in which dust heavily contaminated with animal faeces can gain ready access to theatre air.

Wet environmental sites   A great variety of bacterial pathogens—both gram positive and gram negative—can persist for several days at wet environmental sites contaminated from a human source; from there they may be transferred by contact to other patients. Most of them tend to die out slowly or to be overgrown by other less exacting organisms, generally gram-negative bacilli. Organisms that multiply at moist situations may have come from a non-human source or from a patient who had acquired it from an 'earlier' non-human source or from another patient. Conditions in hospitals that favour the multiplication of these organisms outside the human body were summarized on p. 202.

Routes

Aerial transfer   The airborne dissemination of infective particles was discussed in Chapter 9; the upper part of Table 58.3 summarizes ways in which it may be responsible for micro-organisms reaching a site of implantation into the patient. The size of the particles determines the range of their dispersion when they are liberated into the air of rooms and the depth of their penetration into the respiratory tract; it also sets a limit to the number of organisms per droplet.

Organisms of human origin reach the air either as droplets from the respiratory tract or as dry particles from the body surface. The mouth is the chief source of respiratory droplets, which vary greatly in size. Very small droplets evaporate within seconds, and the resulting *droplet nuclei* remain suspended in the air for sufficiently long for the surviving organisms to be distributed over considerable distances by movements of the air; when inhaled, the majority of them escape entrapment in the upper respiratory passages and reach the lung. Large salivary droplets remain suspended for only a few seconds and are usually deposited in the upper respiratory tract, but each one may contain hundreds or thousands of organisms (see Chapters 9 and 51). Epidemiological evidence (Chapter 59) suggests that respiratory infection with group A streptococci occurs only over very short distances (3 m or less) and is thus mediated by large salivary particles.

Face masks are intended to prevent the infection of patients by organisms expelled from the respiratory tract of members of the hospital staff. There is little agreement about the extent to which they do this and the circumstances in which they should be worn. Their use can be justified in situations in which droplets might be expelled directly into an open wound or on to sterile equipment, but in few others. Some air is generally expelled laterally around the edges of the mask, so that many of the smaller particles escape. Masks are more effective in preventing the inspiration than the expiration of pathogenic organisms; they are thus of considerable value in protecting hospital staff from highly pathogenic organisms transmitted by the respiratory route. The older types of reusable cotton masks—generally inefficient and commonly misused—have been replaced by single-use, non-wettable products. For a critical assessment of the use of masks, and for methods of testing their efficiency, see Rogers (1980).

Dry infective particles generated from the body surface, when of a size equivalent to that of a 15-25 μm sphere, stay in the air sufficiently long for infection to be transmitted from one end of a hospital ward to the other. Most of the skin scales liberated during physical activity are of this size (Chapter 60), as are many of the particles of dried nasal secretion and wound exudate released from clothing and dressings. In burned patients, however, most of the particles shed are said to be rather smaller (3-6 μm) (Hambraeus 1973*a*).

Particles from the skin pass readily through the weave of ordinary fabrics, including the usual types of protective garment worn by members of the hospital staff (Duguid and Wallace 1948, Hare and Thomas 1956, Charnley and Eftekhar 1969; see also p. 209). Organisms that have contaminated the outside of protective clothing from contact with patients may pass inwards through the fabric and contaminate the clothing beneath (Speers *et al.* 1969), from which they may subsequently be transferred to freshly donned outer garments (Hambraeus 1973*b*) and eventually become resuspended in the air.

There is uncertainty about the relative importance of room air and contaminated objects (including hands) in the transmission of gram-positive organisms from human sources (see, for example, Chapter 60). The risk of airborne infection would be expected to be greatest when large areas of tissue are exposed for a long time, as in the operating theatre (p. 208) and in burns wards (p. 210). Direct aerial contamination of

Table 58.3   Common routes of infection in hospitals

Route	Origin	Site of implantation	Examples
1. *Aerial: from persons* (a) Small particles (*ca* 5μm); droplets→droplet nuclei; long range (room to room)	Mouth	Respiratory tract (including lung)	Tuberculosis, measles, chickenpox
(b) Large particles (> 100 μm); liquid droplets; short range (< 3 m)	Mouth	Respiratory tract (upper only)	Streptococcal respiratory; and probably meningococcal and *Haem. influenzae* infection
(c) Medium-sized particles (15–25 μm); dry (skin scales, pus, secretions); moderate range (same room)	Nose, skin, wound exudate	Nose, wound, skin	Staphylococcal and streptococcal sepsis; cutaneous mycoses
2. *Aerial: from inanimate sources* Wide range of particle sizes (only small particles infect at a distance)	(a) In circuit of respiratory machine (b) Nebulizer (c) Air-conditioning plant (d) Shower-bath spray	} Respiratory tract	Various enterobacterial and *Ps. aeruginosa* infections *Legionella* pneumonia
3. *Contact: from persons* Usually indirect, via hands or contaminated objects	(a) Respiratory secretions (b) Wound exudate, skin (c) Urine	} Wound, skin, urinary tract	Streptococcal, staphylococcal, enterobacterial and *Ps. aeruginosa* sepsis
	(d) Faeces	Mouth	Enterobacterial and viral diarrhoea
4. *Contact: from inanimate sources** (a) 'Wet': fluids, contaminate equipment, food and medicaments	Soil, water, etc. (subsequent multiplication)	Injection, instillation, by mouth, or (in very susceptible patients), surface contamination	Sepsis by enterobacteria, and pseudomonads and other 'non-fermenters'
(b) 'Dry': equipment (often remotely contaminated)	As above (but no subsequent multiplication)	Injection, wound contamination	Tetanus
5. *Direct transfer: from persons or animals*	Blood, blood products, tissue	Injection, wound contamination, tissue implantation	Hepatitis B, malaria, etc.; various latent viruses

*Independent of contamination from a human source.

wounds in post-operative wards is probably of less importance, at least for wounds that can be adequately covered, but the indirect consequences of the aerial transmission of gram-positive cocci in hospital wards may be considerable. The constant inhalation of *Staph. aureus* in ward air is an important determinant of nasal colonization by antibiotic-resistant strains. Among adult patients the risk of acquiring a particular strain has been shown to be proportional to the amount of aerial exposure to it (Chapter 60). In the initial colonization of newborn babies, which usually takes place on the skin, contact infection is of much greater significance than aerial infection.

Organisms from non–human sources are most often disseminated aerially when they are liberated as liquid droplets generated mechanically from fluids. The droplets may be of various sizes; when small enough to reach the lungs in inhaled air, they are an important cause of respiratory colonization and sometimes of pneumonia in susceptible patients. The organisms responsible are generally gram-negative aerobic bacilli that have multiplied in the fluid.

Respiratory apparatus that delivers humidified air is responsible for many of these infections (Hoffman and Finberg 1955, Bassett *et al.* 1965, Phillips and Spencer 1965, Reinarz *et al.* 1965, Hovig 1981); the organism has usually multiplied in the fluid reservoir or elsewhere in the moist interior of the apparatus. Dissemination is favoured by the inclusion of a nebulizer in the system.

Few gram-negative bacilli are readily transmitted through the general atmosphere of rooms at normal levels of humidity. The legionellae form an exception. In hospitals, as in other buildings, they appear to survive drying—possibly within free-living amoebae (Chapter 43)—and may infect at a considerable distance from the point of generation of the liquid aerosol, even occasionally causing disease in persons who did not enter the hospital but merely passed near the air-discharge ducts (Dondero *et al.* 1980). Dissemination in hospitals, as in other buildings (Chapter 73), appears to be either (1) from an air-conditioning system into the general air supply, or (2) from atomized water in shower-baths. In both circumstances the organism seems to have multiplied within the hospital: in the cooling tower or evaporative condensers of the air-conditioning system (Marks *et al.* 1979, Dondero *et al.* 1980) or in the water storage and distribution system (Tobin *et al.* 1980, 1981, Fisher-Hoch *et al.* 1981). However, the organisms can often be found in situations in hospitals in which cases of clinical infection have not been recognized (Tobin *et al.* 1981). (For legionellae in nebulizer fluids see Gorman *et al.* 1980.) The circumstances in which epidemics develop have not yet been defined (see Fallon 1980). Patients with lowered cell-mediated immunity have been the chief sufferers in some incidents; in others, healthy members of the staff and visitors have suffered severe clinical attacks.

Contact transfer   Patient-to-patient transfer by the contact route is usually indirect, and is mediated by contaminated objects. The sorts of organisms transmitted on solid objects will depend upon the total time of exposure of the organisms to dry conditions and to some extent upon the size of the initial inoculum. If, as is the case in hospital wards, transmission usually takes place within a few hours, not only gram-positive cocci but also gram-negative aerobic bacilli will be transferred, particularly from heavily contaminated objects. Contact transfer from objects contaminated by deposition of organisms from the air will be almost exclusively of gram-positive organisms.

Aseptic practice in hospitals is designed to prevent contact infection (1) by sterilizing or disinfecting instruments, utensils and other items of medical equipment, and (2) by decontaminating the hands of members of the staff at appropriate times (below). The outer clothing of staff members may become contaminated externally by contact with patients or their bedding. Organisms from the skin or undergarments of the wearer may also reach the surface; these may include some derived from patients with whom the wearer had earlier contact (p. 203). Contact transfer to other patients is facilitated when parts of the garment become wet. It may be prevented by donning a fresh disposable or easily decontaminated plastic apron for each close contact with a patient.

Contact infections from 'wet' inanimate sources, usually caused by nutritionally unexacting gram-negative bacilli, occur frequently in the modern hospital. We have referred (p. 204) to the conditions under which these organisms may be inhaled by patients. They also may be introduced, often in large numbers, by injection (p. 211), or by instillation into wounds or normally sterile areas such as the conjunctiva and the urinary tract. Outbreaks of severe disease in immu-

nodeficient patients, including premature babies, are sometimes associated with the surface application of a contaminated fluid. It may then be difficult to assess the relative importance of invasion through the undamaged skin or mucous membrane and infections associated with common procedures such as venepuncture or cannulation. Fluid medicaments for oral use, and topical applications, are frequently contaminated with gram-negative organisms, particularly *Ps. aeruginosa*, during manufacture, storage or repeated use (Report 1971, Baird *et al.* 1976) and have on occasion been shown to transmit infection (Shooter *et al.* 1969, Baird and Shooter 1976). (For salmonella infections from oral medicaments see p. 214, and for gut colonization from food, Chapter 61.)

Infection from the hands   The *resident flora* (Chapter 8) of the skin of the hands is a primary source and its *transient flora* an intermediate reservoir of infection. Broadly speaking, the former consists of organisms that are not removed by washing with soap and water and the latter of organisms that can be so removed.

On the well kept hands of surgeons, and of persons usually employed in experiments on hand decontamination, the resident flora seldom includes organisms that cause septic infection in healthy persons, though a small proportion consistently yield *Staph. aureus*, usually in small numbers (Chapter 60). The rest of the organisms on the 'normal' recently washed hand are coagulase-negative staphylococci and diphtheroids that seldom cause significant disease unless implanted into tissues on a foreign body, or in immunosuppressed patients. On hands subject to frequent cuts or abrasions, overhydration, or irritation from exposure to strong disinfectants, heavy and prolonged colonization, not only with *Staph. aureus*, but also with *Ps. aeruginosa*, klebsiellae and other gram-negative aerobes, and with *Candida*, frequently develops, and hand washing causes only a small and transient reduction in their numbers. Inflammatory lesions on the hands—staphylococcal boils and areas of infected eczema, chronic nail-bed infections with *Ps. aeruginosa* and *Candida*, and herpetic whitlows—are rich sources of infecting organisms. (For the hand carriage of gram-negative organisms see Bruun and Solberg 1973 and Chapter 61.)

An important method of preventing septic infections is to avoid contact between the hands of the staff and the patient's tissues by employing 'no touch' techniques for wound dressing and similar procedures whenever practicable. When tissues must be handled, gloves are worn to minimize the access of resident flora to them. However, the organisms inside the glove tend to increase progressively in numbers with time and may gain access to the wound if the glove is torn, through the invisible holes that appear in some 20 per cent of gloves during a major operation, or by the seepage of moisture from the hands to the sleeves of the operating gown.

The traditional 'surgical scrub' for several minutes with soap and water is intended to remove transient organisms and to reduce temporarily the number of resident organisms on the surface of the skin. Its effect is enhanced if a detergent solution containing a mild disinfectant, such as hexachlorophane, chlorhexidine or povidone iodine is used; this usually results in a 50–90 per cent reduction in the count. Several such treatments cause a progressive decrease to an equilibrium level which varies with the disinfectant used, and in the case of 4 per cent chlorhexidine-detergent solution represents a reduction of over 99 per cent of the count (Lowbury and Lilly 1973). Further treatment with aqueous solutions causes no further lowering of the count, but if at this stage the skin is rubbed for 2 min with alcoholic chlorhexidine, or even with 95 per cent ethanol, there is a further 100-fold reduction (Lowbury and Lilly 1973, Lilly *et al.* 1979). However, repetition of the two-stage process in reverse order— repeated application of ethanol followed by an application of the aqueous solution—does not cause a corresponding reduction. A single treatment with *ca* 10 ml of 0.5 per cent alcoholic chlorhexidine, rubbed on to the hands until they are dry, reduces the skin count by 98 per cent (Lowbury *et al.* 1974). The antiseptic on the skin has good residual activity, and will prevent the build-up of organisms inside the glove even during long operations.

Washing with water is the usual means of removing transient flora from the hands of members of the hospital staff. If performed carefully for a period of *ca* 30 sec it will remove 98 per cent of recently deposited organisms, but it is considerably less effective against organisms rubbed into the skin, removing only some 70 per cent of them (Lilly and Lowbury 1978). In either circumstance, washing in 70 per cent ethanol destroys or removes over 99 per cent of the organisms; it gives slightly better results than washing in aqueous chlorhexidine-detergent solution. Another argument in favour of using either ethanol or alcoholic chlorhexidine is that it reduces temporarily the number of organisms on the skin of hand carriers, which is little influenced by washing with soap and water.

If hygienic hand decontamination is regularly practised between attending successively to patients its frequency may well exceed 50 per 8-hr shift in some hospital departments. Unless the preparations used are carefully chosen, hands may become irritated and the carrier rate for potentially pathogenic organisms may rise (Ojajärvi *et al.* 1977); the dehyrdrating effect of repeatedly applied alcohol may be counteracted, for example, by including some glycerol in the preparation. Washed hands may become recontaminated with gram-negative organisms from soap-dishes (Jarvis *et al.* 1979), bulk supplies of liquid detergents and hand-care lotions (Morse and Schoenbeck 1968). In some 'poor' countries, hand-washing facilities are scarce in hospital wards; dipping the hands in 70 per cent alcohol between attending to successive patients is a useful alternative. Even when wash-basins are freely available, however, they may not be used as often as they should; in one survey in a number of intensive-care units (Albert and Condie 1981), members of the staff were observed to wash their hands after only 25–50 per cent of contacts with patients.

(For a general account of the effects of antimicrobial substances on the skin flora see Ayliffe 1980.)

Direct transfer Pathogens may be transferred directly from one person to another in blood or blood products by injection or wound contamination, or in transplanted tissue or organs. By far the commonest example of this is hospital-acquired hepatitis B infection (p. 214), but occasional examples of transfusion-related malaria, syphilis and other infections are on record. The transfer of latent viruses to immunosuppressed patients is one of the hazards of organ or bone-marrow transplantation. (For infections from porcine xenografts, see p. 202.)

Important clinical types of hospital-acquired infection

Septicaemia

In Chapter 57 we discussed the natural history of septicaemic infections and outlined the changes that had taken place in their aetiology in the last 40 years. Most studies of this subject were initiated in the laboratory and were in fact retrospective investigations of patients in whom bacteria had been isolated from the blood. In such studies it is often difficult to decide to what extent the organism isolated was responsible for the symptoms recorded or the outcome of the incident. The frequency with which bacteraemia is detected will depend upon the assiduity with which it is sought, particularly in patients suffering from terminal illnesses, and the extent to which laboratory results are submitted to critical scrutiny. Here we need only consider the circumstances in which bacteraemia is detected in hospital patients.

According to Maki (1981), hospital-acquired bacteraemia occurred in 1 in 200 patients admitted to hospitals in the USA. In over one-half of the patients a local source of the infection was not identified. Of identified sources, intravenous cannulation or a similar procedure was the commonest. When bacteraemia was associated with a local infection, this was most often in a surgical wound, the urinary tract or the lung, in that order of frequency. In the years 1971–80, of 70 epidemics of bacteraemia recorded, 29 were associated with intravenous therapy. The organisms responsible for epidemics of bacteraemia were, in order of frequency: pseudomonads other than *Ps. aeruginosa*, *Klebsiella*, *Enterobacter*, *Serratia*, *Ps. aeruginosa* and *Esch. coli*. Sporadic cases, on the other hand, were most often caused by *Esch. coli*, *Staph. aureus*, *Klebsiella*, group D streptococci, *Enterobacter*, *Ps. aeruginosa* and anaerobes, in that order. (See also Svanbom 1979.) The proportion of bacteraemic infections in which two or more organisms were isolated from the blood is said to have increased from 6 per cent in 1970 to 13 per cent in 1975 (Kiani *et al.* 1979). Polymicrobial bacteraemia is particularly associated with malignancy and immunosuppressive chemotherapy, and with a higher mortality than bacteraemia in which a single organism is isolated.

Post-operative wound sepsis

The frequency with which sepsis develops after operations through apparently uninfected tissues varies widely. In one series of nearly 3000 non-emergency operations in 21 British hospitals (Report 1960), the total clinical sepsis rate was 10 per cent, but figures for individual hospitals varied from 5 to 22 per cent. Further investigation (Lidwell 1961) showed that these differences were almost entirely accounted for by factors individual to the patients and to the types of operation performed; age over 60 years, long duration of the operation, large incision, and the insertion of a drain were each independently associated with increased risk of sepsis. Moreover, certain types of operation in which the sepsis rate was high were ones in which a viscus with an abundant bacterial flora had been incised—the category now often referred to as 'clean contaminated' operations (see Jeffrey and Sklaroff 1958). In a recent survey in Sweden, Bengtsson and co-workers (1979) give sepsis rates of 5 per cent for strictly 'clean', 20 per cent for 'clean contaminated' and 35 per cent for 'dirty' or 'infected' operations. Operations at 'clean contaminated' sites, e.g. appendicectomy, are attended by a much lower sepsis rate when planned than when performed in an emergency (Kippax and Thomas 1966). Sepsis rates for colorectal operations without chemoprophylaxis range from 35 to 66 per cent (see Raahave *et al.* 1981). Not surprisingly, a relation between the bacterial content of the wound at the end of the operation and the risk of sepsis has been repeatedly demonstrated, for example, by Bruun (1970) and Davidson and co-workers (1971*a*) (see also Raahave 1979). Similar operations in comparable age groups have distinctly higher sepsis rates in males than in females (Ayliffe *et al.* 1977).

Bacterial causes

Staph. aureus continues to be the most common organism isolated from septic wounds after strictly 'clean' operations (Chapter 60). In the British survey (Report

1960), which included a proportion of 'clean contaminated' sites, *Staph. aureus* was isolated alone from 45 per cent, and with other organisms from a further 15 per cent (see also Report 1964). In the survey of Bengtsson and co-workers (1979) the percentages of *Staph. aureus* and of gram-negative aerobic bacilli were respectively: for sepsis in strictly 'clean' operations, over 40 and *ca* 20, and in 'clean contaminated' operations *ca* 30 and over 60.

There are few large-scale studies of the importance of gram-negative anaerobic bacilli as causes of post-operative sepsis (but see Leigh 1981). Indeed, this is difficult to estimate by bacteriological means because these organisms are seldom present in pure culture (see Chapter 57). Wound infection with group A streptococci (Chapter 59) is not very common nowadays; it tends to occur in distinct ward outbreaks that are easy to recognize and control. Outbreaks of wound diphtheria have been recorded (for example by Bensted 1936) but not recently.

Colonization of wounds without signs of local inflammation is at least as common as wound sepsis. In susceptible patients colonization may be a prelude to septicaemia or even endocarditis. The role in wounds of a number of organisms that do not cause a characteristic acute local lesion is difficult to define. Wound colonization by group B streptococci, for example, is commonly without ill effect, but on rare occasions septicaemia follows. Coagulase-negative staphylococci are often the only organisms to be found in wounds showing mild superficial infection. They are generally ignored, but *Staph. epidermidis* appears to cause a definite superficial lesion in chest wounds, and this may be followed by endocarditis associated with a heart-valve prosthesis (Chapter 60). *Staph. epidermidis* and several other skin organisms that gain entry to wounds sometimes give rise to chronic suppuration around implanted foreign bodies many weeks or months later. Whyte and his colleagues (1981) isolated *Staph. epidermidis* 23 times, *Propionibacterium* spp. 12 times and gram-positive anaerobic cocci seven times from 51 joint implants removed because of signs of inflammation (see also Petrini *et al.* 1979, Lidwell *et al.* 1983). Skinner and co-workers (1978) record a number of infections with *Propionibacterium* spp. after the insertion of shunt devices or other neurosurgical operations.

Clostridial infections of surgical wounds are infrequent but often disastrous in their effects. Outbreaks of *post-operative tetanus* are now very rare in economically developed countries. The few that occurred in Britain betwen 1946 and 1957 (see Lowbury 1963) appear to have resulted from airborne contamination of theatre equipment either by outside air or from animal hair in plaster work exposed during alterations to the building. *Post-operative gas gangrene*, usually caused by *Cl. perfringens*, occurs in hospital in a variety of circumstances. The great majority of cases occur sporadically and appear to arise by self-infection from the gut contents. Local gangrene of the abdominal wall or perineum is associated with operations on the bowel or gall-bladder. After 'clean' operations it is very rare except after amputation through the thigh in patients with impaired arterial blood-supply to the leg, and after the insertion of an inanimate implant to the hip joint or femur (Ayliffe and Lowbury 1969, Parker 1969). Drewitt and his colleagues (1972) drew attention to the frequency with which *Cl. perfringens* can be isolated from the skin of the upper part of the thigh. Although *Cl. perfringens* can usually be found in the dust of operating theatres (Lowbury 1963), infection from external sources at the time of operation appears to be rare. However, Parker (1969) refers to one epidemic associated with an operating theatre to which a clinical case of gas gangrene had been taken at a time when the dressings autoclave was faulty; several cases of gas gangrene developed after hip operations performed within the next few days. Pearson and his colleagues (1980) attributed an outbreak to the application of elastic bandages contaminated with *Cl. perfringens* to amputation stumps of legs with arterial insufficiency. It should be noted that infections with *Cl. perfringens* after neurosurgical operations lead to intracranial abscesses and not gas gangrene; septicaemic infections with clostridia, which are not uncommon in terminally ill patients, are also very seldom complicated by gas gangrene.

Infection in the operating theatre

Aerial infection The rituals of aseptic surgery as practised in the earlier part of the twentieth century were designed mainly to prevent organisms from reaching the wound on contaminated objects. At about the time of the Second World War the isolation of potential pathogens from the air led to renewed interest in this as a source of wound contamination. Mechanical ventilation systems were installed partly for this reason and partly to improve the physical comfort of the theatre staff. Early systems extracted air from the theatre and discharged it externally ('negative-pressure' ventilation), but it was soon apparent that this often resulted in organisms such as antibiotic-resistant strains of *Staph. aureus* being drawn into the theatre from elsewhere in the hospital. In consequence, a requirement for positive-pressure ventilation with filtered air came to be accepted (Lidwell and Williams 1960). In association with this (Report 1962*a*), the theatre suite was 'zoned' and supplies of air to the various rooms was so adjusted that air always passed from the more to the less 'clean' areas ('plenum' ventilation).

The objectives of ventilation in operating suites are: (1) to prevent the ingress of airborne organisms to the part of the theatre in which an operation is in progress from elsewhere in the operating suite and from outside

it; and (2) to reduce by dilution the number of airborne bacteria-carrying particles generated within the suite. Recommendations for ventilation systems in British operating theatres (Report 1972*a*) were that incoming air should pass through a filter of average pore diameter 5 μm at a rate that ensured 20 air changes per hr in the operating room. If the filters are functioning properly the air entering the theatre will contain no more than 35 bacteria-carrying particles per m³ and *Cl. perfringens* and *Staph. aureus* will be absent from 30 m³ of air. The bacteria-carrying particles in the air of such an operating room will nearly all have been generated from the bodies of persons in the room. If excessive physical activity by members of the staff is controlled, air counts during operations in a properly ventilated theatre seldom exceed 180 per m³. Strict comparisons are difficult to make, but many workers believe that attaining this standard led to a significant reduction in the rate of sepsis in general surgery (see Blowers 1963), though contrary views have been expressed, for example, by Bengtsson and his colleagues (1979).

In recent years, means of securing very much greater reductions in the air count in the neighbourhood of the surgical wound have become available. Much greater volumes of air and finer filters are employed. Charnley (1970) placed the wound and the surgical team in a small enclosure in which over 300 air changes an hour could be made, and others used various forms of unidirectional ('laminar-flow') ventilation. Surgical isolators, in which the wound is effectively separated from the air and from contact with the operating team, have also been developed (McLauchlan *et al.* 1974). 'Ultra-clean' air systems give air counts up to 100 times less than those in conventionally ventilated theatres. In a careful comparison of two adjacent operating rooms, one ventilated conventionally and the other unidirectionally, Whyte and his colleagues (1982) observed total air counts averaging respectively 413 and 4 per m³; washings from comparable wounds just before suturing gave average counts respectively of 103 and 3 per wound. Calculations from these data led to the conclusion that only a minority of the organisms found in the wounds had fallen directly into them from the air; the rest had presumably fallen elsewhere and later been transferred into the wound by contact. The value of 'ultra-clean' methods was convincingly demonstrated by Lidwell and his colleagues (1982); in a large controlled trial on joint-replacement operations performed either in conventional or 'ultra-clean' conditions by the same surgical teams, a significant reduction (over two-fold) was observed in the latter. When special precautions were also taken to prevent dissemination of organisms from the staff in the 'ultra-clean' situation, the reduction was even greater—nearly a seven-fold reduction when 'body-exhaust' suits were worn (see below). There is little doubt that the use of 'ultra-clean' air systems can be justified for operations in which the risk of endogenous infection is slight and at which it is practicable to take effective measures to prevent the access of extraneous organisms by routes other than the air. Their use in general surgery has yet to be evaluated.

Protective clothing worn in the operating theatre does little to prevent the aerial dispersion of organisms from the body-surface of staff members (p. 203). This can be reduced by wearing garments made of fabrics that cannot be penetrated by skin scales and fit closely at the neck, wrists and ankles. When made of closely woven cotton they are often hot and uncomfortable; there are a number of non-woven alternatives, some of them intended to be disposable (see Mackintosh 1982). Experimental tests show that some of these newer garments are very effective (Lidwell and Mackintosh 1978, Lidwell *et al.* 1978, Mitchell *et al.* 1978) though not always very comfortable. Charnley and Eftekhar (1969) describe a system of 'body-exhaust' ventilation for use with gowns of tightly woven material. (For face masks see p. 203.)

Contact infection In addition to surgical instruments and other equipment and the hands of the surgical team (p. 205), the skin of the patient is an important source of infection. Substances favoured for preoperative decontamination of the operation site are chlorhexidine, iodine or an iodophor, each in 70 per cent ethanol. They are most effective when applied liberally over a wide area and rubbed in vigorously for several minutes, preferably with the gloved hand, a second glove being donned for this (Lowbury and Lilly 1975). For elective, high-risk operations, several previous applications of an aqueous detergent-antiseptic solution are desirable, beginning on the day before the operation. When special efforts to remove clostridial spores are indicated, as on hands with ingrained dirt or on the thigh and nearby areas, the application for half-an-hour of a compress of 10 per cent aqueous povidine iodine is recommended. Pre-operative shaving of the skin, if essential, should be deferred until the day of the operation to minimize the colonization of abrasions by bacteria, or a depilatory cream may be used. Mucous membranes are difficult to decontaminate; aqueous solutions of disinfectants or creams are generally used. (For more practical details see Lowbury *et al.* 1981.)

The importance of bacteria on 'general' surfaces in operating theatres is difficult to estimate. Contact infection from them is preventible by strict application of 'no-touch' techniques. Re-suspension of dust from them will contribute to the content of organisms in the air; for this reason, high standards of physical cleanliness of horizontal surfaces, maintained by frequent mopping or wet dusting, the rigid exclusion of all clothing and bedding emanating from wards, and the avoidance of unnecessary and particularly of violent movements in the theatre, are of importance. Chemical disinfection of floors is indicated only for areas obviously contaminated with blood or pus. (For the effects of a strict zoning system on the contamination of floors see Hambraeus *et al.* 1978.)

General discussions of infection in operating theatres are to be found in the following books or papers: Walter (1948), Altemeier and co-workers (1976), Hambraeus and Laurell (1980), Lidwell (1981), Lowbury and co-workers (1981), Ayliffe and Lowbury (1982).

Infection acquired post-operatively

Opinions differ widely about the relative proportion of wound infections that are acquired at operation and subsequently. This question has been most extensively studied in relation to *Staph. aureus* infection (Chapter 60). The highly discrepant conclusions reached can be attributed not only to differences in the local conditions in the theatres and wards studied and in the operations performed, but also to the criteria used by the investigators to determine the place of infection. Thus, failure to detect the infecting organism in any member of the theatre staff is often taken as evidence that the infection was acquired in the ward, and finding a possible source in the ward may lead to the assumption that the organism entered the wound after operation. Uncritical interpretation of incomplete evidence, particularly when a full bacteriological examination of the patient was not made immediately before the operation, easily leads to an overestimate of the frequency with which infection is acquired post-operatively.

Patients in post-operative wards are heavily exposed to micro-organisms from many sources, but the access of the organisms to the depths of the wound is often limited. If the wound has been closed in the theatre, most infections are superficial and limited, and are caused by organisms from the surrounding skin. If a drain is left in the wound, however, organisms can readily pass along the static column of fluid in it to reach the depths of the wound (Casey 1971; for methods of closed drainage see Lowbury *et al.* 1981, p. 168). In wounds that cannot be closed, and those that require repeated dressing, the risks of infection are considerable. The great success that attended the introduction of 'no-touch' techniques for war wounds (Williams *et al.* 1945) suggests that contact-transfer of organisms on hands, or on sterilized but subsequently recontaminated instruments, was a major cause of this type of infection. The use of a fresh or recently decontaminated plastic apron by persons performing wound dressings would nowadays form part of the 'no-touch' routine.

The evidence that mechanical ventilation and physical subdivision of surgical wards reduces the frequency of wound infection is uncertain (see Ayliffe and Lowbury 1982). The effect of these measures on aerial exposure to staphylococci may well be neutralized by failure to prevent redispersal of the organisms from nurses' clothing (Lidwell *et al.* 1975). The use of appropriately ventilated cubicles for the containment of heavy aerial dispersers of gram-positive pathogens (p. 218) is fully justified, but separately ventilated dressing-rooms in general-surgery wards are probably not (Ayliffe and Lowbury 1982).

Infections of burns

These resemble more closely the infections of war wounds than those of surgical incisions. The area of tissue exposed to infection is large, and bacteria from the skin of the patient are usually multiplying actively in the lesion at the time of admission to hospital. To these are soon added bacteria from other patients and from inanimate sources in the hospital. Study of the bacteriology of burns began soon after special units for their treatment were established and it became clear that added infection occurred frequently. In addition to *Str. pyogenes* (Cruickshank 1935), Colebrook and his colleagues (1948) noted many infections with *Ps. aeruginosa* (Chapter 61), *Proteus* spp. and *Staph. aureus*. Dressings applied to burns often became soaked with exudate in which abundant bacterial multiplication occurred (Colebrook and Hood 1948), providing a reservoir from which other patients became infected. Bacteriological observations led these workers to conclude that the aerial route of infection was important, so that, in addition to the use of 'no-touch' dressing techniques, they installed artificial ventilation in the burn-dressing station. Some reduction was observed in the frequency of streptococcal infection, but adequate control of this was attained only when chemoprophylaxis with penicillin was added.

Thereafter, local antibiotic prophylaxis was for some years the main means of controlling infection in burns, and the agents used had to be changed periodically as new resistances appeared in the prevalent pathogens (Lowbury 1960, Lowbury *et al.* 1964, 1969). After 1960, interest was mainly concentrated on *Ps. aeruginosa* infection, and the serious consequences of this were subsequently much ameliorated by means other than the use of antibiotics (see Chapter 61). In the meantime, a number of other gram-negative bacilli with wide patterns of antibiotic resistance—strains of *Esch. coli*, *Proteus* spp., klebsiellae, *Enterobacter* spp. and *Acinetobacter* spp.—were becoming prevalent and causing occasional serious infections (Davis *et al.* 1969, Roe and Lowbury 1972, Fujita *et al.* 1982). Infection with gram-negative bacilli is favoured by treatment regimens in which burns are kept covered with moist dressings, particularly when antibiotic prophylaxis is used. Treatment by exposure to a dry atmosphere and the sparing use of antibiotics gives lower infection rates with gram-negative organisms but this is associated with a considerable frequency of *Str. pyogenes* infection (Wormald 1970).

The clinical consequences of infection in burns depend very much on the infecting organism. *Ps. aeruginosa* is responsible for most of the lethal infections. Colonization with *Staph. aureus* is almost universal and generally symptomless, but local purulent infections, secondary abscesses and a few cases of septicaemia are seen; scarlatiniform rashes (Chapter 60) have been recorded (Colebrook *et al.* 1948). *Str. pyogenes* may cause febrile illnesses, scarlet fever, and occasional septicaemia; perhaps the most serious consequence of this infection is that skin grafts frequently die.

Under experimental conditions, complete elimination of spread by contact greatly reduced the frequency of *Ps. aeruginosa* infection even when the aerial route was not blocked; but interruption of both contact and aerial spread did not prevent the acquisition of *Staph. aureus* (Lowbury *et al.* 1971). The difficulty of controlling infection in a burns unit with separately ventilated cubicles and high standards of nursing was attributed by Hambraeus (1973b) to lack of success in preventing the spread of organisms on the clothing of staff (p. 203; for attempts to do this, see Hambraeus and Ransjo 1977, Ransjo 1978, 1979).

Infections of the urinary tract

These form the largest single group of hospital-acquired infections, and account for between one-quarter and one-half of the total (p. 200; see also Turck and Stamm 1981). If only the bladder is infected the consequences are often slight and transient, but many of the patients develop pyelonephritis and a significant proportion—variously estimated as 0.1–0.8—septicaemia (see Jepsen *et al.* 1982). The causative organisms are usually gram-negative aerobic bacilli, but *Esch. coli* does not predominate among them as it does in urinary-tract infections in the general population (Chapter 57). For example, Jepsen and his colleagues (1982) state that, of 233 infections acquired in medical wards, 32 per cent were caused by *Esch. coli*, a similar percentage by other enterobacteria (*Proteus* and members of the *Klebsiella-Enterobacter-Serratia* groups), and 12 per cent by *Ps. aeruginosa*; only 18 per cent were caused by gram-positive cocci (faecal streptococci 12 per cent; staphylococci 6 per cent). Infection occurs predominantly by the ascending route and in most cases follows catheterization or other forms of instrumentation of or operation on the urinary tract. Prevalence rates among female medical patients were reported by Jepsen and co-workers (1982) as 38 per cent of those who had recently been catheterized and 4 per cent of those who had not; among males the corresponding percentages were 27 and 2.

Categories of patients in whom urinary-tract infection occurs frequently are: (1) those under investigation or receiving surgical treatment for diseases of the urinary tract; (2) those in childbirth or undergoing operations on the lower abdomen, in whom temporary catheterization is usual; (3) seriously ill patients who are catheterized, usually for longer periods of time, as part of an intensive-care 'package'; and (4) incontinent elderly patients, and patients with spinal injury or disease, in whom catheterization is permanent or semi-permanent.

At one time it was assumed that hospital-acquired

urinary-tract infection, like that in the general population, was autogenous. Studies in the 1950s revealed, however, that certain strains of klebsiellae (Ørskov 1952, 1954) and other coliform bacilli, *Ps. aeruginosa* and *Staph. aureus* (Gillespie 1956, Dutton and Ralston 1957) were endemic causes of the infection in particular urological wards. The proportions of each organism tended to differ from hospital to hospital and over extended periods of time in the one ward (Gillespie 1956). The infecting organisms were found in large numbers on the skin and bedding of the patients, in urinals, bedpans and sinks, and on the hands of the nurses (Ørskov 1954, Pyrah *et al.* 1955, Macleod 1958). It seemed therefore that infected urine was the main reservoir of infection. More recent studies (Chapter 61) give a rather different picture. Many hospital populations have now become widely colonized by 'hospital' strains of, for example, klebsiellae, which are responsible for a variety of infections including those of the urinary tract. In these circumstances it is very difficult to distinguish between self-infection and cross-infection. Hospital-acquired *Esch. coli* infections appear to be mainly autogenous (Bettelheim *et al.* 1971).

Attempts to prevent urinary-tract infections associated with urological and gynaecological surgery have thrown light on the means by which the organisms gain entry to the bladder (Miller *et al.* 1958, 1960a,b, Gillespie *et al.* 1960, 1964, 1967). There appear to be three main routes: (1) introduction into the bladder on instruments, such as cystoscopes, that contain heat-sensitive components and are difficult to sterilize chemically; (2) passage up the lumen of the catheter, which can be prevented by closed drainage of the bladder; and (3) transfer of organisms up the urethra outside the catheter, especially in females, which can be minimized by the use of antibacterial creams and by immobilizing the catheter. Strict application of measures to prevent the ingress of organisms by all three routes is strikingly successful in preventing infection for short periods of time—up to 3 days (Slade 1980). Platt and his colleagues (1983), in a controlled study of non-surgical patients undergoing short-term catheterization (mean duration 3 days), showed that the effective maintenance of closed drainage led to a 2.7-fold reduction in the infection rate in patients not receiving antibiotics; this was associated with a highly significant reduction in the total mortality rate. Schaberg and co-workers (1976) concluded from a study of outbreaks of urinary-tract infection caused by *Klebsiella*, *Serratia* and *Proteus* strains that the source of the infecting organism was usually another patient and that transmission was more often by the hands of attendants than by contaminated equipment; important risk factors included prolonged catheterization, prior antibiotic treatment and the repeated irrigation of catheters, particularly with antibiotic-containing fluids.

Permanent catheterization in aged patients is almost invariably associated with bacteriuria, usually with a mixed flora. The use of antibiotics locally or systemically serves merely to select more resistant organisms without eliminating infection. Many of the patients are symptomless for most of the time, but infection may be exacerbated, often with serious or fatal consequences, when the catheter is disturbed; it is perhaps justifiable to give brief antibiotic cover when changing catheters, but an effective agent may be difficult to choose. The permanent catheterization of incontinent aged patients has been called into question (Brandberg *et al.* 1980, Seeberg *et al.* 1982); if the catheter is removed and absorbent pads are used to soak up the urine, more than half of the bacteriuric infections undergo spontaneous cure or can be eliminated by a single course of antibiotic treatment.

Sepsis associated with injections and infusions

Contaminated re-usable syringes were at one time an important means of spreading infection (staphylococcal and streptococcal sepsis, tetanus and gas gangrene, local abscesses caused by *Myco. tuberculosis* and other mycobacteria, viral hepatitis, even occasionally syphilis and malaria; see Report 1962b). Even with the pre-sterilized, single-use syringes now in general use in hospitals, giving injections is not without risk of infection. The main sources of it are now the substance given, the skin of the patient and the hands of the injector. The organisms responsible are generally gram-negative aerobic bacilli, which may have multiplied in multiple-dose vials of drugs, in 'sterile' water used to make up solutions for injection, and in soaps, hand-care preparations, and weak aqueous solutions of disinfectants.

The transfusion of blood heavily contaminated with gram-negative bacilli leads to a pyrexial reaction with severe hypotensive shock, which is often fatal; Borden and Hall (1951) correctly attributed this to bacterial endotoxin (Chapter 57). The organisms responsible are cold-tolerant strains of fluorescent pseudomonads, other non-fermenting gram-negative bacilli or enterobacters; after 2–3 weeks at refrigerator temperature they may reach concentrations of 10^8–10^{10} per ml in stored blood (Braude *et al.* 1955, McEntegart 1956). Other contaminated blood products—albumen, and platelet and leucocyte concentrates—may cause similar complications.

Gram-negative bacilli that can grow in fluids with a minimal nutrient content (normal saline, glucose saline) may after storage at ambient temperature reach numbers (10^5–10^6 per ml) sufficient to cause serious disease when the fluid is given intravenously; growth often ceases before the fluid has become turbid. A number of epidemics associated with contaminated batches of intravenous fluid, some prepared in hospitals and some commercially, have been recorded

(Phillips *et al.* 1972, Meers *et al.* 1973, Maki *et al.* 1976) and it is very likely that others have escaped notice. The organisms responsible included pseudo-monads of the *cepacia* group, *Erwinia herbicola* (*Enterobacter agglomerans*), and klebsiellae. Most of the affected patients suffered a septicaemic illness, some-times with endotoxic shock, and endocarditis sub-sequently developed in a few. Such mishaps may fol-low a failure of initial sterilization, but they appear more often to result from recontamination through faulty or ill designed bottle closures.

Intravenous administration of initially sterile fluids is often associated with the appearance of bacteraemic infection, particularly when the infusion is long con-tinued (Moran *et al.* 1965, Smits and Freedman 1967). Organisms may gain access to the fluid reservoir or the delivery system by a variety of means (see Holmes and Allwood 1979). The number of organisms found in the fluid is usually quite small, gram-positive cocci and *Bacillus* spp. predominate, and heavy contamination with gram-negative bacilli, though it has been re-corded, is distinctly uncommon (Maki *et al.* 1974). When infusion is prolonged, organisms from the patient's skin may also gain entry to the blood stream; colonization of the site of insertion of the needle or cannula is very common, sometimes with local signs of inflammation.

The main hazard of prolonged intravenous cannu-lation is 'catheter sepsis' in which micro-organisms multiply in fibrin clot adhering to the tip of the catheter and signs of acute septicaemia appear; this condition rapidly subsides when the catheter is re-moved. The organisms grown from the tips of catheters are various; coagulase-negative staphylo-cocci predominate, but *Staph. aureus*, coliform bacilli and streptococci and diphtheroids are quite often found (for references see Keighley and Burdon 1979). Opinions differ as to the relative importance of the delivery system, the patient's skin, and other endogen-ous foci as their source. The risk of 'catheter sepsis' is particularly high in 'central venous catheterization', when the end of the catheter lies in a great vein. When used to monitor central venous pressure, extraneous organisms may gain entry from the air or from con-taminated transducer fluid. The administration of nu-tritive substances ('total parenteral nutrition') is generally through a central venous catheter and may be continued for many weeks. Some observers have noted a predominance of systemic infections with *Candida albicans* in association with total parenteral nu-trition (Curry and Quie 1971). This may be related to the ability of this organism, but not of a number of bacteria, to grow rapidly in certain protein hydroly-sates administered to patients (Gelbart *et al.* 1973). Nevertheless, bacterial infections in these patients appear to arise in the same way as those in other forms of long-term intravenous therapy (Powell-Tuck *et al.* 1979).

(For a general account of infections associated with intravenous infusions, see Phillips *et al.* 1976, and for preventive routines, Keighley and Burdon 1979 and Lowbury *et al.* 1981.)

Infections of the respiratory tract

Infections of the lower respiratory tract form, accord-ing to various estimates, 10 to 40 per cent of recorded hospital-acquired infections (p. 200). Most of them occur in elderly patients suffering from other serious medical complaints, in whom pneumonia is difficult to distinguish from exacerbations of chronic bronchial disease and from the pulmonary effects of congestive heart failure; indeed, the indefinite phrase 'infection of the lower respiratory tract' is widely used. Many of the patients have received antibiotics before pulmon-ary signs appeared; the sputum usually gives a heavy growth of gram-negative aerobic bacilli, and the com-mon causes of bacterial pneumonia, such as the pneumococcus, are seldom isolated. At necropsy, vari-ous gram-negative bacilli, and somewhat less often *Staph. aureus*, usually predominate in the lungs (see, for example, Tanner *et al.* 1969), but it is very difficult to decide how often they are the primary cause of the disease. According to Pierce and his colleagues (1966), a disease described as 'gram-negative bacillary necro-tizing pneumonia' can be recognized *post mortem* by its characteristic histological appearances—alveolar septal necrosis with necrosis of the walls of small ar-teries. They state that it was present in some 8 per cent of 522 patients whose lungs were examined at necropsy (see also Pierce and Sanford 1974, LaForce 1981). The disease has been associated with over-growth in the lungs of various gram-negative bacilli, including *Esch. coli*, members of the *Klebsiella-Enterobacter-Serratia* group, *Proteus*, pseudomonads and several other non-fermenting organisms. The illness is difficult to diagnose in life, and bacteriological examination of the sputum or the lung is seldom helpful.

Colonization of the lower respiratory tract by gram-negative aerobic bacilli is particularly common in patients undergoing mechanical ventilation, whether by tracheostomy or by endotracheal tube, and signs of infection of the lower respiratory tract appear in some of them. The respiratory apparatus is sometimes the source of these organisms (p. 204), but nowadays this possibility can often be excluded. In other cases, perhaps the majority, the organisms appear to have passed down the trachea outside the endotracheal tube by the seepage of contaminated oropharyngeal or even gastric secretions (see Sander-son 1983). Infection is thus often autogenous, but the organisms can also be passed from the respiratory tract of one patient to that of another by means of contaminated tracheal-aspiration tubes or on the hands of nurses. Secretions from the mouth or upper respiratory tract may be aspirated into the lungs of drowsy or unconscious patients and give rise to pneumonia, in which the lung flora is generally mixed and often includes non-sporing anaerobes (Chapter 62). In immunodeficient patients a number of other organisms cause pneumonia: various pathogenic fungi, moulds, *Candida, Toxoplasma, Pneumocystis*, cytomegalovirus, and varicella-zoster and herpes simplex viruses. Sporadic cases of legionella pneumonia tend to appear in patients with deficient cell-mediated immunity, but personal susceptibility appears not to be of great importance in hospital outbreaks of this disease (p. 205).

Viral respiratory-tract infections spread in hospitals as in other institutions. Hospital epidemics of influenza A are usually accompanied by a wave of pneumonia deaths in elderly patients. In paediatric wards, respiratory syncytial virus (RSV) and influenza A and para-influenza viruses readily spread among small children. Of these, RSV is of the greatest importance as a cause of acute bronchiolitis and pneumonia. It may cause clinical deterioration in children with such illnesses as cystic fibrosis or congenital heart disease. Studies in Newcastle-upon-Tyne (Gardner *et al.* 1973, Martin *et al.* 1978) and Rochester, New York (Hall *et al.* 1975a,b) have demonstrated high infection rates (30–60 per cent) in childrens' wards during epidemics. Sufferers remain infectious for several weeks. The spread of infection is only partly controlled by the isolation of patients in cubicles, perhaps because members of the staff frequently contract mild infections. (See also Hall and Douglas 1981.)

Diarrhoeal diseases

Most hospital outbreaks of bacterial food poisoning, bacillary dysentery, and various other forms of bacterial or viral diarrhoea resemble those that occur in other institutions; information about these will be found elsewhere in this book (respectively in Chapters 72, 69 and 71). Many of the incidents of salmonella infection in hospitals have somewhat different characters, which will be considered below, together with the diarrhoeal complications of antibiotic treatment.

Salmonella infection

In addition to causing outbreaks of food-borne diarrhoeal disease in hospitals, salmonellae often spread widely from person to person. Endemic prevalences among small children have been recognized for many years (see for example, Rubbo 1948, Mackerras and Mackerras 1949). These often continued for months or years, sometimes with spread from hospital to hospital and into the patients' families. Cases of all degrees of severity were seen, and symptomless excretion by other patients and staff was often widespread. Mortality rates were variable, from nil to 30 per cent or more. Systemic infections, notably meningitis among the younger infants, were a feature of some outbreaks. Experience in Britain was that until towards the end of the 1950s, hospital outbreaks not associated with food were confined almost entirely to young children. More recently, however, a number of outbreaks have been reported in which large numbers of adults were affected and the organism appeared to spread from patient to patient (for example, Datta and Pridie 1960, Sanford *et al.* 1969). Less than one-third of outbreaks of salmonella infection in hospitals in England and Wales in 1974–77 were thought to be food-borne (Report 1980*b*).

After 1960, many of the salmonellae responsible for hospital-acquired infections possessed additional antibiotic resistances. In Chapter 37 we noted that in the late 1970s strains with very broad spectra of resistance, often determined by a single plasmid, had given rise to extensive epidemics in hospitals in many countries. It is not clear whether, as has been suggested, these strains possessed exceptional virulence; mortality rates among children in some of the pre-1960 epidemics were also very high. Little is known about the way in which antibiotic resistance favours the spread of salmonellae in hospitals, if indeed it does. Antibiotic treatment is singularly ineffective in eliminating salmonellae from the gut flora even when they are sensitive *in vitro* to the agent given. Some experimental evidence (Chapter 37) suggests that antibiotics favour infection by small doses of salmonellae by suppressing the anaerobes of the gut flora that normally exert an interfering action on salmonellae. The additional advantage postulated for resistant strains in this situation (Hirsch *et al.* 1965, Adler *et al.* 1970) has yet to be demonstrated. The main clinical significance of antibiotic resistance is in restricting the range of agents available for the treatment of systemic salmonella infections.

Salmonellae of serotypes that under normal circumstances seldom give rise to serious systemic infections do this much more often not only in babies but also in elderly patients with a major underlying disease, particularly a malignant neoplasm (Han *et al.* 1967); reactivation of a latent infection sometimes appears to have played a part. (For further discussion of the determinants for systemic salmonella infection see Chapter 68.)

When salmonella infection is endemic in hospital wards the organisms are isolated from a variety of inanimate sites: from dust (Bate and James 1958), towels (Rubbo 1948), wash basins (Mackerras and Mackerras 1949), and from a water-bath used to warm bottles of milk (Epstein *et al.* 1951); it would be reasonable to assume, however, that they usually reach the patient on the hands of staff members. Outbreaks in maternity departments usually begin when a mother with diarrhoea infects her child at parturition (Watt and Carlton 1945, Abramson 1947, Neter 1950); other babies in the nursery subsequently acquire the disease (see Rowe *et al.* 1969, Report 1980*b*). Secondary case-to-case spread of infection may follow an initial food-borne outbreak in a hospital (Steere *et al.* 1975) but is by no means inevitable.

Other modes of infection are more specifically related to hospital procedures. The organism may be introduced orally in feeds given by gastric tube (Kunz and Ouchterlony 1955) or on inadequately disinfected endoscopes (Chmel and Armstrong 1976, Beecham *et al.* 1979); or rectally by means of barium enemas (Steinbach *et al.* 1960) or rectal thermometers (Im *et al.* 1981). Other sources of infection include stored human milk (Ryder *et al.* 1977), special dietary supplements (Steere *et al.* 1975), and apparatus used for the resuscitation of the newborn (Rubenstein and Fowler 1955, Ip *et al.* 1976). Infection from transfused blood products has been reported (Rhame *et al.* 1973). Various medicinal products of animal origin—thyroid powder (Kallings *et al.* 1966), cochineal (Lang *et al.* 1967) and pancreatin (Glencross 1972, Rowe and Hall 1975)—have been responsible for outbreaks.

Yersinial infection

Hospital outbreaks of infection with *Yersinia enterocolitica* have been reported in which infection appears to have been transmitted from person to person (Toivanen *et al.* 1973, Kist *et al.* 1980): a number of the victims were members of the staff. (For systemic infections in predisposed hospital patients, see Chapter 55.)

Diarrhoea associated with the administration of antibiotics

Patients receiving antibiotics often suffer from diarrhoea but this is usually mild and short-lived. Two serious forms of antibiotic-associated diarrhoea have been attributed to the effects of the agent on the bowel flora: (1) that said to result from the overgrowth of antibiotic-resistant strains of *Staph. aureus* and to be caused by enterotoxin B (Chapter 60), and (2) *pseudomembranous enterocolitis*, due to the production of a toxin or toxins by *Cl. difficile* (Chapter 63).

The disease associated with staphylococci was reported mainly in the years 1955–65, at a time when the clostridial disease had not been recognized. Thus the suspicion remains that some of the earlier cases were caused by clostridia, as were most of the severe cases of antibiotic-associated diarrhoea investigated appropriately in the late 1970s. However, the staphylococcal disease was described as an acute but short-lived cholera-like diarrhoea, whereas *Cl. difficile* causes a necrotizing disease often involving both the large and the small intestine and prolonged in its effects. Many different antibiotics have been held responsible for severe diarrhoea, but the staphylococcal disease tended to be associated with the use of broad-spectrum agents such as tetracycline and neomycin, and pseudomembranous enterocolitis with lincomycin and erythromycin. The situation is further complicated by the fact that pseudomembranous enterocolitis has long been recognized as an occasional complication of major intra-abdominal surgery (Pettet *et al.* 1954) in the absence of antibiotic treatment; and a somewhat similar disease of premature babies—*neonatal necrotizing enteritis* (p. 216)—also appears not to be related to the use of antibiotics.

In the staphylococcal disease, a strain of *Staph. aureus* resistant to the antibiotic being given often became the main aerobic component of the faecal flora; it was usually a strain endemically present in the ward in which the patient was being treated. The way in which antibiotic administration leads to pseudomembranous enterocolitis is not understood. Toxin production in the gut by *Cl. difficile* is well established as the cause of the lesion whether or not the disease was associated with antibiotic administration (Chapter 63), but not apparently in the neonatal disease. *Cl. difficile*, though commonly present in the faeces of normal infants, is seldom found in healthy adults. Strains isolated from patients with post-antibiotic enterocolitis are in many cases sensitive to the antibiotic previously administered to the patient (George *et al.* 1978, Dzink and Bartlett 1980, Shuttleworth *et al.* 1980). This suggests that the relevant action of the antibiotic may be on some other constituent of the flora that normally exerts interference against *Cl. difficile*. In-vitro evidence has also been advanced that antibiotics may stimulate the production of toxins by *Cl. difficile* (Onderdonk *et al.* 1979, Honda *et al.* 1983).

Unaccountable fluctuations have been recorded in the frequency of enterocolitis whether or not this was associated with antibiotic administration (Pettet *et al.* 1954, Swartzberg *et al.* 1977). Greenfield and his colleagues (1981) saw eight cases in two adjacent hospital wards within the space of 11 days; the faeces of all the patients contained *Cl. difficile* toxin in large amount but only four of the patients had received antibiotics. This suggests that pseudomembranous enterocolitis may be infectious; *Cl. difficile* is certainly widespread in the immediate environment of patients with the disease (Kim *et al.* 1981).

Hepatitis B

Hospital infections with hepatitis B first became a matter of serious concern in the late 1960s, when special units were created for the treatment of chronic renal failure by haemodialysis and kidney transplantation. Before this time they had been recognized as an occasional consequence of transfusing human blood products, injecting pharmaceutical substances that contained human serum or plasma, and the inadequate sterilization of multiple-use syringes. Twelve outbreaks of hepatitis, in which there were 357 cases and 18 deaths, were identified retrospectively in British renal units in the years 1965–71 (Report 1972*b*). In a prospective study of 20 units in 1968–69 (Polakoff *et al.* 1972), outbreaks of hepatitis B were detected in three and possibly four of them; and in 1970, when systematic serological testing had been instituted, the infection was detected in 5.6 per cent of patients and

0.3 per cent of the staff in 28 units. Although the severity of infection varied considerably from outbreak to outbreak, infections in patients were often more mild and atypical than those among the staff (Report 1972b). Nevertheless, many of the carrier patients were very highly infectious; this is attributed to immunosuppression often associated with renal failure or artificially induced in patients undergoing kidney transplantation.

The infectious agent of hepatitis B has been detected in most body fluids and secretions, but blood is believed to be the main source of infection in renal units. The infection may be introduced into the unit in transfusion blood, but the admission of a carrier patient is the more usual first event (Polakoff *et al.* 1972). Other patients appear usually to become infected by the injection of blood-contaminated material during renal dialysis or a similar procedure. Earlier types of dialysis machine provided ample opportunities for this (Report 1968b, 1975); disposable dialysers are now available but are often re-used in the interests of economy. When setting up or dismantling dialysis systems it is difficult to avoid contaminating neighbouring surfaces with blood (Report 1975) so that contact transfer of infection to other patients is very likely. Accidental injuries from contaminated needles and other sharp objects are an important means of infection in members of the staff. Contamination with blood of other recent skin injuries, and of the conjunctiva and mucous membranes cannot be excluded.

The spread of hepatitis in British renal units was effectively controlled (Report 1974b, 1976) by the implementation of the recommendations of the Rosenheim Committee (Report 1972b), the main elements in which were as follows. (1) Exclusion of all infected persons from renal units (serological screening of all patients before entry, and of patients and staff in the unit periodically) and either housing them in a separate isolation section not far from the main unit or discharging them to continue dialysis at home. (2) Early discharge of uninfected patients to home dialysis, and kidney transplantation at the first opportunity. (3) Minimal use of blood transfusion; and screening for the hepatitis B agent of all the blood used. (4) Maximal use of disposable equipment and its subsequent destruction by burning; heat sterilization whenever possible of re-used items; disposable dialysers at least for infected patients. Detailed specifications for nursing and other precedures designed to take account of the exceptional resistance of the agent to heat and disinfectants (Chapter 101) have since been worked out (see, for example, Lowbury *et al.* 1981). The success of such preventive measures is possible only when the results of a sensitive test for the presence of the agent in blood are readily and promptly available (see Chapter 101).

Patients in other hospital departments who are suspected of hepatitis-B infection are nursed with precautions similar to those used in renal units ('stool-urine-needle' isolation, p. 219) while laboratory tests for the agent are being made, and subsequently if there are pressing reasons for keeping an infected patient in the ward.

There is a small but definite risk that a carrier of the hepatitis-B agent on the hospital staff may infect patients. Most recorded incidents of this have followed procedures in which the staff member may have sustained trauma to the hands and deposited a small quantity of blood in the patient's tissues, such as in dental operations and other operations in which the fingers are liable to be pricked (Report 1980c). The frequency with which hepatitis B is transmitted in this way is said to be reduced by suitable modifications of surgical technique (Carl *et al.* 1982). Most healthy hepatitis-B carriers appear to have only small amounts of the agent in their blood, but one doctor with chronic active hepatitis is believed to have infected patients while performing venepuncture (Grob *et al.* 1981).

(For further information about hepatitis B, and about passive and active immunization against it, see Chapter 101.)

Neonatal infection

According to the broad definition now employed (p. 193), all neonatal infections in babies born in hospital are deemed to have been 'acquired' there, but some of them are caused by organisms that reach the baby from the mother at or before the time of birth. This is true of 'early-onset' systemic sepsis caused by *Esch. coli* (Chapter 61), group-B streptococci (Chapter 59) and *Listeria monocytogenes* (Chapter 50), as well as congenital infections with cytomegalovirus and *Toxoplasma*, and ophthalmia due to gonococci and chlamydiae. Serious bacterial infections—septicaemia or meningitis or both—in the first week of life are very much more common in premature babies than in those born at full term (Remington and Klein 1976). Until around 1970, nearly all babies with a birth weight of less than 1000 g died, as did one-half of those weighing 1000–1500 g; nowadays, survival rates respectively of 35–50 and over 85 per cent are to be expected (Hurley and de Louvois 1981). Thus, the number of 'early-onset' systemic bacterial infections seen in hospitals has increased considerably in recent years. In addition, the measures employed to increase the survival rate in small babies, which include infusion of fluids, assisted respiration, and the lavish use of broad-spectrum antibiotics, as well as the performance of extensive surgical operations on those with congenital abnormalities, have added greatly to the risks that infection will be acquired subsequently. Sepsis rates in neonatal intensive-care or 'special-care' units are among the highest to be found anywhere in hospitals (p. 200).

Infections appearing after the first few days of life are predominantly caused by gram-negative aerobic bacilli, including a variety of enterobacteria, *Ps. aeruginosa*, and other 'non-fermenters'; their epidemiology resembles that of infections in other groups of patients receiving intensive medical care. *Flavobacterium meningosepticum* and *Citrobacter koseri*, however, appear to cause meningitis only in the newborn (Chapter 61). When it was the practice to keep most newborn babies in hospital for over 7 days, staphylococcal skin infections of both the vesicular and the pustular type were common and often occurred in epidemics associated with the spread of a single strain of *Staph. aureus*. Severe pustular skin lesions—often with systemic complications—were particularly prominent in the 1950s and early 1960s and affected healthy full-term as well as premature babies. Some of the staphylococci responsible for these outbreaks were also exceptionally virulent for adults in contact with the infants. Trivial skin lesions from which *Staph. aureus* can be isolated continue to be common in babies born in hospitals, but severe infections are now less often seen. The extent to which this can be attributed to the early discharge of healthy babies from hospital, better hygiene in nurseries for the newborn or the disappearance of the more virulent and communicable strains of *Staph. aureus* (Chapter 60) is uncertain.

In the last 20 years *neonatal necrotizing enterocolitis* (Chapter 63) has come to be recognized as an important hazard to the lives of premature babies in 'special-care' units. It appears in about 3 per cent of them and has a mortality rate of some 30 per cent. The fact that cases sometimes occur in clusters suggests that it may be a communicable disease (Book *et al.* 1977), but no one infecting organism, and certainly not *Cl. difficile*, has been incriminated. Little is known for certain about the aetiology of the disease (see Thomas 1982). Hypoxia, leading to ischaemia of the intestine, is said by some to cause increased permeability of the mucosa to bacteria or their products (Lake and Walker 1977). The disease does not begin until oral feeding has started; this suggests that the postulated causative agent is ingested. Lawrence and co-workers (1982) point out that the hygienic precautions taken in 'special-care' units tend to lead to homogeneity in the gut flora, in which only one or a very few bacterial strains may multiply unchecked during the first few days of life. They draw a parallel with the fatal enteritis that develops in young germ-free experimental animals when a pure culture of certain bacterial strains, but not of others, is given by the mouth.

Other diarrhoeal diseases of the newborn—caused by salmonellae, *Esch. coli* and viruses—are discussed elsewhere in this volume (see p. 213 and Chapter 71). General accounts of neonatal infections, and references to their less common bacterial causes, are given by McCracken and Shinefield (1966), Overall (1970), Remington and Klein (1976), Hurley *et al.* (1979), Siegel and McCracken (1981), and Hurley (1982).

Prevention

The most obvious means of preventing infections in hospitals is by *hygienic measures* that deny pathogenic micro-organisms access to the patient, or at least to body-sites from which they can easily invade the tissues. The sources of organisms in the immediate environment of the patient are so numerous and the routes of transmission are so many and complex (pp. 200–206) that the precautions needed to prevent all possibility of infection from extraneous sources would be very elaborate indeed; and a considerable proportion of the patients would continue to suffer infections with organisms from their own body flora. In designing a programme of hygienic precautions for use in a particular hospital we are therefore forced to be selective, with the objective of reducing the chances of infection as much as possible within the constraints of a finite budget and a not unlimited supply of trained staff. The 'mix' of precautions should include as many as possible for which there is good evidence of efficacy, and exclude inessential elements, particularly when these are expensive or time-consuming and would divert resources from more useful ends (Report 1968*a*). Full weight should be given to the frequency of the infection to be prevented by each measure and the severity of its consequences. The measures recommended must be practicable in existing circumstances, clearly specified in writing, repeatedly explained to those with the responsibility of implementing them, and subject to a reasonable amount of day to day monitoring.

The *prophylactic use of antimicrobial agents* to destroy organisms after they have reached the tissues provides an apparent alternative to hygienic precautions. There are circumstances in which we are forced to adopt this procedure (p. 219); we then have to seek ways of minimizing its untoward consequences, particularly for other patients in the hospital. The use of antimicrobial prophylaxis should never be looked upon as an excuse for relaxing hygienic standards.

In addition to strictly microbiological measures, it is important to limit as far as possible disturbances to the patients' defence mechanisms arising from procedures to which they are subjected, for example, by 'good' surgical technique, by avoiding unnecessary 'invasive' methods for diagnosis and treatment, and by limiting therapeutic immunosuppression to the minimum necessary.

Choice of hygienic precautions

First we must consider how the efficacy of measures designed to exclude micro-organisms from susceptible sites are to be assessed, (Table 58.4).

Table 58.4 Types of evidence for the possible efficacy of hygienic precautions

---

Clinical
Reduction of the sepsis rate in a controlled trial

Bacteriological: reduction of
(1) colonization
    (a) at a possible site of invasion,
    (b) elsewhere;
(2) access of pathogens (or an index organism) to a possible site of invasion
    (a) in patients,
    (b) in an experimental model;
(3) intrusion of pathogens (or an index organism) into the patient's immediate environment

---

Clinical evidence

The investigation of an individual epidemic may provide clear evidence that the infection spread in a particular way, e.g. from a staphylococcal disperser known not to have touched the patients, or from a bacterially contaminated material injected into patients who subsequently became ill. Such 'anecdotal' information may draw attention to unsuspected means of infection, but the frequency of such incidents and the efficacy of measures designed to prevent them remain uncertain. The ultimate justification for a hygienic precaution is evidence that its introduction lowers the rate of clinical infection. Infection-control programmes are now so 'overloaded' that evidence of this sort should be obtained before expensive new measures are added.

Early hypotheses about sources and routes of infection, based on clinical observations (p. 193), led to the introduction of various 'packages' of highly successful preventive methods directed mainly—despite Lister's brief use of antiseptic aerial sprays (see Williams 1956)—at preventing contact infection from hands and other objects. The diseases then under investigation—puerperal fever and post-operative sepsis—were associated with high infection rates, and the fact that the efficacy of the measures introduced could be easily demonstrated by uncontrolled studies indicated that a single route of transmission predominated. Later in the nineteenth century, clinical studies in childrens' wards (p. 218) showed the value of methods that would now be described as 'bed isolation' or 'barrier nursing' in preventing the spread of certain infectious diseases, e.g. diphtheria and scarlet fever, but not of others, e.g. measles and chickenpox.

The conditions under which precautions must now be evaluated are somewhat different. Many of the more obvious sources of infection have already been eliminated, and a plethora of additional precautions has been introduced, but frequently in a patchy manner. Patients are more often predisposed to infection by pre-existing disease and are subjected to a greater variety of procedures that may result in infection. Only strictly controlled trials of specific procedures in vir-

tually identical patients can be expected to give useful information. Randomly selected test and control groups must be checked for the distribution of known predisposing factors; and operations on members of the two groups must be performed contemporaneously and by the same surgical team. (For recent examples of well controlled trials, see Lidwell *et al.* 1982, and Platt *et al.* 1983).

On p. 200 we referred to some of the difficulties of securing test and control groups of sufficient size, particularly when the expected infection rate is not very high. In most of the diseases under investigation, including post-operative wound sepsis, the infecting organism may reach the patient's tissues by one of several routes; the precaution may 'block' only one of these and only a proportional reduction of the sepsis rate is to be expected. To secure groups of sufficient size it may be necessary to include patients from a number of hospitals with test and control patients in each. It is then necessary to persuade all the clinicians of the ethical propriety of not adding other precautions in which they believe. Even if this cannot be done, however, it may be possible to make allowance for these differences and even sometimes to make a rough assessment of the value of the additional precaution. Thus, in the study of 'ultra-clean' operative precautions by Lidwell and his colleagues (1982) some evidence was obtained of the efficacy of chemoprophylaxis in both test and control groups.

When infection is an unusual event, prevention of transmission by one route may be expected to cause a fall in the sepsis rate proportional to the importance of the route. When the sepsis rate is high this may not always be so, as shown by the studies of Gillespie and his colleagues (1960) on the prevention of urinary-tract infection after prostatectomy. Measures designed to prevent infection by three putative routes were introduced successively (A = closed bladder drainage; B = improved sterilization of catheters and cystoscopes; C = continuous decontamination of the urethral meatus). In one experiment, A reduced the infection rate from 76 per cent to 55 per cent, A + B to 19 per cent and A + B + C to 6 per cent. In another, in a different hospital, the measures were introduced in reverse order; C + B gave a reduction from 92 per cent only to 62 per cent, but the addition of A led to a further reduction to 8 per cent. Had each measure been tested alone its importance might have been underestimated; had all been introduced together an unnecessary precaution might have been included.

The effect on sepsis rates of measures designed to reduce contact infection has been clearly demonstrated under a variety of circumstances. Although there is good bacteriological evidence that some gram-positive organisms are dispersed aerially (p. 203), a lowering of sepsis rates associated with reduced aerial exposure has not until recently been demonstrated convincingly. This has now been done (p. 208), but only with 'ultra-clean' air and in conditions in which infection by other routes was minimal.

Properly controlled clinical trials are expensive and yield results slowly; nevertheless infection-control measures will be placed on a firm scientific base only when many more of them have been made. It would, however, be unreasonable to expect that all preventive measures could be evaluated in this way. Less direct evidence must be used to reach reasonable assessments of the remainder.

Bacteriological evidence

Bacteriological investigations have the advantage that results can be accumulated rapidly. They are thus a valuable means of eliminating preventive measures that are not worth testing clinically. If a precaution is not expensive and does not add materially to the work-load we may be prepared to recommend its use on bacteriological evidence alone.

Evidence that a precaution reduces the frequency of colonization by pathogens is of significance if the site colonized is one from which invasion normally occurs. Thus, studies of the colonization of burns (for example, Lowbury et al. 1971), or of the skin in newborn babies with *Staph. aureus* (Mortimer et al. 1966) provide valuable indications of the relative importance of preventing contact and aerial infection under particular circumstances. On the other hand, studies of colonization at some other sites are less easy to interpret; as we shall see (Chapter 60), the effects of ventilation and segregation on the nasal carrier rate of *Staph. aureus* by patients in hospital wards is to change the types of staphylococci carried rather than the carrier rate. Studies of the access of pathogens to a site of potential invasion appear to be of considerable relevance; the examination of 'wound washouts' performed just before closure appear to reflect fairly accurately the likelihood of subsequent post-operative sepsis and the organisms that will cause it (Davidson et al. 1971a, Kelly 1980, Whyte et al. 1982), particularly if quantitative methods are used. Experimental models of nursing procedures, performed with artificially contaminated objects, have been used to measure the risks of contamination (for example by M'tero and co-workers 1981). Studies of the intrusion of pathogens from identifiable sources into the immediate environment of patients, e.g. their clothing or bedding, can be adapted to the study of the details of nursing procedure (Lidwell et al. 1974) but are perhaps relevant only to the protection of highly susceptible patients. Total bacterial counts on 'general' ward surfaces, though widely used in the past to assess the 'cleanliness' of hospitals, appear to be of little significance in relation to risks of infection.

There is broad agreement (Report 1968a, Ayliffe et al. 1969, Daschner et al. 1978, Parker 1978, Eickhoff 1981) that the value of certain classes of preventive measure, all of them intended to prevent contact infection, can be taken as proved. These include (1) 'aseptic' practices: the sterilization or decontamination of all objects—instruments, utensils, apparatus and substances—likely to come into contact with parts of the patient's body from which invasion is possible; (2) *elimination of infection by manual contact*: avoidance

as far as possible of contact between the bare hands and exposed tissue; appropriate hand decontamination when this is impracticable, and between attendance on successive patients (p. 205); and (3) *protection of artificially created sites of bacterial invasion*: in addition to the surgical wound, these include: the catheterized urinary bladder (p. 210), drainage tubes (p. 209), sites of intravenous cannulation (p. 212), and the artificially ventilated respiratory tract (p. 212). The value of certain other classes of measure, though not yet fully established experimentally, or so established only in some circumstances, may reasonably be inferred from the bacteriological evidence. These include decontamination of the skin of the operation site (p. 209), and the protection of the exposed wound from airborne organisms, the value of which may be confined to circumstances in which the risk of infection from other sources is low. There is also a strong case for various isolation procedures (below) as a means of either limiting the spread of infection from known sources or protecting very susceptible persons; the difficulties here are in allocating priorities to various sorts of patient. There are, however, a number of widely used precautions for which there is little clinical or bacteriological evidence. Among these are attempts to eliminate micro-organisms from the 'general' hospital environment by the wholesale application of chemical disinfectants to floors, walls, sinks or lavatories, by spraying them into the air, or by ultraviolet irradiation.

Practical applications

The detailed application of these principles is beyond the scope of this chapter. For guidance the reader is referred to books or monographs on the subject, including those of Williams et al. (1966), Parker (1978), Bennett and Brachman (1979), Report (1979), Castle (1980), Lowbury et al. (1981) and Ayliffe et al. (1982), and to individual papers presented at conferences (Brachman and Eickhoff 1971, Daschner 1978, Report 1981a). References on the prevention of surgical wound infection were given on p. 209. For sterilization and disinfection, see Chapter 4, Kelsey (1972), and Russell, Hugo and Ayliffe (1982), and for the use of chemical disinfectants in hospitals, Maurer (1978). Methods of sterilization or disinfection appropriate for various items of equipment, and for the performance of specific procedures on patients, approved for use in hospitals in one British region, are set out authoritatively by Lowbury and his colleagues (1981). The book by Ayliffe, Collins and Taylor (1982) provides rational explanations of these and other preventive measures suitable for non-microbiologists. For the administrative aspects of infection control, see Report (1959), Parker (1978) and Lowbury et al. (1981) and for guidance on the investigation and control of individual outbreaks of infection, Ayliffe et al. (1982).

Isolation

Source isolation The isolation of infectious persons—was first practised in the nineteenth century as a nursing technique ('bed isolation'; p. 217) to prevent contact spread of infection. A physical barrier between beds was often added, but this seldom extended from floor to ceiling and was generally open at one end; it was looked upon more as a reminder to the staff ('barrier nursing') than as a means of preventing aerial infection (Grancher 1900). Later, separate rooms or 'cubicles' with closable doors were provided in many infectious-disease hospitals (see Williams 1956) to house patients with diseases believed to be spread aerially. It was realized that micro-organisms might escape from such rooms when the door was open, and single rooms with doors leading directly to the open air were built in some hospitals (see Parker *et al.* 1965). McKendrick and Emond (1976) showed that, when aerial spread through doorways of single rooms was minimized and nursing standards were high, very low rates of cross-infection could be attained, e.g. only 0.2 per 1000 days of exposure to measles.

Source isolation in single rooms is widely recommended for heavy dispersers of pyogenic cocci in general hospitals, preferably with extract ventilation discharging contaminated air to the outside and an air-lock vestibule the two doors of which cannot be open at the same time. An estimate made in 1965 was that such an arrangement was required for nearly 10 per cent of patients in general surgical wards (Report 1965*b*). Many hospitals cannot provide this, and single unventilated rooms are commonly used.

Attempts have been made to rationalize the use of available resources by defining categories of isolation suitable for different types of patient (Kunin and Henley 1967). In one such scheme (Report 1974*a*), three categories were recognized: (1) 'strict' isolation; (2) 'standard' isolation; and (3) 'stool-urine-needle' isolation, each for a separate list of diseases and for stated periods of time, and with specific requirements for nursing procedures. The last of these categories included, in addition to the normal precautions against contact infection, measures intended to prevent the spread of hepatitis B. Single rooms were considered advisable for all categories of isolation, with a separate lavatory, and a hand-washing basin in the anteroom. When these are not available in sufficient numbers, all the other precautions should be applied with particular care, and a system of priorities for isolation may have to be used. Recommendations for 'standard' and 'strict' isolation did not differ greatly (see also Lowbury *et al.* 1981) and were probably simply an expression of such priorities. 'Strict' isolation as defined cannot be relied upon to prevent the spread of the most infectious diseases, provision for which is best made in separate hospitals.

Protective isolation   Attempts are now being made to protect from infection certain classes of patient with severe immunodeficiency of limited duration. Systems of various degrees of elaboration have been proposed, from single-bed rooms with positive-pressure or plenum ventilation (Bagshawe 1964) to plastic isolators, 'air curtains' and laminar-flow ventilation (Lowbury *et al.* 1971, Lidwell and Towers 1972, Trexler *et al.* 1975). In addition, the patients' food is rendered substantially germ-free, and attempts may be made to suppress completely the flora of the alimentary tract, skin and mucous membranes by the oral or topical application of antimicrobial agents. Selwyn (1980) gives a critical assessment of the available methods.

(For a general discussion of isolation methods see Bagshawe, Blowers and Lidwell 1978.)

The use of antimicrobial agents and the control of infection

The selection of resistant strains of bacteria by these agents and their frequent transfer between patients by cross-infection have a mutually reinforcing effect. Thus, infection control is an important means of delaying the spread of resistant strains in hospitals; and the limitation of antibiotic use should, by reducing the frequency of massive overgrowth of resistant strains in the flora of treated patients, decrease the hazards of infection in other patients.

Surveys of antibiotic use in hospitals (see Report 1982 for references) suggest that between one-third and two-thirds of all courses given are either unnecessary or badly chosen, that is to say, unlikely to achieve their intended object or inordinately selective for resistant flora. Good antibiotic prescribing for *therapeutic purposes*, though much dependent on the clinical skills of the individual physician (Moss *et al.* 1981), can be fostered by collective decisions by the hospital staff to adhere to agreed guidelines (an 'antibiotic policy'), but this is unlikely to be effective unless good and up-to-date microbiological and pharmacological advice is available to the physician at all times (Report 1982).

About one-third of all courses of antimicrobial agents are given *prophylactically*. Until quite recently gross misuse for this purpose was rife (Achong *et al.* 1977, Report 1977, Moss *et al.* 1981) but a more general application of principles that are now becoming widely accepted would greatly improve the situation.

There is little doubt that appropriate antimicrobial prophylaxis reduces the frequency of sepsis after *surgical operations* on the intestinal and the infected biliary and urinary tracts, and after 'clean contaminated' gynaecological operations (Keighley and Burdon 1979, Report 1982). 'Cover' may be given either for non-sporing anaerobes only or for these organisms and gram-negative aerobic bacilli; the choice of agents for particular operations is still debatable. Prophylaxis is not held to be justifiable for most 'clean' surgical procedures. For amputation of ischaemic limbs (p. 208), however, prophylaxis with benzylpenicillin is essential. There is evidence to support its use, but with broader 'cover' for gram-positive organisms, when a foreign body is implanted surgically.

Surgical prophylaxis is likely to be effective only when an initially 'clean' site has been recently contaminated. Studies of experimental wounds (Burke 1961, Alexander and Altemeier 1965, Stone *et al.* 1976) indicate that the critical period for successful prophylaxis is the 4 hr immediately after the implantation of the organisms in the wound. The aim is to provide high blood and tissue levels of the agent at this time (Bernard and Cole 1964). Prophylaxis begun more than a few hours before operation is likely to defeat its objective by causing overgrowth of resistant bacteria at sites from which the wound may become contaminated. Prophylactic agents should be given in full dosage by a systemic route (except metronidazole, which may be given by rectal suppository) for a period not exceeding 24 hr.

There are few clear indications for *non-surgical prophylaxis* in hospital patients. It is widely practised, and often for long periods of time, for example in patients in coma or with respiratory or cardiac failure. There are many consequent ill effects, including superinfections in the patient and resistant infections in other patients in the same department. These risks may be considered acceptable for patients with severe immunodeficiency of short duration, but the consequences for other patients must be adequately guarded against by isolation.

The legitimate uses of antimicrobial agents are now so numerous that total consumption of them, at least in some hospital departments, will continue to be high. For example, in a urological department in a university hospital, in which a strictly controlled 'antibiotic policy' was followed, antimicrobial agents were given to 31 per cent of all patients (31 per cent therapeutic; 43 per cent prophylactic) and on 32 per cent of patient-days (Casewell *et al.* 1981). Under such circumstances, untoward consequences should be minimized by giving appropriate agents for as short a time as possible, maintaining strict hygienic precautions, and monitoring the appearance and controlling the spread of each new resistant bacterial strain.

References

Abramson, H. (1947) *Amer. J. Dis. Child.* **74**, 570.

Achong, M. R., Wood, J., Theal, H. K., Goldberg, R. and Thompson, D. A. (1977) *Lancet* **ii**, 1118.

Adler, J. L., Anderson, R. L., Boring, J. R. III and Nahmias, A. J. (1970) *J. Pediat.* **77**, 970.

Albert, R. K. and Condie, F. (1981) *New Engl. J. Med.* **304**, 1465.

Alexander, J. W. and Altemeier, W. A. (1965) *Surg. Gynec. Obstet.* **120**, 263.

Allen, J. C. [Ed.] (1976) *Infection and the Compromised Host.* Williams and Wilkins, Baltimore.

Allen, J. R., Hightower, A. W., Martin, S. M. and Dixon, R. E. (1981) *Amer. J. Med.* **70**, 389.

Allison, V. D. and Brown, W. A. (1937) *J. Hyg., Camb.* **37**, 153.

Almquist, E. (1891) *Z. Hyg. InfektKr.* **10**, 253.

Altemeier, W. A., Burke, J. F., Pruitt, B. A. and Sandusky, W. R. (1976) *Manual on Control of Infection in Surgical Patients.* Lippincott, Philadelphia.

Ayliffe, G. A. J. (1980) *J. Hosp. Infect.* **1**, 111.

Ayliffe, G. A. J., Brightwell, K. M., Collins, B. J. and Lowbury, E. J. L. (1969) *Lancet* **ii**, 1117.

Ayliffe, G. A. J., Brightwell, K. M., Collins, B. J., Lowbury, E. J. L., Goonatilake, P. C. L. and Etheridge, R. A. (1977) *J. Hyg., Camb.* **79**, 299.

Ayliffe, G. A. J., Collins, B. J. and Lowbury, E. J. L. (1966) *Brit. med. J.* **ii**, 442.

Ayliffe, G. A. J., Collins, B. J., Lowbury, E. J. L., Babb, J. R. and Lilly, H. A. (1967) *J. Hyg., Camb.* **65**, 515.

Ayliffe, G. A. J., Collins, B. J. and Taylor, L. J. (1982) *Hospital-acquired Infection; Principles and Prevention.* Wright PSG, Bristol.

Ayliffe, G. A. J. and Lowbury, E. J. L. (1969) *Brit. med. J.* **ii**, 337; (1982) *J. Hosp. Infect.* **3**, 217.

Bagshawe, K. D. (1964) *Brit. med. J.*, **ii**, 871.

Bagshawe, K. D., Blowers, R. and Lidwell, O. M. (1978) *Brit. med. J.* **ii**, 609, 634, 744, 808, 879.

Baird, R. M., Brown, W. R. L. and Shooter, R. A. (1976) *Brit. med. J.* **i**, 511.

Baird, R. M. and Shooter, R. A. (1976) *Brit. med. J.* **ii**, 349.

Bassett, D. C. J., Thompson, S. A. S. and Page, B. (1965) *Lancet* **i**, 781.

Bate, J. G. and James, U. (1958) *Lancet* **ii**, 713.

Beecham, H. J. III, Cohen, M. L. and Parkin, W. E. (1979) *J. Amer. med. Ass.* **241**, 1013.

Bengtsson, S., Hambraeus, A. and Laurell, G. (1979) *J. Hyg., Camb.* **83**, 41.

Bennett, J. V. and Brachman, P. S. (Eds.) (1979) *Hospital Infections.* Little, Brown and Co., Boston.

Bensted, H. J. (1936) *J. R. Army med. Cps* **67**, 295.

Bernander, S., Hambraeus, A., Myrbäck, K.-E., Nyström, B. and Sundelöf, B. (1978) *Scand. J. infect. Dis.* **10**, 66.

Bernard, H. R. and Cole, W. R. (1964) *Surgery* **56**, 151.

Bettelheim, K. A., Dulake, C. and Taylor, J. (1971) *J. clin. Path.* **24**, 442.

Blowers, R. (1963) In: *Infection in Hospitals: Epidemiology and Control*, p. 199. Ed. by R. E. O. Williams and R. A. Shooter. Blackwell, Oxford.

Bodey, G. P., Buckley, M., Sathe, Y. S. and Freireich, E. J. (1966) *Ann. intern. Med.* **64**, 328.

Bodey, G. P. and Rodriguez, V. (Eds.) (1979) *Hospital-associated Infections in the Compromised Host.* Marcel Dekker, New York.

Book, L. S., Overall, J. C., Herbst, J. J., Britt, M. R., Epstein, B. and Jung, A. L. (1977) *New Engl. J. Med.* **297**, 984.

Borden, C. W. and Hall, W. H. (1951) *New Engl. J. Med.* **245**, 760.

Brachman, P. S. and Eickhoff, T. C. (Eds) (1971) *Proc. int. Conf. on Nosocomial Infections, 1970.* American Hospital Association, Chicago.

Brandberg, Å., Seeberg, S., Bergström, G. and Nordquist, P. (1980) *J. Hosp. Infect.* **1**, 245.

Braude, A. I., Carey, F. J. and Siemienski, J. (1955) *J. clin. Invest.* **34**, 311.

Bruce, D. L. and Wingard, D. W. (1971) *Anaesthesiology* **34**, 271.

Bruun, J. N. (1970) *Acta med. scand.* suppl. 514.

Bruun, J. N. and Solberg, C. O. (1973) *Brit. med. J.* **ii**, 580.

Burke, J. F. (1961) *Surgery* **50**, 161.

Burke, J. F. and Hildick-Smith, G. Y. (Eds.) (1978) *The Infection-prone Hospital Patient.* Little, Brown and Co., Boston.

Carl, M., Blakey D. L., Francis, D. P. and Maynard, J. E. (1982) *Lancet* **i**, 1982.

Casewell, M. W., Pugh, S. and Dalton, M. T. (1981) *J. Hosp. Infect.* **2**, 55.

Casey, B. H. (1971) *Med. J. Aust.* **ii**, 718.

Castle, M. (1980) *Hospital Infection Control: Principles and Practice.* Wiley, New York.

Charnley, J. (1970) In *Aerobiology: Proceedings of the Third International Symposium*, p. 191. Ed. by I. H. Silver. Academic Press, London.

Charnley, J. and Eftekhar, N. (1969) *Lancet* **i**, 172.

Chmel, H. and Armstrong, D. (1976) *Amer. J. Med.* **60**, 203.

Clarke, S. K. R., Dalgleish, P. G., Parry, E. W. and Gillespie, W. A. (1954) *Lancet* **ii**, 211.

Cohen, J. (1983) *Proc. R. Soc. Med.* **76**, 508.

Colebrook, L., Duncan, J. M. and Ross, W. P. D. (1948) *Lancet* **i**, 893.

Colebrook, L. and Hood, A. M. (1948) *Lancet* **ii**, 682.

Cruickshank, R. (1935) *J. Path. Bact.* **41**, 367; (1944) *Brit. med. Bull.* **2**, 272.

Curry, C. R. and Quie, P. G. (1971) *New Engl. J. Med.* **285**, 1221.

Daschner, F. (Ed.) (1978) *Proven and Unproven Methods in Hospital Infection Control.* Fischer, Stuttgart; (1981*a*) *Münch. med. Wschr.* **123**, 658; (1981*b*) *Dtsch. med. Wschr.* **106**, 101.

Daschner, F., Borneff, G. G., Jackson, G. G. and Parker, M. T. (1978) In; *Proven and Unproven Methods in Hospital Infection Control*, p. 105. Ed. by F. Daschner. Fischer, Stuttgart.

Datta, N. and Pridie, R. B. (1960) *J. Hyg., Camb.* **58**, 229.

Datta, N., Richards, H. and Datta, C. (1981) *Lancet* **i**, 1181.

Davidson, A. I. G., Clark C. and Smith, G. (1971*b*) *Brit. J. Surg.* **58**, 333.

Davidson, A. I. G., Smith, G. and Smylie, H. G. (1971*a*) *Brit. J. Surg.* **58**, 326.

Davies, T. W. and Cottingham, J. (1979) *J. Infect.* **1**, 329.

Davis, B., Lilly, H. A. and Lowbury, E. J. L. (1969) *J. clin. Path.* **22**, 634.

Devenish, A. E. and Miles, A. A. (1939) *Lancet* **i**, 1089.

Dondero, T. J. Jr *et al.* (1980) *New Engl. J. Med.* **302**, 365.

Drewitt, S. E., Payne, D. J. H., Tuke, W. and Verdon, P. E. (1972) *Lancet* **i**, 1172.

Duguid, J. P. and Wallace, A. T. (1948) *Lancet* **ii**, 845.

Dutton, A. A. C. and Ralston, M. (1957) *Lancet* **i**, 115.

Dzink, J. and Bartlett, J. G. (1980) *Antimicrob. Agents Chemother.* **17**, 695.

Eickhoff, T. C. (1981) *Amer. J. Med.* **70**, 381.

Epstein, H. C., Hochwald, A. and Ashe, R. (1951) *J. Pediat.* **38**, 723.

Fallon, R. J. (1980) *J. Hosp. Infect.* **1**, 299.

Favero, M. S., McDade, J. J., Robertson, J. A., Hoffman, R. K. and Edwards, R. W. (1968) *J. appl. Bact.* **31**, 336.

Fisher-Hoch, S. P. *et al.* (1981) *Lancet* **i**, 932.

Fleming, A. and Tytler, W. H. (1923) In: *History of the Great War; Medical Service; Pathology*, p. 119. HMSO, London.

Forlenza, S. W., Newman, M. G., Lipsey, A. L., Siegel, S. E. and Blachman, U. (1980) *Lancet* **i**, 567.

Fraser, D. W. (1981) *Amer. J. Med.* **70**, 432.

Freeman, J., Rosner, B. A. and McGowan, J. E. Jr (1979) *J. infect. Dis.* **140**, 732.

Freid, M. A. and Vosti. K. L. (1968) *Arch. intern. Med.* **121**, 418.

Fujita, K., Lilly, H. A. and Ayliffe, G. A. J. (1982) *J. Hosp. Infect.* **3**, 29.

Gardner, P. S., Court, S. D. M., Brocklebank, J. T., Downham, M. A. P. S. and Weightman, D. (1973) *Brit. med. J.* **iii**, 571.

Gelbart, S. M., Reinhardt, G. F. and Greenlee, H. B. (1973) *Appl. Microbiol.* **26**, 874.

George, W. L., Sutter, V. L. and Finegold, S. M. (1978) *Curr. Microbiol.* **1**, 55.

Gill, V. J., Manning, C., Lamson, M., Woltering, P. and Pizzo, P. A. (1981) *J. clin. Microbiol.* **13**, 472.

Gillespie, W. A. (1956) *Proc. R. Soc. Med.* **49**, 1045.

Gillespie, W. A., Lennon, C. G., Linton, K. B. and Phippen, G. A. (1967) *Brit. med. J.* **ii**, 90.

Gillespie, W. A., Lennon, C. G., Linton, K. B. and Slade, N. (1964) *Brit. med. J.* **ii**, 423.

Gillespie, W. E., Linton, K. B., Miller, A. and Slade, N. (1960) *J. clin. Path.* **13**, 187.

Glencross, E. J. G. (1972) *Brit. med. J.* **ii**, 376.

Goffin, Y., Primo, G., Legrand, J. and Beers, D. van (1981) *Lancet* **i**, 1421.

Gorman, G. W. *et al.* (1980) *Ann. intern. Med.* **93**, 572.

Graevenitz, A. von and Sall, T. (Eds.) (1975) *Pathogenic Microbes from Atypical Clinical Sources.* Marcel Dekker, New York.

Grancher, J. (1900). See Williams, R. E. O. (1956).

Greenfield, C., Burroughs, A., Szawathowski, M., Bass, N., Noone, P. and Pounder, R. (1981) *Lancet* **i**, 1981.

Grob, P. J., Bischof, B. and Naeff, F. (1981) *Lancet* **ii**, 1218.

Haley, R. W., Schaberg, D. R., Allmen, S. D. von and McGowan, J. E. Jr (1980) *J. infect. Dis.* **141**, 248.

Haley. R. W. *et al.* (1981) *Amer. J. Med.* **70**, 947.

Hall, C. B. and Douglas, R. G. Jr (1981) *Amer. J. Dis. Child.* **135**, 512.

Hall, C. B., Douglas, G. Jr and Geiman, J. M. (1975*a*) *J. infect. Dis.* **132**, 151.

Hall, C. B., Douglas, R. G., Geiman, J. M. and Messner, M. K. (1975*b*) *New Engl. J. Med.* **293**, 1343.

Hambraeus, A. (1973*a*) *J. Hyg., Camb.* **71**, 787; (1973*b*) *Ibid.* **71**, 799.

Hambraeus, A., Bengtsson, S. and Laurell, G. (1978) *J. Hyg., Camb.* **80**, 169.

Hambraeus, A. and Laurell, G. (1980) *J. Hosp. Infect.* **1**, 15.

Hambraeus, A. and Ransjö, U. (1977) *J. Hyg., Camb.* **79**, 193.

Han, T., Sokal, J. E. and Neter, E. (1967) *New Engl. J. Med.* **276**, 1045.

Hare, R. and Thomas, C. G. A. (1956) *Brit. med. J.* **ii**, 840.

Hirsch, W., Sapiro-Hirsch, R., Berger, A., Winter S. T., Mayer, G. and Merzbach, D. (1965) *Lancet* **ii**, 828.

Hoffman, M. A. and Finberg, L. (1955) *J. Pediat.* **46**, 626.

Holmes, C. J. and Allwood, M. C. (1979) *J. appl. Bact.* **46**, 247.

Honda, T., Hernandez, I., Katoh, T. and Miwatani, T. (1983) *Lancet* **i**, 655.

Hovig, B. (1981) *J. Hosp. Infect.* **2**, 301.

Hurley, R. (1982) *J. Hosp. Infect.* **3**, 323.

Hurley, R. and Louvois, J. de (1980) *Proc. R. Soc. Med.* **73**, 770; (1981) *J. clin. Path.* **34**, 271.

Hurley, R., Louvois, J. de and Drasar, F. (Eds.) (1979) *Perinatal and Neonatal Infections. J. antimicrobial Chemother.* **5**, suppl. A.

Hutinel, V. (1894) See Williams, R. E. O. (1956).

Im, S. W. K., Chow, K. and Chau, P. Y. (1981) *J. Hosp. Infect.* **2**, 171.

Ip, H. M. H., Sin, W. K., Chau, P. Y., Tse, D. and Teoh-Chan, C. H. (1976) *J. Hyg., Camb.* **77**, 307.

Jarvis, J. D., Wynne, C. D., Enwright, L. and Williams, J. D. (1979) *J. clin. Path.* **32**, 732.

Jeffrey, J. S. and Sklaroff, S. A. (1958) *Lancet* **i**, 365.

Jepsen, O. B. and Mortensen, N. (1980) *J. Hosp. Infect.* **1**, 237.

Jepsen, O. B. *et al.* (1982) *J. Hosp. Infect.* **3**, 241.

Johanson, W. G., Pierce, A. K. and Sanford, J. P. (1969) *New Engl. J. Med.* **281**, 1137.

Kallings, L. O., Ringertz, O., Silverstolpe, L. and Ernerfeldt, F. (1966) *Acta pharm. suecica* **3**, 219.

Keighley, M. R. B. and Burdon, D. W. (1979) *Antimicrobial Prophylaxis in Surgery.* Pitman Medical, Tunbridge Wells.

Kelly, M. J. (1980) *Ann. roy. Coll. Surg. Engl.* **62**, 52.

Kelsey, J. C. (1972) *Lancet* **ii**, 1301.

Kiani, D., Quinn, E. L., Burch, K. H., Madhavan, T., Saravolatz, L. D. and Neblett, T. R. (1979) *J. Amer. med. Ass.* **242**, 1044.

Kim, K.-H. *et al.* (1981) *J. infect. Dis.* **143**, 42.

Kippax, P. W. and Thomas, E. T. (1966) *Lancet* **ii**, 1297.

Kist, M., Langmaack, H. and Just, M. (1980) *Dtsch. med. Wschr.* **195**, 185.

Kunin, C. M. and Henley, R. W. Jr (1967) *J. Amer. med. Ass.* **200**, 295.

Kunz, L. J. and Ouchterlony, O. T. G. (1955) *New Engl. J. Med.* **253**, 761.

LaForce, F. M. (1981) *Amer. J. Med.* **70**, 664.

Lake, A. M. and Walker, W. A. (1977) *Clin. Gastroenterol.* **6**, 463.

Lalonde, G. and Hand, R. (1980) *Canad. med. Ass. J.* **122**, 316.

Lang, D. J., Kunz, L. J., Martin, A. R., Schroeder, S. A. and Thomson, L. A. (1967) *New Engl. J. Med.* **276**, 829.

Lawrence, G., Bates, J. and Gaul, A. (1982) *Lancet* **i**, 137.

Leigh, D. A. (1981) *J. Hosp. Infect.* **2**, 207.

Lidwell, O. M. (1961) *J. Hyg., Camb.*, **59**, 259; (1963) In *Infection in Hospitals: Epidemiology and Control*, p. 43. Ed. by R. E. O. Williams and R. A. Shooter. Blackwell, Oxford; (1981) *Amer. J. Med.* **70**, 693.

Lidwell, O. M., Brock, B., Shooter, R. A., Cooke, E. M. and Thomas, G. E. (1975) *J. Hyg., Camb.* **75**, 445.

Lidwell, O. M. and Lowbury, E. J. L. (1950) *J. Hyg., Camb.* **48**, 21.

Lidwell, O. M., Lowbury, E. J. L., Whyte, W., Blowers, R., Stanley, S. J. and Lowe, D. (1982) *Brit. med. J.* **285**, 10; (1983) *J. Hosp. Infect.* **4**, 19.

Lidwell, O. M. and Mackintosh, C. A. (1978) *J. Hyg., Camb.* **81**, 433.

Lidwell, O. M., Mackintosh, C. A. and Towers, A. G. (1978) *J. Hyg., Camb.* **81**, 453.

Lidwell, O. M. and Towers, A. G. (1972) *Lancet* **i**, 347.

Lidwell, O. M., Towers, A. G., Ballard, J. and Gladstone, B. (1974) *J. appl. Bact.* **37**, 649.

Lidwell, O. M. and Williams, R. E. O. (1960) *J. Hyg., Camb.* **58**, 449.

Lilly, H. A. and Lowbury, E. J. L. (1978) *J. clin. Path.* **31**, 919.

Lilly, H. A., Lowbury, E. J. L. and Wilkins, M. D. (1979) *J. clin. Path.* **32**, 382.

Louria, D. B. and Kaminski, T. (1962) *Amer. Rev. resp. Dis.* **85**, 649.

Lowbury, E. J. L. (1960) *Brit. med. J.* **i**, 994; (1963) In *Infection in Hospitals: Epidemiology and Control*, p. 171.

Ed. by R. E. O. Williams and R. A. Shooter. Blackwell, Oxford; (1969) *Brit. J. Derm.* **81**, suppl 1, 55.

Lowbury, E. J. L., Ayliffe, G. A. J., Geddes, A. N. and Williams, J. D. (Eds.) (1981) *Control of Hospital Infection*, 2nd edn. Chapman and Hall, London.

Lowbury, E. J. L., Babb, J. R., Brown, V. I. and Collins, B. J. (1964) *J. Hyg., Camb.* **62**, 221.

Lowbury, E. J. L., Babb, J. R. and Ford, P. M. (1971) *J. Hyg., Camb.* **69**, 529.

Lowbury, E. J. L. and Fox, J. (1953) *J. Hyg., Camb.* **51**, 203.

Lowbury, E. J. L., Kidson, A., Lilly, H. A., Ayliffe, G. A. J. and Jones, R. J. (1969) *Lancet* **ii**, 448.

Lowbury, E. J. L. and Lilly, H. A. (1973) *Brit. med. J.* **i**, 510; (1975) *Lancet* **ii**, 153.

Lowbury, E. J. L., Lilly, H. A. and Ayliffe, G. A. J. (1974) *Brit. med. J.* **iii**, 369.

McCracken, G. H. and Shinefield, H. R. (1966) *Amer. J. Dis. Child.* **112**, 33.

McEntegart, M. G. (1956) *Lancet* **ii**, 909.

McKendrick, G. D. W. and Emond, R. T. D. (1976) *J. Hyg., Camb.* **76**, 23.

Mackerras, I. M. and Mackerras, M. J. (1949) *J. Hyg. Camb.* **47**, 166.

Mackintosh, C. A. (1982) *J. Hosp. Infect.* **3**, 5.

McKissock, W., Wright, J. and Miles, A. A. (1941) *Brit. med. J.* **ii**, 375.

McLauchlan, J., Pilcher, M. F., Trexler, P. C. and Whalley, R. C. (1974) *Brit. med. J.* **i**, 322.

McLeod, J. W. (1958) *Lancet* **i**, 394.

Maddocks, J. L. and May, J. R. (1969) *Lancet* **i**, 793.

Maki, D. G. (1981) *Amer. J. Med.* **70**, 719.

Maki, D. G., Anderson, R. L. and Shulman, J. A. (1974) *Appl. Microbiol.* **28**, 778.

Maki, D. G., Rhame, F. S., Mackel, D. C. and Bennett, J. V. (1976) *Amer. J. Med.* **60**, 471.

Marks, J. S. *et al.* (1979) *Ann. intern. Med.* **90**, 565.

Martin, A. J., Gardner, P. S. and McQuillin, J. (1978) *Lancet* **ii**, 1035.

Maurer, I. M. (1978) *Hospital Hygiene*, 2nd edn. Arnold, London.

Medici, M. A. and Gatti, R. A. (1978). See Burke, J. F. and Hildick-Smith, G. Y., p. 71.

Meers, P. D., Calder, M. W., Mazhar, M. M. and Lawrie, G. M. (1973) *Lancet* **ii**, 1189.

Miles, A. A. *et al.* (1940) *Brit. med. J.* **ii**, 855, 895.

Miller, A., Gillespie, W. A., Linton, K. B., Slade, N. and Mitchell, J. P. (1958) *Lancet* **ii**, 608; (1960b) *Ibid.* **ii**, 886.

Miller, A., Linton, K. B., Gillespie, W. A., Slade, N. and Mitchell, J. P. (1960a) *Lancet* **i**, 310.

Mitchell, N. J., Evans, D. S. and Kerr, A. (1978) *Brit. med. J.* **i**, 696.

Mitchell, R. G. (1983) In *Oxford Textbook of Medicine*, Vol. 1, 5.363. Ed. by D. J. Wetherall, J. G. G. Ledingham and D. A. Warrell. Oxford University Press, Oxford.

Moran, J. M., Atwood, R. P. and Rowe, M. I. (1965) *New Engl. J. Med.* **272**, 554.

Morse, L. J. and Schonbeck, L. E. (1968) *New Engl. J. Med.* **278**, 376.

Mortimer, E. A. Jr, Wolinsky, E., Gonzaga, A. J. and Rammelkamp, C. H. Jr (1966) *Brit. med. J.* **i**, 319.

Moss, F., McNicol, M. W., McSwiggan, D. A. and Miller, D. L. (1981) *Lancet* **i**, 349, 407, 461.

M'tero, S. S., Sayed, M. and Tyrrell, D. A. J. (1981) *J. Hosp. Infect.* **2**, 317.

Neter, E. (1950) *Amer. J. publ. Hlth* **40**, 929.

Ørskov, I. (1952) *Acta path. microbiol. scand.* Suppl. 93, 259; (1954) *Ibid.* **35**, 194.

Ojajärvi, J., Mäkelä, P. and Rantasalo, I. (1977) *J. Hyg., Camb.* **79**, 107.

Okell, C. C. and Elliott, S. D. (1936) *Lancet* **ii**, 836.

Onderdonk, A. B., Lowe, B. R. and Bartlett, J. G. (1979) *Appl. environm. Microbiol.* **38**, 637.

Overall, J. C. Jr (1970) *J. Pediat.* **76**, 499.

Park, S. K., Brody, J. L., Wallace, H. A. and Blakemore, W. S. (1971) *Lancet* **i**, 53.

Parker, M. T. (1969) *Brit. med. J.* **iii**, 671; (1978) (Ed.) *Hospital-acquired Infections: Guidelines to Laboratory Methods.* (*World Hlth Org. Reg. Publ. Europ. Ser.* No. 4.) Copenhagen.

Parker, M. T., John, M., Emond, R. T. D. and Machacek, K. A. (1965) *Brit. med. J.* **i**, 1101.

Pearson, T. A., Braine, H. G. and Rathbun, H. K. (1977) *J. Amer. med. Ass.* **238**, 1737.

Pearson, R. D., Valenti, W. M. and Steigbigel, R. T. (1980) *J. Amer. med. Ass.* **244**, 1128.

Petrini, B., Welin-Berger, T. and Nord, C.-E. (1979) *Med. Microbiol. Immunol.* **167**, 155.

Pettet, J. D., Baggenstoss, A. H., Dearing, W. H. and Judd, E. S. Jr (1954) *Surg. Gynec. Obstet.* **98**, 546.

Pettit, F. and Lowbury, E. J. L. (1968) *J. Hyg., Camb.* **66**, 393.

Phillips, I., Eykyn, S. and Laker, M. (1972) *Lancet* **i**, 1258.

Phillips, I., Meers, P. D. and D'Arcy, P. F. (1976) *Microbiological Hazards in Infusion Therapy.* MTP Press, Lancaster.

Phillips, I. and Spencer, G. (1965) *Lancet* **ii**, 1325.

Phillips, I., Warren, C., Harrison, J. M., Sharples, P., Ball, L. C. and Parker, M. T. (1976) *J. med. Microbiol* **9**, 393.

Pierce, A. K., Edmonson, E. B., McGee, G., Ketchersid, J., Loudon, R. G. and Sanford, J. P. (1966) *Amer. Rev. resp. Dis.* **94**, 309.

Pierce, A. K. and Sanford, J. P. (1974) *Amer. J. resp. Dis.* **110**, 647.

Platt, R., Polk, B. F., Murdock, B. and Rosner, B. (1983) *Lancet* **i**, 893.

Polakoff, S., Cossart, Y. E. and Tillett, H. E. (1972) *Brit. med. J.* **iii**, 94.

Pollack, M., Charache, P., Nieman, R. E., Jett, M. P., Reinhardt, J. A. and Hardy, P. H. Jr (1972) *Lancet* **ii**, 668.

Powell-Tuck, J., Lennard-Jones, J. E., Lowes, J. A., Danso, K. T. and Shaw, E. J. (1979) *J. clin. Path.* **32**, 549.

Pyrah, L. N., Goldie, W., Parsons, F. M. and Raper, F. P. (1955) *Lancet* **ii**, 314.

Raahave, D. (1979) *Bacterial Densities in Operation Wounds.* FADL's Forlag, Copenhagen.

Raahave, D., Hansen, O. H., Carstensen, H. E. and Friis-Møller, A. (1981) *Acta chir. scand.* **147**, 215.

Ransjö, U. (1978) *Scand. J. infect. Dis.*, suppl. 11; (1979) *J. Hyg., Camb.* **82**, 369.

Ransjö, U. and Hambraeus, A. (1982) *J. Hosp. Infect.* **3**, 81.

Reinarz, J. A., Pierce, A. K., Mays, B. B. and Sandford, J. P. (1965) *J. clin. Invest.* **44**, 831.

Remington, J.. S. and Klein, J. O. (1976) *Infectious Diseases of the Fetus and Newborn Infant.* Saunders, Philadelphia, London and Toronto.

Report (1959) *Staphylococcal Infections in Hospitals.* HMSO London; (1960) *Lancet* **ii**, 659; (1962a) *Lancet* **ii**, 945; (1962b) *The Sterilization, Use and Care of Syringes* (Med. Res. Counc. Memo. No. 41). HMSO, London; (1964) *Ann.* *Surg.* **160**, suppl. 2; (1965a) *J. Hyg., Camb.* **63**, 457; (1965b) *Lancet* **ii**, 895; (1966) *Brit. med. J.* **i**, 313; (1968a) *Lancet* **i**, 705, 763, 831; (1968b) *Brit. med. J.* **iii**, 454; (1971) *Pharm. J.* **207**, 96; (1972a) *Ventilation in Operating Suites.* HMSO, London; (1972b) *Hepatitis and the Treatment of Chronic Renal Failure.* HMSO, London; (1974a) *Brit. med. J.* **ii**, 41; (1974b) *Brit. med. J.* **iv**, 751; (1975) *J. Hyg., Camb.* **74**, 133; (1976) *Brit. med. J.* **ii**, 1579; (1977) *Lancet* **i**, 1351; (1979) *Infection Control in the Hospital*, 4th edn. American Hospital Association, Chicago; (1980a) *Amer. J. Epidem.* **111**, 465; (1980b) *J. Hosp.Infect.* **1**, 307; (1980c) *Lancet* **i**, 1; (1981a) *Amer. J. Med.* **70**, 379, 631, 899; (1981b) *J. Hosp. Infect.* **2** suppl.; (1982) *World Hlth Org.*, no. WHO/BVI/PHA/ANT/82.1.

Rhame, F. S., Root, R. K., MacLowry, J. D., Dadisman, T. A. and Bennett, J. V. (1973) *Ann. intern. Med.* **78**, 633.

Roe, E. and Lowbury, E. J. L. (1972) *J. clin. Path.* **25**, 176.

Rogers, K. B. (1951) *J. Hyg., Camb.* **49**, 497; (1980) *J. Hosp. Infect.* **2**, 1.

Rose, H. D. and Schreier, J. (1968) *Amer. J. med. Sci.* **255**, 228.

Rowe, B., Giles, C. and Brown, G. L. (1969) *Brit. med. J.* **iii**, 561.

Rowe, B. and Hall, M. L. M. (1975) *Brit. med. J.* **iv**, 51.

Rubbo, S. D. (1948) *J. Hyg., Camb.* **46**, 158.

Rubenstein, A. D. and Fowler, R. N. (1975) *Amer. J. publ. Hlth* **45**, 1109.

Rubin, R. H. and Young, L. S. (Eds) (1981) *Clinical Approach to Infection in the Compromised Host.* Plenum, New York.

Russell, A. D., Hugo, W. B. and Ayliffe, G. A. J. (Eds) (1982) *Principles and Practice of Disinfection, Preservation and Sterilization.* Blackwell, Oxford.

Ryder, R. W., Crosbie-Ritchie, A., McDonough, B. and Hall, W. J. III (1977) *J. Amer. med. Ass.* **238**, 1533.

Sanderson, P. J. (1983) *J. Hosp. Infect.* **4**, 15.

Sanford, D. A., Leslie, D. A., McKeon, J. A., Crone, P. P. and Hobbs, B. C. (1969) *J. Hyg., Camb.* **67**, 75.

Schaberg, D. R., Rubens, C. E., Alford, R. H., Farrar, W. E., Schaffer, W. and McGee, Z.A. (1981) *Amer. J. Med.* **70**, 445.

Schaberg, D. R., Weinstein, R. A. and Stamm, W. E. (1976) *J. infect. Dis.* **133**, 363.

Seeberg, S., Brandberg, Å., Bergström, G. and Nordquist, P. (1982) *J. Hosp.Infect.* **3**, 159.

Selwyn, S. (1980) *J. Hosp. Infect.* **1**, 5.

Shooter, R. A. et al. (1969) *Lancet* **i**, 1227.

Shuttleworth, R., Taylor, M.and Jones, D. M. (1980) *J. clin. Path.* **33**, 1002.

Siegel, J. D. and McCracken, G. H. Jr (1981) *New Engl. J. med.* **304**, 642.

Skinner, P. R., Taylor, A. J. and Coakham, H. (1978) *J. clin. Path.* **31**, 1085.

Slade, N. (1980) *Proc. R. Soc. Med.* **73**, 739.

Smits, H. and Freedman, L. R. (1967) *New Engl. J. Med.* **276**, 1229.

Speers, R. Jr, Shooter, R. A., Gaya, H., Patel, N. and Hewitt, J. H. (1969) *Lancet* **ii**, 233.

Spooner, E. T. C. (1941) *J. Hyg., Camb.* **41**, 320.

Stamm, W. E., Weinstein, R. A. and Dixon, R. E. (1981) *Amer. J. Med.* **70**, 393.

Steere, A. C. et al. (1975) *Lancet* **i**, 319.

Steinbach, H. L., Rousseau, R., McCormack, K. R. and Jawetz, E. (1960) *J. Amer. med. Ass.* **174**, 1207.

Stone, H. H., Hooper, C. A., Kolb, L. D., Geheber, C. E. and Dawkins, E. J. (1976) *Ann. Surg.* **184,** 443.

Svanbom, M. (1979) *Scand. J. infect. Dis.* **11,** 187.

Swartzberg, J. E., Maresca, R. M. and Remington, J. S. (1977) *J. infect. Dis.* **135,** S99.

Tanner, E. I., Gray, J. D., Rebello, P. V. N. and Gamble, D. R. (1969) *J. Hyg., Camb.* **67,** 477.

Thomas, D. F. M. (1982) *Proc. Roy. Soc. Med.* **75,** 838.

Tobin, J. O'H., Swann, R. A. and Bartlett, C. L. R. (1981) *Brit. med. J.* **282,** 515.

Tobin, J. O'H. *et al.* (1980) *Lancet* **ii,** 118.

Toivanen, P., Toivanen, A., Olkkonen, L. and Aantaa, S. (1973) *Lancet* **i,** 801.

Trexler, P. C., Spiers, A. S. D. and Gaya, H. (1975) *Brit. med. J.* **iv,** 549.

Turck, M. and Stamm, W. (1981) *Amer. J. Med.* **70,** 651.

Tyras, D. H., Kaiser, G. C., Barner, H. B., Laskowski, L. F. and Marr, J. J. (1978) *J. thorac. cardiovasc. Surg.* **75,** 331.

Valenti, W. M., Trudell, R. G. and Bentley, D. W. (1978) *New Engl. J. Med.* **298,** 1108.

Walter, C. W. (1948) *The Aseptic Treatment of Wounds.* MacMillan, New York.

Wassermann, M. (1901) *Virchows Arch.* **165,** 342.

Watt, J. and Carlton, E. (1945) *Publ. Hlth Rep., Wash.* **60,** 734.

Weinstein, R. A. and Stamm, W. E. (1977) *Lancet* **ii,** 862.

Wenzel, R. P., Osterman, C. A. and Hunting, K. J. (1976*b*) *Amer. J. Epidem.* **104,** 645

Wenzel, R. P., Osterman, C. A., Hunting, K. J. and Gwaltney, J. M. Jr (1976*a*) *Amer. J. Epiderm.* **103,** 251.

Whyte, W., Hodgson, R. and Tinkler, J. (1982) *J. Hosp. Infect.* **3,** 123.

Whyte, W., Hodgson, R., Tinkler, J. and Graham, J. (1981) *J. Hosp. Infect.* **2,** 219.

Williams, R. E. O. (1956) *Bull. Hyg.* **31,** 965.

Williams, R. E. O., Blowers, R., Garrod, L. P. and Shooter, R. A. (1966) *Hospital Infection, Causes and Prevention,* 2nd edn. Lloyd-Luke, London.

Williams, R. E. O., Clayton-Cooper, B., Howat, T. W. and Miles, A. A. (1945) *Brit. J. Surg.* **32,** 425.

Wormald, P. J. (1970) *J. Hyg., Camb.* **68,** 633.

Young, L. S. (1981) *Amer. J. Med.* **70,** 398.

Young, V. M., Meyers, W. F., Moody, M. R. and Schimpff, S. C. (1981) *Amer. J. Med.* **70,** 646.

59

Streptococcal diseases

M. T. Parker

Introductory

In this chapter we have space to give adequate consideration only to some of the more important streptococcal diseases of man; others will be discussed elsewhere in the book (see Table 59.1). At the end of the chapter we give a brief summary of the streptococcal diseases of animals.

Table 59.1 Streptococcal diseases of man

Streptococcus (Lancefield group)	Diseases
pyogenes (A)	*Local infection:* of respiratory tract, wounds, skin, etc. *Systemic infection* *Other acute diseases:* scarlet fever; erysipelas *Late sequelae:* rheumatic fever; acute glomerulonephritis
equisimilis (C) *sp.* (G)	Local and systemic infection
agalactiae (B)	Septicaemia and meningitis
suis, serotype 2 ('R')	Meningitis
milleri (usually ungroupable)*	Abscesses in internal organs; septicaemia; endocarditis (see also Chapter 57)
pneumoniae	Pneumonia (Chapter 67); meningitis (Chapter 65); septicaemia; otitis media
faecalis (D)	Local infection; septicaemia; endocarditis (Chapter 57)
sanguis *mitior* *mutans* *bovis* (D)	Endocarditis (Chapter 57)

* Some strains A, C, G, or F.
  For *Str. mutans* and dental disease, see Chapter 57.
Note. Almost any streptococcus may occasionally cause endocarditis.

The greater part of the chapter is devoted to diseases caused by *Streptococcus pyogenes* (the group A streptococcus); these occur almost exclusively in man, for whom it is an important and remarkably versatile pathogen. The group A streptococcus is responsible not only for several types of acute local inflammation and occasionally for a highly fatal septicaemic disease; it also causes certain acute non-suppurative manifestations—scarlet fever and erysipelas—and late sequelae—rheumatic fever and post-streptococcal glomerulonephritis—that are almost unique in human and animal pathology. Other β-haemolytic streptococci, members of groups B, C and G, cause acute inflammatory diseases in man as well as in animals, but the human infections are seldom acquired from an animal source. Occasional human infections with *Str. suis* and with the *zooepidemicus* species of group C streptococci appear to be almost the only zoonotic streptococcal infections.

We shall describe the septic infections by *Str. milleri* in this Chapter, and those due to pneumococci and faecal streptococci elsewhere (see Table 59.1). Several other streptococci that rarely cause local sepsis are important causes of subacute endocarditis; this is discussed in Chapter 57, as is the role of streptococci in dental disease.

Diseases due to *Streptococcus pyogenes*

Acute infections of the respiratory tract

Streptococci were recognized as causes of sepsis in the 1880s (Chapter 29) but their role as respiratory pathogens was not understood clearly until many years later. At that time, scarlet fever was a common and serious disease the generalized features of which tended to distract attention from the primary streptococcal infection. Streptococci had been isolated from the blood of a proportion of the cases (Klein 1887), but they failed to reproduce the disease when inoculated into animals, and most workers regarded them merely as important secondary invaders. Much later (see Butler 1909) it was realized that cases of sore throat and scarlet fever occurred together in families, and also in community outbreaks, especially those associated with milk, for which it might be supposed that a single infectious agent was responsible. Later, Lancefield (1928) showed that practically all streptococci from both diseases belonged to group A, and the serological studies of Griffith and others (Glover and Griffith 1931, Griffith 1934) demonstrated that a single type of streptococcus, spreading in a closed community such as a school, would produce scarlet fever in some persons and sore throat in others.

Streptococcal infection of the respiratory tract is a disease of temperate and cold climates and occurs mainly in the winter and spring. In Britain we usually refer to streptococcal sore throat as 'tonsillitis', but there is some justification for the American term 'pharyngitis', because in the early acute stage there is widespread inflammation of the mucosa of the pharynx and nasopharynx. The appearances of acute follicular tonsillitis—sore throat, pain on swallowing, swelling and reddening of the tonsils with discrete white spots on their surface, enlargement of the anterior cervical lymph nodes, fever and leucocytosis—are often considered diagnostic of streptococcal disease; but in fact it is difficult to make a reliable clinical diagnosis, because the typical appearances are limited commonly to older children and adolescents, and almost any combination of these symptoms and signs may occur in other infections of the upper respiratory tract (Glezen *et al.* 1967, Kaplan *et al.* 1971, Valkenburg *et al.* 1971).

Streptococcal tonsillitis is rare in infants. Among pre-school children it is not very frequent and its manifestations are usually indefinite (Powers and Boisvert 1944, Alpert *et al.* 1966); infection of the throat may be overshadowed by the appearance of otitis media or cervical adenitis. Among school children and adolescents it is at times very common and may constitute 20–40 per cent of all cases of sore throat; this figure varies however not only with time and place but also according to the criteria used to establish the clinical and the bacteriological diagnosis (Kaplan *et al.* 1971). It is in this age group that the more severe and clinically typical throat infections are seen, and that the non-suppurative sequelae—especially rheumatic fever—occur most often; nevertheless, many cases are mild and most are uncomplicated. When children and young adults are crowded together, for example, in residential schools and military camps, large outbreaks may occur; these affect most severely recent entrants to the group, and are sometimes followed by epidemics of rheumatic fever or nephritis (Rantz *et al.* 1945, 1946). During adult life, attacks of streptococcal tonsillitis become progressively less frequent.

After the initial generalized inflammation of the mucosa of the upper respiratory tract, localization usually takes place in the lymphatic tissues around the pharynx. From here, inflammation may spread to adjacent areas, giving rise to a peritonsillar or retropharyngeal abscess or to otitis media. More commonly, spread is by the lymphatic route to the anterior cervical lymph nodes; occasionally it is along the connective-tissue planes of the neck. Bronchitis and pneumonia are rare, but have been seen in association with measles or influenza. Invasion of the blood stream from the respiratory tract occasionally leads to septicaemia.

Patients suffering from streptococcal tonsillitis often show symptoms and signs of general illness, and these may persist after the acute disease has subsided. In addition to those who subsequently suffer from rheumatic fever, other patients may have pyrexia, a raised erythrocyte-sedimentation rate and abnormal electrocardiographic appearances in the weeks following an attack of tonsillitis. The existence of this so-called 'post-streptococcal state' (Rantz *et al.* 1945, 1947a, Wheeler and Jones 1945), may indicate that the borderline between an uncomplicated attack of tonsillitis and one followed by rheumatic fever is not clear cut.

Little is known about the way in which *Str. pyogenes* infects the respiratory tract; and the part played in the process by the various exotoxins described in Chapter 29 is uncertain. As we have seen (Chapter 29), surface lipoteichoic acid appears to be responsible for the adherence of streptococci to buccal epithelium. There is good reason to believe that the antiphagocytic activity of the M antigen is concerned in the invasion of tissues, because the presence of antibody to it effectively prevents clinical infection in man. Wannamaker and his colleagues (1953a) studied respiratory-tract infections in airmen whose blood had previously been examined for antibody to the prevalent M types; the attack rate in men without antibody to the infecting type was 13 per cent, and in men with the corresponding antibody it was 2 per cent. Once M antibody has appeared in the blood it often persists for many months or years (Lancefield 1959). It may well be that the progressive decrease in the frequency of disease during adult life is due to the accumulation of specific immunity to more and more streptococcal types (see p. 238). Whether the hyaluronic-acid capsule also has significant antiphagocytic activity in natural infections is less certain. Many strains responsible for severe epidemics of respiratory-tract infection are heavily capsulated.

In the temperate zone, the M types most commonly responsible for tonsillitis are nos. 1, 3, 4, 5, 6, 9, 12, and 18, which usually have strongly antigenic M proteins (see Chapter 29) and, with the exception of type 4, fail to form serum opacity factor. In the tropics these types are rarely seen, and few of the strains associated with clinical tonsillitis have yet been characterized. The temperate-zone 'throat' types are rarely isolated from cases of impetigo but often cause wound infection. M types associated with impetigo are seldom the cause of tonsillitis but may colonize the throat (p. 229).

Septic infections

When *Str. pyogenes* gains entry through the skin it may be responsible for one of several types of local lesion, or it may cause only minimal inflammation at the site of implantation but spread rapidly throughout the body. As in streptococcal disease of the respiratory tract, dissemination is usually by the lymphatic route in the first instance. Infection of quite a small puncture wound, for example, may lead to an acute cellulitis of the subcutaneous tissues and the neighbouring fascial planes, or to a rapidly spreading lymphangitis. The lymph nodes become swollen, but seldom appear to prevent the onward passage of the organism, which may reach the blood stream within a few hours. Unless treatment is given, streptococcal septicaemia often leads to death within 24–36 hr. Occasionally the first sign of illness is acute inflammation of a deep seated tissue, for example, a joint; this suggests that a transient bacteraemia sometimes occurs. Acute streptococcal myositis (Barrett and Gresham 1958) may arise in a similar way.

The consequences of *surgical wound infection* vary from trivial local inflammation to acute fatal generalized infection; and if the organisms are introduced into deep tissues, for example, into the peritoneal cavity at the time of operation, an overwhelming infection, sometimes without fever or localizing signs, may develop within a few hours. Nevertheless, symptomless colonization of wounds is not uncommon, particularly when the organism is introduced some time after infliction of the wound. Infection of burns, in addition to causing a variable amount of local inflammation and giving rise to an occasional case of septicaemia, is a frequent cause of failure of skin-grafting (Jackson *et al.* 1951). Many outbreaks of sepsis have followed the inadvertent injection of streptococci subcutaneously or intramuscularly (Allison 1938, Cayton and Morris 1966, Plueckhahn and Banks 1970). This sometimes leads to local abscess formation, but fatal septicaemic infections have been recorded.

Mention must be made of the type of local streptococcal infection sometimes called *necrotizing fasciitis*. This was described by Meleney (1924) as 'streptococcal gangrene', and is perhaps identical with the 'hospital gangrene' so often seen in the nineteenth century. It usually, but by no means always, follows trauma. After an initial cellulitis, dusky areas appear, often with haemorrhagic bullae; these ulcerate, revealing an extensive area of gangrene beneath; necrosis may extend widely along tissue planes, and metastatic abscesses may occur. The fatality rate is high even with penicillin treatment (see Paine *et al.* 1963, Beathard and Guckian 1967, Buchanan and Haserick 1970, Seal and Leppard 1982). This form must be distinguished from *progressive bacterial synergistic gangrene*, also described by Meleney, which spreads more slowly and nearly always follows a surgical operation (Chapter 60).

Acute *cellulitis of the perianal region* (Amren *et al.* 1966) and acute *purulent vaginitis* are diseases of children, and appear to be associated with streptococcal infection of the respiratory tract in the patient or a family contact; several cases of perianal cellulitis may occur in the same family (Hirschfeld 1970).

Puerperal fever

This may be looked upon as a special case of streptococcal wound infection in which the organism invades the placental site and rapidly enters the blood stream. It was once the most common cause of serious infection in the puerperium. Dora Colebrook (1935), in reviewing the results of various workers, noted that, among the fatal cases, the proportion due to haemolytic streptococci varied from 68 to 97 per cent. Lancefield and Hare (1935) showed that most of the streptococci belonged to group A, and a small proportion to groups B, C, D and G. The present frequency of group A streptococcal infection in the puerperium is difficult to estimate because it is common practice to give an antibiotic as soon as fever develops and before a swab has been collected. Fatalities are now rare in hospital practice.

The main features of the epidemiology of puerperal fever were established by careful field investigations long before its bacterial cause had been discovered. Evidence that it was an infectious disease, and that it was often conveyed by the obstetrician or midwife, was first brought forward by Alexander Gordon of Aberdeen in 1795, again in 1843 by Oliver Wendell Holmes in the USA, and most strongly of all in 1861 by Semmelweis in Vienna, but found little acceptance at the time. The causative organism was seen in the lochial exudate by Coze and Feltz in 1869 and in the blood by Pasteur in 1879 (for early references see L. Colebrook 1956), but even after the bacterial cause had been established, and the Listerian revolution in hospital practice had taken place, puerperal fever continued to be an important cause of maternal mortality. The Registrar-General's returns show that from 1880 to 1930 there were about 2000 deaths a year from sepsis associated with childbirth. This figure had been reduced by little more than a fifth by 1935, when the first effective chemotherapeutic agent was introduced; by 1944 the number of deaths had fallen to 16 per cent of the 1930 figure and by 1948 to 6 per cent. We must conclude, therefore, that the conquest of puerperal fever, in Britain at least, was a triumph of chemotherapy rather than of preventive medicine.

Neonatal sepsis

An occasional case of acute septicaemia or meningitis due to *Str. pyogenes* may occur in newborn infants. These infections are usually very acute and are seldom preceded by obvious local sepsis. A series of almost simultaneous investigations (Boissard and Eton 1956, Gray 1956, Kwantes and James 1956, Langewisch 1956) revealed that widespread colonization of the umbilical stump of newborn infants in a nursery might occur with little or no evidence of inflammation, and escape attention until either a baby suffered from a serious infection or an outbreak of puerperal fever occurred in the maternity ward. The rarity of serious disease in the newborn may perhaps be attributed to maternally acquired protective antibodies (Zimmerman and Hill 1969).

Streptococcal impetigo

In this disease, the lesion is a superficial discrete crusted spot seldom exceeding an inch in diameter, which lasts 1–2 weeks and heals spontaneously without leaving a scar. A somewhat more severe type of lesion—termed *streptococcal ecthyma*—consists of an ulcerated area of similar size, often with a raised, oedematous edge. The lesions may appear singly or in groups, and several at different stages of evolution may be present at one time; they are nearly always found on exposed areas of skin. Unlike other forms of streptococcal sepsis, invasion from the impetigenous lesion is limited to a mild lymphadenitis and occasionally a little cellulitis of the surrounding tissues. The lesion causes little pain, there are few if any general symptoms and a scarlatiniform rash rarely appears (see Wannamaker 1970).

Engman (1901) distinguished clearly between streptococcal impetigo, which begins as a tiny vesicle and soon develops a thick 'stuck-on' crust, and staphylococcal impetigo, which begins as a larger vesicle or bulla and after rupture is covered by a thin crust (Chapter 60). Balmain (1926) and Haxthausen (1927) found that the streptococci isolated from the former type of lesion were β-haemolytic, and Evans (1941) showed that they were usually members of group A. Secondary infection of the lesion with *Staph. aureus* is very common; coagulase-negative staphylococci, haemolytic streptococci of other groups, and diphtheria bacilli may also be present in the lesions. Toxigenic diphtheria bacilli may be present as secondary invaders in impetigo. They usually have little or no effect on the appearance of the lesion (Belsey *et al.* 1969) but occasionally give rise to diphtheritic paralysis. Cases of faucial diphtheria may occur among non-immunized contacts (Bray *et al.* 1972).

The serotypes of group A streptococci responsible for impetigo are different from those that cause tonsillitis. In 1955 Parker, Tomlinson and Williams observed that strains from impetigo lesions in Britain very seldom had M antigens that were detectable with the typing sera then available, thus differing from strains isolated from throat infections in the same geographical area. Wannamaker and his colleagues (see Top *et al.* 1967) studied M-untypable strains from impetigo lesions in American Indian children in small communities and identified several hitherto unrecognized M types. This led to an intensive study of strains from skin lesions in geographical areas in which impetigo was prevalent and often resulted in glomerulonephritis, notably in the USA and Central America (Anthony *et al.* 1967*b*, Dillon *et al.* 1967, 1968, 1974, Johnson *et al.* 1968, Parker *et al.* 1968, Potter *et al.* 1968, Kaplan *et al.* 1970*b*, Bassett 1972, Nelson *et al.* 1976). Many new M types were discovered, and it became clear that some of them had a specific association with the skin. These 'skin' types were responsible for impetigo but seldom for tonsillitis, though they were often found in the throat in localities in which they caused impetigo (Dillon *et al.* 1967); among the more common are M types 33, 41, 43, 49, 52, 55, 56, and 60, and M-type 2 strains that have an agglutination pattern in the 8,25,Imp.19 complex (Wannamaker 1970). Most but by no means all of these 'skin' types form opacity factor and have poorly antigenic M proteins (Chapter 29). The distinction between 'skin' and 'throat' types is, however, not absolute, and some types, for example, nos 58 and 65, occupy an intermediate position (Dillon *et al.* 1975, Anthony *et al.* 1976).

In Europe, streptococcal impetigo affects mainly young children living under poor hygenic conditions, and adolescents and adults in mental institutions, prisons and athletic teams. In Britain, it commonly affects workers who handle raw meat, among whom it may cause extensive outbreaks (Fraser *et al.* 1977, Report 1979). Streptococcal impetigo seldom has serious consequences nowadays in Europe, but in many other parts of the world it is often followed by acute glomerulonephritis (p. 235).

Two important determinants for impetigo are poor socioeconomic conditions and a warm, humid climate (Taplin *et al.* 1973); yet it is not in any sense a tropical disease, because epidemics may occur at quite high latitudes during the summer months. In tropical countries, it is essentially a disease of children in poor communities, particularly in country districts. Prevalence rates may be very high; Bassett (1972) found that on average 15 per cent of the children in one rural school in Trinidad were infected throughout the whole year, and calculated that there was at least one new streptococcal infection per child each 10 weeks. The disease was somewhat less common in urban schools. Impetigo, unlike streptococcal tonsillitis, affects mainly pre-school children and tends to decrease in frequency throughout the school (see also Wannamaker 1970). As we shall see (p. 241), the main route of infection is from skin to skin; colonization of the skin with streptococci often precedes the appearance of the lesion. Nevertheless, there is evidence that small traumatic lesions provide a portal of entry for the organism. Lesions occur mostly on unclothed parts of the body and their distribution often appears to correspond with that of insect bites. Scabies lesions may sometimes determine the site and frequency of streptococcal impetigo (Rumao and Sant 1971, Svartman *et al.* 1972). Experiments on human subjects (Leyden *et al.* 1980) showed that group A streptococci applied to the skin surface did not cause a clinical infection even when it was kept moist. When 10^4 organisms or more were rubbed into the scarified skin, however, a lesion appeared; if evaporation from the surface was prevented after the application of the organism the severity of the lesion was enhanced.

Scarlet fever

In Britain in the second half of the nineteenth century scarlet fever was one of the commonest causes of death from infectious diseases in children over the age of one year. In 1861–70, the annual mortality rate from scarlet fever under the age of 15 years in England and Wales was 2617 per million. A steady fall in mortality began long before any effective therapeutic measures were introduced, and by 1921–30 the rate was down to 73 per million. Fatal cases have now ceased to occur, and the disease, though still quite common, is very mild and tends not to be officially notified. The disease has gone through a similar evolution in other countries in Europe and North America (see, for example, Măgureanu and Feller 1966).

Landsteiner and his colleagues (1911) made the initial observation that a disease resembling scarlet fever could be produced in monkeys by inoculating the throat with faucial exudate from patients with the disease. The subsequent proof that scarlet fever is a consequence of streptococcal infection was obtained from a series of observations on man. Krumwiede, Nicoll and Pratt (1914) recorded an instance in which a laboratory worker sucked a culture of streptococci into her mouth and had a typical attack of scarlet fever three days later. Dick and Dick (1921, 1923) produced several attacks of tonsillitis and a single case of scarlet fever by inoculating cultures of haemolytic streptococci into the throat of volunteers; application to the throat of filtrates of cultures was without effect. When a small volume of diluted filtrate was injected intradermally, however, it produced a typical local erythematous reaction in some normal persons but not in others (Dick and Dick 1924*a*, *b*, 1925*a*, *b*). The administration of larger doses to susceptible persons yielded even more demonstrative results. Sometimes they induced a generalized reaction with fever, nausea, vomiting and a transient scarlatiniform rash—so-called *miniature scarlet fever*. The skin reaction did not occur in persons who had had scarlet fever, nor in persons injected with a serum prepared (Moser 1902) by inoculation of horses with scarlatinal streptococci. It thus appeared that a toxic substance produced by haemolytic streptococci caused the symptoms of scarlet fever, but that only a minority of the normal adult population was susceptible to it.

Susceptibility to scarlet fever

Schultz and Charlton (1918) showed that serum from a patient convalescent from scarlet fever, when injected into the skin of a patient displaying a scarlatiniform rash, caused blanching of the rash immediately around the site of injection; serum withdrawn at the height of the attack did not do this (see also Mair 1923). Trask and Blake (1924) demonstrated that the serum of patients acutely ill with scarlet fever produced an erythematous reaction when given intracutaneously to persons who had not had scarlet fever and whose serum gave a negative Schultz-Charlton reaction. It thus appeared that the Dick test—the intradermal injection of diluted filtrate of a broth culture of a scarlatinal strain of streptococcus—was a means of determining susceptibility to scarlet fever, and that a negative Dick test was an indication of antitoxic immunity. Surveys of the population (e.g., Zingher 1924) showed that some 55 per cent of young infants were Dick-negative, that this proportion fell to under 30 per cent by the age of 2 years, and then rose progressively through childhood to reach 77 per cent in the age group 10–15 years and 86 per cent in adults. An analogy was drawn with the Schick test for susceptibility to diphtheria toxin, and the steady rise in the percentage of negatives during childhood was taken as evidence of the acquisition of active antitoxic immunity either by passing through an attack of scarlet fever or by having an inapparent infection. The fall in the proportion of negatives during infancy could not, however, be attributed exclusively to the disappear-

ance of passive immunity derived from the mother, because Cooke (1927, 1928*a*, *b*) showed that the child of a Dick-positive mother might be Dick-negative in early infancy but become Dick-positive some weeks or months later. The blood of 'early' Dick-negative infants did not neutralize erythrogenic toxin, but the blood of an initially Dick-positive infant who subsequently became Dick-negative did. Such observations led Dochez and Stevens (1927) to conclude that scarlet fever was not a result of the direct toxicity of the erythrogenic substance but was an allergic reaction to it.

The allergic concept of scarlet fever was not widely accepted, but has been revived in modified form by Watson and his colleagues (Kim and Watson 1970, Watson and Kim 1970, Schlievert *et al.* 1979) as a result of extensive experiments in animals with purified erythrogenic toxin. It will be recalled (Chapter 29) that the toxin does not cause an erythematous reaction in young experimental animals, but is responsible for fever and death. Watson prefers to call it 'streptococcal pyrogenic exotoxin'; he believes that it is an important factor in the pathogenicity of the streptococci, but that alone it is not responsible for the complete clinical syndrome of scarlet fever. In his opinion, the toxin is combined in the bacterial cell with a heat-stable antigenic material and with hyaluronic acid; streptococcal infection results in a delayed-type hypersensitivity to this, and one of the consequences of this is hyper-reactivity to the erythrogenic toxin. The heat-stable antigen is common to streptococci and other organisms found in the intestinal flora—at least of experimental animals—so that hypersensitivity may perhaps develop in the absence of a streptococcal infection. Scarlet fever, then, is thought by these workers to result from a combination of the direct toxicity of the pyrogenic exotoxin with the secondary toxicity caused by hypersensitivity to the heat-stable part of the antigenic complex. Thus, a person might give a negative Dick reaction for one of two reasons: a very young child may be neither hypersensitive nor immune, and an adult, after repeated streptococcal infections, might be hypersensitive but have antibody to erythrogenic toxin. Second attacks of scarlet fever are uncommon, and may be attributable to failure to form antibody to erythrogenic toxin, to the disappearance of antibody with time, or to successive infections with streptococci that form erythrogenic toxin of a different antigenic type (Chapter 29).

The streptococci isolated from cases of scarlet fever are almost exclusively members of group A; definite evidence that other streptococci can cause the disease is lacking. A disease somewhat resembling scarlet fever may occasionally result from infection with certain strains of *Staph. aureus* (Chapters 30 and 60). For reasons that have not been satisfactorily explained, certain serological types of group A streptococci appear to have a greater ability to cause scarlet fever than others, but these types vary from country to country and at different times. Thus, in Britain, type 4 was for over 20 years isolated more often from scarlet fever than from tonsillitis (Report 1954, 1957). In 1964–65, the commonest type associated with scarlet fever was type 4 in Britain, type 22 in East Germany, and type 1 in Holland and the USSR (Parker 1967). According to Zabriskie (1964) the formation of erythrogenic toxin is determined by the presence in the streptococcus of a temperate phage, but no laboratory studies have been made of the relation of erythrogenicity or the production of pyrogenic exotoxin to serological type.

Most cases of scarlet fever are associated with streptococcal sore throat, but some accompany infections of wounds or burns, or puerperal fever.

Erysipelas

This is an acute, spreading skin lesion, which was formerly common but is rarely encountered today. The lesion is intensely erythematous, with a sharply demarcated but irregular edge, and sometimes with superficial vesicles or bullae. In some cases, particularly of facial erysipelas, the attack may be preceded by a streptococcal infection of the upper respiratory tract (Keefer and Spink 1936, de Waal 1941, Dowsett *et al.* 1975); in others it is associated with an infected abrasion or surgical wound, but in many a primary lesion is not identifiable. The sufferers are usually adults, many of them elderly, and the Dick reaction is usually negative early in the disease. A single attack confers no protection; indeed, some persons suffer many attacks, usually in the same skin area. According to Hektoen (1935) there are no records of epidemics in which some patients suffer from scarlet fever, some from erysipelas, and some from both. Erysipelas is a disease of temperate climates and occurs more often in the winter than the summer. It is unusual for cases to be epidemiologically related, but small epidemics have been reported (see, for example, Dowsett *et al.* 1975). The organism is difficult to isolate from the skin lesions. The suggestion that erysipelas is caused by a special streptococcal toxin has not been substantiated, and the recurrent nature of the disease suggests that it may be a hypersensitivity reaction to streptococcal products.

Rheumatic fever

The term 'rheumatic fever' is applied to a loosely associated group of clinical manifestations—notably acute migratory polyarthritis, carditis, chorea, subcutaneous nodules and erythema marginatum (Jones 1944, Report 1966)—that may follow infection with a group A streptococcus after a latent period of 1–5 weeks (on average $2\frac{1}{2}$ weeks, see Rammelkamp and Stolzer 1961–62). Only in some cases is the heart affected, but damage to it is responsible for most of the serious consequences of rheumatic fever. Myocarditis and endocarditis may cause death in the acute attack, and in many of the survivors there is permanent scarring of the heart valves and myocardium. This in turn may be followed by bacterial endocarditis (see Chapter 57)—or by progressive incompetence of the valves and death from congestive heart failure. Many patients suffer from repeated attacks of rheumatic fever, and once the heart has been affected the

chance of further damage in subsequent attacks is greatly increased (Report 1965, Feinstein and Stern 1967). The clinical manifestations in later attacks tend to mimic those of the first attack (Feinstein and Spagnuolo 1960).

Aetiology

Rheumatic fever in the general population has long been known to be associated with poor living conditions. The fact that rheumatic fever may at times also be common in institutional groups living under good economic conditions suggests that the real predisposing factor is close contact in groups of persons who are susceptible to streptococcal infection. In Europe and North America, rheumatic fever is predominantly a disease of school children and adolescents, but institutional groups of young adults may also be affected; in the tropics, the first attack often occurs before the age of 5 years. Rheumatic fever and rheumatic heart disease are both more common in girls than in boys.

Family studies, including pedigrees of rheumatic families and of identical and non-identical twins, suggest that hereditary factors play a part; but it is often difficult to distinguish between environmental and genetic influences (see Bywaters 1959). According to Spagnuolo and Taranta (1968), there is a tendency for members of the same family to have clinically similar rheumatic attacks. There appears to be a significant association of susceptibility to rheumatic fever with blood group. A number of investigations have shown that susceptibility is greater in non-O persons of the ABO blood-group system, and in persons who do not secrete substances in the ABH system in their saliva (for a summary of these see Haverkorn and Goslings 1969). Glynn and Holborow (1961) suggested that the configuration of the blood-group substance in the saliva in susceptible persons might determine the ability of the streptococci to absorb it and thereby render it antigenic. Studies of the tissue antigens of rheumatic patients and their families have given discordant results (see Read *et al.* 1980).

Evidence from mortality rates and from necropsy studies on children leaves little doubt that, in the economically developed countries of the temperate zone, rheumatic fever in its more serious and easily recognized forms has decreased progressively in incidence since at least the beginning of the twentieth century. This process underwent some acceleration after 1930 and has since continued, but foci of higher incidence persist in certain urban areas of the United States (Gordis *et al.* 1969a, Markowitz 1970) and Britain (Report 1967), and in some institutional groups (Bates 1967). Along with this reduction in frequency, the disease has changed clinically (Perry 1969); death during the acute phase of the disease is now rare, chorea seldom occurs, and carditis occurs in no more than one-half of first attacks. Many cases thus manifest themselves as an uncomplicated arthritis, sometimes of a single joint, and might easily escape recognition. Chronic rheumatic heart disease is very common in economically undeveloped countries in the tropics and subtropics, though until recently in many of them rheumatic fever was rarely diagnosed. The clinical picture that is now being built up there is of a disease that often begins before the age of 5 years, in which severe carditis is common, but chorea, subcutaneous nodules and erythema marginatum are rare; damage to the heart valves progresses rapidly (Shaper 1972, Strasser 1978).

The role of streptococci

The clinical association between tonsillitis and the subsequent appearance of rheumatic fever was recognized by many nineteenth-century physicians. The possibility that acute rheumatism was one of the manifestations of infection with haemolytic streptococci was raised independently by several British workers (Schlesinger 1930, Collis 1931, Glover and Griffith 1931, Sheldon 1931) and by Coburn (1931) in the United States. The starting point of these studies was the observation of an epidemiological relationship, particularly in institutional groups, between haemolytic streptococcal infections of the throat and attacks of rheumatic fever. It was also found that, in individual cases, a recurrence of rheumatic fever was frequently associated with the reappearance of haemolytic streptococci in the throat.

The epidemiology of acute streptococcal infections of the upper respiratory tract among the armed forces in North America during the 1939–45 war provided further evidence of this association. For example, Rantz and his colleagues (1945, 1947*a, b*) observed 15 cases of rheumatic fever among 410 men who had had invasive streptococcal infections, and none among 1100 not so infected. The incidence of rheumatic fever in different social groups (Holmes and Rubbo 1953) and different age groups (Streitfield and Saslaw 1961) is roughly proportional to the streptococcal carrier rates.

The attack rate of rheumatic fever appears to vary widely in different circumstances. We have mentioned outbreaks in military camps in which it was 3–4 per cent of those who suffered from streptococcal tonsillitis—and many more could be quoted (for example, Wannamaker *et al.* 1951). In the general population such outbreaks are rare, except in residential schools and similar institutions, but Zimmerman, Siegel and Steele (1962) describe an epidemic of 18 cases in a small rural community in the USA. The attack rate after sporadic streptococcal infections in families is generally believed to be very much lower. Thus Siegel, Johnson and Stollerman (1961), who studied untreated streptococcal tonsillitis in 608 children in Chicago, observed rheumatic fever in only 0.33 per cent. But most of the Chicago children had a mild tonsillitis, and only 45 per cent of them showed a rise of titre of antistreptolysin O. In most of the American military studies the diagnostic criteria for streptococcal tonsillitis were very strict; at the Warren Air Force Base, for example (Wannamaker *et al.* 1951), 85 per cent of the patients included gave a significant antibody response. In an extensive study of people of all ages in the general population of the Netherlands (Valkenburg *et al.* 1971) the proportion with sore throat, pain on swal-

lowing, group A streptococci in the throat and a significant rise in the titre of anti-streptolysin O that subsequently suffered from rheumatic fever was 0.8 per cent. The difference between attack rates in institutions and in the general population appears therefore to have been somewhat exaggerated but is nevertheless real.

Studies in institutional groups suggest that a severe attack of tonsillitis is more likely than a mild attack to be followed by rheumatic fever. On the other hand many rheumatic attacks in the general population follow mild sore throat. Thus, for example, in 281 first attacks in Baltimore (Gordis *et al.* 1969*b*), there was no history of preceding tonsillitis in 34 per cent, and in a further 32 per cent there was such a history but the patient had not sought medical advice.

One attack of rheumatic fever enormously increases the risk of subsequent attacks, but all patients are not equally liable to recurrences. A high recurrence rate after subsequent streptococcal infection is associated with (1) a history of sore throat and fever, (2) youth, (3) a short interval since the last rheumatic attack, (4) pre-existing heart disease and (5) numerous previous attacks (Spagnuolo *et al.* 1971; see also Taranta 1967).

Rheumatic fever may follow infections by streptococci of a number of different M types. It would be wrong, however, to assume that all types are rheumatogenic. There are accounts of institutional outbreaks of rheumatic fever in northern parts of the USA or in Britain associated with M types 1, 3, 5, 6, 12, 14, 17, 18, and 19 (Maxted 1978), but similar information about sporadic cases is scanty. In the southern states of the USA and in the Caribbean area, where both rheumatic fever and post-streptococcal glomerulonephritis occur in the same communities but hardly ever in the same patient, identifiable M types of 'skin' streptococci responsible for nephritis are rarely isolated from cases of rheumatic fever (Bisno *et al.* 1970, 1977, Potter *et al.* 1972); these usually yield strains that cannot be typed with existing sera. M types known to be associated with outbreaks of rheumatic fever are 'throat' types that do not form opacity factor but elicit a strong antibody response to their cellular protein antigens (Widdowson *et al.* 1974, Maxted 1978).

Patients with rheumatic fever nearly always show serological evidence of recent infection with group A streptococci. Most of them have a raised titre of anti-streptolysin O. It has been suggested (Coburn 1936, Quinn and Liao 1950) that average titres are higher in rheumatic fever than in uncomplicated tonsillitis, and that the titre in rheumatic fever reaches its maximum later and stays up longer. This may signify no more than that a streptococcal infection must be of more than a minimum duration or intensity to give rise to rheumatic fever, or it may suggest that re-infection is a significant determinant factor (Quinn *et al.* 1951, Rantz *et al.* 1951). However, the titre of antistreptolysin O is not raised in some 20 per cent of rheumatic subjects.

A strong antibody response has also been observed to a variety of other antigens of *Str. pyogenes*. These include not only extracellular products such as streptokinase, hyaluronidase, DNAase B, and NADase, but also various cellular antigens such as group polysaccharide, M protein and M-associated protein (see p. 237–8).

Antibody to one or more streptococcal products is to be found, usually in high titre, in the serum of nearly every patient with rheumatic fever, but the spectrum of antibody production varies between individuals; and equally high titres of antibody to these products may be found in cases of uncomplicated streptococcal infection. Perhaps the most constant finding is a raised titre of antibody to M-associated protein, but this, too, may occur in uncomplicated streptococcal infection.

It was suggested that the rheumatic subject may be non-specifically over-responsive to antigenic stimuli, but this possibility has been excluded. Rheumatic children are, however, more responsive than non-rheumatic to streptolysin O (Quinn 1957)— an effect perhaps attributable to secondary responsiveness in children already sensitized to this streptococcal antigen.

The serum protein reacting with the pneumococcal C carbohydrate, which appears in human serum in the acute stage of many infections appears also in acute rheumatic fever, and in many cases of rheumatoid arthritis, and is considered by Anderson and McCarty (1950) to be one of the most sensitive indicators of rheumatic activity; though it is, of course, not a specific index of the disease.

There is further support for the role of *Str. pyogenes* in the fact that the prophylactic administration of sulphonamide or penicillin to rheumatic subjects reduces the incidence both of streptococcal infection of the upper respiratory tract and of rheumatic recurrences, and that penicilllin treatment of manifest streptococcal infection, and mass antibiotic prophylaxis in communities, reduce the frequency of first attacks of the disease (see p. 242). On the other hand, the indirectness of the association between streptococcal infection and rheumatic fever is emphasized by the inefficacy of treatment with massive doses of penicillin, once the rheumatic attack has begun, either in influencing its clinical course or in preventing the subsequent appearance of cardiac damage (Foster *et al.* 1944, Carter *et al.* 1962, Vaisman *et al.* 1965).

Pathogenesis

Under this heading we consider three main views about the pathogenesis of rheumatic fever: that (1) it is a consequence of the persistence of streptococci or their products in the tissues, (2) it results from hypersensitivity to streptococcal antigens, or (3) it is a manifestation of auto-immunity.

(1) **Persistence** The fact that rheumatic fever is not prevented unless penicillin treatment of the preceding streptococcal infection is continued for 10 days suggests that this objective is attained only by complete elimination of the streptococci. There is no evidence, however, that living streptococci are present in early rheumatic lesions (Watson *et al.* 1961). It has been suggested that streptococci may persist in the body in the L phase (Schmitt-Slomska *et al.* 1972), but no definite evidence has yet been obtained that this occurs in rheumatic fever. The length of the latent period before the onset of rheumatic fever makes it unlikely that any of the extracellular toxins are responsible, but cellular constituents might be released slowly into the tissues and cause a delayed reaction. Schwab and his colleagues (Chapter 29) described granulomatous lesions of the skin of the rabbit resulting from the slow release of peptidoglycan from cell-wall fragments; when similar material is injected intraperitoneally into mice, lesions somewhat resembling those of rheumatic carditis appear in the myocardium, and streptococcal antigen can be detected in the areas immediately around them (Ohanian *et al.* 1969). It is therefore possible that the rheumatic lesion may be a cellular reaction to streptococcal constituents that have been translocated in phagocytic cells. Macrophages that have ingested fragments of *Str. pyogenes* cell walls become cytotoxic for cultures of mouse fibroblasts (Smialowicz and Schwab 1977).

(2) **Hypersensitivity** Rantz and his colleagues (1951) believed that rheumatic fever is rare in children under 3 years of age—at least in the USA—because it is a consequence of repeated infections. They pointed out that respiratory infection in very young children tends to be diffuse (Powers and Boisvert 1944) and to be associated with only a small increase in antistreptolysin O titre. In older children, a well defined pharyngitis and a more brisk antibody response occurs in the age groups in which rheumatic fever is most frequent. The early onset of rheumatic fever in some tropical countries may be an indication that multiple streptococcal infections are common in the very young. The latent period between the streptococcal infection and the onset of rheumatic fever is the same in first and subsequent attacks (Rammelkamp and Stolzer 1961–62), as would probably be the case if the initial rheumatic episode represented a second or subsequent exposure to streptococcal antigens. Patients who have suffered one attack of rheumatic fever are unlikely to have a recurrence if they can be kept free from streptococcal infection for at least 5 years (Stollerman 1954).

The lesions of rheumatic fever, unlike those of the primary streptococcal infection, in some respects resemble those of the delayed hypersensitivity reaction in the guinea-pig, both in histology and susceptibility to modification by hormonal treatment. Typical Aschoff bodies are not produced in experimental animals by streptococcal infection, but rheumatic lesions in man are broadly similar to the cardiac lesions induced in rabbits by repeated intracutaneous injections of live *Str. pyogenes* (Murphy and Swift 1949, 1950, Kirchner and Howie 1952) or by producing repeated pharyngeal infection (Glaser *et al.* 1956). Blood lymphocytes from rheumatic patients exhibit cellular immune responses to both cell-wall and cell-membrane fractions of *Str. pyogenes* (Read *et al.* 1974, Sapru *et al.* 1977, Williams *et al.* 1977).

(3) **Auto-immunity** The lesions may be due to the production in response to streptococcal infection of antibodies that react with a component of the host's tissues. Reactions between the sera of patients with rheumatic fever and heart tissue have been repeatedly demonstrated (Brokman *et al.* 1937, Calveti 1945, Polzer and Steffen 1958, Kaplan 1960). The identification by immunofluorescence of deposits of γ-globulin in the myocardium of rheumatic subjects (Kaplan and Dallenbach 1961) suggested that an antigen-antibody reaction had taken place. A connection with the streptococcus was established when Kaplan and Meyeserian (1962) showed that rabbit antisera against group A streptococci contained antibody that reacted with normal cardiac tissue. This antibody causes immunofluorescence of the sarcolemmal membrane of the myofibrils of heart muscle and of the smooth muscle of blood vessels, but it also reacts with skeletal muscle. Kaplan (1963, 1965) concluded that the corresponding antigen was in the cell wall, but Zabriskie and Freimer (1966) identified it in the cell membrane of all group A and some group C strains (see also van der Rijn *et al.* 1977). 'Heart-reactive' antibody is present in large amount in the serum in acute rheumatic fever, but is also found in smaller amount in patients with uncomplicated streptococcal infections. Its distribution thus appears to be rather like that of antibody to M-associated protein (Widdowson *et al.* 1971).

In addition to the 'heart-reactive' antibody that reacts with streptococci, the sera of patients who have suffered myocardial infarction or undergone cardiotomy may contain another antibody against heart tissue (Kaplan *et al.* 1960); whereas the former is absorbable from the serum with streptococcal membranes and with heart muscle, the latter is removed only by heart muscle (Zabriskie *et al.* 1970). Goldstein and his colleagues (1967) described quite a different cross-reaction between streptococci and heart tissue, the antigenic relation being between the streptococcal group polysaccharide and the structural glycoprotein of the heart valve. As we shall see (p. 238), antibody to group polysaccharide persists in the blood of patients with rheumatic fever that has caused permanent valvular damage.

Cross-reactions between streptococci and brain tissue have also been reported (Kingston and Glynn 1971). Sera from cases of rheumatic chorea, and rabbit anti-streptococcal sera, give an immunofluorescent reaction with brain tissue (Husby *et al.* 1976, Kingston and Glynn 1976).

One difficulty in accepting the 'heart-reactive' protein as the determinant for rheumatic fever is the widely accepted view that the Aschoff nodule is a lesion of connective tissue, but Becker and Murphy (1980) advance evidence that the nodules in myocar-

dium, endocardium and heart-valves arise from muscle cells. The multiplicity of cross-reactions between streptococci and mammalian tissues may also seem to detract from their significance, but it is not impossible that they are responsible for different components of this polymorphous disease. Much but not all of the evidence for the cross-reactions was obtained by immunofluorescence; Holm and his colleagues (1979) point out that the non-specific absorption of immunoglobulins by group A streptococci (Chapter 29) may cause false reactions in immunofluorescence tests and advise caution in their interpretation.

In summary, therefore, we may say that only streptococcal infections that give rise to a strong antibody response are followed by rheumatic fever, and that the streptococcus or constituents of it must persist in the tissues for more than a minimum time. We do not know whether this persistence must be at the site of the future rheumatic lesion or can be at a distance; nor do we know whether the tissues are damaged by a streptococcal product, by an antigen–antibody reaction, or by antibody alone. Nor can we say why rheumatic fever develops in only a few patients of those infected with equal severity by a single strain.

Post-streptococcal glomerulonephritis

Aetiology

A clinical association between acute glomerulonephritis and various forms of streptococcal infection—notably tonsillitis and skin sepsis—has been recognized since early in the twentieth century. As with rheumatic fever, an interval of 1–3 weeks elapses between the primary infection and the onset of nephritis. The causative organism is almost invariably *Str. pyogenes*, but occasional cases associated with *Str. zooepidemicus* have been reported (p. 247).

Rammelkamp, Weaver and Dingle (1952) were impressed by the wide variations in the attack rate of nephritis in different outbreaks of streptococcal tonsillitis, and suggested that only some strains of *Str. pyogenes* were nephritogenic. They found that M-type 12 predominated among streptococci isolated both from sporadic cases and in epidemics of nephritis in the USA (see also Rammelkamp and Weaver 1953, Siegel *et al.* 1955). Similar observations have been made in Europe. Members of other M types have seldom been implicated in nephritis associated with infection of the respiratory tract. In some outbreaks of tonsillitis due to type-12 streptococci attack rates in excess of 10 per cent of the number of streptococcal infections have been observed (Stetson *et al.* 1955), but in others nephritis has been rare or absent. In addition to patients with clinically manifest nephritis, many others suffer from a subclinical disease as revealed by the presence of blood, albumen and casts in the urine.

Throughout Europe nephritis has for many years

been associated mainly with an antecedent throat infection, but before 1930 it was observed as a common sequel to impetigo or wound infection (for references see Dillon 1967). In the United States, either respiratory or skin infection may precede nephritis, but in different circumstances; in the winter, outbreaks and sporadic cases associated with type-12 infection of the respiratory tract occur in many parts of the country and in all types of community; in the summer, on the other hand, nephritis associated with impetigo is common in certain populations—particularly of Negroes and American Indians—living under poor hygienic conditions, and is not due to type-12 streptococci. In tropical countries, nephritis nearly always follows skin sepsis; it has been found almost everywhere when it has been sought, and is very prevalent in some areas.

A number of M types have been incriminated as causes of nephritis associated with streptococcal impetigo; these include (1) type 49 (Updyke *et al.* 1955, Maxted *et al.* 1967, Anthony *et al.* 1967*a*) in the USA, the Caribbean area, many parts of Africa and Asia, and occasionally Europe, (2) the type referred to as 'Alabama 2' (Dillon *et al.* 1968), which has the M-antigen 2 but the T-typing pattern 25/Imp. 19 (see Chapter 29) and has been found only in the south-eastern USA, (3) type 55 (Parker *et al.* 1968, Potter *et al.* 1968, Lasch *et al.* 1971, Poon-King *et al.* 1973) in the Caribbean, Africa and the Middle East, (4) type 57 (Parker 1969, Ferrieri *et al.* 1970) in the Caribbean and the USA, and (5) type 60 (Dillon *et al.* 1974) with a distribution similar to that of type 57. It should be noted, however, that in Egypt M-type 55 is said to cause nephritis associated with streptococcal tonsillitis (El Tayeb *et al.* 1979).

Evidence for the nephritogenic role of the various serotypes from skin lesions was obtained from epidemics and large seasonal prevalences in which the type was isolated frequently from cases of nephritis but was found much less often in the skin lesions of patients without nephritis. When investigating sporadic cases of nephritis, it is impossible to be certain that the streptococcus isolated from a skin lesion on admission to hospital is the one that initiated the disease. Multiple infection of the skin, even of the same lesion, with two or more distinct streptococci, occurs frequently; in Trinidad, a study of serial swabs from skin lesions indicated that as many as one-third of the serotypes isolated were different from those present in the lesions of the same patient two weeks before (Parker *et al.* 1968, Bassett 1972). Several serotypes other than the ones we have mentioned have been isolated from the skin lesions of small groups of nephritis cases, but the evidence that they have nephritogenic properties is as yet insufficient.

The epidemiological conditions under which nephritis is associated with impetigo vary. In some areas, such as Alabama, there is a regular summer peak of nephritis associated with the seasonal prevalence of impetigo, and the relative importance of different streptococcal serotypes varies from year to year (Dillon 1967); in the isolated American Indian reservation at Red Lake, Minnesota, impetigo occurs each summer, but nephritis occurs only in those years when a nephritogenic serotype is present (Anthony *et al.* 1966, Ferreiri *et al.* 1970); in Trinidad, where impetigo is common at all times, nephritis appears irregularly in large non-seasonal epidemics caused by a succession of types: type 55 in 1964,

type 49 in 1965, type 60 in 1967, type 57 in 1968, type 55 again in 1971, and types 49 and 60 again in 1977-78 (Poon-King *et al.* 1979). The genesis of epidemics appears to be determined, at different times and in different places, by the frequency of the primary skin lesion, the presence of a nephritogenic strain, and the amount of type-specific immunity in the child population.

The incidence of post-streptoccal nephritis varies within a wide range. As we have seen, the attack rate after tonsillitis due to type 12 strains sometimes exceeded 10 per cent in the 1950s. In Britain, where this type has for many years formed 10-20 per cent of all group A isolations, post-streptococcal nephritis is now quite uncommon. Extensive surveys in the general population in Holland (Valkenburg *et al.* 1971) gave an annual incidence rate of 20 per 100 000, and an attack rate of 1 per cent in persons who had suffered from bacteriologically confirmed streptococcal tonsillitis. In the Trinidad epidemic of 1964-65, about one child in 300 in the whole population had clinical nephritis (McDowall *et al.* 1970). Anthony and his colleagues (1969) describe an epidemic in an isolated community in which one-quarter of all children with skin infections due to type 49, but only 5 per cent of those with throat infections with this type, suffered from acute nephritis or unexplained haematuria.

The mortality rate among patients admitted to hospital with acute nephritis seldom exceeds 3 per cent, and the long-term prognosis for survivors is good, at least in children. Thus, none of the 61 children who had clinical nephritis at Red Lake in 1954 showed progressive renal disease 10 years later (Perlman *et al.* 1965); and Lieberman and Donnell (1965), who studied retrospectively the case records of 485 children with acute nephritis, concluded that physical abnormalities were rare two years later. In a prospective investigation in which the children were examined by renal biopsy during the acute phase and three years afterwards, Dodge and his colleagues (1972) observed complete healing in 39 of 41 patients who appeared on histological evidence to be suffering from first attacks of nephritis; in the remaining two progressive renal disease developed. Of six other children whose kidneys showed evidence of a previous renal lesion when first examined, only three made a complete recovery. In these cases, however, it is impossible to be sure that the original lesion was post-streptococcal. A study of 722 Trinidadian patients (Potter *et al.* 1982), nearly all of them children at the time of the original attack of glomerulonephritis 12-17 years before, showed that only 3.6 per cent of them had chronic renal disease. Unlike rheumatic fever, one attack of nephritis does not predispose to further attacks. Roy, Wall and Etteldorf (1969), in Tennessee, observed 12 second attacks in 509 children and found them to be clinically indistinguishable from first attacks; renal biopsy from 10 patients showed evidence of healing in each case. Nephritis is very much less common in adults than in children, but it is said that the long-term prognosis is considerably less favourable (Schacht *et al.* 1976).

Pathogenesis

We do not know for certain how the streptococcus causes acute glomerulonephritis. Many workers have produced renal lesions in experimental animals by the injection of streptococci or streptococcal products, but these lesions differ somewhat from those of the natural disease. Most workers favour the view that damage to the kidney is of immunological rather than of directly toxic origin. The histological appearances of the acute disease in man closely resemble those of experimentally produced serum-sickness nephritis in the rabbit, in which there occurs a deposition of antigen-antibody complexes in the glomeruli. Studies by electronmicroscopy of the glomeruli in the human disease reveal electron-dense deposits along the basement membrane and in the mesangium; these contain γ-globulin and complement, but attempts to identify antigenic components of the streptococcus gave uncertain results (Andres *et al.* 1966, Michael *et al.* 1966). The suggested hypothesis is that soluble streptococcal antigen–antibody complexes are formed during the latent period, that some of these are deposited in the glomeruli, and that complement is bound by the complex and an inflammatory response follows.

The implantation of a Millipore chamber containing group A streptococci in the peritoneal cavity of the rat (Lindberg *et al.* 1967) is followed by an early transient proteinuria. If the streptococcus is a nephritogenic type-12 strain, a greater degree of proteinuria appears after a latent period—at about the same time as type-specific antibody appears in the blood—and γ-globulin, complement and M antigen can then be detected on the glomerular basement membrane, but no significant histological changes appear (see Vosti *et al.* 1970). Lindberg and Vosti (1969) found that globulin eluted from the homogenized kidney reacted specifically with the nephritogenic streptococcus, and they implicated the type-12 M antigen and its corresponding antibody in the reaction.
Treser and his colleagues (1969; see Lange *et al.* 1976) have identified a protein ('endostreptosin') in cell-membrane fractions of all group A and some group C streptococci that is unrelated to M antigen and other identifiable products and is said to be detectable by immunofluorescent staining in the glomeruli in the early stages of nephritis. More recently, an extracellular protein has been described (Villareal *et al.* 1979) that is formed by some members of the 'nephritogenic' M types but not by others. This has been detected by similar methods in the nephritic glomerulus.
An alternative view of the pathogenesis of glomerulonephritis is that a component of the membrane of the streptococcus is immunologically cross-reactive with a component of the glomerulus (Markowitz and Lange 1964), and that both are glycoproteins of low molecular weight (see Markowitz *et al.* 1971).
The view that nephritis results from an antigen–antibody reaction is supported by the observation that the amount of complement in the blood is reduced during the acute stage of the disease. This is attributable to the disappearance of the C3 component of complement (West *et al.* 1964). Reduction in blood content of the C3 component has been widely used as a diagnostic test for post-streptococcal glomerulonephritis

(Kohler and Ten Bensel 1969, Derrick *et al.* 1970). According to Onyewotu and Mee (1978) there is a significant correlation between the amount of soluble immune complexes in the blood and the severity of the disease as judged by the blood-urea level.

Anaphylactoid purpura   Despite the clinical impression that anaphylactoid purpura might in some cases be a sequel of streptococcal infection, Ayoub and Hoyer (1969), who studied 33 patients with this disease, found that the titres of antistreptolysin O, anti-DNAase B and anti-NADase were similar to those in controls.

The antibody response to *Str. pyogenes*

Antibodies to many different constituents of the organism are formed as a result of infection. Here we shall describe their production in various types of infection, but shall defer consideration of the diagnostic use of tests for them until later in this chapter (p. 242).

Antibody to extracellular substances

The extracellular products formed by *Str. pyogenes* (Chapter 29) include streptolysin O, streptolysin S, erythrogenic toxin, proteinase, streptokinase, hyaluronidase, the DNAases and NADase; of these streptolysin S is not antigenic and proteinase is poorly antigenic. We have already given our reasons for believing that the Dick test may not in all cases be a measure of antibody to erythrogenic toxin; in any event, it is difficult to standardize and not suitable for quantitative measurements. Antibody to the remaining five substances can be detected by testing for neutralization of their specific biological activity. Special precautions must be taken to distinguish between antibody and non-specific inhibitors in the serum.

Antistreptolysin O (ASO)   The test for this antibody has for many years been the standard method for the retrospective diagnosis of infection with *Str. pyogenes*. Dilutions of serum are mixed with one unit of reduced streptolysin O, the potency of which is controlled by making parallel tests on the International Standard Serum (Spaun *et al.* 1961) or a local standard based on this. Serum and lysin are allowed to act at 4° (Gooder 1961); a suspension of human or sheep erythrocytes is added; and the end-point of inhibition of lysis is determined visually (Rantz and Randall 1945) or spectrophotometrically (Liao 1951) after incubation at 37°. The dilution of the serum at the end-point gives the number of units of ASO per ml. Fasth (1974) describes a plate method for measuring ASO by reversed radial immunodiffusion.

Streptococci of groups C and G also form streptolysin O, and infections with them may cause the production of ASO, but the immunological relation between streptolysin O and the oxygen-labile lysins of clostridia and pneumococci does not lead to significant cross-reactions. Free cholesterol inhibits the action of streptolysin O (Hewitt and Todd 1939), but the normal esterified or protein-bound cholesterol of serum does not (Watson *et al.* 1972). Bacterially contaminated sera may inhibit streptolysin O non-specifically and so give false positive reactions in the ASO test (Hewitt and Todd 1939); the inhibitor is probably either free cholesterol or partly hydrolysed lipoprotein released by bacterial action (Watson and Kerr 1975). Similar non-specific inhibitors may be present in the sera of patients with jaundice, nephrosis, rheumatoid arthritis, and staphylococcal and occasionally other bacterial infections (see Watson and Kerr 1978 for references). The inhibitor may be removed or rendered inactive by treating the serum with dextran sulphate (Hällén 1963), Rivanol (Mihalco and Mitrică 1965), digitonin (Badin and Barillec 1970), or a polyene antibiotic such as amphotericin B (Watson and Kerr 1978). Very high titres of true ASO are occasionally seen in association with monoclonal hypergammaglobulinaemia (Waldenström *et al.* 1964, Winblad 1966).

Anti-hyaluronidase Many strains of *Str. pyogenes* appear to produce little hyaluronidase *in vitro*, but antibody to the enzyme is formed in most infections in man. Tests for it depend on the detection of the hyaluronic acid remaining after incubation with a standard amount of streptococcal hyaluronidase and dilutions of serum. An insoluble complex of hyaluronic acid and protein is formed at acid pH and may be detected turbidometrically or by the formation of a gelatinous precipitate in alcohol (the 'mucin-clot prevention test'). A modification of the latter test (Di Caprio *et al.* 1952) has been used widely. The serum must be heated at 56° for 30 min to destroy non-specific inhibitor of hyaluronidase (Friou 1949).

Anti-DNAase B *Str. pyogenes* produces four serologically distinct DNAases of which the B enzyme is the most regularly formed. Antibody to it is measured by incubating dilutions of heated serum with a standard amount of DNAase B, and then testing for unneutralized enzyme by adding DNA and incubating again; undigested DNA is detected by precipitating with alcohol (Ayoub and Wannamaker 1962). Alternatively, a complex of DNA and methyl green is used; this is decolorized when the DNA is digested (Nelson *et al.* 1968). The use of commercially manufactured kits has eliminated most of the difficulties in standardizing the test.

Anti-NADase Antibody to NADase (Chapter 29) is measured by its ability to neutralize the action of a standard amount of the enzyme on the pure substrate; undigested NAD is estimated spectrophotometrically (Kellner *et al.* 1958, Bernhard and Stollerman 1959). The value of the test is limited by the fact that members of a number of serotypes do not form the enzyme.

Antistreptokinase This antibody can be measured by determining the dilution of serum that prevents the lysis of fibrin clot by a standard dose of streptokinase. The test is not recommended for general use, partly because it is difficult to standardize and partly because antibody to streptokinase is formed less often in streptococcal infection than is antibody to other extracellular products (see Wannamaker and Ayoub 1960).

We shall not attempt to summarize the extensive literature on the production of antibodies to the extracellular antigens, but direct our readers to the comprehensive paper by Wannamaker and Ayoub (1960). In streptococcal infections, except impetigo, the patterns of antibody response to streptolysin O, hyaluronidase, DNAase B and NADase are similar. With this exception, and if each antibody is considered separately, between 50 and 80 per cent of all patients who suffer from clinical infection have a raised titre. Titres in rheumatic fever tend to be somewhat higher than in uncomplicated tonsillitis, but the difference is not clear cut, and almost as many patients fail to give a significant response to each antigen. But because low titres of individual antibodies occur in a random fashion, the use of two or three tests will increase the proportion of patients giving a significant antibody response to one or more antigens almost to 100 per cent. In streptococcal pyoderma the situation is different in that there is little or no ASO or anti-NADase response, even when the infection is complicated by glomerulonephritis; but titres of anti-DNAase B and anti-hyaluronidase are usually raised, and are on average higher than those seen in throat infections (Potter *et al.* 1968, Kaplan *et al.* 1970*a*). Titres in glomerulonephritis do not differ significantly from those in uncomplicated streptococcal infection.

In infections of the respiratory tract the titre of ASO usually begins to rise towards the end of the first week of illness and increases during the next 2–4 weeks. It then falls slowly, and in the absence of re-infection tends to approach its original value in 6–12 months; but there is much individual variation in this respect. As we have seen, the titre tends to rise for a longer period and stay up longer if rheumatic fever develops, and the response is on average poorer in infants than in older children and adults. Early treatment with penicillin in adequate dosage and for a sufficient period of time greatly lessens both the frequency and the magnitude of the antibody response (Wannamaker *et al.* 1951).

The Streptozyme test This is a passive haemagglutination reaction in which erythrocytes sensitized with a crude preparation of the extracellular products of a group A streptococcus are mixed with serum on a slide. The fact that it detects antibodies to a number of extracellular antigens is said to remove the necessity of performing two or three conventional tests (Klein and Jones 1971). It is simple to perform, and different batches of reagent give reasonably uniform results. The test is sensitive and specific, giving a positive result in nearly all cases of clinical infection whether of the throat or of the skin (Bisno and Ofek 1974, Lütticken *et al.* 1976), and is thus a useful means of screening sera; it also serves to eliminate false positive ASO results.

Anti-Streptozyme titres begin to rise before titres of antibody to the known extracellular products. They reach an early peak at the end of the first week and then fall, rising again along with the ASO and anti-DNAase titres (Kaplan and Wannamaker 1975). According to Bisno and his colleagues (1976), the material responsible for the early rise in titre can be separated from the other extracellular products.

Antibody to cellular constituents

Group polysaccharide   The group A polysaccharide may be adsorbed to tanned erythrocytes (Schmidt and Moore 1965), or it may be esterified and then adsorbed to untreated cells (Goldstein and Caravano 1967); both are suitable for indirect haemagglutination tests. Antibody may also be measured by radioimmunoassay (Dudding and Ayoub 1968).

By each of these methods, raised titres of antibody can be detected in patients suffering from acute streptococcal infection (Goedvolk-de Groot *et al.* 1974, Kaplan *et al.* 1974). Dudding and Ayoub (1968) found that the high titre of antibody present in the early stages of rheumatic fever decreased progressively in the next few months unless the heart valves had been damaged, when the titre remained high, often for many years after the last rheumatic attack. Ayoub and Dudding (1969) examined the blood of patients with bacterial endocarditis secondary to rheumatic fever or to congenital valvular disease and found the antibody only in the former; this suggested that it was unlikely to be autoantibody against valvular tissue (see p. 234).

Surface protein antigens   Attention has been focused on the production of antibody to M protein because it is directed against an important virulence factor of *Str. pyogenes*. As with antibodies to other antigens that have not been obtained in a state of chemical purity, the only reliable method of detecting M antibody is by specific inhibition of its biological activity. It can be measured (1) by passive protection tests in mice, if a mouse-virulent strain of appropriate type is available, (2) by the long-chaining test, (3) by the direct or the indirect bactericidal test, or (4) by a phagocytic test with mouse peritoneal leucocytes (see Chapter 29). Most of the information we have was obtained by means of the indirect bactericidal test.

Studies made in outbreaks of severe tonsillitis in young adults suggest that M antibody is formed regularly after untreated infections (Rothbard *et al.* 1948, Denny *et al.* 1957) but appears much later than antibody to the extracellular substances. Usually it can be detected within one month of the onset of illness, but instances of its appearance later have been quoted. Siegel and his colleagues (1961), who studied children suffering from mild tonsillitis due to types 3 and 12, detected M antibody in the sera of 70 per cent of those who did not receive penicillin treatment; the factor most closely correlated with the formation of M antibody was the duration of carriage of the streptococcus. In patients with impetigo (Potter *et al.* 1971) M antibody appears more frequently after infection with some serotypes than with others. Widowson and her colleagues (1974) studied the antibody response in institutional outbreaks of tonsillitis due to a number of serotypes and concluded that M antibody appeared rather infrequently after infection with serotypes that form the opacity factor. In general, infection with serotypes that do not form this factor result more

often in the appearance of M antibody, but there are considerable differences between individual strains in this respect. A similar inverse relation between the presence of opacity factor in the strain and the antigenicity of the M protein is observed in rabbits.

Early and adequate penicillin treatment suppresses the formation of M antibody (Daikos and Weinstein 1951, Denny *et al.* 1957, Siegel *et al.* 1961). Lancefield (1959) studied sera from a few patients who had been bled serially over periods of 10 to 32 years and concluded that 'bactericidal' antibody persisted for many years in at least half of the subjects. On the other hand, Bergner-Rabinowitz and her colleagues (1971), who used mouse peritoneal leucocytes to study the antibody response to type 55 after impetigo with nephritis, observed that titres had usually fallen to a low figure 7 months after infection.

The bactericidal test is very time-consuming, and studies of the distribution of antibody to many types in large populations of normal persons have so far not been made. Unpublished information available to us indicates that adults possess 'bactericidal' antibody against only a minority of the common serotypes. But patients may have significant type-specific immunity in the absence of the corresponding circulating antibody. A single injection of a preparation of streptococcal cell walls or of purified M antigen, which seldom gives rise to a serum-antibody response in controls, may cause a secondary-type response in persons who earlier had been infected with the respective type of streptococcus (Potter 1965).

Other serological tests for M antibody have given confusing results. Purified M antigen has been used in agglutination tests after adsorption on to erythrocytes or to inert particles or in radioimmunoassay (Chapter 29). Widespread cross-reactions between types were observed, but these could usually be removed by absorption of the serum with a heterologous strain of *Str. pyogenes*. Doubtless many of them were due to the presence of M-associated protein in the preparation of antigen.

The formation of antibody to M-associated protein (anti-MAP; Chapter 29) is of some interest. Little antibody is found in healthy persons, and the titre usually does not exceed 20. High titres ($\geqq 60$) are almost invariable in rheumatic fever but are uncommon in glomerulonephritis (Widdowson *et al.* 1971). After tonsillitis the antibody response is variable; in general there is a direct correlation between the production of M antibody and high titres of anti-MAP in the patients, and an inverse correlation of both with the formation of opacity factor by the infecting streptococcus (Widdowson *et al.* 1974). Patients with uncomplicated tonsillitis due to strains that caused rheumatic fever in other persons very frequently developed high anti-MAP titres as well as forming the corresponding M antibody. Patients infected with serotypes that form opacity factor sometimes produce neutralizing antibody to it (Maxted *et al.* 1973).

Epidemiology

Infections of the respiratory tract

Infection in streptococcal tonsillitis is mainly from the upper respiratory tract of another person. During the acute stage of the disease, the organism is widely distributed throughout the pharynx, nasopharynx and nasal cavities; at this stage it can usually be grown in large numbers from both throat and nose swabs. In the early stages of convalescence the numbers decrease unless there is a suppurative complication in the throat or nasal sinuses. Observations made in pre-antibiotic times indicated that the organism usually persisted in the throat for many weeks, though they tended to disappear much more quickly from the nose (see, for example, Brown and Allison 1935). More recent information about the bacteriological findings in untreated tonsillitis is difficult to obtain.

Treatment with penicillin usually causes a rapid reduction in the number of streptococci in the upper respiratory tract; nose and throat swabs may become negative within a few days. Even when a 10-day course is given, however, throat carriage persists in some 20 per cent of patients, and repetition of the treatment does not accelerate the rate of clearance. It has been suggested (Kundsin and Miller 1964, Markowitz *et al.* 1967) that the presence of penicillinase-forming staphylococci in the throat favours the persistence of streptococci, but Quie and his colleagues (1966) were unable to confirm this (see also Rosenstein *et al.* 1968).

Hamburger (1944) found *Str. pyogenes*, in numbers ranging from 10^3 to 10^6 per ml, in the saliva of 65 per cent of patients with tonsillitis or scarlet fever and a positive throat swab; but there was little correspondence between the numbers isolated from the two specimens. The frequency of salivary carriage fell by one-half during the first fortnight of convalescence. Ross (1971*a*), however, isolated the organism from over 90 per cent not only of children with tonsillitis but also of symptomless throat carriers; although the mean count was higher for the cases than for the carriers, there was no clear line of demarcation between the two groups.

The frequency with which *Str. pyogenes* can be isolated from the throat of normal persons not in contact with epidemic streptococcal infection of the respiratory tract varies widely with age, season and country, and doubtless other factors, including the bacteriological techniques employed. Rates for adults (excluding hospital staff) of 2 to 8 per cent have been reported. Among children the rates are more variable but seldom as low as in adults; they usually lie within the range 10–25 per cent but may be even higher (Quinn *et al.* 1957). The study of Holmes and Williams (1954) of about 2400 London children revealed some of the variables that may be concealed within an average carrier rate. The total rate was 21 per cent, but within individual units, such as schools and day nurseries, rates of 0 to 79 per cent were observed. Carriage at age 3–4 years was higher than at any other age; and the rate for children with tonsils was 28 per cent, compared with 8 per cent for children without tonsils. Ross (1971*b*), who found a carriage rate of 10 per cent in Edinburgh school children, observed seasonal variation in the range 4 to 21 per cent, with the highest rates between October and March; carriage was most frequent at ages 5–7

years and occurred more often when the tonsils were present than when they were absent. In the nose, *Str. pyogenes* is much less common than in the throat, and rates as low as 1 per cent have been quoted. Holmes and Williams (1954) found 4 per cent of nasal carriers among London children, but rates in individual units were as high as 42 per cent. Most nose carriers are also throat carriers, and the streptococci from the two sites usually belong to the same serotype. In tropical areas where tonsillitis is uncommon and impetigo is prevalent, the nasal carrier rate often exceeds the throat carrier rate (Bassett 1972).

Carrier rates in institutions in which an outbreak of streptococcal tonsillitis is in progress are extremely variable. There may be few cases even in the presence of a high carrier rate, and many cases when the carrier rate is low (Hamburger 1944, Hartley *et al.* 1945, Rubenstein and Foley 1945).

Despite the long persistence of the organism in the upper respiratory tract after an attack of tonsillitis, the infectivity of patients decreases rapidly during convalescence. When scarlet fever was habitually treated in hospital and some 50 per cent of patients were throat carriers when they were discharged, only about 2 per cent gave rise to further cases among home contacts. Brown and Allison (1935) found that heavy carriers were more likely to prove infective than light carriers, and that patients discharged during the 5th, 6th and 7th weeks were more infective than those discharged earlier. This observation was interpreted as suggesting that the primary infecting strain loses most of its infectivity by the 4th week, and that the cross-infecting strain is still infective at the time of the patient's discharge. The idea that streptococci carried by convalescents underwent a loss of virulence received support from accounts of the loss of M antigen by strains carried for many weeks after an acute attack; for reasons set out in the 6th Edition of this book, these can no longer be accepted. In our experience, the isolation of true-M-negative variants from patients is exceedingly uncommon.

Another explanation of the early loss of infectivity during convalescence follows from the observations of Hamburger and his colleagues (1945, 1949) that the environment around nasal carriers is much more heavily contaminated with *Str. pyogenes* than the environment around throat carriers. The convalescent throat carrier may therefore have a low infectivity because he does not disseminate many organisms; and the high infectivity of acute cases may be related to the frequency of nasal carriage at this stage of the disease. It must be emphasized, however, that some carriers may remain infective for many weeks or months (Griffith 1934, Boissard and Fry 1944). This is seldom so if the organism is present only in the throat. Many of the long-term sources are nasal carriers, who include patients with sinusitis or with infected excoriations of the external nares. Patients with a discharging ear or extensive skin lesions secondarily infected with *Str. pyogenes* may occasionally be responsible for outbreaks of streptococcal tonsillitis (Loosli *et al.* 1950).

In hospital wards housing cases and carriers, the streptococci are present in the dust and the air (White 1936) and on the bedclothes of infected patients. Quantitative studies suggested that the organisms in the air were mainly derived from dust and bedclothes (Hamburger *et al.* 1944). Studies of nasal carriers (Hamburger and Green 1946) revealed that the main route of dissemination was not directly into the air but to the hands and the handkerchief and so to the clothing and bedclothes, whence they were released into the air by bodily movement and bedmaking. Airborne streptococci that settle on surfaces from the air may be resuspended as a result of ward activities such as sweeping. Streptococci may persist for weeks in dust, but there is some evidence that the survivors have a reduced ability to infect by the respiratory route. Thus, it was observed that volunteers could be readily infected by inoculating the throat with moist secretions containing group A streptococci, but that insufflation into the nasopharynx of similar numbers in dust from naturally contaminated blankets did not lead to infection (Rammelkamp *et al.* 1958). According to Srisuparbh and Sawyer (1972), this loss of infectivity is attributable to damage to the hyaluronic acid capsule.

It is generally believed that transmission by the respiratory route is favoured by close contact; and the observations of Wannamaker (1954) suggest that as the distance between the beds is increased beyond 8 ft the rate of infection falls off progressively. This is inconsistent with the view that streptococci in airborne droplet nuclei, or resuspended in the air from the dust, are the main source of infection. It suggests rather that large droplets expelled from the respiratory tract, which might be inhaled by close contacts before they have become dry or sedimented from the air, are of greater importance. This conclusion receives support from the failure—in most but not in all cases—of dust suppression to prevent the spread of streptococcal infection in hospital wards (Williams 1960). But the high infectivity of nasal carriers is undoubted, and streptococci are not dispersed directly into the air from the nose. This discrepancy might be resolved if it could be shown that nasal carriage and heavy salivary carriage were closely associated, particularly in acutely ill persons, but, as we have seen, the evidence on this point is conflicting.

Foodborne tonsillitis   Streptococci may reach the upper respiratory tract in food. Numerous milk-borne outbreaks of tonsillitis are on record, though they have become very rare in the last 30 years. It was believed at one time that the usual source of infection of the milk was the cough spray or ear discharge (see Henningsen and Ernst 1939) of a worker on the farm or in the dairy, but the evidence is now more in favour of infection of the udder of the cow itself. Most cases of streptococcal mastitis are due to *Str. agalactiae*; but in occasional cases *Str. pyogenes* is responsible. In such instances the milk may contain enormous numbers of streptococci that prove highly infective for man. Milk that is contaminated directly from a human case or carrier is liable to be less infective because *Str. pyogenes* does not usually multiply to any considerable extent in milk at room temperature (Pullinger and Kemp 1937). Several outbreaks have been recorded in which infection has been traced to the diseased udder of a cow, and in some a human carrier has been found from whom the cow presumably derived its infection. Stebbins, Ingraham and Reed (1937) were able to trace infection in 6 of 7 streptococcal outbreaks of sore throat to mastitis in one of the animals supplying the milk. In the remaining outbreak the milk had been infected from a human source, having been bottled and capped by a worker suffering from an acute sore throat. Occasional outbreaks due to food other than milk are on record (Hamburger *et al.* 1945, Reynell 1948, Boissard and Fry 1955, Farber and Korff 1958, Hill *et al.* 1969b).

Septic infections

A wound may become infected from the respiratory tract of the patient, a medical attendant, or another patient. The streptococcus may reach the wound by the airborne route. There is no reason to believe that dried streptococci have lost their infectivity for wounds; in addition, therefore, to large droplets expelled from the respiratory tract, small dry particles dispersed from the body surface of a carrier, or resuspended from the dust, may convey streptococci to wounds by the airborne route. Outbreaks of postoperative wound infection and puerperal fever have occasionally been caused by medical attendants—including anaesthetists—who were carriers of *Str. pyogenes* only on the perianal skin (Di Caprio *et al.* 1966, McIntyre 1968, Schaffner *et al.* 1969, Gryska and O'Dea 1970), or even in the vagina (Stamm *et al.* 1978). It should not be concluded, however, that all wounds are infected by the airborne route; contaminated hands and objects also probably play an important part.

Puerperal fever   The streptococcus reaches the blood stream from the uterus, and in most cases appears to have been introduced into the genital tract during labour or early in the puerperium. Group A streptococci are rarely found in the vagina before labour (Hare and Colebrook 1934), on the perineal or perianal skin during pregnancy (Colebrook *et al.* 1935), or in the faeces (Hare and Maxted 1935). It therefore seemed likely that the organism came from the nose or throat of the patient or of some other person in the vicinity. Smith (1931, 1933), studying 49 cases of puerperal fever caused by *Str. pyogenes*, found an antigenically identical strain in the throat or nose of 8 of the patients and of 31 of the contacts, thus accounting for 39 out of 49 of the series. Dora Colebrook (1935), studying 63 cases, obtained corresponding figures of 24 and 39 respectively, though in some of the latter group the organism was isolated from the nasopharynx of both the contact and the patient herself, rendering it difficult to decide in which direction the transfer had occurred. One important additional source of infection—the babies themselves—was unsuspected in the days when puerperal infection was the subject of intensive investigation, and may still be overlooked (see p. 228).

Impetigo The available evidence suggests that the streptococci that cause impetigo are derived mainly from skin lesions in other persons (Bassett 1972). Organisms first detected in a skin lesion may subsequently be found in the nose and less often in the throat; spread in the reverse direction is unusual. Little is known about the routes by which the organism passes from one person to another. As we have seen (p. 229), lesions often appear at the site of a minor injury to the skin; but the observations of Dudding and his colleagues (1970) suggest that the infecting streptococcus may often colonize the skin up to 2 weeks before the impetigo lesion appears, though the number present is usually small (Brown *et al.* 1975). This sequence of events is the reverse of that which usually occurs in surgical wound infection. We do not know what changes in the skin permit colonization by *Str. pyogenes* or which elements in the complex of local overcrowding, absence of washing facilities, and unsatisfactory personal habits are responsible for the frequency of streptococcal skin infection.

There is evidence to suggest that, in some geographical areas, *Str. pyogenes* may be spread by non-biting insects, particularly by 'eye gnats' of the genus *Hippelates* (Taplin *et al.* 1965, Bassett 1967). These small flies feed on moist skin lesions, become contaminated with streptococci, and may convey the organisms mechanically from one lesion to another. According to Bassett (1970), this may contribute to the very rapid spread of individual serotypes through the child population of Trinidad.

Bacteriological diagnosis

A clinical diagnosis of infection with *Str. pyogenes* can seldom be made with certainty, and confirmation must be obtained by identifying the organism in the lesion or by demonstrating streptococcal antibody in the blood (see p. 242). If a bacteriological diagnosis is made early in the disease and prompt antibiotic treatment is given, the antibody response may be slight. We are thus seldom in a position to observe an antibody response in paired sera of patients from whom the organism was isolated in the acute stage of the disease. The isolation of the organism from a patient with the 'right' symptoms provides us with reasonable grounds for instituting specific treatment. But the mere presence of *Str. pyogenes* at the site of the supposed infection does not provide absolute proof that it is the cause. At least 10 per cent of children carry the organism in the throat; if they suffer from tonsillitis due to another cause they may erroneously be considered to be suffering from streptococcal disease. In practice, therefore, we must accept a certain amount of over-diagnosis if it is our aim to treat as many streptococcal infections as possible at the stage at which serious complications can be prevented. On the other hand, in any outbreak, some of the patients from whom the organism can be isolated exhibit few or no

symptoms but give an antibody response. We are unlikely, therefore, to succeed in recognizing all the streptococcal infections by any practicable system of surveillance.

Cultural methods For throat swabs, culture on blood agar is sufficient, but there is no general agreement about the optimal composition of the medium. On horse-blood agar, *Str. pyogenes* gives clear β-haemolysis; but so do the haemolytic haemophili, and it is necessary to be able to distinguish between the two by means of their colonial appearances. This problem does not arise on sheep-blood agar (Schaub *et al.* 1958), but the streptococci often give poor haemolysis when incubated aerobically on this medium. Anaerobic incubation improves haemolysis, particularly on sheep-blood agar, and also decreases the growth of other organisms such as staphylococci. The clarity of zones of haemolysis is improved by incorporating blood only in the upper layer of the medium. Isolation from other sites, such as the nose, the skin, and wound swabs, can also often be made on blood agar, particularly when it is incubated anaerobically. Overgrowth by other organisms is prevented by the use of various selective media (Chapter 29).

Throat cultures from cases of untreated tonsillitis usually yield much heavier growths of group A streptococci than do throat cultures from carriers. Semi-quantitative reporting of the results of cultures that have been seeded in a standard way may therefore be helpful. For example, Bell and Smith (1976) found that 71 per cent of positive swabs from patients with sore throat, but only 10 per cent from healthy carriers, yielded more than 200 colonies on primary plates.

In Chapter 20 we discussed the taking of throat swabs and the necessity to ensure that the swab was free from inhibitory substances. To aid the survival of streptococci during transit to the laboratory, swabs should if possible be kept at 4° and desiccated. A desiccant substance may be added to the swab container (Redys *et al.* 1968) or the swab may be rubbed on a strip of filter paper (Hollinger *et al.* 1960) which is then dried and may be sent by post. According to Ross (1977), coating swabs with serum or albumin does not improve viability in transit if the swab is free from inhibitory substances. Efficient detachment of the streptococci from dry swabs is of importance; on arrival in the laboratory, swabs should be agitated in broth before plating. Survival is less than optimal on swabs that remain moist for a long time during transit, particularly at a high atmospheric temperature. (See Chapter 29 for selective and enrichment media for streptococci, and for grouping methods.)

Several methods have been advocated for the rapid identification of *Str. pyogenes*. Direct immunofluorescent staining with sera directed against the group polysaccharide (Moody *et al.* 1958) was in vogue in the 1960s; swabs must first be incubated in broth for several hours, and really specific sera are difficult to prepare. Immunofluorescence may also be used for the rapid identification of colonies on primary plates, but this can be done more easily by co-agglutination (Chapter 29). A microtechnique for the extraction of group antigen from throat scrapings and its subsequent identification by co-agglutination has been proposed (El Kholy *et al.* 1978).

Serological methods   The main practical use of serological tests is for the retrospective diagnosis of infection in patients thought to be suffering from a late complication such as rheumatic fever or glomerulonephritis. We shall therefore not expect to demonstrate a rising titre of antibody, but must compare results obtained on a single specimen with the supposed 'normal value' for corresponding members of the general population. This tends to vary with age, geographical location and the frequency of streptococcal infection in the area. Because some of our 'normal' subjects will almost certainly have experienced a recent streptococcal infection, the best we can do is to make use of a figure for the 'upper limit of normal' defined as a titre below which a stated percentage of titres will fall. Table 59.2 shows figures for the upper limit of normal of the ASO, anti-hyaluronidase, anti-DNAase B and anti-DPNase tests collected by Wannamaker and Ayoub (1960), mainly from sources in the USA, in which this percentage is set at 80 (see also Klein *et al.* 1971). It also gives comparable figures for the lower limit of titres to be expected in patients in the early acute stage of rheumatic fever. We would merely add that the upper limit of normal for infants aged 6 months to 2 years is probably below 100 units for ASO and 60 for anti-DNAase B.

Such comparisons are of limited value. When used to support a diagnosis of, for example, rheumatic fever, they tell us only that the patient has had a recent attack of streptococcal infection; when used to exclude the diagnosis, estimation of a single antibody is insufficient evidence, but a negative result in two or three tests is highly conclusive. Chorea may not develop for several months after the initial streptococcal infection, and by this time the ASO titre may be normal. Nevertheless, when several antibody tests are used, evidence of previous streptococcal infection can be obtained in many cases, including a number in which there were no other rheumatic manifestations (Ayoub and Wannamaker 1966).

Of the many antibody tests that are available (see p. 236) there is little doubt that the ASO—which is the most widely used and the most easily standardized— yields as much information as any other test, except in streptococcal impetigo. If another is to be added our choice would be the anti-DNAase B test or the anti-hyaluronidase test.

Antimicrobial treatment

The therapeutic and prophylactic uses of antimicrobial agents against streptococci cannot be clearly separated. We treat patients who are suffering from an acute streptococcal infection for the purpose of ameliorating the clinical illness, lessening their infectivity, and preventing the subsequent appearance of a rheumatic attack (so-called 'primary prevention' of rheumatic fever); and we give drugs prophylactically to patients who have had a rheumatic attack to prevent a recurrence (*individual prophylaxis* or 'secondary prevention' of rheumatic fever) and to groups of healthy persons to limit the spread of infection in communities (*mass prophylaxis*).

When the sulphonamides were first introduced they proved valuable in the treatment of invasive infections such as puerperal fever, but not of tonsillitis, and they did not prevent the subsequent appearance of rheumatic fever. On the other hand, they were effective for individual prophylaxis in rheumatic subjects (e.g. Bywaters and Thomas 1958). At present some 5 per cent of strains isolated in Britain are sulphonamide resistant.

Penicillin is highly effective in the treatment of all acute infections, and penicillin resistance has not yet been observed in *Str. pyogenes*. Treatment for as little as 3–5 days will cut short the effect of severe attacks of tonsillitis and prevent local septic complications such as otitis media, but only if treatment is continued for 10 days will it be effective in preventing rheumatic attacks. As we have seen (p. 239), penicillin treatment fails to eliminate streptococci completely in some 20 per cent of cases. Whether it can be relied upon to prevent glomerulonephritis is uncertain (see Weinstein and Le Frock 1971). Systemic penicillin is highly effective in the treatment of streptococcal impetigo (Dillon 1970). Its use in prophylaxis will be discussed in the next section.

If the use of penicillin is contraindicated, a macrolide or lincomycin, given for a similar length of time, appears to be

Table 59.2   Antibody titres in the normal population, and in cases of early acute rheumatic fever (modified from Wannamaker and Ayoub 1960)

Antibody	Upper limits for the normal population*		Lower limits for rheumatic fever†
	5–12 years	young adults	
Anti-streptolysin O	333	200	250
Anti-hyaluronidase	110	80	300
Anti-DNAase B	. . .	80‡	320
Anti-DPNase	. . .	130‡	175

* 80 per cent of subjects have titres at this level or lower.

† 80 per cent of subjects have titres at this level or higher.

‡ Includes a few children.

clinically effective and to eliminate the streptococci from the throat in about the same proportion of cases as does penicillin. Resistance to these antibiotics is unusual in the general population, at least in Britain (for further details see Chapter 29). Resistance to tetracycline is so common in *Str. pyogenes* that this antibiotic should never be used for the treatment of streptococcal infections.

(For information about the appropriate dosage of antibiotics see Report 1966.)

Prevention

General hygienic measures, such as ensuring adequate ventilation and the suppression of dust, appear to be ineffective in controlling outbreaks of streptococcal infection of the respiratory tract in institutions. The isolation of ill persons, though a rational step, will seldom exert a decisive influence, either because dangerous carriers are left at large or because the chain of infection is not broken quickly enough. Attempts to eradicate an actively spreading infection by mass swabbing and segregation of carriers often fails because further persons become infected before the bacteriological tests have been completed; and when the carrier rate is very high, segregation becomes an impossibility.

Nowadays, therefore, it is usual to employ some form of chemotherapy or chemoprophylaxis. The prompt treatment of clinical cases, with isolation for 2–3 days if possible, is a minimal measure. Boissard and Fry (1966) advocate the rapid detection of heavy dispersers of streptococci among the contacts by the examination of swabs of the nose and the labio-dental sulcus; nasal carriers are treated by the insufflation of sulphonamide, and salivary carriers with penicillin lozenges, but are not isolated; this, combined with segregation and treatment of clinical cases, they claim will often cut short epidemics in schools; but controlled studies have not been made.

Many workers favour some form of *mass prophylaxis* with penicillin.

The administration of oral penicillin for 10–14 days to the whole population of a large military camp dramatically cuts short an epidemic of streptococcal infection (Wannamaker *et al.* 1953*b*), but cases often begin to occur again when prophylaxis is stopped, and may have returned to their original frequency within 3 weeks. Seal (1955) advocated giving intermittent courses of 500 000 units daily for 10 days whenever an epidemic occurred, and reported an effective control of rheumatic fever in military camps by this means. Others (*e.g.* Chancey *et al.* 1955) favour a single intramuscular injection of 600 000 units of the long-acting benzathine penicillin, which gives almost complete protection against the acquisition of *Str. pyogenes* for 3 weeks. Finally, Frank, Stollerman and Miller (1965), who had found that intermittent mass prophylaxis gave imperfect control of the infection, instituted a single injection of benzathine penicillin to all new entrants to military camps at any season of the year, and claimed a greater than tenfold reduction in the morbidity from streptococcal infection. Subjects with a history of penicillin allergy were not treated. In 9 years' experience of this prophylactic regimen on a total of 315 000 men, drug reactions of the serum-sickness type were seen in less than 1 per cent of subjects and no death from anaphylaxis was recorded.

Despite the undoubted short-term efficacy of mass prophylaxis it is difficult to define precisely the circumstances in which it should be employed. In situations in which the risk of rheumatic fever is predictable, such as large training camps for military recruits, there is a case for the institution of prophylaxis as soon as an increase in the frequency of streptococcal tonsillitis is detected; but the advantage of this course of action will be lost unless prophylaxis is repeated early in every outbreak. One effect of mass prophylaxis is to prevent the acquisition of type-specific immunity; it follows that its occasional use may be merely to delay the onset of an outbreak. Furthermore, there is probably little to be gained from beginning it several weeks after the onset of an outbreak. Mass prophylaxis is therefore unlikely to be effective in, for example, boarding schools or in sections of the general population unless (1) the population is under continuous surveillance and the outbreak is detected early, and (2) the period of risk is of limited duration (see Zimmerman *et al.* 1966, Report 1968). In small residential institutions it might be advantageous to treat all new entrants with penicillin, or only the carriers on admission, but the effect of this has not been investigated.

The prevention of rheumatic fever   Rheumatic fever would be prevented if early and adequate antibiotic therapy was given to all patients with acute streptococcal infections. A number of community programmes for the primary prevention of rheumatic fever have been established but it is still too early to assess their effectiveness. On the other hand, the value of *individual prophylaxis* for the prevention of recurrences in rheumatic subjects is now well documented. The prophylactic agent must be given continuously; the usual recurrence rate of 40–70 per cent in untreated patients can be lowered by this means to about 4 per cent. The agents recommended are daily oral sulphadiazine, daily oral penicillin (preferably the phenoxymethyl derivative) and 3-weekly intramuscular injection of benzathine benzylpenicillin (for dosage see Report 1966). The results obtained with intramuscular injections of long-acting penicillin are in general superior to the results of oral prophylaxis (Wood *et al.* 1964, Feinstein *et al.* 1968), probably because the latter is often not continuous (Gordis *et al.* 1969*c*). It is not yet clear how long prophylaxis should be continued. In the absence of damage to the heart a period of 5 years, or until the child leaves school, whichever is the longer, should be adequate (Report 1960, Johnson *et al.* 1964, Feinstein *et al.* 1966), but in its presence it is perhaps advisable to extend the period by several years.

(For the organization of preventive programmes in developing countries, see Strasser *et al.* 1981, and for a general review, Krause 1975.)

Vaccine prophylaxis

Subcutaneous injection into human volunteers of a vaccine of purified M protein, with aluminium hydroxide as adjuvant, resulted in the appearance of M antibody in the serum and afforded highly significant protection against challenge with a virulent strain of the same type when this was applied to the tonsils on a swab (Fox *et al.* 1973, Wittner *et al.* 1979). The M proteins were of types 1, 3 and 12, all types known to give a good antibody response. A surprising finding was that the same vaccines given by intranasal spray gave equally good protection, though little M antibody appeared in the blood.

An effective polyvalent M-protein vaccine would provide an attractive alternative to chemoprophylaxis for preventing the serious sequelae of *Str. pyogenes* infection. It should now be possible to make a vaccine that is unlikely to elicit the formation of 'heart-reactive' antibody (see Chapter 29). It is doubtful, however, whether it is feasible to produce a polyvalent vaccine with a sufficiently wide coverage of prevalent M types, and with protective activity against many of the less antigenic M proteins.

Further information about diseases caused by *Str. pyogenes* will be found in papers in the following volumes: Wannamaker and Matsen (1972), Skinner and Quesnel (1978), Parker (1979*b*), Read and Zabriskie (1980) and Holm and Christensen (1982).

Infections of man due to *Str. agalactiae* (the group B streptococcus)

The recognition that the group B streptococcus is an important pathogen for man came late—nearly 80 years after it was described as a cause of bovine mastitis (Chapters 29 and 57). The first serious human infection was described by Colebrook and Purdie in 1937; in the next few years a number of others were recorded, mainly cases of puerperal septicaemia or endocarditis, but only three cases of neonatal infection were reported before 1958 (for early references, see Parker 1977). Between this date and 1968, 141 systemic infections were recorded, 82 of them in the newborn. This gave rise to the view that neonatal group-B streptococcal infection was a 'new' disease, but this is far from certain. Many bacteriologists were unfamiliar with the organism until quite recently, and many of the cases occurred in premature babies who would probably not have survived in earlier years. It is difficult to believe, however, that neonatal meningitis, which often affects full-term babies, would have escaped recognition until 1958 unless its frequency had increased considerably.

Infections of the newborn

Several workers (Quirante and Cassady 1972, Baker and Barrett 1973, Franciosi *et al.* 1973) observed that there were two distinct forms of serious neonatal infection: (1) *early-onset disease* in the first week of life, always septicaemic and sometimes meningitic, might be septicaemic or meningitic, usually affected infants with a low birth-weight, was often associated with a long interval between rupture of the membranes and delivery, and had a poor prognosis; and (2) *late-onset disease*, which occurred after the 7th day, was almost invariably meningitic, affected apparently healthy babies after a normal delivery, and had a better prognosis if treatment was given promptly. In early-onset cases group B streptococci of the infecting serotype were nearly always present in the vagina of the mother; in late-onset cases this was seldom so. The clinical appearances in early-onset disease are mainly of shock or respiratory distress; these tend to overshadow the signs of meningitis when this is present. When the respiratory tract is affected, the disease closely resembles hyaline-membrane disease (Ablow *et al.* 1976), and indeed the respiratory passages are filled with a membranous exudate in which enormous numbers of streptococci are embedded (Katzenstein *et al.* 1976); a definite bronchopneumonia is sometimes present.

Early-onset disease may be present at birth; more often it develops within the next 24 hr, and nearly always in the first 5 days of life. Late-onset disease is most frequent between the 10th and 20th days, but occurs with decreasing frequency through the later neonatal period into infancy and early childhood. Many cases of early-onset disease escape detection unless post-mortem culture is regularly practised, and it appears that at least 70 per cent of all neonatal systemic infections are of the early-onset type. Estimates of the frequency of serious neonatal group-B streptococcal disease vary; most of them lie between 1 and 3 per 1000 live births (see Parker 1977), but figures as high as 6 per 1000 have been recorded (Pass *et al.* 1979). However, most of these figures have wide confidence limits, and a current British survey of over 200 000 births gives a figure of only 0.3 per 1000 live births (Mayon-White 1982). Purulent localization of group B streptococci elsewhere than in the meninges is uncommon in the newborn, but may occur in bones or joints; necrotizing fasciitis and bullous skin eruptions have been described.

Infections in adults

Systemic disease is usually a septicaemia; less than 10 per cent of cases are of meningitis; endocarditis develops in some 20 per cent of cases and purulent localization—usually in a joint or the peritoneal cavity—in about 15 per cent. It may follow local sepsis at a variety of sites: the female genital tract, surgical wounds, extensive non-bacterial skin lesions, and the urinary tract. Some patients are diabetics or have lymphoreticular diseases. Most systemic infections in adults occur in patients aged more than 45 years, and males are affected as often as females (Parker and Stringer 1979). Group B streptococci are often isolated from superficial lesions, whether or not signs of inflammation are present, and are usually present in mixed culture. According to Mhalu (1977), they are responsible for about 1 per cent of urinary-tract infections, nearly all in females.

Distribution of serotypes

With a few exceptions, observers have reported a considerable excess of type III strains and a deficiency of type II strains in meningitis, in comparison with other systemic diseases (Franciosi *et al.* 1973, Baker and Barrett 1974, but see Butter and de Moor 1967). In the USA, the respective percentages for strains isolated from cerebrospinal fluid and blood in neonates were: for type III, 87 and 49; for type II, <1 and 10 (Wilkinson 1978*a*). In Britain, the corresponding percentages for meningitis and for all other systemic infections were: for type III 57 and 37; for type II, 1.3 and 18 (Parker 1979*a*). In general, the type distribution in septicaemic infections corresponds with that for carrier sites and superficial lesions.

Carriage

The presence of group B streptococci in the vagina has attracted most attention. Reported vaginal carrier rates vary widely even when selective and enrichment media are used. For women in late pregnancy, figures varying from 5 to 25 per cent or more have been reported (see Parker 1977). The exact site swabbed influences the result, swabs from the lower part of the vagina giving considerably higher carriage rates than high vaginal or cervical swabs (Kexel and Beck 1965). There are probably racial differences in carriage rates; thus, American women of Mexican descent carry the organism only half as often as those of non-Mexican descent (Anthony *et al.* 1978). Despite somewhat divergent reports, it appears that carriage rates in the different stages of pregnancy do not differ much. However, repeated swabbing gives considerably higher total carriage rates than a single examination; according to Anthony and his colleagues (1978), less than 40 per cent of women who harboured the organism during pregnancy gave positive cultures on three or more of the four occasions on which they were examined. Counts of the number of group B streptococci on vaginal swabs vary widely, from less than 10 to more than 10^5 (Schauf and Hlaing 1976).

Carriage rates in non-pregnant women have often been reported to be higher than in pregnant women. This is true not only of nurses in maternity departments but also in women unconnected with hospitals (Baker and Barrett 1973, Franciosi *et al.* 1973, Finch *et al.* 1976). Several reports suggest that women seen in venereal-disease departments have even higher carriage rates than other non-pregnant women, but there is considerable overlap between rates for the two categories in different geographical locations, and sexual promiscuity does not appear to be related to carriage in otherwise comparable groups (Baker *et al.* 1977*a*).

The frequent intermittency of carriage in the vagina suggests that this may not be the primary site of carriage. Group B streptococci can often be isolated from urethral, perianal and rectal swabs in women and men (Christensen *et al.* 1976, 1978, Badri *et al.* 1977, Mitchell *et al.* 1977). Gastro-intestinal carriage appears to be confined mainly to the ano-rectal area (Islam and Thomas 1980, Easmon *et al.* 1981). The pattern of cross-contamination between the genital, urinary and ano-rectal sites appears to be complex.

Less than 10 per cent of adults carry the organism in the throat, and there is little correspondence in serotype between strains isolated from respiratory and abdominal sites in the same person (Franciosi *et al.* 1973, Christensen and Christensen 1978).

Epidemiology of neonatal infection

Babies in whom early-onset disease develops almost certainly acquire the infecting strain from the mother's vagina. Occasional cases occur in babies delivered by Caesarian section, but it is generally held that the organism is acquired during passage through the birth canal. Some authors believe, however, that the importance of intrauterine infection has been underestimated. They point to the frequency with which the baby is already ill at birth or becomes ill within a few hours of delivery, and believe that in these cases the organism had penetrated the apparently unruptured membranes (see Baker 1978).

When the vaginal swab of the mother is positive at the time of delivery, 70 per cent of the infants yield group B streptococci from the body surface 18 hr after birth; when it is negative, the percentage is only 10 (Baker and Barrett 1973). The heaviness of initial contamination of the baby is directly proportional to the profuseness of carriage by the mother (Ancona *et al.* 1980). The organism may be detected on the skin or in the upper respiratory tract, but can be isolated most often from the external auditory meatus (Ferrieri *et al.* 1977).

In the late-onset disease, the mother's vagina can usually be eliminated as a source of the organism. Prospective bacteriological studies of babies in whom this infection subsequently develops have not been made, but several instances of epidemiologically related infections are on record (Winterbauer *et al.* 1966, Steere *et al.* 1975, Boyer *et al.* 1980). It is also now clear that postnatal acquisition of the organism by babies from non-maternal sources in hospital is considerably more common than had been thought (Aber *et al.* 1976, Paredes *et al.* 1977, Duben *et al.* 1978). The evidence suggests that infection is usually transferred from baby to baby in the course of nursing procedures, and is less often acquired from members of the staff; this has since been confirmed by phage typing (Anthony *et al.* 1979, Boyer *et al.* 1980).

Predisposition to systemic infection in the newborn

Why low birth-weight predisposes to the early-onset disease is uncertain, but it is generally assumed that the cellular defences of the immature fetus against group B streptococci are deficient. A long interval between rupture of the membranes and delivery is thought to allow time for the organism to multiply in the remaining fluid surrounding the fetus and so to increase the infecting dose. However, a considerable proportion of sufferers from early-onset disease were born after a short labour. If the organism often invades the amniotic cavity through intact membranes, this might be a significant cause of premature labour.

Evidence is accumulating that a deficiency in type-specific antibody in the mother's blood may be an important predisposing factor. Most of this comes from the study of late-onset meningitis, for which no other predisposing cause has yet been identified. Baker and Kasper (1976; see also Baker *et al.* 1977*b*, 1978) compared the content of antibody against 'native' type III polysaccharide (Chapter 29) in the sera of mothers whose babies had late-onset meningitis due to type III organisms with its content in the sera of mothers who were vaginal carriers of the organism but whose babies had remained healthy. In the first group, the antibody was seldom present in more than minimal amount, but in the second, many of the mothers had considerably higher levels of antibody. Most babies that suffered non-fatal attacks of the disease formed this antibody. In similar tests for antibody against the incomplete 'core' antigen, no difference could be shown between the two groups of mothers, and the babies showed little antibody response (Baker *et al.* 1979). Adults with low levels of 'native' antibody respond well to vaccination with purified preparations of the corresponding antigen.

The conclusion that only antibody against the 'native' antigen confers type-specific protection may be questioned. In Rebecca Lancefield's original demonstration of protection in mice (Chapter 29) the hot-acid extracts used to prepare antisera presumably contained little 'native' antigen. Group B streptococci of type III and pneumococci of capsular type 14 contain the same 'core' antigen; and antisera against pneumococci of this type are bactericidal for type III group B streptococci (Fischer *et al.* 1979). Until these doubts are resolved it will not be possible to predict the likely effects of vaccinating antibody-deficient mothers during pregnancy, or indeed to decide which mothers are deficient in the essential antibody. The case for vaccination would be strengthened if it was shown that antibody deficiency was a determinant for early- as well as for late-onset neonatal disease.

Bacteriological diagnosis

According to Cumming and Ross (1979), group B streptococci survive rather better on plain swabs than in fluid transport media. They can usually be isolated from clinical specimens on blood agar, but selective and enrichment media should be used in surveys of carriage; Waitkins (1982) recommends the use in selective media of (mg per l): neomycin 32 (rather than gentamicin 8) and nalidixic acid 15 (see also Chapter 29). The CAMP test may be used for presumptive identification, but co-agglutination tests with growth from primary plates (Chapter 29) provide a more convenient alternative.

The following methods have been used for the rapid detection of group antigen in blood, body fluids or even pharyngeal exudates in patients suspected of having systemic disease: counter-current immunoelectrophoresis (Henning and Tenstam 1973, Fenton and Harper 1978, Slack and Mayon-White 1978); agglutination of latex particles sensitized with group antibody (Webb and Baker 1980); immunofluorescent staining of organisms from blood-culture bottles (Romero and Wilkinson 1974); and co-agglutination tests on primary cultures in enrichment broth (Szilagyi *et al.* 1978). The urine is said to be the best source of antigen in patients with systemic disease (Ingram *et al.* 1982; see also Ingram *et al.* 1980).

Prevention

There is little agreement about how early-onset neonatal disease can be prevented. Attempts to eliminate maternal sources of infection by antibiotic treatment before labour are unlikely to be uniformly successful. The intermittency of vaginal carriage makes it difficult to detect all the mothers whose babies are at risk from this source; vaginal swabbing on a single occasion in late pregnancy is insufficient (Ferrieri *et al.* 1977), but swabbing of the vagina, urethra and perianal region at this time might be more effective. An alternative is to prevent invasion of the baby by giving antibiotic prophylaxis immediately after birth. Steigman and his colleagues (1978) claim that a single dose of 50 000 units of benzylpenicillin given intramuscularly prevents all serious neonatal group-B streptococcal infections, but no controls were included in their study. In a controlled study of over 30 000 babies (Siegel *et al.* 1982), single-dose prophylaxis was associated with a significant reduction in the frequency of systemic infection with group B streptococci but not in the total mortality rate for all systemic infections. Thus, the general use of antibiotic prophylaxis may well not be justifiable. Its selective use, for low birth-weight babies, or for the babies of carrier mothers if these could be reliably detected, is perhaps worthy of more serious investigation. It should be remembered, however, that a number of the babies with early-onset disease are gravely ill at birth.

The present indications are that late-onset disease follows the post-natal acquisition of group B streptococci. The general lack of success in controlling the acquisition of *Staph. aureus* in the neonatal period by modest improvements in hygienic practice suggests that this, too, will not be easy to prevent. This provides another argument against nurseries for normal babies in maternity departments, and for the meticulous use of 'no touch' techniques in special-care baby units. Whether the vaccination of antibody-deficient mothers during pregnancy (see p. 246) is a practicable means of preventing group B streptococcal disease in the newborn remains to be seen.

Treatment

Systemic group-B streptococcal infections are nearly always fatal if untreated. With early and vigorous treatment, the mortality rate in late-onset meningitis is about 10 per cent, but a number of the survivors suffer neurological sequelae; in early-onset disease it is 30–50 per cent. The drug of choice is probably benzylpenicillin, but ampicillin is often given, in the initial stages at least, because treatment must be begun before a bacteriological diagnosis has been made. Although the minimum inhibitory concentration of benzylpenicillin is well under 0.1 mg per 1 in conventional tests, it may be 2–4 mg per 1 in tests with an inoculum of 10^8 (Baker 1977). Relapses and recrudescences are common in meningitis even when large doses of penicillin are given. Experiments in mice suggest that mixtures of ampicillin or benzylpenicillin with gentamicin would eradicate infection more effectively than a penicillin alone (Deveikis *et al.* 1977). Chloramphenicol is said to inhibit the bactericidal action of ampicillin against group B streptococci *in vitro* (Weeks *et al.* 1981). According to Shigeoka and her colleagues (1978), transfusions of fresh whole blood reduce the mortality in early-onset disease if the blood contains opsonins for the infecting strain.

For general accounts of group B streptococcal disease, see Baker (1977), Jelinková (1977), and Wilkinson (1978*b*).

Other pyogenic infections in man

Group C and group G streptococci

Streptococci of groups C and G are found in the throat of a small proportion of healthy persons—normally not exceeding 2–3 per cent for each group—in Europe and North America, but carrier rates may be considerably higher in some parts of the tropics (see, for example, Ogunbi *et al.* 1974). When streptococci of either group are isolated from the throat of a single patient with tonsillitis it is very difficult to be sure whether it is responsible for the illness. There is little doubt, however, that in suitable circumstances group G streptococci can cause tonsillitis. One carefully investigated food-borne epidemic of this, in which the patients had rising titres of ASO, was reported by Hill and his colleagues (1969*a*). Streptococci of groups C and G are quite often found in minor septic lesions, often in pure culture. They are both occasionally responsible for septicaemia, sometimes with endocarditis.

The group C streptococci found in man nearly always belong to the *equisimilis* species. Rarely, septic infections with *Str. zooepidemicus* may occur in persons closely associated with domestic animals and sore throat in drinkers of unpasteurized milk (for references see Barnham *et al.* 1983). Acute glomerulonephritis has been reported in two such groups of patients (Duca *et al.* 1969, Barnham *et al.* 1983).

Str. suis, serotype 2

Human infections with this organism are confined to persons who are in contact with pigs or handle raw meat: farmers, workers in abattoirs or meat-processing factories, retail butchers, and occasionally housewives. In 36 reported infections in the Netherlands, Denmark, Britain and France, 33 patients suffered from meningitis; the mortality rate was 8 per cent; and nearly half of the survivors had residual deafness or giddiness (Zanen and Engel 1979); this was not attributable to treatment with aminoglycosides.

Str. milleri

During the last 40 years there have been scattered accounts of septic infections caused by β-haemolytic streptococci, usually belonging to groups F, G or C, that can now be recognized as *Str. milleri*, though most members of this species are non-haemolytic and ungroupable (Chapter 29). Since the recognition of the species as a whole, it has been shown to be a fairly common cause of purulent and septicaemic infections. Guthof (1956) found it in abscesses around the jaw, and Lütticken and his colleagues (1978) later identified it repeatedly in cervical 'actinomycosis'. It often causes abscesses in internal organs and purulent exudates in body cavities (Parker and Ball 1976, 1979): abscesses in the frontal lobe of the brain and localized meningitis; pleural empyema and lung abscesses; and it is the streptococcus most often isolated from liver abscesses and is often present in purulent peritonitis and subphrenic abscesses (see Chapter 57 for these, and for the possible role of *Str. milleri* in appendicitis). It is also responsible for septicaemic illnesses—usually less acute than those due to other pyogenic streptococci—

and for endocarditis, but this develops rarely in patients with visceral abscesses (Parker and Ball 1979).

Str. milleri is commonly present, and in large numbers, in the human oral cavity, especially the gingival sulcus (Mèjare and Edwardsson 1975), the throat, the bowel (Unsworth 1980) and the vagina. It has also been isolated from various animals (Hahn and Tolle 1979) including cattle, dogs and guinea-pigs.

Streptococcal diseases of domestic animals

In Chapter 57 we discussed streptococcal mastitis of cattle, and mentioned the streptococci that occasionally cause endocarditis in pigs, cattle and sheep; and in Chapter 29 there are many references to diseases of animals caused by streptococci. Here we shall merely summarize the streptococcal diseases that are common in a few domestic species.

Pigs Serotype 1 of *Str. suis* causes septicaemia, meningitis and arthritis in pigs aged 2-6 weeks, and serotype 2 causes meningitis in weaned pigs aged 10-14 weeks. The serotype 2 disease has increased considerably in frequency in Britain since 1974, probably because of the growth of the practice of transferring weaned pigs from farm to farm, and of overcrowding in post-weaning houses (Windsor 1977). Vaccines of purified type 2 polysaccharide (Elliott *et al.* 1980) are under trial (Chapter 29). *Str. lentus* is a frequent cause of cervical adenitis in young pigs in North America, but this disease is not often seen in Britain. *Str. equisimilis* and streptococci of group L cause septic arthritis in young pigs (see Ross 1972).

Horses *Str. equi* causes an acute infection of the pharynx and nasopharynx that leads to the formation of abscesses in the cervical lymph nodes; in a few cases the form of laryngeal paralysis known as 'strangles' follows this. The disease may occur in epidemics, affects young horses, and is said to lead to acute carditis. A killed bacterial vaccine made from a single strain of the organism gives good protection (Bryans and Moore 1972). *Str. zooepidemicus* is prevalent in the respiratory tract of healthy horses, and is found in a variety of septic lesions, many of which are secondary to trauma. Septicaemia may follow umbilical infection in foals.

Sheep Septicaemia and polyarthritis of lambs have been associated with an organism resembling *Str. dysgalactiae*, and pneumonia with *Str. zooepidemicus*.

For a general account of streptococcal diseases of animals, see Wilson and Salt (1978).

References

Aber, R.C., Allen, N., Howell, J.T., Wilkinson, H.W. and Facklam, R.R. (1976) *Pediatrics, Springfield* **58**, 346.
Ablow, R.C. *et al.* (1976) *New Engl. J. Med.* **294**, 65.
Allison, V.D. (1938) *Lancet.* **i**, 1067.
Alpert, J.J., Pickering, M.R. and Warren, R.J. (1966) *Pediatrics, Springfield* **38**, 663.
Amren, D.P., Anderson, A.S. and Wannamaker, L.W. (1966) *Amer. J. Dis. Child.* **112**, 546.

Ancona, R.J., Ferrieri, P. and Williams, P.P. (1980) *J. med. Microbiol.* **13**, 273.
Anderson, H.C. and McCarty, M. (1950) *Amer. J. Med.* **8**, 445.
Andres, G.A., Accini, L., Hsu, K.C., Zabriskie, J.B. and Seegal, B.C. (1966) *J. exp. Med.* **123**, 399.
Anthony, B.F., Kaplan, E.L., Briese, F.W., Chapman, S.S. and Wannamaker, L.W. (1968) *J. Lab. clin. Med.* **72**, 850.
Anthony, B.F., Kaplan, E.L., Chapman, S.S., Quie, P.G. and Wannamaker, L.W. (1967*a*) *Lancet* **ii**, 787.
Anthony, B.F., Kaplan, E.L., Wannamaker, L.W., Briese, F.W. and Chapman, S.S. (1969) *J. clin. Invest.* **48**, 1697.
Anthony, B.F., Kaplan, E.L., Wannamaker, L.W. and Chapman, S.S. (1976) *Amer. J. Epidem.* **104**, 652.
Anthony, B.F., Okada, D.M. and Hobel, C.J. (1978) *J. infect. Dis.* **137**, 524; (1979) *J. Pediat.* **95**, 431.
Anthony, B.F., Perlman, L.V. and Wannamaker, L.W. (1967*b*) *Pediatrics, Springfield* **39**, 263.
Ayoub, E.M. and Dudding, B.A. (1969) *J. Lab. clin. Med.* **74**, 747.
Ayoub, E.M. and Hoyer, J. (1969) *J. Pediat.* **75**, 193.
Ayoub, E.M. and Wannamaker, L.W. (1962) *Pediatrics, Springfield* **29**, 527; (1966) *Ibid.* **38**, 946.
Badin, J. and Barillec, A. (1970) *J. Lab. clin. Med.* **75**, 975.
Badri, M.S. *et al.* (1977) *J. infect. Dis.* **135**, 308.
Baker, C.J. (1977) *J. infect. Dis.* **136**, 137; (1978) *J. Pediat.* **93**, 124.
Baker, C.J. and Barrett, F.F. (1973) *J. Pediat.* **83**, 919; (1974) *J. Amer. med. Ass.* **230**, 1158.
Baker, C.J., Edwards, M.S. and Kasper, D.L. (1978) *J. clin. Invest.* **61**, 1107.
Baker, C.J. and Kasper, D.L. (1976) *New Engl. J. Med.* **294**, 753.
Baker, C.J., Kasper, D.L. and Schiffman, G. (1979) In: *Pathogenic Streptococci*, p. 150. Ed. by M.T. Parker. Reedbooks, Chertsey, Surrey.
Baker, C.J. *et al.* (1977*a*) *J. infect. Dis.* **135**, 392; (1977*b*) *J. clin. Invest.* **59**, 810.
Balmain, A.R. (1926) *Lancet* **ii**, 484.
Barnham, M., Thornton, T.J. and Lange, K. (1983) *Lancet* **i**, 945.
Barrett, A.M. and Gresham, G.A. (1958) *Lancet* **i**, 347.
Bassett, D.C.J. (1967) *Lancet* **i**, 503; (1970) *Trans. R. Soc. trop. Med. Hyg.* **64**, 138; (1972) *Brit. J. Derm.* **86**, Suppl. 8, 55.
Bates, M.M. (1967) *Mon. Bull. Minist. Hlth Lab. Serv.* **26**, 132.
Beathard, G.A. and Guckian, J.C. (1967) *Arch. intern. Med.* **120**, 63.

Becker, C.G. and Murphy, G.E. (1980) In: *Streptococcal Diseases and the Immune Response*, p. 23. Ed. by S.E. Read and J.B. Zabriskie. Academic Press, New York.

Bell, S.M. and Smith, D.D. (1976) *Lancet* **ii**, 61.

Belsey, M.A., Sinclair, M., Roder, M.R. and Le Blanc, D.R. (1969) *New Engl. J. Med.* **280**, 135.

Bergner-Rabinowitz, S., Ofek, I., Davies, M.A. and Rabinowitz, K. (1971) *J. infect. Dis.* **124**, 488.

Bernhard, G.C. and Stollerman, G.H. (1959) *J. clin. Invest.* **38**, 1942.

Bisno, A.L. and Ofek, I. (1974) *Amer. J. Dis. Child.* **127**, 676.

Bisno, A.L., Ofek, I. and Beachey, E.H. (1976) *Infect. Immun.* **13**, 407.

Bisno, A.L., Pearce, I.A. and Stollerman, G.H. (1977) *J. infect. Dis.* **136**, 278.

Bisno, A.L., Pearce, I.A., Wall, H.P., Moody, M.D. and Stollerman, G.H. (1970) *New Engl. J. Med.* **283**, 561.

Boissard, J.M. and Eton, B. (1956) *Brit. med. J.* **ii**, 574.

Boissard, J.M. and Fry, R.M. (1944) *Mon. Bull. Minist. Hlth Lab. Serv.* **3**, 160; (1955) *J. appl. Bact.* **18**, 478; (1966) *J. Hyg., Camb.* **64**, 221.

Boyer, K.M., Vogel, L.C., Gotoff, S.P., Gadzala, C.A., Stringer, J. and Maxted, W.R. (1980) *Amer. J. Dis. Child.* **134**, 964.

Bray, J.P., Burt, E.G., Potter, E.V., Poon-King, T. and Earle, D.P. (1972) *J. infect. Dis.* **126**, 34.

Brokman, H., Brill, J. and Frendzel, J. (1937) *Klin. Wschr.* **16**, 502.

Brown, J., Wannamaker, L.W. and Ferrieri, P. (1975) *J. med. Microbiol.* **8**, 503.

Brown, W.A. and Allison, V.D. (1935) *J. Hyg., Camb.* **35**, 283.

Bryans, J.T. and Moore, B.O. (1972) In: *Streptococci and Streptococcal Diseases*, p. 327. Ed. by L.W. Wannamaker and J.M. Matsen. Academic Press, New York.

Buchanan, C.S. and Haserick, J.R. (1970) *Arch. Derm.* **101**, 664.

Butler, W. (1909) *Proc. R. Soc. Med.* **2**, *Sect. Epidem.* 69.

Butter, M.N.W. and Moor, C.E. de (1967) *Antonie v. Leeuwenhoek* **33**, 439.

Bywaters, E.G.L. (1959) *Rheumatic Fever: Epidemiology and Prevention*. Blackwell, Oxford.

Bywaters, E.G.L. and Thomas, G.T. (1958) *Brit. med. J.* **ii**, 350.

Calveti, P.A. (1945) *Proc. Soc. exp. Biol., N.Y.* **60**, 379.

Carter, M.E., Bywaters, E.G.L. and Thomas, G.T. (1962) *Brit. med. J.* **i**, 965.

Cayton, H.R. and Morris, C.A. (1966) *Mon. Bull. Minist. Hlth Lab. Serv.* **25**, 87.

Chancey, R.L. *et al.* (1955) *Amer. J. med. Sci.* **229**, 165.

Christensen, K.K. and Christensen, P. (1978) *Scand. J. infect. Dis.* **10**, 209.

Christensen, K.K., Christensen, P., Jaldorf, F. and Pettersson, L. (1978) *Scand. J. infect. Dis.* **10**, 291.

Christensen, K.K., Ripa, T., Agrup, G. and Christensen, P. (1976) *Scand. J. infect. Dis.* **8**, 75.

Coburn, A.F. (1931) *The Factor of Infection in the Rheumatic State*. Williams and Wilkins Co., Baltimore; (1936) *Lancet* **ii**, 1025.

Colebrook, D.C. (1935) *Spec. Rep. Ser. med. Res. Coun., Lond.* No. 205.

Colebrook, L. (1956) *Brit. med. J.* **i**, 247.

Colebrook, L., Maxted, W.R. and Johns, A.M. (1935) *J. Path. Bact.* **41**, 521.

Colebrook, L. and Purdie, A.W. (1937) *Lancet* **ii**, 1237.

Collis, W.R.F. (1931) *Lancet* **i**, 1941.

Cooke, J.V. (1927) *Amer. J. Dis. Child.* **34**, 969; (1928*a*) *Ibid.* **35**, 762; (1928*b*) *Ibid.* **35**, 772.

Cumming, C.G. and Ross, P.W. (1979) *J. clin. Path.* **32**, 1066.

Daikos, G. and Weinstein, L. (1951) *Proc. Soc. exp. Biol., N.Y.* **78**, 160.

Denny, F.W., Perry, W.D. and Wannamaker, L.W. (1957) *J. clin. Invest.* **36**, 1092.

Derrick, C.W., Reeves, M.S. and Dillon, H.C. Jr (1970) *J. clin. Invest.* **49**, 1178.

Deveikis, A., Schauf, V., Mizen, M. and Riff, L. (1977) *Antimicrob. Agents Chemother.* **11**, 817.

Di Caprio, J.M., Rantz, L.A. and Randall, E. (1952) *Arch. intern. Med.* **89**, 374.

Di Caprio, J.M., Roberts, C.E. and Sherris, J.C. (1966) *Lancet* **ii**, 1007.

Dick, G.F. and Dick, G.H. (1921) *J. Amer. med. Ass.* **77**, 782; (1923) *Ibid.* **81**, 1166; (1924*a*) *Ibid.* **82**, 265; (1924*b*) *Ibid.* **83**, 84; (1925*a*) *Ibid.* **84**, 802; (1925*b*) *Ibid.* **84**, 1477.

Dillon, H.C. (1967) *Annu. Rev. Med.* **18**, 207; (1970) *J. Pediat.* **76**, 676.

Dillon, H.C., Derrick, C.W. and Dillon, M.S. (1974) *J. infect. Dis.* **130**, 257.

Dillon, H.C. Jr, Derrick, C.W. and Gooch, P.E. (1975) *J. gen. Microbiol.* **91**, 119.

Dillon, H.C., Moody, M.D., Maxted, W.R. and Parker, M.T. (1967) *Amer. J. Epidemiol.* **86**, 710.

Dillon, H.C., Reeves, M.S. and Maxted, W.R. (1968) *Lancet* **i**, 543.

Dochez, A.R. and Stevens, F.A. (1927) *J. exp. Med.* **46**, 487.

Dodge, W.F. *et al.* (1972) *New Engl. J. Med.* **286**, 273.

Dowsett, E.G., Herson, R.N., Maxted, W.R. and Widdowson, J.P. (1975) *Brit. med. J.* **i**, 500.

Duben, J., Jelinková, J. and Neubauer, M. (1978) *Zbl Bakt.* **A242**, 168.

Duca, E. *et al.* (1969) *J. Hyg., Camb.* **67**, 691.

Dudding, B.A. and Ayoub, E.M. (1968) *J. exp. Med.* **128**, 1081.

Dudding, B.A., Burnett, J.W., Chapman, S.S. and Wannamaker, L.W. (1970) *J. Hyg., Camb.* **68**, 19.

Easmon, C.S.F., Tanna, A., Munday, P. and Dawson, S. (1981) *J. clin. Path.* **34**, 921.

El Kholy, A., Facklam, R., Sabri, G. and Rotta, J. (1978) *J. clin. Microbiol.* **8**, 725.

Elliott, S.D., Clifton-Hadley, F. and Tai, J. (1980) *J. Hyg., Camb.* **85**, 275.

El Tayeb, S.H.M., Wannamaker, L.W., Nasr, E.M.M. and El Salaam, E.A. (1979) In: *Pathogenic Streptococci*, p. 131. Ed. by M.T. Parker. Reedbooks, Chertsey, Surrey.

Engman, M.F. (1901) *J. cutan. Dis.* **19**, 180.

Evans, P. (1941) *Lancet* **i**, 737.

Farber, R.E. and Korff, F.A. (1958) *Publ. Hlth Rep., Wash.* **73**, 203.

Fasth, A. (1974) *Acta path. microbiol. scand.* **B82**, 715.

Feinstein, A.R. and Spagnuolo, M. (1960) *New Engl. J. Med.* **262**, 533.

Feinstein, A.R., Spagnuolo, M., Jonas, S., Kloth, H., Tursky, E. and Levitt, M. (1968) *J. Amer. med. Ass.* **206**, 565.

Feinstein, A.R., Spagnuolo, M., Jonas, S., Levitt, M. and Tursky, E. (1966) *J. Amer. med. Ass.* **197**, 949.

Feinstein, A.R. and Stern, E.K. (1967) *J. chron. Dis.* **20**, 13.

Fenton, L.J. and Harper, M.H. (1978) *J. clin. Microbiol.* **8**, 500.

Ferrieri, P., Cleary, P.P. and Seeds, A.E. (1977) *J. med. Microbiol.* **10**, 103.

Ferrieri, P., Dajani, A.S., Chapman, S.S., Jensen, J.B. and Wannamaker, L.W. (1970) *New Engl. J. Med.* **283**, 832.

Finch, R.G., French, G.L. and Phillips, I. (1976) *Brit. med. J.* **ii**, 1245.

Fischer, G.W., Lowell, G.H., Crumrine, M.H. and Wilson, S.R. (1979) *Lancet* **i**, 75.

Foster, E.P. *et al.* (1944) *J. Amer. med. Ass.* **126**, 281.

Fox, E.N., Waldman, R.H., Wittner, M.K., Mauceri, A.A. and Dorfman, A. (1973) *J. clin. Invest.* **52**, 1885.

Franciosi, R.A., Knostman, J.D. and Zimmermann, R.A. (1973) *J. Pediat.* **82**, 707.

Frank, P.F., Stollerman, G.H. and Miller, L.F. (1965) *J. Amer. med. Ass.* **193**, 775.

Fraser, C.A.M., Ball, L.C., Morris, C.A. and Noah, N.D. (1977) *J. Hyg., Camb.* **78**, 283.

Friou, G.J. (1949) *J. infect. Dis.* **84**, 240.

Glaser, R.J. *et al.* (1956) *J. exp. Med.* **103**, 173.

Glezen, W.P., Clyde, W.A. Jr. Senior, R.J., Sheaffer, C.I. and Denny, F.W. (1967) *J. Amer. med. Ass.* **202**, 455.

Glover, J.A. and Griffith, F. (1931) *Brit. med. J.* **ii**, 521.

Glynn, L.E. and Holborow, E.J. (1961) *Arthr. Rheum.* **4**, 203.

Goedvolk-de Groot, L.E., Michel-Bensink, N., Es-Boon, M.M. van, Vonno, A.H. van and Michel, M.F. (1974) *J. clin. Path.* **27**, 891.

Goldstein, I. and Caravano, R. (1967) *Proc. Soc. exp. Biol., N.Y.* **124**, 1209.

Goldstein, I., Halpern, B. and Robert, L. (1967) *Nature, Lond.* **213**, 44.

Gooder, H. (1961) *Bull. World Hlth Org.* **25**, 173.

Gordis, L., Lilienfeld, A. and Rodriguez, R. (1969a) *J. chron. Dis.* **21**, 645, 655; (1969b) *Publ. Hlth Rep., Wash.* **84**, 333.

Gordis, L., Markowitz, M. and Lilienfeld, A.M. (1969c) *Pediatrics, Springfield* **43**, 173.

Gray, J.D.A. (1956) *Lancet* **ii**, 132.

Griffith, F. (1934) *J. Hyg., Camb.* **34**, 542.

Gryska, P.F. and O'Dea, A.E. (1970) *J. Amer. med. Ass.* **213**, 1189.

Guthof, O. (1956) *Zbl. Bakt.* **166**, 553.

Hahn, G. and Tolle, A. (1979) *Zbl. Bakt.* I Abt. Orig. **A244**, 427.

Hällén, J. (1963) *Acta path. microbiol. scand.* **66**, 93.

Hamburger, M. (1944) *J. infect. Dis.* **75**, 58, 71.

Hamburger, M. and Green, M.J. (1946) *J. infect. Dis.* **79**, 33.

Hamburger, M., Green, M.J. and Hamburger, V.G. (1945) *J. infect. Dis.* **77**, 68, 96.

Hamburger, M., Lemon, H.M. and Platzer, R.F. (1949) *Amer. J. Hyg.* **49**, 140.

Hamburger, M., Puck, T.T., Hamburger, V.G. and Johnson, M.A. (1944) *J. infect. Dis.* **75**, 79.

Hare, R. and Colebrook, L. (1934) *J. Path. Bact.* **39**, 429.

Hare, R. and Maxted, W.R. (1935) *J. Path. Bact.* **41**, 513.

Hartley, G.J., Enders, J.F., Mueller, J.H. and Schoenbach, E.B. (1945) *J. clin. Invest.* **24**, 92.

Haverkorn, M.J. and Goslings, W.R.O. (1969) *Amer. J. hum. Genet.* **21**, 360.

Haxthausen, H. (1927) *Ann. Derm. Syph., Paris* **8**, 201.

Hektoen, L. (1935) *J. Amer. med. Ass.* **105**, 1.

Henning, C. and Tenstam, J. (1973) *Scand. J. infect. Dis.* **5**, 313.

Henningsen, E.J. and Ernst, J. (1939) *J. Hyg., Camb* **39**, 51.

Hewitt, L.F. and Todd, E.W. (1939) *J. Path. Bact.* **49**, 45.

Hill, H.R., Caldwell, G.G., Wilson, E., Hager, D. and Zimmerman, R.A. (1969a) *Lancet* **ii**, 371.

Hill, H.R., Zimmerman, R.A., Reid, G.V.K., Wilson, E. and Kilton, R.M. (1969b) *New Engl. J. Med.* **280**, 917.

Hirschfeld, A.J. (1970) *Pediatrics, Springfield* **46**, 799.

Hollinger, N.F. *et al.* (1960) *Publ. Hlth Rep., Wash.* **75**, 251.

Holm, S.E. and Christensen, P. (Eds) (1982) *Basic Concepts of Streptococci and Streptococcal Diseases.* Reedbooks, Chertsey, Surrey.

Holm, S.E., Christensen, P. and Schalén, C. (1979) In: *Pathogenic Streptococci*, p. 72. Ed. by M.T. Parker. Reedbooks, Chertsey, Surrey.

Holmes, M.C. and Rubbo, S.D. (1953) *J. Hyg., Camb.* **51**, 450.

Holmes, M.C. and Williams, R.E.O. (1954) *J. Hyg., Camb.* **52**, 165.

Husby, G., Rijn, I. van der, Zabriskie, J.B., Abdin, Z.H. and Williams, R.C. (1976) *J. exp. Med.* **144**, 1094.

Ingram, D.L., Pendergrass, E.L., Bromberger, P.I., Thullen, J.D., Yoder, C.D. and Collier, A.M. (1980) *Amer. J. Dis. Child.* **134**, 754.

Ingram, D.L., Suggs, D.M. and Pearson, A.W. (1982) *J. clin. Microbiol.* **16**, 656.

Islam, A.K.M.S. and Thomas, E. (1980) *J. clin. Path.* **33**, 1006.

Jackson, D.M., Lowbury, E.J.L. and Topley, E. (1951) *Lancet* **ii**, 705.

Jelinková, J. (1977) *Curr. Topics Microbiol. Immunol.* **76**, 127.

Johnson, E.E., Stollerman, G.H. and Grossman, B.J. (1964) *J. Amer. med. Ass.* **190**, 407.

Johnson, J.C., Baskin, R.C., Beachey, E.H. and Stollerman, G.H. (1968) *J. Immunol.* **101**, 187.

Jones, T.D. (1944) *J. Amer. med Ass.* **126**, 481.

Kaplan, E.L., Anthony, B.F., Chapman, S.S., Ayoub, E.M. and Wannamaker, L.W. (1970a) *J. clin. Invest.* **49**, 1405.

Kaplan, E.L., Anthony, B.F., Chapman, S.S. and Wannamaker, L.W. (1970b) *Amer. J. Med.* **48**, 9.

Kaplan, E.L., Ferrieri, P. and Wannamaker, L.W. (1974) *J. Pediat.* **84**, 21.

Kaplan, E.L., Top, F.H. Jr, Dudding, B.A. and Wannamaker, L.W. (1971) *J. infect. Dis.* **123**, 490.

Kaplan, E.L. and Wannamaker, L.W. (1975) *J. Lab. clin. Med.* **86**, 91.

Kaplan, M.H. (1960) *Ann. N.Y. Acad. Sci.* **86**, 974; (1963) *J. Immunol.* **90**, 594; (1965) *Ann. N.Y. Acad. Sci.* **124**, 904.

Kaplan, M.H. and Dallenbach, F.D. (1961) *J. exp. Med.* **113**, 1.

Kaplan, M.H. and Meyeserian, M. (1962) *Lancet* **i**, 706.

Kaplan, M.H., Meyeserian, M. and Kushner, I. (1960) *J. exp. Med.* **113**, 17.

Katzenstein, A.L., Davis, C. and Braude, A. (1976) *J. infect. Dis.* **133**, 430.

Keefer, C.S. and Spink, W.W. (1936) *J. clin. Invest.* **15**, 17, 21.

Kellner, A., Freeman, E.B. and Carlson, A.S. (1958) *J. exp. Med.* **108**, 299.

Kexel, G. and Beck, K.J. (1965) *Geburtsh. Frauenheilk.* **25**, 1078.

Kim, Y.B. and Watson, D.W. (1970) *J. exp. Med.* **131**, 611.

Kingston, D. and Glynn, L.E. (1971) *Immunology* **21**, 1003; (1976) *Brit. J. exp. Path.* **57**, 114.

Kirschner, L. and Howie, J.B. (1952) *J. Path. Bact.* **64**, 367.

Klein, E. (1887) *Proc. roy. Soc.* **42**, 158.

Klein, G.C., Baker, C.N. and Jones, W.L. (1971) *Appl. Microbiol.* **21**, 999.

Klein, G.C. and Jones, W.L. (1971) *Appl. Microbiol.* **21**, 257.

Kohler, P.F. and Ten Bensel, R. (1969) *Clin. exp. Immunol.* **4**, 191.

Krause, R.M. (1975) *J. infect. Dis.* **131**, 592.

Krumwiede, C., Nicoll, M. and Pratt, J.S. (1914) *Arch. intern. Med.* **13**, 909.

Kundsin, R.B. and Miller, J.M. (1964) *New Engl. J. Med.* **271**, 1395.

Kwantes, W. and James, J.R.E. (1956) *Brit. med. J.* **ii**, 576.

Lancefield, R.C. (1928) *J. exp. Med.* **47**, 91, 469, 481, 843, 857; (1959) *Ibid.* **110**, 271.

Lancefield, R.C. and Hare, R. (1935) *J. exp. Med.* **61**, 335.

Landsteiner, K., Levaditi, C. and Prasek, E. (1911) *Ann. Inst. Pasteur* **25**, 754.

Lange, K., Ahmed, U., Kleinberger, H. and Treser, G. (1976) *Clin. Nephrol.* **5**, 207.

Langewisch, W.H. (1956) *Pediatrics, Springfield* **18**, 438.

Lasch, E.E., Frankel, V., Vardy, P.A., Bergner-Rabinowitz, S., Ofek, I. and Rabinowitz, K. (1971) *J. infect. Dis.* **124**, 141.

Leyden, J.J., Stewart, R. and Kligman, A.M. (1980) *J. invest. Derm.* **75**, 196.

Liao, S.J. (1951) *J. Lab. clin. Med.* **38**, 648.

Lieberman, E. and Donnell, G.N. (1965) *Amer. J. Dis. Child.* **109**, 398.

Lindberg, L.H. and Vosti, K.L. (1969) *Science* **166**, 1032.

Lindberg, L.H., Vosti, K.L. and Raffel, S. (1967) *J. Immunol.* **98**, 123.

Loosli, C.G., Smith, M.H.D., Cline, J. and Nelson, L. (1950) *J. Lab. clin. Med.* **36**, 342.

Lütticken, R., Wannamaker, L.W., Kluitmann, G., Neugebauer, M. and Pulverer, G. (1976) *Dtsch. med. Wschr.* **101**, 958.

Lütticken, R., Wendorff, V., Lütticken, D., Johnson, E.A. and Wannamaker, L.W. (1978) *J. med. Microbiol.* **11**, 419.

McDowall, M.F., Ramkissoon, R. and Bassett, D.C.J. (1970) *Clin. Pediat.* **9**, 580.

McIntyre, D.M. (1968) *Amer. J. Obstet. Gynec.* **101**, 308.

Măgureanu, E. and Feller, H. (1966) *Arch. roum. Path. exp. Microbiol.* **25**, 1041.

Mair, W. (1923) *Lancet* **ii**, 1390.

Markowitz, A.S., Horn, D., Aseron, C., Novak, R. and Battifora, H.A. (1971) *J. Immunol.* **107**, 504.

Markowitz, A.S. and Lange, C.F. (1964) *J. Immunol.* **92**, 565.

Markowitz, M. (1970) *Circulation* **41**, 1077.

Markowitz, M. *et al.* (1967) *J. Pediat.* **71**, 132.

Maxted, W.R. (1978) In: *Streptococci*, p. 107. Ed. by F.A. Skinner and L.B. Quesnel. Academic Press, London.

Maxted, W.R., Fraser, C.A.M. and Parker, M.T. (1967) *Lancet* **i**, 641.

Maxted, W.R., Widdowson, J.P. and Fraser, C.A.M. (1973) *J. Hyg., Camb.* **71**, 35.

Mayon-White, R.T. (1982) In: *Basic Concepts of Streptococci and Streptococcal Diseases*, p. 305. Ed. by S.E. Holm and P. Christensen. Reedbooks, Chertsey, Surrey.

Mejàre, B. and Edwardsson, S. (1975) *Arch. oral Biol.* **20**, 757.

Meleney, F.L. (1924) *Arch. Surg.* **9**, 317.

Mhalu, F.S. (1977) *Post Grad. med. J.* **53**, 216.

Michael, A.F., Drummond, K.N, Good, R.A. and Vernier, R.L. (1966) *J. clin. Invest.* **45**, 237.

Mihalco, F. and Mitrică, N. (1965) *Arch. roum. Path. exp. Microbiol.* **24**, 985.

Mitchell, R.G., Guillebaud, J. and Day, D.G. (1977) *J. clin. Path.* **30**, 1021.

Moody, M.D., Ellis, C.D. and Updyke, E.L. (1958) *J. Bact.* **75**, 553.

Moser (1902) *Wien. klin. Wschr.* **15**, 1053.

Murphy, G.E. and Swift, H.F. (1949) *J. exp. Med.* **89**, 687; (1950) *Ibid.* **91**, 485.

Nelson, J., Ayoub, E.M. and Wannamaker, L.W. (1968) *J. Lab. clin. Med.* **71**, 867.

Nelson, K.E., Bisno, A.L., Waytz, P., Brunt, J., Moses, V.K. and Haque, R.-U. (1976) *Amer. J. Epidem.* **103**, 270.

Ogunbi, O., Lasi, Q. and Lawal, S.F. (1974) In: *Streptococcal Disease and the Community*, p. 282. Ed. by M. Haverkorn. Excerpta Medica, Amsterdam.

Ohanian, S.H., Schwab, J.H. and Cromartie, W.J. (1969) *Proc. Soc. exp. Biol.*, N.Y. **129**, 37.

Onyewotu, I.I. and Mee, J. (1978) *J. clin. Path.* **31**, 817.

Paine, T.F., Novick, R.P. and Hall, W.H. (1963) *Arch. intern. Med.* **112**, 936.

Paredes, A., Wong, P., Mason, E.O. Jr, Taber, L. and Barrett, F.F. (1977) *Pediatrics, Springfield* **59**, 679.

Parker, M.T. (1967) *Bull. World Hlth Org.* **37**, 513; (1969) *Brit. J. Derm.* **81**, Suppl., 37; (1977) *Post Grad. med. J.* **53**, 598; (1979a) *J. antimicrob. Chemother.* **5** (suppl. A), 27; (1979b) [Ed.] *Pathogenic Streptococci.* Reedbooks, Chertsey, Surrey.

Parker, M.T. and Ball, L.C. (1976) *J. med. Microbiol.* **9**, 275; (1979) In: *Pathogenic Streptococci*, p. 234. Ed. by M.T. Parker. Reedbooks, Chertsey, Surrey.

Parker, M.T., Bassett, D.C.J., Maxted, W.R. and Arneaud, J.D. (1968) *J. Hyg., Camb.* **66**, 657.

Parker, M.T. and Stringer, J. (1979) In: *Pathogenic Streptococci*, p. 171. Ed. by M.T. Parker. Reedbooks, Chertsey, Surrey.

Parker, M.T., Tomlinson, A.J.H. and Williams, R.E.O. (1955) *J. Hyg., Camb.* **53**, 458.

Pass, M.A., Gray, B.M., Khare, S. and Dillon, H.C. Jr (1979). In: *Pathogenic Streptococci*, p. 175. Ed. by M.T. Parker. Reedbooks, Chertsey, Surrey.

Perlman, L.V., Herdman, R.C., Kleinman, H. and Vernier, R.L. (1965) *J. Amer. med. Ass.* **194**, 63.

Perry, C.B. (1969) *Ann. rheum. Dis.* **28**, 471.

Plueckhahn, V.D. and Banks, J. (1970) *Med. J. Aust.* **i**, 405.

Polzer, K. and Steffen, C. (1958) *Klin. Wschr.* **36**, 211.

Poon-King, T. *et al.* (1973) *Lancet* **i**, 475; (1979) In: *Pathogenic Streptococci*, p. 124. Ed. by M.T. Parker. Reedbooks, Chertsey, Surrey.

Potter, E.V. (1965) *J. Lab. clin. Med.* **65**, 40.

Potter, E.V., Lipschultz, S.A., Abdih, S., Poon-King, T. and Earle, D.P. (1982) *New Engl. J. Med.* **307**, 725.

Potter, E.V., Moran, A.F., Poon-King, T. and Earle, D.P. (1968) *J. Lab. clin. Med.* **71**, 126.

Potter, E.V., Svartman, M., Burt, E.G., Finklea, J.F., Poon-King, T. and Earle, D.P. (1972) *J. infect. Dis.* **125**, 619.

Potter, E.V. *et al.* (1971) *J. clin. Invest.* **50**, 1197.

Powers, G.F. and Boisvert, P.L. (1944) *J. Pediat.* **25**, 481.

Pullinger, E.J. and Kemp, A.E. (1937) *J. Hyg., Camb.* **37**, 527.

Quie, P.G., Pierce, H.C. and Wannamaker, L.W. (1966) *Pediatrics, Springfield* **37**, 467.

Quinn, R.W. (1957) *J. clin. Invest.* **27**, 471.

Quinn, R.W., Denny, F.W. and Riley, H.D. (1957) *Amer. J. publ. Hlth* **47**, 995.

Quinn, R.W. and Liao, S.J. (1950) *J. clin. Invest.* **29**, 1156.

Quinn, R.W., Liao, S.J. and Quinn, J.P. (1951) *Amer. J. Hyg.* **54**, 331.

Quirante, J. and Cassady, G. (1972) *Clin. Res.* **20**, 43.

Rammelkamp, C.H., Morris, A.J., Catanzaro, F.J., Wannamaker, L.W., Chamovitz, R. and Marple, E.C. (1958) *J. Hyg., Camb.* **56**, 280.

Rammelkamp, C.H. and Stolzer, B.L. (1961–62) *Yale J. Biol. Med.* **34**, 386.

Rammelkamp, C.H. Jr and Weaver, R.S. (1953) *J. clin. Invest.* **32**, 345.

Rammelkamp, C.H., Weaver, R.S. and Dingle, J.H. (1952) *Trans. Ass. Amer. Phycns* **65**, 168.

Rantz, L.A., Boisvert, P.J. and Spink, W.W. (1945) *Arch. intern. Med.* **76**, 131; (1947*a*) *Ibid.* **79**, 401.

Rantz, L.A., Maroney, M. and di Caprio, J.M. (1951) *Arch. intern. Med.* **87**, 360.

Rantz, L.A. and Randall, E. (1945) *Proc. Soc. exp. Biol., N.Y.* **59**, 22.

Rantz, L.A., Rantz, H.H., Boisvert, P.J. and Spink, W.W. (1946) *Arch. intern. Med.* **77**, 121.

Rantz, L.A., Spink, W.W. and Boisvert, P.J. (1947*b*) *Arch. intern. Med.* **79**, 272.

Read, S.E., Fischetti, V.A., Untermohlen, V., Falk, R. and Zabriskie, J.B. (1974) *J. clin. Invest.* **54**, 439.

Read, S.E., Poon-King, T., Reid, H.F.M. and Zabriskie, J.B. (1980). In: *Streptococcal Diseases and the Immune Response*, p. 347. Ed. by S.E. Read and J.B. Zabriskie. Academic Press, New York.

Read, S.E. and Zabriskie, J.B. (Eds) (1980) *Streptococcal Diseases and the Immune Response*. Academic Press, New York.

Redys, J.J., Hibbard, E.W. and Borman, E.K. (1968) *Publ. Hlth Rep., Wash.* **83**, 143.

Report (1954) *Mon. Bull. Minist. Hlth Lab. Serv.* **13**, 171; (1957) *Ibid.* **16**, 584; (1960) *Prevention of Initial Attacks and Recurrences of Rheumatic Fever.* Min. of Hlth, Lond.; (1965) *Brit. med. J.* **ii**, 607; (1966) *WHO Tech. Rep. Ser.* No. 342; (1967) *Rheumatic Fever in Scotland*, HMSO Edinburgh; (1968) *WHO Tech. Rep. Ser.* No. 394; (1979) In: *Pathogenic Streptococci*, p. 117. Ed. by M.T. Parker. Reedbooks, Chertsey, Surrey.

Reynell, P.C. (1948) *J. Hyg., Camb.* **46**, 148.

Rijn, I. van der, Zabriskie, J.B. and McCarty, M. (1977) *J. exp. Med.* **146**, 579.

Romero, R. and Wilkinson, H.W. (1974) *Appl. Microbiol.* **28**, 199.

Rosenstein, B.J. *et al.* (1968) *J. Pediat.* **73**, 513.

Ross, P.W. (1971*a*) *J. Hyg., Camb.* **69**, 347; (1971*b*) *Hlth Bull., Edinburgh* **29**, 108; (1977) *J. med. Microbiol.* **10**, 69.

Ross, R.F. (1972) In: *Streptococci and Streptococcal Diseases*, p. 339. Ed. by L.W. Wannamaker and J.M. Matsen. Academic Press, New York.

Rothbard, S., Watson, R.F., Swift, H.F. and Wilson, A.T. (1948) *Arch. intern. Med.* **82**, 229.

Roy, S., Wall, H.P. and Etteldorf, J.N. (1969) *J. Pediat.* **75**, 758.

Rubenstein, A.D. and Foley, G.E. (1945) *Amer. J. publ. Hlth* **35**, 905.

Rumao, L.M. and Sant, M.V. (1971) *Indian J. med. Sci.* **25**, 100.

Sapru, R.P., Ganguly, N.K., Sharma, S., Chandani, R.E. and Gupta, A.K. (1977) *Brit. med. J.* **ii**, 422.

Schacht, R.G., Gluck, M.C., Gallo, G.R. and Baldwin, D.S. (1976) *New Engl. J. Med.* **295**, 977.

Schaffner, W., Lefkowitz, L.B., Goodman, J.S. and Koenig, M.G. (1969) *New Engl. J. Med.* **280**, 1224.

Schaub, I.G., Mazeika, I., Lee, R., Dunn, M.T., Lachaine, R.A. and Price, W.H. (1958) *Amer. J. Hyg.* **67**, 46.

Schauf, V. and Hlaing, V. (1976) *Obstet. and Gynec.* **47**, 719.

Schlesinger, B. (1930) *Arch. Dis. Childh.* **5**, 411.

Schlievert, P.M., Bettin, K.M. and Watson, D.W. (1979) *Infect. Immun.* **26**, 467.

Schmidt, W.C. and Moore, D.J. (1965) *J. exp. Med.* **121**, 793.

Schmitt-Slomska, J., Boué, A. and Caravano, R. (1972) *Infect. Immun.* **5**, 389.

Schultz and Charlton (1918) *Z. Kinderheilk.* **17**, 328.

Seal, D.V. and Leppard, B. (1982) *Trans. R. Soc. trop. Med. Hyg.* **76**, 392.

Seal, J.R. (1955) *Amer. J. publ. Hlth* **45**, 662.

Shaper, A.G. (1972) *Brit. med. J.* **iii**, 683, 743, 805, **iv**, 32.

Sheldon, W.P.H. (1931) *Lancet* **i**, 1337.

Shigeoka, A.O., Hall, R.T. and Hill, H.R. (1978) *Lancet* **i**, 636.

Siegel, A.C., Johnson, E.E. and Stollerman, G.H. (1961) *New Engl. J. Med.* **265**, 559, 566.

Siegel, A.C., Rammelkamp, C.H. Jr and Griffeath, H.I. (1955) *Pediatrics, Springfield* **15**, 33.

Siegel, J.D., McCracken, G.H. Jr, Threlkeld, N., DePasse, B.M. and Rosenfeld, C.R. (1982) *Lancet* **ii**, 1426.

Skinner, F.A. and Quesnel, L.B. (Eds) (1978) *Streptococci.* Academic Press, London.

Slack, M.P.E. and Mayon-White, R.T. (1978) *Arch. Dis. Childh.* **53**, 540.

Smialowicz, R.J. and Schwab, J.H. (1977) *Infect. Immun.* **17**, 599.

Smith, J. (1931) *Causation and Source of Infection in Puerperal Fever.* HMSO, London; (1933) *J. Obstet. Gynec.* **40**, 991.

Spagnuolo, M., Pasternak, B. and Taranta, A. (1971) *New Engl. J. Med.* **285**, 641.

Spagnuolo, M. and Taranta, A. (1968) *New Engl. J. Med.* **278**, 183.

Spaun, J., Bentzon, M.W., Larsen, S.O. and Hewitt, L.F. (1961) *Bull. World Hlth Org.* **24**, 271.

Srisuparbh, K. and Sawyer, W.D. (1972) *Infect. Immun.* **5**, 176.

Stamm, W.E., Feeley, J.C. and Facklam, R.R. (1978) *J. infect. Dis.* **138**, 287.

Stebbins, E.L., Ingraham, H.S. and Reed, E.A. (1937) *Amer. J. publ. Hlth* **27**, 1259.

Steere, A.C. *et al.* (1975) *J. Pediat.* **87**, 784.

Steigman, A.J., Bottone, E.J. and Hanna, B.A. (1978) *Pediatrics, Springfield* **62**, 842.

Stetson, C.A., Rammelkamp, C.H., Krause, R.M., Kohen, R.J. and Perry W.D. (1955) *Medicine, Baltimore* **34**, 431.

Stollerman, G.H. (1954) *Amer. J. Med.* **17**, 757.

Strasser, T. (1978) *Chron. World Hlth Org.* **32**, 18.

Strasser, T. *et al.* (1981) *Bull. World Hlth Org.* **59**, 285.

Streitfield, M.M. and Saslaw, M.S. (1961) *J. infect. Dis.* **108**, 270.

Svartman, M., Potter, E.V., Finklea, J.F., Poon-King, T. and Earle, D.P. (1972) *Lancet* **i**, 249.

Szilagyi, G., Mayer, E. and Eidelman, A.I. (1978) *J. clin. Microbiol.* **8**, 410.

Taplin, D., Lansdell, L., Allen, A.M., Rodriguez, R. and Cortes, A. (1973) *Lancet* **i**, 501.

Taplin, D., Zaias, N. and Rebell, G. (1965) *Arch. environm. Hlth* **11**, 546.

Taranta, A. (1967) *Annu. Rev. Med.* **18**, 159.

Top, F.H. Jr, Wannamaker, L.W., Maxted, W.R. and Anthony, B.F. (1967) *J. exp. Med.* **126**, 667.

Trask, J.D. and Blake, F.G. (1924) *J. exp. Med.* **40**, 381.

Treser, G. *et al.* (1969) *Science* **163**, 676.

Unsworth, P.F. (1980) *J. Hyg., Camb.* **85**, 153.

Updyke, E.L., Moore, M.S. and Conroy, E. (1955) *Science* **121**, 171.

Vaisman, S. *et al.* (1965) *J. Amer. med. Ass.* **194**, 1284.

Valkenburg, H.A., Haverkorn, M.J., Goslings, W.R.O., Lorrier, J.C., Moor, C.E. de and Maxted, W.R. (1971) *J. infect. Dis.* **124**, 348.

Villareal, H. Jr, Rijn, I. van der, Fischetti, V.A., Mahabir, R.N. and Zabriskie, J.B. (1979) In: *Pathogenic Streptococci*, p. 135. Ed. by M.T. Parker. Reedbooks, Chertsey, Surrey.

Vosti, K.L., Lindberg, L.H., Kosek, J.C. and Raffel, S. (1970) *J. infect. Dis.* **122**, 249.

Waal, H.L de. (1941) *J. Hyg., Camb.* **41**, 65.

Waitkins, S.A. (1982) *Med. Lab. Sci.* **39**, 185.

Waldenström, J., Winblad, S., Hällén, J. and Liungman, S. (1964) *Acta med. scand.* **176**, 619.

Wannamaker, L.W. (1954) In: *Streptococcal Infections*, p. 157. Columbia Univ. Press, New York; (1970) *New Engl. J. Med.* **282**, 23, 78.

Wannamaker, L.W. and Ayoub, E.M. (1960) *Circulation* **21**, 598.

Wannamaker, L.W., Denny, F.W., Perry, W.D., Siegel, A.C. and Rammelkamp, C.H. (1953a) *Amer. J. Dis. Child.* **86**, 347.

Wannamaker, L.W. and Matsen, J.M. (Eds) (1972) *Streptococci and Streptococcal Diseases.* Academic Press, New York and London.

Wannamaker, L.W. *et al.* (1951) *Amer. J. Med.* **10**, 673; (1953b) *New Engl. J. Med.* **249**, 1.

Watson, D.W. and Kim, Y.B. (1970) In: *Microbial Toxins 3*, 173. Ed. by T.C. Montie, S. Kadis and S.J. Ajl. Academic Press, New York.

Watson, K.C. and Kerr, E.J.C. (1975) *J. med. Microbiol.* **8**, 465; (1978) *J. clin. Path.* **31**, 230.

Watson, K.C., Rose, T.P. and Kerr, E.J.C. (1972) *J. clin. Path.* **25**, 885.

Watson, R.F., Hirst, G.K. and Lancefield, R.C. (1961) *Arthr. and Rheum.* **4**, 74.

Webb, B.J. and Baker, C.J. (1980) *J. clin. Microbiol.* **12**, 442.

Weeks, J.L., Mason, E.O. Jr and Baker, C.J. (1981) *Antimicrob. Agents Chemother.* **20**, 281.

Weinstein, L. and Le Frock, J. (1971) *J. infect. Dis.* **124**, 229.

West, C.D., Northway, J.D. and Davis, N.C. (1964) *J. clin. Invest.* **43**, 1507.

Wheeler, S.M. and Jones, T.D. (1945) *Amer. J. med. Sci.* **209**, 58.

White, E. (1936) *Lancet* **i**, 941.

Widdowson, J.P., Maxted, W.R., Notley, C.M. and Pinney, A.M. (1974) *J. med. Microbiol.* **7**, 483.

Widdowson, J.P., Maxted, W.R. and Pinney, A.M. (1971) *J. Hyg., Camb.* **69**, 553.

Wilkinson, H.W. (1978a) *J. clin. Microbiol.* **7**, 176; (1978b) *Annu. Rev. Microbiol.* **32**, 41.

Williams, R.C. Jr, Zabriskie, J.B., Mahros, F., Hassaballa, F. and Abdin, Z.H. (1977) *Clin. exp. Immunol.* **27**, 135.

Williams, R.E.O. (1960) *Annu. Rev. Microbiol.* **14**, 43.

Wilson, C.D. and Salt, G.F.H. (1978) In: *Streptococci*, p. 143. Ed. by F.A. Skinner and L.B. Quesnel. Academic Press, London.

Winblad, S. (1966) *Acta path. microbiol. scand.* **66**, 93.

Windsor, R.S. (1977) *Vet. Rec.* **101**, 378.

Winterbauer, R.H., Fortuine, R. and Eickhoff, T.C. (1966) *Pediatrics, Springfield* **38**, 661.

Wittner, M.K. *et al.* (1979) In: *Pathogenic Streptococci*, p. 33. Ed. by M.T. Parker. Reedbooks, Chertsey, Surrey.

Wood, H.F., Feinstein, A.R., Taranta, A., Epstein, J.A. and Simpson, R. (1964) *Ann. intern. Med.* **60**, Suppl. 5, 31.

Zabriskie, J.B. (1964) *J. exp. Med.* **119**, 89.

Zabriskie, J.B. and Freimer, E.H. (1966) *J. exp. Med.* **124**, 661.

Zabriskie, J.B., Hsu, K.C. and Seegal, B.C. (1970) *Clin. exp. Immunol.* **7**, 147.

Zanen, H.C. and Engel, H.W.B. (1979) In: *Pathogenic Streptococci*, p. 232. Ed. by M.T. Parker. Reedbooks, Chertsey, Surrey.

Zimmerman, R.A., Cross, W.M., Miller, D.R. and Sciple, G.W. (1966) *J. Pediat.* **69**, 40.

Zimmerman, R.A. and Hill, H.R. (1969) *Pediatrics, Springfield* **43**, 809.

Zimmerman, R.A., Siegel, A.C. and Steele, C.P. (1962) *Pediatrics, Springfield* **30**, 712.

Zingher, A. (1924) *J. Amer. med. Ass.* **83**, 432.

60

Staphylococcal diseases

M.T. Parker

Introductory

The staphylococci are parasites of the body surface of animals, and their ability to cause disease appears to be only an incidental character. Almost all of them have some ability to multiply in the tissues, but for most of the time and in most individual hosts they are denied the oportunity to do so by the natural defence mechanisms of the body. Nevertheless, staphylococcal diseases are very common because the conditions that permit the organism to gain access to the tissues are numerous and frequent.

The most important pathogen in the group is *Staphylococcus aureus;* broadly speaking, it causes two forms of disease. (1) *Acute inflammation,* which usually begins at or near the point of entry of the organisms into the tissue. It is in most instances mild and localized but at times spreads widely by direct extension and occasionally leads to generalized infection. (2) *Acute toxæmia*, which results from the absorption of extracellular products formed by staphylococci multiplying in a local lesion, at a carrier site, or even outside the body. *Exfoliation of the skin* is a result of the formation, in local lesions and possibly at carrier sites, of epidermolytic toxin. *Staphylococcal diarrhoea*—usually accompanied by vomiting—is caused by the absorption of enterotoxin formed during the multiplication of the organism in food (staphylococcal food poisoning; Chapter 72) or in the lumen of the gut (staphylococcal post-antibiotic diarrhoea). *Toxic shock syndrome* (Todd *et al.* 1978) is a recently described form of generalized staphylococcal toxaemia associated with the local multiplication of *Staph. aureus* at carrier sites or in tissues.

As we have seen (Chapter 30), *Staph. aureus* comprises a number of varieties that are adapted to parasitize particular animal hosts. The diseases of various animal species appear mainly to be caused by the staphylococci of the animal affected, but a few exceptions have been recorded. Thus, strains with the characters of human *Staph. aureus* have occasionally colonized dogs and goats in animal hospitals, and have then been the source of clinical infections in the human attendants (Pagano *et al.* 1960, Live and Nichols 1961, Poole and Baker 1966).

The ability of *Staph. aureus* to gain access to the tissues and to cause septic lesions appears to be determined as much by host factors as by the character of the infecting strain. The organism may enter either by penetrating apparently unbroken skin, or through a break in the skin due to accidental trauma or surgical operation, or through a mucous membrane damaged by viral infection or other pathological process, or even a normally sterile cavity such as the bladder.

Although, as we shall see, one of the main determinants for staphylococcal sepsis is a local or general lowering of host resistance, there is evidence that strains of *Staph. aureus* differ in their ability to invade the tissues and establish a local lesion.

In what we may describe as normal circumstances, most staphylococcal infections are self-infections, and are sporadic. This is usually so with minor skin sepsis in man, and there is evidence that sporadic staphylococcal disease also occurs in some wild animals (McDiarmid 1955, Osebold and Gray 1960, Hagen 1963). In certain artificial circumstances, however, staphylococcal sepsis may be a communicable disease. The most familiar example of this is human infection in hospitals, which arises from the creation of populations of susceptible patients in which surgical and medical treatment provide ample opportunities for the infection of many persons by a single strain. Certain communicable staphylococcal diseases of domestic animals—for example, bovine mastitis—are similarly consequences of the techniques of animal husbandry (Chapter 57).

The coagulase-negative staphylococci are much less often responsible for disease. The ability of *Staph. epidermidis* to cause local sepsis in superficial tissues is a matter of debate, but it certainly gives rise to endocarditis or chronic septicaemia in certain categories of predisposed patients. Some strains of *Staph. saprophyticus* are a common cause of urinary-tract infection in previously healthy women.

In this chapter we shall consider mainly diseases due to *Staph. aureus*, and shall use the term 'staphylococcal diseases' without qualification to refer to them. At the end of the chapter we shall make a brief reference to diseases due to coagulase-negative staphylococci and micrococci, and to the anaerobic cocci.

Staph. aureus infections

Diseases of the skin and subcutaneous tissues

Staph. aureus is responsible for two distinct forms of skin disease, pustular and exfoliative, of which the former is by far the more common.

Pustular diseases

The characteristic pustular lesion is the boil, a subepidermal collection of pus, often around the root of a hair-follicle. A variety of different strains of *Staph. aureus* cause boils, but a comparison of the phage-typing patterns of strains from these lesions with those from nasal swabs of normal persons suggests that some strains are more likely to be responsible than others (Williams and Jevons 1961). The ability to cause boils is not associated with particular phage groups, but appears to be a strain-specific character that may sometimes be associated with a particular

phage-typing pattern. Thus, the 52, 52A, 80, 81 complex of phage-group I strains (Chapter 30), when prevalent in hospitals in the 1950's and 1960s, showed exceptional ability to cause boils not only in patients but also in members of the hospital staff (Rountree 1978); certain phage-group III strains that were frequently responsible for other forms of sepsis in hospitals at this time were rarely found in boils. In phage-group II, boil-producing strains are seldom lysed only by phage 71. Stàphylococci from boils are nearly always egg-yolk positive (Alder *et al.* 1953), but not all egg-yolk-positive strains appear to be producers of boils.

It has been estimated that 5-9 per cent of the population have one or more minor skin infections a year, most of which are boils (Gould and Cruickshank 1957, Kay 1962). About one-third of the patients have two or more lesions in succession, and over 10 per cent have recurrent lesions spread over several months, or even years (Roodyn 1960). The type of staphylococcus found in the lesion corresponds in general with that found in the patient's nasal swab (Hobbs *et al.* 1947, Valentine and Hall-Smith 1952, Tulloch 1954). The correspondence is almost complete for styes and sycosis barbae, but the infecting strain is absent from the nose of between a third and a half of patients with boils (Roodyn 1954, Kay 1962). Kay found that nasal carriage of the infecting strain occurred more often when the lesion was on the face than when it occurred elsewhere on the body. What proportion of the infections in persons who are not nasal carriers come from other carrier sites, such as the perineum (see p. 264), and what proportion are acquired from associates it is hard to say, but intrafamilial spread of infection is certainly very common (Roodyn 1954, 1960); on half of the occasions on which one person has a boil the infecting staphylococcus can also be isolated from another member of the family (Kay 1962). In only 13 per cent of cases is it absent from the nose of both the patient and his family associates (see also Miller 1962, Kundsin 1966).

In the 1950s, sepsis of the skin appeared to be more common among family contacts of persons recently discharged from hospital, and especially of babies recently delivered in hospital, than in the rest of the general population (Ravenholt *et al.* 1957, Hurst and Grossman 1958, Galbraith 1960), but this was not the experience in later studies (e.g. Oliver *et al.* 1964, Blowers *et al.* 1967). The apparent discrepancy may perhaps be explained by the decreasing prevalence of staphylococci of the 52, 52A, 80, 81 complex in the years after 1960 (see also Johnson *et al.* 1960).

Numerous host factors are probably concerned in the development of boils, for example pressure and minor trauma, as in beat disease of miners (see Atkins and Marks 1952)—an occupational furunculosis and chronic cellulitis due to *Staph. aureus*. The part played by mechanical factors in experimental staphylococcal infections of man and of laboratory animals is discussed in Chapter 30 and on p. 260. The liability of diabetics to suffer from boils appears to form part of a general increase in susceptibility to bacterial infection, reinforced by the higher than normal staphylococcal carrier rate in patients suffering from the disease (Smith *et al.* 1966). As we shall see, skin sepsis of all sorts due to *Staph. aureus* is very common in infants, whether or not they are born in hospital. Boils are more frequent in adolescents than in older persons (Miller 1962).

Exfoliative diseases

These are characterized by stripping of the superficial layers of the skin from the underlying tissues by the action of epidermolytic toxins. According to Parker and his colleagues (1955, Parker 1958) most of the strains responsible belonged to phage-group II and were lysed only by phage 71; they never gave rise to opacity in egg yolk or split Tween 80, but many of them formed a characteristic bacteriocine (Chapter 30). Later reports, supported by laboratory tests for the specific toxin, suggested that about one-quarter of the strains did not belong to phage-group II and had a variety of other phage-typing patterns (see de Azavedo and Arbuthnott 1981).

Exfoliative staphylococcal lesions have been variously named according to their extent, distribution and severity. They include bullous impetigo (Engman 1901) and pemphi-

gus of the newborn (Almquist 1891), in which the lesions are localized and consist of flat blebs which contain fluid that is at first clear but becomes turbid in 2–3 days. There is little underlying inflammatory reaction; when the vesicles rupture a thin reddish crust appears, and the lesion heals without scarring. Staphylococcal impetigo is usually a disease of children but may affect adults living under poor hygienic conditions. Staphylococci with the characters described above can be isolated in large numbers, and usually in pure culture, from impetiginous lesions, and it is assumed that these result from the local production of epidermolytic toxin.

In babies and young children, but very rarely in adults, the lesion may be more extensive and continuous, affecting the skin of complete regions and sometimes of the whole body surface. This was described first by Ritter von Rittershain (1878)—hence the name Ritter's disease—and later by Lyell (1956, 1967) as the 'scalded skin syndrome' or toxic epidermal necrolysis (see also Lyell *et al.* 1969). The onset is abrupt, with generalized erythema closely resembling that of scarlet fever; within 1–2 days the skin becomes wrinkled and peels off on light stroking (Nikolsky's sign). Then large flaccid bullae appear and extensive areas of skin are exfoliated. There are signs of acute toxaemia, and death may occur unless the correct antimicrobial treatment is given. Staphylococci similar to those found in impetigo are also responsible for these diseases (Howells and Jones 1961, Parker and Williams 1961, Benson *et al.* 1962, Lyell *et al.* 1969) but may not be isolated in large numbers from the lesions. This was explained when Melish and Glasgow (1970) showed in baby mice that epidermolytic toxin could act at a distance from the site of multiplication of the staphylococcus.

Several authors had earlier described the association of a *scarlatiniform rash* with staphylococcal infection (Stevens 1927, Aranow and Wood 1942, Simpson 1953, Negro *et al.* 1956, Dunnet and Schallibaum 1960), and an erythrogenic substance had been demonstrated in cultures of *Staph. aureus* from cases of 'staphylococcal scarlet fever'. Melish and Glasgow (1971) reported scarlatiniform rashes in 11 young children. These were distinguished from scarlet fever mainly by the absence of an enanthem, and by the presence of a transient Nikolsky's sign in 9 of the cases. Thus, exfoliative skin diseases of varying extent and severity may result from epidermolytic toxin formed at a site remote from the skin lesions. More recent reports suggest, however, that epidermolytic toxin may not be the only staphylococcal product to cause scarlatiniform skin lesions. They are also common in toxic-shock syndrome (p. 259); though extensive exfoliation does not occur in the acute phase of this disease, peeling of the extremities during convalescence is usual. However, the staphylococcal strains associated with toxic-shock syndrome do not produce epidermolytic toxin.

Staph. aureus often infects skin lesions of non-microbial origin, such as atopic dermatitis and psoriasis, as well as the lesions of chickenpox, smallpox and scabies, but the consequences are seldom serious; in many cases colonization is symptomless. Like group A streptococci (see Chapter 59), *Staph. aureus* is occasionally responsible for *necrotic fasciitis*, in which there is widespread destruction of subcutaneous connective tissue and subsequent ulceration of overlying skin. The disease is acute and the patient exhibits severe toxaemia. It is a rare complication of superficial traumatic lesions, which may have been trivial, or less often of a surgical operation.

Sepsis in newborn infants   In addition to pemphigus and Ritter's disease, newborn infants suffer from a variety of other staphylococcal skin lesions. These are due to many different strains of *Staph. aureus*. The milder lesions are often nondescript in appearance and commonly follow minor trauma to the skin. Severe lesions may be more like boils, with much underlying cellulitis.

It has long been recognized that neonatal skin infections in hospital tend to occur in epidemics, and many of these are associated with single staphylococcal strains. It must be remembered, however, that apparent epidemics, in which a number of strains participate, may be attributable to a deterioration in nursery technique. Moreover, investigations made in the absence of epidemics have shown that minor skin sepsis is often very common in maternity hospitals (see Williams *et al.* 1966) and may affect 10–25 per cent of infants without causing alarm. If lesions of all degrees of severity are counted, the sepsis rate in infants born at home may be of the order of 10 per cent (Williams 1961). It seems likely, therefore, that many recorded epidemics were incidents in which the lesions had become more severe rather than more frequent.

Male infants are more likely than female to become carriers of *Staph. aureus*, and male carriers suffer from staphylococcal lesions more often than do female carriers (Thompson *et al.* 1966). The pustular type of skin infection is sometimes accompanied by serious lesions elsewhere in the body, including deep abscesses, osteomyelitis and pneumonia. These complications were particularly common in epidemics caused by staphylococci belonging to the 52, 52A, 80, 81 complex (Rountree and Freeman 1955, Beavan and Burry 1956, Timbury *et al.* 1958). They may develop in babies in whom skin sepsis was not detected. Epidemics of *cervical adenitis* in babies aged 2–4 months have been described (Ayliffe *et al.* 1972, Boyce *et al.* 1976). These may be associated with birth in a particular hospital during a limited period of time and caused by a single strain of *Staph. aureus* that was not responsible for sepsis in the newborn. Neonatal pemphigus and Ritter's disease are very seldom followed by abscess formation.

The development of *acute mastitis* by nursing mothers is influenced by local predisposing causes, of which engorgement of the breasts is the most important. The infection is commoner in hospital than in domiciliary midwifery, though it is often not recognized until after the mother has left hospital. The causative organism is usually transmitted to the mother from the nasopharynx of her infant (Duncan and Walker 1942, Ravenholt *et al.* 1957, Wysham *et al.* 1957);

and epidemics of breast abscess are usually a consequence of the presence of a virulent strain of *Staph aureus* in the nursery. Infants of either sex occasionally suffer from neonatal breast abscess. (For mastitis in cattle see Chapter 57.)

Skin sepsis of animals   Many species of domestic animal suffer from staphylococcal skin infections, which are usually secondary to some form of trauma. Chronic abscesses of the skin, subcutaneous tissues and spermatic cord of horses—often referred to as *botriomycosis*—are usually post-traumatic staphylococcal infections. Staphylococcal infection in poultry and other birds, which usually takes the form of a septicaemia with purulent arthritis, may also be an inoculation infection resulting from pecking injuries or trauma to the feet (Smith 1959).

Sepsis in wounds and burns   *Staph. aureus* is the commonest cause of sepsis in small accidental wounds (Williams and Miles 1949) and is responsible for between one-third and one-half of all sepsis following 'clean' surgical operations (see Chapter 58). The frequency of clinical infection varies widely in different types of operation, and the consequences range from a mild reddening of the edges of the wound to extensive suppuration, septicaemia and metastatic abscesses. Symptomless colonization of wounds is at least as frequent as clinical sepsis. *Staph. aureus* can be isolated from most burns, but it often does not delay healing or lead to the rejection of grafts.

Infections of the urinary tract   Cystitis and pyelonephritis due to *Staph. aureus* occur rarely outside hospital, but may follow catheterization or operations on the bladder or prostate in hospital patients (Chapter 58). They form only a very small proportion of hospital-acquired urinary-tract infections but often have serious consequences; according to Demuth and co-workers (1979), some 15 per cent of them are complicated by bacteraemia. Haematogenous spread to the kidney or perinephric tissue may occur in pyaemic infections.

Staphylococcal pneumonia

Staph. aureus may invade the lungs from the blood stream, giving rise to the formation of abscesses; more often it causes a primary pneumonia. This occurs mainly in the following classes of patient: (1) young infants, especially when colonized by a particularly virulent staphylococcal strain, (2) healthy young adults, secondary to influenzal infection, and (3) adults suffering from other serious diseases.

Staphylococcal pneumonia of infants tends to affect the premature or sickly, but epidemics may occur among groups of healthy newborn infants in hospital; they were a feature of outbreaks of sepsis due to staphylococci of the 52, 52A, 80, 81 complex (Beavan and Burry 1956, Disney *et al.* 1956). The immediate mortality is high, and survivors may suffer from empyema, lung abscess or pneumocystocoele. *Post influenzal pneumonia* is the main cause of death in healthy young people after influenza-A infection (Slot 1950, Hers *et al.* 1958, Oswald *et al.* 1958, Report 1958). The onset of disease is very rapid, the sputum is profuse, watery and evenly blood-stained, and death often occurs in 1–2 days. The staphylococci are present in enormous numbers in the lung and resemble in antibiotic resistance and phage-typing patterns the strains prevalent in the general population. Influenzal pneumonia appears to result from a massive invasion of the lung by the strain of *Staph. aureus* carried by the victim and is a consequence of extensive damage to the mucosa of the lower respiratory tract by the virus. Staphylococcal pneumonia is rare in young adults in the absence of influenza and appears to be a less important cause of fatal influenza in elderly than in young adults.

The isolation of *Staph. aureus* is a poor guide to the frequency of staphylococcal pneumonia in debilitated adult hospital patients. Weiss and Flippin (1963) concluded that only 1 in 30 of isolations of the organism from the sputum on admission to hospital and 1 in 6 'acquisitions' in hospital were of clinical significance. Nevertheless, clear evidence of staphylococcal pneumonia was found at necropsy in 6 per cent of a large series of patients dying in hospital in Britain, though between one-half and three-quarters of them would not have been expected to have survived had this infection not occurred (Report 1966).

Chronic staphylococcal lung infection is very common in patients with cystic fibrosis, but is probably of less clinical significance than infection with *Pseudomonas aeruginosa*. Nevertheless, its frequency tends to be underestimated because the organism may be difficult to isolate unless the sputum is liquefied and a selective medium is employed. In patients who have received long-term antimicrobial treatment the organisms may be colonially atypical and may include thymidine-dependent strains (Sparham *et al.* 1978).

Osteomyelitis *Staph. aureus* is by far the commonest cause of acute osteomyelitis (Chapter 57), which is usually accompanied by bacteraemia.

Generalized infections

Staph. aureus may spread from a superficial lesion through the tissue planes, or along the lymphatics to the regional lymph nodes, but it more often causes serious disease when it spreads by the blood stream. This may occasionally follow a trivial local lesion in an apparently healthy person. Indeed, the well-attested fact that staphylococcal osteomyelitis seldom follows a recognized local lesion but may be precipitated by mild injury to the bone suggests that transient staphylococcal bacteraemia may be quite common. Abscesses of the kidney ('renal carbuncle') also result from deposition of staphylococci from the blood. They develop in the renal cortex, do not communicate with the renal collecting system, and tend to spread to the adjacent perinephric tissues. Nowadays they are less common than formerly except in association with the intravenous self-administration of narcotics.

Staphylococcal septicaemia, pyaemia and endocarditis

Acute septicaemia may occur in association with local suppuration, for example, wound sepsis, pneumonia or osteomyelitis, or it may occur spontaneously. It often develops in patients who are predisposed to it by a serious underlying disease, such as neoplasia, liver disease, diabetes, rheumatic arthritis, or certain extensive skin diseases, or by septic complications of procedures for their relief, notably surgical operation and intravenous cannulation (Smith and Vickers 1960, Keene *et al.* 1961, Cluff *et al.* 1968, Iannini and Crossley 1976). The special liability of patients with severe rheumatoid arthritis to suffer from staphylococcal septicaemia, often with pyoarthrosis, is only partly attributable to treatment with steroids (Kellgren *et al.* 1958, de Andrade and Tribe 1962).

In many cases, endocarditis develops. According to Nolan and Beaty (1976), endocarditis occurs rather infrequently in cases of septicaemia in which there is an identifiable primary staphylococcal infection. When the primary focus is one that can be easily removed, e.g. an infected cannula site, the septicaemia usually responds to quite a short course of antimicrobial treatment (Iannini and Crossley 1976). Septicaemia without an obvious primary source is more commonly associated with endocarditis and with the subsequent development of pyaemic abscesses elsewhere in the body (Nolan and Beaty 1976) and is much more difficult to eradicate. *Staph. aureus* is one of the important causes of endocarditis associated with the intravenous self-administration of drugs (Chapter 57; see also Tuazon *et al.* 1975, Sklaver *et al.* 1978).

Fulminating attacks of staphylococcal septicaemia may be associated with thrombocytopaenia, disseminated intravascular coagulation, glomerulonephritis, and occasionally symmetrical peripheral gangrene (Rahal *et al.* 1968, Murray *et al.* 1977) which may be attributable to activation of the alternative complement pathway (O'Connor *et al.* 1978). *Staph. aureus* is often present in considerable numbers in the urine in cases of staphylococcal septicaemia even in the absence of renal abscess (Lee *et al.* 1978).

Tick pyaemia in lambs This is another example of a special susceptibility to generalized staphylococcal infection. Infected lambs may die from septicaemia a few days after birth; those that escape suffer from chronic abscesses of the joints and liver; the meninges may also be invaded. The disease occurs only in geographical areas in which sheep are infested with the tick *Ixodes ricinus*. It is apparently caused by the *Staph. aureus* strains normally present on the skin of lambs (Foggie 1947), whether or not they are tick-infested (Watson 1964*a*, *b*, 1965). There is no evidence that the tick acts as a vector of staphylococci, though the organisms may enter the tissues through its bite. According to Foggie (1956), the profound neutropaenia that accompanies tick-borne fever in lambs is the predisposing factor.

Tropical pyomyositis

This is a generalized staphylococcal disease that affects apparently healthy young persons. Large abscesses develop in voluntary muscles and are often multiple. The staphylococci are presumed to reach the muscle through the blood stream, but the factors that determine the site of abscess formation are unknown. Tiny areas of myonecrosis have been seen elsewhere in the musculature of sufferers, but the cause of these is uncertain (Taylor *et al.* 1976*b*, Smith *et al.* 1978). Scattered cases have been reported from many parts of the tropics, and the disease is particularly common in Central Africa and New Guinea (see Shepherd 1983).

Foster (1965), who investigated the disease in Uganda, found that nearly every case was due to *Staph. aureus*, that 60 per cent of the strains isolated from lesions belonged to phage-group II and that three-quarters of the phage-group II strains had the typing pattern 3A/3B/3C/55/71. Strains with this pattern were much less often found in other types of lesion, but were responsible for a number of cases of furunculosis in the same population. It does not appear, however, that pyomyositis is often a complication of skin sepsis.

Staphylococcal toxic-shock syndrome

Todd and his colleagues (1978) described under this name a characteristic syndrome of high fever, headache, confusion, conjunctival reddening, subcutaneous oedema, vomiting and diarrhoea, and profound hypotensive shock in older children and adolescents. In the more severe cases, acute renal failure, disseminated intravascular coagulation, peripheral gangrene, and even death sometimes occurred. All the patients had a scarlatiniform rash, and fine desquamation of the hands and feet often occurred during convalescence. Patients examined bacteriologically yielded *Staph. aureus* from one or more sites; a few of them had a local septic lesion due to this organism, but blood cultures gave negative results. The strains isolated gave negative serological tests for epidermolytic toxins A and B.

In 1979, cases of a similar illness in young women began to be reported with increasing frequency in the USA. They showed a striking association with menstruation and the use of intravaginal tampons, particularly when these were employed continuously and changed infrequently (Davis *et al.* 1980, Shands *et al.* 1980). Two or more attacks, each associated with menstruation and the use of tampons, were reported by a number of women. The vaginal swabs of sufferers almost invariably yielded *Staph. aureus*, which was seldom resistant to antibiotics other than penicillin. The risk of acquiring the disease appeared to be mainly associated with the use of highly absorbent tampons and particularly with one brand of them. According to Tierno and colleagues (1983), these tampons contained a form of carboxymethylcellulose that is liable to enzymic degradation by vaginal organisms to yield glucose, and tampon material treated with β-glucosidase was able to support the growth of *Staph. aureus*. These workers believe that the role of the tampon may be to obstruct the menstrual flow and when digested to provide a readily available source of nutrient for the rapid growth of the staphylococci.

Davis and co-workers (1980) calculated that the incidence of toxic-shock syndrome in Wisconsin in 1979–80 was as high as 6 per 10 000 menstruating women per year, but the disease appears not to have been very common outside North America (see, for example, de Saxe *et al.* 1982). Growing familiarity with the clinical appearances of toxic-shock syndrome has led to the recognition of many more cases not associated with menstruation. Reingold and colleagues (1982) identified 54 in the USA in 1981. They were associated with a variety of staphylococcal inflammatory conditions: of the skin or in wounds, of the breast or the lung, and occasionally with puerperal infection or 'primary' bacteraemia.

Schlievert and co-workers (1979) drew attention to resemblances between the effects of staphylococcal pyrogenic exotoxins in experimental animals (Chapter 30) and the clinical manifestations of toxic-shock syndrome. Subsequently they showed that strains from this disease formed the C type of toxin (Schlievert *et al.* 1981). They also observed that this substance caused a 50 000-fold enhancement of the action of enterobacterial endotoxin on rabbits (Schlievert 1982). Cohen and Falkow (1981) identified two immunologically distinct proteins in whole-cell preparations of about 80 per cent of strains from cases of toxic-shock syndrome but of only 25 per cent of control strains. Bergdoll and co-workers (1981) described the separation and purification of a low-molecular-weight protein with the general characters of an enterotoxin from a toxic-shock strain. This protein caused vomiting when given orally to monkeys and was named enterotoxin F. Its presence was detected immunologically in 94 per cent of *Staph. aureus* strains from the vagina of cases of toxic shock but in only about 10 per cent of strains from other sources. Patients who suffered from toxic-shock syndrome seldom had high titres of antibody to enterotoxin F in their acute-phase sera or produced more antibody later. Indeed, high titres of antibody were significantly more common in matched control sera that in those of cases of toxic-shock syndrome. The relation between the substances described by these authors is still uncertain. It has been suggested (Schutzer *et al.* 1983) that the presence of a temperate phage may determine the ability of staphylococci to produce the toxic agent, but the evidence adduced is not conclusive. According to Barbour (1981) strains of *Staph. aureus* from the vagina of toxic-shock patients are significantly less haemolytic on sheep blood agar than other vaginal isolates of this organism.

It is unlikely that toxic-shock syndrome is a 'new' staphylococcal disease, but its association with menstruation is certainly a new manifestation of it. The possibility that staphylococcal overgrowth in the vagina is a consequence of antibiotic administration has been excluded. If the symptoms are attributable to the local production of enterotoxin F, this is the first clear example in which the systemic effects of an enterotoxin overshadow its action on the gut. It will be recalled (Chapter 30) that other enterotoxins have systemic effects on experimental animals resembling those of the endotoxin of gram-negative bacilli. 'Scarlet fever' has been described as a consequence of the overgrowth of resistant strains of *Staph. aureus* in patients treated with broad-spectrum antibiotics (Hazen *et al.* 1951), and Hallander and Körlof (1967) suggested that the severe systemic effects of surgical wound infections caused by a methicillin-resistant phage-group III strain could be attributed to its production of enterotoxin B. These findings should redirect attention to the extra-intestinal effects of enterotoxins in staphylococcal infection and carriage. (For more information about toxic-shock syndrome see Report 1982.)

Diarrhoea and vomiting

It has been known for many years that diarrhoea and vomiting may follow the ingestion of food in which *Staph. aureus* has multipled and formed enterotoxin (Chapters 30 and 72), but only since 1950 has acute diarrhoea due to the multiplication of the organism in the gut been recognized. Soon after the introduction of the tetracyclines, reports appeared of severe choleriform diarrhoea in patients undergoing treatment with one of these antibiotics (Jackson *et al.* 1951, Janbon *et al.* 1952, Terplan *et al.* 1953). The normal coliform flora of the faeces had been replaced in most cases by a *Staph. aureus* strain that was resistant to several antibiotics including tetracycline. Microscopic examination of the faeces usually revealed the presence of numerous clumps of gram-positive cocci and few other organisms (Matthias *et al.* 1957). Severe dehydration and death often followed. Prompt treatment of the patient with an antibiotic to which the faecal strain of *Staph. aureus* was sensitive resulted in speedy amelioration of the symptoms (Dearing and Heilman 1953).

Similar diarrhoeal attacks were observed after the administration of several other antibiotics, including chloramphenicol, neomycin and penicillin-streptomycin mixtures; they may follow systemic as well as oral treatment (Fairlie and Kendall 1953, Lunsgaard-Hansen *et al.* 1960, Tinsdale *et al.* 1960, Meyer *et al.* 1965). There was a tendency for cases to occur in outbreaks associated with the presence of a particular strain of *Staph. aureus* in the affected ward (Brodie *et al.* 1955). In the early 1960s, many cases were associated with the lavish use of neomycin for the preoperative preparation of the bowel. A few years later, the disease began to decrease in frequency, and it is rarely reported nowadays.

According to Dack (1956), nearly all of the staphylococcal strains isolated from cases of post-antibiotic diarrhoea formed enterotoxin, and enterotoxin B was looked upon as the causative agent. Most of the strains responsible, in our experience, were multiple-antibiotic resistant members of phage-group III. Some two-thirds of the strains of this description isolated from British hospital patients in the early 1960s formed enterotoxin B, but many of them also formed enterotoxin A. However, there is no direct proof that enterotoxin is responsible for the diarrhoea associated with the overgrowth of staphylococci in the gut.

In any event, it is unlikely that the post-antibiotic diarrhoea seen between 1950 and 1965 was all caused by staphylococci. The subsequent discovery that a cytotoxic agent formed by *Clostridium difficile* might cause enterocolitis, both as a complication of antibiotic treatment and also spontaneously in debilitated hospital patients (Chapter 58), may appear to cast doubt on the role of staphylococci in post-antibiotic diarrhoea. However, staphylococcal overgrowth in the gut was associated with short-lived but often severe attacks of diarrhoea but did not cause the necrosis of the bowel tissue characteristic of the clostridial disease.

As we have seen, diarrhoea is a common manifestation of toxic-shock syndrome, now attributed to enterotoxin F, but in this disease it tends to be overshadowed by other systemic disturbances.

Pathogenesis of the local inflammatory lesion

The intradermal or subcutaneous injection of *Staph. aureus* into human subjects or adult experimental animals does not result in the appearance of a local lesion unless the infecting dose is very large. Much smaller inocula are sufficient, in man, guinea-pigs and mice but not in rabbits, if the organisms are introduced on a foreign body; in rabbits, other forms of skin injury, notably burning, have a similar effect. The conclusions reached from various experiments in man and animals (see Chapter 30 and Goshi *et al.* 1961*a,b*) indicate that the following local conditions favour the appearance of a staphylococcal lesion in the skin or subcutaneous tissues: (1) vasoconstriction, (2) necrosis, (3) acute inflammation due to another micro-organism, (4) a hypersensitivity reaction, and (5) the presence of a foreign body. Increased susceptibility due to acute inflammation and to a local hypersensitivity reaction persists for only a short time and is much less than that caused by a burn. Anti-inflammatory agents, including the endotoxin of gram-negative bacilli (Conti *et al.* 1961), and some but not all corticosteroids (Agarwal 1967*b*), enhance the severity of the lesion.

Comparison of the lesions in mice after introducing *Staph. aureus* (1) subcutaneously in a fluid medium, and (2) on a plug of cotton dust (Noble 1965) showed that the effect of the foreign body was to slow down the rate of exudation of fluid and migration of leucocytes into the lesion (Agarwal 1967*a*), so permitting the cocci to multiply. In man, acute inflammation also results from placing small numbers of

Staph. aureus on the skin after the outer epidermal layers have been stripped off by the repeated application and removal of adhesive tape (Marples and Kligman 1971); the conditions necessary for the production of a severe lesion are that (1) the area is kept moist, (2) serum exudes from the surface, (3) leucocytes do not migrate into the exudate before the lesion is established, and (4) few competing organisms are present. Extensive spreading cellulitis appears regularly if a few hundred cocci are applied soon after the skin is stripped. If the inoculation is delayed for 24 hr the lesion is less severe and more localized; if for 48 hr it is minimal or absent. Protection of the staphylococcus from the cellular defences of the body, and delay in mobilizing them for only a short time, thus appear to be important determinants of wound infection.

We know little about how *Staph. aureus* penetrates the apparently unbroken skin to cause a boil. The anatomy of the lesion suggests that it often begins around the root of a hair follicle, but experimental infections have not been produced except by the application of numbers of cocci considerably in excess of those normally found on the skin.

The fact that almost all *Staph. aureus* strains that cause boils form a lipase active on Tween 80 (Chapter 30) and produce opacity in egg-yolk broth suggests that the ability to attack certain skin lipids is a determinant for the formation of boils. This hypothesis has been tested by preparing, by lysogenic conversion, Tween-negative variants of boil-producing strains, and comparing the ability of the two organisms to cause lesions when instilled into the hair follicles of the pig (Jessen and Bülow 1967). It is unlikely, however, that the production of a Tween-splitting lipase is the only determinant for boil production, because some Tween-positive phage-group III strains, which often cause postoperative wound sepsis, rarely if ever cause boils.

Local multiplication of staphylococci on the surface of the uninjured skin occurs under conditions of high humidity (Foster and Hutt 1960), and is favoured by preliminary treatment of the skin with ethanol, which presumably acts by removing the competitive bacterial flora (Gurmohan Singh *et al.* 1971). This may lead to the appearance of a lesion, which is a papulo-vesicular eruption rather than a boil; the staphylococci are confined to the superficial layers of the epidermis; and necrosis, oedema and haemorrhages appear in the underlying tissue. According to Gurmohan Singh and his colleagues this lesion is a toxic dermatitis due to staphylococcal extracellular products. Similar lesions form spontaneously under occlusive dressings on the skin of carriers of neomycin-resistant *Staph. aureus* if they are treated with neomycin (Marples and Kligman 1969).

Phagocytosis

Various serum factors (opsonins) are necessary for the phagocytosis of staphylococci. Most human sera contain antibacterial antibodies, the action of which is enhanced by complement. In addition, complement is activated by staphylococci through the alternative pathway and has primary opsonic activity independent of the presence of specific antibody. However, opsonization by the classical pathway is rapid and

efficient; by the alternative pathway it is slower. (For a summary of the requirements for opsonization see Quie *et al.* 1981.) After phagocytosis in the presence of serum, there is an initial rapid fall in the number of viable *Staph. aureus*, though a small number survive; under these conditions, coagulase-negative staphylococci are usually completely destroyed.

The increased susceptibility to *Staph. aureus* infections of patients with various abnormalities of the phagocytic process indicates that the cellular element of the early inflammatory response may be of primary importance in preventing the development of septic lesions. These defects in phagocytic function have been reviewed by Quie and his colleagues (1974, 1981). Absence of antibody and abnormalities of the metabolism of complement, leading to poor opsonization, are occasional causes of increased susceptibility to staphylococcal infection, but the most striking effects result from deficiencies in the response of the leucocytes to chemotactic stimulation or in their bactericidal function.

Job's syndrome (Hill *et al.* 1974*a*) is a rare disease in which recurrent, severe 'cold' staphylococcal abscesses and chronic eczema are associated with a very high serum IgE. The leucocytes of these patients exhibit a profoundly depressed leucotaxic response. Depression of leucotaxis has also been reported in juvenile rheumatoid arthritis and in diabetes mellitus (Mowat and Baum 1971, Quie *et al.* 1974). However, according to Sood and her colleagues (1975), both the accumulation of leucocytes and the exudation of fluid into the lesion are delayed in alloxan-diabetic mice given *Staph. aureus* subcutaneously.

In *chronic granulomatous disease* (Quie *et al.* 1967), phagocytosis occurs normally but does not lead to the usual burst of metabolic activity that results in the intracellular accumulation of H_2O_2 and is necessary for the bactericidal action of the myeloperoxidase-H_2O_2 system. As a result, patients suffer from repeated chronic septic infections—notably of the cervical glands and lungs—caused by catalase-forming bacteria such as staphylococci and aerobic gram-negative bacilli. Catalase-negative organisms themselves form sufficient H_2O_2 to activate the bactericidal system in the leucocytes. Several rare diseases in which there are other deficiencies in the bactericidal activity of the leucocytes and an increased susceptibility to staphylococcal sepsis have been described. There is also some evidence that defects in both leucotaxis and intracellular killing may develop in very severe bacterial infections.

Bacteristasis by serum   Coagulase-positive staphylococci have a much greater ability than coagulase-negative staphylococci to grow in normal human serum (Chapter 30). However, Ehrenkranz and his colleagues (1971) showed that strains of *Staph. aureus* vary widely in their ability to grow in serum, and that sera differ in their power to inhibit the growth of both coagulase-positive and coagulase-negative strains. The importance of this in the pathogenesis of staphylococcal infection is uncertain.

Cellular virulence factors

Some staphylococcal strains form substances that delay the mobilization of local defence mechanisms. The bacterial body has aggressive activity, so that the addition of dead organisms to a living inoculum favours the production of a lesion (Fisher 1963, Gow *et al.* 1963); this effect is attributable to heat- and acid-resistant material in the cell wall. Agarwal (1967*b*) showed that strains of *Staph. aureus* differed in their ability, when injected subcutaneously into mice with cotton dust, to cause a necro-purulent lesion, and that aggressive action in this experimental model was associated with ability to delay the inflammatory response. In further experiments, Hill (1968) obtained similar lesions by injecting a small amount of cell-wall material from a strain with 'aggressive' properties with a sub-infective dose of cocci. The active material was present in the cell-wall residue after extraction with deoxycholate and appeared to be a complex of peptidoglycan with protein. Residues from 'aggressive' but not from 'non-aggressive' *Staph. aureus* strains inhibited the production of oedema around a subcutaneously implanted plug of cotton dust, and the migration of leucocytes (Weksler and Hill 1969). Active immunization of mice with the aggressin gave protection against subcutaneous infection with staphylococcal strains that formed this substance (Hill 1969).

The capsular polysaccharides of *Staph. aureus* (Chapter 30), when formed in large amount, have antiphagocytic action, and capsulated strains are exceptionally virulent when given intraperitoneally to mice. Examination by conventional methods suggests that few strains are heavily capsulated; the significance of the smaller amounts of polysaccharide said to be present on the surface of most *Staph. aureus* strains in initiating infection is doubtful. Easmon (1980) showed that a capsulated strain had poor ability to cause local lesions when given subcutaneously to mice with cotton dust; and, strains that produced the cell-wall aggressin and caused severe local lesions in the skin were considerably less virulent than the capsulated strain by the intraperitoneal route. Dossett and his colleagues (1969) concluded that protein A had antiphagocytic activity and that resistance to phagocytosis could be correlated with the amount formed. There is also some suggestion that protein A may act as an aggressin by forming complexes with gamma-globulin and causing a local hypersensitivity reaction (Gustafson *et al.* 1968). Hale and Smith (1945) considered that the production of free coagulase favoured the establishment of the staphylococcal lesion by causing the deposition of fibrin around the cocci and so protecting them from phagocytosis; but doubt has been cast upon this (Cawdery *et al.* 1969).

We noted in Chapter 30 that continuous passage of *Staph. aureus* in rabbits greatly enhanced its virulence

for rabbits by the intrathoracic route. This was associated with an increase in the resistance of the organisms to intracellular killing by rabbit polymorphonuclear leucocytes, to the bactericidal substances released by lysed leucocytes, and to the bactericidal action of rabbit serum, but not with inhibition of phagocytosis (Adlam *et al.* 1970).

The role of extracellular toxins

It would be reasonable to think that damage to tissues by the extracellular toxins of *Staph. aureus* might contribute to the production of septic lesions, but few of these substances appear to be formed by all the strains that cause severe disease in their respective natural hosts (Chapter 30). On at least two occasions (Kellaway *et al.* 1928, Olin and Lithander 1948), the injection into human patients of medicaments in which *Staph. aureus* had multiplied led within a few hours to vomiting, high fever, cyanosis, convulsions and, in a number of cases, death. Staphylococcal α-lysin, which acts on many types of tissue cell (Chapter 30) may well be thought to have caused these acute lethal effects, but this could not be established definitely at the time (see Wilson 1967). There is evidence (below) that this toxin may contribute to the necrosis and perhaps to the delayed mobilization of leucocytes, in the local staphylococcal lesion, though the leucocidal action of the α-lysin *in vitro* has not yet been convincingly separated from that of the Panton-Valentine leucocidin. α-Lysin-negative variants cause smaller local lesions when implanted subcutaneously along with a foreign body, and do not cause necrosis (Taubler *et al.* 1963, van der Vijer *et al.* 1975).

In the rabbit, the presence of antibody to α-lysin is associated with resistance to staphylococcal infection in burns but not in the normal subcutaneous tissue (Goshi *et al.* 1961*b*); in the mouse, it reduces the amount of necrosis after the subcutaneous implantation of the organisms in cotton dust (Agarwal 1967*b*); but in neither animal is it certain that this was the only staphylococcal antibody that had been formed. Immunization with purified α-toxoid results in the rabbit in a more rapid infiltration of staphylococcal skin lesions with leucocytes, and in the inhibition of necrosis and of the multiplication of the organism (Goshi *et al.* 1963*a*). But, in both man and the rabbit, it also leads to a dermal hypersensitivity, which is proportional to the α-antitoxin titre (Goshi *et al.* 1963*b*, Smith *et al.* 1963). Immunization with purified α-lysin does not protect lactating mice against the development of mastitis after the intramammary infusion of *Staph. aureus*, but it does prevent the lethal haemorrhagic form of the disease caused by certain highly virulent strains (Adlam *et al.* 1977). Doubt is cast on the role of α-lysin by the fact that it is not formed by all *Staph. aureus* strains that cause septic lesions. Admittedly, in-vitro tests are not an entirely reliable indication of the ability of a strain to form α-lysin or any other toxin (Gladstone and Glencross 1960), but there is general agreement that α-lysin is seldom produced by animal strains of undoubted pathogenicity. Animal staphylococci more often form β-lysin, but no definite part in pathogenesis can be assigned to this (see Adlam *et al.* 1977). Staphylokinase might conceivably aid the dissemination of staphylococci by causing dislodgement of infected clots in veins, but it is rarely formed by β-lysin-producing strains.

Staphylococcal extracellular products may enhance the chemotactic activity of leucocytes (Russell *et al.* 1975); the most striking effect on polymorphonuclear cells was observed when staphylococci or their products had first been incubated with plasma, but monocytes exhibited a direct chemotactic effect. Although high doses of α-lysin damage polymorphs, small doses enhance their ability to take up staphylococci and kill them (Gemmell *et al.* 1982).

Hypersensitivity

Panton and Valentine (1929) gave repeated small doses of *Staph. aureus* intracutaneously to rabbits, and observed an increase in the severity of the cutaneous lesions after the later injections. Bøe (1946) confirmed this, and claimed to have desensitized rabbits by the intravenous injection of culture filtrates of staphylococci. Vaccination of rabbits with washed, heat-killed staphylococci also results in an increase in their susceptibility to subsequent infection by the intracutaneous route (Johnson *et al.* 1961). Johanovský (1956, 1958*b*) 'immunized' rabbits with staphylococcal toxoid or by the intravenous injection of a live but somewhat attenuated strain of *Staph. aureus*. The rabbits developed delayed cutaneous hypersentitivity to staphylococci; also, when subsequently given a lethal intravenous injection of *Staph. aureus*, the organisms were more rapidly cleared from the blood than in the controls, but the survival time was shortened. The state of delayed hypersensitivity, and the increased susceptibility to lethal infection, could be transferred to other rabbits by injections of spleen cells or peritoneal leucocytes. Delayed cutaneous hypersensitivity to an extract of disintegrated *Staph. aureus* develops in the mouse after repeated injection of this material into the foot-pad, and can be transferred by means of spleen cells but not of plasma (Taubler 1968). *In vitro*, staphylococcal extracts inhibit the migration of macrophages from explants of the spleen of mice previously infected with staphylococci (Taubler and Mudd 1968).

After repeated infections induced by the subcutaneous injection of *Staph. aureus* with cotton dust, mice showed a delayed hypersensitivity to staphylococcal cell walls that was mediated by T lymphocytes (Easmon and Glynn 1975). The mice were hypersensitive to the peptidoglycan but not to the teichoic acid or protein A of *Staph. aureus* and not to cell-wall constituents of other species of staphylococcus (Easmon and Glynn 1978). The sera of hypersensitive mice protected uninfected mice against the dermonecrotic effect of subsequent challenge by the subcutaneous route. However, the transfer of spleen cells led to an increased susceptibility and the appearance of severe dermonecrotic lesions. When both serum and spleen cells were transferred, the effects were similar to those of giving serum alone. Thus, the harmful effects of delayed hypersensitivity were expressed only in the absence of a humoral response (Easmon and Glynn 1975). Subsequent experiments (Easmon and Glynn 1977, 1979) gave evidence of a humoral factor in infected mice that suppressed the delayed hypersentitivity response, but its effects were overridden by repeated infections. This suggested (Easmon 1980) that cell-mediated hypersensitivity was likely to influence the severity of natural disease only when staphylococci or their cellular constituents persisted in the tissues, for example, when the patient had a defect in the mechanism of phagocytosis or intracellular killing of staphylococci (p. 261). It must be remembered, however, that the serum-antibody response to staphylococcal products in man is often remarkably poor.

Sources and routes of infection

The majority of staphylococcal infections arise from endogenous sources. This is true not only of most of the sporadic skin infections and minor wound sepsis in the general population and of post-influenzal pneumonia, but also of a considerable proportion of the surgical wound infections and nearly all of the neonatal sepsis in the hospital population. In certain circumstances, however, *Staph. aureus* from another person may reach a susceptible site in the patient and cause a lesion; sometimes it is introduced by direct contact and at others indirectly through the air or on a contaminated object. This occurs mainly in hospitals and in relation to surgical wounds.

We therefore have to consider staphylococcal carriage from two standpoints: (1) as a determining factor for clinical infection in the carrier, and (2) as a source of infection or colonization of other persons. The state of staphylococcal carriage or non-carriage is a fairly stable character except in the first few weeks of life and in persons exposed to antibiotics.

Carriage

The carriage of *Staph. aureus* by healthy persons was first described by Hallman (1937) and has since been studied intensively (see Williams 1963). Between 35 and 50 per cent of normal adults carry the organism in the *anterior nares* at any one moment, and it can usually be found elsewhere on the skin of nasal carriers if several swabs are examined and a sensitive method of culture is used (Williams 1946). Lower nasal carriage rates (7–25 per cent) have been recorded in some primitive peoples (see, for example, Rountree *et al.* 1967), often in association with high carriage rates of enterobacteria. There is some evidence for a familial predisposition to nasal carriage (Noble *et al.* 1967); a study of identical and non-identical twins (Hoeksma and Winkler 1963) suggests that genetic influences may be responsible for this.

Age influences the frequency of nasal carriage. The highest rates are seen in young infants; by the age of 2 weeks, figures of 60–70 per cent of those born at home and 80–100 per cent of those born in hospital have been recorded. The rate declines to about 20 per cent by the end of infancy and rises again to reach the adult level by the age of 5–8 years. In elderly persons it declines to around 20–25 per cent.

Repeated swabbing of the same population of normal individuals yields cumulative nasal carriage rates of 60–90 per cent; 20–35 per cent of persons are persistent carriers, 30–70 per cent are intermittent or occasional carriers and 10–40 per cent are never carriers (Williams 1963). Persistent carriers usually harbour the same strain for many months or even for years. Some intermittent carriers are in fact examples of short-term persistent carriage of a single strain, which is then lost; but others are truly intermittent carriers of the same strain over a long period of time. Intermittent or occasional nasal carriage may occur in persons who live in close contact with a persistent carrier (Kay 1963) or who are heavy carriers of the same strain elsewhere on the body surface.

The population of *Staph. aureus* in the nose of carriers thus tends to be stable, and non-carriers exhibit considerable powers of resistance to colonization. Bacterial interference exerted by the existing nasal flora appears to be an important factor in maintaining the stability of both carriage and non-carriage of *Staph. aureus*. As we shall see (p. 271), instillation into the nose of the newborn infant of a large dose of *Staph. aureus* strain 502A usually leads to colonization, and as long as this persists the implantation of other staphylococcal strains is prevented. Studies on the artificial colonization of the nose of adults have established that all strains of *Staph. aureus*, when present in the nose in large numbers, are effective in preventing colonization by other strains; that treatment with antibiotics to which the resident strain is sensitive renders the nose susceptible to colonization by other strains; and that persistent non-carriers are less susceptible to colonization than persons whose nasal strain of *Staph. aureus* has been suppressed by antibiotic treatment, but become highly susceptible if given an antibiotic such as oxacillin (Boris *et al.* 1964, Budd *et al.* 1965, Ehrenkranz 1966, Shinefield *et al.* 1966). Furthermore, after the local application of lysostaphin to the nose, which removes *Staph. aureus* but has less effect on the rest of the flora, natural recolonization with *Staph. aureus* occurs more slowly than after treatment with an antibiotic that acts on all elements in the nasal flora (Martin and White 1968). It thus appears that the total flora of the nose exerts a non-specific inhibitory effect on colonization with new strains.

Information about carriage of *Staph. aureus* in the *throat* of healthy persons is remarkably discrepant, and rates varying between 4 and 64 per cent have been quoted (see Williams 1963). *Faecal carriage rates* in normal infants correspond fairly closely to nasal carriage rates, and the number of organisms present is rather variable; the faeces of some 20 per cent of healthy adults yield small numbers of *Staph. aureus* on a single examination, and nasal carriers are more often faecal carriers than are nasal non-carriers (Matthias *et al.* 1957).

Skin carriage rates of 10–20 per cent are found in most areas of the body except the hands, where up to 40 per cent of swabs may yield *Staph. aureus*. In general, the numbers present on the skin are less than in the nose, but there are exceptions. Undoubtedly, most *Staph. aureus* on the skin are 'transients', in that they have reached the skin from the nose by contact and can be removed by washing. The skin organisms usually correspond in phage-typing patterns to the nasal strain, and their elimination from the nose by antibiotic treatment results in cessation of skin carriage (Solberg 1965). However, 'independent' skin carriage also occurs and is sometimes heavy. Some 5–10 per cent of persons appear from a single examination to carry *Staph. aureus* only on the skin, though when examined repeatedly many of them are found to be occasional nasal carriers. By far the commonest site of 'independent' skin carriage is the perineum; 10–20 per cent of persons, about half of whom are not nasal carriers, are perineal carriers (Hare and Ridley 1958, Ridley 1959, Bøe *et al.* 1964). Patients with chronic skin diseases, particularly atopic eczema or psoriasis, are often very heavy skin carriers of *Staph. aureus*. Not only are they liable to be rich sources of airborne staphylococci but they are at increased risk of self-infection when they undergo surgical operations. In psoriasis the organisms are present in large numbers on the normal skin as well as in the lesions (Noble and Savin 1968).

Airborne dissemination

Few staphylococci are disseminated directly into the air from the respiratory tract (Duguid and Wallace 1948, Hare and Thomas 1956). Much larger numbers reach the environment from elsewhere on the body surface, especially during physical activity. Carriers vary greatly in their ability to disseminate organisms in this manner; in hospitals, a small minority of carrier patients contribute a disproportionately large percentage of the staphylococci found in the air (Shooter *et al.* 1958, Noble 1962). Most but by no means all profuse dispersers are heavy nasal carriers (White *et al.* 1964, Solberg 1965); on the other hand, some heavy nasal carriers disperse few staphylococci into the environment. Heavy perineal carriers almost invariably disperse large numbers of *Staph. aureus* (Ridley 1959, Solberg 1965). Many of the more profuse dispersers are not suffering from staphylococcal disease (Noble 1962); but certain classes of patient—notably those with infected skin lesions, pressure sores, profuse wound discharge, staphylococcal pneumonia or diarrhoea, and some moribund patients (Barber and Dutton 1958, Hare and Cooke 1961, White 1961, Alder and Gillespie 1964, Noble and Davies 1965, Selwyn 1965)—may shed very large numbers of organisms. This is far from true, however, of all patients suffering from staphylococcal disease. Thus, Thom and White (1962) observed little dispersal by patients during the surgical incision of minor staphylococcal lesions, but heavy contamination of the air when the soiled dressings were disturbed. Treatment of a carrier with an antibiotic to which his strain of *Staph. aureus* is resistant may increase the profuseness of carriage and the heaviness of dispersion of the organism (Ehrenkranz 1964).

Staphylococci accumulate rapidly on the clothes and bedding of the disperser, and are disseminated on airborne particles when these are disturbed. At one time it was assumed that the particles that carried the staphylococci were themselves derived from blankets or clothing. It now appears, however, that most of them are desquamated cells from the skin of carriers (Davies and Noble 1962, 1963). Normal persons shed skin scales at varying rates, but the rate of dispersal of organisms into the air is less closely related to this than to the density of the bacterial population on the skin (Noble and Davies 1965, Noble *et al.* 1976). Among healthy persons, heavy dispersion of *Staph. aureus* is much more common in males than females (Hill *et al.* 1974b, Mitchell and Gamble 1974). During exercise, larger numbers of *Staph. aureus* are released from the skin of the lower than of the upper part of the body (Bethune *et al.* 1965), even when their ultimate source is the nose rather than the perineum. The convectional upward movement of skin scales in the layer of air immediately adjacent to the body surface (Lewis *et al.* 1969; see Chapter 9) is therefore probably of less significance in causing dispersion than is the frictional detachment of scales during physical activity. The observation of May and Pomeroy (1973) that organisms are dispersed as frequently during exercise by naked as by clothed subjects suggests that friction between skin and clothing is less important than friction between adjacent skin surfaces.

The median diameter of airborne particles carrying *Staph. aureus* is about 12 μm, with a range of 4–24 μm (Noble *et al.* 1963). These particles remain suspended in the air sufficiently long to be carried for considerable distances on convectional air currents before settling on surfaces or being inhaled by patients; they also pass easily through ordinary clothing, including surgical gowns. Staphylococci in dust have greater powers of survival than have gram-negative bacilli (Lidwell and Lowbury 1950, Lowbury and Fox 1953, McDade and Hall 1963, 1964), particularly at low relative humidity. About half of them survive the initial drying process, and thereafter their death rate is exponential, some 10 per cent surviving the first 24 hr. Little information is available about their infectivity in the dry state (see Williams 1966), but epidemiological experience suggests that few infections from environmental sources occur more than a day or so after the removal of the infectious person from the vicinity. Contaminated dust may be resuspended in the air by activities such as bed-making and dry sweeping of floors, but the effect of walking over a dusty floor is much less (see Ayliffe *et al.* 1967). Staphylococci in dust on walls and ceilings are not easily dispersed into the air.

Dissemination by contact

Less is known about the transmission of staphylococci by contact than by the airborne route, because suitable methods of studying it quantitatively have yet to be developed and evaluated (Favero *et al.* 1968). The undoubted reduction in wound-sepsis rates that followed the introduction of 'no-touch' dressing techniques (see Williams 1971), and several clear instances of wound infection by hand-carriers of *Staph. aureus*, indicate the importance of this route as a means of

transferring staphylococci from septic lesions or carrier sites to uninfected wounds (see Chapter 58). Direct contamination of the clothing of hospital staff by contact with patients (Speers *et al.* 1969) may also be of significance, both as an immediate source of infection for other patients and as a means of dispersal of the organism into the air at a distance from the infected patient.

Acquisition at carrier sites

The neonatal period Newborn babies are very susceptible to colonization by *Staph. aureus*, which is the usual consequence of their first exposure to the organism; clinical sepsis seldom develops without prior colonization. Newborn infants usually acquire *Staph. aureus* first on the skin and later in the nose (Hurst 1960); often the umbilicus (Jellard 1957), and sometimes the circumcised penis (Messinger *et al.* 1963), become heavily colonized at an early stage. Furthermore, if colonization of the skin is prevented by the application of hexachlorophane (p. 271), the nasal carrier rate is much reduced (Gezon *et al.* 1964). In the nursery, a newborn baby usually acquires its first strain of staphylococcus from one of the older infants (Parker and Kennedy 1949, Rountree and Barbour 1950). Sometimes a single strain predominates among the babies, but at other times there may be several; serial observations often show a pattern of successive waves of colonization with fresh strains (Baldwin *et al.* 1957). The introduction of a new strain by a member of the staff may be observed from time to time (Baldwin *et al.* 1957), but the staphylococci carried by members of the staff are commonly quite different from those prevalent among the babies (Parker and Kennedy 1949, Wysham *et al.* 1957). Few of the staphylococci acquired by babies in nurseries come from their mothers.

There is evidence that *Staph. aureus* may be disseminated in nurseries both by the airborne route and by contact. The organism can often be isolated in large numbers from surfaces, bedding and the air, but it can also be found on the hands of some 12 per cent of nurses who have touched a carrier baby (Love *et al.* 1963). In a series of well conducted trials, Rammelkamp and his colleagues (Wolinsky *et al.* 1960, Mortimer *et al.* 1962, 1966) compared the rate of colonization of infants from an index source in the same room when they were attended to (1) with unwashed hands, (2) with effectively washed hands, (3) with gloved hands, and (4) without direct manual contact with the source of infection. Briefly, contact with unwashed hands caused five times as many acquisitions as airborne infection, but if the hands were carefully cleansed or were gloved infection from the two sources was about equal in frequency. Thus, though infection was more 'easily' spread by contact than through the air, hand-washing alone probably cannot be relied upon to prevent infection. Transmission by fomites has also been demonstrated (Gonzaga *et al.* 1964), but only with heavily contaminated clothing.

Later acquisitions   After the immediate neonatal period, changes in the carrier state occur much less frequently. As we have seen (page 256), there is some spread of *Staph. aureus* to carrier sites within the family, but in general the established flora of the nose prevents the acquisition of new strains unless antibiotic treatment is given or the subject enters hospital. After 1945, the strains of *Staph. aureus* carried by hospital patients were more often resistant to antibiotics than those carried by members of the non-hospital population. Frequent contact with antibiotics led to the elimination of sensitive organisms from the flora of many of the patients and their replacement by resistant strains of *Staph. aureus* acquired by cross-infection from other persons in the hospital. Thus, continuous chains of transmission led to the endemic prevalence of certain strains of *Staph. aureus* in the hospital population as a whole.

When penicillinase-forming strains of *Staph. aureus* first became prevalent, Barber (1947*a*, *b*) showed that this was due to their selection and subsequent spread from patient to patient in the hospital. Patients admitted to hospital tended to acquire the resistant strains prevalent in the ward. This occurred most frequently in patients being treated with penicillin, but a number of patients who had not received the antibiotic, and many members of the hospital staff, also became carriers of the resistant strains (Barber and Rozwadowska-Dowzenko 1948). The observation of Gould (1958) that the inanimate environment of the treated patients became contaminated with penicillin provided a possible explanation of this.

The successive introduction of other antibiotics was followed by the appearance in hospitals of staphylococci resistant to them. Resistance to each new antibiotic appeared almost exclusively in penicillin-resistant strains, and there was a tendency for more and more resistances to accumulate in a few strains, which then spread widely by cross-infection. The acquisition of a resistant strain was commoner in patients who had received one of the antibiotics to which it was resistant than in those who had not (Knight and Holzer 1954, Williams *et al.* 1959, Berntsen and McDermott 1960) but was by no means confined to them. Also, antibiotic treatment of non-carriers of *Staph. aureus* increased the rate of acquisition of resistant strains at least as much as did treatment of carriers of sensitive strains (Noble *et al.* 1964). This confirms the view that elimination of organisms other than *Staph. aureus* increases susceptibility to colonization.

Heavy exposure to antibiotics increases the rate of loss of sensitive strains of *Staph. aureus* from carrier sites, and effective isolation of patients decreases the rate of acquisition of resistant strains (Parker *et al.* 1965, Lidwell *et al.* 1966). The net effect of this on the total carrier rate of *Staph. aureus* is variable (Lidwell *et al.* 1970), and depends upon the antibiotic regimen in the hospital and the extent to which the individual patient is exposed to staphylococci dispersed from other patients.

The staphylococci carried by hospital patients are a mixture of those present on admission and those acquired subsequently. Strains of either category may cause clinical disease by self-infection, but a considerable minority of hospital-acquired staphylococcal lesions develop in non-carriers who acquired the infecting organism from an extraneous source. In the 1950s, the highly antibiotic-resistant strains then endemic in hospitals were responsible for over one-half of hospital-acquired staphylococcal sepsis, but this proportion subsequently fell considerably (p. 268).

We are still uncertain about the route by which *Staph. aureus* reaches the carrier sites of adult hospital patients; few studies have been made of the relationship in time of acquisition of the skin and in the nose, and clear evidence that skin carriage precedes nasal carriage has not been produced. It would be difficult to imagine how *Staph. aureus* could often reach the nose from an extraneous source by direct contact; on the other hand, the patient is constantly breathing in air in which the organisms are present. Several studies (Lidwell *et al.* 1966, 1970, 1971) have shown that there is a direct quantitative relationship between the rate of acquisition in the nose of particular classes of staphylococcal strains and the amount of airborne exposure to them (see also Williams 1966).

The characters of 'hospital' strains

The strains that established themselves in hospitals in the 1950s and 1960s were in general resistant to antibiotics in common use at the time. There is little evidence that, as a class, they differed from less resistant strains found outside hospital either in pathogenicity or communicability, but they certainly showed considerable variability among themselves in these respects.

Antibiotic resistance

Penicillinase-forming strains existed in small numbers before penicillin was used therapeutically (Chapter 30). The rapid increase in their prevalence in hospitals in the mid-1940s was a consequence of their spread by cross-infection. Soon afterwards, resistance to antibiotics other than penicillin—which in most instances had not been detected before the corresponding agent was taken into use—appeared in a minority of the penicillinase-formers. These strains, generally termed 'multiple-antibiotic resistant', in their turn spread widely in hospitals. The spectrum of resistance of individual strains became progressively wider, and at various times included almost any combination of penicillinase formation with resistance to streptomycin and various other aminoglycosides, tetracycline and minocycline, chloramphenicol, the macrolides, lincomycin, bacitracin, fucidin and trimethoprim, as well as a non-enzymic resistance to penicillins and cephalosporins ('methicillin resistance').

Other penicillin-resistant strains, however, failed to develop multiple-antibiotic resistance and came to play a progressively smaller part in hospital infection except in nurseries for the newborn. The frequency of penicillinase production by staphylococcal strains in the general population rose very much more slowly, and did not reach 50 per cent until 1960 (Ashley and Brindle 1960, Munch-Petersen and Boundy 1962); it is now usually well in excess of 80 per cent. Strains resistant to antibiotics other than penicillin are still uncommon except in persons who have had a recent contact with a hospital.

The genetic determinants for resistance to various antimicrobial agents are—with a few exceptions—distinct, and comprise a series of plasmid-borne and chromosomal genes that appear to have been acquired separately by the 'hospital' strains (Chapter 30). Their accumulation by a small number of strains, and the fact that nearly all of them form penicillinase, must therefore be explained on ecological rather than genetic grounds.

As we have seen (Chapter 30), those of the multiple-antibiotic resistant strains that spread widely usually formed large amounts of penicillinase—and, particularly, large amounts of extracellular penicillinase. It thus appeared that a really effective mechanism for dealing with high concentrations of penicillin was a minimum requirement of the survival of a strain in the hospital population sufficiently long for it to acquire the genetic determinants for other resistances. On the other hand, weak penicillinase production appeared to confer some selective advantage on staphylococci in the general population.

'*Methicillin resistance*', a broad-spectrum tolerance for all penicillins and cephalosporins, probably existed before the penicillinase-resistant penicillins were introduced, but there was no means of detecting it in strains that also produced penicillinase. It became quite common in some parts of the world in the later 1960s, but its prevalence at that time was not closely related to the use of penicillinase-resistant penicillins or cephalosporins. The fact that its effects were fully expressed only at temperatures well below 37° suggested that it might confer a selective advantage to staphylococci on the body surface under conditions in which penicillinase production provides inadequate protection against conventional penicillins; for example, it might permit colonization by small numbers of staphylococci implanted on a surface site in a patient being treated with a penicillin (Parker and Hewitt 1970). Indeed, one of the highest prevalence-rates for 'methicillin resistance' was reported from a hospital in which large doses of benzylpenicillin were given prophylactically to a considerable proportion of the patients but in which methicillin usage was negligible (Siboni and Poulsen 1968). 'Methicillin resistance' is found almost exclusively in strains that also form large amounts of penicillinase, but the two characters have separate genetic determinants.

When a new antibiotic came into use, resistance to it sometimes appeared promptly, but in several instances there was a delay of several years. For example, neomycin resistance was not reported until this antibiotic had been widely used for over a decade. In Britain, it was first identified in strains with the phage-typing pattern 84/85, which then became widely distributed in hospitals. It was confined to these strains for several years, after which it appeared in rapid succession in other phage-group III strains (Jevons and Parker 1964, Parker *et al.* 1974). The appearance of resistance to gentamicin was also delayed for over 10 years, but in 1976

several clearly distinguishable resistant strains appeared almost simultaneously in different parts of the country (for references, see Shanson 1981). This suggests that the initial mutation to resistance to certain antibiotics, or its transfer to *Staph. aureus* from another organism, is a very rare event, but that once it has been acquired by certain strains of *Staph. aureus* it may be rapidly transferred to other strains.

In general, therefore, selection and cross-infection were the main factors responsible for building up and maintaining populations of resistant staphylococci, and gene transfer appeared to play only a secondary role in the process.

There are, however, a few exceptions to this. Mutants resistant to fucidin and novobiocin can be found in most large populations of *Staph. aureus*, and resistance therefore tends to appear in the infecting strain during the treatment of the individual patient with these antibiotics. These resistances, therefore, may be scattered in distribution and sometimes do not form part of the usual broad patterns of multiple antibiotic resistance. Mutation to the constitutive (high-level) type of resistance to macrolides—often with lincomycin resistance—occurs at high frequency only in strains with the inducible type of macrolide resistance, which are almost always resistant to several other unrelated antibiotics.

Phage-typing patterns

From 1947 onwards, phage-typing patterns were widely used to characterize the antibiotic-resistant strains that were prevalent in hospitals; indeed they provided valuable help in following their spread. Later it became apparent that phage-typing patterns might give misleading results when used to study the course of long continued endemic prevalences of particular strains.

In Chapter 30 we mentioned the 80/81 strain, which gave rise to a number of variants with different patterns of reaction with phages 52, 52A, 80 and 81 by the loss or gain of prophages. These changes in pattern concealed the continued prevalence of what in all other respects was the same strain. A few years later, a series of 'new' phage-group III strains that were untypable with existing phages appeared in hospitals in various parts of the world. Some of them were shown to have arisen by lysogenization of existing strains (Jevons and Parker 1964, Jevons *et al.* 1966, Jessen *et al.* 1969, Rosendal and Bülow 1971). The new strain was in several instances resistant to an antibiotic to which its putative parent was sensitive, but the acquisition of resistance and the change in phage-typing pattern appear to have been separate events. Thus, the appearance of 'new' strains of *Staph. aureus* in hospitals has been observed on a number of occasions in the last 30 years. Several of these strains were initially recognized by their insusceptibility to lysis by phages of the current basic typing set. Their untypability was usually a result of prophage immunity. The frequency with which lysogenization occurs in staphylococcal strains that are endemic in hospitals may be attributed to the instability of the nasal flora in hospital patients, which provides many opportunities for the co-existence of different staphylococcal strains in biological 'mixed culture'. The most recent of the 'new' hospital staphylococci, the members of phage-group V, are an exception to this rule, because their untypability is attributable to the possession of a very exclusive restriction-modification system (Asheshov *et al.* 1977).

For a general account of the relations between phage-typing pattern and antibiotic resistance in *Staph. aureus*, see Parker (1983).

Pathogenicity and transmissibility

The behaviour of individual staphylococcal strains in hospitals provides evidence that they differ among themselves not only in the frequency with which they cause septic infection, but also in the nature of the resulting lesion. Moreover, some spread more easily among the hospital population than do others, and this is not always attributable to the possession of a wider spectrum of resistance to antibiotics.

The suggestion that there are 'epidemic strains' of *Staph. aureus* was first made in pre-antibiotic days in connection with neonatal sepsis. It now seems likely that attention was drawn to a number of these strains by their ability to cause the exfoliative type of skin lesion. The early antibiotic-resistant 'hospital' strains were not particularly virulent, but in 1952 a strain that caused lesions of remarkable severity appeared in Australia and soon became generally prevalent throughout the country (Isbister *et al.* 1954, Rountree and Freeman 1955, Rountree 1978). This 80/81 strain caused severe skin pustules in newborn babies, their mothers and the nursing staff, wound infections, abscesses in deep tissues and septicaemic disease in hospital patients of all ages, and epidemics of neonatal pneumonia. In addition, it caused widespread nasal colonization and spread extensively in the general community, at one time being responsible for one-half of all staphylococcal sepsis seen in general practice (Johnson *et al.* 1960). All 80/81 staphylococci isolated in Australia up to the end of 1954 were resistant only to penicillin, though other multiple-antibiotic resistant strains had been prevalent in the country for several years. Between 1953 and 1955, 80/81 strains with similar pathogenicity became established in a number of other countries, but these were generally multiple-antibiotic resistant. It seems likely that all the 80/81 staphylococci arose from a common source (Asheshov and Winkler 1966), and it may have been justifiable to use the term 'epidemic' to describe their initial dissemination. Eventually, however, they and their lysogenized derivatives became established endemically in hospitals in most countries in the world. After several years, they began to decrease in frequency, and were rarely seen in Britain after 1970 (Parker *et al.* 1974).

Before, during and after the 80/81 incident, many other strains, also with characteristic phage-typing patterns, assumed importance as causes of sepsis in particular districts, countries or even continents, but they rarely exhibited an equal ability to cause severe lesions in otherwise healthy persons. After 1959, numerous 'new' phage-group III strains, lysed only by phages in the 83A, 84, 85, 86, 88, 89, 93 series and with very broad spectra of resistance to antibiotics, became prevalent in various parts of the world (see Jevons *et al.* 1966, Jessen *et al.* 1969, Kryński *et al.* 1976, Rosendal *et al.* 1976). Although capable of causing severe wound sepsis, pneumonia and septicaemia, particularly in debilitated patients, they rarely gave rise to boils either in patients or in

members of the hospital or became a major cause of neonatal sepsis.

It thus appears that 'hospital staphylococci' may differ in virulence both quantitatively and qualitatively, and that transmissibility is a character separate from virulence. This view receives support from the observation that some strains spread from patient to patient without causing sepsis (Barber *et al.* 1953, Shooter *et al.* 1958) and that others seldom colonize fresh patients even when disseminated profusely into the air (Noble 1962).

Recent history of the 'hospital staphylococci'

After 1960 the relative importance of multiple-antibiotic resistant strains as causes of staphylococcal infections in hospitals began to decrease, but the extent and timing of this change was variable. It was first noted in 1959-62 by Bulger and Sherris (1968) in the USA. In several London hospitals the fall began in or around 1964 (Parker *et al.* 1974); in Denmark it took place several years later (Rosendal *et al.* 1977).

The main pattern of decline in the London area was a progressive fall in the size of endemic prevalences caused by established multiple-antibiotic resistant strains rather than their complete disappearance (Parker *et al.* 1974). At the same time, strains with increasingly wide spectra of resistance continued to appear (see Shanson 1981), but they gave rise to only small incidents of infection. By 1975-6, only some 20 per cent of infections in London hospitals were caused by multiple-antibiotic resistant strains and staphylococci resistant only to penicillin were responsible for one-half (Parker 1983). The more sensitive strains caused many small groups of epidemiologically related infections. The only identifiable 'new strains' to assume increasing importance over a wide geographical area were phage-group V strains lysed by phages 94 and 96 (Marraro and Mitchell 1975), which were seldom resistant to antibiotics other than penicillin and sometimes tetracycline. Identifiable 'new' phage-group III strains were all limited in distribution.

It is difficult to account for these widespread changes. The total usage of antimicrobial substances had not decreased. Individual workers attributed their success in controlling the spread of resistant strains to the operation of an antibiotic 'policy' or to the restriction of the use of particular agents (see Shanson 1981), but the changes were not confined to single hospitals. Considerable efforts had been made at the time to improve hygiene in many hospitals, but it was not easy to attribute the changes to any specific preventive measures. In the early 1970s hospital doctors were concerned about other apparently more pressing questions, such as infections with gram-negative bacilli and hepatitis B (Chapter 58), and they tended to consider staphylococcal infection a problem of the past. However, though multiple-antibiotic resistant strains were limited in distribution in the individual hospital, they continued to cause serious infections in predisposed patients. This situation was tolerable as long as it was easy to find an appropriate antibiotic to treat these infections, but within a few years this became increasingly difficult. Gentamicin resistance was much more widespread and soon began to be associated with methicillin resistance in strains with very wide spectra of resistance to other agents. A number of local outbreaks caused by such strains have now been reported in Britain, the USA and Australia (Crossley *et al.* 1979, Graham *et al.* 1980, Price *et al.* 1980, Shanson and McSwiggan 1980, King *et al.* 1981, Linnemann *et al.* 1982; see also *Medical Journal of Australia* 1982). Some of these outbreaks were extensive and long continued, with evidence of inter-hospital spread. The strains responsible generally appeared to belong to phage-group III but were not easy to characterize by phage typing. They did not appear to be exceptionally virulent, but a number of deaths occurred in infections in susceptible patients. Methicillin resistance, hitherto rarely seen in the USA, is now common in staphylococci in a number of large hospitals (Haley *et al.* 1982). Reports from Eastern Europe suggest some recrudescence of infection by members of the 52, 52A, 80, 81 complex in the mid-1970s (see, for example, Rische *et al.* 1980), but the strains concerned did not have exceptionally wide patterns of resistance. The future of multiple-antibiotic resistant *Staph. aureus* strains as causes of hospital-acquired infection is difficult to predict.

Sources and routes of infection of wounds

It remains to be seen how and when the infecting staphylococcus reaches the wound. For wounds infected outside hospital, the patient is probably the main source of infection (Williams and Miles 1949) and the organism is introduced into the wound by contact. For surgical wound infection, the potential sources are more numerous and the routes more complex. Opinions differ as to the relative importance of the operating room and the ward as the place of infection (see Chapter 58). Indeed this probably varies from hospital to hospital (see, for example, Shooter *et al.* 1958, Bassett *et al.* 1963), and at different times in the same hospital.

The same appears to be true of the sources from which the infecting organism comes and the routes by which it reaches the wound. Estimates of the relative proportions of self-infections and cross-infections vary widely. Thus, Williams and his colleagues (1959) reported wound-sepsis rates of 7.1 per cent in carriers of *Staph. aureus* and only 2 per cent in non-carriers; in about half of the cases, sepsis was due to the strain carried in the nose (for somewhat similar findings see Calia *et al.* 1969). Other workers have not found this relation between nasal carriage and sepsis (Bassett *et al.* 1963, Moore and Gardner 1963); it is perhaps significant that, in at least some of the hospitals in which self-infection does not appear to be frequent, a large proportion of the infections were acquired in the operating theatre. In self-infection, it is usually very difficult to decide whether the organism entered the wound at the time of operation or subsequently. Infection can come from the patient's own skin when it is inadequately sterilized (Harrison and Cruickshank 1952), or when the edges of the incision are recontaminated during the operation (Black *et al.* 1962; see also McNeill *et al.* 1961). Calia and his colleagues (1969),

who showed that self-infection was more frequent in heavy than in sparse nasal carriers, and in nose carriers who were also skin carriers than in those who were not, were unable to isolate the infecting strain from strips of skin excised from the margin of the wound at the beginning of the operation, or from swabs taken just before the wound was closed. This suggests that recontamination of the wound site and multiplication of the organism in moist exudate during the early post-operative period may be of significance.

Exogenous infection in the operating theatre may be from the surgeon (Devenish and Miles 1939, McDonald and Timbury 1957, Shooter *et al.* 1957) or from an assistant who touched the wound or the instruments (Penikett *et al.* 1958). There are, however, several instances in which a member of the theatre staff who was not in direct contact with the wound, but who was a heavy disperser of staphylococci into the air, gave rise to an outbreak of post-operative sepsis (Walter *et al.* 1963, Ayliffe and Collins 1967, Payne 1967). It is unlikely that such incidents are responsible for more than a small proportion of wound infections. Airborne infections from more remote sources—even from air sucked into the theatre from other parts of the hospital (Blowers *et al.* 1955, Shooter *et al.* 1956)—were not uncommon before effective operating-theatre ventilation became the rule.

It is impossible to give a comprehensive picture of the sources and routes of infection in surgical wards in general, and few attempts have been made to do this even for a single hospital or department. Lindbom (1970) investigated 111 staphylococcal wound infections in a department of thoracic surgery during six years, and concluded that 33 per cent were self-infections—20 per cent with organisms present in the patients on admission and 13 per cent with organisms acquired pre-operatively; 57 per cent were exogenous—11 per cent acquired at operation, 39 per cent in the intensive care ward, and 7 per cent in the general ward; the source of the remainder could not be traced.

(For further discussion of the relative importance of infection in the operating theatre and in the post-operative wards, and of aerial and contact infections, see Chapter 58.)

Diagnosis

A diagnosis of staphylococcal sepsis can be made from the laboratory only when the material examined has come from a closed lesion. The presence of *Staph. aureus* in wound exudate and in the secretions from the respiratory tract must be interpreted in the light of the clinical circumstances.

Serological diagnosis

The sera of normal persons usually contain small amounts of antibody against several components of

Staph. aureus, and the antibody response to infection is remarkably poor (see Oeding *et al.* 1981). Serological methods play no part in the diagnosis of most localized infections; they are used for the detection of certain chronic deep-seated infections, such as osteomyelitis of the spine, and may be of some assistance in the clinical assessment of septicaemic infections, but the results are often difficult to interpret.

Antibody to α-*lysin* is of no significance unless present in excess of 2 units per ml (for references, see Towers and Gladstone 1958). Titres in staphylococcal osteomyelitis may reach 15 units per ml or more, but as many as 50–80 per cent of all cases do not give a significant antibody response (Lack 1957), and false-positive results may occur in bone tuberculosis (Taylor *et al.* 1975). Microscopic tests for antibody against *Panton-Valentine leucocidin* (Towers and Gladstone 1958) are said to give more positive results in staphylococcal osteomyelitis but are not convenient for routine use; the more sensitive indirect haemagglutination test (Towers 1961) may give false-positive results, especially in bone tuberculosis (Taylor *et al.* 1975). Tests for antibody to γ-*lysin* (Taylor and Plommet 1973) are said to give few false-positive reactions, and may give a positive result in some cases of staphylococcal osteomyelitis in which the anti-α-lysin titre is not raised; according to Taylor and his colleagues (1975), some 80 per cent of cases of staphylococcal bone infection can be recognized if the anti-α and anti-γ tests are used in combination. Antibody to *staphylococcal nuclease* is said to appear only in infections with *Staph. aureus* (Taylor *et al.* 1976*a*); tests for this may give a positive result in some cases of osteomyelitis in which the anti-α and anti-γ titres are not raised.

Antibody to the *ribitol teichoic acid* of the cell wall of *Staph. aureus* can be detected in most sera by a highly sensitive method, for example, enzyme-linked immunosorbent assay. The gel-diffusion precipitation test is very much less sensitive, but gives positive results in many cases of bone infection and in practically all cases of endocarditis due to *Staph. aureus* (Crowder and White 1972, Nagel *et al.* 1975). According to Tuazon and her colleagues (1978), the appearance of sufficient antibody to be detected by the gel-diffusion method serves to distinguish between bacteraemic patients who have endocarditis or metastatic abscesses elsewhere in the body and those who do not; this is an important guide to treatment (below). A few healthy persons and a greater proportion of sufferers from systemic lupus erythematosus or yersinial arthritis have significant titres of antibody (Larinkari *et al.* 1983). Counterimmunoelectrophoresis is a method of intermediate sensitivity. (For antibodies to staphylococcal peptidoglycan, see Verbrugh *et al.* 1981.)

Treatment

Whether or not to give specific treatment for a particular staphylococcal infection—and the choice between surgical and antimicrobial treatment—are matters for clinical judgement. On the one hand, antibiotic treatment of closed septic lesions is unlikely to succeed unless given very early, and adequate surgical drainage of the lesion may render it unnecessary; on the other, prompt administration of an appropriate antibiotic may make surgical intervention unnecessary. It is essential to give early and vigorous antibiotic treatment for staphylococcal osteomyelitis, post-influenzal pneumonia, widespread exfoliative disease, toxic-shock syndrome, post-antibiotic diarrhoea, and all bacteraemic infections. Treatment should be continued for 4–6 weeks if bacteraemia is not associated with a focus of primary infection that can be easily removed or effectively treated, and if there is evidence of endocarditis or pyaemic dissemination. Once the decision has been taken to use an antimicrobial agent, the help of the laboratory in choosing the appropriate one becomes necessary. Wherever possible, the choice should be based on knowledge of the in-vitro sensitivity of the causative organism to the therapeutically effective drugs. For most purposes, tests of the bacteristatic action of the drug on the organism—measurement of its minimum inhibitory concentration, or inference about the minimum inhibitory concentration from the results of disk-diffusion tests—are adequate, but for the eradication of generalized infections, particularly where there is evidence of endocarditis, tests for the bactericidal action of drugs or of drug combinations are essential. (For further information see Garrod *et al.* 1981.) If treatment has to be begun before the results of the sensitivity tests are available, the laboratory should be able to give guidance based upon up-to-date knowledge about the resistance to particular drugs of the strains of *Staph. aureus* prevalent in the local community.

Prevention

Most of the measures we take to prevent staphylococcal infections in hospital form part of a general programme for the prevention of septic infections (Chapter 58). Briefly, our main objectives should be (1) to prevent the access of *Staph. aureus* to susceptible sites from which they can invade the tissues; (2) to lessen the chance that organisms that do reach such sites will cause sepsis; and (3) to reduce as far as possible the number of human sources of *Staph. aureus* in the immediate neighbourhood of the patient.

Many of the precautions we discussed in Chapter 58 were designed to achieve the first of these objectives. Eliminating some of the predisposing factors detailed earlier in this chapter, without which *Staph. aureus* seldom causes serious infection, may also be helpful. There is little doubt, for example, that careful surgical technique lessens the risk of staphylococcal wound sepsis. It is now generally accepted that the prophylactic use of antibacterial agents may be a useful means of preventing the multiplication of bacteria implanted into wounds (Chapter 58). However, this must be done under carefully defined conditions if the ill effects of the practice are not to outweigh its advantages. The frequency of carriage of

Staph. aureus by normal persons makes it impracticable, save under conditions of exceptional risk, to ensure that a patient will not be exposed to the organism in hospital.

It is seldom possible to insist that members of the hospital staff should be non-carriers, except perhaps for a few engaged in special tasks. General attempts to abolish nasal carriage, in patients or staff, by mass chemoprophylaxis, either systemic or local, have not proved very effective and have often created problems with antibiotic resistance (see Williams *et al.* 1966). On the other hand, the removal and treatment— preferably outside the hospital—of individual members of the staff who have actually caused infections is justified. The topical application of creams containing gentamicin, neomycin or vancomycin usually frees the nose from *Staph. aureus* strains that are sensitive to the agent employed, but in some cases the organism reappears later (Williams *et al.* 1967).

Mention must be made of the use of *hexachlorophane* for the suppression of *Staph. aureus* on the skin. It acts predominantly on gram-positive organisms and has a cumulative effect when applied repeatedly. Its regular application, in a detergent washing fluid, a cream or a powder, is a most effective means of preventing skin colonization in newborn infants and so preventing neonatal staphylococcal sepsis (for references see Williams *et al.* 1966). Evidence of the toxicity of hexachlorophane led to a great reduction in its use. Many maternity hospitals then reported an increase in the frequency of neonatal staphylococcal disease (Dixon *et al.* 1973, Kaslow *et al.* 1973, Scopes *et al.* 1974). The value of hexachlorophane had been established during the years when staphylococci of the 52, 52A, 80, 81 complex were widespread in hospitals. By the time hexachlorophane came under suspicion these strains were seldom found in maternity hospitals (see, for example, Light *et al.* 1975). The increase in sepsis that followed discontinuance of its use appears in most instances to have been only moderate. (For a review of the hazards of hexachlorophane to newborn infants, of the circumstances in which it may be used safely, and for alternatives to it, see *Lancet* 1982.)

Lysostaphin (Chapter 30) has a powerful and selective action on staphylococci. Although unsuitable for parenteral use, it is an effective means of eliminating *Staph. aureus*, and to a lesser extent other staphylococci, from the nose when applied topically (Martin and White 1967, Harris *et al.* 1968). Because it does not remove all organisms, the patients appear to remain somewhat resistant to recolonization by *Staph. aureus*.

The value of *bacterial interference* as a means of preventing colonization by *Staph. aureus* is now well established. Shinefield and his colleagues (1963) showed that instillation of large inocula of *Staph. aureus* strain 502A into the nose of newborn infants prevented subsequent colonization by other strains of the same organism. This was done in a number of nurseries in which outbreaks of serious infection—many due to members of the 52, 52A, 80, 81 complex—were in progress and resulted in a considerable reduction in the frequency of sepsis (see also Light *et al.* 1967). The 502A strain was chosen because it was believed to be somewhat 'avirulent'; subsequent experience has tended to confirm this, but the organism has certainly given rise to a number of mild purulent lesions, and at least on one occasion to a fatal septicaemia (Houck *et al.* 1972).

Another use of bacterial interference is to prevent recurrent furunculosis in individuals and families. Carriage of the infecting staphylococcus is suppressed temporarily by antibiotic treatment before the 502A strain is instilled into the nares, and sometimes also on to the skin. In a number of instances this has resulted in an interruption, for periods of a year or more, in the succession of furuncles (Drutz *et al.* 1966, Boris *et al.* 1968, Strauss *et al.* 1969). Any strain of *Staph. aureus* present in sufficiently large numbers will prevent subsequent colonization by another; this is true not only in the nose but also in experimental burns (Anthony and Wannamaker 1967). Coagulase-negative staphylococci also interfere with a subsequently implanted strain of *Staph. aureus* in burns (Wickman 1970).

Active immunization

Toxic filtrates of *Staph. aureus* may be rendered innocuous by treatment with formalin. A course of injections of toxoid results in a high concentration of α-antitoxin in the blood, both in animals and in man. There is little convincing evidence, however, that it influences the course of chronic or recurrent infections. Staphylococcal toxoid also contains variable amounts of P-V leucocidin and several workers have observed a prophylactic effect which they attribute to the anti-leucocidin formed in response to immunization with toxoid.

Johanovský (1958a) injected pregnant women with a toxoid that contained P-V leucocidin; later he estimated the α-antitoxin and the antileucocidin in the cord blood of immunized and control patients. There was no relation between the α-antitoxin titre and the neonatal sepsis rate. In the unimmunized controls a high anti-leucocidin titre was associated with a reduced risk of neonatal sepsis and of breast abscess. Few infections occurred in the immunized group. A later and more extensive trial (Šebek *et al.* 1959) showed a five-fold reduction of serious staphylococcal infections in the babies, and a two-fold reduction of maternal mastitis, in the immunized group. Bänffer (1962) obtained rather similar results (see also Bänffer and Franken 1967). It appears, therefore, that staphylococcal toxoid rich in P–V leucocidin may have some value for the 'prospective' protection of susceptible persons likely to be exposed to staphylococcal infection, but it is of little or no value for the prevention of recurrent skin sepsis.

Attempts to produce antibacterial immunity to staphylococcal infections by vaccination have given even less convincing results. Staphylococcal vaccines have been widely used in man in attempts to prevent recurrent skin infection, but, as with staphylococcal toxoids, little convincing evidence of their efficacy has been produced. This may be due in part to the role of hypersensitivity in chronic furunculosis. (For immunization against staphylococcal mastitis see Chapter 57.)

Infections due to coagulase-negative staphylococci and micrococci

We have seen (Chapter 30) that these organisms form part of the normal flora of the skin. They are often found in swabs from wounds and various skin lesions, and from the conjunctiva, but it is difficult to decide whether their presence is of clinical significance. However, the role of certain of them as causes of other specific types of infection is now firmly established.

Urinary-tract infection

Mild infections, usually caused by *Staph. epidermidis* biotype 1 of Baird-Parker, may follow operations on or instrumental manipulation of the lower urinary tract (Mitchell 1964). This organism very rarely causes urinary-tract infection in healthy persons who have not been subjected to such procedures.

A quite distinct staphylococcus—a novobiocin-resistant member of *Staph. saprophyticus* biotype 3 of Baird-Parker (see Chapter 57)—causes spontaneous acute urinary-tract infection in previously healthy persons. With a few exceptions these are women aged 16–25 years; the disease appears soon after the woman becomes sexually active and closely resembles urinary-tract infection caused by *Escherichia coli* in its symptomatology. Many of the patients show evidence of pyelonephritis, but the disease is usually self-limiting. The staphylococcus is believed to enter the urinary tract by the ascending route. Indistinguishable organisms form a small and rather inconstant proportion (1–20 per cent) of the staphylococci on the arms and legs of about one-half of normal adults, but are rarely found in the axilla or the nares (Kloos and Musselwhite 1975). They may occasionally be found, though usually in very small numbers, in the urethra of normal persons of either sex.

It is likely that novobiocin-resistant strains of *Staph. saprophyticus* biotype 3 form only a small proportion of the staphylococci introduced mechanically into the urethra by physiological means, and that they have a specific ability to invade by this route. They multiply in urine no more rapidly than other staphylococci (Anderson *et al.* 1976). Namavar and his colleagues (1978) observed that novobiocin-resistant strains of *Staph. saprophyticus* biotype 3 from urinary-tract infections were consistently more virulent for baby mice by the intracerebral route than were strains of *Staph. epidermidis* biotype 1 or *Staph. saprophyticus* biotypes 1 or 2. The same was true of novobiocin-resistant but not of novobiocin-sensitive strains of *Staph. saprophyticus* biotype 3 from the skin. Hovelius and Mårdh (1979) made the even more significant observation that the urinary pathogen had the property, almost unique among staphylococci, of forming a trypsin-sensitive, D-mannose-resistant haemagglutinin for sheep erythrocytes; it is also said to adhere to epithelial cells from the human urinary tract (Colleen *et al.* 1979).

Systemic infections

Endocarditis

This may occur in two quite distinct circumstances: (1) spontaneously in persons whose heart valves have suffered previous damage; and (2) after an open operation on the heart (Quinn *et al.* 1966). In the first, endocarditis is usually of the subacute form, the blood culture is often only intermittently positive, and the staphylococcus may be sensitive to antibiotics and can be eradicated by intensive treatment. Coagulase-negative staphylococci are among the less common causes of endocarditis of the chronically damaged heart valve and in persons who use narcotic drugs intravenously (see Chapter 57 and Simberkoff 1977). Post-cardiotomy endocarditis is usually of the acute form, the blood culture is in most cases regularly positive, the staphylococcus is nearly always resistant to penicillin and to several other antibiotics, and the prognosis is poor. Coagulase-negative staphylococci are responsible for about one-quarter of endocarditis developing within 6 months of operation and for nearly as many of the cases that appear subsequently (Karchmer and Swartz 1977).

The causative organisms in both spontaneous and postcardiotomy endocarditis are usually described as *Staph. epidermidis*; when further examination has been made, most of them have been found to belong to Baird-Parker's biotype 1. Marples and Richardson (1980) have described a true micrococcus responsible for a few cases of endocarditis; it is uniformly resistant to methicillin and its cultural characters do not conform to those of any recognized species in the genus.

The operations on the heart after which endocarditis develops are usually those in which a valve has been replaced by a rigid prosthesis (Amoury *et al.* 1966). The most consistent lesion in fatal cases develops immediately adjacent to the site of attachment of the prosthesis; micro-abscesses may form and extend into the surrounding heart muscle; vegetations on the prosthesis are a less regular feature (Arnett and Roberts 1977). The disease thus appears to be a true infection of cardiac tissue.

Cases of post-cardiotomy endocarditis generally occur sporadically, but a few apparent epidemics have been described. Two investigations have revealed evidence of groups of cases caused by a single strain of *Staph. epidermidis*. In one of these (Blouse *et al.* 1978), it was concluded that two separate incidents, caused by different strains and separated by several months, had taken place. At the time of each incident, the respective infecting strain was prevalent in nasal swabs of members of the staff. It was concluded that the operating team was the main source of infection, which may have reached the patients by the aerial or the contact route, or via the bypass pump. In a different investigation (Marples *et al.* 1978), an outbreak caused by a highly resistant strain was identified. This strain was found on only one occasion during the several months of the outbreak in a member of the theatre staff and was not identified preoperatively in any of the patients. Evidence was advanced that in this department infections were acquired in the postoperative ward.

'Ventricular shunt' infections

These are febrile illnesses characterized by splenomegaly and anaemia that develop in children after the insertion of a valve to drain off cerebrospinal fluid in cases of hydrocephalus (Callaghan *et al.* 1961), usually but not always into the cardiac atrium. Coagulase-negative staphylococci—nearly always *Staph. epidermidis* biotype 1 (Holt 1969)—are responsible for about one-half and *Staph. aureus* for one-quarter of the infections; the organism is usually present in the blood and less often in the cerebrospinal fluid (Schoenbaum *et al.* 1975).

This disease differs from post-cardiotomy infections in that a fibrinous vegetation is regularly present on the prosthesis, and the heart valve is unaffected even when the tip of the atrial catheter is in contact with it. Over two-thirds of infections develop within 2 months of the insertion of the valve. Infection can seldom be eradicated by means of antibiotics; if the infected prosthesis is first removed this can usually be done without great difficulty.

Other infections In long-term intravenous cannulation, a fibrinous mass packed with bacteria may appear on the tip of the plastic catheter, and the patient shows signs of septicaemia (Chapter 58). The organisms are often coagulase-negative staphylococci. According to Christensen and co-workers (1982), 63 per cent of strains of *Staph. epidermidis* from catheter-tips form a polysaccharide slime in carbohydrate-containing media, a character seldom seen in strains from other sources. Slime-forming strains formed macrocolonies on catheters *in vitro*. *Staph. epidermidis* is the commonest cause of peritonitis in patients undergoing continuous ambulatory peritoneal dialysis for renal failure, and is responsible for over 40 per cent of infections (Gokal *et al.* 1982). (For a general review of infections caused by coagulase-negative staphylococci and micrococci, see Parker 1981.)

Infections due to anaerobic gram-positive cocci

Anaerobic cocci are in general much less often present in purulent lesions, and in the blood stream, than are anaerobic gram-negative bacilli. They tend to be associated with sepsis of the respiratory and female genital tracts rather than with other abdominal sepsis (Bartlett *et al.* 1974, Sanderson *et al.* 1979). When present, they are seldom in pure culture, and it is therefore often difficult to be sure when they are the primary cause of infection. There is little doubt that they are an important cause of puerperal septicaemia, and that they give rise, either alone or in conjunction with aerobic bacilli, to characteristic forms of wound infection. They are generally sensitive to penicillin and to most other antibiotics used for the treatment of septic infections, but they are so often found in association with gram-negative anaerobes that an agent, such as metronidazole, to which both groups of organisms are sensitive, is generally used.

Puerperal infections

In pre-antibiotic days, the anaerobic cocci were only a little less important than the haemolytic streptococci as a cause of severe or fatal puerperal infections. Colebrook and Hare (1933), for instance, recorded that among 100 positive blood cultures from unselected cases, 60 yielded *Streptococcus pyogenes*, 38 anaerobic 'streptococci' and 2 both types of organism. Although the infections were of a less fulminating type than those due to *Str. pyogenes*, the case-fatality rate from infection with anaerobic cocci was high—about 40 per cent (Colebrook, 1930). Mixed infections were common, two or more types of anaerobic cocci being isolated from the blood of 50 per cent of the cases (Colebrook and Hare 1933). The anaerobic coccus most commonly associated with puerperal infection is now named *Peptostreptococcus anaerobius* (or *putridus*) (Chapter 30).

Wound infections

Anaerobic cocci, usually 'streptococci', are found in a number of local infections of man, and are often associated with putrid suppuration (see Sandusky *et al.* 1942). Anaerobic cocci occur in post-operative wounds (Pulaski *et al.* 1941) and in war wounds, especially the larger wounds (Gardner 1946). In a few of the war wounds infected with anaerobic cocci (MacLennan 1943*a, b*), the muscle becomes massively infected, gangrenous, and gassy. This condition of *anaerobic 'streptococcal' myositis*, is separable from clostridial gas gangrene (Chapter 63) on clinical grounds and by the detection of large numbers of polymorphonuclear leucocytes and gram-positive cocci in chains in the wound exudates. (See also Hayward and Pilcher 1945.) The treatment is radical excision, supported by chemotherapy.

A rare disease known as *bacterial synergistic gangrene* (Meleney *et al.* 1945) usually develops in association with an operation wound, especially of the abdominal wall. It is a slowly spreading and ultimately ulcerating gangrene in which an anaerobic coccus is found in association with an aerobe, which may be *Staph. aureus* or a gram-negative bacillus. The disease can be reproduced experimentally only if the anaerobe and the aerobe are injected into the skin together. It must be distinguished from necrotic fasciitis (page 257), an acute disease in which symptoms of toxaemia are more prominent. The usual treatment is radical excision, but massive chemotherapy and hyperbaric oxygen together may be successful (Clarke 1947, Grainger *et al.* 1967).

Peptococcus magnus has been isolated in pure culture from deep abscesses on a number of occasions (Lambe *et al.* 1974, Taylor *et al.* 1979, Bourgault *et al.* 1980). A peptococcus that does not belong to any of the recognized species but was originally designated *M. abscedens-ovis* (Chapter 30) is said to be responsible for abscesses and lymphadenitis in sheep and goats.

For general accounts of clinical infections due to anaerobic cocci, see Lambe and co-workers (1974) and Finegold (1977).

References

Adlam, C., Pearce, J.H. and Smith, H. (1970) *J. med. Microbiol.* **3**, 147, 157.

Adlam, C., Ward, P.D., McCartney, A.C., Arbuthnott, J.P. and Thorley, C.M. (1977) *Infect. Immun.* **17**, 250.

Agarwal, D.S. (1967a) *Brit. J. exp. Path.* **48**, 436; (1967b) *Ibid.* **48**, 469, 483.

Alder, V.G. and Gillespie, W.A. (1964) *Lancet* **ii**, 1356.

Alder, V.G., Gillespie, W.A. and Herdan, G. (1953) *J. Path. Bact.* **66**, 205.

Almquist, E. (1891) *Z. Hyg. InfektKr.* **10**, 253.

Amoury, R.A., Bowman, F.O. Jr, and Malm, R.M. (1966) *J. thor. cardiovasc. Surg.* **51**, 36.

Anderson, J. D., Forshaw, H.L., Adams, M.A., Gillespie, W.A. and Sellin, M.A. (1976) *J. med. Microbiol.* **9**, 317.

Andrade, J.R. de and Tribe, C.R. (1962) *Brit. med. J.* **i**, 1516.

Anthony, B.F. and Wannamaker, L.W. (1967) *J. exp. Med.* **125**, 319.

Aranow, H. Jr and Wood, W.B. Jr. (1942) *J. Amer. med. Ass.* **119**, 1491.

Arnett, E.N. and Roberts, W.C. (1977) In: *Infections of Prosthetic Heart Valves and Vascular Grafts*, p. 17. Ed. by R.J. Duma. University Park Press, Baltimore.

Asheshov, E.H., Coe, A.W. and Porthouse, A. (1977) *J. med. Microbiol.* **10**, 171.

Asheshov, E.H. and Winkler, K.C. (1966) *Nature, Lond.* **209**, 638.

Ashley, D.J.B. and Brindle, M.J. (1960) *J. clin. Path.* **13**, 336.

Atkins, J.B. and Marks, J. (1952) *Brit. J. indust. Med.* **9**, 296.

Ayliffe, G.A.J., Brightwell, K.M., Ball, P.M. and Derrington, M.M. (1972) *Lancet* **ii**, 479.

Ayliffe, G.A.J. and Collins, B.J. (1967) *J. clin. Path.* **20**, 195.

Ayliffe, G.A.J., Collins, B.J. and Lowbury, E.J. (1967) *J. Hyg., Camb.* **65**, 515.

Azavedo, J. de and Arbuthnott, J.P. (1981) *J. med. Microbiol.* **14**, 341.

Baldwin, J.N., Rheins, M.S., Sylvester, R.F. and Schaffer, T.E. (1957) *Amer. J. Dis. Child.* **94**, 107.

Bänffer, J.R.J. (1962) *Brit. med. J.* **ii**, 1224.

Bänffer, J.R.J. and Franken, J.F. (1967) *Path. et Microbiol., Basel*, **30**, 166.

Barber, M. (1947a) *J. Path. Bact.*, **59**, 373; (1947b) *Brit. med. J.* **ii**, 863.

Barber, M. and Dutton, A.A.C. (1958) *Lancet* **ii**, 64.

Barber, M. and Rozwadowska-Dowzenko, M. (1948) *Lancet* **ii**, 641.

Barber, M., Wilson, B.D.R., Rippon, J.E. and Williams, R.E.O. (1953) *J. Obstet. Gynaec.* **60**, 476.

Barbour, A.G. (1981) *Infect. Immun.* **33**, 442.

Bartlett, J.G., Gorbach, S.L., Thadepalli, H. and Finegold, S.M. (1974) *Lancet* **i**, 338.

Bassett, H.F.M., Ferguson, W.G., Hoffman, E., Walton, M., Blowers, R. and Conn, C.A. (1963) *J. Hyg., Camb.* **61**, 83.

Beavan, D.W. and Burry, A.F. (1956) *Lancet* **ii**, 211.

Benson, P.F., Rankin, G.L.S. and Rippey, J.J. (1962) *Lancet* **i**, 999.

Bergdoll, M.S., Crass, B.A., Reiser, R.F., Robbins, R.N. and Davis, J.P. (1981) *Lancet* **i**, 1017.

Berntsen, C.A. and McDermott, W. (1960) *New Engl. J. Med.* **262**, 637.

Bethune, D.W.R., Blowers, R., Parker, M. and Pask, E.A. (1965) *Lancet* **i**, 480.

Black, T., Lynch, P.F. and Summers, M.M. (1962) *Lancet* **i**, 612.

Blouse, L.E., Lathrop, G.D., Kolonel, L.N. and Brockett, R.M. (1978) *Zbl Bakt.* **A241**, 119.

Blowers, R., Hodgkin, K. and Sklaroff, S. (1967) *Brit. med. J.* **ii**, 642.

Blowers, R., Mason, G.A., Wallace, K.R. and Walton, M. (1955) *Lancet* **ii**, 786.

Bøe, J. (1946) *Acta derm-venereol., Stockh.* **26**, 111.

Bøe, J., Solberg, C.O., Vogelsang, T.M. and Wormnes, A. (1964) *Brit. med. J.* **ii**, 280.

Boris, M. Sellers, T.F., Eichenwald, H.F., Ribble, J.C. and Shinefield, H.R. (1964) *Amer. J. Dis. Child.* **108**, 252.

Boris, M., Shinefield, H.R., Romano, P., McCarthy, D.P. and Florman, A.L. (1968) *Amer. J. Dis. Child.* **115**, 521.

Bourgault, A.-M., Rosenblatt, J.E. and Fitzgerald, R.H. (1980) *Ann. intern. Med.* **93**, 244.

Boyce, J.M., Garner, J.S., Twenge, J.A., Shipley, J.M. and Dixon, R.E. (1976) *Pediatrics, Springfield* **57**, 854.

Brodie, J., Jamieson, W. and Sommerville, T. (1955) *Lancet* **ii**, 223.

Budd, M.A., Boring, J.R. and Brachman, P.S. (1965) *Antimicrob. Agents Chemother. 1964* p. 681.

Bulger, R.J. and Sherris, J.C. (1968) *Ann. intern. Med.* **69**, 1099.

Calia, F.M., Wolinsky, E., Mortimer, E.A. Jr, Abrams, J.S. and Rammelkamp, C.H. Jr (1969) *J. Hyg., Camb.* **67**, 49.

Callaghan, R.P., Cohen, S.J. and Stewart, G.T. (1961) *Brit. med. J.* **i**, 860.

Cawdery, M., Foster, W.D., Hawgood, B.C. and Taylor, C. (1969) *Brit. J. exp. Path.* **50**, 408.

Christensen, G.D., Simpson, W.A., Bisno, A.L. and Beachey, E.H. (1982) *Infect. Immun.* **37**, 318.

Clarke, S.H.C. (1947) *Lancet* **i**, 748.

Cluff, L.E., Reynolds, R.C., Page, D.L. and Breckinridge, J.L. (1968) *Ann. intern. Med.* **69**, 859.

Cohen, M.L. and Falkow, S. (1981) *Science, N. Y.* **211**, 842.

Colebrook, L. (1930) *Brit. med. J.* **ii**, 134.

Colebrook, L. and Hare R. (1933) *J. Obstet. Gynaec.* **40**, 609.

Colleen, S., Hovelius, B., Wieslander, Å. and Mårdh, P.-A. (1979) *Acta path. microbiol. scand.* **B87**, 321.

Conti, C.R., Cluff, L.E. and Scheder, E.P. (1961) *J. exp. Med.* **113**, 845.

Crossley, K., Landesman, B. and Zaske, D. (1979) *J. infect. Dis.* **139**, 280.

Crowder, J.G. and White, A. (1972) *Ann. intern. Med.* **77**, 87.

Dack, G.M. (1956) *Amer. J. Surg.* **92**, 765.

Davies, R.R. and Noble, W.C. (1962) *Lancet* **ii**, 1295; (1963) *Ibid.* **i**, 1111.

Davis, J.P., Chesney, P.J., Wand, P.J. and La Venture, M. (1980) *New Engl. J. Med.* **303**, 1429.

Dearing, W.H. and Heilman, F.R. (1953) *Proc. Mayo. Clin.* **28**, 121.

Demuth, P.J., Gerding, D.N. and Crossley, K. (1979) *Arch. intern. Med.* **139**, 78.

Devenish, E.A. and Miles, A.A. (1939) *Lancet* **i**, 1089.

Disney, M.E., Wolff, J and Wood, B.S.B. (1956) *Lancet* **i**, 767.

Dixon, R.E., Kaslow, R.A., Mallison, G.F. and Bennett, J.V. (1973) *Pediatrics, Springfield* **51**, 413.

Dossett, J.H., Kronvall, G., Williams, R.C. Jr and Quie, P.G. (1969) *J. Immunol.* **103**, 1405.

Drutz, D.J., Way, M.H. van, Shaffner, W. and Koenig, M.G. (1966) *New Engl. J. Med.* **275**, 1161.

Duguid, J.P. and Wallace, A.T. (1948) *Lancet* **ii**, 845.

Duncan, J.T. and Walker, J. (1942) *J. Hyg., Camb.* **42**, 474.

Dunnet, W.N. and Schallibaum, E.M. (1960) *Lancet* **ii**, 1227.

Easmon, C.S.F. (1980) *J. med. Microbiol.* **13**, 495.

Easmon, C.S.F. and Glynn, A.A. (1975) *Immunology* **29**, 75; (1977) *Ibid.* **33**, 767; (1978) *Infect. Immun.* **19**, 341; (1979) *Immunology* **38**, 103.

Ehrenkranz, N.J. (1964) *New Engl. J. Med.*, **271**, 225; (1966) *J. Immunol.* **96**, 509.

Ehrenkranz, N.J., Elliott, D.F. and Zarco, R. (1971) *Infect. Immun.* **3**, 664.

Engman, M.F. (1901) *J. cutan. Dis.* **19**, 180.

Fairlie, C.W. and Kendall, R.E. (1953) *J. Amer. med. Ass.* **153**, 90.

Favero, M.S., McDade, J.J., Robertson, J.A., Hoffman, R.K. and Edwards, R.W. (1968) *J. appl. Bact.* **31**, 336.

Finegold, S.M. (1977) *Anaerobic Bacteria in Human Disease.* Academic Press, New York.

Fisher, S. (1963) *J. infect. Dis.* **113**, 213.

Foggie, A. (1947) *J. comp. Path.* **57**, 245; (1956) *Ibid.* **66**, 278.

Foster, W.D. (1965) *J. Hyg., Camb.* **63**, 517.

Foster, W.D. and Hutt, M.S.R. (1960) *Lancet* **ii**, 1373.

Galbraith, N.S. (1960) *Proc. R. Soc. Med.* **53**, 253.

Gardner, C.E. (1946) *Arch. Surg.* **53**, 387.

Garrod, L.P., Lambert H.P. and O'Grady, F. (1981) *Antibiotic and Chemotherapy*, 5th ed. Livingstone, Edinburgh and London.

Gemmell, C.G., Peterson, P.K., Schmeling, D.J. and Quie, P.G. (1982) *Infect. Immun.* **38**, 975.

Gezon, H.M., Thompson, D.J., Rogers, K.D., Hatch, T.F. and Taylor, P.M. (1964) *New Engl. J. Med.* **270**, 379.

Gladstone, G.P. and Glencross, E.J.G. (1960) *Brit. J. exp. Path.* **41**, 313.

Gokal, R. *et al.* (1982) *Lancet* **ii**, 1388.

Gonzaga, A.J., Mortimer, E.A. Jr, Wolinsky, E. and Rammelkamp, C.H. Jr (1964) *J. Amer. med. Ass.* **189**, 711.

Goshi, K. Cluff, L.E. and Johnson, J.E. (1961*b*) *J. exp. Med.* **113**, 259.

Goshi, K., Cluff, L.E., Johnson, J.E. and Conti, C.R. (1961*a*) *J. exp. Med.* **113**, 249.

Goshi, K., Cluff, L.E. and Norman, P.S. (1963*a*) *Bull. Johns Hopk. Hosp.* **112**, 31.

Goshi, K., Smith, E.W., Cluff, L.E. and Norman, P.S. (1963*b*) *Bull. Johns Hopk. Hosp.* **113**, 183.

Gould, J.C. (1958) *Lancet* **i**, 489.

Gould, J.C. and Cruickshank, J.D. (1957) *Lancet* **ii**, 1157.

Gow, T.L., Sweeney, F.J., Witmer, C.M. and Wise, R.I. (1963) *J. Bact.* **86**, 611.

Graham, D.R., Correa-Villasenor, A., Anderson, R.L., Vollman, J.H. and Baine, W.B. (1980) *J. Pediat.* **97**, 972.

Grainger, R.W., MacKenzie, D.A. and McLachlin, A.D. (1967) *Canad. J. Surg.* **10**, 439.

Gurmohan Singh, Marples, R.R. and Kligman, A.M. (1971) *J. invest. Derm.* **57**, 149.

Gustafson, G.T., Stålenheim, G., Forsgren, A. and Sjöquist, J. (1968) *J. Immunol.* **100**, 530.

Hagen, K.W. (1963) *J. Amer. vet. med. Ass.* **142**, 1421.

Hale, J.H. and Smith, W. (1945) *Brit, J. exp. Path.* **26**, 209.

Haley, R.W. *et al.* (1982) *Ann. intern. Med.* **97**, 297.

Hallander, H.O. and Körlof, B. (1967) *Acta path. microbiol, scand.* **71**, 359.

Hallman, F.A. (1937) *Proc. Soc. exp. Biol., N.Y.* **36**, 789.

Hare, R. and Cooke, E.M. (1961) *Brit, med. J.* **ii**, 333.

Hare, R. and Ridley, M. (1958) *Brit. med. J.* **i**, 69.

Hare, R. and Thomas, C.G.A. (1956) *Brit. med. J.* **ii**, 840.

Harris, R.L., Nunnery, A.W. and Riley, H.D. (1968) *Antimicrob. Agents Chemother.*, 1967, p. 110.

Harrison, G.K. and Cruickshank, D.B. (1952) *Lancet* **i**, 288.

Hayward, N.J. and Pilcher, R. (1945) *Lancet,* **ii**, 560.

Hazen, L.N., Jackson, G.G., Chang, S.-M., Place, E.H. and Finland, M. (1951) *J. Pediat.* **39**, 1.

Hers, J.F.P., Masurel, N. and Mulder, J. (1958) *Lancet* **ii**, 1141.

Hill, H.R. *et al.* (1974*a*) *Lancet* **ii**, 617.

Hill, J., Howell, A. and Blowers, R. (1974*b*) *Lancet* **ii**, 1131.

Hill, M.J. (1968) *J. med. Microbiol.*, **1**, 33; (1969) *Ibid.* **2**, 1.

Hobbs, B.C., Carruthers, H.L. and Gough, J. (1947) *Lancet* **ii**, 572.

Hoeksma, A. and Winkler, K.C. (1963) *Acta leidensia* **32**, 123.

Holt, R.J. (1969) *J. clin. Path.* **22**, 475.

Houck, P.W., Nelson, J.D. and Kay, J.L. (1972) *Amer. J. Dis. Child.* **123**, 45.

Hovelius, B. and Mårdh, P.-A. (1979) *Acta. path. microbiol. scand.* **B87**, 45.

Howells, C.H.L. and Jones, H.E. (1961) *Arch. Dis. Childh.* **36**, 214.

Hurst, V. (1960) *Pediatrics, Springfield* **25**, 11.

Hurst, V. and Grossman, M. (1958) *Calif. Med.* **89**, 107.

Iannini, P.B. and Crossley, K. (1976) *Ann. intern. Med.* **84**, 558.

Isbister, C., Durie, E.B., Rountree, P.M. and Freeman, B.M. (1954) *Med. J. Aust.* **ii**, 897.

Jackson, G.G., Haight, T.H., Kass, E.H., Womack, C.R., Gocke, T.M. and Finland, M. (1951) *Ann. intern. Med.* **35**, 1175.

Janbon, L., Bertrand, J., Roox, J. and Salvaing, J. (1952) *Bull. Acad. nat. Méd.* **136**, 59.

Jellard, J. (1957) *Brit. med. J.* **i**, 925.

Jessen, O. and Bülow, P. (1967) *Acta path. microbiol. scand.*, Suppl. 187, p. 48.

Jessen, O., Rosendal, K., Bülow, P., Faber, V. and Eriksen, K.R. (1969) *New Engl. J. Med.* **281**, 627.

Jevons, M.P., John, M. and Parker, M.T. (1966) *J. clin. Path.* **19**, 305.

Jevons, M.P. and Parker, M.T. (1964) *J. clin. Path.* **17**, 243.

Johanovský, J. (1956) *Čas. Lék. čes.* **95**, 455, 459; (1958*a*) *Nature, Lond.* **182**, 1454; (1958*b*) *Z. ImmunForsch.* **116**, 318.

Johnson, A., Rountree, P.M., Smith, K., Stanley, N.F. and Anderson, K. (1960) *Natl Hlth med. Res. Coun., Canberra, Spec. Rep. Ser.*, No. 10.

Johnson, J.E., Cluff, L.E. and Goshi, K. (1961) *J. exp. Med.* **113**, 235.

Karchmer, A.W. and Swartz, M.N. (1977) In: *Infective Endocarditis*, p. 58. Ed. by E.L. Kaplan and A.V. Taranta. American Heart Association, Dallas.

Kaslow, R.A. *et al.* (1973) *Pediatrics, Springfield* **51**, 418.

Kay, C.R. (1962) *Brit. med. J.* **i**, 1048; (1963) *J. Coll. gen. Pract.* **6**, 47.

Keene, W.R., Minchew, B.H. and Cluff, L.E. (1961) *New Engl. J. Med.* **265**, 1128.

Kellaway, C.H., MacCallum, P. and Terbutt, A.H. (1928) *Report of the Commission to the Government of the Commonwealth of Australia*. Melbourne.

Kellgren, J.H., Ball, J., Fairbrother, R.W. and Barnes, K.L. (1958) *Brit. med. J.* **i**, 1193.

King, K., Brady, L.M. and Harkness, J.L. (1981) *Lancet* **ii**, 698.

Kloos, W.E. and Musselwhite, M.S. (1975) *Appl. Microbiol.* **30**, 381.

Knight, V. and Holzer, A.R. (1954) *J. clin. Invest.* **33**, 1196.

Kryński, S., Galiński, J. and Becla, E. (1976) *Zbl. Bakt,* I Abt., suppl. 5, p. 1003.

Kundsin, R.B. (1966) *Clin. Med.* **73**, 27.

Lack, C.H. (1957) *Proc. R. Soc. Med.,* **50**, 625.

Lambe, D.W., Vroon, D.H. and Rietz, C.W. (1974) In: *Anaerobic Bacteria, Role in Disease*, p. 585. Ed. by A. Balows, R.M. Dehaan, V.R. Dowell and L.B. Guze. Thomas, Springfield.

Lancet (1982) **i**, 87.

Larinkari, U. *et al.* (1983) *J. med. Microbiol.* **16**, 45.

Lee, B.K., Crossley, K. and Gerding, D.N. (1978) *Amer. J. Med.* **65**, 303.

Lewis, H.E., Foster, A.R., Mullan, B.J., Cox, R.N. and Clark, R.P. (1969) *Lancet* **i**, 1273.

Lidwell, O.M., Davies, J., Payne, R.W., Newman, P. and Williams, R.E.O. (1971) *J. Hyg., Camb.* **69**, 113.

Lidwell, O.M. and Lowbury, E.J. (1950) *J. Hyg., Camb.* **48**, 6, 21, 28.

Lidwell, O.M. *et al.* (1966) *J. Hyg., Camb.* **64**, 321; (1970) *Ibid.* **68**, 417.

Light, I.J., Atherton, H.D. and Sutherland, J.M. (1975) *J. infect. Dis.* **131**, 281.

Light, I.J., Walton, R.L., Sutherland, J.M., Shinefield, H.R. and Brackvogel, V. (1967) *Amer. J. Dis. Child.* **113**, 291.

Lindbom, G. (1970) *Acta Univ. Upsal.* No. 87.

Linnemann, C.C.Jr, Mason, M., Moore, P., Korfhagen, T.R. and Staneck, J.L. (1982) *Amer. J. Epidemiol.* **115**, 941.

Live, I. and Nichols, A.C. (1961) *J. infect. Dis.* **108**, 195.

Love, I.G., Gezon, H.M., Thompson, D.J., Rogers, K.D. and Hatch, T.F. (1963) *Pediatrics, Springfield* **32**, 956.

Lowbury, E.J.L. and Fox, J. (1953) *J. Hyg., Camb.* **51**, 203.

Lunsgaard-Hansen, P., Senn, A., Roos, B. and Waller, U. (1960) *J. Amer. med. Ass.* **173**, 1008.

Lyell, A. (1956) *Brit. J. Derm.,* **68**, 355; (1967) *Ibid.* **79**, 662.

Lyell, A., Dick, H.M. and Alexander, J. O'D. (1969) *Lancet* **i**, 787.

McDade, J.J. and Hall, L.B. (1963) *Amer. J. Hyg.* **77**, 98; (1964) *Ibid.* **80**, 183, 192.

McDiarmid, A. (1955) *J. comp. Path.* **65**, 17.

McDonald, S. and Timbury, M.G. (1957) *Lancet* **ii**, 863.

MacLennan, J.D. (1943*a*) *Lancet* **i**, 582; (1943*b*) *Ibid.* **ii**, 63, 94, 123.

McNeill, I.F., Porter, I.A. and Green, C.A. (1961) *Brit. med. J.* **ii**, 798.

Marples, R.R., Hone, R., Notley, C.M., Richardson, J.F. and Crees-Morris, J.A. (1978) *Zbl. Bakt.* **A241**, 140.

Marples, R.R. and Kligman, A.M. (1969) *J. invest. Derm.* **53**, 11; (1971) In: *Epidermal Wound Healing*, p. 241. Ed. by H.I. Maibach and D.T. Rovee. Year Book Medical Publishers, Chicago.

Marples, R.R. and Richardson, J.F. (1980) *J. med. Microbiol.* **13**, 355.

Marraro, R.V. and Mitchell, J.L. (1975) *J. clin. Microbiol.* **12**, 180.

Martin, R.R. and White, A. (1967) *J. Lab. clin. Med.* **70**, 1; (1968) *Ibid.* **71**, 791.

Matthias, J.Q., Shooter, R.A. and Williams, R.E.O. (1957) *Lancet* **i**, 1173.

May, K.R. and Pomeroy, N.P. (1973) In *Airborne Transmission and Airborne Infection*, p. 426. Ed. by J.F.P. Hers and K.C. Winkler. Oosthoek Publishing Co., Utrecht.

Medical Journal of Australia (1982) **i**, 448.

Meleney, F.L., Friedman, S.T. and Harvey, H.D. (1945) *Surgery* **18**, 243.

Melish, M.E. and Glasgow, L.A. (1970) *New Engl. J. Med.* **282**, 1114; (1971) *J. Pediat.* **78**, 958.

Messinger, H.B., Bonestell, A.E., Blum, H.L., Browne, A.S., Dingley, M. and Flett, J. (1963) *Amer. J. Hyg.* **78**, 310.

Meyer, W., Winter, K.-H. and Kienitz, M. (1965) *Dtsch. Gesundheitswes.* **31**, 1429.

Miller, D.L. (1962) *J. Hyg., Camb.* **60**, 467.

Mitchell, N.J. and Gamble, D.R. (1974) *Lancet* **ii**, 1133.

Mitchell, R.G. (1964) *J. clin. Path.* **17**, 105.

Moore, B. and Gardner, A.M.N. (1963) *J. Hyg., Camb.* **61**, 95.

Mortimer, E.A. Jr, Lipsitz, P.J., Wolinsky, E., Gonzaga, A.J. and Rammelkamp, C.H. (1962) *Amer. J. Dis. Child.* **104**, 289.

Mortimer, E.A. Jr, Wolinsky, E., Gonzaga, A.J. and Rammelkamp, C.H. Jr (1966) *Brit med. J.* **ii**, 319.

Mowat, A.G. and Baum, J. (1971) *New Engl. J. Med.* **284**, 621.

Munch-Petersen, E. and Boundy, C. (1962) *Bull. World Hlth Org.* **26**, 241.

Murray, H.W., Tuazon, C.U. and Sheagren, J.N. (1977) *Arch. intern. Med.* **137**, 844.

Nagel, J.G., Tuazon, C.V., Cardella, T.A. and Sheagren, J.N. (1975) *Ann. intern. Med.* **82**, 13.

Namavar, F., Graaff, J. de, With, C. de and MacLaren, D.M. (1978) *J. med. Microbiol.* **11**, 243.

Negro, R.C., Gentile-Ramos, I. and Galiana, J. (1956) *An. Fac. Med. Montevideo* **41**, 263.

Noble, W.C. (1962) *J. clin. Path.,* **15**, 552; (1965) *Brit. J. exp. Path.* **46**, 254.

Noble, W.C. and Davies, R.R. (1965) *J. clin. Path.* **18**, 16.

Noble, W.C., Habbema, J.D.F., Furth, R. van, Smith, I. and Raay, C. de (1976) *J. med. Microbiol.* **9**, 53.

Noble, W.C., Lidwell, O.M. and Kingston, D. (1963) *J. Hyg., Camb.* **61**, 385.

Noble, W.C. and Savin, J.A. (1968) *Brit med. J.* **i**, 417.

Noble, W.C., Valkenburg, H.A. and Wolters, C.H.L. (1967) *J. Hyg., Camb.* **65**, 567.

Noble, W.C., Williams, R.E.O., Jevons, M.P. and Shooter, R.A. (1964) *J. clin. Path.* **17**, 79.

Nolan, C.M. and Beaty, H.N. (1976) *Amer. J. Med.* **60**, 495.

O'Connor, D.T., Weisman, M.H. and Fierer, J. (1978) *Clin. exp. Immunol.* **34**, 179.

Oeding, P., Natås, O.B. and Fleurette, J. (1981) In: *The Staphylococci*, p. 94. Ed. by A. Macdonald and G. Smith. Aberdeen University Press, Aberdeen.

Olin, G. and Lithander, A. (1948) *Acta path. microbiol. scand.* **25**, 52.

Oliver, V.L., Sargent, C.A., Damann, G.L.A. and Albrecht, R.M. (1964) *Amer. J. Hyg.* **79**, 302.

Osebold, J.W. and Gray, D.M. (1960) *J. infect. Dis.* **106**, 91.

Oswald, N.C., Shooter, R.A. and Curwen, M.P. (1958) *Brit. med. J.* **ii**, 1305.

Pagano, J.S., Farrer, S.M., Plotkin, S.A., Brachman, P.S., Fekety, F.R. and Pidcoe, V. (1960) *Science* **131**, 927.

Panton, P.N. and Valentine, F.C.O. (1929) *Brit. J. exp. Path.* **10**, 257.

Parker, M.T. (1958) *J. Hyg., Camb.* **56**, 238; (1981) In: *The Staphylococci*, p. 156. Ed. by A. Macdonald and G. Smith. Aberdeen University Press, Aberdeen; (1983) In: *Staphylococci and Staphylococcal Infections*, vol. 1, p. 33. Ed. by C.S.F. Easmon and C. Adlam. Academic Press, London.

Parker, M.T., Asheshov, E.H., Hewitt, J.H., Nakhla, L.S. and Brock, B.M. (1974) *Ann. N.Y. Acad. Sci.* **236**, 466.

Parker, M.T. and Hewitt, J.H. (1970) *Lancet* **i**, 800.

Parker, M.T., John, M., Emond, R.T.D. and Machacek, K.A. (1965) *Brit. med. J.* **ii**, 1101.

Parker, M.T. and Kennedy, J. (1949) *J. Hyg., Camb.* **47**, 213.

Parker, M.T., Tomlinson, A.J.H. and Williams, R.E.O. (1955) *J. Hyg., Camb.* **53**, 458.

Parker, M.T. and Williams, R.E.O. (1961) *Acta paediat., Stockh.* **50**, 101.

Payne, R.W. (1967) *Brit. med. J.* **iii**, 17.

Penikett, E.K.J., Knox, R. and Liddell, J. (1958) *Brit. med. J.* **i**, 812.

Poole, P.M. and Baker, J.R. (1966) *Mon. Bull. Minist. Hlth Lab. Serv.* **25**, 116.

Price, E.H., Brain, A. and Dickson, J.A.S. (1980) *J. Hosp. Infect.* **1**, 221.

Pulaski, E.J., Meleney, F.L. and Spaeth, W.L.C. (1941) *Surg. Gynec. Obstet.* **72**, 982.

Quie, P.G., Hill, H.R. and Davis, A.T. (1974) *Ann. N.Y. Acad. Sci.* **236**, 233.

Quie, P.G., Verhoef, J., Kim, Y. and Peterson, P.K. (1981) In: *The Staphylococci*, p. 83. Ed. by A. Macdonald and G. Smith. Aberdeen University Press, Aberdeen.

Quie, P.G., White, J.G., Holmes, G. and Good, R.A. (1967) *J. clin. Invest.* **46**, 668.

Quinn, E.L., Cox, F. and Drake, E.H. (1966) *J. Amer. med., Ass.* **196**, 815.

Rahal, J.J. Jr, MacMahon, H.E. and Weinstein, L. (1968) *Ann. intern. Med.* **69**, 35.

Ravenholt, R.T., Wright, P. and Mulhern, M. (1957) *New Engl. J. Med.* **257**, 789.

Reingold, R.L., Dan, B.B., Shands, K.N. and Broome, C.V. (1982) *Lancet* **i**, 1.

Report (1958) *Brit. med. J.* **i**, 915; (1966) *Ibid.* **i**, 313: (1982) *Ann. intern. Med.* **96**, 831.

Ridley, M. (1959) *Brit. med. J.* **i**, 270.

Rische, H., Witte, W. and Hummel, R. (1980) *Z. ges. Hyg.* **26**, 729.

Rittershain, G., Ritter von (1878) *ZentZeit. Kinderheilkde* **2**, 3.

Roodyn, L. (1954) *Brit. med. J.* **ii**, 1322; (1960) *J. Hyg., Camb.* **58**, 1, 11.

Rosendal, K. and Bülow, P. (1971) *Acta path. microbiol. scand.* **B 79**, 377.

Rosendal, K., Bülow, P., Bentzon, M.W. and Eriksen, K.R. (1976) *Acta path. microbiol. scand.* **B85**, 143.

Rosendal, K., Jessen, O., Benzon, M.W. and Bülow, P. (1977) *Acta path. microbiol. scand.* **B85**, 143.

Rountree, P.M. (1978) *Med. J. Aust.* **ii**, 543.

Rountree, P.M. and Barbour, R.G.H. (1950) *Med. J. Aust.* **i**, 525.

Rountree, P.M., Beard, M.A., Arter, W. and Woolcock, A.J. (1967) *Med. J. Aust.* **i**, 967.

Rountree, P.M. and Freeman, B.M. (1955) *Med. J. Aust.* **ii**, 157.

Russell, R.J., Wilkinson, P.C., McInroy, R.J., McKay, S., McCartney, A.C. and Arbuthnott, J.P. (1975) *J. med. Microbiol.* **8**, 433.

Sanderson, P.J., Wren, M.W.D. and Baldwin, A.W.F. (1979) *J. clin. Path.* **32**, 143.

Sandusky, W.R., Pulaski, E.J., Johnson, B.A. and Meleney, F.L. (1942) *Surg. Gynec. Obstet.* **75**, 145.

Saxe, M.J. de, Wieneke, A., Azavedo, J. de and Arbuthnott, J.P. (1982) *Brit med. J.* **284**, 1641.

Schlievert, P.M. (1982) *Infect. Immun.* **36**, 123.

Schlievert, P.M., Schoettle, D.J. and Watson, D.W. (1979) *Infect. Immun.* **23**, 609.

Schlievert, P.M., Shands, K.N., Dan, B.B., Schmid, G.P. and Nishimura, R.D. (1981) *J. infect. Dis.* **143**, 509.

Schoenbaum, S.C., Gardner, P. and Shillito, J. (1975) *J. infect. Dis.* **131**, 543.

Schutzer, S.E., Fischetti V.A. and Zabriskie, J.B. (1983) *Science* **220**, 316.

Scopes, J.W., Eykyn, S. and Phillips, I. (1974) *Lancet* **ii**, 1392.

Šebek, V., Schubert, J. and Johanovský, J. (1959) *Čas. Lék čes.* **98**, 1181.

Selwyn, S. (1965) *J. Hyg., Camb.* **63**, 59.

Shands, K.N. *et al.* (1980) *New Engl. J. Med.* **303**, 1436.

Shanson, D.C. (1981) *J. Hosp. Infect.* **2**, 11.

Shanson, D.C. and McSwiggan, D.A. (1980) *J. Hosp. Infect.* **1**, 171.

Shepherd, J.J. (1983) *Lancet* **ii**, 1240.

Shinefield, H.R., Ribble, J.C., Boris, M. and Eichenwald, H.F. (1963) *Amer. J. Dis. Child.* **105**, 646 (and 5 papers on succeeding pages).

Shinefield, H.R. Wilsey, J.D., Ribble, J.C., Boris, M., Eichenwald, H.F. and Dittmar, C.I. (1966) *Amer. J. Dis. Child.* **111**, 11.

Shooter, R.A., Griffiths, J.D., Cook, J. and Williams, R.E.O. (1957) *Brit. med. J.* **i**, 433.

Shooter, R.A., Taylor, G.W., Ellis, G. and Ross, J.P. (1956) *Surg. Gynec. Obstet.* **103**, 257.

Shooter, R.A. *et al.* (1958) *Brit. med. J.* **i**, 607.

Siboni, K. and Poulsen, E.D. (1968) *Danish med. Bull.* **15**, 161.

Simberkoff, M.S. (1977) In: *Infective Endocarditis*, p. 46. Ed. by E.L. Kaplan and A.V. Taranta. American Heart Association, Dallas.

Simpson, J. (1953) *Med. Offr* **89**, 85.

Sklaver, A.R., Hoffman, T.A. and Greenman, R.L. (1978) *Sth Med. J.* **71**, 638.

Slot, W.J.B. (1950) *Ned. Tidschr. Geneesk.* **94**, 3438.

Smith, E.W., Goshi, K., Norman, P.S. and Cluff, L.E. (1963) *Bull. Johns Hopk. Hosp.* **113**, 247.

Smith, H.W. (1959) In: *Infectious Diseases of Animals*, vol. 2, p. 557. Ed. by A.W. Stableforth and I.A. Galloway. Butterworth, London.

Smith, I.M. and Vickers, A.B. (1960) *Lancet* **i**, 1318.

Smith, J.A., O'Connor, J.J. and Willis, A.T. (1966) *Lancet* **ii**, 776.

Smith, P.G., Pike, M.C., Taylor, E. and Taylor, J.F. (1978) *Trans. R. Soc. trop. Med. Hyg.* **72,** 46.

Solberg, C.O. (1965) *Acta med. scand.* **178,** Suppl. 436,

Sood, U., Agarwal, D.S. and Aurora, A.L. (1975) *Indian. J. med. Res.* **63,** 1564.

Sparham, P.D., Lobban, D.I. and Speller, D.C.E. (1978) *J. clin. Path.* **31,** 913.

Speers, R. Jr, Shooter, R.A., Gaya, H., Patel, N. and Hewitt, J.H. (1969) *Lancet* **ii,** 233.

Stevens, F.A. (1927) *J. Amer. med. Ass.* **88,** 1956.

Strauss, W.G., Maibach, H.I. and Shinefield, H.R. (1969) *J. Amer. med. Ass.* **208,** 861.

Taubler, J.H. (1968) *J. Immunol.* **101,** 546.

Taubler, J.H., Kapral, F.A. and Mudd, S. (1963) *J. Bact.* **86,** 51.

Taubler, J.H. and Mudd, S. (1968) *J. Immunol.* **101,** 550.

Taylor, A.G., Cook, J., Fincham, W.J. and Millard, F.J.C. (1975) *J. clin. Path.* **28,** 284.

Taylor, A.G., Fincham, W.J. and Cook, J. (1976a) *Zbl Bakt.* I Abt, suppl. 5, 911.

Taylor, A.G., Fincham, W.J., Golding, M.A. and Cook, J. (1979) *J. clin. Path.* **32,** 61.

Taylor, A.G. and Plommet, M. (1973) *J. clin. Path.* **26,** 409.

Taylor, J.F., Fluck, D. and Fluck, D. (1976b) *J. clin. Path.* **29,** 1081.

Terplan, K., Paine, J.R., Sheffer, J., Egan,R. and Lansky, H. (1953) *Gastroenterology* **24,** 476.

Thom, B.T. and White, R.G. (1962) *J. clin. Path.* **15,** 559.

Thompson, D.J., Gezon, H.M., Rogers, K.D., Yee, R.B. and Hatch, T.F. (1966) *Amer. J. Epidem.* **84,** 314.

Tierno, P.M., Hanna, B.A. and Davies, M.B. (1983) *Lancet,* **i,** 615.

Timbury, M.C., Wilson, T.S., Hutchinson, J.G.P. and Govan, A.D.T. (1958) *Lancet* **ii,** 1081.

Tinsdale, W.A., Fenster, L.F. and Klatskin, G. (1960) *New Engl. J. Med.* **263,** 1014.

Todd, J., Fishaut, M., Kapral, F. and Welch, T. (1978) *Lancet* **ii,** 1116.

Towers, A.G. (1961) *J. clin. Path.* **14,** 161.

Towers, A.G. and Gladstone, G.P. (1958) *Lancet* **ii,** 1192.

Tuazon, C.U., Cardella, T.A. and Sheagren, J.N. (1975) *Arch. intern. Med.* **135,** 1555.

Tuazon, C.U., Sheagren, J.N., Choa, M.S., Marcus, D. and Curtin, J.A. (1978) *J. infect. Dis.* **137,** 57.

Tulloch, L.G. (1954) *Brit. med. J.* **ii** 912.

Valentine, F.C.O. and Hall-Smith, S.P. (1952) *Lancet* **ii** 351.

Verbrugh, H.A., Peters, R., Rozenberg-Arska, M., Peterson, P.K. and Verhoef, J. (1981) *J. infect. Dis.* **144,** 1.

Vijver, J.C.M. van der, Es-Boon, M.M. van and Michel, M.F. (1975) *J. med. Microbiol.* **8,** 265, 279.

Walter, C.W., Kundsin, R.B. and Brubaker, M.M. (1963) *J. Amer. med. Ass.* **186,** 908.

Watson, W.A. (1964a) *Vet. Rec.* **76,** 743; (1964b) *Ibid.* **76,** 793; (1965) *Ibid.* **77,** 477.

Weiss, W. and Flippin, H. (1963) *Amer. J. med. Sci.* **245,** 440.

Weksler, B.B. and Hill, M.J. (1969) *J. Bact.* **98,** 1030.

White, A. (1961) *J. clin. Invest.* **40,** 23.

White, A., Smith, J. and Varga, D.T. (1964) *Arch. intern. Med.* **114,** 651.

Wickman, K. (1970) *Acta path. microbiol. scand.* **B 78,** 15.

Williams, J.D., Waltho, C.A., Ayliffe, G.A.J. and Lowbury, E.J.L. (1967) *Lancet* **ii,** 390.

Williams, R.E.O. (1946) *J. Path. Bact.* **58,** 259; (1961) *Lancet* **ii,** 173; (1963) *Bact. Rev.* **27,** 56; (1966) *Ibid.* **30,** 660; (1971) In: *Proc. int. Conf. on Nosocomial Infections,* p. 1. Ed. by P.S. Brachman and T.C. Eickhoff. American Hospital. Association, Chicago.

Williams, R.E.O., Blowers, R., Garrod, L.P. and Shooter, R.A. (1966) *Hospital Infection: Causes and Prevention,* 2nd ed. Lloyd-Luke, London.

Williams, R.E.O. and Jevons, M.P. (1961) *Zbl. Bakt* **181,** 349.

Williams, R.E.O. and Miles, A.A. (1949) *Spec. Rep. Ser. med. Res. Coun.* No. 266, London.

Williams, R.E.O. *et al.* (1959) *Brit med. J.* **ii,** 658.

Wilson, G.S. (1967) *The Hazards of Immunization,* pp. 80, 84. Athlone Press, London.

Wolinsky, E., Lipstitz, P.J., Mortimer, E.A. and Rammelkamp, C.H. Jr (1960) *Lancet* **ii,** 620.

Wysham, D.N., Mulhern, M.E., Navarre, G.C., La Veck, G.D., Kennan, A.L. and Giedt, W.R. (1957) *New Engl. J. Med.* **257,** 295.

61

Septic infections due to gram-negative aerobic bacilli

M. T. Parker

Introductory

In this chapter we shall consider diseases caused by gram-negative aerobic bacilli other than those dealt with fully elsewhere in this volume: *Actinobacillus* (Chapter 49), *Pasteurella*, *Francisella* and *Yersinia* (Chapter 55), *Brucella* (Chapter 56), *Haemophilus* (Chapter 67), *Salmonella* (Chapters 58, 68 and 72), and *Legionella* (Chapter 73). The general pattern of septic infections caused by gram-negative aerobic bacilli in hospital patients was outlined in Chapter 58.

The pseudomonads

Several of the organisms now brought together in the genus *Pseudomonas* (Chapter 31) have for many years been recognized as important pathogens of man and other animals. The genus also includes a number of plant pathogens, at least two of which may also cause disease in man. In this chapter we shall consider separately the diseases caused by each pathogen, having first drawn attention to certain resemblances and differences between them.

Ps. aeruginosa causes a wide variety of septic infections in man and other vertebrates, and is also pathogenic for some insects and plants. Transmission of infection from one animal species to another has rarely been observed. The organism is world-wide in distribution and multiplies freely in the inanimate environment. It causes local infection of the skin and wounds, and abscesses in the internal organs, and may give rise to acute septicaemic infections. Serious disease in man is rare unless resistance to infection is lowered or the organism is introduced directly into the tissues or into certain normally sterile areas of the body.

Ps. pseudomallei (Whitmore's bacillus) causes melioidosis in man and certain animals but, as with *Ps. aeruginosa*, the disease appears to occur independently in each animal species. Melioidosis resembles glanders, and is characterized by the appearance of multiple abscesses in the internal organs, including the lungs. It is confined to certain areas of the tropics, where Whitmore's bacillus is a common inhabitant of the soil. Serious infection may occur in previously healthy persons.

Ps. mallei causes glanders, a disease of equine animals, from which it may on occasion be transmitted to man or other mammals. It is an obligate animal parasite and was formerly world-wide in distribution but is now found in some countries only. It causes abscesses in lympatic glands and internal organs, including the lungs. Healthy persons who come into contact with infected animals may suffer from serious disease.

A few other pseudomonads are pathogenic. Members of the *cepacia* group are plant pathogens and common inhabitants of the soil. In man they may give rise to superficial skin infections and are also a cause of serious disease when introduced directly into the tissues, or in immunosuppressed patients. The fluorescent pseudomonads other than *Ps. aeruginosa* are soil and water organisms and include a number of strains that are pathogenic for plants. They rarely cause disease in man unless injected intravenously in large doses, for example in a contaminated blood product. A number of other pseudomonads that live in water, soil and other moist situations may from time to time be found on mucous membranes or in open lesions in patients; occasionally they cause bacteraemic infections in highly susceptible persons.

Pseudomonas aeruginosa

By the middle of the nineteenth century it had been recognized that the agent responsible for 'blue pus' was transmissible and it was suspected that it might be a living agent. Sédillot (1850) showed that a blue-green discoloration developed on dressings contaminated with pus and then removed from the wound, and Fordos (1860) extracted pyocyanin from the dressings and obtained it in crystalline form (for early references see Bulloch 1929). Between 1870 and 1890, *Ps. aeruginosa* had been obtained in pure culture and characterized (see Chapter 31); by the end of the century it had been recognized as the cause of a highly fatal septicaemic disease in man (Williams and Cameron 1896, Hitchmann and Kreibich 1897, Brill and Libman 1899). During the next forty years little attention was paid to the organism. The reawakened interest in it since 1950 or so appears not to be due to a change in the behaviour of the organism, but to changed circumstances in hospitals.

Diseases in man

Infections in healthy persons are nearly always trivial and superficial, and are usually associated with warm, moist environments. Chronic paronychia, sometimes with staining of the nail with pyocyanin ('green nail syndrome'), tends to affect domestic workers; and follicular pustules may develop in bathers in heated water contaminated with *Ps. aeruginosa* (Jacobson *et al.* 1976). In the tropics, a soggy dermatosis of the toe webs is common, and in severe cases may be associated with general swelling of the feet ('immersion foot'). When the human skin is kept in a superhydrated state by means of an impermeable dressing, a vesiculopapular rash develops only when *Ps. aeruginosa* multiplies under the dressing (Hojyo-Tomoka *et al.* 1973), but the organisms do not penetrate the skin.

Acute otitis externa (Hall *et al.* 1968) is often caused by *Ps. aeruginosa*. It may occur in epidemics associated with swimming-pools (Reid and Porter 1981), and among 'saturation' divers (Alcock 1977), who appear to become infected in their pressurized and humid living quarters. Necrotizing external otitis (Evans and Richards 1973, Doroghazi *et al.* 1981) is an invasive disease of the tissues around the ear that occurs in patients with poorly controlled diabetes mellitus. Puncture wounds of the foot may occasionally be followed by osteomyelitis of a neighbouring bone (MacKinnon 1975).

Serious infections occur in the following circumstances: (1) the patient has a lowered resistance to infection; (2) the organism is introduced directly into the body, thus by-passing the natural defence mechanisms; and (3) the subject is a newborn or especially a premature infant. Such patients are usually in hospital and opportunities for infection with *Ps. aeruginosa* are greater here than elsewhere.

Lowered resistance to infection with *Ps. aeruginosa* is seen in extreme form in patients receiving drug treatment for cancer or leukaemia, or suffering from certain diseases of the reticuloendothelial system. In leukaemia, susceptibility to *Ps. aeruginosa* is determined by the degree of absolute granulocytopaenia (Sickles *et al.* 1973). Lesser degrees of lowered resistance may occur in other diseases, but these cannot easily be measured; and it is often difficult to disentangle their effects from those of other adverse circumstances in hospital, or indeed in severely ill patients to separate the consequences of infection from the effects of underlying disease.

In many patients, *Ps. aeruginosa* enters the body through a wound; this usually leads only to the local formation of pus, but the blood stream may be invaded in predisposed patients. Local infection of burns often leads to the rejection of skin grafts (Jackson *et al.* 1951); in extensive burns, septicaemia is very common and is the main cause of death in patients who have survived for several days after the accident (Markley *et al.* 1957, Rabin *et al.* 1961).

The virulence of individual strains of *Ps. aeruginosa* for the experimentally burned rat (Chapter 31) is said to be directly proportional to their ability to resist killing by normal rat phagocytes (McEuen *et al.* 1976a). The increased susceptibility of the burned animal to invasion is attributable to the decreased ability of its phagocytes to kill the pseudomonad; burning leads to depletion of the store of granulocytes in the bone marrow (Eurenius and Brouse 1973) and to the disappearance from the serum of a heat-labile factor or factors necessary for the bactericidal action of the blood granulocytes on the organisms (Alexander and Wixson 1970, McEuen *et al.* 1976b). When the body is invaded by *Ps. aeruginosa*, granulocyte production is further depressed (Newsome and Eurenius 1973). According to Montie and his colleagues (1982) non-flagellate variants have a reduced virulence for the burned mouse.

Serious infections of the eye (Stanley 1947, Spencer 1953, Forkner 1960) which, once established, almost invariably result in the loss of sight, may follow quite small injuries if a contaminated fluid is instilled into a wound or the conjunctival sac. They have been recorded not only after operations on the eye (see Ayliffe *et al.* 1966) but also after minor abrasions of the cornea, and even in the absence of known trauma, for example, when contaminated fluid is used for the storage of contact lenses (Golden *et al.* 1971). Serious eye lesions may also develop in the absence of obvious trauma in newborn infants infected by the respiratory route (Drewett *et al.* 1972). Meningitis has frequently followed the introduction of the organism at lumbar puncture or operation on the brain or spinal cord (Forkner 1960).

It is, however, not always necessary for the organism to be implanted directly into tissue; introduction into a normally sterile body cavity, such as the bladder or the lower respiratory tract, often causes sepsis. Infection of the urinary tract follows catheterization, cystoscopy or operations on the bladder (see Chapter 58). The organism is often introduced into the bronchi in the course of mechanically assisted respiration or anaesthesia, and may subsequently be isolated from the sputum or the tracheostomy wound.

Neonatal infections

The particular susceptibility of the newborn to systemic infections with *Ps. aeruginosa* has long been recognized, as has the frequent presence of purulent thrombophlebitis of the umbilical vessels (Wassermann 1901) in fatal cases.

Many authors believe, however, that neonatal pseudomonas infection has become more common since 1950 and attribute this to the maintenance of high humidity in nurseries for the newborn, to the frequent contamination of various articles of equipment, such as resuscitators, to the use of mechanically assisted ventilation for the treatment of respiratory distress, or an increase in the surgical treatment of severe congenital defects (Hoffman and Finberg 1955, Asay and Koch 1960, Cooper 1967, Barson 1971). Most nurseries for the newborn remain free from *Ps. aeruginosa* for long periods of time. When infection is introduced it often gives rise to a carrier epidemic in which many infants are colonized on the skin—including the umbilical stump—and in the gut; many of them remain well, but a few may suffer from septicaemia or meningitis or both (Neter and Weintraub 1955, Bassett *et al.* 1965). *Noma neonatorum* is a highly fatal gangrenous disease of the perioral and perianal tissues in premature infants in which *Ps. aeruginosa* is nearly always present in the local lesions and usually also in the blood (Ghosal *et al.* 1978); but, for the role of non-sporing anaerobes in this disease, see Chapter 62. Epidemics of otitis media of the newborn (Kwantes 1960, Victorin 1967) have been attributed to infection from contaminated bath water.

Systemic infections

In adult hospital patients these are seen mainly in elderly persons with serious underlying illnesses, notably malignancy (Flick and Cluff 1976, Baltch and Griffin 1977); younger patients are often under treatment for leukaemia. Systemic infections are often preceded by colonization of the respiratory tract with the causative strain. *Ps. aeruginosa* is one of the less common causes of endocarditis, but considerable numbers of cases now occur in association with operations on the heart and the intravenous self-administration of drugs (Archer *et al.* 1974).

In very serious bacteraemic infections, particularly those that follow the infection of large burns, there may be leucopaenia (Brill and Libman 1899) and terminal hypothermia. Characteristic skin lesions may appear ('ecthyma gangrenosum'; Hitchmann and Kreibich 1897), beginning as red macules or vesicles, which become bullous, then black and haemorrhagic, and finally ulcerate. In more chronic cases, nodular subcutaneous swellings appear (Markley *et al.* 1957). Ecthyma gangrenosum is a local haemorrhagic necrosis with little cellular infiltration. Fraenkel (1906) showed that similar lesions occur throughout the body, and that these result from focal colonization of the walls of small vessels with bacteria, leading to vasculitis and local necrosis (see also Teplitz 1965). This has been attributed to the action of bacterial protease on the elastin of the vessel wall (Chapter 31). In recent years, however, greater attention has been given to the role of lethal toxins than of focal embolic lesions as a cause of death in *Ps. aeruginosa* infection, and particularly to that of the exotoxin A of Liu (Chapter 31). This toxin is formed, for example, in the experimentally infected burn wound and can subsequently be detected in the blood; its specific action on the tissues of visceral organs can be demonstrated in experimental animals (Saelinger *et al.* 1977). In bacteraemic infections in man, survival is associated with infection by strains that do not form exotoxin A, and with low concentrations of this toxin in the blood (Cross *et al.* 1980).

Infection of the respiratory tract

Colonization of the respiratory tract frequently occurs in hospital patients, especially those with serious preexisting disease who have received antibiotics and mechanically assisted ventilation. Some of the patients show signs of pulmonary consolidation, but it is often difficult to attribute these definitely to *Ps. aeruginosa* unless bacteraemia is present or frank suppuration is detected *post mortem* (Iannini *et al.* 1974). We noted (Chapter 58) that the histological appearances in the lungs of patients with 'gram-negative necrotizing pneumonia' attributed by some workers to *Ps. aeruginosa* and to various other gram-negative aerobes, are said to resemble those described by Fraenkel (1906) in various organs in systemic pseudomonas infection. Johnson and co-workers (1982) suggest that the elastase of *Ps. aeruginosa* aids invasion of the lung by destroying a proteinase inhibitor present in bronchial mucus which normally protects lung tissue against dissolution by the proteolytic enzymes of bacteria and of leucocytes.

Persistent colonization of the respiratory tract with *Ps. aeruginosa* is common in patients with bronchitis and bronchiectasis who are treated with antibiotics. Antibiotic-destroying enzymes formed by the pseudomonad may prevent the elimination of a primary pathogen in the respiratory tract (Maddocks and May 1969). It should be noted that *Ps. aeruginosa* strains from such infections are seldom mucoid (see below), and that serum precipitins against the organism are seldom formed (Burns 1973).

Infection in cystic fibrosis Since children with cystic fibrosis of the pancreas have been kept alive by antibiotic treatment, an increasingly large proportion of them have acquired *Ps. aeruginosa* in the respiratory tract (Garrard *et al.* 1951). These patients have viscous sputum and suffer from recurrent bronchial infections due to a variety of organisms, including *Staphylococcus aureus*, pneumococci and *Haemophilus influenzae* (Burns and May 1968); later they become colonized with *Ps. aeruginosa*. Over three-quarters of the *Ps. aeruginosa* strains isolated from patients with cystic fibrosis are of the mucoid colonial form (Doggett *et al.* 1964, 1966, Høiby 1975). Patients are first colonized by non-mucoid strains but these later become mucoid or are replaced by mucoid strains (Doggett *et al.* 1971, Høiby 1974). Individual centres for the treatment of cystic fibrosis tend to have a predominant endemic mucoid strain (Høiby and Rosendal 1980) which appears to be acquired by the patient from contaminated equipment in the centre rather than by person-to-person spread (Kelly *et al.* 1982, Zimakoff *et al.* 1983). Chronic colonization with a mucoid strain that develops before puberty is usually associated with progressive deterioration of pulmonary function (Høiby *et al.* 1977).

According to Doggett and his colleagues (1971) the mucoid material from cystic-fibrosis strains differs from that of other mucoid strains in containing guluronic as well as mannuronic acid. Mucoid strains are difficult to type; they are often polyagglutinable by O-typing sera, and the slime prevents penetration by phages and bacteriocines. Mucoid strains resist phagocytosis (Schwarzmann and Boring 1971).

Persistent colonization of the respiratory tract by mucoid strains is associated with the presence of high titres of antibody against *Ps. aeruginosa*. The number of distinct lines of precipitation in gel against a single strain of the organism increases as the disease progresses, and a high total number indicates a poor prognosis (Høiby 1977). IgA titres in sputum sometimes exceed those in serum (Schiøtz *et al*. 1979). There is no evidence that antibody plays any part in eliminating the organism. In advanced infections the lymphocytes develop a specific unresponsiveness to stimulation by *Ps. aeruginosa* (Sorenson *et al*. 1977).

Diseases in animals

There are numerous accounts of infections in domestic animals (for a full review see Lusis and Soltys 1971). These include septicaemia, pneumonia, abscesses of the lungs, liver and other organs, mastitis, infections of the reproductive tract, and various local abscesses. Disease has been recorded in cattle, buffalo, sheep, goats, horses, pigs, dogs, cats, mink, chinchillas, laboratory mice, guinea-pigs and poultry. Mastitis has been described in cattle, sheep and goats (Chapter 57). Outbreaks of septicaemia and pneumonia occur on mink farms. Spontaneous septicaemia is common among irradiated mice. *Ps. aeruginosa* is said to cause disease in grasshoppers, but in general it appears to affect mainly domesticated or captive animals.

Diagnosis

The isolation and identification of *Ps. aeruginosa* from lesions seldom presents much difficulty. If the colonial appearances are typical and pyocyanin is formed, the organism can be recognized immediately and confirmatory tests are hardly necessary; but only about 80 per cent of strains are of colonial type 1, and pigment production may not be apparent on the general-purpose media used for primary culture. It is therefore necessary to be familiar with the other colonial forms of the organism and to subculture all suspected colonies to a medium on which pyocyanin production is optimal (Chapter 31). Some 90 per cent of all cultures will then produce the characteristic pigment.

Further tests will be required for the identification of unpigmented strains. Oxidation of glucose, a positive oxidase test, and the production of ammonia from arginine will establish a presumption that the organism is a fluorescent pseudomonad, and this can be confirmed in most cases by demonstrating that it produces a fluorescent yellow-green pigment (Lowbury *et al*. 1962). Final identification may be made by demonstrating growth in three successive subcultures at 42°, but this may take too long. Alternative tests applicable to organisms known to be fluorescent pseudomonads are oxidation and slime production in gluconate at 37°, production of red colonies on tetrazolium agar at

37°, caseinolysis (Brown and Foster 1970) and collagenase production. However, some non-pigmented strains that grow at 42° but give a negative gluconate test are difficult to classify (see, for example, Oberhofer 1981). In our experience, lysis by *Ps. aeruginosa* phages, production of pyocines active on strains of the same species, and, with rare exceptions, agglutinability by O antisera of the Habs series, are rapid and reliable means of identification. (See also Chapter 31.)

With heavily contaminated material the use of selective and indicator media may be necessary. Many of these include cetrimide as a selective agent (Lowbury and Collins 1955). Brown and Lowbury (1965) used cetrimide in a medium designed to favour the production of the fluorescent pigment, incubated the plates strictly at 37° and then examined them under ultraviolet light. Under these conditions, most of the fluorescent colonies were of *Ps. aeruginosa*. The addition of nalidixic acid to this medium (Tinne *et al*. 1967) reduced the number of other colonies appearing on the plates. Alternatively, nitrofurantoin may be added to cetrimide agar (Thom *et al*. 1971), or an agar medium containing novobiocin, penicillin and cyclohexamide may be used (Sands and Rovira 1970). Enrichment media include an asparagine, ethyl alcohol, salts solution (Drake 1966) and cetrimide broth (Shooter *et al*. 1966). Infection in burns may be recognized immediately by observing fluorescence when the lesion is inspected under ultraviolet light (Polk *et al*. 1969). Growth in blood culture is significantly delayed if air is excluded (Knepper and Anthony 1973). (For methods of enumerating *Ps. aeruginosa* in samples from inanimate sources, see Carson *et al*. 1975.)

Typing methods

Ps. aeruginosa may be typed serologically or by means of pyocines or phages (Chapter 31). Typing by O agglutination or by one of the 'active' pyocine-typing methods (see Chapter 20) has proved satisfactory as a routine method of identifying large prevalences of individual strains in hospitals but not for detailed studies of sporadic infections, for which there are usually many possible sources to be excluded.

When used alone, the O-serological and pyocine methods lack discrimination because a few types are very prevalent (Wahba 1965*a*, Govan and Gillies 1969). The results of O typing are highly reproducible; those of the 'active' pyocine-typing methods are somewhat less so, and variations of a single reaction are not uncommon. Some of these would result in the allocation of different isolations of the same strain to different 'types'. If single-reaction differences are ignored, discrimination is considerably reduced. In the various phage-typing methods very many different patterns of lysis can be recognized, but reproducibility is often poor, particularly with some strains. If pattern differences of up to two 'strong reactions' are ignored, fairly reliable results are obtained, but discrimination is low. In our opinion, phage typing should not be used alone in epidemiological investigations.

'Passive' or 'reversed' pyocine-typing methods have been less widely used but have certain advantages. High-titre preparations of pyocines can be prepared by induction with mitomycin C, freed from phage by ultraviolet irradiation and used as standard reagents (Rampling *et al.* 1975). 'Reversed' typing is useful for characterizing mucoid strains (Williams and Govan 1973), which are often untypable by other methods.

Combined O and H serotyping (Pitt 1981) is both reproducible and discriminatory. Alternatives include various combined systems in which O serology is used to make primary divisions and pyocine typing (Wahba 1965*a*, Csiszár and Lányi 1970) or phage typing (Meitert and Meitert 1966, Asheshov 1974) to make further subdivisions in the more common types. Since the number of strains to be compared by the secondary typing method is thus greatly reduced, it is possible to make allowances for known levels of non-reproducibility and still obtain an acceptable level of discrimination. (For reviews of typing methods, see Asheshov 1974, Bergan and Norris 1978, Brokopp and Farmer 1979 and Pitt 1980.)

Epidemiology

Distribution

Although *Ps. aeruginosa* can survive for long periods of time in moist situations, and is regularly present in sewage (Ringen and Drake 1952), there is doubt whether it is truly a free-living organism. It is generally believed to be absent from natural waters that are not subject to faecal pollution, but it may occasionally be found in surface waters that do not yield *Escherichia coli* (Reitler and Seligmann 1957). Its presence in soil is variously reported to be rare (Ringen and Drake 1952, Hoadley *et al.* 1968) or very common (Green *et al.* 1975). It is said to multiply on vegetation in horticultural establishments when the humidity is high (Green *et al.* 1975). Faecal carriage by wild animals is rare (Mushin and Ashburner 1964, Hoadley *et al.* 1968); in domestic and captive animals it is variable in frequency, being common among calves on some farms and in laboratory animals in some breeding establishments (Matthews and Fitzsimmons 1964, Mushin and Ashburner 1964, Hoadley and McCoy 1968).

Estimates of the frequency of *faecal carriage in man* vary widely. In healthy persons not associated with hospitals, it appears to be between 6 and 12 per cent (Shooter *et al.* 1966, Sutter *et al.* 1967, Hoadley and McCoy 1968, Stoodley and Thom 1970); most persons excrete less than 10 000 per g of faeces. Higher carriage rates have been observed in the stools of hospital out-patients with diarrhoea (21 per cent; Stoodley and

Thom 1970) and in patients very soon after admission to hospital (20 per cent; Shooter *et al.* 1969). The rate tends to increase during stay in hospital; serial examinations show that nearly one-third of patients excrete the organism at some time whilst in hospital (Shooter *et al.* 1966, 1969) and a number of them excrete several different strains; nevertheless high carrier rates of individual strains, such as are sometimes seen with *Klebsiella*, are uncommon except occasionally in newborn babies. Implantation of this organism in the gut of healthy persons occurs rarely even when as many as 10^6 organisms are given by the mouth unless an antibiotic such as ampicillin is taken orally (Buck and Cooke 1969; see also Gaya *et al.* 1970).

The organism is found on the skin or in the upper respiratory tract of some 20 per cent of hospital patients (Grogan 1966). The skin-carriage rate tends to rise during stay in hospital, and is highest in the groin.

Survival and multiplication in the environment

Survival in liquids is much better than in the dry state. Emmanouilidou-Arseni and Koumentakou (1964) found that, when a standard inoculum was placed in sterile distilled water and stored at temperatures between 6° and 37°, organisms could be recovered for at least 10 months; an equal number of organisms dried on filter paper and stored at room temperature had disappeared within 3 months. Nevertheless, the ability of dried organisms to survive is considerable (Wahba 1965*b*); *Ps. aeruginosa* could, in fact, be isolated from some 5 per cent of samples of hospital dust (Darrell and Wahba 1964). Lowbury and Fox (1953) compared the survival on glass slides of *Ps. aeruginosa* with that of staphylococci and streptococci, and found that the main difference was in the first 1–2 hr; in this time, about 99 per cent of the former and only 50 per cent of the latter died, but subsequently the death-rates were similar. Pettit and Lowbury (1968) found that survivors after drying were more resistant to a second drying than were organisms that had not been so treated or had been subcultured after the first drying. Clearly, therefore, dry surfaces will remain contaminated for a very long time if the original inoculum is large. Hurst and Sutter (1966) found 3000 *Ps. aeruginosa* per unit volume of washings from the floor of a recently evacuated burns ward, but the number had fallen to 22 per unit volume 5 weeks later. The organism could not be isolated from washings in the 8th week, but pieces of eschar picked up from the floor at that time contained millions. For similar reasons, though it is unusual to isolate *Ps. aeruginosa* from the air of general hospital wards, small numbers are often found in the air of burns wards (Lowbury 1954).

Nevertheless, there is little doubt that moist objects and fluids are more important sources of *Ps. aeruginosa* than are dry objects, and this is attributable to the ability of the organisms to multiply over a wide range of temperatures—though not, like some other pseudomonads, at refrigerator temperature—in the presence of minimal nutrients (see Chapter 31).

Ps. aeruginosa has often been found in large numbers in hospital supplies of *distilled water*; its multiplication is believed to be made possible by the presence of traces of carbon compounds in the fluid or its container (Botzenhart and Röpke 1971). Favero and his colleagues (1971) observed an increase from 4000 to 10^6 organisms per ml in hospital distilled water within 24 hr at 25°, and found that *Ps. aeruginosa* taken directly from the water multiplied more readily in distilled water than did the same organisms after growth overnight in laboratory culture medium. Organisms that had multiplied in water also exhibited an increased resistance to many disinfectants (Carson *et al.* 1972). These observations partly explain why *Ps. aeruginosa* is so often to be found in large numbers in fluids, creams, ointments and liquid medicaments that may be applied to or ingested by patients (Shooter *et al.* 1969, Baird *et al.* 1976). In addition, the organism can also persist and multiply in badly chosen, incorrectly compounded or improperly stored disinfectant solutions (Maurer 1969). Undue reliance has been placed on the disinfectant properties of quaternary ammonium compounds, to which *Ps. aeruginosa* is partly resistant (Chapter 31); these may become completely inactive if improperly stored (Lowbury 1951), and can then act as a source of carbon and energy for the growth of some strains (Adair *et al.* 1969). Phenolic compounds are also irregular and uncertain in their action on *Ps. aeruginosa* (Bean and Farrell 1967, Simmons and Gardner 1969).

Sanitary appliances such as sinks (Ayliffe *et al.* 1974) and baths in hospitals are often contaminated with *Ps. aeruginosa*, and in this respect differ from similar appliances in the home (Whitby and Rampling 1972). Contamination is persistent and almost impossible to eradicate by the use of chemical disinfectants. The organisms tend to be of types isolated from the patients but may persist for months after all infected persons have left the ward. *Ps. aeruginosa* is often present on floor cloths and mops and in cleaning fluids.

Epidemiology of hospital-acquired infection

Ps. aeruginosa infections are unevenly distributed in hospitals, being infrequent at most times in many general wards but tending to be concentrated in departments that accommodate specially susceptible patients. Three main patterns of infection are discernible.

1. Infection from common environmental sources Numerous groups of patients have been infected from fluids injected intravenously or intrathecally, instilled into wounds or into the bladder, or nebulized or other-wise introduced into the lower respiratory tract (see Bassett 1971*a*). Medical and surgical equipment that is difficult if not impossible to sterilize because of its construction—notably respiratory machines, incubators, resuscitators, humidifiers, apparatus for cardiac catheterization—present serious and ever-changing technical problems.

2. Endemic infection In some hospital departments there are concentrations of patients who are susceptible to infection with *Ps. aeruginosa* and who, when infected, provide sources from which other patients may acquire the organism. This is the case, for example, in burns wards, intensive-care units, and urological wards (see Chapter 58) in which individual strains of *Ps. aeruginosa* tend to become endemic.

The fact that treatment is standardized and repetitive, and entails a good deal of handling by the nurses, provides opportunity for transfer of organisms from patient to patient either on the hands or on shared equipment. Lowbury and Fox (1954) showed that the hands of one-third of the nursing staff in a burns department were contaminated with *Ps. aeruginosa*, usually of types endemic in the ward. Most of these organisms are removed from the hands by washing with soap and water (Lowbury 1969), but a few persist even after rinsing in disinfectant. Some persons are persistent hand carriers of *Ps. aeruginosa*. Endemic infection is nearly always associated with widespread contamination of moist sites in the ward, but the evidence suggests that these are of less importance than imperfectly washed hands as sources of the organism (Lowbury *et al.* 1970, Wormald 1970).

3. Sporadic infection This is seen in most types of hospital departments, and there is evidence that many cases are self-infections (Darrell and Wahba 1964, Shooter *et al.* 1966).

On admission to hospital, ill patients have a higher faecal carrier rate than do normal persons, and many of the patients acquire further strains while in hospital. Groups of patients in a hospital ward at about the same time may acquire the same strain (Shooter *et al.* 1966). Attempts to relate these acquisitions to organisms present in the ward environment are seldom successful, but several clear instances have been observed of the acquisition in the gut of a strain known to have been ingested in the food or in a liquid medicament (Shooter *et al.* 1969). A significant reduction in the acquisition rate can be brought about by improving the hygiene of the hospital kitchen (Parker 1971). Faecal colonization is often associated with the presence of the same organism in the upper respiratory tract or on the skin. It is unlikely that colonization by new strains would happen often unless a considerable proportion of the patients had received antibiotics.

Infection in nurseries for newborn infants Outbreaks of infection among newborn infants have often been shown to be from a common environmental source (see Bassett 1971*a*). Infection may be either by the respiratory or the cutaneous route. In some outbreaks, however, a continuing source of infection in the environment cannot be found (Jellard and Churcher 1967), or the epidemic may continue after the original environmental source has been dealt with. Probably in this case infection is spread from baby to baby on the nurses' hands.

Epidemiology of infection in animals

As we have seen, faecal carriage of *Ps. aeruginosa* is infrequent in animals except when they are in captivity or domesticated; it is often confined to young animals. Contaminated food and drinking water are thought to have been responsible for outbreaks of infection in chinchillas (Larrivee and Elvehjem 1954), mink (Trautwein *et al.* 1962) and laboratory animals (Flynn 1963). In cattle, faecal carriage in particular herds has been attributed to the presence of the organism in marshy pastures (Hoadley and McCoy 1968), and pseudomonas mastitis to its multiplication in the water used for washing the cows' udders (Curtis 1969). For further information see Lusis and Soltys (1971).

Antibody formation

Type-specific agglutinating antibody appears in the serum of burned patients who are infected with *Ps. aeruginosa* (Fox and Lowbury 1953*a*). Precipitating antibody can be detected in gel-diffusion tests with extracts of the organism; it is present in the serum of nearly all patients who are recovering from bacteraemia but not of those with localized lesions or with fatal bacteraemia; it likewise appears to be type specific (Young *et al.* 1970). Type-specific anti-lipopolysaccharide antibody can also be detected by means of a passive haemagglutination reaction (Crowder *et al.* 1974, Shigeta *et al.* 1978). Antibody to exotoxin A may be measured by neutralization of toxicity or by radioimmunoassay. It is present in small amount in the sera of some normal persons and in larger amount in those of colonized and infected patients. Considerably higher titres are found in patients who recovered from serious *Ps. aeruginosa* infections than in those who succumbed, but a similar correspondence is found between high titre of antibody to lipopolysaccharide and a non-fatal outcome (Pollack *et al.* 1976, Pollack and Young 1979, Cross *et al.* 1980). The subcutaneous administration of a polyvalent dissolved vaccine (p. 287) that is effective in preventing invasive infections in burned patients leads to the production of haemagglutinating antibody against each of the constituent types of *Ps. aeruginosa*. Mouse-protective antibody is also produced, giving peak titres 1 week after the first dose and tending to disappear very soon after the last dose (Jones 1979). (See p. 283 for the production of antibody in cystic fibrosis.)

Prevention and treatment

The hygienic measures appropriate for the prevention of pseudomonas infection in hospital patients will be apparent from our discussion of its epidemiology. It must be admitted, however, that some types of highly susceptible patient have proved extremely difficult to protect from infection. The poor prognosis in systemic disease due to *Ps, aeruginosa*, and the indifferent results obtained from antibiotic therapy, have stimulated interest in the possible uses of vaccines and sera for the prevention or treatment of infection.

Active immunization

Vaccination with killed organisms gives protection to rats and mice against lethal infection by the homologous strain of *Ps. aeruginosa* (Walker *et al.* 1964, Markley and Smallman 1968). Similar results were obtained with vaccines of various sub-cellular fractions of *Ps. aeruginosa* (Alexander *et al.* 1966, Alms and Bass 1967, Jones 1968, 1969). Active immunity against intraperitoneal infection in the mouse appears to be type-specific. According to Fisher and his colleagues (1969) the 'immunotypes' so defined do not correspond to the O groups, but this conclusion conflicts with the results Bass and McCoy (1971) obtained in passive protection tests (see also Hanessian *et al.* 1971).

Feller (1967) gave a vaccine prepared from a single strain of *Ps. aeruginosa* to recently burned patients and claimed to have brought about a reduction in the frequency of pseudomonas septicaemia and death. Even better results were obtained when the patients also received hyperimmune plasma prepared by vaccinating human volunteers (see Feller and Pierson 1968). Alexander and his colleagues (1969, 1971) vaccinated burned patients with a polyvalent dissolved vaccine containing representatives of Fisher's seven 'immunotypes' and observed a five-fold reduction in mortality from pseudomonas infection. The development of early immunity after vaccination was soon confirmed in animals. Jones (1971) gave a single injection of a culture filtrate of *Ps. aeruginosa* to burned and to unburned mice that were challenged by intraperitoneal injection of the homologous organism at various times afterwards. Protection was evident from the first day in burned mice and from the third day in unburned mice and persisted until the 14th day. Serum from mice vaccinated 3 days earlier protected other mice against intraperitoneal challenge with the organism (Jones *et al.* 1971), but agglutinating antibody and bactericidal activity were not detectable in it. These findings suggested that early protection was due to the production of antibody rather than to endotoxin tolerance. However, it was not type specific, though its coverage of heterologous types of *Ps. aeruginosa* was variable (Jones 1972). Jones and his colleagues (1976) developed a polyvalent vaccine containing surface antigens obtained by a mild extraction process from the living cells of 16 serologically distinct *Ps. aeruginosa* strains (Miler *et al.* 1977). Three doses of this were given at intervals of a week, beginning as soon as possible after burning. In controlled trials in burned patients in India, the use of this vaccine led to the disappearance of bacteraemic *Ps. aeruginosa* infections and to a reduction in the total mortality rate from 41 per cent to 7 per cent in adult patients and from 21 per cent to 5 per cent in children (Jones *et al.* 1979).

Passive immunization

Rabbit antiserum against *Ps. aeruginosa*, and human antiserum containing agglutinating antibody, confer type-specific protection on mice subsequently infected by various routes (Millican and Rust 1960, Jones and Lowbury 1965, Jones *et al.* 1966, Bass and McCoy 1971). Pooled human γ-globulin is said to prevent death in mice given intraperitoneal injections of *Ps. aeruginosa* (Rosenthal *et al.* 1957, Fisher and Manning 1958), and to have some therapeutic action in man, especially when given in combination with an antibiotic (Fisher 1957, Waisbren and Lepley 1962). Antitoxin against purified exotoxin A protects burned mice against lethal infections with toxigenic strains (Pavloskis *et al.* 1977). The sera of animals immunized with polyvalent vaccine (Miller *et al.* 1977) contain mouse-protective antibody; immunoglobulins from the sera of volunteers given this vaccine confer protection on burned patients of a similar order to that given by active immunization (Jones *et al.* 1980).

Chemotherapy and chemoprophylaxis

The number of antimicrobial agents available for the treatment of serious infections is limited and their efficacy is difficult to evaluate (Andriole 1979). The affected patients are clinically heterogeneous and usually suffer from grave underlying diseases; few controlled comparisons of two or more drugs have been made. Polymyxins seem to be of limited value in invasive infections. The antibiotics most widely used are (1) carbenicillin and related synthetic penicillins and (2) gentamicin and some other aminoglycosides. The enhanced bactericidal action observed *in vitro* with combinations of carbenicillin and gentamicin (Sonne and Jawetz 1969) was confirmed in studies of experimental cystitis in the rat (Koníčková and Prát 1971). Other members of the two groups of antibiotics may show similar synergism.

The minimal inhibitory concentration of carbenicillin for many strains of *Ps. aeruginosa* is near to or above the maximum attainable in the blood; nevertheless, Brumfitt and his colleagues (1967) were successful in the treatment of 40 per cent of severe infections,

mainly of the urinary tract, and Jones and Lowbury (1967) obtained good results with it in invasive infections in burned patients. Penicillins that usually kill *Ps. aeruginosa* at lower concentrations include ticarcillin, azlocillin and piperacillin. Some cephalosporins, for example cefsulodin, inhibit it strongly *in vitro*, but their efficacy in natural infections has yet to be assessed. The aminoglycosides gentamicin, tobramycin and amikacin usually inhibit it at concentrations of less than 2 mg per 1.

Resistance to carbenicillin and to aminoglycosides has become increasingly common in hospitals in recent years but is still far from general. A plasmid-determined β-lactamase (Chapter 31) found in both *Ps. aeruginosa* and various enterobacteria destroys carbenicillin and ticarcillin; its action on piperacillin is variable but on the whole less strong (Milne and Waterworth 1978). A less common β-lactamase (Livermore *et al.* 1982) has greater action on some of the anti-pseudomonal cephalosporins than on carbenicillin. Resistance to gentamicin and similar aminoglycosides (Chapter 31) may be non-enzymic or may be mediated by antibiotic-destroying enzymes. Non-enzymic resistance may be only partial but is often sufficiently strong to be of clinical significance. It reduces susceptibility to all aminoglycosides and may appear in a strain during the course of treatment, especially if the antibiotic is used topically. Enzymic resistance is total but often narrower in spectrum. Thus gentamicin-resistant strains may be sensitive to tobramycin or amikacin or to both, but strains resistant to either of these antibiotics are invariably gentamicin resistant. Resistance to carbenicillin or to aminoglycosides tends to be patchy in distribution; it is most common in departments in which the corresponding agent is much used and there are many opportunities for the spread of resistant organisms between patients (Lowbury *et al.* 1972, Holder 1976, Meyer *et al.* 1976, Keys and Washington 1977).

Clinical experience suggests that transfusions of fresh whole blood or of leucocytes, and injections of normal immunoglobulin, may be useful adjuncts to chemotherapy in systemic *Ps. aeruginosa* infections.

Prolonged and repeated courses of systemic carbenicillin, gentamicin, or both, are used for the treatment of cystic fibrosis patients with chronic *Ps. aeruginosa* infections (Friis 1979), but the effects of this are only temporary. Continuous use of an aerosol spray of the two drugs leads to clinical improvement, but the infection is not eliminated (Hodson *et al.* 1981). Opinions differ as to whether mucoid strains are on the whole more or less sensitive to antibiotics than non-mucoid strains (Govan and Fyfe 1978, Demko and Thomassen 1980), and individual colonies in primary cultures tend to differ in antibiotic susceptibility. The exopolysaccharide of mucoid strains impedes the diffusion of aminoglycosides but not of carbenicillin (Slack and Nicholls 1981), and in any event concentrations of systemically administered aminoglycosides are low in the sputum (Mombelli *et al.* 1981).

The local application of antibiotics may prevent colonization of burns by *Ps. aeruginosa*, but the use of carbenicillin (Lowbury *et al.* 1972) and gentamicin (Holder 1976) is contraindicated because of the frequency with which resistant strains may appear. The topical use of polymyxins does not have this disadvantage, and they are suitable applications for burns (Cason and Lowbury 1960) and the mucous membranes of immunosuppressed patients (Storring *et al.* 1977). Several chemical agents reduce colonization with *Ps. aeruginosa*: mafenide cream (Lindberg *et al.* 1965); compresses of 0.5 per cent silver nitrate (Moyer *et al.* 1965), silver sulphadiazine, a mixture of silver nitrate and chlorhexidine (Lowbury *et al.* 1976), and iodophors. Mafenide, silver salts and iodophors may cause metabolic disturbances if applied to extensive burned areas, and silver sulphadiazine encourages the spread of sulphonamide-resistant klebsiella strains that have plasmid-borne resistance to several antibiotics (Bridges and Lowbury 1977). According to Monafo and his colleagues (1976), cerium nitrate is effective and of low toxicity, and permits the growth only of gram-positive organisms; a mixture of the cerium and the silver salt has been advocated.

Pseudomonas pseudomallei: melioidosis

When Whitmore and Krishnaswami (1912) first encountered human cases of melioidosis in Rangoon they were impressed by its resemblance to glanders. The course of the disease was acute, the clinical appearances were those of septicaemia or pyaemia, and abscesses or caseous deposits were found *post mortem* in the lungs, liver, spleen and other internal organs. The slender, irregularly staining, gram-negative bacillus found in the lesions resembled the glanders bacillus morphologically but was motile; and when injected into guinea-pigs it caused a disease rather like glanders. Whitmore (1913) therefore named the organism *Bacillus pseudomallei*. Stanton and Fletcher (1921, 1925) observed similar human infections in Malaysia and called the disease 'melioidosis'. Subsequently, chronic infections were described; in some there were multiple abscesses in viscera and bones (Mayer and Finlayson 1944, Gutner and Fisher 1948), but in others there was a single abscess which sometimes healed without specific chemotherapy (Peck and Zwanenburg 1947). Melioidosis also appears as a chronic pneumonitis with fever, wasting and haemoptysis, which is difficult to distinguish from pulmonary tuberculosis (Everett and Nelson 1975).

The frequency with which antibodies against *Ps. pseudomallei* are found in the indigenous human population of south-east Asia suggests that infection may be quite common. Nigg (1963) found complement-fixing antibody in the serum of 8 per cent of healthy male Thais; and in Malaysia, Strauss and his colleagues (1969*b*) found antibody by indirect haemagglutination in the serum of between 2 and 16 per cent of various groups of adult males; the highest percentage was found in the population of rice-growing areas and the lowest in forest aborigines.

Epidemiology

The disease is exclusively one of warm climates, and most of the reports of infection in man have been from south-east Asia, notably Burma, Malaysia, Thailand and Viet-Nam. A few cases have been reported from the Philippines and Guam. Another focus of human infection is in Queensland and the Northern Territories of Australia, and a few cases have been described in Central America (see Biegeleisen *et al.* 1964).

Whitmore's patients in Rangoon (1913) were indigent males, many of whom were taking morphine by injection. A number of subsequent reports have described a close association of melioidosis with serious pre-existing disease and with admission to hospital for its treatment. Thus in Viet-Nam, Roques and Dauphin (1943) described four cases in patients who had recently undergone surgical operations and had been in hospital for several weeks when the symptoms of acute melioidosis appeared; Alain and his colleagues (1949) reported a number of similar cases. In Australia (Rimington 1962, Crotty *et al.* 1963), 4 of 8 infections were in diabetics. On the other hand, the disease was also often observed in soldiers on active service in the French (Bres 1958) and the American army (Spotnitz *et al.* 1967) in Viet-Nam, and many authors (see Chambon 1955, Bres 1958, Thin *et al.* 1970) drew attention to the frequency with which the patients had been immersed in or contaminated with muddy water in the course of hunting parties, automobile accidents, military operations or alcoholic stupor. The frequency of respiratory infections in American soldiers in Viet-Nam has, however, been attributed to the inhalation of contaminated dust raised by helicopter rotors. Transmission between human subjects is very rare.

In animals, infection with *Ps. pseudomallei* is distributed irregularly in tropical areas and sometimes gives rise to outbreaks of clinical disease: in laboratory animals in Malaysia (Stanton and Fletcher 1925); in wild rats in Queensland (Cook 1962) but not in Saigon (Alain *et al.* 1949); in sheep in Queensland (Lewis and Olds 1952); in pigs in Malaysia (Omar *et al.* 1962), Madagascar (Girard 1936) and Upper Volta (Dodin *et al.* 1974); and in sheep, goats and pigs in the Netherlands Antilles (Sutmöller *et al.* 1957). The areas in which animal disease occurs are all, with the exception of those in Africa, ones in which human disease has been reported; there is, however, remarkably little evidence that human disease comes from animal sources.

Whitmore's bacillus is a normal inhabitant of soil

and water in south-east Asia. Stanton and Fletcher (1932) observed that it would survive in soil for 27 days, and Vaucel (1937) and Chambon (1955) in Viet-Nam isolated it from naturally contaminated water and mud. In an extensive survey of soil and surface waters in Malaysia (Strauss *et al.* 1969*a*) it was found in 14–33 per cent of samples from rice fields and newly planted fields of oil palm, but in only 1–3 per cent of samples from forested areas (see also Thin *et al.* 1971). It is apparently common in water and soil in Upper Volta (Dodin *et al.* 1974).

Most cases of melioidosis in healthy persons can be related to an accidental circumstance in which organisms from their natural habitat are introduced into the body by the cutaneous or the respiratory route. However, the disease also occurs in hospital patients and in others who habitually receive injections. The contamination of fluids for injection and of hospital equipment with *Ps. pseudomallei* in the tropics merits further investigation. (See also Ashdown 1979*a*).

Bacteriological diagnosis

Diagnosis of the disease in life is not always easy. The organism may be grown from the blood or from pus, but its presence is sometimes unexpectedly scanty. It grows on plain agar and even better on glycerol agar. Useful selective media include 3 per cent glycerol agar containing 1 in 200 000 crystal violet (Miller *et al.* 1948*a*) and MacConkey's agar containing colistin S 20 mg per l (Farkas-Himsley 1968). For isolation from soil and water, a susceptible animal should be used; Ellison and his colleagues (1969) recommend the intraperitoneal injection of the material into weanling hamsters and culture of the organs of those that die on crystal violet glycerol agar. Whitmore's bacillus can be detected in experimentally contaminated soil, or in the tissues of infected animals, by immunofluorescence (Moody *et al.* 1956, Thomason *et al.* 1956), but this method does not distinguish it from *Ps. mallei*.

Serological tests Bacterial agglutination tests are not of much value, but an *indirect haemagglutination test* and a *complement-fixation test* have proved useful. In the former, a heat-stable and presumably polysaccharide antigen is adsorbed on to erythrocytes; Nigg (1963) used a phenol–water extract of bacteria, but most other workers used a mallein-like material (*melioidin*) obtained by boiling the supernate of an old broth culture (Boyden 1950, Ileri 1965, Strauss *et al.* 1969*b*). In the complement-fixation test, the antigen is an unheated crude aqueous extract of disintegrated bacteria (Nigg and Johnston 1961). In acute melioidosis in man, both types of antibody begin to appear towards the end of the first week of illness. From the 2nd to the 5th week about 90 per cent of serum specimens are positive by both the indirect haemagglutination and the complement-fixation test, and if the two tests are used a serological diagnosis can be made in nearly every case (Alexander *et al.* 1970). Neither test distinguishes between glanders and melioidosis, but with this exception the specificity of the haemagglutination test is excel-

lent. Low titre cross-reactions occur in the complement-fixation test with sera prepared against some serotypes of *Ps. aeruginosa* and sometimes in patients infected with this organism. Titres of 1 in 40 in the haemagglutination test and 1 in 8 in the complement-fixation test are strongly suggestive of melioidosis. According to Ashdown (1981*a*, *b*), the indirect immunofluorescence test for specific IgM is the most reliable means of detecting active cases of melioidosis serologically. An *intradermal melioidin test* has been used to identify subclinical cases of melioidosis in goats (Olds and Lewis 1954).

Treatment

As stated in Chapter 31, most strains of Whitmore's bacillus are sensitive *in vitro* to sulphadiazine, co-trimoxazole (see Bassett 1971*c*), tetracycline, chloramphenicol, novobiocin and rifampicin. In acute infections, however, the response to therapy may be poor, and treatment should always be continued for at least 30 days. The drugs most favoured are tetracycline, chloramphenicol and novobiocin, and combinations of two drugs are often used. In-vitro tests (Calabi 1973) suggest that combinations of novobiocin and tetracycline are synergic in their bactericidal action on Whitmore's bacillus. Aminoglycosides are of doubtful value (Ashdown 1979*b*). Treatment should, where possible, be controlled by quantitative sensitivity tests on the infecting organism (see Malizia *et al.* 1969, Eikhoff *et al.* 1970). (The following general reviews of melioidosis should be consulted: de Moor *et al.* 1932, Stanton and Fletcher 1932, Redfearn *et al.* 1966, Howe *et al.* 1971).

Pseudomonas mallei: glanders

Numerous investigations during the nineteenth century indicated that glanders was infectious, and that the disease in man was the same as that in the horse. The causative organism, however, was not isolated until 1882, when Loeffler and Schütz (Loeffler 1886) succeeded in culturing it from a horse that had died of glanders. (For a history of glanders, see Wilkinson 1981.)

The disease in animals

Glanders is primarily a disease of equine animals, but goats, sheep, dogs and cats sometimes contract it naturally; it has also occurred in zoological gardens among carnivora fed on infected horse flesh.

There are two clinical types of disease in horses and asses—glanders and farcy. In *glanders*, which may be acute or chronic, the lungs are almost invariably affected. Rounded, greyish, firm nodules, 0.5 to 1 cm in diameter, appear in small numbers. When recent, they have a dirty white centre and a dark red, or sometimes yellow, gelatinous periphery. In the centre is thick pus composed of polymorphonuclear cells. In older nodules, the greyish centre is surrounded by dry crumbling material. A zone of epithelioid and giant cells lies outside the central necrotic area, and surrounding the whole is a layer of fibrous tissue; occasionally calcification may occur. Nodular and later ulcerative lesions of the nasal mucosa and trachea, and subcutaneous abscesses, are also frequent in glanders. In acute cases there may be pneumonic infiltration of the lungs, and nodules are distributed throughout the spleen, liver and other internal organs. In the ass and the mule the disease is almost invariably acute, death occurring in 3–4 weeks.

In *farcy*, swellings appear in the skin or subcutaneous tissues, particularly of the limbs and flanks, which break down and ulcerate. The lymphatic vessels leading from these swellings become firm and enlarged, standing out beneath the skin as hard cords. These can be traced to the lymphatic nodes, which likewise become prominent.

Glanders was at one time common in most parts of the world, but has decreased steadily in frequency during the twentieth century. By 1939 it had been eliminated from most parts of Western Europe, the USA and Canada. It is still seen in Asia, Africa and the Middle East, though less often than formerly.

In Britain, the disease used to be especially prevalent in places where large numbers of horses were kept in close contact; once infection had been introduced it spread rapidly. Apart from the clinical cases there was an even larger number of carriers, which served to perpetuate the disease and to render its control extremely difficult. Months may elapse between infection and the onset of symptoms; and in many animals the disease may be permanently dormant or undergo spontaneous cure.

Mode of infection   Though the primary lesion often appears in the lung, attempts to produce chronic pulmonary glanders by infection via the respiratory tract fail; intratracheal and intranasal injection of the organism result in acute disease. On the other hand, typical chronic pulmonary lesions follow infection by the mouth (M'Fadyean 1904, Bonome 1906). Animals infected in this way, however, usually also show inflammatory lesions in the intestinal wall and mesenteric lymph nodes, but these are uncommon in the natural disease. It is thus impossible to say whether the infectious material reaches the lungs directly or via the pharyngeal mucosa or the intestine. In any event, the nasal and pulmonary discharges, and sometimes the urine and faeces, of animals suffering from glanders are infectious. In nearly every case of farcy, glanders nodules are present in the lungs at necropsy (M'Fadyean 1904).

Glanders may be reproduced in horses, asses, and mules by feeding with cultures of the bacillus, and by subcutaneous inoculation. Sheep and goats are easily infected, but cows and pigs are absolutely resistant.

The disease in man

Glanders was a rare disease in man even when it was common in horses (Robins 1906). It may be acute or chronic, and it may be localized chiefly in the respiratory organs, or in the skin and subcutaneous tissues. In acute glanders there is generally fever, a mucopurulent discharge from the nose, and prostration out of all proportion to the clinical signs (Bernstein and Carling 1909); a generalized pustular eruption is very frequent. Death occurs in a week or 10 days unless specific treatment is given. In the chronic disease there may be coryza; there may be multiple subcutaneous and intramuscular abscesses, often associated with enlargement of the lymphatic glands and vessels; nodules may form in the mucosa of the respiratory and alimentary tracts, and may break down and ulcerate; necrotic foci may appear in the bones, and nodular lesions in the viscera. The disease may remain active for months or even years; without treatment it is usually but not invariably fatal. Sometimes after apparent recovery the disease may break out again; latent periods up to 10 years have been observed (Bernstein and Carling 1909). The incubation period varies from a few hours to several weeks.

The disease attacks chiefly those who come into close contact with horses. Infection results most frequently from contamination of a scratch or wound, but primary infection of the nasal mucosa may occur. Several cases of glanders have been reported in laboratory workers; indeed, probably no organism, with the possible exception of the tularaemia bacillus, is as dangerous to work with as the glanders bacillus. In one laboratory, several members of the staff became ill a few days after the breaking of a centrifuge tube. In another, which was specifically designed for work with this organism (Howe and Miller 1947), one-half of the workers were infected within a year (see also Bernstein and Carling 1909). It is therefore rather surprising that the disease is not more common among those who come into contact with glandered animals.

Diagnosis

Cultures should be put up on nutrient agar or glycerol agar. If the material is contaminated, it may be incubated for 3 hr at 37° in physiological saline containing 1000 units of penicillin per ml and then plated on to agar containing 1 in 200 000 crystal violet (Miller *et al.* 1948*a*). A more sensitive method, however, is to inject the material, preferably after preliminary incubation with penicillin, subcutaneously into a guinea-pig or hamster and recover the bacillus from the enlarged glands. Not all strains of *Ps. mallei* are virulent on isolation; sometimes two or three passages have to be made before the typical disease develops (Dudgeon *et al.* 1918). As a rule material from the acute disease in horses proves virulent, but from the chronic disease it is generally avirulent.

The mallein test This is used for the diagnosis of latent or chronic glanders in horses. Mallein is an autoclaved whole culture of *Ps. mallei*. In the *subcutaneous test*, 2.5 ml of unconcentrated mallein is given; a positive result is characterized by a local swelling and a febrile reaction, but interpretation is sometimes difficult and antibodies may appear in the blood and cause confusion in subsequent serological tests. Tentimes concentrated mallein may be rubbed into the shaved and scarified skin (the *cutaneous test*) or into the inner surface of the eyelid (the *conjunctival test*), but nowadays the *intradermal palpebral test* is most often used; 0.1 ml of concentrated mallein is injected into the cutaneous surface of the eyelid near its edge. A positive reaction is characterized by tender oedematous swelling of the eyelid, with congestion of the conjunctiva and a mucous discharge from the eye, which begins 9–10 hr after injection and reaches its height between the 24th and 36th hr. Intradermal and conjunctival tests are apparently without effect on the agglutinin titre (Lovell 1935). The mallein test is a fairly specific and sensitive method of diagnosis, but occasional false positive results occur and the test may be negative in acute or advanced cases. If the result of the ophthalmic test is negative or doubtful it may be repeated 24 hr later, when a more brisk reaction may occur. The mallein test has seldom been employed in man.

Serological tests Bacterial agglutination and precipitation tests are unreliable. Indirect haemagglutination (Boyden 1950) and complement-fixation tests (Reinhardt 1919, Poppe 1919) are said to give better results. Hole and Coombs (1947) found that the *conglutinating complement-absorption test* was more specific than the ordinary complement-fixation test for the diagnosis of experimental glanders in ponies. It may prove useful for testing the sera of mules, asses and pregnant mares, which are often anticomplementary.

The serum of normal human subjects almost invariably contains agglutinins for *Ps. mallei* in a titre ranging from 10 to 320. In glanders the titre sometimes rises as high as 5120 (Galtier 1881, Collins 1908, Dudgeon *et al.* 1918, Cravitz and Miller 1950) but may be extremely low (Gabriélidès and Remlinger 1902, Bernstein and Carling 1909). The complement-fixation reaction is sometimes of value (Watson 1923, Sabolotny 1926).

Immunity, prophylaxis and treatment

Horses may recover spontaneously from glanders (Nocard 1901). Only a small proportion of healthy horses that give a positive mallein test develop clinical signs of the disease (M'Fadyean 1900); many of the remainder subsequently lose their reaction to mallein, but this is not necessarily proof of cure (M'Fadyean 1901, Bonome 1906). Spontaneous recovery from glanders does not, however, leave the animals with a high degree of immunity (M'Fadyean 1900, Nocard 1901, Mohler and Eichhorn 1914). Despite apparent success in producing active immunity in guinea-pigs (Babes 1892, Nicolle 1906) attempts to render horses immune by injections of mallein or killed vaccines were unsuccessful (M'Fadyean 1901, Mohler and Eichhorn 1914).

For the control of glanders in horses the best results have followed a procedure such as that laid down in the British Glanders or Farcy Order of 1907. Briefly this provides that every animal with clinical evidence of glanders, and every animal giving a positive mallein test, is slaughtered. The diseased carcases are buried or suitably destroyed, and thorough disinfection of the infected premises is carried out. Contacts are tested by mallein, and all positive reactors are slaughtered.

Sulphadiazine is bacteristatic for *Ps. mallei* in therapeutically attainable concentrations (Miller *et al.* 1948b) and is effective in the treatment of experimentally infected hamsters. Howe and Miller (1947) obtained good results with it in the treatment of six cases of human glanders. (See also Cravitz and Miller 1950, Ansari and Minou 1951.) *Ps. mallei* is resistant to penicillin but is said to be sensitive to streptomycin, chloramphenicol and the tetracyclines. The results of experimental chemotherapy with these drugs are, however, contradictory, and little information is available about the value of antibiotics in the treatment of natural infections.

Other pseudomonads

The *cepacia* group

These organisms, which include strains described as *Ps. cepacia*, *Ps. thomasi* and *Ps. picketti* (Chapter 31), have caused many outbreaks of infection in hospital patients associated with contaminated fluids: of urinary-tract infection (Mitchell and Hayward 1966) and wound sepsis (Bassett *et al.* 1970) attributable to contamination of irrigation fluids, and of septicaemia after the intravenous infusion of aqueous solutions containing large numbers of the organisms (Phillips *et al.* 1972). Endocarditis may develop in septicaemic infections so acquired, and also in heroin addicts (Noreiga *et al.* 1975).

Organisms of this group are commonly found in tap water (Bassett *et al.* 1970) and readily recontaminate and multiply in distilled water in pharmacies; in the epidemic reported by Phillips and his colleagues (1972), organisms from distilled water gained entry to bottles of autoclaved fluids during the cooling process. However, many outbreaks have been made possible by the ability of *Ps. cepacia* and related organisms to multiply in fluids containing certain disinfectants; these include not only quaternary ammonium compounds but chlorhexidine (Mitchell and Hayward

1966, Bassett *et al.* 1970, Hardy *et al.* 1970, Phillips *et al.* 1971). *Ps. cepacia* will grow in use-dilutions of aqueous chlorhexidine and of Savlon (cetrimide and chlorhexidine). The action of chlorhexidine on it is highly dependent on pH (Bassett 1971*b*); the advice of the manufacturer to add a small percentage of alcohol to diluted solutions of chlorhexidine-containing disinfectants should be followed.

Ps. cepacia also causes lesions of the feet in persons who have been immersed for long periods of time in swamp water in the sub-tropics (Taplin *et al.* 1971). The skin of the toe-web becomes macerated and hyperkeratotic, and the lesions resemble those caused by *Ps. aeruginosa* in similar circumstances.

Ps. cepacia is rather less resistant to cetrimide than is *Ps. aeruginosa*. It is resistant to colistin, gentamicin and carbenicillin but sensitive to cotrimoxazole (Taplin *et al.* 1971).

Less frequently encountered pseudomonads

Fluorescent pseudomonads other than *Ps. aeruginosa* cause disease rarely unless they are injected intravenously in large numbers. This may happen when fluids for intravenous infusion, especially blood and leucocyte concentrates, become contaminated with a cold-tolerant strain and are subsequently kept for several days in the refrigerator (see Chapter 58).

A number of other pseudomonads are found from time to time in pathological material, but the significance of this is very difficult to assess. On the evidence that it has been isolated by a number of workers in pure culture from the meninges and other closed sites, and from the blood stream, *Ps. maltophilia* should probably be looked upon as an occasional pathogen, at least in patients with serious underlying diseases (for references see Zuravleff and Yu 1982). Single isolations have been made under similar circumstances of members of several other species: *Ps. paucimobilis* (Hadjiroussou *et al.* 1979); *Ps. testosteroni* (Atkinson *et al.* 1975); and '*Ps.*' *putrefaciens* (Vandepitte and Debois 1978). (See also Gilardi 1972.) The possibility of 'pseudo-septicaemia' due to the contamination with pseudomonads of the apparatus used for blood collection, or of the hands of the operator, should always be borne in mind (Semel *et al.* 1978).

Escherichia coli

Esch. coli forms the bulk of the aerobic bacterial flora of the gut in man and many other mammals (Chapter 8). It may invade the body and give rise to various common septic diseases, but this is in general conditional upon one of a number of predisposing conditions in the host. Predisposition may be local and aid invasion by a particular route, for example, the urinary tract; or it may be general, as in the newborn, who are prone to systemic infections without a preceding local inflammation. However, not all strains of *Esch. coli* have an equal ability to cause septic diseases even in predisposed persons. As we saw (Chapter 34), the distribution of O serogroups among strains from extra-intestinal septic infections is quite different from that among strains that cause diarrhoeal diseases; and strains from septic infections tend to have surface polysaccharides (K antigens) of the L and A classes. Several other laboratory markers for the ability to cause particular septic infections have been described.

Although *Esch. coli* has considerable ability to survive in the environment of man and animals, particularly in moist places, it seldom multiplies significantly in fluids with a minimal content of organic matter, as do pseudomonads and a number of other enterobacteria. It is therefore not a cause of explosive 'common-source' epidemics in hospitals. Individual strains of *Esch. coli* do not spread widely among hospital patients except occasionally in the newborn. Many patients suffer from *Esch. coli* infections in hospital, but these are mainly self-infections.

(Methods for typing *Esch. coli* were described in Chapter 34.)

Urinary-tract infections in man

Esch. coli is by far the commonest cause of urinary-tract infections in the general population. Briefly (see Chapter 57) the infections occur mainly in females, except in infancy, and become progressively more frequent in successive age-groups. Infection is often symptomless and thus detectable only by examination of the urine. Clinically apparent infections may be confined to the bladder (cystitis) or extend up into the renal pelvis (pyelonephritis). The organism enters the urinary tract by the ascending route; its establishment in the kidney is favoured by stasis of the urinary flow arising from minor abnormalities of the renal pelvis or from pregnancy.

A number of phenetic characters are more common in *Esch. coli* strains from urinary-tract infections than in unselected faecal strains. As long ago as 1945 Vahlne found that members of certain O serogroups predominated in the urinary strains. Many other workers have since made similar observations; among the more prevalent O groups all of them included group 4, almost all of them group 6, many of them groups 2 and 75, and some of them groups 1, 7, and 18a,c (Ujváry 1958, Vosti *et al.* 1964, Grüneberg and Bettelheim 1969, Wong and Bettelheim 1976, Brooks *et al.* 1980). Nearly all of these groups are among those that cause parenteral rather than diarrhoeal disease (Chapter 34), all have O antigens that migrate towards the cathode, and most of them have surface or capsular acid-polysaccharide K antigens (see Chapter 34). However, differences were observed in the distribution of O groups in urinary-tract infections in different geographical areas and in the same area at different times. It should also be noted that a significant minority of strains responsible for urinary-tract infections have incomplete O antigens and exhibit various degrees of roughness.

Glynn and his colleagues (1971) reported that strains of *Esch. coli* from cases of pyelonephritis formed more K antigen than did strains from cases of cystitis or from the faeces of normal persons (see also Kaijser 1973), but only some K antigens appear to be associated with the ability to infect the urinary tract (Kaijser *et al.* 1977). The earlier suggestion (Glynn *et al.* 1971) that K antigens prevented the complement-dependent bactericidal action of normal serum on *Esch. coli* has since been disputed. Serum sensitivity occurs not only in rough and semi-rough strains, but also in smooth members of some O groups associated with urinary-tract infection (Taylor 1974). Whether K antigens are important inhibitors of phagocytosis is still uncertain (see p. 294).

Strains of *Esch. coli* from clinical urinary-tract infections are said to adhere more strongly to epithelial cells from the human urinary tract than do faecal strains. This ability has been associated with the presence of fimbriae of a type responsible for mannose-resistant agglutination of human erythrocytes (Edén *et al.* 1977, Edén and Hannson 1978). These must be distinguished from the common fimbriae responsible for mannose-sensitive haemagglutination that are found on most enterobacteria. According to Källenius and co-workers (1981*a*), the 'mannose-resistant' fimbriae become attached to antigens of the P blood group on the surface of urinary epithelial cells; and there is an association between the P_1 blood group and susceptibility to urinary-tract infection. However, Harber and his colleagues (1982) state that strains of *Esch. coli* from urinary-tract infections do not adhere to the cells when first isolated but become adherent only after serial subculture in the laboratory. P-fimbriate strains are said to be isolated much more often from cases of pyelonephritis than of cystitis (Källenius *et al.* 1981*b*). Thus, adhesion of this type may not be responsible for the initial establishment of the organism in the urinary tract, but it may play some role in the invasion of the renal pelvis.

Other characters that are said to be more common in strains from urinary-tract infections than in faecal strains from healthy persons include: the production of α-haemolysin (Chapter 34), which is often associated with the formation of a cytotoxin demonstrable by its effect when injected intradermally into rabbits and on chick-embryo fibroblasts (Cooke and Ewins

1975); the acidification of dulcitol and salicin, and the presence of the H antigen 1 (Brooks *et al.* 1980, 1981, Nimmich *et al.* 1980).

Among hospital patients, urinary-tract infections with *Esch. coli* usually follow catheterization, but these are outnumbered by infections with other organisms. They are mainly sporadic self-infections except in the newborn. A number of small epidemics of spontaneous urinary-tract infection, often with bacteraemia and caused by organisms of a single serotype, have been recorded in neonatal nurseries (Porter and Giles 1956, Kenny *et al.* 1962, Sweet and Wolinsky 1964, Balassanian and Wolinsky 1968).

In acute pyelonephritis, but not in cystitis, antibody against the O antigen of the infecting strain usually appears in the blood (Chapter 57). It may be accompanied by antibody against other constituents of the cell and its appendages (see Hanson *et al.* 1981). When the kidney is infected, antibody-coated *Esch. coli* are often present in the urine (Chapter 57).

Other local septic infections

Coliform bacteria are found in many wounds, whether accidental or surgical, and until recently most of them were *Esch. coli*. Most but by no means all of these infections follow trauma to or operations on the gut, the gall-bladder or the female genitalia. In one large American series (Report 1964), *Esch. coli* was present in 22 per cent of all post-operative septic infections. In some frankly septic wounds it is present in pure culture, but in most of those after transperitoneal operations it occurs together with anaerobes, especially *Bacteroides fragilis*. The same is true of septic lesions associated with the bowel or pelvic organs. Many of these infections are susceptible to treatment with metronidazole, but for the reasons given in Chapter 57 this does not exclude the possibility that enterobacteria and gram-negative non-sporing anaerobes act synergically in intra-abdominal sepsis. Kelly (1978) demonstrated that subinfective doses of *Esch. coli* and *Bact. fragilis* when given together caused clinical sepsis in experimental wounds in the guinea-pig. One important consequence of intra-abdominal infection with *Esch. coli* is that instrumentation or other manipulations of inflamed organs may lead to bacteraemia and fatal endotoxic shock (see below).

Bovine mastitis *Esch. coli* became a common cause of bovine mastitis in the years following the widespread use of penicillin for the treatment of streptococcal mastitis. It appears seldom to give rise to epidemics, and the disease is caused by strains of a wide range of O serogroups (Linton *et al.* 1979).

Systemic infections

The sudden discharge of large numbers of *Esch. coli* into the blood stream may result from instrumental interference with an intra-abdominal organ. Rapid invasion of the pelvic veins is soon followed by vascular collapse and other signs of endotoxic shock (see Chapter 57). A similar condition can be produced in granulocytopaenic rabbits by feeding selected strains of *Esch. coli* and causing mild trauma to the rectum (Braude *et al.* 1969). Septicaemia of a less fulminating nature, in which signs of endotoxic shock are usually less prominent, may occur as a result of a variety of local infections.

Systemic infections in the young

In man　*Esch. coli* is the commonest cause of neonatal meningitis and septicaemia. Colony counts in the blood of septicaemic babies are very high, exceeding 1000 per ml in about one-third of cases (Dietzman *et al.* 1974). Systemic invasion occurs without a preceding local inflammation. Infections are usually sporadic, but small outbreaks in hospital nurseries have been described (Bacon *et al.* 1975, Czirók *et al.* 1977, Headings and Overall 1977), as have epidemics of pyelonephritis with septicaemia (p. 293).

Esch. coli is the commonest cause of *neonatal meningitis*. The strains responsible belong to a variety of O serogroups but most of them possess the K antigen 1 (Chapter 34): according to Robbins and his colleagues (1974), 84 per cent of them, in contrast with 34 per cent of strains from neonatal septicaemia without meningitis and 11 per cent from meningitis in adults. Neonatal meningeal infections with K1 strains have a fatally rate of about 30 per cent; with other *Esch. coli* strains, adequately treated meningitis is seldom fatal (McCracken *et al.* 1974). The presence of large amounts of K1 antigen in the cerebrospinal fluid indicates a poor prognosis; its persistence for more than 2 days in survivors is associated with permanent neurological damage (McCracken *et al.* 1974). The fact that the K1 antigen had a close similarity to the group B polysaccharide of *Neisseria meningitidis* appeared at first sight to be of significance, but it is now known that the K92 antigen of *Esch. coli*, which is not particularly associated with meningitis, has a similar antigenic relation to the group C polysaccharide of *Neiss. meningitidis* (Glode *et al.* 1977); cross-reactions between antigen K7 and *Str. pneumoniae* type 3, and between antigen K100 and *Haemophilus influenzae* type b have also been described.

In meningitis due to K1 strains the infecting organism is usually derived from the mother and colonizes the baby's intestinal tract (Sarff *et al.* 1975). Some babies acquire the organism from other sources in nurseries, and there is a tendency for the carriage rate of K1 strains to rise in those who are retained for several weeks in special-care units (Peter and Nelson 1978). About 10 per cent of normal adults carry K1 strains, but considerably higher frequencies have been recorded among mothers in maternity departments.

In domestic animals　Acute septicaemia occurs in young animals of several species, notably cattle, sheep and chickens. The disease in newborn calves is caused by strains of *Esch. coli* that belong to a number of serotypes but are quite distinct from those that cause diarrhoea. It affects animals that are deficient in IgG because they have been deprived of colostrum or for other reasons. Strains of *Esch. coli* isolated from the blood of calves will multiply in the blood of IgG-deficient calves but not in the blood of calves with a normal IgG content (Smith 1962). When a strain of serotype O78:K80—a serotype often responsible for bacteraemic infection—is given experimentally to IgG-deficient calves either intravenously or by the mouth an acute and rapidly fatal septicaemia develops (Smith and Halls 1968). Calves with normal amounts of IgG in the serum are resistant to oral challenge with the strain; after intravenous challenge most of the bacteria are taken up by cells of the reticulo-endothelial system and the calves survive. The O78:K80 strain disappears from IgG-containing blood more quickly *in vivo* than *in vitro*, and certain other *Esch. coli* strains that grow in this blood are nevertheless rapidly eliminated *in vivo* after intravenous inoculation (Smith and Halls 1968). This suggests that removal by the reticulo-endothelial system is of greater importance than the serum bactericidal effect. The main protective mechanism thus appears to be phagocytosis by reticulo-endothelial cells in the presence of colostrum-derived antibody.

Pathogenesis

It is difficult to generalize about the characters of *Esch. coli* strains that cause extra-intestinal infections because these seem to be to some extent specific for individual types of disease. There is little evidence that the O antigens play a significant part in the process of invading the tissues, but the distribution of O serogroups in strains isolated from septic infections suggests that other virulence factors may be O-group associated. This may be to some extent true of the ability to survive in normal serum or to resist phagocytosis by blood leucocytes. However, serum resistance does not appear to be essential for the establishment of infection in the urinary tract. Evidence has been presented that the K antigens are the main anti-phagocytic factors in *Esch. coli* (Wolberg and de Witt 1969, van Dijk *et al.* 1977), but Björksten and his colleagues (1976) could not confirm this. In a study of strains from systemic infections in babies, they observed wide variations in the requirements for opsonization; a considerable proportion, including a number of K1 strains, activated complement by the alternative pathway and were well opsonized in the presence of normal serum. The special significance of the K1 antigen, or of the virulence factor closely related to it, appears to be confined to invasion of the blood stream, and subsequently of the meninges, in babies. Virulence for calves seems to be associated with the ability to resist phagocytosis by fixed cells in the reticulo-endothelial system in the absence of maternally derived immunoglobulins.

There is little doubt that the release of endotoxin into the blood is an important cause of death once septicaemia has become established. Some experimental evidence suggests that immunization against common core constituents of the lipopolysaccharide protects leucopaenic rabbits against the lethal action of *Esch. coli* (Johns *et al.* 1977).

Two other virulence factors have been demonstrated by their ability to cause increased death rates in experimental animals; both are plasmid-determined (Smith 1974, Smith and Huggins 1976). The plasmid coding for one of these virulence factors determines the production of colicine V, but it is unlikely that the colicine is directly responsible for the virulence (Chapter 34). The majority of strains from bacteraemia in calves, lambs and chickens (Smith and Huggins 1976) but only some 10 per cent from similar infections in man (Minshew *et al.* 1978) form colicine V. The colicine-producing strains belong to a variety of O serogroups. Strains carrying the plasmid give rise to more bacteraemic infections than the corresponding plasmid-free strains when given intravenously or intramuscularly to chickens, intravenously to mice, or intravenously or orally to colostrum-deprived calves (Smith and Huggins 1976). Increased virulence is not associated with the production of an extracellular toxin.

The second virulence factor, coded for by the plasmid *vir*, is an extracellular toxin that causes increased death rates in chickens, rabbits and mice. It is non-dialysable, sensitive to heat and acid, and causes liver damage and acute death. The *vir* plasmid was found in a strain from a bacteraemic infection in a lamb; it appears to be uncommon.

Antibiotic resistance

In Britain, strains of *Esch. coli* from the faeces of members of the human population outside hospitals are in the main sensitive to antibiotics, but at least one-half of all persons carry some resistant organisms (Linton *et al.* 1974), though usually in small numbers. The more common resistances of clinical significance are to sulphonamides, ampicillin and tetracycline. If one of these agents, particularly tetracycline, is given, resistant strains of *Esch. coli* rapidly become predominant in the faeces; these strains are resistant to the antibiotic given and often also to others. Some of them disappear soon after treatment ceases, but others persist in considerable numbers for many months in the absence of further antibiotic treatment (Hartley and Richmond 1975). About 1 per cent of the general public has a predominantly resistant faecal flora of

Esch. coli at any moment (Richmond and Linton 1980). Resistant organisms predominate in the faeces of a considerably greater proportion of hospital patients, and resistance patterns are generally wider than those seen in the general population.

Resistant strains are very much more common in the faeces of certain domestic animals, notably calves (Howe and Linton 1976), pigs (Linton 1977) and poultry (Linton *et al.* 1977*a*), than in man. This is generally attributed to the use of antibiotics for growth promotion and for prophylaxis and treatment in these animals (Linton 1981). Strains isolated from adult cattle are much less often resistant. Meat from calves, pigs and poultry becomes heavily contaminated with resistant organisms in the abattoir. Identifiable resistant strains from poultry have been detected in the faeces of workers who handled the carcasses and persisted there in considerable numbers for at least 10 days in the absence of antibiotic administration (Linton *et al.* 1977*b*). There is as yet no definite evidence that resistant strains of *Esch. coli* acquired from animals cause disease in man, but the possibility exists that resistance plasmids from animal strains may be transferred to the indigenous bowel flora of man. However, the evidence suggests that the selective action of antibiotics—particularly tetracycline—administered to man, either in hospitals or outside them, is sufficient to account for the observed frequencies of resistant *Esch. coli* in the respective populations (Richmond and Linton 1980).

The sensitivity of *Esch. coli* strains responsible for septic infections will thus depend upon whether the patient had received antibiotics or been admitted to hospital before the onset of the disease, but resistance to commonly used agents in strains from the non-hospital population shows considerable geographical variation even within a single country. Agents for treatment should therefore whenever possible be chosen according to the results of in-vitro tests. In infections of the urinary tract arising outside hospital in Britain, up to one-third of the causative strains may be resistant to ampicillin or sulphonamides, but resistance to trimethoprim, nalidixic acid and nitrofurantoin is usually considerably less common (Brunton and Heggie 1979, Garrod *et al.* 1981). (For the treatment of urinary-tract infections, see Chapter 57.)

Escherichia vulneris Brenner and his colleagues (1982) have described a new group of coliform organisms that they believe, on grounds of DNA-relatedness, should be placed in the genus *Escherichia*. Nearly all of the 61 strains they studied had been isolated from wounds in human patients in the USA. Clinical microbiologists are unlikely to confuse this organism with *Esch. coli*. Two-thirds of the strains formed a yellow pigment; they all fermented malonate but failed to form indole or to decarboxylate ornithine. They differed from *Erwinia herbicola* ('*Enterobacter agglomerans*') in nearly always decarboxylating lysine, arginine, or both.

Klebsiella, *Enterobacter* and *Serratia*

Until about 30 years ago members of these genera were looked upon as infrequent causes of disease. Biochemically atypical klebsiella strains were known to be responsible for occasional cases of pneumonia and for rhinoscleroma in man, and a few systemic infections with klebsiellae and pigmented strains of *Serratia* had been described. In 1933, Burke-Gaffney had observed that many chronic urinary-tract infections were caused not by *Esch. coli* but by 'aerogenes-like' organisms that would nowadays be classified as *Klebsiella* or *Enterobacter*. The role of the *aerogenes* variety of *Klebsiella* as a cause of hospital-acquired urinary-tract infection, and its ability to spread widely in hospitals, was first recognized by I. Ørskov (1952, 1954). Retrospective surveillance of bacteraemia in the Boston General Hospital (Chapter 57; Table 57.1) indicated that cases of septicaemia caused by *Klebsiella* and *Enterobacter* first appeared in significant numbers in the late 1940s. In a study in this hospital in 1964 (see Steinhauer *et al.* 1966), 205 clinical infections with *Klebsiella* and 39 with *Enterobacter* were identified. The recognition that *Serratia* was an important cause of hospital-acquired infection followed the demonstration by Davis and her colleagues (1957) that the pigmented strains formed only a small proportion of the genus, but it appears to have become a widespread cause of serious infections only in the mid-1960s.

Nomenclature of the causative organisms

The names now approved by taxonomists for species in the genus *Klebsiella* would be most confusing if used in clinical bacteriology. Briefly (Chapter 34), the name *Kl. pneumoniae* is applied to all members of the genus that fail to liquefy gelatin or form indole; slow liquefiers of gelatin, which usually form indole, are known as *Kl. oxytoca*. Most of the clinically important members of the genus belong to the species *Kl. pneumoniae* so defined. However, these include not only the common and biochemically typical strains that clinical microbiologists in Britain habitually call *Kl. aerogenes* but also the biochemically atypical strains found mainly in the human respiratory tract. In this chapter, therefore, we shall not use the specific name *Kl. pneumoniae*, and shall refer to the biochemical varieties of the organism as *aerogenes*, *rhinoscleromatis*, *ozaenae* and *pneumoniae* strains of the genus *Klebsiella*, corresponding to the subspecies of similar name described in Chapter 34. Although not at present recognized by the taxonomists, the group of respiratory strains described as *Kl. edwardsi* by Cowan and his colleagues (1960) will be mentioned separately; we do not, however, imply that they form a separate species. It should also be noted that *Kl. oxytoca*, though accorded specific status on genetic grounds, closely resembles

the *aerogenes* strains biochemically and is seldom identified separately by clinical bacteriologists.

Most pathogenic members of the genus *Enterobacter* belong to the species *aerogenes* or *cloacae*. The organism often referred to in the American literature as *Ent. agglomerans* will be mentioned later in this chapter (p. 302) under the name *Erwinia herbicola*. In *Serratia* the main pathogen is *Ser. marcescens*, but strains of the species *liquefaciens*, *odorifera*, and occasionally *plymuthica* may be encountered.

Typing methods Recent interest in the epidemiology of klebsiella and serratia infections in hospitals has led to demands for facilities to type large numbers of strains of these organisms. The available methods were outlined in Chapter 34. For klebsiella strains of the *aerogenes* variety, capsular serotyping became the method of choice when typing sera became widely available. The number of recognizable serotypes is large, and the typing system is therefore sufficiently discriminating for most purposes. If a single serotype becomes locally prevalent, attempts to subdivide it, for example, by susceptibility to bacteriocines, may be helpful; but the results of such 'passive' bacteriocine-typing methods are only moderately reproducible (Simoons-Smit *et al.* 1983) and 'active' bacteriocine typing of klebsiellae is not practicable.

Typing of *Serratia* presents difficulties of the sort encountered with *Ps. aeruginosa* (p. 283). O serotyping, if used alone, is a poor means of distinguishing between strains because the number of types is small. Bacteriocine typing is usually performed by testing for susceptibility to bacteriocines of the A class (Chapter 34). It has been used as an alternative to O serotyping, but small differences in the pattern of susceptibility on replicate testing are rather common (Anderhub *et al.* 1977, Traub 1978) and limit the usefulness of the method. According to Pitt and his colleagues (1980), subdivision of the more common O types either by the detection of H antigens or by phage typing gives adequate discrimination for most epidemiological studies (see also Pitt 1982).

Klebsiella

Infections of the respiratory tract in man

Pneumonia The 'pneumobacillus' described by Friedländer (1883) as a cause of fatal pneumonia is difficult to place in modern classifications of *Klebsiella*. There is little doubt, however, about the existence of the characteristic type of pneumonia associated with the name of this author, though it is not common. Friedländer's pneumonia is usually an acute lobar consolidation, often affecting an upper lobe and tending to proceed to abscess formation, but sometimes the disease is chronic from the start (Limson *et al.* 1956). Cases are sporadic and usually develop outside hospital in elderly males who are suffering from other debilitating diseases (Wylie and Kirschner 1950). About one-half of the infections are accompanied by bacteraemia and the mortality rate is high.

The causative organism is in most cases a biochemically atypical klebsiella that belongs to capsular type 1, 3, 4 or 5. Foster and Bragg (1962) concluded that the klebsiellae from such cases usually corresponded not to the biotype now called *pneumoniae* but to the organism described as *Kl. edwardsi* var. *edwardsi* (Cowan *et al.* 1960; see Chapter 34). Of eight strains of this organism from sputum, all were from cases of acute pneumonia, five of which were fatal and two of which led to the formation of lung abscesses. Steinhauer and co-workers (1966) found only 14 members of capsular types 1, 3, 4 and 5 among 205 strains of *Klebsiella* from clinical infections; 11 of them were from cases of pneumonia.

Biochemically typical *aerogenes* organisms, nearly all of which belong to capsular types other than 1, 3, 4 and 5, are commonly found in sputum, particularly of hospital patients who have received antibiotics. As we shall see, colonization of the upper respiratory tract with klebsiellae is now very common in hospital patients, and if the patient is producing sputum this is likely to contain the organism. Cooke and her colleagues (1979) found klebsiellae in 31 per cent of specimens of sputum from hospital patients submitted to the laboratory. Most such isolations are probably of no clinical significance. There is some evidence that klebsiellae of the *aerogenes* type may be responsible occasionally for pneumonia, but this is very difficult to establish in the individual case. Thus, Steinhauer and his colleagues (1966) observed that, of 16 patients with 'broncho-pneumonia' who had klebsiellae of capsular types other than 1, 3, 4 or 5 in the sputum, 5 had bacteraemia and 4 of these died; but evidence of this sort does not establish with certainty that the klebsiella was the cause of the pulmonary lesion.

It has been suggested that the presence of klebsiellae in the sputum in cases of chronic bronchial disease, especially bronchiectasis, may be of clinical significance. Burns (1968) found that such patients usually had serum precipitins to an 'enterobacterial' antigen, and associated this with the presence of organisms of the *ozaenae* biotype in the sputum. Fallon (1973) reported that over one-half of klebsiella strains from exacerbations of chronic bronchitis belonged to the *ozaenae* biotype. This cannot be attributed to the administration of antibiotics because these organisms are usually sensitive to broad-spectrum agents.

Rhinoscleroma

This is a chronic disease of the upper respiratory tract that occurs endemically in Poland, Hungary, South-West Russia, Central America, Egypt and Indonesia. It is commoner in women than in men and usually appears in early adult life. Multiple cases often occur in families at long intervals of time. The disease is associated with prolonged residence under overcrowded and insanitary conditions (Murrell 1966, Krasnilikov and Izraitel 1966). In the Lublin province of Poland it occurs most often in agricultural communities in marshy districts (Semczuk *et al.* 1968).

Lesions develop in the nose, larynx, throat and trachea, according to Klonowski (1968) in 86, 69, 47 and 7 per cent of cases respectively. They are granulomatous infiltrations of the submucosa, the mucosa remaining intact. The fibrous granulations contain many plasma cells and smaller numbers of characteristic vacuolated, foamy macrophages known as Mickulicz's cells (Cunning and du Guerry 1942). Nasal respiration is often impaired and thick mucus dries to form crusts. Extensive scarring may occur.

As first shown by von Frisch (1882) the highly characteristic *rhinoscleromatis* variety of *Klebsiella* can usually be obtained from the lesions. It may be difficult to isolate from swabs, and culture of excised tissue may be necessary. According to Klonowski (1968) the organism can be isolated from about 90 per cent of cases. Many attempts to reproduce the disease failed, but Steffen and Smith (1961) and Godoy (1966) were able to set up a chronic granulomatous pneumonia in mice by the intranasal instillation of freshly isolated organisms.

Complement-fixing antibodies against the *rhinoscleromatis* organism are present in the serum of the majority of patients (Tomášek 1925, Kouwenaar *et al.* 1934, Levine and Hoyt 1948, Krasnilikov 1968). According to Klonowski (1968), 75 per cent of patients initially gave a positive result in this test; this percentage rose to 95 during treatment with streptomycin but eventually fell to 40. Other serological tests include agglutination with capsulated *rhinoscleromatis* organisms, indirect haemagglutination with capsular polysaccharide, and the detection of incomplete antibody by Coombs's method (for references see Klonowski 1968).

Treatment with streptomycin halts the disease in 60 per cent of cases, but only if total dosage exceeds 120 g (Klonowski 1968).

Ozaena

This name was given to a chronic disease characterized by atrophy of the nasal mucosa and the formation of greenish-yellow crusts with a foul and particularly penetrating odour. Klebsiellae of the *ozaenae* biotype have been isolated from the nasal secretions on a number of occasions, but the claim that they are the cause of the disease has not been substantiated. The disease is seldom reported nowadays; it may well have been a foetid sinusitis caused by gram-negative non-sporing anaerobes. (See also Perez's bacillus in Chapter 35.)

Other infections in man

With the exceptions noted above, the klebsiella strains responsible for septic infections are generally considered to belong to the *aerogenes* biotype. It must be admitted, however, that biochemical tests to exclude the phenetically similar *Kl. oxytoca* are by no means always employed in clinical bacteriology laboratories. Some workers, for example Davis and Matsen (1974) and Edmondson and his colleagues (1980), have recorded substantial proportions of indole-positive strains among klebsiellae isolated from clinical sources. The role of *Kl. oxytoca* as a human pathogen merits further study.

Most klebsiella infections seen in hospital patients are of the urinary tract—69 per cent in the series from the Boston General Hospital (Steinhauer *et al.* 1966). These showed the expected association with pre-existing disease of the urinary tract (55 per cent) and catheterization (71 per cent), and some 80 per cent of infections were judged to have been acquired in hospital. Infections of wounds and cannulation sites amounted to no more than 4 per cent of the total. However, hospital epidemics of klebsiella wound infection, and of meningitis after intracranial operations (Price and Sleigh 1970), are on record.

In units for the care of sick babies the spread of a single klebsiella strain may lead to serious illnesses, including septicaemia and meningitis (Hill *et al.* 1974). On the other hand, Riser and her colleagues (1980), who isolated klebsiellae from 98 babies in a special-care unit, considered that only 14 babies suffered illness definitely attributable to this and observed only one systemic infection.

Systemic infections

Septicaemia is a common complication of klebsiella infections in hospital patients. Steinhauer and co-workers (1966) obtained positive blood cultures in 23 per cent of all clinical infections. The mortality rate was 56 per cent; one-half of the deaths took place within 48 hr of the taking of the first positive blood culture. Endotoxic shock is a prominent feature of septicaemia associated with chronic urinary-tract infection. Bacteraemia, not usually preceded by a local septic infection, occurs frequently in patients under treatment for cancer, especially if they are neutropaenic (Umsawasdi *et al.* 1973). Metastatic abscess formation is infrequent, but occasional cases of meningitis, intra-abdominal abscesses and empyema are seen.

Capsular polysaccharide can be detected by counter-current immunoelectrophoresis in about one-half of bacteraemic infections, but the reaction is stronger in patients with a major purulent lesion, e.g. in Friedländer-type pneumonia and meningitis, than in patients with a simple bacteraemia (Pollack 1976). Capsular material, though non-toxic, enhances mortality when given intraperitoneally along with non-capsulate organisms such as streptococci and salmonellae (Kato *et al.* 1976*a*). This is attributed to a neutral polysaccharide component rather than to the acidic type-specific antigen. The effect of the neutral polysaccharide is said to be to delay the emigration of leucocytes into the peritoneal cavity (Kato *et al.* 1976*b*).

Epidemiology of hospital-acquired infection

Although *aerogenes*-like klebsiellae seem to have caused infections in patients with chronic urinary obstruction in pre-antibiotic days, their emergence as an important cause of infections in hospitals was clearly related to antibiotic use. This happened early because the *aerogenes* organisms were from the beginning naturally resistant to currently available antibiotics. They have since retained this advantage by acquiring resistance to new antibiotics to which they were initially sensitive. The 'respiratory' biotypes of klebsiellae, on the other hand, were much less often resistant to broad-spectrum agents and have shown little or no tendency to spread in hospitals.

Earlier investigations suggested that urinary-tract infection was commonly spread from patient to patient by indirect contact. Urine appeared to be the main source of infection, and transmission was on contaminated equipment or the hands of staff; catheterization, cystoscopy and continuous bladder drainage provided the occasions for this (Chapter 58). In the 1960s, a number of common-source epidemics came to light in which the klebsiellae had multiplied in various fluids and were subsequently introduced into the blood stream, the respiratory tract or the bladder. On the whole, however, klebsiellae appear to have been responsible for fewer outbreaks of this nature than have *Ps. aeruginosa* or *Serr. marcescens*. More recently, increasing attention has been given to the faeces as a reservoir of klebsiellae in hospital patients. Selden and his colleagues (1971) isolated them from the faeces of 25 per cent of patients in hospital wards but of only 2 per cent of the hospital staff (see also Montgomerie *et al.* 1970). Acquisition of the organisms could be related to the giving of antibiotics, and patients who acquired them in hospital were at significantly greater risk of infection than non-carriers or patients who were already colonized when admitted to hospital (Selden *et al.* 1971). In addition to providing opportunities for self-infection, high faecal carrier rates increased the number of sources from which other patients might become infected by the contact route.

Faecal carriage is usually accompanied by the presence of the organism in the nasopharynx and on the skin of the hands and groin. Like *Ps. aeruginosa*, klebsiellae are often found on the hands of nurses after contact with infected or carrier patients; however, according to Casewell and Phillips (1977) they are much less easily removed by the washing with soap.

Long-term studies in wards for the chronic sick and in intensive-care units reveal a complicated situation in which many distinguishable strains of klebsiellae are present, some appearing sporadically, others infecting a few patients, and a few becoming established endemically (see, for example, Casewell and Phillips 1978b). Many of the strains have acquired a formidable array of additional resistances; in recent years wide dissemination of individual strains has been associated with resistance to gentamicin, which first appeared in the mid-1970s (Casewell *et al.* 1977, Rennie and Duncan 1977, Curie *et al.* 1978). Casewell and Talsania (1981) found that most of the gentamicin-resistant strains in British hospitals were also resistant to tobramycin and at least 10 other antibiotics. Resistance to gentamicin is often mediated by an antibiotic-destroying enzyme and is determined by a variety of transferable plasmids that also code for resistance to various other antibiotics; however, in some strains it is non-enzymic and not transferable (Hardy *et al.* 1980).

It should not necessarily be concluded that all the *aerogenes* organisms that infect hospital patients come from a human source. Similar organisms are common in surface waters and other moist situations in the external environment, where they appear to lead an existence independent of faecal contamination (Chapter 9). It has therefore been suggested that klebsiellae ingested in food may contribute to the intestinal colonization of patients who are receiving antibiotics (Montgomerie *et al.* 1970). The organisms are widely distributed in hospital kitchens, and foods prepared there may contain considerable numbers of klebsiellae (Cooke *et al.* 1980). Casewell and Phillips (1978a) drew attention to the frequency of such contamination of liquids used for nasogastric feeding and observed an association between the serotypes of klebsiellae fed to patients and those that appeared in their faeces. However, the predominant klebsiellae found in natural waters and on vegetables tend to differ biochemically from those that cause human infections (Naemura and Seidler 1978). Edmondson and his colleagues (1980) confirmed this, and found that strains isolated in hospital kitchens and from hospital food were usually of the 'human' rather than the 'environmental' type.

Treatment It is often difficult to select an appropriate antibiotic for the treatment of serious infections with *aerogenes*-type klebsiellae in hospital patients. In 1981, Casewell and Talsania found that nearly all strains were still susceptible *in vitro* to cefotaxime, cefuroxime, cefoxitin and amikacin.

Disease in animals

Numerous outbreaks of mastitis due to klebsiellae have been recorded in herds of milking cattle (Buntain and Field 1953, Barnes 1954, Hinze 1956, Easterbrooks and Plastridge 1956, White 1957). In some cases the disease is chronic, but in others the animals are acutely ill with fever, weakness and respiratory symptoms, and may die. The disease can be reproduced by intramammary instillation of the organism. Calves fed the milk from infected cows may suffer a fatal pneumonia with septicaemia. A variety of capsular types have been associated with bovine mastitis (Braman *et al.* 1973) (See also Chapter 57.) Klebsiella mastitis also occurs in pigs.

Capsular type 2 strains have been isolated from metritis in mares (Edwards 1928) and a paralytic tick-borne disease in moose (Wallace *et al.* 1933). Flamm (1957) observed an epidemic of neck abscesses and cervical adenitis in laboratory mice; cases of pneumonia and disseminated abscess formation were seen. An epidemic of pneumonia, fibrinous pleurisy and occasional cases of peritonitis due to type 3 organisms in a guinea-pig colony was described by Dennig and Eidmann (1960).

Enterobacter

Members of this genus cause some infections in hospital patients, but these are much less common than klebsiella infections. They cause inflammation mainly in the urinary tract, usually under circumstances rather similar to those of klebsiella urinary-tract infections. On the whole, *Ent. cloacae* seems to be rather more pathogenic than *Ent. aerogenes* for man; thus, bacteraemia occurred in 5 of 14 urinary-tract infections with *Ent. cloacae* but only in 1 in 11 such infections with *Ent. aerogenes* (Steinhauer *et al.* 1966). Mayhall and co-workers (1979) describe an epidemic of bacteraemia caused by *Ent. cloacae* in a burns ward. Like *Klebsiella* and *Serratia*, strains of *Enterobacter* grow in fluids with minimal nutrients; *Ent. cloacae* caused many cases of bacteraemia when present, along with *Erwinia herbicola* in a contaminated batch of fluid that was given intravenously to patients (Report 1971). Pigmented enterobacter strains, now called *Ent. sakazakii*, have occasionally been found in the blood stream (Monroe and Tift, 1979) and have caused a few cases of neonatal meningitis (Muytgens *et al.* 1983).

Serratia

Before 1960 almost all accounts of *Serratia* as a pathogen were concerned with pigmented strains. Soon after the genus had been defined biochemically it became clear that non-pigmented members of it were quite prevalent among hospital patients. Clayton and von Graevenitz (1966) isolated serratiae 181 times in one hospital in the space of a year, but obtained little evidence that they were significant pathogens. On the other hand, Stenderup and co-workers (1966) described six fatal septicaemia infections within a few months in a unit for premature babies, and Dodson (1968) reported many cases of post-operative septicaemia. Altemeier and his colleagues (1969), in a 7-year survey of serratia infections, reported a sharp increase in their number between 1965 and 1966. Most of the infections developed in patients who had received antibiotics; the organism had been introduced via the urinary tract, sites of cannulation, or—in patients on mechanically assisted ventilation—the respiratory tract (see also Cabrera 1969). Similar reports appeared with increasing frequency in subsequent years.

Serratia infections occur predominantly in hospital patients and usually follow the transfer of the pathogen to the site of invasion in the course of a medical or nursing procedure. The consequences of this depend not only on the site contaminated but on the susceptibility of the recipient. Thus widespread colonization of healthy newborn babies may not result in any clinical illnesses (McCormack and Kunin 1966), but infections acquired in neonatal intensive-care units are often highly fatal. As with klebsiellae, difficulties are encountered in assessing the clinical significance of *Serratia* in the sputum.

Systemic infections

These are quite common and highly fatal (Wilfert *et al.* 1968); they are usually associated with antigenaemia (Crowder *et al.* 1971) and often with clinical signs of endotoxic shock (Vic-Dupont *et al.* 1969).

Some strains of *Serratia* isolated from clinical specimens are sensitive to the bactericidal action of unheated normal serum, but strains that cause bacteraemia are invariably serum resistant (Simberkoff *et al.* 1976*a*). Serum-resistant strains are ingested and killed by human leucocytes in the presence of fresh normal serum, which contains complement-dependent opsonins, but this serum does not protect mice against lethal infection with virulent strains. Vaccination of mice with heat-killed organisms confers an immunity that is O-type specific and appears to be associated with the presence of complement-independent opsonins; similar opsonins appear in the sera of some patients convalescent from septicaemic serratia infections (Simberkoff *et al.* 1976*b*).

Serratia organisms are an occasional cause of endocarditis in hospital patients; in some geographical regions they are said to be responsible for a considerable proportion of cases of endocarditis in intravenous users of narcotics (Mills and Drew 1976).

Epidemiology of hospital-acquired infection

The ability of *Serratia* to multiply in fluids with a minimal content of organic nutrients and at ambient temperature accounts for the prominence of common-source epidemics in which the organisms are introduced in such fluids into the blood stream (Rabinowitz and Schriffrin 1952, Donowitz *et al.* 1979, Ehrenkranz *et al.* 1980), the respiratory tract (Ringrose *et al.* 1968, Sanders *et al.* 1970), or the bladder (Cabrera 1969). Cold-tolerant strains may multiply in bottled blood and give rise to fatal transfusion reactions (Black *et al.* 1967). According to Farmer and his colleagues (1976) about one-half of all incidents of infection in hospitals are common-source outbreaks caused by strains of normal antibiotic susceptibility and the rest are more long-continued endemic prevalences in which infection is spread from patient to patient and the causative strain has additional antibiotic resistances.

Spread of infection from patient to patient appears to be by indirect contact, mediated either by contaminated apparatus or the hands of members of the staff. The hands may have been recently contaminated by contact with patients (Maki *et al.* 1973) or ward utensils (Rutala *et al.* 1981), but some members of the staff may be persistent hand carriers (Mutton *et al.* 1981).

Faecal carriage of *Serratia* is uncommon in the general population, but some hospital patients become carriers. However, this appears to occur much less often than with klebsiellae, and self-infection in the colonized patient is correspondingly less important.

Antibiotic treatment   Serratia strains are all resistant to cephalosporins. Some are moderately sensitive to ampicillin and carbenicillin, but many possess a plasmid-borne β-lactamase that renders them highly resistant to both antibiotics. Many are sensitive to cefoxitin. Although resistant *in vitro* to polymyxins, these agents are said to exhibit synergy with certain other drugs (see below). Until recently, the aminoglycosides were the agents of first choice for the treatment of serious infections, but resistance to gentamicin and amikacin had become prevalent in the USA by the mid-1970s, though not apparently in Britain (Gray *et al.* 1978). Cotrimoxazole, colistin and cotrimoxazole, and colistin and rifampicin are possible alternatives to aminoglycosides. (For a general account of serratia infections, see Yu 1979.)

Proteus

Pr. mirabilis, and less often *Pr. vulgaris*, are common causes of urinary-tract infections and various systemic infections in man.

Pr. vulgaris appears not to be a uniform species. Three main biochemical varieties have been recognized (Hickman *et al.* 1982): (1) indole negative, salicin and aesculin not attacked; (2) indole positive, acid from salicin and aesculin hydrolysed; and (3) indole positive, salicin and aesculin tests negative. The name

Pr. penneri is now proposed for group 1, all strains of which are said to be resistant to chloramphenicol; its natural pathogenicity has not yet been recorded.

Infections of the urinary tract

In the general population, *Proteus* is an infrequent cause of urinary-tract infection except in male children. It is a prominent cause of cystitis in previously healthy boys up to the age of puberty (Saxena and Bassett 1975, Hallett *et al.* 1976). The illness is usually benign but may be recurrent. The organism is present in the preputial secretions of sufferers but in only about 10 per cent of age-matched controls (Hallett *et al.* 1976). A few infections occur in adults, especially in elderly males with urinary obstruction. However, infection in adults is most often seen in patients in urological wards after surgical operations or instrumentation (Chapter 58). Single strains of *Proteus* are sometimes endemic in such wards.

The urine becomes alkaline in proteus infection and, if the kidney is involved, renal calculi often form; this is attributed to the local production of urease by the organism. Urease production assists the organism in invading the renal pelvis (Braude and Siemienski 1960, MacLaren 1969); the release of alkali damages the urinary epithelium and inactivates complement. Treatment of rats with inhibitors of urease prevents the development of pyelonephritis after *Pr. mirabilis* has been instilled into the bladder (Musher *et al.* 1975). According to Edén and co-workers (1980), *Pr. mirabilis* differs from *Esch. coli* strains from urinary-tract infections in that it adheres to squamous- but not to transitional-epithelial cells.

Systemic infections Septicaemia, usually with meningitis, may occur in epidemics among newborn babies in hospital (Shortland-Webb 1968, Burke *et al.* 1971); it is often preceded by umbilical colonization. With this exception, septicaemia occurs only in patients with serious underlying diseases or as a complication of operations on the urinary or biliary tracts, or occasionally of parturition (Lewis and Fekety 1969, Adler *et al.* 1971). According to Larsson and Olling (1977) strains from the blood stream are more often resistant to the bactericidal action of normal human serum than are strains from the urinary tract.

Other hospital-acquired infections Miscellaneous infections, mainly of operation wounds and bedsores, occur sporadically in most hospital wards and are usually considered to be self-infections from the gut flora (Story 1954, Adler *et al.* 1971). Endemic prevalences of individual strains have been reported (Larrson *et al.* 1978, Chow *et al.* 1979) and variously attributed to contact-spread by staff or to faecal colonization of patients. In recent years they have been increasingly associated with the spread of aminoglycoside-resistant strains (see Chapter 35, which also contains further information about the antibiotic sensitivity of proteus strains). Although reportedly transmitted on contaminated apparatus and utensils, outbreaks of proteus infection associated with the multiplication of the organism in fluids seldom appear to occur.

Morganella morgani is an infrequent cause of clinical disease. Occasional small groups of related infections in hospital patients have been reported (Tucci and Isenberg 1981, Williams *et al* 1983). According to Senior (1983) it grows more slowly in urine than does *Pr. mirabilis* and forms less alkali.

Providencia

Of the three species of *Providencia* only two, *Prov. stuarti* and *Prov. rettgeri*, are important causes of disease in man. Infections are almost confined to hospital patients, are mainly of the urinary tract, and are particularly associated with the use of indwelling catheters (Solberg and Matsen 1971, Fischer 1972). Infection of wounds and burns may occur; septicaemia is infrequent but may sometimes contribute to death. Long-term prevalences of providencia infections, especially in wards for the chronic sick, have been reported with increasing frequency since 1970; some of these are caused by single strains, but in others many different strains are present (Kocka *et al.* 1980, Toni *et al.* 1980). It seems, therefore, that most individual strains have limited ability to spread among patients. Strains of *Providencia* are often found in the gut flora, and their pattern of resistance to antimicrobial agents is always very broad (Chapter 35); they are therefore easily selected in hospital departments in which antibiotics are used lavishly. Transferable R factors for carbenicillin and various aminoglycosides are now prevalent in many hospitals (see Chapter 35 and McHale *et al.* 1981).

Other members of the Enterobacteriaceae

Citrobacter

Members of the two main species, *Cit. freundi* and *Cit. koseri*, are commonly found in human faeces though seldom as the predominant aerobic organism. They are isolated from time to time from clinical specimens, notably of sputum and urine (Hodges *et al.* 1978), but the significance of this is often difficult to evaluate.

The one clear instance of pathogenicity is the ability of *Cit. koseri* to cause meningitis in newborn babies (McKay and Smith 1958, Gross *et al.* 1973; see also Chapter 34). The disease affects mainly premature infants in the first 2 weeks of life. Cases tend to occur in small groups, each associated with a particular nursery but widely spaced in time. The mortality rate is high, and brain abscesses develop in three-quarters of the cases (Graham and Band 1981). In one special-care baby unit in which *Cit. koseri* of a single serotype caused three cases of meningitis in $1\frac{1}{2}$ years, faecal carriage of the organism was frequent and often profuse in healthy contact babies (Ribiero *et al.* 1976); it was estimated that meningitis developed in only 4.6 per cent of the colonized babies.

Citrobacter strains are resistant to ampicillin; *Cit. koseri* is regularly resistant to carbenicillin but sometimes sensitive to cephalosporins; *Cit. freundi* is carbenicillin sensitive but cephalosporin resistant (Chapter 34).

Edwardsiella

These organisms are particularly associated with the intestinal tract of cold-blooded vertebrates and have been found on a number of occasions in human faeces in South-East Asia (see Chapter 34). They have been reported as an uncommon cause of meningitis, septicaemia, or both, in man (Okubadejo and Alausa 1968, Sonnenworth and Kallus 1968, Jordan and Hadley 1969, Sachs *et al.* 1974).

Erwinia

Gram-negative bacilli that ferment sugars and form a yellow non-diffusible pigment are occasionally found in clinical material. Some of them are *Ent. sakazakii* or *Esch. vulneris* (p. 295); others—usually anaerogenic— appear to belong to the poorly characterized genus *Erwinia*, which includes a number of plant pathogens.

The organism we describe in Chapter 34 as *Erw. herbicola* ('*Ent. agglomerans*') is occasionally isolated from the human upper respiratory tract or urine, or from swabs of superficial lesions (Meyers *et al.* 1972), and from various animals (Muraschi *et al.* 1965, Lev *et al.* 1969). Similar organisms are common in water and soil and may be found on surfaces and in moist places in hospitals. Such serious human infections as have been described appear to have arisen from heavy exposure of patients by the intravenous or the respiratory route. Fluids for intravenous use have become contaminated with *Erw. herbicola* on several occasions (Lapage *et al.* 1973). In one widespread epidemic associated with the intravenous injection of a fluid containing *Erw. herbicola* and *Ent. cloacae* in large numbers (Report 1971), patients had septicaemic illnesses in which either or both organisms were present in the blood. They suffered intermittent pyrexia but symptoms of endotoxic shock were not prominent. Ansorg and his colleagues (1974) reported a fatal septicaemic illness in a newborn baby attributed to the contamination of respiratory apparatus with *Erw. herbicola*.

Other aerobic gram-negative bacteria

Vibrio

Certain of the halophilic vibrios (Chapter 27) cause septic rather than diarrhoeal diseases. The most im-

portant of these are the organisms often referred to as lactose-positive or L+ vibrios, for which the name *Vib. vulnificus* has now been proposed. They cause two distinct diseases (Blake *et al.* 1979): (1) wound infection, often clinically severe and accompanied by fever, in previously healthy persons who have recently sustained a penetrating injury that has been contaminated with sea-water; and (2) septicaemic disease of acute onset with initial pyrexia but often proceeding rapidly to collapse and death. An initial focus of infection cannot be detected in the septicaemic disease, but ecchymoses, bullae and ulceration of the skin may develop during its course. It occurs nearly always in patients with a serious underlying disease, usually of the liver, or in alcoholics. Although the vibrio is sensitive to a wide range of antibiotics, including benzylpenicillin (Hollis *et al.* 1976), over half of the septicaemic patients die despite intensive treatment (Blake *et al.* 1979). Both types of disease affect mainly middle-aged or elderly males in the summer. It has been suggested that the septicaemic infection results from the ingestion of the vibrio, possibly in raw oysters (Hollis *et al.* 1976).

Occasional local infections, particularly of the ear in bathers and fishermen, have been attributed to *Vib. alginolyticus* (see Ayres and Barrow 1978 for references). Another species, *Vib. damsela*, has been isolated from a number of inflamed wounds in otherwise healthy persons in whom the wound had been exposed to salt or brackish water at the time of the injury (Morris *et al.* 1982). Rubin and his co-workers (1981) describe a case of meningitis in a baby caused by a cholera-like vibrio of a serogroup other than 1, i.e. a NAG vibrio (Chapter 27). (For a review of septic infections caused by vibrios see Blake *et al.* 1980).

Aeromonas

Aero. hydrophila causes systemic diseases in cold-blooded animals, such as red leg in frogs (Kulp and Borden 1942) and ascites in carp and other fish (Schäperklaus 1939, Wunder and Dombrowski 1953). In 1954, Hill, Caselitz and Moody isolated it from a fatal case of acute disseminated myositis in an adult human patient; a number of other systemic infections have since been recorded in man.

The organism grows over a wide range of temperatures (4–45°) and is almost universally present in surface waters both fresh and brackish (Hazen *et al.* 1978). It is found from time to time in material collected from mucous membranes and superficial lesions (von Graevenitz and Mensch 1968, McCracken and Barkley 1972) and in human faeces. Systemic infections are uncommon in man and usually occur in persons suffering from severe underlying diseases, notably in patients under treatment for cancer or leukaemia (Ketover *et al.* 1973) or with cirrhosis of the liver (McCracken and Barkley 1972). Disseminated myonecrosis and skin lesions resembling ecthyma gangrenosa develop in some cases. Meningitis and solitary abscesses in internal organs have been described. Wound infections, sometimes clinically severe, have occasionally been attributed to the contamination of recently inflicted wounds with fresh or sea-water (Phillips *et al.* 1974, Hanson *et al.* 1977, Fulghum *et. al.* 1978).

Daily and co-workers (1981) state that strains from lesions tend to differ from free-living strains in having a lower median lethal dose for mice and in possessing various virulence factors; these strains they would place in a separate species, *Aer. sobria*. *Aer. hydrophila* is resistant to penicillin, ampicillin and carbenicillin, usually resistant to cephalothin, and sometimes resistant to gentamicin.

Aer. salmonicida causes an economically important disease called furunculosis in salmon and trout (see Chapter 27), but does not cause disease in warm-blooded animals.

Plesiomonas shigelloides This organism is found quite often in human and animal faeces. It is said to cause ulcerative stomatitis in snakes but has rarely been incriminated as a cause of septic disease in man (see, however, Appelbaum *et al.* 1978).

Chromobacterium

All pathogenic chromobacteria belong to the mesophilic species *Chr. violaceum*. Infections have been recorded in man, cattle, pigs and monkeys (see Sneath 1960, Groves *et al.* 1969). Human infections usually take the form of an acute septicaemia with the formation of multiple abscesses in visceral organs; they are usually fatal. In some cases the first sign is a granulomatous skin lesion. The disease is very uncommon; only about 20 cases have been recorded. Local abscess formation without systemic spread (Victorica *et al.* 1974) is even more rare. Nearly all recorded cases occurred in South-East Asia or the south-eastern USA. As stated in Chapter 32, the organism is commonly present in soil and water in warm countries. Although it is usually sensitive to chloramphenicol, tetracycline and aminoglycosides, treatment of the systemic disease is seldom successful.

Flavobacterium

This genus is composed mainly of saprophytes (Chapter 32) but includes one member with pathogenic ability. *Fl. meningosepticum* has given rise to a number of epidemics of meningitis in newborn babies in hospitals (Brody *et al.* 1958, Vandepitte *et al.* 1958, Cabrera and Davis 1961, George *et al.* 1961, Seligmann *et al.* 1963, Plotkin and McKitrick 1966). Not all of these epidemics were fully investigated, but it was noted (Cabrera and Davis 1961, Seligmann *et al.* 1963) that many of the healthy contacts became temporarily colonized in

the nose; and in one epidemic (Plotkin and McKitrick 1966) the organism appeared to have multiplied in a bottle of saline used for irrigation of babies' eyes. In a detailed study of a nursery in which seven sporadic cases of meningitis due to various serotypes of *Fl. meningosepticum* had occurred in the previous 5 years (Thong *et al.* 1981), the organisms were isolated very frequently from suction apparatus, baths, and aqueous solutions of chlorhexidine, and on one occasion from distilled water from the dispensary. Coyle-Gilchrist and her colleagues (1976) also found *Fl. meningosepticum* on a number of occasions in bottles of chlorhexidine solution kept at use-dilution in hospital wards. The distribution of the organism in the inanimate environment of babies suggests that contamination of skin and mucous membranes, possibly followed by colonization, is the normal means of infection; but the possibility that contamination of fluids for injection or the hands of the injector played a significant part in these incidents cannot be excluded.

The inadvertent injection of large doses of the organism intravenously into adult patients (Olsen *et al.* 1965, Olsen 1967) resulted in short-lived febrile illnesses from which recovery was uneventful. However, a few sporadic bacteraemic or meningitic infections have occurred in adults suffering from severe underlying non-bacterial diseases.

Fl. meningosepticum exhibits a wide spectrum of resistance to antibiotics (Chapter 32). In-vitro tests suggest cotrimoxazole, clindamycin and rifampicin as possible agents for the treatment of serious infections.

Acinetobacter

These organisms, whether of the saccharolytic or the asaccharolytic variety (Chapter 32), are commonly found in soil and water, and also as a constitutent of the skin flora—particularly of the axilla and groin— in about one-quarter of healthy persons. They are often present in superficial lesions, in which they are usually of no significance (Alami and Riley 1966, Rosenthal and Freundlich 1977). However, they do upon occasion give rise to serious infections. These occur mainly in hospital patients and are usually septicaemic or meningitic illnesses, sometimes with a petechial rash and other signs resembling those of the Waterhouse-Friederichsen syndrome (Daly *et al.* 1962, Burrows and King, 1966, Donald and Doak 1967). The isolation of *Acinetobacter* from the sputum in association with pulmonary consolidation has led to a number of reports that the organism causes pneumonia both in hospital patients (Glew *et al.* 1977) and in the general community (Cordes *et al.* 1981).

Septicaemic infections are often associated with prolonged intravenous cannulation (Daly *et al.* 1962). Although they generally appear sporadically, epidemics associated with particular hospital wards are on record. In one of these affecting patients receiving intravenous fluids, the source of infection was bedside humidifiers in which the organism multiplied (Smith and Massanari 1977); aerial dispersion of the organism from the humidifier was demonstrated, and was thought to have been the means of contaminating the sites of cannulation. Respiratory apparatus may also disseminate the organism (Buxton *et al.* 1978, Cunha *et al.* 1980). Hand carriage of the organism by hospital staff, which is frequent and sometimes persistent, may be an important means of dissemination in some ward epidemics (Buxton *et al.* 1978, French *et al.* 1980).

The choice of antibiotics for the treatment of serious infections is often very limited (Chapter 32); some strains now seen in hospitals are resistant to as many as 18 antibiotics (French *et al.* 1980).

'Achromobacter xylosoxidans' McGuckin and co-workers (1982) record an outbreak of bacteraemia affecting 10 patients and attributed to the intravenous injection of a contaminated solution.

References

Adair, F. W., Geftic, S. G. and Gelzer, J. (1969) *Appl. Microbiol.* **18**, 299.

Adler, J. L., Burke, J. P., Martin, D. F. and Finland, M. (1971) *Ann. intern. Med.* **75**, 517, 531.

Alain, M., Saint-Etienne, J. and Reynes, V. (1949) *Méd. trop.* **9**, 119.

Alami, S. Y. and Riley, H. D. (1966) *Amer. J. med. Sci.* **252**, 537.

Alcock, S. R. (1977) *J. Hyg., Camb.* **78**, 395.

Alexander, A. D., Huxsoll, D. L., Warner, A. R., Shepler, V. and Dorsey, A. (1970) *Appl. Microbiol.* **20**, 825.

Alexander, J. W., Brown, W., Walker, H., Mason, A. D. and Moncrieff, J. A. (1966) *Surg. Gynec. Obstet.* **123**, 965.

Alexander, J. W., Fisher, M. W. and MacMillan, B. G. (1971) *Arch. Surg.* **102**, 1.

Alexander, J. W., Fisher, M. W., MacMillan, B. G. and Altemeier, W. A. (1969) *Arch. Surg.* **99**, 249.

Alexander, J. W. and Wixson, D. (1970) *Surg. Gynec. Obstet.* **130**, 431.

Alms, T. H. and Bass, J. A. (1967) *J. infect. Dis.* **117**, 249, 257.

Altemeier, W. A., Culbertson, W. R., Fullen, W. D. and McDonough, J. J. (1969) *Arch. Surg.* **99**, 232.

Anderhub, B., Pitt, T. L., Erdman, Y. J. and Willcox, W. R. (1977) *J. Hyg., Camb.* **79**, 89.

Andriole, V. T. (1979) *J. Lab. clin. Med.* **94**, 196.

Ansari, M. and Minou, M. (1951) *Ann. Inst. Pasteur* **81**, 98.

Ansorg, R., Thomssen, R. and Stubbe, P. (1974) *Med. Microbiol. Immunol.* **159**, 161.

Appelbaum, P. C., Bowen, A. J., Adhikari, M., Robins-Browne, R. M. and Koornhof, H. J. (1978) *J. Pediat.* **92**, 676.

Archer, G., Fekety, F. R. and Supena, R. (1974) *Amer. Heart J.* **88**, 570.

Asay, L. D. and Koch, R. (1960) *New Engl. J. Med.* **262**, 1062.

Ashdown, L. R. (1979a) *Rev. infect. Dis.*, **1**, 891; (1979b) *Med. J. Aust.* **i**, 284; (1981a) *Pathology* **13**, 597; (1981b) *J. clin. Microbiol.* **14**, 361.

Asheshov, E. H. (1974) In: *Proc. 6th Nat. Cong. Bact.*, p. 9. Ed. by A. Arseni, Athens.

Atkinson, B. E., Smith, D. L. and Lockwood, W. R. (1975) *Ann. intern. Med.* **83**, 369.

Ayliffe, G. A. J., Babb, J. R., Collins, B. J. Lowbury, E. J. L. and Newsom, S. W. B. (1974) *Lancet* **ii**, 578.

Ayliffe, G. A. J., Barry, D. R., Lowbury, E. J. L., Roper-Hall, M. J., and Walker, W. M. (1966) *Lancet* **i**, 1113.

Ayres, P. A. and Barrow, G. I. (1978) *J. Hyg., Camb.* **80**, 281.

Babes, A. (1892) *Arch. Méd. exp.* **4**, 450.

Bacon, C. J., Kenna, A. P., Ingham, H. R., Gross, R. J. and Rowe, B. (1975) *Lancet* **ii**, 1091.

Baird, R. M., Brown, W. R. L. and Shooter, R. A. (1976) *Brit. med. J.* **i**, 511.

Balassanian, N. and Wolinsky, E. (1968) *Pediatrics, Springfield* **41**, 463.

Baltch, A. L. and Griffin, P. E. (1977) *Amer. J. med. Sci.* **274**, 119.

Barnes, L. E. (1954) *J. Amer. vet. med. Ass.* **125**, 50.

Barson, A. J. (1971) *Arch. Dis. Childh.* **46**, 245.

Bass, J. A. and McCoy, J. C. (1971) *Infect. Immun.* **3**, 51.

Bassett, D. C. J. (1971a) *Proc. R. Soc. Med.* **64**, 980; (1971b) *J. clin. Path.* **24**, 708; (1971c) *Ibid.* **24**, 798.

Bassett, D. C. J., Stokes, K. J. and Thomas, W. R. G. (1970) *Lancet*, **1**, 1188.

Bassett, D. C. J., Thompson, S. A. S. and Page, B. (1965) *Lancet*, **i**, 781.

Bean, H. S. and Farrell, R. C. (1967) *J. Pharm., Lond.* **19**, Suppl. 183S.

Bergan, T. and Norris, J. R. (Eds) (1978) *Methods in Microbiology*, vol. 10. Academic Press, London.

Bernstein, J. M. and Carling, E. R. (1909) *Brit. med. J.* **i**, 319.

Biegeleisen, J. Z., Mosquera, R. and Cherry, W. B. (1964) *Amer. J. trop. Med. Hyg.* **13**, 89.

Björkstén, B., Bortolussi, R., Gothefors, L. and Quie, P. G. (1976) *J. Pediat.* **89**, 892.

Black, W. A., Pollock, A. and Batchelor, E. L. (1967) *J. clin. Path.* **20**, 883.

Blake, P. A., Merson, M. H., Weaver, R. E., Hollis, D. G. and Heublein, P. C. (1979) *New Engl. J. Med.* **300**, 1.

Blake, P. A., Weaver, R. E. and Hollis, D. G. (1980) *Annu. Rev. Microbiol.* **34**, 341.

Bonome, A. (1906) *Zbl. Bakt.* **38**, 97.

Botzenhart, K. and Röpke, S. (1971) *Arch. Hyg., Berl.* **154**, 509.

Boyden, S. V. (1950) *Proc. Soc. exp. Biol. Med.* **73**, 289.

Braman, S. K., Eberhart, R. J., Asbury, M. A. and Hermann, G. J. (1973) *J. Amer. vet. med. Ass.* **162**, 109.

Braude, A. I., Douglas, H. and Jones, J. (1969) *J. Bact.* **98**, 979.

Braude, A. I. and Siemienski, J. (1960) *J. Bact.* **80**, 171.

Brenner, D. J., McWhorter, A. C., Knutson, J. K. L. and Steigerwalt, A. G. (1982) *J. clin. Microbiol.* **15**, 1133.

Bres, P. (1958) *Méd. trop.*, *Marseille* **18**, 694.

Bridges, K. and Lowbury, E. J. L. (1977) *J. clin. Path.* **30**, 160.

Brill, N. E. and Libman, E. (1899) *Amer. J. med. Sci.* **118**, 153.

Brody, J. A., Moore, H. and King, E. O. (1958) *Amer. J. Dis. Child.* **96**, 1.

Brokopp, C. D. and Farmer, J. J. III (1979) In: *Pseudomonas aeruginosa: Clinical Manifestations of Infection and Current Therapy*, p. 89. Ed. by R. G. Doggett. Academic Press, New York.

Brooks, H. J. L., Benseman, B. A., Peck, J. and Bettelheim, K. A. (1981) *J. Hyg., Camb.* **87**, 53.

Brooks, H. J. L., O'Grady, F., McSherry, M. A. and Cattell, W. R. (1980) *J. med. Microbiol.* **13**, 57.

Brown, M. R. W. and Foster, J. H. S. (1970) *J. clin. Path.* **23**, 172.

Brown, V. I. and Lowbury, E. J. L. (1965) *J. clin. Path.* **18**, 752.

Brumfitt, W., Percival, A. and Leigh, D. A. (1967) *Lancet* **i**, 1289.

Brunton, W. A. T. and Heggie, D. (1979) *J. Infect.* **1**, 183.

Buck, A. C. and Cooke, E. M. (1969) *J. med. Microbiol.* **2**, 521.

Bulloch, W. (1929) In: *A System of Bacteriology in Relation to Medicine*, vol. 4, p. 326. Med. Res. Council, London.

Buntain, D. and Field, H. I. (1953) *Vet. Rec.* **65**, 91.

Burke, J. P., Ingall, D., Klein, J. O., Gezon, H. M. and Finland, M. (1971) *New Engl. J. Med.* **284**, 115.

Burke-Gaffney, H. J. O'D. (1933) *J. Hyg., Camb.* **33**, 510.

Burns, M. W. (1968) *Lancet* **i**, 383; (1973) *Brit. med. J.* **iii**, 382.

Burns, M. W. and May, J. R. (1968) *Lancet*, **i**, 270.

Burrows, S. and King, E. O. (1966) *Amer. J. clin. Path.* **46**, 234.

Buxton, A. E., Anderson, R. L., Werdegar, D. and Atlas, E. (1978) *Amer. J. Med.* **65**, 507.

Cabrera, H. A. (1969) *Arch. intern. Med.* **123**, 650.

Cabrera, H. A. and Davis, G. H. (1961) *Amer. J. Dis. Child.* **101**, 289.

Calabi, O. (1973) *J. med. Microbiol.*, **6**, 293.

Carson, L. A., Favero, M. S., Bond, W. W. and Petersen, N. J. (1972) *Appl. Microbiol.* **23**, 863.

Carson, L. A., Petersen, N. J., Favero, M. S., Doto, I. L., Collins, D. E. and Levin, M. A. (1975) *Appl. Microbiol.* **30**, 935.

Casewell, M. W., Dalton, M. T., Webster, M. and Phillips, I. (1977) *Lancet* **ii**, 444.

Casewell, M. W. and Phillips, I. (1977) *Brit. med. J.* **ii**, 1315; (1978a) *J. clin. Path.* **31**, 845; (1978b) *J. Hyg., Camb.* **80**, 295.

Casewell, M. W. and Talsania, H. G. (1981) *J. antimicrob. Chemother.* **7**, 237.

Cason, J. S. and Lowbury, E. J. L. (1960) *Lancet* **ii**, 501.

Chambon, L. (1955) *Ann. Inst. Pasteur* **89**, 229.

Chow, A. W., Taylor, P. R., Yoshikawa, T. T. and Guze, L. B. (1979) *J. infect. Dis.* **139**, 621.

Clayton, E. and Graevenitz, A. von (1966) *J. Amer. med. Ass.* **197**, 1059.

Collins, K. R. (1908) *J. infect. Dis.* **5**, 401.

Cook, I. (1962) *Med. J. Aust.* **ii**, 627.

Cooke, E. M., Brayson, J. C., Edmondson, A. S. and Hall, D. (1979) *J. Hyg., Camb.* **82**, 473.

Cooke, E. M. and Ewins, S. P. (1975) *J. med. Microbiol.* **8**, 107.

Cooke, E. M., Sazegar, T., Edmondson, A. S., Brayson, J. C. and Hall, D. (1980) *J. Hyg., Camb.* **84**, 97.

Cooper, R. G. (1967) *Med. J. Aust.* **i**, 527.

Cordes, L. G. *et al.* (1981) *Ann. intern. Med.* **95**, 688.

Cowan, S. T., Steel, K. J., Shaw, C. and Duguid, J. P. (1960) *J. gen. Microbiol.* **23**, 601.

Coyle-Gilchrist, M. M., Crewe, P. and Roberts, G. (1976) *J. clin. Path.* **29**, 824.

Cravitz, L. and Miller, W. R. (1950) *J. infect. Dis.* **86**, 53.

Cross, A. S., Sadoff, J. C., Inglewski, B. H. and Sokal, P. A. (1980) *J. infect. Dis.* **142**, 538.

Crotty, J. M., Bromwich, A. F. and Quinn, J. V. (1963) *Med. J. Aust.* **i**, 274.

Crowder, J. G., Devlin, H. B., Fisher, M. and White, A. (1974) *J. Lab. clin. Med.* **83**, 853.

Crowder, J. G., Gilkey, G. H. and White, A. C. (1971) *Arch. intern. Med.* **128**, 247.

Csiszár, K. and Lányi, B. (1970) *Acta microbiol. hung.* **17**, 361.

Cunha, B. A., Klimek, J. J., Gracewski, J., McLaughlin, J. C. and Quintiliani, R. (1980) *Post Grad. med. J.* **56**, 169.

Cunning, D. S. and Guerry, P. du (1942) *Arch. Otolaryng., Chicago* **36**, 662.

Curie, K., Speller, D. C. E., Simpson, R. A., Stephens, M. and Cooke, D. I. (1978) *J. Hyg., Camb.* **80**, 115.

Curtis, P. E. (1969) *Vet. Rec.* **84**, 476.

Czirók, É., Milch, H., Madár, J. and Semjén, G. (1977) *Acta microbiol. hung.* **24**, 115.

Daily, O. P. *et al.* (1981) *J. clin. Microbiol.* **13**, 769.

Daly, A. K., Postic, B. and Kass, E. H. (1962) *Arch. intern. Med.* **110**, 580.

Darrell, J. H. and Wahba, A. H. (1964) *J. clin. Path.* **17**, 236.

Davis, B. R., Ewing, W. H. and Reavis, R. W. (1957) *Int. Bull. bact. Nomencl.* **7**, 151.

Davis, T. J. and Matsen, J. M. (1974) *J. infect. Dis.* **130**, 402.

Demko, C. A. and Thomassen, M. J. (1980) *Curr. Microbiol.* **4**, 69.

Dennig, H. K. and Eidmann, E. (1960) *Berl. Münch. tierärztl. Wschr.* **73**, 273.

Dietzman, D. E., Fischer, G. W. and Schoenknecht, F. D. (1974) *J. Pediat.* **85**, 128.

Dijk, W. C. van, Verbrugh, H. A., Peters, R., Tol, M. E. van der, Peterson, P. K. and Verhoef, J. (1977) *J. med. Microbiol.* **10**, 123.

Dodin, A., Ferry, R., Moussa, A., Sanson, R. and Guénolé, A. (1974) *Bull. Soc. Path. exot.* **67**, 121.

Dodson, W. H. (1968) *Arch. intern. Med.* **121**, 145.

Doggett, R. G., Harrison, G. M. and Carter, R. E. Jr. (1971) *Lancet* **i**, 236.

Doggett, R. G., Harrison, G. M., Stilwell, R. N. and Wallis, E. S. (1966) *J. Pediat.* **68**, 215.

Doggett, R. G., Harrison, G. M. and Wallis, E. S. (1964) *J. Bact.* **87**, 427.

Donald, W. D. and Doak, W. M. (1967) *J. Amer. med. Ass.* **200**, 287.

Donowitz, L. G., Marsik, F. J., Hoyt, J. W. and Wenzel, R. P. (1979) *J. Amer. med. Ass.* **242**, 1749.

Doroghazi, R. M., Nadol, J. B. Jr, Hyslop, N. E. Jr, Baker, A. S. and Axelrod, L. (1981) *Amer. J. Med.* **71**, 603.

Drake, C. H. (1966) *Hlth Lab. Sci.* **3**, 10.

Drewett, S. E., Payne, D. J. H., Tuke, W. and Verdon, P. E. (1972) *Lancet* **i**, 946.

Dudgeon, L. S., Symonds, S. L. and Wilkin, A. (1918) *J. comp. Path.* **31**, 43.

Easterbrooks, H. L. and Plastridge, W. N. (1956) *J. Amer. vet. med. Ass.* **128**, 502.

Edén, C. S., Eriksson, B. and Hanson, L. Å. (1977) *Infect. Immun.* **18**, 767.

Edén, C. S. and Hannson, H. A. (1978) *Infect. Immun.* **21**, 229.

Edén, S. C., Larsson, P. and Lomberg, H. (1980) *Infect. Immun.* **27**, 804.

Edmondson, A. S., Cooke, E. M., Wilcock, A. P. D. and Shinebaum, R. (1980) *J. med. Microbiol.* **13**, 541.

Edwards, P. R. (1928) *J. Bact.* **15**, 245.

Ehrenkranz, N. J., Bolyard, E. A., Wiener, M. and Cleary, T. J. (1980) *Lancet* **ii**, 1289.

Eikhoff, T. C., Bennett, J. V., Hayes, P. S. and Feeley, J. (1970) *J. infect. Dis.* **121**, 95.

Ellison, D. W., Baker, H. J. and Mariappan, M. (1969) *Amer. J. trop. Med. Hyg.* **18**, 694.

Emmanouilidou-Arseni, A. and Koumentakou, I. (1964) *J. Bact.* **87**, 1253.

Eurenius, K. and Brouse, R. O. (1973) *Amer. J. clin. Path.* **60**, 337.

Evans, I. T. G. and Richards, S. H. (1973) *J. Laryng.* **87**, 13.

Everett, E. D. and Nelson, R. A. (1975) *Amer. Rev. resp. Dis.* **112**, 331.

Fallon, R. J. (1973) *J. clin. Path.* **26**, 523.

Farkas-Himsley, H. (1968) *Amer. J. clin. Path.* **49**, 850.

Farmer, J. J. III *et al.* (1976) *Lancet* **ii**, 455.

Favero, M. S., Carson, L. A., Bond, W. W. and Petersen, N. J. (1971) *Science*, **173**, 836.

Feller, I. (1967) *J. Trauma* **7**, 93.

Feller, I. and Pierson, C. (1968) *Arch. Surg.* **97**, 225.

Fischer, R. (1972) *Z. ges. Hyg.* **18**, 133.

Fisher, M. W. (1957) *Antibiot. and Chemother.* **7**, 315.

Fisher, M. W., Devlin, H. B. and Gnabasik, F. J. (1969) *J. Bact.* **98**, 835.

Fisher, M. W. and Manning, M. C. (1958) *J. Immunol.* **81**, 29.

Flamm, H. (1957) *Schweiz. Z. allg. Path.* **20**, 23.

Flick, M. R. and Cluff, L. E. (1976) *Amer. J. Med.* **60**, 501.

Flynn, R. J. (1963) *Lab. Anim. Care* **13**, 1.

Forkner, C. E. (1960) *Pseudomonas aeruginosa Infections.* Grune and Stratton, New York.

Foster, W. D. and Bragg, J. (1962) *J. clin. Path.* **15**, 478.

Fox, J. E. and Lowbury, E. J. L. (1953) *J. Path. Bact.* **65**, 519, 533.

Fraenkel, E. (1906) *Virchows Arch.* **183**, 405.

French, G. L., Casewell, N. W., Roncorni, A. J., Knight, S. and Phillips, I. (1980) *J. Hosp. Infect.* **1**, 125.

Friedländer, C. (1883) *Fortschr. Med.* **1**, 715.

Friis, B. (1979) *Scand. J. infect. Dis.* **11**, 211.

Frisch, A. von (1882) *Wien. med. Wschr.* **32**, 969.

Fulghum, D. D., Linton, W. R. Jr and Taplin, D. (1978) *Sth. med. J., Bgham, Ala.* **71**, 739.

Gabriélidès and Remlinger (1902) *C. R. Soc. Biol., Paris* **54**, 1147.

Garrard, S. D., Richmond, J. B. and Hirsch, M. M. (1951) *Pediatrics, Springfield* **8**, 482.

Garrod, L. P., Lambert, H. P. and O'Grady, F. (1981) *Antibiotic and Chemotherapy*, 5th edn, p. 374, Churchill Livingstone, Edinburgh.

Gaya, H., Adnitt, P. I. and Turner, P. (1970) *Brit. med. J.* **iii**, 624.

George, R. M., Cochran, C. P. and Wheeler, W. E. (1961) *Amer. J. Dis. Child.* **101**, 296.

Ghosal, S. P., Sen Gupta, P. C., Mukherjee, A. K., Choudhury, M., Dutta, N. and Sarkar, A. K. (1978) *Lancet* **ii**, 289.

Gilardi, G. L. (1972) *Ann. intern. Med.* **77**, 211.

Girard, G. (1936) *Bull. Soc. Path. exot.* **29**, 712.

Glew, R. H., Moellering, R. C. and Kunz, L. J. (1977) *Medicine* **56**, 79.

Glode, M. P., Robbins, J. B., Liu, T.-Y., Gotschlich, E. C., Ørskov, I. and Ørskov, F. (1977) *J. infect. Dis.* **135**, 94.

Glynn, A. A., Brumfitt, W. and Howard, C. J. (1971) *Lancet* **i**, 514.

Godoy, G. A. (1966) *Rev. lat.-am. Microbiol. Parasitol.* **8**, 119.

Golden, B., Fingerman, L. H. and Allen, H. F. (1971) *Arch. Ophthal., N.Y.* **85**, 543.

Govan, J. R. W. and Fyfe, J. A. M. (1978) *J. antimicrob. Chemother.* **4**, 233.

Govan, J. R. W. and Gillies, R. R. (1969) *J. med. Microbiol.* **2**, 17.

Graevenitz, A. von and Mensch, A. H. (1968) *New Engl. J. Med.* **278**, 245.

Graham, D. R. and Band, J. D. (1981) *J. Amer. med. Ass.* **245**, 1923.

Gray, J., McGhie, D. and Ball, A. P. (1978) *J. antimicrob. Chemother.* **4**, 551.

Green, S. K., Schroth, M. N., Cho, J. J., Kominos, S. D. and Vitanza-Jack, V. B. (1975) *Appl. Microbiol.* **28**, 987.

Grogan, J. B. (1966) *J. Trauma* **6**, 639.

Gross, R. J., Rowe, B. and Easton, J. A. (1973) *J. clin. Path.* **26**, 138.

Groves, M. G., Strauss, J. M., Abbas, J. and Davis, C. E. (1969) *J. infect. Dis.* **120**, 605.

Grüneberg, R. N. and Bettelheim, K. A. (1969) *J. med. Microbiol.* **2**, 219.

Gutner, L. B. and Fisher, M. W. (1948) *Ann. intern. Med.* **28**, 1157.

Hadjiroussou, V., Holmes, B., Bullas, J. and Pinning, C. A. (1979) *J. clin. Path.* **32**, 953.

Hall, J. H., Callaway, J. L., Tindall, J. P. and Smith, J. G. (1968) *Arch. Derm.* **97**, 312.

Hallett, R. J., Pead, L. and Maskell, R. (1976) *Lancet* **ii**, 1107.

Hanessian, S., Regan, W., Watson, D. and Haskell, T. H. (1971) *Nature New Biol., Lond.* **229**, 209.

Hanson, L. Å., Fasth, A., Jodal, U., Kaijser, B. and Edén, C. S. (1981) *J. clin. Path.* **34**, 695.

Hanson, P. G., Standridge, J., Jarrett, F. and Maki, D. G. (1977) *J. Amer. med. Ass.* **238**, 1053.

Harber, M. J., Chick, S., Mackenzie, R. and Asscher, A. W. (1982) *Lancet* **i**, 586.

Hardy, D. J., Legeai, R. J. and O'Callaghan, R. J. (1980) *Antimicrob. Agents Chemother.* **18**, 542.

Hardy, P. C., Ederer, G. M. and Matsen, J. M. (1970) *New Engl. J. Med.* **282**, 33.

Hartley, C. L. and Richmond, M. H. (1975) *Brit. med. J.* **iv**, 71.

Hazen, T. C., Fliermans, C. B., Hirsch, R. P. and Esch, G. W. (1978) *Appl. envir. Microbiol.* **36**, 731.

Headings, D. L. and Overall, J. C. Jr (1977) *J. Pediat.* **90**, 99.

Hickman, F. W., Steigerwalt, A. G., Farmer, J. J. III and Brenner, D. J. (1982) *J. clin. Microbiol.* **15**, 1097.

Hill, H. R., Hunt, C. E. and Matsen, J. M. (1974) *J. Pediat.* **85**, 415.

Hill, K. R., Caselitz, F.-H. and Moody, L. M. (1954) *W. Indian med. J.* **3**, 9.

Hinze, P. M. (1956) *Vet. Med.* **51**, 257.

Hitchmann, R. and Kreibich, K. (1897) *Wien. klin. Wschr.* **10**, 1093.

Hoadley, A. W. and McCoy, E. (1968) *Cornell Vet.* **58**, 354.

Hoadley, A. W., McCoy, E. and Rohlich, G. A. (1968) *Arch. Hyg., Berl.* **152**, 328.

Hodges, G. R., Degener, C. E. and Barnes, W. G. (1978) *Amer. J. clin. Path.* **70**, 37.

Hodson, M. E., Penketh, A. R. L. and Batten, J. C. (1981) *Lancet* **ii**, 1137.

Hoffman, M. A. and Finberg, L. (1955) *J. Pediat.* **46**, 626.

Høiby, N. (1974) *Acta path. microbiol. scand.* **B82,** 551; (1975) *Ibid.* **B83,** 549; (1977) *Ibid.* **C,** suppl. 262.

Høiby, N., Flensborg, E. W., Beck, B., Friis, B., Jacobsen, S. V. and Jacobsen, L. (1977) *Scand. J. resp. Dis.* **58,** 65.

Høiby, N. and Rosendal, K. (1980) *Acta path. microbiol. scand.* **B88,** 125.

Hojyo-Tomoka, M. T., Marples, R. R. and Kligman, A. M. (1973) *Arch. Derm.* **107,** 723.

Holder, I. A. (1976) *J. antimicrob. Chemother.* **2,** 309.

Hole, N. H. and Coombs, R. R. A. (1947) *J. Hyg., Camb.* **45,** 497.

Hollis, D. G., Weaver, R. E., Baker, C. N. and Thornsberry, C. (1976) *J. clin. Microbiol.* **3,** 425.

Howe, C. and Miller, W. R. (1947) *Ann. intern. Med.* **26,** 93.

Howe, C., Sampath, A. and Spotnitz, M. (1971) *J. infect. Dis.* **124,** 598.

Howe, K. and Linton, A. H. (1976) *J. appl. Bact.* **40,** 317.

Hurst, V. and Sutter, V. L. (1966) *J. infect. Dis.* **116,** 151.

Iannini, P. B., Claffey, T. and Quintiliani, R. (1974) *J. Amer. med. Ass.* **230,** 558.

Ileri, S. Z. (1965) *Brit. vet. J.* **121,** 164.

Jackson, D. M., Lowbury, E. J. L. and Topley, E. (1951) *Lancet,* **ii,** 137.

Jacobson, J. A., Hoadley, A. W. and Farmer, J. J. III. (1976) *Amer. J. publ. Hlth* **66,** 1092.

Jellard, C. H. and Churcher, G. M. (1967) *J. Hyg., Camb.* **65,** 219.

Johns, M. A., Bruins, S. C. and McCabe, W. R. (1977) *Infect. Immun.* **17,** 9.

Johnson, D. A., Carter-Hamm, B. and Dralle, W. M. (1982) *Amer. Rev. resp. Dis.* **126,** 1070.

Jones, R. J. (1968) *Brit. J. exp. Path.* **49,** 411; (1969) *J. Hyg., Camb.* **67,** 241; (1971) *Brit. J. exp. Path.* **52,** 100; (1972) *J. Hyg. Camb.* **70,** 343; (1979) *Ibid.* **82,** 453.

Jones, R. J., Jackson, D. McG. and Lowbury, E. J. L. (1966) *Brit. J. plastic Surg.* **19,** 43.

Jones, R. J., Lilly, H. A. and Lowbury, E. J. L. (1971) *Brit. J. exp. Path.* **52,** 264.

Jones, R. J. and Lowbury, E. J. L. (1965) *Lancet,* **ii,** 623; (1967) *Brit. med. J.* **iii,** 79.

Jones, R. J., Roe, E. A. and Gupta, J. L. (1979) *Lancet* **ii,** 977; (1980) *Ibid.* **ii,** 1263.

Jones, R. J., Roe, E. A., Lowbury, E. J. L., Miler, J. J. and Spilsbury, J. F. (1976) *J. Hyg., Camb.* **76,** 429.

Jordan, G. W. and Hadley, W. K. (1969) *Ann. intern. Med.* **70,** 283.

Kaijser, B. (1973) *J. infect. Dis.* **127,** 670.

Kaijser, B., Hanson, L. Å., Jodal, U., Lidin-Janson, G. and Robbins, J. B. (1977) *Lancet* **i,** 663.

Källenius, G., Svenson, S. B., Möllby, R., Cedergren, B., Haltberg, H. and Winberg, J. (1981a) *Lancet* **ii,** 604.

Källenius, G. *et al.* (1981b) *Lancet* **ii,** 1369.

Kato, N., Kato, O. and Nakashima, I. (1976a) *Jap. J. Microbiol.* **20,** 163; (1976b) *Ibid.* **20,** 415.

Kelly, M. J. (1978) *J. med. Microbiol.* **11,** 513.

Kelly, N. M., Fitzgerald, M. X., Tempany, E., O'Boyle, C., Falkiner, F. R. and Keane, C. T. (1982) *Lancet* **ii,** 688.

Kenny, J. F., Medearis, D. N., Drachman, R. H., Gibson, L. E. and Klein, S. W. (1962) *Amer. J. Dis. Child.* **104,** 461.

Ketover, B. P., Young, L. S. and Armstrong, D. (1973) *J. infect. Dis.* **127,** 284.

Keys, T. F. and Washington, J. A. II (1977) *Mayo. Clin. Proc.* **52,** 797.

Klonowski, S. (1968) *Arch. Immunol. Ther. exp.* **16,** 338.

Knepper, J. G. and Anthony, B. F. (1973) *Lancet* **ii,** 285.

Kocka, F. E., Srinivasan, S., Mowjood, M. and Kantor, H. S. (1980) *J. clin. Microbiol.* **11,** 167.

Koníčková, L. and Prát, V. (1971) *J. clin. Path.* **24,** 113.

Kouwenaar, W., Maasland, J. H. and Wolff, J. W. (1934) *Geneesk Tijdschr. Ned.-Ind.* **74,** 1447.

Krasnilikov, A. P. (1968) *Arch. Immun. Ther. exp.* **16,** 474.

Krasnilikov, A. P. and Izraitel, N. A. (1966) *J. Hyg. Epidem., Praha* **10,** 145.

Kulp, W. L. and Borden, D. G. (1942) *J. Bact.* **44,** 673.

Kwantes, W. (1960) *Mon. Bull. Minist. Hlth Lab. Serv.* **19,** 169.

Lapage, S. P., Johnson, R. and Holmes, B. (1973) *Lancet* **ii,** 284.

Larrivee, G. P. and Elvehjem, C. A. (1954) *J. Amer. vet. med. Ass.* **124,** 447.

Larsson, P., Andersson, H. E. and Norlén, B. (1978) *Infection* **6,** 105.

Larsson, P. and Olling, S. (1977) *Med. Microbiol. Immunol.* **163,** 77.

Lev, M., Alexander, R. H. and Sobel, H. J. (1969) *J. appl. Bact.* **32,** 429.

Levine, M. G. and Hoyt, R. E. (1948) *Arch. Otolaryng., Chicago* **47,** 438.

Lewis, F. A. and Olds, R. J. (1952) *Aust. vet. J.* **28,** 145.

Lewis, J. and Fekety, F. R. (1969) *Johns Hopk. med. J.* **124,** 151.

Limson, B. M., Romansky, M. J. and Shea, J. G. (1956) *Ann. intern. Med.* **44,** 1070.

Lindberg, R. B., Moncrief, J. A., Switzer, W. E., Order, S. E. and Mills, W. (1965) *J. Trauma* **5,** 601.

Linton, A. H. (1977) *Vet. Rec.* **100,** 354; (1981) *Ibid.* **108,** 328.

Linton, A. H., Howe, K., Bennett, P. M., Richmond, M. H. and Whiteside, E. J. (1977b) *J. appl. Bact.* **43,** 465.

Linton, A. H., Howe, K., Hartley, C. L., Clements, A. M. and Richmond, M. H. (1977a) *J. appl. Bact.* **42,** 365.

Linton, A. H., Howe, K., Sojka, W. J. and Wray, C. (1979) *J. appl. Bact.* **46,** 585.

Linton, K. B., Richmond, M. H., Bevan, M. and Gillespie, W. A. (1974) *J. med. Microbiol.* **7,** 91.

Livermore, D. M., Williams, R. J., Lindridge, M. A., Slack, R. C. B. and Williams, J. D. (1982) *Lancet* **i,** 1466.

Loeffler (1886) *Arb. ReichsgesundhAmt.* **1,** 141.

Lovell, R. (1935) *J. R. Army vet. Cps* **6,** 69.

Lowbury, E. J. L. (1951) *Brit. J. indust. Med.,* **8,** 22; (1954) *Lancet* **i,** 292; (1969) *Brit. J. Derm.* **81,** Suppl. 1, 55.

Lowbury, E. J. L., Babb, J. R., Bridges, K. and Jackson, D. M. (1976) *Brit. med. J.* **i,** 493.

Lowbury, E. J. L., Babb, J. R. and Roe, E. (1972) *Lancet* **ii,** 941.

Lowbury, E. J. L. and Collins, A. G. (1955) *J. clin. Path.* **8,** 47.

Lowbury, E. J. L. and Fox, J. (1953) *J. Hyg., Camb.* **51,** 203; (1954) *Ibid.* **52,** 403.

Lowbury, E. J. L., Lilly, H. A. and Wilkins, M. D. (1962) *J. clin. Path.* **15,** 339.

Lowbury, E. J. L., Thom, B. T., Lilly, H. A., Babb, J. R. and Whittall, K. (1970) *J. med. Microbiol.* **3,** 39.

Lusis, P. I. and Soltys, M. A. (1971) *Vet. Bull.* **41,** 169.

McCormack, R. C. and Kunin, C. M. (1966) *Pediatrics* **37,** 750.

McCracken, A. W. and Barkley, R. (1972) *J. clin. Path.* **25,** 970.

McCracken, G. H. Jr *et al.* (1974) *Lancet* **ii**, 246.

McEuen, D. D., Blair, P., Delbene, V. E. and Eurenius, K. (1976a) *Infect. Immun.* **13**, 1360.

McEuen, D. D., Gerber, G. C., Blair, P. and Eurenius, K. (1976b) *Infect. Immun.* **14**, 399.

M'Fadyean, J. (1900) *J. comp. Path.* **13**, 55; (1901) *Ibid.* **14**, 265; (1904) *Ibid.* **17**, 295.

McGuckin, M. B., Thorpe, R. J., Koch, K. M., Alavi, A., Staum, M. and Abrutyn, E. (1982) *Amer. J. Epidem.* **115**, 785.

McHale, P. J., Keane, C. T. and Dougan, G. (1981) *J. clin. Microbiol.* **13**, 1099.

McKay, E. and Smith, J. (1958) *Arch. Dis. Childh.* **33**, 358.

MacKinnon, A. E. (1975) *Post Grad. med. J.* **51**, 591.

MacLaren, D. M. (1969) *J. Path. Bact.* **97**, 43.

Maddocks, J. L. and May, J. R. (1969) *Lancet* **i**, 793.

Maki, D. G., Hennekens, G. C., Phillips, C. W., Shaw, W. V. and Bennett, J. V. (1973) *J. infect. Dis.* **128**, 579.

Malizia, W. F., West, G. A., Brundage, W. G. and Walden, D. C. (1969) *Hlth Lab. Sci.* **6**, 27.

Markley, K., Gurmendi, G., Chavez, P. K. and Bazan, A. (1957) *Ann. Surg.* **145**, 175.

Markley, K. and Smallman, E. (1968) *J. Bact.* **96**, 867.

Matthews, P. R. J. and Fitzsimmons, W. M. (1964) *Res. vet. Sci.* **5**, 171.

Maurer, I. M. (1969) *Pharm. J.* **203**, 529.

Mayer, J. H. and Finlayson, M. H. (1944) *S. Afr. med. J.* **18**, 109.

Mayhall, C. G., Lamb, V. A., Gayle, W. E. Jr and Haynes, B. W. Jr (1979) *J. infect. Dis.* **139**, 166.

Meitert, T. and Meitert, E. (1966) *Arch. roum. Path. exp. Microbiol.* **25**, 427.

Meyer, R. D., Lewis, R. P., Halter, J. and White, M. (1976) *Lancet* **i**, 580.

Meyers, B. R., Bottone, E., Hirschman, S. Z. and Schneierson, S. S. (1972) *Ann. intern. Med.* **76**, 9.

Miler, J. M., Spilsbury, J. F., Jones, R. J., Roe, E. A. and Lowbury, E. J. L. (1977) *J. med. Microbiol.* **10**, 19.

Miller, W. R., Pannell, L., Cravitz, L., Tanner, W. A. and Ingalls, M. S. (1948a) *J. Bact.* **55**, 115.

Miller, W. R., Pannell, L. and Ingalls, M. S. (1948b) *Amer. J. Hyg.* **47**, 205.

Millican, R. C. and Rust, J. D. (1960) *J. infect. Dis.* **107**, 389.

Mills, J. and Drew, D. (1976) *Ann. intern. Med.* **84**, 29.

Milne, S. E. and Waterworth, P. M. (1978) *J. antimicrob. Chemother.* **4**, 247.

Minshew, B. H., Jorgensen, J., Counts, G. W. and Falkow, S. (1978) *Infect. Immun.* **20**, 50.

Mitchell, R. G. and Hayward, A. C. (1966) *Lancet,* **i**, 793.

Mohler and Eichhorn (1914) *J. comp. Path.* **27**, 183.

Mombelli, G., Coppens, L., Thys, J. P. and Klastersky, J. (1981) *Antimicrob. Agents Chemother.* **19**, 72.

Monafo, W. W. *et al.* (1976) *Surgery* **80**, 465.

Monroe, P. W. and Tift, W. L. (1979) *J. clin. Microbiol.* **10**, 850.

Montgomerie, J. Z., Doak, P. B., Taylor, D. E. M., North, J. D. K. and Martin, W. J. (1970) *Lancet* **ii**, 787.

Montie, T. C., Doyle-Huntzinger, D., Craven, R. C. and Holder, I. A. (1982) *Infect. Immun.* **38**, 1296.

Moody, M. D., Goldman, M. and Thomason, B. M. (1956) *J. Bact.* **72**, 357.

Moor, C. E. de, Soekarnen and Walle, N. van der. (1932) *Geneesk. Tijdschr. Ned.-Ind.* **24**, 1618.

Morris, J. G. Jr *et al.* (1982) *Lancet* **i**, 1294.

Moyer, C. A., Brentano, L., Gravens, D. L., Mergraf, H. W. and Monafo, W. W. (1965) *Arch. Surg., Chicago* **90**, 812.

Muraschi, T. F., Friend, M. and Bolles, D. (1965) *Appl. Microbiol.* **13**, 128.

Murrell, T. G. C. (1966) *Trans. R. Soc. trop. Med. Hyg.* **60**, 681.

Musher, D. M., Griffith, D. P., Yawn, D. and Rossen, R. D. (1975) *J. infect. Dis.* **131**, 177.

Mushin, R. and Ashburner, F. M. (1964) *J. appl. Bact.* **27**, 392.

Mutton, K. J., Brady, L. M. and Harkness, J. L. (1981) *J. Hosp. Infect.* **2**, 85.

Muytgens, H. L. *et al.* (1983) *J. clin. Microbiol.* **18**, 115.

Naemura, L. G. and Seidler, R. J. (1978) *Appl. envir. Microbiol.* **35**, 392.

Neter, E. and Weintraub, D. H. (1955) *J. Pediat.* **46**, 280.

Newsome, T. W. and Eurenius, K. (1973) *Surg. Gynec. Obstet.* **136**, 375.

Nicolle (1906) *Ann. Inst. Pasteur* **20**, 625, 698, 801.

Nigg, C. (1963) *J. Immunol.* **91**, 18.

Nigg, C. and Johnston, M. M. (1961) *J. Bact.* **82**, 159.

Nimmich, W., Naumann, G., Budde, E. and Straube, E. (1980) *Zbl. Bakt.,* **A247**, 35.

Nocard, E. (1901) *J. comp. Path.* **13**, 80.

Noriega, E. R., Rubinstein, E., Simberkoff, M. S. and Rahal, J. J. (1975) *Amer. J. Med.* **59**, 29.

Oberhofer, T. R. (1981) *J. clin. Microbiol.* **14**, 492.

Ørskov, I. (1952) *Acta path. microbiol. scand.* Suppl. 93, 259; (1954) *Ibid.* **35**, 194.

Okubadejo, O. A. and Alausa, K. O. (1968) *Brit. med. J.* **iii**, 357.

Olds, R. J. and Lewis, F. A. (1954) *Aust. vet. J.* **30**, 253.

Olsen, H. (1967) *Danish med. Bull.* **14**, 6.

Olsen, H., Frederiksen, W. C. and Siboni, K. E. (1965) *Lancet* **ii**, 1204.

Omar, A. R., Cheah Kok Kheong and Mahendranathan, T. (1962) *Brit. vet. J.* **118**, 421.

Parker, M. T. (1971) In: *Proc. int. Conference on Nosocomial Infections,* p. 35. Ed. by P.S. Brachman and T.C. Eikhoff. Amer. Hosp. Ass., Chicago.

Pavloskis, O. R., Pollack, M., Callaghan, L.T. III and Inglewski, B.H. (1977) *Infect. Immun.* **18**, 596.

Peck, C. R. and Zwanenburg, T. (1947) *Brit. med. J.* **i**, 337.

Peter, G. and Nelson, J. S. (1978) *J. Pediat.* **93**, 866.

Pettit, F. and Lowbury, E.J.L. (1968) *J. Hyg., Camb.* **66**, 393.

Phillips, I., Eykyn, S., Curtis, M. A. and Snell, J.J.S. (1971) *Lancet* **i**, 375.

Phillips, I., Eykyn, S. and Laker, M. (1972) *Lancet* **i**, 1258.

Phillips, J.A., Bernhardt, H.E. and Rosenthal, S.G. (1974) *Pediatrics, Springfield* **53**, 110.

Pitt, T.L. (1980) *J. Hosp. Infect.* **1**, 193; (1981) *J. med. Microbiol.* **14**, 261; (1982) *J. Hosp. Infect.* **3**, 9.

Pitt, T.L. Erdman, Y.J. and Bucher, C. (1980) *J. Hyg., Camb.* **84**, 269.

Plotkin, S.A. and McKitrick, J.C. (1966) *J. Amer. med. Ass.* **198**, 662.

Polk, H.C., Ward, C.G., Clarkson, J.G. and Taplin, D. (1969) *Arch. Surg.* **98**, 292.

Pollack, M. (1976) *Infect. Immun.* **13**, 1543.

Pollack, M., Callahan, L.T. III and Taylor, N.S. (1976) *Infect. Immun.* **14**, 942.

Pollack, M. and Young, L.S. (1979) *J. clin. Invest.* **63**, 276.

Poppe (1919) *Berl. tierärztl. Wschr.* **35**, 173.

Porter, K.A. and Giles, H.McC. (1956) *Arch. Dis. Childh.* **31**, 303.

Price, D.J.E. and Sleigh, J.D. (1970) *Lancet*, **ii**, 1213.

Rabin, E.R., Graber, C.D., Vogel, E.H., Finkelstein, R.A. and Tumbusch, W.A. (1961) *New. Engl. J. Med.* **265**, 1225.

Rabinowitz, K. and Schiffrin, R. (1952) *Acta med. orient.* **11**, 181.

Rampling, A., Whitby, J.L. and Wildy, P. (1975) *J. med. Microbiol.* **8**, 531.

Redfearn, M.S., Palleroni, N.J. and Stanier, R.Y. (1966) *J. gen. Microbiol.* **43**, 293.

Reid, T.M.S. and Porter, I.A. (1981) *J. Hyg., Camb.* **86**, 357.

Reinhardt, R. (1919) *Berl. tierärztl. Wschr.* **35**, 453, 465.

Reitler, R. and Seligmann, R. (1957) *J. appl. Bact.* **20**, 145.

Rennie, R.P. and Duncan, I.B.R. (1977) *Antimicrob. Agents Chemother.* **11**, 179.

Report (1964) *Ann. Surg.* **160**, Suppl. 2; (1971) *Morbid. Mortal.* **20**, suppl. to no. 9.

Ribiero, C.D., Davis, P. and Jones, D.M. (1976) *J. clin. Path.* **29**, 1094.

Richmond, M.H. and Linton, K.B. (1980) *J. antimicrob. Chemother.* **6**, 33.

Rimington, R.A. (1962) *Med. J. Aust.* **i**, 50.

Ringen, L.M. and Drake, C.H. (1952) *J. Bact.* **64**, 841.

Ringrose, R.E., McKown, B., Felton, F.G., Barclay, B.O., Muchmore, H.G. and Rhoades, E.R. (1968) *Ann. intern. Med.* **69**, 719.

Riser, E., Noone, P. and Howard, F.M. (1980) *J. clin. Path.* **33**, 400.

Robbins, J.B., McCracken, G.H. Jr, Gotschlich, E.C., Ørskov, F., Ørskov, I. and Hanson, L.A. (1974) *New Engl. J. Med.* **290**, 1216.

Robins, G.D. (1906) *Stud. R. Victoria Hosp., Montreal* **2**, No. 1.

Roques and Dauphin (1943) *Rev. méd. Franc. extr. Orient* **21**, 267.

Rosenthal, S.L. and Freundlich, L.F. (1977) *Hlth Lab. Sci.* **14**, 194.

Rosenthal, S.M., Millican, R.C. and Rust, J. (1957) *Proc. Soc. exp. Biol., N.Y.* **94**, 214.

Rubin, L.G., Altman, E., Epple, L.K. and Yolken, R.H. (1981) *J. Pediat.* **98**, 940.

Rutala, W.A., Kennedy, V.A., Loflin, H.B. and Sarubbi, F.A. Jr (1981) *Amer. J. Med.* **70**, 659.

Sabolotny, S.S. (1926) *Zbl. Bakt.* **97**, 168.

Sachs, J.M., Pacin, M. and Counts, G.W. (1974) *Amer. J. Dis. Child.* **128**, 387.

Saelinger, C.B., Snell, K. and Holder, I.A. (1977) *J. infect. Dis.* **136**, 555.

Sanders, C.V. Jr, Luby, J.P., Johanson, W.G., Barnett, J.A. and Sanford, J.P. (1970) *Ann. intern. Med.* **73**, 15.

Sands, D.C. and Rovira, A.D. (1970) *Appl. Microbiol.* **20**, 513.

Sarff, L.D. *et al.* (1975) *Lancet*, **i**, 1099.

Saxena, S.R. and Bassett, D.C.J. (1975) *Arch. Dis. Childh.* **50**, 899.

Schäperclaus, W. (1939) *Z. Fisch.* **37**, 6.

Schiøtz, P.O., Høiby, N., Permin, H. and Wiik, A. (1979) *Acta path. microbiol. scand.* **C87**, 229.

Schwarzmann, S. and Boring, J. III (1971) *Infect. Immun.* **3**, 762.

Selden, R., Lee, S., Wang, W.L.L., Bennett, J.V. and Eickhoff, T.C. (1971) *Ann. intern. Med.* **74**, 657.

Seligmann, R., Komarov, M. and Reitler, R. (1963) *Brit. med. J.* **ii**, 1528.

Semczuk, B., Klonowski, S., Dylewski, B. and Hencner, Z. (1968) *Arch. Immunol. Ther. exp.* **16**, 895.

Semel, J.D., Trenholme, G.M., Harris, A.A., Jupa, J.E. and Levin, S. (1978) *Amer. J. Med.* **64**, 403.

Senior, B.W. (1983) *J. med. Microbiol.* **16**, 317.

Shigeta, S., Yasunaga, Y. and Ogata, M. (1978) *J. clin. Microbiol.* **8**, 489.

Shooter, R.A. *et al.* (1966) *Lancet* **ii**, 1331; (1969) *Ibid.* **i**, 1227.

Shortland-Webb, W.R. (1968) *J. clin. Path.* **21**, 422.

Sickles, E.A., Young, V.M., Greene, W.H. and Wiernik, R.H. (1973) *Ann. intern. Med.* **79**, 528.

Simberkoff, M.S., Moldover, N.H. and Rahal, J.J. Jr (1976*b*) *J. infect. Dis.* **134**, 348.

Simberkoff, M.S., Ricupero, I. and Rahal, J.J. Jr (1976*a*) *J. Lab. clin. Med.* **87**, 206.

Simmons, N.A. and Gardner, D.A. (1969) *Brit. med. J.* **ii**, 668.

Simoons-Smit, A.M., Verweij-van Vught, A.M.J.J., Kanis, I.Y.R. and Maclaren, D.M. (1983) *J. Hyg., Camb.* **90**, 461.

Slack, M.P.E. and Nichols, W.W. (1981) *Lancet* **ii**, 502.

Smith, H.W. (1962) *J. Path. Bact.* **84**, 147; (1974) *J. gen. Microbiol.* **83**, 95.

Smith, H.W. and Halls, S. (1968) *J. med. Microbiol.* **1**, 61.

Smith, H.W. and Huggins, M.B. (1976) *J. gen. Microbiol.* **92**, 335.

Smith, P.W. and Massanari, R.M. (1977) *J. Amer. med. Ass.* **237**, 795.

Sneath, P.H.A. (1960) *Iowa St. J. Sci.* **34**, 243.

Solberg, C.O. and Matsen, J.M. (1971) *Amer. J. Med.* **50**, 241.

Sonne, M. and Jawetz, E. (1969) *Appl. Microbiol.* **17**, 893.

Sonnenworth, A.C. and Kallus, B.A. (1968) *Amer. J. clin. Path.* **49**, 92.

Sorensen, R.U., Stern, R.C. and Polmar, S.H. (1977) *Infect. Immun.* **18**, 735.

Spencer, W.H. (1953) *Calif. Med.* **79**, 438.

Spotnitz, M., Rudnitzky, J. and Rambaud, J.J. (1967) *J. Amer. med. Ass.* **202**, 950.

Stanley, M.M. (1947) *Amer. J. Med.* **2**, 253, 347.

Stanton, A.T. and Fletcher, W. (1921) *Trans. 4th Congr. Far East Ass. trop. Med.* **2**, 196; (1925) *J. Hyg., Camb.* **23**, 347; (1932) *Studies Inst. med. Res., F.M.S.* No. 21.

Steffen, T.N. and Smith, I.M. (1961) *Ann. Otol. Rhinol. Laryngol.* **70**, 935.

Steinhauer, B.W., Eickhoff, T.C., Kislak, J.W. and Finland, M. (1966) *Ann. intern. Med.* **65**, 1180.

Stenderup, A., Faergeman, O. and Ingerslev, M. (1966) *Acta path. microbiol. scand.* **68**, 157.

Stoodley, B.J. and Thom, B.T. (1970) *J. med. Microbiol.* **3**, 367.

Storring, R.A., Jameson, B. and McElwain, T.J. (1977) *Lancet* **ii**, 837.

Story, P. (1954) *J. Path. Bact.* **68**, 55.

Strauss, J.M., Alexander, A.D., Rapmund, G., Gan, E. and Dorsey, A.E. (1969*b*) *Amer. J. trop. Med. Hyg.* **18**, 703.

Strauss, J.M., Groves, M.G., Mariappan, M. and Ellison, D.W. (1969*a*) *Amer. J. trop. Med. Hyg.* **18**, 698.

Sutmöller, P., Kraneveld, F.C. and Schaaf, A. van der (1957) *J. Amer. vet. med. Ass.* **130**, 415.

Sutter, V.L., Hurst, V. and Lane, C.W. (1967) *Hlth Lab. Sci.* **4,** 245.

Sweet, A.Y. and Wolinsky, E. (1964) *Pediatrics, Springfield* **33,** 865.

Taplin, D., Bassett, D.C.J. and Mertz, P.M. (1971) *Lancet,* **ii,** 568.

Taylor, P.W. (1974) *J. clin. Path.* **27,** 626.

Teplitz, C. (1965) *Arch. Path.* **80,** 297.

Thin, R.N.T., Brown, M., Stewart, J.B. and Garrett, C.J. (1970) *Quart. J. Med.,* NS, **39,** 115.

Thin, R.N.T., Groves, M., Rapmund, G. and Mariappan, M. (1971) *Singapore med. J.* **12,** 181.

Thom, A.R., Stephens, M.E., Gillespie, W.A. and Alder, V.G. (1971) *J. appl. Bact.* **34,** 611.

Thomason, B.M., Moody, M.D. and Goldman, M. (1956) *J. Bact.* **72,** 362.

Thong, M.L., Puthucheary, S.D. and Lee, E.L. (1981) *J. clin. Path.* **34,** 429.

Tinne, J.E., Gordon, A.M., Bain, W.N. and Mackey, W.A. (1967) *Brit. med. J.* **iv,** 313.

Tomášek, V. (1925) *Zbl. Bakt. Ref.* **79,** 564.

Toni, M., Casewell, M.W. and Schito, G.C. (1980) *J. antimicrob. Chemother.* **6,** 527.

Traub, W.H. (1978) *Zbl. Bakt.* I Abt. Orig., **A240,** 57.

Trautwein, G., Humboldt, C.F. and Nielsen, S.W. (1962) *J. Amer. med. Ass.* **140,** 701.

Tucci, V. and Isenberg, H.D. (1981) *J. clin. Microbiol.* **14,** 563.

Ujváry, G. (1958) *Zbl. Bakt.* **170,** 394.

Umsawasdi, T., Middleman, E.A., Luna, M. and Bodey, G.P. (1973) *Amer. J. med. Sci.* **265,** 473.

Vahlne, G. (1945) *Serological Typing of Colon Bacteria.* Gleerupka Univ.—Bokhandeln, Lund.

Vandepitte, J., Beeckmans, C. and Buttiaux, R. (1958) *Ann. Soc. belge Méd. trop.* **38,** 563.

Vandepitte, J. and Debois, J. (1978) *J. clin. Microbiol.* **7,** 70.

Vaucel, M. (1937) *Bull. Soc. Path. exot.* **30,** 10.

Vic-Dupont, V. *et al.* (1969) *Ann. Méd. interne, Paris* **120,** 395.

Victorica, B., Baer, H. and Ayoub, E.M. (1974) *J. Amer. med. Ass.* **230,** 578.

Victorin, L. (1967) *Acta. paed. scand.* **56,** 344.

Vosti, K. L., Goldberg, L. M., Monto, A. S. and Rantz, L. A. (1964) *J. clin. Invest.* **43,** 2377.

Wahba, A. H. (1965a) *Brit. med. J.* **i,** 86; (1965b) *Arch. Hyg., Berl.* **149,** 187.

Waisbren, B. A. and Lepley, D. (1962) *Arch. intern. Med.* **109,** 712.

Walker, H. L., Mason, A. D. and Raulston, G.L. (1964) *Ann. Surg.* **160,** 297.

Wallace, G. I., Cahn, A. R. and Thomas, L. J. (1933) *J. infect. Dis.* **53,** 386.

Wassermann, M. (1901) *Virchows Arch.* **165,** 342.

Watson, E. A. (1923) *J. Amer. vet. med. Ass.* **64,** 146.

Whitby, J. L. and Rampling, A. (1972) *Lancet* **i,** 15.

White, F. (1957) *Vet. Rec.* **69,** 566.

Whitmore, A. (1913) *J. Hyg., Camb.* **13,** 1.

Whitmore, A. and Krishnaswami, C. S. (1912) *Indian med. Gaz.* **47,** 262.

Wilfert, J. N., Barrett, F. F. and Kass, E. H. (1968) *New Engl. J. Med.* **279,** 286.

Wilkinson, L. (1981) *Med. Hist.* **25,** 363.

Williams, E. P. and Cameron, K. (1896) *J. Path. Bact.* **3,** 344.

Williams, E. W., Hawkey, P. M., Penner, J. L., Senior, B. W. and Barton, L. J. (1983) *J. clin. Microbiol.* **18,** 5.

Williams, R. J. and Govan, J. R. W. (1973) *J. med. Microbiol.* **6,** 409.

Wolberg, G. and Witt, C. W. de (1969) *J. Bact.* **100,** 730.

Wong, W. T. and Bettelheim, K. A. (1976) *Zbl Bakt.,* **A236,** 481.

Wormald, P. J. (1970) *J. Hyg., Camb.* **68,** 633.

Wunder, W. and Dombrowski, H. (1953) *Z. Fisch.* **2** (n.s.), 327.

Wylie, R. H. and Kirschner, P. A. (1950) *Amer. Rev. Tuberc.* **61,** 465.

Young, L. S., Yu, B. H. and Armstrong, D. (1970) *Infect. Immun.* **2,** 495.

Yu, V. L. (1979) *New Engl. J. Med.* **300,** 887.

Zimakoff, J., Høiby, N., Rosendal, K. and Guilbert, J. P. (1983) *J. Hosp. Infect.* **4,** 31.

Zuravleff, J. J. and Yu, V. L. (1982) *Rev. infect. Dis.* **4,** 1236.

62

Infections due to gram-negative non-sporing anaerobic bacilli

Brian I. Duerden

Introductory

Gram-negative, non-sporing, anaerobic bacilli of the bacteroides-fusobacterium group have been recognized as significant pathogens since the late nineteenth century. Miller described fusiform bacteria in patients with ulcerative stomatitis in 1883 (Finegold 1977), and Loeffler (1884) first associated 'Bacillus necrophorus' with calf diphtheria. Plaut (1894) and Vincent (1896) independently described a fusiform organism frequently associated with a spirochaete (*Treponema vincenti*) in necrotic ulcerative lesions of the throat that resembled diphtheria; this disease was later known as *Plaut-Vincent's angina*. Vincent found similar organisms in gangrenous lesions. Veillon and Zuber (1897, 1898) isolated non-sporing anaerobic bacilli from appendicitis, pyogenic arthritis, and abscesses of the lung, brain and pelvis. They called these organisms *Bacillus ramosus*, *B. serpens*, *B. fragilis*, *B. fusiformis* and *B. furcosus*; *B. fusiformis* was apparently identical with the fusiform bacillus described by Vincent. Subsequently Veillon and his colleagues showed that gram-negative anaerobic bacilli played a part in septicaemia, empyema, and infections of the middle ear and female genital tract, including puerperal infections. Improved anaerobic methods have revealed similar organisms in many human and animal infections.

Many infections with gram-negative anaerobic bacilli are endogenous, the source of infection being the normal flora of the patient's own mucous membranes. Many arise close to the mucous membranes of the mouth, lower gastro-intestinal tract, and genital tract; they often follow surgical or accidental injury, especially in debilitated patients. There is usually extensive tissue damage, cell necrosis or abscess formation, or both, with foul-smelling pus and sometimes gas production. The lesions arise where dead or damaged tissue with a poor blood supply produces a low Eh; a foreign body may provide a nidus for infection.

Patients debilitated by malignant disease, and those whose tissues are rendered hypoxic by vascular disease are commonly affected; they are placed further at risk by the thrombophlebitis that often results from the infection.

Bacteroides or *Fusobacterium* spp. are isolated in pure culture from only a minority of infections. The majority yield one or more anaerobic species together with organisms that are aerobic and facultatively anaerobic. The gram-negative anaerobic bacilli may often be the most significant component; mixed post-operative sepsis can be both treated and prevented by agents active only against anaerobes (Willis *et al.* 1975, 1976, 1977). Moreover, '*B. melaninogenicus*' strains that were probably *B. asaccharolyticus* were found to cause necrosis in experimental infection only when mixed with aerobes, which were often of weak pathogenicity themselves (Weiss 1943, Socransky and Gibbons 1965). Synergy between aerobic and anaerobic bacteria may occur. It is probable that the aerobic species contribute to a reduction in the Eh of the tissue, and supply growth factors for the anaerobic species (Hite *et al.* 1949, Kelly 1980). (For a fuller description of the Bacteroidaceae, see Chapter 26.)

Pathogenic species of Bacteroidaceae

Because infection is usually endogenous, the bacterial species present might be expected to reflect the constitution of the normal flora at the site from which infection originated. However, the isolation rates of *Bacteroides* and *Fusobacterium* spp. from infections and from normal flora differ significantly (Duerden 1980*a*, *b*). The more reliable studies suggest that a small number of species represent a large proportion of the pathogenic isolates; this indicates that certain species of Bacteroidaceae have particular pathogenic potential. *B. fragilis* is the commonest species isolated from human infections. The other species of the fragilis group, although more common than *B. fragilis* itself in the faecal flora, are isolated much less often. They include *B. thetaiotaomicron*, which is isolated four times less commonly than *B. fragilis* (Werner and Pulverer 1971, Werner 1974, Holland *et al.* 1977, Duerden 1980*a*). The remaining species in the fragilis group are seldom isolated from infections—and then usually as part of a mixed flora derived from gross faecal soiling.

Pigmented *Bacteroides* spp., referred to as '*B. melaninogenicus*', have often been identified in human infections, though much less frequently than *B. fragilis* (Burdon 1928, 1932, Heinrich and Pulverer 1960). They are rarely found in pure culture. Precise identification of pigmented species is not to be found in the earlier reports; however, Heinrich and Pulverer (1960) regarded *B. melaninogenicus* as asaccharolytic; and *B. asaccharolyticus* was the commonest pigmented

species and the second most common species in Duerden's (1980*a*) series.

F. necrophorum was once fairly common in serious infections in man. It is sensitive to penicillin and many other antibacterial agents and has been encountered much less frequently since the advent of antibiotics. Although now overshadowed by *B. fragilis*, it can cause serious or fatal disease, including lung abscesses, septicaemia, and infections of the central nervous system (Finegold 1977).

B. fragilis

B. fragilis accounts for more than 50 per cent of all significant isolates of Bacteroidaceae (Werner and Pulverer 1971, Werner 1974, Holland *et al.* 1977, Duerden 1980*a*). Polk and Kasper (1977) and Duerden (1980*a*) found that *B. fragilis* represented 63 and 75 per cent respectively of the clinically significant isolates of the fragilis group. Infections with *B. fragilis* often originate from the lower gastro-intestinal tract and, when the source of infection is the appendix, colon or rectum, about 60 per cent of all significant isolates and more than 75 per cent of the fragilis-group isolates consist of *B. fragilis*. In the normal faecal flora most *Bacteroides* strains belong to the fragilis group, but *B. vulgatus* and *B. thetaiotaomicron* are present in the largest numbers, *B. fragilis* being greatly outnumbered. This indicates that members of the fragilis group differ in pathogenicity. *B. fragilis* is commoner than any other species in the small number of infections in which a gram-negative anaerobic bacillus is present alone. *B. fragilis* is also responsible for most serious bacteroides infections such as bacteraemias, serious wound infections, intraperitoneal abscesses, and peritonitis (Polk and Kasper 1977, Duerden 1980*a*).

The pathogenicity of *B. fragilis* is related principally to its cell-surface properties and polysaccharide capsule (see Chapter 26). All clinical isolates of *B. fragilis* are encapsulated, but the capsule is lost during repeated subculture on artificial media. A smaller and possibly unrelated capsule has been described in other members of the fragilis group (Babb and Cummins 1978, Burt *et al.* 1978, Lindberg *et al.* 1979). Only encapsulated strains of *B. fragilis* produce intraperitoneal abscesses in rats. Partly purified capsular material alone produces sterile abscesses. Mixtures of capsular material or killed capsulated strains with non-capsulated strains also produce abscesses (Onderdonk *et al.* 1977). The capsulated strains of *B. fragilis* adhere better than non-capsulated strains to rat peritoneal mesothelium; and adherence may be important in pathogenesis (Onderdonk *et al.* 1978). The capsule of *B. fragilis* also confers resistance to phagocytosis. Clinical isolates of *B. fragilis* are ingested and killed by polymorphonuclear leucocytes only after opsonization by serum factors (Casciato *et al.* 1975, Bjornson *et al.* 1976). *B. fragilis* may inhibit the phagocytosis of aerobic, or of aerobic and facultatively anaerobic organisms in mixed cultures, explaining perhaps why treatment of mixed infections with agents active against anaerobes alone is often successful (Ingham *et al.* 1977*b*).

B. fragilis produces a number of extra-cellular or membrane-associated enzymes that may act as toxins or aggressins and thus contribute to virulence. They include collagenase and other proteases, fibrinolysin, haemolysin, neuraminidase, phosphatase, DNAase, hyaluronidase, chondroitin sulphatase, and heparinase (Rudek and Haque 1976). These enzymes are active against many components of mammalian tissue and may cause the tissue damage and necrosis of bacteroides infections, the capsule already having assisted in the initiation of infection. Heparinase activity may be responsible in part for the local thrombophlebitis that often develops and gives rise to septic emboli and metastatic abscesses.

Melaninogenicus-oralis group

Species in the melaninogenicus-oralis group that are part of the normal flora of the gingival crevice and the vagina also cause infections related to these sites. *B. melaninogenicus* plays an important role in oral infections, including periodontal disease and gingivitis, peri-apical abscesses, and soft-tissue infections of the head and neck. Acute ulcerative gingivitis (AUG; *Vincent's stomatitis*) has been accepted generally as a synergistic infection with *Treponema vincenti* and a *Fusobacterium* sp.; but there is evidence that oral *Bacteroides* spp., in particular *B. melaninogenicus*, are also essential and may be chiefly responsible for the gingival damage that occurs. Tissue destruction, especially in the gums, may result from the action of a variety of exoenzymes or toxins produced by *B. melaninogenicus*. Most strains are strongly proteolytic and show vigorous collagenase and fibrinolytic activity (Gibbons and MacDonald 1961, Kaufman *et al.* 1972, Hardie 1974). As in other bacteroides infections, thrombosis of local vessels occurs commonly.

Many strains of *B. melaninogenicus* and *B. oralis* hydrolyse dextran (Staat *et al.* 1973, Holbrook and McMillan 1977); this may be significant in the development of dental plaque and periodontal disease. The predominant bacteria in early plaque are dextran-producing streptococci; colonization by *Bacteroides* spp. of the melaninogenicus-oralis group follows and leads to periodontal disease (Hardie 1974; see also Chapter 57). *B. melaninogenicus* and *B. oralis* are also important pathogens in chronic lung infections such as lung abscess, and in cerebral abscesses secondary to lung infections or to chronic otitis media and mastoiditis.

B. melaninogenicus is usually found in mixed infections, in which it appears to be the most important

pathogenic component. Some of the evidence for this was obtained from strains that were defined solely on pigmentation and may have been *B. asaccharolyticus*. In experimental infections, *B. melaninogenicus* produces severe lesions when mixed with organisms of weak pathogenicity, whereas *B. melaninogenicus* alone or mixed infections without it do not result in tissue damage (Weiss 1943, Socransky and Gibbons 1965). Most natural human infections are mixed.

Other species in the melaninogenicus-oralis group are less often implicated in serious infection. *B. ruminicola* subsp. *brevis* is found in a few infections of, or originating from, the mouth and gastro-intestinal tract; and *B. bivius*, a normal vaginal commensal, may cause uterine and pelvic infection with abscess formation and bacteraemia in the puerperium or after surgical treatment of the female reproductive tract.

B. asaccharolyticus

The pigmented, non-fermentative *B. asaccharolyticus*—an important pathogen—is a commensal of the lower gastro-intestinal tract. It is commonly isolated from infections of, or originating from, the appendix, colon, rectum, and perianal area, where it is usually present with *B. fragilis*. It may also be of pathogenic significance in the ulcers and gangrene of patients with peripheral vascular disease and diabetes (Peromet *et al.* 1973, Duerden 1980*a*). *B. asaccharolyticus* is a demanding organism and is rarely isolated in pure culture.

Its pathogenicity may be due in part to a small capsule and extracellular slime. The characteristic production of foul necrotic tissue is probably due to extracellular enzymes or toxins and, in particular, to vigorous proteolytic activity.

B. gingivalis

Oral strains of pigmented asaccharolytic *Bacteroides* are now assigned to a separate species—*B. gingivalis*—that is found irregularly and in only small numbers in the normal gingival flora but is implicated in the pathogenesis of periodontal disease. It can be isolated in large numbers from periodontal pockets in patients with destructive periodontitis (Coykendall *et al.* 1980).

B. ureolyticus (formerly *B. corrodens*)

The strictly anaerobic *B. ureolyticus* was first isolated from abscesses, principally of the buccal region (Eiken 1958); and most subsequent reports have described strains isolated from infections, usually mixed (Khairat 1967, Schröter and Stawru 1970). *B. ureolyticus* is not a major component of the normal flora. It may be of pathogenic significance in ulcerative or gangrenous lesions of the groin, perineum and scrotum, and in some cases of paronychia.

F. necrophorum

Fusiform organisms described under a variety of names but now recognized as *F. necrophorum* are responsible for certain necrotic lesions in animals, including calf diphtheria (Loeffler 1884), labial necrosis in rabbits (Schmorl 1891), and foot-rot of sheep and cattle (Langworth 1977). It also causes severe necrotic infections in man, including lung abscesses with septicaemia (Cohen 1932), hepatic abscess (Beaver *et al.* 1934), puerperal fever (Harris and Brown 1927), tonsillar and pharyngeal infections (Hansen 1950), chronic otitis media and mastoiditis with septicaemia and brain abscess, and lesions associated with ulcerative colitis (Dack *et al.* 1936).

Pathogenicity appears to be related to endotoxin and exotoxins. The cell-wall lipopolysaccharide has strong endotoxic properties, which resemble those of enterobacterial endotoxin. It is lethal for mice, rabbits and chicken embryos, and produces localized and generalized Shwartzman reactions in rabbits (see Chapter 26). *F. necrophorum* produces several potent exotoxins or exoenzymes; these include a lipase, haemolysin, DNAase, protease, a cytoplasmic toxin and a leucocidin that inhibits phagocytosis by polymorphs (Roberts 1967, 1970, Langworth 1977).

Pure cultures of *F. necrophorum* are pathogenic for rabbits and mice. Intraperitoneal or intravenous inoculation of rabbits causes death with intense inflammation, exudate and coagulative necrosis; subcutaneous injection into the lower lip gives rise to labial necrosis and death in 4–12 days. Subcutaneous inoculation of mice produces spreading necrotic lesions, and death after about 12 days.

Infections in man due to Bacteroidaceae

Gram-negative anaerobic bacilli were recognized long ago as the causative agents of foul-smelling necrotic lesions (Vincent 1896, Veillon and Zuber 1898); but many reports have seriously underestimated the incidence of infection in man, because poor anaerobic techniques were used. Dack (1940) isolated non-sporing anaerobes from less than 4 per cent of 5180 specimens from patients in a surgical unit. Stokes (1958) found anaerobes in 10.5 per cent of 4737 unselected specimens that yielded bacterial growth. Subsequent improvements in anaerobic techniques led to increased awareness of the pathogenicity of the Bacteroidaceae. Martin (1971) isolated anaerobes from 35 per cent of specimens received in the Mayo Clinical Laboratory; they represented 49.3 per cent of all bacteria isolated. In a subsequent study, anaerobes were isolated from 49 per cent of culture-positive specimens (Martin 1974). Anaerobes were found in 58.5 per cent of bacteria-positive cultures from specimens other than stool, urine, sputum and blood (Holland *et al.* 1977); 70 per cent were *Bacteroides* spp. Leigh (1974) showed that, with improved techniques, the isolation rate of *Bacteroides* spp. from wound infections after abdominal operations rose from 17 per cent in 1970–72 to 81 per cent in 1973; the rate of significant post-operative infection remained the same. Increased clinical and laboratory awareness of anaerobes, good anaerobic techniques, and quick transportation of appropriate specimens to the laboratory have shown that gram-negative anaerobic bacilli are common pathogens. The source of these organisms is often the mucous membranes of the mouth, lower gastro-intestinal tract, and female genital tract; and the species isolated vary with the site of infection. Many originate from the bacteroides component of the faecal flora and are associated with the large intestine, but all body systems are susceptible to infection with *Bacteroides* and *Fusobacterium* spp.

Abdominal and perineal infections

Anaerobic bacteria, especially *Bacteroides* spp., are the predominant pathogens in infections associated with surgery, injury, perforation, or other underlying abnormality of the gastro-intestinal tract (Gorbach *et al.* 1974, Gorbach and Bartlett 1974). Infection with *Bacteroides* spp. is most common after surgical treatment or perforation of the large intestine; it is less common after surgical treatment of the upper gastro-intestinal tract, where contamination with anaerobes from the normal gut flora is less.

Two groups of *Bacteroides* spp. cause infections associated with the oesophagus, stomach and small intestine—(a) the fragilis group derived from the faecal flora, and (b) the melaninogenicus-oralis group from the oral flora. The normal stomach and upper small intestine generally contain few bacteria, but they become colonized by faecal bacteria, including those of the fragilis group, when their normal anatomy and physiology are altered (Drasar and Hill 1974). *B. fragilis* is a significant cause of post-operative wound infection, peritonitis, and the intra-abdominal abscesses that follow perforation or surgical operations. Normal gastric acidity provides a barrier between the bacterial populations of the mouth and the large intestine. The normal flora of the oesophagus is derived from the mouth, but most organisms are destroyed when they enter the stomach. Any disturbance of acid production results in colonization of the stomach by oral bacteria, including the melaninogenicus-oralis group of *Bacteroides*. *B. melaninogenicus* subsp. *melaninogenicus* and subsp.

intermedius and *B. oralis* are found in wound infections, peritonitis and empyemas after surgical treatment or perforation of the oesophagus and stomach (Duerden 1980*a*), and also in peritonitis secondary to perforation of duodenal ulcers. These infections are usually mixed, and the *Bacteroides* spp. are present in association with other oral organisms such as non- and α-haemolytic streptococci and anaerobic cocci.

The most common bacteroides infections are wound infections, intra-abdominal abscesses, and peritonitis associated with the appendix or large intestine. Finegold (1977) found that anaerobes, mainly *Bacteroides* spp., were associated with 86 per cent of intra-abdominal infections. *Bacteroides* spp. are the most significant pathogens in the majority of wound infections after surgical treatment of the appendix or large intestine (Leigh 1974, 1976, Swenson *et al*. 1974, Willis *et al*. 1975, 1976, 1977). Similarly, *Bacteroides* spp. are almost always associated with peritonitis and intra-abdominal abscesses (sub-phrenic, pelvic and paracolic for example), resulting from perforation of the large intestine.

The importance of anaerobes as the cause of peritonitis and abscesses in appendicitis was recognized by Veillon and Zuber (1898); gram-negative anaerobic bacilli were the predominant pathogens in 21 of their 22 cases of suppurative appendicitis and appendix abscess. Bornstein and his colleagues (1964) found that 25 per cent of bacteroides infections were associated with the appendix. Altemeier (1938) isolated gram-negative anaerobic bacilli from 96 of 100 cases of peritonitis associated with acute perforated appendicitis. Felner and Dowell (1971) found that 10 per cent of a series of 250 cases of bacteroides bacteraemia originated from appendiceal lesions. Leigh and his colleagues (1974) isolated *Bacteroides* spp. from 78 per cent of 322 swabs taken at appendicectomy, and from more than 90 per cent of the infectious complications of those operations. Willis (1977) found that the incidence of post-appendicectomy sepsis varied from 4 per cent for normal appendices to 77 per cent for gangrenous or perforated appendices.

Similar infections result from surgical treatment or perforation of the colon. Diverticulosis is a common non-infectious disease but *Bacteroides* spp. are almost invariably present in the infections that arise when diverticulitis leads to diverticular and paracolic abscesses and peritonitis. These infections frequently give rise to bacteraemia; 6–10 per cent of bacteroides bacteraemias originate from diverticulitis (Felner and Dowell 1971, Wilson *et al*. 1972). There is also a close association between abdominal infections with *Bacteroides* spp. and carcinoma of the colon or rectum. Peritonitis, intra-abdominal abscesses and wound infections are common complications of the surgical treatment of these tumours. Bacteroides bacteraemia, peritonitis, or abscess formation may on occasion be the first manifestation of the malignancy (Finegold *et*

al. 1971*a*). Perirectal and perianal abscesses frequently yield mixtures of anaerobes in which *Bacteroides* spp. predominate (see Finegold 1977).

B. fragilis is the predominant pathogen in 60–70 per cent of abdominal infections. *B. asaccharolyticus* is the next most common species but is never isolated in pure culture. *B. thetaiotaomicron* is about four times less common than *B. fragilis*; other members of the fragilis group are isolated infrequently (Werner and Pulverer 1971, Werner 1974, Holland *et al*. 1977, Polk and Kasper 1977, Duerden 1980*a*). *B. fragilis* is found in at least 60 per cent of perianal and perirectal abscesses; *B. asaccharolyticus* is important but less common and rarely present alone. *B. ureolyticus* is rarely isolated from abdominal infections but sometimes occurs in perianal abscesses and other perineal infections (Henriksen 1948, Duerden 1980*a*).

Liver and biliary tract infections

Liver abscesses are often caused by anaerobes, particularly *Bacteroides* spp. (Sabbaj *et al*. 1972). The source of infection is usually the faecal flora, and the route the portal venous system. Colorectal malignancy, ulcerative colitis, Crohn's disease, and intraperitoneal abscesses following perforation of the large intestine are common underlying diseases. Consequently the causative organism is usually *B. fragilis*; *F. necrophorum* occurs occasionally. *Bacteroides* spp. may be found in amoebic abscesses (Legrand and Axisa 1905, May *et al*. 1967) and hydatid cysts of the liver (Deve and Guerbet 1907, Chabrol *et al*. 1948); and they may cause the first clinical signs of disease.

Bacteroides bacteraemia is a common complication of liver abscess and may help in establishing a clinical and bacteriological diagnosis. Bacteraemia was present in a series of five cases, diagnosed only at autopsy, that might have been diagnosed bacteriologically during life (Sabbaj *et al*. 1972). Liver abscess and bacteraemia due to *B. fragilis* sometimes occur in liver transplant patients (Fulginiti *et al*. 1968).

Faecal bacteroides are responsible for only a minority of cases of cholangitis and cholecystitis. They have been isolated from the bile and the gall-bladder wall of patients with bilary obstruction.

Oral and dental infections

The mouth, particularly the gingival crevice, is colonized by *Bacteroides* and *Fusobacterium* spp. after the eruption of the deciduous teeth (see Chapter 26). Most infections are derived from the gingival flora. Gram-negative anaerobes are the commonest cause of suppurative infection in the mouth.

Bacteroides spp. of the melaninogenicus-oralis group and *Fusobacterium* spp. are frequently implicated, along with α-haemolytic streptococci, in infections of the dental pulp and root canals (Möller 1966).

Such infections may result in dento-alveolar abscesses with purulent necrosis of the surrounding tissue and bone. When chronic, the abscesses are often referred to as *pyogenic granulomas*. They may affect adjacent teeth and cause osteomyelitis of the jaw. Large abscesses may discharge spontaneously via fistulae to the exterior or into the oral or nasal cavity (Moore and Russell 1972, Hardie 1974). Reliable studies indicate that the most significant organisms are *B. melaninogenicus*, *Fusobacterium* spp., and *Leptotrichia buccalis*; oral streptococci are also often found. Gram-negative anaerobic bacilli may have a synergistic role in actinomycotic abscesses (Hardie 1974). *B. melaninogenicus* and *Fusobacterium* spp. are frequently present with *Actinomyces israeli* (Pulverer 1958).

The treatment of dento-alveolar infection often necessitates tooth extraction. The post-extraction syndrome, *dry socket*, is the result of infection, leading to localized osteomyelitis and necrosis of the surrounding tissue. Such infections sometimes occur in previously clean sockets, and are painful and foul smelling. The presence of fusiform bacilli and spirochaetes may simulate Vincent's angina; other oral anaerobes are also present. *B. melaninogenicus* and *Fusobacterium* spp., especially *F. nucleatum*, have a particularly important role in the osteomyelitis and cellulitis that sometimes occur in the jaw and surrounding tissues as a result of compound fractures, tooth extraction, and dento-alveolar abscess (Heinrich and Pulverer 1960).

Acute ulcerative gingivitis—otherwise known as *AUG, Vincent's angina, Plaut-Vincent's infection, trench mouth, and fusospirochaetosis*—was one of the first anaerobic infections to be recognized (Plaut 1894, Vincent 1896). It is associated with poor environmental conditions, malnutrition, debility, and poor oral hygiene, such as are common in wartime, and is characterized by pain, haemorrhage, a foul odour, destruction of the interdental papillae, inflammation, recession of the gingival margin, and the formation of a pseudomembrane. It constitutes one form of periodontal disease, the most common cause of tooth loss in developed countries. It is now generally accepted that AUG is a synergistic infection with a fusobacterium and a large coarse spirochaete. The fusiform organism is usually *Leptotrichia buccalis*—previously known as Vincent's bacillus (see Chapter 26)—but *F. nucleatum* has also been implicated (Willis 1977); the spirochaete is *Treponema vincenti*. These organisms are present in the exudate and pseudomembrane, and diagnosis is made by direct microscopy. There is evidence that oral *Bacteroides* spp., in particular *B. melaninogenicus*, are essential components of the infection, and may be the main cause of the tissue damage and necrosis (Kaufman *et al.* 1972, Hardie 1974, Coykendall *et al.* 1980).

The bacteraemia that often follows extraction of the teeth may lead to endocarditis caused by *Bacteroides*

spp., *Fusobacterium* spp., or streptococci (Okell and Elliot 1935, Pressman and Bender 1944, Khairat 1966). Likewise, the lungs may suffer from the effect of dental sepsis, the result being necrotizing pneumonia, empyema, and abscess formation caused by these same organisms, occasionally including *B. fragilis*. Dento-alveolar and periodontal infection may spread directly or by the blood stream to cause retropharyngeal abscesses, osteomyelitis, extradural and subdural abscesses, and brain abscesses, with or without meningitis. Brain abscesses derived from oral sources usually have a mixed bacterial flora. *Actinomyces* spp. and streptococci are common components, but *B. melaninogenicus*, *B. oralis* or *Fusobacterium* spp. are almost always present; *B. fragilis* occurs occasionally.

Infections of the female genito-urinary tract

Gram-negative anaerobic bacilli are part of the normal vaginal flora (see Chapter 26) and are the prinicpal pathogens in the serious infections of the female genital tract that sometimes follow surgical interference or accompany other diseases (Ledger *et al.* 1971, Gorbach and Bartlett 1974, Willis 1977). *Bacteroides* spp. are implicated in infections of the uterus and pelvis, such as endometritis, pyometra, parametritis, salpingitis, tubal, tubo-ovarian and ovarian abscesses, pelvic cellulitis and abscesses, chorio-amnionitis, intra-uterine sepsis, post-abortal and puerperal infections, and wound infections and abscesses incidental to gynaecological surgical treatment; they are also found in vulvovaginal infections such as abscesses of the vaginal wall, vulva, and Bartholin's and Skene's glands. Many infections may be complicated by peritonitis, intra-abdominal abscesses, pelvic thrombophlebitis, bacteraemia with metastatic abscesses, and cellulitis or abscess formation in the perineum, groin or abdominal wall. *Bacteroides* spp. commonly play a part in infections of pregnancy, especially during the puerperium and after abortion. Slow rupture of the membranes, prolonged labour, and post-partum haemorrhage are predisposing factors.

Most cases of septic abortion are due to anaerobes (Rotheram and Schick 1969). Pearson and Anderson (1970) estimated that 10–15 per cent of cases were associated with bacteroides infection, but later studies suggested an incidence of more than 70 per cent (Thadepalli *et al.* 1973, Rotheram 1974). Patients suffer from fever, uterine tenderness, and a foul-smelling cervical discharge. Most respond well to uterine evacuation. The complications of septic abortion include broad-ligament abscesses, ovarian and pelvic abscesses, peritonitis, and pelvic thrombophlebitis; bacteraemia and metastatic abscesses may follow, though less often now than in the pre-antibiotic era. Similar lesions occur in puerperal infections, caused as a rule by *Bacteroides* spp., often in association with anaerobic or microaerophilic cocci (Pearson and Anderson 1970, Thadepalli *et al.* 1973). Pearson and Anderson (1970) found that bacteria from the normal vaginal flora caused amnionitis in about 10 per cent of deliveries; most such infections were apparently insignificant, but there seemed to be an association with perinatal mortality (Pearson and Anderson 1967). In more severe infections there is also an endometritis, which may spread to produce parametritis, pelvic abscess, and peritonitis. These infections produce fever, lower abdominal tenderness, and a foul lochial discharge.

Anaerobes often play a part in infections of the non-pregnant uterus, such as pyometra or endometritis. Such infections are associated with malignancy, especially after irradiation of the pelvic organs, and with cervical cauterization or stenosis (Burdon 1928, Carter *et al.* 1951, Finegold 1977). Intrauterine contraceptives have on rare occasions been associated with endometritis, and adnexal and pelvic infections, caused by anaerobes. *Bacteroides* spp. occur in non-gonococcal tubo-ovarian abscesses and pelvic inflammatory disease (Eschenbach *et al.* 1975). Anaerobes are common causes of wound infections and of pelvic abscesses after operations on the uterus, adnexa, and vagina (Ledger *et al.* 1971, Swenson *et al.* 1973, 1974, Thadepalli *et al.* 1973).

Bacteroides spp. are important pathogens in superficial infections of the female genitalia (Pearson and Anderson 1970); they are the most common cause of bartholinitis, and are found in necrotizing, gangrenous infections, often referred to as *noma*, that resemble fusospirochaetal infections in the mouth. The role of *Bacteroides* spp. in patients with non-specific vaginitis and a foul vaginal discharge is debatable. Some workers have found an increased number of *Bacteroides* organisms in specimens from these patients (Müller *et al.* 1967, Lindner *et al.* 1978); and a good response to metronidazole was obtained in the absence of *Trichomonas vaginalis* (Finegold 1977).

It has been assumed that *Bacteroides* spp. from the vaginal commensal flora reach the normally sterile uterus, fallopian tubes and pelvis by the ascending route as a result of surgical interference, trauma, or dysfunction (Galask *et al.* 1976). *B. melaninogenicus* and non-pigmented members of the melaninogenicus-oralis group, especially *B. bivius*, which are the predominant *Bacteroides* spp. in the vaginal flora (see Chapter 26), may cause severe infection; but *B. fragilis*, which is not part of the normal vaginal flora, has also been isolated from pelvic and gynaecological infections. Werner and his colleagues (1978) isolated clinically important anaerobes from 42 per cent of women *post partum—B. oralis* from 25 out of 56 patients, *B. fragilis* from 19, and *B. melaninogenicus* from five. Duerden (1980*a*) found *B. fragilis* to be the commonest gram-negative anaerobic pathogen in infections of the female genital tract, whereas *B. bivius* and *B. melaninogenicus* were present in only 30 per

cent. This is further evidence for the specific pathogenicity of *B. fragilis. Fusobacterium* spp. and spirochaetes are sometimes found in the lesions that resemble Vincent's infection.

Urinary tract infection Enterobacteria from the faecal flora are the common cause of infections of the urinary tract in women. Although *Bacteroides* spp. outnumber enterobacteria in the faecal flora by a factor of 10^3 (Drasar and Hill 1974), and form part of the normal urethral flora, they are not responsible for primary infections of the urinary tract. Possibly the Eh and the dissolved O_2 concentration in urine are too high for anaerobes to become established. However, *Bacteroides* spp. may enter the urine and behave as pathogens in obstructive pyonephrosis, renal or peri-renal abscesses, para- or peri-urethral cellulitis or abscesses, cowperitis, prostatitis, and prostatic abscesses.

Infections of the respiratory tract

Gram-negative anaerobic bacilli can infect all sites in the respiratory tract. Infections of the tonsils and pharynx are less common now than in the earlier part of the twentieth century. Plaut-Vincent's angina (see p. 316) has long been known to cause painful ulcerating lesions of the throat and tonsils (Plaut 1894, Vincent 1896). A severe tonsillar infection with peritonsillar cellulitis and abscess formation caused by *F. necrophorum* (*Bacteroides funduliformis*) was recognized later; it frequently led to bacteraemia, and sometimes to death. It has been seen less commonly since the introduction of antibiotics. Oral *Bacteroides* spp., *B. melaninogenicus* in particular, and *Fusobacterium* spp. are found in other peritonsillar and retropharyngeal abscesses, and may destroy the surrounding tissues, or cause bacteraemia and metastatic lesions (Hansen 1950).

In the pre-antibiotic era *F. necrophorum* was a common cause of acute otitis media, mastoiditis and sinusitis. Anaerobes are now rarely isolated in acute cases, but in the chronic forms of these diseases *Bacteroides* and *Fusobacterium* spp. are often found. Though Rist (1901) considered anaerobes to be the cause of all cases of chronic otitis media, a mixed flora is usually present. Gram-negative anaerobes often predominate and may give rise to serious complications. Thus, Guillemot (1899) and Rist (1901) reported cases of pulmonary gangrene secondary to chronic otitis media. Local osteomyelitis, soft tissue abscesses, and bacteraemia with widespread metastatic infection are also not uncommon. The most serious complication is brain abscess, sometimes with meningitis, subdural or extradural empyema, and intracranial thrombophlebitis; *Bacteroides* spp. are the most important components of the mixture of organisms usually isolated. Of the species associated with chronic otitis media and its complications, *B. melaninogenicus, B. oralis* and some *Fusobacterium* spp. are derived from the normal flora of the mouth and upper respiratory tract; *B. fragilis* is also often concerned and *F. necrophorum* occurs in fulminating bacteraemic infections.

Anaerobic infections of the lung and pleural space are sufficiently common in America to suggest that anaerobes are second only to the pneumococcus as causes of pneumonia in hospital patients (Bartlett and Finegold 1974), but this is not true of Great Britain. This discrepancy may be real or due to differences in laboratory methods, such as the more widespread use of anaerobic cultural techniques, or of aspiration techniques for obtaining specimens (Finegold 1977). The source of infection is the patient's mouth and upper respiratory tract. Sputum specimens are unsuitable for diagnosis because they are contaminated with the oropharyngeal flora. Satisfactory specimens from cases of empyema can be obtained by aspiration, but for specimens of lung parenchyma either per-cutaneous transtracheal aspiration (Bartlett *et al.* 1973) or needle biopsy (Beerens and Tahon-Castel 1965) is necessary. In Great Britain, sputum samples are still commonly used. Collection of specimens by the fibreoptic bronchoscope has been described (Wong *et al.* 1974).

Guillemot and his colleagues (1904) found that empyema fluid from patients with putrid pleurisy contained few aerobes but many anaerobes, which resembled those of the normal oral flora. Aspiration and pneumonia lead to empyema. The incidence of lung abscesses and empyema is increased by the presence of periodontal disease, peritonsillar infections, and chronic otitis media or sinusitis, and is low in edentulous patients. Pharyngeal infections with *F. necrophorum* may be accompanied by bacteraemia and pulmonary complications. All these, however, together with aspiration pneumonia, are less common now than formerly.

Anaerobic pleuropulmonary infection can be divided into four categories—aspiration pneumonia, necrotizing pneumonia, lung abscess, and empyema (Bartlett and Finegold 1972). There is some overlapping, but the pathogenesis and source of infection are the same for each category. Anaerobic aspiration pneumonia without necrosis or abscess formation is easily overlooked because of the absence of foul pus and tissue necrosis. It is an acute illness which responds rapidly to appropriate treatment (Bartlett *et al.* 1974*a*).

Necrotizing pneumonia is characterized by suppuration, with areas of necrosis and cavity formation. It usually affects one segment or lobe but may spread rapidly through the lung, destroying the tissue and leaving only putrid sloughs; the x-ray findings are characteristic and the disease is associated with a high mortality. It was called *pulmonary gangrene* by the early workers, who recognized the role of anaerobes in its pathogenesis (Guillemot *et al.* 1904, Rona 1905).

Lung abscess has also been recognized as a predominantly anaerobic infection for many years. The mechanism of infection is aspiration, but the infection is localized. It is often a complication of a tumour or of enlarged lymph nodes; the lung becomes infected by organisms from the upper respiratory tract. There is usually a low grade fever with anaemia and weight loss; tissue destruction is seen on x-ray as a cavity partly filled with fluid. The copious sputum is putrid and the patient has halitosis. Cerebral abscess is a serious complication of lung abscess, resulting from pulmonary thrombophlebitis, bacteraemia, and septic emboli. Empyema is almost invariably associated with underlying parenchymal disease, but is occasionally secondary to subphrenic abscesses. Anaerobes are the predominant pathogens (Bartlett *et al.* 1974*a*). The empyema fluid is usually purulent and often foul; it may be loculated and difficult to drain, but drainage is an essential part of the treatment.

Anaerobic pleuropulmonary infections are invariably mixed, but gram-negative anaerobic bacilli are usually the major pathogens. They are also responsible in part for the progressive damage and destruction that occur in bronchiectatic cavities. *B. melaninogenicus*, *B. oralis*, and *F. nucleatum*, which are part of the oral flora, are the commonest species isolated. *B. fragilis* is found in a significant minority of cases of simple aspiration pneumonia, lung abscess, and bronchiectasis and, more frequently, in empyema; it is less common in necrotizing pneumonia, in which *B. melaninogenicus* and *F. nucleatum* predominate. *F. necrophorum* is less common now than formerly but is occasionally responsible for severe pleuropulmonary infections with abscess formation, empyema, and bacteraemia (Bartlett *et al.* 1974*b*).

Infections of the central nervous system

Anaerobes, particularly gram-negative anaerobic bacilli, are the commonest cause of brain abscesses, especially those originating from chronic otitis media, mastoiditis, and sinusitis (Heineman and Braude 1963, Ingham *et al.* 1977*a*). They are found less often in other infections of the central nervous system. Reports of isolated cases were reviewed by Finegold (1977).

Brain abscesses due to *Bacteroides* and *Fusobacterium* spp. may arise from infection at adjacent sites, or by metastatic spread from suppurative lesions elsewhere. Cholesteatoma is an important predisposing factor. Abscesses secondary to chronic otitis media and mastoiditis—the most important primary causes—usually affect the temporal lobes of the cerebrum, the cerebellum being the second most common site. Infection occurs by direct extension, often with localized osteomyelitis; thrombosis or thrombophlebitis of the lateral sinus is a recognized complication. Abscesses of the frontal lobe occasionally occur as a result of chronic sinusitis. Dental infections, oral infections, and abscesses may also give rise to brain abscesses, either by direct extension or via the blood stream. Abscesses of the frontal lobe occasionally follow penetrating injuries of the orbit and sinuses.

Metastatic brain abscesses are not uncommon complications of anaerobic pleuropulmonary infections. They tend to be multiple and usually occur at the junction of the grey and white matter. The frontal, parietal and occipital lobes are most commonly affected; metastatic abscesses of the temporal lobe and cerebellum are rare. It was once thought that metastatic spread from pleuropulmonary infections occurred via the spinal venous system, with the assistance of the increased intrathoracic pressure produced by coughing, but serial arteriographic studies show that the areas first affected are those supplied by the middle cerebral artery, especially on the right side (Prolo and Hanbery 1968). This indicates that the infection spreads through the arterial system by septic emboli from pulmonary thrombophlebitis. Heineman and Braude (1963) believe that previous cerebral haemorrhage or infarction may predispose to brain abscess.

The *Bacteroides* and *Fusobacterium* spp. isolated from brain abscesses vary with the source of infection. The melaninogenicus-oralis group and fusobacteria, especially *F. nucleatum*, are frequently isolated from abscesses secondary to otitis media and sinusitis, but the most common species is *B. fragilis*. Brain abscesses arising by direct extension usually yield more than one pathogen; *Bacteroides* and *Fusobacterium* spp. are the commonest and most significant organisms, but facultatively or strictly anaerobic gram-positive cocci such as *Streptococcus milleri*, and enterobacteria such as *Proteus* spp., are often found (Ingham *et al.* 1977*a*). In metastatic brain abscesses complex mixtures of organisms are uncommon. *Fusobacterium* spp. are generally the cause of abscesses secondary to pleuropulmonary infections; *B. fragilis* is rare. Organisms of the melaninogenicus-oralis group and *Fusobacterium* spp. are often present in metastatic abscesses secondary to dental or oral infection, but *Actinomyces* spp. are the commonest cause; mixed infections with *A. israeli* and *B. melaninogenicus* have been described.

Meningitis is occasionally caused by anaerobes; most such cases are associated with extradural, subdural, or brain abscesses, but some with chronic otitis media and mastoiditis. Anaerobes from the sinuses, mouth, and lungs are rarely concerned unless access to the meninges has been made by trauma or surgical interference. *Fusobacterium* spp., especially *F. necrophorum*, are the commonest anaerobic pathogens in meningitis. *B. melaninogenicus*, *B. oralis* and *B. fragilis* may also be concerned.

Subdural empyema usually originates from sinusitis, often of the frontal sinus; trauma, surgical interference, and otitis media play only a small part. Anaerobic and microaerophilic cocci are the commonest

species in infections originating from the frontal sinus, but *Bacteroides* spp., including *B. fragilis*, and *Fusobacterium* spp., are frequently implicated. The pathogenesis and bacteriology of extradural empyema—a less dangerous disease—are the same as for brain abscesses and meningitis.

In the pre-antibiotic era suppurative thrombophlebitis of the venous sinuses was a common complication of anaerobic infections of the middle ear, mastoid, sinuses and throat, and was associated with meningitis and brain abscess.

Skin and soft tissue infections

Gram-negative anaerobic bacilli from mucous membranes cause wound infection after surgical treatment or accidental injury. Human and animal bites may be infected with oral *Bacteroides* spp. and fusobacteria (Linscheid and Dobyns 1975). Other lesions include cellulitis, gangrene, and abscesses. Infections are generally mixed, but anaerobes play a major part in the tissue destruction and necrosis. *Bacteroides* and *Fusobacterium* spp. play a role in Meleney's *synergistic gangrene* and *necrotizing fasciitis*.

Necrotic and gangrenous lesions of the face and neck are often caused by anaerobes from the oral flora. Many such infections are associated with periodontal disease, dental abscesses, or tooth extraction. *Cancrum oris* in poorly nourished and debilitated children (Emslie 1963) starts at a mucocutaneous junction and spreads to large areas of the face. It is one manifestation of the disease *noma*. It occurs most commonly around the mouth but may also affect the nose, ear, vulva, prepuce, or anus. The destructive process, which is associated with a foul odour, also affects periosteum and bone, with sequestrum formation. The gums and cheeks are destroyed, the teeth fall out and there may be almost complete destruction of the face before death; without antimicrobial therapy, the mortality was 70–100 per cent (Blumer and MacFarlane 1901). Predisposing factors include systemic disease, chronic infections and infestations, malnutrition, and poor oral hygiene. Noma is usually regarded as a fusospirochaetal synergistic infection, but there is evidence that *B. melaninogenicus* may play an important role (Kaufman *et al.* 1972).

A similar disease known as *Fournier's gangrene* or *necrotizing dermogenital infection* affects the perineum, groin, vulva, scrotum, and penis. It may be a manifestation of noma, usually taking the form of a spreading anaerobic cellulitis and necrotizing fasciitis. It may follow surgical operations on the groin, perineum or lower abdomen and begin as a wound infection; other cases begin as minor lesions of the genitalia. In males, a tight prepuce and poor genital hygiene predispose to balanitis that may spread to surrounding areas. The necrotic lesions extend rapidly and may cover huge areas of the abdominal wall, loin, buttocks, and thighs. Without antibiotic therapy and extensive débridement the mortality is high. Infection is always mixed. Anaerobic or microaerophilic cocci, *B. fragilis*, *B. melaninogenicus*, *B. asaccharolyticus*, *B. ureolyticus* and *F. nucleatum* have all been isolated.

Elsewhere in the body, *Bacteroides* and *Fusobacterium* spp. are found in large numbers in infected pilonidal cysts or abscesses, and sebaceous cysts (Bornstein *et al.* 1964, Pien *et al.* 1972). The pus is usually foul and also often contains low-grade pathogens such as micrococci and diphtheroids. Paronychia is caused by *B. melaninogenicus* and *Fusobacterium* spp. often of oral origin, and sometimes by *B. ureolyticus*.

In varicose or decubitus ulcers, and in gangrenous diabetic ulcers, the superficial areas are colonized by bacteria that are aerobic and facultatively anaerobic, but heavy growth of *Bacteroides* and *Fusobacterium* spp. regularly occurs when specimens taken from the depth of the lesions are cultivated. Some lesions progress to osteomyelitis, and several workers have reported bacteroides bacteraemia, sometimes fatal, associated with decubitus ulcers (Felner and Dowell 1971). Various gram-negative anaerobic bacilli have been isolated, but several reports have recognized the special role of *B. asaccharolyticus*, or strains of '*B. melaninogenicus*' that were probably *B. asaccharolyticus* (Peromet *et al.* 1973, Duerden 1980*a*).

Breast abscesses that occur in the puerperium are usually caused by *Staphylococcus aureus*, but *B. melaninogenicus*, *Fusobacterium* spp., and occasionally *B. fragilis* have been isolated, particularly from recurrent abscesses of the subareolar region (Pearson 1967).

Bone and joint infections

Osteomyelitis due to anaerobes arises either by spread from an adjacent lesion or by haematogenous infection. The bones most commonly infected are the mastoid and ethmoid; the infection arises from chronic otitis media, mastoiditis, and sinusitis. The maxilla and mandible become infected as a consequence of dental or periodontal disease, and sometimes as the result of tooth extraction or other trauma. The causative organisms are *B. melaninogenicus*, *B. oralis*, *Fusobacterium* spp., and occasionally *B. fragilis*.

Elsewhere in the body, *B. melaninogenicus*, *B. asaccharolyticus*, and *B. fragilis* cause chronic osteomyelitis with the formation of sequestra and sinuses containing foul pus. Osteomyelitis of the foot and leg often arises by spread from adjacent lesions in soft tissue, especially in diabetics. There appears to be an association between bacteroides infections and diabetes, alcoholism, or drug addiction (Ziment *et al.* 1967, 1969, Pearson and Harvey 1971).

A few reports, most from the pre-antibiotic era, describe purulent arthritis due to haematogenous infection with *F. necrophorum*.

Bacteraemia and cardiovascular infections

Bacteroides spp. are reported to cause 0.2 to 10 per cent of all bacteraemias, depending upon the type of patient studied. Transient and insignificant bacteraemia due to *Bacteroides* spp. often follows dental manipulation in patients with periodontal disease (Rogosa *et al.* 1960, Francis and de Vries 1968); it may also result from the insertion of instruments into the gastro-intestinal and genito-urinary tracts (Le Frock *et al.* 1973, 1975, Chow and Guze 1974). Significant bacteroides bacteraemia may, however, indicate serious underlying diseases (Felner and Dowell 1971, Chow and Guze 1974), usually intra-abdominal abscesses and peritonitis, less often septic abortion and puerperal infection, and occasionally pleuropulmonary infections, brain abscess, chronic otitis media, and decubitus or varicose ulcers. Bacteroides bacteraemia is sometimes the first manifestation of a malignant tumour of the colon, rectum or cervix. In the pre-antibiotic era, necrotic tonsillitis caused by *F. necrophorum* often resulted in bacteraemia. Septic thrombophlebitis is the immediate source of many bacteroides bacteraemias.

B. fragilis is the usual cause of bacteraemia arising from intra-abdominal sepsis; *Fusobacterium* spp., especially *F. necrophorum*, are most often responsible when the bacteraemia arises from oral or upper respiratory tract infection. Aerobic and facultatively anaerobic organisms are often also present in the blood (Hermans and Washington 1970, Wilson *et al.* 1972). *B. fragilis* and *Escherichia coli* often occur together, and in such instances demonstration of the anaerobe may necessitate the use of a selective medium (von Graevenitz and Sabella 1971).

Mortality associated with bacteroides bacteraemia depends on the underlying cause; it varies from 1–2 per cent in patients with septic abortion or puerperal sepsis (Smith *et al.* 1970) to at least 70 per cent in patients with malignant tumours (Kagnoff *et al.* 1972).

Abscesses resulting from bacteraemia are less common now than in the pre-antibiotic era. Bacteraemic shock sometimes occurs in bacteroides bacteraemia, though less commonly than in bacteraemia due to enterobacteria (Wilson *et al.* 1967), despite the lack of 2-keto-3-deoxyoctanate and heptose in bacteroides endotoxin. Disseminated intravascular coagulation sometimes occurs (Attar *et al.* 1966, Yoshikawa *et al.* 1971). Endotoxin can be detected by the *Limulus* lysate assay (Sonnenwirth *et al.* 1972).

Endocarditis due to *Bacteroides* spp. is rare. Masri and Grieco (1972) reviewed 27 cases and found pre-existing heart damage in 60 per cent. Most patients had underlying bacteroides infections, and the mortality was 36 per cent. Felner (1974) reported similar findings in a review of 22 cases.

Diagnosis of bacteroides infections

The clinical history and nature of the lesion are often valuable guides. Diagnosis is confirmed by isolating *Bacteroides* or *Fusobacterium* spp. with or without other bacteria. The most reliable specimens are pus or exudate from the depths of an open lesion or from a closed lesion. Specimens such as sputum, being contaminated with the normal flora of mucous membranes, are unsuitable; and cotton-wool swabs often give poor results (Collee *et al.* 1974). Most pathogens are fairly aerotolerant (Tally *et al.* 1975), but they may be killed in transit by prolonged exposure to air, or by desiccation, which can be prevented by the use of a transport medium. Specimens should be transported to the laboratory quickly. Attebery and Finegold (1969) recommended the use of tubes containing an atmosphere of O_2-free CO_2. Bartlett and his colleagues (1976) reported good results with pus held in closed containers at room temperature for up to 48 hours. Direct plating of material and immediate anaerobic incubation at the patient's bedside is likely to give the best results.

Direct microscopy may show the pleomorphic gram-negative forms that typify *B. fragilis* and the pointed filaments of fusobacteria. The presence of large numbers of fusobacteria and coarse spirochaetes may suggest Vincent's angina, or other fusospirochaetal infections. Immunofluorescent methods enable *B. fragilis* and *B. melaninogenicus* to be identified rapidly in smears (Stauffer *et al.* 1975). Direct examination of specimens of pus or exudates by gas liquid chromatography is also an aid to rapid diagnosis (Gorbach *et al.* 1976, Phillips *et al.* 1976). Species that are aerobic and facultatively anaerobic produce only acetic acid; the demonstration of other short-chain fatty acids shows that anaerobes are present, but the method does not provide a specific diagnosis.

Bacteroides and *Fusobacterium* spp. can be isolated by conventional anaerobic techniques (Collee *et al.* 1972). The addition of 5–10 per cent CO_2 to the anaerobic atmosphere is often helpful (Watt 1973, Stalons *et al.* 1974). The roll-tube (Hungate 1950, 1969) and anaerobic-cabinet methods (Aranki *et al.* 1969)—designed specifically for very strict anaerobes—are not essential (Rosenblatt *et al.* 1973, Watt *et al.* 1974).

B. fragilis grows well on blood agar. Some of the less common *Bacteroides* and *Fusobacterium* spp. require an enriched nutrient medium such as BM medium, which contains trypticase, proteose peptone and yeast extract (Williams *et al.* 1975). The addition of menadione (vitamin K) provides an essential growth factor for some strains of *B. melaninogenicus* (Lev 1959), and the use of lysed-blood agar assists in the early recognition of pigmented species (Finegold *et al.* 1971*b*). The media should either be freshly prepared or stored in an anaerobic atmosphere.

Selective media usually contain high concentrations of aminoglycoside antibiotics; for *Bacteroides* and *Fusobacterium* spp. neomycin—a useful selective agent for clostridia—is less suitable than kanamycin 75µg/ml (Loesche *et al.* 1971). Nalidixic acid has proved useful (Wren 1978).

Members of the fragilis group of *Bacteroides* produce recognizable colonies after incubation for 18–24 hours, but the more demanding species, including *B. ureolyticus* and most fusobacteria, require longer incubation. Primary cultures should therefore be incubated for at least 3 days, and the anaerobic conditions should not be interrupted until 2 full days have elapsed (Wren 1980). An anaerobic cabinet enables cultures to be examined without exposure to air.

Specific identification may be made by conventional bacteriological tests (Holdeman *et al.* 1977, Duerden *et al.* 1980) or by sets of tests prepared commercially (Moore *et al.* 1975, Stargel *et al.* 1976, Tharagonnet *et al.* 1977).

Serological response to bacteroides infections

Antibodies that react with purified lipopolysaccharide of *B. fragilis*, *B. melaninogenicus* and *F. nucleatum* are present in the serum of most normal human subjects; they are not present in umbilical-cord serum taken at birth (see Chapter 26). An immune response can be demonstrated in patients with *B. fragilis* septicaemia or deep abscesses by means of agglutination, precipitation and immunofluorescent techniques (Danielsson *et al.* 1974, Rissing *et al.* 1974, Lambe *et al.* 1975). Such tests are not used in the diagnosis of bacteroides infections because of the heterogeneity of the surface antigens, especially in members of the fragilis group (see Chapter 26).

Agglutinating antibodies against *F. necrophorum*, unlike antitoxin (Garcia *et al.* 1973), are not protective. The main determinant of pathogenicity in *B. fragilis* is probably the polysaccharide capsule (see Chapter 26); the capsular antigen is the same for all strains. The antibody response is best demonstrated when the antigen is associated with the outer membrane (Kasper *et al.* 1977). Low concentrations of antibody to the capsular polysaccharide are present in normal subjects, but there is a large increase in women with pelvic inflammatory disease due to *B. fragilis* (Kasper *et al.* 1978). These antibodies may prove useful in the diagnosis of deep infection with *B. fragilis*. Poxton (1979) described a species-specific outer-membrane antigen. It is unlikely that serological methods will play a major part in the diagnosis of acute infections with *Bacteroides* spp., but they may prove useful in the diagnosis and management of chronic infections and inflammatory bowel diseases.

(For reviews of anaerobic infections in man see Balows and his colleagues 1974, Willis 1977, Finegold 1977.)

Animal infections caused by gram-negative anaerobic bacilli

Infections occur in cattle, sheep, goats, pigs, rabbits, cats and wild animals, and cause serious economic loss. The organism most commonly responsible is *F. necrophorum*, which produces tissue necrosis and abscess formation, accompanied by a putrid odour. Initiation of infection usually requires a minor injury to the skin or mucosa. Bacteraemia may occur and result in widespread metastatic abscesses. Unlike most infections in man, animal infections usually occur as primary diseases rather than as secondary complications.

Calf diphtheria

This disease, first associated with 'Bacillus necrophorus' (*F. necrophorum*) by Loeffler in 1884, is a necrotizing infection of the oral mucosa, tongue, pharynx and larynx. It occurs in calves of all ages and in cattle up to the age of two or more years. There is a false membrane, which may extend from the throat down into the trachea, but which more commonly occurs in patches, firmly adherent to underlying, circumscribed and indurated swellings. Initiation of infection probably requires some minor damage to the mucosa; the organism then proliferates in the damaged mucosa and spreads to the submucosa and muscle. The disease may be fatal. Necrotic material sometimes obstructs the larynx and causes asphyxiation. Smaller amounts of necrotic material may be aspirated and cause lung abscesses or pneumonia (Langworth 1977).

   F. necrophorum can be isolated in culture and seen in smears prepared from the lesions; it forms characteristic long wavy rows of bacilli in the deeper layers of the membrane. The disease has not been reproduced by the introduction of *F. necrophorum* or infected tissue into the throats of healthy calves (Jensen and Mackey 1974). Untreated animals usually die within 5 days from asphyxia, broncho-pneumonia or toxaemia.

Liver abscess

Liver abscesses or *necrotic focal hepatitis* caused by *F. necrophorum* result in serious economic losses in cattle. The disease also occurs in sheep but rarely affects other farm animals. The abscesses are often multiple. *F. necrophorum* is the predominant organism seen in smears prepared from the lesions, and is consistently isolated from the pus (Feldman *et al.* 1936, Madin 1949). The encapsulated abscesses are usually discovered at slaughter. Most cattle in areas where there is a high incidence of abscesses have antibodies to *F. necrophorum* (Feldman *et al.* 1936). The factors that predispose to abscess formation have not been clearly established. *F. necrophorum* is thought to reach the liver via the portal circulation from the rumen owing to some alteration in the mucosal protective mechanism (Madin 1949). Smith (1944) suggested a relation between rumen ulcers and liver abscesses. Abscesses are more common in animals on artificial feeds or concentrates than in those on pasture. Jensen and Mackey (1974) noted that the change from grazing to a high-concentrate diet resulted in increased acid production by the rumen flora which sometimes led to ulceration. Metallic foreign bodies in the bovine stomach may also cause ulceration. Liver abscesses have been reproduced experimentally in cattle and sheep by the intraportal injection of *F. necrophorum* (Jensen *et al.* 1954). (See Chapter 53 for gram-negative anaerobes in summer mastitis.)

Foot-rot

Foot-rot is an infection of the hooves and surrounding tissues in ungulates, and is a major cause of lameness in cattle and sheep (Jensen and Mackey 1974); it also affects goats, pigs and deer. The disease is a bacterial dermatitis; infection begins with superficial necrosis of the skin of the interdigital space and spreads rapidly to the dermis, adjacent skin, and hoof margin; there is a foul discharge of pus and necrotic tissue, and the horn may separate. The disease may be mild or debilitating, and may require surgical interference.

   F. necrophorum is implicated in most forms of foot-rot (Langworth 1977) but is only one constituent of a mixed infection; the disease may be another example of synergistic infection. *F. necrophorum* may be the primary pathogen in ovine foot-abscess. In combination with *Corynebacterium pyogenes*, it causes ovine interdigital dermatitis (Roberts 1967); this takes the form of red and swollen interdigital skin covered by a film of necrotic tissue. Roberts (1967) showed that *F. necrophorum* produced a leucocidal toxin that allowed *C. pyogenes* to become established, and *C. pyogenes* produced a filtrable, heat-labile, non-dialysable macromolecule that stimulated the growth and pathogenicity of *F. necrophorum*. Ovine interdigital dermatitis can be reproduced by placing pads soaked with *F. necrophorum* on scarified interdigital skin (Parsonson *et al.* 1967). *F. necrophorum* is also implicated in foot-rot of cattle, possibly in association with several other species. Berg and Loan (1975) isolated *F. necrophorum* and *B. melaninogenicus* from eight cases of bovine foot-rot. The application of mixed inocula of the two species to scarified interdigital skin produced typical lesions of foot-rot.

Bacteroides nodosus is regarded by some workers as the cause of foot-rot in sheep and goats (Egerton and Parsonson 1966, Wilkinson *et al.* 1970). In combination with *F. necrophorum* it causes an infection of the hoof and underlying tissues in cattle, which may result in the detachment of the hoof. It has been isolated from some cases of foot-rot in cattle in Australia (Egerton and Parsonson 1966, Wilkinson *et al.* 1970), Holland (Toussaint Raven and Cornelisse 1971) and Great Britain (Thorley and Calder 1977).

Predisposing factors for the development of foot-rot in any species of animal are unhygienic conditions such as damp soil or litter underfoot and prior injury to the foot. Experimental infections can be established only after scarification or similar minor injury to the interdigital skin (Parsonson *et al.* 1967, Berg and Loan 1975). *F. necrophorum* can survive in damp soil for several months (Garcia *et al.* 1971), but material from affected feet does not remain infective under saprophytic conditions for more than a week. The disease may be eradicated from a farm by segregation and treatment of infected animals, and removal of uninfected animals to pasture where no infection has occurred in the previous 2–3 weeks. Good foot hygiene must be maintained (see Beveridge 1941, Ensor 1957). Treatment consists of surgical toilet, topical application of antimicrobial substances, and sometimes antibiotics injected parenterally. Egerton and Roberts (1971) found that a vaccine composed of *B. nodosus* killed by formalin and emulsified in Freund's incomplete adjuvant gave substantial protection against experimental infections; it gave partial protection under field conditions (Egerton and Morgan 1972).

Labial necrosis of rabbits

Schmorl (1891) first described this disease in an outbreak among his laboratory rabbits. It is caused by *F. necrophorum* and the disease is characterized by a dark bluish-red swelling of the underlip that spreads along the ventral aspect of the jaw and throat, and in about 8 days reaches the base of the neck. By the 5th day there is a watery nasal discharge and the animal is febrile. It dies in an emaciated condition with considerable dyspnoea. At post-mortem examination the underlip is seen to be converted into a yellowish-white, compact, bacon-like, necrotic mass, which may extend to the bone. The cervical glands are swollen, juicy and greyish-red; sometimes they show small caseous foci. There is a bloody, slightly turbid exudate in the pleural and pericardial cavities; sometimes there are a few pneumonic areas in the lung. The viscera appear normal. *F. necrophorum* can be isolated from the necrotic tissue and from the exudates.

Subcutaneous abscesses in cats

Anaerobic bacteria, in particular *Bacteroides* spp., are the commonest cause of subcutaneous abscesses in cats. These abscesses are often encountered in small-animal veterinary practice and are the result of bites sustained in fights. A mixture of bacterial species is isolated from the foul pus of most abscesses but the majority of strains are anaerobes. Love and her colleagues (1979) found that 66 per cent of their anaerobic isolates were gram-negative bacilli. They isolated 48 strains of *Bacteroides* and 32 of *Fusobacterium* from 36 abscesses; the commonest species were *B. fragilis*, *B. asaccharolyticus*, *F. nucleatum*, *B. melaninogenicus* and *F. necrophorum*.

References

Altemeier, W. A. (1938) *Ann. Surg.* **107**, 517.

Aranki, A., Syed, S. A., Kenney, E. B. and Freter, R. (1969) *Appl. Microbiol.* **17**, 568.

Attar, S. *et al.* (1966) *Ann. Surg.* **164**, 41.

Attebery, H. R. and Finegold, S. M. (1969) *Appl. Microbiol.* **18**, 558.

Babb, J. L. and Cummins, C. S. (1978) *Infect. Immun.* **19**, 1088.

Balows, A., Haan, R. M. de, Dowell, V. R. Jr and Guze, L. B. (1974) *Anaerobic Bacteria: Role in Disease.* Charles C Thomas, Springfield, Ill.

Bartlett, J. G. and Finegold, S. M. (1972) *Medicine, Baltimore* **51**, 413; (1974) *Amer. Rev. resp. Dis.* **110**, 56.

Bartlett, J. G., Gorbach, S. L., Tally, F. P. and Finegold, S. M. (1974a) *Amer. Rev. resp. Dis.* **109**, 510.

Bartlett, J. G., Rosenblatt, J. E. and Finegold, S. M. (1973) *Ann. intern. Med.* **79**, 535.

Bartlett, J. G., Sullivan-Gigler, N., Louis, T. J. and Gorbach, S. L. (1976) *J. clin. Microbiol.* **3**, 133.

Bartlett, J. G., Sutter, V. L. and Finegold, S. M. (1974b). In: *Anaerobic Bacteria: Role in Disease*, p. 327. Charles C Thomas, Springfield, Ill.

Beaver, D. C., Henthorne, J. C. and Macy, J. W. (1934) *Arch. Path.* **17**, 493.

Beerens, H. and Tahon-Castel, M. (1965) *Infections Humaines à Bactéries Anaérobies Non-Toxigénes.* Présses Acad. Eur., Bruxelles.

Berg, J. N. and Loan, R. W. (1975) *Amer. J. vet. Res.* **36**, 1115.

Beveridge, W. I. B. (1941) *Counc. sci. industr. Res. Aust.* Bull. No. 140.

Bjornson, A. B., Altemeier, W. A. and Bjornson, H. S. (1976) *Infect. Immun.* **14**, 843.

Blumer, G. and MacFarlane, A. (1901) *Amer. J. med. Sci.* **122**, 527.

Bornstein, D. L., Weinberg, A. N., Swartz, M. N. and Kunz, L. J. (1964) *Medicine, Baltimore* **43**, 207.

Burdon, K. L. (1928) *J. infect. Dis.* **42**, 161; (1932) *Proc. Soc. exp. Biol., N.Y.* **29**, 1144.

Burt, S., Meldrum, S., Woods, D. R. and Jones, D. T. (1978) *Appl. environ. Microbiol.* **35**, 439.

Carter, B., Jones, C. P., Ross, R. A. and Thomas, W. L. (1951) *Amer. J. Obstet. Gynec.* **62**, 693.

Casciato, D. A., Rosenblatt, J. E., Goldberg, L. S. and Bluestone, R. (1975) *Infect. Immun.* **11**, 337.

Chabrol, E., Sterboul and Fallot (1948) *Bull. Acad. nat. Méd.* **132**, 562.

Chow, A. W. and Guze, L. B. (1974) *Medicine, Baltimore* **53**, 93.

Cohen, J. (1932) *Arch. Surg.* **24**, 171.

Collee, J. G., Watt, B., Brown, R. and Johnstone, S. (1974) *J. Hyg., Camb.* **72**, 339.

Collee, J. G., Watt, B., Fowler, E. G. and Brown, R. (1972) *J. appl. Bact.* **35**, 71.

Coykendall, A. L., Kaczmarek, F. S. and Slots, J. (1980) *Int. J. syst. Bact.* **30**, 559.

Dack, G. M. (1940) *Bact. Rev.* **4**, 227.

Dack, G. M., Dragstedt, L. R. and Heinz, T. E. (1936) *J. Amer. med. Ass.* **106**, 7.

Danielsson, D., Lambe, D. W. Jr and Persson, S. (1974) In: *Anaerobic Bacteria: Role in Disease*, p. 173. Charles C Thomas, Springfield, Ill.

Deve, F. and Guerbet, M. (1907) *C. R. Soc. Biol., Paris* **63**, 305.

Drasar, B. S. and Hill, M. J. (1974) *Human Intestinal Flora*. Academic Press, London.

Duerden, B. I. (1980a) *J. Hyg., Camb.* **84**, 301; (1980b) *J. med. Microbiol.* **13**, 69, 79, 89.

Duerden, B. I., Collee, J. G., Brown, R., Deacon, A. G. and Holbrook, W. P. (1980) *J. med. Microbiol.* **13**, 231.

Egerton, J. R. and Morgan, I. R. (1972) *Vet. Rec.* **91**, 453.

Egerton, J. R. and Parsonson, I. M. (1966) *Aust. vet. J.* **42**, 425.

Egerton, J. R. and Roberts, D. S. (1971) *J. comp. Path.* **81**, 179.

Eiken, M. (1958) *Acta path. microbiol. scand.* **43**, 404.

Emslie, R. D. (1963) *Dent. Pract.* **13**, 481.

Ensor, C. R. (1957) *N.Z.J. Agric.* **94**, 218.

Eschenbach, D. A. *et al.* (1975) *New Engl. J. Med.* **293**, 166.

Feldman, W. H., Hester, H. R. and Wherry, F. P. (1936) *J. infect. Dis.* **59**, 159.

Felner, J. M. (1974) In: *Anaerobic Bacteria: Role in Disease*, p. 345. Charles C Thomas, Springfield, Ill.

Felner, J. M. and Dowell, V. R. Jr (1971) *Amer. J. Med.* **50**, 787.

Finegold, S. M. (1977) *Anaerobic Bacteria in Human Disease*. Academic Press, New York.

Finegold, S. M., Marsh, V. H. and Bartlett, J. G. (1971a) *Proc. Int. Conf. Nosocomial Infections* 1970, 123.

Finegold, S. M., Sugihara, P. T. and Sutter, V. L. (1971b) In: *Isolation of Anaerobes*. Ed. by D. S. Shapton and R. G. Board. Academic Press, London.

Francis, L. E. and Vries, J. A. de (1968) *Dent. Clin. N. Amer.* p. 243.

Fulginiti, V. A. *et al.* (1968) *New Engl. J. Med.* **279**, 619.

Galask, R. P., Larsen, B. and Ohm, M. J. (1976) *Clin. Obstet. Gynec.* **19**, 61.

Garcia, M. M., Dorward, W. J., Alexander, D. C., Magwood, S. E. and McKay, K. A. (1973) *Canad. J. comp. Med.* **38**, 222.

Garcia, M. M., Neil, D. H. and McKay, K. A. (1971) *Appl. Microbiol.* **21**, 809.

Gibbons, R. J. and MacDonald, J. B. (1961) *J. Bact.* **81**, 614.

Gorbach, S. L. and Bartlett, J. G. (1974) *New Engl. J. Med.* **290**, 1177, 1237, 1289.

Gorbach, S. L., Mayhew, J. W., Bartlett, J. G., Thadepalli, H. and Onderdonk, A. B. (1976) *J. clin. Invest.* **57**, 478.

Gorbach, S. L., Thadepalli, H. and Norsen, J. (1974) In: *Anaerobic Bacteria: Role in Disease*, p. 399. Charles C Thomas, Springfield, Ill.

Graevenitz, A. von and Sabella, W. (1971) *J. Med.* **2**, 185.

Guillemot, L. D. (1899) *Recherches sur la Gangrène Pulmonaire*. Thèses de Paris.

Guillemot, L. D., Halle, J. and Rist, E. (1904) *Arch. Méd. exp.*, **16**, 571.

Hansen, A. (1950) *Nogle Undersogelsen over Gramnegative Anaerobe IkkeSpore Daunende Bacterier Isolerede fra Peritonsilloere Abscesser bos Mennesker*. Munksgaard, Copenhagen.

Hardie, J. M. (1974) In: *Infection with Non-Sporing Anaerobic Bacteria*. Churchill Livingstone, Edinburgh.

Harris, J. W. and Brown, J. H. (1927) *Bull. Johns Hopk. Hosp.* **40**, 203.

Heineman, H. S. and Braude, A. I. (1963) *Amer. J. Med.* **35**, 682.

Heinrich, S. and Pulverer, G. (1960) *Z. Hyg. InfektKr.* **146**, 331.

Henriksen, S. D. (1948) *Acta path. microbiol. scand.* **25**, 368.

Hermans, P. E. and Washington, J. A. II (1970) *Ann. intern. Med.* **73**, 387.

Hite, K. E., Locke, M. and Hesseltine, H. C. (1949) *J. infect. Dis.* **84**, 1.

Holbrook, W. P. and McMillan, C. (1977) *J. appl. Bact.* **43**, 369.

Holdeman, L. V., Cato, E. P. and Moore, W. E. C. (1977) *Anaerobe Laboratory Manual*, 4th edn. Blacksburg, Virginia.

Holland, J. W., Hill, E. O. and Altemeier, W. A. (1977) *J. clin. Microbiol.* **5**, 20.

Hungate, R. E. (1950) *Bact. Rev.* **14**, 1; (1969) In: *Methods in Microbiology*, vol. 3B. Ed. by J. R. Norris and D. W. Ribbons. Academic Press, London.

Ingham, H. R., Selkon, J. B. and Roxby, C. M. (1977a) *Brit. med. J.* **ii**, 991.

Ingham, H. R., Sisson, P. R., Tharagonnet, D., Selkon, J. B. and Codd, A. A. (1977b) *Lancet* **ii**, 1252.

Jensen, R., Flint, J. C. and Griner, L. A. (1954) *Amer. J. vet. Res.* **54**, 5.

Jensen, R. and Mackey, D. R. (1974) *Diseases of Feedlot Cattle*. Lea and Febiger, Philadelphia.

Kagnoff, M. F., Armstrong, D. and Blevins, A. (1972) *Cancer* **29**, 245.

Kasper, D. L., Eschenbach, D. A., Hayes, M. E. and Holmes, K. K. (1978) *J. infect. Dis.* **138**, 74.

Kasper, D. L., Onderdonk, A. B. and Bartlett, J. G. (1977) *J. infect. Dis.* **136**, 789.

Kaufman, E. J., Mashimo, P. A., Hausmann, E., Hanks, C. T. and Ellison, S. A. (1972) *Arch. oral Biol.* **17**, 577.

Kelly, M. J. (1980) *Ann. R. Coll. Surg. Engl.* **62**, 52.

Khairat, O. (1966) *J. dent. Res.* **45**, 1191; (1967) *J. Path. Bact.* **94**, 29.

Lambe, D. W. Jr, Danielsson, D., Vroon, D. H. and Carver, R. K. (1975) *J. infect. Dis.* **131**, 499.

Langworth, B. F. (1977) *Bact. Rev.* **41**, 373.

Ledger, W. J., Sweet, R. L. and Headington, J. T. (1971) *Surg. Gynec. Obstet.* **133**, 837.

Le Frock, J. L., Ellis, C. A., Klainer, A. S. and Weinstein, L. (1975) *Arch. intern. Med.* **135**, 835.

Le Frock, J. L., Ellis, C. A., Turchik, J. B. and Weinstein, L. (1973) *New Engl. J. Med.* **289**, 467.

Legrand, H. and Axisa, E. (1905) *Dtsch. med. Wschr.* **31**, 1959.

Leigh, D. A. (1974) *Brit. med. J.* **iii**, 225; (1976) In: *Selected Topics in Clinical Bacteriology*. Ed. by J. de Louvois. Baillière Tindall, London.

Leigh, D. A., Simmons, K. and Norman, E. (1974) *J. clin. Path.* **27**, 997.

Lev, M. (1959) *J. gen. Microbiol.* **20**, 697.

Lindberg, A. A., Berthold, P., Nord, C. E. and Weintraub, A. (1979) *Med. Microbiol. Immunol.* **167**, 29.

Lindner, J. G. E. M., Plantema, F. H. F. and Hoogkamp-Korstanje, J. A. A. (1978) *J. med. Microbiol.* **11**, 233.

Linscheid, R. L. and Dobyns, J. H. (1975) *Orthoped. Clin. N. Amer.* **6**, 1063.

Loeffler, F. (1884) *Mitt. ReichsgesundhAmt.* **2**, 421.

Loesche, W. J., Hockett, R. and Syed, S. A. (1971) *Arch. oral. Biol.* **16**, 813.

Love, D. N., Jones, R. F., Bailey, M. and Johnson, R. S. (1979) *J. med. Microbiol.* **12**, 207.

Madin, S. H. (1949) *Vet. Med.* **44**, 248.

Martin, W. J. (1971) *Appl. Microbiol.* **22**, 1168; (1974) *Mayo Clin. Proc.* **49**, 300.

Masri, A. F. and Grieco, M. H. (1972) *Amer. J. med. Sci.* **263**, 357.

May, R. P., Lehmann, J. D. and Sanford, J. P. (1967) *Arch. intern. Med.* **119**, 69.

Möller, Å, J. R. (1966) *Odont. Tijdschr.* **74**, Suppl., 365.

Moore, H. B., Sutter, V. L. and Finegold, S. M. (1975) *J. clin. Microbiol.* **1**, 15.

Moore, R. J. and Russell, C. (1972) *Dent. Pract.* **22**, 390.

Müller, W. A., Holtorff, J. and Blaschke-Hellmessen, R. (1967) *Arch. Hyg., Berl.* **151**, 609.

Okell, C. C. and Elliott, S. D. (1935) *Lancet* **ii**, 869.

Onderdonk, A. B., Kasper, D. L., Cisneros, R. L. and Bartlett, J. G. (1977) *J. infect. Dis.* **136**, 82.

Onderdonk, A. B., Moon, N. E., Kasper, D. L. and Bartlett, J. G. (1978) *Infect. Immun.* **19**, 1083.

Parsonson, I. M., Egerton, J. R. and Roberts, D. S. (1967) *J. comp. Path.* **77**, 309.

Pearson, H. E. (1967) *Surg. Gynec. Obstet.* **125**, 1.

Pearson, H. E. and Anderson, G. V. (1967) *Obstet. Gynec.* **30**, 486; (1970) *Ibid.* **35**, 31; (1970) *Amer. J. Obstet. Gynec.* **107**, 1264.

Pearson, H. E. and Harvey, J. P. Jr (1971) *Surg. Gynec. Obstet.* **132**, 876.

Peromet, M., Labbe, M., Yourassowsky, E. and Schoutens, E. (1973) *Acta clinica belgica* **28**, 117.

Phillips, K. D., Tearle, P. V. and Willis, A. T. (1976) *J. clin. Path.* **29**, 428.

Pien, F. D., Thompson, R. L. and Martin, W. J. (1972) *Mayo Clin. Proc.* **47**, 251.

Plaut, H. C. (1894) *Dtsch. med. Wschr.* **20**, 920.

Polk, B. F. and Kasper, D. L. (1977) *Ann. intern. Med.* **86**, 569.

Poxton, I. R. (1979) *J. clin. Path.* **32**, 294.

Pressmann, R. S. and Bender, I. B. (1944) *Arch. intern. Med.* **74**, 346.

Prolo, D. J. and Hanbery, J. W. (1968) *Arch. Surg., Chicago* **96**, 58.

Pulverer, G. (1958) *Z. Hyg. InfecktKr.* **145**, 293.

Rissing, J. P., Crowder, J. G., Smith, J. W. and White, A. (1974) *J. infect. Dis.* **130**, 70.

Rist, E. (1901) *Zbl. Bakt., I. Abt. Orig.* **30**, 287.

Roberts, D. S. (1967) *Brit. J. exp. Path.* **48**, 665; (1970) *J. comp. Path.* **80**, 247.

Rogosa, M. *et al.* (1960) *J. Amer. dent. Ass.* **60**, 171.

Rona, S. (1905) *Arch. Derm. Syph., Berl.* **74**, 171.

Rosenblatt, J. E., Fallon, A. and Finegold, S. M. (1973) *Appl. Microbiol.* **25**, 77.

Rotheram, E. B. Jr (1974) In: *Anaerobic Bacteria: Role in Disease*, p. 369. Charles C Thomas, Springfield, Ill.

Rotheram, E. B. Jr and Schick, S. F. (1969) *Amer. J. Med.* **46**, 80.

Rudek, W. and Haque, R. U. (1976) *J. clin. Microbiol.* **4**, 458.

Sabbaj, J., Sutter, V. L. and Finegold, S. M. (1972) *Ann. intern. Med.* **77**, 629.

Schmorl, G. (1891) *Dtsch. Z. Thiermed.* **17**, 375.

Schröter, G. and Stawru, J. (1970) *Z. med. Mikrobiol.* **155**, 241.

Smith, H. A. (1944) *Amer. J. vet. Res.* **5**, 234.

Smith, J. W., Southern, P. M. Jr and Lehmann, J. D. (1970) *Obstet. Gynec.* **35**, 704.

Socransky, S. S. and Gibbons, R. J. (1965) *J. infect. Dis.* **115**, 247.

Sonnenwirth, A. C., Yin, E. T., Sarmiento, E. M. and Wesler, S. (1972) *Amer. J. clin. Nutr.* **25**, 1452.

Staat, R. H., Gawronski, T. H. and Schachtele, C. F. (1973) *Infect. Immun.* **8**, 1009.

Stalons, D., Thornsberry, C. and Dowell, V. R. Jr (1974) *Appl. Microbiol.* **27**, 1098.

Stargel, M. D., Thompson, F. S., Phillips, S. E., Lombard, G. L. and Dowell, V. R. Jr (1976) *J. clin. Microbiol.* **3**, 291.

Stauffer, L. R., Hill, E. O., Holland, J. W. and Altemeier, W. A. (1975) *J. clin. Microbiol.* **2**, 337.

Stokes, E. J. (1958) *Lancet* **i**, 668.

Swenson, R. M., Lorber, B., Michaelson, T. C. and Spaulding, E. H. (1974) *Arch. Surg., Chicago* **109**, 398.

Swenson, R. M., Michaelson, T. C., Daly, M. J. and Spaulding, E. H. (1973) *Obstet. Gynec.* **42**, 538.

Tally, F. P., Stewart, P. R., Sutter, V. L. and Rosenblatt, J. E. (1975) *J. clin. Microbiol.* **1**, 161.

Thadepalli, H., Gorbach, S. L. and Keith, L. (1973) *Amer. J. Obstet. Gynec.* **117**, 134.

Tharagonnet, D., Sisson, P. R., Roxby, C. M., Ingham, H. R. and Selkon, J. B. (1977) *J. clin. Path.* **30**, 505.

Thorley, C. M. and Calder, H. A. McC. (1977) *Vet. Rec.* **100**, 387.

Toussaint Raven, E. and Cornelisse, J. L. (1971) *Vet. med. Rev. No.* 2/3, 223.

Veillon, A. and Zuber, A. (1897) *C. R. Soc. Biol., Paris* **49**, 253; (1898) *Arch. Méd. exp. Anat. Pathol.* **10**, 517.

Vincent, H. (1896) *Ann. Inst. Pasteur* **10**, 488.

Watt, B. (1973) *J. med. Microbiol.* **6**, 307.

Watt, B., Collee, J. G. and Brown, R. (1974) *J. med. Microbiol.* **7**, 315.

Weiss, C. (1943) *Surgery* **13**, 683.

Werner, H. (1974) *Arzneimittel-Forsch.* **24**, 340.

Werner, H., Lang, N., Kraseman, C., Tolkmitt, G. and Feddern, R. (1978) *Curr. med. Res. Opin.* **5**, Suppl. 2, 52.

Werner, H. and Pulverer, G. (1971) *Dtsch. med. Wschr.* **96**, 1324.

Wilkinson, F. C., Egerton, J. R. and Dickson, J. (1970) *Aust. vet. J.* **46**, 382.

Williams, R. A. D., Bowden, G. H., Hardie, J. M. and Shah, H. (1975) *Int. J. syst. Bact.* **25**, 298.

Willis, A. T. (1977) *Anaerobic Bacteriology: Clinical and Laboratory Practice*, 3rd edn. Butterworths, London.

Willis, A. T. *et al.* (1975) *J. antimicrob. Chemother.* **1**, 393; (1976) *Brit. med. J.* **i**, 318; (1977) *Ibid.* **i**, 607.

Wilson, R. F., Chiscano, A. D., Quadros, E. and Tarver, M. (1967) *Anesth. Analg., Cleveland* **46**, 751.

Wilson, W. R., Martin, W. J., Wilkowski, C. J. and Washington, J. A. II (1972) *Mayo Clin. Proc.* **47**, 639.

Wong, G. A., Hoeprich, P. D., Barry, A. L., Pierce, T. H. and Rausch, D. C. (1974) *Clin. Res.* **22**, 127A.

Wren, M. W. D. (1978) *Med. Lab. Sci.* **35**, 371; (1980) *J. med. Microbiol.* **13**, 257.

Yoshikawa, T. T., Tanaka, K. R. and Guze, L. B. (1971) *Medicine, Baltimore* **50**, 237.

Ziment, I., Davis, A. and Finegold, S. M. (1969) *Arthr. Rheum.* **12**, 627.

Ziment, I., Miller, L. G. and Finegold, S. M. (1967) *Antimicrob. Agents Chemother.* **7**, 77.

63

Gas gangrene and other clostridial infections of man and animals

J. W. G. Smith and Geoffrey Smith

Introductory

Gas gangrene in man

In the pre-antiseptic days the formation of sloughs, the occurrence of massive necrosis, and the onset of gangrene, so frequently followed upon any extensive operative procedure that they were regarded as more or less inevitable features of surgical practice. With the advent of antiseptic surgery, and the consequent healing of wounds by first intention, these processes became less and less common, till eventually even the names by which they were described took on an archaic air. It was therefore not unnatural that when gas gangrene appeared in the early days of the war of 1914–18, many surgeons were confronted with it for the first time. It was the older surgeons, with experience of pre-Listerian methods, who recognized its true nature and pointed out the impossibility of preventing it by the current aseptic methods. Nevertheless, though the disease was old, its study had to be taken up anew. The problems were attacked with such success that by the end of the war gas gangrene had ceased to be of more than minor importance.

In 1914 the incidence of gas gangrene amongst the British Expeditionary Force amounted to over 12 per cent of the number wounded; of these 20–25 per cent died. By 1918, owing to earlier evacuation of the wounded, excision of wounds, the employment of an antiseptic technique, and other measures, the incidence had fallen to under 1 per cent (Report 1919). Indeed, in the base hospitals in 1918, amongst a total of about 25 000 wounded reported on by Bowlby (1919), only 84 patients contracted serious or massive gas gangrene—an incidence of 0.34 per cent. In the French and German armies the incidence appears to have been rather higher. The incidence of gas gangrene in the 1939–45 war varied with the theatre of fighting. Among US troops it ranged from 0 to 45 per 1000 (Smith 1949), the higher figures usually being associated with land fighting, and, presumably, with contaminated soil (see Smith and Gardner 1949). Among the British forces it ranged, according to the theatre of war, from 3 to 8 per 1000 (MacLennan 1962).

The rarity of gas gangrene in wounded soldiers in Vietnam has led to the suggestion that it might be regarded as virtually extinct as a disease of warfare (Unsworth 1973). However, in peace-time the disease still occurs but is uncommon, most frequently developing after severe compound fractures (see King 1931) and during the puerperium, particularly after illegal abortions (see Hill 1936, Butler 1945). The number of cases occurring in developed countries in the 1970s was probably of the order of 0.1–1.0 per million population per year, based on the number of patients treated in referral centres in California (Hart *et al.* 1975) and Holland (Roding *et al.* 1972), serving populations of 30 million and 15 million respectively.

We can only outline the original bacteriological investigations of gas gangrene; for details the reader is referred to the monograph of Weinberg and Séguin (1918) and to the Report of the Anaerobic Committee (see Report 1919). The more recent work on gas gangrene is extensively reviewed by MacLennan (1962) (see also Willis 1969).

Bacteriology

Examination of the wound of a man dying of gas gangrene reveals a number of organisms, aerobic and anaerobic; it is uncommon to meet with only one species of bacterium. Thus in 91 cases, Weinberg and Séguin (1918) found anaerobes in 24, and anaerobes plus aerobes in 67; in only 10 of the former group was a single species isolated—a monomicrobial infection. The diversity of flora was so great that no fewer than 38 different combinations of anaerobic bacteria occurred in their 91 cases; but from the results obtained at this time by the French and English workers, it was clear that certain anaerobes, alone or in combination, were primarily responsible for gas gangrene (Table 63.1).

Numerous other anaerobic bacilli were also found, such as *Cl. histolyticum*, *Cl. aerofoetidum*, *Cl. putrificum*, *Cl. bifermentans*, and *Cl. tertium*. Of the aerobes commonly encountered the majority consisted of diplo-streptococci and organisms belonging to the *Proteus* group (Weinberg and Séguin 1918). In 37 of Weinberg and Séguin's cases only one species of anaerobe was isolated; of these 29 were *Cl. perfringens*,

5 *Cl. novyi*, 1 *Cl. septicum*, 1 *Cl. fallax*, and 1 *Cl. aerofoetidum*.

The suggestion that each of these organisms was capable by itself of giving rise to gas gangrene was borne out by animal experiments. With the possible exception of *Cl. sporogenes*, these five organisms were the only ones that could reproduce in animals a disease simulating the picture in man. With regard to *Cl. sporogenes* it is not clear whether it is pathogenic by itself. The work of the English bacteriologists suggested that it played a subsidiary but important role in assisting *Cl. perfringens*, *Cl. novyi*, and *Cl. septicum* in their attack on the tissues, but that by itself it was harmless. On the other hand, Weinberg and Séguin regarded it as potentially though weakly pathogenic.

From his study of wound infections in the British Army campaigns in North Africa, 1940 to 1942, MacLennan (1943, 1944a) recognized three clinical manifestations of clostridial infection, namely, simple contamination which occurred at some stage in the history of 20–30 per cent of all the wounds examined, 'anaerobic cellulitis' developing in about 5 per cent of contaminated wounds, and 'clostridial myositis', the true gas gangrene, developing in about 1.5 per cent of wounds containing anaerobes, and in about 0.34 per cent of all wounds.

The clostridia most frequent in 259 contaminated wounds were *Cl. tertium* (46.7 per cent), *Cl. butyricum* (34.3 per cent), *Cl. perfringens* (23.5 per cent) and *Cl. sporogenes* (20.1 per cent). Unidentified clostridia were present in 13.2 per cent; the incidence of *Cl. novyi* was 5.8 per cent; and no *Cl. septicum* was found. In anaerobic cellulitis, characterized by foul, sero-purulent infection of the depths and crevices of the wound, without progressive involvement of muscle, the predominant organisms were proteolytic and non-toxigenic clostridia, and *Cl. perfringens*. In 146 cases of true gas gangrene, the pathogenic clostridia were as follows: *Cl. perfringens* (57 per cent), *Cl. novyi* (37.7 per cent), *Cl. septicum* (20.0 per cent), *Cl. tetani* (13.1 per cent), *Cl. histolyticum* (6.9 per cent), *Cl. bifermentans* (4.8 per cent) and *Cl. fallax* (1.4 per cent). As the fighting moved from desert to more cultivated country, the incidence of gas gangrene was doubled and the percentage of *Cl. perfringens* infection increased. The frequency of the pathogenic clostridia when they were present singly in wounds was as follows:

Table 63.1   The percentage frequency of different anaerobes in gas gangrene (Report 1919)

	Weinberg and Séguin 91 Cases	McIntosh 41 Cases	McIntosh 52 Cases	Henry 50 Cases
Cl. perfringens	77	43.9	67.3	80
Cl. novyi	34	—	4.0	10
Cl. sporogenes	27	36.5	38.7	—
Cl. fallax	16.5	—	—	6
Cl. septicum	13	19.5	16.3	16

Cl. perfringens 34.3 per cent, *Cl. novyi* 17.1 per cent, *Cl. septicum* 4.8 per cent, *Cl. histolyticum* and *Cl. bifermentans*, each 0.7 per cent. In other cases more than one species was isolated. Anaerobic streptococci (see Chapter 29) were found in 8.8 per cent of cases. Other surveys in the 1939–45 war and later revealed substantially the same distribution of pathogenic clostridia (Zeissler 1944, Lindberg *et al.* 1955), excepting that Stock (1947) and Smith and George (1946) recorded an incidence of 4 per cent or less for *Cl. septicum*. In India, Dhayagude and Purandare (1949) found no *Cl. novyi* in their patients with gas gangrene, and 20 and 36 per cent of *Cl. histolyticum* and *Cl. septicum* respectively. On the whole, the incidence of the various clostridia in gas gangrene corresponds to their prevalence in soil (MacLennan 1962).

In civilian practice in peace-time the comparatively uncommon sporadic cases of postoperative gas gangrene are caused as a rule by strains of *Cl. perfringens* derived from the patient's own bowel. They occur particularly after operations on the leg in elderly patients whose arteries are obstructed (Ayliffe and Lowbury 1969, Parker 1969).

It should be noted that *Cl. botulinum* has been identified in wounds (Hall 1945), in some cases giving rise to clinical botulism (Davis *et al.* 1951, Thomas *et al.* 1951; see also Chapter 72).

Summing up, we may say that gas gangrene may be caused by a number of different anaerobic bacteria, usually clostridia, but sometimes by anaerobic streptococci. Usually the clostridia are present either in combination with each other or with aerobic organisms. The three most important anaerobes are *Cl. perfringens*, *Cl. novyi*, and *Cl. septicum*; of less importance are *Cl. bifermentans*, *Cl. fallax*, *Cl. histolyticum*, and *Cl. sporogenes*.

The presence of pathogenic clostridia in a wound, however, is not necessarily indicative of gas gangrene infection. Most commonly they appear to be simple contaminants of wounded tissues. In war wounds they are frequently associated with cellulitis and only infrequently with true gas gangrene, *i.e.* clostridial myositis; in civil wounds the association with gas gangrene is infrequent and with cellulitis even more so (Mac-Lennan 1962).

Reproduction of gas gangrene in animals

The experimental diseases caused by *Cl. perfringens*, *Cl. septicum*, and *Cl. novyi* are described in Chapter 42. Each of these organisms synthesizes a variety of soluble biologically active substances, including one or more frank exotoxins; and we must consider how far these substances are responsible for the manifestations of natural gas gangrene.

Mode of development of gas gangrene in man

Initiation of the disease

Broth cultures of *Cl. perfringens*, *Cl. septicum* and *Cl. novy* will each, on injection into a guinea-pig, give rise to a disease in many respects simulating gas gangrene; whereas bacilli washed free of the toxic substances present in the broth are non-pathogenic except in large doses. But small doses of toxin-free bacilli and non-lethal doses of culture filtrate together are intensely virulent. Thus guinea-pigs surviving a certain dose of washed *Cl. perfringens* intramuscularly succumbed within 19 hours to less than one-thousandth of the dose when it was mixed with 0.5 ml of culture filtrate (de Kruif and Bollman 1917). The broth filtrate enables the bacilli to proliferate in the tissues. Without it the bacilli are partly lysed by the tissue fluids and partly taken up by phagocytes; with it neither lysis nor phagocytosis occurs, but the organisms divide, produce fresh toxin, and ultimately kill the animal.

Subsequent work has elaborated, but not changed, this simple picture of gas gangrene as a progressive disease, in which various toxins and other debilitating substances break down the tissue of defences so that otherwise harmless, but potentially toxigenic, bacilli are able to multiply.

Many substances are present in broth filtrates that 'activate' the infection. Young cultures of *Cl. perfringens* contain a non-toxic antigen which activates washed bacilli, and whose aggressive effect is neutralized by the corresponding antibody (Fredette and Frappier 1946). As we describe in Chapter 42, more mature culture filtrates of *Cl. perfringens* type A contain among other substances the α and θ toxins, collagenase (κ), hyaluronidase (μ), deoxyribonuclease (ν) and a fibrinolysin. Two other antigenic substances are reported, whose action is neutralized by antitoxic sera; one having the same action as enterobacterial endotoxins in hypersensitizing blood vessels to the action of adrenaline, the other inhibiting phagocytosis (Ganley *et al.* 1955). The lesions in naturally infected human muscle (Kettle 1919, Govan 1946) are simulated by the injection of toxic filtrates into animals, and are characterized by oedema, necrosis, and capillary and venous thrombosis. Proteolysis, lipolysis and liberation of lipid phosphorus, and tissue disintegration, partly referable to the bacterial collagenase, lecithinase and hyaluronidase, are also evident (Frazer *et al.* 1945, Robb-Smith 1945). The importance of these substances in initiating and maintaining the infection is not clear, but initiation appears to be related to the tissue damage they produce. Grossly damaging substances like calcium chloride and silicic acid, such as are contained in soils that get into wounds, are also effective activators (Bullock and Cramer 1919). It is generally assumed that the growth of the anaerobe is facilitated by the lowering of the O-R potential in foci of necrosis like those produced by these exotoxins or simple chemicals (see Fildes 1927, Russell 1927, Oakley 1954, Chapter 64). A similar effect is also produced by interruption of the blood supply as a result of trauma or pressure, foreign bodies like fragments of clothing, or in tissue made necrotic by extensive trauma. It is possible that the concomitant secondarily infecting organisms in the gas-gangrenous wound, whether aerobes or anaerobes, induce similar changes in the O-R potential, either as the result of their local proliferation, or by destroying the host tissues. But perhaps the activation that is most important from the point of view of surgical prophylaxis and treatment is the opportunity provided by the mechanical injury that results in crushed tissues and bones, effusion of blood and interruption of the main blood supply.

It should be noted that a transient tissue anoxia may enable the gas gangrene organisms to gain a foothold. An intracutaneous dose of adrenaline that blanches the skin of the guinea-pig and lowers the O-R potential for as little as two hours, leaving no sign of tissue damage when its effect has worn off, is sufficient to increase the susceptibility of the injected area to washed *Cl. perfringens* over ten thousand-fold (Evans *et al.* 1948).

The role of clostridial toxins in the established disease

The evidence for the participation of the biologically active products of clostridia in gas gangrene is of three kinds: (*a*) association of pathogenicity with capacity to produce these active substances; (*b*) mimicry of the disease by the substances on injection into susceptible tissue; and (*c*) protective or curative effect of antibodies to the substances.

As regards (*a*), there is some association, which is by no means clear cut, between α toxin production and pathogenicity among strains of *Cl. perfringens* type A; and hyaluronidase is produced by about half of the toxigenic strains (Robertson and Keppie 1941, McClean *et al.* 1943, Keppie and Robertson 1944, Evans *et al.* 1945*a*, Kass *et al.* 1945). Collagenase production, like α toxigenicity, is also associated with virulence in *Cl. perfringens* (Evans 1947*a*). Simple association, however, is not proof of a causal linkage. The mimicry of the natural lesion by toxic filtrates containing several active components, noted above, is a better proof; but when more stringent immunological evidence (*c*) is sought, the result is meagre. Evans (1943*a*, *b*, 1947*b*) tested the protection afforded to guinea-pigs by antibodies against α and θ toxin, hyaluronidase and collagenase; and found that of the four, only α antitoxin was protective. The efficacy of active immunization by α toxoid reinforces this view of the predominating importance of α toxin, and the subsidiary part, if any, played by the hyaluronidase and collagenase in the genesis of gas gangrene.

Smith (1949) commented on the possible role of collagenase in damaging uninfected muscle and in supplying, by proteolysis, amino acids and peptides for bacterial growth.

It is possible too that hyaluronidase, originating either in *Cl. perfringens*, or, as McClean and Rogers (1944) suggest, in accompanying pyogenic cocci of the wound, may facilitate spread of the infection. Hyaluronidase *per se* cannot penetrate the intercellular spaces to any large extent (Hechter 1946), but infected fluid under pressure of the oedematous lesion would move more readily under its influence. The relief of such pressure, and its consequences, may be one of the reasons for the success of early surgical treatment in gas gangrene. The effect of such relief is well illustrated by Evans, who in the guinea-pig made a simple incision in 4-hour-old *Cl. perfringens* lesions of the thigh muscles, and recorded that 13 of 18 animals survived, whereas 18 of 18 control animals not so treated died within 3 days (Hartley and Evans 1946).

The role of the lecithinase in the muscle lesions is more readily understood than in the toxaemia of gas gangrene. Type A filtrates have an acute action on heart muscle (see Nicholson 1934–5, Kellaway *et al.* 1940); the isolated heart perfused with α-toxin is killed, with the liberation of phosphoryl choline (see Wright 1950), suggesting a direct action of lecithinase. Moreover, not only can α toxin be detoxified by mixing with lecithin before injection (Wright and Hopkins 1946), but emulsions of lecithin, extracted from human animal tissue, injected into the blood of dogs and mice, protect them against the toxaemia (Zamecnik *et al.* 1945, see also Gordon *et al.* 1954). On the other hand, MacLennan and Macfarlane (1945, Macfarlane and MacLennan 1945) question the conclusion that the generalized toxaemia is directly due to α toxin, mainly on the grounds that even the early establishment in the blood of high concentrations of antitoxin does not protect against toxaemia; that α toxin is rapidly fixed to muscle (see also Wright and Hopkins 1946); that intravascular α toxin, though it induces haemolysis, does not necessarily produce toxaemia; and that little or no free toxin is ever detectable in the blood or tissues of intoxicated man or animals (see also Balch and Ganley 1957, Linsey 1959). They suggest that the toxaemia is due to products of tissue injury secondary to local intoxication, similar to those produced, for example, in traumatic shock. (See also Zamecnik *et al.* 1947, Berg *et al.* 1951, Berg and Levinson 1955, 1957, 1959.) The absence of detectable α toxin in lesions or in the blood during natural or experimental gas gangrene does not rule out its participation in either the local or the general disease. Indeed in the local disease, the myonecrosis, with disruption of sarcolemma and fragmentation of muscle fibres (Robb-Smith 1945), and the oedema are compatible with intoxication. It is, moreover, evident that α toxin may have a profound effect on the membranes of cells or cell particles (Macfarlane and Datta 1954). A plausible role can also be postulated for the hyaluronidase, deoxyribonuclease, collagenase and fibrinolysin. Proof of such roles is nevertheless difficult. Aikat and Dible (1956), for example, found no good evidence that *Cl. perfringens* collagenase is responsible for the observed damage to connective tissue elements, which they attribute mainly to the α toxin and the hyaluronidase. In *Cl. novyi* infections they conclude that the α toxin produces the disruptive oedema and the muscle changes, and the ε toxin the lipolysis; and they find no evidence of a collagenase or a hyaluronidase effect in either *Cl. novyi* or *Cl. septicum* myositis (Aikat and Dible 1960).

Experiments on rabbits infected with *Cl. perfringens*, alone or in combination with *Cl. sporogenes*, indicate that the resulting disease occurs in two stages: in the first the capillaries are destroyed by the α-lecithinase in the toxin; and in the second the bacteria invade the bloodstream and give rise to a general bacteraemia. The local lesion provides a reservoir both of toxin and of bacteria (Katitch *et al.* 1964).

The separation of the toxic components of *Cl. novyi* and *Cl. septicum* has not yet progressed to the point where their in-vivo effects can be profitably analysed, even to the limited extent realized with those in *Cl. perfringens*. Nevertheless, the protective effect in experimental animals of active immunization with toxoids made from the α toxins of these three clostridia and the inefficacy of passive antitoxin in the presence of persisting regions of infected muscle (see below) is good immunological evidence for the participation of the toxins of *Cl. perfringens*, *Cl. novyi* and *Cl. septicum* in the genesis of gas gangrene; but only so far as the local myonecrosis is concerned. The 'toxaemic' events in the natural disease are largely unaccounted for. (For a discussion of the systemic effects of *Cl. perfringens* α toxin, see MacLennan 1962.)

Type A antitoxins also contain an antibody-like factor bactericidal for *Cl. perfringens* in tissue fluids that is neither an agglutinin nor one of the recognized antitoxins, though it is absorbable by suspensions of *Cl. perfringens*; it appears to be responsible for the protective effect of antitoxin in experimental infection in the chick embryo, guinea-pig, and mouse. It is noteworthy, in the light of the generally recognized inefficacy of antitoxin in animals with established myonecrosis, that the bactericidal effect of the factor is most evident at relatively high O-R potentials (Bullen and Cushnie 1962).

Diagnosis

It cannot be too strongly emphasized that the diagnosis of gas gangrene must primarily be made upon clinical grounds. As Table 63.1 shows, the demonstration of one of the pathogenic clostridia in a wound, and especially of *Cl. perfringens*, is by itself of little pathognomonic significance. The presence of gas does not necessarily indicate clostridial infection, since gas formation may occur in anaerobic cellulitis and in gangrenous infections caused by bacteria other than clostridia (see Bessman and Wagner 1975). The technique of anaerobic bacteriology is at present too laborious and time-consuming to permit much in the way of confirmatory evidence that will be of value in a disease whose incubation period may be as short as 4 hours, and is on the average 1–2 days. Direct examination, however, of the wound material may yield valuable information, especially for the diagnosis of anaerobic streptococcal myositis, and should never be omitted (see MacLennan 1943, Memorandum 1943, Hayward 1945). Butler (1945) considers that the presence in pathological material of *capsulated* perfringiform bacilli is almost diagnostic of pathogenic *Cl. perfringens* in obstetric infection.

McClean and his colleagues (McClean *et al.* 1943, McClean and Rogers 1943, 1944) devised a simple test for hyaluronidase in wounded tissue, but the consistently positive results they obtained in guinea-pigs infected with *Cl. perfringens*, *Cl. septicum* and with some strains of *Cl. novyi* were not obtained in gas gangrenous patients (MacLennan 1944*b*); perhaps because the clostridial hyaluronidases were destroyed by other bacteria present, or neutralized by the prophylactic antitoxins given to the patients.

Full bacteriological examination must include, besides examination of films, direct plating, in order to assess the relative numbers of the various bacilli present; and seeding into cooked meat broth for enrichment purposes, with subsequent plating. The plates include those with indicator substances for various toxins, with half the surface treated with appropriate antitoxins for a serological identification of the indicator effect. The following scheme, based on that of Hayward (1945), is recommended by Willis (1964).

(*a*) Inoculate aerobic and anaerobic fresh blood agar; aerobic and anaerobic heated blood agar; anaerobic lactose egg yolk milk agar with and without neomycin, half of each plate being spread with a mixture of *Cl. perfringens* type A and *Cl. novyi* type A antitoxins (Willis and Hobbs 1959); and aerobic lactose milk agar. On the Willis and Hobbs medium, a presumptive identification may be made after 1 and 2 days' incubation, according to the Nagler reactions and their inhibition by antitoxins, and the evidence of pearly layer formation, lactose fermentation and proteolysis (Table 63.2).

(*b*) Inoculate into four cooked meat broths, and heat respectively at 100° for 5, 10, 15, and 20 minutes. Plate as above after 24 and 48 hours, especially for species, like *Cl. novyi* and *Cl. septicum*, with highly heat-resistant spores.

(*c*) Inoculate a cooked meat broth, and plate as above after 1, 2, 4, and 7 days, both directly and after heating a portion to 80° for 10 minutes.

The tendency of clostridia to form spreading colonies on the surface of solid media may be counteracted by the use of nutrient gels containing up to 8 per cent of agar (Miles and Hayward 1943), which permit the ready isolation of pure cultures. Methods of distinguishing the various types among the clostridial species are discussed in Chapter 42. For methods of isolation, see also McClung and Toabe (1947) and Willis (1964).

Pathogenic may be separated from non-pathogenic organisms by animal inoculation. Moreover the individual species of pathogenic bacteria may be separated by injection into protected animals. Thus, when a mixture of *Cl. perfringens*, *Cl. septicum* and *Cl. novyi* is inoculated into a guinea-pig that has received a dose of antiserum prepared against the latter two organisms, then *Cl. perfringens* alone will proliferate and invade the blood stream.

Blood cultures during life are frequently positive (Weinberg and Séguin 1918). *Cl. perfringens* and *Cl. septicum* are especially liable to invade the blood; *Cl. novyi*, which acts chiefly by virtue of its toxin, often remains confined to the local lesion.

Table 63.2   Reactions of clostridial colonies on lactose egg yolk milk agar (after Willis and Hobbs 1959)

| *Clostridium* | Reaction | | | | |
| | Opalescence | | Pearly layer | Lactose fermentation | Proteolysis |
	Produced	Inhibited			
perfringens, types A–F	+	+	–	+	–
bifermentans, sordelli	+	+	–	–	+
botulinum, types A, B, C, E	+	–	+	–	(+)*
sporogenes	+	–	+	–	+
novyi type A	+	+	+	–	–
types B, D	+	–	–	–	–
type C	–	–	–	–	–
histolyticum	–	–	–	–	+
septicum, chauvoei, sphenoides, tertium,					
butyricum, fallax	–		–	+	–
tetani, cochlearium, capitovale, tetanomorphum	–		–	–	–

* Proteolytic and non-proteolytic variants common in certain types (Dolman and Murakami 1961).

Prophylaxis and treatment

Passive immunization

Towards the end of the war of 1914–18, antitoxic and antibacterial sera were prepared against *Cl. perfringens, Cl. novyi*, and *Cl. septicum*. No extensive trial was possible, but the results of prophylaxis in a small series of badly wounded soldiers, and of treatment in a number of soldiers suffering from gas gangrene, with mixtures of these sera suggested that both procedures were beneficial (Weinberg and Séguin 1918, Report 1919).

In experimental gas gangrene, both antitoxins and antibacterial sera containing O antibody are protective. But since none of the chief gas-gangrene clostridia is antigenically homogeneous, the efficacy of an antibacterial serum is confined to infections with a given serological type (see Robertson and Felix 1930, Henderson 1934, 1935). It is clearly impracticable to maintain antibacterial sera specific for the multiplicity of serological types. On the other hand the toxic antigens in culture filtrates of these clostridia, though they vary in the content of different components, are to a large extent species-specific, and for routine therapy a mixture of antitoxins can therefore be used. The antitoxin appears to diminish the toxaemia, and thus, by lifting the burden of intoxication from the tissues, makes more effective the antibacterial mechanisms of the host (see, *e.g.*, Stewart 1943*a*, Evans 1945*b*). (See also Boyd *et al*. 1972.)

Several methods have been employed for the specification of gas-gangrene antitoxins. References to earlier work on this problem will be found in the 4th edition of this book. International antitoxin standards, with appropriate unitages, are available for antitoxins against *Cl. perfringens* type A, type B and type D; *Cl. septicum, Cl. novyi, Cl. histolyticum* and *Cl. sordelli* (Report 1961).

The two chief methods used for titrating the antitoxin content are intramuscular or intravenous inoculation of mice, and intracutaneous inoculation of guinea-pigs or rabbits, with toxin-antitoxin mixtures (see, *e.g.*, British Pharmacopoeia 1958). Other methods include flocculation tests analogous to the Ramon titration of diphtheria antitoxin, and, with *Cl. perfringens*, neutralization of the lecithinase activity of the α toxin. No international action has yet been taken to define these antitoxic sera in terms of the multiplicity of toxins; it may be assumed that they are mainly active against the chief lethal toxin.

During the 1939–45 war, the dose of polyvalent antitoxin recommended for prophylaxis in the British Army was 9000 units of *perfringens* antitoxin, 4500 units of *septicum* antitoxin and 3000 units of *novyi* antitoxin, to be given intravenously if possible, otherwise intramuscularly. For treatment, three times this dose was recommended, to be given intravenously and repeated 4–6 hourly, according to the response of the patient (see Memorandum 1943). The value of prophylaxis in seriously wounded soldiers cannot at present be estimated. The evaluation of antitoxin therapy of established gas gangrene is complicated by the wide variations in the severity of the disease encountered in the field, in the surgical treatment, and in the type of chemotherapy, local and systemic, often combined with antitoxin treatment.

The British figures for the African and Italian campaigns (Macfarlane 1943, MacLennan 1943, 1944*a*, MacLennan and Macfarlane 1944, Macfarlane 1945) suggest that the death rates in established gas gangrene were significantly lowered by large total doses of antitoxin (50 000 units or more). Thus in one series 84 per cent of 25 untreated patients died but only 51.5 per cent of 114 antitoxin-treated patients. The antitoxin was effective only in patients receiving adequate surgical treatment. In another series of 75 patients with gas gangrene, the mean incubation period of those given prophylactic antitoxin was 68 hours, and in those not given antitoxin, 33 hours. Neither antitoxin nor chemotherapy appears to have much effect on the case-fatality rates when excision of the affected part is incomplete. Occasional striking successes for antitoxin are recorded (Patterson *et al.* 1945).

With the adoption of hyperbaric oxygen treatment (see below), the use of gas gangrene antitoxin has become uncommon, owing to the danger of anaphylaxis associated with its injection and to doubts about its value. Roding and his colleagues (1972) considered antitoxin unnecessary, since 31 of their 130 cases developed gas gangrene despite having received antitoxin. However, this evidence is inconclusive since it was not known how many patients had been given antitoxin and did not develop gas gangrene. The clinical and experimental evidence of the therapeutic value of antitoxin justifies consideration of its use in severe cases, especially when hyperbaric oxygen is not available (Weinstein and Barza 1973).

Chemoprophylaxis and chemotherapy

For the prevention of gas gangrene in dirty lacerated wounds, and before operation in patients whose arterial supply to the part is impaired, penicillin is the antibiotic of choice. Benzylpenicillin, to which all clostridia are sensitive, should be given in a dose of 500 000 units six-hourly; patients for whom this drug is unsuitable should receive erythromycin (Parker 1969). The efficacy of penicillin varies with the mode of infection, the clostridial species and strain concerned, and the route of administration. Strains of *Cl. novyi*, in particular, that are readily inhibited by antibiotics *in vitro* may fail to respond *in vivo* (Schallehn 1970).

Table 63.3 Effect of different treatments in dogs experimentally infected with *Cl. perfringens* (Demello, Haglin and Hitchcock 1973)

Treatment	No. of survivors/ No. treated
None	0/70
Surgery alone	0/12
HBO alone	0/12
Antibiotic alone	10/20
HBO and surgery	0/13
HBO and antibiotic	11/20
Surgery and antibiotic	14/20
HBO and surgery and antibiotic	19/20

HBO = hyperbaric oxygen treatment.

Cephalosporins, chloramphenicol, clindamycin, erythromycin and metronidazole are also active against *Cl. perfringens*, but a proportion of the strains show some resistance to tetracycline (Dornbusch *et al.* 1975, Schwarztman *et al.* 1977). It has been suggested that antimicrobials are not needed in the treatment of gas gangrene unless a mixed infection is present (Roding *et al.* 1972). However, it is difficult to discount findings such as those of Demello, Haglin, and Hitchcock (1973), who treated experimentally infected dogs by means of hyperbaric oxygen, with or without conservative surgical débridement or antibiotic, with the results shown in Table 63.3.

The part played by antitoxin and by chemotherapeutic agents in the treatment of gas gangrene is difficult to assess. In the 1939–45 war the sulphonamides appeared to be effective as adjuvants to both surgical and antitoxin treatment (see MacLennan 1943). Jeffrey and Thomson (1944) recorded a striking arrest of clostridial myositis by penicillin, though in some cases the arrest was followed by a fatal intoxication, apparently due to absorption of substances from the retained and degenerating muscle (but see Conway 1946).

In both the prevention and the treatment of gas gangrene penicillin should be combined with antitoxin. For prophylactic purposes a dose of mixed antitoxins may be given consisting of 10 000 units each of *Cl. perfringens* and *Cl. novyi* and 5000 units of *Cl. septicum* antitoxin, but antitoxin appears rarely to be used for prophylaxis nowadays, and infrequently in treatment (Roding *et al.* 1972, Deveridge and Unsworth 1973, Darke *et al.* 1977).

In general, prophylaxis is successful, but treatment, especially when delayed for a few hours after infection, usually results in little more than a short prolongation of the time to death. Neither prophylactic nor therapeutic treatment with drugs and antitoxins is likely to be of much help without radical surgical interference; the efficacy of these substances declines sharply within a few hours of the onset of infection (see Altemaier *et al.* 1951, Garrod 1958, MacLennan 1962).

Hyperbaric oxygen

Hyperbaric oxygen has become widely used in the treatment of gas gangrene (Boerema and Brummelkamp 1960, Roding *et al.* 1972). Its effectiveness has not been examined by controlled trial but clinical observations, supported by experimental evidence (Boerema and Brummelkamp 1960), have led to acceptance of its therapeutic value. Roding and his colleagues (1972), over a twelve-year period in Holland, treated 130 patients with 29 deaths (22.3 per cent) of which only 14 were attributed to gas gangrene. The death rate in cases due to trauma, mostly road traffic accidents, was only 12.2 per cent, compared with 45.2 per cent in post-operation cases. These workers emphasized the fact that the use of hyperbaric oxygen had led to the saving of limbs in many patients who otherwise would have suffered amputations. They found that among patients who survived for 48 hours after the initiation of hyperbaric treatment, the disease was almost certainly effectively controlled, although death from other causes might supervene.

Darke and co-workers (1977) described the results in 88 patients treated in the Regional Hyperbaric Oxygen Unit at Whipps Cross Hospital in England during 1964–73. Among patients with non-uterine clostridial wound infection, 14 of 35 (40 per cent) in whom gas was detected died, compared with 5 of 23 patients (22 per cent) in whom gas was undetected. The series included 22 patients with non-clostridial gangrene, usually with gas formation, of whom 8 (30 per cent) died; the response of this group to hyperbaric oxygen was poor. Holland and his colleagues (1975) in a ten-year period treated 49 patients with a regimen which included hyperbaric oxygen. Of 26 patients in whom the infection was confined to the extremities only two died, but when the trunk was involved 11 of 23 patients died. In wounded soldiers in world war II, before hyperbaric oxygen was used in treatment, the mortality rate in cases of gas gangrene was 30 per cent (Matheson 1968), but it is difficult to compare the results in civilian cases generally with those in war injuries.

Treatment usually consists in exposure to oxygen at 2.5–3 atmospheres pressure for 1.5–2 hours as soon as possible after admission, essential resuscitation measures, and the intravenous administration of penicillin. A single-patient chamber filled with pressurized oxygen may be used, or a larger chamber which will also accommodate medical staff. The larger chambers are pressurized with air and the patient is given oxygen by means of a face mask. Surgical débridement is usually delayed until after the first hyperbaric treatment. Subsequent treatments in the hyperbaric chamber are given at about 8-hour intervals for the first 24 hours and then at 12-hour intervals. A broad-spectrum antibiotic may be given in addition to penicillin, since many infections are mixed (Hart *et al.* 1975, Darke *et al.* 1977, Skiles *et al.* 1978). The development of uraemia is a serious complication of gas gangrene and may require treatment with peritoneal or renal dialysis. Of 18 patients with uraemia 9 died in the series reported by Darke and his colleagues (1977).

Hyperbaric oxygen acts directly on the clostridia to impair growth (Hopkinson and Towers 1963) and also to prevent the formation by *Cl. perfringens* of α toxin (Unnik 1965, Kaye 1966). Vegetative bacilli are also killed. Hill and Osterhout (1972) implanted agar disks seeded with approximately 1000 viable *Cl. perfringens* subcutaneously in mice. In untreated mice the organisms multiplied in the disks about 10 000-fold over a 48-hour period, without causing illness, whereas few viable organisms could be recovered from the disks in mice exposed four times to 3 atmospheres of oxygen for 90 minutes. The inclusion of blood or of mixed, fresh mouse muscle in the disk protected the clostridia and permitted growth to occur. Hyperbaric oxygen may also promote the viability of tissues to which the blood supply is impaired. (See also Table 63.3.)

Active immunization

Formol and alum-precipitated toxoids induce in guinea-pigs, rabbits and sheep protective antitoxic immunity to experimental infection with *Cl. perfringens*, *Cl. novyi*, and *Cl. septicum* (Penfold and Tolhurst 1937, Kesterman and Vogt 1940, Penfold *et al.* 1941, Kolmer 1942, Stewart 1942, Robertson and Keppie 1943, Waters and Moloney 1949, Boyd *et al.* 1972). As with other toxoids, a stimulating dose some time after the primary injections is necessary for high degrees of immunity. In man, the antitoxic response to these toxoids is good (Penfold and Tolhurst 1938, Stewart 1943*b*, Adams 1947, Bernheimer 1947, Tytell *et al.* 1947); but at present there are no data from which the value of prophylactic immunization can be estimated. It should be noted that in the guinea-pig, the response to a stimulating dose was not demonstrable for 3–10 days (Robertson and Keppie 1943). Immunization of man by means of a stimulating dose at the time of wounding is not, therefore, likely to affect the outcome of a disease of which the incubation period is commonly less than 2 days. (See also MacLennan 1962.)

Ito (1968, 1970) pointed out that the α toxin of *Cl. perfringens* was unusually sensitive to formalin. By careful control of the conditions, however, he prepared from purified toxin a toxoid that was highly immunogenic for guinea-pigs. Incidentally, this toxoid proved to be of greater protective potency than one made from crude α toxin. Again, working with guinea-pigs, Klose and Schallehn (1967, 1969) found that an alum-precipitated toxoid prepared from *Cl. novyi* given in three doses within 4 weeks led to a good lasting immunity; toxoid without alum was unsatisfactory.

Other clostridial infections of man

The active participation of toxigenic clostridia, especially of *Cl. perfringens*, in the severe and sometimes fatal 'toxaemia' that accompanies acute obstruction of the intestine or acute appendicitis has long been a matter of debate. Williams (1926) obtained evidence suggesting that the toxaemia was due to the formation of *Cl. perfringens* toxin in the lumen of the obstructed small intestine, and that antitoxin was beneficial in the treatment of such cases. Paralytic ileus following acute inflammation in the abdominal cavity appeared to have the same effect as mechanical obstruction in the genesis of toxaemia; clostridial infection of the inflamed tissue was not observed. Weinberg (Weinberg *et al.* 1928*b*, Weinberg 1929), however, observed that clostridia could be obtained from 30 per cent of acute lesions of the appendix (subsequent observers have recorded higher figures); and he tested the effect of antisera prepared against the gas-gangrene organisms in the treatment of conditions associated with an anaerobic invasion of the tissues (Weinberg and Howard 1927, Weinberg *et al.* 1928*a*). At most, it can be said that such antisera appear to decrease the case-fatality rates, especially when used in conjunction with surgical treatment (for later references see Priestly and McCormack 1936).

Bower, Burns and Mengle (1938) induced a fatal peritonitis in dogs by ligation of the appendiceal vessels, followed by a purgative dose of castor oil; *Cl. perfringens* antitoxin reduced the mortality, and prolonged the lives of animals that eventually succumbed to the infection. Normal horse serum was likewise protective, but to a lesser extent—an effect attributed to the natural antitoxins in the serum. Prophylactic immunization with *Cl. perfringens* toxin also decreased the death rate from this experimental disease (Mengle *et al.* 1938). The active participation of *Cl. perfringens* may also be inferred from the fact that an antitoxin response is demonstrable in patients with appendicitis. Thus, Bower, Mengle and Paxson (1938) found that 22.2 per cent of patients with acute appendicitis, 46.6 per cent of those with active or quiescent pelvic peritonitis, and 69.0 per cent of those with septic peritonitis after perforation, had more than 1 unit of *Cl. perfringens* antitoxin per ml of circulating blood. The corresponding figure in control patients was 0.0 per cent.

There is little evidence that clostridia, as distinct from other intestinal bacilli, have any significant role in the development of haemorrhagic shock (Fine 1952).

A case of necrotic hepatitis due to *Cl. novyi*, resembling braxy of sheep (see below), is reported in man (Mollaret *et al.* 1948).

Intestinal diseases

We describe below a number of well recognized intestinal infections of animals due to clostridia. No such disease in man was recognized until 1946, when a severe and often fatal enteritis was described in Germany, characterized by a diffuse sloughing enteritis of the jejunum, ileum, and colon; and named *enteritis necroticans* ('Darmbrand'). Zeissler and Rassfeld-Sternberg (1949) isolated from the lesion a type of *Cl. perfringens*, subsequently designated type F (see Chapter 42), in which α, β, and γ toxins predominated. Type F is found in a small proportion of normal stools, and is said to induce enteritis in guinea-pigs when given intraduodenally, but not by mouth (Schütz 1949). In man, the organism is apparently ingested with food. The disease attacks mainly infants and persons over 30.

A similar disease—a necrotizing jejunitis called locally *pig-bel*—is met with among the natives of New Guinea after feasting on pork (Murrell *et al.* 1966). It is caused by the β toxin of *Cl. perfringens* type C—a type that differs from type F mainly in being thermolabile. Untreated cases have a fatality rate of 44–85 per cent. A formol toxoid vaccine prepared from *Cl. perfringens* type-C cultures has been found in a controlled clinical trial to be protective (Lawrence

et al. 1979). From animal experiments there is little doubt that *Cl. perfringens* type A plays an important part in death from intestinal obstruction (Weipers *et al.* 1964, Yale and Balish 1971).

A variety of *Cl. perfringens* type A is responsible for outbreaks of food poisoning in man (see Chapter 72). *Cl. perfringens* type D, commonly a cause of enterotoxaemia in sheep, has also been found in human patients with diarrhoea (Kohn and Warrack 1955), and in the intestine of a patient dying of acute intestinal obstruction (Gleeson-White and Bullen 1955).

Neonatal necrotizing enterocolitis is a rare illness in which the aetiological factors include injury to the intestinal mucosa, alteration of intestinal function and the presence of bacteria. A variety of bacterial species have been implicated, including *Escherichia coli*, klebsiellae, pseudomonads, and clostridia (Kliegman 1979; and see also Brown *et al.* 1980). In twelve cases encountered in a London hospital over a six-week period, *Cl. butyricum* was identified from the blood of nine of the patients by means of gas-liquid chromatography and also by culture in seven (Howard *et al.* 1977).

The pathological changes in haemorrhagic fever—a disease attacking US troops in the Korean war—in some respects resemble those of the animal clostridial enterotoxaemias; in convalescents from the disease the serum concentration of iota antitoxin was higher than in controls, indicating a possible association of the disease with *Cl. perfringens* type E (Marshall and Anslow 1955).

Pseudomembranous enterocolitis This is a rare disease, liable to prove fatal in its fulminating form. Most cases arise as a complication of antimicrobial treatment, especially in the seriously ill patient after abdominal surgery, but it sometimes follows intestinal surgery or vascular collapse in patients who have not had antibiotics, as in the cases described in the early reports (Finney 1893, Penner and Berheim 1939). The disease became commoner in the 1970s in association with the use of ampicillin and, especially, lincomycin and clindamycin.

Cl. difficile is evidently responsible for many cases of pseudomembranous colitis, and for diarrhoea and colitis without pseudomembrane formation following antimicrobial treatment. Larson and his colleagues (1977) detected a cytopathic agent in faecal filtrates from five cases. The activity could not be transferred serially in tissue culture and was shown to be due to a toxin, first identified as that of *Cl. sordelli*, since it was neutralized by the corresponding commercial antitoxin (Larson and Price 1977, Rifkin *et al.* 1977); but later study showed that the organism was *Cl. difficile*, the cytopathic toxin of which cross-reacts with that of *Cl. sordelli* (Bartlett *et al.* 1978a, George *et al.* 1978). The toxin can be demonstrated in almost all affected patients, and is much less frequent in the faeces of patients with enteritis not associated with pseudo-membrane formation.

The role of *Cl. difficile* in human disease is supported by experimental work, which has established that it causes the fatal caecitis which develops in guinea-pigs and hamsters treated with penicillin or clindamycin (Bartlett *et al.* 1977, Bartlett *et al.* 1978*b*, Larson *et al.* 1978). The caecal contents of affected animals contain the organism and a filtrable toxin which reproduces the disease when injected into the caecum of healthy animals; it can also be produced by injecting pure cultures of the organism, which can be re-isolated from affected animals (Bartlett *et al.* 1977, Larson *et al.* 1978). The illness can be prevented, in hamsters injected with clindamycin, by passive immunization with *Cl. sordelli* antitoxin (Allo *et al.* 1979) and by treatment with vancomycin, to which *Cl. difficile* is sensitive (Bartlett *et al.* 1977).

The course of events is thought to result from the overgrowth of antibiotic-resistant *Cl. difficile* after suppression of competing gut flora by antimicrobial treatment. The mucosa of the colon and terminal ileum is attacked by *Cl. difficile* toxin, becoming necrotic with the formation, in fulminating cases, of an exudative membrane which resembles that seen in diphtheria. The pseudomembrane is formed in multiple, friable plaques a few centimetres in diameter, attached to the mucosal surface; they may become confluent. Microscopically, the membrane appears as a fibrinous exudate containing white blood cells, epithelial cells and mucin, and the underlying intestinal submucosa shows a varying degree of necrosis and inflammatory reaction. The consequent illness may be characterized by watery diarrhoea, abdominal pain, and fever. Diagnosis depends on observing the typical appearance on sigmoidoscopy, supplemented by biopsy findings, but negative results do not exclude the diagnosis owing to the patchy distribution of membrane in many cases. Toxicity for tissue cultures is detectable in faecal filtrates (see George *et al.* 1979). The effect is neutralized by *Cl. sordelli* antitoxin, and the organism can usually be cultured from stools by taking advantage of the selective action of cycloserine and cefoxitin (George *et al.* 1979, Willey and Bartlett 1979), or kanamycin and nalidixic acid (Keighley *et al.* 1978*a*).

The incidence of the disease is uncertain. Published reports show a wide variation between studies and suggest that episodes of locally increased prevalence occur. Miller and Jick (1977) reported no cases among 26 294 hospitalized patients of whom 34 per cent received at least one antibiotic, and including 163 who had had clindamycin. Tedesco and his colleagues reported that of 200 patients treated with clindamycin, 42 (21 per cent) subsequently suffered from diarrhoea. Proctoscopy of 33 of these patients revealed the features of pseudomembranous colitis in 20, suggesting an incidence of 10 per cent after clindamycin treatment, but none of these cases became seriously ill

(Tedesco *et al.* 1974). Other reports suggest that severe illness from this cause occurs in about 0.01–0.3 per cent of patients given clindamycin, but less severe forms of the disease may be commoner, not being detected unless sigmoidoscopy and biopsy are performed and, when necessary, repeated. Keighley and his colleagues (1978*a*) studied diarrhoea after gastrointestinal operations. Among 119 patients who received antibiotics, 46 (39 per cent) subsequently had diarrhoea, nine of them with toxigenic *Cl. difficile* in the stools. The patients had received lincomycin, clindamycin, kanamycin, or metronidazole. In 122 patients who did not have antibiotics, only 12 (10 per cent) had diarrhoea, and from only one of these could toxigenic *Cl. difficile* be isolated.

Though the part played by *Cl. difficile* in the causation of pseudomembranous colitis is now well established, the role of the cytopathic toxin is less certain. *Cl. difficile* has been shown also to produce an enterotoxin (Taylor *et al.* 1980) which is immunologically distinct from the cytopathic toxin (Libby *et al.* 1982). The possibility that the enterotoxic factor may be responsible for, or contribute to, the colitis is suggested by the observation of Burdon and his colleagues (Burdon *et al.* 1978) that only 10 of 27 strains isolated from 8 cases in adults produced cytopathic toxin, and also by the fact that *Cl. difficile* and cytopathic toxin may be found in the faeces of nearly 50 per cent of healthy infants (Larson *et al.* 1978, Rietra *et al.* 1978). The high isolation rate in the newborn may be a consequence of vaginal carriage: *Cl. difficile* was cultured from vaginal specimens in 18 per cent of women attending family planning clinics, and 72 per cent of women attending venereal disease clinics (Hafiz *et al.* 1975).

Treatment depends on supportive measures, including steroids and fluid replacement, and withdrawal of the antibiotics which the patient may be receiving. *Cl. difficile* is usually resistant to many antibiotics, but sensitive to vancomycin and metronidazole. There is evidence from both animal work (Bartlett *et al.* 1977) and man (Keighley *et al.* 1978*b*) that vancomycin is effective, although relapses may occur as long as two to three weeks after treatment has been stopped. The relapses may again require vancomycin but have also been reported to respond to treatment with intestinal binding agents. A response to oral metronidazole has been reported in a small number of cases (Pashby *et al.* 1979), but it may be less reliable than vancomycin since, unlike the latter, it does not attain high concentrations in the colon.

Cl. difficile may possibly play a part in chronic inflammatory bowel disease. Although it was not detected in patients with cancer of the large intestine or ulcerative colitis (Larson *et al.* 1978), *Cl. difficile* and toxin were found in stool samples of patients with an exacerbation of ulcerative colitis or with diarrhoea associated with Crohn's disease, immune deficiency, and chronic active hepatitis (Bolton *et al.* 1980, La-Mont and Trnka 1980). Clinical improvement, in most cases following treatment with vancomycin or metronidazole, was associated with disappearance of the toxin from faecal samples.

Clostridial infections of animals

Blackleg and gas gangrene

Synonyms: blackquarter, quarter-evil.
French: *Charbon symptomatique*. German: *Rauschbrand*.

Blackleg is a disease attacking cattle, sheep, and less often swine (Meyer 1915). It is characterized by the appearance of a crepitant fluctuating swelling, generally on one of the quarters, followed rapidly by death. Calves are resistant for the first 6 months of life; they then become susceptible and remain so till about 2 years of age. The disease is commoner in the summer than in the winter, and occurs chiefly in animals in good bodily condition kept on permanent pasturage. It is sometimes associated with particular farms or pastures. In sheep it usually follows shearing, docking, parturition, and castration. In cattle it more often appears spontaneously, without obvious wounding; here the route of infection may be intestinal. The production of active infection from spores lying latent in the tissues may occur as a result of bruising. Arloing, Cornevin and Thomas (1881) first showed that the disease could be transmitted to cattle by inoculation of infective material. From the lesions they isolated an anaerobic spore-bearing bacillus—*Cl. chauvoei*. The washed bacilli are non-infective, but they can be activated by a small dose of broth filtrate, or by substances such as lactic acid, potassium chloride, or sand (Nocard and Roux 1887, Roux 1888, Leclainche and Vallée 1900). Local damage by an injected enterotoxaemia vaccine is recorded as a predisposing factor in an outbreak of sheep (Albiston 1937).

The infection in essentials is, like human gas gangrene, a clostridial myonecrosis. Blackleg is generally caused by *Cl. chauvoei*, occasionally by *Cl. septicum* and more rarely by *Cl. perfringens* and *Cl. novyi* (see, *e.g.*, Weinberg and Mihailesco 1929, Roberts and McEwen 1931, Vallée 1940, Jamieson 1950, Iyer 1952). An outbreak of *Cl. novyi* infection in cattle is reported by Byrne and Armstrong (1948). *Cl. chauvoei* blackleg in pigs was described by Sterne and Edwards (1955), but *Cl. septicum* is more often responsible. The toxin of *Cl. chauvoei* is less powerful than that produced by many other pathogenic clostridia, but the organism multiplies profusely in the affected muscle and is found in many other sites at death.

Both *Cl. chauvoei* and *Cl. septicum* can be identified in smears of muscle or in anaerobic cultures by immunofluorescence (Sterne and Batty 1975).

(For a review of anaerobic infections see the monographs by Willis 1972, Finegold 1977, Brown and Sweet 1980.)

Cl. chauvoei can also be distinguished from *Cl. septicum* by its ability to ferment sucrose but not salicin, and its usual failure to show long chains of organisms in liver impression smears from experimentally infected guinea-pigs. Mice and guinea-pigs can be protected from *Cl. chauvoei* infection by antiserum to *Cl. chauvoei* or *Cl. septicum*. *Cl. chauvoei* antiserum, on the other hand, will not protect against the specific α toxin of *Cl. septicum*. As in other clostridial diseases diagnosis becomes difficult or impossible if the carcass has begun to putrefy. *Cl. septicum* is a frequent and rapid post-mortem invader; *Cl. chauvoei* invades more slowly.

Vaccination has been successfully practised for many years. Heated cultures or material from infected tissues were among the earlier vaccines used (see Roux 1888, Cornevin 1895, Nitta 1919, Dalling 1925). A formolized broth culture of *Cl. chauvoei* is more effective than non-formolized broth filtrate (Leclainche and Vallée 1925, McEwen 1926, Roberts 1946). Heated washed cells and solutions of heat-stable bacillary antigen (Henderson 1932, Mason 1936) are also effective. Formolized whole cultures precipitated with alum (see Buddle 1954*a*) are commonly employed. The immunity, which is both antibacterial and antitoxic, is high for about one year, after which it declines (Scheuber 1944). Vaccines are equally successful when combined with a *Cl. septicum* vaccine against braxy (Coackley and Weston 1957). Most clostridial vaccines contain adjuvants, of which aluminium hydroxide gel is the commonest, followed by potassium aluminium sulphate (potash alum); aluminium phosphate and other substances are sometimes used.

Favourable therapeutic results in sheep are recorded for large doses of antiserum to formolized whole culture, prepared in horses. Penicillin in large doses is also curative in the experimental (Buddle 1952) and the natural disease (Kealy and McKinley 1953).

Braxy

Braxy is the term used for a disease of sheep occurring mainly in the north-western part of Europe. It is characterized by inflammation of the fourth stomach, and is due to *Cl. septicum*. It has been confused with the so-called German braxy, or *Bradsot*, which is characteristically an infectious necrotic hepatitis of sheep, and occasionally cattle, due to *Cl. novyi* (see Black disease, below).

The classical or northern braxy occurs on the western coast of Norway, in Iceland, the Faroes, and Scotland. Sheep are most susceptible during their first year. The disease breaks out in the late autumn and early winter months when the animals descend from the hills to the pastures of the lowlands. It is often associated with severe frost, and the ingestion of frozen food. In Scotland the losses from braxy in some districts amounted to as much as 35 per cent of the first year's sheep (Gaiger 1924).

Clinically, in the usual acute form of the disease, death occurs before any symptoms have been noticed; in the subacute form there is weakness, followed by coma, dyspnoea, and death. The duration of the disease is a matter of a few hours. Death, which results from toxaemia, is preceded by bacteraemia. *Post mortem*, there is severe inflammation of the fourth stomach and duodenum, with oedema and haemorrhage—sometimes necrosis—of the mucosa and submucous tissue. The peritoneal fluid is often turbid and in excess; it may contain detectable amounts of the specific lethal α toxin. There are degenerative changes in the visceral organs, and an acute parenchymatous nephritis.

The causative organism was described by Nielsen in 1888, who found it in large numbers in the abomasal wall. It was identified as a clostridium by C. O. Jensen in 1896, and as *Cl. septicum* by Gaiger (1922). The mode of infection is obscure. Experimental feeding with pure cultures of the bacillus apparently fails to reproduce the disease; Jensen was successful only 5 times in 1,545 attempts (Gaiger 1924). Subcutaneous injection produces blackleg; *post mortem*, lesions may be present in the abomasum simulating those of the natural disease (Hamilton 1902, 1906). There is no evidence, however, that natural infection occurs by this route. Experimental work on guinea-pigs by Borthwick (1934) suggests that intestinal stasis and exposure of the animal to cold are important factors in determining the invasion of the tissues by *Cl. septicum*. Though the feeding of cultures to normal guinea-pigs rarely sets up infection, inhibition of peristalsis by narcotine, or preliminary exposure to cold, resulted in the development of a fatal infection in a fairly high proportion of the animals. It may be added that Gelev (1964) in Bulgaria isolated an aerobic pleomorphic gram-negative bacillus from a sheep dying of braxy, and with it reproduced the disease experimentally. The diagnosis of braxy is made by methods similar to those used for blackleg.

Gaiger (1922) obtained promising results by vaccinating with a sterile filtrate of a culture of *Cl. septicum*. Of 10 340 hoggs (1st year sheep) vaccinated, 2.39 per cent died of braxy, whereas among 3800 control animals the mortality from braxy was 9 per cent.

Dalling (1925, Dalling *et al.* 1925, 1926) used a combined vaccine against blackleg and braxy in sheep,

consisting of a culture filtrate of *Cl. chauvoei* and a mixture of *Cl. septicum* toxin and antitoxin. Among 549 inoculated sheep the mortality from braxy was 0.36 per cent; among 464 uninoculated sheep, it was 15.5 per cent. Using a formolized whole culture vaccine, Gordon (1934*a*) observed an 0.8 per cent mortality in 3588 immunized hoggs, as against 8.4 per cent in 1886 control animals on the same farms (see also Dungal 1932).

Vaccination of lambs in early autumn is recommended. Vaccines consisting of formolized cultures, with or without alum, and of alum-precipitated toxoid are available.

Black disease

Synonyms: Infectious necrotic hepatitis; German braxy (bradsot).

This disease occurs in Australia, particularly Victoria, New South Wales, and Tasmania, in Britain and America, and in certain parts of Germany, where it is termed braxy. In Germany two types have been recognized: (1) meadow braxy, affecting mainly ewes at pasture in well watered valleys; and (2) stall braxy, affecting chiefly young castrated lambs that are being fattened for market (Miessner *et al.* 1931). In Australia the meadow type alone seems to be common. Well nourished animals are affected. Clinically the course in meadow braxy is usually very acute, and the animals often die from toxaemia before suspicion of illness has been aroused, but in stall braxy symptoms may be present for 12 hours before death. At death there is subcutaneous venous engorgement, which gives the skin a dark colour—hence the term black disease— amber-coloured fluid in the pleural, pericardial, and peritoneal cavities, and a number of characteristic necrotic areas in the liver. Unlike true braxy the abomasum is slightly or not at all affected, and no obvious degenerative changes are present in the visceral organs. The liver lesions are most typical of the disease. Greenish bile-stained areas of necrosis, about 1 cm in diameter, containing cavities filled with blood clot and immature liver flukes, are usually seen in small numbers. In addition to these non-specific foci, there are a few characteristic, irregularly circular, necrotic areas, 2–3 cm in diameter, sharply defined, yellowish-white in colour, and surrounded by a wide zone of venous congestion. Bacteriologically the liver can be shown to contain large anaerobic bacilli, referred to by the Germans as *Cl. gigas*, which are *Cl. novyi* (*oedematiens*) type B. The effusions in the serous cavities may contain demonstrable amounts of α and β toxin.

According to Roberts, Güven and Worrall (1970) the lesions are caused by the α toxin of type B. The β toxin of type D is responsible for the haemorrhagic lesions of red water disease of cattle (see below).

The pathogenesis of the disease was elucidated by

Dodd (1918, 1921), Albiston (1927), and particularly Turner (1930), who sought for predisposing factors which, as Fildes had demonstrated in the case of infections with *Cl. tetani* (Chapter 64), would increase the reducing intensity of the tissues so as to favour the germination of the spores of *Cl. novyi*. Turner demonstrated that in black disease the necessary conditions for growth of the causative organism occurred in the necrotic areas produced by invasion of the liver with wandering immature liver flukes of the species *Fasciola hepatica*. Many sheep were found to be latently infected with the spores of *Cl. novyi*, but not until suitable necrotic foci were established by invasion with liver flukes did the infection become active. Jamieson (1949) showed that in northern Scotland *Cl. novyi* type B was present in about 1 per cent of livers of healthy sheep and cattle in areas free from black disease, and in about 17 per cent of livers of healthy animals in black disease areas. Latent disease in infected guinea-pigs could be activated by the oral administration of cercariae of *F. hepatica*.

The causative organism, a fastidious anaerobe, can be identified in smears from liver, exudates, or cultures by immunofluorescence (Batty and Walker 1964). The α and β toxins of *Cl. novyi* type B may be demonstrable in exudates or extracts of necrotic liver by neutralization tests made intradermally in guinea-pigs or intravenously in mice. The lecithinase activity of β toxin may also be of help in diagnosis. *Cl. novyi* type D produces β but not α toxin.

Turner prepared a formolized broth vaccine which gave very satisfactory results in field trials. The mortality among sheep receiving three doses was less than half that among control animals on the same farm. Vaccination of over 270 000 sheep in Victoria, Australia, during the years 1933-36 was accompanied by a reduction in the annual mortality due to black disease from 4-15 per cent to 0.02-0.05 per cent (Wardle 1936; see also Rose 1936). According to Oxer (1937) alum-precipitated vaccine is a more effective immunizing antigen.

Red water disease of cattle   *Cl. novyi* infection is also associated with a fatal infectious disease of cattle and sometimes of sheep in the western States of the USA and in Central America, characterized by haemoglobinuria and infarction in the liver (see Vawter and Records 1945). The organism was originally described as *Cl. haemolyticum*; it is included (Chapter 42) among the varieties of *Cl. novyi* as type D. Infestation by liver flukes may be a factor determining infection. Cattle have been protected for a year or more by whole culture vaccines adsorbed by aluminium hydroxide (see Smith 1957).

Osteomyelitis of buffaloes   The buffaloes of Indonesia are said to suffer from a non-fatal osteomyelitis, which particularly affects the humerus and femur. The causative organism is indistinguishable from *Cl. novyi*, except that it is non-pathogenic to laboratory animals. The pathogenesis of the disease is obscure; and inoculation of the buffalo with the organism is not followed by osteomyelitis (Kraneveld 1930, Kraneveld and Djaenoedin 1933).

In **pigs** *Cl. novyi* may produce a rapidly fatal septicaemic disease. The liver at necropsy looks like spongy chocolate.

In America and South Africa an infection known as 'big head' occurs in sheep, especially young rams, as a result of fighting; its cause is *Cl. novyi* type A. Rare cases of necrotic hepatitis in the horse due to *Cl. novyi* have been reported. *Cl. sordelli* has been known to cause necrotic hepatitis in sheep and cattle, and infections resulting from the use of contaminated hypodermic syringes.

Infections due to *Cl. perfringens*: lamb dysentery; 'struck'; infectious enterotoxaemia or pulpy kidney disease; other infections

A number of acute toxaemic diseases have been described in different parts of the world, mostly affecting sheep, in which the lesions appear to result from an intestinal infection with a variety of *Cl. perfringens*. The pathogenesis of these diseases is still incompletely understood, but considerable progress has been made in their control by vaccination and serotherapy.

Lamb dysentery

This is a disease that takes a heavy toll of life among the lambs during their first 2 weeks of life (Gaiger and Dalling 1921). It is particularly prevalent in the Border counties of England and Scotland, and in North Wales. It may be associated with the ingestion of excessive quantities of milk. Pathologically it is characterized by an enteritis varying from a mild congestion of the intestinal mucosa to an extensive necrosis and ulceration of the small and large intestine. On the surface of the inflamed mucosa and ulcerated areas a bacillus is found in large numbers (Dalling *et al.* 1925, Dalling 1928*b*, 1931-32). It is now classified as *Cl. perfringens* type B. The organisms are confined mainly to the intestine. The body tissues are not usually invaded, death being due to the type B toxins—chiefly the β toxin—absorbed from the gut. Dysentery associated with *Cl. perfringens* type B has also been reported in foals (Mason and Robinson 1938). A similar disease in North American lambs, due to a variety of *Cl. perfringens* type C, is reported by Griner and Johnson (1954).

Diagnosis is confirmed by neutralization tests—made intravenously in mice or intradermally in guinea-pigs—with filtrates of bowel contents (see Sterne and Batty 1975). In lamb dysentery the filtrates contain mainly β, but also ε and α toxins.

The newborn lambs may be protected passively, either by antitoxin from the colostrum of immunized ewes, or by direct injection of antitoxin. Pregnant ewes have been successfully immunized with an over-neutralized toxin-antitoxin mixture, but formolized cultures are both effective and simpler to prepare. In one field experiment with over 700 animals in each group, the mortality among lambs born of uninoculated ewes was 13.09 per cent; that of ewes given toxin-antitoxin mixture, 1.07 per cent; and that of ewes given formolized culture, 1.12 per cent (see Dalling 1926, 1928*a*, Dalling *et al.* 1928, 1929, Mason *et al.* 1930). The vaccine should contain type D as well as type B, to ensure an adequate concentration of ε toxoid, and should be tested for its capacity to induce both β and ε antitoxins. Vaccines are often precipitated by potash alum. Ewes are given one dose in the autumn, and one a fortnight before lambing. (See also Jansen 1961.)

Antitoxin given to the newborn lamb is protective. Thus on 15 farms the mortality amongst 1122 lambs injected with serum was 0.44 per cent and amongst 1241 control lambs on the same farms it was 17.16 per cent (Dalling 1928*b*). A dose containing 6000 units of β and 600 units of ε antitoxin is recommended.

Struck

This is the local name given to a disease of sheep and lambs occurring during the late winter and spring months on the Romney Marsh in Kent. Occasional outbreaks of a similar disease have been reported in other parts of the world. Clinically the symptoms are of short duration, and the case fatality is high. In fresh carcasses the main lesions are those of an acute enteritis, sometimes with ulceration, and peritonitis, but in animals that have been dead for some hours the subcutaneous and muscular tissues present a picture similar to that found in gas gangrene. Blood-tinged peritoneal, pleural, and pericardial effusions are often present. McEwen (1930, McEwen and Roberts 1931) showed that the disease was essentially a toxaemia consequent on the multiplication in the intestinal canal, and perhaps the body tissues, of an organism resembling *Cl. perfringens*. Originally named *Cl. paludis*, the bacillus is now classified as *Cl. perfringens* type C. Both calves (Griner and Bracken 1953) and piglets (Field and Gibson 1955, Köhler 1978) are recorded as suffering from an analogous type-C infection. The toxaemic effects are largely due to the β toxin, which is demonstrable in the tissues and bowel contents of infected animals; though Buddle (1954*b*) found type C strains which produced little β toxin associated with a struck-like disease in New Zealand sheep. The diagnosis of struck is confirmed by neutralization tests.

McEwen (1935) successfully immunized Romney Marsh sheep with formolized culture filtrates from type C strains. Lamb dysentery vaccines, which contain β toxoid, provide an alternative immunogen for protection against struck.

Infectious enterotoxaemia

This is a clostridial disease of sheep described by Bennetts (1932) in Western Australia, and apparently due to the absorption of toxin from the small intestine. The organism, for which the name *Cl. ovitoxicum* was suggested, is now classified as *Cl. perfringens* type D. The same organism appears to be responsible for the disease known as **pulpy kidney disease of lambs**, which has been studied in New Zealand by Gill (1932) and in Wales by Montgomerie and Dalling (1933) and Montgomerie and Rowlands (1933, 1934). Gordon (1934*b*), who examined filtrates of intestinal contents from lambs, sheep, and horses that had apparently died of acute toxaemia, found a toxin neutralizable by an antitoxin to *Cl. perfringens* type D—suggesting that this organism plays a part in a number of diseases of sheep, lambs, and perhaps horses. Type D enterotoxaemia also occurs in calves (Griner *et al.* 1956).

The main clinical and pathological findings in this disease appear to be the result of intoxication by the ε toxin (Nicholson 1934–35, Gordon *et al.* 1940, Kellaway *et al.* 1940, Quesada 1963). At post-mortem examination the stomach is full, the mucosa of the stomach and intestine is congested and ecchymotic; and there may be small necrotic foci in the liver, congestion of the lungs, serous or bloody fluid in the peritoneum and pericardium, and early disintegration of the renal cortex after death. The disease must be distinguished from *acidosis*, due to fermentation in the stomach of wheat or other rich starchy food with consequent passage of lactic acid into the circulation. In acidosis of this type the pH of the rumen may fall to 4.1, but the glucose content of the blood remains normal: in enterotoxaemia the blood glucose is raised, but there is no change in the pH of the rumen (Bullen 1963). In both enterotoxaemia and acidosis death may occur in a day or two. The effect of ε toxin is to increase the blood concentrations of catecholamine, cyclic adenosine 3′,5′-monophosphate, glucose, pyruvate, lactate, and α-ketoglutarate; the hyperglycaemia, which is commonly accompanied by glycosuria, results from the rapid mobilization of hepatic glycogen (Gardner 1973*a*, *b*, Buxton 1978, Worthington *et al.* 1979). There is severe vascular endothelial damage, haemoconcentration and, in the brain, oedema and sometimes focal malacia (Buxton *et al.* 1978). In 1935 Bosworth and Glover observed that the lethal power of the toxin of *Cl. perfringens* type D, but not of types A, B or C, was increased by mixing with an intestinal filtrate of a normal sheep, guinea-pig, or rabbit; though there was no increase in the amount of antitoxin required to neutralize the toxic effect. The observation that toxin is activated by a variety of proteolytic enzymes, including those of *Cl. perfringens* itself, led Turner and Rodwell (1943) to postulate an ε prototoxin, convertible to ε toxin by tryptic enzymes (see Chapter 42).

The disease is associated with overeating or with continued feeding on rich diets. Roberts (1938) concluded from his investigations that in milk-fed lambs the acidity of the abomasal contents normally inhibited growth of *Cl. perfringens*, and that this restriction disappeared with depression of the acidity by the large amounts of casein taken in with abnormally large feeds; in consequence proliferation of the organism and toxin formation occurred in the intestine. In older sheep, *Cl. perfringens* type D is often present in the intestines (Bullen 1952). Using animals with permanent fistulae in the rumen, duodenum, and ileum, Bullen and his colleagues observed that spores are largely destroyed in the rumen, but those reaching the intestine intact multiply rapidly there for several hours, producing substantial amounts of toxin, and then disappear. The ε toxin formed increases the permeability of the intestinal mucosa to large-molecular proteins, including, presumably, the toxin itself. In normal animals the acidity in the rumen, and the rapid movement of the intestinal contents, preclude any accumulation of toxin in the intestine. After large feeds, when the rumen acidity is depressed and unfermented starch passes into the small intestine, the conditions there may be suitable for formation of toxin, in concentrations high enough and persistent enough to produce toxaemia (Bullen *et al.* 1953, Bullen and Batty 1957, Bullen and Scarisbrick 1957).

Diagnosis is confirmed by neutralization tests in mice or guinea-pigs to identify the toxin in samples such as bowel contents. Trypsinization may be necessary to convert ε prototoxin to toxin. In areas where the disease is common many sheep will be protected, by circulating antitoxin, from ε toxin formed in the intestine. The detection of glycosuria may be helpful in diagnosis.

Protection of lambs is afforded by ingestion of colostrum from ewes vaccinated during pregnancy (Oxer 1936, Dayus 1938, Batty *et al.* 1954, Smith and Matsuoka 1959) or by giving high-titre ε antitoxin to animals at risk (Montgomerie and Rowlands 1933). Formolized cultures confer active protection on a proportion of sheep so immunized (Bennetts 1936, Rose 1936; see also Baldwin *et al.* 1945, Bythell and Parker 1946). The incidence of pulpy kidney is highest in lambs aged 3–6 weeks, so that active immunization must start early. Alum-precipitated toxoid, given to lambs under 3 days old, and 1 month later, induces a good antitoxin response (Batty *et al.* 1954). It is noteworthy that trypsin activation of young culture filtrates, which contain mainly ε prototoxin, increases the immunogenicity of toxoid prepared from them (Thomson and Batty 1953, Smith and Matsuoka 1954).

The source of infection is presumably *Cl. perfringens* type D in the soil, though it is found there only occasionally, type A or non-toxic *Cl. perfringens* being the predominating varieties (see Taylor and Gordon 1940). However, it is known that type D strains readily lose their power to produce ε toxin on subculture (Borthwick 1937), and it is possible that many of the type A strains found in soils and the intestinal content of animals are degraded type D strains.

Parish (1961) described an **enterotoxaemia in fowls**,

associated with the presence in the intestine of a toxin neutralized by types B and C antitoxins. In the acute form haemorrhagic necrosis of the small intestine and degenerative changes in the liver were seen; in the chronic form, degeneration in liver, spleen, and bone marrow, and diffuse regions of non-haemorrhagic degeneration in the small intestine. The causal organism appeared to be *Cl. perfringens* type F, the strain that was isolated producing large amounts of the ν antigen (deoxyribonuclease). The fact that the disease could be reproduced, though inconsistently, by feeding the strain to birds given $CaCO_3$, $NaHCO_3$ and opium suggests some dysfunction of the alimentary tract as a predisposing cause. There is evidence that *Cl. perfringens* type A can produce necrotic enteritis in chickens (see Balauca 1976, Prescott *et al.* 1978).

In **rabbit** colonies enterotoxaemia can be caused by *Cl. perfringens* type E (Baskerville *et al.* 1980).

For further information on diagnosis, and on vaccines and antisera, see Sterne and Batty (1975) and Report (1976). For **botulism** in animals see Chapter 72.

References

Adams, M. H. (1947) *J. Immunol.* **56**, 323.

Aikat, B. K. and Dible, J. H. (1956) *J. Path. Bact.* **71**, 461; (1960) *Ibid.* **79**, 227.

Albiston, H. E. (1927) *Aust. J. exp. Biol. med. Sci.* **4**, 113; (1937) *Aust. vet. J.* **13**, 245.

Allo, M., Silva, J., Fekety, R., Rifkin, G. D. and Waskin, H. (1979) *Gastroenterology* **76**, 351.

Altemeier, W. A., McMurrin, J. A. and Alt, L. P. (1951) *J. Amer. med. Ass.* **145**, 440.

Arloing, Cornevin and Thomas (1881) *C. R. Acad. Sci.* **92**, 1246.

Ayliffe, G. A. J. and Lowbury, E. J. L. (1969) *Brit. med. J.* **ii**, 333.

Balauca, N. (1976) *Arch. exp. VetMed.* **30**, 903.

Balch, H. H. and Ganley, O. H. (1957) *Ann. Surg.* **146**, 86.

Baldwin, E. M., Frederick, L. D. and Ray, J. D. (1945) *Amer. J. vet. Res.* **9**, 296.

Bartlett, J. G., Chang, T. W., Gurwith, M., Gorbach, S. L. and Onderdonk, A. B. (1978*b*) *New Engl. J. Med.* **298**, 531.

Bartlett, J. G., Chang, T. W. and Onderdonk, A. B. (1978*a*) *Lancet*, **i**, 338.

Bartlett, J. G., Onderdonk, A. B., Cisneros, R. L. and Kasper, D. L. (1977) *J. infect. Dis.* **136**, 701.

Baskerville, M., Wood, M. and Seamer, J. H. (1980) *Vet. Rec.* **107**, 18.

Batty, I., Thomson, A. and Hepple, J. R. (1954) *Vet. Rec.* **66**, 249.

Batty, I. and Walker, P. D. (1964) *J. Path. Bact.* **88**, 327.

Bennetts, H. W. (1932) *Aust. Coun. sci. industr. Res. Bull.* No. 57; (1936) *Aust. vet. J.* **12**, 196.

Berg, M. and Levinson, S. A. (1955) *Arch. Path.* **59**, 656; (1957) *Ibid.* **64**, 633; (1959) *Ibid.* **68**, 83.

Berg, M., Levinson, S. A. and Wang, K. J. (1951) *Arch. Path.* **51**, 137.

Bernheimer, A. W. (1947) *J. Immunol.* **56**, 317.

Bessman, A. N. and Wagner, W. (1975) *J. Amer. med. Ass.* **233**, 9, 958.

Boerema, I. and Brummelkamp, W.H. (1960) *Ned. Tijdschr. Geneesk.* **104**, 2548.

Bolton, R. P., Sherriff, R. J. and Read, A. E. (1980) *Lancet* **i**, 383.

Borthwick, G. R. (1934) *Brit. J. exp. Path* **15**, 153; (1937) *Ibid.* **18**, 475.

Bosworth, T. J. and Glover, R. E. (1935) *Proc. R. Soc. Med.* **28**, 1004.

Bower, J. O., Burns, J. C. and Mengle, H. A. (1938) *Arch. Surg.* **37**, 751.

Bower, J. O., Mengle, H. A. and Paxson, N. F. (1938) *J. Immunol.* **34**, 185.

Bowlby, A. (1919) *Brit. med. J.* **i**, 205.

Boyd, N. A., Thomson, R. O. and Walker, P. D. (1972) *J. med. Microbiol.* **5**, 459, 467.

British Pharmacopoeia (1958) Pharmaceutical Press, London.

Brown, E. G., Ainbender, E., Henley, W. L. and Hodes, H. L. (1980) *See* Brown and Sweet (1980), p. 69.

Brown, E. G. and Sweet, A.Y. [Eds] (1980) *Neonatal Necrotizing Enterocolitis.* Grune and Stratton, New York.

Buddle, M. B. (1952) *N.Z. vet. J.* **1**, 13; (1954a) *N.Z. J. Sci. Tech.* **35**, 395; (1954b) *J. comp. Path.* **64**, 217.

Bullen, J. J. (1952) *J. Path. Bact.* **64**, 201; (1963) *Bull. Off. int. Épiz.* **59**, 1453.

Bullen, J. J. and Batty, I. (1957) *J. Path. Bact.* **73**, 511.

Bullen, J. J. and Cushnie, G. H. (1962) *J. Path. Bact.* **84**, 177.

Bullen, J. J. and Scarisbrick, R. (1957) *J. Path. Bact.* **73**, 485.

Bullen, J. J., Scarisbrick, R. and Maddock, A. (1953) *J. Path. Bact.* **65**, 209.

Bullock, W. E. and Cramer, W. (1919) *Proc. roy. Soc., B* **90**, 513.

Burdon, D. W. *et al.* (1978) *New Engl. J. Med.* **299**, 48.

Butler, H.M. (1945) *Surg. Gynec. Obstet.* **81**, 475.

Buxton, D. (1978) *J. med. Microbiol.* **11**, 293.

Buxton, D., Linklater, K. A. and Dyson, D. A. (1978) *Vet. Rec.* **102**, 241.

Byrne, J. L. and Armstrong, J. H. O. (1948) *Canad. J. comp. Med.* **12**, 155.

Bythell, D. W. P. and Parker, W. H. (1946) *Vet. Rec.* **58**, 367.

Coackley, W. and Weston, S. J. (1957) *J. comp. Path.* **67**, 157.

Conway, H. (1946) *Surgery* **19**, 553.

Cornevin (1895) *J. comp. Path.* **8**, 233.

Dalling, T. (1925) *J. Path. Bact.* **28**, 536; (1926) *J. comp. Path* **39**, 148; (1928a) *Vet. Rec.* **8**, 841; (1928b) *Ann. Congr. Nat. Vet. Med. Ass. G.B.I.*; (1931–32) *Proc. R. Soc. Med.* **25**, 807.

Dalling, T., Allen, H. R. and Mason, J. H. (1925) *Vet. Rec.* **5**, 561; (1926) *Ibid.* **6**, 505.

Dalling, T., Mason, J. H. and Gordon, W. S. (1928) *Vet. J.* **84**, 640; (1929) *Vet. Rec.* **9**, 902.

Darke, S. G., King, A. M. and Slack, W. K. (1977) *Brit. J. Surg.* **64**, 104.

Davis, J. B., Mattman, L. H. and Wiley, M. (1951) *J. Amer. med. Ass.* **146**, 646.

Dayus, C. V. (1938) *Aust. vet. J.* **14**, 251.

Demello, F. J., Haglin, J. J. and Hitchcock, C. R. (1973) *Surgery* **73**, 936.

Deveridge, R. J. and Unsworth, I. P. (1973) *Med. J. Aust.* **1**, 1106.

Dhayagude, R. G. and Purandare, N. M. (1949) *Indian J. Med. Res.* **37**, 283.

Dodd, S. (1918) *J. comp. Path.* **31**, 1; (1921) *Ibid.* **34**, 1.

Dolman, C. E. and Murakami, L. (1961) *J. infect. Dis.* **109**, 107.

Dornbusch, K., Nord, C. E. and Dahlback, A. (1975) *Scand. J. infect. Dis.* **7**, 127.

Dungal, N. (1932) *J. comp. Path.* **45**, 313.

Evans, D. G. (1943a) *Brit. J. exp. Path.* **24**, 81; (1943b) *J. Path. Bact.* **55**, 427; (1945a) *Ibid.* **57**, 75; (1945b) *Brit. J. exp. Path.* **26**, 104; (1947a) *Ibid.* **28**, 24; (1947b) *J. gen. Microbiol.* **1**, 378.

Evans, D. G., Miles, A. A. and Niven, J. F. (1948) *Brit. J. exp. Path.* **29**, 20.

Field, H. I. and Gibson, E. A. (1955) *Vet. Rec.* **67**, 31.

Fildes, P. (1927) *Brit. J. exp. Path* **8**, 387.

Fine, J. (1952) *Shock and Circulatory Homeostasis*, p. 140. Josiah Macy Foundation, New York.

Finegold, S. M. (1977) *Anaerobic Bacteria in Human Disease.* University of California.

Finney, J. M. T. (1893) *Bull. Johns Hopk. Hosp.* **4**, 53.

Frazer, A. C. *et al.* (1945) *Lancet* **i**, 457.

Fredette, V. and Frappier, A. (1946) *Rev. Canad. Biol.* **5**, 436.

Gaiger, S. H. (1922) *J. comp. Path.* **35**, 191, 235; (1924) *Ibid.* **37**, 163.

Gaiger, S. H. and Dalling, T. (1921) *J. comp. Path.* **34**, 79.

Ganley, O. H., Merchant, D. J. and Bohr, D. F. (1955) *J. exp. Med.* **101**, 605.

Gardner, D. E. (1973a) *J. comp. Path.* **83**, 499; (1973b) *J. comp. Path.* **83**, 525.

Garrod, L. P. (1958) *J. roy. Army med. Corps* **104**, 209.

Gelev, I. (1964) *Zbl. Bakt.* **194**, 222.

George, R. H. *et al.* (1978) *Brit. med. J.* **1**, 695.

George, W. L. *et al.* (1979) *J. clin. Microbiol.* **9**, 214.

Gill, D. A. (1932) *N.Z. J. Agric.* **45**, 332.

Gleeson-White, M. H. and Bullen, J. J. (1955) *Lancet* **i**, 384.

Gordon, J., Turner, G. C. and Dmochowski, L. (1954) *J. Path. Bact.* **67**, 605.

Gordon, W. S. (1934a) *Vet. Rec.* **14**, 1; (1934b) *Ibid.* **14**, 1016.

Gordon, W. S., Stewart, J., Holman, H. H. and Taylor, A. W. (1940) *J. Path. Bact.* **50**, 251.

Govan, A. D. T. (1946) *J. Path. Bact.* **58**, 423.

Griner, L. A., Aichelman, W. W. and Brown, G. D. (1956) *J. Amer. vet. med. Ass.* **129**, 375.

Griner, L. A. and Bracken, K. (1953) *J. Amer. vet. med. Ass.* **122**, 99.

Griner, L. A. and Johnson, H. W. (1954) *J. Amer. vet. med. Ass.* **125**, 125.

Hafiz, S., McEntegart, M. G., Morton, R. S. and Waitkins, S. A. (1975) *Lancet* **i**, 420.

Hall, I. C. (1945) *J. Bact.* **50**, 213.

Hamilton (1902) *Trans. Highland and Agricultural Soc. of Scotland*; (1906) Rept Departmental Committee of Bd Agric. on Louping-ill and Braxy. London.

Hart, G. B., Cave, R. H., Goodman, D. B., O'Reilly, R. R. and Broussard, N. D. (1975) *Milit. Med.* **140/7**, 461.

Hartley, P. and Evans, D. G. (1946) *Proc. R. Soc. Med.* **39**, 295.

Hayward, N. J. (1945) *Proc. Ass. clin. Path., Lond.* **1**, 5.

Hechter, O. (1946) *Science* **104**, 409.

Henderson, D. W. (1932) *Brit. J. exp. Path.* **13**, 421; (1934) *Ibid.* **15**, 166; (1935) *Ibid.* **16**, 393.

Hill, A. M. (1936) *J. Obstet. Gynaec.* **43**, 201.

Hill, G. B. and Osterhout, S. (1972) *J. infect. Dis.* **125**, 26.

Holland, J. A., Hill, G. B., Wolfe, W. G., Osterhout, S., Saltzman, H. A. and Brown, I. W. (1975) *Surgery* **77**, 75.

Hopkinson, W. I. and Towers, A. G. (1963) *Lancet* **ii**, 1361.

Howard, F. M., Bradley, J. M., Flynn, D. M., Noone, P. and Szawatkowski, M. (1977) *Lancet* ii, 1099.

Ito, A. (1968) *Jap. J. med. Sci. Biol.* **21**, 379; (1970) *Ibid.* **23**, 21.

Iyer, S. V. (1952) *Indian vet. J.* **29**, 27.

Jamieson, S. (1949) *J. Path. Bact.* **61**, 389; (1950) *Vet. Rec.* **62**, 772.

Jansen, B. C. (1961) *Onderstepoort J. vet. Res.* **28**, 495.

Jensen, C. O. (1896) See *Kolle and Wassermann's Handbuch der pathogenen Mikro-organismen* 1 Aufl. 1902, **2**, 624.

Jeffrey, J. S. and Thomson, S. (1944) *Brit. J. Surg.* **32**, 159.

Kass, E. H., Lichstein, H. C. and Waisbren, B. A. (1945) *Proc. Soc. exp. Biol., N.Y.* **58**, 172.

Katitch, R., Voukitchevitch, Z., Djoukitch, B., Miljkovitch, V. and Tchavitch, B. (1964) *Rev. Immunol.* **28**, 63.

Kaye, D. (1966) *Proc. Soc. exp. Biol., N.Y.* **124**, 360.

Kealy, J. K. and McKinley, M. G. (1953) *Irish vet. J.* **7**, 234.

Keighley, M. R. B. *et al.* (1978a) *Lancet* ii, 1165; (1978b) *Brit. med. J.* ii, 1667.

Kellaway, C. H., Trethewie, E. R. and Turner, A. W. (1940) *Aust. J. exp. Biol. med. Sci.* **18**, 225, 253.

Keppie, J. and Robertson, M. (1944) *J. Path. Bact.* **56**, 123.

Kesterman, E. and Vogt, K.-E. (1940) *Klin. Wschr.* **19**, 1009.

Kettle, E. H. (1919) *See* Report (1919).

King, W. E. (1931) *Amer. J. Surg.* **14**, 460.

Kliegman, R. M. (1979) *Ped. Clin. N. Amer.* **26**, 327.

Klose, F. and Schallehn, G. (1967) *Z. ImmunForsch.* **134**, 350; (1969) *Ibid.* **138**, 480.

Köhler, B. (1978) *Arch. exp. VetMed.* **32**, 841.

Kohn, J. and Warrack, G. H. (1955) *Lancet.* i, 385.

Kolmer, J. A. (1942) *J. Immunol.* **43**, 289.

Kraneveld, F. C. (1930) *Ned.-Ind. Bl. Diergeneesk.* **42**, 564.

Kraneveld, F. C. and Djaenoedin, R. (1933) *Ned.-Ind. Bl. Diergeneesk.* **45**, 80.

Kruif, P. H. de and Bollman, J. L. (1917) *J. infect. Dis.* **21**, 588.

LaMont, J. T. and Trnka, Y. M. (1980) *Lancet* i, 381.

Larson, H. E., Parry, J. V., Price, A. B., Davies, D. R., Dolby, J. and Tyrrell, D. A. J. (1977) *Brit. med. J.* i, 1246.

Larson, H. E. and Price, A. B. (1977) *Lancet* ii, 1312.

Larson, H. E., Price, A. B., Honour, P. and Borriello, S. P. (1978) *Lancet* i, 1063.

Lawrence, G., Shann, F., Freestone, D. S. and Walker, P. D. (1979) *Lancet* i, 227.

Leclainche, E. and Vallée, H. (1900) *Ann. Inst. Pasteur* **14**, 202; (1925) *C. R. Soc. Biol.* **92**, 1273.

Libby, J. M., Jornter, B. S. and Wilkins, T. D. (1982) *Infect. Immun.* **36**, 822.

Lindberg, R. R. *et al.* (1955) *Ann. Surg.* **141**, 369.

Linsey, D. (1959) *Amer. J. Surg.* **97**, 592.

McClean, D. and Rogers, H. J. (1943) *Lancet* i, 707; (1944) *Ibid.* ii, 434.

McClean, D., Rogers, H. J. and Williams, B. W. (1943) *Lancet* i, 355.

McClung, L. S. and Toabe, R. (1947) *J. Bact.* **53**, 139.

McEwen, A. D. (1926) *J. comp. Path.* **39**, 253; (1930) *Ibid.* **43**, 1; (1935) *J. S.-E. agric. Coll.*, Wye No. 35, p. 45.

McEwen, A. D. and Roberts, R. S. (1931) *J. comp. Path.* **44**, 26.

Macfarlane, M. G. (1943) *Brit. med. J.* ii, 636; (1945) *Ibid.* i, 803.

Macfarlane, M. G. and Datta, N. (1954) *Brit. J. exp. Path* **35**, 202.

Macfarlane, R. G. and MacLennan, J. D. (1945) *Lancet* ii, 328.

MacLennan, J. D. (1943) *Lancet* i, 582; *Ibid.* ii, 63, 94, 123; (1944a) *Ibid.* i, 203; (1944b) *Ibid.* ii, 433; (1962) *Bact. Rev.* **26**, 177.

MacLennan, J. D. and Macfarlane, R. G. (1944) *Brit. med. J.* i, 683; (1945) *Lancet* ii, 301.

Marshall, J. D. and Anslow, R. O. (1955) *Proc. Soc. exp. Biol., N.Y.* **90**, 265.

Mason, J. H. (1936) *Onderstepoort J. vet. Sci.* **7**, 433.

Mason, J. H., Dalling, T. and Gordon, W. S. (1930) *J. Path. Bact.* **33**, 783.

Mason, J. H. and Robinson, E. M. (1938) *Onderstepoort J. vet. Sci.* **11**, 333.

Matheson, J. M. (1968) *Ann. R. Coll. Surg. Engl.* **42**, 347.

Memorandum (1943) *War Memorandum Med. Res. Coun., Lond.* No. 2, 2nd ed.

Mengle, H. A., Paxson, N. F. and Bower, J. O. (1938) *J. Immunol.* **34**, 429.

Meyer, K. F. (1915) *J. infect. Dis.* **17**, 458.

Miessner, H., Meyn, A. and Schoop, G. (1931) *Zbl. Bakt.* **120**, 257.

Miles, A. A. and Hayward, N. J. (1943) *Lancet* ii, 116.

Miller, R. R. and Jick, H. (1977) *Clin. Pharmacol. Therap.* **22**, 1.

Mollaret, P., Prévot, A. R. and Guéniot, M. (1948) *Ann. Inst. Pasteur* **75**, 195.

Montgomerie, R. F. and Dalling, T. (1933) *Vet. J.* **89**, 223.

Montgomerie, R. F. and Rowlands, W. T. (1933) *Vet. J.* **89**, 388; (1934) *Ibid.* **90**, 399.

Murrell, T. G. C., Roth, L., Samels, J. and Walker, P. D. (1966) *Lancet* i, 217.

Nicholson, J. A. (1934–5) *4th Rep. Director, Inst. anim. Path., Camb.* p. 212.

Nielsen, I. (1888) *Tideskrift f. Veterinärer.* See *Mh. prakt. Thierheilk.*, 1897, **8**, 55.

Nitta, N. (1919) *J. comp. Path.* **32**, 122.

Nocard and Roux (1887) *Ann. Inst. Pasteur* **1**, 257.

Oakley, C. L. (1954) *Brit. med. Bull.* **10**, 52.

Oxer, D. T. (1936) *Aust. vet. J.* **12**, 54; (1937) *Ibid.* **13**, 3.

Parish, W. E. (1961) *J. comp. Path.* **71**, 377, 394, 405.

Parker, M. T. (1969) *Brit. med. J.* iii, 671.

Pashby, N. L., Bolton, R. P. and Sherrif, R. J. (1979) *Brit. med. J.* i, 1605.

Patterson, T. C., Keating, C. and Clegg, H. W. (1945) *Brit. J. Surg.* **33**, 74.

Penfold, W. J. and Tolhurst, J. C. (1937) *Med. J. Aust.* i, 982; (1938) *Ibid.* i, 604.

Penfold, W. J., Tolhurst, J. C. and Wilson, D. (1941) *J. Path. Bact.* **52**, 187.

Penner, A. and Berheim, A. I. (1939) *Arch. Path.* **27**, 966.

Prescott, J. F., Sivendra, R. and Barnum, D. A. (1978) *Canad. vet. J.* **19**, 181.

Priestly, J. T. and McCormack, C. J. (1936) *Surg. Gynec. Obstet.* **63**, 675.

Quesada, A. (1963) *Bull. Off. int. Épiz.* **59**, 1313.

Report (1919) *Spec. Rep. Ser. med. Res. Coun., London.* No. 39; (1961) *World Hlth Org. techn. Rep. Ser.* **222**, 28; (1976) Joint OIE-IABS Symposium on Clostridial Products in Veterinary Medicine. *Developments in Biological Standardization*, vol. 32. S. Karger, Basel.

Rietra, P. J. G., Slaterus, K. W. and Zanen, H. C. (1978) *Lancet* ii, 319.

Rifkin, G. D., Fekety, F. R., Silva, J. and Sack, R. B. (1977) *Lancet* **ii**, 1103.

Robb-Smith, A. H. T. (1945) *Lancet* **ii**, 362.

Roberts, R. S. (1938) *Vet. Rec.* **50**, 591; (1946) *J. comp. Path.* **61**, 150.

Roberts, R. S. Güven, S. and Worrall, E. E. (1970) *J. comp. Path.* **80**, 9.

Roberts, R. S. and McEwen, A. D. (1931) *J. comp. Path.* **44**, 180.

Robertson, M. and Felix, A. (1930) *Brit. J. exp. Path.* **11**, 14.

Robertson, M. and Keppie, J. (1941) *J. Path. Bact.* **53**, 95; (1943) *Lancet* **ii**, 311.

Roding, B., Groeneveld, P. H. A. and Boerema, I. (1972) *Surg. Gynec. Obstet.* **134**, 579.

Rose, A. L. (1936) *Aust. vet. J.* **12**, 185.

Roux, E. (1888) *Ann. Inst. Pasteur* **2**, 49.

Russell, D. S. (1927) *Brit. J. exp. Path.* **8**, 377.

Schallehn, G. (1970) *Zbl. Bakt.* **215**, 348.

Scheuber, J. R. (1944) *Onderstepoort J. vet. Sci.* **19**, 17.

Schütz, F. (1949) *Zbl. Bakt.* **154**, 197.

Schwartzman, J. D., Barth Reller, L. and Wang, W. L. (1977) *Antimicrob. Agents Chemother.* **11**, 695.

Skiles, M. S., Covert, G. K. and Fletcher, S. H. (1978) *Surg. Gynec. Obstet.* **147**, 65.

Smith, L. DS. (1949) *Bact. Rev.* **13**, 233; (1957) *Proc. 60th Ann. Mtg U.S. live Stk. sanit. Ass.* p. 135.

Smith, L. DS. and Gardner, M. V. (1949) *J. Bact.* **58**, 407.

Smith, L. DS. and George, R. L. (1946) *J. Bact.* **51**, 271.

Smith, L. DS. and Matsuoka, T. (1954) *J. Amer. vet. med. Ass.* **15**, 361; (1959) *Amer. J. vet. Res.* **20**, 91.

Sterne, M. and Batty, I. (1975) *Pathogenic Clostridia.* Butterworth, London and Boston.

Sterne, M. and Edwards, J. B. (1955) *Vet. Rec.* **67**, 314.

Stewart, S. E. (1942) *War Med., Chicago* **2**, 87; (1943a) *Publ. Hlth Rep., Wash.* **58**, 1277; (1943b) *War Med., Chicago* **3**, 508.

Stock, A. H. (1947) *J. Bact.* **54**, 169.

Taylor, A. W. and Gordon, W. S. (1940) *J. Path. Bact.* **50**, 271.

Taylor, N. S., Thorne, G. M. and Bartlett, J. G. (1980) *Clin. Res.* **28**, 285A.

Tedesco, F. J., Barton, R. W. and Alpers, D. J. (1974) *Ann. intern. Med.* **81**, 429.

Thomas, C. G., Keleher, M. F. and McKee, A. P. (1951) *Arch. Path.* **51**, 623.

Thomson, A. and Batty, I. (1953) *Vet. Rec.* **65**, 659.

Turner, A. W. (1930) *Aust. Coun. sci. industr. Res.* Bull. No. 46.

Turner, A. W. and Rodwell, A. W. (1943) *Aust. J. exp. Biol. med. Sci.* **21**, 17, 27.

Tytell, A. A. *et al.* (1947) *J. Immunol.* **55**, 233.

Unnik, A. J. M. van (1965) *Antonie v. Leeuwenhoek* **31**, 181.

Unsworth, I. P. (1973) *Med. J. Aust.* **i**, 1077.

Vallée, H. (1940) *C. R. Soc. Biol.* **133**, 51.

Vawter, L. R. and Records, E. (1945) *Bacillary Haemoglobinuria of Cattle and Sheep.* University of Nevada, Reno.

Wardle, R. N. (1936) *Aust. vet. J.* **12**, 189.

Waters, G. G. and Moloney, P. J. (1949) *Canad. J. Res.* **27**, 171.

Weinberg, M. (1929) *Bull. Inst. Past.* **27**, 529.

Weinberg, M. and Howard, A. F. (1927) *C. R. Soc. Biol.* **97**, 221.

Weinberg, M. and Mihailesco, M. (1929) *Ann. Inst. Pasteur* **43**, 1408.

Weinberg, M., Prévot, A. R., Davesne, J. and Renard, C. (1928a) *C. R. Soc. Biol.* **98**, 749, 752; (1928b) *Ann. Inst. Pasteur* **42**, 1167.

Weinberg, M. and Séguin, P. (1918) *La Gangrène Gazeuse.* Paris.

Weinstein, L. and Barza, M. A. (1973) *New Engl. J. Med.* **289**, 1129.

Weipers, W. L., Harper, E. M. and Warrack, G. H. (1964) *J. Path. Bact.* **87**, 279.

Willey, S. H. and Bartlett, J. G. (1979) *J. clin. Microbiol.* **10**, 880.

Williams, B. W. (1926) *Brit. J. Surg.* **14**, 295.

Willis, A. T. (1964) *Anaerobic Bacteriology in Clinical Medicine*, 2nd edn. Butterworth, London; (1969) *Clostridia of Wound Infection.* Butterworth, London; (1972) Anaerobic Infections. *Publ. Hlth Lab. Serv., Monogr. Ser.*, No. 3.

Willis, A. T. and Hobbs, G. (1959) *J. Path. Bact.* **77**, 511.

Worthington, R. W., Bertschinger, H. J. and Mülders, M. S. G. (1979) *J. med. Microbiol* **12**, 497.

Wright, G. P. (1950) *Proc. R. Soc. Med.* **43**, 886.

Wright, G. P. and Hopkins, S. J. (1946) *J. Path. Bact.* **58**, 573.

Yale, C. E. and Balish, E. (1971) *Infect. Immun.* **3**, 481.

Zamecnik, P. C., Folch, J. and Brewster, L. E. (1945) *Proc. Soc. exp. Biol., N.Y.* **60**, 33.

Zamecnik, P. C., Nathanson, I. T. and Aub, J. A. (1947) *J. clin. Invest.* **26**, 394.

Zeissler, J. (1944) *Vorsicht bei der bakteriologischen Diagnostik der Gasödeme.* Hamburg; (1949) *Zbl. Bakt.* **154**, 200.

Zeissler, J. and Rassfeld-Sternberg, L. (1949) *Brit. med. J.* **i**, 267.

64

Tetanus

J. W. G. Smith

Introductory

This disease was described by Hippocrates, Aretaeus, and others, but its nature remained obscure till Carle and Rattone in 1884 demonstrated its transmissibility to animals. They injected a number of rabbits with a suspension of an acne pustule, which had been the starting point of a fatal attack of tetanus. The injections, which were made into the sciatic nerve or into the back muscles, were followed in 2 or 3 days by tetanus. The disease could be transmitted to fresh animals by injection of a suspension of nerve tissue.

In the same year Nicolaier (1884, 1886), working at Göttingen, found that the inoculation of earth into mice, guinea-pigs or rabbits was frequently followed by a disease closely simulating human tetanus. In the pus at the site of inoculation, besides cocci and other organisms, he noticed long, thin bacilli. Though unable to isolate the organism in pure culture, he grew it in deep coagulated serum for seven generations, and reproduced the disease by injection of the last culture. In animals dying of the experimental disease Nicolaier was unable to find the bacilli microscopically except in the local lesion and, occasionally, in the sciatic nerve sheath and spinal cord. Their limited distribution led him to suggest that the organisms multiplied locally and produced a strychnine-like poison which, on absorption, reproduced the disease. In 1886 Rosenbach observed a similar bacillus with a round terminal spore in a human case of tetanus, and the pus proved infective to animals. The final demonstration of the aetiological role of the tetanus bacillus was furnished in 1889 by Kitasato, who isolated it in pure culture from pus. This he accomplished by heating the pus to 80° for 45 to 60 minutes to destroy non-sporing organisms, plating out on gelatin, and incubating in an atmosphere of hydrogen. The inoculation of pure cultures into animals was successful in reproducing the disease.

Epidemiology

Tetanus is an infective disease resulting generally from the contamination of a wound or raw surface, and, more rarely, from the parenteral injection of substances contaminated with tetanus bacilli. It is characterized by a series of tonic reflex muscular spasms superimposed upon muscular rigidity. In mild cases the spasms may not develop, and occasionally, in neonatal tetanus, they occur in the absence of muscular rigidity. The wounds most commonly implicated are those of the legs, followed by those of hands, arms and neck; in as many as 20 per cent of cases there may be no history of injury nor a detectable wound (Press 1948, Patel, Mehta and Goodluck 1965). The frequency with which the masseter muscles are affected has given rise to the popular term 'lockjaw'. In newborn infants the cut surface of the umbilical cord may afford an entrance for the bacilli, giving rise to *tetanus neonatorum*, and in women rare cases of tetanus occur shortly after childbirth, as the result of infection in the genital tract; both forms of tetanus have a high case-fatality rate. A few cases are attributable to growth of tetanus bacilli in an ear affected by chronic otitis media. Infections of the face or head give rise to a peculiar form—cephalic tetanus—characterized by facial paralysis and dysphagia. Visceral tetanus is an uncommon form, in which infection appears to originate from the bacilli in the intestinal tract.

Incidence

The incidence of tetanus varies greatly in different countries. In parts of Asia, Africa and South America, where much of the disease is in the form of neonatal tetanus, it is a major cause of death (see Veronesi 1965, Vaishnava and Goyal 1966); in Europe, the United States and other developed parts of the world, it is comparatively uncommon. Bytchenko (1972) estimated that throughout the world tetanus was responsible for 423 000 deaths each year, of which probably over 70 per cent were neonatal. In England and Wales it now occurs less frequently than formerly. For example, the mean annual number of civilian deaths during the ten years 1915–24 was 158: during 1970–1979 total cases notified numbered only 194 with 112 deaths (Galbraith *et al.* 1981). Among the small number of cases reported in recent years older women figure prominently, often having received a trivial wound whilst gardening. In the United States numerous cases used to follow the 4th July celebrations, most of them from blank cartridge wounds (Smith 1908). Writing in 1957, Axnick and Alexander gave the annual number of cases in the USA as 500 and the deaths about 300. By 1970, however, the cases had fallen to about 260 a year, with a case-fatality rate of 58 per cent (Buchanan *et al.* 1970, Report 1970, Blake and Feldman 1975). The incidence is higher in the non-white than white population of the USA, probably owing to greater exposure and lower immunization rates. Men and women appear to be equally susceptible, the distribution of cases between the sexes depending largely on occupational habits and the active immunization of men in the armed forces. In the developing countries *tetanus neonatorum* may be responsible for 20–30 per cent of the cases and 70 per cent of deaths; this form of tetanus is extremely rare in countries where good hygiene is practised in midwifery. In agricultural countries the rural population suffers more than the urban, and more in the summer than the winter. Infection of the legs and feet are prone to occur in heavily manured areas (see, e.g. Bezjak and Fališevac 1955, Möse 1955, Rigby 1960, Manzanillo *et al.* 1964). Those addicted to narcotic drugs are at risk from tetanus; of 142 cases in New York in 1955–65, 102 were in addicts (Cherubin 1967).

During the war of 1914–18 the total number of British wounded on all fronts was officially reported as 2 032 142; of these 2385 contracted tetanus—an incidence of 1.17 per 1000 wounded. The ratio was

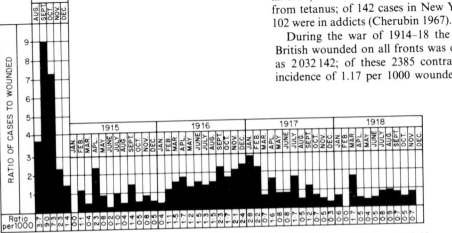

Fig. 64.1 The ratio per 1000 of cases of tetanus to wounded in the British armed forces 1914–18 (after Bruce.)

much higher at the beginning of the war (see Fig. 64.1). Among British troops on the western front in the 1939–45 war, Boyd (1946) recorded an incidence of 0.06 per 1000. This great reduction was largely attributable to prophylactic immunization (see p. 359).

The mortality rate from tetanus throughout the world appears to be declining (Bytchenko 1972) for a number of reasons which vary in importance in different countries—improved living standards, urbanization, better hygiene especially in relation to childbirth, increasing use of active immunization and more effective treatment (see Adams *et al.* 1969, Fraser 1972, Ebisawa and Fukutomi 1979).

Incubation period

As Sartwell (1950) found with certain other diseases, the incubation period of tetanus, when plotted logarithmically, shows a normal distribution (Tateno 1963). When it can be calculated accurately, as after a wound, it is commonly 7–10 days, but greater variation occurs, and it may range from 2–30 days. It tends to be shorter after head and face wounds than after leg and foot wounds; and to be 7 days or less in *tetanus neonatorum* (Patel and Mehta 1963). In about 10–30 per cent of cases the disease comes on without obvious cause, so that the length of the incubation period remains doubtful. Prophylactic antitoxin treatment after a wound, when it does not prevent the disease completely, may increase the length of the incubation period. Thus in the 1914–18 war the incubation period in patients treated in hospitals in Britain lengthened from 11.8 days in 1914 to 50 days in 1918–19 (Bruce 1920).

Case fatality

This depends on a number of factors such as age, sex, general physique, type and severity of wound, length of incubation period, period of onset (see below), receipt of prophylactic antitoxin, the stage at which treatment is begun, the time interval between first symptoms and admission to hospital, and the standard of available treatment—especially in severe cases, which demand curarization and artificial respiration (Cole 1940, Adams *et al.* 1969, Smythe *et al.* 1974, Armitage and Clifford 1978). Observations made on large series of patients suggest that the type of injury or its site does not influence the mortality, except in uterine infections and *tetanus neonatorum*, in which the case-fatality rate in the absence of modern treatment methods may be as high as 90 per cent (Adams and Morton 1955, Patel *et al.* 1965, Adams 1968). The death rate, apart from that in the newborn, tends to be low in children, and to rise later. In England and Wales, the United States and Japan, 50 per cent or so of the deaths are in persons aged 45 years or over; this may be due partly to the failure of elderly persons to

have received immunization. There is good evidence that the shorter the incubation period, the worse is the prognosis (Table 64.1) (see Bruce 1920); but, according to Cole (1940) and numerous subsequent observers, the period of onset is the more reliable index, this period being defined as the time between appearance of the first symptom of the disease and the first reflex spasms i.e. convulsive spasms superimposed on the underlying tonic rigidity. From their analysis of the data on 1385 patients treated in India with antitoxin but without curarization, Armitage and Clifford (1978) conclude that the probability of death is separately related to the period of onset, the interval between first symptoms and admission to hospital—long

Table 64.1 Tetanus mortality in relation to incubation period

Type of tetanus	Number of cases	Incubation period	Case fatality rate, per cent
Adult	2497	0–7 days	52.5
		More than 7 days	32.2
Neonatal	956	0–8 days	60.7
		More than 8 days	19.0

(See Patel *et al.* 1965, Adams 1968.)

periods being associated with better prognosis—and the severity of illness on admission. The case-fatality rate was 51 per cent when the period of onset was less than 10 hours, and 15 per cent when it was greater than 72 hours. Tetanus in patients with reflex spasms but whose illness was judged on admission to be mild had a fatality rate of 26 per cent; when the illness was severe it was 73 per cent. In patients who do not have reflex spasms the prognosis is more favourable—the fatality rate was only 18 per cent in the cases studied by Armitage and Clifford (1978).

In pre-serum days the case fatality was about 85 per cent (Bruce 1920). Bruce (1915, 1916, 1917*b*, 1919, 1920) reported on 1458 cases of tetanus treated in British hospitals during the war of 1914–18. After 1914 almost every patient had received one or more prophylactic injections of antiserum; the majority had also been treated with it therapeutically. The mortality from tetanus in these cases is given in Table 64.2.

Table 64.2 Fatality of cases of tetanus in the war of 1914–1918*

Year	Number of cases	Died of tetanus	Recovered	Fatality, per cent
1914	182	101	81	55.5
1915	138	78	60	56.5
1916	451	170	281	37.7
1917	376	76	294	20.5
1918	291	73	204	26.4
1919	20	3	16	15.8

*Cases (21) dying from causes other than tetanus have been excluded.

Conybeare (1959), on the basis of the Hospital In-Patient Enquiry, estimated the mortality rate in 1953–56 as 15–30 per cent. In certain parts of the United States it was also thought to be about 15–30 per cent (Axnick and Alexander 1957), but on the basis of notified cases it was 58 per cent in 1970–71 (Blake and Feldman 1975). In Japan during 1947–60 it was as high as 75 per cent (Tateno *et al.* 1961). With full modern treatment, including paralysis by curare, positive pressure ventilation, and antitoxin therapy, it can be reduced to 10 per cent (Smythe *et al.* 1974, Edmonson and Flowers 1979).

Distribution of tetanus bacilli

The recorded prevalence of tetanus bacilli in various parts of the environment depends, of course, on the sensitivity of the method of detection used. Sanada and Nishida (1965) found that the lower the degree of pre-heating of a soil suspension the greater was the isolation rate of tetanus bacilli and the higher the toxicity of the strains isolated. The most toxic strains were cultivated from soil heated to 60° for 10 minutes, and the least toxic from soil heated to 100° for 30 minutes.

Soil and dust

Nicolaier (1884) demonstrated the presence of tetanus bacilli in 12 out of 18 samples of earth; some of the specimens— from Leipzig and Berlin—had been kept for years. Grünbaum and Hertwig (1959), by mouse inoculation, found the bacilli in 220 of 400 samples of Berlin street dust. In Britain, Fildes (1925) found them in 45 per cent of 73 samples from waste and cultivated ground. In France, percentages of 50 and 64 were recorded. In Baltimore, Gilles (1937*a*) demonstrated toxigenic strains in 9 of 63 samples of street dust; and Oster-tag (1942) in 21 of 100 samples. The bacilli were found in the dust of British surgical operating theatres (Robinson *et al.* 1946, Sevitt 1949), and with some consistency both in the air of a city hospital and in that of operating theatres ventilated by a system that took in air from the hospital corridors (Lowbury and Lilly 1958). Contamination by factory and warehouse dust of fibres, especially unbleached wool fibres, used in surgical dressings constitutes a special hazard (see, e.g., Report 1959). The bacilli also occur in raw catgut used for surgical suture (Cuboni 1957).

Animal faeces

Toledo (Toledo and Veillon 1891) demonstrated the presence of *Clostridium tetani* in horse and cow dung. Noble (1915) found it in 18 per cent of 61 samples of horse faeces, but not in 20 samples of cow faeces. Other figures include 30 per cent of 23 samples of guinea-pig faeces (Sanfelice 1893) and 17 per cent of 200 samples from London horses (Fildes 1925). Kerrin (1929), in an examination of between 21 and 141 faecal samples from each of several species, obtained the following approximate percentages of positive results: horse, 15; cow, 5; sheep, 27; dog, 46; rat, 37; and hen, 18. About half the strains were atoxic. The presence of the bacillus in

faeces, and consequently in manure, may account for the greater prevalence of the organism in cultivated, than in uncultivated, soils.

Human faeces

Pizzini (1898) demonstrated the presence of *Cl. tetani* in human faeces; he found it in 3 out of 10 samples from ostlers, and in 2 out of 90 samples from peasants. From these results, it has been generally assumed that contact with horses strongly predisposes to the carrier condition in man, but we doubt if this assumption is justified. In Great Britain, Tulloch (1919–20) found the bacillus in 5 of 31 specimens of human faeces, and Fildes (1925) only twice in 200 specimens. Others have failed to find it in 304 (Kerrin 1928, 1929), 50 (Scheu-nemann 1931) and 92 (Bandmann 1953) samples. On the other hand, Bauer and Meyer (1926) in California found it in 24.6 per cent of 487 specimens of faeces. Bandmann (1953) summarized earlier investigations, in which the human faecal carrier rates lay between 0 and 40 per cent (see also Lowbury and Lilly 1958).

It is uncertain whether the tetanus bacillus multiplies in the human and animal intestine or is an organism of passage, having been ingested in food; the evidence is inconclusive. Other clostridia are able to thrive in the intestine, and there are reports that small amounts of antitoxin may be found in the serum of some animals and in man (see p. 352). Even if proliferation, and toxin release, does occur in the alimentary tract, tetanus, unlike botulism, does not arise as a result of absorption from the intestine.

Reproduction of the disease in animals

Tetanus can be reproduced by the injection of pure cultures or of toxin into mice, rats, guinea-pigs, rabbits, goats, sheep, horses, monkeys and other animals.

Cats and dogs are more resistant; birds and cold-blooded animals are highly resistant. The most susceptible animal, calculated on the amount of toxin per gram of body weight necessary to prove fatal on injection, is the horse. This species is about 12 times as susceptible as the mouse; the guinea-pig is 6 times and the monkey 4 times as susceptible as the mouse (von Lingelsheim 1912, Sherrington 1917). On the other hand, the rabbit is twice, the dog 50 times, the cat 600 and the hen 30 000 times as resistant as the mouse (Kitasato 1891, von Lingelsheim 1912). These figures are approximations only (for other estimates, see Wright 1955). Toxin is equally effective by the intramuscular and subcutaneous routes; much smaller doses are effective on injection into nerve trunks or the spinal cord (see, e.g., Roux and Borrel 1898). (For descriptions of the experimental disease in laboratory animals, see Chapter 42.)

Mode of infection in tetanus

Vaillard and Rouget in 1892 found that when tetanus cultures were heated to 65–67° for half an hour to destroy the vegetative bacilli and the toxin, the toxin-free spores remaining could be injected in large numbers into a guinea-pig without giving rise to the disease. The spores did not germinate in the tissues, but were rapidly taken up by the phagocytes; in 2 or 3 days they were completely ingested. When, however, the spores were protected from the phagocytes by being wrapped in filter paper, they germinated and gave rise to fatal tetanus. The same results were obtained by injury at the site of injection sufficient to cause necrosis or effusion of blood; and by injecting a second bacterial species, particularly an aerobic one. Spores germinated in damaged tissue, but not in clean, aseptic wounds. These observations on the activation of otherwise innocuous toxin-free spores were extended by several workers. Francis (1924) found that tetanus occurred in guinea-pigs inoculated simultaneously with washed spores and either staphylococcal culture or quinine. When the spores were given 9–30 days before the adjuvant, only 2 of 20 animals died. Clearly, spores can remain dormant in the tissues, including the phagocytes (Vaillard and Rouget 1892), for considerable periods.

Dormancy was evident in soldiers wounded during the 1914–18 war. The organism was frequently found in the wounds of men with no symptoms of the disease. Thus of 100 soldiers without tetanus, Tulloch (1919–20) found *Cl. tetani* in 19. Tetanus sometimes did not develop for weeks or months after a wound had healed, and might then appear suddenly after an operation, perhaps on another part of the body. There are instances of tetanus developing up to 14 years after the presumed introduction of the spores into the tissues, and Bonney, Box and MacLennan (1938) described one case in which tetanus bacilli were isolated from the uterine tissue ten years after a pelvic operation which had at the time been followed by a typical attack of tetanus.

Tulloch (1919–20) activated toxin-free spores *in vivo* by injecting them with tissue poisons like lactic acid, saponin, and trimethylamine; toxic filtrates of *Clostridium perfringens* and *Clostridium septicum* were particularly effective activators.

Bullock (Gye) and Cramer (1919) made the important discovery that the injection of small quantities of ionizable calcium salts together with toxin-free spores invariably led to the development of tetanus. They obtained the same result when washed bacilli—but not spores—and the salt were injected into different sites. The lesion occurred at the site of injection of the salt. There the organisms were present in large numbers and were not ingested by phagocytes; at the other injection site the bacilli underwent lysis and rapid phagocytosis. Phagocytosis of spores could not, however, be the sole cause of non-germination, because it was always incomplete. The spores germinated only when necrotic areas were produced by the injection of soil or calcium chloride. The calcium ions did not alter the virulence of the bacteria or the potency of the toxin, but damaged the tissues at the site of injection, breaking down their immunity (kataphylaxis).

The effect of soil was lost if its content of calcium salts was removed by precipitation. Bullock and Cramer concluded that the ionizable calcium salts of soil were an important factor in promoting tetanus in soil-contaminated wounds. They noted that calcium salts were more abundant in cultivated than in waste soil and that tetanus was far commoner at the beginning of the war, when the fighting was mainly on the highly cultivated soil of Flanders, than towards the end, when the soil had largely become waste. As we relate in Chapter 42, the effect of many of the agents that activate tetanus spores is referable to the induction in the tissues of an anaerobic focus of low oxidation-reduction (O-R) potential. It appears that spores germinate in tissues in which, by reasons of the necrosis, the O-R potential is decreased to a sufficiently low point (Fildes 1927, Russell 1927). This view is consistent with the inability of tetanus spores to germinate in a medium at pH 7.0–7.6 unless the potential is decreased to Eh + 0.1 volt or less. The Eh of the subcutaneous tissue of the guinea-pig is higher than this, but when it is lowered by holding the animals in an atmosphere containing only 7 per cent instead of the normal 21 per cent of oxygen, spores will often germinate *in vivo* under the influence of only mild activating agents (Knight and Fildes 1930, Campbell and Fildes 1931.

Growth of the bacilli in a necrotic, anaerobic lesion may be rapid: from lesions produced in mice by the injection of about 4000 spores in calcium chloride solution, of which about 1000 remained at the injection site, Smith and MacIver (1974) recovered 1.7×10^5 colony-forming units after 18 hours and 6.8×10^6 after 24 hours. They estimated that 2 MLD of toxin were probably formed as early as 7 hours after the spore challenge. The growing bacilli may increase the extent of the anaerobic focus; in guinea-pigs Smith and MacIver (1969) found that anaerobic lesions produced by calcium chloride became about ten times larger in volume when spores also were injected.

We may conclude that when tetanus spores are introduced into a wound by contamination with soil, horse faeces, or other material, their fate depends largely on the presence or absence of certain accessory factors. Many of these have been described—notably trauma, haemorrhage, tissue necrosis, and foreign bodies; chemicals such as lactic acid, saponin, trimethylamine, colloidal silicic acid, and ionizable calcium salts; substances such as the toxins of *Cl. septicum* and *Cl. perfringens*; and infection by other microbes. The presence of a suitable accessory factor enables the spores to germinate and multiply in the tissues; it seems probable that this action is dependent on the production by injury or by tissue debilitants of a region of necrosis with a sufficiently low oxidation-reduction potential to permit the spores to germinate. A similar local debilitating action by gelatin, vaccine lymph, antitoxic sera, and bacterial vaccines is probably the reason for the tetanus that follows the injection of these substances (Smith 1908). In post-operative tetanus the spores may be derived either from infected catgut (see Smith 1908, Mackie 1928, Bulloch 1929, Mackie *et al.* 1929, Savolainen 1950), imperfectly sterilized instruments or dressings, plasters and talcum powders, or some other source such as the intestinal or the respiratory tract or the contaminated air of the operating room (see Carbone and Perrero 1895, Motzfeldt 1912, Wright 1930, Tremewan 1946, Murray and Denton 1949, Mackay-Scollay 1959).

Absorption and mode of action of tetanus toxin

The means by which tetanus toxin, elaborated in a small, localized wound, gives rise to clinical tetanus has been studied for over ninety years. Evidence has slowly accumulated that the main site of action of the neurotoxin is the central nervous system to which the toxin gains access by ascending the motor nerves (see review by Mellanby and Green 1981).

Toxin does not appear to spread by the blood stream because, to produce fatal tetanus in experimental animals, more toxin is required by the intravenous than by the subcutaneous route; and with a given dose of toxin, the incubation period can be longer after the intravenous injection. In the rabbit, toxin given intravenously can later be found in blood, kidneys, and to a slight extent in the lumbar part of the spinal cord; toxin given into the sciatic nerve is readily detected in the cord (Bruschettini 1890, 1892). Gumprecht reported in 1894 that local tetanus was stopped by division of the motor nerve. Likewise it does not develop in a denervated limb (Couermont and Doyon 1899). It can be produced by the injection of a very small and otherwise ineffective dose of toxin directly into the sciatic nerve (Tizzoni and Cattani 1890, Meyer and Ransom 1903). Toxin is more effective given straight into the sciatic nerve than into the forepaw; but even when given into the forepaw, tetanus fails to develop when the second cervical nerve on that side is cut (Marie 1897).

Marie and Morax (1902) demonstrated the presence of toxin in the sciatic nerve of guinea-pigs some 30 minutes after toxin was injected into the foot, but only when the peripheral portion of the nerve was intact. Ascent of toxin may be blocked by antitoxin or by sclerosis of the motor nerve. Meyer and Ransom (1903) injected antitoxin into the right sciatic nerve of a rabbit. When toxin was injected subcutaneously into both hind legs, tetanus developed only in the left leg. Sclerosis of the nerve by tincture of iodine prevents the spasticity that follows *distal* but not *proximal* injection of toxin (Teale and Embleton 1920, Baylis *et al.* 1952a). As a rule, the smaller the animal and the less the distance the toxin has to travel along the nerves, the shorter is the incubation period. In the mouse, for example, it is about 12 hours, in the rabbit 18 to 36 hours, in the dog 36 to 48 hours and in the horse 5 days (Meyer and Ransom 1903). The dose of toxin administered is also a determining factor. The larger this is, the shorter is the incubation period. Wright and his colleagues (1950) induced strabismus, torticollis, salivation, and bradycardia by the careful injection of toxin into either the facial, vagus, or hypoglossal nerves—an observation best explained by the central action of toxin on neighbouring bulbar nuclei. Local tetanus in a limb can be abolished by barbiturate anaesthesia, probably because it acts within the central nervous system (Wright *et al.* 1952). Schellenberg and Matzke (1958) reported that in parabiotic rats local tetanus could be produced in one of the rats when toxin was injected into the hind limb of the other, provided a crossed regenerated nerve was present.

Transmission of toxin to the central nervous system along *motor* nerves was suggested as early as 1890 by the work of Bruschettini (1890, 1892) who observed that toxin injected into the muscle in rabbits could be recovered from the motor nerves supplying the surrounding area. Meyer and Ransom (1903), by injecting toxin directly into the anterior horn cells, produced tetanus within 3 hours, the incubation period being much longer when the toxin was given in the limb. When the toxin is injected into the posterior roots between the ganglion and the cord, tetanus, as it is generally understood, does not occur. Moreover, when toxin is injected into a part of the body devoid of motor nerves, such as the testes or peritoneum, a peculiar form of splanchnic tetanus develops, characterized by a longer incubation period, a more rapid illness, and an absence of muscular spasms (von Lingelsheim 1912).

The development of tetanus in one limb, after injection of toxin into it, is usually followed by local tetanus of the opposite limb (Sawamura 1909), suggesting spread of toxin from the anterior horn of one side to that of the other in the segment of cord affected. With larger doses of toxin, there is a gradual ascent of spasticity, first of the muscles of the back, then of the forelimbs and neck. That this is due to movement of toxin up the spinal cord is evident from the results of transection of the cord, whereby the spasm is limited to musculature enervated below the level of transection (Meyer and Ransom 1903, Firor *et al.* 1940, Friedemann *et al.* 1941, Baylis *et al.* 1952b). Kryzhanovsky (1966) found that muscle electrical activity in an affected limb increased only when toxin reached the anterior horn cells. In various animal species, including the donkey, in which, as in man, descending tetanus (see p. 351) occurs, Kryzhanovsky and his colleagues (see Kryzhanovsky 1965, 1966, 1967, 1973) studied the spread of toxin at intervals during the course of experimental intoxication by excising pieces of nerve and titrating their contained toxin in mice. In an extensive series of experiments they demonstrated that toxin ascended the nerves, both in ascending and descending tetanus, reaching the grey matter of the ventral horns of the spinal cord by way of the ventral roots.

Meyer and Ransom (1903) suggested that the part of the nerve through which toxin was transmitted was the axis cylinder. This view became generally accepted, despite a lack of evidence (Wright 1955). The axon route was later thought to be improbable, on the grounds that a large protein molecule like that of tetanus toxin could not diffuse in an axon as fast as the known rates of toxin movement demand, and was thought to be excluded by the inhibiting effect of sclerosing agents that leave the axon intact. The ascent of nerves by substances such as India ink, dyes, and isotopically labelled albumin led Wright (1953, 1955) to propose that toxin passed in the interstitial spaces of nerve, propelled by the pressures resulting from muscular contraction. Studies with tritium-labelled toxin (Fedinec and Matzke 1959, Fedinec 1967) also indicated the endoneural tissue spaces as the probable route of transmission. In later work, in which toxin was identified by means of horseradish peroxidase through a specific antibody bridge, and its activity confirmed as far as possible by injecting dissected parts of the nerve into mice, Fedinec (1972) showed that toxin was present in nerve stripped of epineurium, and in the perineurium, but not in the endoneurium (see also King and Fedinec 1973, 1974). However, more recent work with toxin labelled with ^{125}I has provided support for the earlier axonal hypothesis. Price and his colleagues (1975, 1977) used autoradiographic techniques to follow the passage of ^{125}I-labelled tetanospasmin in mice and rats after intramuscular and intraneural injection. The nerves were crushed proximal to the injection site to cause the accumulation of labelled toxin transported to the site of the crush blockage and, in consequence, greater radioactivity in the routes of migration. After intramuscular injection, radioactive label was found to be concentrated at the axon terminals of the neuromuscular junctions and in the axon, with little or none in the endo-, peri- or epineurium. Label was not found in the nerve at the site of the crush after intraneural injection of toxin at a site distal to the crush, suggesting that toxin is taken up only through the neuromuscular nerve endings. The authors pointed out that their findings accorded with recent evidence that intra-axonal transport is an important route for the movement of molecules along nerves. Erdmann and his colleagues (1975) left the nerves uncrushed and were able to demonstrate labelled toxin in the axons at the level of the ventral roots. Green and co-workers (1977) compared the diameter of labelled and unlabelled motor neurons in the ventral roots of cats previously given labelled toxin intramuscularly, and found that the diameter of the labelled fibres corresponded to that of the population of alpha motor fibres. Horseradish peroxidase, on the other hand, accumulates more in gamma than in alpha fibres in cat motor neurons (see Strick *et al.* 1976), and the reason for the association of toxin with only alpha fibres is not known. Evidence of the importance of the axonal route is therefore strong, although it has been criticized on the grounds that the fate of the labelled toxin may not be identical with that of native toxin, and because the significance of the toxin found by previous workers in the perineurium is unexplained (Zacks and Sheff 1976). However, most of the labelled toxin in the cord appears to be intact, active toxin; Habermann and his colleagues (1977) extracted the toxin from the spinal cord of rats and cats and found that 85 per cent of the radioactivity was bound specifically by antitoxin and had common precipitin lines in gel-diffusion tests with labelled and unlabelled toxins.

Tetanus in man is usually descending in character, that is, stiffness begins in the head and neck and spreads down the body. A similar form of tetanus is produced in the experimental animal by intravenous injection of toxin and it is presumed that in man toxin is transmitted *via* the blood stream to the bulbar motor nuclei. The means by which the blood-borne toxin reaches the central nervous system has been controversial. Wright (1955) suggested that local capillary permeability, in the *areae postremae* in the fourth ventricle, for example, permitted transport of toxin into the brain, but Fedinec (1967) in experiments with tritium-labelled toxin found no supportive evidence. An alternative explanation depends upon the uptake of toxin from the blood and tissue fluids by motor-nerve endings all over the body (Stöckel *et al.* 1975); the onset of symptoms first in the head and neck is then accounted for by the shorter length of the cephalic nerves. The work of Kryzhanovsky and his colleagues, already referred to, provided much evidence to support the latter hypothesis, and Habermann and Dimpfel (1973) found that labelled toxin was not detected in the central nervous system of rats until 6 hours after intravenous injection, by which time the concentration in the plasma had fallen sharply. Moreover, the pontine and post-pontine brain stem, from which nerves V–XII arise, were much richer in toxin than the region from which the smaller nerves III and IV arise. Stöckel and his colleagues (1975) provided evidence against direct blood-brain passage of toxin by the observation that, in rats given an injection of labelled toxin intramuscularly, toxin appeared in the blood stream and capillaries in the whole spinal cord, but no label could be detected in the nervous tissue adjacent to these blood vessels.

Wassermann and Takaki reported in 1898 that toxin was absorbed and rendered non-toxic by a suspension of brain tissue but not by liver, kidney, or spleen. The toxin receptor in the nervous system was eventually identified (van Heyningen 1959, van Heyningen and Miller 1961) as a ganglioside which, extracted from beef brain, could fix toxin at an estimated ratio of 1 mole of toxin to 2 moles of ganglioside. It was later reported that it was the heavy chain of the toxin molecule which became bound to the ganglioside (van Heyningen 1976). Tetanus toxin also binds to preparations of synaptosomal membranes, which have been shown to contain gangliosides (Habermann 1976). Interestingly, toxin also binds to the plasma membrane of thyroid cells, probably through the ganglioside receptors for thyrotropic hormone, which competitively inhibits toxin fixation by the membrane (Ledley *et al.* 1977); this may account for the state resembling acute thyroid overactivity which sometimes is seen in cases of human tetanus. Cultured neuronal cells from the central nervous system also bind toxin, apparently through long-chain ganglioside receptors (Dimpfel *et al.* 1977, Mirsky *et al.* 1978). Two binding sites have been distinguished on differentiating mouse-neuroblastoma cells in culture. One accepts toxin, but not toxoid, with a resulting contraction of neurone-like processes formed by the cells in culture; this effect is not inhibited by treatment of the cells with neuraminidase or β-galactosidase, both of which modify membrane gangliosides. The second receptor binds both toxin and toxoid but with no visible morphological effect; ganglioside will inhibit the binding of toxoid. The receptor through which toxin mediates its neurological effects may not, therefore, be the ganglioside (Zimmerman and Piffaretti 1977).

The mechanism by which tetanus toxin affects the functioning of the central nervous system is unknown. It appears to depress selectively the inhibitory impulses from pre-synaptic fibres which normally act upon the motor neurones, thereby permitting the unchecked activity of the poly-synaptic reflexes; this results in tetanus (see Davies *et al.* 1954, Brooks *et al.* 1957, Wilson *et al.* 1960, Bizzini 1977). The blocking of post-synaptic inhibition of motor neurones probably occurs pre-synaptically (Curtis and de Groat 1968), by interfering with transmitter release, but the molecular process responsible has not yet been elucidated, despite a growing understanding of the molecular basis of neurotransmission (see reviews by Bizzini 1977, van Heyningen 1980, Mellanby and Green 1981).

In addition to its principal effect on the central nervous system, tetanus toxin may also cause flaccid paralysis by affecting neuromuscular transmission (Abel 1934). Indeed, when toxin is prevented from reaching the central nervous system in the rabbit, paralytic signs develop (Miyasaki *et al.* 1967). Ambache and his colleagues (1948*a, b*) induced non-spastic paralysis of the iris by the injection of crude toxin into the rabbit eye. The affected iris would not respond to stimulation of the oculomotor nerve but contracted in response to the local injection of acetylcholine. The effect was probably due to tetanospasmin, as a non-spasmogenic fraction left after absorption of toxin with ganglioside had no activity on the iris (Mellanby *et al.* 1968). It was deduced that the toxin impeded the release of acetylcholine at the local nerve endings. Toxin causes flaccid paralysis in the goldfish, an animal as sensitive as the mouse to tetanus toxin—apparently as a result of blockage of neuromuscular transmission pre-synaptically; miniature junction potentials in nerve-muscle preparations from pectoral fins partly paralysed with toxin were reduced in frequency but not in amplitude, implying that post-synaptic sensitivity to acetylcholine was unaffected (Diamond and Mellanby 1971, Mellanby and Thompson 1972). Tachycardia and disturbances of heart rhythm, sometimes observed in tetanus patients, have been attributed to sympathetic overactivity (Vakil *et al.* 1965, Kerr *et al.* 1968, Adams *et al.* 1969). There is no evidence, however, that this effect is due to the action of toxin on the sympathetic nervous system.

In summary, local and 'ascending' tetanus from locally injected toxin is produced by toxin absorbed at the motor nerve endings, and carried to the anterior horn cells, where, by interfering at the synaptic junctions with the inhibitory neurones governing the generation of motor impulses, it permits motor hyperactivity and hence spasticity in the appropriate muscles. In 'descending' tetanus the toxin appears to reach the brain stem from the blood stream also by ascending the motor nerves, the onset of symptoms in the head and neck being accounted for by the shorter length of the nerves supplying those areas. The mode of absorption of toxin, of its carriage up the nerve trunk and spinal cord, of its passage from the blood to the brain, or of its action on the susceptible nerve cells, is not known with any precision, but current evidence favours absorption *via* ganglioside receptors and passage along the motor axons. Though a local action of toxin on myoneural junctions may occur, central action of the toxin appears to account for most of the manifestations of natural and experimental tetanus.

Immunity to tetanus

Natural immunity

Mammals vary in their susceptibility to tetanus; some, such as the mouse, the monkey, and the horse, are highly susceptible; others, such as the dog and cat, much less so. Most birds and cold-blooded animals are extremely resistant. Examination of the blood of susceptible and partly susceptible animals shows that, though most cattle contain more than 1/500 unit of antitoxin per ml, no antitoxin is present in the blood of dogs, pigs, monkeys, or rodents. Sheep and goats may contain small quantities (Coleman and Meyer 1926, Coleman 1931, Ramon and Lemétayer 1934, 1935), and also horses in some countries (R. Veronesi, *pers. commun.*). There are a few reports of small amounts of antitoxin being detected in human sera in India and Brazil (Vakil *et al.* 1968, Veronesi *et al.* 1975, D'Sa *et al.* 1978, Ray *et al.* 1978). However, the antitoxin was detected in these studies by in-vitro haemagglutination rather than neutralization. There is no evidence (see Lahiri 1939) that these small amounts of antitoxin provide any natural immunity in man. The frequent presence of antitoxin in the blood of ruminants may be causally associated with the comparative resistance of these animals to tetanus. It is suggested that in ruminants tetanus bacilli multiply and form toxin in the digestive reservoirs which precede the true stomach, and that the toxin so formed, partly modified perhaps by the products of bacterial fermentation, is absorbed and gives rise to antitoxin. Examination of the blood of naturally resistant species has proved it to be devoid of antitoxin (Vaillard 1892). In some, it may be due to a failure of toxin to reach susceptible nerve tissue. Thus, whereas in the susceptible rabbit a large intravenous dose of toxin disappears from the blood within a day (Marie 1897), it persists in the blood of the insusceptible hen for several days (Vaillard 1892).

Acquired immunity

One of the most important advances in the progress of bacteriology was the discovery (Behring and Kitasato 1890, Kitasato 1891, 1892, Behring 1892) that the injection of small amounts of a culture filtrate of tetanus bacilli into an animal stimulated the production of an antibody which was able to neutralize the toxin. This antibody protected not only against injections of toxin, but also against inoculation with tetanus bacilli.

Antiserum acts not by interfering with the germination of spores, or with the multiplication of the bacilli in the tissues, but by neutralizing the toxin as it is formed. When antitoxin is present in the circulating blood and lymph, it will neutralize the toxin before it has time to enter the nerves. When the antitoxin is not given until after the formation of toxin has begun, it may be too late to prevent the absorption by the nerves.

It follows that unless antitoxin comes into intimate contact with free toxin, it will be useless. Toxin in the nerve trunks or the central nervous system is largely inaccessible to antitoxin in the blood and tissue fluids, and very large amounts of antitoxin in the blood stream are required in order to save the life of an animal once the toxin has gained access to the central nervous system; very shortly afterwards antitoxin is quite ineffective. Roux and Borrel (1898) showed that an actively immunized rabbit, capable of withstanding large doses of toxin subcutaneously or intravenously, succumbed as readily to the intracerebral injection of toxin as a normal rabbit: thus we have the anomaly of an animal whose blood contains sufficient antitoxin to save the life of hundreds of mice itself dying of tetanus.

Two stages in the action of toxin—adsorption to the nerve cells, and fixation—have been distinguished (Abel *et al.* 1938, Abel and Chalian 1938). Fixation coincided with the appearance of tetanic symptoms. In the sheep, antitoxin given intravenously after demonstrable absorption of lethal amounts of toxin by the nervous tissue, but before symptoms of descending tetanus had appeared, had a curative action. In animals previously given injections of toxin, antitoxin injected intravenously and intracerebrally can prevent death, even when given after the onset of signs (Sherrington 1917, Smith 1965, Kryzhanovsky 1973), but signs progress for some time after the antitoxin has been injected, and antitoxin is incapable of reversing the effects of toxin (Webster and Laurence 1963). Toxin that has entered the central nervous system may be regarded as 'free' and susceptible to neutralization by antitoxin that gains access to it, or 'fixed' and insusceptible to neutralization (Webster and Laurence 1963, Habermann 1972). Toxin bound *in vitro* by brain extract can be neutralized by antitoxin (Kryzhanovsky 1973).

Diagnosis

The bacteriological diagnosis of tetanus is often difficult, not least because the site of infection may be very small.

Microscopical examination should be made where possible of the pus or necrotic material in the wound, and special note taken of bacilli with round, terminal spores. The presence of 'drumstick' bacilli is not, however, pathognomonic of infection with *Cl. tetani* (see, e.g., Boyd and MacLennan 1942). The organisms may be present in such small numbers that they are altogether overlooked.

Cultural methods are of more importance. The pus or wound scrapings, or tissue taken from a necrotic focus in the wound, should be plated on blood agar for anaerobic incubation and seeded into one or more bottles of cooked-meat medium. One bottle should be incubated unheated and the remainder after heating to 80° in a waterbath for various times from 5 to 20 minutes with the aim of killing non-sporing bacilli and leaving tetanus spores unharmed. Enrichment of the wound material in broth containing 1 in 100 000 crystal violet is said to be effective (Gilles 1937b), and ascorbic acid polymyxin thyoglycollate broth has also been used for this purpose (Wetzler *et al.* 1956). The inoculated liquid media are incubated at 37° and subcultured on half of a blood agar plate daily for at least four days. The plates are incubated anaerobically and examined for the swarming edge of growth of *Cl. tetani*. The incubated cooked-meat broths should also be heated at 80° and samples plated at short time intervals. A blood agar plate containing 4 per cent agar to minimize swarming, one half of which has been treated with antitoxin, may be seeded on each half with growth assumed

to be *Cl. tetani* and incubated for two days at 37°. Colonies haemolytic on the untreated half, but not so on the antitoxin half, are almost certainly toxigenic *Cl. tetani* (Lowbury and Lilly 1958).

The isolation of toxigenic bacilli is confirmed by demonstrating that they produce tetanospasmin. Two mice, one unprotected and the other protected by an injection of 1000 units of tetanus antitoxin, are given a subcutaneous injection in the hind leg of 0.5 ml of a 48-hour cooked-meat broth culture of the isolate. The protected mouse remains well; the unprotected animal will show typical ascending tetanus.

Prophylaxis and treatment

For the prevention and treatment of tetanus both general and specific measures are required. Hygienic precautions, especially in midwifery, are important, and proper cleansing of the wound with removal of foreign bodies and of all necrotic material should be undertaken. When suitably situated, the wound may be completely excised. Once symptoms of the disease have appeared, nursing care and supportive treatment are of prime importance, supplemented when necessary by anti-convulsive therapy and, in severe cases, by neuromuscular block with curare accompanied by tracheostomy and positive pressure ventilation (see Ellis 1963, Laurence and Webster 1963, Smythe *et al.* 1974). Death may occur not only from exhaustion, respiratory infection, circulatory failure, or spasm of the glottis, but also in all probability from the effect of destructive lesions in the medulla leading to sudden respiratory failure (Baker 1942).

Specific prophylactic and therapeutic measures include the use of tetanus antitoxin, tetanus toxoid, and antibiotics.

Preparation and specification of antitoxin

Owing to its exceptional potency, tetanus toxin cannot readily be used in the *preparation* of antitoxin. Many substances have been used to detoxify the toxin, without destroying its immunogenicity.

Behring and Kitasato (1890) and Kitasato (1891) successfully immunized rabbits and mice with toxin modified by iodine trichloride; the serum of the immune animals contained toxin-neutralizing antibodies. Antitoxin in large quantities was prepared by this method in sheep and horses (Behring 1892). It protected mice from infection with spores of *Cl. tetani* when given before, at the same time as, and even 24 hours after the injection of spores. The therapeutic dose required was very much greater than the prophylactic dose. Thus in one experiment the effective dose of serum was 0.001 ml when given 15 hours before, 0.1 ml when given at the same time as, and 0.4 ml when given 24 hours after the infection (Kitasato 1892). Vaillard (1891, 1892) confirmed in rabbits the immunogenicity of iodized toxin, and the efficacy of passive protection of mice by antitoxin. The effect of prophylactic antitoxin lasted for a few days, but in his hands therapeutic antitoxin was not curative when given a few hours after tetanus was manifest.

Antitoxin was later produced on a large scale by the injection of horses with toxin-antitoxin mixtures (Buxton and Glenny 1921). The toxin was over-neutralized, and injections were given at intervals of a few days over a period of some months. Detoxification with formaldehyde (Descombey 1924) is now widely used. Usually about 0.4 per cent of formol is added to the toxin, and the mixture is incubated at 37–39° till it is sufficiently detoxified. About 20 ml of formolized toxin should prove non-toxic to the guinea-pig on subcutaneous injection (Hosoya *et al.* 1931, Wilcox 1934). The toxoid is commonly adsorbed with an aluminium salt to increase its antigenic efficacy (see Glenny *et al.* 1926). After inducing substantial immunity with formol toxoid, immunization of the horse may be completed with crude toxin, to yield serum containing as many as 1000 units of antitoxin per ml.

Human anti-tetanus immunoglobulin has replaced horse serum antitoxin for tetanus prophylaxis in many countries, since it has the considerable advantages of a longer half-life (see page 355), and freedom from the risks of adverse reactions (see page 356) and of early elimination (see page 355). It is obtained by separating the IgG fraction from the plasma of immunized volunteers, usually by cold-ethanol fractionation, and is commonly provided at a concentration of 250 units of antitoxin in an injection volume of 1.0 ml. The product appears to be free from the risk of transmitting hepatitis (Rubbo and Suri 1962, Rubinstein 1962, McComb and Dwyer 1963).

The *specification* of the potency of tetanus antitoxin, whether of animal or human origin, is made in terms of a stable, standard antitoxin to which is assigned an arbitrary figure expressed as units.

In 1928, the Permanent Commission on Standardization of the League of Nations recommended the US National Institutes of Health (NIH) standard (Rosenau and Anderson 1908) as the international standard preparation. Although this recommendation was adopted, the international unit was made equal not to the established NIH unit but to one-half this unit, with the result that for many years the international unit (1928) was in conflict with the NIH unit. This discrepancy was resolved in 1950 by the Expert Committee on Biological Standardization of the World Health Organization, which doubled the size of the international unit. The international unit (1950) is now equal to the NIH unit (Report 1950) and, as defined in 1969 by the WHO Expert Committee on Biological Standardization, is the activity contained in 0.033 84 mg of the second international standard for tetanus antitoxin (Spaun and Lyng 1970).

Antitoxin titrations are carried out by comparing the amount needed to protect mice against a fixed test dose of toxin with the amount of standard antitoxin needed to give the same degree of protection. A convenient test dose of toxin is the smallest quantity which, mixed with one international unit of the standard antitoxin (or a fraction of one unit), causes clearly recognizable paralysis within four days but does not cause significant suffering—mild but definite paralysis of the hind leg is suitable for this purpose. This Lp dose of toxin is preferable to the L + dose, i.e., that which kills the animal by the fourth day, as it minimizes the illness to which the mice are subjected (Mussett and Sheffield 1976, British Pharmacopoeia 1980).

Comparative assays in a number of institutions show that, provided one specimen of toxin is employed in the assay of antitoxin in terms of the standard, similar results are obtained by different workers. The toxic filtrates commonly used for assay, however, vary greatly with the strain of *Cl. tetani* and the medium selected for its growth. Even when the greatest care is taken in the selection of the test toxin, in the technique of administration, and in the choice of strain, there are qualitative as well as quantitative differences among antitoxins that may vitiate the assay (Report 1938, Smith 1938, Ipsen 1940–41, Hornibrook 1952). The relative potency of two antitoxins may vary with the species of animal used in the test (Smith 1943–44) or with the route of administration of the toxin-antitoxin mixtures (Friedemann and Hollander 1943). The change of potency ratio with change of the biological system used for an assay indicates a heterogeneity of some kind in the materials used. In this case we may postulate heterogeneity of the antibodies, either in reactivity (see, e.g., Cinader and Weitz 1953) or in specificity. The specificity of the antibody may vary because molecules of toxin vary, or because the antibodies are specific only for portions of the toxin molecule. Nagel and Cohen (1973) described four different antigenic determinants on both toxin and formol toxoid, and all of the corresponding four antitoxins were capable of neutralizing toxin *in vivo*. Antibodies to tetanus toxin are able to neutralize even though they do not bind to the ganglioside-binding site of the toxin (Kryzhanovsky 1973).

The potency of unrefined antitoxins may be measured by flocculation methods analogous to the Ramon titration of diphtheria antitoxin (see, e.g., Goldie *et al.* 1942), but proteolytically refined antitoxins may not flocculate satisfactorily.

Antitoxin may also be assayed *in vitro* by agglutination of tanned red cells coated with toxoid (Fulthorpe 1957, 1958), by the enzyme-linked immunosorbent assay (Voller *et al.* 1976) and, in screening assays, by immunodiffusion methods (Eldridge and Entwhistle 1975, Winsnes 1979). Immuno-electrophoresis methods (Wiseman and Gascoigne 1977) have been used to select donors for the preparation of human tetanus antitoxin. However, such tests do not always precisely reflect neutralizing activity, probably owing to avidity effects as well as to the existence of different antigenic determinants on the toxin molecule.

Properties of tetanus antitoxin

Tetanus antitoxin prepared in the horse, or sometimes in other domestic animals such as the sheep or cow, is used in the form of native serum or, more commonly, proteolytically refined globulins. The proteolytic treatment splits the immunoglobulin molecules, the Fc fragments being then separated by heat precipitation, leaving the Fab fragments, which carry the combining sites, in solution (Pope 1963, Porter 1963). The Fab fragments are precipitated with ammonium sulphate and resuspended, at an appropriate concentration, to provide the refined antitoxin preparation. This product is less antigenic than untreated preparations, and is associated with fewer allergic reactions.

The passive immunity conferred by an injection of antitoxin is of limited duration. The half-life of refined horse antitoxin in man is about 7–12 days (Barr and Sachs 1955,

Turner *et al.* 1958, Smolens *et al.* 1961), but owing to the development of an antibody response to the heterologous protein 'rapid immune elimination' may succeed normal catabolic elimination and cut short the duration of passive immunity. An antibody response to the horse globulin by no means always occurs. Reisman and his colleagues (1961) detected anti-horse serum antibody in only 27 of 71 patients, 15 of whom had a high titre of antibodies before the injection of horse antitoxin. Suri and Rubbo (1961) observed immune elimination in 5 of 30 patients, and some of those who showed no accelerated elimination had previously received horse antitoxin. Nevertheless, the occurrence of immune elimination is a drawback to the use of heterologous antitoxins, especially when this process begins immediately owing to the presence of circulating anti-horse antibodies as a result of earlier exposure (Godfrey *et al.* 1960).

The half-life of homologous human antitoxin in man is longer than that of heterologous antitoxin—approximately 28 days; consequently, a given period of passive protection can be secured with a smaller dose. Although immunoglobulin allotypes occur, immune elimination does not appear to have been reported.

The minimum concentration of antitoxin in the serum that is adequate to provide immunity is uncertain but a figure of 0.01 unit per ml is often accepted (McComb and Dwyer 1963, Adams *et al.* 1969, Smith 1969). Thus, when a population of women immunized during pregnancy had a mean serum antitoxin of 0.01 unit per ml, *tetanus neonatorum* in their babies was prevented (MacLennan *et al.* 1965). Similar studies in Colombia suggested that the immunity threshold may be less than 0.01 unit per ml (Newell *et al.* 1971), but mild tetanus has been described in patients known to have had concentrations greater than 0.01 unit per ml in their serum (Goulon *et al.* 1972, Berger *et al.* 1978). Moreover, although the neutralizing capacity of antitoxin is considerable—250 units has the capacity to neutralize *in vitro* about 30 million minimal lethal doses for the mouse—passive immunity is less protective than active immunity (d'Antona and Valensin 1937, see page 359). Probably 0.01 unit per ml of serum should be regarded as the minimum protective level in passive immunity, but for patients with severe, soiled wounds, prophylactic doses of antitoxin larger than usual should be considered.

Antitoxin prophylaxis

Passive immunization with horse antitoxin, coupled with surgical débridement of the wound, has long been the standard method for prevention of tetanus. It was used extensively for the wounded in the first world war and, for those who have not been actively immunized, is still being used in many countries where human antitoxin is not available. The aim is to provide a protective amount of antitoxin in the serum of the injured person until toxin production in a wound is likely to have ceased. A dose of 1500 units of refined horse antitoxin, subcutaneously or intramuscularly, as soon as possible after wounding is recommended for this purpose (Adams *et al.* 1969), though Edsall (1967) and others (see Eckmann 1967) would raise this to 3000 or, if more than 48 hours after the injury, to 6000 units. A dose of 1500 units should provide, in the

absence of immune elimination, a peak content of about 0.2–0.5 unit per ml of serum two days after injection, falling to 0.01–0.05 unit at three weeks. A dose of 250 units of human antitoxin should provide protection for about 4 weeks in adults (Rubbo 1966), including the severely injured; but in extensively burned patients this dose may be insufficient (Lowbury *et al.* 1978). The suggested indications for passive immunization of injured persons who are not known to be actively immune are: wounds over six hours old; infected wounds, wounds through skin contaminated with soil, street mud or dust, animal faeces and the like; deep wounds, including puncture wounds; wounds with devitalizing tissue damage; wounds that cannot be closed (see Parish *et al.* 1957, Smith *et al.* 1975). In severe wounding, and especially burns, the systemic dose may be doubled or trebled. A portion of the dose may be injected round the wound. The systemic dose should be repeated every 4 weeks until healing is assured.

Animal experiments clearly establish the value of prophylactic antitoxin. In man the evidence is indirect and often equivocal in that it is not always possible to exclude other causes for the effects attributed to antitoxin.

(1) The prophylactic injection of antitoxin within a few hours of the receipt of a wound appears to diminish greatly the chances of tetanus. The incidence during the war of 1914–18 is recorded in Fig. 64.1, p. 346. The sudden fall in November 1914 might well have been due to the prophylactic use of antitetanic serum that was introduced about the middle of the preceding month. The rise towards the end of 1916 may have been fictitious; Bruce ascribes it to improved diagnosis, particularly of local tetanus, which was largely overlooked at the beginning of the war. The subsequent fall in 1917 and 1918 may perhaps be related to the more effective prophylactic serum treatment which commenced in June 1917, but more probably to the practice of early excision of wounds which was introduced at about the same time. These changes in incidence can provide only suggestive evidence of the value of antitoxin, as they may have been caused by other factors such as the nature of the terrain over which battles were fought, or the efficiency of the field surgical services.

(2) When it does not prevent the development of tetanus, antitoxin appears to lengthen the incubation period. The average incubation period in soldiers of the 1914–18 war treated in hospitals in Britain rose from 11.8 days in 1914 to 48.0 days in 1917; and in a direct comparison of men given and not given prophylactic antitoxin, the average periods were respectively 45.5 and 10.9 days. Since the passive immunity conferred by a single dose of antitoxin lasts only some 2–3 weeks, it is possible that late developing tetanus in some of the immunized men was due to tetanus bacilli in the wound becoming effectively toxigenic in the absence of immunity; this might have been prevented by repeated doses of antitoxin.

(3) Tetanus developing in persons who have received a prophylactic injection appears to be less fatal than in uninoculated persons (Bruce 1920) (Table 64.3). Moreover, the average fatality among the inoculated during the war fell more rapidly than among the uninoculated. However, the

Table 64.3 Showing the fatality of tetanus in patients with and without antitoxin prophylaxis (Bruce 1920)

Prophylactic serum	Number of cases	Died	Case fatality, per cent
Given	899	203	22.6
Not given	559	298	53.3

treatment of those who were given antiserum may have differed in other respects from that of patients who were not given antiserum.

(4) Among all sufferers from tetanus, the percentage of patients in whom the disease became generalized fell from 98.9 in 1914 to 83.5 in 1918 (Bruce 1920). This may be attributed partly to the more general and effective use of antitoxin prophylactically, and partly to improved diagnosis of local tetanus.

Suggestive evidence is also provided by Bazy (1914). Of 200 soldiers with similar wounds, 100 were not given antitoxin, for reasons Bazy did not specify, and 18 of these patients developed tetanus. In the 100 who were given antitoxin only one case occurred, on the day after injection.

Later records of effective antitoxin prophylaxis are scanty and can be considered as no more than suggestive. Thus Wildegans (1940) observed in the Polish campaign of 1939 that the incidence of tetanus among Polish soldiers, over half of whom had received no prophylactic antitoxin, was 66 per 10 000, as compared with one of 3.6 per 10 000 in the German soldiers, the majority of whom had been passively immunized (see also Boyd 1946). From a study of the records of a series of civil cases of similar severity, Bianchi (1962) estimated a fatality rate of 55.7 per cent in 80 patients receiving less than 3000 units of antitoxin, and of 32.4 per cent in 34 receiving 3000 units or more.

Lucas and Willis (1965) reported that, when horse serum was replaced by antibiotic prophylaxis for injured patients attending a casualty department in Ibadan, Nigeria, the incidence of tetanus among them rose from 2.3 to 7.8 cases per 10 000 patients, and the rate fell to 2.4 cases when there was a return to the use of antiserum.

Field evidence of the value of human antitoxin prophylaxis does not appear to have been reported.

Reactions to antitoxic serum

Passive immunization with heterologous antibody has the disadvantage that persons receiving it for the first time may become sensitized to immediate and delayed types of allergic reactions upon subsequent injection of the same proteins; and, in persons already sensitized, of causing similar reactions. The incidence of local and general reactions to horse antitoxin has been variously estimated at from 5 to 50 per cent, the variation reflecting differences in the criteria used for defining a reaction, the thoroughness of follow-up, the purity of the antitoxin used, and the differences in the populations studied. In Britain, it has been reported to be 8.7 per cent after a first injection and 10.5 per cent after re-injection (Binns 1961). This difference is small and may reflect the difficulties of establishing an accurate history of previous exposure to horse serum.

In developing countries, where fewer people might be expected to have received horse serum injections, the reported incidence of reactions is less (Patel *et al.* 1965).

Before injecting the full dose of heterologous antitoxin, it is wise to test the patient's reactivity. The type of test that provides the best indicator of a potential reaction to the full dose is uncertain. An intradermal, conjunctival or subcutaneous test can be used. Laurent and Parish (1958) distrust the intradermal reaction. They have often injected a full dose of serum without ill effect when the test has been positive, and they have known of fatal cases of anaphylactic shock when it has been negative. They regard the reaction to a subcutaneous trial injection of 0.2 ml of serum as giving a more certain indication of the possibility of the reaction most to be feared—acute anaphylaxis. Cox, Knowelden and Sharrard (1963) found no relation between a local reaction to a subcutaneous trial dose and subsequent reactions to the full dose, but Binns (1961) found that a local reaction within an hour of injecting a trial dose was associated with an increased risk of a later general reaction. Presumably, the occurrence of a local reaction to a skin, conjunctival or subcutaneous test is evidence of some form of immunologically sensitive state and should be regarded as an indication that a reaction might follow injection of the full dose.

The amount of antitoxin to be given in a subcutaneous trial dose requires thought, since fatal anaphylaxis has been reported after the injection of even 0.01 ml of serum (Buff 1960). Too small a dose, however, may fail to reveal the latent hypersensitivity. A compromise dose of 0.05 ml has been suggested (Adams *et al.* 1969), but in patients with a history of allergy this should be preceded by a trial of 0.05 ml of a 1 in 10 dilution.

The use of human antitoxin in place of horse serum has largely overcome the danger of serious reactions to passive tetanus prophylaxis. Mild local reactions sometimes follow intramuscular injection, and a few instances of severe generalized reactions have been reported (Glaser and Wyss-Souffront 1961, MacKenzie and Vlahcevic 1974).

It should be added that active immunization should be offered to all patients given horse serum antitoxin and is advisable for those given human antitoxin (see page 362, simultaneous active and passive immunization).

Therapeutic antitoxin

Although many years have elapsed since the introduction of antitoxin therapy, its efficacy has been examined by means of controlled clinical trials only in recent years. The results of antitoxin treatment during the war of 1914–18 (see Bruce 1915, 1916, 1917*a, b*,

1919, Dean 1917, Leishman and Smallman 1917) gave no clear answer.

Experiments with small laboratory animals have shown that, although antitoxin is capable of preventing death when it is given shortly after the onset of symptoms, any further delay in its administration makes it difficult or impossible to interfere with the progress of the disease. On monkeys, however—chiefly *Macacus rhesus* and *Callithrix*—Sherrington (1917) obtained unequivocal results. Batches of 25 were given injections of 8 MLD of toxin into the gastrocnemius muscle. From 47 to 78 hours later, when the early symptoms of tetanus had appeared, he injected 20 000 units of antitoxin per kg body-weight, giving the serum by a different route in each batch. All animals receiving no antitoxin died; the fatality amongst the inoculated was as shown in Table 64.4.

Table 64.4

Route by which serum given	Deaths	Recoveries	Fatality, per cent
Subcutaneous	23	2	92
Intramuscular	22	3	88
Intravenous	18	7	72
Lumbar intrathecal	11	14	44
Bulbar intrathecal	7	13	35
Cerebral subdural	10	0	100

This experiment showed that antitoxin, when given by the intrathecal and to a lesser extent by the intravenous route, was of undoubted therapeutic value. However, determination of its value in human tetanus, and of the optimal dose to be injected, is extremely difficult. Clinical reports are highly contradictory. The reason for this appears to be mainly that the severity and fatality rate of tetanus are influenced by so many factors that, in order to obtain reliable results, comparisons must be made between groups as identical as possible, of adequate size, and in properly designed therapeutic trials.

Brown and his colleagues (1960) were among the first to bring fairly strong experimental evidence in favour of the therapeutic value of antitoxin in man. They applied sequential analysis to a series of pairs of patients, in each of which one patient received 200 000 units, the other none. In this way the withholding of antitoxin from the controls could be stopped as soon as treatment proved to be significantly beneficial. The result was a clear-cut difference ($P = <0.05$) between a fatality rate of 49 per cent in the treated and of 76 per cent in the untreated. The effect of 200 000 units of antitoxin was compared with 500 000 units in 270 patients (Lucas *et al.* 1965), 50 000 units in 293 patients (Vakil *et al.* 1963), 20 000 units in 300 patients (Vakil *et al.* 1964) and with 10 000 units in 796 patients (Vakil *et al.* 1968); in none of these studies was the larger dose shown to be preferable to the smaller. As a result the authors recommended a dose of 10 000 units. These findings have not been accepted everywhere as providing proof that antitoxin is beneficial: the patients in the pairs were selected not by matching but by random allocation; the numbers in the first trial were very small—41 in the treated group and 38 in the controls; and it may therefore be objected that the results were influenced too much by adventitious factors. Sequential analysis is a good method when the universe of treated and untreated is homogeneous, but with a heterogeneous universe, such as

that of tetanus patients, there is a danger that a significant difference may be reached as the result of chance rather than of the mode of treatment. Creech and his colleagues (1957), for example, report that in a series of 52 cases having a fatality rate of 42 per cent, 19 consecutive cases were treated without a single death.

To overcome this difficulty either large numbers must be used, or the patients must be allocated to groups of like severity. This is what Patel and his colleagues (1964) did in an attempt to determine whether a dose of 20 000 units was better than one of 10 000. They used a matched pair sequential analysis, the patients being classified into five prognostic grades. Altogether 886 cases were studied. The results showed no significant difference between the two doses. The same conclusion was drawn by De (1966) in a study of 621 cases. Patel and Mehta (1963), who observed 2007 cases divided into five prognostic groups, gave antitoxin in doses ranging from 20 000 to 240 000 units. No virtue was found in the higher dosage (Table 64.5). None of these last three trials,

Table 64.5 Fatality rate in relation to dosage of tetanus antitoxin (ATS) (after Patel and Mehta 1963)

ATS units	Number of cases	Fatal cases	Fatality, per cent
240 000	55	18	32.7
120 000	243	84	34.6
60 000	1266	487	38.5
20 000	151	31	20.5
Total	1715	620	36.2

of course, proved the value of serum treatment, because no control group was included. Another trial, planned differently, in which 470 patients were randomly allocated to four groups, was reported on by Vaishnava and his colleagues (1966). Group 1 received no antitoxin and served as a control. The other three groups received different doses of antitoxin. The results given in Table 64.6 showed virtually no difference

Table 64.6 Fatality rate in relation to dosage of tetanus antitoxin (ATS) (after Vaishnava *et al.* 1966)

ATS units	Number of cases	Fatal cases	Fatality, per cent
None	103	47	45.6
10 000	127	52	40.9
30 000	133	66	49.6
60 000	107	51	47.7
Total	470	216	46.0

in fatality rate between the treated groups and the control group or between the treated groups themselves. Nor was any difference observed in the course of the disease or in the development of complications. The authors therefore concluded that antitoxin had little value in the treatment of tetanus.

We are left, it will be realized, with no conclusive proof of the therapeutic usefulness of antitoxin. There is fairly strong evidence that large doses have no advantage over small ones, and in fact may be detrimental to the general condition of the patient. Nor is this conclusion surprising. If all that can be expected of therapeutic antitoxin is to neutralize toxin before it is 'fixed' to the nervous tissue, then a single dose of reasonable size, such as 10 000 units, should be sufficient. Nevertheless, this aim is an entirely reasonable one. The prevention of further intoxication of the central nervous system might be expected to make a difference between life and death in a small proportion of patients—those who, when treatment is begun, have not yet taken up into the nervous system a lethal dose of toxin but in whom toxin is still present in the tissues. Moreover, it is difficult to discount the therapeutic effect of antitoxin in experimental animals. It is possible that doses of less than 10 000 units would be equally effective, but their value has not been studied in controlled trials. In patients with wounds that cannot be properly cleansed, a second dose 2–3 weeks later may be advisable, but otherwise there seems to be no reason for giving more than one injection.

Human antitoxin is increasingly used for tetanus treatment in place of horse antitoxin. In therapy it is advantageous to inject the antitoxin intravenously in order to secure a high level of antitoxin in the serum as quickly as possible. After intramuscular injection a peak level of horse antitoxin is not reached till 48 hours and of human antitoxin not until 2–5 days (Turner *et al.* 1958, Rubinstein 1962, Piringer *et al.* 1964). The difference may possibly be accounted for by the slower catabolism of the homologous protein. A disadvantage of the intravenous use of human antitoxin, however, is the occurrence in some patients of adverse reactions, including circulatory collapse, an effect which may be due to the anticomplementary effect of aggregates of immunoglobulin (Barundun *et al.* 1962) or to contamination of the preparation with vasoactive peptides (Alving *et al.* 1979). Preparations suitable for intravenous injection are, however, available in some countries (Editorial 1973) and the ordinary material has been given safely by slow infusion in an intravenous drip (Reed *et al.* 1973, Schwander and Wegman 1973).

Although a smaller dose of human than of animal antitoxin may be used in prophylaxis, this may not be true in treatment, where the main effect sought is probably the rapid neutralization of toxin that is free in the tissues before it is taken up into the central nervous system. Since the neutralizing effect of human and animal antitoxins, unit for unit, is the same, the therapeutic effect might be expected to be the same, a view supported by experiments in mice (Smith and Schallibaum 1970). On the other hand, as mentioned above, the therapeutic value of less than 10 000 units has not been examined in controlled clinical trials; small doses of human antitoxin may, therefore, perhaps be sufficient. After the initial neutralization of all free toxin in a tetanus patient, toxin subsequently formed in the wound could presumably be dealt with adequately by a small dose of homologous

antitoxin, such as 250 units, which should provide protection for about 4 weeks.

Summing up, we conclude that antitoxin probably exerts some therapeutic effect, but that this is mainly limited to the neutralization of toxin that has not yet reached the central nervous system. For this purpose a single dose of 10 000 units should be injected intravenously at the earliest possible moment. Larger doses of antitoxin are not called for, and only when the surgical treatment of the wound is unsatisfactory is there any need to give a second injection later. Human antitoxin should be used, when it is available, in the same dose as that recommended for horse serum antitoxin. It too should be given intravenously and, with the preparations usually available, slowly.

Intrathecal antitoxin

The experiments of Sherrington (1917) in monkeys (Table 64.4) clearly showed that antitoxin had a greater therapeutic effect when it was given intrathecally than by other routes.

Similar findings have been reported by others (Roux and Borrel 1898, Park and Nicoll 1914, Firor 1940, Friedemann and Traub 1949), but some workers (Permin 1914, Florey and Fildes 1927, Fedinec and Matzke 1959, Habermann and Erdmann 1978) found no advantage in the use of intrathecal antitoxin. However, failure to demonstrate an advantage for the intrathecal route might be due to a failure to treat at the optimum time (Friedemann *et al.* 1939). Smith (1966) found that intracerebral antitoxin was able to protect mice 6 hours after the subcutaneous injection of toxin, when the same dose of antitoxin intravenously was ineffective (see also Kryzhanovsky 1973). At two hours after the toxin injection the two routes were equally effective and at 12 hours equally ineffective. The intrathecal route might be expected to have an advantage only in a small proportion of patients treated at a critical stage in the disease, when a lethal dose of toxin had passed from the blood stream into the central nervous system but not become 'fixed' and inaccessible to neutralization.

Despite a few reports of its effectiveness in man (Paterson 1930, Nabarro 1932, Yodh 1932, Sanders *et al.* 1977) the intrathecal route has been little used, because an intrathecal injection may provoke convulsions (Dietrich 1940, Spaeth 1941), and because horse serum can be a meningeal irritant (Andrewes 1917, Flexner and Amoss 1917). However, the adoption of treatment by sedation and curarization, and the availability of human antitoxin, has permitted the question to be re-explored. Vakil and his colleagues (1979) and Sedaghatian (1979) found no benefit when intrathecal antitoxin was given to patients in addition to intravenous antitoxin. These studies both concerned severe tetanus, but in early tetanus Gupta and his colleagues (1980) reported only one fatality in 49 patients given intrathecal antitoxin compared with 10 in 48 patients given intramuscular antitoxin.

Active immunization

Since the demonstration of the antitoxigenic value of toxoid (formaldehyde-treated toxin; Descombey

1924) and of toxoid precipitated with alum (Bergey 1934), aluminium hydroxide or phosphate (Tasman and van Ramshorst 1952), or calcium phosphate (Relyveld *et al.* 1969), these agents have been widely tested in man. Owing to the low incidence of tetanus in peace-time, the efficacy of active immunization has been for the most part judged by the antitoxin response in the inoculated subject, but evidence from the 1939-45 war showed that it was highly effective in preventing tetanus.

British troops in the 1939-45 war were given two doses of 1 ml of toxoid at an interval of six weeks and a reinforcing dose nine months later; this was extended in 1942 to an annual reinforcing dose. On wounding, the men received also 3000 units of antitoxin, repeated weekly in those not actively immunized, until healing was apparent. During 1939-40, although many of the wounded failed to get prophylactic antitoxin, there were in France no cases in over 16 000 actively immunized British soldiers; 8 cases occurred in 18 000 who had not been immunized (Bensted 1940). In the same area, 80 cases of tetanus occurred in the German land forces, who had only antitoxin prophylaxis; but none in the German air forces, who had been actively immunized (Long and Sartwell 1947).

In the United States army three doses of 1 ml of toxoid were given at intervals of 3 weeks, and a reinforcing dose a year later or before military operations. A reinforcing dose was also given at the time of wounding. In the United States army on all fronts there were 0.044 cases per 10 000 wounded, as compared with 0.6 per 10 000 in the 1914-18 war (Scheibel 1955). Among the 22 million members of the American Forces there were only about 50 cases of clinical tetanus; and among 2.5 million wounded only about 12, of whom 8 had not received all their toxoid injections (Long and Sartwell 1947). In contrast, the Axis powers, who did not practise routine active immunization, had about 50 000 cases (Robles *et al.* 1967). The practice of relying for specific protection solely on a reinforcing dose of toxoid at the time of wounding has been criticized on the grounds that the antitoxin response takes 2–5 days to develop, which may be too long when the incubation period is less than the average of 7–10 days; and that gross shock, chill, fatigue or malnutrition may inhibit the antibody response. It is probable, however, that in most properly immunized persons the antitoxin content of the blood before reinforcement is sufficient for protection; and the results in armies where active immunization was compulsory and no prophylactic antitoxin was used clearly justify the procedure. The improvement on the results in 1914-18 cannot be attributed to fighting on sparsely contaminated terrain. Thus Glenn (1946) in Manila recorded that, among over 1100 non-immunized civilians admitted to hospital for treatment in the same area as that occupied by the US troops, 156 cases of tetanus occurred. Long (1948) cited an incidence of 1 per 100 wounded in the non-immunized Japanese army during 1940–44; though here again differences in the efficacy of non-specific prophylaxis must also be considered.

A reduction in the incidence of tetanus associated with the introduction of active immunization is reported for a number of civilian communities (see Scheibel 1955, Edsall 1959, Petrella 1960, Ebisawa and Fukutomi 1979); this is, at least in part, attributable to immunization.

Tetanus toxoid

Formol toxoid is produced from the tetanus toxin released by the bacilli in liquid culture in large capacity fermenters (Hepple 1968). The choice of medium and of the strain of *Cl. tetani* used is important, both to ensure a high yield of toxin and freedom from protein in the final product, which might lead to hypersensitivity reactions (Mueller and Miller 1940, 1954, Latham *et al.* 1962, Nielsen 1967, Bizzini *et al.* 1969). The crude toxin, or the toxoid, may be purified before being adsorbed with a suitable adjuvant (Latham *et al.* 1965). It is often used combined with diphtheria and pertussis vaccines, and in some countries with killed poliomyelitis vaccines.

Immunization course

The basic course for immunization adopted in many countries consists of three spaced injections. Although a slow antitoxin response usually occurs after one injection of adsorbed toxoid and two injections suffice to induce a satisfactory immunity (Newell *et al.* 1966, MacLennan *et al.* 1973, Ruben *et al.* 1978, Black, Huber and Curlin 1980, Breman *et al.* 1981), the third injection is usually advised to ensure a long lasting immunity. The intervals recommended in Britain, for example, are 4–6 weeks between the first and second injections and 6–12 months between the second and third. The course adopted for immunization of infants in some countries—the USA and Canada for example—comprises four injections of combined diphtheria, tetanus and pertussis vaccine, three doses being given at intervals of 4–8 weeks and the fourth dose about 1 year after the third (Report 1977, National Advisory Committee on Immunization 1979).

There is little doubt that the injection into man of two to three doses of toxoid at properly spaced intervals raises the antitoxin content of the blood within a few weeks to prophylactic concentrations. There is good reason to suppose that a given blood content indicates a higher degree of immunity in the actively than in the passively immunized person. Thus d'Antona and Valensin (1937) found that actively immunized guinea-pigs with blood concentrations of 0.012 to 0.05 unit per ml withstood 200 MLD of toxin; passively immunized animals with 0.2 to 0.5 units per ml either died or had severe local tetanus. The difference may in part be due to a better distribution of antitoxin in the actively immunized animal (see also Sneath *et al.* 1937, Bergey *et al.* 1939, Bernard and Servant 1957). The toxin produced in the test infection may in these circumstances induce a secondary antitoxin response (Jones and Jamieson 1936, Wolters and Dehmel 1940, Zuger *et al.* 1942), but the effect is variable. Moreover, the response is slow, and may therefore be too late to affect the issue. Ipsen (1961), however, found in rabbits that immunity to toxin rose within a few hours of a secondary stimulus by toxoid, although a rise in serum antitoxin was not evident until the third day. Clearly the lag in detectable antitoxin response does not necessarily reflect absence of increase in total body immunity. In this connection it is noteworthy that in both animals and man the antigenic stimulus of an attack of tetanus is insufficient to induce either a primary antitoxin response, or secondary responsiveness to toxoid (see, e.g., Ramon and Lemétayer 1935, Wolters and Dehmel 1942, Turner *et al.* 1958, Tasman 1959). The same observation has been made in relation to diphtheria.

The duration of immunity provided by the basic course of immunization can be long lasting (Trinca 1967), and may be lifelong (Rubbo 1966), though necessarily becoming weaker. The duration is likely to be greater after the use of toxoids of high Lf and aluminium adjuvant content (Trinca 1965, White *et al.* 1969, MacLennan *et al.* 1973), but these advantages are liable to be offset to some extent by an increased incidence of side-effects of vaccination (see below). Since toxoids in use in different countries may differ in these attributes, as well as in the number and spacing of the injections, results from different studies cannot be strictly compared, but numerous observations have shown that even after 10–20 years the antitoxin content of the blood serum is usually above that generally accepted as indicative of immunity, namely 0.01 per unit per ml (Scheibel 1955, Eckmann 1959, Edsall 1959, Gottlieb *et al.* 1964, Scheibel *et al.* 1966, Trinca 1967).

According to Gottlieb and his colleagues (1964) circulating antitoxin falls gradually after primary immunization for about ten years and then remains more or less stationary. However, there is variability in the response to primary and secondary immunization in different subjects (see Ramon and Zoeller 1933, Evans 1943, MacLennan *et al.* 1973). Thus, among the 191 children studied by Scheibel and her colleagues (1966) about 12 years after primary immunization with three doses of adsorbed toxoid, although most were immune with a mean serum antitoxin of 0.45 unit per ml, 4.2 per cent had less than 0.01 unit per ml. Poor response has been associated with a particular immunoglobulin allotypic marker (Schanfield *et al.* 1979). The proportion of unsatisfactory responders is liable to be greater after the use of plain toxoid. Thus White and his colleagues (1969) found that 12 of 91 adults immunized with three spaced doses of plain toxoid had less than 0.01 unit of antitoxin in their sera 2–4 years after immunization, compared with only 1 of 80 subjects who had had adsorbed toxoid.

The response of the immune mechanism to a reinforcing dose of toxoid in a subject conditioned by primary immunization is striking. Within 3–4 days the antitoxin content of the serum rises and within 10 days often reaches a titre far higher than after primary immunization (Miller *et al.* 1949, Bigler 1951, Regamey and Schlegel 1951, Ellis 1963, Gottlieb *et al.* 1964, Robles *et al.* 1967). Regamey and Schlegel's observations are reproduced in Table 64.7. The response to a reinforcing dose is said to vary directly with the pre-injection titre (Gottlieb *et al.* 1964). There appears to be no difference in the rapidity with which an antitoxin response can be detected after a reinforcing dose of plain or adsorbed toxoid (Trinca 1965).

In view of these considerations there seems little justification for giving repeated reinforcing injections as a routine. At the most, one every ten years should suffice to keep the antibody content well above the minimal protective concentration. A reinforcing dose may be required after an injury that may be followed by tetanus, but in view of the increasing number of reports of adverse reactions a reinforcing dose is not generally advisable within one year (Report 1972a) or five years (Report 1972b, Smith *et al.* 1975) of a previous injection.

Reactions to tetanus toxoid

Purified fluid tetanus toxoid, prepared with protein-free media, is one of the least irritating of vaccines. The primary series of injections seldom causes trouble. In a study of 209 casualty patients at Sheffield only five delayed reactions were noted—three local and two general (Cox *et al.* 1963). But with repeated reinforcing doses allergy is liable to develop and increase in intensity till injections are followed by severe local and sometimes general reactions (Schneider 1964, Edsall *et al.* 1967). Very occasionally a reaction of the anaphylactic type occurs in a sensitized subject (Fardon 1967). It is wise to watch all persons for 10 minutes after injection. Adsorbed toxoid is more irritant. Mahoney and his colleagues (1967), for example, reported 16 per cent of delayed local reactions after its use. White and his colleagues (1973) reported that among a factory population local reactions occurred in 0.3 per cent of adults after the first injection of the primary course of adsorbed toxoid, in 2.7 per cent after

Table 64.7 Showing the duration of responsiveness to reinforcing doses of tetanus toxoid induced by three primary injections (after Regamey and Schlegel 1951)

Years since primary inoculations	Number of subjects tested	Mean number of antitoxin units per ml of serum at the stated time (days) after reinforcing dose			
		0	4	6	8
1	26	0.70	1.27	9.9	18.9
2–3	10	0.52*	2.55	39.6	69.4
4–7	6	0.55	1.32	17.6	63.4
8–9	10	0.22	0.74	25.1	86.0
10	8	0.28	0.63	16.5	90.0

*Minimum individual serum antitoxin in this group, 0.007 5; in all other groups minima were 0.035 or more.

the second dose, and in 7.5 per cent after the third. The rate was 1.6 per cent after post-injury reinforcing doses. Reactions are more frequently seen in females than males, and are more likely to follow subcutaneous than intramuscular injections (Relihan 1969). Side-effects may have become commoner owing to the increasing use of active immunization, especially of reinforcing doses after injury (Collier *et al.* 1979). Although adsorbed toxoid is more liable to cause reactions, it is more effective, both for primary immunization and for reinforcement of existing immunity. The frequency of reactions is greater when vaccines containing a higher Lf content are used. As with diphtheria (see Chapter 53), low-dose adsorbed toxoid containing, for example, 1 Lf of antigen, has been found to minimize reactions to reinforcing doses while stimulating a satisfactory antibody response (McComb and Levine 1961, Trinca 1963). Care should be taken to keep tetanus toxoid at a lowish temperature; if kept for long at 37° it may regain part of its toxicity (Akama *et al.* 1971). Local reactions to purified toxoid are probably the result of an immunologically mediated reaction to the toxoid antigen itself rather than to impurities in the vaccine. Thus, reactors usually have a high titre of serum antitoxin, or respond rapidly to the vaccine (Edsall *et al.* 1967). Positive skin tests, with the characters of delayed type hypersensitivity, to highly purified toxoid may be found in reactors who do not respond to control tests with purified medium (White *et al.* 1973); and IgE responses occur to reinforcing doses of toxoid (Nagel *et al.* 1977). However, adjuvants contribute to local reactions, and merthiolate sensitivity sometimes plays a part (Hansson and Möller 1971). Toxoid has been reported to increase anti-A and anti-B antibodies, owing to the presence of traces of blood-group antigens in the vaccine. Immunization of pregnant women with the aim of preventing *tetanus neonatorum* may, therefore, increase the risk of haemolytic disease in the newborn, although no published reports of such a side-effect have been made (Gupte and Bahtia 1979).

Effect of simultaneous immunization with other antigens

It is convenient to mix toxoid with enteric vaccines, pertussis vaccine, diphtheria toxoid or other antigens. The response to toxoid is reported by Maclean and Holt (1940), Miller and Saito (1942), Greenberg and Fleming (1947, 1948), and Regamey and Schlegel (1951) to be enhanced by the presence of pertussis vaccine. Scheibel (1944) recorded a depression of the antigenic potency of both diphtheria and tetanus toxoids (adsorbed by $Al(OH)_3$) when they were injected together into guinea-pigs. Barr and Llewellyn-Jones (1953) observed a depression of the response when toxoid and enteric vaccines were given in certain proportions to guinea-pigs. This occurred especially in animals that had had primary inoculations with the enteric vaccine and in which the secondary response to the vaccine 'crowded out' the response to toxoid. A similar crowding out of primary responses to tetanus toxoid by secondary responses to diphtheria toxoid was observed in children, though not to the extent of diminishing the secondary responsiveness to later injections of tetanus toxoid (Chen *et al.* 1956, 1957). In man, enteric vaccines are reported to have either no effect (Hegyessey *et al.* 1956), or a definite adjuvant effect, on toxoid administered with them (Ikić 1958). Multiple immunizing antigens should not be given without considering the nature and proportions of the differ-

ent antigens in the mixture and the immune state of the subject. The frequency of reactions to the diphtheria component of combined diphtheria and tetanus toxoids has led, in the USA, to the use in adults of a combined preparation containing the normal amount of tetanus toxoid and a reduced amount of diphtheria toxoid (Report 1972*b*). A combined preparation of enteric vaccines with fluid tetanus toxoid has also been found effective in stimulating a satisfactory antitoxin response when given in an intradermal dose of 0.1 ml containing 2 Lf (Barr *et al.* 1959), though the response to the enteric components has not been subjected to clinical trial of the protective effect.

Formerly active immunization was confined mainly to the fighting services. The wisdom of immunizing the whole population is now realized, however, and most countries include tetanus toxoid in the vaccines routinely offered to children. The immunization of adults should especially be directed to those at increased risk, such as agricultural, industrial, and road workers, persons with chronic ulcers of the lower limbs, and injured patients. Despite such efforts many people will remain unprotected (Crossley *et al.* 1979). In Britain, older women who did not serve in the forces during the second world war are often susceptible, and some of the small number of cases that have occurred in recent years are in this group, often as a result of gardening injuries (Report 1971, Atrakchi and Wilson 1977).

Pregnant women should be immunized, to protect the infant against *tetanus neonatorum* by placentally transmitted maternal antitoxin rather than the mother against puerperal tetanus (MacLennan *et al.* 1965, Newell *et al.* 1966). Two spaced doses of adsorbed toxoid should be given during pregnancy, if possible before the sixth month (Report 1967). Good results have been reported with a single dose of calcium phosphate-adsorbed vaccine containing 30 Lf of toxoid (Kielmann and Vohra 1977).

With certain reservations (see below), the actively immunized person when wounded may safely be treated with a reinforcing dose of toxoid, in addition to receiving surgical care of the wound. The reliability of this procedure depends on knowing with certainty that the wounded person has undergone an effective course of immunization. Though each immunized person should have a record card, it is not likely to be available in many of the emergencies in which prophylaxis is essential; in such circumstances the patient has to be regarded as non-immunized, and the need for passive antitoxin considered.

In treatment of the wounded, prophylactic antitoxin—preferably in the form of human immunoglobulin—may be needed for those who have a wound thought liable to a risk of tetanus (see page 355) and (a) are not immunized or have had only a single dose of toxoid previously, (b) are immunized but have had no reinforcing dose for ten years, (c) whose immunization state is unknown or uncertain, or (d) are severely shocked or wounded. In other cases a dose of toxoid is all that is necessary, but even this is not needed for patients who have had a reinforcing dose of toxoid within the preceding year (Report 1972*a*) or five years (Report 1972*b*, Smith *et al.* 1975).

Simultaneous active and passive immunization

The simultaneous injection of tetanus toxoid and antitoxin is of value to begin the active immunization in a wounded person. Where horse serum antitoxin is used, active immunization will also ensure that the patient will not need horse serum on the occasion of a future injury.

The antitoxin may interfere with the response of the body to the toxoid—the so-called blanketing effect. This has been demonstrated in the guinea-pig (see Ramon and Zoeller 1933, Wolters and Dehmel 1951, Smith 1964*a*), but in man, provided adsorbed toxoid is used, the objection is more of theoretical than of practical interest. Though the serum may delay somewhat the response to toxoid given simultaneously, the ultimate immunity after the second and third doses appears to be the same as after the administration of toxoid alone (Bergentz and Philipson 1958, Tasman and Huygen 1962, Fulthorpe 1965, Züst 1967). There is now ample evidence to support the soundness and wisdom of the procedure both with horse and human antitoxin. In practice, various doses of toxoid and antitoxin are given. Toxoid, adsorbed on to aluminium hydroxide or aluminium phosphate, is generally injected in a dose of 5–20 Lf into one arm, and 1000–1500 units of refined horse tetanus antitoxin or 250 units of human tetanus immunoglobulin are injected into the other (Suri and Rubbo 1961, McComb and Dwyer 1963, Smith *et al.* 1963, Levine *et al.* 1966, Züst 1967). The subsequent injections of toxoid are given at the usual intervals. Ideally, simultaneous immunization should ensure a continuous immunity—passive replaced by active—from the time of injection, but this is difficult to secure. Cohen and Leussink (1973) examined the effect of different adsorbed toxoids in small groups of subjects given 250 units of human antitoxin in a different limb. Vaccine that contained 5 Lf of toxoid resulted, 3 weeks after simultaneous immunization, in titres of antitoxin higher than those produced by vaccine containing 10 Lf; this was attributed to the capacity of the toxoid to bind the passive antitoxin. In patients suffering from tetanus and treated with a large dose of antitoxin, active immunization should be deferred for 3 or 4 weeks.

The assay of tetanus toxoid

The specification of fluid tetanus toxoid for use in man is based on the antigenic response in guinea-pigs. British regulations, for instance, require that not fewer than nine guinea-pigs are given injections of five times the human dose; the vaccine satisfies the requirements if, after 6 weeks, at least two-thirds of the guinea-pigs contain 0.05 or more unit of antitoxin per ml of blood serum (British Pharmacopoeia 1980). In many countries plain fluid toxoid is no longer used. The European Pharmacopoeia, for example, includes a monograph only for adsorbed toxoid, the potency of which is controlled in terms of International Units by comparison with the International Standard for Tetanus Toxoid (adsorbed). The vaccine is required to possess a minimum potency of not less than 40 units per human dose determined by comparing its ability to immunize mice or guinea-pigs with that of the International Standard (van Ramshorst and Sundaresan 1972, Mussett and Sheffield 1973, European Pharmacopoeia 1977). The potency of adsorbed toxoids determined in mice or guinea-pigs has been found to reflect immunizing potency in man (Ikić 1960, Pittman *et al.* 1970, van Ramshorst *et al.* 1973). Potency assay of fluid toxoid in mice is less consistent than that of adsorbed toxoid (Cohen *et al.* 1959, Scheibel *et al.* 1968).

The Lf value of a toxoid preparation is not a reliable guide to its antigenicity, mainly because flocculation zones are multiple, and single zones can normally be obtained only with absorbed sera (see Ramon and Descombey 1926, Glenny and Stevens 1938, Lahiri 1940, Koerber 1943, Moloney and Hennessy 1944, Regamey 1947, Nagel and Cohen 1973). With a suitable reference preparation the Lf value can be of use, however, in controlling the purity of tetanus vaccine (Spaun and Lyng 1970, Sheffield *et al.* 1979). The European Pharmacopoeia (1977) and the World Health Organization Requirements (1964) specify a minimum acceptable purity for the toxoid of 500 Lf per mg of protein nitrogen. A purity of over 1000 Lf per mg of protein nitrogen can readily be secured by salting out with ammonium sulphate.

Chemoprophylaxis

Experimental tetanus in animals can be controlled with sulphonamides (Mayer 1938, 1939, Evans *et al.* 1945) and with antibiotics (Novak *et al.* 1949, Bliss *et al.* 1950, Taylor and Novak 1951, Lennert-Petersen 1954, Anwar and Turner 1956, McDonald *et al.* 1960). The frequency of side-effects after the use of horse antitoxin stimulated the introduction of chemoprophylaxis in man (Filler and Ellerbeck 1960, Cox *et al.* 1963). Sharrard (1965) at Sheffield, for example, replaced horse antitoxin by antibiotic prophylaxis, except in patients with wounds more than 24 hours old sustained under conditions in which tetanus contamination was very likely or in which the inclusion of a foreign body in the wound was suspected. Combined with careful surgical treatment, this policy was pursued for some years without any increase in the number of tetanus cases admitted to the Royal Infirmary. The procedure was widely used in Britain and elsewhere, but with the general availability of human antitoxin the need for chemoprophylaxis has become less.

There is no question that the replacement of antitoxin by penicillin or erythromycin (Lowbury *et al.* 1978) will save a great many patients from unpleasant serum reactions, but whether it will prove as effective prophylactically is doubtful. Tetanus bacilli are inhibited by a concentration of penicillin of 0.05 unit per ml, but are not completely killed even by one of 2 units per ml. Experiments on mice showed that unless penicillin was given within 6 hours of infection it could not prevent tetanus; and that unless the concentration of penicillin in the blood serum was maintained at 0.1 unit per ml or over until the wound was healed, the resistant spores might germinate after the concentration had fallen below this figure and lead to fatal tetanus. Antitoxin, on the other hand, was effective when given even 20 hours after infection (Smith 1964*b*). In the presence of penicillinase-producing staphylococci, penicillin failed to prevent tetanus even when given at the time of infection. In practice, several cases of tetanus have occurred after antibiotic prophylaxis, and some cases after prophylactic treatment with antitoxin; though Ellis (1964) at Leeds stated that not a single patient out of about 100 000 treated with antitoxin during the previous 15 years had contracted tetanus. Ellis further pointed out that satisfactory surgical treatment of a wound, for example by excision, was not always possible. He instanced penetrating wounds of the hand or foot in which excision would lead to unsightly and possibly disabling scars.

Working with horses, Corréa and Tavares (1967) reported that, of six animals infected with tetanus spores along with activating aerobic organisms and given 18 000 units of antitoxin subcutaneously at the time of infection or 12 hours later, not one contracted tetanus. On the other hand, of two horses given 7 200 000 units of penicillin G benzathine 12 hours before infection, and of five given the same dose at the time of infection, every one contracted tetanus.

In man an informative comparison was made by Lucas and Willis (1965) at the University College Hospital of Ibadan during the 4-year period 1959 to 1962. During the first part of the period, when serum prophylaxis was used, the incidence of tetanus in the treated patients was 2.3 per 10 000; during the second period when antitoxin was replaced by penicillin it was 7.8 per 10 000; and during the third period, when antitoxin was reverted to it was 2.4 per 10 000. Sequential comparisons of this sort are never entirely satisfactory, and the total number of cases of tetanus in the series during the 4 years was only twenty. Chance may therefore have played a part, but the results nevertheless point to the presumptive superiority of antitoxin prophylaxis. Side-effects provide the main drawback to the use of horse antitoxin, but it must be remembered that antibiotics are also troublesome and sometimes dangerous in this respect. It might well be that, provided adequate surgical treatment of the wound was practicable and that antibiotic treatment was started before significant toxin production had occurred and continued until the wound had healed, antibiotic prophylaxis would prove to be satisfactory. In mice Smith and MacIver (1969) found that the amount of antitoxin required to protect was far less when penicillin was injected simultaneously, and there may be value in supplementing antitoxin prophylaxis with antibiotic in patients with badly soiled wounds. Many such patients might in any case be given antibiotics to prevent infections by bacteria other than *Cl. tetani*. Combined chemotherapy and serotherapy may well be the best treatment available—penicillin to prevent multiplication of the organisms and serum to neutralize any toxin formed.

Chemotherapy Once clinical signs of tetanus have appeared, it is doubtful whether antibiotics can do much to influence the course of events. Antibiotics should, however, be capable of preventing further multiplication of the bacilli. Antibiotic treatment may, of course, also be indicated for dealing with sepsis in the wound and for prophylaxis and treatment of respiratory infection, especially in curarized patients.

Tetanus in animals

The natural disease attacks horses more frequently than other animals, and is especially common in warm and tropical countries. In Northern Europe the incidence is low; thus during 1899–1908 in the Prussian Army the annual average number of equine cases was only 0.6 per 1000 (Hutyra and Marek 1912). Tetanus occurs less frequently in cattle, sheep and goats, though it is not uncommon in ewes after parturition and in lambs after docking and castration. It is less common in the pig and uncommon in the dog. Infection occurs by contamination of wounds of skin, or of raw surfaces; nails in the hoof, castration, tail docking, harness galling, bone fractures, and imperfect treatment of the navel at birth, seem to be responsible for the majority of cases.

Prophylactic injection of antitoxin is recommended in operations on the horse. Nocard (see Mohler and Eichhorn 1911) reported that in 2727 horses given serum after various operations not a single case of tetanus occurred, whereas during the same period 259 cases occurred in horses that were not inoculated (number not stated). A dose of not less than 3000 units is recommended. The passive immunity so conveyed lasts longer in the horse than in other animals because the antibodies are homologous, and are therefore destroyed more slowly (Mohler and Eichhorn 1911). The recommended prophylactic dose for pigs, calves and sheep is not less than 500 units, and for piglets, lambs and dogs, not less than 250 units.

Buxton and Glenny (1921) immunized horses actively by three injections at intervals of 3 days of a toxin-antitoxin mixture. A month later the animals withstood 2000 guinea-pig MLD of crude toxin (see also Glenny, Hamp and Stevens 1932).

Tetanus toxoid is an effective prophylactic in the horse (Ramon and Descombey 1925) and other domestic animals. Two doses of 1–2 ml of purified toxoid adsorbed with aluminium phosphate, four weeks or more apart, provide a satisfactory immune response without side-effects; reactions to less pure toxoids may be troublesome in thoroughbred horses (Kerry *et al.* 1976). Lambs and sheep respond satisfactorily to tetanus toxoid (Adey and Kennedy 1939), which may be given as a component of a multiple clostridial vaccine adsorbed with alum (Sterne *et al.* 1962); it has also been given in Freund's incomplete adjuvant, the water-in-oil emulsion being injected intraperitoneally (Thompson *et al.* 1969).

In horses, the therapeutic effect of antitoxin is dubious (Mohler and Eichhorn 1911). A dose of not less than 10 000 units followed by daily injections of 1–3 million units has been recommended.

References

Abel, J. J. (1934) *Science* **79**, 63, 121.
Abel, J. J. and Chalian, W. (1938) *Johns Hopk. Hosp. Bull.* **62**, 610.

Abel, J. J., Firor, W. M. and Chalian, W. (1938) *Johns Hopk. Hosp. Bull.* **63**, 373.

Adams, E. B. (1968) *S. Afr. med. J.* **42**, 739.

Adams, E. B., Laurence, D. R. and Smith, J. W. G. (1969) *Tetanus*. Blackwell Scientific Publications. Oxford.

Adams, J. Q. and Morton, R. F. (1955) *Amer. J. Obstet. Gynec.* **69**, 169.

Adey, C. W. and Kennedy, M. (1939) *Aust. vet. J.* **15**, 205.

Akama, K., Ito, A., Yamamoto, A. and Sadahiro, S. (1971) *Jap. J. med. Sci. Biol.* **24**, 181.

Alving, B. M., Tankersley, D. L., Mason, B. L., Rossi, F., Aronson, D. L. and Finlayson, J. S. (1979) *Thromb. and Haem.* **42**, 253. (Abstract)

Ambache, N., Morgan, R. J., and Payling-Wright, G. (1948*a*) *J. Physiol.* **107**, 45; (1948*b*) *Brit. J. exp. Path.* **29**, 408.

Andrewes, F. W. (1917) *Lancet* **i**, 682.

Anwar, A. A. and Turner, T. B. (1956) *Johns Hopk. Hosp. Bull.* **98**, 85.

Armitage, P. and Clifford, R. (1978) *J. infect. Dis.* **138**, 1.

Atrakchi, S. A. and Wilson, D. H. (1977) *Brit. med. J.* **i**, 179.

Axnick, N. W. and Alexander, E. R. (1957) *Amer. J. publ. Hlth* **47**, 1493.

Baker, A. B. (1942) *J. Neuropath. exp. Neurol.* **1**, 394.

Bandmann, F. (1953) *Z. Hyg. InfektKr.* **136**, 559.

Barr, M. and Llewellyn-Jones, M. (1953) *Brit. J. exp. Path.* **34**, 12.

Barr, M. and Sachs, A. (1955) *Report Army Pathology Advisory Committee*. W.O. Code No. 1162, War Office, London.

Barr, M., Sayers, M. H. P. and Stamm, W. P. (1959) *Lancet* **i**, 816.

Barundun, S., Kistler, P., Jeunet, F. and Isliker, H. (1962) *Vox Sang.* **7**, 157.

Bauer, J. H. and Meyer, K. F. (1926) *J. infect. Dis.* **38**, 295.

Baylis, J. H. *et al.* (1952*b*) *J. Path. Bact.* **64**, 47.

Baylis, J. H., Mackintosh, J., Morgan, R. S. and Wright, G. P. (1952*a*) *J. Path. Bact.* **64**, 33.

Bazy, M. (1914) *C. R. Acad. Sci., Paris* **159**, 794.

Behring (1892) *Z. Hyg. InfektKr.* **12**, 45.

Behring and Kitasato, S. (1890) *Dtsch. med. Wschr.* **16**, 1113.

Bensted, H. J. (1940) *Lancet* **ii**, 788.

Bergentz, S. E. and Philipson, L. (1958) *Acta chir. scand.* **116**, 58.

Berger, S. A., Cherubin, C. E., Nelson, S. and Levine, L. (1978) *J. Amer. med. Ass.* **240**, 769.

Bergey, D. H. (1934) *J. infect. Dis.* **55**, 72.

Bergey, D. H., Brown, C. P. and Etris, S. (1939) *Amer. J. publ. Hlth* **29**, 334.

Bernard, J. G. and Servant, P. (1957) *Rev. Immunol., Paris* **21**, 188.

Bezjak, B. and Fališevac, J. (1955) *Archiv. Hig. Rada., Zagreb* **6**, 115.

Bianchi, G. (1962) *Helv. med. Acta* **1**, 1.

Bigler, J. A. (1951) *Amer. J. Dis. Child.* **81**, 226.

Binns, P. M. (1961) *Brit. J. prev. soc. Med.* **15**, 180.

Bizzini, B. (1977) The Specificity and Action of Animal, Bacterial and Plant Toxins. In: *Receptors and Recognition. Ser. B.* Vol. 1, p. 1. Ed. by P. Cuatrecasas. Chapman and Hall, London.

Bizzini, B., Turpin, A. and Raynaud, M. (1969) *Ann. Inst. Pasteur* **116**, 686.

Black, R. E., Huber, D. H. and Curlin, G. T. (1980) *Bull. World Hlth Org.* **58**, 927.

Blake, P. A. and Feldman, R. A. (1975) *J. infect. Dis.* **131**, 745.

Bliss, E. A., Warth, P. T. and Chandler, C. A. (1950) *Ann. N.Y. Acad. Sci.* **53**, 277.

Bonney, V., Box, C. and MacLennan, J. (1938) *Brit. med. J.* **ii**, 10.

Boyd, J. S. K. (1946) *Lancet* **ii**, 113.

Boyd, J. S. K. and MacLennan, J. D. (1942) *Lancet* **ii**, 745.

Breman, J. G., Wright, G. G., Levine, L., Latham, W. C. and Compaoré, K. P. (1981) *Bull. World Hlth Org.* **59**, 745.

British Pharmacopoeia (1980), II, 879, H.M.S.O., London.

Brooks, V. B., Curtis, D. R. and Eccles, J. C. (1957) *J. Physiol.* **135**, 655.

Brown, A. *et al.* (1960) *Lancet* **ii**, 227.

Bruce, D. (1915) *Lancet* **ii**, 901; (1916) *Ibid.* **ii**, 929; (1917*a*) *Brit. med. J.* **i**, 118; (1917*b*) *Lancet* **i**, 986, **ii**, 411, 925; (1919) *Ibid.* **i**, 311; (1920) *J. Hyg., Camb.* **19**, 1.

Bruschettini, A. (1890) *La Riform. med.* **6**, 1346; (1892) *Ibid.* **8**, 256.

Buchanan, T. M., Brooks, G. F., Martin, S. and Bennett, J. V. (1970) *J. infect. Dis.* **122**, 564.

Buff, B. H. (1960) *J. Amer. med. Ass.* **174**, 1200.

Bulloch, W. (1929) *Spec. Rep. Ser. med. Res. Coun., Lond.* No. 138.

Bullock, W. E. and Cramer, W. (1919) *Proc. roy. Soc. B* **90**, 513.

Buxton, J. B. and Glenny, A. T. (1921) *Lancet* **ii**, 1109.

Bytchenko, B. (1972) In: *Third International Conference on Tetanus*, San Paulo, 1970, p. 17. Pan American Health Org., Publ. no. 253, Washington.

Campbell, J. A. and Fildes, P. (1931) *Brit. J. exp. Path.* **12**, 77.

Carbone, T. and Perrero, E. (1895) *Zbl. Bakt.* **18**, 193.

Carle and Rattone (1884) *G. Accad. Med. Torino*, 3rd Ser. **32**, 174.

Chen, B. L. *et al.* (1956) *J. Immunol.* **77**, 144; (1957) *Ibid.* **79**, 393.

Cherubin, C. E. (1967) *Arch. environm. Hlth* **14**, 802.

Cinader, B. and Weitz. R. (1953) *J. Hyg., Camb.* **51**, 293.

Cohen, H. and Leussink, A. B. (1973) *J. biol. Stand.* **1**, 313.

Cohen, H., Ramshorst, J. D. van and Tasman, A. (1959) *Bull. World Hlth Org.* **20**, 1133.

Cole, L. (1940) *Lancet* **i**, 164.

Coleman, G. E. (1931) *Amer. J. Hyg.* **14**, 515.

Coleman, G. E. and Meyer, K. F. (1926) *J. infect. Dis.* **39**, 332.

Collier, L. H., Polakoff, S. and Mortimer, J. (1979) *Lancet* **i**, 1364.

Conybeare, E. T. (1959) *Proc. roy. Soc. Med.* **52**, 112.

Corréa, A. and Tavares, J. (1967) *Rev. Inst. Med. trop., S. Paulo* **9**, 309.

Couermont, J. and Doyon, M. (1899) In: *Le Tetanos*, p. 53. Baillière, Tindall and Cox, Paris.

Cox, C. A., Knowelden, J. and Sharrard, W. J. W. (1963) *Brit. med. J.* **ii**, 1360.

Creech, O., Glover, A. and Ochsner, A. (1957) *Ann. Surg.* **146**, 369.

Crossley, K., Irvine, P., Warren, J. B., Lee, B. K. and Mead, K. (1979) *J. Amer. med. Ass.* **242**, 2298.

Cuboni, E. (1957) *Boll. Ist. sieroterap.* **36**, 1.

Curtis, D. R. and Groat, W. C. de (1968) *Brain Res.* **10**, 208.

D'Antona, D. and Valensin, M. (1937) *Rev. Immunol., Paris* **3**, 437.

Davies, J. R. *et al.* (1954) *Arch. int. Physiol.* **62**, 248.

De, S. (1966) *Calcutta med. J.* **63**, 302.

Dean, H. R. (1917) *Lancet* i, 673.

Descombey, P. (1924) *C. R. Soc. Biol.* **91**, 239.

Diamond, J. and Mellanby, J. M. (1971) *J. Physiol.* **215**, 727.

Dietrich, H. F. (1940) *Amer. J. Dis. Child.* **59**, 693.

Dimpfel, W., Huang, R. T. C. and Habermann, E. (1977) *J. Neurochem.* **29**, 329.

D'Sa, J. A., Dastur, F. D., Awatramani, V. P., Dixit, S. K. and Nair, K. G. (1978) *J. ass. Phys. Ind.* **26**, 891.

Ebisawa, I. and Fukutomi, K. (1979) *Jap. J. exp. Med.* **49**, 131.

Eckmann, L. (1959) *Schweiz. med. Wschr.* **89**, 311; (1967) In: *Principles on Tetanus.* Proc. int. Conf. Tetanus, p. 576. Huber, Berne.

Editorial (1973) *Med. J. Aust.* **ii**, 303.

Edmonson, R. S. and Flowers, M. W. (1979) *Brit. med. J.* **i**, 1401.

Edsall, G. (1959) *J. Amer. med. Ass.* **171**, 417; (1967) *Arch. environm. Hlth* **15**, 473.

Edsall, G., Elliott, M. W., Peebles, T. C., Levine, L. and Eldred, M. C. (1967) *J. Amer. med. Ass.* **202**, 17.

Eldridge, P. L. and Entwhistle, C. C. (1975) *Vox Sang.* **28**, 62.

Ellis, M. (1963) *Brit. med. J.* **i**, 1123; (1964) *Ibid.* **i**, 1438.

Erdmann, G., Weigand, H. and Wellhöner, H. H. (1975) *Naunyn-Schmeideberg's Arch. Pharmacol.* **290**, 357.

European Pharmacopoeia (1977). Suppl. to Vol III, p. 174. Maisonneuve S.A., France.

Evans, D. G. (1941) *Lancet* **ii**, 628; (1943) *Ibid.* **ii**, 316.

Evans, D. G., Fuller, A. T. and Walker, J. (1945) *Lancet* **ii**, 336.

Fardon, D. E. (1967) *J. Amer. med. Ass.* **199**, 125.

Fedinec, A. A. (1967) In: *Principles on Tetanus.* Proc. int. Conf. Tetanus, p. 169. Huber, Berne; (1972) In: Proc. 3rd int. Conf. Tetanus, p. 63. *Pan American Hlth Org. Sci. Publ.* No. 253.

Fedinec, A. A. and Matzke, H. A. (1959) *J. exp. Med.* **110**, 1023.

Fildes, P. (1925) *Brit. J. exp. Path.* **6**, 62; (1927) *Ibid.* **8**, 387.

Filler, R. M. and Ellerbeck, W. (1960) *J. Amer. med. Ass.* **174**, 1.

Firor, W. M. (1940) *Arch. Surg.* **41**, 299.

Firor, W. M., Lamont, A. and Shumacker, H. B. (1940) *Ann. Surg.* **111**, 246.

Flexner, S. and Amoss, H. L. (1917) *J. exp. Med.* **25**, 525.

Florey, H. and Fildes, P. (1927) *Brit. J. exp. Path.* **8**, 393.

Francis, E. (1924) *Bull. U.S. hyg. Lab.* No. 95.

Fraser, D. W. (1972) *Amer. J. Epidem.* **96**, 306.

Friedemann, U. and Hollander, A. (1943) *J. Immunol.* **40**, 325.

Friedemann, U., Hollander, A. and Tarlov, I. M. (1941) *J. Immunol.* **40**, 325.

Friedemann, U. and Traub, F. B. (1949) *J. Immunol.* **63**, 23.

Friedemann, U., Zuger, B. and Hollander, A. (1939) *J. Immunol.* **36**, 473, 485.

Fulthorpe, A. J. (1957) *J. Hyg., Camb.* **55**, 382; (1958) *Ibid.* **56**, 183; (1965) *Ibid.* **63**, 243.

Galbraith, N. S., Forbes, P. and Tillett, H. (1981) *J. Infect.* **3**, 181.

Gilles, E. C. (1937a) *J. Amer. med. Ass.* **109**, 484; (1937b) *Amer. J. Hyg.* **26**, 394.

Glaser, J. and Wyss-Souffront, W. A. (1961) *Pediatrics, Springfield* **28**, 367.

Glenn, F. (1946) *Ann. Surg.* **124**, 1030.

Glenny, A. T. *et al.* (1926) *J. Path. Bact.* **29**, 31.

Glenny, A. T., Hamp, A. G. and Stevens, M. F. (1932) *Vet. J.* **88**, 90.

Glenny, A. T. and Stevens, M. F. (1938) *J. R. Army med. Cps* **70**, 308.

Godfrey, M. P., Parsons, V. and Rawstron, J. R. (1960) *Lancet* **ii**, 1229.

Goldie, H., Parsons, C. H. and Bowers, M. S. (1942) *J. infect. Dis.* **71**, 212.

Gottlieb, S., McLaughlin, F. X., Levine, L., Latham, W. C. and Edsall, G. (1964) *Amer. J. publ. Hlth* **54**, 961.

Goulon, M., Girard, D., Grosbius, S., Desormeau, J. -P. and Capponi, M. F. (1972) *Nouv. Presse méd.* **1**, 3049.

Green, J., Erdmann, G. and Wellhöner, H. H. (1977) *Nature, Lond.* **265**, 370.

Greenberg, L. and Fleming, D. G. (1947) *Canad. J. publ. Hlth* **38**, 279; (1948) *Ibid.* **39**, 131.

Grünbaum, B. and Hertwig, F. (1959) *Z. f.d.g. Hyg. Grenzgebiete* **5**, 303.

Gumprecht, F. (1894) *Dtsch. med. Wschr.* **20**, 546.

Gupta, P. S., Kapoor, R., Goyal, S., Batra, V. K. and Jain, B. K. (1980) *Lancet* **ii**, 439.

Gupte, S. C. and Bhatia, H. M. (1979) *Indian J. med. Res.* **70**, 221.

Habermann, E. (1972) *Naunyn-Schmiedeberg's Arch. Pharmacol.* **272**, 75; (1976) *Ibid.* **293**, 1.

Habermann, E. and Dimpfel, W. (1973) *Naunyn-Schmiedeberg's Arch. Pharmacol.* **276**, 327.

Habermann, E. and Erdmann, G. (1978) *Toxicon* **16**, 611.

Habermann, E., Wellhöner, H. H. and Räker, K. O. (1977) *Naunyn-Schmiedeberg's Arch. Pharmacol.* **299**, 187.

Hansson, H. and Möller, H. (1971) *Acta Allerg. Kbh.* **26**, 150.

Hegyessey, G., Bozsoky, S. and Schulek, E. (1956) *Brit. J. exp. Path.* **37**, 300.

Hepple, J. R. (1968) *Chem. & Ind.* May 25, 670.

Heyningen, S. van (1976) *F.E.B.S. Letters* **68**, 5; (1980) *Pharmacol. and Therap.* **11**, 141.

Heyningen, W. E. van (1959) *J. gen. Microbiol.* **20**, 310.

Heyningen, W. E. van and Miller, P. A. (1961) *J. gen. Microbiol.* **24**, 107.

Hornibrook, J. W. (1952) *J. Lab. clin. Med.* **40**, 58.

Hosoya, S., Takada, M. and Terao, S. (1931) *Jap. J. exp. Med.* **9**, 33.

Hutyra, F. and Marek, J. (1912) *Special Pathology and Therapeutics of the Diseases of Domestic Animals.* London.

Ikić, D. (1958) *Acta med. jugoslav.* **12**, 179; (1960) *Proc. int. Symp. microbiol. Standardization* p. 415.

Ipsen, J. (1940–41) *Bull. Hlth Org., L.o.N.* **9**, 447, 452; (1961) *J. Immunol.* **86**, 50.

Jones, F. G. and Jamieson, A. (1936) *J. Bact.* **32**, 33.

Kerr, J. M., Corbett, J. L., Prys-Roberts, C., Crampton Smith, A. and Spalding, J. M. K. (1968) *Lancet* **ii**, 236.

Kerrin, J. C. (1928) *Brit. J. exp. Path.* **9**, 69; (1929) *Ibid.* **10**, 370.

Kerry, J., Thompson, R. O., Epps, H. B. G. and Foster, W. H. (1976) *Vlaams diergeneesk. Tijdschr.* **45**, 333.

Kielmann, A. A. and Vohra, S. R. (1977) *Indian J. med. Res.* **66**, 906.

King, L. E. and Fedinec, A. A. (1973) *Acta Neuropath. (Berlin)* **24**, 244; (1974) *Naunyn-Schmiedeberg's Arch. Pharmacol.* **281**, 391.

Kitasato, S. (1889) *Z. Hyg. InfektKr.* **7**, 225; (1891) *Ibid.* **10**, 267; (1892) *Ibid.* **12**, 256.

Knight, B. C. J. G. and Fildes, P. (1930) *Biochem. J.* **24**, 1496.

Koerber, W. L. (1943) *J. Immunol.* **46**, 391.

Koerber, W. L. and Mook, G. E. (1943) *J. Immunol.* **46**, 411.

Kryzhanovsky, G. N. (1965) In: *Recent Advances in the Pharmacology of Toxins* **9**, 105. Proc. 2nd int. pharmacol. meeting, Prague. Pergamon, Oxford; (1966) *Tetanus*. State Publishing House, Moscow; (1967) In: *Principles on Tetanus.* Proc. int. Conf. Tetanus, p. 155. Huber, Berne; (1973) *Naunyn-Schmiedeberg's Arch. Pharmacol.* **276**, 247.

Lahiri, D. C. (1939) *Indian. J. med. Res.* **27**, 581; (1940) *Ibid.* **27**, 651.

Latham, W., Bent, D. F. and Levine, L. (1962) *Appl. Microbiol.* **10**, 146.

Latham, W. C. *et al.* (1965) *J. Immunol.* **95**, 487.

Laurence, D. R. and Webster, R. A. (1963) *Clin. Pharmacol. Ther.* **4**, 36.

Laurent, L. J. M. and Parish, H. J. (1952) *Brit. med. J.* **i**, 1214; (1958) *Lancet* **ii**, 376.

Ledley, F. D., Lee, G., Kohn, L. D., Habig, W. H. and Hardegree, M. C. (1977) *J. biol. Chem.* **252**, 4049.

Leishman, W. B. and Smallman, A. B. (1917) *Lancet* **i**, 131.

Lennert-Petersen, O. (1954) *Acta. path. microbiol. scand.* **35**, 591.

Levine, L., McComb, J. A., Dwyer, R. C. and Latham, W. C. (1966) *New Engl. J. Med.* **274**, 186.

Lingelsheim, von (1912) See *Kolle and Wassermann's Handbuch der pathogen Mikro-organismen* 2te Aufl. 1912–13, **4**, 737.

Long, A. P. (1948) *Amer. J. publ. Hlth* **38**, 485.

Long, A. P. and Sartwell, P. E. (1947) *Bull. U.S. Army med. Dept* **7**, 371.

Lowbury, E. J. L., Kidson, A., Lilly, H. A., Wilkins, M. D. and Jackson, D. M. (1978) *J. Hyg., Camb.* **80**, 267.

Lowbury, E. J. L. and Lilly, H. A. (1958) *Brit. med. J.* **ii**, 1334.

Lucas, A. O. *et al.* (1965) *Clin. Pharmacol. Ther.* **6**, 592.

Lucas, A. O. and Willis, A. J. P. (1965) *Brit. med. J.* **ii**, 1333.

McComb, J. A. and Dwyer, R. C. (1963) *New Engl. J. Med.* **268**, 857.

McComb, J. A. and Levine, L. (1961) *New Engl. J. Med.* **265**, 1152.

McDonald, R. T., Chaikoff, L. and Truant, J. P. (1960) *Surg. Gynec. Obstet.* **110**, 702.

Mackay-Scollay, E. M. (1959) *Proc. R. Soc. Med.* **52**, 110.

Mackenzie, D. L. and Vlahcevic, L. R. (1974) *New Engl. J. Med.* **290**, 749.

Mackie, T. J. (1928) *An Inquiry into Post-operative Tetanus—A Report to the Scottish Board of Health.* H.M.S.O., Edinburgh.

Mackie, T. J., McLachan, D. G. S. and Anderson, E. J. M. (1929) *Certain Factors that Promote the Development of the Tetanus Bacillus in the Tissues, with Special Reference to Post-operative Tetanus—an Experimental Inquiry. A Report to the Department of Health for Scotland.* H.M.S.O., Edinburgh.

Maclean, I. H. and Holt, L. B. (1940) *Lancet* **ii**, 581.

MacLennan, R., Levine, L., Newell, K. W. and Edsall, G. (1973) *Bull. World Hlth Org.* **49**, 615.

MacLennan, R., Schofield, F. D., Pittman, M., Hardegree, M. C. and Barile, M. F. (1965) *Bull. World Hlth Org.* **32**, 683.

Mahoney, L. J., Aprile, M. A. and Moloney, P. J. (1967) *Canad. med. Ass. J.* **96**, 1401.

Manzanillo, G., Bianco, G. and Leonardi, L. (1964) *Ann. Sanit. publ.* **25**, 829.

Marie, A. (1897) *Ann. Inst. Pasteur* **11**, 591; (1898) *Ibid.* **12**, 91.

Marie, A. and Morax, V. (1902) *Ann. Inst. Pasteur* **16**, 818.

Mayer, R-L. (1938) *Bull. Acad. Med., Paris* **120**, 277; (1939) *C. R. Soc. Biol., Paris* **130**, 1560.

Mellanby, J. and Green, J. (1981) *Neuroscience* **6**, 281.

Mellanby, J., Pope, D. and Ambache, N. (1968) *J. gen. Microbiol.* **50**, 479.

Mellanby, J. and Thompson, P. A. (1972) *J. Physiol.* **224**, 407.

Meyer, H. and Ransom, F. (1903) *Arch. Path. Pharm.* **49**, 369.

Miller, J. J., Ryan, M. L. and Beard, R. R. (1949) *Pediatrics, Springfield* **3**, 64.

Miller, J. J. and Saito, T. M. (1942) *J. Pediat.* **21**, 31.

Mirsky, R., Wendon, L. M. B., Black, P., Stolkin, C. and Bray, D. (1978) *Brain Res.* **148**, 251.

Miyasaki, S. *et al.* (1967) *Jap. J. exp. Med.* **37**, 217.

Mohler, J. R. and Eichhorn, A. (1911) *28th Ann. Rep. Bur. Anim. Ind.* p. 185.

Moloney, P. J. and Hennessy, J. N. (1944) *J. Immunol.* **48**, 345.

Möse, J. R. (1955) *Arch. Hyg. Bakt.* **139**, 137.

Motzfeldt, K. (1912) *Zbl. Bakt.* **65**, 60.

Mueller, J. M. and Miller, P. A. (1940) *J. Immunol.* **40**, 21; (1954) *J. Bact.* **67**, 271.

Murray, E. G. D. and Denton, G. D. (1949) *Canad. med. Ass. J.* **60**, 1.

Mussett, M. V. and Sheffield, F. (1973) *J. biol. Stand.* **1**, 259; (1976) *Ibid.* **4**, 141.

Nabarro, D. (1932) *Lancet* **i**, 450.

Nagel, J. and Cohen, H. (1973) *J. Immunol.* **110**, 1388.

Nagel, J., Svec, D., Waters, T. and Fineman, P. (1977) *J. Immunol.* **118**, 334.

National Advisory Committee on Immunization (1979) *A Guide to Immunization for Canadians.* Laboratory Centre for Disease Control, Canada.

Newell, K. W., Dueñas Lehmann, A., LeBlanc, D. R. and Garces Osorio, N. (1966) *Bull. World Hlth Org.* **35**, 863.

Newell, K. W., LeBlanc, D. R., Edsall, G., Levine, L., Christensen, H., Montouri, M. H. and Ramirez, N. (1971) *Bull. World Hlth Org.* **45**, 773.

Nicolaier, A. (1884) *Dtsch. med. Wschr.* **10**, 842; (1886) *Jber. Fortschr. path. Mikroorg.* **2**, 270.

Nielsen, P. A. (1967) *Appl. Microbiol.* **15**, 453.

Noble, W. (1915) *J. infect. Dis.* **16**, 132.

Novak, M., Goldin, M. and Taylor, W. I. (1949) *Proc. Soc. exp. Biol. Med.* **70**, 573.

Ostertag, H. (1942) *Z. Hyg. InfektKr.* **123**, 698.

Parish, H. J., Laurent, L. J. M. and Moynihan, N. H. (1957) *Brit. med. J.* **i**, 639.

Park, W. H. and Nicoll, M. (1914) *J. Amer. med. Ass.* **63**, 235.

Patel, J. C. and Mehta, B. C. (1963) *Indian J. med. Sci.* **17**, 791.

Patel, J. C., Mehta, B. C. and Goodluck, P. L. (1965) *Proc. 1st int. Conf. Tetanus, Bombay, 1963,* p. 1.

Patel, J. C., Mehta, B. C. and Modi, K. N. (1965) *Proc. 1st int. Conf. Tetanus, Bombay, 1963,* p. 181.

Patel, J. C., Mehta, B. C., Modi, K. N. and Lotlikar, K. (1964) _Indian J. med. Sci._ **18**, 697.

Paterson, A. E. (1930) _Med. J. Aust._ **1**, 832.

Permin, C. (1914) _Mitt. Grenzgab. Med. Chir._ **27**, 1.

Petrella, A. (1960) _Acta microbiol., Budapest_ **7**, 65.

Piringer, E., Auserwald, W., Kuisewetter, E. and Olegnik, I. (1964) _Progr. immunobiol. Standard._ **2**, 52.

Pittman, M. _et al._ (1970) _Bull. World Hlth Org._ **43**, 469.

Pizzini, L. (1898) _Zbl. Bakt._ **24**, 890.

Pope, C. G. (1963) _Brit. med. Bull._ **19**, 230.

Porter, R. R. (1963) _Brit. med. Bull._ **19**, 197.

Press, E. (1948) _New Engl. J. Med._ **239**, 50.

Price, D. L., Griffin, J. W. and Peck, K. (1977) _Brain Res._ **121**, 379.

Price, D. L., Griffin, J., Young, A., Peck, K. and Stocks, A. (1975) _Science_ **188**, 945.

Ramon, G. and Descombey, P. (1925) _C. R. Soc. Biol._ **93**, 508; (1926) _Ibid._ **95**, 434.

Ramon, G. and Lemétayer, E. (1934) _C. R. Soc. Biol._ **116**, 275; (1935) _Rev. Immunol., Paris_ **1**, 209.

Ramon, G. and Zoeller, C. (1933) _C. R. Soc. Biol._ **112**, 347.

Ramshorst, J. D. van, Cohen, H., Levine, L. and Edsall, G. (1973) _J. biol. Standard._ **1**, 215.

Ramshorst, J. D. van and Sundaresan, T. K. (1972) _Bull. World Hlth Org._ **46**, 53.

Ray, S. N. _et al._ (1978) _Indian J. med. Res._ **68**, 901.

Reed, W. D. _et al._ (1973) _Lancet_ **ii**, 1347.

Regamey, R. H. (1947) _Schweiz. Z. allg. Path. Bakt._ **10**, 492.

Regamey, R. H. and Schlegel, H. J. (1951) _Schweiz. Z. allg. Path. Bakt._ **14**, 550.

Reisman, R. E., Rose, N. R., Witebsky, E. and Arbesman, C. E. (1961) _J. Allergy_ **32**, 531.

Relihan, M. (1969) _J. Irish med. Ass._ **62**, 430.

Relyveld, E. H. _et al._ (1969) _Ann. Inst. Pasteur_ **116**, 300.

Report. (1938) _Bull. Hlth Org., L.o.N._ **7**, 713; (1950) _World Hlth Org. tech. Rep. Ser._ No. 2; (1959) _Brit. med. J._ **i**, 1150; (1967) In: _Principles on Tetanus._ Proc. int. Conf. Tetanus, p. 576, Huber, Berne; (1970) _World Hlth Statist. Annu. for 1967._ WHO, Geneva; (1971) _Brit. med. J._ **ii**, 372; (1972a) _Immunization against Infectious Disease._ Dept Hlth Soc. Security, HMSO., London; (1972b) _Morbid. and Mortal._ **21**, 5; (1977) _Ibid._ **26**, 401.

Rigby, E. P. (1960) _East Afr. med. J._ **37**, 1.

Robinson, D. T., McLeod, J. W. and Downie, A. W. (1946) _Lancet_ **i**, 152.

Robles, N. L., Walske, B. R. and Personeus, G. (1967) _Amer. J. Surg._ **114**, 627.

Rosenau, M. J. and Anderson, J. F. (1908) _Bull. U.S. hyg. Lab._ No. 43.

Rosenbach (1886) _Arch klin. Chir._ **34**, 306.

Roux, E. and Borrel, A. (1898) _Ann. Inst. Pasteur_ **12**, 225.

Rubbo, S. D. (1966) _Lancet_ **ii**, 449.

Rubbo, S. D. and Suri, J. C. (1962) _Brit. med. J._ **ii**, 79.

Ruben, F. L., Nagel, J. and Fineman, P. (1978) _Amer. J. Epidem._ **108**, 145.

Rubinstein, H. M. (1962) _Amer. J. Hyg._ **76**, 276.

Russell, D. S. (1927) _Brit. J. exp. Path._ **8**, 377.

Sanada, I. and Nishida, S. (1965) _J. Bact._ **89**, 626.

Sanders, R. K. M., Joseph, R., Martyn, B. and Peacock, M. L. (1977) _Lancet_ **i**, 974.

Sanfelice F. (1893) _Z. Hyg. InfektKr._ **14**, 339.

Sartwell, P. E. (1950) _Amer. J. Hyg._ **51**, 310.

Savolainen, T. (1950) _Ann. med. exp. Biol. fenn._ **28**, 55.

Sawamura, S. (1909) _Arb. Inst. InfektKr., Bern._ **4**, 1.

Schanfield, M. S., Wells, J. V. and Fudenberg, H. H. (1979) _J. Immunogenetics_ **6**, 311.

Scheibel, I. (1944) _Acta path. microbiol. scand._ **21**, 130; (1955) _Bull. World Hlth Org._ **13**, 381.

Scheibel, I., Bentzon, M. W., Christensen, P. E. and Biering, A. (1966) _Acta path. microbiol. scand._ **67**, 380.

Scheibel, I., Chen, B.-L., Bentzon, M. W. and Zia, K. (1968) _Acta path. microbiol. scand._ **73**, 115.

Schellenberg, D. B. and Matzke, H. A. (1958) _J. Immunol._ **80**, 367.

Scheunemann, K. (1931) _Arch. Hyg._ **105**, 287.

Schneider, C. H. (1964) _Med. J. Aust._ **ii**, 303.

Schwander, D. and Wegman, A. (1973) _Schweiz. med. Wschr._ **103**, 1184.

Sedaghatian, M. R. (1979) _Arch. Dis. Child._ **54**, 623.

Sevitt, S. (1949) _Lancet_ **ii**, 1075.

Sharrard, W. J. W. (1965) _Proc. R. Soc. Med._ **58**, 221.

Sheffield, F., Baldwin, P. W., Knight, P. A. and Richer, P. R. (1979) _J. biol. Stand._ **7**, 301.

Sherrington, C. S. (1917) _Lancet_ **ii**, 964.

Smith, J. W. G. (1964a) _J. Hyg., Camb._ **62**, 379; (1964b) _Brit. med. J._ **ii**, 1293; (1965) _Proc. R. Soc. Med._ **58**, 226; (1966) _Brit. J. exp. Path._ **47**, 17; (1969) _Brit. med. Bull._ **25**, 177.

Smith, J. W. G., Evans, D. G., Jones, D. A., Gear, M. W. L., Cunliffe, A. C. and Barr, M. (1963) _Brit. med. J._ **i**, 237.

Smith, J. W. G., Laurence, D. R. and Evans, D. G. (1975) _Brit. med. J._ **ii**, 453.

Smith, J. W. G. and MacIver, A. G. (1969) _J. med. Microbiol._ **2**, 385; (1974) _J. med. Microbiol._ **7**, 497.

Smith, J. W. G. and Schallibaum, E. M. (1970) _Brit. J. exp. Path._ **51**, 73.

Smith, M. L. (1938) _Bull. Hlth Org., L.o.N._ **7**, 739; (1943-44) _Quart. Bull. Hlth Org., L.o.N._ **10**, 104.

Smith, T. (1908) _Trans. Chicago path. Soc._ **7**, No. 4.

Smolens, J. _et al._ (1961) _J. Pediatrics_ **59**, 899.

Smythe, P. M., Bowie, M. D. and Voss, T. J. V. (1974) _Brit. med. J._ **i**, 223.

Sneath, P. A. T., Kerslake, E. G. and Scruby, F. (1937) _Amer. J. Hyg._ **25**, 464.

Spaeth, R. (1941) _Arch. intern. Med._ **68**, 1133.

Spaun, J. and Lyng, J. (1970) _Bull. World Hlth Org._ **42**, 523.

Sterne, M., Batty, I., Thompson, A. and Robinson, J. M. (1962) _Vet. Rec._ **74**, 909.

Stöckel, K., Schwab, M. and Thoenen, H. (1975) _Brain Res._ **99**, 1.

Strick, P. L., Burke, R. E., Kanda, K., Kim, C. C. and Walmsley, B. (1976) _Brain Res._ **113**, 582.

Suri, J. C. and Rubbo, S. D. (1961) _J. Hyg., Camb._ **59**, 29.

Tasman, A. (1959) _Ann. Inst. Pasteur_ **97**, 835.

Tasman, A. and Huygen, F. J. A. (1962) _Bull. World Hlth Org._ **26**, 397.

Tasman, A. and Ramshorst, J. D. van (1952) _Ned. Tijdschr. Leeuwenhoek_ **18**, 357.

Tateno, I. (1963) _Jap. J. exp. Med._ **33**, 149.

Tateno, I., Suzuki, S. and Kitamoto, O. (1961) _Jap. J. exp. Med._ **31**, 365.

Taylor, W. E. and Novak, M. (1951) _Ann. Surg._ **133**, 44.

Teale, F. H. and Embleton, D. (1920) _J. Path. Bact._ **23**, 50.

Thompson, R. D. _et al._ (1969) _Vet. Rec._ **85**, 81.

Tizzoni, G. and Cattani, G. (1890) _Arch. exp. Path. Pharmak._ **27**, 432.

Toledo, S. and Veillon (1891) _Zbl. Bakt._ **9**, 18.

Tremewan, H. C. (1946) _N.Z. med. J._ **45**, 312.

Trinca, J. C. (1963) *Med. J. Aust.* **ii**, 389; (1965) *Ibid.* **ii**, 116; (1967) *Med. J. Aust.* **ii**, 153.

Trinca, J. C. and Reid, J. C. (1967) *Lancet* **i**, 76.

Tulloch, W. J. (1919–20) *J. Hyg., Camb.* **18**, 103.

Turner, T. B., Velasco-Joven, E. A. and Prudovsky, S. (1958) *Johns Hopk. Hosp. Bull.* **102**, 71.

Vaillard, L. (1891) *C. R. Soc. Biol.* **3**, 462; (1892) *Ann Inst. Pasteur* **6**, 224.

Vaillard, L. and Rouget, J. (1892) *Ann. Inst. Pasteur* **6**, 224.

Vaishnava, H. and Goyal, R. K. (1966) *Brit. med. J.* **ii**, 466.

Vaishnava, H., Goyal, R. K., Neogy, C. N. and Mathur, G. P. (1966) *Lancet* **ii**, 1371.

Vakil, B. J. *et al.* (1965) *Proc. 1st int. Conf. Tetanus, Bombay, 1963*, p. 255.

Vakil, B. J., Armitage, P., Clifford, R. E. and Laurence, D. R. (1979) *Trans. R. Soc. trop. Med. Hyg.* **73**, 579.

Vakil, B. J., Tulpule, T. H., Armitage, P. and Laurence, D. R. (1963) *Clin. Pharmacol. Therap.* **4**, 182; (1964) *Ibid.* **5**, 695; (1968) *Ibid.* **9**, 465.

Vakil, B. J., Tulpule, T. H., Rao, S. S. and Borker, M. B. (1968) *Indian J. med. Res.* **56**, 1188.

Veronesi, R. (1965) *J. Hyg. Epidem. Microbiol. Immunol.* **9**, 421.

Veronesi, R. *et al.* (1975) *Proc. 4th int. Conf. Tetanus, Dakar*, p. 613.

Voller, A., Bidwell, D. E. and Bartlett, A. (1976) In: *Protides of the Biological Fluids*, p. 751. 24th Colloquium, Bruges, 1975. Pergamon Press, Oxford.

Wassermann, A. and Takaki, T. (1898) *Berl. klin. Wschr.* **35**, 5.

Webster, R. A. and Laurence, D. R. (1963) *J. Path. Bact.* **86**, 413.

Wetzler, T. F., Marshall, J. D. and Cardella, M. A. (1956) *Amer. J. clin. Path.* **26**, 418.

White, W. G. *et al.* (1973) *J. Hyg., Camb.* **71**, 283.

White, W. G., Gall, D., Barnes, G. M., Barker, E., Griffith, A. H. and Smith, J. W. G. (1969) *Lancet*, **ii**, 95.

Wilcox, H. L. (1934) *J. Immunol.* **27**, 195.

Wildegans, H. (1940) *Dtsch. med. Wschr.* **66**, 869.

Wilson, V. J., Diecke, F. P. J. and Talbot, W. H. (1960) *J. Neurophysiol.* **23**, 659.

Winsnes, R. (1979) *Acta path. microbiol. scand. B* **87**, 191.

Wiseman, J. C. and Gascoigne, M. E. (1977) *J. clin. Path.* **30**, 177.

Wolters, K. L. and Dehmel, H. (1940) *Z. Hyg. InfektKr.* **122**, 603; (1942) *Ibid.* **124**, 326; (1951) *Ibid.* **132**, 582.

World Health Organization (1964) *Tech. Rep. Ser.* No. 293, p. 25.

Wright, E. A., Morgan, R. S. and Wright, G. P. (1950) *J. Path. Bact.* **62**, 569; (1952) *Lancet* **ii**, 316.

Wright, G. P. (1953) *Proc. R. Soc. Med.* **46**, 319; (1955) *Pharmacol. Rev.* **7**, 413.

Wright, H. D. (1930) *Univ. Coll. Hosp. Mag.* **15**, 64.

Yodh, B. B. (1932) *Brit. med. J.* **ii**, 589.

Zacks, S. I. and Sheff, M. F. (1976) *Science* **192**, 158.

Zimmerman, J. M. and Piffaretti, J.-Cl. (1977) *Naunyn-Schmiedeberg's Arch. Pharmacol.* **296**, 271.

Zuger, B., Greenwald, C. K. and Gerber, H. (1942) *J. Immunol.* **44**, 309.

Züst, B. (1967) *Z. ImmunForsch.* **132**, 261.

65

Bacterial meningitis

Graham Wilson

Introductory

Meningitis is an inflammatory affection of the membranes surrounding the brain and spinal cord, which occurs as either a primary disease or secondarily to disease in some other part of the body. It may be caused by a multitude of different microbes—in fact by practically every known pathogenic organism, including a number of viruses (see Volume 4) and occasional fungi and protozoa. In the United States the commonest form of bacterial meningitis, excluding tuberculous, is influenzal, followed in order by meningococcal, pneumococcal, streptococcal B, and listerial (Report 1979). In this chapter we shall deal at some length with these forms, and list at the end references to those that are less common.

The incidence of the various forms varies greatly with age. In the perinatal period premature infants and those with some anatomical defect such as a meningocele are often attacked. The organisms responsible are mainly enterobacteria and are probably derived from the infected genital tract of the mother (Berman and Banker 1966). Over 80 per cent of cases of meningitis due to these organisms occur during the first year of life.

Between 3 months and 5 years of age, meningococcal, influenzal (*Haemophilus influenzae*) and pneumococcal cases prevail. Tuberculous meningitis is commonest between 1 and 6 years. Of cases caused by *Listeria monocytogenes* practically all occur under 1 year and over 50 years of age; and of those caused by leptospirae all occur between 5 and 50 years of age. During the period 10 to 20 years there is a high incidence of viral meningitis caused chiefly by the mumps, herpes, coxsackie and echo viruses.

In England and Wales during non-epidemic times about 70 per cent of cases of meningitis are of bacterial and 30 per cent of viral origin. Of those caused by bacteria, 84 per cent are accounted for by the meningococcus, the pneumococcus, and *Haemophilus influenzae*; and, except for tuberculous and leptospiral meningitis, cases of all forms are commonest in children under 5 years of age (Bevan-Jones and Miller 1967).

Cerebrospinal meningitis

Disease caused by the meningococcus ranges from an inapparent infection of the nasopharynx to the fulminating Waterhouse–Friderichsen syndrome characterized by circulatory shock and collapse proving fatal within a few hours. In between come a number of different clinical manifestations. Of these the commonest is an acute purulent meningitis, affecting particularly the base of the brain. Alternatively the organism may produce a simple rhinopharyngitis; or a generalized infection accompanied by septicaemia without meningitis; or a slow intermittent fever lasting for weeks or months; or occasionally a rheumatic type of illness characterized by endocarditis (see Banks 1948, Goeters 1954). The common form of the disease has an incubation period of 2 to 10 days, and is marked clinically by a sudden onset with fever, intense headache, nausea and often vomiting, painful stiffness of the neck, retraction of the head, spinal rigidity, and a petechial rash. Death, if it occurs, is usually within the first five days. Among the complications are deafness, optic neuritis, polyarthritis, and hydrocephalus. Post-mortem examination of a case of acute meningitis reveals an inflamed pia-arachnoid membrane, a purulent exudate at the base of the brain and over the vertex, sometimes extending down the cord or invading the ventricles. In the chronic form the ventricles are often distended with turbid fluid, the convolutions of the brain are flattened, and hydrocephalus may be present. In the fulminant form there is little or no meningitis, but there are extensive haemorrhages into the adrenal glands and skin.

The disease is caused by the meningococcus, *Neisseria meningitidis*. At least nine serological groups of this organism are recognized, A, B, C, D, X, Y, Z, W135 and 29E, of which A and B are the commonest. At one time great confusion existed over the part played in the causation of cerebrospinal meningitis by Jaeger's coccus (see Chapter 28), but it was ultimately established that this organism was not aetiologically related to the disease (for references, see 5th edition, p. 1736).

Epidemiology

Morbidity and mortality

Two of the most striking features of a cerebrospinal fever epidemic are the low morbidity and the high case fatality. As in poliomyelitis, only a very small proportion of those who become infected contract the clinical disease. The case-fatality rate before the introduction of sulphonamide treatment ranged from 40 to 90 per cent, but is now only about 10 per cent.

In 1966 over 42 000 cases were reported in the world, of which more than half were in Africa. During the present century some countries in Africa, particularly the Sudan, Northern Nigeria, and Ghana, have suffered devastating epidemics accompanied by a high case-fatality rate (see Waddy 1957, Archibald 1962). Lapeyssonnie (1963) defined a 'meningitis belt' stretching across Africa 8 to 16 degrees north of the equator, lying between the Sahara to the north and the equatorial forest to the south, in which endemic disease is punctuated from time to time by epidemics sweeping from east to west and causing at their height a morbidity of 20 per 1000 of the population. In this belt the dry season lasts from October to April, the temperature getting higher and higher and the absolute and relative humidity lower and lower. The epidemic reaches its peak in March and April and declines rapidly when the rains come. Infection seems to be favoured by the overcrowding that occurs in the huts at night and by the dry unhealthy state of the nasal mucosa which can no longer act as an effective barrier to the invading organisms.

This incidence, occurring as it does at the height of the hot season, is very different from that in temperate climates, in which the maximum prevalence occurs during the cold weather of late winter and spring. Waddy (1957) explains this paradox by pointing out that in both regions the absolute humidity during the epidemic period is low. This explanation is supported to some extent by Simpson (1958), who ascribes the high incidence of colds during the winter months in temperate climates to the dryness of the indoor air caused by artificial heating. This leads to crusting of the nasal mucosa, increasing viscidity of the mucus coat, damage to the underlying epithelial cells, and interference with ciliary action.

It has already been mentioned that meningococcal meningitis has its highest incidence between 3 months and 5 years of age, though in Africa Lapeyssonnie (1963) says that cases are most frequent between 10 and 15 years. Throughout the world the case-fatality rate during the years 1960–64 was said to be (Report 1969):

Under 1 year	35.9 per million live births
1–4 years	7.6 per million population
5–14 years	1.0 per million population

Miners and soldiers, especially young recruits, are the two occupational groups on which the disease falls most heavily (Dopter 1921). Goeters (1954) ascribed the high incidence in miners partly to overcrowding in the homes and partly to the sudden large changes in temperature and atmospheric pressure to which they are subject. Recruits to the Services often come from rural areas where thay have had little opportunity of becoming latently immunized with the common flora of the nasopharynx, and are subjected to a vigorous training characterized by an unaccustomed degree of physical exertion.

Stressing the predisposing effect of fatigue, Dopter tells an impressive story of a party of recruits who made a long and very tiring march to join their regiment at Versailles. On reaching their destination cerebrospinal fever broke out, and of a total strength of 153 men no fewer than 79 were attacked.

History and incidence

The first definite knowledge that we have of cerebrospinal fever is due to Vieusseux, who in 1805 described an outbreak of the disease at Geneva. In the following year it appeared in Medfield, Massachusetts, and during 1806–07 it attacked the Prussian Army. Since then it has invaded most countries in the world, rising at one time to epidemic proportions, subsiding at other times to remain dormant for long periods, but inevitably reappearing in outbreaks of greater or less severity. The disease has, in fact, shown a gradual increase, both in its geographical range and in the number of persons it has attacked. Thus, according to Hirsch (1886) from 1805 to 1830 the disease was most prevalent in the United States; from 1837 to 1850 France was attacked most severely; from 1854 to 1874 epidemics occurred both in Europe and in America; and from 1875 onwards not only these two continents but Asia, Africa, and Australia have been invaded (Low 1916).

The African epidemics, which were caused by group A meningococci, are not the only ones that have occurred in recent years. For example, an epidemic lasting four years caused by group B organisms attacked new recruits at Fort Ord, California (Brown and Condit 1965); and another, due to groups C and A, affected mainly children and young adults at São Paulo, Brazil (Report 1974*a*).

In Great Britain cases have occurred since 1830. During the first world war the disease broke out in epidemic form among the Canadian Expeditionary Force (Reece 1916) and rapidly spread to both the civil and the military population. There was another outbreak during the second world war, causing in 1940 no fewer than 11 162 cases and 2459 deaths. Thereafter the disease declined till in the decade 1961–70 cases numbered about 500 and deaths about 50 a year.

Mode of spread of the disease

The disease is endemic in large towns, from which, at least in the post-basic form, it rarely disappears. Every winter and spring a few fresh cases occur—sporadic cases affecting children, less frequently young adults— which appear at widely separated points having no apparent connection with each other. In villages the endemic state is practically unknown.

Every now and then an epidemic breaks out. As a rule it starts insidiously by the occurrence of one or two cases at intervals of a week or more; it then begins to advance by the formation of small multiple foci in families, schools, barracks, or gaols. Often it works itself out in one focus before passing on to another. Thus, at Strassburg (Dopter 1921) in 1840 the 7th line regiment was attacked in October; three months later it spread to the 69th regiment, of which two companies were quartered with a part of the 7th. In January the 29th line regiment and the 11th artillery were attacked; in February the 34th line regiment and the 1st artillery; and in March, 6 months after the commencement of the outbreak, it invaded the pontonniers.

In a town the cases are not aggregated together, but occur widely scattered, as if there were no causal relation between them. Thus in Hamburg between 1880 and 1885 the 180 cases of cerebrospinal fever that occurred were distributed over 131 streets. The distribution in a country is similar. For example, in the 1966 epidemic in the Netherlands 957 communities were affected, yet only 217 had more than one case (Severin *et al.* 1969).

The spread of the epidemic is irregular and capricious: often groups, which from their situation appear certain to be attacked, escape, whereas others, situated remotely from the primary focus, are attacked. Similarly with regard to time; the successive outbreaks of an epidemic are irregular; there are paroxysms of intensity followed by remissions or intermissions. The decline of the epidemic takes place slowly, fewer and fewer foci remaining, till finally the endemic state is regained (Dopter 1921).

It is remarkable what a predilection the disease has for the military population; frequently garrisons have been attacked without a single case among the neighbouring civilians. Of 75 epidemics in France, collected by Hirsch (1886), 39 were confined to the troops.

Carriers

Long before the discovery of the causal organism, it had been noticed that overcrowding favoured the development of the disease. The reason for this was not known. At a post-mortem examination of a case of cerebrospinal fever, Weigert noticed the presence of a purulent rhinitis, and on the basis of this observation von Strumpell suggested that the nasopharynx might serve as a portal of entry for the infective material.

Weichselbaum (1887) in his original paper likewise mentioned that one of his cases had a purulent sinusitis, and expressed the opinion that the organisms might find their way to the meninges via the nose. Further observations showed that meningococci were often present in the nasopharynx not only of patients suffering from the disease—about 60 per cent of cases in the first week—but also in contacts and even non-contacts (Albrecht and Ghon 1901, von Lingelsheim 1905, Kutscher 1906, Ostermann 1906, Scott 1916). The risk of household contacts developing the disease is estimated at 0.4–0.5 per cent (Report 1974b, 1979) and of classroom contacts at 0.3 per cent (Jacobson *et al.* 1976). According to Vedros and Hottle (1970), antibodies are present in about 75 per cent of carriers, having been formed presumably in response to repeated exposures to meningococci throughout life (Rake 1935). These form the basis of maternal immunity, since experience has shown that meningitis occurs only in persons devoid of antibodies (Goldschneider *et al.* 1969).

During the war of 1914–18 observations of much interest were made on the carrier rate. By swabbing large numbers of the military population in camps and depots, it was found that before an outbreak of cerebrospinal fever the proportion of carriers of the meningococcus increased steadily. The normal carrier rate among troops was recorded as 2–4 per cent, but preceding an epidemic it rose to 20–30 per cent. Soon after it had passed the 20 per cent mark, isolated cases of meningitis began to appear, and as the epidemic gained a foothold the carrier rate likewise rose, sometimes to as high as 88 per cent.

Investigating the cause of this 'warning rise' in the carrier rate, Glover (1920) was led to suspect a relation between it and overcrowding in the sleeping huts. These huts were at the best poorly ventilated (Eagleton 1919–20), and during the stress of war the mobilization standard had been overstepped, so that beds, instead of being separated by 1 ft 4 in, were practically touching each other. Glover (1920) noticed that the carriers in a given hut tended to be aggregated together; three type II carriers were in adjacent beds, two type I carriers, and so on. This pointed strongly to the direct transmission of the meningococcus from one man to another sleeping in the next bed. Finding that the spraying capacity during normal sleep was not more than about 3 ft, Glover tried the effect of spacing out the beds in the hope that the infection would be diminished. The results obtained were in accordance with expectation. The effect of the distance between the beds was not confined to the carrier rate. At Caterham depot, where there was severe overcrowding, an outbreak of cerebrospinal fever had occurred during each winter of the war, but after the adoption of the spacing-out policy in 1917–18, not a single case occurred.

Because of the far-reaching conclusions on the relation between overcrowding and the carrier rate, and the importance of detecting the 'warning rise', Glover's work received a great deal of attention.

In the Detroit epidemic of 1928–29, Norton and Baisley (1931) found no association between the degree of overcrowding in the home and the contact carrier rate. During the outbreak of 1931 at Aldershot, Armstrong and his colleagues (1931), from a study of the position of carriers in dormitory barrack rooms, were unable to obtain any evidence that infection occurred mainly at night. The carriers were scattered quite irregularly without any particular relation to the position of the beds. Our own nasopharyngeal surveys of the civilian population carried out before the second world war showed us that the carrier rate in institutions might be as high as 20 per cent and over, without any outbreak of cerebrospinal fever occurring. Rake (1934) made the same observation. Perhaps the most striking figures, however, are afforded by Dudley and Brennan (1934). Working at the Chatham naval hospital, they found that between January 1932 and March 1933 there were 11 cases of cerebrospinal meningitis with a carrier rate of about 13 per cent. During the period March 1933 to May 1934, the carrier rate was 54 per cent, yet not a single case of meningtitis occurred. During the same period at the Royal Naval Hospital, Portsmouth, there were 6 cases of meningitis with a carrier rate of only 5 per cent. Analysis of the distribution of carriers at Chatham showed no constant relation between the density of the population and the carrier rate. The senior ratings with the most spacious sleeping accommodation had as high a carrier rate—60 per cent— as the recruits with the worst sleeping quarters. Phair, Schoenbach and Root (1944) carried out a survey at an army camp in which cerebrospinal fever was prevalent. There were no cases, however, in the unit they examined. A group of 99 men was swabbed 28 times in 68 days; no fewer than 91 of these men were found on one or more occasions to be harbouring meningococci in the nasopharynx. The average carrier rate was 41 per cent.

The behaviour of the carrier rate may be quite anomalous. Farrell and Dahl (1966), for example, found that at one training centre for recruits the average carrier rate, which on arrival was about 16 per cent, actually fell during the 6-week stay to below 10 per cent. The recruits were housed in barracks of four different types, but no relation was found between the carrier rate and the type of barrack, the rainfall, the temperature, or the humidity. No cases of meningitis occurred at this centre.

These observations lend no support to the suggestion that there is a direct relation between the carrier rate and the incidence of meningitis. No predictive value can be attached to the carrier rate, at any rate in small populations. Nevertheless, the prevalence of carriers and the number of cases of meningitis in the population at large are associated; and carrier rates are higher in epidemic than in inter-epidemic periods.

It is reasonable to suppose that overcrowding increases the frequency of exchange of nasopharyngeal flora, but it is clear that it by no means always leads to a rise in the carrier rate or to the development of cerebrospinal meningitis. Weight must be attached to the degree of virulence of the meningococci on the one hand and the resistance of the population at risk on the other. During epidemic times the organisms, as judged by the mouse test (Cohen 1936, Branham 1940), tend to be more virulent than in between epidemics, and group A organisms appear to be more virulent, as a rule, than those belonging to other groups. The fact that both the carrier rate and the incidence of meningitis are lower in seasoned troops than in raw recruits points to the part played by latent immunization in resistance to the disease. Aycock and Mueller (1950), who review the evidence, conclude that the prevalence of cerebrospinal fever cannot be regarded as a simple function of the carrier rate. It seems to be more closely associated with variations affecting the host.

Routes of infection

There is now no doubt that the meningococcus gains access to the body via the nasopharynx. It is unusual to find it in the nose itself or in the saliva, where, as Gordon (see Report 1917) showed, the salivary streptococci exert an inimical action on its growth; it may be isolated from the throat, but the most important site is undoubtedly the nasopharynx. It is here that it multiplies, and sometimes sets up a rhinopharyngitis.

Whereas there is now general agreement that the nasopharynx is the portal of entry, it is still uncertain by what route the organism gains access to the meninges. There are two main views: the first, supported by Netter and Debré (1911) and more recently by Fairbrother (1947), postulates the direct transmission of the organism from the nose to the meninges; the second, upheld by several workers in Germany, America and England (Westenhoeffer 1906, Elser and Huntoon 1909, Herrick 1918, Baeslack *et al.* 1918, Rolleston 1919, Banks 1948), is that it is carried to the meninges by the blood stream. The arguments in favour and against each of these two routes were set out in full in the 6th edition (pp. 1885–1887), and there is no need to repeat them here. Probably most workers favour the haematogenous route, but there is sufficient evidence to render it probable that direct spread from the nose to the meninges cannot be excluded. Herpes simplex virus, for example, has been demonstrated in the olfactory nerve system and the temporal lobe of the brain (Ojeda 1980).

Diagnosis

In inter-epidemic times the clinical diagnosis of cerebrospinal meningitis is by no means always easy to make. It may be confused with influenzal, leptospiral or viral meningitis, and in the tropics with cerebral malaria. Even during an epidemic not every case of meningitis is necessarily caused by the meningococcus. In Africa as many as 10 per cent of cases may be due to other organisms, particularly the pneumococcus and *Moraxella duplex* (Waddy 1957, Lapeyssonnie 1963). The correct diagnosis can be made with assurance only in the laboratory.

In cerebrospinal meningitis the causative organism is always present in the cerebrospinal fluid and quite often in the nasopharynx (see Report 1920). For diagnostic purposes, therefore, it is essential to perform a lumbar puncture, and desirable to take a nasopharyngeal swab. In the early stages of the disease, and in cases treated by chemotherapy, however, the organisms may be difficult or impossible to demonstrate in the spinal fluid, even in the presence of a polymorphonuclear or lymphocytic pleocytosis. In such cases, and those in which the cell count is normal and the blood culture negative, a second lumbar puncture should be performed (Smales and Rutter 1979). The meningococcus is a delicate organism intolerant of heat, cold, and sunlight, and soon undergoes autolysis in spinal fluid after its withdrawal by lumbar puncture. It is therefore desirable to put up cultures or to inoculate tubes of transport medium (Amies 1967, Gästrin *et al.* 1968) at the bedside or, when this is impossible, to transmit the material for examination to the laboratory in an insulated container at a temperature of about 37°.

Examination of the spinal fluid

The spinal fluid should be withdrawn and divided into three portions, which should be treated as follows:

(1) Examination of centrifuged deposit for cells and organisms, and of the supernatant fluid for specific precipitinogens The fluid is centrifuged, the deposit stained by Gram, and examined microscopically for the presence of the oval gram-negative diplococci, most of which are situated within the polymorphonuclear cells. Dissolved antigenic material of polysaccharide nature is tested for by layering a few drops of the supernatant fluid on to a similar quantity of meningococcal antiserum in a Dreyer's tube. Usually a fine white opaque disc of precipitate becomes apparent in a few minutes. If it does not, the tube is incubated in a water-bath for 1–2 hours at 37°. Both Rake (1933*b*) and Maegraith (1935) obtained very satisfactory results with this method (see also Alexander and Rake 1937). In Maegraith's series of 120 cases, a positive reaction occurred in 116, whereas of 180 control fluids, all but 2 were negative. No other method affords such a rapid diagnosis.

(2) Immediate cultivation   As soon as the spinal fluid is withdrawn, it should be streaked in quantities of 1 ml over two or three plates of chocolate blood agar, ascitic agar, Loeffler's serum, or Gordon's trypagar (Gordon *et al.* 1916). Alternatively it may be inoculated in a quantity of 5 ml or more into 0.25 per cent glucose broth and, after incubation overnight, plated on a solid medium. Growth is often improved by the addition of 5–10 per cent CO_2 to the atmosphere. Colonies should be apparent within 18 hours. If made up of gram-negative cocci, their identity should be determined by the fluorescent antibody technique (Biegeleisen *et al.* 1965), and by agglutination with specific antisera. For this purpose the growth should be suspended in a small quantity of saline, heated to 65° for 1 hour, standardized to 500 million per ml, and tested against the group sera; incubation should be at 50° for an hour.

Slide agglutination gives an immediate answer, but must always be checked by the tube method (Bell 1920*b*). Alternatively the spinal fluid may be seeded directly on to agar plates containing specific group antisera as described by Petrie (1932), incubated, and examined for haloes. This method enables large numbers of strains to be rapidly identified (Kuzemenská *et al.* 1977, Craven *et al.* 1979). When there is insufficient growth on the primary plates to allow of either method of agglutination, subcultures must be made on to fresh plates, and the growth examined after 24 hours. The fact must not be forgotten that one variety of organism found in meningitis, *Neisseria flavescens*, may produce a golden-yellow pigment (Branham 1930), and that an antigenic relation exists between *Escherichia coli* 07: K1 and *N. meningitidis* group B (Grados and Ewing 1970).

Not all strains are readily agglutinable by the available type sera. Strains from sporadic cases are usually less readily agglutinated than those isolated during an epidemic. Much depends upon the strains selected for the preparation of the agglutinating sera (see Griffith 1917, Scott 1917, Bell 1920*a*, Murray 1929, Branham 1932, Rake 1933*a*).

It is advisable to test the fermentation reactions of the organisms isolated. Colonies should be inoculated into glucose, maltose, and sucrose peptone water containing 10 per cent of ascitic fluid or blood serum. The meningococcus produces acid in glucose and maltose, but not in sucrose; *N. flavescens* is without action on any of the sugars. The final identification of any strain must be made on the basis of morphological, cultural, fermentation, and serological reactions.

(3) Incubation of the spinal fluid with subsequent subcultivation   It is always wise to incubate 5 ml of the fluid at 37° in a sterile tube, in case direct plating proves negative. Sometimes the organisms grow in the fluid itself when they fail to do so in cultures made at the time of lumbar puncture. As soon as turbidity appears, the fluid should be subcultured on to suitable media, and the resulting colonies identified in the usual way.

The frequency with which positive cultures are to be expected depends partly on the stage of the disease and partly on the carefulness of one's technique. Flack (see Report 1917), in an examination of 55 specimens of cerebrospinal fluid, obtained the following results:

	No.	Culture positive (per cent)
In clear fluids	15	33
In slightly turbid fluids	13	61
In turbid fluids	27	96

In the early stages of the disease before the spinal fluid is purulent, the proportion of positive cultures is much lower than in the later stages. In chronic cases the fluid, after being frankly purulent, may clear again, and the organisms become increasingly difficult to isolate. Ultimately they disappear and can no longer be demonstrated.

Cultivation from the nasopharynx

At the time of lumbar puncture a nasopharyngeal swab should be taken. There are two advantages in this practice. First, a positive culture may be obtained from the nasopharynx, especially in the early stages of the disease, when the spinal culture is negative. Second, the nasopharyngeal swab may yield a heavier growth than the spinal plate, and so enable the serological grouping of the organisms to be carried out more rapidly. Since the nasopharyngeal and spinal strains are almost invariably of the same type, there is little danger of error in this procedure.

For examining the nasopharynx, West's swab may be used; this consists of a wire curved at one end, fitted with a pledget of cotton-wool, and enclosed in a glass tube. As soon as the swab has been taken it should be streaked over a plate of ascitic or chocolate blood agar or of Thayer and Martin's medium (1966). The plates must be incubated immediately, or kept in a warm container. The following day they are examined for the typical lenticular colonies of meningococci. The further examination of these colonies should be performed in the same way as that of the spinal colonies. When the swab cannot be cultured at once, it should be stored in a tube of transport medium. Whatever type of swab is used, care should be taken to avoid its becoming contaminated with saliva; otherwise the meningococci may be overgrown by salivary streptococci.

Examination of the blood

In view of the not infrequent occurrence of meningococcal septicaemia it is advisable in every suspected case of cerebrospinal fever to make a blood culture.

For this purpose 10 ml of blood should be withdrawn from the median basilic vein and inoculated into 100 ml of a digest broth containing 0.05 per cent Liquoid. Another 10 ml should be distributed in 1 ml quantities over a series of ten ascitic or chocolate agar plates. Positive results may be expected in 25 per cent of cases examined during the first week of the disease.

Examination of petechial skin lesions

The presence of meningococci in petechial spots was demonstrated by Muir (1919), using the cultural method. In a considerable proportion of cases the organisms can be found microscopically in smears. McLean and Caffey (1931) were successful by this means in 14 out of 18 cases in children, and Tompkins (1943) in the Army in 39 out of 48 cases. The petechial spot—macules are unsatisfactory—should be punctured with a Hagedorn needle, while the surrounding skin is pinched so as to prevent the access of capillary blood to the lesion. Small drops of blood should be smeared on to a slide and stained with Giemsa. Typical intracellular diplococci with the adjacent surfaces flattened can be detected in this way, and confirmed by a Gram stain. They can often be found when a blood culture is negative.

Rapid diagnosis of meningitis

The standard methods just described are seldom suited for the rapid diagnosis of early cases and cases already treated by chemotherapy. During recent years several methods have been introduced to obviate this defect.

(1) Countercurrent immuno-electrophoresis (CIE)

This is the most widely used technique for the rapid diagnosis of bacterial meningitis, especially for that of influenzal or pneumococcal origin. For meningococcal meningitis its value is impaired by the lack of any suitable antiserum against organisms of group B, which may be responsible for as many as half the cases (Kaplan and Feigin 1979). Even with the other groups antisera vary in their reliability, so that the results of the test must not be regarded as more than presumptive (Finch and Wilkinson 1979). According to McCracken (1976), it is no more sensitive than a gram smear, but has the advantage of showing the species to which the organisms belong. The polysaccharides that take part in the test are of high molecular weight (Greenwood and Whittle 1974). They are found in the spinal fluid and, in severe cases, in the blood serum. Together with standard bacteriological methods, the CIE test enabled 84.8 per cent of cases to be correctly diagnosed (Whittle *et al.* 1975). High concentrations of polysaccharides were often accompanied by complications, such as arthritis, vasculitis and neurological damage, and pointed to a bad prognosis. The results of the CIE test are available within about one hour.

(See also Denis and his colleagues 1979, at Dakar, who examined 700 cases of purulent meningitis by electroimmunodiffusion.)

(2) Other rapid diagnostic tests

The *latex agglutination test* has been used successfully and found to be of greater sensitivity than countercurrent immuno-electrophoresis (Leinonen and Käyhty 1978). Endotoxins are produced by gram-negative organisms and can be detected by the *Limulus lysate assay*. This is a rapid test and may be positive even when the patient is undergoing antibiotic treatment (McCracken 1976, Jorgensen and Lee 1978). Special care, however, has to be taken to keep all glassware, laboratory equipment, and water used in the assay free from endotoxin (Kaplan and Feigin 1979).

A positive *nitroblue-tetrazolium test* on the cerebrospinal fluid is generally indicative of bacterial meningitis but does not distinguish between the different causative agents (Kölmel and Egri 1980).

In meningitis the *lactate* content of the cerebrospinal fluid is increased, presumably as the result of anaerobic glycolysis induced by diminished cerebral blood flow and oxygen supply. It can be measured by gas liquid chromatography (Ferguson and Tearle 1977, Hoban 1978) or by an enzyme test (Kleine *et al.* 1979). Examining spinal fluid from various types of meningitis, Gästrin, Briem and Rombo (1979) found the mean content of lactate in the spinal fluid of cases of meningococcal meningitis to be 16.5 ± 8.2 mmol/litre as against the normal content of 1.3 ± 0.54 mmol/litre. In the experience of Kleine and his colleagues (1979) a lactate content of 3.5 mmol/litre is not found except in bacterial meningitis; such a content, together with a leucocyte count in the spinal fluid of over 800/ml, practically rules out a diagnosis of cerebral tumour or thrombosis.

Post-mortem examination

In the examination of bodies at autopsy, it is important to take cultures within 12 hours of death if possible; the meningococcus dies out rapidly, and though it may be found microscopically in smears from the meninges, cultures often prove to be sterile. For cultural purposes, it is best to use the pus on the meninges; when there is no pus visible in this situation, the interior of the lateral ventricles should be carefully examined; occasionally some may be found in the posterior horn. In cases of doubt the spinal cord should be removed, and a careful search made, especially of the posterior aspect.

Prophylaxis and treatment

We have seen that the meningococcus lives as a parasite in the nasopharynx. In non-epidemic times the meningococcus is present in 2–8 per cent of healthy civilians; the actual proportion will depend to a certain extent on the time of year.

Infection of normal persons occurs from contact with cases or carriers. It might therefore be supposed that meningitis could be prevented by detecting and isolating carriers and treating them till they have become negative. Such a course, however, is impracticable; the cost and work would be altogether too great. Moreover, there is no entirely satisfactory method of ridding the nasopharynx of meningococci. Attempts at nasal disinfection by inhalation of a 1 per cent zinc sulphate or 0.1 per cent acriflavine spray were made in the first world war without any conspicuous success. More hopeful results were obtained in the second world war by use of the sulphonamides. Fairbrother (1940) found that both convalescents and healthy carriers treated with sulphapyridine cleared up rapidly. Controlled observations on servicemen in the United States showed that the carrier rate fell precipitately when sulphadiazine was given two or three times a day (Kuhns *et al.* 1943, Phair *et al.* 1944; for details see 6th edition, p. 1892).

During the course of an epidemic in Nigeria, Vollum and Griffiths (1962) gave one to four 1 gram doses of sulphadimidine snuff to about 99 per cent of a population numbering 100 000. Treatment was completed in two days. A dramatic reduction in the incidence of the disease is said to have followed, whereas in similar districts left untreated it continued to rise. Though Archibald (1962) had no success with this method, and though Vollum and Griffiths' results have been widely discredited, its rationale appears to be fundamentally sound. The clearance of nasal carriers of diphtheria bacilli and of streptococci with sulphapyridine snuff is well attested, and there seems no reason why meningococcal carriers should not yield to the same treatment. Controlled field trials of this method should certainly be carried out when occasion presents itself.

Since these observations were made, many strains of meningococci have become resistant to the sulphonamides, so that these drugs are of very doubtful value in prophylaxis (Bristow *et al.* 1965, Feldman 1965). For the present, there is little call for antibiotic prophylaxis, except for those who are in close contact with cases of meningitis, such as members of the household and persons exposed to the patient's oral secretions (Jacobson *et al.* 1976). Only in exceptional circumstances should it be needed for contacts with healthy carriers. Systemic treatment of carriers with antibiotics is ineffective (Bristow *et al.* 1965).

Mass chemoprophylaxis by the systemic route is not without its dangers. Toxic reactions may occur; sensitization may be induced; and strains that are still sensitive may become resistant. As a temporary expedient in a limited area for conserving the manpower of an army, or for controlling the disease in a closed community, it may occasionally be of value; but its application on a large scale is to be deprecated. The method of treating the whole population simultaneously with sulphadimidine snuff in order to suppress the circulation of meningococci in the community would be ideal if it was found to be effective; but until further observations are made, its usefulness must remain in doubt.

Prophylactic vaccination

Prophylactic vaccination is still in the experimental stage. As already mentioned, evidence points to the protective effect of antibodies in the blood (Goldschneider *et al.* 1969, Report 1971). These may be against the group-specific capsular polysaccharides, or the type-specific protein antigens, the latter of which are shared by a number of serogroups (Fallon 1979). Early attempts, such as those by Riding and Corkill (1932) in the Sudan, to produce a satisfactory vaccine were a failure; but more hopeful results were obtained in the trial carried out by Gold and Artenstein (1971) on Army recruits in the United States. Every fifth recruit was injected with a group C polysaccharide vaccine, prepared according to the method of Gotschlich, Liu and Artenstein (1969). During the succeeding 8 weeks 2/28 390 vaccinated subjects and 77/114 964 controls contracted group C meningococcal disease, indicating a diminution of 89.5 per cent among the vaccinated. The group C vaccine was specific; it did not protect against disease due to group B meningococci.

In Finland, Mäkelä and her colleagues (1975) vaccinated recruits with a group A polysaccharide vaccine. All but 1 per cent of these formed detectable antibodies. During the next nine months only one of the 16 458 vaccinated subjects suffered from group A meningitis as opposed to eight of the 20 748 unvaccinated controls, indicating approximately a protective effect of 89 per cent. Similarly favourable results were recorded from the use of a highly purified capsular polysaccharide vaccine against groups A, C, and Y (Farquhar *et al.* 1978; see also Gotschlich *et al.* 1978). Immunity is directed against only the meningococcal groups contained in the vaccine. According to Gold (1978), the response to an A plus C vaccine is much better in adults than in infants.

The vaccine is costly, protection is of short duration, and as epidemics are of irregular occurrence, routine vaccination in a country such as Africa, is impracticable. Given early, however, in the course of an epidemic, it should help to limit the spread of disease (Binkin and Band 1982).

Serum treatment

At one time specific antiserum was widely used in treatment, but it has now been abandoned in favour of the antibiotics. Though antiserum treatment was first introduced in Germany by Jochmann (1906), it was Flexner (1907b, 1912, 1913) and Flexner and Jobling (1908a, b) in the United States of America who were mainly responsible for its development and for the early assessment of its results. There is no justification here for describing the mode of preparation, titration, and use of the antiserum or for analysing its effect in practice; those who are interested may be referred to pp. 1444–8 of the 3rd edition of this book. Suffice it to say that, in patients treated in the early stage of the disease with sufficient quantities of a good univalent type-1 antiserum, the case-fatality rate was reduced by 50 per cent or more. Even the most favourable results, however, of antiserum treatment cannot compare with those obtained by chemotherapeutic means.

Chemotherapy

Treatment with the sulphonamides is excellent so long as the organisms are sensitive. Archibald (1962) recommended the administration of 2–2.5 grams of sulphadimidine or sulphamezathine three times a day orally or by injection; and Lapeyssonnie (1963) in Africa a single injection of a repository preparation such as sulphamethoxy-pyridazine. A high proportion of strains, particularly of groups B and C, are now resistant, and recourse has to be had to the antibiotics. Levin and Painter (1966) gave penicillin G in large doses by the intravenous infusion of an aqueous solution at 1–2-hour intervals in order to obtain a high concentration in the cerebrospinal fluid. Such treatment must be continued for at least seven days. Patients who are sensitive to penicillin and are infected with sulphonamide-resistant strains may be treated with chloramphenicol (Feldman 1965), rifampicin (Fallon 1979) or cephaloridine (Benner and Hoefrich 1972). Macfarlane and his colleagues (1979) in Nigeria found that a single injection of a long-acting oily preparation of chloramphenicol gave as good results as a 5-day course of crystalline and procaine penicillin (see also Wali et al. 1979). Cortisone therapy is recommended by Kaiser (1978), among numerous others, though not for patients exhibiting the Waterhouse-Friderichsen syndrome. As the supply of glucose to the brain is diminished in meningitis, Menkes (1979) suggests including glucose infusion in antibiotic treatment. Whatever method of treatment is used, the fatality rate is higher in infants and in patients over 60 years of age than at other times of life, when it is lowered to between 3 and 6 per cent. The earlier treatment is instituted the more favourable are the results. Treatment of the carrier state is most unsatisfactory; no drugs can eliminate it (see Vedros and Hottle 1970).

Pneumococcal meningitis

This form of meningitis occurs usually as a complication of middle-ear disease, with or without mastoiditis. It is seldom associated with lobar pneumonia (Spink 1978). Sometimes a primary pneumococcal meningitis is seen. The pneumococci are present in large numbers in the cerebrospinal fluid, which is turbid and full of pus cells; the cocci may be recognized by their appearance as gram-positive, lanceolate, capsulated, and mostly extra-cellular diplococci. Cultures are best taken on to blood agar or ascitic agar. The type of coccus may be determined in the usual way. In untreated cases the disease is almost uniformly fatal. Treatment with univalent type serum given preferably by the combined intravenous and intrathecal routes, together with sulphadiazine, may be of some value (Finland et al. 1938); but penicillin treatment is far more effective. Patients, however, who have recovered suffer only too often from residual disturbances such as loss of vision or hearing, and severe brain damage.

Streptococcal meningitis

Streptococcal meningitis is most frequently due to extension from middle-ear disease, especially in children, and to perforated wounds of the skull. The cerebrospinal fluid is turbid, and contains large numbers of pus cells and of streptococci, which are usually of the β-haemolytic type. Before the sulphonamide era recovery was regarded as exceptional, though Watson-Williams (1938) points out that early mastoidectomy, combined with repeated lumbar puncture and the administration of antistreptococcal serum, may result in a substantial proportion of recoveries. In practice, sulphonamide or penicillin therapy should be instituted as early as possible, along with appropriate surgical measures. Of recent years infections of the new-born with group-B haemolytic streptococci have become frequent; indeed they are said to be one of the commonest causes of early septicaemia and meningitis in the neonatal period (Galbraith et al. 1980). The type of organism differs from *Str. agalactiae*, which infects the cow's udder.

Tuberculous meningitis

Tuberculous meningitis is probably always secondary to a tuberculous focus in some other part of the body, but, as it is often difficult to determine the site of the primary lesion, even at autopsy, the disease not infrequently appears to be primary. Osler (Osler and McCrae 1920) states that it may occur during the 1st year of life, but is commonest between the 2nd and 5th years. It is uncommon in adults, except as a terminal infection in pulmonary tuberculosis. Delage and Dusseault (1979) in Canada, reviewing 1412 cases, found that only 18 were in adults. The laboratory diagnosis is often difficult. Lumbar puncture reveals a fluid which may be limpid, and contain only mononuclear cells. The protein content is raised; the sugar, and usually the chloride, content are lowered. The cell count generally ranges between 100 and 1000 per mm^3. Lymphocytes predominate but are accompanied by polymorphonuclear and often some plasma cells. When the fluid is allowed to stand at room temperature, a clot not infrequently forms; this should be spread on a slide, and stained by Ziehl-Neelsen. If no clot forms, the fluid should be centrifuged at high speed for half an hour, the deposit spread on albuminized slides, or fixed with methyl alcohol, and stained. with Ziehl-Neelsen. It is often only after prolonged search that a group of tubercle bacilli is found. As well as being examined microscopically, the clot or centrifuged deposit should be injected intramuscularly into a guinea-pig, which is killed after 4–8 weeks (see Chapter 51). Sometimes it is possible to culture the organisms when none is visible microscopically; for this purpose the deposit should be distributed over three or four Dorset egg tubes, and incubated for a month. Harvey (1952), who examined cerebrospinal fluid from 150 cases by all 3 methods, found tubercle bacilli in the first sample by microscopical means in 103 (69 per cent), by culture in 103 (69 per cent), and by guinea-pig inoculation in 100 (67 per cent). The combination of microscopical and chemical examination, in conjunction with the clinical findings, should enable a correct diagnosis to be made in over 90 per cent of the cases. In England and Wales up to the end of the second world war the bovine type of tubercle bacillus was responsible for about 25 per cent of the cases at all ages, but with the rapid growth of the Attested Herds scheme for the eradication of tuberculosis from cattle, and with the great increase in the pasteurization of milk for human consumption, meningitis due to this type of organism soon became uncommon.

At one time the disease was uniformly fatal, but under modern antibiotic and chemotherapeutic treatment this is no longer true. Streptomycin is mainly used, combined with PAS or isoniazid. McKenzie and his colleagues (1979) recommend isoniazid, rifampicin, and streptomycin, sometimes with ethambutol to replace para-aminosalicylic acid. Successful results can be expected only when treatment is begun early and continued for several months. Failure may be due to the acquisition of drug resistance by the organisms during this long course of therapy. In a series of 1491 cases in Canada treated mainly during the 1970s, 35 per cent proved fatal (Delage and Dusseault 1979); but in a series of 52 cases in Scotland treated intensively—13 being operated on surgically and 43 receiving corticosteroid therapy—the fatality rate was only 15 per cent (Kennedy and Fallon 1979). Cerebral sequelae may be expected in 10 to 25 per cent of those who recover.

Haemophilus influenzae meningitis

This form is met with sporadically; it is not commoner during an influenza epidemic. Most cases appear to be primary. It was first described by Slawyk (1899), and has since been recognized by a number of workers. Neal and her colleagues (1934) met with 111 cases of this type among a total of 2727 cases of meningitis. The incidence is highest in childhood. It is rare under 2 months or over 6 years of age (Fothergill and Wright 1933). About 60–80 per cent of cases occur during the first 2 years of life. The risk of contracting the disease in close contacts is said to be 2.3 per cent (Glode *et al.* 1980). In the USA the disease has increased during the last 50 years or so, and is now much commoner than meningococcal meningitis. It is responsible for something like 10 000 cases annually (Robbins 1975). The disease closely simulates cerebrospinal meningitis. It usually lasts 10–20 days, and when untreated has a fatality rate of 95 per cent. The cerebrospinal fluid is under pressure, is cloudy or purulent, has a high cell count, especially of polymorphs, and shows a variable increase in protein and often a moderate decrease in sugar. Gram-negative bacilli are present with the typical Pfeiffer morphology, but sometimes accompanied by long thread forms. In the early stage of the disease they are intracellular, but later they become free in the fluid. On suitable media they can be cultivated from the spinal fluid, and often from the blood, and from the purulent joint exudates that occur not infrequently as a complication of this disease. Their nutritional requirements and their indole production are typical of *H. influenzae*, but they are distinct in being capsulated, in producing characteristic colonies on Levinthal's medium, in forming a type-specific polysaccharide, and in being considerably more virulent for laboratory animals; they correspond to Pittman's type b (see Pittman 1931, Mulder 1939). Occasional strains, however, of the respiratory type are encountered (Gordon, Woodcock and Zinnemann 1944). Inoculated intraperitoneally into rabbits, guinea-pigs, rats, and mice, they often give rise to a fatal septicaemia and to purulent effusions in the serous cavities. Diagnosis may be made in the laboratory by microscopical examination of the spinal fluid, aided when practicable by immunofluorescence (Page *et al.* 1961), by the capsule-swelling reaction with specific antiserum in hanging-drop preparations of the spinal fluid, by culture, and by demonstration of the specific polysaccharide in the spinal fluid either by a precipitin or a haemagglutination test (Warburton, Keogh and Williams 1949). Rapid diagnostic tests include counterimmuno-electrophoresis (CIE), latex agglutination, and the *Limulus* lysate assay. Care has to be taken in the CIE test, because the antisera to type b may cross-react with K1 strains of *Esch. coli* and with the polysaccharide antigens of types 6, 15, 29 and 35 pneumococci (Kaplan and Feigin 1979). The test is valuable in that the influenzal polysaccharide can be detected not only in the spinal fluid but in pericardial and joint fluids when pericarditis or septic arthritis is present, and in the urine. For treatment, chloramphenicol is probably the most reliable drug (McGowan *et al.* 1974). Next to it comes ampicillin. Though treatment may be life-saving, about 30 per cent of the survivors are left with residual defects of one sort or another. Rifampicin is sometimes used prophylactically in contacts.

Non-suppurative (aseptic) meningitis

We may refer here to a form of meningitis that follows contamination of the cerebrospinal fluid during lumbar puncture and spinal anaesthesia. It is usually of a low-grade type—sometimes referred to as serous meningitis—but fatal cases occur from time to time. The disease is caused by organisms which under ordinary conditions are non-pathogenic, such as *Pseudomonas* and '*Achromobacter*'. Contamination is often due to the use of so-called sterile water for rinsing syringes. To prevent this totally unnecessary disease, strict asepsis should be practised. All apparatus used for lumbar puncture should be sterilized by heat— preferably dry heat—and water for rinsing and cooling purposes, except from sealed autoclaved bottles used on one occasion only, should be avoided (see Smith and Smith 1941, Garrod 1946, Report 1948).

Other forms of meningitis

Occasionally meningitis may be caused by organisms, such as *Bacillus anthracis*, *Esch. coli*, Friedländer's bacillus (Gordon and Norton 1930), *Salmonella typhi*, *Salm. enteritidis* (Stevenson and Wills 1933), *Salm.* spp. (Burton *et al.* 1977), *Salm. paratyphi B* (Patterson 1942), *Pseudomonas mallei*, *Brucella suis* (Hansmann and Schencken 1932, Hartley *et al.* 1934), *Br. abortus* (Magoffin *et al.* 1949), the gonococcus (Strumia and Kohlhas 1933, Branham *et al.* 1938), *Diplococcus mucosus* (Cowan 1938, Bray and Cruickshank 1943, Christie and Cook 1947), *Neisseria lactamica* (Kuzemenská *et al.* 1976, Lauer and Fisher 1976, Holten *et al.* 1978), *Mima polymorpha* (Olafsson *et al.* 1958), *Listeria monocytogenes* (see Chapter 23), *Gaffkya tetragena* (Peron *et al.* 1958), *Flavobacterium meningosepticum* (King 1959, Cabrera and Davis 1961, Madruga *et al.* 1970), *Neisseria subflava* (Lewin and Hughes 1966), *Streptococcus agalactiae* (Kexel and Schönbohm 1965), *Str. suis*, *Clostridium perfringens* (Willis and Jacobs 1934), *Bacteroides convexus* (Werner 1964), *Bacteroides fragilis* (Warner *et al.* 1979), *Nocardia asteroides* (Gilligan *et al.* 1962), *Actinomyces bovis* (Henry 1910), *Leptospira icterohaemorrhagiae* (Marie and Gabriel 1935, Mollaret and Erber 1935), *Leptospira canicola* (Chapter 74), and *Enterobacter sakazaki* (Muytjens *et al.* 1983). Apart from cases of meningitis secondary to infection with *Entamoeba his-* *tolytica* or *Histoplasma capsulatum* (Plouffe and Fass 1979), primary cases of amoebic meningitis have been described, caused by *Acanthamoeba* and *Naegleria fowleri* (see Galbraith *et al.* 1980). It may be added that the presence of specific polysaccharides in the cerebrospinal and other body fluids, and of C-reactive protein in the patient's blood, distinguishes bacterial from viral meningitis (Mäkelä 1982)

(For a brief review of various forms of meningitis at different ages, see Grumbach and Kikuth 1958, Turner 1959, Paul 1964, Berman and Banker 1966, Spink 1978, Report 1979; for meningitis caused by filtrable viruses, see Volume 4; and for a general review of meningitis, see Mims 1977, McGee and Kaiser 1979.)

References

Albrecht, H. and Ghon, A. (1901) *Wien. klin. Wschr.* **14**, 984.

Alexander, H. E. and Rake, G. (1937) *J. exp. Med.* **65**, 317.

Amies, C. R. (1967) *Canad. J. publ. Hlth* **58**, 296.

Archibald, H. M. (1962) *J. trop. Med. Hyg.* **65**, 196.

Armstrong, C., Fotheringham, J. B., Hood, A., Little, C. J. H. and Thompson, T. O. (1931) *J. R. Army med. Cps* **57**, 321.

Aycock, W. L. and Mueller, J. H. (1950) *Bact. Rev.* **14**, 115.

Baeslack, F. W., Bunce, A. H., Brunelle, G. C., Fleming, J. S., Klugh, G. F., McClean, E. H. and Salomon, A. V. (1918) *J. Amer. med. Ass.* **70**, 684.

Banks, H. S. (1948) *Lancet*, **ii**, 635, 677.

Bell, A. S. G. (1920a) *Spec. Rep. Ser. med. Res. Coun., Lond.* No. 50, p. 57; (1920b) *Ibid.* No. 50, p. 63.

Benner, E. J. and Hoefrich, P. D. (1972) In: *Infectious Diseases*, p. 931. Ed. by P. D. Hoefrich. Harper and Row, Hagerstown, Md.

Berman, P. H. and Banker, P. Q. (1966) *Pediatrics, Springfield* **38**, 6.

Bevan-Jones, H. and Miller, D. L. (1967) *Mon. Bull. Minist. Hlth* **26**, 22.

Biegeleisen, J. Z., Mitchell, M. S., Marcus, B. B., Rhoden, D. L. and Blumberg, R. W. (1965) *J. Lab. clin. Med.* **65**, 976.

Binkin, N. and Band, J. (1982) *Lancet* **ii**, 315.

Branham, S. E. (1930) *Publ. Hlth Rep., Wash.* **45**, 845; (1932) *J. Immunol.* **23**, 49; (1940) *Bact. Rev.* **4**, 59.

Branham, S. E., Mitchell, R. H. and Brainin, W. (1938) *J. Amer. med. Ass.* **110**, 1804.

Bray, P. T. and Cruickshank, J. C. (1943) *Brit. med. J.* **i**, 601.

Bristow, W. M., Peenen, P. F. D. van and Volk, R. (1965) *Amer. J. publ. Hlth* **55**, 1039.

Brown, J. W. and Condit, P. K. (1965) *Calif. Med.* **102**, 171.

Burton, B. K., Marr, T. J., Traisman, H. S. and Davis, A. T. (1977) *Amer. J. Dis. Child.* **131**, 1031.

Cabrera, H. A. and Davis, G. H. (1961) *Amer. J. Dis. Child.* **101**, 289.

Christie, A. L. M. and Cook, G. T. (1947) *J. Hyg., Camb.* **45**, 149.

Cohen S. M. (1936) *J. Immunol.* **30**, 203.

Cowan, S. T. (1938) *Lancet* **ii**, 1052.

Craven, D. E., Frasch, C. E., Mocca, L. F., Rose, F. B. and Gonzalez, R. (1979) *J. clin. Microbiol.* **10**, 302.

Delage, G. and Dusseault, M. (1979) *Canad. med. Ass. J.* **120**, 305.

Denis, F., Chiron, J. P., M'Boup, S., Cadoz, M. and Mar, I. D. (1979) *Méd. Mal. infect.* **9**, 353.

Dopter, C. (1921) *L'Infection Meningococcique.* Paris.

Dudley, S. F. and Brennan, J. R. (1934) *J. Hyg., Camb.* **34**, 525.

Eagleton, A. J. (1919-20) *J. Hyg., Camb.* **18**, 264.

Elser, W. J. and Huntoon, F. M. (1909) *J. med. Res.* **20**, 371.

Fairbrother, R. W. (1940) *Brit. med. J.* **ii**, 859; (1947) *J. clin. Path.* **1**, 10.

Fallon, R. J. (1979) In *Recent Advances in Infection*, p. 77. Ed. by D. Reeves and A. Geddes. Churchill Livingstone, Edinburgh.

Farquhar, J. D. *et al.* (1978) *Proc. Soc. exp. Biol. Med.* **157**, 79.

Farrell, D. G. and Dahl, E. V. (1966) *J. Amer. med. Ass.* **198**, 1189.

Feldman, H. A. (1965) *J. Amer. med. Ass.* **196**, 391.

Ferguson, I. R. and Tearle, P. V. (1977) *J. clin. Path.* **30**, 1163.

Finch, C. A. and Wilkinson, H. W. (1979) *J. clin. Microbiol.* **10**, 519.

Finland, M., Brown, J. W. and Rauh, A. E. (1938) *New Engl. J. Med.* **218**, 1033.

Flexner, S. (1907a) *J. exp. Med.* **9**, 142; (1907b) *Ibid.* **9**, 168; (1912) *J. State Med.* **20**, 257; (1913) *J. exp. Med.* **17**, 553.

Flexner, S. and Jobling, J. W. (1908a) *J. exp. Med.* **10**, 141; (1908b) *Ibid.* **10**, 690.

Fothergill, L. D. and Wright, J. (1933) *J. Immunol.* **24**, 273.

Galbraith, N. S., Forbes, P. and Mayon-White, R. T. (1980) *Brit. med. J.* **ii**, 427.

Garrod, L. P. (1946) *Brit. med. Bull.* **4**, 106.

Gästrin, B., Briem, H. and Rombo, L. (1979) *J. infect. Dis.* **139**, 529.

Gästrin, B., Kallings, L. O. and Marcetic, A. (1968) *Acta path. microbiol. scand.* **74**, 371.

Gilligan, B. S., Williams, I. and Perceval, A. K. (1962) *Med. J. Aust.* **ii**, 747.

Glode, M. P., Daum, R. S., Goldmann, D. A. and Leclair, J. (1980) *Brit. med. J.* **i**, 899.

Glover, J. A. (1920) *Spec. Rep. Ser. med. Res. Coun., Lond.* No. 50, p. 133.

Goeters, W. (1954) *Ergebn. Hyg. Bakt. ImmunForsch.* **28**, 1.

Gold, R. (1978) *Infect. Dis. Rev.* **5**, 93.

Gold, R. and Artenstein, M. S. (1971) *Bull. World Hlth Org.* **45**, 279.

Goldschneider, I., Gotschlich, E. C. and Artenstein, M. S. (1969) *J. exp. Med.* **129**, 1307, 1327.

Gordon, J. E. and Norton, J. E. (1930) *J. prev. Med., Baltimore* **4**, 339.

Gordon, J. Woodcock, H. E. de C. and Zinnemann, K. (1944) *Brit. med. J.* **i**, 779.

Gordon, M. H., Hine, T. G. M. and Flack, M. (1916) *Brit. med. J.* **ii**, 678.

Gotschlich, E. C., Austrian, R., Cvjetanović, B. and Robbins, J. B. (1978) *Bull. World Hlth Org.* **56**, 509.

Gotschlich, E. C., Liu, T. Y. and Artenstein, M. S. (1969) *J. exp. med.* **129**, 1349, 1367.

Grados, O. and Ewing, W. H. (1970) *J. infect. Dis.* **122**, 100.

Greenwood, B. M. and Whittle, H. C. (1974) *Clin. exp. Immunol.* **16**, 413.

Griffith, F. (1917) *Rep. loc. Govt Bd, New Ser.* No. 111, p. 52.

Grumbach, A. and Kikuth, W. (1958) *Die Infektionskrankheiten des Menschen und ihre Erreger*, G. Thieme Verlag, Stuttgart.

Hansmann, G. H. and Schencken, J. R. (1932) *Amer. J. Path.* **8**, 435.

Hartley, G. A., Millice, G. S. and Jordan, P. H. (1934) *J. Amer. med. Ass.* **103**, 251.

Harvey, R. W. S. (1952) *Brit. med. J.* **ii**, 360.

Henry, H. (1910) *J. Path. Bact.* **14**, 164.

Herrick, W. W. (1918) *J. Amer. med. Ass.* **71**, 612.

Hirsch, A. (1886) *Handbook of Geographical and Historical Pathology*, Trans. New Sydenham Soc., Vol. 3, p. 547. London.

Hoban, S. (1978) *Canad. J. med. Technol.* **40**, 171.

Holten, E., Bratlid, D. and Bøvre, K. (1978) *Scand. J. infect. Dis.* **10**, 36.

Jacobson, J. A., Camargos, P. A. M., Ferreira, J. T. and McCormick, J. B. (1976) *Amer. J. Epidem.* **104**, 552.

Jochmann, G. (1906) *Dtsch. med. Wschr.* **32**, 789.

Jorgensen, J. H. and Lee, J. C. (1978) *J. clin. Microbiol.* **7**, 12.

Kaiser, D. H. (1978) *Dtsch. med. Wschr.* **103**, 1642.

Kaplan, S. L. and Feigin, R. D. (1979) In: *Rapid Diagnosis in Infectious Disease.* Ed. by M. W. Rytel. CRC Press, Boca Raton, Florida.

Kennedy, D. H. and Fallon, R. J. (1979) *J. Amer. med. Ass.* **241**, 264.

Kexel, G. and Schönbohm, S. (1965) *Dtsch. med. Wschr.* **90**, 258.

King, E. O. (1959) *Amer. J. clin. Path.* **31**, 241.

Kleine, T. O. *et al.* (1979) *Dtsch. med. Wschr.* **104**, 553.

Kölmel, H. W. and Egri, T. (1980) *Infection* **4**, 142.

Kuhns, D. M., Nelson, C. T., Feldman, H. A. and Kuhn, L. R. (1943) *J. Amer. med. Ass.* **123**, 335.

Kutscher, K. (1906) *Dtsch. med. Wschr.* **32**, 1071.

Kuzemenská, P., Burian, V., Frasch, C. E. and Švandová, E. (1977) *Bull World Hlth Org.* **55**, 659.

Kuzemenská, P. *et al.* (1976) *Zbl. Bakt., I. Abt. Orig.* **A236**, 559.

Lapeyssonnie, L. (1963) *Bull. World Hlth Org.* **28**, Suppl., pp. 1-114.

Lauer, B. A. and Fisher, C. E. (1976) *Amer. J. Dis. Child.* **130**, 198.

Leinonen, M. and Käyhty, H. (1978) *J. clin. Path.* **31**, 1172.

Levin, S. and Painter, M. B. (1966) *Ann. intern. Med.* **64**, 1049.

Lewin, R. A. and Hughes, W. T. (1966) *J. Amer. med. Ass.* **195**, 821.

Lingelsheim, von (1905) *Dtsch. med. Wschr.* **31**, 1017.

Low, R. B. (1916) *Rep. loc. Govt Bd, New Ser.* No. 110, p. 115.

McCracken, G. H. (1976) *J. Pediat.* **88**, Pt 1, 706.

Macfarlane, J. T. *et al.* (1979) *Trans. R. Soc. trop. Med. Hyg.* **73**, 693.

McGee, Z. A. and Kaiser, A. B. (1979) In: *Principles and Practice of Infectious Diseases*, p. 738. Ed by G. L. Mandell, R. G. Douglas and J. E. Bennett. John Wiley and Sons, New York.

McGowan, J. E. *et al.* (1974) *J. infect. Dis.* **130**, 119.

McKenzie, M. S., Burckart, G. J. and Ch'ien, L. T. (1979) *Clin. Pediat.* **18**, 75, 78, 82.

McLean, S. and Caffey, J. (1931) *Amer. J. Dis. Child.* **42**, 1053.

Madruga, M., Zanon, U., Pereira, G. M. N. and Galvão, A. C. (1970) *J. infect. Dis.* **121**, 328.

Maegraith, B. G. (1935) *Lancet* **i**, 545.

Magoffin, R. L., Kabler, P., Spink, W. W. and Fleming, D. (1949) *Publ. Hlth Rep., Wash.* **64**, 1021.

Mäkelä, P. H. (1982) *Scand. J. infect. Dis., Suppl.* No. 36, p. 111.

Mäkelä, P. H. *et al.* (1975) *Lancet* **ii**, 883.

Marie, J. and Gabriel, P. (1935) *Bull. Mém. Soc. méd. Hôp., Paris* **51**, 1454.

Menkes, J. H. (1979) *Lancet* ii, 559.

Mims, C. (1977) *The Pathogenesis of Infectious Disease.* Academic Press, London.

Mollaret, P. and Erber, B. (1935) *Bull. Mém. Soc. méd. Hôp. Paris* **51**, 1632.

Muir, R. (1919) *J. R. Army med. Cps* **33**, 404.

Mulder, J. (1939) *J. Path. Bact.* **48**, 175.

Murray, E. G. D. (1929) *Spec. Rep. Ser. med. Res. Coun., Lond.* No. 124, p. 111.

Muytjens, H. L. *et al.* (1983) *J. clin. Microbiol.* **18**, 115.

Neal, J. B., Jackson, H. W. and Appelbaum, E. (1934) *J. Amer. med. Ass.* **102**, 513.

Netter, A. and Debré, R. (1911) *La Méningite Cérébro-spinale.* Paris.

Norton, J. F. and Baisley, I. E. (1931) *J. prev. Med.* **5**, 357.

Ojeda, V. J. (1980) *Pathology (Sydney)* **12**, 429.

Olafsson, M., Lee, Y. C. and Abernethy, T. J. (1958) *New Engl. J. Med.* **258**, 465.

Osler, W. and McCrae, T. (1920) *The Principles and Practice of Medicine*, 9th edn. London.

Ostermann, A. (1906) *Dtsch. med. Wschr.* **32**, 414.

Page, R. H., Caldroney, G. L. and Stulberg, C. S. (1961) *Amer. J. Dis. Child.* **101**, 155.

Patterson, W. H. (1942) *Lancet* **i**, 678.

Paul, H. (1964) *The Control of Diseases*, 2nd edn. Churchill Livingstone, Edinburgh.

Peron, M. L., Fielding, R. T. and Nelson, H. M. (1958) *J. Pediat.* **53**, 484.

Petrie, G. F. (1932) *Brit. J. exp. Path.* **13**, 380.

Phair, J. J., Schoenbach, E. B. and Root, C. M. (1944) *Amer. J. publ. Hlth* **34**, 148.

Pittman, M. (1931) *J. exp. Med.* **53**, 471.

Plouffe, J. F. and Fass, R. J. (1979) *Ann intern. Med.* **92**, 189.

Rake, G. (1933a) *J. exp. Med.* **57**, 561; (1933b) *Ibid.* **58**, 375; (1934) *Ibid.* **59**, 553; (1935) *Ibid.* **61**, 545.

Reece (1916) *Rep. loc. Govt Bd, New Ser.* No. 110.

Report (1917) *Spec. Rep. Ser. med. Res. Coun., Lond.* No. 3; (1920) *Ibid.* No. 50; (1948) *Mon. Bull. Minist. Hlth Lab.*

Serv. **7**, 23; (1969) *WHO Chron.* **23**, 47; (1971) *J. Amer. med. Ass.* **216**, 1419; (1974a) *Morbid. and Mortal.* **23**, 349; (1974b) *J. Amer. med. Ass.* **235**, 261; (1979) *Morbid. and Mortal.* **28**, 277.

Riding, D. and Corkill, N. K. (1932) *J. Hyg., Camb.* **32**, 258.

Robbins, J. B. (1975) In: *Microbiology,* p. 400. Ed. by D. Schlesinger. *Amer. Soc. Microbiol.*

Rolleston, H. (1919) *Lancet* **i**, 541, 593, 645.

Scott, W. M. (1916) *Rep. loc. Govt Bd, New Ser.* No. 110, p. 56; (1917) *Ibid.* No 114, p. 111.

Severin, W. P. J., Ruys, A. C., Bijkerk, H. and Butter, M. N. W. (1969) *Zbl. Bakt.* **210**, 364.

Simpson, R. E. H. (1958) *Roy. Soc. Hlth J.* **78**, 593.

Slawyk (1899) *Z. Hyg. InfektKr.* **32**, 443.

Smales, O. R. C. and Rutter, N. (1979) *Brit. med. J.* **i**, 588.

Smith, W. and Smith, M. M. (1941) *Lancet* **ii**, 783.

Spink, W. W. (1978) *Infectious Diseases.* Dawson & Son Ltd, Folkestone.

Stevenson, F. N. and Wills, L. K. (1933) *Lancet* **ii**, 1084.

Strumia, M. M. and Kohlhas, J. J. (1933) *J. infect. Dis.* **53**, 212.

Thayer, J. D. and Martin, J. E. (1966) *Publ. Hlth Rep., Wash.* **81**, 559.

Tompkins, V. N. (1943) *J. Amer. med. Ass.* **123**, 31.

Turner, E. K. (1959) *Med. J. Aust.* **ii**, 806.

Vedros, N. A. and Hottle, G. A. (1970) In: *Infectious Agents and Host Reactions*, p. 271. Ed. S. Mudd. W. B. Saunders Co., Philadelphia.

Vollum, R. L. and Griffiths, P. W. W. (1962) *J. clin. Path.* **15**, 50.

Waddy, B. B. (1957) *J. trop. Med. Hyg.* **60**, 179, 218.

Wali, S. S. *et al.* (1979) *Trans. R. Soc. trop. Med. Hyg.* **73**, 698.

Warburton, M. F., Keogh, E. V. and Williams, S. W. (1949) *Med. J. Aust.* **i**, 135.

Warner, J. F., Perkins, R. L. and Cordero, L. (1979) *Arch. intern. Med.* **139**, 167.

Watson-Williams, E. (1938) *Lancet* **i**, 807.

Weichselbaum, A. (1887) *Fortschr. Med.* **5**, 573, 620.

Werner, H. (1964) *Zbl. Bakt.* **194**, 203.

Westenhoeffer (1906) *Berl. klin. Wschr.* **43**, 1267, 1313.

Whittle, H. C. *et al.* (1975) *Amer. J. Med.* **58**, 823.

Willis, A. T. and Jacobs, S. I. (1964) *J. Path. Bact.* **88**, 312.

66

Gonorrhoea

A. E. Wilkinson

Introductory

That gonorrhoea has a history stretching back into antiquity we know, but we are entirely ignorant of the time and place of its first appearance. There is little to suggest that the disease has altered in its nature during the passage of time; it appears to run its course now in the same way as it has done for hundreds of years. It therefore differs from many diseases in which the clinical picture alters with the centuries. The explanation of this may lie in the absence of both natural and acquired immunity in man.

The causal agent, the gonococcus (*Neisseria gonorrhoeae*), was discovered by Neisser in 1879, who found it in 35 cases of gonorrhoea which had lasted for varying lengths of time, from 3 days to 13 weeks. He failed to find it in non-gonorrhoeal pus, such as that accompanying chancres and buboes or in simple vaginal discharges; he demonstrated its presence in seven cases of ophthalmia neonatorum, and in two cases of adult ophthalmia. Though unable through illness to complete his investigation, he placed the aetiological role of the gonococcus on a footing which has never since been seriously challenged. It was left, however, to Bumm (1885*a*, *b*) to cultivate the organism, and by inoculation experiments on human subjects to demonstrate its pathogenicity in pure cultures. The history of gonorrhoea has been reviewed by Morton (1977).

Gonorrhoea in adults

Gonorrhoea is an acute infectious disease, generally characterized by primary invasion of the genito-urinary tract, and sometimes complicated by secondary disturbances of greater or less severity.

By far the commonest method of infection is by sexual intercourse. In men the anterior urethra is first attacked. After an incubation period of 3 to 5 days, a mucoid discharge appears, which rapidly becomes purulent. About ten per cent of infections in men may be symptomless, although the patient may remain infectious for some time (Handsfield *et al.* 1974). At first some gonococci may be extracellular, but soon they are taken up by the pus cells, which ingest them in large numbers; as many as 100 may be counted in a single cell.

In patients not successfully treated at this stage, gonococci may spread to the posterior urethra and prostate; this was largely responsible for the chronicity of the disease before the advent of antibiotics. Gonococcal proctitis may result from homosexual intercourse. Pharyngeal infection with gonococci is being increasingly recognized in patients who have had orogenital sexual contacts, especially in homosexuals. It may produce a pharyngitis but is often symptomless (Brø-Jorgensen and Jensen 1973).

In women, the cervix is first invaded and later the urethra. Uncomplicated infections may produce few symptoms and up to 50 per cent of patients may be unaware that they are infected. Involvement of the ano-rectal region, usually symptomless, occurs in about 40 per cent of infections in women (Scott and Stone 1966, Schmale *et al.* 1969).

Complications of gonorrhoea in adults

Although infection may remain localized, it may spread by direct continuity and involve glands draining into the genito-urinary tract, producing local complications. Much less frequently it may invade the blood stream and produce metastatic lesions. In males, affection of Littré's glands and the sub-epithelial tissues may progress to fibrosis and stricture formation or to the development of a periurethral abscess. The prostate, seminal vesicles, epididymes and Cowper's glands may be invaded with the production of inflammatory lesions. These complications of gonorrhoea are very much less frequent since the introduction of rapidly effective treatment by antibiotics. In women, the main local complications are due to the involvement of the glands of Skene and Bartholin or to spread of infection upward to the Fallopian tubes. In both sexes infection may be spread to the eye by direct contact, producing an acute conjunctivitis.

Spread through the blood stream occurs in about 1 per cent of patients with gonorrhoea (Holmes *et al.* 1971). It occurs more often in women than in men and is often associated with pregnancy or recent menstruation. There is mild fever with flitting polyarthritis or tenosynovitis. Single large joints may be affected; an acute purulent exudate collects but most workers have been able to isolate gonococci from this in only about half the cases. In the early stages blood cultures may be positive, but by the time arthritis has developed they are usually found to be negative. A sparse papulo-pustular rash may develop, usually over the extremities (Abu-Nassar *et al.* 1963, Förstrom *et al.* 1972); this is thought to be a reaction resulting from an embolus. It is difficult to culture gonococci from these skin lesions but Tronca and his colleagues (1974) found gonococci in 20 of 35 smears from lesions by direct immunofluorescence. Endocarditis, pericarditis and meningitis are rare manifestations. Strains isolated from disseminated infections have some properties which contrast with those from infections which remain localized; they may cause disseminated infections in sexual partners, are usually very sensitive to penicillin (Wiesner *et al.* 1973), belong to auxotypes with requirements for arginine, hypoxanthine and uracil (Knapp and Holmes 1975), and tend to resist the bactericidal action of normal human serum. Disseminated infections have been reviewed by Holmes, Counts and Beaty (1971) and Handsfield (1977).

Gonorrhoea in children

Ophthalmia neonatorum results from contamination during birth, but less than 30 per cent of cases are due to gonococcal infection. This produces a very rapid, acute conjunctivitis which unless promptly treated may irretrievably damage the cornea and lead to impaired vision or blindness. It has become much less common since the introduction of Credé's preventive treatment. The meningococcus can also produce a conjunctivitis in children, but the commonest cause of ophthalmia neonatorum appears to be a chlamydial infection.

Formerly **vulvovaginitis** was often seen as an institutional disease transmitted by infected towels or linen. An outbreak of severe infection in children associated with arthritis and proctitis was described by Cowperman (1927); transmission was probably by inadequately sterilized rectal thermometers. Today, vulvovaginitis is most often seen as a 'family' infection, often from the child sharing a bed with infected parents and coming into contact with soiled bedding. It may also result from sexual assault; both these modes are illustrated in the cases described by Shore and Winkelstein (1971) in an Alaskan community. Nongonococcal vulvovaginitis is much commoner than the gonococcal form; Sersiron and Roiron (1967) found only 21 infections with gonorrhoea among 872 girls with leucorrhoea; most of these were in children between 2 and 5 years of age. Other organisms, including neisseriae other than the gonococcus, may be present; nine such strains were found in this series, and full identification is necessary if a correct diagnosis is to be made. This is especially true of cases in which sexual assault may have occurred. (For review see Litt, Edberg and Finberg 1974.)

The gonococcus attacks columnar rather than squamous epithelium. Lactic acid produced by lactobacilli from glycogen in the stratified squamous epithelium protects the vagina against gonococci in adults. Before puberty the vagina and vulva are lined by an immature squamous epithelium which lacks glycogen and so are not protected against infection. Vulvovaginitis and conjunctivitis are commoner in children than in adults, whereas complications involving the internal genito-urinary organs are more frequently seen in adults. Acute pelvic peritonitis may, however, occur in children. The mucosa of the urethra and rectum are susceptible throughout life. Both in adult gonorrhoea in the female and in vulvovaginitis of children the rectum is stated to be frequently affected—in 20-30 per cent of the cases.

The pathogenesis of gonorrhoea

Changes in the urethral mucosa appear within three to four days after infection. Gonococci rapidly reach the subepithelial connective tissue, in which the capillaries become dilated and there is an infiltration of polymorphonuclear leucocytes and smaller numbers of mononuclear cells. The overlying epithelium is shed in a patchy manner and eventually replaced by an ingrowth of layers of squamous epithelial cells. If the subepithelial inflammatory changes are sufficiently severe to lead to the production of much fibrous tissue, later contraction of this may cause the formation of a stricture or occlusion of the Fallopian tubes. For a review of the histopathology of gonorrhoea see Harkness (1948).

Studies with the electronmicroscope have shown that gonococci become attached to the columnar epithelial cells by their pili (Ward and Watt 1972, Ward, Watt and Robertson 1974). This attachment must be rapid and sufficiently firm to prevent gonococci being washed off by the flow of urine, as urination immediately after intercourse does not appear to lessen the risk of infection (Bernfeld 1972). Although T1 and T2 colonies of gonococci are pilated *in vitro*, pili are rarely seen on electronmicroscopy of urethral pus (Novotny, Short and Walker 1975, Arco, Bullard and Duncan 1976); possibly the pilus protein may not always be extruded as filaments from gonococci *in vivo* but may form part of the outer wall of the organism. Once attached to the surface of epithelial cells the gonococci are taken into the cell by phagocytosis and can be seen lying within membrane-lined vesicles in the cytoplasm. Penetration between epithelial cells is not thought to occur. After passing through the epithelial cells gonococci are set free into the subepithelial connective tissue where they become exposed to the action of natural antibody and complement. This leads to a polymorphonuclear response with phagocytosis of the gonococci which is characteristic of the disease. Whether some organisms survive phagocytosis and are protected against the action of penicillin is not yet certain. Swanson and his colleagues (1975) have described a surface protein, leucocyte-association factor, which is thought to be more closely associated with the interaction of gonococci with leucocytes than the possession of pili by the organisms. Capsules have been seen on freshly isolated gonococci (James and Swanson 1977, Richardson and Sadoff 1977) and these may perhaps affect the susceptibility to phagocytosis. Recent studies on the interaction between gonococci and leucocytes have been reviewed by Roberts (1977). Novotny and his colleagues (1977) have described clusters of dividing gonococci surrounded by the debris of degenerated phagocytic cells in gonococcal pus. These 'infectious units' appear to attach to epithelial cells, and the covering of material from the host may protect the enclosed gonococci from phagocytosis and humoral defence mechanisms (Novotny and Short 1977). Toxic material from gonococci may also play a part in pathogenesis. In organ cultures of human Fallopian tubes both the supernatant fluid from cultures infected with gonococci and purified gonococcal lipopolysaccharide produce loss of ciliary activity with eventual sloughing of the ciliated epithelium (Gregg *et al.* 1981, McGee, Johnson and Taylor-Robinson 1981).

The response of the host is affected by the nature of the infecting strain and ranges from an acute inflammation to an infection so mild that it passes unnoticed by the patient, who is nevertheless infectious for others. Strains with requirements for arginine, hypoxanthine and uracil (AHU⁻ strains) are often found in symptomless infections (Crawford, Knapp, Hale and Holmes 1977), but invasive strains which disseminate through the blood stream and produce little local reaction are frequently of this auxotype (Knapp and Holmes 1975), although Thompson and his colleagues (1978) found a requirement for proline to be more common. These invasive strains are extremely sensitive to penicillin and have a surface receptor for an IgG blocking antibody which protects them from the bactericidal action of normal human serum and complement (McCutchan *et al.* 1978). They have been reported to have an outer membrane protein which is rarely found in strains that produce localized infection, and which is independent of the auxotype (Hildebrandt and Buchanan 1978).

For general reviews of the interaction of gonococci and epithelial surfaces see Ward and Watt (1975) and Watt and Lambden (1978).

Reproduction of the disease in man

Bumm (1885*a*) produced infection in man by injecting a pure culture into the urethra. Subsequent workers have had varied success; Mahoney and his colleagues (1946) had unpredictable results with cultures but were more successful in reproducing the disease in male volunteers with gonococcal pus. These variations are explained by the observations of Kellogg and his colleagues (1963), who found, like Atkin (1925), that, when freshly isolated, gonococci were largely of two colonial forms, types 1 and 2; these were virulent when inoculated into volunteers. On repeated subculture, two further types of colony, 3 and 4, replaced the original types and were avirulent. The virulence of a strain could be maintained by selective subculture of type 1 colonies, by the passage of old strains in the allantoic fluid of fertile hens' eggs (Walsh *et al.* 1963), or by passage in the exudate collecting in plastic chambers implanted subcutaneously into guinea-pigs (Veale *et al.* 1975). Ward, Watt and Glynn (1970) have shown that gonococci growing *in vivo* have a protective factor, possibly a surface antigen, which confers resistance to natural antibody and complement; this is lost on subculture. Freshly isolated strains of colonial types 1 and 2 carry pili, but avirulent types 3 and 4 do not. Pili may be associated with the ability of the organism to effect a lodgement on a mucosal surface (Ward and Watt 1972). Pilated gonococci attach more readily to human epithelial cells, red blood cells and sperm, and are more resistant to phagocytosis than gonococci without pili; they also show twitching motility which may facilitate their spread over a mucosal surface. Besides the Kellogg colonial types, variations in the colour, opacity and consistency of gonococcal colonies have been described, some of which are linked with the presence of particular outer membrane proteins. These have been reviewed by Swanson (1977).

Although humoral and local antibodies are produced during infection, the titre tends to be low, especially as treatment is usually given early in the disease. One attack does not confer immunity; indeed repeated infections are all too common, partly because of poor antibody response and partly because of antigenic diversity among gonococci.

Diagnosis

The diagnosis of gonorrhoea rests on the demonstration of the causative organism. In acute infections in the male and female the finding of typical clusters of gram-negative intracellular diplococci in smears of exudate affords a presumptive diagnosis (Fig. 66.1).

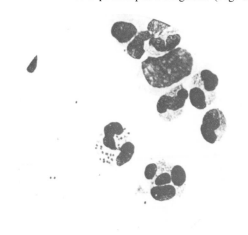

Fig. 66.1 *Neisseria gonorrhoeae* in film of pus from acute gonorrhoea (× 1000).

Single stains, such as methylene blue, should not be used. A definitive diagnosis can be made only by isolation in culture followed by identification tests. Cultural methods are superior to the examination of stained smears. Gram-stained smears will detect about 95 per cent of cases in males but only about 60 per cent of infections in females. Both smears and cultures should be taken, as occasionally the smear may be positive but the culture negative. In men, specimens should be taken from the urethra and in homosexuals from the rectal mucosa. If the urethral discharge is scanty or absent, specimens should be collected early

in the morning before the patient has passed urine. In females, the cervix and urethra should always be tested and, when possible, the rectum, as symptomless involvement of the last site is common. High vaginal swabs should not be used. Koch (1948) stated that cervical cultures were most likely to be positive during the oestrogenic phase of the menstrual cycle when the cervical mucus is almost neutral in reaction, but Falk and Krook (1967) found no such relation. Repeated tests may be required, but Jephcott and Raschid (1978) found that 99 per cent of cases could be diagnosed on two sets of tests.

For diagnostic cultivation the chocolate blood agar medium described by McLeod and his colleagues (1934) gives good results. The selective media described by Thayer and Martin (1964, 1966), Martin and others (1967), Martin and Lester (1971) and Faur and co-workers (1973) have proved a major advance. These consist of a supplemented base to which vancomycin 3 μg/ml, colistimethate sodium 12.5 μg/ml and nystatin 12.5 units/ml are added to suppress overgrowth of gonococci by other bacteria or yeasts. Other neisseriae except the meningococcus are inhibited, as are most strains of *Acinetobacter*. *Proteus* is not inhibited, but the addition of trimethoprim at a concentration of 3 to 8 μg/ml effects this (Seth 1970). These selective media greatly facilitate the isolation of gonococci from heavily contaminated sites such as the rectum. However, a minority of strains, mainly those sensitive to penicillin, may be inhibited by the concentration of vancomycin present in these media (Reyn 1969, Cross *et al.* 1971). Ødegaard and his colleagues (1975) recommend substituting lincomycin 0.5 μg/ml for vancomycin. Faur and her co-workers (1978), on the other hand, found less contamination and obtained more positive cultures with vancomycin 2 μg/ml than with lincomycin 1–4 μg/ml. For comparative studies on the efficiency of isolation and transport media see Taylor and Phillips (1980). Whatever medium is selected, the minimum amount of agar necessary to give it stability should be used. Cultures should be incubated at 36° in a moist atmosphere in a sealed container with added CO_2; these requirements can be met by placing a lighted candle in a petri dish containing a little water in the container before sealing. The need for CO_2 seems to be stricter for strains on primary isolation; subcultures are less demanding. Platt (1976) found that 39 per cent of 120 primary isolates would grow only in the presence of CO_2; Chapel, Smeltzer and Dassel (1976) reported that a delay of more than two hours before incubating primary cultures in the presence of CO_2 resulted in the loss of some strains. Earl and his colleagues (1976) stated that the requirement for CO_2 could be met by incorporating $NaHCO_3$ 0.024 mol/l in the medium. After incubation for 24 to 48 hours oxidase reagent (freshly prepared 1 per cent tetramethyl-*p*-phenylene-diamine hydrochloride) is added to suspected colonies.

Those of gonococci turn first pink and then deepen to purple, and can be picked off for microscopical examination and subculture for identification tests. Alternatively, a portion of the colony can be picked off with a platinum loop and smeared on filter paper moistened with the oxidase reagent; a positive result is shown by the development of a purple colour within ten seconds. Before being discarded as negative, plates should be flooded with the reagent; this should not be allowed to come into contact with the skin as sensitization is possible. The test is not specific for gonococci; it is given by all the neisseriae and by gram-negative rods of several genera. Occasionally colonies giving a positive oxidase reaction and consisting of gram-negative cocci such as *Mima polymorpha* (Neugebauer and Lucas 1959, Svihus *et al*. 1961), *Moraxella* (Geizer 1970), or other neisseriae (Wilkinson 1952, Roiron 1957) are isolated from the genito-urinary tract. These are not likely to be confused with the gonococcus if properly investigated, and are usually inhibited by the selective media described above.

The gonococcus is susceptible to drying; and when direct plating is not practicable specimens should be sent to the laboratory on swabs in Amies's (1967) modification of Stuart's transport medium. Wooden applicator sticks and cotton-wool may be toxic; swabs should be boiled in Sørensen's buffer, pH 7.4, and impregnated with finely divided charcoal to neutralize toxic substances, probably fatty acids, which are present in some batches of agar (Moffett *et al*. 1948). Reyn, Korner and Bentzon (1960) recovered gonococci from artificially infected swabs in Stuart's medium after 72 hours' storage at 23 to 25°. Stuart, Toshach and Patsula (1954) found that the transport medium allowed the isolation of gonococci from clinical material in 90 per cent of cases after storage for 24 hours; beyond this time the proportion of isolations decreased. Gästrin and his colleagues (1968) compared various transport media for gonococci and other organisms. Other media have been devised which permit the combined transport and culture of gonococci, such as Transgrow (Martin and Lester 1971), Jembec-Neigon (Jephcott *et al*. 1976), and Till-U-Test (Donald 1980). After inoculation, these should be incubated overnight before being sent to the laboratory.

(For a review of diagnosis see Young 1981).

Tests for the identification of gonococci

Isolates from the urogenital tract which produce typical colonies of oxidase-positive gram-negative diplococci on selective media almost always prove to be *N. gonorrhoeae*; the margin of error is probably less than 2 per cent. Confirmatory tests are desirable, and are essential on isolates from extragenital or unusual sites, from children, and in cases which may involve medico-legal issues.

Carbohydrate degradation tests   (See Chapter 28.) These may depend on changes in the substrate during growth of the organism, or they may be rapid methods in which a very heavy inoculum of bacteria containing pre-formed enzymes is added to lightly buffered solutions of sugars with an indicator to show any change in pH. In the latter, acid production can usually be detected after incubation for 2–3 hours. Comparative studies of various methods have been reported by Shtibel and Toma (1978). The serum-free medium described by Flynn and Waitkins (1972) can be recommended for conventional sugar degradation tests.

Identification by immunofluorescence staining   This method was described by Deacon and his colleagues (1960) and Deacon (1961). Antisera should be prepared against a wide variety of local strains. Methods for preparing, absorbing and conjugating antisera were described by Peacock (1970). With properly absorbed conjugates, gonococci from young (18-hour) cultures show brilliant fluorescence at the periphery of the cell. Other neisseriae either do not fluoresce or show a very faint uniform staining. The method is rapid and the results compare favourably with the more laborious sugar degradation tests (Olcén *et al*. 1978).

Immunofluorescence staining can also be used to detect gonococci in fixed smears of secretions. Background staining of cells, particularly polymorphs, may interfere unless counterstains such as naphthalene black are used. Conjugates may react non-specifically with strains of *Staphylococcus aureus* that contain protein A (Danielsson 1965*b*, Lind 1968, Lind and Mansa 1968). This can be blocked by diluting the conjugate in normal rabbit or human serum or in an antiserum against protein A-containing *Staph. aureus*. The method has been found more sensitive than gram-stained films for the detection of gonococci (Danielsson 1963, 1965*a*, Lind 1969, Thin 1970, Barteluk *et al*. 1974). Dead gonococci may still fluoresce; the test, should not therefore, be used to control the results of treatment. It is too laborious for routine use, but is of value in the examination of skin lesions or joint fluid, from which cultures are frequently negative.

Co-agglutination test   Danielsson and Kronvall (1974) showed that protein A-containing *Staph. aureus*, sensitized with specific anti-gonococcal antibody, would co-agglutinate when mixed with a suspension of gonococci. The method is much quicker than that of immunofluorescence, and with properly absorbed antisera is both sensitive and specific. Favourable comparisons with immunofluorescence and sugar degradation tests have been reported by Danielsson, Olcén and Sandström (1977), Lewis and Martin (1980), Rufli (1980), and Johnston (1981). Difficulty may be experienced with some strains which are autoagglutinable or have undergone mucoid degeneration. This can usually be overcome by treating suspensions from young cultures with trypsin (Menck 1976), or by boiling the suspension and testing the supernate after centrifugation.

Gonococci have been classified into three serogroups, WI, WII and WIII, by the co-agglutination technique. These groups vary in their geographical distribution and their sensitivity to antibiotics (Sandström and Danielsson 1980, Bygdeman 1981, Bygdeman Danielsson and Sandström 1981).

Serological diagnosis of gonorrhoea

In the past, chronic or complicated gonorrhoea was common and serological tests were widely used as an aid to diagnosis. With the introduction of the sulphonamides and later penicillin, infections, especially in the male, are cut short before much stimulus to antibody formation is possible. Because of this and of improved selective media for the isolation of the gonococcus, serological tests have been relegated to a minor place and abandoned in many centres. The recrudescence of gonorrhoea in recent years has led to the view that this may have been premature and that serological tests may still have a place, particularly as an aid to the detection of latent infections in the female, which form a reservoir of infection.

Complement-fixation and other tests   Following the original observations of Müller and Oppenheim (1906) *complement-fixation tests* with suspensions or crude extracts of gonococci were used for many years. These were not very sensitive in uncomplicated infections and gave non-specific results. Experience with the test today is reviewed by Young, Hendriksen and McMillan (1980). Much recent work has been directed to the use of defined gonococcal antigens in radioimmunoassay, ELISA and haemagglutination techniques. These have included *pili* (Buchanan *et al.* 1973, Oates *et al.* 1977, Reiman and Lind 1977, Young and Low 1981) and *outer membrane protein* antigens (Glynn and Ison 1978).

Antibodies of IgG, IgM and IgA classes are produced during gonococcal infection; their detection in serum as an aid to diagnosis has been studied by immunofluorescence methods (Welch and O'Reilly 1973, Wilkinson 1975, Gaafar 1976, McMillan, McNeillage and Young 1979*a*, McMillan *et al.* 1979*b*) and by ELISA tests (Ison and Glynn 1979). Other approaches have been directed to the detection of local (secretory) IgA antibody in secretions (Kearns *et al.* 1973, O'Reilly, Lee and Welch 1976; McMillan, McNeillage and Young 1979*a*, McMillan *et al.* 1980) or gonococcal antigens in urine (Thornley *et al.* 1979). An enzyme, 1,2-propanediol oxidoreductase, is produced in large amounts by gonococci and to a much lesser degree by other neisseriae and *Acinetobacter* but not by other organisms found in the genital tract. A fluorimetric assay for its detection has been described by Takeguchi and his colleagues (1980); it is a sensitive

method for detecting present, rather than past infection, but its specificity in a low-risk group of women was only 75 per cent.

Although many of these new methods are much more sensitive than the original complement-fixation test and may give useful results in high-risk groups, they are not entirely satisfactory. They do not distinguish present from past infections and therefore give an appreciable proportion of false positive results. Some of these may be due to cross-reactions with antibodies to meningococci. Toshach (1978), for example, obtained positive serological results in flocculation and immunofluorescence tests for gonorrhoea in 35 per cent of meningococcal carriers who had no clinical or bacteriological evidence of gonorrhoea. This limits the value of these tests as screening methods for detecting unsuspected latent infections. They may be of use in the investigation of patients with suspected complications of gonorrhoea, such as salpingitis or disseminated infection, if a rise in titre can be demonstrated, but their role is subordinate to the isolation of gonococci in culture. (For reviews of work on serological tests see Sandström and Danielsson 1975, and Holmes *et al.* 1978.)

Prevention and treatment

Since the middle of the 1950s there has been a worldwide increase in the incidence of gonorrhoea. In 1980 no fewer than 60 824 new patients with gonorrhoea attended clinics in the United Kingdom. Factors which have contributed to this resurgence are the social changes of a permissive society, the introduction of oral contraceptives, an influx of immigrants, an increasing number of infections among homosexuals, and the emergence of strains of gonococci with diminished sensitivity to antibiotics. Control methods must be based on the prompt detection and treatment of patients with the disease and the energetic tracing of their contacts, both the primary (source) and secondary contacts who have subsequently been exposed to infection. This must be reinforced by wider educational measures in personal and social hygiene.

Vaccination

The isolation of the major components of the outer membrane of the gonococcus has aroused interest in their use as possible vaccines. Diena and his colleagues (1978) found that immunization with lipopolysaccharide protected mice against intracerebral challenge. In experiments in chambers implanted subcutaneously in guinea-pigs immunization with outer membrane proteins or pili was found to give some protection against challenge with the homologous strain and to a lesser degree against heterologous strains (Buchanan *et al.* 1977, Buchanan and Arco 1978). However, Turner and Novotny (1976), who used the same system, concluded that protection after immunization or in animals in which infection in the chambers had died out naturally did not depend on the production of antibody against pili, and that not all auxotypes could infect these chambers (Novotny *et al.* 1978). These models provide conditions far removed from the disease in man, but Arco, Duncan and their colleagues (1976) reported that chimpanzees immunized with formolized type 2 gonococci became refractory to reinfection of the urethra or pharynx; in some animals this persisted for up to two years. There have been few experiments on man. A vaccine produced from autolysed gonococci gave no protection in a field trial, although it stimulated a good antibody response (Greenberg *et al.* 1974). Brinton and his colleagues (1978) immunized a small number of volunteers with purified pili from a single strain and found that this decidedly increased the number of gonococci of the same strain needed to produce urethral infection on challenge. A similar vaccine produced minimal side effects; the antibody that was produced blocked attachment of the vaccine strain and to a lesser degree of heterologous strains to human buccal epithelial cells (Tramont *et al.* 1981). The prospects for the development of a vaccine do not seem hopeless, but to be of value it will have to confer a solid immunity against a wide variety of strains and prevent the establishment of symptomless infection. For a general review see Robbins (1977).

Chemotherapy

Treatment was revolutionized by the introduction of the sulphonamides. Resistant strains soon emerged which, however, were sensitive to relatively small doses of penicillin. In about 1957, strains of diminished sensitivity to penicillin also appeared, associated with failure of treatment at the dosage then employed (Curtis and Wilkinson 1958, Reyn *et al.* 1958). These strains have increased in frequency throughout the world. Until recently the upper limit of resistance rarely exceeded 0.5–1.0 μg/ml, although more resistant strains were prevalent in the Far East. Shtibel (1980), however, has reported the isolation of strains in Canada that had an MIC of 50 units/ml but which did not produce β-lactamase. Such penicillin-insensitive strains apparently arose through chromosomal mutations. They often show cross-resistance to streptomycin, tetracycline, erythromycin, rifampicin, and chloramphenicol. Genetic factors determining their resistance were reviewed by Sparling (1977). Surveys of trends in antibiotic resistance of gonococci in the United States were reported by Reynolds and his colleagues (1979) and in England by Seth, Kolator and Wilkinson (1979).

Strains of gonococci which were truly resistant to penicillin through the production of a β-lactamase were first reported by Ashford, Golash and Hemming (1976) in American servicemen returning from the Far East and by Phillips (1976) in a patient from West Africa. A localized outbreak due to similar strains occurred in Liverpool in 1976 (Percival *et al.* 1976). β-Lactamase-producing gonococci are now widely disseminated throughout the world and have been reported from at least 27 countries. Except in areas of high prevalence (the Philippines, Singapore, Thailand, and probably West Africa) they have generally produced only sporadic infections, usually imported from abroad. Of 104 infections with these strains seen in the United Kingdom in 1979, 29.8 per cent were contracted in the Far East, 17.3 per cent in West Africa and 39.4 per cent in the United Kingdom (Johnston, Kolator and Seth 1981). They now appear, however, to be well established in Holland and accounted for about 2.5 per cent of all infections with gonorrhoea (Nayyar *et al.* 1980). These strains are fully virulent and can produce the whole clinical spectrum of gonococcal infection.

β-Lactamase production is mediated by R plasmids, which are different in the Far Eastern and West African strains; the two varieties appear to have arisen independently but almost simultaneously. Many Far Eastern strains also have a conjugative plasmid that can transfer resistance to other gonococci, and experimentally to *Escherichia coli* or other neisseriae (Perine *et al.* 1977). The R plasmids show similarities to the TEM β-lactamase of enterobacteria and ampicillin-resistant *Haemophilus influenzae* and may have originated from these. (For the epidemiological behaviour of these strains see Arya *et al.* 1978, Perine *et al.* 1979, and Rajam *et al.* 1981.)

Except for β-lactamase-producing strains, penicillin remains the drug of choice in a dosage of 2.4–4.8 mega units of aqueous procaine penicillin G preceded by 1 g probenecid to increase the concentration in the blood. Oral or slow-release penicillins should not be used. Ampicillin 3.5 g plus 1 g probenecid also gives high rates of cure. When penicillin cannot be given, tetracycline, kanamycin or cotrimoxazole are suitable alternatives. Infections with β-lactamase-producing strains can be treated with spectinomycin, cefuroxime, cefotaxime, gentamicin or kanamycin, though strains doubly resistant to penicillin and spectinomycin have now been met with. Whatever treatment is used, full bacteriological tests of cure are essential, and should include serological tests for concomitant infection with syphilis. In men who have contracted both gonorrhoea and non-specific urethritis symptoms of the latter may appear after the gonococcal infection has responded to penicillin.

For general reviews of gonorrhoea, its incidence, epidemiology, diagnosis and treatment the reader is referred to the monographs by Morton (1977), Roberts (1977) and Report (1978). (For non-gonococcal urethritis [NGU], see Chapters 76 and 78.)

References

Abu-Nassar, H., Hill, N., Fred, H. L. and Yow, E. M. (1963) *Arch. intern. Med.* **112**, 731.

Amies, C. R. (1967) *Canad. J. publ. Hlth* **58**, 296.

Arco, R. J., Bullard, J. C. and Duncan, W. P. (1976) *Brit. J. vener. Dis.* **52**, 316.

Arco, R. J., Duncan, W. P., Brown, W. J., Peacock, W. L. and Tomizawsa, T. (1976) *J. infect. Dis.* **133**, 441.

Arya, O. P., Rees, E., Percival, A., Alergant, C. D. *et al.* (1978) *Brit. J. vener. Dis.* **54**, 28.

Ashford, W. A., Golash, R. G. and Hemming, V. G. (1976) *Lancet* ii, 657.

Atkin, E. E. (1925) *Brit. J. exp. Path.* **6**, 235.

Barteluk, R., Lavender, G. W., Wilkinson, A. E. and Rodin, P. (1974) *Brit. J. vener. Dis.* **50**, 195.

Bernfeld, W. K. (1972) *Brit. med. J.* **iv**, 173.

Brinton, C. C., Bryan, J., Dillon, J.-A., Guerina, N. *et al.* (1978). In: *Immunobiology of Neisseria gonorrhoeae*. Ed. by G. F. Brooks, E. C. Gotschlich, K. K. Holmes, W. D. Sawyer and F. H. Young. American Society for Microbiology, Washington, D.C.

Brø-Jørgensen, A. and Jensen, T. (1973) *Brit. J. vener. Dis.* **49**, 491.

Buchanan, T. M. and Arco, R. J. (1978) *J. infect. Dis.* **135**, 879.

Buchanan, T. M., Pearce, W. A., Schoolnik, G. K. and Arco, R. J. (1977) *J. infect. Dis.* **136** (suppl.), 132.

Buchanan, T. M., Swanson, J., Holmes, K. K., Kraus, S. J. and Gotschlich, E. C. (1973) *J. clin. Invest.* **52**, 2896.

Bumm, E. (1885*a*) *Dtsch. med. Wschr.* **11**, 508; (1885*b*) *Ibid.* **11**, 910.

Bygdeman, S. (1981) *Acta path. microbiol. scand. B* **89**, 227.

Bygdeman, S., Danielsson, D. and Sandström, E. (1981) *Acta derm-venereol., Stockh.* **61**, 423.

Chapel, T. A., Smeltzer, M. and Dassel, R. (1976) *Hlth Lab. Sci.* **13**, 45.

Cowperman, M. B. (1927) *Amer. J. Dis. Child.* **33**, 932.

Crawford, G., Knapp, J. S., Hale, J. and Holmes, K. K. (1977) *Science* **196**, 1352.

Cross, R. C., Hoger, M. B., Neibauer, R., Pasternack, B. and Brady, F. J. (1971) *HSMHA Hlth Rep.* **86**, 990.

Curtis, F, R. and Wilkinson, A. E. (1958) *Brit. J. vener. Dis.* **34**, 70.

Danielsson, D. (1963) *Acta derm-venereol., Stockh.* **43**, 511; (1965*a*) *Ibid.* **45**, 74; (1965*b*) **45**, 61.

Danielsson, D. and Kronvall, G. (1974) *Appl. Microbiol.* **27**, 368.

Danielsson, D., Olcén, P. and Sandström, E. (1977) In: *Gonorrhoea; Epidemiology and Pathogenesis*. Ed. F. A. Skinner, P. D. Walker and H. Smith. Academic Press, London.

Deacon, W. E. (1961) *Bull. World Hlth Org.* **24**, 349.

Deacon, W. E., Peacock, W. L., Freeman, E. M., Harris, A. and Bunch, W. L. (1960) *Publ. Hlth Rep., Wash.* **75**, 125.

Diena, B. B., Ashton, F. E., Ryan, A. and Wallace, R. (1978) *Canad. J. Microbiol.* **24**, 117.

Donald, W. H. (1980) *Brit. J. vener. Dis.* **56**, 81.

Earl, R. G., Dennison, D., Whadford, V., Hayes, J. and Baugh, C. L. (1976) *J. Amer. vener. Dis. Ass.* **3**, 40.

Falk, V. and Krook, G. (1967) *Acta derm-venereol., Stockh.* **47**, 190.

Faur, Y. C., Weisburd, M. H. and Wilson, M. E. (1978) *Hlth Lab. Sci.* **15**, 22.

Faur, Y. C., Weisburd, M. H., Wilson, M. E. and May, P. S. (1973) *Hlth Lab. Sci.* **10**, 44.

Flynn, J. and Waitkins, S. (1972) *J. clin. Path.* **25**, 525.

Førstrom, L., Mustakallio, K. K., Sivonen, A. and Kousa, M. (1972) *Ann. clin. Res.* **4**, 49.

Gaafar, H. A. (1976) *J. clin. Microbiol.* **4**, 423.

Gästrin, B., Kallings, L. O. and Marcetic, A. (1968) *Acta path. microbiol. scand.* **74**, 371.

Geizer, E. (1970) *Zbl. Bakt.* **214**, 75.

Glynn, A. A. and Ison, C. (1978) *Brit. J. vener. Dis.* **54**, 97.

Greenberg, L. *et al.* (1974) *Canad. J. publ. Hlth* **65**, 29.

Gregg, C. R., Melly, M. A., Hellerqvist, C. G., Coniglio, J. C. and McGee, Z. A. (1981) *J. infect. Dis.* **143**, 432.

Handsfield, H. H. (1977). In: *The Gonococcus*. Ed. by R. B. Roberts. John Wiley and Sons, New York.

Handsfield, H. H., Lipman, J. O., Harnisch, J. P., Tronca, E. and Holmes, K. K. (1974) *New Engl. J. Med.* **290**, 117.

Harkness, A. H. (1948) *Brit. J. vener. Dis.* **24**, 137.

Hildebrandt, J. F. and Buchanan, T. M. (1978) In: *Immunobiology of Neisseria gonorrhoea*. Ed by G. F. Brooks, E. C. Gotschlich, K. K. Holmes, W. D. Sawyer and F. E. Young. American Society for Microbiology, Washington, D.C.

Holmes, K. K., Buchanan, T. M., Adam, J. L. and Eschenbach, D. A. (1978) In: *Immunobiology of Neisseria gonorrhoeae*. Ed. by G. F. Brooks, E. C. Gotschlich, K. K. Holmes, W. D. Sawyer and F. E. Young. American Society for Microbiology, Washington, D.C.

Holmes, K. K., Counts, G. W. and Beaty, H. N. (1971) *Ann. intern. Med.* **74**, 979.

Holmes, K. K., Wiesner, P. J. and Pedersen, A. H. B. (1971) *Ann. intern. Med.* **75**, 470.

Ison, C. and Glynn, A. A. (1979) *Lancet*, ii, 1165.

James, J. F. and Swanson, J. (1977) *J. exp. Med.* **145**, 1082.

Jephcott, A. E., Bhattacharyya, M. N. and Jackson, D. H. (1976) *Brit. J. vener. Dis.* **52**, 250.

Jephcott, A. E. and Raschid, S. (1978) *Brit. J. vener. Dis.* **54**, 155.

Johnston, N. A. (1981) *Brit. J. vener. Dis.* **57**, 315.

Johnston, N. A., Kolator, B. and Seth, A. D. (1981) *Lancet* i, 263.

Kearns, D. H., O'Reilly, R. J., Lee, L. and Welch, B. G. (1973) *J. infect. Dis.* **127**, 99.

Kellogg, D. S., Peacock, W. I., Deacon, W. E., Brown, L. and Pirkle, C. I. (1963) *J. Bact.* **85**, 1274.

Knapp, J. S. and Holmes, K. K. (1975) *J. infect. Dis.* **132**, 204.

Koch, M. L. (1948) *J. Bact.* **56**, 83.

Lewis, J. S. and Martin, J. E. (1980) *J. clin. Microbiol.* **11**, 153.

Lind, I. (1968) *Acta path. microbiol. scand.* **73**, 624; (1969) *Ibid.* **76**, 279.

Lind, I. and Mansa, B. (1968) *Acta path. microbiol. scand.* **73**, 637.

Litt, I. F., Edberg, S. C. and Finberg, L. (1974) *J. Pediat.* **85**, 595.

McCutchan, J. A., Katzenstein, D., Norqvist, D., Chikani, G., Wunderlich, A. and Braude, A. I. (1978) *J. Immunol.* **121**, 1884.

McGee, Z. A., Johnson, A. P. and Taylor-Robinson, D. (1981) *J. infect. Dis.* **143**, 413.

McLeod, J. W., Coates, J. C., Happold, F. C., Priestley, D. P. and Wheatley, B. (1934) *J. Path. Bact.* **39**, 221.

McMillan, A., McNeillage, G. and Young, H. (1979*a*) *J. infect. Dis.* **140**, 89.

McMillan, A., McNeillage, G., Young, H. and Bain, S. R. (1979*b*) *Brit. J. vener. Dis.* **55**, 5; (1980) *Ibid.* **56**, 223.

Mahoney, J. F., Slyke, C. J. van, Cutler, J. C. and Blum, H. L. (1946) *Amer. J. Syph.* **30**, 1.

Martin, J. E., Billings, T. E., Hackney, J. F. and Thayer, J. D. (1967) *Publ. Hlth Rep., Wash.* **82**, 361.

Martin, J. E. and Lester, A. (1971) *HSMHA Hlth Rep.* **86**, 30.

Menck, H. (1976) *Acta path. microbiol. scand. B* **84**, 139.

Moffett, M., Young, J. L. and Stuart, R. D. (1948) *Brit. med J.* **ii**, 421.

Morton, R. S. (1977) *Gonorrhoea.* W. B. Saunders Ltd, London.

Müller, R. and Oppenheim, M. (1906) *Wien. klin. Wschr.* **19**, 894.

Nayyar, K. C., Michel, M. F. and Stolz, E. (1980) *Brit. J. vener. Dis.* **56**, 249.

Neisser, A. (1879) *Zbl. med. Wiss.* **17**, 497; (1894) *Dtsch. med. Wschr.* **20**, 335.

Neugebauer, D. L. and Lucas, R. N. (1959) *Amer. J. clin. Path.* **32**, 364.

Novotny, P. *et al.* (1977) *J. med. Microbiol.* **10**, 347.

Novotny, P., Broughton, E. S., Cownley, K., Hughes, M. and Turner, W. H. (1978) *Brit. J. vener. Dis.* **54**, 88.

Novotny, P. and Short, J. A. (1977) In: *Gonorrhoea; Epidemiology and Pathogenesis.* Ed. by F. A. Skinner, P. D. Walker and H. Smith. Academic Press, London.

Novotny, P., Short, J. A. and Walker, P. D. (1975) *J. med. Microbiol.* **8**, 413.

Oates, S. A., Falkler, W. A., Joseph, J. M. and Warfel, L. E. (1977) *J. clin. Microbiol.* **5**, 26.

Ødegaard, K., Solberg, O., Lind, J., Myhre, G. and Nyland, B. (1975) *Acta path. microbiol. scand. B* **83**, 301.

Olcén, P., Danielsson, D. and Kjellander, J. (1978) *Acta path. microbiol. scand. B* **86**, 327.

O'Reilly, R. J., Lee, L. and Welch, B. G. (1976) *J. infect. Dis.* **133**, 113.

Peacock, W. L. (1970) *Publ. Hlth Rep., Wash.* **85**, 733.

Percival, A., Corkhill, J. E., Arya, O. P., Rowlands, J. *et al.* (1976) *Lancet* **ii**, 1379.

Perine, P. L., Morton, R. S., Piot, P., Siegel, M. S. and Antal, G. M. (1979) *Sex. transm. Dis.* **6**, 152.

Perine, P. L., Thornsberry, C., Schalla, W., Biddle, J. *et al.* (1977) *Lancet* **ii**, 993.

Phillips, I. (1976) *Lancet* **ii**, 656.

Platt, D. J. (1976) *J. clin. Microbiol.* **4**, 129.

Rajam, N. S., Thirumoorthy, T. and Tan, N. J. (1981) *Brit. J. vener. Dis.* **57**, 158.

Reiman, K. and Lind, I. (1977) *Acta path. microbiol. scand. C* **85**, 115.

Report (1978) *Neisseria gonorrhoeae and gonococcal infections. Technical Report Series 616.* World Health Organization, Geneva.

Reyn, A. (1969) *Bull. World Hlth Org.* **40**, 245.

Reyn, A., Korner, B. and Bentzon, M. W. (1958) *Brit. J. vener. Dis.* **34**, 227; (1960) *Ibid.* **36**, 243.

Reynolds, G. H., Zaidi, A. A., Thornsberry, C., Guinan, M. *et al.* (1979) *Sex. transm. Dis.* **6**, 103.

Richardson, W. P. and Sadoff, J. C. (1977) *Infect. Immun.* **15**, 663.

Robbins, J. B. (1977) *Brit. J. vener. Dis.* **53**, 170.

Roberts, R. B. (1977) *The Gonococcus.* John Wiley and Sons, New York.

Roiron, V. (1957) *Rev. du Practicien* **7**, 2513.

Rufli, T. (1980) *Brit. J. vener. Dis.* **56**, 144.

Sandström, E. and Danielsson, D. (1975) In: *Genital Infections and their Complications.* Ed. by D. Danielsson, L. Juhlin and P.-A. Mårdh. Almqvist and Wiksell, Stockholm; (1980) *Acta path. microbiol. scand. B* **88**, 27.

Schmale, J. D., Martin, J. E. and Domescik, G. (1969) *J. Amer. med. Ass.* **210**, 312.

Scott, J. and Stone, A. H. (1966) *Brit. J. vener. Dis.* **42**, 103.

Sersiron, D. and Roiron, V. (1967) *Brit. J. vener. Dis.* **43**, 33.

Seth, A. (1970) *Brit. J. vener. Dis.* **46**, 201.

Seth, A. D., Kolator, B. and Wilkinson, A. E. (1979) *Brit. J. vener. Dis.* **55**, 325.

Shore, W. B. and Winkelstein, J. A. (1971) *J. Pediat.* **79**, 661.

Shtibel, R. (1980) *Lancet* **ii**, 39.

Shtibel, R. and Toma, S. (1978) *Canad. J. Microbiol.* **24**, 177.

Sparling, P. F. (1977) In: *The Gonococcus.* Ed. by R. B. Roberts. John Wiley and Sons, New York.

Stuart, R. D., Toshach, S. R. and Patsula, T. M. (1954) *Canad. J. publ. Hlth* **45**, 73.

Svihus, R. H., Lucerno, E. M., Mikolajczyk, R. J. and Carter, E. E. (1961) *J. Amer. med. Ass.* **177**, 121.

Swanson, J. (1977) In: *The Gonococcus.* Ed. by R. B. Roberts. John Wiley and Sons, New York.

Swanson, J., Sparks, E., Young D. and King, G. (1975) *Infect. Immun.* **11**, 1352.

Takeguchi, M. M. *et al.* (1980) *Brit. J. vener. Dis.* **56**, 304.

Taylor, E. and Phillips, I. (1980) *Brit. J. vener. Dis.* **56**, 390.

Thayer, J. D. and Martin, J. E. (1964) *Publ. Hlth Rep., Wash.* **79**, 49; (1966) *Ibid.* **81**, 559.

Thin, R. N. T. (1970) *Brit. J. vener. Dis.* **46**, 27.

Thompson, S. E. *et al.* (1978) *Sex. transm. Dis.* **5**, 127.

Thornley, M., Wilson, D. V., Harmaeche, R. D. de, Oates, J. K. and Coombs, R. R. A. (1979) *J. med. Microbiol.* **12**, 161.

Toschach, S. (1978) *Canad. J. publ. Hlth* **69**, 127.

Tramont, E. C. *et al.* (1981) *J. clin. Invest.* **68**, 881.

Tronca, E., Handsfield, H. H., Wiesner, P. J. and Holmes, K. K. (1974) *J. infect. Dis.* **129**, 583.

Turner, W. H. and Novotny, P. (1976) *J. gen. Microbiol.* **92**, 224.

Veale, D. R., Smith, H., Witt, K. A. and Marshall, R. B. (1975) *J. med. Microbiol.* **8**, 325.

Walsh, M. J., Brown, B. C., Brown, L. and Pirkle, C. I. (1963) *J. Bact.* **86**, 478.

Ward, M. E. and Watt, P. J. (1972) *J. infect. Dis.* **126**, 601; (1975) In: *Genital Infections and their Complications.* Ed. by D. Danielsson, L. Juhlin and P.-A. Mårdh. Almqvist and Wiksell, Stockholm.

Ward, M. E., Watt, P. J. and Glynn, A. A. (1970) *Nature, Lond.* **227**, 382.

Ward, M. E., Watt, P. J. and Robertson, J. N. (1974) *J. infect. Dis.* **129**, 650.

Watt, P. J. and Lambden, P. R. (1978) In: *Modern Topics in Infection.* Ed. by J. D. Williams. Heinemann, London.

Welch, B. G. and O'Reilly, R. J. (1973) *J. infect. Dis.* **127**, 69.

Wiesner, P. J., Handsfield, H. H. and Holmes, K. K. (1973) *New Engl. J. Med.* **288**, 1221.

Wilkinson, A. E. (1952) *Brit. J. vener. Dis.* **28**, 24; (1962) *Ibid.* **38**, 145; (1975) *Ibid.* **51**, 28.

Young, H. (1981) In: *Recent Advances in Sexually Transmitted Diseases 2.* Ed. by J. R. W. Harris. Churchill-Livingstone, Edinburgh.

Young, H., Henricksen, C. and McMillan, A. (1980) *Med. Lab. Sci.* **37**, 165.

Young, H. and Low, A. C. (1981) *Med. Lab. Sci.* **38**, 41.

67

Bacterial infections of the respiratory tract

J. W. G. Smith

General review of respiratory infection

Excluding for the moment the causative organisms of well recognized diseases such as whooping cough, primary pneumococcal pneumonia and influenza, the range of microbial pathogens now known to be capable of infecting the respiratory tract is wide. Despite this new knowledge, particularly in virology, the cause of individual episodes of respiratory infection often remains unknown. Thus, sputum examination and culture are often unhelpful or unreliable. The value of serological diagnosis is limited by the impracticability of including in routine tests more than a small number of antigens, representing the commoner of the many known respiratory pathogens. Probably there are also causative agents that still remain to be recognized.

The aetiology and symptomatology of respiratory diseases vary with age, season, the type of population at risk, and other factors. No classification so far devised is entirely satisfactory. An anatomical one, such as into rhinitis, pharyngitis, laryngitis, tracheitis, bronchitis, and pneumonitis, suffers from the disadvantage that often more than one part of the respiratory tract is affected at the same time. Practically all of this disease falls into the communicable category, being caused by infection with bacteria, viruses, chlamydiae, or mycoplasmas. An aetiological classification, however, is limited by the difficulty of determining the cause in individual patients. One pathogen may give rise to a wide range of clinical and pathological effects; and many of the illnesses produced by different pathogens are alike, although broad differences in the pattern of illness they cause can often be discerned.

Most workers prefer a symptomatological classification (Dingle 1948). Stuart-Harris (1968) recognized seven syndromes (excluding influenza): (1) common cold; (2) feverish cold; (3) sore throat; (4) croup or laryngo-tracheo-bronchitis; (5) bronchitis; (6) bronchiolitis; and (7) segmental, lobar, broncho- or atypical pneumonia. The classification of acute respiratory illness seen in general practice adopted by Hope-Simpson and Miller (1973) was into: common cold, pharyngitis, primary otitis media, laryngitis, tracheitis, acute bronchitis, and pneumonia, together with influenza to describe illness in which systemic symptoms outweighed respiratory ones. For a series of 7590 children admitted to hospital the categories initially used were similar, except that tonsillitis was included and tracheitis excluded. Later, however, they were simplified to three main groups: (1) upper respiratory infection syndrome, comprising colds, pharyngitis, tonsillitis, and otitis media; (2) middle respiratory infection, i.e. croup; and (3) lower respiratory infection comprising acute bronchitis, acute bronchiolitis, pneumonia, and influenza.

The laboratory findings in Great Britain (Miller 1973), together with those at Tecumseh, USA (Monto and Cavallaro 1971, Monto and Ullman 1974), show that a suspected pathogen is isolated from about a quarter of the specimens taken from patients with acute respiratory illness. Most of the isolates are viruses, but β-haemolytic streptococci constitute between 5 and 25 per cent, depending on age and the nature of the illness. The common cold is caused mainly by rhinoviruses, less often by parainfluenza, adeno-, and respiratory syncytial (RS) viruses. In acute otitis media β-haemolytic streptococci, adeno-, and RS viruses predominate, but some cases are caused by *Haemophilus influenzae* and pneumococci. Group-A haemolytic streptococci often cause pharyngitis and tonsillitis—diseases that may also be associated with adeno-, entero-, rhino-, and parainfluenza viruses, as well as *Herpesvirus hominis*. Parainfluenza viruses tend to predominate in croup. In lower respiratory infection RS virus is an important pathogen, especially in very young children, but rhino- and adeno- viruses are also responsible for some cases. Influenzal illness may be caused by parainfluenza, rhino-, RS, and Coxsackie A viruses as well as influenza viruses. Bacteria, particularly the pneumococcus, *Haemophilus* spp. and *Staphylococcus aureus*, are concerned in lower respiratory tract infection, either as (1) primary causes of acute illness, such as pneumococcal pneumonia and Legionnaire's disease; or (2) as invaders of lung damaged by viral or other agencies; or (3) where the defences are impaired by circumstances such as old age, cardiac disease, or steroid therapy (Kripke 1961, Dowling and Lefkowitz 1963, Parrott *et al.* 1963, Higgins *et al.* 1964, Report 1965, Sterner *et al.* 1967, Brandt *et al.* 1969, Poole and Tobin 1973).

Anaerobic bacteria give rise to pleuropulmonary disease, such as lung abscess, necrotizing pneumonia, and empyema (Tillotson 1976; see also Chapter 62).

In the northern hemisphere acute respiratory disease is commonest in the January–March quarter and least common in the July–September quarter, though this seasonal periodicity is liable to be disturbed by epidemic influenza.

Monto and Ullman (1974) surveyed a large random sample of the 7500 residents of Tecumseh, USA, over the period 1965–69. They estimated that the average number of respiratory illnesses was 3.0 per person per year. The highest rate (6.1 per year) was in infants under 1 year of age, falling steadily to 1.3 per year in adults over 60 years, except for a peak of 2.8 per year in the 20 to 29-year age group. Similar figures have been reported in New Zealand (Jennings *et al.* 1978), and Western Europe (McDonald 1963). In these studies the common cold has accounted for at least half the total illnesses. Infection may recur and become chronic with the development of chronic suppurative otitis media, sinusitis, bronchitis, bronchiectasis, or asthma (Court 1968). The mortality in advanced countries in 1974–5 was 0.575 per 1000, i.e. about 6.3 per cent of all deaths. The rates were higher in less advanced countries. The highest death rates from acute respiratory infection in 1973–4 were in the age group 65 and over, in which rates greater than 300 per 1000 were recorded in twelve advanced countries; and in infants under one year of age among whom death rates in excess of 50 per 1000 were recorded in eleven such countries (Cockburn 1979). Influenza outbreaks are responsible for many winter deaths, the numbers of which can be estimated by calculations of excess mortality, i.e. the number of deaths in excess of those that would have been expected in the absence of influenza. In England and Wales the mean excess winter mortality associated with the small influenza outbreaks of 1970 to 1975 was 10 147, but the estimate was 25 406 for the more severe outbreak of 1969 to 1970 (Clifford *et al.* 1977).

The remainder of the present chapter will be devoted to respiratory illness caused by bacteria. Streptococcal sore throat and otitis media have already been described in Chapter 59 and staphylococcal infections in Chapter 60. The viral infections are dealt with in Volume 4. For information on respiratory disease, especially that caused by viruses, see Report (1963), Cooney and co-workers (1972), Miller (1973), and Monto and Ullman (1974).

Whooping cough

Whooping cough (pertussis) is an infectious disease of childhood caused by a small gram-negative bacillus described by Bordet and Gengou and now known as *Bordetella pertussis* (see Chapter 39). After an incubation period, usually of 7–10 days, there is an initial catarrhal stage, followed in 1–2 weeks by a paroxysmal stage in which the early irritating cough becomes more severe and spasmodic and assumes the character from

Table 67.1　Age distribution of whooping-cough in selected years—England and Wales

Age in years	Notifications in			Deaths in			Case-fatality rate % in		
	1950	1976	1978	1950	1976	1978	1950	1976	1978
0–	13 045	478	6 676	277	1	10	2.12	0.21	0.15
1–	88 292	2 097	39 662	106	2	2	0.12	0.10	0.01
5–	51 579	1 026	14 102	8	0	0	0.02	0.00	0.00
10 and over	4 836	265	4 247	3	0	0	0.06	0.00	0.00
Total	157 758	3 907	65 956	394	3	12	0.25	0.08	0.02

which the disease is named. In its typical form whooping cough lasts about six weeks, but the spasmodic cough may continue for three months or more. Apart from pulmonary sequelae, such as bronchitis and pneumonia, the disease may be complicated by convulsions and encephalitis, and may, according to White and his colleagues (1964), leave behind it some degree of mental impairment. Under the poor environmental conditions that existed in the early years of the century, whooping cough in Great Britain, as in many other western countries, was a fatal disease killing some 10 000 children annually.

In advanced countries, both morbidity and mortality have declined greatly during recent years. For example, in England and Wales the number of annual notifications fell from a peak of 169 000 in 1951 to an average of 8500 in the years 1971 to 1976. During this period there were only 58 deaths. Indeed the case-fatality rate has been about 1 per 1000 over the past 25 years compared with 11 per 1000 during the war years 1939–45. However, in parts of Africa, India, and South America whooping cough is still a serious disease, with a fatality rate among hospital patients of 15 per cent or more (Morley *et al.* 1966). The decline in incidence in developed countries is attributed to improved social conditions and nutrition, and to the effects of vaccination. In England and Wales, where the vaccination rate had fallen from 79 per cent in 1974 to about 31 per cent in 1978, a resurgence of the disease began in the autumn of 1977 and was responsible for 66 000 notified cases in 1978 (Table 67.1), the largest number since 1957; no corresponding epidemic was recorded in any other country in Europe in which high vaccination rates had been maintained.

All age groups are susceptible, but 80 per cent of clinical cases occur in children under the age of ten. The disease is most severe in young babies, in whom there is an inverse relation between age and severity. The disease is most fatal in early infancy; and there seems to be no benefit from maternal immunity. Of the total deaths in England and Wales, more than 90 per cent occur in the first year of life (Table 67.1), and 75 per cent within the first six months. In girls whooping cough is rather more prevalent and fatal than in boys. One attack leaves behind it a substantial degree of immunity, second attacks being uncommon. The

disease may occur at any time of the year but is most frequent in late winter and spring.

The role of *Bord. pertussis* in whooping cough has been demonstrated experimentally in human beings and monkeys. Macdonald and Macdonald (1933) produced the disease in two children with a pure culture, and Shibley and Hoelscher (1934) were similarly successful in the chimpanzee. A mild disease can be produced in rhesus monkeys (*Macaca mulatta*) by intranasal or intratracheal inoculation, and the organism can be recovered from the lungs if the animals are killed within the following two weeks (North *et al.* 1940). A respiratory infection, usually non-fatal, can be produced in mice by intranasal inoculation. (Experimental infections in laboratory animals are described in Chapter 39.)

Bord. pertussis is present in the respiratory tract during the first two weeks of the disease. After this it becomes difficult to demonstrate. It can be isolated from the lungs at autopsy, and in histological sections is seen in dense masses between the cilia of the epithelial cells (Pittman 1970). Adhesion to cilia, which probably plays a part in pathogenicity, can be demonstrated *in vitro* in cultured tracheal rings by means of scanning electronmicroscopy (Sato *et al.* 1978).

A closely similar organism, *Bordetella parapertussis*, gives rise to a generally milder illness of short duration, but is much less common than *Bord. pertussis*. Eldering and Kendrick (1952) in the USA found that it caused only 2 per cent of cases diagnosed as whooping cough; sometimes both *Bord. pertussis* and *parapertussis* were present together (see also Linnemann and Berry 1977). In a special survey in England and Wales only 0.6 per cent of cases were attributed to *Bord. parapertussis* (Report 1969). Lautrop (1958), however, in Denmark showed that during the years 1950 to 1957 it occurred with about the same frequency as *Bord. pertussis*. In Czechoslovakia illness caused by *Bord. parapertussis* is said to assume epidemic proportions (Vysoka 1958). In Mexico *Bord. parapertussis* is said to be as common as *Bord. pertussis* (Miravete and de la Mora 1966); and in Moscow a frequent cause of clinical and subclinical infection (Neimark *et al.* 1961). The two organisms do not immunize against each other. *Bordetella bronchiseptica* is regarded by some workers as an occasional cause of whooping cough (Lautrop and Lacey 1960), and viruses, particularly adenoviruses, by others (Connor 1970, Pereira and Candeias 1971, Feldman *et al.* 1972).

Natural infection occurs mainly by direct contact with patients suffering from the disease. The organisms gain access by inhalation of material expelled in the cough spray, especially that produced during the early part of the illness. According to Lawson (1933) about 70 per cent of infections are passed on during the catarrhal stage, and about 18 per cent during the first week of the paroxysmal stage. Some infections may be contracted indirectly through fomites. Ocklitz and Milleck (1967), for example, found that in dried droplets of a casamino acid solution *Bord. pertussis* could survive for 3 to 5 days on plastic, woollen, and dress materials. Infection of infants and pre-school children occurs to some extent from their older brothers and sisters, and from adults (Nelson 1978), but less commonly than might be expected. In a survey made by the London County Council, for example, only 9 per cent of cases in children under five years of age could be traced to contact with infected siblings. However, 61 per cent of cases in children of 5 to 14 years originated from siblings. Much of the infection of the pre-school children must have been by casual contact with children in the late incubation period or early catarrhal stage before the true nature of the illness had revealed itself (Report 1951*a*). Healthy carriers of *Bord. pertussis* do not appear to exist (Kristensen 1933, Linnemann *et al.* 1968); this cannot be regarded as certain, however, as carriage of organisms in a form other than phase I would not be recognized by the available culture methods. On the other hand, it is probable that mild and subclinical cases occur in which the diagnosis of pertussis is never made (Mishulow *et al.* 1942).

Diagnosis

The clinical diagnosis of whooping cough is often difficult. Confirmation may be sought from the laboratory; but when suspicion of the true nature of the disease is not aroused for a fortnight or so the organisms may be difficult to isolate. During the catarrhal stage there is usually a leucopaenia, but this changes during the paroxysmal stage to a leucocytosis with an absolute and relative lymphocytosis. Discharges from the patient should be examined as soon as possible. If this cannot be done, the material should be added to a suitable transport medium containing starch, yeast extract, charcoal, and penicillin (Kendrick 1969).

Microscopical

A rapid diagnosis may be made by the fluorescent antibody technique. Whitaker and co-workers (1960) examined nasal smears by this method, and obtained confirmation of a clinical diagnosis of pertussis in 78 per cent of 128 patients. On the other hand Strachiloff and his colleagues (1966) were successful in only 29

per cent. The exact proportion will depend, among other things, on the stage of the disease at which the material is taken. Non-specific fluorescence is often confusing, and may lead to false reactions. The technique when carefully controlled has about the same degree of sensitivity as that of culture (Linnemann *et al.* 1968).

Cultural

Three methods are available. (1) In the *cough-plate* method the patient is induced to cough directly on to a plate of freshly made Bordet-Gengou medium. This is examined for the characteristic pearly colonies after 48–72 hours' incubation (see Chapter 39). As a rule, the organism can be demonstrated in 70–90 per cent of cases in the catarrhal stage, in 40–60 per cent in the second week, diminishing to a negligible proportion after 4 or 5 weeks (for references see 4th edition, p. 1874). (2) To avoid some of the inconveniences of the cough-plate method, Maclean (1937) introduced the *post-nasal* or *nasopharyngeal swab*. Cultures are liable to be overgrown, but contaminants can be largely suppressed by streaking the swab on to a Bordet-Gengou plate over which a few drops of a suitable strength of penicillin have been spread, or by using a selective medium (see Chapter 39). (3) Superior to the nasopharyngeal swab is the *pernasal swab* (Bradford and Slavin 1940, Cockburn and Holt 1948). This consists of a thin wisp of wool wound round the end of a length of fine flexible nichrome wire. It is passed through the anterior nares till it reaches the pharynx. By this method the organism can be cultivated from 60–80 per cent of cases during the catarrhal stage.

Serological tests

The presence of specific *complement-fixing antibodies* in the blood was demonstrated by Bordet and Gengou (1906, 1907). Their observations were confirmed by Winholt (1915) and by Chievitz and Meyer (1916), who showed that the reaction first became positive in the 3rd or 4th week, and rose to a maximum about the 8th or 10th week. It usually becomes negative again in the course of a few months. In infants and children over six months of age complement-fixing antibodies are present in a high proportion of patients; under six months the proportion is much lower (Bradstreet *et al.* 1972). *Agglutinins*, best detected by a slide agglutination test with a strain containing serotype 1 antigen, appear in much the same proportion of patients as do complement-fixing antibodies (Report 1970). Haemagglutinins may also be demonstrated; the antigen used consists of an extract of *Bord. pertussis* adsorbed on to tanned sheep red blood corpuscles (Schubert *et al.* 1961). Although haemagglutination appears to be the most sensitive of the serological tests, a combination of tests improves the likelihood of diagnosis (Combined Scottish Study 1970, Macaulay 1979).

A *skin test* to distinguish susceptible from immune subjects was introduced by Flosdorf and his colleagues (1943). The antigen, which is believed to be the agglutinogen, is prepared by aqueous extraction of the frozen and thawed organisms or by acid extraction of a fresh culture. The test was reported on favourably by Smolens and Mudd (1943) and by Sauer and Markley (1946) in the United States, but experience of it in Great Britain has proved less satisfactory.

It will be seen therefore that in the early stage of the disease, before the characteristic paroxysmal cough has developed, diagnosis in the laboratory should be made by fluorescence microscopy of the nasopharyngeal secretion or by cultivation by the cough-plate or pernasal swab method. It is at this stage that a bacteriological method of diagnosis is most helpful. In those cases in which the diagnosis remains in doubt for 3–4 weeks both cultural and serological tests may be employed. When a laboratory diagnostic method is sought in a case of more than 4 weeks' standing a combination of serological tests may be used. (For the ELISA test, see Mertsola *et al.* 1983.)

Prophylaxis

General measures

Patients should be separated from susceptible children and excluded from school and other public places until three weeks after the onset of the paroxysmal stage. Though the spasmodic cough may not have ceased by then, the patient is no longer infectious. Special care should be taken to protect babies and young infants from exposure to infection.

Vaccination

The place of pertussis vaccine in prophylaxis depends on the protective value and safety of the vaccine, and the prevalence and severity of the disease.

Vaccine efficacy

The early trials of pertussis vaccine from those of Madsen (1933) onwards yielded conflicting results (for references see 5th edition p. 2012). In some of the trials no control group was included, and many others were unsatisfactory for one reason or another. To satisfy the most rigorous conditions a field trial was carried out in England by the Whooping-cough Immunization Committee of the Medical Research Council (Report 1951*b*).

Infants and children of 6–18 months of age whose parents agreed for them to take part in the trial were divided by random sampling into two groups of equal size. One was given three doses, at monthly intervals, of pertussis vaccine, the other of an anti-catarrhal vaccine. The pertussis vaccine included five batches from three different manufacturers. Each child was visited frequently by a special nurse, and swabs were taken for bacteriological diagnosis from clinical cases of whooping cough. Altogether 10 trials were carried out in five separate areas. A total of 3801 children injected with pertussis vaccine and 3757 injected with the control vaccine were followed up for a period of 2–3 years. The average attack rate in all the trials taken together was 1.45 per 1000 child-months of observation in the vaccinated and 6.3 in the control group. The best comparative index proved to be the attack rate among children under 14 years of age exposed to risk of infection in their own homes. These rates were 18.2 per cent in the vaccinated and 87.3 per cent in the control group. The vaccines varied in their protective power. With the best an attack rate among home exposures of only 7.3 per cent was recorded; with the least effective the corresponding figure was 30.4 per cent.

This trial showed without doubt that substantial protection against whooping cough could be conferred by vaccination, and that different vaccines varied in their protective potency.

Further trials were carried out, but by a modified method. Once pertussis vaccine had been shown to confer a high degree of protection against the disease, it was no longer considered justifiable to include a control group of children. Instead, the efficacy of a given vaccine was judged by its ability in relation to other pertussis vaccines to reduce the attack rate among home exposures. By this means it was shown that vaccines prepared in liquid medium, combined vaccines, and a vaccine made with the antigenic fraction extracted by Pillemer and his colleagues (1954) were as efficacious as the American reference vaccine prepared from organisms grown on solid medium made up according to the formula of the Michigan Department of Health.

In the whole series of trials conducted by the Medical Research Council 19 separate vaccines were tested in over 36 000 children, each of whom was observed for 2–3 years. The results showed that suitable vaccines produced a high degree of protection as judged by a reduction in the attack rate among home contacts, and a diminution in the severity and duration of the disease in those who were not completely protected. Different vaccines varied greatly in their protective action; the poorest and best were associated with attack rates in home contacts of 87 per cent and 4 per cent, respectively (Report 1956*a*, 1959*b*).

Vaccine potency is expressed in international units, and is measured by comparison with the international reference standard for pertussis vaccine in terms of the ability to immunize mice against intracerebral challenge. The second international standard for pertussis vaccine has an assigned potency of 46 IU per ampoule (see Seagroatt and Sheffield 1981). A minimum estimated potency of 4 units per human dose, and a maximum number of organisms, measured in opacity units, has been internationally agreed upon (Report 1964). It is usual for manufacturers to ensure that the vaccine contains strains of *Bord. pertussis* carrying the three most important agglutinogens (Griffith 1978). What is it then that constitutes a good vaccine? Of this we are still ignorant, and our knowledge remains empirical. Phase I strains are essential, but their properties vary considerably (Standfast 1951), and it is impossible to say whether any particular antigenic property is closely associated with protective power (see Chapter 39). According to Pittman (1979) the exotoxin is non-immunogenic, and no test has yet been devised for measuring it.

Potency tests

Numerous attempts have been made to assess the protective potency of whooping-cough vaccines in the laboratory. The intranasal method of infection in mice (Burnet and Timmins 1937), which results in the multiplication of the organisms in the lungs with the production of pneumonia, has been recommended, chiefly by Australian workers (North *et al.* 1941, Gray 1947, 1949), but the Medical Research Council's trial showed that it did not arrange vaccines in the order of their protective potency for human beings (Standfast 1958). The most satisfactory test is the Kendrick method, in which immunized mice are challenged by the intracerebral injection of live bacilli (Kendrick *et al.* 1947, 1949). The results are subject to a wide degree of variation unless large numbers of animals are used (Irwin and Standfast 1957); but judged by field trials in children the test does distinguish between vaccines differing substantially in their protective power (Standfast 1958).

Agglutinins, complement-fixing antibodies or opsonocytophagic antibodies may be studied in children after vaccination. The production of agglutinins corresponds more closely than the production of any other antibody with the protective power of the vaccine (Report 1957). According to Pittman (1970), the toxicity of pertussis vaccines is best measured by the mouse-weight gain test. However, the test is difficult to reproduce, and the evidence that it gives results related to minor side-effects in vaccinated children is incomplete (Cohen *et al.* 1967, Hilton and Burland 1969, Perkins *et al.* 1969).

A survey of the efficacy of vaccines used in the United Kingdom for some years prior to 1968 revealed little difference in the attack rates (56 and 67 per cent) among vaccinated and unvaccinated children exposed to infection in the home. The poor protection has been attributed by some to the use of vaccines with a potency of less than 4 mouse-protective units per human dose, and by others to an absence of agglutinogen 3 in the vaccines. Preston (1965) pointed out that the prevalent strains contained agglutinogens 1 and 3, not 1 and 2 as they had done previously. However, the role of agglutinogens in protection is controversial (see Chapter 39). The possible deficiencies in the vaccines were rectified, and recent reports suggest that vaccines now have a protective effect of the order of 70 per cent or more against home exposure (Bassili and Stewart 1976, Noah 1976, Malleson and Bennett 1977, Jenkinson 1978, Church 1979). A few reports have been less encouraging (Stewart 1977, Ditchburn 1979).

Vaccines are prepared in the liquid medium of Cohen and Wheeler (1946) or on solid media such as those of Bordet and Gengou or Cohen and Wheeler solidified with agar. They consist of suspensions in saline, with or without a suitable adjuvant such as aluminium hydroxide. They may be given alone or combined with diphtheria and tetanus toxoids. Experience has shown that potency is best maintained in vaccines inactivated by heating at 56° for 30 minutes and preserved with merthiolate.

Three doses of vaccine are usually recommended, at suitable intervals. Because the disease is most serious in children under the age of six months, and because effective passive immunity to pertussis is not acquired from the mother, the course of active immunization is often begun at the age of three months or earlier. Combined diphtheria, tetanus and pertussis (DTP) vaccines are used in most countries, but in others quadruple vaccines, which also contain killed poliomyelitis vaccine, types I, II and III, are used. (For a discussion of some of the factors that have to be taken into account in formulating combined vaccines see 6th edition p. 2157.)

Reactions and complications

Pertussis vaccine, whether given alone or in combination with other vaccines, is liable to give rise to reactions, mostly of a minor nature. They include sore arm, irritability, and pyrexia. Their incidence is reduced slightly by the incorporation of an aluminium adjuvant in the vaccine (Burland *et al.* 1968, Butler *et al.* 1969). Serious reactions include persistent screaming beginning about two hours after the injection, shock and collapse, vomiting lasting 24 hours or so, and convulsions developing within three days of vaccination.

The incidence of serious reactions is uncertain and has been variously estimated by different observers at about 1 in 5000 to 1 in 50 000 injections (Byers and Moll 1948, Köng 1953, Malmgrem *et al.* 1960, Ström 1960, Jeavons and Bower 1964, Dane *et al.* 1966, Haire *et al.* 1967, Ehrengut 1974). The wide range reflects the difficulty of closely following up large numbers of vaccinated infants, the use of different criteria for defining a reaction, the care with which the contra-indications are observed and, possibly, variation between different vaccines. Moreover, estimates need to take into account the natural incidence of similar events. A first convulsion, febrile or non-febrile, may occur naturally in approximately one child in 400 in the age group 3–9 months (van den Berg and Yerushalmy 1969); it follows that one child in 10 000 given three doses of vaccine between the ages of 3 and 9 months might by chance have a first convulsion within three days of an injection. In Holland the incidence of convulsions occurring within 3 days of vaccination in children given four doses of DTP-polio vaccine containing 16 opacity units of *Bord. pertussis* per dose was estimated by Hannick and Cohen (1978) to be about 1 in 2700 children. The incidence of shock was about the same. A prospective survey in California compared reactions in infants who had DTP and DT vaccine; the findings in 2300 DTP recipients and 89 DT recipients suggested that the pertussis component was associated with persistent crying in 1 in 25 children, convulsions of brief duration in 1 in 500, and collapse in 1 in 500—all unusually high rates (Baraff and Cherry 1978). Of particular concern is the possible occurrence of reactions leading to permanent brain damage (Kulenkampf *et al.* 1974) or death. The belief that such occurrences are not infrequent has led to the view expressed by some workers that pertussis vaccination is no longer justified in developed countries (Ehrengut 1974, Bassili and Stewart 1976, Stewart 1977, 1978,

1979). The accompanying public debate in Britain was responsible for a decline in the proportion of infants vaccinated against pertussis from about 79 per cent in 1974 to about 31 per cent in 1978. The frequency of brain damage following pertussis vaccination is, however, unknown. In Holland, encephalopathy has been estimated, on the basis of careful but unfortunately uncontrolled observations, to occur in 1 in 400 000 vaccinated children (Hannick and Cohen 1978). In the United Kingdom estimates as high as 1 in 10 000 to 1 in 60 000 children have been made (Stewart 1977, 1978, 1979); their validity, however, has been seriously questioned (Report 1977, Griffith 1978, 1979, Stuart-Harris 1978). Probably the most reliable estimate is given by a study of 1000 cases of encephalopathy in children admitted to hospital in England, Wales, and Scotland in 1976–79. Their immunization histories were compared with those of 1955 control children matched for age, sex, and area of residence. Thirty-five cases of encephalopathy occurred in previously normal children within 7 days of immunization with DTP vaccine. The attributable risk of encephalopathy was estimated to be 1 in 110 000 injections. A year later 21 of the 35 cases had recovered, apparently completely, but two had died and nine had mental retardation. The risk of encephalopathy with sequelae was estimated to be 1 in 310 000. The confidence limits of both estimates were wide (Miller, Ross, Alderslade, Bellman and Rawson 1981). This study did not include cases in which convulsions lasted for less than 30 minutes (Ehrengut 1981, but see also Miller *et al.* 1982).

The frequency of encephalopathy following pertussis vaccination is believed to be higher in children with a close family history of epilepsy or other central nervous system disease, and in those with a history of convulsions, neurological defects, or reactions to previous doses of vaccine (Byers and Moll 1948, Heilig 1957, Report 1977). The importance of allowing these suspicions to serve as contra-indications to vaccination is doubtful (Griffith 1978).

The encephalopathy following the use of pertussis vaccine does not seem to be of the same nature as the allergic demyelinating encephalitis that occurs after the use of smallpox and certain other vaccines. It appears rather to be due to some toxic constituent of the whooping-cough bacillus itself (see review by Wilson 1967). This conclusion is supported by the work of Kantchourine and Kapitonova (1969), who found that encephalopathy could be produced in guinea-pigs by pertussis bacilli mixed with vaseline without the addition of any nervous tissue. No other gram-negative organism has been shown to do this. Further support is afforded by Pittman, who regards the active factor as an exotoxin and defines its properties, in a valuable review on the subject (Pittman 1979; see also Pittman *et al.* 1980).

Provocation disease

A different type of complication is that of *provocation disease*.

Attention to *provocation poliomyelitis* was drawn by McCloskey (1950, 1951) in Australia, and later by workers in Great Britain. The special committee set up by the Medical Research Council studied 355 cases of the paralytic disease that developed after inoculation against diphtheria or whooping cough in England and Wales. A clear relation was found between inoculation and the onset of poliomyelitis

within 4 weeks, generally starting in the inoculated limb. The risk was greatest after combined diphtheria-pertussis vaccines, with or without alum, and least with plain diphtheria toxoid and pertussis vaccine given separately (Report 1956b). The pathogenesis of the provocation disease is obscure, but presumably the vaccine activates a latent viral infection (Raettig 1959a, b). Vaccination against many diseases including pertussis may be dangerous during the incubation period. It may activate either the homologous organism—as Raška and his colleagues (1957) found with pertussis vaccine injected intranasally into mice—or a heterologous organism such as the virus of poliomyelitis. Those who are inclined to doubt this will do well to read the contributions of Raettig (1959a, b) to the experimental study of provocation disease. Where poliomyelitis is prevalent it may be advisable to immunize against this disease before undertaking vaccination with diphtheria, tetanus and pertussis antigens.

Prevalence of whooping cough

In England and Wales vaccination of children started first in 1937 on a small scale and then slowly increased. Its use for all infants was officially recommended in 1952, but it was not until 1956 that 90 per cent of local authorities were using the vaccine (Griffith 1974). Examination of Fig. 67.1 shows that the case-fatality rate had been falling since 1942. The subsequent fall was presumably a continuation of this, aided perhaps by the use of antibiotics for the respiratory complications. The influence that vaccination may have had is uncertain. The morbidity began to fall at about the time that vaccination became general, and continued to fall until 1977, apart from the interruption every three years or so of epidemics, each smaller than the last. The rapid fall in morbidity between 1954 and 1962 was presumably ascribable, in some measure at least, to the effect of vaccination. This inference is supported by the similar falls that occurred after the introduction of vaccination, though at different dates, in both the United States and Denmark. Moreover, in the United Kingdom in 1978 an epidemic which began a year earlier produced the highest number of notifications since 1956. This reappearance of the disease must be attributed to the decline in use of the vaccine, and suggests that the herd immunity presumably induced by vaccination was lost. A similar increase in the prevalence of whooping cough occurred in Japan after a fall in the acceptance rate of pertussis vaccine (Kanai 1980). Were vaccination to be discontinued, the disease would probably re-emerge as a common illness in children.

Severity of whooping cough

As may be seen from the decline in the case-fatality rate of whooping cough in England and Wales, the disease has become less severe in developed countries.

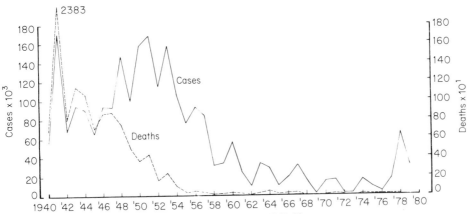

Fig. 67.1 Pertussis: cases and deaths in England and Wales, 1940–80

However, from a survey of 8092 cases notified to area health authorities in England and Wales in the 1974–75 winter, Miller and Fletcher (1976) concluded that the disease was still fairly severe; 10 per cent of patients were admitted to hospital and ten deaths were reported. Convulsions occurred in 0.4 per cent of the cases. Severity, in terms of hospital admission rate, was greater in infants under the age of six months; 60 per cent of 545 notified cases in this age group were admitted, half being regarded as seriously ill (see also Islur *et al.* 1975). Trollfors (1979) concluded from a study of 59 infected children aged 2 to 26 weeks admitted to hospital in Sweden that the disease was usually mild. Further information will probably be forthcoming from studies of the United Kingdom epidemic of 1978–80, in which the case-fatality rate was only 0.02 per cent (Table 67.1). Information is lacking on the respiratory and other sequelae of whooping cough as it occurs today, but whether the lowered case-fatality rate represents a milder disease or reflects improved treatment is uncertain. On present evidence the disease appears to be sufficiently severe to justify the use of vaccination.

Vaccination policy

A decision as to whether pertussis vaccine should be given to an individual infant, or recommended for a national vaccination programme, has to be reached on the basis of incomplete data (Miller *et al.* 1982). A precise measure of the efficacy or safety of the vaccine is not available. The current severity of whooping cough is not known, and its prevalence in the absence of vaccination cannot with certainty be forecast. However, the disease is liable to reappear when vaccination rates fall. In most babies the vaccine is harmless, apart from minor local reactions; and provided contra-indications to vaccination are observed, serious reactions are probably rare. When infection does occur, *Bord. pertussis* may still cause severe protracted illness which may endanger life and have respiratory and other sequelae, especially in those under the age of six months. Such events are likely to be minimized by modern therapeutic methods, including the use of antibiotics and intensive care techniques. It must also be remembered that the severe complication of encephalitis can occur in the natural disease.

On the whole, the balance of evidence favours routine vaccination of infants. It would shift still further in this direction if vaccines of greater purity and safety were developed. On the contrary, if the disease became milder, or other methods of prophylaxis or effective means of treatment became available, the balance would shift in the opposite direction.

Treatment

Because about 90 per cent of deaths from whooping cough occur before the age of one year, mostly in the first six months of life, vaccination can confer little benefit on infants who are at greatest risk of severe infection. Indirectly they may benefit by the protection given to older siblings from whom the infant may contract the disease, and the herd immunity possibly induced by widespread vaccination. Antisera prepared against the exotoxin of *Bord. pertussis* appear to have little effect on the course of the disease (see Pittman *et al.* 1979). Similarly, there is no evidence to show that gamma-globulin from the blood of vaccinated donors has any prophylactic effect in children exposed to infection (Morris and McDonald 1957). Chloramphenicol given in the very early stage of the disease may be beneficial, but its use is too dangerous for routine practice. The clinical course of whooping cough appears to be uninfluenced by other antibiotics. Erythromycin may eliminate the organism from the nasopharynx and thus indirectly protect contacts (Bass *et al.* 1969, Nelson 1969, Baraff *et al.* 1978); and alone (Altemeier and Ayoub 1977) or in conjunction with cotrimoxazole (Adcock *et al.* 1972) may be of prophylactic value in those exposed to infection.

Bronchitis

Bronchitis is an inflammatory disease of the bronchi that may be acute or chronic. Certain atmospheric conditions, particularly smoke and fog, predispose to the disease and perpetuate it. Cigarette smoking is an important contributory factor. In many cases allergy to pollen, moulds, bacteria, and other sensitizing agents seems to be responsible. In Britain chronic bronchitis is a serious disease. It is three times as common in men as in women; it has a bad prognosis, and it accounts for 4 per cent of all deaths. It cannot be regarded as an infectious disease, nor do bacteria appear to be primarily responsible for it. Their role is a secondary one, serving to maintain and aggravate the inflammatory lesions in the bronchial mucosa. It is difficult to determine which organisms in a mixed flora are responsible. Bacteria may be irregularly distributed in sputum, and the predominant organism is not necessarily the significant one (Turk and May 1967). Several specimens of sputum should be taken from each patient and should preferably be homogenized before culture. Clinical improvement of the patient and disappearance of a given organism as the result of treatment, followed by a reversal of these two events when treatment is discontinued, points to the organism's aetiological importance.

The normal bronchus is sterile. In chronic bronchitis and in bronchiectasis the two most frequent infecting organisms are the non-capsulated types of *Haemophilus influenzae* and the pneumococcus (Mulder *et al.* 1952, May 1954, Allibone *et al.* 1956, Lees and McNaught 1959, Holdaway and Turk 1967). The influenza bacillus predominates in purulent sputum, is much less frequent in mucoid sputum, and is seldom found in bronchial asthma. For isolation of the influenza bacillus Hovig and Aandahl (1969) recommend the use of a chocolate agar medium containing 300 µg bacitracin per ml. Fallon and Brown (1967) reported the isolation of capsulated strains, types b and f, from 38 of 50 chronic bronchitic patients with purulent sputum. The presence of precipitins in the serum of 69 per cent of patients suffering from mucopurulent or obstructive bronchitis as against only 25 per cent of patients suffering from simple bronchitis, led Burns and May (1967) to conclude that *H. influenzae* was not a primary infecting agent, but occurred only when the disease had progressed beyond the stage of mucus hypersecretion alone.

Bronchiectasis may follow pneumonia, pertussis, measles, asthma, or other respiratory disturbance. Owing probably to the use of tetracyclines and other antibiotics it is becoming comparatively uncommon, both in Britain and in the United States (Field 1969). On serological grounds there is reason to believe that klebsiellae are not infrequently responsible for the chronic destructive lesions that are seen in this disease (Burns 1968).

In treatment, whether given prophylactically at the beginning of winter or therapeutically when bronchitis has developed, the tetracyclines, ampicillin, or cotrimoxazole are of value. They lead to an amelioration of the symptoms and reduction in the amount and purulence of the sputum without, however, permanently suppressing the characteristic flora (see Edwards *et al.* 1957). Prophylactic injections of influenza vaccine have not been found to affect the attack rate (Report 1959*a*).

The acute *epidemic bronchiolitis of infants* is of virus origin and is considered in Chapter 98. (For the role of viruses in the acute bronchitis of children, see Horn *et al.* 1979.)

Primary lobar pneumonia

Primary lobar pneumonia is a severe pyrexial illness characterized by sudden onset with a rigor and high fever, consolidation of one or more lobes of the lung, and in non-fatal cases recovery usually by crisis after 7–10 days. Bacteraemia is often present during the initial phase of the infection, and is a constant feature in severe and fatal cases. In spite of the efficacy of modern treatment in diminishing the death rate, the present incidence in the United States is said to be about as high as before the antibiotic era; Austrian (1968) estimated that there were 700 000 cases a year with an average fatality rate of 10 per cent. In England and Wales in 1976, the estimated number of hospital admissions for pneumococcal pneumonia was 11 300 and there were 2917 deaths. These figures do not enable the case-fatality rate to be calculated, but they indicate that primary pneumonia was responsible for not more than 0.5 per cent of all deaths. Fatality is highest in infants and old people.

Bacteriologically, 80–90 per cent of cases are caused by the pneumococcus; the remainder mostly by *Streptococcus pyogenes*, *Staphylococcus aureus*, *Haemophilus influenzae*, and *Klebsiella pneumoniae* (Wallman *et al.* 1955, Goldstein *et al.* 1967, Austrian 1968, Basiliere *et al.* 1968). Of the 80 or so serotypes of pneumococci some—Danish serotypes 1, 2, 3, 4, 5, 6, 7, 8, 9, 12, 14, 18, 19, 20, 23, and 25—are found mainly in association with disease, whereas the others are present in the nose and throat of healthy carriers.

In the USA from 1967 to 1978, the types associated with bacteraemia were, in order of frequency, Danish serotypes 8, 4, 1, 14, 3, 7, 12, 6, 18, 9, 19, 23, and 5, which together accounted for 80 per cent of infections (Austrian *et al.* 1976, Austrian 1978). In Great Britain, the ten commonest types isolated from serious infections in 1969–1977 were all represented in the USA listing, although the rank order differed. Results from other countries show various differences from the British or American patterns (Austrian 1978, Parker 1978, Henrichsen 1979, Greenwood *et al.* 1980), including the occurrence of types which are not currently prevalent in Great Britain or the USA, such as type 2 in South Africa.

The incidence of the different types varies to some extent from year to year, place to place, and with age. Some types are more fatal than others, type 3 being the most fatal of all. According to Austrian (1978) the fatality rate in type 1 cases of bacteraemic pneumonia treated with penicillin is 6 per cent, but in type 3 cases 48 per cent. Some types tend to be associated with disease in parts of the body other than the lung, such as the middle ear, conjunctiva, pleura, or meninges. Healthy carriage in the throat is quite common, being found in 40 or 50 per cent of persons at a single sampling. The same type may be carried for years. Multiple carriage is not infrequent; as many as seven types may be recorded simultaneously in the same person. Types responsible for serious disease in children tend to be less common in adults. (For references see 5th edition, p. 2021.)

Though primary pneumonia must be regarded as an infectious disease, the various aetiological factors concerned are by no means clear. Outbreaks in young recruits may occur, but even in these it is usually not known how far case-to-case and carrier-to-case infections are primarily responsible, or how far unfavourable environmental or physiological conditions may have lowered the resistance of carriers. The prevalence of the disease in newly joined South African diamond miners suggests the importance of exposure to serotypes of which they have had no previous experience. Increased susceptibility is observed in patients who have had a splenectomy, or suffer from sickle-cell anaemia, or alcoholism.

Not much light is thrown on this problem by study of the experimental disease in animals. The work of Blake and Cecil and of numerous others showed that lobar pneumonia could be produced in monkeys by inoculating small numbers of pneumococci into the trachea, and in dogs by inoculating the organisms suspended in mucin into a terminal bronchus (for references see 5th edition, pp. 2018–20). Gaskell (1927) found that the type of disease in rabbits after intratracheal inoculation depended largely on the virulence of the strain used, ranging, as virulence decreased, from a rapidly fatal septicaemia without pulmonary changes, through lobar and lobular pneumonia, and through lesions confined almost entirely to the bronchi and bronchioles, to a complete absence of reaction.

The evidence culled from observations on the natural disease in man and the experimental disease in animals leaves little doubt that infection occurs by the respiratory route. The original focus is generally near the periphery of the lung, and the infecting organisms seem to be carried in oedema fluid from one alveolus to another, passing directly through the pores of Kohn (Loeschcke 1931). The inflammatory reaction in the lung progresses through its successive stages, frequently accompanied by a bacteraemia that may or may not increase in severity. At some time between the 4th and 7th days in favourable cases specific antibodies appear in the blood. Simultaneously, or soon after, the temperature falls by crisis, the pneumococci disappear from the blood, and the clinical condition of the patient greatly improves; abnormal signs in the lungs may persist for several days or even weeks. There are good grounds for believing that the degree and persistence of the bacteraemia, determined as they must be by the number of organisms that enter the blood from the infected tissues, afford the surest index to death or recovery. The grave prognostic significance of a positive blood culture is, in fact, one of the best attested observations on the disease.

Diagnosis

In primary pneumococcal pneumonia the organism can be isolated from the sputum by culture on blood agar, and is quite often demonstrable in the blood. Its main differential properties are the characteristic colonies on blood agar, diffuse growth in serum broth, sensitivity to optochin, and solubility in bile (see Chapter 29). Examination for haemolysis, fermentation, and antibiotic sensitivity is of little practical value. Pneumococcal antibodies are not found until the 4th or 5th day of illness, and the detection by a precipitation test of type-specific polysaccharide in the urine (Dochez and Avery 1917) is seldom successful.

More sensitive serological methods, such as radioimmunoassay (Schiffman and Austrian 1971) and enzyme-linked immunosorbent assays (Berntsson *et al.* 1978) may be of value in confirming the diagnosis in patients treated with antibiotics. Identification of capsular antigen in the sputum or serum by means of counter-current immunoelectrophoresis (Dorff *et al.* 1971) increases the rate of detection of pneumococcal infection in patients with radiological evidence of pneumonia to 72 per cent; this compares with only 12 per cent when sputum culture alone is used to detect the presence of pneumococci (Dulake 1979). Attempts have been made to improve the diagnostic value of sputum culture in lobar pneumonia and other chest infections, by such means as the selection of purulent material from the sputum, either directly or by washing; by homogenization; and by quantitative culture in order to detect organisms present in excess of 10^6 per ml (Louria 1962, Guckian and Christensen 1978). Culture of aspirates collected by means of a needle inserted into the trachea through the skin is more reliable than sputum culture (Bartlett and Finegold 1978), but the procedure is not without occasional complications.

When typing of the pneumococcus is required, a specimen of the sputum is injected into the peritoneal cavity of a mouse. The animal is killed after 6 to 8 hours, and the sediment and supernatant fluid of the peritoneal washings are examined by agglutination and precipitation tests with type-specific antisera. A quicker method is to mix the agglutinating suspension with antiserum and methylene blue on a slide and to note the swelling of the capsule that results from the action of the antiserum, as well as the agglutination itself. This *capsule swelling (Quellung) reaction*, which was first described by Neufeld (1902), can be applied to specimens of sputum direct (see Sabin 1929, 1933, Armstrong 1932, Beckler and MacLeod 1934). In experienced hands, the agreement of the results of typing by the direct swelling reaction and the indirect reaction after mouse inoculation ranges between 75 and 90 per cent.

Skin reactions elicited in patients by intradermal injection of the purified polysaccharide have been studied (see 6th edition, p. 2163).

Prophylaxis and treatment

General measures, such as good ventilation, avoidance of overcrowding, the control of coughing or sneezing except into a handkerchief, and of spitting except into a receptacle that can be sterilized, may be of some value in prevention. Isolation of the patient is unlikely to be of much value prophylactically, owing to the presence of healthy carriers.

The *serum treatment* of pneumococcal pneumonia, although of value, has passed into disuse. Those interested are referred to the 4th edition (pp. 1886–87) or the 5th edition (pp. 2022–23) of this book.

Vaccination   *Vaccines*, prepared from whole pneumococci, were first studied in field trials among gold miners in South Africa by Almroth Wright shortly before the first world war (Wright *et al.* 1914). Although these trials gave evidence of a protective effect, both this and later work was generally considered to be inconclusive (see 6th edition p. 2164). Further progress depended upon the observation that capsular polysaccharide was immunogenic in man (Frances and Tillett 1930). MacLeod and his coworkers (1945) later showed that vaccines prepared from the polysaccharide capsular antigen of four

prevalent pneumococcal types induced type-specific protection in recruits to the US Armed Forces. With the advent of chemotherapy, interest in vaccination disappeared, until it was recognized in the USA that, despite the undoubted value of penicillin, the pneumococcus remained a pathogen to be reckoned with (Austrian and Gold 1964, Austrian 1977). As we have seen (p. 400), only a small number of serotypes is responsible for most cases of serious infection in a country or region; and a vaccine prepared from about 14 serotypes should be capable of providing immunity against about 80 per cent of serious infections. Field trials among groups particularly susceptible to pneumococcal pneumonia, such as miners in South Africa (Smit *et al.* 1977), children with sickle-cell disease, patients who have had their spleens removed (Amman *et al.* 1977), and normal adult natives of Papua New Guinea (Riley *et al.* 1977), have demonstrated the presence of a 78–85 per cent protection against the serotypes included in polyvalent vaccines. A single dose of vaccine, containing 50 micrograms of the purified capsular polysaccharide of each of 12–14 serotypes, is able to induce in most recipients over about 2 years of age a greater than four-fold increase in antibody, measured as antibody nitrogen by radioimmunoassay, to most of the capsular antigens included in the vaccine (Austrian *et al.* 1976, Amman *et al.* 1977, Weibel *et al.* 1977).

Local reactions to the vaccines, sometimes accompanied by fever, are seen in as many as 80 per cent of recipients, but are rarely severe. The antibody response to a single injection of pneumococcal antigen is sustained, but follow-up of recipients of modern vaccines for longer than 3–4 years has not yet been reported (Heidelberger *et al.* 1950, Hilleman *et al.* 1979). Re-vaccination within this period is unnecessary, and is also inadvisable owing to an increased frequency of local reactions (Borgono *et al.* 1978). Children under the age of two years have been found to respond poorly to some of the serotypes in polyvalent pneumococcal vaccines.

Chemotherapy   This has been generally used for the treatment of pneumococcal pneumonia ever since Evans and Gaisford (1938) demonstrated the efficacy of sulphonamides. Pneumococci normally are sensitive to all the usual antibiotics so that penicillins usually suffice for most therapeutic purposes. Nevertheless, pneumococcal pneumonia is still liable to prove fatal despite treatment (Austrian and Gold 1964). Strains resistant to penicillin and other antibiotics have been described in recent years (Howes and Mitchell 1976, Dixon *et al.* 1977).

Broncho-pneumonia

Broncho-pneumonia is a loose term referring to a patchy consolidation of the lungs as opposed to the lobar or segmental consolidation of lobar pneumonia. It is often bilateral. The inflammatory process may affect the walls of the bronchi and bronchioles, leading to an interstitial pneumonia with consequent incomplete resolution and subsequent fibrosis. The onset is often insidious. The disease may be primary, as in infants, but is more usually secondary to bronchitis, to one of the specific fevers, to gastro-enteritis, or to chronic illness such as diabetes, nephritis, anaemia, or carcinoma. It may follow surgical operations, especially those on the abdomen, which render coughing painful and predispose to atelectasis, and is a hazard to patients with a tracheostomy. Or again, it may be due to aspiration of food or vomit into the trachea from one cause or another. Cases may be sporadic or occur in outbreaks such as are seen among recruits during an influenza epidemic. In influenzal pneumonia the lesions of the bronchi are often severe; haemorrhages into the alveoli, the bronchi, or the peribronchiolar tissue are frequently encountered, and localized areas of necrosis or suppuration are not uncommon (see Opie *et al.* 1921). The frequency of bronchopneumonia naturally depends on the prevalence of the predisposing causes, and the case fatality on such factors as age, previous health, and the nature of the invading organism. In England and Wales the deaths from all forms of the disease amount to about 5 per cent of the total deaths registered—a figure similar to that for chronic bronchitis. (For a beautifully illustrated study of influenzal pneumonia, see Mulder and Hers 1972.)

The *bacteriology* of the disease varies greatly from time to time and from place to place. For example, whereas Smillie and Duerschner (1947) in the United States found that in terminal broncho-pneumonia the chief organisms were pneumococci, influenza bacilli, and β-haemolytic streptococci, Tanner and her colleagues (1969) in England found that only 3–8 per cent of cases of this type could be accounted for by these three organisms; in 34 per cent of their cases *Escherichia coli* was isolated from the lungs at autopsy and in 31 per cent *Staphylococcus aureus*. As in bronchitis, determination of the responsible organism may be difficult. Culture of lung tissue *post mortem* is the most reliable method. The trachea and larger bronchi are subject to ante- or post-mortem invasion in persons who die suddenly from causes other than respiratory disease (Smillie and Duerschner 1947).

In *primary broncho-pneumonia of infants* pneumococci are the organisms most frequently incriminated. In broncho-pneumonia *secondary* to the acute fevers, pneumococci, staphylococci, and influenza bacilli commonly occur. The pneumococci tend to be of the commensal types that lead a harmless existence in the upper part of the respiratory tract rather than the well recognized virulent types that predominate in lobar pneumonia.

Anaerobes of the *Bacteroides* and *Fusobacterium* genera may be implicated in cases of aspiration pneumonia and empyema, and in some cases of chronic bronchitis, but their detection may require culture of material obtained by transtracheal aspiration or bronchoscopy (Bartlett *et al.* 1974,

Ries *et al.* 1974, Gardiner *et al.* 1977; see also Chapter 62).

Staphylococcus aureus may give rise to primary or secondary broncho-pneumonia which, as the organisms are often resistant to antibiotics, has a high case-fatality rate. The lesions in the lungs consist of multiple abscesses appearing on x-ray examination as cavities. In patients that recover these cavities disappear and the lung takes on a normal appearance again. In the hyperacute form that is sometimes seen during outbreaks of influenza, the patient may be dead within 24 hours before the characteristic lesions have time to develop. The infection of the lung is often associated with a generalized staphylococcal septicaemia following osteomyelitis or abscesses in other parts of the body.

Friedländer's bacillus, *Klebsiella aerogenes*, is not a common cause of broncho-pneumonia, but is capable of producing an acute suppurative and usually fatal disease (see Solomon 1937, 1940, Hyde and Hyde 1943). The sputum is characterized by extreme viscosity. The disease may become chronic and last for weeks or months.

Other organisms of the gram-negative group, such as *Escherichia coli* and species of *Pseudomonas*, *Proteus*, and *Achromobacter* are less often met with (Fetzer *et al.* 1967, Tillotson and Lerner 1967). They are most frequently seen in broncho-pneumonia of middle-aged patients suffering from alcoholism, diabetes, or cardiac, pulmonary, or other debilitating disease.

In the broncho-pneumonia that follows *influenza* and plays so large a part in determining the case-fatality rate, various organisms are encountered, mainly *Haemophilus influenzae*, pneumococci, *Streptococcus pyogenes*, and *Staphylococcus aureus*. The view that Pfeiffer's influenza bacillus may give rise to post-influenzal broncho-pneumonia is considered by Turk and May (1967) to be unsupported by firm evidence. This doubt is difficult to understand in the face of the overwhelming evidence produced by such competent observers as Fildes and McIntosh (1920) and McIntosh (1922), who found Pfeiffer's bacillus at post-portem examination of the lungs in 80 per cent or more of patients dying during the autumn wave of pandemic influenza in 1918 and the following wave at the beginning of 1919. The lower in the respiratory tract the examination was made, the larger were the numbers and the purer the state in which the organism was found. Writing before the discovery of the influenza virus, Andrewes (1920) stated that, whether Pfeiffer's bacillus was a primary or a secondary cause of the disease, there could be no doubt that it played a part of vast importance, and was the chief cause of the haemorrhagic oedema and the localized haemorrhagic areas in the lungs that formed such a conspicuous feature of the 1918–19 epidemic. (See also Mulder and Hers 1972.)

As a rule, little can be done to prevent the development of broncho-pneumonia in a patient rendered susceptible by some predisposing disease. Such a patient should be protected from the secondary infections that are liable to occur under unfavourable conditions. There is ample evidence that, for example, overcrowding in hospital wards or military camps may seriously increase the mortality during an outbreak of influenza or measles (see Memorandum 1944). In streptococcal and pneumococcal infections sulphonamides and penicillin will usually be effective in treatment, but in staphylococcal infections and those caused by Pfeiffer's influenza bacillus the tetracyclines, erythromycin, or even chloramphenicol must be given. The disease is often rapidly fatal and no time must be lost in beginning treatment. The form caused by Friedländer's bacillus is peculiarly dangerous; treatment with an aminoglycoside antibiotic may be effective.

Mention should perhaps be made of the *interstitial plasma-cell pneumonia* of infants. This disease, which attacks premature infants particularly, starts with coryza and passes on to a terminal asphyxia or recovery. The case-fatality rate is thought to be about 20 per cent. *Post mortem*, the lungs are heavy and airless. Histologically the interalveolar septa are seen to be wide and filled with plasma cells. The disease is caused by a parasite—*Pneumocystis carini*—which is common in the lungs of animals. It forms cysts or vacuoles in the alveolar exudate and can be demonstrated by Giemsa's stain (see Reye and Seldam 1956, Gadjusek 1957, 1976). Both cotrimoxazole and pentamidine are of some therapeutic value (Hughes *et al.* 1978).

Another disease that may attack infants and immuno-suppressed patients with serious underlying disease (Dutz *et al.* 1976, Walzer *et al.* 1976) is the so called *giant-cell pneumonia*. This appears to be a non-specific reaction to various infective agents, notably measles virus. Microscopically, the lungs at post-mortem examination show an interstitial pneumonia, characterized by the presence of multinucleated giant cells containing intranuclear and intracytoplasmic eosinophilic inclusion bodies. (For further information see Kripke 1961.)

Respiratory infections in the lower animals

Except for *Bordetella bronchiseptica* infection, which is dealt with below, the numerous respiratory infections of animals are described in other chapters. Some animal pneumonias, for example contagious bovine pleuropneumonia (see Chapter 78), are specifically caused by a single agent. Many others, such as chronic respiratory disease of chickens (see Chapter 78) and shipping fever of cattle (see Chapter 55), are more frequently caused by infection with two or more agents that may include bacteria, mycoplasmas (see Chapter 78), and viruses; often in such diseases one agent appears to predominate. Some bacterial pathogens, such as *Pasteurella haemolytica* biotype A (see Chapter 55), show a strong predilection for the lung; others, such as streptococci, corynebacteria and *Fusobacterium necrophorum* are found no more frequently in diseased lung than in lesions elsewhere. Stress factors such as fatigue, overcrowding, and fear or excitement are widely believed to play a part in the pathogenesis of many respiratory infections. High concentrations of atmospheric ammonia in animal enclosures may lead to an increase in disease. Some pneumonias, such as enzootic pneumonia of sheep (see Chapter 55), are difficult to reproduce experimentally, even by the endobronchial injection of homogenized lung tissue

from natural cases. The prevalence of pneumonias, particularly those with a multiple aetiology, has increased as a result of modern intensive systems of animal husbandry; such pneumonias often tend to be chronic and to lead to loss of productivity.

Respiratory infections due to *Bordetella bronchiseptica*

In 1911 M'Gowan isolated *Bord. bronchiseptica* from the respiratory tract of dogs and several other species. He regarded the organism as a primary cause of canine distemper—a view later disproved (see Chapter 98). *Bord. bronchiseptica*, however, is now known to be a frequent cause of infectious pneumonia, not only in guinea-pigs (Smith 1913), but also in other laboratory rodents, rabbits, pigs, horses, dogs, cats, and turkeys. Infection is often chronic, and may be pure or mixed. In puppies the organism may act as a primary or secondary pathogen of the respiratory tract (Wright *et al.* 1973); it is one of the causes of *kennel cough*. *Bord. bronchiseptica* has been isolated from the upper respiratory tract of several wildlife species (Farrington and Jorgenson 1976). Injected intraperitoneally into guinea-pigs in a dose of 0.5 ml of a 24-hour broth culture it causes death in 24 to 48 hours; *post mortem* there are small haemorrhages on the peritoneum, and a viscid translucent exudate forming pseudomembranes on the liver, spleen, and the less mobile parts of the intestine; the bacilli are easily isolated from the peritoneal cavity, but with difficulty from the blood, liver, or lungs. Subcutaneous injection produces a local lesion only; feeding and inhalation are without effect. The organism injected intraperitoneally is non-pathogenic for mice. It rapidly loses its virulence in culture.

Bord. bronchiseptica is thought to be the main cause of *infectious atrophic rhinitis* of pigs, a disease whose aetiology is nonetheless incompletely understood. Other organisms such as *Pasteurella multocida* may sometimes play a part (see Chapter 55). Lesions of the disease are seen in many countries, mainly in pigs aged 2 to 5 months. An initial catarrhal rhinitis is followed later by hypoplasia of the nasal turbinate bones, especially the inferior scroll of the ventral turbinate. The lesion of turbinate atrophy usually occurs only in pigs originally infected within the first few weeks of life. Deformities of the snout sometimes occur, and the disease, when severe, leads to unthriftiness. Numerous workers have reproduced rhinitis and turbinate atrophy in young pigs by the intranasal administration of virulent strains of *Bord. bronchiseptica* from swine or other animal species (Cross and Claflin 1962, Ross *et al.* 1967, Brassinne *et al.* 1976). The organism is widespread in pig populations. Of 844 pigs from central, eastern, or southern England examined at abattoirs in 1978-79, 50 per cent harboured it in the nasal cavity, 10 per cent had severe lesions of turbinate atrophy, and 15 per cent had moderate lesions

(Cameron *et al.* 1980). In a study based on heat-stable O antigens and heat-labile K antigens, Pedersen (1975) found that the porcine strains examined were all of a single antigenic type (O 1,2; K 1,2); isolates from other animal species showed some variation. The organism is susceptible to sulphonamides, but resistant strains are not uncommon. (For a detailed account of atrophic rhinitis see Switzer and Farrington 1975; and for vaccination of pigs with *Bord. bronchiseptica* see Giles and Smith 1983.)

References

Adcock, K. J., Reddy, S., Okubadejo, O. A. and Montefiore, D. (1972) *Arch. Dis. Childh.* **47**, 311.

Allibone, E. G., Allison, P. R. and Zinnemann, K. (1956) *Brit. med. J.* **i**, 1457.

Altemeier, W. A. and Ayoub, E. M. (1977) *Pediatrics, Springfield* **59**, 623.

Ammann, A. J., Addiego, J., Warg, D. W., Lubin, B., Smith, W. B. and Mentzer, W. C. (1977) *New Engl. J. Med.* **297**, 897.

Andrewes, F. W. (1920) *Rep. publ. Hlth med. Subj. No. 4* p. 110. Minist. Hlth, London.

Armstrong, R. R. (1932) *Brit. med. J.* **i**, 187.

Austrian, R. (1968) *J. clin. Path.* **21**, Suppl., 93; (1977) *J. infect. Dis.* **186**, Suppl. 38; (1978) In: *Pathogenic Streptococci. Proc. VIIIth Int. Symp. on Streptococcal Disease, Oxford, 1978.* p. 187. Ed. by M. T. Parker. Reedbooks, Chertsey, Surrey.

Austrian, R. and Gold, J. (1964) *Ann. intern. Med.* 60, 759.

Austrian, R. *et al.* (1976) *Trans. Ass. Amer. Phys.* **89**, 184.

Baraff, L. J. and Cherry, J. D. (1978) In: *Int. Symp. on Pertussis*, p. 291. Ed. by C. R. Manclark and J. C. Hill. U.S. Dept Hlth Educ. and Welf., Bethesda, U.S.A.

Baraff, L. J., Wilkins, J. and Wehole, P. F. (1978) *Pediatrics, Springfield* **61**, 224.

Bartlett, J. G. and Finegold, S. M. (1978) *Amer. Rev. resp. Dis.* **117**, 1019.

Bartlett, J. G., Gorbach, S. L. and Finegold, S. M. (1974) *Amer. J. Med.* **56**, 202.

Basiliere, J. L., Bistrong, H. W. and Spence, W. F. (1968) *Amer. J. Med.* **44**, 580.

Bass, J. W., Klenk, E. L., Kotheimer, J. B., Linnemann, C. C. and Smith, M. H. D. (1969) *J. Pediat.* **75**, 768.

Bassili, W. R. and Stewart, G. T. (1976) *Lancet* **i**, 471.

Beckler, E. and MacLeod, P. (1934) *J. clin. Invest.* **13**, 901.

Berg, B. J. van den and Yerushalmy, J. (1969) *Pediat. Res.* **3**, 298.

Berntsson, E., Broholm, K. A. and Kaijser, B. (1978) *Scand. J. infect. Dis.* **10**, 177.

Bordet, J. and Gengou, O. (1906) *Ann. Inst. Pasteur* **20**, 731; (1907) *Ibid.* **21**, 720.

Borgono, J. M. *et al.* (1978) *Proc. Soc. exp. Biol. Med.* **157**, 148.

Bradford, W. L. and Slavin, B. (1940) *Proc. Soc. exp. Biol., N.Y.* **43**, 590.

Bradstreet, C. M. P., Tannahill, A. J. and Edwards, J. M. B. (1972) *J. Hyg., Camb.* **70**, 75.

Brandt, C. D. *et al.* (1969) *Amer. J. Epidem.* **90**, 484.

Brassinne, M., Dewaele, A. and Gouffaux, M. (1976) *Res. vet. Sci.* **20**, 162.

Burland, W. L., Sutcliffe, W. M., Joyce, M. A., Hilton, M. L. and Muggleton, P. W. (1968) *Med. Offr* **119**, 17.

Burnet, F. M. and Timmins, C. (1937) *Brit. J. exp. Path.* **18**, 83.

Burns, M. W. (1968) *Lancet* **i**, 383.

Burns, M. W. and May, J. R. (1967) *Lancet* **i**, 354.

Butler, N. R., Voyce, M. A., Burland, W. R. and Hilton, M. L. (1969) *Brit. med. J.* **i**, 663.

Byers, R. K. and Moll, F. C. (1948) *Pediatrics, Springfield* **1**, 437.

Cameron, R. D. A., Giles, C. J. and Smith, I. M. (1980) *Vet. Rec.* **107**, 146.

Chievitz, I. and Meyer, A. H. (1916) *Ann. Inst. Pasteur* **30**, 503.

Church, M. A. (1979) *Lancet* **ii**, 118.

Clifford, R. E., Smith J. W. G., Tillett, H. E. and Wherry, P. J. (1977) *Int. J. Epidemiol.* **6**, 1151.

Cockburn, W. C. (1979) *J. Infect.* **1**, Suppl. 2, 3.

Cockburn, W. C. and Holt, H. D. (1948) *Mon. Bull. Minist. Hlth Lab. Serv.* **7**, 156.

Cohen, H., Ramshorst, J. D. van and Drion, E. F. (1967) *Symp. Ser. Immunobiol. Standard.* **10**, 53.

Cohen, S. M. and Wheeler, M. W. (1946) *Amer. J. publ. Hlth* **36**, 371.

Combined Scottish Study (1970) *Brit. med. J.* **iv**, 637.

Connor, J. D. (1970) *New Engl. J. Med.* **283**, 390.

Cooney, M. K., Hall C. E. and Fox, J. P. (1972) *Amer. J. Epidemiol.* **96**, 285.

Court, S. D. M. (1968) *J. clin. Path.* **21**, Suppl. 30.

Cross, R. F. and Claflin, R. M. (1962) *J. Amer. vet. med. Ass.* **141**, 1467.

Dane, D. S., Haire, M. and Dick, G. (1966) *J. Hyg., Camb.* **64**, 475.

Dingle, J. H. (1948) *J. Amer. med. Ass.* **136**, 1084.

Ditchburn, R. K. (1979) *Brit. med. J.* **i**, 1601.

Dixon, J. M. S., Lipinski, A. E. and Graham, M. E. P. (1977) *Canad. med. Ass. J.* **117**, 1159.

Dochez, A. R. and Avery, O. T. (1917) *J. exp. Med.* **26**, 477.

Dorff, G. J., Coonrod, J. D. and Rytel, M. W. (1971) *Lancet* **i**, 578.

Dowling, H. F. and Lefkowitz, L. B. (1963) *Amer. Rev. resp. Dis.* **88**, 61.

Dulake, C. (1979) *J. Infect.* **1**, Suppl. 2, 45.

Dutz, W., Post, C., Vessel, K. and Kohout, E. (1976) *Nat. Cancer Inst. Monogr.* **43**, 31.

Edwards, G., Buckley, A. R., Fear, E. C., Williamson, G. M. and Zinnemann, K. (1957) *Brit. med. J.* **ii**, 259.

Ehrengut, W. (1974) *Dtsch. med. Wschr.* **99**, 2273; (1981) *Brit. med. J.* **ii**, 494.

Eldering, G. and Kendrick, P. L. (1952) *Amer. J. publ. Hlth* **42**, 27.

Evans, D. G. and Gaisford, W. F. (1938) *Lancet* **ii**, 14.

Fallon, R. J. and Brown, W. (1967) *Lancet* **i**, 614.

Farrington, D. O. and Jorgenson, R. D. (1976) *J. Wildlife Dis.* **12**, 523.

Feldman, G. V., Macaulay, D., Abbott, J. D., Craddock-Watson, J. E. and Tobin, J. O'H. (1972) *Lancet* **i**, 379.

Fetzer, A. E., Werner, A. S. and Hagstrom, J. W. C. (1967) *Amer. Rev. resp. Dis.* **96**, 1121.

Field, C. E. (1969) *Arch. Dis. Childh.* **44**, 551.

Fildes, P. and McIntosh, J. (1920) *Brit. J. exp. Path.* **1**, 119, 159.

Florsdorf, E. W., Felton, H., Bondi, A. and McGuinness, A. C. (1943) *Amer. J. med. Sci.* **206**, 421.

Francis, T. and Tillett, W. S. (1930) *J. exp. Med.* **52**, 573.

Gadjusek, D. C. (1957) *Pediatrics, Springfield* **19**, 543; (1976) *Nat. Cancer Inst. Monogr.* 43, 1.

Gardiner, I. T., Blenkharn, I., Stradling, P. and Darrell, J. H. (1977) *Brit. J. Dis. Chest* **71**, 277.

Gaskell, J. F. (1927) *Lancet* **ii**, 951.

Giles, C. J. and Smith, I. M. (1983) *Vet. Bull.* **53**, 327.

Goldstein, E., Daly, A. K. and Seamans, C. (1967) *Ann. intern. Med.* **66**, 35.

Gray, D. F. (1947) *J. Path. Bact.* **59**, 235; (1949) *J. Immunol.* **61**, 35.

Greenwood, B. M. *et al.* (1980) *Lancet* **i**, 360.

Griffith, A. H. (1974) *Proc. R. Soc. Med.* **67**, 16; (1978) *Brit. med. J.* **i**, 809; (1979) *Scot. med. J.* **24**, 42.

Guckian, J. C. and Christensen, W. D. (1978) *Amer. Rev. resp. Dis.* **118**, 997.

Haire, M., Dane, D. S. and Dick, G. (1967) *Med. Offr* **117**, 55.

Hannick, C. A. and Cohen, H. (1978) In: *Int. Symp. on Pertussis*, p. 279. Ed. by C. R. Manclark and J. C. Hill. US Dept Hlth, Educ. and Welf., Bethesda, U.S.A.

Heidelberger, M., Dilapi, M. M., Siegel, M. and Walter, A. W. (1950) *J. Immunol.* **65**, 535.

Heilig, W. R. (1957) *Minn. Med.* **40**, 542.

Henrichsen, J. (1979) *J. Infect.* **1**, Suppl. 2, 31.

Higgins, P. G., Boston, D. G. and Ellis, E. M. (1964) *Mon. Bull. Minist. Hlth Lab. Serv.* **23**, 93.

Hilleman, M. R., McLean, A. A., Vella, P. P., Weibel, R. E. and Woodhour, A. F. (1979) *J. Infect.* **1**, Suppl. 2, 73.

Hilton, M. L. and Burland, W. L. (1969) *Symp. Ser. Immunobiol. Standard.* **13**, 150.

Holdaway, M. D. and Turk, D. C. (1967) *Lancet* **i**, 358.

Hope-Simpson, R. E. and Miller, D. L. (1973) *Postgrad. med. J.* **49**, 763.

Horn, M. E., Reed, S. E. and Taylor, P. (1979) *Arch. Dis. Childh.* **54**, 587.

Hovig, B. and Aandahl, E. H. (1969) *Acta. path. microbiol. scand.* **77**, 676.

Howes, J. U. and Mitchell, R. G. (1976) *Brit. med. J.* **i**, 996.

Hughes, W. T., Feldman, S., Chandhary, S. C., Ossi, M. J., Cox, F. and Sanyal K. (1978) *J. Pediat.* **92**, 285.

Hyde, L. and Hyde, B. (1943) *Amer. J. med. Sci.* **205**, 660.

Irwin, J. O. and Standfast, A. F. B. (1957) *J. Hyg., Camb.* **55**, 50.

Islur, J., Anglin, C. S. and Middleton, P. J. (1975) *Clin. Pediat.* **14**, 171.

Jeavons, P. M. and Bower, B. D. (1964) *Infantile Spasms*, Clinics in Developmental Medicine, No. 15. Heinemann, London.

Jenkinson, D. (1978) *Brit. med. J.* **iii**, 557.

Jennings, L. C., McDiarmid, R. D. and Miles, J. A. (1978) *J. Hyg., Camb.* **81**, 49.

Kanai, K. (1980) *Jap. J. med. Sci. Biol.* **33**, 107.

Kantchourine, A. K. and Kapitonova, M. E. (1969) *Ann. Inst. Pasteur* **117**, 378.

Kendrick, P. L. (1969) *Publ. Hlth Lab.* **27**, 85.

Kendrick, P. L., Elderling, G., Dixon, M. K. and Misner, J. (1947) *Amer. J. publ Hlth* **37**, 803.

Kendrick, P. L., Updyke, E. L. and Eldering, G. (1949) *Amer. J. publ. Hlth* **39**, 179.

Köng, E. (1953) *Helv. pediat. Acta* **8**, 90.

Kripke, S. S. (1961) *Amer. J. Dis. Child.* **102**, 123.

Kristensen, B. (1933) *J. Amer. med. Ass.* **101**, 204.

Kulenkampff, M., Schwartzman, J. S. and Wilson, J. (1974) *Arch. Dis. Childh.* **49**, 46.

Lautrop, H. (1958) *Acta path. microbiol. scand.* **43**, 255.

Lautrop, H. and Lacey, B. W. (1960) *Bull. World Hlth Org.* **23**, 15.

Lawson, G. M. (1933) *Amer. J. Dis. Child.* **46**, 1454.

Lees, A. W. and McNaught, W. (1959) *Lancet* **ii**, 1112, 1115.

Linnemann, C. C., Bass, J. W. and Smith, M. H. D. (1968) *Amer. J. Epidemiol.* **88**, 422.

Linnemann, C. C. and Berry, E. B. (1977) *Amer. J. Dis. Child.* **131**, 560.

Loeschcke, H. (1931) *Beitr. path. Anat.* **86**, 201.

Louria, D. B. (1962) *J. Amer. med. Ass.* **182**, 1082.

Macaulay, M. E. (1979) *J. Hyg., Camb.* **83**, 95.

McCloskey, B. P. (1950) *Lancet* **i**, 659; (1951) *Med. J. Aust.* **i**, 613.

McDade, J. E. *et al.* (1977) *New Engl. J. Med.* **297**, 1197.

Macdonald, H. and Macdonald, E. J. (1933) *J. infect. Dis.* **53**, 328.

McDonald, J. C. (1963) *Amer. Rev. resp. Dis.* **88**, No. 33 Pt 2, p. 35.

M'Gowan (1911) *J. Path. Bact.* **15**, 372.

McIntosh, J. (1922) *Spec. Rep. Ser. med. Res. Coun., Lond.*, No. 63.

Maclean, I. H. (1937) *J. Path. Bact.* **45**, 472.

MacLeod, C. M., Hodges, R. G., Heidelberger, M. and Berhnard, W. G. (1945) *J. exp. Med.* **82**, 445.

Madsen, T. (1933) *J. Amer. med. Ass.* **101**, 187.

Malleson, P. N. and Bennett, J. C. (1977) *Lancet* **i**, 237.

Malmgren, B., Vahlquist, B. and Zetterström, R. (1960) *Brit. med. J.* **ii**, 1800.

May, J. R. (1954) *Lancet* **ii**, 839.

Memorandum (1944) *War Memorandum No. 11 Med. Res. Coun., Lond.*

Mertsola, J., Ruuskanen, O., Kuronen, T. and Viljanen, M. K. (1983) *J. infect. Dis.* **147**, 252.

Miller, D. L. (1973) *Postgrad. med. J.* **49**, 749.

Miller, D. L., Alderslade, R. and Ross, E. M. (1982) *Epidem. Rev.* **4**, 1.

Miller, D. L. and Fletcher, W. B. (1976) *Brit. med. J.* **i**, 117.

Miller, D. L. and Ross, E. M. (1978) *Brit. med. J.* **ii**, 992.

Miller, D. L., Ross, E. M., Alderslade, R., Bellman, M. H. and Rawson, N. S. B. (1981) *Brit. med. J.* **282**, 1595.

Miravete, A. P. and Mora, S. P. de la (1966) *Revta. Invest. Salud. publ.* **26**, 235.

Mishulow, L., Siegel, M., Leifer, L. and Berkey, S. R. (1942) *Amer. J. Dis. Child.* **63**, 875.

Monto, A. S. and Cavallaro, J. J. (1971) *Amer. J. Epidemiol.* **94**, 280.

Monto, A. S. and Ullman, B. M. (1974) *J. Amer. med. Ass.* **227**, 164.

Morley, D., Woodland, M. and Martin, W. J. (1966) *Trop. geogr. Med.* **18**, 169.

Morris, D. and McDonald, J. C. (1957) *Arch. Dis. Childh.* **32**, 236.

Mulder, J., Goslings, W. R. O., Plas, M. C. van der and Lopes Cardozo, P. (1952) *Acta med. scand.* **143**, 32.

Mulder, J. and Hers, J. F. P. (1972) *Influenza*. Wolters-Noordhoff Publishing, Groningen.

Neimark, F. M., Lugovaya, L. V. and Belova, N. (1961) *Microbiol. Epidemiol. and Immunobiol.* **32**, 49.

Nelson, J. D. (1969) *Pediatrics, Springfield* **44**, 474; (1978) *Amer. J. Dis. Child.* **132**, 371.

Neufeld, F. (1902) *Z. Hyg. InfektKr.* **40**, 54.

Noah, N. (1976) *Brit. med. J.* **i**, 128.

North, E. A., Anderson, G. and Graydon, J. J. (1941) *Med. J. Aust.* **ii**, 589.

North, E. A., Keogh, E. V., Christie, R. and Anderson, G. (1940) *Aust. J. exp. Biol. med. Sci.* **18**, 125.

Ocklitz, H. W. and Milleck, J. (1967) *Zbl. Bakt.* **203**, 79.

Opie, E. L., Blake, F. G., Small, J. L. and Rivers, T. M. (1921) *Epidemic Respiratory Diseases*, London.

Parker, M. T. (1978) In: *Pathogenic Streptococci.* Proc. VIIIth Int. Symp. on Streptococci and Streptococcal Disease, Oxford, 1978, p. 191. Ed. by M. T. Parker. Reedbooks, Chertsey, Surrey.

Parrott, R. H., Vargosko, A. J., Kim, H. W. and Chanock, R. M. (1963) *Amer. Rev. resp. Dis.* **88**, No. 3 Pt 2, p. 73.

Pedersen, K. B. (1975) *Acta path. microbiol. scand.*, B **83**, 590.

Pereira, M. S. and Candeias, J. A. N. (1971) *J. Hyg.*, *Camb.* **69**, 399.

Perkins, F. T., Sheffield, F., Miller, C. L. and Skegg, J. L. (1969) *Symp. Ser. Immunobiol. Standard.* **13**, 141.

Pillemer, L., Blum, L. and Lepow, I. H. (1954) *Lancet* **i**, 1257.

Pittman, M. (1952) *J. Immunol.* **69**, 201; (1970) In: *Infectious Agents and Host Reactions*, p. 239. Ed. by S. Mudd. W. B. Saunders Co., Philadelphia; (1979) *Rev. inf. Dis.* **1**, 401.

Pittman, M., Furman, B. L. and Wardlaw, A. C. (1980) *J. infect. Dis.* **142**, 56.

Pittman, M., Gardner, R. A. and Marshall, J. F. (1979) *J. biol. Standard.* **7**, 261.

Poole, P. M. and Tobin, J. O'H (1973) *Postgrad. med. J.* **49**, 778.

Preston, N. W. (1965) *Brit. med. J.* **ii**, 11.

Raettig, H. (1959a) *Zbl. Bakt.* **175**, 236, 245, 618; (1959b) *Ibid.* **176**, 346.

Raška, K., Kratochilova, V. and Jelinek, J. (1957) *Brit. J. exp. Path.* **38**, 217.

Report (1951a) *Ann. Rep. M.O.H. Lond. Cty Coun. for 1951* p. 24; (1951b) *Brit. med. J.* **i**, 1464; (1956a) *Brit. med. J.* **ii**, 454; (1956b) *Lancet* **ii**, 1223; (1957) *Brit. med. J.* **i**, 1336; (1959a) *Ibid.* **ii**, 909; (1959b) *Brit. med. J.* **i**, 994; (1963) Conference on Newer Respiratory Disease Viruses, *Amer. Rev. resp. Dis.* **88**, No. 3, Pt 2, pp. 1–419; (1964) *World. Hlth Org. Tech. Rep. Ser.* No. 274, Geneva; (1965) *Brit. med. J.* **ii**, 319; (1969) *Brit. med. J.* **iv**, 329; (1970) *Brit. med. J.* **iv**, 637; (1977) *Whooping Cough Vaccination*, Department of Health and Social Security, H.M.S.O., London.

Reye, R. D. K. and Seldam, R. E. J. ten (1956) *J. Path. Bact.* **72**, 451.

Ries, M. F., Levison, M. E. and Kaye, D. (1974) *Arch. intern. Med.* **133**, 453.

Riley, I. D. *et al.* (1977) *Lancet* **i**, 1338.

Ross, R. F., Switzer, W. P. and Duncan, J. R. (1967) *Canad. J. comp. Med.* **31**, 53.

Sabin, A. B. (1929) *Amer. J. publ. Hlth* **19**, 1148; (1933) *J. Amer. med. Ass.* **100**, 1584.

Sato, Y., Izumiya, M. A. and Sato, H. (1978) In: *Int. Symp. on Pertussis*, p. 512. Ed. by C. R. Manclark and J. C. Hill. U.S. Dept Hlth Educ. and Welf., Bethesda, U.S.A.

Sauer, L. W. and Markley, E. D. (1946) *J. Amer. med. Ass.* **131**, 967.

Schiffman, G. and Austrian, R. (1971) *Fed. Proc.* **30**, 658.

Schubert, J. H., Eleff, M. G. and Hermann, G. J. (1961) *Amer. J. publ. Hlth* **51**, 441.

Seagroatt, V. and Sheffield, F. (1981) *J. biol. Standard.* **9**, 351.

Shibley, G. S. and Hoelscher, H. (1934) *J. exp. Med.* **60**, 403.

Smillie, W. G. and Duerschner, D. R. (1947) *Amer. J. Hyg.* **45**, 1, 13.

Smit, P., Oberholzer, D., Hayden-Smith, S., Koornhof, H. J. and Hilleman, M. R. (1977) *J. Amer. med. Ass.* **238**, 2613.

Smith, T. (1913) *J. med. Res.* **29**, 291.

Smolens, J. and Mudd, S. (1943) *J. Immunol.* **47**, 155.

Solomon, S. (1937) *J. Amer. med. Ass.* **108**, 937; (1940) *Ibid.* **115**, 1527.

Standfast, A. F. B. (1951) *J. gen. Microbiol.* **5**, 531; (1958) *Immunology* **1**, 135.

Sterner, G. *et al.* (1967) *Acta med. scand.* **183**, 805.

Stewart, G. T. (1977) *Lancet* **i**, 234; (1978) In: *Int. Symp. on Pertussis*, p. 262. Ed. by C. R. Manclark and J. C. Hill. U.S. Dept Hlth, Educ. and Welf., Bethesda, U.S.A.; (1979) *J. Epidem. Comm. Hlth* **33**, 150.

Strachiloff, D., Mebel, S., Mohr, J. and Weiczorek, H. (1966) *Zbl. Bakt.* **201**, 235.

Ström, J. (1960) *Brit. med. J.* **ii**, 1184.

Stuart-Harris, C. H. (1968) *J. clin. Path.* **21**, Suppl. p. 1; (1978) In: *Int. Symp. on Pertussis*, p. 256. Ed. by C. R. Manclark and J. C. Hill. U.S. Dept Hlth Educ. and Welf., Bethesda, U.S.A.

Switzer, W. P. and Farrington, D. O. (1975) In: *Diseases of Swine*, 4th edn, p. 687. Ed. by H. W. Dunne and A. D. Leman. Iowa State University Press, Ames.

Tanner, E. I., Gray, J. D., Rebello, P. V. N. and Gamble, D. R. (1969) *J. Hyg.*, *Camb.* **67**, 447.

Tillotson, J. R. (1976) *Infect. Dis. Rev.* **4**, 1.

Tillotson, J. R. and Lerner, A. M. (1967) *New Engl. J. Med.* **277**, 115.

Tilton, R. C. (1979) *Ann. intern. Med.* **90**, 697.

Trollfors, B. (1979) *Acta. paediat. scand.* **68**, 323.

Turk, D. C. and May, J. R. (1967) *Haemophilus influenzae. Its Clinical Importance.* Engl. Univ. Press Ltd, London.

Vysoka, B. (1958) *J. Hyg. Epidem.*, *Prague* **2**, 196.

Wallman, I. S., Godfrey, R. C. and Watson, J. R. H. (1955) *Brit. med. J.* **ii**, 1423.

Walzer, P. D., Perl, D. P., Krogstad, D. J., Rawson, P. G. and Schultz, M. G. (1976) *Nat. Cancer Inst. Monogr.* **43**, 55.

Weibel, R. E., Vella, S. P., MacLean, A. A., Woodhour, A. F., Davidson, W. L. and Hilleman, M. R. (1977) *Proc. Soc. exp. Biol. Med.* **156**, 144.

Whitaker, J. A., Donaldson, P. and Nelson, J. D. (1960) *Amer. J. Dis. Child.* **100**, 678.

White, R., Finberg, L. and Tramer, A. (1964) *Pediatrics*, *Springfield* **33**, Pt 1, p. 705.

Wilson, G. S. (1967) *The Hazards of Immunization*, p. 195, Athlone Press, London.

Winholt, W. (1915) *J. infect. Dis.* **16**, 389.

Wright, A. E., Morgan, W. P., Colebrook, L. and Dodgson, R. W. (1914) *Lancet* **i**, 187.

Wright, N. G., Thompson, H., Taylor, O. and Cornwell, H. T. C. (1973) *Vet. Rec.* **93**, 486.

68

Enteric infections: typhoid and paratyphoid fever

M. T. Parker

Introductory

The symptomatology and post-mortem appearances of typhoid fever were first accurately described by French workers such as Louis and Chomel in the early years of the nineteenth century (see Gay 1918). Bretonneau in 1823 called the disease 'dothienentérite' (Gr. dothien, a boil, and enteron, intestine), on account of the swollen Peyer's patches in the small gut. The difference between typhus and typhoid fevers was clearly recognized by Gerhard (1836–7) in the Philadelphia outbreak of typhus in 1836 and later by Schoenlein (1839) who referred to the two diseases as 'typhus exanthematicus' and 'typhus abdominalis'. As noted by Creighton (1894), the final recognition of enteric fever in Britain resulted from the elaborate analysis of the symptoms of the different types of continued fever carried out by Sir William Jenner between 1849 and 1851.

The most striking contribution to our knowledge of the natural history of typhoid fever, before the opening of the bacteriological era, was that made by William Budd (1856, 1873) of North Tawton in Devon. He insisted on its spread by contagion, on the reproduction of the specific poison within the living body, on the excretion of the infective material in the faeces, on the spread of the disease through the family by the hands of those who waited on the sick, on the part played by the contamination of water and milk in its epidemic spread, and on the destruction of the *materies morbi* by suitable disinfectants. His writings afford an admirable example of the value of accurate observations by a medical practitioner carried out in the spirit of the field naturalist.

With the description of the typhoid bacillus by Eberth in 1880, and its isolation by Gaffky in 1884, it became possible to attack the problems of enteric infection by bacteriological methods. The subsequent isolation and study of the various species of paratyphoid bacilli showed that enteric fever, though a clinical syndrome, might result from infection with several distinct but nearly related species.

Enteric fever is in essence a septicaemia and is the common manifestation of infection with *Salmonella typhi, S. paratyphi A, S. paratyphi B* and *S. sendai*. We may recall that several other salmonellae which cause septicaemia also frequently give rise to pyaemic lesions in the internal organs. The chief of these are *S. choleraesuis, S. paratyphi C, S. blegdam, S. enteritidis* var. *chaco* and *S. dublin*. The distinction between *salmonella septicaemia*, or enteric fever, and *salmonella pyaemia* (Bornstein 1943) is useful, but is by no means absolute, and some of the usual effects of typhoid fever are due to the secondary localization of bacteria in individual organs. Most members of the *Salmonella* group, other than those already mentioned, give rise to acute gastro-enteritis (Chapter 72), but may occasionally cause a septicaemic or pyaemic illness.

Bacteriology

During an attack of typhoid fever the causative bacilli can be isolated from the faeces (Pfeiffer 1885), the urine (Hueppe 1886), the rose spots of the eruption (Neuhaus 1886), and the blood (Vilchur 1887). Occasionally they may be cultivated from the faeces and rarely from the blood of the so-called *precocious carrier* during the incubation period of the disease (Conradi 1907, Mayer 1910), from subclinical cases and from perfectly healthy persons exposed to infection— so-called *symptomless excreters* (Billet *et al.* 1910). Specific agglutinins appear in the blood during the course of most attacks of typhoid fever (Widal and Sicard 1896, Grünbaum 1896). An instructive picture of the usual course of events may be obtained by charting, for the successive weeks of the disease, the average frequency with which *S. typhi* can be isolated from the blood, or from the faeces, and the frequency with which the serum gives the diagnostic agglutinin reaction. In preparing Fig. 68.1 along these lines, we have employed for the faeces the figures obtained by A.C. Jones (1953) in the Oswestry outbreak, for the blood those of Coleman and Buxton (1907), and for the agglutinins those recorded by Park and Williams (1925).

As indicated in the figure, the onset of symptoms is associated with a bacteraemia. From the 1st week onwards the frequency with which *S. typhi* can be isolated from the blood falls. By the end of the 3rd week it can be found in about half the cases; after the 4th week its isolation is infrequent. It must be remembered that we are dealing with a frequency chart, not with the course of events in any individual case, and that the frequencies noted for the results of blood culture after the 3rd week refer to patients who were febrile at that period. In any given case, it will usually be found that the decline of fever is associated with the disappearance of bacteria from the blood. In severe cases the bacteraemia may not reach its peak till the 3rd week of the disease (Shaw and Mackay 1951); and in cases that terminate fatally as the result of the primary infection, apart from the secondary accidents of haemorrhage and perforation, its intensity usually

increases until death (Jochmann 1914). Typhoid bacilli are most easily isolated from the faeces between the 3rd and the 5th week of illness. Early workers were able to isolate the organism from the faeces of little more than 50 per cent of cases during the 1st week of disease, but with the more sensitive methods now available it can nearly always be detected at this stage. There is no doubt, however, that the number of organisms in the faeces increases greatly from the 1st to the 3rd week of illness. The significant point is the decreasing frequency of detectable bacteraemia, and the increasing frequency of *S. typhi* in the intestine, during the first 2 to 3 weeks. This clearly suggests that the main line of invasion during this period is from the blood stream to the intestine, not the reverse. As Fig. 68.1 indicates, the bacilli do not disappear from the

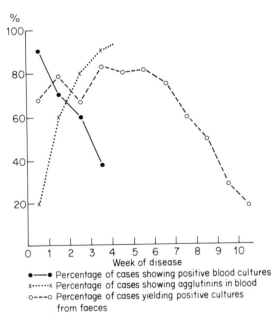

●——● Percentage of cases showing positive blood cultures
x······x Percentage of cases showing agglutinins in blood
o---o Percentage of cases yielding positive cultures
       from faeces

Fig. 68.1 Bacteriological findings in typhoid fever (see text for sources of the data)

intestine so quickly or so completely as from the blood. Many patients are still excreting typhoid bacilli at a time when positive blood cultures can no longer be obtained, and some continue excreting well into convalescence or beyond it.

Our figures for the frequency of agglutinins in the blood serum of the patient are based mainly on the results obtained with formolized broth cultures, which react with the H (flagellar) agglutinins (Dreyer *et al.* 1915; Dreyer and Walker 1916, Dreyer and Inman 1917). During the 1st week about 20 per cent of the cases show the presence of these agglutinins. The curve then rises sharply, crossing the blood-culture curve just before the end of the 2nd week, and still rising attains a value of 90 per cent or over by the 4th week. It remains at a high level for some weeks.

In enteric fever due to *S. paratyphi B* we see the same general picture, with some differences in detail. The blood culture is nearly always positive in the 1st week of illness, but septicaemia and the associated pyrexia are usually of shorter duration in paratyphoid than in typhoid fever. Faecal excretion is regular and profuse during the early stages of the illness, and 80–100 per cent of all specimens collected during the first fortnight contain paratyphoid B bacilli (Glass and Wright 1937, Rubinstein *et al.* 1944, George *et al.* 1953). Clearance of the faeces during convalescence is rather more rapid in paratyphoid than in typhoid fever (see p. 420). In some paratyphoid infections the fever may be quite short-lived, and the blood cultures only transiently positive. Faecal excretion is usually as regular and prolonged in these mild attacks as in typical enteric illnesses.

Course of the infection

Our knowledge of the epidemiology of enteric fever makes it clear that the bacilli enter the body by the alimentary tract. Having passed through the stomach, they penetrate the mucosa of the intestine without undergoing significant multiplication in the gut lumen. Their presence in large numbers in the faeces during the 2nd and 3rd weeks of an attack of typhoid fever is due in large part to bacteria re-entering the lumen from the biliary tract or from the intestinal lesions that develop later in the disease. The observation of Hornick and his colleagues (1970) that human subjects who gargled with and expectorated cultures of *S. typhi* did not become infected indicates that the organism does not enter through the pharyngeal mucosa. However, occasional accidental infections of hospital patients by the rectal route (Chapter 58) indicate that penetration through the mucosa of the lower bowel is possible.

The sequence of events in natural typhoid infection is probably similar to that observed by Ørskov, Jensen and Kobayashi (1928) in experimental infection of the mouse *per os* with *S. typhimurium*: the destruction in the intestinal tract of the greater part of the ingested bacteria, as evidenced by a transitory excretion of bacilli followed by a period during which none could be recovered from the faeces; the passage

of some bacilli to the mesenteric glands, and their further passage to the blood stream, probably via the thoracic duct; a transitory bacteraemia, rapidly brought to an end by the removal of the bacilli by the reticulo-endothelial cells, particularly those of the liver and the spleen; a phase during which active proliferation is proceeding in the liver and spleen, as evidenced by the recovery of increasing numbers of bacilli from these organs, but during which the blood remains sterile; a phase of secondary bacteraemia, associated with a generalization of bacteria throughout the tissues, and a secondary invasion of the intestine. Thus, the incubation period in typhoid fever in man corresponds to the phases of the infection that precede the establishment of a sustained bacteraemia. Naylor (1983) suggests that a short incubation period indicates that the blood stream is invaded directly from the lymphoid tissue associated with the intestine; if the local cellular immune response prevents this, some organisms nevertheless reach lymphoid tissues elsewhere, multiply in them, and invade the blood stream later (p. 410). Such a conception receives support from the occasional isolation of typhoid bacilli from the faeces and from the blood during the incubation period of the disease.

If cultures are taken from various situations during a post-mortem examination, the distribution of typhoid bacilli is usually found to be as follows. In the intestine itself almost pure cultures are frequently obtained from the duodenum and upper part of the jejunum. At lower levels the bacilli become less and less numerous in proportion to the other intestinal bacteria (von Drigalski 1904). Outside the intestine, *S. typhi* is almost constantly present in the spleen, the enlarged mesenteric glands, the bone marrow, and the gall-bladder; but, as was pointed out by Levy and Gaehtgens (1908), other lymph nodes are frequently unaffected.

The typhoid bacillus can be isolated from the gall-bladder in most fatal cases of the disease (von Fütterer 1888, Gilbert and Girode 1890, Chiari 1894, Pratt 1901); it may persist here for many years after convalescence (Droba 1889), and is occasionally found—sometimes but not always in the presence of cholelithiasis—in the absence of any history of typhoid fever (Dudgeon 1908, see also Ledingham and Arkwright 1912, Gay 1918).

The gall-bladder of the rabbit becomes infected within a few minutes of an intravenous injection of typhoid bacilli (Blachstein 1891, Blumenthal 1910), and may remain so for many months (Morgan 1911, Johnston 1912, Gay and Claypole 1913, Weinfurter 1915); gall-stones sometimes form (Richardson 1899). It is uncertain whether the bacilli reach the gall-bladder from the blood by being excreted from the liver capillaries into the bile canaliculi (Doerr 1905), through the capillaries in the wall of the gall-bladder (Koch 1909), or by both routes (Gay 1918).

The characteristic lesions in Peyer's patches appear to result from an initial hyperplasia of the endothelial cells in response to the bacterial invasion, followed by necrosis and sloughing. Perforation and haemorrhage are in the nature of occasional associated accidents. The gall-bladder is one of the most frequent sites of persistent infection; and is, from the epidemiological point of view, the most important, because of the ease with which bacilli escape thence to the intestine, to be excreted in the faeces; but it is by no means the only situation in which the bacilli may remain latent over long periods of time. The well known typhoid periostitis affords an example of a lesion developing long after convalescence from the original illness.

Relapses may occur during convalescence, particularly in patients treated with antibiotics. They are usually milder and of shorter duration than the original attack, and the fatality rate is low (Gay 1918). They are undoubtedly due to a re-invasion of the blood stream from the tissues in which typhoid bacilli are still proliferating at the time when the bacteraemic phase of the primary attack is brought to a close.

In both typhoid and paratyphoid fever the *incubation period* is subject to considerable variation. In typhoid fever it is usually 8–15 days, seldom as short as 5 days and sometimes as long as 30 or even 35 days; in paratyphoid fever it tends to be less and may be so short as to lead to a suspicion of food poisoning. The typical symptoms of either disease may be preceded by acute gastro-enteritis coming on shortly after consumption of the infected water or food (see Savage and White 1925, Robertson 1936, Warren 1941). Whether these symptoms represent an early reaction of the tissues to the infecting organism or are attributable to a coincident epidemic of food poisoning may be difficult to determine. A short episode of diarrhoea and vomiting sometimes occurs in the first day or two of an attack of typhoid fever (Colon 1976), though constipation rather than diarrhoea characterizes the later stages of the disease. In paratyphoid fever, the symptoms may resemble those of typhoid infection or the disease may be a diarrhoea with or without fever.

Experience with *experimental infection in human subjects* (see Hornick *et al.* 1970) has confirmed and extended our knowledge of the pathogenesis of typhoid fever. The organism was given by the mouth to healthy adults with no history of typhoid infection, and an illness identical in every respect with the natural disease resulted. The chance of contracting typhoid fever was directly proportional, and the length of the incubation period inversely proportional, to the size of the dose (Table 68.1); but the clinical course of the disease was unrelated to the number of organisms ingested. With the Quailes strain, the dose necessary to cause a typhoid-like illness in 50 per cent of subjects was *ca* 10^7 organisms, but a few volunteers remained well after ingesting 100 times this number. With doses of the order of 10^7 organisms of a strain of average virulence, most of the volunteers who escaped typhoid fever became symptomless excreters; many of them formed antibodies against *S. typhi*; and a few had a transient pyrexia with a positive blood culture. The pattern of faecal excretion in the absence of symptoms resembled that in typhoid fever. The infectivity of two other Vi-positive strains was similar to that of the Quailes strain. Vi-negative strains, though rather less infective (Table 68.1), caused clinically typical disease in some of the volunteers and symptomless infection in others.

Naylor (1983) commenting on the results obtained by Hornick and his colleagues (1970) pointed out that, in infections with a constant dose of the Quailes strain (10^5 organisms), which gave an attack rate of about 35 per cent, the incubation periods did not have a normal distribution but showed evidence of three peaks the modal points of which were respectively at 7, 21 and 28 days, the first peak being by far the largest. This he attributed to the presence of subpopulations of persons differing in susceptibility, pos-

Table 68.1 Virulence of typhoid bacilli for man[1]

Strain	Presence of Vi antigen	Dose given	Number of persons challenged	Number (and percentage) with typhoid fever[2]	infected[3]	Incubation period in days: median (range)
Quailes	+	10^9	42	40(95)		5 (3–32)
		10^8	9	8(89)		
		10^7	32	16(50)		7.5 (4–56)
		10^5	116	32(28)		9 (6–33)
		10^3	14	0(0)		
Quailes	+	10^7	30	16	28	
Zermatt	+	10^7	11	6	10	
Ty2V	+	10^7	6	2	5	
	Total with Vi antigen	10^7	47	24(51)	43(91)	
O-901	−	10^7	20	6	12	
Ty2W	−	10^7	19	4	14	
	Total Vi negative	10^7	39	10(26)	26(67)	

[1] Modified from Hornick *et al.* (1970).
[2] Temperature of 39.4° or higher for >36 hr.
[3] One or more of the following (i) typhoid fever (2) low-grade fever (3) significant serological response (4) positive blood culture (5) excretion of *S. typhi* in stools for >5 days.

sibly as a result of genetic factors. He then examined the distribution of incubation periods in a number of water- and food-borne epidemics of typhoid fever in which the dates of exposure and of the onset of infection had been accurately recorded. In the five water-borne epidemics the incubation periods formed a single rather flat peak, with modes at 18–21 days. Of ten food-borne outbreaks, five had short incubation periods (modes 6–9 days) and individual dates of onset closely spaced around the mode, and two resembled the water-borne epidemics in having long incubation periods (modes 20 days) and a single flat peak spanning 14 days or more. In three of the food-borne epidemics, however, the distribution of incubation periods was bimodal (modes respectively at 15 and 29 days, 7 and 21 days, and 10 and 20 days). He concluded that the factors determining the incubation period were the virulence of the strain, the infecting dose, and the susceptibility of individual members of the exposed population. Low-dose infections, whether from food or water, usually had a long incubation period, a single-peaked distribution and a low attack rate. High-dose infections from foods—presumably those in which the organism had multiplied before being ingested—had a short incubation period with closely spaced dates of onset; in some the attack rate was very high. The food-borne outbreaks in which the distribution of incubation periods was bimodal were probably intermediate-dose outbreaks in which individual susceptibility determined the date of onset in those who developed clinical infection.

The intravenous injection of the *lipopolysaccharide endotoxin* (Chapter 33) of the typhoid bacillus into healthy volunteers causes a febrile reaction with symptoms rather like those of enteric fever, but repeated doses lead to an increasing tolerance. Nevertheless, experimental typhoid fever in the tolerant volunteer is clinically typical, and repeated doses of endotoxin during an attack of typhoid fever evoke progressively smaller responses without influencing the course of the disease (Hornick *et al.* 1970). Butler and his colleagues (1978) found some evidence of disseminated intravascular coagulation in typhoid fever but concluded that it was not the main cause of the toxaemia or of bleeding from the intestine. They pointed out that the *Limulus* test for endotoxinaemia was uniformly negative in typhoid fever, and that the number of typhoid bacilli in the blood was very much smaller (< 10 to 90 per ml) than in the 'gram-negative' septicaemias caused, for example, by coliform bacilli. It is therefore unlikely that endotoxin is directly responsible for the general symptoms of typhoid fever. More probably the local action of endotoxin on tissue cells near to sites of bacterial multiplication leads to the release of endogenous pyrogens by leucocytes (see Hornick and Greisman 1978).

Untreated typhoid fever lasts on average 4 weeks and then resolves by lysis. By this time high serum titres of specific antibody are often present, but this is by no means always so (p. 414). The development of *cell-mediated immunity* is probably the main means by which the infection is overcome. In studies by means of the leucocyte migration-inhibition test, cell-mediated immunity to *S. typhi* can be demonstrated in most patients suffering from typhoid fever of normal severity, but it may be weak or absent in complicated or potentially fatal infections (Sarma *et al.* 1977). In typhoid fever the cell-mediated-immunity response bears little relation to the serum-antibody titres, and in healthy persons given injections of killed vaccine it is weak and inconstant (Nyerges *et al.* 1976, Nath *et al.* 1977, Sarma *et al.* 1977)

(For further information about the clinical picture in typhoid and paratyphoid fevers see Adams 1983.)

Diagnosis

The methods of diagnosis that are available and their relative value at different stages of the disease are indicated by the time-relations shown in Fig. 68.1. Though at one time reliance was placed chiefly on a serological method of diagnosis—the Widal reaction—modern practice lays far more emphasis on the cultural isolation of the causative organism from the blood, faeces or urine. Serological tests can afford only indirect evidence of infection, and their interpretation may be rendered difficult by past exposure to infection, previous immunization with TAB vaccine, or early antibiotic treatment. Only if the causative organism is isolated can it be tested for susceptibility to the drug that is to be used for treatment.

Cultural methods

The method of choice is **blood culture**. This has the great advantage over culture from the faeces, urine or bile of showing not only that the patient is infected with the bacillus but that the infection is active and is almost certainly responsible for the disease from which he is suffering.

At least 10 ml of blood should be withdrawn, since the number of organisms in the blood, particularly in mild and recovering cases, may be quite small, even as few as one per ml (Watson 1955*a*). For preference it should be inoculated directly into a large volume (150–200 ml) of medium. Cultures should be incubated at 37° and plated out daily. They should not be discarded as negative for at least 11 days (Shaw and Mackay 1951). According to Felix (1924b), the blood of cases of typhoid fever often contains substances bactericidal for the typhoid bacillus. For this reason, the volume of medium in which whole blood is incubated should be as large as possible. Certain substances, such as liquoid and bile salt, inhibit this bactericidal effect. If whole blood has to be transported to the laboratory before inoculation into medium, liquoid is therefore the anticoagulant of choice. For a similar reason, 0.5 per cent sodium taurocholate broth and pure ox bile have been recommended as blood culture media for typhoid bacilli (see Kaye *et al.* 1966), but even in bile-salt broth considerable dilution (at least 6-fold) is desirable (Watson 1978). Mikhail and co-workers (1983) describe a rapid method for the presumptive identification of *S. typhi* in blood cultures by co-agglutination with a suspension of *Staph. aureus* sensitized with O and Vi antibodies.

Blood clot from which serum has been removed often gives a positive result when a similar volume of whole blood yields no growth (Felix 1924b). Thomas, Watson and Hewstone (1954) confirmed this, but noted that there was often a considerable time lag in the growth of typhoid bacilli in clot cultures. If the clot was dissolved with streptokinase, not only did growth occur more rapidly but the total number of positive cultures was increased. Clot cultures with streptokinase gave approximately twice as many positive results as whole blood cultures with a 3–4 fold dilution in taurocholate broth. Further experiments suggested that there was a bactericidal factor not only in serum but in unlysed clots (Watson 1954, 1955a, b). Watson (1978) recommends the addition of 100 units of streptokinase to bile-salt broth for the culture of both clots and whole blood.

Bone-marrow culture (Ling *et al.* 1948) may give a positive result when blood culture fails, particularly in patients admitted to hospital while on antibiotic treatment. Gilman and his colleagues (1975a) state that the diagnosis would have been missed in 24 of 62 cases of typhoid fever had they not undertaken cultures of bone marrow and of skin biopsies from rose spots as well as blood cultures. It is unusual to find the causative organism in the **urine** during the first few weeks of an attack of enteric fever unless it is also present in the faeces; nevertheless, occasional exceptions occur (see Report 1941) and it is advisable to examine the urine as well as the faeces at any stage of the disease. Typhoid bacilli are generally more numerous in the **bile** than in the faeces; the passage of a duodenal tube would be too distressing for a patient with enteric fever, but swallowing a weighted nylon string in a gelatin capsule and withdrawing it after a suitable interval is said to be an effective alternative (Gilman and Hornick 1976).

Cultures of the faeces should be put up at the same time as the blood culture. Numerous selective and enrichment media are available. Few of these are equally good for the growth of all salmonella serotypes or indeed for all strains of the same serotype. It is therefore unwise to rely upon a single medium.

Early *enrichment media* for salmonellae contained various dyes, notably brilliant green. Tetrathionate broth (Muller 1925, Schäfer 1935) was a much better enrichment medium for most salmonellae, but had the disadvantage of allowing the growth of *Proteus* and giving a poor isolation rate of typhoid bacilli. Kauffmann's (1930–31, 1935–36) modification of Muller's medium contained brilliant green and bile. The brilliant green helped to check the growth of *Proteus*. A better means of suppressing this organism is to use a tetrathionate medium containing no excess of thiosulphate (Knox *et al.* 1942). Other workers recommended the addition of novobiocin (Jeffries 1959) and sodium lauryl sulphate (Jameson 1961) to inhibit *Proteus*. For most purposes, including the isolation of the typhoid bacilli, the best medium is probably sodium selenite broth (Leifson 1936), but it should be noted that *S. choleraesuis* is inhibited in this and in ordinary tetrathionate broth; according to Chung and Frost (1970) it grows well in tetrathionate medium made with half the normal amount of iodine. Harvey and Price (1964) recommend the use of double-strength selenite broth for the isolation of *S. typhi* from large volumes of heavily contaminated fluid material. Some workers prefer Rappaport's medium, which contains hypertonic magnesium chloride and malachite green (Rappaport *et al.* 1956), to selenite broth for the isolation of salmonellae other than the typhoid bacillus, and for the latter others advocate the use of strontium selenite broth (Iveson and Mackay-Scollay 1969). The incubation of sodium selenite broth at 43° (Harvey and Thomson 1953) favours the growth of salmonellae other than *S. typhi* in cultures of heavily contaminated material.

Solid *selective media* containing one or other of the triphenylmethane group of dyes have the advantage that they suppress the growth of *Proteus*, and are therefore of value when used in conjunction with tetrathionate broth. The simplest of these, brilliant green agar, is favoured by veterinary and food bacteriologists because it allows the growth of a wide range of salmonellae, including the pullorum bacillus and *S. choleraesuis*. It requires careful standardization and is not good for typhoid bacilli. Brilliant green MacConkey agar (Wilson and Darling 1918) is strongly recommended by Harvey (1956) for the isolation of salmonellae other than the typhoid bacillus. Other dye-containing media include brilliant green eosin agar (Teague and Clurman 1916), brilliant green eosin methylene blue agar (Knox *et al.* 1942) and acid fuchsin brilliant green MacConkey agar (Hoyle 1943).

One of the best selective media is the bismuth sulphite brilliant green agar of Wilson and Blair (1927, 1931). It is a particularly good medium for the isolation of *S. typhi*, though most other salmonellae grow well on it. Although it is less selective than many other media, salmonellae form on it characteristic black colonies surrounded by a metallic sheen. It is a difficult medium to prepare, and is rather unstable, but a good dehydrated medium can be obtained commercially. It should never be used as the only selective medium for the examination of faeces, since some strains of salmonellae do not form black colonies on it. The other medium that is widely used in clinical bacteriology is Leifson's (1935) deoxycholate citrate agar, which permits the growth of shigellae as well as of most salmonellae, including the typhoid bacillus. Unfortunately, many other gram-negative bacilli grow well on deoxycholate citrate agar; for some purposes, therefore, it is an advantage to incubate it anaerobically with 10 per cent CO_2 (Ryan 1972); *Pseudomonas aeruginosa* is suppressed, *Proteus* is partly inhibited and salmonellae form characteristic colonies with jet-black centres. To obtain optimal growth of the typhoid bacillus the lactose should be omitted. For the isolation of salmonellae from sewage and foods see respectively p. 421 and Chapter 72.

Most British bacteriologists favour the direct inoculation of Leifson's deoxycholate citrate agar and Wilson and Blair's medium, and enrichment cultures in selenite broth with subsequent plating on the same two solid media. Others would include enrichment in one of the modified tetrathionate broths, or the use of one of the less selective dye-containing media such as brilliant green MacConkey agar. If only one selective medium can be used it should be deoxycholate citrate agar, and if only one enrichment medium it should be selenite broth.

The isolation of typhoid or paratyphoid bacilli from a specimen of faeces depends more on the proportion of these bacteria to the rest of the enterobacterial flora in the particular specimen examined than on any detail of technique. No reliance can be placed on a single negative result, whatever method of examination is employed; and in carriers or convalescents, where the excretion of these organisms may be intermittent, many specimens may have to be examined before a positive result is obtained.

For the cultivation of *urine* two or three 10-ml quantities should be inoculated into liquid enrichment media and plated out after incubation on to the selective media already mentioned. *Bile*, duodenal fluid and the distal bile-stained portion of a nylon string withdrawn from the duodenum are treated similarly.

Great care should be exercised over the collection of specimens of faeces and urine for examination. In hospitals imperfectly sterilized bedpans and urinals afford a frequent source of contamination, and it is much better to provide every patient from whom a specimen is required with a special grease-proof cardboard container into which the excreta can be passed directly (see Holt *et al.* 1942). Alternatively a rectal swab may be taken, though this is less satisfactory than faeces when the bacilli are scanty (Shaughnessy *et al.* 1948, McCall *et al.* 1966). The faeces should be examined as fresh as possible. If this is impracticable, they may be collected into buffered glycerol saline (Sachs 1939) or Stewart's transport medium (see Ewing *et al.* 1966). An alternative method of preserving salmonellae, including the typhoid bacillus, is to place several drops of liquid faeces or faecal emulsion on a sheet of filter paper and allow it to dry in air (Dold and Ketterer 1943, 1947). The paper is placed in a cellophane envelope for dispatch to the laboratory. Many of the intestinal bacteria die during the drying, but the pathogens are rather more resistant than other enterobacteria. In the laboratory, a saline suspension of the material from the filter paper is plated heavily on a selective medium. Good results have been obtained after periods of up to 13 days (Joe 1950, 1956, Bailey and Bynoe 1953).

Colonies suspected of being salmonellae may be tentatively identified by slide agglutination and confirmed by biochemical and tube-agglutination reactions. Failure to appreciate that organisms subcultured from colonies on selective media are frequently impure is one of the commonest pitfalls of enteric bacteriology. Other pitfalls include lack of recognition that some salmonellae, especially the typhoid bacillus, may appear to be non-motile for one or two subcultures after they have been grown on a selective medium, and that the presence of Vi antigen may render a typhoid bacillus inagglutinable by *S. typhi* O serum. The occurrence of small-colony variants must also be remembered.

Serological methods

Agglutination reactions (the Widal test)

These are usually performed in tubes or in wells in plastic plates; H agglutination is best seen in Dreyer's tubes and O and Vi agglutination in round-bottomed tubes or wells.

Serial dilutions of sera are made, to give a starting dilution, after the addition of the bacterial suspension, of 1 in 20 for the H, 1 in 50 for the O and 1 in 5 for the Vi-agglutination test. H-agglutination tests should be incubated in a 50° water-bath for 2 hr, and read after being left for 3 hr at room temperature. O- and Vi-agglutination tests should be incubated for 2 hr at 37° and left in the cold room overnight before being read. The suspensions to be employed must vary to some extent with the prevalence of different infecting organisms. In Europe, for example, nearly all cases of enteric fever are caused by *S. typhi* or *S. paratyphi B*. In Asia and Africa, however, *S. paratyphi A* is met with, and in the Middle East and Asia *S. paratyphi C*. Routine tests in Europe should therefore be made with suspensions containing typhoid H, typhoid O, paratyphoid

B phase 1 H and paratyphoid B O antigens; in addition, a suspension containing the common phase-2 salmonella H antigens is useful for detecting antibodies against the second phase of *S. paratyphi B* and some of the 'food poisoning' salmonellae that occasionally cause enteric fever. If the presence of *S. paratyphi A* or *C* is suspected the corresponding H and O suspension should be used. (The use of the Vi agglutination test for the detection of typhoid carriers is considered on p. 421).

To obtain reliable results, bacterial suspensions of uniform agglutinability are essential (Dreyer *et al.* 1915, Gardner 1937), the technique must be standardized, and end-points must be read consistently. One important error in O agglutination is the use of too high temperatures of incubation, which tend to reduce agglutination by low-titre though not by high-titre sera. Experience with the Vi-antibody test, for which it is impossible to produce a stable suspension of uniform agglutinability, made it clear that bacterial antigens do not provide a satisfactory standard for agglutination tests. The standard must be a serum, which can be dried and preserved without alteration for a long time (Felix 1950). Not only does the use of a standard serum remove difficulties due to varying agglutinability of suspensions, but it eliminates the subjective element from the reading of the result. This is important in both the Vi and the O agglutination tests, in which the end-points are rather diffuse. Tests are put up at the same time with the standard serum and the patient's serum. The titre of the patient's serum corresponds to the tube showing agglutination equal in amount to that given by the standard serum at the dilution which has been designated as its Standard Agglutinating Titre. Provisional International Standard Agglutinating Sera for O, H and Vi antibodies are available (Felix 1950, Felix and Bensted 1954), and may be used for the preparation of local standard sera. A full titration of the standard serum should be carried out with every routine test for typhoid Vi antibodies. Once an experienced worker has become familiar with the degree of agglutination given by the standard sera with particular batches of H and O suspensions he will not find it necessary to include them in his tests more often than weekly or fortnightly.

Before discussing the interpretation of the agglutination test, we must consider some variables that affect the result.

Frequency distribution of agglutinins in the population The serum of a proportion of persons in any country contains antibodies capable of reacting to a variable titre in the Widal test. The frequency distribution of H antibodies, O antibodies, or both has been studied by numerous authors (for example, Rosher and Fielden 1922, Smith *et al.* 1930, Gardner and Stubington 1932, Giglioli 1933a, Tabet 1940, Mackenzie and Taylor 1945, Hughes 1955, Collard *et al.* 1959, Schubert *et al.* 1959, Levine *et al.* 1978, Pang

and Puthucheary 1983). In the absence of previous inoculation with TAB vaccine, the frequency of H agglutinins in a population reflects its experience of salmonellae with the corresponding antigens—either in the form of enteric fever or of latent infection—and therefore varies widely from country to country. Where enteric infection has for many years been uncommon—as in Britain—few sera contain detectable H agglutinins against the phase 1 H antigens of the enteric bacilli, but a small proportion—not more than one or two per cent—have titres ranging from 20 to 320 or more. A rather larger percentage may have the phase 2 H antigen of *S. paratyphi B* and *S. typhimurium*, in some cases resulting from past attacks of salmonella food poisoning. In countries in which enteric fever is more common, H antibodies against the prevalent strains may be found in up to one-quarter of all sera. The frequency and concentration of O agglutinins, on the other hand, varies much less in different parts of the world. A low titre of antibody against *S. typhi* O or *S. paratyphi B* O is found in a considerable proportion of any population; titres of 25 are to be expected in some 20 per cent, of 50 in 2–3 per cent, and of 100 in up to 1 per cent of persons.

This agrees with our knowledge of the distribution of salmonella antigens and of the duration of the antibody response to them. The H antigens are rare outside the *Salmonella* group and the corresponding antibodies persist for many years; the O antigens are widely distributed not only among salmonella serotypes but also among other enterobacteria, and the antibody response is more transitory. Though some O antibodies may result from experience of salmonella infection, others may well not.

The stage of the disease H agglutinins usually begin to appear towards the end of the 1st week of an attack of enteric fever, increase to a maximum during the 3rd week, and persist for months or years afterwards. Thus, a low titre (20–40) early in the disease may be of some significance in countries in which the corresponding enteric infection has for many years been rare and when previous TAB inoculation can be excluded, but in many parts of the world such titres must be ignored (see, for example Levine *et al.* 1978). There is great variation in the titre reached at a particular stage of the disease (Felix 1924a, b, Gardner and Stubington 1932, Jones 1951), but a value of 80 or more will be found by the 3rd week in most patients. In some, however—according to Brodie (1977) as many as 15 per cent in the Aberdeen typhoid outbreak—H agglutinins may be absent at all stages of the disease.

O agglutination titres in excess of 100 are usual at some stage in most cases of typhoid fever, but they may appear quite late in the disease (Brodie 1977); in paratyphoid fever they are found in less than 50 per cent of cases (Gardner and Stubington 1932, Watkins 1959). In typhoid fever antibody to the Vi antigen appears irregularly; in some patients it may reach a titre of 80–160, but in others there is no antibody response. Occasionally Vi antibodies are formed in the absence of H and O agglutinins (Report 1941, 1942*a*).

It may be noted that a patient suffering from paratyphoid fever may form O agglutinins against *S. typhi*. If the infecting strain of *S. paratyphi B* is mainly in phase 2, specific H antibody may not be formed; thus it is essential to include a phase-2 suspension in the Widal test.

The effect of previous inoculation   H agglutinins may persist in titres of 50 to 800 for years after TAB inoculation (Smith 1932, Wyllie 1932, Giglioli 1933*b*, Report 1942*b*). O agglutinins are also formed in response to TAB inoculation but do not reach a high titre or persist for more than a few weeks or months. Downie and his colleagues (Report 1942*b*) found that only some 5 per cent of vaccinated men had titres exceeding 100 in the 3 months after vaccination, and that these had all disappeared 1–3 years later (see also Beattie and Elliot 1937, MacKenzie and Taylor 1945, Wilson 1945, Schubert *et al.* 1959).

The effect of antibiotic treatment   The early treatment of cases of typhoid fever with chloramphenicol has a profound effect on the antibody response. If agglutinins have not appeared when treatment is begun they are unlikely to do so subsequently; if they are already present, no further rise in titre is to be expected (De Blasi 1950, Subrahmanyan 1952, Seeliger and Vorlaender 1953, El-Rooby and Gohar 1956).

Use of the Widal test

The results of the test are unlikely to be helpful in the 1st week of disease or in patients given early treatment with a relevant antibiotic. In those who have received TAB vaccine at any time, only a high titre of O or Vi agglutinins is of any significance. In any event, interpretation of the results will be difficult unless information is available about the frequency distribution of titres of the various agglutinins in the population from which the patient came.

Before effective antimicrobial treatment of enteric fever was possible, rapid diagnosis was of little advantage in the individual patient. The demonstration of a rising titre of H or O antibodies in serial specimens of serum, the first collected as early as possible in the disease, was then a more or less acceptable alternative to cultural diagnosis, particularly in endemic areas in which epidemiological investigation of sporadic infections was not practicable. Nowadays, the Widal reaction is useful only insofar as the examination of a

single serum from a recently admitted patient helps to make a prompt diagnosis; opinions differ on its value for this purpose. Brodie (1977), in a non-endemic area, found that the frequency of weak and delayed H and O antibody responses severely limited the usefulness of the test. Levine and his colleagues (1978) in Central America found H titres of 80 or O titres of 40 so often in the normal population that the examination of a single serum was of little use except in children. On the other hand, Pang and Puthucheary (1983) in Malaysia obtained considerably more reliable results if only H and O titres of 160 or more were considered significant; over 90 per cent of bacteriologically confirmed cases of typhoid fever but only 3 per cent of persons suffering from other fevers had such a titre to one or both antigens (see also Senewiratne and Senewiratne 1977). It seems likely that many of the patients in endemic areas who give a rapid and strong antibody response have had earlier exposure to the typhoid bacillus.

Other serological tests   Salmonella O antibodies can be detected by indirect haemagglutination tests (Neter 1956) but the results appear to offer little advantage over bacterial agglutination. Precipitation reactions, detected by counterimmunoelectrophoresis, between the patient's serum and veronal-buffer extracts of *S. typhi*, are said to be more sensitive and specific than the Widal test (Tsang and Chau 1981). Tsang and his colleagues (1981) used solid-phase radioimmunoassay to detect serum antibodies against lipopolysaccharide and against a protein fraction of *S. typhi*. Antibody to the two components was formed in similar amounts by typhoid patients, but the titres of anti-lipopolysaccharide appeared to bear little relation to the O-agglutination titres. The immunoglobulins were of the IgG, IgA and IgM classes. Similar methods were used to detect antibodies in faeces (Chau *et al.* 1981); in typhoid fever these were of the IgM and IgA classes, but in typhoid carriers only IgA antibody was found. Enzyme-linked immunosorbent assay is another sensitive method of measuring salmonella O antibodies (Karlsson *et al.* 1980, Lange *et al.* 1980); if combined with determination of the class of the immunoglobulins it may provide a means of identifying recently formed antibody.

Barrett and his colleagues (1982) used enzyme-linked immunosorbent assay to detect Vi *antigen* in the urine. It was present, in the absence of typhoid bacilli, in a number of typhoid-fever patients; these workers consider the test to be a rapid and reliable means of establishing a diagnosis.

Epidemiology

The main conclusions that emerge from the extensive literature bearing on the epidemiology of enteric fever are that the only source of typhoid bacilli, and the main source of paratyphoid bacilli, is the infected person; and that the disease can be prevented by blocking the various routes by which the bacilli pass from the intestine or urinary tract of one person to the mouth of another. The dramatic fall in the incidence of enteric fever that took place in most countries in Western Europe, North America and Australia from 1880 onwards was the result of putting this knowledge into practice. Between 1901 and 1951 the standardized mortality ratio for England and Wales decreased over 200-fold (Table 68.2) as a result of improved sanita-

Table 68.2 Enteric fever: England and Wales. Standardized mortality ratios (all ages) (Base years 1950–52 taken as 100)

Year	Ratio
1901–10	23 581
1911–20	8 926
1921–30	2 729
1931–39	1 180
1940–49	387
1950	83 ⎫
1951	115 ⎬100
1952	102 ⎭
1960	17
1971	0

Modified from Report (1961*a*, 1972).

tion; since that time the added effect of chemotherapy has been to reduce the annual mortality for enteric fever almost to vanishing point. As Gay (1918) pointed out, the incidence in western countries in the mid-nineteenth century was greatest in the cities, where it varied with the density of the population; but after this time the sanitation of the large cities improved rapidly and steadily, with the result that the disease became relatively commoner in small towns and villages which had not made corresponding improvements in the disposal of excreta and the provision of a pure water supply.

By 1939, the annual incidence of enteric fever over a large and continuous area of Northern and Western Europe was less than 5 cases per 100 000 population, and the rates in adjacent areas were rapidly falling towards this level. In consequence of the disorganization of community life and the large movements of people that occurred in the second world war, this trend was dramatically reversed in Germany, Austria, Holland, Finland and Eastern Europe (Stowman 1947–48), where incidence rates rose five to ten-fold. By 1954, however, pre-war levels had been reached again, and by 1960 a rate of less than 1 per 100 000

had become the rule in most years in Britain, Norway, Denmark and Holland. Somewhat higher rates continue to be experienced in countries bordering the Mediterranean. Indigenous typhoid infection has for some years been rare in Britain. Of the 200 or so infections reported annually in England and Wales in recent years, nearly 90 per cent were contracted abroad, most in the Indian subcontinent. Among the minority of patients who acquired the infection in Britain, over one-half belonged to households of Asian origin. In the USA, where the incidence of infection is even lower than in Britain (see Cvjetanović *et al.* 1978), a rather smaller proportion was acquired abroad—in 1975–76, for example, only about one-third (Ryder and Blake 1979). However, a few years earlier, a large epidemic of typhoid fever in Mexico resulted in a sharp but temporary increase in the number of infections in travellers returning to the USA (Baine *et al.* 1977). Typhoid fever is very common in most countries in Asia, Africa and Central and South America, but its true incidence is difficult to estimate. In these countries it is a disease mainly of urban slums and of rural communities.

In western industrialized countries during the twentieth century, the highest incidence of enteric fever was in children over the age of 5 years, in adolescents, and in young adults. The fatality rate, however, tended to rise steadily with increasing age (Godfrey 1928). Ashcroft (1964) has pointed out that the situation is different in areas of high endemicity; where hygiene is very primitive, infection is almost universal, but recognizable typhoid fever is not common. The first infection of infants is often clinically atypical and may go unrecognized, but the resulting immunity prevents appearance of overt disease in older children and adults. As conditions of hygiene improve, the date of primary infection may be postponed to an age at which the disease is recognized and its incidence is then highest in school children. (For an account of typhoid fever in children see Scragg *et al.* 1969.) With increasing urbanization and the provision of common supplies of water which may occasionally become contaminated, epidemics begin to occur and the incidence of diagnosed infection tends to increase in persons of all ages.

It is usually said that most typhoid infections result in typical illness, except in very young children, and that true symptomless excretion is rare. Most of the observations on which these conclusions are based were made during the last 35 years in western counries in which typhoid epidemics are usually food-borne and the attack rate is high, but even in some of these outbreaks a proportion of the infections were symptomless (Jones 1951). Six per cent of persons infected from meat products in the Aberdeen outbreak had no symptoms (Walker 1965). Whether the proportion of symptomless excreters is higher when the infecting dose is smaller, for example, in water-borne outbreaks, we do not know. In food-borne paratyphoid B infection, about half of all infections are symptomless, and many of the illnesses experienced are atypical, and could easily be confused with influenza or bacterial food poisoning. Indeed, some strains of *S. paratyphi* B cause not a continued fever, but a gastro-enteritis which may or may not be accompanied by fever.

Infections with the typhoid bacillus are almost entirely confined to man. There have been few isolations of *S. typhi* from animals, though Brygoo (1974) reported a number from fruit bats in Madagascar. In any event, human infection from an animal source has never been reliably established. The paratyphoid B bacillus is like-wise a human pathogen but has been isolated from other mammals on numerous occasions; and infected cattle have been responsible for several milk- and meat-borne outbreaks of disease in man. There is also some evidence that cattle may acquire paratyphoid B infection from the contamination of pasture with human excreta or sewage (Smith and Thomas 1966, George *et al.* 1972). Paratyphoid A infection appears to be entirely human in origin.

The main routes of epidemic spread differ notably in typhoid and paratyphoid B fever in that, though both diseases are carried by food, typhoid fever is often borne by water and shell-fish, whereas these vehicles are rare in the genesis of paratyphoid fever (Savage 1942).

Water-borne infection

The part played by water in the spread of typhoid fever was first established by the painstaking observations of William Budd (1856, 1873) in this country, and has been confirmed by experience in many parts of the world. For references to classical examples of water-borne typhoid epidemics see Scott (1934) and earlier editions of this book.

Among the most frequent causes of contamination of water are cross-connection of a main with a polluted water supply, seepage of surface water or sewage into a gravity conduit, surface contamination of shallow wells, and inadequate filtration or chlorination (Gorman and Wolman 1939). It may be noted that the evidence inculpating a particular water supply is almost always circumstantial. The pollution which gives rise to an epidemic is often temporary or intermittent, and has usually ceased by the time the water supply falls under suspicion. The evidence is, none the less, convincing enough. Water-borne outbreaks are generally characterized by a typically explosive onset. The curve for the whole epidemic is often characteristically skew, the primary cases due to direct infection from the water supply being followed by secondary crops of contact cases. Less often, particularly when infection of the water is slight and intermittent, instead of an explosive outburst, there is a series of single cases or small groups of cases occurring over a considerable period of time, and affecting only a small proportion of the consumers. Suspicion of the water should always arise if building operations have been proceeding on a gathering ground, or if extensions, repairs or other alterations have been going on in the supply services. The incubation period in water-borne infection is generally long (p. 411), and the attack rate is generally lower than that seen in many food-borne epidemics.

Water-borne outbreaks of paratyphoid B infection are uncommon, but a few small ones are on record (Jones *et al.* 1942, Savage 1942).

Although the public water supplies in many countries are now above suspicion, some water-borne infection persists. Occasional cases and small outbreaks follow the drinking of stream water by campers, and a few sporadic infections have probably resulted from bathing or playing in sewage-polluted water (Martin 1947, Steiniger 1953, Stevenson 1953). The danger to bathers is apparently very slight, and seems to occur mainly in fresh water, despite the enormous amount of sewage discharged directly into the sea and the ease with which salmonellae can often be isolated from sea-water (Report 1959*b*).

Food-borne infection

In many food-borne epidemics of enteric infection the source of the organism was polluted water. Not only typhoid, but also paratyphoid B infection, may be spread in this way if there is opportunity for the organisms to multiply in the food. Paratyphoid B bacilli have gained access to milk from polluted rivers or streams on several occasions (Wallace and Mackenzie 1947, Thomas *et al.* 1948). The washing of cooked crabs in heavily polluted harbour water may also have given rise to paratyphoid fever (Glatzel 1944). There is some evidence that housewives may contract typhoid fever by handling contaminated fish before it is cooked (Hompesch 1953) and that typhoid bacilli may be disseminated on salad vegetables which have been manured with sewage (Harmsen 1954). *Canned food* may be contaminated if the cans are cooled in polluted water. There have been at least five such outbreaks in Britain (W.B. Moore 1950, Couper *et al.* 1956, Ash *et al.* 1964, Report 1964*a*, Anderson and Hobbs 1973). In the most recent of these, in Aberdeen in 1964, there were over 500 infections, some from the originally contaminated corned beef, but many more from other cooked meats that had become cross-contaminated from it in the shop (see Howie 1968). Untreated river water had been used for cooling the filled and sealed cans after heat sterilization. The typhoid bacilli are believed to have entered the can through a minute gap in the seam and to have multiplied without causing 'blowing'. Meers and Goode (1965) showed that the typhoid bacillus multiplied actively in corned beef, and that multiplication under anaerobic conditions is enhanced by a concentration of nitrate similar to that added to this meat; moreover, cans inoculated with a mixture of *S. typhi* and *Escherichia coli* failed to 'blow.'

The technical problems that complicate the isolation of enteric bacilli from sewage are discussed on page 421.

Shell-fish, particularly oysters, are another important source of infection in typhoid fever, though rarely incriminated in paratyphoid fever. They are often bred or fattened in the sewage-polluted waters of tidal estuaries, and they are usually consumed uncooked. Numerous epidemics are on record which can be traced to this source (see Gay 1918, Belin 1934), and sporadic cases have also been common. Nowadays oysters are a rare source of typhoid fever in Britain (but see Report 1959a).

Milk follows close on water as an important source of explosive epidemics. Numerous references to milk-borne epidemics of enteric fever will be found in earlier editions of this book.

The organisms may gain access to the milk from the hands of an infected person or from contamination of milking utensils. Enteric bacilli multiply in milk at ordinary atmospheric temperatures, so that even a trivial contamination may cause an explosive outbreak; but sporadic infections, spread over a considerable period of time, may result if the milk is contaminated during the process of distribution (see Bradley 1943, Bekenn and Edwards 1944). There is no evidence to suggest that cows excrete typhoid bacilli in their milk (Scott and Minett 1947) but, as we have seen, they may excrete paratyphoid B bacilli.

Apart from liquid milk, *other dairy products* or substitutes such as cream, artificial cream, synthetic cream, custard and ice-cream, are frequent vehicles for the enteric group of organisms, particularly paratyphoid bacilli. Of the 40 paratyphoid outbreaks analysed by Savage (1942), 8 were due to liquid milk, 16 to cream and cream cakes, 2 to ice-cream, 1 to trifle, 1 to cream cheese and 2 to other foods. *Ice-cream* contaminated by a urinary carrier caused the large outbreak of typhoid fever at Aberystwyth in 1946 (Evans 1947). Several outbreaks of typhoid fever in Canada and the United States of America have been ascribed to the consumption of unripened hard *cheese* made from raw milk handled by chronic carriers (Bowman 1942, Gauthier and Foley 1943, Menzies 1944, Fabian 1947). *Butter* has been identified as the vehicle of infection in a number of outbreaks of enteric fever in Austria and in France (Studeny 1947, Boyer and Tissier 1950).

Numerous other foods, including meat products, have been responsible for outbreaks of enteric fever in which there was evidence that contamination had taken place from a human source during preparation of the food. In a few instances, however, meat from an infected animal may have been responsible for outbreaks due to *S. paratyphi B* (Hemmes 1951, Nicol 1956, Smith and Thomas 1966).

Bakery products containing synthetic cream were for many years the commonest cause of outbreaks of paratyphoid B in

Britain, but repeated attempts to isolate the causative organism from the cream before it reached the bakery were unsuccessful. The work of Newell, Hobbs and their colleagues (Newell 1955a, b, Newell et al. 1955, Hobbs and Smith 1955) revealed that paratyphoid B bacilli, together with other salmonellae, were often present in *dried and frozen egg products* imported from China and widely used in the bakery industry. Most of the egg was used in the manufacture of products that were later heated to sterilizing temperatures, but in the meantime the paratyphoid bacilli in the egg mix had been transferred to other materials, including synthetic cream, that were added to the cakes after baking. The heat treatment of dried and frozen egg products has now virtually eliminated this source of paratyphoid infection.

Wilson and Mackenzie (1955) described a widespread epidemic of typhoid fever in Australia associated with *dried coconut* from New Guinea. Cases of typhoid fever were associated with infections due to certain other serotypes of *Salmonella*. These serotypes were also found in the coconut accompanied by a few typhoid bacilli. *S. paratyphi B* was at one time often present in coconut exported from Sri Lanka (Kovacs 1959, Galbraith et al. 1960), but it appears to have been responsible for few infections in recipient countries.

Infection by contact

There is no doubt that enteric infection can be transmitted from one person to another without the interposition of water or food, but the frequency of this is difficult to estimate. When typhoid infection spreads within the family the source is usually a chronic carrier—nearly always a middle-aged or elderly female (p. 420). Thus contamination of food within the household is difficult to exclude. Many chronic carriers only rarely transmit infection to other members of their family and often only after many years; this suggests that they are not very infectious and that transmission is associated with rare lapses in food-handling practice. Paratyphoid B infection is thought to spread in families more frequently and often to originate from a symptomless excreter or an ambulant case, but in retrospective investigations of such incidents it is often difficult to exclude the possibility that several persons were infected from an unsuspected food source.

Hospital-acquired infections provide a number of convincing examples of person-to-person spread. In endemic typhoid infection in mental hospitals the source is usually a chronic carrier. Typhoid patients nursed in open hospital wards seldom transmit infection, even under poor hygienic conditions, but Ayliffe and his colleagues (1979) record a small outbreak in a maternity department in which the source was a parturient woman with typhoid fever. (See Chapter 58 for the spread of typhoid infection in hospitals by means of contaminated equipment.) With paratyphoid B infection, on the other hand, person-to-person spread in hospitals is more common, particularly among young children; newborn babies appear to acquire this infection readily, though it is frequently mild or even symptomless (Foley 1947, Jones and Pantin 1956).

From time to time *laboratory workers* contract typhoid fever. This hazard is not confined to those handling infected clinical specimens. According to Blaser and co-workers (1980) 24 cases of laboratory-acquired typhoid fever were identified in the USA in less than three years; these constituted over 10 per cent of all indigenous typhoid infections in the country during this time. Only three of the infections were derived from clinical specimens, the rest being from cultures used for quality-control or teaching purposes. A considerable proportion of those infected had not knowingly handled the cultures. Hickman and his colleagues (1982) advocate the use of strains of *S. typhi* of reduced virulence for teaching and for the evaluation of techniques.

Infection from flies

There is little doubt that flies play a part in transferring enteric bacilli from excreta to food in some countries, but this is difficult to estimate quantitatively. Reed, Vaughan and Shakespeare (1899) drew attention to the clear association between the prevalence of typhoid fever among the American troops in the Spanish-American War and the exposure of excreta, and suggested that flies played an important part in the spread of the disease.

Experimental studies (Firth and Horrocks 1902, Ficker 1903, Graham-Smith 1910) showed that external contamination of flies with *S. typhi* was of short duration but that the organism persisted in the intestinal canal for some days. Faichnie (1909) examined flies caught in infected areas and isolated typhoid bacilli on several occasions from the crushed bodies of the insects after the exterior had been sterilized by flaming (see also de la Paz 1939).

The typhoid and paratyphoid carrier

In 1900, Horton-Smith recorded the case of a urinary carrier of *S. typhi*, but the thesis that chronic carriers were the main source of the persistence of enteric infection in the community was first established by the work of a Commission in Germany, under the auspices of Robert Koch (Frosch 1903, von Drigalski 1904; see Ledingham and Arkwright 1912). To von Drigalski we owe the clear demonstration that the typhoid bacillus may lead a prolonged existence in the human body and be excreted in the faeces.

It is necessary to define our terms. There is now fairly general agreement on the use of the word 'temporary' or 'transient' for carriers who excrete the bacillus for not more than a year, and the word 'chronic' for carriers of over a year's standing. In order to make the distinction more clear cut, we prefer to use the terms *chronic carrier* and *temporary excreter*. Temporary excreters are of two sorts: *convalescent excreters* who have had a clinically recognizable attack of the disease, and *symptomless excreters* who have

not. There is some evidence to suggest that persons may harbour the infecting bacilli in the tissues for a long time without excreting them in the urine or faeces. Though the term 'carrier' is strictly applicable to such persons, it would in practice be misleading to use it in this sense without some qualification, such as 'closed'.

Faecal carriers

In considering the frequency of carriers we must realize that our technique of demonstrating typhoid and paratyphoid bacilli, especially in the faeces, is far from perfect. With the great technical improvements that have resulted in recent years from the introduction of more selective media, the conviction has been growing that intermittency of excretion is less frequent than used to be thought, particularly in paratyphoid infection (see Holt and Wright 1942). This may well be true of chronic carriers (see, for example, Merselis *et al.* 1964), but periods of intermittency appear to be quite common among symptomless excreters of *S. typhi*. Thus, according to Jones (1953), only 52 per cent of 173 samples of faeces from nine symptomless excreters examined over periods ranging from 2–16 weeks were positive, and periods of intermittency of up to 4 weeks were observed. On the other hand, the cultural results might have been better if the method of preliminary dilution of the faeces recommended by Thomson (1954) had been applied.

Thomson found that, with both typhoid and paratyphoid carriers, direct inoculation of a drop of faecal suspension, diluted 1 in 1000 or so, on to a suitable selective medium yielded far more positive results than the usual method of previous enrichment in tetrathionate or selenite broth. Incidentally, his quantitative examinations showed what enormous numbers of typhoid or paratyphoid bacilli may be excreted by chronic carriers. Figures of over one million per gram were common, and in one instance as many as 12 000 million per gram were found.

Though the methods now in use permit the detection of small numbers of organisms in the faeces more readily than before, it may be doubted whether their application will lead to any substantial modification of our estimate of the proportion of patients with enteric fever who become chronic carriers. We shall deal first with the typhoid carrier. The proportion of patients excreting typhoid bacilli falls slowly during the first few weeks of convalescence, and then more rapidly between the 7th and 12th weeks. Most of those still positive after 3 months are excreting large numbers of organisms, and it is therefore safe to accept the opinion of the early workers that most, though not all, of those still excreting after 3 months will become chronic carriers. Lentz (1905) found that 4.5 per cent of 400 typhoid convalescents excreted *S. typhi* for more than 10 weeks and 3 per cent for longer than 13 months. The figures given by Ames and Robins (1943) are 4.3 per cent at the 12th week and 2.9 per cent at 1 year. Other estimates of the proportion of patients ultimately becoming chronic carriers include the following: Kayser (1907) 3 per cent, Brückner (1910) 4 per cent, Gill (1927) 9 per cent, Gray (1938) 3 per cent, Vogelsang and Bøe (1948) 3 per cent, Morzycki (1949) 2 per cent. A further point of interest is the influence of age and sex on the chance of becoming a typhoid carrier. The figures given in Table 68.3 were obtained

Table 68.3 Percentage of patients with typhoid fever who became carriers (Ames and Robins 1943)

Age (years)	Male	Female
0–10	0.6	—
—20	0.4	0.2
—30	2.1	2.1
—40	2.8	6.2
—50	3.5	16.4
—60	9.1	11.5
60+	6.2	9.4
All ages	2.1	3.8
	2.9	

by Ames and Robins (1943), and refer to 90 carriers resulting from 3130 cases of typhoid fever in New York State in the years 1930–39. During childhood and early adult life the chance of becoming a chronic carrier increases slowly. Women are much more likely than men to become carriers in middle age. Many chronic carriers give no clear history of an attack of typhoid fever.

Our information on the frequency of typhoid carriers among the population at large is somewhat scanty. Several investigations have been made in which specimens of faeces from large samples of the population have been examined for *S. typhi*. The following is a selection of the percentages of chronic carriers found: in Germany 0.9 per cent by Klinger (1906), 0.4 per cent by Minelli (1906) and 0.3 per cent by Prigge (1909); in the United States, 0.3 per cent (Rosenau *et al.* 1909) and 3.6 per cent (Welch *et al.* 1925). Most of the communities studied were ones in which enteric fever had been present for years. Indirect attempts to assess the carrier rate, based on notification rates, or mortality rates, during the preceding decades, or upon the proportion of persons with a history of enteric fever, give very much lower figures. Ames and Robins (1943) calculated that there were 42 carriers per 100 000 population (0.04 per cent) in New York State. Other estimates include one of 48 per 100 000 for Massachusetts (Anderson *et al.* 1936) and 288 per 100 000 for Mississippi (Gray 1938). We can safely conclude that the frequency of chronic carriers in most counries in Western Europe is now very low—of the order of 2 per 100 000 or less—except in localities in which a significant proportion of the population are immigrants from or frequent visitors to high-incidence areas.

Turning now to the paratyphoid carrier, we find a number of accounts of the rate at which *S. paratyphi B* disappears from the faeces during convalescence (Glass and Wright 1937, Davies *et al.* 1940, Gell and Knox 1942, Holt *et al.* 1942, Vogelsang and Bøe 1948, Kennedy and Payne 1950, George *et al.* 1953). Clearance begins towards the end of the 3rd week of illness—3 to 4 weeks earlier than in typhoid fever—and proceeds rapidly; about half of the patients have ceased to excrete the organism by the 5th to 7th week, and all but 10 per cent by the 10th to 11th week. Thereafter a slow fall continues to the 20th week, at which time about 4 per cent of patients still have positive faeces. Kennedy and Payne (1950) found no difference in the rate of clearance of cases and symptomless excreters. Vogelsang and Bøe (1948), who examined 1027 cases of paratyphoid B fever, found that 1.9 per cent became chronic carriers, which is significantly lower than their figure for typhoid fever (3.3 per cent). There is little exact information on the duration of excretion of paratyphoid A bacilli, but the observations of Bumke (1926) suggest that these organisms disappear from the faeces at a considerably greater rate than typhoid bacilli.

Urinary carriers

Urinary excretion is common in the later part of the febrile period and during convalescence but has usually disappeared within 6–8 weeks. Occasionally it occurs for the first time in late convalescence. In most countries, however, chronic urinary carriers are much less common than faecal carriers. Vogelsang and Bøe (1948) found that only one of 32 chronic enteric carriers was excreting the organism in the urine. Urinary carriage is thought to be associated with the presence of a chronic lesion in the kidney. It is frequently unilateral, and occasional cures have followed a nephrectomy (El-Sadr 1953). In countries, such as Egypt, in which urinary schistosomiasis is common, many urinary carriers of enteric bacilli are found, and may constitute up to 3 per cent of the population in some areas (Archer *et al.* 1952). Urinary carriage develops in about 10 per cent of patients with chronic urinary schistosomiasis who contract typhoid fever, and in about 50 per cent of those with paratyphoid A fever (Hathout *et al.* 1966). Carriage is persistent and often associated with intermittent bacteraemia (Farid *et al.* 1970). Successful anti-schistosomal treatment may lead to the cessation of carriage. Lo Verde and co-workers (1980) suggest an even more intimate association between salmonellae and *Schistosoma*, having observed the adherence of the bacteria to the surface of the worm; this they attribute to bacterial pili.

The literature contains numerous and well authenticated records of the danger of the chronic carrier. For accounts of such classical instances as the case of the Strassburg Master-Baker's Wife, the Folkestone

Milker, and 'Typhoid Mary,' the student is referred to the monograph of Ledingham and Arkwright (1912), and to papers by Soper (1939) and Jones and Pantin (1961).

Laboratory methods for detecting carriers

The only reliable means of detecting carriers of enteric bacilli is by isolating the organisms from the faeces or the urine. The cultural methods for this have been described (p. 412). Because carriage may be intermittent, many examinations may sometimes be necessary before a positive result is obtained. It is almost impossible to specify a number of negative cultures of the faeces that will completely exclude carriage. If a number of specimens have given negative results and the suspicion of carriage remains, the duodenal bile should be examined, preferably by means of a nylon string (see p. 412 and Gilman *et al.* 1979).

Serological methods have been widely used to pick out persons who are likely to be carriers from larger numbers of subjects. The carrier usually shows the presence of H and O antibodies in low titre, but greater value attaches to the demonstration of Vi antibody (Felix 1938*a*). Vi agglutinins are found in the serum of some 85 per cent of chronic carriers at a titre of 5 or more (Eliot 1940, Eliot and Cameron 1941, Klein 1943, Pijper and Crocker 1943), but their presence is in no sense diagnostic of the carrier state. The frequency with which randomly collected sera, or sera from non-carriers, have been reported to have titres of 5 or more is highly variable. In England percentages of 3.7 (Mackenzie and Taylor 1945) and 1.2 (Report 1961*b*) have been recorded, but in areas of high incidence of typhoid infection percentages of 15 and more are usual (for example, see Joe *et al.* 1957). Thus, the use of this test as a means of screening suspects is clearly impracticable in many parts of the world (Bokkenheuser *et al.* 1964). In any event, not all carriers have Vi agglutinins in their sera, so that the complete omission of cultural examination of the faeces and urine is not justifiable.

To demonstrate Vi agglutinins, a suspension of a Vi-rich strain of typhoid bacillus (Felix 1938*b*) is mixed with serial dilutions of the serum in round-bottomed tubes or wells and incubated at 37° for 2 hr. Readings are made after a further 22 hr in the refrigerator. Carefully standardized suspensions of heat-killed organisms—which have a 'shelf-life' of no more than 2 months—are used. A full titration of a control serum must be included with each batch of tests. End-points are difficult to determine in sera with low titres of agglutinin, and considerable experience in reading the results is necessary.

Several other methods of detecting Vi antibody have been reported, of which the *indirect haemagglutination test* is the most widely used. Spaun (1951) and Corvazier (1952) coated red cells by treating them with aqueous or saline extracts of cultures of a Vi-positive strain of *S. typhi*, and Landy and Lamb (1953) used purified Vi antigen. Nolan and his colleagues (1980) sensitized sheep red cells with a purified Vi antigen from a strain of *Citrobacter*. Haemagglutination tests are a very sensitive and quite specific means of detecting typhoid carriers. Chau and Tsang (1982) used an antigen prepared by the same method and obtained positive results in tests of sera from 14 of 15 carriers but only with 3 per cent of controls among Hong Kong Chinese. Lanata and co-workers (1983) in Chile obtained positive results (titres of 160 or more) in 75 per cent of 36 carriers, in 8 per cent of 388 non-carrier women, and in 3 per cent of 59 randomly selected healthy subjects. Chau and Tsang (1982) compared the results of passive-haemagglutination and counter-immunoelectrophoresis-precipitation tests with the same purified antigen. The two tests had a similar sensitivity but the precipitation tests were rather more specific. Barrett and his colleagues (1983) obtained rather similar results by indirect haemagglutination and enzyme-linked immunosorbent assay of antibody against purified Vi antigen.

In *urinary carriers* who are suffering from schistosomiasis, H antibodies are often found in the urine, usually to a titre 1/5–1/20 of that in the blood serum. Serological examination of the urine has therefore been used in Egypt in searching for carriers among prospective food handlers (see Archer *et al.* 1952, Naylor and Caldwell 1953).

The introduction by B. Moore (1948, 1950) of the *sewer swab technique* provided a most useful means of circumscribing the search for human carriers. By following the contamination of the sewage with enteric bacilli back from the main sewer to tributary sewers, and thence to the drains from individual houses, it is often possible to demonstrate the actual house in which the carrier is living (see also Lendon and Mackenzie 1951, Moore *et al.* 1952, Harvey and Phillips 1955, Kwantes and Speedy 1955).

It is more difficult to isolate enteric bacilli, particularly *S. typhi*, from sewage than from faeces. Enrichment in selenite broth, preferably by addition to an equal quantity of double-strength medium, followed by plating on Wilson and Blair's medium, is the most generally favoured method. According to Lendon and MacKenzie (1951) several serial tenfold dilutions of fluid expressed from sewer swabs should be cultured in enrichment medium.

The epidemiological significance of the carrier state, whether transient or permanent, is much enhanced by making use of *bacteriophage typing* (see Chapter 37). For the value of the method reference may be made to papers by Cruickshank (1947), Martin (1947), Felix (1951) and Felix and Anderson (1951) in Great Britain, Crocker (1947) in South Africa, Desranleau (1947) in Canada, Henderson and Ferguson (1949) in the United States, and Joe (1949) in Indonesia. Information on the geographical distribution of phage types of *S. typhi*, *S. paratyphi A* and *S. paratyphi B* is given in Report (1973, 1982). For reviews of the carrier problem in general see the monographs by Ledingham and Arkwright (1912), Browning and his colleagues (1933) and Vogelsang and Bøe (1948).

Prevention of enteric fever

We may attempt to prevent enteric fever by detecting and eradicating its sources, by blocking its routes of transmission, or by immunizing persons liable to be infected.

Control of carriers

Little can be done during the acute illness to prevent the development of the carrier state, but every effort should be made to recognize it in convalescents. Most of those who are excreting the organism 3 months after the onset of illness will become chronic carriers. It is reasonable to encourage them to submit to chemotherapy at this stage, though success cannot be guaranteed. Persons who have excreted the organism for over a year very seldom cease spontaneously to do so, and at this stage cholecystectomy must be considered. Some 70-80 per cent of faecal carriers of typhoid and paratyphoid bacilli are permanently cured by cholecystectomy (see Browning *et al.* 1933, Vogelsang and Bøe 1948, Vogelsang 1964). The bacilli may be found in the faeces for a time after operation but eventually disappear (Littman *et al.* 1949). When removal of the gall-bladder proves a failure, there is in some cases reason to believe that the organism persists in the bile passages (Erlik and Reitler 1960, Vogelsang 1964). Chronic typhoid cholangitis is said to be particularly common among Hong King Chinese (McFadzean 1966). Before recommending cholecystectomy, it is wise to consider the age and occupation of the carrier, to seek evidence of chronic cholecystitis and of cholangitis, and to exhaust the possibilities of chemotherapy.

The titre of Vi agglutinins in the blood during convalescence is often taken as a guide to the expected duration of excretion of the organism. Before discharge from hospital the patient should have his blood examined for Vi agglutinins; these will normally be found in about half the convalescent cases. If they are absent at a titre of 5 nothing further need be done; the patient may still be excreting typhoid bacilli in the faeces or urine but is unlikely to continue to do so for long. If, on the other hand, Vi agglutinins are present, a second serum examination should be made 3 months later. Disappearance of the Vi agglutinins or a considerable fall in titre at this stage may be regarded as satisfactory. Persistence of the original titre, however, or a rise in titre, should lead to the thorough examination of the faeces and urine for typhoid bacilli; six or more tests at intervals should be performed, if necessary, over a period of 3 weeks. If these organisms are found, the patient should be regarded as a persistent carrier; if not, a further examination for Vi agglutinins is desirable after another three months (Report 1945).

The names of all known typhoid carriers, whether detected in convalescence or as a result of investigations of outbreaks of infection, and the Vi-phage types of the organisms carried, should be recorded centrally. The carriers themselves should be kept under supervision by local health officers, who should ensure that the sewage from their dwellings is safely disposed of, give them appropriate instruction in personal hygiene, and warn them against taking employment in the manufacture or handling of comestible substances. However, few food- or water-borne outbreaks are caused by known chronic carriers. There is general agreement that all new technical employees of water undertakings should be examined bacteriologically, but opinions differ about the value of routine screening of workers in the catering trade. Use may be made of cultures of the sewage effluent (p. 421) from large food factories as an alternative to the examination of individual faecal specimens.

Hygienic measures

A detailed account of appropriate measures is beyond the scope of this book. Briefly, the most important are those designed to ensure that excreta are safely disposed of, that a safe water supply is provided, and that food is handled hygienically. From an examination of the available evidence, Cvjetanović and his colleagues (1978) concluded that exposure to infection in endemic regions could be halved by the construction of privies and the provision of uncontaminated water. This would reduce the rate of generation of new enteric carriers and lead with the passage of years to a steady and continuous fall in the number of sources of infection and a consequent decline in the incidence of clinical disease. This is what is believed to have happened in many western countries over the last century.

Prophylactic immunization

Antityphoid vaccination in man was initiated by Wright in 1897 (see Wright and Semple 1897, Wright 1902, 1908)—though the observations of Pfeiffer and Kolle (1896) have been held by some to establish a claim to priority—and has since been tried under field conditions on a large scale.

Vaccination by parenteral routes

The first vaccines to be used extensively were heat-killed suspensions of laboratory strains and were given subcutaneously. Trials in British troops stationed abroad appeared to show a five-fold reduction in the incidence of enteric fever in vaccinated subjects (Report 1913; see also Boyd 1966). According to Cockburn (1955), the design of this and a number of similar trials was faulty and the results cannot be accepted without question. As a result of the work of Bensted, Findlay and Perry, and of Felix (Chapter 37), the British Army later adopted a vaccine prepared from freshly selected and fully mouse-virulent strains, killed by heat and preserved with phenol, but this vaccine did not provoke the appearance of Vi antibody when given subcutaneously. Felix (1941) therefore advocated the use of a vaccine in which the organism was killed with 75 per cent alcohol and preserved in 25 per cent alcohol.

More recently numerous field trials of typhoid vaccines have been performed with great care to ensure that the inoculated and control groups were strictly alike. The common features of these trials were as follows: participants were mainly school children living in endemic areas; all received injections, in the experimental groups of various typhoid vaccines and in the control groups of an unrelated organism or antigen; they were kept under surveillance, and appropriate bacteriological tests were carried out on patients with febrile illnesses; a diagnosis of enteric fever was made only when the blood culture was positive. The first trial, in Jugoslavia (Report 1957, 1962), showed that two doses, at a 3-week interval, of heat-killed phenolized vaccine were followed during the next year by a fall in the average annual attack rate per 10 000 from 19.2 to 6.1, but that two doses of alcoholized vaccine made from the same strain resulted in an attack rate of 14.1. The next trial, in Guiana (Report 1964b) compared a heat-killed phenolized vaccine (L) with one killed with acetone with the object of preserving the Vi antigen (K). The attack rate per 10 000, for $3\frac{1}{2}$ years' exposure, was 41 in the control group receiving tetanus toxoid, 11 in the group given vaccine L, and 2 in the group given vaccine K. The apparent superiority of the acetone-killed vaccine was in accordance with its behaviour in Topley's studies of infection of mice with *S. typhimurium* (Raistrick and Topley 1934).

Further studies of the vaccines L and K, which had been established as international reference preparations (Spaun and Uemura 1964), were made in Jugoslavia, Guiana, Poland and the USSR (Cvjetanović and Uemura 1965). These indicated that both vaccines were effective and that K was superior to L. The percentage effectiveness (Table 68.4) of the L vaccine in various trials ranged from 51 to 73 per cent and of the K vaccine from 79 to 94 per cent. Moreover, the duration of protection by vaccine K was greater than had been expected; after two doses there was virtually no loss of effectiveness in 3 years. Ashcroft and his colleagues (1967) in Guiana concluded that protection was undiminished for at least 5 years, and that a single dose was as effective as two. It was emphasized, however, that the latter statement may hold true only in areas of high endemicity. The efficacy of these vaccines was confirmed by the protection they afforded to persons infected experimentally with typhoid bacilli. With an inoculum of 10^5 organisms *per os* of a strain that gave a 40 per cent attack rate in unvaccinated subjects, the percentage effectiveness of the L vaccine was 67

Table 68.4   Effectiveness of acetone-killed (K) and heat-killed phenolized (L) typhoid vaccines; results of field trials[1]

Country (and year)	Vaccine group K		Vaccine group L		Control group		Percentage effectiveness of vaccine[2] (and 95 per cent confidence limits)	
	No. persons in trial	No. cases of typhoid	No. persons in trial	No. cases of typhoid	No. persons in trial	No. cases of typhoid	K	L
Jugoslavia (1960–63)	5 028	16	5 068	37	5 039	75	79 (63–88%)	51 (26–68%)
Guiana (1960–64)	24 046	6	23 431	26	24 241	99	94 (86–98%)	73 (58–83%)
Poland (1961–63)	81 534	4	...	...	83 734	31	87 (62–97%)	...
USSR (1962–63)	...	...	36 112	13	36 999	50	...	73 (50–87%)

[1] Modified from Cvjetanović and Uemura (1965).
[2] Percentage effectiveness $= 100(b-a)/b$ where $a=$ incidence in vaccinated group and $b=$ incidence in control group.

per cent and of the K vaccine 75 per cent (Hornick and Woodward 1967, Woodward 1980). These studies were made in a non-endemic area and provided some indirect evidence of the duration of protection after routine vaccination in persons who were seldom naturally exposed to typhoid bacilli. Among the control subjects in these experiments, those whose only known previous exposure to typhoid antigens was routine TAB vaccination on military service—on average 12 years before—the attack rate from experimental infection was 20 per cent, but 48 per cent of those without this experience developed clinical infection (Woodward 1980).

Little is known about the efficacy of the paratyphoid A component of the TAB vaccine. Trials of paratyphoid B vaccine in the USSR (Hejfic *et al.* 1968) showed a reduction in incidence of over 70 per cent for 30 months after two doses and a similar reduction, but only for 12 months, after a single dose. Opinion now favours the use of a vaccine composed only of typhoid bacilli because paratyphoid B infection is seldom of much importance in the geographical areas in which typhoid infection is prevalent and the vaccine is likely to be used.

The results of properly controlled trials and of experimental infections in man have thus shown conclusively that killed vaccines, which had been used empirically for over 60 years, confer considerable protection against typhoid fever. Their use in persons likely to be exposed frequently to infection is therefore to be recommended. The protection afforded is, however, far from absolute. Numerous outbreaks have occurred in well vaccinated communities as a result of heavy exposure to infection, particularly when foodborne (for references see Miller *et al.* 1951). The observations of Hornick and Woodward (1967) that vaccination gave considerable protection against experimental infection with 10^5 typhoid bacilli but none against 10^7 organisms provide an explanation for this.

Whether to carry out 'emergency' vaccination of those exposed to infection in water- or milk-borne outbreaks, or as a consequence of natural or man-made disasters, is questionable. Vaccination during the latter part of the incubation period may, instead of protecting, induce the opposite effect, namely the so-called *provocation typhoid*. There is reason to believe that this risk is substantial (Topley 1938); when typhoid fever begins within a few days of an injection of TAB vaccine the onset tends to be sudden, and the course is in most cases severe. Whether vaccination also increases the incidence of enteric fever by converting a symptomless into a clinical attack is not easy to determine, but there is ample evidence that giving salmonella vaccines to naturally or experimentally infected mice greatly increases the frequency of fatal bacteraemia (Geks 1949, Raettig 1959*a, b, c,* 1966). These results confirm the observations made by Jürgens (1927) and Stroebe (1928) on human beings in the

Hanover outbreak of typhoid fever in 1926, by Raettig (1950) in the Greifswald outbreak of 1945, and by various other observers in Germany and the United States. The findings of these workers and the whole subject of provocation disease were reviewed by Wilson (1967). Faced with an epidemic, therefore, our decision whether or not to vaccinate will depend on our estimate of the rapidity with which the source of infection can be brought under control, and of our ability to keep under close observation, and when necessary to treat at an early stage, those who are exposed to risk. In general, when there is reason to believe that numerous symptomless excreters and patients in the incubation period of the disease are present in the population, it is wise to refrain from vaccination.

Typhoid vaccine is usually given subcutaneously to adults in two doses of 500 and 1000 million organisms at an interval of 2–4 weeks. Since it often causes a severe local and systemic reaction, it is wise when reinforcing injections are given at a later date to test the patient's reactivity by injecting 25 or 50 million organisms. Siler and Dunham (1939) of the American Army pointed out that the reaction was less when a smaller dose, 100 million organisms, was given intradermally. This dose appears to provide as strong an antibody response as the larger dose injected subcutaneously (see Miles 1958, Barr *et al.* 1959) but nothing is known about its protective effect. Alcoholized vaccine is unsuitable for intradermal use. The use of jet injection in mass vaccination campaigns is said to lead to more local reactions than injection by syringe (Edwards *et al.* 1974).

Other killed vaccines have been tried on a small scale. In general, alcoholized vaccines seemed to be somewhat inferior to those of the L and K types, and more drastic treatment of the bacilli in attempts to 'detoxify' the vaccine led to loss of activity (Cvjetanović and Uemura 1965). Vaccines of purified Vi antigen (Levin *et al.* 1975) are less toxic than cellular vaccines and lead to the formation of the corresponding antibody, but they give no protection against experimental infection in man (Woodward 1980).

Extensive investigations have been made of the antibody response to various killed typhoid vaccines, and numerous laboratory tests have been proposed for assessing their *potency* (for references see Cvjetanović and Uemura 1965, Pittman and Bohner 1966), but it is not possible by either means to reach firm conclusions on the degree of protection conferred by a vaccine on man.

In animals, live attenuated vaccines given subcutaneously often afford greater protection against infection by the oral route than do killed vaccines; this has been demonstrated with *S. gallinarum* in chickens (Smith 1956*a, b*), with *S. dublin* in calves, and with *S. choleraesuis* in pigs (Smith 1965). Many of the vaccines that give good protection are in the rough state, but not all rough strains are effective; and the cross-protection between different salmonella serotypes suggests that neither O nor H antigens are concerned in the immunity (Smith and Halls 1966, Rankin *et al.* 1967, Knivett and Stevens 1971). Multiplication of the vaccine strain in the animal body is necessary for the production of immunity. A *gal E* mutant (see below) of *S. typhimurium* given subcutaneously to calves protected them against oral challenge with both *S. typhimurium* and *S. dublin* (Wray *et al.* 1977).

Oral vaccination

In 1902, Wright (1904) made a small trial of a killed *S. typhi* vaccine *per os*. Two years later, Carroll, in the US Army, gave an oral vaccine to 12 soldiers, but it was imperfectly sterilized and at least seven of them developed typhoid fever (see Tigertt 1959). Oral vaccination with dead organisms has since been advocated from time to time, but evidence for its efficacy is scanty. In a recent trial (see Woodward 1980), the administration of a commercially available oral vaccine of acetone-inactivated organisms did not provoke any antibody response and conferred only minimal protection against experimental infection in man.

The success of live attenuated vaccines given parenterally in protecting animals against salmonella infection (above) stimulated a search for strains of *S. typhi* of low virulence by the oral route. The first to be studied was a streptomycin-dependent mutant; when freshly harvested and given by the mouth it had a 66–78 per cent effectiveness in preventing typhoid fever in human subjects challenged with a dose of 10^5 organisms (Levine *et al.* 1976). However, this vaccine lost its protective effect when freeze-dried. In 1975, Germanier and Fürer described a stable double mutant of the virulent but Vi-negative strain Ty 2 that lacks the enzyme UDP-galactose-4-epimerase. This *gal E* mutant, designated Ty 21a, proved to be safe to administer and stable in the human body (Gilman *et al.* 1977); when given orally (twice a week, 5–8 doses in all, each of 10^{10}–10^{11} organisms) its effectiveness in preventing experimental typhoid fever in man was 87 per cent. The vaccine strain could be isolated from the stools on the first day after administration but seldom thereafter. A large field trial of the Ty 21a vaccine was made in schoolchildren in Egypt (Wahdan *et al.* 1980, 1982; see also Sutton and Merson 1983). Some 19 000 children were given three doses of the vaccine at 2-day intervals (freeze-dried material reconstituted in sucrose-phosphate buffer) 10 min after a tablet of sodium citrate had been chewed. A similar number of children received a placebo. The incidence of confirmed cases of typhoid fever in the next 3 years was: in the control group, 4.9 per 10 000 per year; in the vaccinated group, 0.2 per 10 000 per year. The efficacy of the vaccine was thus 96 per cent.

Deployment of preventive measures

If the hygienic improvements that have taken place in western countries could be reproduced elsewhere, it is safe to predict that enteric fever would eventually disappear. Whether a similar or more rapid reduction in the incidence of the disease could be brought about by mass vaccination, and what the relative costs of the

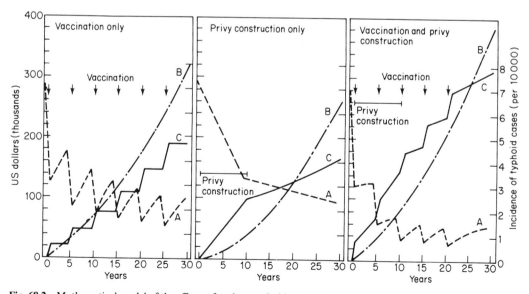

Fig. 68.2 Mathematical model of the effects of various typhoid-control programmes on the incidence of the disease, and on the cumulative costs and benefits, in an endemic area. A = incidence; B = cumulative benefits; C = cumulative costs. (Reproduced, with permission, from Cvjetanović *et al.* 1978, *Bull. World Hlth Org.* **56**, suppl. no. 1, p. 60.)

two operations would be, are matters of considerable importance. Cvjetanović, Grab and Uemura (1978) examined this question by means of a complex mathematical model, incorporating in this current estimates of the effects of the numerous contributory factors, such as the size of the infecting dose, the number of infected persons in the community, the ability of the organisms to survive outside the human body, the susceptibility of the population to infection, the efficacy and cost of various preventive measures, and many more. They applied this to situations in which (1) mass vaccination with a current killed vaccine was performed at 5-yearly intervals, (2) hygienic improvements—privy construction and the provision of safe

water—were introduced and maintained, and (3) both of these measures were applied. Representative predictions are given in Fig. 68.2, which shows the incidence of typhoid fever, the cost of the operation, and the financial benefits from it over a 30-year period. Studies of this sort, particularly if they take account of the degree and duration of the immunity to be expected from oral vaccination, may provide valuable guidance to national and international health authorities. However, it will always be difficult to make allowance for the beneficial effects of the hygienic measures on other diseases, or of unpredictable factors, such as war and natural disasters, which would tilt the balance in favour of active immunization.

Chemotherapy in enteric fever

Effective chemotherapy for enteric fever first became possible with chloramphenicol. When given in proper dosage early in the disease, treatment with this antibiotic leads in most patients to a rapid clinical cure (Woodward *et al.* 1948); but relapses occur frequently unless the drug is given for 10–14 days after the temperature has become normal (Smadel *et al.* 1949, Good and Mackenzie 1950). The action of the drug is bacteristatic rather than bactericidal. It is said that relapses occur most frequently in cases treated early in the disease. Many workers have reported that typhoid bacilli disappear rapidly from the blood when chloramphenicol is given, but Watson (1955c) often obtained positive results by clot culture with streptokinase during treatment. He believes that intracellular organisms may survive the action of chloramphenicol both in the blood stream and in the tissues (Watson 1955d, 1957). Chloramphenicol is also effective in the treatment of paratyphoid A and B fevers (Curtin 1949, Marmion 1952, Reimann and Lian 1955).

There is no evidence that chloramphenicol has any effect on the rate of clearance from the intestine, or on the number of patients who will eventually become chronic carriers (Smadel 1950, O'Connor 1958). In experimental typhoid in man (Hornick *et al.* 1970), treatment always aborted the illness but did not influence the pattern of faecal excretion. No illness occurred if treatment was begun 24 hr after the ingestion of *S. typhi* and continued for 28 days; but the patients showed a significant antibody response, suggesting that the organisms had very quickly become intracellular and so could not be eradicated by the antibiotic. The frequency of chloramphenicol resistance in *S. typhi* in some parts of the world (Chapter 37), and the liability of this antibiotic to cause bone-marrow damage, and to precipitate haemolytic crises in persons with the Mediterranean type of glucose 6-phosphate dehydrogenase deficiency, are serious drawbacks to its use.

Typhoid fever responds slowly to ampicillin, but amoxycillin gives results that, according to some workers, for example Pillay and his colleagues (1975) and Scragg (1976), are at least as good as those obtained with chloramphenicol. Co-trimoxazole is said by some observers, for example Kamat (1970), to be highly effective, but Scragg and Rubridge (1971) report a number of treatment failures with it. Gilman and his colleagues (1975b) found that, in infections with antibiotic-sensitive strains of *S. typhi*, the duration of fever was shorter in patients treated with co-trimoxazole than in those given amoxycillin; in patients infected with a strain that had plasmid-encoded resistance to chloramphenicol and sulphonamide, however, this advantage to the co-trimoxazole-treated patients was no longer apparent. Ball and his colleagues (1979) report very good results from the use of pivmecillinam, but others have found this drug to be relatively ineffective. Oxolinic acid has little therapeutic action though *S. typhi* is sensitive to it *in vitro* (Sanford *et al.* 1976). Furazolidone is safe and quite effective, but it acts rather slowly (Kamat *et al.* 1969).

Early treatment with a number of antimicrobial agents reduces mortality, and the frequency of serious complications. There is, however, little comparative information about relapse rates, the duration of excretion, and the frequency of chronic carriage after treatment with the various agents.

S. typhi may be resistant not only to chloramphenicol and sulphonamides (Chapter 37) but also occasionally to ampicillin and, more recently, to trimethoprim (see Ward *et al.* 1982). First-choice treatment in typhoid fever must be made on the basis of the known frequency of resistance to potentially useful drugs in the area in which the infection was contracted, and should be confirmed as soon as possible by testing the organism from the patient. If the infection fails to respond quickly, later isolates of the organism must be examined to exclude the possibility that it has subsequently acquired resistance—not only to the agent given but also to others—by gene transfer from another organism in the patient's flora (Kobayashi *et al.* 1979, Datta *et al.* 1981).

Treatment of carriers Chronic faecal carriers have sometimes been treated successfully with very large doses of benzylpenicillin and sulphonamide, or of the penicillin alone if probenecid was also given (see the 6th Edition of this book for references). Vigorous treatment with ampicillin eliminates carriage in over three-quarters of cases if given continuously by the oral route for 3 months (Christie 1964), or parenterally with probenecid for 6 weeks (Münnich *et al.* 1965-66). Successes with 8-hourly intravenous ampicillin for a fortnight have even been reported even in carriers who had chronic cholecystitis with gallstones (Scioli *et al.* 1972). Not all patients are willing to complete such courses of treatment, and hypersensitivity to the drug may cause trouble, but chemotherapy is now a serious alternative to cholecystectomy. Treatment of chronic carriers with co-trimoxazole is often successful (Kassur and Kociszewski 1975). Although chloramphenicol is without effect on faecal carriage it is said to eliminate excretion by about two-thirds of urinary carriers (Kennedy and Millar 1951, Lewin *et al.* 1951, Miller and Floyd 1954). According to Farid and his colleagues (1978), patients with chronic urinary-tract infection and recurrent bacteraemia due to *S. typhi* or *S. paratyphi A* respond dramatically to treatment with amoxycillin.

Other generalized salmonella infections in man

Generalized infections due to salmonellae other than the typhoid and paratyphoid bacilli include some which are clinically indistinguishable from enteric fever; others present a picture of pyaemia with localization in various internal organs. Certain serotypes, such as *S. choleraesuis*, *S. paratyphi C*, *S. blegdam*, *S. moscow*, and *S. enteritidis* var. *chaco*, seldom if ever cause a gastro-enteritis, but give rise to pyaemic and localized suppuration in a large proportion of those affected (Chapter 37).

Most other salmonellae give rise to occasional typhoidal or septic infections—for reasons that are imperfectly understood. One factor of importance is certainly the age of the patient. The severity of salmonella infection, and the resultant fatality rate, are high at the two extremes of life. Many old people who die of salmonella infection are suffering from some other serious illness. Cardiovascular insufficiency often leads to fatal circulatory disturbances, and a number of patients are diabetics or have cancer. Generalized salmonella infection may develop in patients undergoing chemotherapy or irradiation for neoplastic disease. There is a tendency for the organisms to localize in damaged tissues, such as a neoplasm, the site of an intramuscular injection, or an inanimate implant. These events often appear to arise from reactivation of a latent salmonella infection (see Chapter 58).

Some babies die from the biochemical consequences of their gastro-enteritis, but many more have severe septicaemic illnesses in which there is extra-intestinal localization of infection with collections of pus in previously healthy tissue and especially in the meninges (Schiff 1938, Guthrie and Montgomery 1939, Curbelo and Cruz 1941, Hormaeche and Peluffo 1941, Leeder 1956). Most of these infections are caused by salmonellae that produce a simple gastro-enteritis in older persons.

The association between generalized salmonella infection and relapsing or typhus fevers has been attributed to their being transmitted together by the bite of the louse (Chapter 37). Salmonella septicaemia—usually due to *S. typhimurium*—occurs in over one-third of all cases of bartonellosis and is probably the main cause of death in this disease (Cuadra 1956). The septicaemia usually occurs late in the febrile stage; this suggests that the salmonella is not introduced by the bite of the sandfly but that the bartonellosis activates latent salmonella infection. Patients with the sickle-cell trait have a greatly increased susceptibility to salmonella osteomyelitis but not to osteomyelitis due to other organisms (Hodges and Holt 1951, Smith 1953, Vandepitte *et al.* 1953, Hook *et al.* 1957, Silver *et al.* 1957, Roberts and Hilburg 1958). Symptoms of osteomyelitis often appear after a haemolytic crisis. Kaye and his colleagues (1965, 1967) report that infection with *Plasmodium berghei* enhances salmonella infection in mice and conclude that erythrophagocytosis by reticulo-endothelial cells interferes with their capacity to kill salmonellae. Hand and King (1978) report, however, that the serum of sickle-cell patients is often deficient in heat-labile factors required for the opsonization of *S. typhimurium*. Ortiz-Neu and co-workers (1978) observed predisposing circumstances in one-half of 46 salmonella infections of bone or joints, notably haemoglobinopathies, lymphoma, and antecedent local trauma or surgical operation. Infections of joints generally responded well to chemotherapy, but one-half of the bone infections became chronic despite treatment.

Enteric infection in animals

It is impossible to consider in any detail the numerous varieties of enteric infection which are known to occur among the lower animals under natural conditions. References to the relation of the different species or types, within the *Salmonella* group, to various forms of animal infection will be found in Chapter 37. It may be noted that the bacterial species that cause enteric infection in man are seldom natural pathogens of animals, and that the species and types which produce enteric infection in swine, rats, mice, etc., usually give rise to acute gastro-enteritis when ingested by man (see Chapter 72).

In hog cholera, which is caused by a virus (see Volume 4), *S. choleraesuis* may give rise to intestinal complications.

Bacillary white diarrhoea of chicks and fowl typhoid

S. gallinarum causes fowl typhoid, an enteric infection of adult chickens, and its *pullorum* variety causes bacillary white diarrhoea of baby chickens. (See Chapter 37.)

Bacillary white diarrhoea is of special interest, as it affords an example of congenital infection via the egg. The organism multiplies within the egg and causes a high mortality before hatching. The few surviving chicks disseminate the organism to the rest of the brood, and the outbreak usually assumes epidemic proportions within a few days or weeks. The organism is excreted in the faeces, and is also spread in dried dust and fluff. It is probably acquired by inhalation as well as by ingestion. Susceptibility to the acute form of the disease is confined to the first few weeks of life. The average duration of the disease is 2 or 3 days; and the case fatality is high—on average about 70 per cent. The survivors, and probably other chicks that have never suffered from the typical disease, frequently have a persistent ovarian infection and perpetuate the disease from one breeding season to another.

In *fowl typhoid*, a disease of young adult chickens, the mortality rate is variable, and many of the survivors become chronic faecal carriers. Common symptoms of the disease are listlessness, thirst and greenish diarrhoea. In birds that have died from acute fowl typhoid the liver and spleen may be enlarged, but the appearances are not characteristic. When death occurs later in the disease there are many large necrotic foci in the heart muscle, and discrete lesions in the intestine not unlike typhoid ulcers (Smith 1955*a*).

The problem of controlling bacillary white diarrhoea resolves itself into the elimination of the carrier hens, and so preventing perpetuation of the disease from one season to the next. This has been rendered possible by the application of the O-agglutination test with suspensions of *S. gallinarum* var. *pullorum*. Agglutinins may be sought in a tube test with serum, or in a rapid slide test with whole blood. Opinions on the relative merit of the two methods differ (see Buxton 1959). The agglutination test with pullorum antigen also serves to detect chronic fowl typhoid, but control of this disease by the elimination of carriers and improved hygiene has not been entirely successful. Many attempts to produce active immunity have therefore been made. As we have seen (p. 424), live vaccines made from attenuated strains—even from rough strains—have a stronger and more lasting effect than killed vaccines; their value has been confirmed in field trials (Harbourne 1957).

No chemotherapeutic treatment of either bacillary white diarrhoea or fowl typhoid which does not prevent the development of chronic infection is of much practical value. The effects of the sulphonamides, streptomycin, chloramphenicol and neomycin are disappointing. The most valuable drug so far described is furazolidone (Grumbles *et al.* 1954, Smith 1954*a, b*, 1955*b*). If it is given immediately after infection with *S. gallinarum* or its *pullorum* variety, not only is acute infection prevented, but very few birds become chronic carriers. When treatment is delayed for several days there is still a considerable reduction in mortality, but many birds remain persistently infected. It is doubtful whether chronic infection is often eliminated by treatment with furazolidone (Gordon and Tucker 1955, 1957, Smith 1955*b*, Wilson 1956), but there is general agreement that eggs laid by carrier hens during treatment and for some time afterwards are seldom infected.

(For information about other salmonella infections in birds and domestic animals see Chapter 72.)

References

Adams, E.B. (1983) In: *Oxford Textbook of Medicine*, vol. 1, p. 5.183. Ed. by D.J. Weatherall, J.G.G. Ledingham and D.A. Warrell. Oxford University Press, Oxford.

Ames, W.R. and Robins, M. (1943) *Amer. J. publ. Hlth.* **33**, 221.

Anderson, E.S. and Hobbs, B.C. (1973) *Israel J. med. Sci.* **9**, 162.

Anderson, G.W., Hamblen, A.D. and Smith, H.M (1936) *Amer. J. publ. Hlth* **26**, 396.

Archer, G.T.L., Goffe, A.P. and Ritchie, A. (1952) *J.R. Army med. Cps* **98**, 40, 125, 189, 237, 334.

Ash, I., McKendrick, G.D.W., Robertson, M.H. and Hughes, H.L. (1964) *Brit. med. J.* **i**, 1474.

Ashcroft, M.T. (1964) *J. trop. Med. Hyg.* **67**, 185.

Ashcroft, M.T., Singh, B., Nicholson, C.C., Ritchie, J.M., Sobryan, E. and Williams, F. (1967) *Lancet* **ii**, 1056.

Ayliffe, G.A.J., Geddes, A.M., Pearson, J.E. and Williams, T.C. (1979) *J. Hyg., Camb.* **82,** 353.

Bailey, W.R. and Bynoe, E.T. (1953) *Canad. J. publ. Hlth* **44,** 468.

Baine, W.B., Farmer, J.J. III, Gangarosa, E.J., Hermann, G.T., Thornsberry, C. and Rice, P.A. (1977) *J. infect. Dis.* **135,** 649.

Ball, A.P., Farrell, I.D., Gillett, A.P., Geddes, A.N., Clarke, P.D. and Ellis, C.J. (1979) *J.Infect.* **1,** 353.

Barr, M., Sayers, M.H.P. and Stamm, W.P. (1959) *Lancet* **i,** 816.

Barrett, T.J., Blake, P.A., Brown, S.L., Hoffman, K., Llort, J.M. and Feeley, J.C. (1983) *J. clin. Microbiol.* **17,** 625.

Barrett, T.J., Snyder, J.D., Blake, P.A. and Feeley, J.C. (1982) *J. clin. Microbiol.* **15,** 235.

Beattie, C.P. and Elliot, J.S. (1937) *J. Hyg., Camb.* **37,** 36.

Bekenn, C.H.G. and Edwards, J.L. (1944) *Mon. Bull. Minist. Hlth Lab. Serv.* **3,** 60.

Belin, V.M. (1934) *Coquillages et Fièvres Typhoides. Un Point d'Histoire Contemporaine.* Les Presses universitaires de France, Paris.

Billet, le Bihan, Thèrault, Lamandé, Lutrot and Louis. (1910) *Arch. Méd. Pharm. milit.* **55,** 259.

Blachstein (1891) *Johns Hopk. Hosp. Bull.* **2,** 96.

Blaser, M.J., Hickman, F.W., Farmer, J.J. III, Brenner, D.J., Balows, A. and Feldman, R.A. (1980) *J. infect. Dis.* **142,** 934.

Blasi, R. de (1950) *Boll. Ist. sieroter. milan.* **29,** 11.

Blumenthal, E. (1910) *Zbl. Bakt.* **55,** 341.

Bokkenheuser, V., Smit, P. and Richardson, N. (1964) *Amer. J. publ. Hlth* **54,** 1507.

Bornstein, S. (1943) *J. Immunol.* **46,** 439.

Bowman, M. (1942) *Canad. publ. Hlth. J.* **33,** 541.

Boyd, J. (1966) *J.R. Army med. Cps* **112,** 4.

Boyer, J. and Tissier, M. (1950) *Presse méd.* **58,** 634.

Bradley, W.H. (1943) *Brit. med. J.* **i,** 438.

Brodie, J. (1977) *J. Hyg., Camb.* **79,** 161

Browning, C.H., with Coulthard, H.L., Cruickshank, R., Guthrie, K.J. and Smith, R.P. (1933) *Spec. Rep. Ser. med. Res. Coun., Lond.* No. 179.

Brückner (1910) *Arb. ReichsgesundhAmt.* **33,** 435.

Brygoo, E.R. (1974) *Ann. Inst. Pasteur, Madagascar* **42,** 239.

Budd, W. (1856) *Lancet,* **ii,** 618, 694; (1873) *Typhoid Fever.* London.

Bumke, E. (1926) *Z. Hyg. InfektKr.* **105,** 342.

Butler, T., Bell, W.R., Levin, J., Linh, N.N. and Arnold, K. (1978) *Arch. intern. Med.* **138,** 407.

Buxton, A. (1959) In: *Infectious Diseases of Animals,* **2,** p. 481. Ed. by A.W. Stableforth and I.A. Galloway. Butterworth, London.

Chau, P.Y. and Tsang, R.S.W. (1982) *J. Hyg., Camb.* **89,** 261.

Chau, P.Y., Tsang, R.S.W., Lam, S.K., La Brooy, J.T. and Rowley, D. (1981) *Clin. exp. Immunol.* **46,** 515.

Chiari, H. (1894) *Zbl. Bakt.* **15,** 648.

Christie, A.B. (1964) *Brit. med. J.,* **i,** 1609.

Chung, G.T. and Frost, A.J. (1970) *J. appl. Bact.* **33,** 449.

Cockburn, W.C. (1955) *J.R. Army med. Cps* **101,** 171.

Coleman, W. and Buxton, B.H. (1907) *Amer. J. med. Sci.* **133,** 896.

Collard, P., Sen, R. and Montefiore, D. (1959) *J. Hyg., Camb.* **57,** 427.

Colon, A.R. (1976) *Sth. med. J., Bgham, Ala.* **69,** 914.

Conradi (1907) *Klin. Jb.* **17,** 273.

Corvazier, P. (1952) *Ann. Inst. Pasteur* **83,** 173.

Couper, W.R.M., Newell, K.W. and Payne, D.J.H. (1956) *Lancet* **i,** 1057.

Creighton, C. (1894) *A History of Epidemics in Britain.* Cambridge University Press.

Crocker, C.G. (1947) *J. Hyg., Camb.* **45,** 118.

Cruickshank, J.C. (1947) *Mon. Bull. Minist. Hlth Lab. Serv.* **6,** (April), 88.

Cuadra, C.M. (1956) *Texas Rep. Biol. Med.* **14,** 97.

Curbelo, A. and Cruz, J.A.M. (1941) *Arch. de Méd. inf.* **10,** 170. Quoted by Black, P.H., Kunz, L.J. and Swartz, M.N. (1960) *New Engl. J. Med.* **262,** 864.

Curtin, M. (1949) *Brit. med. J.* **ii,** 1504.

Cvjetanović, B., Grab, B. and Uemura, K. (1978) *Bull. World Hlth Org.* **56,** suppl. no. 1.

Cvjetanović, B and Uemura, K. (1965) *Bull. World Hlth Org.* **32,** 29.

Datta, N., Richards, H. and Datta, C. (1981) *Lancet* **i,** 1181.

Davies, I.G., Cooper, K.E., Wiseman, J. and Davies, J.M. (1940) *Lancet* **ii,** 778.

Desranleau, J-M. (1947) *Canad. J. publ. Hlth* **38,** 343.

Doerr, R. (1905) *Zbl. Bakt.* **39,** 624.

Dold, H. and Ketterer, M. (1943) *Z. Hyg. InfektKr.* **125,** 215; (1947) *Ibid.* **127,** 326.

Dreyer, G. and Inman, A.C. (1917) *Lancet* **i,** 365.

Dreyer, G. and Walker, E.W.A. (1916) *Lancet* **ii,** 419.

Dreyer, G., Walker, E.W.A. and Gibson, A.G. (1915) *Lancet* **i,** 324.

Drigalski, von (1904) *Zbl. Bakt.* **35,** 776.

Droba (1889) *Wien. klin. Wschr.* **12,** 1141.

Dudgeon, L.S. (1908) *Lancet* **ii,** 1651.

Eberth, C.J. (1880) *Virchows Arch.* **81,** 58.

Edwards, E.A., Johnson, D.P., Pierce, W.E. and Peckinpaugh, R.O. (1974) *Bull. World Hlth Org.* **51,** 501.

El-Rooby, A. and Gohar, M.A. (1956) *J. trop. Med. Hyg.* **59,** 47.

El-Sadr, A.R. (1953) *J, Egypt. med. Ass.* **36,** 499.

Eliot, C.P. (1940) *Amer. J. Hyg.* **31,** B, 8.

Eliot, C.P. and Cameron, W.R. (1941) *Amer. J. publ. Hlth* **31,** 599.

Erlik, D. and Reitler, R. (1960) *Lancet* **i,** 1216.

Evans, D.I. (1947) *Med. Offr* **77,** 39.

Ewing, W.H., McWhorter, A.C. and Montague, T.S. (1966) *Publ. Hlth Lab.* **24,** 63.

Fabian, F.W. (1947) *Amer. J. publ. Hlth* **37,** 987.

Faichnie, N. (1909) *J.R. Army med. Cps* **13,** 580, 672.

Farid, Z., Bassily, S., Mikhail, I.A., Edman, D.C., Hassan, A. and Miner, W.F. (1978) *J. infect. Dis.* **132,** 698.

Farid, Z. *et al.* (1970) *J. trop. Med. Hyg.* **73,** 153.

Felix, A. (1924a) *Z. ImmunForsch.,* **39,** 127; (1924b) *J. Immunol.* **9,** 115; (1938a) *Lancet* **ii,** 738; (1938b) *J. Hyg., Camb.* **38,** 750; (1941) *Brit. med. J.* **i,** 391; (1950) *Bull. World Hlth Org.* **2,** 643; (1951) *Brit. med. Bull.* **7,** 153.

Felix, A. and Anderson, E.S. (1951) *J. Hyg., Camb.* **49,** 349.

Felix, A. and Bensted, H.J. (1954) *Bull. World Hlth Org.* **10,** 919.

Ficker, M. (1903) *Arch. Hyg.* **46,** 274.

Firth, R.H. and Horrocks, W.H. (1902) *Brit. med. J.* **ii,** 936.

Foley, A.R. (1947) *Canad. J. publ. Hlth* **38,** 73.

Frosch (1903) *Festschr. 60 Geburtst. R. Koch.* p. 691. Jena.

Fütterer, von (1888) *Münch. med. Wschr.* **35,** 316.

Gaffky (1884) *Mitt. ReichsgesundhAmt.* **2,** 372.

Galbraith, N.S., Hobbs, B.C., Smith, M.E. and Tomlinson, A.J.H. (1960) *Mon.Bull. Minist. Hlth Lab. Serv.* **19,** 99.

Gardner, A.D. (1937) *J. Hyg., Camb.* **37**, 124.

Gardner, A.D. and Stubington, E.F. (1932) *J.Hyg., Camb.* **32**, 516.

Gauthier, J. and Foley, A.R. (1943) *Canad. publ. Hlth J.* **34**, 543.

Gay, F.P. (1918) *Typhoid Fever.* The Macmillan Co., New York.

Gay, F.P. and Claypole, E.J. (1913) *Arch. intern. Med.* **12**, 621.

Geks, F.J. (1949) *Z. Hyg. InfektKr.* **129**, 215.

Gell, P.G.H. and Knox, R. (1942) *Publ. Hlth, Lond.* **56**, 8.

George, J.T.A., Wallace, J.G., Morrison, H.R. and Harbourne, J.F. (1972) *Brit. med. J.* **iii**, 208.

George, T.C.R., Harvey, R.W.S. and Thomson, S. (1953) *J. Hyg., Camb.* **51**, 532.

Gerhard, W.W. (1836–37) *Amer. J. med. Sci.* **19**, 289.

Germanier, E. and Fürer, E. (1975) *J. infect. Dis.* **131**, 553.

Giglioli, G. (1933a) *J. Hyg., Camb.* **33**, 379; (1933b) *Ibid.* **33**, 387.

Gilbert, A. and Girode, J. (1890) *Sem. méd.* **10**, 481.

Gill, D.G. (1927) *J. Amer. med. Ass.* **89**, 1198.

Gilman, R.H. and Hornick, R.B. (1976) *J. clin. Microbiol.* **3**, 456.

Gilman, R.H., Hornick, R.B., Woodward, W.E., DuPont, H.L., Snyder, M.J., Levine, M.M. and Libanoti, J.P. (1977) *J. infect. Dis.* **136**, 717.

Gilman, R.H., Islam, S., Rabbani, S. and Ghosh, H. (1979) *Lancet* **i**, 795.

Gilman, R.H., Terminel, M., Levine, M.M., Hernandez-Mendoza, P. and Hornick, R.B. (1975a) *Lancet* **i**, 1211.

Gilman, R.H. *et al.* (1975b) *J. infect. Dis.* **132**, 630.

Glass, V. and Wright, H.D. (1937) *J. Path. Bact.* **45**, 431.

Glatzel, E. (1944) *Z. Hyg. InfektKr.* **126**, 81.

Godfrey, E.S. (1928) *Amer. J. publ. Hlth* **18**, 616.

Good, R.A. and Mackenzie, R.D. (1950) *Lancet* **i**, 611.

Gordon, R.F. and Tucker, J. (1955) *Vet. Rec.* **67**, 116; (1957) *Brit. vet. J.* **113**, 99.

Gorman, A.E. and Wolman, A. (1939) *J. Amer. Wat. Wks Ass.* **31**, 225.

Graham-Smith, G.S. (1910) *Rep. loc. Govt Bd publ. Hlth* **16**, 9.

Gray, A.L. (1938) *Amer. J. publ. Hlth* **28**, 1415.

Grumbles, L.C., Wills, F.K. and Boney, W.A. (1954) *J. Amer. vet. med. Ass.* **124**, 217.

Grünbaum, A.S. (1896) *Lancet* **ii**, 806, 1747.

Guthrie, K.J. and Montgomery, G.L. (1939) *J. Path. Bact.* **49**, 393.

Hand, W.L. and King, N.L. (1978) *Amer. J. Med.* **64**, 388.

Harbourne, J.F. (1957) *Vet. Rec.* **69**, 1102.

Harmsen, H. (1954) *Städthyg.* **5**, 54.

Harvey, R.W.S. (1956) *Mon. Bull. Minist. Hlth Lab. Serv.* **15**, 118.

Harvey, R.W.S. and Phillips, W.P. (1955) *Lancet* **ii**, 137.

Harvey, R.W.S. and Price, T.H. (1964) *Mon. Bull. Minist. Hlth Lab. Serv.* **23**, 233.

Harvey, R.W.S. and Thomson, S. (1953) *Mon. Bull. Minist. Hlth Lab. Serv.* **12**, 149.

Hathout, S.E.D., El-Ghaffar, Y.A., Awny, A.Y. and Hassan, K. (1966) *Amer. J. trop. Med. Hyg.* **15**, 156.

Hejfic, L.B., Levina, L.A., Kuz'minova, M.L., Salmin, L.V., Slavina, A.M. and Vasil'eva, A.V. (1968) *Bull. World Hlth Org.* **38**, 907.

Hemmes, G.D. (1951) *Tijd. Diergeneesk.* **76**, 258.

Henderson, N.D. and Ferguson, W.W. (1949) *J. Lab. clin. Med.* **34**, 739.

Hickman, F.W., Rhoden, D.L., Esaias, A.O., Baron, L.S., Brenner, D.J. and Farmer, J.J. III (1982) *J. clin. Microbiol.* **15**, 1085.

Hobbs, B.C. and Smith, M.E. (1955) *J. appl. Bact.* **18**, 471.

Hodges, F.J. and Holt, J.F. (1951) *Yearbook of Radiology*, p. 89. Year Book Publications, Chicago.

Holt, H.D., Vaughan, A.C.T. and Wright, H.D. (1942) *Lancet* **i**, 133.

Holt, H.D. and Wright, H.D. (1942) *J. Path. Bact.* **54**, 247.

Hompesch, H. (1953) *Z. Hyg. InfektKr.* **137**, 541.

Hook, E.W., Campbell, C.G., Weens, H.S. and Cooper, G.R. (1957) *New Engl. J. Med.* **257**, 403.

Hormaeche, E. and Peluffo, C.A. (1941) *Puerto Rico J. publ. Hlth trop. Med.* **17**, 99.

Hornick, R.B. and Greisman, S. (1978) *Arch. intern. Med.* **138**, 357.

Hornick, R.B., Greisman, S.E., Woodward, T.E., DuPont, H.L., Dawkins, A.T. and Snyder, M.J. (1970) *New Engl. J. Med.* **283**, 686, 739.

Hornick, R.B. and Woodward, T.E. (1967) *Trans. Amer. clin. climatol. Ass.* **78**, 70.

Horton-Smith, P. (1900) *Brit. med. J.* **i**, 827.

Howie, J.W. (1968) *J. appl. Bact.* **31**, 171.

Hoyle, L. (1943) *Mon. Bull. Minist. Hlth Lab. Serv.* **2**, 26.

Hueppe (1886) *Fortschr. Med.* **4**, 447.

Hughes, M.H. (1955) *J. Hyg., Camb.* **53**, 368.

Iveson, J.B. and Mackay-Scollay, E.M. (1969) *J. Hyg., Camb.* **67**, 457.

Jameson, J.E. (1961) *J. Hyg., Camb.* **59**, 1.

Jeffries, L. (1959) *J. clin. Path.* **12**, 568.

Jochmann (1914) *Lehrbuch der Infektionskrankheiten.* Berlin.

Joe, L.K. (1949) *Docum. neerl. indones. Morb. trop.* **1**, 145; (1950) *Ned. Tijdschr. Geneesk.* **94**, 1246; (1956) *Amer. J. trop. Med. Hyg.* **5**, 133.

Joe, L.K., Wiratmajda, N.S., Hardjowardojo, S.D., Kartanegara, S. and Harmiati, S. (1957) *Docum. Med. geograph. trop.* **9**, 27.

Johnston, J.A. (1912) *J. med. Res.* **27**, 177.

Jones, A.C. (1951) *J. Hyg., Camb.* **49**, 335; (1953) *Pers. comm.*

Jones, D.J., Gell, P.G.H. and Knox, R. (1942) *Lancet* **ii**, 362.

Jones, D.M.M. and Pantin, C.G. (1956) *J. clin. Path.* **9**, 128; (1961) *Mon. Bull. Minist. Hlth Lab. Serv.* **20**, 20.

Jürgens (1927) *Med. Klin.* **23**, 1012, 1053.

Kamat, G.R., Bulchandani, D. and Katrak, P.H. (1969) *J. Ass. Physns India* **17**, 615.

Kamat, S.A. (1970) *Brit. med. J.* **iii**, 320.

Karlsson, K., Carlsson, H.E., Neringer, R. and Lindberg, A.A. (1980) *Scand. J. infect. Dis.* **12**, 41

Kassur, B. and Kociszewski, J. (1975) *Przegl. epidem.* **29**, 301.

Kauffmann, F. (1930–31) *Zbl. Bakt.* **119**, 148; (1935–36) *Z. Hyg. InfektKr.* **117**, 26.

Kaye, D., Gill, F.A. and Hook, E.W. (1967) *Amer. J. med. Sci.* **254**, 205.

Kaye, D., Merselis, J.G. Jr and Hook, E.W. (1965) *Proc. Soc. exp. Biol., N.Y.* **120**, 810.

Kaye, D., Palmieri, M. and Rocha, H. (1966) *J. Bact.* **91**, 945.

Kayser, H. (1907) *Arb. ReichsgesundhAmt.* **25**, 223.

Kennedy, J. and Payne, A. M-M. (1950) *Mon. Bull. Minist. Hlth Lab. Serv.* **9**, 168.

Kennedy, J.M. and Millar, E.L.M. (1951) *Lancet* **i**, 92.

Klein, M. (1943) *J. infect. Dis.* **72**, 49.

Klinger, P. (1906) *Arb. ReichsgesundhAmt.* **25**, 223.

Knivett, V.A. and Stevens, W.K. (1971) *J. Hyg., Camb.* **69**, 233.

Knox, R., Gell, P.G.H. and Pollock, M.R. (1942) *J. Path. Bact.* **54**, 469.

Kobayashi, K., Kitaura, T., Goto, N. and Nakaya, R. (1979) *Microbiol. Immunol.* **23**, 423.

Koch, J. (1909) *Z. Hyg. InfektKr.* **62**, 1.

Kovacs, N. (1959) *Med. J. Aust.* **i**, 557.

Kwantes, W. and Speedy, W.J.Y. (1955) *Mon. Bull. Minist. Hlth Lab. Serv.* **14**, 120.

Lanata, C.F. *et al.* (1983) *Lancet* **ii**, 441.

Landy, M. and Lamb, E. (1953) *Proc. Soc. exp. Biol., N.Y.* **82**, 593.

Lange, S., Elwing, H., Larsson, P. and Nygren, H. (1980) *J. clin. Microbiol.* **12**, 637.

Ledingham, J.C.G. and Arkwright, J.A. (1912) *The Carrier Problem in Infectious Diseases.* London.

Leeder, F.S. (1956) *Ann. N.Y. Acad. Sci.* **66**, Article 1, 54.

Leifson, E. (1935) *J. Path. Bact.* **40**, 581; (1936) *Amer. J. Hyg.* **24**, 423.

Lendon, N.C. and Mackenzie, R.D. (1951) *Mon. Bull. Minist. Hlth Lab. Serv.* **10**, 23.

Lentz (1905) *Klin. Jb.* **14**, 475.

Levin, D.M., Wong, K.H., Reynolds, H.Y., Sutton, A. and Northrup, R.S. (1975) *Infect. Immun.* **12**, 1290.

Levine, M.M., Grados, O., Gilman, R.H., Woodward, W.E., Solis-Plaza, R. and Waldman, W. (1978) *Amer. J. trop. Med. Hyg.* **27**, 795.

Levine, M.M. *et al.* (1976) *J. infect. Dis.* **133**, 424.

Levy, E. and Gaehtgens, W. (1908) *Arb. ReichsgesundhAmt.* **28**, 168.

Lewin, W., Bersohn, I., Gaylis, B. and Mundel, B. (1951) *S. Afr. med. J.* **25**, 621.

Ling, C.C., Liu, J. and Chen, T.Y. (1948) *Chin. med. J.* **66**, 66.

Littman, A., Vaichulis, J.A. and Ivy, A.C. (1949) *J. Lab. clin. Med.* **34**, 549.

Lo Verde, P.T., Amento, C. and Higashi, G.I. (1980) *J. infect. Dis.* **141**, 177.

McCall, C.E., Martin, W.T. and Boring, J.R. (1966) *J. Hyg., Camb.* **64**, 261.

McFadzean, A.J.S. (1966) *Brit. med. J.* **i**, 1567.

Mackenzie, E.F.W. and Taylor, E.W. (1945) *J. Hyg., Camb.* **44**, 31.

Marmion, D.E. (1952) *Trans. R. Soc. trop. Med. Hyg.* **46**, 619.

Martin, P.H. (1947) *Mon. Bull. Minist. Hlth Lab. Serv.* **6**, 148.

Mayer, G. (1910) *Zbl. Bakt.* **53**, 234.

Meers, P.D. and Goode, D. (1965) *Lancet* **i**, 426.

Menzies, D.B. (1944) *Canad. publ. Hlth J.* **35**, 431.

Merselis, J.G., Kaye, D., Connolly, C.S. and Hook, E.W. (1964) *Amer. J. trop. Med. Hyg.* **13**, 425.

Mikhail, I.A., Sanborn, W.R. and Sippel, J.E. (1983) *J. clin. Microbiol.* **17**, 564.

Miles, A.A. (1958) *J. R. Army med. Cps* **104**, 89.

Miller, W.S., Clark, D.L. and Dierkhising, O.C. (1951) *Amer. J. trop. Med.* **31**, 535.

Miller, W.S. and Floyd, T.M. (1954) *Lancet* **i**, 343.

Minelli, S. (1906) *Zbl. Bakt.* **41**, 406.

Moore, B. (1948) *Mon. Bull. Minist. Hlth Lab. Serv.* **7**, 241; (1950) *Ibid.* **9**, 72.

Moore, B., Perry, E.L. and Chard, S.T. (1952) *J. Hyg., Camb.* **50**, 137.

Moore, W.B. (1950) *J.R. sanit. Inst.* **70**, 93.

Morgan, H. de R. (1911) *J. Hyg., Camb.* **11**, 202.

Morzycki, J. (1949) *Bull. Inst. mar. trop. Med. Gdansk* **2**, 41.

Muller, L. (1925) *C. R. Soc. Biol.* **93**, 433.

Münnich, D., Békési, I. and Uri, J. (1965–66) *Chemotherapia, Basel* **10**, 253.

Nath, T.R., Malaviya, A.N., Kumar, R., Balakrishnan, K. and Singh, B.P. (1977) *Clin. exp. Immunol.* **30**, 38.

Naylor, G.R.E. (1983) *Lancet* **i**, 864.

Naylor, G.R.E. and Caldwell, R.A. (1953) *J. Hyg., Camb.* **51**, 245.

Neter, E. (1956) *Bact. Rev.* **20**, 166.

Neuhaus (1886) *Berl. klin. Wschr.* **23**, 89.

Newell, K.W. (1955a) *Mon. Bull. Minist. Hlth Lab. Serv.* **14**, 146; (1955b) *J. appl. Bact.* **18**, 462.

Newell, K.W., Hobbs, B.C. and Wallace, E.J.G. (1955) *Brit. med. J.* **ii**, 1296.

Nicol, C.G.M. (1956) *Mon. Bull. Minist. Hlth Lab. Serv.* **15**, 240.

Nolan, C.M., Feeley, J.C., White, P.C. Jr, Hambie, E.A., Brown, S.L. and Wong, K.-H. (1980) *J. clin. Microbiol.* **12**, 22.

Nyerges, G., Szerdahelyi, F., Tankó, S., Bognár, S. and Funk, O. (1976) *Acta microbiol. hung.* **23**, 293.

O'Connor, M.E. (1958) *Publ. Hlth Rep., Wash.* **73**, 1039.

Ørskov, J., Jensen, K.A. and Kobayashi, K. (1928) *Z. ImmunForsch.* **55**, 34.

Ortiz-Neu, C., Marr, J.S., Cherubin, C.E. and Neu, H.C. (1978) *J. infect. Dis.* **138**, 820.

Pang, T. and Puthucheary, S.D. (1983) *J. clin. Path.* **36**, 471.

Park, W.H. and Williams, A.W. (1925) *Pathogenic Organisms.* New York.

Paz, G.C. de la (1939) *Bull. Hyg., Lond.* **14**, 635.

Pfeiffer, R. (1885) *Dtsch. med. Wschr.* **11**, 500.

Pfeiffer, R. and Kolle, W. (1896) *Dtsch. med. Wschr.* **22**, 735.

Pijper, A. and Crocker, C.G. (1943) *J. Hyg., Camb.* **43**, 201.

Pillay, N., Adams, E.B. and North-Coombes, D. (1975) *Lancet* **ii**, 333.

Pittman, M. and Bohner, H.J. (1966) *J. Bact.* **91**, 1713.

Pratt (1901) *Amer. J. med. Sci.* **122**, 584.

Prigge, R. (1909) *Klin. Jb.* **22**, 245.

Raettig, H. (1950) *Zbl. Bakt.* **155**, 239; (1959a) *Ibid.*, **174**, 192; (1959b) *Ibid.* **175**, 236, 245; (1959c) *Ibid.* **175**, 618; (1966) *Ibid.* **201**, 253.

Raistrick, H. and Topley, W.W.C. (1934) *Brit. J. exp. Path.* **15**, 113.

Ranklin, J.D., Taylor, R.J. and Newman, G. (1967) *Vet. Rec.* **80**, 720.

Rappaport, F., Konforti, N. and Navon, B. (1956) *J. clin. Path.* **9**, 261.

Reed, Vaughan and Shakespeare. (1899) *Abs. Rep. Typhoid Fever in U.S. military Camps—Spanish War*, 1898. Wash.

Reimann, H.A. and Lian, P.T. (1955) *Arch. intern. Med.* **96**, 777.

Report. (1913) *Antityphoid Com., G.B.* London; (1941) *Mon. Bull. Emerg. publ. Hlth Lab. Serv.* **1**, Dec., p. 6; (1942a) *Ibid.* **1**, Jan., p. 10; (1942b) *Ibid.* **1**, Dec., p. 7; (1945) *Ibid.* **4**, 224; (1957) *Bull. World Hlth Org.* **16**, 897; (1959a) *Ann. Rep. for 1958, C.M.O., Min. Hlth*, London; (1959b) *J.*

Hyg., Camb. **57**, 345; (1961*a*) *Ann. Rep. for 1960, C.M.O., Min. Hlth,* London; (1961*b*) *J. Hyg., Camb.* **59**, 231; (1962) *Bull. World. Hlth Org.* **26**, 357; (1964*a*) *Scott. Home Hlth Dep.,* Comd 2542. HMSO, Edinburgh; (1964*b*) *Bull. World Hlth Org.* **30**, 631; (1972) *Ann. Rep. for 1971, C.M.O., Min. Hlth,* London; (1973) *J. Hyg., Camb.* **71**, 59; (1982) *Ibid.,* **88**, 231.

Richardson, M.W. (1899) *J. Boston Soc. med. Sci.* **3**, 79.

Roberts, A.R. and Hilburg, L.E. (1958) *J. Paediatrics* **52**, 170.

Robertson, E.C. (1936) *Canad. publ. Hlth J.* **27**, 37.

Rosenau, M.J., Lumsden, L.L. and Kastle, J.H. (1909) *Bull. U.S. hyg. Lab.* No. 52.

Rosher, A.B. and Fielden, H.A. (1922) *Lancet* **i**, 1088.

Rubinstein, A.D., Feemster, R.F. and Smith, H.M. (1944) *Amer. J. publ. Hlth* **34**, 841.

Ryan, W.J. (1972) *J. med. Microbiol.* **5**, 533.

Ryder, R.W. and Blake, P.A. (1979) *J. infect. Dis.* **139**, 124.

Sachs, A. (1939) *J. R. Army med. Cps* **73**, 235.

Sanford, J.P., Linh, N.N., Kutscher, E., Arnold, K. and Gould, K. (1976) *Antimicrob. Agents Chemother.* **9**, 387.

Sarma, V.N.B., Malavaya, A.N., Kumar, R., Ghai, O.P. and Bakhtary, M.M. (1977) *Clin. exp. Immunol.* **28**, 35.

Savage, W.G. (1942) *J. Hyg., Camb.* **42**, 393.

Savage, W.G. and White, P.B. (1925) *J. Hyg., Camb.* **24**, 37.

Schäfer, W. (1935) *Zbl. Bakt.* **133**, 458.

Schiff, F. (1938) *J. Amer. med. Ass.,* **111**, 2458.

Schoenlein, J.L. (1839) *Allgemeine und specielle Pathologie und Therapie.* Freiburg.

Schubert, J.H., Edwards, P.R. and Ramsey, C.H. (1959) *J. Bact.* **77**, 648.

Scioli, C., Fiorentino, S. and Sasso, G. (1972) *J. infect. Dis.* **125**, 170.

Scott, H.H. (1934) *Some Notable Epidemics.* Arnold, London.

Scott, W.M. and Minett, F.C. (1947) *J. Hyg., Camb.* **45**, 159.

Scragg, J.N. (1976) *Brit. med. J.* **ii**, 1031.

Scragg, J.N. and Rubridge, C.J. (1971) *Brit. med. J.* **iii**, 738.

Scragg, J.N., Rubridge, C.J. and Wallace, H.L. (1969) *Arch. Dis. Childh.* **44**, 18.

Seeliger, H. and Vorlaender, K.O. (1953) *Z. ImmunForsch.* **110**, 128.

Senewiratne, B. and Senewiratne, K. (1977) *Gastroenterology* **77**, 233.

Shaughnessy, H.J., Friewer, F. and Snyder, A. (1948) *Amer. J. publ. Hlth* **38**, 670.

Shaw, A.B. and Mackay, H.A.F. (1951) *J. Hyg., Camb.* **49**, 299, 315.

Siler, J.F. and Dunham, G.C. (1939) *Amer. J. publ. Hlth* **29**, 95.

Silver, H.K., Simon, J.L. and Clement, D.H. (1957) *Pediatrics, Springfield* **20**, 439.

Smadel, J.E. (1950) *Trans. R. Soc. trop. Med. Hyg.* **43**, 555.

Smadel, J.E., Woodward, T.E. and Bailey, C.A. (1949) *J. Amer. med. Ass.* **141**, 129.

Smith, A.J.K. and Thomas, K.L. (1966) *Mon. Bull. Minist. Hlth Lab. Serv.* **25**, 219.

Smith, H.W. (1954*a*) *Vet. Rec.* **66**, 215; (1954*b*) *Ibid.* **66**, 493; (1955*a*) *J. comp. Path.* **65**, 37; (1955*b*) *Ibid.* **65**, 55; (1956*a*) *J. Hyg., Camb.* **54**, 419; (1956*b*) *Ibid.* **54**, 433; (1965) *J. Hyg., Camb.* **63**, 117.

Smith, H.W. and Halls, S. (1966) *J. Hyg., Camb.* **64**, 357.

Smith, J. (1932) *J. Hyg., Camb.* **32**, 143.

Smith, M.M., McVie, M.H. and Newbold, E. (1930) *J. Hyg., Camb.* **30**, 55.

Smith, W.S. (1953) *Ohio med. J.* **49**, 642.

Soper, G.W. (1939) *Bull. N.Y. Acad. Med.* **15**, 698.

Spaun, J. (1951) *Acta path. microbiol. scand.* **29**, 416.

Spaun, J. and Uemura, K. (1964) *Bull. World. Hlth Org.* **31**, 761.

Steiniger, F. (1953) *Zbl. Bakt.* **160**, 80.

Stevenson, A.H. (1953) *Amer. J. publ. Hlth* **43**, 529.

Stowman, K. (1947–48) *Epid. vit. Stat. Rep., World Hlth Org.* **1**, 166.

Stroebe, F. (1928) *Z. klin. Med.* **108**, 752.

Studeny, O. (1947) *Wien. med. Wschr.* **97**, 347.

Subrahmanyan, P. (1952) *J. Indian med. Ass.* **22**, 99.

Sutton, R.G.A. and Merson, M.H. (1983) *Lancet* **i**, 523.

Tabet, F. (1940) *Lab. med. Prog.,* Cairo **1**, 13.

Teague, O. and Clurman, A.W. (1916) *J. infect. Dis.* **18**, 647.

Thomas, J.C., Watson, K.C. and Hewstone, A.S. (1954) *J. clin. Path.* **7**, 50.

Thomas, W.E., Stephens, T.H., King, G.J.G. and Thomson, S. (1948) *Lancet* **ii**, 270.

Thomson, S. (1954) *J. Hyg., Camb.* **52**, 67.

Tigertt, W.D. (1959) *Milit. Med.* **124**, 342.

Topley, W.W.C. (1938) *Lancet* **i**, 181.

Tsang, R.S.W. and Chau, P.Y. (1981) *Brit. med. J.* **282**, 1505.

Tsang, R.S.W., Chau, P.Y., Lam, S.K., La Brooy, J.T. and Rowley, D. (1981) *Clin. exp. Immunol.* **46**, 508.

Vandepitte, J., Colaert, J., Lambotte-Legrand, C. and Perin, F. (1953) *Ann. Soc. belge trop. Méd.* **33**, 511.

Vilchur (1887) *Etiology and Clinical Bacteriology of Typhoid Fever.* St. Petersburgh.

Vogelsang, T.M. (1964) *J. Hyg., Camb.* **62**, 443.

Vogelsang, T.M. and Bøe, J. (1948) *J. Hyg., Camb.* **46**, 252.

Wahdan, M.H., Sérié, C., Cerisier, Y., Sallam, S. and Germanier, R. (1982) *J. infect. Dis.* **145**, 292.

Wahdan, M.H. *et al.* (1980) *Bull. World Hlth Org.* **58**, 469.

Walker, W. (1965) *Scot. med. J.* **10**, 466.

Wallace, W.S. and Mackenzie, R.D. (1947) *Mon. Bull. Minist. Hlth Lab. Serv.* **6**, 32.

Ward, L.R., Rowe, B. and Threlfall, E.J. (1982) *Lancet* **ii**, 705.

Warren, S.H. (1941) *Publ. Hlth, Lond.* **54**, 139.

Watkins, J.F. (1959) *J. R. Army med. Cps* **105**, 167.

Watson, K.C. (1954) *J. clin. Path.* **7**, 305; (1955*a*) *J. Lab. clin. Med.* **46**, 128; (1955*b*) *J. clin. Path.* **8**, 52; (1955*c*) *Ibid.* **8**, 55; (1955*d*) *J. Lab. clin. Med.* **45**, 97; (1957) *Amer. J. trop. Med. Hyg.* **6**, 72; (1978) *J. clin. Microbiol.* **7**, 122.

Weinfurter, F. (1915) *Zbl. Bakt.* **75**, 379.

Welch, S.W., Dehler, S.A. and Havens, L.C. (1925) *J. Amer. med. Ass.* **85**, 1036.

Widal, G.F.I. and Sicard, A. (1896) *Bull. Soc. Méd. Paris,* 3rd Ser. **13**, 681.

Wilson, G.S. (1967) *The Hazards of Immunization,* p. 265. Athlone Press, London.

Wilson, J.E. (1956) *Vet. Rec.* **68**, 748.

Wilson, J.F. (1945) *J. Hyg., Camb.* **44**, 129.

Wilson, M.M. and Mackenzie, E.F. (1955) *J. appl. Bact.* **18**, 510.

Wilson, W.J. and Blair, E.M.McV. (1927) *J. Hyg., Camb.* **26**, 374; (1931) *Ibid.* **31**, 138.

Wilson, W.J. and Darling, G. (1918) *Lancet* **ii**, 105.

Woodward, T.E., Smadel, J.E., Ley, H.L., Green, R. and Mankikar, D.S. (1948) *Ann. intern. Med.* **29**, 131.

Woodward, W.E. (1980) *Trans. R. Soc. trop. Med. Hyg.* **74,** 553.

Wray, C., Sojka, W.J., Morris, J.A. and Morgan, W.J.B. (1977) *J. Hyg., Camb.* **79,** 17.

Wright, A.E. (1902) *Lancet,* **ii,** 651; (1904) *A Short Treatise on Antityphoid Inoculation.* Constable, Westminster; (1908) *Zbl. Bakt.* **46,** 188.

Wright, A.E. and Semple, D. (1897) *Brit. med. J.* **i,** 256.

Wyllie, J. (1932) *J. Hyg., Camb.* **32,** 375.

69

Bacillary dysentery

M.T. Parker

Introductory

Dysentery is generally defined as an inflammation of the bowel with the passage of blood in the stools. Until the last quarter of the nineteenth century it was regarded as a single, well defined disease. In 1875 Loesch demonstrated the presence of parasitic amoebae in the stools of dysenteric patients, and in 1883 Robert Koch observed these amoebae in the intestinal wall in patients dying of dysentery in Egypt. In 1898-1901, Shiga in Japan, Flexner in the Philippines and Kruse in Germany brought evidence to show that another form of dysentery was caused by bacteria that now form the genus *Shigella* (for references, see Chapter 36). This disease is called bacillary dysentery. It is an acute diarrhoeal disease characterized in the more severe infections by the presence of blood and mucus in the stools. These symptoms result from acute inflammation of the mucosa and submucosa of the large

intestine, often with ulceration.

As described in Chapter 36, there are four subgroups or species of shigellae: (1) subgroup 1 (*Sh. dysenteriae*), with 10 serotypes, of which serotype 1 is traditionally known as Shiga's bacillus, serotype 2 is often called Schmitz's bacillus, and the rest are generally designated by their serotype numbers; (2) subgroup 2 (*Sh. flexneri*), with six serotypes and various subtypes; (3) subgroup 3 (*Sh. boydi*), with 14 serotypes; and (4) subgroup 4 (*Sh. sonnei* or Sonne's bacillus), with a single serotype. All of these organisms cause bacillary dysentery.

The shigellae are closely related to *Escherichia coli*, certain strains of which—often described as '*entero-invasive*'—are known to cause a disease clinically very similar to bacillary dysentery. This we shall discuss briefly at the end of the present chapter (p. 442).

Dysentery due to *Shigella* species

Pathogenesis and clinical manifestations

Bacillary dysentery is a local disease mainly of the large bowel in which the causative organism is usually

confined to the gut wall and the mesenteric lymph glands. Its effects are chiefly on the lower part of the colon, and range from acute inflammation to shallow ulceration, which may be extensive. Virulent strains of shigellae penetrate the mucosa and multiply beneath the lamina propria (Chapter 36). The ability of shigellae to invade epithelial cells is associated with the presence of a 140-megadalton plasmid (Sansonetti *et al.* 1981, 1982). In the bowel wall, shigellae cause necrosis of cells. This may in large part be due to the action of the exotoxin first detected in strains of Shiga's bacillus but formed, though apparently in smaller amount, by other shigellae. This substance—often called 'Shiga toxin'—is cytotoxic in tissue cultures and inhibits intracellular protein synthesis.

The incubation period is short—about 48 hr as a rule. In the typical case there is diarrhoea—often bloody—and tenesmus. The severity of the disease varies widely; a considerable proportion of the patients suffer from a simple diarrhoea, but in all but the mildest infections there is usually some constitutional disturbance. In severe infections, toxaemia is profound. Koster and his colleagues (1978) observed acute haemolytic anaemia, thrombocytopaenia, oliguria and sometimes kidney failure in a number of severe bacillary dysentery infections in children in Bangladesh. Bacterial endotoxin could be detected by means of the *Limulus* test (Chapter 57) in the blood of one-half of the patients. It seems likely that this so-called '*haemolytic-uraemic syndrome*' is attributable to the systemic effects of endotoxin released from the gut wall. In severe bacillary dysentery there is often leucocytosis exceeding 50 000 cells per mm³ with a gross preponderance of granulocytes, and immune complexes can often be detected in the blood. Scragg and his colleagues (1978) detected shigellae in the blood of some 2 per cent of black South African children admitted to hospital with bacillary dysentery; bacteraemia was said to be even more common in some epidemics of severe infection with Shiga's bacillus.

In typical cases of severe bacillary dysentery toxaemia overshadows dehydration. Nevertheless, in some patients diarrhoea is profuse and watery and there is considerable fluid loss. This is sometimes the predominant clinical picture in food-borne outbreaks, particularly those caused by Sonne's bacillus (Rohleder 1938, Report 1942*a*). The mode of action of shigellae on the small bowel is not understood. It will be recalled (Chapter 36) that Shiga toxin causes the dilatation with fluid of ligated loops of rabbit ileum. In experimental oral infection of monkeys with Flexner's bacillus (Rout *et al.* 1975) some animals suffer dysentery, some watery diarrhoea and some both. However, mutants of Shiga's bacillus that form toxin but do not invade the mucosa of the colon fail to cause disease. There are no macroscopic lesions in the small intestine in human bacillary dysentery and shigellae have not been seen to invade the jejunum in experimental dysentery. Thus the role of the toxin in causing fluid loss from the small intestine is uncertain.

Some convalescents from bacillary dysentery subsequently suffer chronic relapsing diarrhoea, but this is uncommon except in infections with Shiga's bacillus. Entirely symptomless infections are frequent, particularly in areas of high endemicity (p. 439). In severe shigella infections, conjunctivitis is sometimes seen early in the second week and iritis a few days later; arthritis and neuritis are likewise sometimes met with (Manson-Bahr 1942). Vulvovaginitis has been recorded (McGinness and Telling 1950).

The case-fatality rate varies widely according to circumstances and age. In the first 3 months of life it tends to be high; it decreases gradually until about the age of 3 years. It is negligible in most western countries; in Britain it has not exceeded 0.1 per cent for many years. In some parts of the world, however, as many as one-quarter of malnourished children with bacillary dysentery in hospital may die (see, for example, Scragg *et al.* 1978). Deaths from bacillary dysentery are particularly common in times of famine and in some epidemics, particularly those caused by Shiga's bacillus. Males are said to have a uniformly higher fatality rate than females at all ages (Glover 1949).

Epidemiology

Bacillary dysentery attracts most attention when it appears in epidemics, but it also occurs endemically in most parts of the world. It is commoner in countries with a warm than a temperate climate, and in 'poor' than in 'affluent' countries. In adults, it has traditionally been associated with the overcrowding and bad hygienic conditions encountered in times of war and other disasters, and in gaols and mental institutions. Among groups of young children, however, it spreads easily under apparently good hygienic conditions, though it is a major cause of death in the young only under poor socio-economic circumstances.

The careful study carried out by Bojlén (1934) in Denmark showed that in that country endemic dysentery was widespread. Similar conclusions were reached in the United States as the result of surveys by McGinnes and his colleagues (1936) in Virginia, and Watt, Hardy and DeCapito (1942) in New Mexico, Georgia and Puerto Rico. A survey of Egyptian villages (Higgins *et al.* 1955*a*) suggested that each child suffered from at least one attack of bacillary dysentery every year, and that shigella infection was responsible for 35 per cent of all diarrhoeal episodes. The real incidence of the disease is difficult to assess, since in most countries only the severer cases are notified.

In England and Wales notifications, which were moderately high after the 1st world war, fell to a minimum of 345 in 1925 (see Table 69.1). After that they rose gradually to a peak of 4170 in 1938, fell steeply in the following year, increased to a yearly average of 7000 between 1941 and 1943, and reached a peak in the epidemic of 1945 when 16 273 cases were notified. For a few years after the end of the 2nd world war notifications returned to a lower level, but resumed their upward trend in 1950, and in 1956 exceeded 49 000. Between 1950 and 1970 there were wide fluctuations; the annual incidence, however, was never less than 10 000 cases. Since 1970 the annual notification rates have been lower, ranging from 3000 to 5000 in most years.

Table 69.1   Dysentery in England and Wales

Year	Notifications	Deaths
1919	1 657	435
1925	345	135
1935	1 177	95
1938	4 170	112
1940	2 860	185
1945	16 273	165
1947	3 802	81
1950	17 286	65
1954	31 858	39
1956	49 009	33
1958	38 107	32
1960	43 285	36
1970	10 765	7

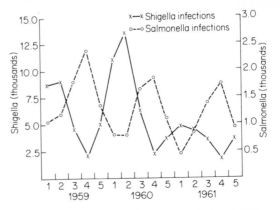

Fig. 69.1   Seasonal incidence of shigella infections in Britain 1959–61 (Sonne and Flexner infections only). Number of infected persons—including symptomless excreters—reported to the Public Health Laboratory Service by laboratories in the United Kingdom, in 10-week periods (excluding weeks 51, 52 and 53). The corresponding figures for infections with salmonellae—other than *Salm. typhi* and *Salm. paratyphi B*—are also shown.

The increased availability of laboratory facilities in England and Wales after 1945 certainly contributed to the high rates of notification in the 1950s, but there is good evidence that mild shigella infections increased considerably at that time. The more recent fall in the notification rate may not indicate a corresponding change in the incidence of infection. The fall in the size of the child population, the great reduction in the number of day nurseries under the direct supervision of local health authorities, and decisions no longer to investigate epidemics of mild diarrhoea in schools may each have contributed to the present low rates of notification.

Young children are more liable than older persons to acquire shigella infections and when infected to suffer from clinical disease (Charles and Warren 1929, Bojlén 1934, Hardy and Watt 1945*a*, Shaw 1953, Floyd *et al.* 1956*a, b*, Scott 1959). Infection is usually rather less common in the first 6 months of life than in the next 2 years (Hardy and Watt 1945*a, b*, Floyd *et al.* 1956*a*), but the illnesses which do occur in young infants are often severe (Haltalin 1967). Four out of every five children under 9 years of age, but only one out of two older persons, who become infected suffer from diarrhoea (Shaw 1953). In early life the attack rate is higher in males than females, but after about 20 years of age the incidence is reversed.

In many countries dysentery is a disease of the summer months. Hardy and Watt (1948) found this to be so in New Mexico and Georgia but not in New York. In Great Britain before 1940 the months of peak incidence were the late summer and early autumn (Blacklock and Guthrie 1937, Carter 1937), but in the years 1941–8 a late winter prevalence appeared (Glover

1949). In subsequent years the February–April peak was further accentuated and the summer prevalence almost entirely disapppeared (Taylor 1957). This alteration in the seasonal incidence of dysentery wås accompanied by other changes in the behaviour of the disease and its mode of spread (see Fig. 69.1 and p. 437).

In some tropical countries the effect of season is more complex. According to Rogers (1913), for example, bacillary dysentery in India was at a minimum in the cool months of January and February. With the onset of the hot weather in March there was a small rise, followed, however, by a fall during the very hot months of May and June. The maximum increase followed closely the beginning of the heavy monsoon rains in July, and the incidence reached its highest point in August and September. With the cessation of the rains late in October and the fall of temperature, the incidence steadily declined.

It is of interest to note that the months when dysentery is most common in sub-tropical and tropical countries are also the months when *flies* are most abundant.

The suggestion that *flies* might carry infection mechanically was first put forward in pre-bacteriological days by Snow (1855). A friend of his had observed that, when an infusion of quassia had been placed in a room for poisoning flies, the bread and butter sometimes tasted bitter. This suggestion has been abundantly confirmed for bacillary dysentery by later workers. For example, Manson-Bahr (see Bahr 1912), working in Fiji, found that dysentery was most prevalent when the fly population was at its maximum. Shiga's bacillus was actually isolated from the lower

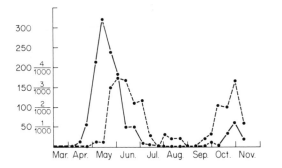

Fig. 69.2 The relationship between the prevalence of flies and the incidence of dysentery in the Salonika Expeditionary Force. Continuous curve = number of flies trapped *per diem*; interrupted curve = incidence of dysentery cases per 1000 *per diem*. (After Taylor.)

intestinal tract of two flies caught on the bed of a patient suffering from acute dysentery. During the 1st world war Taylor (1919) noted that in the Salonika Expeditionary Force the maximum incidence of dysentery occurred when flies were most abundant (Fig. 69.2). Experiments made to test the ability of flies to carry dysentery bacilli showed that these insects, after feeding on milk cultures or on dysenteric faeces, might remain infective for 24 hr. Of 1670 flies caught under natural conditions, 1 from a hospital kitchen was carrying Shiga's bacillus, and 8 were carrying Flexner's bacillus (see also Manson-Bahr 1919). Though dysentery bacilli may be found occasionally in the intestinal contents of these insects, they do not appear to survive for more than a few days. The probability is that flies act mainly by carrying infective material on their feet rather than in their intestine.

The incidence of shigella infection in certain areas in the southern United States was lowered temporarily by spraying DDT (Watt and Lindsay 1948, Lindsay *et al.* 1953). Later, the flies became resistant to the insecticide, but the provision of bore-hole latrines, which denied them access to human excreta, again reduced the incidence of dysentery (McCabe and Haines 1957). It is unlikely that flies play much part in the transmission of either the epidemic or the endemic disease in countries with proper arrangements for the disposal of sewage. In Great Britain, as has been pointed out, dysentery has for some years been more a winter than a summer disease.

Sources and spread of infection

Epidemics of infection may occasionally result from the contamination of milk or ice-cream but are more often attributable to the eating of raw or ready-cooked foods, which are liable to contamination from fingers or by flies. Numerous water-borne epidemics are on record (see, for example, Wade 1922, Report 1938, Kinnaman and Beelman 1944, Drachman *et al.* 1960). In several instances the causative organism has been isolated from the water itself (Green and Macleod 1943, Ross and Gillespie 1952, Baine *et al.* 1975). Merson and co-workers (1975) describe an outbreak on a cruise-liner in which the ship's water supply was implicated. In Indonesia, dysentery was said to be predominantly a water-borne disease (Fairlie and Boyd 1943). Swimming in contaminated river water may occasionally lead to shigella infection (Rosenberg *et al.* 1976a). Black and his colleagues (1978) traced 72 food-borne and 38 water-borne outbreaks of bacillary dysentery in the USA in the years 1961–75; salads, and small, poorly controlled water supplies were respectively the commonest vehicles. It is unlikely, however, that spread by food or water makes more than a minor contribution to the total of shigella infections in Britain or most western countries.

Infection is derived as a rule from cases, particularly ambulant cases, of the disease, from healthy convalescents, from symptomless carriers—usually temporary—and probably far less often, except in mental institutions, from chronic carriers. The organisms, which are excreted in the faeces, may gain access to food through the imperfectly cleansed fingers of the patient or carrier; or they may pass from one person to another by contact with inanimate articles.

The available evidence suggests that most of the shigella infection that occurs nowadays in Great Britain is spread from person to person. It usually occurs during the cold weather, and is more common in the industrial conurbations than in the small towns and rural districts (Bradley and Richmond 1956, Taylor 1957). The annual winter peak of incidence represents the sum of a number of local prevalences. These often begin, early in the winter, with an outbreak in a primary school, which is followed by many secondary infections among the siblings who attend other schools. In due course, further school outbreaks occur, and the disease may spread to surrounding neighbourhoods where the process is repeated. Secondary epidemics may occur in hospitals and nurseries. Only a few of those affected are seriously ill, and many are quite symptomless. It is not unusual to find that a quarter or even half of the apparently healthy contacts in a school or nursery are excreting shigellae. There is reason to believe, however, that the completely healthy excreter is a much less serious source of infection than the clinical case (Davies 1952, Ludkin 1955, Ross 1957). Weissman and co-workers (1974) in the USA draw attention to the importance of pre-school children as spreaders of Sonne dysentery; this they attribute to their high suceptibility to infection and the frequency with which diarrhoea develops when they are infected. In a study of diarrhoea in day-care centres for young children, Pickering and his colleagues (1981) identified 15 epidemics; of these, five were caused by shigellae, two by rotaviruses, one by *Giardia* and seven apparently by more than one agent.

It remains to be explained how the organism passes from one child to another. The most likely possibility is that the hands become contaminated from objects soiled with the faeces of ambulant patients. Hutchinson (1956) investigated the spread of infection by children with Sonne infection. She found that objects such as clothes, toys, furniture and dust were seldom contaminated, but that *Sh. sonnei* could be isolated frequently from places and objects associated with the disposal of faeces, and particularly from the seats of water closets. Part of the contamination arose from splashing which occurred when a lavatory bowl containing liquid faeces was flushed. Sonne's bacillus survived for many days on lavatory seats, particularly in conditions of low temperature, high humidity and subdued lighting. These observations are particularly illuminating in that they provide a possible explanation of the winter prevalence of dysentery.

Earlier in this century dysentery was endemic in many mental hospitals in Europe (Kruse 1901, Mott 1901–2). Bojlén (1934) found that, in contrast to endemic dysentery in the general population, it was characterized by a high case-fatality rate, ranging from about 10 to 20 per cent. Relapses, and probably reinfections, were very common, and as many as 50 per cent of the inmates might suffer from chronic dysentery. Persistent carriers were met with, and together with the numerous patients suffering or recovering from dysentery, they provided a continual source of infective material which was responsible for the further propagation of the disease. In the large mental hospitals of New York State, Hardy (1945) found that outbreaks were non-explosive and of long duration. Most of the clinical cases occurred early in the outbreaks; later, infection was maintained chiefly by symptomless excreters (see also DuPont *et al.* 1970).

The ease with which dysentery bacilli may be spread from person to person by contact suggests that the minimum infective dose must be low. In our experience, laboratory infections with Flexner and Sonne bacilli are common.

Monkeys are naturally susceptible to shigella infection, and have been found infected with most known serotypes. Their clinical condition varies from a rapidly fatal dysentery to symptomless infection. The commonest infecting organisms are *Sh. flexneri* types 4a and 4b; the latter is rarely found in man. Human infections have occasionally been acquired from monkeys (Carpenter and Sandiford 1952).

Bacteriological findings in different parts of the world

In the tropical and sub-tropical countries where dysentery is very common, a large number of different shi-

gella serotypes are present in endemic infections, and it is unusual for any one to predominate. Even at times of high incidence it is seldom that all the infections are caused by a single strain, though sometimes one may predominate. In small localized outbreaks, particularly in institutions, a single type may be found. The proportion of infections due to individual serotypes varies considerably from country to country, and in one country at different times, but the figures quoted by Boyd (1936, 1946) for the British Army in India in 1932–4 and in the Middle East in 1940–3 are fairly characteristic: Shiga's bacillus 10–20 per cent, Schmitz's bacillus 5–10 per cent, other members of subgroup 1 less than 3 per cent, *Sh. flexneri* about 60 per cent, *Sh. boydi* about 5 per cent, and *Sh. sonnei* a little less than 10 per cent. The great variety of different serotypes to be found in quite small populations is well illustrated by the surveys carried out by Floyd (1954, 1956) on Egyptian children. Among 291 positive cultures, obtained during one year from a group of 75 children in three villages, Floyd found Shiga's bacillus and Schmitz's bacillus, several other members of subgroup 1, 7 different Flexner types and subtypes, 4 Boyd types, and *Sh. sonnei*. In the course of three years' work in the same district he isolated every then known serotype of *Sh. boydi*.

In sub-tropical North America there is a good deal of endemic dysentery and a considerable diversity of shigella serotypes. At most times Flexner strains predominate. In the 1970s the northward spread of Shiga's bacillus in Central America led to the importation of many infections into the USA by returning travellers but little subsequent spread there. In the USA as a whole, *Sh. sonnei* replaced *Sh. flexneri* as the most important cause of dysentery after 1960 (Reller *et al.* 1970; see also Rosenberg *et al.* 1976*b*). More recently (Report 1981*b*, 1982*b*) some two-thirds of all isolations in the USA were of *Sh. sonnei* and one-quarter of *Sh. flexneri*.

In England during 1926–39, several Flexner serotypes and *Sh. sonnei* were endemic, Shiga's bacillus was no longer to be found, and Schmitz's bacillus was becoming uncommon. Dysentery occurred most frequently in the late summer, and was commoner in the country than the towns. Though more infections were diagnosed bacteriologically during the years 1939–45, *Sh. flexneri* became less and less common. By the end of the 2nd world war Sonne's bacillus was responsible for 87 per cent of all infections. After 1950 this figure rose to 95–98 per cent. *Sh. sonnei* is now also the predominant cause of dysentery in most part of Western Europe. Nearly all of the rather small number of infections with *Sh. dysenteriae*, *Sh. flexneri* and *Sh. boydi* identified nowadays in England and Wales are in persons recently returned from abroad (Gross *et al.* 1979). However, Flexner dysentery continued until quite recently to be as common as Sonne dysentery in some parts of Scotland (Hunponu-Wusu 1970).

Shiga's bacillus

This organism is now seldom met with except in countries with a warm climate or in persons returning from them, though it caused scattered epidemics in Northern Europe during and soon after the 2nd world war (Imelik and Imelik 1947, vor den Esche 1953). It has a reputation for causing a more severe disease than that produced by the other dysentery bacilli. This was certainly justified for infections in troops in the Middle East in 1914–18 but not, in our experience, in India during the 2nd world war. More recently, large epidemics of Shiga dysentery, with very high mortality rates, have been seen in many countries. The first of these was in Somalia in 1963–4 (Cahill *et al.* 1966). In 1968 Shiga dysentery reappeared in epidemic form in Central America after a long period of infrequency. The epidemic began in Guatemala and subsequently spread to El Salvador, Honduras, Nicaragua and Mexico (Gangarosa *et al.* 1970, Gangarosa 1971). It was estimated that 112 000 cases, with 8300 deaths, occurred in 10 months in Guatemala alone. The epidemic lasted about 3 years and then ceased spontaneously. Outbreaks of infection of a similar nature have since occurred in Bangladesh, Sri Lanka and parts of Central Africa (for references see Chapter 36 and Frost *et al.* 1981). All of these incidents were characterized by infections of exceptional severity, high attack rates in the affected population and the presence of resistance to several antibiotics in the causative strains.

In severe attacks of Shiga dysentery the abdominal symptoms are particularly distressing and there is much toxaemia. Recovery is slow, patients often remain debilitated for months, and diarrhoea tends to recur on resuming physical activity (Fletcher and Mackinnon 1919). Dysenteric arthritis (Graham 1919, Report 1919) occurs almost only in infections with Shiga's bacillus.

Schmitz's bacillus

This organism was first encountered in Eastern Europe in a winter epidemic of diarrhoea among prisoners of war (Schmitz 1917); the morbidity rate was high but deaths were few. Dysentery due to this organism was endemic in Britain from the end of the 1st world war but was never very common. Several epidemics were seen in mental hospitals (Evans 1938, Report 1942*b*, Shera 1943), but no infections have arisen in the general population for a number of years. Elsewhere in the world it is an infrequent cause of dysentery.

Sonne's bacillus

There is every reason to believe that *Sh. sonnei* was endemic in Europe and America before Sonne's description in 1915 drew attention to it. In an investigation of dysentery in Norway, Thjøtta (1919) found that 25 out of 65 strains isolated conformed to the Sonne type; and in Denmark Bojlén (1934) found it to be as common as Flexner's bacillus in the causation of endemic and institutional dysentery. In Great Britain the organism has been recognized since about 1925 (see Channon 1926, Kerrin and Cruickshank 1926), and from 1930 onwards it was probably as common as Flexner's bacillus (Carter 1937). The two diseases occurred together endemically, and local prevalences of infection in which both played a part were common. There was little difference between the clinical manifestations of Sonne and Flexner dysentery (Blacklock and Guthrie 1937). At this time, as well as endemic infections and institutional outbreaks, there were numerous milk-borne outbreaks of Sonne dysentery (*e.g.* Bowes 1938, Faulds 1942, 1943), but they became progressively less common, and only six outbreaks of dysentery due to any foodstuff were recorded in the decade 1945–54 (Taylor 1957). Since 1945 the epidemiology of dysentery in England has been essentially that of Sonne dysentery. Between then and 1960 there was an irregular upward trend in notifications associated with a falling mortality, so that many now consider the disease as little more than a nuisance. Occasional severe infections occur, and in infants may take the acute toxic form known in Japan as *Ekiri*.

Dysentery carriers

Patients recovering from an acute attack of dysentery may continue to excrete the bacilli in their stools. Generally this excretion lasts for only a few weeks, but a small proportion become persistent carriers.

Fletcher and Mackinnon (1919), who studied large numbers of dysentery convalescents, found that carriers of Shiga's bacillus nearly always remained infected for a number of months, often suffered from chronic dysentery, and were subject to frequent clinical relapses. Carriage of Flexner's bacillus, on the other hand, was less often prolonged and more often intermittent, and the patients usually remained in moderately good health. Bojlén (1934) observed many persistent carriers of *Sh. flexneri* in Danish mental institutions, but it is the experience of most workers that carriage for more than a year is exceptional.

The use of good selective media such as deoxycholate citrate or SS agar has established that healthy carriers are quite common in some communities. Hardy, Watt and DeCapito (1942), for example, found that in an endemic area as many as 3.8 per cent of persons who gave no history of illness during the preceding year were carriers of the dysentery bacillus. Higgins, Floyd and Kader (1955*a*) found that Egyptian village children suffered from 2.8 fresh shigella infections a year, half of them without symptoms. In Egyptian adults, only 13 per cent of all shigella infections were accompanied by diarrhoea (Floyd *et al.* 1956*b*). In England and Wales 0.4 per cent of healthy children aged 0–5 years were found to excrete *Sh. sonnei* (Report 1959). Even when good selective media are used it is not uncommon to observe intermittency of excretion in some convalescents and symptomless carriers (Cruickshank and Swyer 1940, Report 1942*c*). Watt, Hardy and DeCapito (1942), working with SS and deoxycholate citrate media, found that 11 per cent of Flexner and 7 per cent of Sonne carriers excreted the bacillus for over ten weeks. There is no difference between the rate of clearance in convalescent cases and symptomless excreters of *Sh. sonnei*, but excretion is rather more prolonged in young children than in older persons (Shaw 1953, Scott 1959).

Diagnosis

Blood and urine cultures are usually sterile. In the bacteriological diagnosis of dysentery most stress should be laid on the isolation of the specific organism from the faeces, less on the presence of agglutinins in the patient's serum. Before the introduction of deoxycholate media great care had to be exercised in the plating of washed fragments of mucus in order to isolate shigellae from acute cases of dysentery. With modern methods of culture such precautions are unnecessary, but it must be remembered that shigellae survive less well than do salmonellae in the faeces, particularly at high atmospheric temperatures. Stools should preferably be transported by hand to the laboratory and plated without delay. If this is impossible, the mucus or faeces should be collected into a buffered 30 per cent glycerol saline solution, which prevents the dysentery bacilli from being destroyed by the acid produced during the growth of other organisms. The dried filter-paper method (Chapter 68) is less satisfactory for shigellae than for salmonellae (Bailey and Bynoe 1953, Kirsche and Prévot 1953). Rectal swabs are often of value in acute cases, or when special care has to be exercised to avoid substitution, or, in hospitals, in order to avoid contamination of the sample by collection in imperfectly sterilized bed-pans. They should always be cultured at once.

Microscopical examination of the mucus in the early stage of acute bacillary dysentery shows a striking degree of cellularity, 90 per cent or more of the cells, apart from red blood corpuscles, being polymorphonuclears. In the later stages the number of cells decreases, the pus cells become degenerate and proportionately fewer, the red blood corpuscles are scanty or absent, mononuclear cells increase, and the exudate, in coming to resemble that found in amoebic dysentery, loses its diagnostic significance.

Cultures should be made on Leifson's (1935) deoxycholate citrate agar or on the so-called SS medium, both of which are very much superior, especially in the examination of convalescents and contacts, to Mac-

Conkey's or Endo's medium (see Hardy *et al.* 1939, 1942, Hynes 1942). Deoxycholate citrate plates are examined after overnight incubation for colonies resembling shigellae. Many workers carry out a preliminary identification by slide agglutination from the growth on the primary plates. In any event, full biochemical and serological tests must be performed on subcultures that have been checked for purity. Serological typing is done by slide agglutination with absorbed sera.

Brilliant green media, including Wilson and Blair's medium, and tetrathionate broth are unsuitable for the growth of shigellae. *Sh. sonnei* and *Sh. flexneri* type 6 survive in selenite broth (Armstrong 1954, Hughes 1955), but other shigellae do not. Park and his colleagues (1977) describe an enrichment medium for shigellae that contains 4-chloro-2-cyclopentyl-β-D-galactopyranoside; lactose-fermenting organisms form from it an aglycone that is toxic for them but not for the shigellae. For further information about the identification of dysentery bacilli, see Carpenter (1968) and Chapter 36.

Serological methods *Agglutinins* may appear in the patient's serum between the 6th and the 12th day, but sometimes not until later. Their detection is not a reliable means of making a diagnosis of dysentery, because normal human serum may contain agglutinins, particularly in endemic areas (Watt and DeCapito 1945), and because a negative result does not exclude the possibility of a recent infection.

The *indirect haemagglutination test*, carried out with red blood cells treated with boiled supernates of shigella broth cultures or with partly purified lipopolysaccharide, has proved a more sensitive and specific test for antibody (Neter and Walker 1954, Neter and Dunphy 1957, Klečková-Aldová 1958, Putožnik *et al.* 1958, Havlík *et al.* 1959, Neter *et al.* 1962, Gotoff *et al.* 1963, Mata *et al.* 1970, Patton *et al.* 1976). The results in Shiga, Flexner and Sonne infections are said to be specific, and polyvalent antigens can be prepared. Raised titres can often be detected in the first week of illness, and sometimes also in symptomless infections; but by no means all patients with dysentery show an increase in antibody.

Serological methods are now little used in the diagnosis of dysentery but may occasionally be of value in the retrospective investigation of outbreaks.

Typing Much attention has been devoted to the perfection of *typing methods* for the subdivision of certain shigella organisms. These methods are useful for tracing the source of infections in communities where a single serotype is predominant, and are therefore particularly appropriate for the investigation of Sonne dysentery. The bacteriophage and colicine methods were described in Chapter 36. (See Helgason and Old 1981, and Old *et al.* 1981, for the use of resistotyping.)

Prophylaxis and treatment

Prophylaxis

The general prophylaxis of bacillary dysentery is so similar to that of typhoid fever that there is no need to consider it in detail. Particular attention should be paid to the extermination of flies, to the protection of food from them and from unwashed hands, and to the safe disposal of human excreta. Persons who have had dysentery should not act as cooks, or have anything to do with the preparation or serving of food, until at least three samples of faeces, taken at intervals of not less than two days, have proved negative for dysentery bacilli; but the examination of clearance samples should not be begun until after clinical recovery, and until at least three days after specific antimicrobial treatment has ceased (see Report 1970).

If an outbreak occurs in an institution, it is of course necessary to exclude the possibility of food- or water-borne spread by careful epidemiological enquiry and to examine the faeces of those who prepare or handle the food. However, the discovery of excreters of the causative organism on the staff may be misleading; often they are also victims of the outbreak. Nevertheless it may be necessary to remove them temporarily from their duties to exclude the possibility of the further spread of infection. In most outbreaks in hospitals, schools and nurseries the infection is not food-borne, and attention should be concentrated on preventing infection by indirect contact via hands and other contaminated objects. Care should be taken in disinfecting bed-pans, lavatory seats, door handles and water-flushing devices. Crockery and feeding utensils should be sterilized by heat after use. All persons must be instructed to wash their hands in a disinfectant solution after visiting the lavatory, whether defaecation takes place or not, and also before eating.

The value of exclusion and isolation in the control of dysentery in day nurseries and schools is not so clear. There is little doubt that early detection of the first case in a day nursery, and its prompt exclusion, is the most important single measure for the prevention of epidemics (Report 1955), but this requires good bacteriological facilities and considerable vigilance on the part of the staff. By the time that several cases have occurred, control by exclusion is unlikely to be successful even if the symptomless excreters are sought and removed. Epidemics in day schools are seldom detected early enough to warrant an attempt to exclude all infected persons. There is evidence that, under good hygienic conditions, the symptomless excreter is less infectious than the clinical case. The ravages of dysentery in schools may therefore be mitigated somewhat by the prompt exclusion of all persons with diarrhoea until their symptoms have ceased, together with a strict control of hand-washing and of the hygiene of the lavatories. It is doubtful whether a more radical policy will give much better results. Those interested in the control of Sonne dysentery in institutions should consult the review by Bradley and his colleagues (1958).

Vaccination

The value of vaccination in the prophylaxis of dysen-

tery is still undecided, but the chances of success now appear somewhat more hopeful than they were a few years ago.

There is no evidence that the parenteral administration of killed vaccines contributes to the prevention of the natural disease (Hardy *et al.* 1948, Higgins *et al.* 1955*b*). In monkeys (Formal *et al.* 1967*b*) subcutaneous vaccination with heat- and acetone-killed vaccines and even with living, virulent organisms of *Sh. flexneri* type 2a gave no protection against oral challenge with the same organism.

On the other hand, Mel and his colleagues (1965*a*) found that, in mice kept on Freter's regimen (Chapter 36), the oral administration of a live vaccine conferred protection against subsequent homologous challenge by the same route. This led them to investigate the possibility of oral immunization in man with attenuated strains of dysentery bacilli. They found (Mel *et al.* 1965*b*) that a streptomycin-dependent mutant of *Sh. flexneri* type 2a very seldom reverted to streptomycin independence and could safely be given to volunteers. In a controlled trial on soldiers there was statistically significant evidence of type-specific protection against clinical dysentery, but the carrier rate was unaffected (Mel *et al.* 1965*c*). Five doses of vaccine were given over a period of 2 weeks; if the dosage was suitably graded even mild diarrhoeal reactions were unusual. In a later trial, a combined vaccine of two Flexner serotypes gave protection against both organisms (Mel *et al.* 1968). Studies of oral vaccines of streptomycin-dependent organisms by another group of workers (Levine *et al.* 1974, 1976) confirmed their safety but gave less clear evidence of their efficacy.

Other types of attenuated living vaccine have been used. These include (1) spontaneous mutants that have lost the ability to penetrate the intestinal mucosa; (2) *Shigella–Esch. coli* hybrids that can penetrate the mucosa and enter the lamina propria but are unable to multiply there; and (3) hybrids of (1) with *Esch. coli* (Formal *et al.* 1965, 1966, 1967*a*, DuPont *et al.* 1972). When given orally to man in doses of 10^{10}, vaccine (1) gives rise to virulent mutants; and vaccine (2) is itself virulent for man in doses exceeding 10^8. But the mutant-*Esch. coli* hybrid vaccine (3) can be given safely in doses of $5–10 \times 10^{10}$ (DuPont *et al.* 1972). Vaccines (1) and (2) rendered monkeys resistant to oral challenge with the homologous strain, but only vaccine (2) did this after a single dose. Istrati and his colleagues (1967, 1973; see also Meitert *et al.* 1973) used a spontaneous non-pathogenic mutant of *Sh. flexneri* serotype 2a as an oral vaccine without ill effect. They claim modest success in preventing dysentery in institutionalized children, as a means of shortening the period of excretion in dysentery convalescents, and as an aid to the resolution of the rectosigmoid lesions in chronic dysentery (see also Sultanov *et al.* 1981). Yet another approach has been explored by Formal and his colleagues (1981), who transferred the plasmid that codes for the production of the form-1 antigen of *Sh. sonnei* into the *gal E* mutant of *Salm. typhi* that is used as a live vaccine (Chapter 68). Giving this modified strain to mice conferred protection on them against both *Sh. sonnei* and *Salm. typhi*.

Treatment

In earlier editions of this book we discussed the value of a number of antimicrobial agents in the treatment of bacillary dysentery. The evidence was far from conclusive but suggested that a number of them—sulphadiazine, streptomycin, chloramphenicol, the tetracyclines and ampicillin—caused some amelioration of the clinical condition and reduced somewhat the period of excretion if the causative organism was sensitive *in vitro* to the agent given. The effects were, however, far from dramatic. It therefore seemed justifiable to give one or other of these drugs to patients with severe toxic attacks of bacillary dysentery but not to the majority of more mildly affected patients.

Resistance to these drugs is now so prevalent worldwide that the choice of a suitable agent often presents great difficulties. For example, the highly virulent strains of Shiga's bacillus that have spread widely in Central America, Asia and Africa were all resistant to sulphonamides, streptomycin and tetracycline and nearly all of them to chloramphenicol and ampicillin as well (Frost *et al.* 1981). Gross and his colleagues (1981) examined 2370 strains of shigellae other than *Sh. sonnei* isolated in the United Kingdom in the period 1974-8—nearly all of them from infections acquired overseas—and found that 76 per cent were resistant to sulphonamides, 58 per cent to streptomycin, 36 per cent to tetracycline, 13 per cent to chloramphenicol and 12 per cent to ampicillin; resistance to cephaloridine, gentamicin, neomycin, nalidixic acid, furazolidone and trimethoprim was uncommon. As we saw (Chapter 36), British strains of *Sh. sonnei*

have for some years been predominantly resistant to ampicillin, sulphonamides and streptomycin, and often resistant to tetracycline; most of them are still sensitive to neomycin and trimethoprim.

Treatment, when required, should ideally be based upon the results of sensitivity tests on the causative shigella, but these may not be readily available in many parts of the world; in any event it is desirable to begin treatment immediately in severe infections. The usual practice is to select an agent to which the prevalent shigellae in the area have been shown to be sensitive. Often the choice is limited, and it may be further restricted by the cost of the newer antibiotics. Thus, in large epidemics of severe infection, the shigella strain may be heavily exposed to a single agent and rapidly acquire resistance to it. In an epidemic of infection with Shiga's bacillus in Zaire (Frost *et al.* 1982), the use of tetracycline was discontinued in favour of co-trimoxazole when it was found that the strain was resistant to this drug and to sulphonamides, streptomycin, chloramphenicol and ampicillin; within a few months 42 per cent of the organisms causing infection were resistant to trimethoprim.

Among the agents to which shigellae are at present generally sensitive, neomycin (or kanamycin) appears not to be very effective (Haltalin *et al.* 1968). Nalidixic acid (Lewis 1967) and its analogue oxolinic acid (Gordon *et al.* 1976) are possible alternatives, but resistant variants are said to appear rapidly *in vitro* and in patients. The use of rifampicin (Naveh *et al.* 1977) on a large scale is probably contraindicated in countries in which it is relied upon for the treatment of leprosy.

Prospects for the chemotherapy of bacillary dysentery are unlikely to improve unless the present indiscriminate use of antimicrobial agents can be halted (Report 1978, 1981*a*). Measures to combat malnourishment, to relieve dehydration (Report 1983), and to limit the effects of bacterial toxaemia are at present the most valuable ways of saving the lives of small children with severe bacillary dysentery.

Dysentery due to *Escherichia coli*

Some strains of *Esch. coli* cause a diarrhoeal disease that closely resembles shigella infection. Many of them were first recognized because they were biochemically atypical (Chapter 34), being non-motile, anaerogenic or fermenting lactose late and thus appearing at first sight to be shigellae. It is now recognized, however, that strains of *Esch. coli* that are typical in these respects may cause dysentery-like disease; the most constant abnormality is that they usually fail to form lysine decarboxylase. Like shigellae, these strains give rise to keratoconjunctivitis when instilled into the conjunctival sac of guinea-pigs and penetrate and cause cytopathic changes in HeLa cells (Chapter 36). They

are referred to as *enteroinvasive* strains (Chapter 71).

The ability to cause dysentery-like disease is confined to members of some O serogroups of *Esch. coli*: 28a,c, 112a,c, 124, 136, 143, 144, 152, 164, and 167 (Sakazaki *et al.* 1967, Trabulsi *et al.* 1967, Ogawa *et al.* 1968, Trabulsi and de Toledo 1969, Ørskov *et al.* 1972, Report 1982*c*). The O antigens of these groups tend to cross-react with those of certain *Shigella* serotypes; that of *Esch. coli* O124 is identical with the O antigen of *Sh. dysenteriae* serotype 3. Strains of *Esch. coli* O124 have been responsible for outbreaks of diarrhoea in schools and hospitals in Britain (Hobbs *et al.* 1949, Rowe *et al.* 1974), for an extensive food-borne epidemic associated with cheese (Marier *et al.* 1973), and for many water-borne epidemics in Eastern Europe (Kétyi *et al.* 1958, Lányi *et al.* 1959, Kubinyi 1965, Hanny and Horvath 1966, Bezjak *et al.* 1972).

In general the symptoms of the diseases caused by enteroinvasive *Esch. coli* strains are similar to those of shigella infection but are less severe and of rather shorter duration.

It should be noted that some strains of *Esch. coli* that are not enteroinvasive appear to cause the passage of bloody stools. *Haemorrhagic colitis* (Report 1982*a*) is a sporadic disease of children associated with strains of serogroup O157 and possibly with members of other serogroups. These organisms do not form heat-labile or heat-stable enterotoxin but are said to produce a cytotoxin active on Vero cells (Chapter 71). O'Brien and her colleagues (1983) report that this toxin is neutralized by antiserum against purified Shiga toxin. Karmali and co-workers (1983) described a haemorrhagic-uraemic syndrome in a number of children who had recently suffered bloody diarrhoea associated with infection by *Esch. coli* strains of several serogroups (including O157) that formed a toxin active on Vero cells (see Chapter 71).

References

Armstrong, E.C. (1954) *Mon. Bull. Minist. Hlth Lab. Serv.* **13,** 70.

Bahr, P.H. (1912) *Dysentery in Fiji during the Year 1910.* Rep. Lond. School trop. Med.

Bailey, W.R. and Bynoe, E.T. (1953) *Canad. J. publ. Hlth* **44,** 468.

Baine, W.B. *et al.* (1975) *Amer. J. Epidem.* **101,** 323.

Bezjak, B., Dragăs, Z. and Jagić, H. (1972) *Zbl. Bakt.* **A219,** 302.

Black, R.E., Craun, G.F. and Blake, P.A. (1978) *Amer. J. Epidem.* **108,** 47.

Blacklock, J.W.S. and Guthrie, K.J. (1937) *J. Path. Bact.* **45,** 79.

Bojlén, K. (1934) *Dysentery in Denmark.* Bianco Lunos Bogtrykkeri A/S, Copenhagen.

Bowes, G.K. (1938) *Brit. med. J.* **i,** 1092.

Boyd, J.S.K. (1936) *J. R. Army med. Cps* **66,** 1; (1946) *J. Path. Bact.* **58,** 237.

Bradley, W.H. and Richmond, A.E. (1956) *Mon. Bull. Minist. Hlth Lab. Serv.* **15,** 2.

Bradley, W.H., Richmond, A.E., Shaw, C.H. and Taylor, I. (1958) *Med. Offr* **99,** 175.

Cahill, K.M., Davies, J.A. and Johnson, R. (1966) *Amer. J. trop. Med. Hyg.* **15,** 52.

Carpenter, K.P. (1968) *Identification of Shigella.* Assoc. clin. Path. Broadsheet No. 68, London.

Carpenter, K.P. and Sandiford, B.R. (1952) *Brit. med. J.* **i,** 142.

Carter, H.S. (1937) *J. Path. Bact.* **45,** 446.

Channon, H.A. (1926) *J. Path. Bact.* **29,** 496.

Charles, J.A. and Warren, S.H. (1929) *Lancet* **ii,** 626.

Cruickshank, R. and Swyer, R. (1940) *Lancet,* **ii,** 803.

Davies, J.B.M. (1952) *Brit. med. J.* **ii,** 191.

Drachman, R.H. *et al.* (1960) *Amer. J. Hyg.* **72,** 321.

DuPont, H.L., Hornick, R.B., Snyder, M.J., Libonati, J.P., Formal, S.B. and Gangarosa, E.J. (1972) *J. infect. Dis.* **125,** 5.

DuPont, H.L. *et al.* (1970) *Amer. J. Epidem.* **92,** 172.

Esche, P. vor dem (1953) *Arch. Hyg., Berl.* **137,** 20.

Evans, A.C. (1938) *Lancet,* **ii,** 187.

Fairley, N.H. and Boyd, J.S.K. (1943) *Trans. R. Soc. trop. Med. Hyg.* **36,** 253.

Faulds, J.S. (1942) *Mon. Bull. Emerg. publ. Hlth Lab. Serv.* **1,** July, p. 6; Sept., p. 5; (1943) *Ibid.* **2,** 143.

Fletcher, W. and Mackinnon, D.L. (1919) *Spec. Rep. Ser. med. Res. Coun., Lond.* No. 29.

Floyd, T.M. (1954) *Amer. J. trop. Med. Hyg.* **3,** 294; (1956) *J. Bact.* **71,** 525.

Floyd, T.M., Blagg, J.W. and Kader, M.A. (1956*b*) *Amer. J. trop. Med. Hyg.* **5,** 812.

Floyd, T.M., Higgins, A.R. and Kader, M.A. (1956*a*) *Amer. J. trop. Med. Hyg.* **5,** 119.

Formal, S.B., Baron, L.S., Kopecko, D.J., Washington, O., Powell, C. and Life, C.A. (1981) *Infect. Immun.* **34,** 746.

Formal, S.B., Kent, T.H., May, H.C., Palmer, A., Falkow, S. and LaBrec, E.H. (1966) *J. Bact.* **92,** 17.

Formal, S.B., LaBrec, E.H., Kent, T.H., May, H.C., Lowenthal, J.P. and Berman, S. (1967*a*) *Proc. Soc. exp. Biol., N.Y.* **124,** 284.

Formal, S.B., LaBrec, E.H., Palmer, A. and Falkow, S. (1965) *J. Bact.* **90,** 63.

Formal, S.B., Maenza, R.M., Austin, S. and LaBrec, E.H. (1967*b*) *Proc. Soc. exp. Biol., N.Y.* **125,** 347.

Frost, J.A., Rowe, B. and Vandepitte, J. (1982) *Lancet* **i,** 963.

Frost, J.A., Rowe, B., Vandepitte, J. and Threlfall, E.J. (1981) *Lancet* **ii,** 1074.

Gangarosa, E.J. (1971) *Salud publ. Méx.* **13,** 301.

Gangarosa, E.J., Perera, D.R., Mata, L.J., Mendizábal-Morris, C., Guzmán, G. and Reller, L.B. (1970) *J. infect. Dis.* **122,** 181.

Glover, J.A. (1949) *Mon. Bull. Minist. Hlth Lab. Serv.* **8,** 138.

Gordon, R.C., Stevens, L.I., Edmiston, C.E. Jr and Mohan, K. (1976) *Antimicrob. Agents Chemother.* **10,** 918.

Gotoff, S.P., Lepper, M.H. and Fiedler, M.A. (1963) *Amer. J. Hyg.* **78,** 261.

Graham, G. (1919) *Lancet* **ii,** 1030.

Green, C.A. and Macleod, M.C. (1943) *Brit. med. J.* **ii,** 259.

Gross, R.J., Rowe, B., Cheasty, T. and Thomas, L.V. (1981) *Brit med. J.* **283,** 575.

Gross, R.J., Thomas, L.V. and Rowe, B. (1979) *Brit. med. J.* **ii,** 744.

Haltalin, K.C. (1967) *Amer. J. Dis. Child.* **114,** 603.

Haltalin, K.C., Nelson, J.D., Hinton, L.V., Kusmiesz, H.T. and Sladoje, M. (1968) *J. Pediat.* **72,** 708.

Hanny, I. and Horvath, I. (1966) *Egészségtudomány,* Budapest **10,** 172.

Hardy, A.V. (1945) *Psychiat. Quart.* **19,** 377.

Hardy, A.V., DeCapito, T. and Halbert, S.P. (1948) *Publ. Hlth Rep., Wash.* **63,** 685.

Hardy, A.V. and Watt, J. (1945*a*) *Publ. Hlth Rep., Wash.* **60,** 57; (1945*b*) *Ibid.* **60,** 521; (1948) *Ibid.* **63,** 363.

Hardy, A.V., Watt, J. and DeCapito, T. (1942) *Publ. Hlth Rep., Wash.* **57,** 521.

Hardy, A.V., Watt, J., DeCapito, T.M. and Kolodny, M.H. (1939) *Publ. Hlth Rep., Wash.* **54,** 287.

Havlík, J., Kott, B. and Putožník, V. (1959) *J. clin. Path.* **12,** 440.

Helgason, S. and Old, D.C. (1981) *J. Hyg., Camb.* **87,** 339.

Higgins, A.R., Floyd, T.M. and Kader, M.A. (1955*a*) *Amer. J. trop. Med. Hyg.* **4,** 271; (1955*b*) *Ibid.* **4,** 281.

Hobbs, B.C., Thomas, M.E.M. and Taylor, J. (1949) *Lancet* **ii**, 530.

Hughes, M.H. (1955) *W. Afr. med. J.* N.S. **4**, 73.

Hunponu-Wusu, O.O. (1970) *Hlth Bull., Edinb.* **28**, 36.

Hutchinson, R.I. (1956) *Mon. Bull. Minist. Hlth Lab. Serv.* **15**, 110.

Hynes, M. (1942) *J. Path. Bact.* **54**, 193.

Imelik, B. and Imelik, S. (1947) *Z. Hyg. InfektKr.* **127**, 93.

Istrati, G. *et al.* (1967) *Zbl. Bakt.* **204**, 555; (1973) *Arch. roum. Path. exp. Microbiol.* **32**, 23.

Karmali, M.A., Steele, B.T., Petric, M. and Lim, C. (1983) *Lancet* **i**, 619.

Kerrin, J.C. and Cruickshank, J. (1926) *J. Path. Bact.* **29**, 315.

Kétyi, I., Kneffel, P. and Domján, J. (1958) *Zbl. Bakt.* **170**, 423.

Kinnaman, C.H. and Beelman, F.C. (1944) *Amer. J. publ. Hlth* **34**, 948.

Kirsche, P. and Prévot, M. (1953) *Bull. Soc. Path. exot.* **46**, 491.

Klečková-Aldová, E. (1958) *Čs. Epidem.* **7**, 188.

Koster, F. *et al.* (1978) *New Engl. J. Med.* **298**, 927.

Kruse, W. (1901) *Dtsch. med. Wschr.* **27**, 370, 386.

Kubinyi, L. (1965) *Egészségtudomány, Budapest* **9**, 84.

Lányi, B., Szita, J., Ringelhann, B. and Korách, K. (1959) *Acta microbiol. acad. sci. hung.* **6**, 77.

Leifson, E. (1935) *J. Path. Bact.* **40**, 581.

Levine, M.M., Gangarosa, E.J., Barrow, W.B. and Weiss, C.F. (1976) *Amer. J. Epidem.* **104**, 88.

Levine, M.M., Gangarosa, E.J., Werner, M. and Morris, G.K. (1974) *J. Pediat.* **84**, 803.

Lewis, M.J. (1967) *Lancet* **ii**, 953.

Lindsay, D.R., Stewart, W.H. and Watt, J. (1953) *Publ. Hlth Rep., Wash.* **68**, 361.

Ludkin, S. (1955) *Mon. Bull. Minist. Hlth Lab. Serv.* **14**, 126.

McCabe, L.J. and Haines, T.W. (1957) *Publ. Hlth Rep., Wash.* **72**, 921.

McGinness, G.F., McLean, A.L., Spindle, F. and Maxcy, K.F. (1936) *Amer. J. Hyg.* **24**, 552.

McGinnes, W.J. and Telling, R.C. (1950) *Brit. med. J.* **ii**, 1424.

Manson-Bahr, P. (1919) *J. R. Army med. Cps* **33**, 117; (1942) *Brit. med. J.* **ii**, 346, 374.

Marier, R., Wells, J.G., Swanson, R.C., Callaghan, W. and Mehlman, I.J. (1973) *Lancet* **ii**, 1376.

Mata, L.J., Gangarosa, E.J., Cácares, A., Perera, D.R. and Mejicanos, M.L. (1970) *J. infect. Dis.* **122**, 170.

Meitert, T. *et al.* (1973) *Arch. roum. Path. exp. Microbiol,* **32**, 35.

Mel, D.M., Arsić, B.L., Nicolić, B.D. and Radovanić, M.L. (1968) *Bull. World. Hlth Org.* **39**, 375.

Mel, D.M., Papo, R.G., Terzin, A.L. and Vukšić, L. (1965b) *Bull. World Hlth Org.* **32**, 637.

Mel, D.M., Terzin, A.L. and Vukšić, L. (1965a) *Bull. World Hlth Org.* **32**, 633; (1965c) *Ibid.* **32**, 647.

Merson, M.H. *et al.* (1975) *Amer. J. Epidem.* **101**, 165.

Mott (1901-2) *Trans. epidem. Soc., London,* **21**, 18.

Naveh, Y., Strahovsky, P. and Friedman, A. (1977) *Arch. Dis. Childh.* **52**, 960.

Neter, E. and Dunphy, D. (1957) *Pediatrics, Springfield* **20**, 78.

Neter, E., Harris, A.H. and Drislane, A.M. (1962) *Amer. J. clin. Path.* **37**, 239.

Neter, E. and Walker, J. (1954) *Amer. J. clin. Path.* **24**, 1424.

O'Brien, A.D., Lively, T.A., Chen, M.E., Rothman, S.W. and Formal S.B. (1983) *Lancet* **i**, 102.

Ørskov, F., Ørskov, I. and Furowicz, A.J. (1972) *Acta path. microbiol. scand.* **B80**, 435.

Ogawa, H., Nakamura, A. and Sakazaki, R. (1968) *Jap. J. med. Sci. Biol.* **21**, 333.

Old, D.C., Helgason, S. and Scott, A.C. (1981) *J. Hyg., Camb.* **87**, 257.

Park, C.E., Rayman, M.K. and Stankiewicz, Z.K. (1977) *Canad, J. Microbiol.* **23**, 563.

Patton, C.M., Gangarosa, E.J., Weissman, J.B., Merson, M.H. and Morris, G.K. (1976) *J. clin. Microbiol,* **3**, 143.

Pickering, L.K., Evans, D.G., DuPont, H.L., Vollet, J.J. III and Evans, D.J. Jr (1981) *J. Pediat,* **99**, 51.

Putožník, V., Kott, B. and Havlík, J. (1958) *Čs. Epidem.* **7**, 193.

Reller, L.B., Gangarosa, E.J. and Brachman, P.S. (1970) *Amer. J. Epidem.* **91**, 161.

Report (1919) *Spec. Rep. Ser. med. Res. Coun., Lond.* No. 40; (1938) *J. publ. Hlth Ass., Japan* **14**, June, p. 1; (1942a) *Mon. Bull. Emerg. publ. Hlth Lab. Serv.* **1**, Jan., p. 10; (1942b) *Ibid.* **1**, May, p. 7; (1942c) *Ibid.* **1**, Nov., p. 2; (1955) *Brit. med. J.* **ii**, 939; (1959) *Mon. Bull. Minist. Hlth Lab. Serv.* **18**, 86; (1970) *Publ. Hlth, Lond.,* **84**, 197; (1978) *Tech. Rep. Ser.* No. 624. World Hlth Org., Geneva; (1981a) *World Hlth Org.* No. WHO/BV1/PHA/ANT/82.1; (1981b) *Morbid. Mortal. wkly Rep.* **30**, 462; (1982a) *Ibid.* **31**, 580; (1982b) *Ibid.* **31**, 681; (1982c) *World Hlth Org. No.* WHO/CDD/BE1/82.4; (1983) *The Management of Diarrhoea and the Use of Oral Rehydration Therapy.* World Hlth Org., Geneva.

Rogers, L. (1913) *Dysenteries, their Differentiation and Treatment.* London.

Rohleder, S. (1938) *Bull. Hyg., Lond.* **13**, 865.

Rosenberg, M.L., Hazlet, K.K., Schaefer, J., Wells, J.G. and Pruneda, R.C. (1976a) *J. Amer. med. Ass.* **236**, 1849.

Rosenberg, M.L., Weissman, J.B., Gangarosa, E.J., Reller, L.B. and Beasley, R.P. (1976b) *Amer. J. Epidem.* **104**, 543.

Ross, A.I. (1957) *Mon. Bull. Minist. Hlth Lab. Serv.* **16**, 174.

Ross, A.I. and Gillespie, E.H. (1952) *Mon. Bull. Minist. Hlth Lab. Serv.* **11**, 36.

Rout, W.R., Formal, S.B., Giannella, R.A. and Dammin, G.J. (1975) *Gastroenterology* **68**, 270.

Rowe, B., Gross, R.J. and Allen, H.A. (1974) *Lancet,* **i**, 224.

Sakazaki, R., Tamura, K. and Saito, M. (1967) *Jap. J. med. Sci. Biol.* **20**, 387.

Sansonetti, P.J., Kopecko, D.J. and Formal, S.B. (1981) *Infect. Immun.* **34**, 75; (1982) *Ibid.* **35**, 852.

Schmitz, K.E.F. (1917) *Z. Hyg. InfektKr.* **84**, 449.

Scott, G.A. (1959) *Hlth Bull., Edinb.* **17**, 12.

Scragg, J.N., Rubridge, C.J. and Appelbaum, P.C. (1978) *J. Pediat.* **93**, 796.

Shaw, C.H. (1953) *Mon. Bull. Minist. Hlth Lab. Serv.* **12**, 44.

Shera, A.G. (1943) *Mon. Bull. Emerg. publ. Hlth Lab. Serv.* **2**, 52.

Snow, J. (1855) *On the Mode of Communication of Cholera,* 2nd edn. Churchill, London.

Sonne, C. (1915) *Zbl. Bakt.* **75**, 408.

Sultanov, G.V., Belaya, Y.A., Sadovskaya, T.A., Osmanova, M.M. and Gadzhieva, G.R. (1981) *Zh. Mikrobiol. Epidem. Immunobiol.* no. 10, 33.

Taylor, I. (1957) *Proc. R. Soc. Med.* **50**, 31.

Taylor, J.F. (1919) *Spec. Rep. Ser. med. Res. Coun., Lond.* No. 40.

Thjøtta, T. (1919) *J. Bact.* **4**, 355.

Trabulsi, L.R., Fernandez, M.R. and Zuliani, M.E. (1967) *Rev. Inst. Med. trop. S. Paulo.* **9,** 31.

Trabulsi, L.R. and Toledo, M.R. de (1969) *Rev. Inst. Med. trop. S. Paulo.* **11,** 358.

Wade, T.W. (1922) *Rep. publ. Hlth med. Subj., Minist. Hlth, Lond.* No. 14.

Watt, J. and DeCapito, T.M. (1945) *Publ. Hlth Rep., Wash.* **60,** 642.

Watt, J., Hardy, A.V. and DeCapito, T.M. (1942) *Publ. Hlth Rep., Wash.* **57,** 524.

Watt, J. and Lindsay, D.R. (1948) *Publ. Hlth Rep., Wash.* **63,** 1319.

Weissman, J.B., Schmerler, A., Weiler, P., Filice, G., Godbey, N. and Hansen, I. (1974) *J. Pediat.* **84,** 797.

70

Cholera

Graham Wilson

Introductory

The nomenclature of this disease is unsettled. So long as cholera was believed to be caused solely by the cholera vibrio, it presented no difficulty; but when it was found that disease clinically indistinguishable from cholera could be caused by *Vibrio eltor*, by non-agglutinable (NAG) vibrios, by *Vibrio parahaemolyticus*, and even by such other organisms as *Escherichia coli*, difficulty soon became apparent. For practical purposes cholera may be defined as an acute diarrhoeal disease caused by *Vibrio cholerae* or *V. eltor*. A similar disease caused by NAG vibrios or by *V. parahaemolyticus* may be termed paracholera—on analogy with typhoid and paratyphoid fevers. An acute watery diarrhoea caused by *Esch. coli* (Sack *et al.* 1971) or other organism is best referred to as cholera-like. As Lindenbaum and his colleagues (1965) point out, choleraic disease is a syndrome, the physiological result of extreme dehydration and electrolyte loss. Nomenclature must therefore be somewhat arbitrary (see Felsenfeld 1967).

Just as the nomenclature of the disease is confused, so also is that of the causative organisms. Until international agreement is reached, we shall follow the recommendations tentatively put forward in a report of the World Health Organization (Report 1980).

1. The epidemic strain is referred to as *V. cholerae* O- Group I or *V. cholerae* O I.

2. Strains that have the O I antigen but do not produce enterotoxin and are non-pathogenic are referred to as atypical *V. cholerae* O I.

3. Strains that are similar biochemically to the epidemic strains but do not agglutinate in polyvalent O I antiserum are referred to as non-O I *V. cholerae*. Some of these strains produce a cholera-like enterotoxin.

4. Other vibrios, such as *V. parahaemolyticus*, *V. alginolyticus*, and 'Group F vibrios' are distinct species.

History

Though cholera has been endemic in India for centuries, there is no record of its spread to the rest of the world previous to 1817 (Kirchner 1906). Between 1817 and 1823 it invaded many parts of Asia. The second pandemic, 1826-37, was more widespread. Starting in India, it spread to Russia in 1829, and thence to Poland, Germany, Austria, Sweden, and England. Throughout the years 1832-33 the whole of Europe was ravaged. Four thousand deaths occurred in London alone, and 7000 in Paris. Canada and New York were infected by Irish immigrants fleeing from their native country. The population of Cuba was decimated, and there was a heavy toll of life in Mexico. The third pandemic, 1846-62, again invaded Europe and America. In 1854 the number of deaths in England was 20 000, in Italy 24 000, and in France 140 000. America was infected by way of New Orleans in 1848; thence the disease spread up the Mississippi valley and reached California. In the fourth pandemic, 1864–75, the disease prevailed widely over Asia, Africa, Europe, and America. The fifth pandemic, 1883–96, spread over Egypt, Asia Minor, and Russia; there was a severe outbreak at Hamburg in 1892; several ports in France, Italy, and Spain were affected; and in 1893 no fewer than 287 cases and 135 deaths were reported in England and Wales. The sixth pandemic, which lasted from 1899 to 1923, affected mainly Asia, Egypt, South-East Europe, and European Russia. For several years from 1923 onwards cholera was confined mainly to India and some of the countries to the east. China suffered severely in 1940 and again in 1946, but between 1949 and 1961 did not have a single case. An outbreak, which was rapidly brought under control, occurred in Egypt in 1947; its origin remained a mystery. In 1937 a disease, regarded as paracholera, caused by the El Tor biotype of *V. cholerae*, was observed by de Moor (1938) in the Celebes. It gave rise to small numbers of cases during the next 20 years in the immediately surrounding countries; but in 1961, starting from the island of Sulawesi in Indonesia, it suddenly assumed epidemic proportions and spread rapidly to Java, eastward to the Philippines, westward to India, the Middle East, and eventually to Africa. This is regarded as the 7th pandemic—the first to be caused by the El Tor organisms (see Mackay 1980).

With increasing standards of hygiene and sanitation in the developing countries it was thought that cholera would before long be forced back to its original endemic focus in Lower Bengal (Craster 1913a, Elkington 1916); but the history of the disease during the 60 years since these prognostications were made has led to a more cautious estimate of its ultimate eradication. For instance in 1971 there were reports of no fewer than 100 000 cases in Africa and 70 000 in Asia, mainly of the El Tor variety. (For history and world incidence, see Pollitzer 1959, Kamal 1963, Felsenfeld 1967, Barua 1972, and for history in Great Britain, see Longmate 1966.)

Clinical picture

Like many other bacterial diseases, cholera may range from a symptomless infection to a fulminant attack rapidly proving fatal. In its most characteristic clinical form it starts abruptly with diarrhoea, soon marked by the passage of rice-water stools, and precipitate vomiting. Griping and tenesmus are absent, and the vomiting is unaccompanied by nausea or retching. There follow thirst, suppression of urine, often cramps in the legs and abdomen, hoarseness of speech, sometimes aphonia, progressive weakness, and collapse. The skin is cold and clammy; the face and eyes are sunken; the pulse is rapid and scarcely to be felt; the expression anxious and the body restless. When left untreated, the patient usually dies, often within 24 hours. Consciousness remains to the end. The loss of fluid from the bowel is enormous, amounting to as much as half the body weight in 24 hours and to several times the body weight over a period of days (Benenson 1970). The stools are rich in potassium and bicarbonate, and the serum shows acidosis and saline depletion, including hypokalaemia (Chaudhuri 1971).

Pathology and pathogenesis

Pathologically, cholera is a local disease characterized by an acute enteritis. The organisms grow on the epithelial cells lining the mucosa. They do not invade the blood stream, or even reach the mesenteric lymph nodes; nor do they appear to multiply in the rice-water fluid that fills the intestine (Freter *et al.* 1961). The violence of the gastro-intestinal symptoms is quite out of proportion to the almost normal appearance of the gut *post mortem*. Observations by Gangarosa and his colleagues (1960), made by taking biopsy specimens through a peroral Crosby tube from different parts of the intestine, revealed the presence of acute inflammation manifested by a mononuclear exudate, vascular congestion and hyperplasia of the goblet cells. There was no evidence of excessive epithelial desquamation such as might be caused by a mucolytic enzyme; nor was there any serious leakage of albumen into the gut, as would be expected if the epithelial lining was denuded. As already mentioned in Chapter 27, the fluid in the lumen is isotonic with the blood plasma, though the concentration of bicarbonate and potassium is higher than would be expected (Banwell *et al.* 1970, Chaudhuri 1971). Filtrates of the intestinal contents can be shown to contain enterotoxin (Panse and Dutta 1961, Dutt 1965).

The enterotoxin has been purified and its mode of action determined. It is a protein of 84 000 relative molecular mass (M_r) consisting of two immunologically distinct portions designated A (active) and B (binding). The B portion is composed of non-covalently associated sub-units of about 11 500 M_r, and is responsible for binding the whole toxin to receptors on the host-cell membrane, which contains a glycolipid, the GM_1 ganglioside. This enables the 8000 A portion to penetrate the host cell, where by enzyme action it cleaves NAD (nicotinamide adenosine dinucleotide) and transfers ADP-ribose (adenosine 5'-diphosphate) to the GTP-binding protein (guanosine 5'-triphosphate protein) associated with the host-cell enzyme adenylate cyclase. This ADP ribosylation of GTP-binding protein prevents the breakdown of GTP to GDP (guanosine 5'-diphosphate), and effectively locks adenylate cyclase in its active state. The result is the continuous formation of excessive amounts of cyclic-AMP (adenosine monophosphate). This, in its turn, leads to a rapid sequence of events, as yet un-clear, that cause intestinal epithelial cells to over-secrete electrolytes followed by water—the cholera stool (Report 1980).

The clinical manifestations, including the late renal failure, are consequent upon extreme dehydration and electrolyte depletion of the blood plasma. Judged by immunohistochem-ical examination of the mouse's intestine, the toxin is specif-ically and selectively adsorbed over the entire mucosal sur-face of the villi and crypt areas; it does not penetrate the epithelium (Peterson *et al.* 1972).

Bacteriology

At the Berlin Conference of July 1884 Koch (1886*a*) announced his discovery of the causative organism of cholera—the comma bacillus or *Vibrio cholerae*. During the previous year in Egypt and in India he had examined the faeces of 32 patients during life, and the intestinal contents at autopsy of 62 patients who had succumbed to the disease. In not a single instance had he failed to demonstrate the comma bacillus. The organisms were most numerous in the lower half of the small intestine. In acute cases they were present in almost pure culture, but in cases that had lasted longer, and in whom secondary changes had occurred, the vibrios were few and more difficult to find. In a smear from the rice-water stools or the intestinal con-tents of typical cases, the comma bacilli were arranged with their long axes parallel to one another, presenting a picture similar to that of fish in a stream. The organ-isms were found in the intestinal contents, in the lumen of the glands, and even between the epithelium and the basement membrane of the mucosa. They were apparently confined to the gut; the mesenteric glands and the blood were sterile. As a control, Koch exam-ined the intestinal contents of more than 30 cadavers of patients who had died from non-choleraic—mostly intestinal—diseases, but he failed to demonstrate the vibrio in a single instance. Though unable to repro-duce the disease in lower animals by administration of the vibrio, he came to the conclusion that it was defin-itely the cause of cholera.

This conclusion was challenged by a number of authorities holding one or other of the multitudinous theories of disease current previous to the bacterio-logical era. During the next few years their position was strengthened by the discovery of vibrios in all sorts of situations—cheese, dirty well, river, and sea water, intestinal abscesses of pigs, diarrhoeal faeces of patients with cholera nostras or dysenteric diseases, and other animal and saprophytic sources (see Chap-ter 27). Since many of these vibrios bore a close resem-blance to Koch's comma bacillus, and since their dif-ferentiation by means of the morphological, cultural, and pathogenicity tests that were then available was difficult or even impossible, grave suspicion was cast on the aetiological significance of the cholera vibrio.

Later on, however, with the introduction of a more highly specialized technique—especially the agglutin-ation and complement-fixation reactions and Pfeif-fer's phenomenon—it was possible to show that the majority of these vibrios differed in one or more essen-tial respects from the comma bacillus. Even then the confusion was not altogether cleared up; for vibrios were isolated from persons who had been in contact with cholera patients, and from water supplies in areas where cholera was epidemic, which, though conform-ing to the true cholera type in most respects, yet failed to agglutinate with a specific serum. The serological studies of Gardner and Venkatraman (see Chapter 27) promised an exact definition of the cholera vibrio, but when the El Tor vibrio was found to give rise to a disease indistinguishable from cholera, and when later, as the result of more intensive investigations, several non-agglutinable (NAG) and halophilic vi-brios were shown to be able to do the same, the range of organisms that should be included in the cholera species gave rise to controversy that is even yet not completely settled.

Numerous studies have now revealed the complex-ity of vibrios that may be met with in an outbreak of cholera: cholera vibrios of different phage types (Mukerjee 1962, Sanyal *et al.* 1968); cholera vibrios of different serotypes, for example the Ogawa, Inaba, and Hikojima types being isolated on three different days from the same patient (Gangarosa *et al.* 1967); *V. eltor* (de Moor 1938, Sanyal *et al.* 1968); non-agglu-tinable vibrios having the same H but a different O antigen from *V. cholerae* (Lindenbaum *et al.* 1965, McIntyre *et al.* 1965, Sakazaki *et al.* 1967, Ghosh *et al.* 1970); and the halophilic *V. parahaemolyticus* (Chatterjee *et al.* 1970, Report 1972; see also West *et al.* 1983).

As examples, Sanyal and his colleagues (1968) found that of 58 strains isolated from patients in Cal-cutta and other parts of India suffering from clinical cholera, 17 behaved like cholera vibrios except that some were lysed by phages i, ii or iii, 19 behaved like cholera vibrios except that some gave a positive VP reaction and were not lysed by phage iv, 5 were *V. eltor*, and 17 were inagglutinable. Of 114 strains isolated from patients admitted to the Infectious Di-seases Hospital, Calcutta, in 1970, 58 proved to be *V. cholerae*, 30 *V. eltor*, and 26 *V. parahaemolyticus* (Neogy and Chatterjee 1971).

Bacterial counts on the rice-water stools show that vibrios are present in greatest concentration on the first day of the disease, reaching 100 million or more per ml, but that they rapidly fall till on the third day, if the patient survives, they number 100 per ml or less (Smith *et al.* 1961).

In man there have been numerous cases of laboratory infection, both intentional and uninten-tional, resulting from cholera cultures. Koch (1886*b*) recorded the case of a medical man who attended a bacteriological course in Berlin, and contracted cholera. Since there was no cholera at the time either in Berlin or the rest of Germany, it is practically certain that he contracted his infection in the laboratory. Dr Oergel of Hamburg infected himself with the peritoneal exudate of a guinea-pig that had been inoculated with cholera vibrios; a drop spurted up and entered his mouth. The following day he was ill, and a few days later he died in coma. There was no cholera in Germany at that time. Pfeiffer and Pfuhl (see Kolle 1894) both contracted infection in the laboratory. Pfeiffer, who infected himself at Berlin while making animal experiments with a strain freshly isolated from a cholera case in Hamburg, suffered a sharp attack; vibrios were found in his faeces and persisted till the 33rd day of the disease. Antibodies to *V. cholerae* were demonstrated in his serum. Pfuhl had only a mild attack, vibrios persisting in his faeces for 8 days. Numerous other instances are recorded of similar laboratory infections (Kolle and Schürmann 1912).

Among the intentional cases, the best known are those of von Pettenkofer and Emmerich. Doubting the value of Koch's work, both these observers decided to test the effect of the vibrio on themselves. After a preliminary dose of sodium bicarbonate, they drank some water to which a small amount of fresh cholera culture had been added. Von Pettenkofer suffered from diarrhoea only; but his less fortunate companion developed a severe attack of cholera. As well as having diarrhoea with typical rice-water stools, he passed into a profound state of toxaemia with suppression of urine and the *vox cholerica*. Many attempts to produce artificial infection in man have, however, been fruitless. Metchnikoff (1893) carried out a number of experiments on himself and others, sometimes with positive, but more often with negative results (see Benyajati 1966, Cash *et al.* 1974).

Gastric acidity is probably a major factor in host resistance. When this is neutralized, or low as in patients with hypochlorhydria or after gastrectomy, there is reason to believe that the infecting dose may be as small as 10^2 or 10^3 living vibrios (Report 1980). So far as the aetiology of the disease is concerned, negative experiments are, as Koch pointed out, of little value. We do not know all the factors that are necessary for the reproduction of the disease.

Epidemiology

Thirty years or so before Koch discovered the cholera vibrio, John Snow (1855), the professional London anaesthetist and amateur epidemiologist, had already solved many of the problems presented by the disease. From a careful study of case histories he brought strong evidence to show that infection spread from one person to another. He concluded that the poison was of particulate nature, that it was swallowed accidentally, that it increased in the stomach and bowels,

and that it was voided in the stools. He gave the first properly documented description of a water-borne outbreak—the famous outbreak associated with the Broad Street pump in Golden Square (see Chave 1958); and by a further painstaking study of the distribution of cholera cases among inhabitants in South London supplied by different water undertakings brought convincing evidence of the part played in the dissemination of the disease by water specifically contaminated with human excrement. He noted that soiled linen might spread the infection and that flies might carry the poison mechanically; and he anticipated current practice by his observation that both storage and sedimentation of water led to the disappearance or decomposition of the poison. Finally he laid down rational preventive measures based on personal cleanliness, avoidance of the faecal contamination of food and drink, and destruction of the poison by cooking or other means.

Cholera may be endemic or epidemic. The endemic form is confined mainly to low-lying river deltas, of which the chief are those of the Ganges and Brahmaputra. Other deltas have permitted the cholera vibrio to become established endemically, but so far have never provided it with more than a temporary home. Further requirements are a dense population, a high absolute humidity, and preferably decaying vegetation (Felsenfeld 1967).

The El Tor biotype appears to have a greater endemic tendency and to cause a higher proportion of inapparent or very mild infections than the classical type. Thus Mosley (1970) estimates the ratio of inapparent or very mild infections to the typical severe cases as being 10:1 for the cholera and as high as 100:1 for the El Tor vibrio. This may be related to its longer survival in water, food, and vegetables. Epidemic cholera may occur in any part of the world but is commoner in the northern than in the southern hemisphere. It is a disease chiefly of the warmer climates, and has failed to gain a foothold north of the 50th parallel or south of the 30th parallel of latitude (see Pollitzer 1959). Though Rogers (1928, 1957), who made a life study of the epidemiology of cholera in India, found a close correspondence between the absolute humidity of the atmosphere and the seasonal incidence, the correlation is clearly not complete. At Dacca in East Pakistan, for example, the maximum incidence is in the winter months of November and December, whereas in Calcutta, only 125 miles away, it is in May and June (McCormack *et al.* 1969a, Martin *et al.* 1969). In some countries it precedes, in others it follows the monsoon rains. In general, however, it tends to be most prevalent in hot weather in which the atmospheric humidity is increased by intermittent rains. Coming after a drought when the water level is low, the rain presumably washes infected excreta into the streams and wells at a time, when, owing to the heat, the demand for water is at its greatest. Heavy and prolonged rain has the opposite effect, leading to a fall in the incidence of the disease. (Those who are interested in the effect of atmospheric conditions should consult the careful statistical analyses of Russell and Sundararajan 1928.) The age incidence is variable. In endemic areas children under 5 years of age bear the brunt of the disease; re-infections, as demonstrated by a rise in antibodies, are frequent; and cases on the whole are mild (Martin *et al.* 1969). The epidemic disease, on the other hand, attacks mainly adults, and the cases tend to be severe.

Clinically, cholera due to *V. eltor* does not differ from that due to *V. cholerae*, though epidemiologically it is said to have a lower morbidity and a higher case-fatality rate (Tanamal 1959, Uylangco *et al.* 1962, Wallace *et al.* 1966, Bart *et al.* 1970). The incubation period of the disease ranges from 1 to 5 days with a median at 48 hours. Both the morbidity and the mortality rates vary in different epidemics. The case-fatality rate of the typical severe case left untreated ranges from about 40 to 80 per cent. It is high in infancy, lowest in the 11–20 age group, and increases with age (Rogers 1911). Secondary cases are liable to occur among family contacts; in East Pakistan they were reported in 16 per cent of families (Martin *et al.* 1969).

Mode of spread

There are many ways in which infection can be transmitted from the sick to the healthy person. A study of the epidemiology of cholera indicates, however, that outbreaks can be divided into those dependent on (1) water-borne infection, and (2) case or carrier infection.

We have already referred to the classical *water-borne* outbreak in 1854 described by John Snow. Another example is the Hamburg epidemic of 1892. Hamburg, and two of its suburbs—Altona and Wandsbeck—were provided each with a different water supply. Hamburg drew its water from the river Elbe at a point above the city, and did not filter it. Altona drew its water from the Elbe below Hamburg, after the fluid refuse and faeces of nearly 800 000 people had been poured into it, but took the precaution to install a highly efficient sand-filtration plant. Wandsbeck received its supply from a relatively non-polluted lake, and likewise filtered it. On 16 August 1892, a case of cholera was reported in Hamburg. Others occurred in rapid succession, till by the end of the month there were about 1000 fresh cases a day. When the epidemic came to an end, on 23 October, the figures were: cases 18 000; deaths 8200. The incidence of cholera in Altona and Wandsbeck was in comparison very low. Altona reported 516 cases, but there is reason to believe that in most of these infection was acquired in Hamburg. On the Hamburg-Altona frontier the line of demarcation between the two towns was very irregular, and there was much overlapping of the water supplies. It was observed that in the same street the

houses that were supplied with the Hamburg water were invaded by cholera, whereas those supplied with the Altona water were spared. The Elbe was probably infected originally from the excreta of cholera patients on board the numerous barges anchoring opposite the Hamburg water intake. During the outbreak, cholera vibrios were isolated from the river not far below the mouth of the main Hamburg sewer (Koch 1894). There is here a perfectly clear demonstration of the ability of water to convey the cholera vibrio, and of the efficacy of sand-filtration in protecting a town against a highly polluted source of water. Numerous other outbreaks are on record in which the water supply has become infected and has led to the development of cholera.

Outbreaks originating in this way are generally explosive; the ascending limb of the incidence curve rises very steeply; the epidemic soon reaches its maximum, and with the cleansing of the water begins to fall. The decline is, however, less steep than the ascent, probably because a number of contact infections occur that are not directly dependent on the water-borne infection, and therefore continue to arise even after the water is clean again.

The second method of infection—*case or carrier infection*—is the one by which the disease is spread from place to place, and attains its widespread distribution. Infected material is conveyed from the sick to the healthy person, either by water, food, or infected linen. Milk, raw fruits and vegetables, and other uncooked food are all able to serve as media for the transference of the vibrio. Clothing, especially linen, if kept moist, can retain its infectivity for several days or even weeks. Numerous experiments have been made to test the survival time of the cholera vibrio under natural conditions.

Greig (1913–14*b*) in India stored the rice-water stools of 94 patients in a dark cupboard at room temperature, taking care to prevent evaporation. The average life of the vibrio was 7 to 8 days in the cooler weather, and 1 to 2 days in the hot weather. The longest survival time was 17 days. Gamaléia (1893) found that on linen strips kept moist in a water-saturated atmosphere the vibrios lived for about 5 weeks; when the strips were dried in a desiccator, they died in 17 hours. Dried on silk threads the vibrios rarely live for more than a few hours, but sometimes they may survive for 3 or 4 days (Kitasato 1889). During the Egyptian epidemic of 1947 Gohar and Makkawi (1948) made observations on the survival of cholera vibrios on various articles kept at room temperature and obtained the following results; linen 5 days, wool 4 days, leather 2 days, paper notes 6 hours, rubber 6 hours, coins 3 hours, exterior of dates 3 days, interior of dates 6 hours, vegetables 5 days, honey and treacle less than 3 hours. In water the cholera vibrio remains alive for a variable time depending on such factors as the pH, the salinity and the degree of contamination with other organisms (see Taylor 1941). Houston (1909), who added cultures of the vibrio to raw river water, found that 99.9 per cent of the organisms were dead in a week; none survived longer than 2 weeks. In Nile water contaminated with faeces the vibrio survived for 4 days (Gohar and Makkawi 1948). Some natural waters, for example the Jumna and the Ganges, appear to be unfavourable to its survival (Hankin 1896). The El Tor vibrio is said to survive longer than the cholera vibrio, and to be more often found in night soil (Bart *et al.* 1970). Both organisms are very susceptible to chlorine. Miyaki and his colleagues (1967), who made observations on frozen foods, found that the vibrios sometimes survived in skim milk at −20° for as long as a month. Further information on the length of survival of cholera and El Tor vibrios in various foods and beverages is given by Barua (1970).

Felsenfeld (1965) regards raw water, ice, eating utensils, sweet soft drinks, non-acid sliced fruits, food contaminated after cooking or pasteurization, and fruits and vegetables refreshed with sewage-polluted water and eaten raw as the most important vehicles of infection.

Speaking generally, we may say that heat and desiccation are rapidly destructive, but that under suitable conditions of moisture and temperature the cholera vibrio may survive outside the body for a sufficient length of time to be of epidemiological importance.

More recent work has shown that cholera vibrios may be present in water far away from any source of human contamination. Both in England (Bashford *et al.* 1979) and the United States (Colwell *et al.* 1980) *V. cholerae* O I vibrios have been isolated from brackish water in areas in which faecal contamination could be excluded. Moreover, outbreaks of cholera have occurred in a number of countries in which illness followed the consumption of raw or partly cooked sea-food (Feachem 1981). The same holds true of non-O I strains, some of which produce a toxin that appears to be almost identical with that of the true cholera vibrio (Furniss 1978, Wilson *et al.* 1981). The ability of cholera vibrios to attach themselves to copepods in water may prove to be of epidemiological significance (Huq *et al.* 1983). (For a review on the transmission and control of cholera, see Feacham 1981, 1982.)

Carriers

Convalescent patients usually cease to excrete the organism within a few days. Gilmour (1952), who observed 113 cholera patients in Calcutta until five or more successive daily negative cultures were obtained from the stools, found that approximately 70 per cent were negative in one week after the onset of the disease, 90 per cent in two weeks, and 98 per cent in three weeks. Four patients excreted the vibrio intermittently

for 20, 21, 23 and 25 days respectively. Longer periods of excretion have been recorded but, as Gilmour points out, cross-infection and re-infection cannot always be excluded. Some of the older records suffer from the fact that the identification of the vibrio was not as exact as it is now. In Shanghai, Peterson (1946) found no persistent carriers; 99.8 per cent of the 1949 patients he examined were negative by the end of the second week, and only one patient excreted the vibrios for as long as 17 days. Wallace and his colleagues (1967) in Calcutta, who investigated 34 convalescent patients, found one that excreted El Tor vibrios over a period of 23 weeks, but only after purging with magnesium sulphate. It is true that one carrier of 4 years' duration, a woman whose biliary tract was infected, was reported in the Philippines (Azurin *et al.* 1967a), but this is quite exceptional. Chronic carriers, in fact, are rare and probably play little part in spreading the disease. Though, judging by animal tests, vibrios from carriers are mostly virulent (Sinha *et al.* 1969), they are excreted in very small numbers, far inferior to those excreted by actual patients, and seldom cause infection (Khan 1967). It is probable that the vibrios survive in the *gall-bladder* and from there are excreted into the intestine. Greig (1913) isolated them from the bile of 30 per cent of fatal cases of cholera, and Chatterjee (1939) from 60 per cent (see also Elkington 1916). During life they can be cultivated from the duodenal juice after intravenous injection of cholecystokinin (Pierce *et al.* 1969). Benenson (1970) believes that the reason why the carrier state is commoner and more persistent in infections due to *V. eltor* than to *V. cholerae* is because in El Tor epidemics adults, whose biliary tract is often unhealthy, have been mainly attacked; whereas in endemic areas, where the infection is due to the cholera vibrio, children, whose biliary tract is healthy, are mainly attacked, and child convalescents do not become carriers. Greig (1913–14a) also reported finding vibrios in the urine, but this observation has never been confirmed.

Healthy contacts may become infected and excrete the vibrios without manifesting any sign of the disease. In India Read and Pandit (1941) isolated the organisms from the stools of 7 per cent of close contacts of cholera patients; and in Hong Kong Forbes and his colleagues (1968) observed a contact carrier rate of 5.4 to 16.7 per cent. The duration of excretion in healthy contacts is shorter than in convalescent carriers (see Pollitzer 1959). Occasionally cholera develops in healthy carriers—so-called precocious carriers. Munson (1915), for example, discovered three healthy carriers in a prison in Manila. They were isolated and examined daily. All three of them eventually contracted the disease, two of them after 16 and 17 days respectively; the third, after 18 days, died in 8 hours. As Munson points out, this last man might have travelled half-way round the world scattering his infection and died in a place many thousands of miles from any other source of the disease.

The spread of contact infection may be aided to some extent by *flies*. Cholera vibrios have been isolated from flies taken in infected houses, from flies caught in a post-mortem room in which cholera corpses had been examined, and from the feet of flies caught 17 hours after their experimental contamination (Elkington 1916).

Contact infection is responsible for the usual *chain-spread* of cholera. Cases occur in different localities, often widely separated from each other, exhibiting a regular sequence of infection. With the rapidity of modern transport is is often very difficult to ascertain all the links in a given chain. When it is remembered that infection may be carried and transmitted by apparently healthy persons, the magnitude of this difficulty will be appreciated. An excellent example of

this chain-spread is afforded by the Hamburg epidemic of 1892. This, as we have already stated, was due to an infected water supply; but from Hamburg the disease was carried by infected persons to nearly 300 places in Germany and other countries.

The two methods of infection described may be, and frequently are, combined. In the suburbs of Calcutta, for example, the tanks, which supply a large proportion of the native population with water, are freely open to pollution from cholera cases, so that whenever a tank belonging to a given household becomes contaminated, a small outbreak may result. Since the water in these tanks is used not only for drinking, but for washing, bathing, and sewage disposal, it is not surprising that they are frequently responsible for the conveyance of infection. (See also Cockburn and Cassanos 1960.)

Diagnosis and prophylaxis

Macroscopically, in about half the cases the faeces are of the rice-water type, sometimes flecked with blood; in the other half they are liquid and stained with bile or contain faecal matter or mucus (Chatterjee *et al.* 1958). A flake of mucus, for preference, should be spread on a slide, stained with dilute carbol-fuchsin, and examined microscopically. Cholera vibrios are at their maximum on the first day of the disease; after that their numbers fall rapidly. The earlier the stools are examined therefore the better (Smith *et al.* 1961). If comma bacilli are present in pure or almost pure culture—especially when they present the typical fish-in-stream appearance—a provisional diagnosis of cholera can be made directly. By this means alone Koch (1894) was able to make a diagnosis from about 50 per cent of the specimens sent for examination to the Institute for Infectious Diseases at Berlin. When, however, other bacilli are present, it is advisable to suspend judgment till further examination has been made. Sack and Barua (1964) recommend the use of the fluorescent antibody technique, and Benenson and his colleagues (1964) the dark-ground technique combined with microscopical agglutination; but these methods are suitable only for stools rich in vibrios—10^7 per g or more. As the results are most likely to be positive in cases diagnosable on clinical grounds, microscopical examination is not of great value; it should never be made the basis of more than a presumptive diagnosis.

Cultural examination

For cultural purposes stools are better than rectal swabs; but a rectal catheter is useful for the collection of fluid contents of the intestine. From cadavers some

of the contents of the small intestine and the gallbladder should be selected. When the material cannot be examined at once, it should be transported to the laboratory in a preservative fluid such as the borate saline mixture of Venkatraman and Ramakrishnan (1941), or the more generally useful sodium glycollate saline agar medium of Cary and Blair (1964). For rectal swabs and small quantities of faeces an alkaline trypticase taurocholate tellurite solution is preferable (Barua 1970). (For a comparison of these media see DeWitt *et al.* 1971.) In the absence of preservative fluids, a strip of thick blotting paper may be soaked in the stools and enclosed in a plastic bag, which is then tightly sealed.

On arrival at the laboratory the stools should be cultured as soon as possible. Various methods and numerous media have been described, which it is impossible to recount here (see K. P. Carpenter 1966, Felsenfeld 1967, Gangarosa *et al.* 1968, Barua 1970). Preferably a non-selective medium, such as alkaline peptone agar, should be used in conjunction with a selective medium, such as the alkaline sucrose dextrin basic fuchsin agar of Aronson (1915), the alkaline tellurite lauryl sulphate agar of Felsenfeld and Watanabe (1958) modified later and marketed as 'cholera medium', the TCBS agar of Kobayashi and his colleagues (1963), the sodium taurocholate tellurite gelatin agar medium of Monsur (1961), or the modified deoxycholate citrate agar of Hynes (1942). This last medium has the advantage of allowing salmonellae and shigellae to grow. Besides direct plating, the stools should be inoculated into a tube of alkaline peptone water (pH 9.0). In this medium cholera vibrios grow more rapidly than most other intestinal organisms; they multiply chiefly at the surface, where a thin pellicle soon forms. The culture should be incubated for 3-6 hours, and a loopful from the surface streaked on to a plain and a selective solid medium. A second enrichment in peptone water may be made when vibrios are few (De *et al.* 1968).

On Aronson's medium colonies after 24 hours are 2-3 mm in diameter, convex, translucent, and have a rather poached-egg appearance with a reddish centre and a paler periphery; coliform colonies, when they are present, are usually small and faintly pink. On Felsenfeld and Watanabe's medium colonies are greyish with a dark centre, and are smaller than on Aronson's medium. On TCBS medium sucrose-fermenting colonies are yellow, non-sucrose-fermenting blue. On deoxycholate citrate medium the colonies appear as non-lactose fermenters, more translucent than salmonella colonies and generally smaller.

As described in Chapter 27, colonies are best examined by a low-power stereoscopic microscope under oblique illumination. Colonies on nutrient agar may be recognized after incubation for 4-5 hours, and their identity confirmed by slide agglutination (Barua 1970). Older colonies, when rubbed up in a 0.5 per cent solution of sodium deoxycholate, form characteristic strings. The *string test* is positive with all vibrios except those of the halophilic group (see p. 448). Colonies on TCBS medium may fail to agglutinate.

Suspicious colonies should be rubbed up in saline so as to form a homogeneous suspension, and tested by slide agglutination against a suitable dilution of a serum containing antibodies to both the O antigens of Gardner and Venkatraman's sub-group I. When positive, the slide test should be confirmed by the tube method, and the colony inoculated into peptone water and submitted to the usual series of tests required for its identification (see Chapter 27). A vibrio giving a positive cholera red, a positive oxidase, and a negative VP reaction, fermenting mannose and sucrose but not arabinose, failing to produce a soluble haemolysin for sheep or goat cells, susceptible to phage iv and to polymyxin B, and agglutinable by a specific subgroup O I antiserum may be regarded as a cholera vibrio. A vibrio with similar properties but VP positive, haemolysing sheep or goat cells, agglutinating chicken red cells, susceptible to phage v but not phage iv, and resistant to polymyxin B may be identified as *V. eltor*. Vibrios with much the same properties as the cholera vibrio, but inagglutinable by subgroup O I antiserum, not subject to lysis by phages iv or v, and varying in their reaction to the VP haemolysis, haemagglutinating, and polymyxin sensitivity tests are probably NAG vibrios. *V. parahaemolyticus* differs in being halophilic, growing at 43°, lysing sheep red cells, and failing to ferment sucrose. McIntyre and his colleagues (1965) point out that, in cases of diarrhoea, NAG vibrios are liable to be overlooked by being inoculated on to an inhibitory medium, or to be mistaken for paracolon bacilli by virtue of their late fermentation of lactose.

In searching for cholera vibrios in convalescents and suspected carriers practice varies greatly. Though repeated rectal swabs may prove successful (Forbes *et al.* 1968), most workers prefer to examine the stools, preferably obtained after purging with magnesium sulphate (Gangarosa *et al.* 1966, Pierce *et al.* 1969). Duodenal juice is even better (Pierce *et al.* 1969). When large numbers of persons, such as pilgrims, have to be examined, preliminary screening by examination of stools from ten subjects is recommended by some observers (Zafari *et al.* 1968), though not by others (Felsenfeld 1967). Vibrios from convalescents and carriers may be partly or wholly rough and inagglutinable with a smooth anti-O serum; their identification must therefore be made by other tests.

In the attempted isolation of cholera vibrios from *water*, about 1-5 litres of water should be filtered through a membrane or a kieselguhr pad, which should then be transferred to a bottle containing 50-100 ml of alkaline peptone broth. After incubation for 6-8 hours the culture should be streaked on to plates of plain and selective medium. Further sub-inoculation may be made after 18 hours (Abou-Gareeb 1960, Felsenfeld 1967).

A phage-typing scheme for *V. cholerae* O I has been devised by the Public Health Laboratory at Maidstone (Furniss *et al.* 1978). A number of phages were screened, of which 14 proved to be of value for typing purposes. By these it was found that 1135 strains fell into 25 separate patterns. The scheme should be useful in the epidemiological investigation of cholera outbreaks, and in the search for cholera vibrios in the environment.

Serological examination

Diagnosis of suspected cases of cholera may be aided by the examination of serum—preferably paired samples—for antibodies. By six days or so after the onset of symptoms a fourfold rise in the titre of agglutinins may be expected in over 90 per cent of patients (Benenson, Saad and Paul 1968), and of vibriocidal antibodies in an even higher percentage (Benenson, Saad and Mosley 1968, McCormack *et al.* 1969b). Serological examination is also useful for retrospective diagnosis. Except in carriers, antibodies generally disappear within six months of the attack. It must be pointed out, however, that vibriocidal antibodies are not entirely reliable as indicative of cholera infection; they are sometimes found in healthy persons, even in those from countries free of the disease (Gangarosa *et al.* 1970). Their production is stimulated by *Yersinia enterocolitica* serotype ix (Barua and Watanabe 1972).

Prophylaxis

Cholera is one of the best examples of a disease that

can be controlled by good sanitation and personal hygiene. The preventive measures required are similar to those taken against other infections of the enteric group, except that no provision need be made against *chronic* carriers, because these rarely occur in cholera. Special attention must be paid to the purity of water and ice for human consumption, and of all foods, especially those to be eaten raw; to the disposal of excreta and garbage; to the disinfection of soiled clothing; to proper drainage; to the control of flies; to notification and isolation of cases; to the examination of contacts and pilgrims; to surveillance after an epidemic is over (see Pollitzer 1959, Felsenfeld 1967, Raška 1970); and to health education. Attempts in the past at enforced quarantine have mainly been defeated by human ingenuity; but the International Sanitary Regulations embody a large number of practicable measures on which local legislation can be based.

In countries threatened by cholera a programme should be instituted for the epidemiological surveillance of all cases of diarrhoeal disease. This should enable cholera cases to be detected early and control measures taken. The general use of antibiotics for prophylaxis should be avoided, since resistant strains may develop, thus diminishing their value for treatment.

Immunity and vaccination

The immunity conferred by an attack of cholera is said to be short-lived. In endemic areas repeated infections are common. As a rule these are unattended by clinical symptoms (Report 1969*b*), but genuine second attacks may occur within a few weeks of the first (Woodward 1971). On the other hand, Levine and his colleagues (1981) found that immunity of volunteers to re-infection with 10^6 cholera vibrios persisted for 33–36 months. Little is known, however, about the factors responsible for immunity; and vaccination has, therefore, so far been empirical. Experimental observations have shown that antibodies prepared against dead, and especially live, vibrios will protect infant rabbits infected by the mouth or adult rabbits infected by the ligated gut technique (Jenkin and Rowley 1960, Panse *et al.* 1964). In man, studies in East Pakistan have revealed good correlation between the titre of vibriocidal antibodies and a fall in the attack rate from cholera (Mosley 1969; see also Neoh and Rowley 1970): but further observations seemed to show that the degree of protection could not be related to the vibriocidal antibody titre of the serum (Mosley *et al.* 1972). These antibodies are formed against the endotoxin (toxin 1) in the cell wall of the vibrios; they contain IgM and IgG immunoglobulins and are dependent for their action on complement. Whether they can operate within the lumen of the gut, or whether immunity is determined more by locally secreted IgA immunoglobulin is still a matter for discussion (Freter

1969, Report 1969*b*). The part, if any, played by the choleragenic toxin (toxin 2) in stimulating immunity is likewise unknown. Vaccines lacking the toxic antigenic factor but giving rise to vibriocidal antibodies have proved effective in man (Oseasohn *et al.* 1965), but whether vaccines containing it would be more effective has still to be discovered (Report 1969*b*). Burrows (1970) regards both an antibacterial and an antitoxic immunity as desirable.

There is no need here to describe the early work of Ferran (1885) in Spain and of Haffkine (1895*a*, 1911, 1913) in India on vaccination against cholera (see 5th edition, pp. 1731–2). Kolle (1896*a*, *b*) in Germany was the first to introduce a killed vaccine. His product was used on a fairly large scale by Murata (1904) in Japan with doubtful benefit, and by Savas (1914) in the Grecian Army during the 2nd Balkan war with more promising results (see Greenwood and Yule 1915).

Cholera vaccine is made in various ways. The vibrios may be grown on agar, killed by heat at 56°, and preserved with 0.5 per cent phenol; or they may be grown in broth and killed by incubation for 3 days in the presence of 0.05 per cent formalin; or they may be treated in some other way. The general manufacturing requirements, production control, standardization and potency tests for the vaccine are described in Report (1969*a*) to which reference may be made (see also Feeley 1970). The usual strength is 8000×10^6 vibrios per ml, and the dose 0.25–1.0 ml according to the age of the subject. The vaccine may be prepared with a single strain, for example of the Inaba type, or with 2, 3 or 4 strains of Inaba, Ogawa, El Tor, and NAG vibrios. Injection may be intramuscular or subcutaneous. Two doses at a week's interval call forth a higher content of vibriocidins than one dose (Verwey *et al.* 1969); but field experience shows that in highly endemic areas one dose is almost as good as two (Benenson *et al.* 1968, Report 1968). About 10 per cent of those vaccinated suffer from a local and general reaction. According to Noble (1964), this may be avoided by injecting the vaccine intradermally in two 0.1 ml doses at an interval of 10 days.

Interest in vaccination remained more of less dormant till the rapid and unexpected spread of the El Tor form of cholera after the 2nd world war stimulated investigation into the disease itself. A number of field trials were then made in three of the main endemic areas—in Calcutta by Das Gupta and his colleagues (1967), in East Pakistan by an American group (Oseasohn *et al.* 1965, Benenson *et al.* 1968, Mosley *et al.* 1969, 1970), and in the Philippines by Azurin and his colleagues (1967*a*). The findings were analysed by Cvjetanović and Watanabe (1968). On the whole the results were disappointing. They showed that, compared with a control, cholera vaccine at its best gave about a 40–80 per cent degree of protection during the first 2 or 3 months, falling till after about 6 months it was no longer significant. There was no specific difference between a vaccine made from *V. cholerae* and *V. eltor*, though strain differences in immunogenicity were sometimes apparent. An oil adjuvant vaccine gave better and more lasting protection, but caused too much reaction to be suitable for general use. No evidence was obtained that vaccination influenced the course of the clinical disease once it had developed (McCormack *et al.* 1969*b*). There is some reason to believe that a live vaccine of subnormal virulence given by the mouth might prove more effective in the stimulation of immunity than a killed one. This view is supported by Holmgren (1981), who reviews the whole subject of cholera including its prevention and treatment. Though the ganglioside GM$_1$ is the natural biological receptor for cholera toxin, and might be useful for prophylaxis, Holmgren is of the opinion that a B sub-unit cholera vaccine given by the mouth would be more efficacious. Such a vaccine would stimulate a mucosal immunity with a local secretion of IgA antibody.

In the 1968-9 trial in rural East Pakistan an Inaba, but not an Ogawa, vaccine was found to protect against an Inaba-type epidemic; and the conclusion was therefore drawn that immunity to cholera induced by vaccines was type-specific (Mosley *et al.* 1970). Whether such a result was due to a specific difference in antigenicity or to a non-specific difference in immunogenicity of the strains used, it is impossible to say; and it would be wise to accept Mosley's conclusion with reserve till it is confirmed. If it is true, why, it may be asked, are second attacks of cholera, unlike pneumonia in which type-specific immunity is concerned, so uncommon?

Though vaccination against cholera cannot be disregarded, it is far inferior in its protective value to the practice of environmental sanitation and personal hygiene, to which it should never be looked upon as being more than an adjunct.

According to Felsenfeld and his colleagues (1973) the antibody response to both the cholera vibrio and the yellow fever virus is less when vaccines against these two organisms are given together than when given 3 weeks or more apart.

Treatment

Neither bacteriophage nor serum treatment has proved of therapeutic value (see 5th edition, p. 1733, and Marčuk *et al.* 1971), but chemotherapy provides a useful complement to the main purpose of treatment, which is the replacement of the lost fluid and electrolytes (Felsenfeld 1967, Benenson 1970, Pierce *et al.* 1970, Chaudhuri 1971). The amount and composition of the blood plasma must be restored as rapidly as possible by the intravenous injection of a suitable fluid, such as the Dacca electrolyte solution, which contains almost exactly the electrolyte content of the stools. Its formula is: glucose 20 g/litre, sodium chloride 4.2 g/l, sodium bicarbonate 4.0 g/l, and potassium chloride 1.8 g/l (Nalin *et al.* 1979). In severely shocked or emetic cases the solution should be given intravenously. The amount should equal that of the fluid lost in the stools. As soon as vomiting has ceased, chlorpromazine should be given in a single dose of 1-5 mg/kilo either by the mouth or intramuscularly. This has the effect of reversing the fluid loss and drastically reducing the purging. Alternatively, tetracycline may be substituted for chlorpromazine in a dose of 0.5 grams every six hours for 48 hours. Care, however, should be taken to make sure that the causative vibrios are not resistant to this drug. With such a regimen the case-fatality rate can be cut down from 60 per cent or so to 1 per cent or even under. Special care is needed in the rehydration of children.

For the treatment of *carriers* eight 1-gram doses of streptomycin may be given at hourly intervals on one day only, or 250-mg doses of chloramphenicol at 6-hourly intervals for 5 days (Lindenbaum *et al.* 1967, Forbes *et al.* 1968). Tetracycline is said to clear up the carrier condition within one day (Neogy *et al.* 1968). For the protection of family contacts, 1 gram of tetracycline daily for 5 days is recommended (McCormack *et al.* 1968).

References

Abou-Gareeb, A. H. (1960) *J. Hyg., Camb.* **58**, 21.

Aronson, H. (1915) *Dtsch. med. Wschr.* **41**, 1027.

Azurin, J. C. *et al.* (1967*a*) *Bull. World Hlth Org.* **37**, 703; (1967*b*) *Ibid.*, 745.

Banwell, J. G. *et al.* (1970) *J. clin. Invest.* **49**, 183.

Bart, K. J., Huq, Z., Khan, M. and Mosley, W. H. (1970) *J. infect. Dis.* **121**, Suppl. p. 17.

Barua, D. (1970) In: *Principles and Practice of Cholera Control*, Publ. Hlth Papers, World Hlth Org., No. 40, p. 29; (1972) *Proc. R. Soc. Med.* **65**, 423.

Barua, D. and Watanabe, Y. (1972) *J. Hyg., Camb.* **70**, 161.

Bashford, D. J., Donovan, T. J., Furniss, A. L. and Lee, J. V. (1979) *Lancet* **i**, 436.

Benenson, A. S. (1970) In: *Infectious Agents and Host Reactions*, p. 285. Ed. by S. Mudd. W. B. Saunders Co., Philadelphia.

Benenson, A. S., Islam, M. R. and Greenough, W. B. (1964) *Bull. World Hlth Org.* **30**, 827.

Benenson, A. S., Mosley, W. H., Fahimuddin, M. and Oseasohn, R. O. (1968) *Bull. World Hlth Org.* **38**, 359.

Benenson, A. S., Saad, A. and Mosley, W. H. (1968) *Bull. World Hlth Org.* **38**, 277.

Benenson, A. S., Saad, A. and Paul, M. (1968) *Bull. World Hlth Org.* **38**, 267.

Benyajati, C. (1966) *Brit. med. J.* **i**, 140.

Burrows, W. (1970) *J. infect. Dis.* **121**, Suppl. p. 58.

Carpenter, K. P. (1966) *Mon. Bull Minist. Hlth Lab. Serv.* **25**, 58.

Cary, S. G. and Blair, E. B. (1964) *J. Bact.* **88**, 96.

Cash, R. A. *et al.* (1974) *J. infect. Dis.* **129**, 45.

Chatterjee, B. D., Gorbach, S. L. and Neogy, K. N. (1970) *Bull. World Hlth Org.* **42**, 460.

Chatterjee, H. (1939) *J. Indian med. Ass.* **8**, 449.

Chatterjee, H. N., Basu, D. K. and Chakravarty, P. K. (1958) *J. Hyg. Epidem. Microbiol. Immunol.* **3**, 172.

Chaudhuri, R. N. (1971) *Trop. Doctor* **1**, 5.

Chave, S. P. W. (1958) *Med. History* **11**, 92.

Cockburn, T. A. and Cassanos, J. G. (1960) *Publ. Hlth Rep., Wash.* **75**, 791.

Colwell, R. R. *et al.* (1980) Proc. 15th U.S. and Japan Conf. on Cholera, *Nat. Inst. Hlth Publ.* No. 80-2003, p. 44.

Craster, C. V. (1913a) *J. infect. Dis.* **12**, 472; (1913b) *J. Amer. med. Ass.* **61**, 2210; (1914) *J. exp. Med.* **19**, 581.

Cvjetanović, B. and Watanabe, Y. (1968) Mimeographed Document, *WHO/BD/Cholera/68.17*, World Hlth Org.

Das Gupta, A. *et al.* (1967) *Bull. World Hlth Org.* **37**, 371.

De, S. P., Ghosh, A. K. and Shoumya, P. de (1968) *Indian J. med. Res.* **56**, 1478.

DeWitt, W. E., Gangarosa, E. J., Huq, I. and Zarifi, A. (1971) *Amer. J. trop. Med. Hyg.* **20**, 685.

Dutt, A. R. (1965) *Indian J. med. Res.* **53**, 605.

Elkington, J. S. C. (1916) *Comm. Aust. Quar. Serv.* No. 7.

Feachem, R. G. (1981) *Trop. Dis. Bull.* **78**, 675, 865; (1982) *Ibid.* **79**, 1.

Feeley, J. C. (1970) In: *Principles and Practice of Cholera Control*, Publ. Hlth Papers, World Hlth Org., No. 40, p. 87.

Felsenfeld, O. (1965) *Bull. World Hlth Org.* **33**, 725; (1967) *The Cholera Problem.* Warren H. Green Inc., St Louis, Missouri.

Felsenfeld, O. and Watanabe, Y. (1958) *U.S. Forces med. J.* **9**, 975.

Felsenfeld, O. *et al.* (1973) *Lancet* **i**, 457.

Ferran, J. (1885) *C. R. Acad. Sci.* **100**, 959.

Forbes, G. I., Lockhart, J. D. F., Robertson, M. J. and Allan, W. G. L. (1968) *Bull. World Hlth Org.* **39**, 381.

Freter, R. (1969) *Tex. Rep. Biol. Med.* **27**, Suppl. 1, p. 299.

Freter, R. F., Smith, H. L. and Sweeney, F. J. (1961) *J. infect. Dis.* **109**, 35.

Furniss, A. L. (1978) In: *Modern Topics in Infection*, p. 126. Ed. by J. D. Williams. Heinemann Med. Books Ltd, London.

Furniss, A. L., Lee, J. V. and Donovan, T. J. (1978) *Publ. Hlth Lab. Serv. Monograph Series*, No. 11. H.M.S.O. London.

Gamaléia, N. (1893) *Dtsch. med. Wschr.* **19**, 1350.

Gangarosa, E. J., Beisel, W. R., Benyajati, C., Sprinz, H. and Piyaratn, P. (1960) *Amer. J. trop. Med. Hyg.* **9**, 125.

Gangarosa, E. J., DeWitt, W. E., Feeley, J. C. and Adams, M. R. (1970) *J. infect. Dis.* **121**, Suppl. p. 36.

Gangarosa, E. J., DeWitt, W. E., Huq, I. and Zarifi, A. (1968) *Trans. R. Soc. trop. Med. Hyg.* **62**, 693.

Gangarosa, E. J., Saghari, H., Emile, J. and Siadat, H. (1966) *Bull. World Hlth Org.* **34**, 363.

Gangarosa, E. J., Sanati, A., Saghari, H. and Feeley, J. C. (1967) *Lancet* **i**, 646.

Ghosh, A. K., De, S. P. and Mukerjee, S. (1970) *Ann. Inst. Pasteur* **118**, 41.

Gilmour, C. C. B. (1952) *Bull. World Hlth Org.* **7**, 343.

Gohar, M. A. and Makkawi, M. (1948) *J. trop. Med. Hyg.* **51**, 95.

Greenwood, M. and Yule, G. U. (1915) *Proc. R. Soc. Med.* **8**, Sec. Epidem. and State Med. p. 113.

Greig, E. D. W. (1913) *Indian med. Gaz.* **48**, 8; (1913–14a) *Indian J. med. Res.* **1**, 90; (1913–14b) *Ibid.*, 481.

Haffkine, K. M. (1892) *C. R. Soc. Biol.* **4**, 635, 671; (1895a) *Rep. to Govt India*, Calcutta; (1895b) *Brit. med. J.* **ii**, 1541; (1911) *Epidemiological Notes*, Calcutta; (1913) *Protective Inoculation against Cholera*, Calcutta.

Hankin, M. E. (1896) *Ann. Inst. Pasteur* **10**, 511.

Holmgren, J. (1981) *Nature, Lond.* **292**, 443.

Houston, A. C. (1909) *4th Res. Rep. met. Water Bd.*

Huq, A. *et al.* (1983) *Appl. environm. Microbiol.* **45**, 275.

Hynes, M. (1942) *J. Path. Bact.* **54**, 193.

Jenkin, C. R. and Rowley, D. (1960) *Brit. J. exp. Path.* **41**, 24.

Kamal, A. M. (1963) *Bull World Hlth Org.* **28**, 277.

Khan, A. Q. (1967) *Lancet* **i**, 245.

Kirchner, M. (1906) *Klin. Jb.* **16**, 1.

Kitasato, S. (1889) *Z. Hyg. InfektKr.* **5**, 134.

Kobayashi, T., Enomoto, S., Sakazaki, R. and Kuwahara, S. (1963) *Jap. J. Bact.* **18**, 387.

Koch, R. (1886a) *New Sydenham Soc.* **115**, 327; (1886b) *Ibid.* **115**, 370; (1894) *The Bacteriological Diagnosis of Cholera*, trans. by Duncan. Edinburgh.

Kolle, W. (1894) *Z. Hyg. InfektKr.* **18**, 42; (1896a) *Zbl. Bakt.* **19**, 97; (1896b) *Ibid.* **19**, 217.

Kolle, W. and Schürmann, W. (1912) See *Kolle and Wassermann's Hdb. path. Mikroorg.*, 2te Aufl., 1912–13, **4**, 1.

Levine, M. M. *et al.* (1981) *J. infect. Dis.* **143**, 818.

Lindenbaum, J., Greenough, W. B., Benenson, A. S., Oseasohn, R., Rizvi, S. and Saad, A. (1965) *Lancet* **i**, 1081.

Lindenbaum, J., Greenough, W. B. and Islam, M. R. (1967) *Bull. World Hlth Org.* **36**, 871.

Longmate, N. (1966) *King Cholera. The Biography of a Disease.* Hamish Hamilton, Lond.

McCormack, W. M., Chowdhury, A. M., Jahangir, N., Ahmed, A. B. F. and Mosley, W. H. (1968) *Bull. World Hlth Org.* **38**, 787.

McCormack, W. M., Mosley, W. H., Fahimuddin, M. and Benenson, A. S. (1969a) *Amer. J. Epidem.* **89**, 393.

McCormack, W. M., Rahman, A. S. M. M., Chowdhury, A. K. M. A., Mosley, W. H. and Phillips, R. A. (1969b) *Bull. World Hlth Org.* **40**, 199.

McIntyre, O. R., Feeley, J. C., Greenough, W. B., Benenson, A. S., Hassan, S. I. and Saad, A. (1965) *Amer. J. trop. Med. Hyg.* **14**, 412.

Mackay, D. M. (1980) *Publ. Hlth, Lond.* **94**, 283.

Marčuk, L. M. *et al.* (1971) *Bull. World Hlth Org.* **45**, 77.

Martin, A. R., Mosley, W. H., Sau, B. B., Ahmed, S. and Huq, I. (1969) *Amer. J. Epidem.* **89**, 572.

Metchnikoff, E. (1893) *Ann. Inst. Pasteur* **7**, 562; (1894) *Ibid.* **8**, 529.

Miyaki, K., Iwahara, S., Sato, K., Fujimoto, S. and Aibara, K. (1967) *Bull. World Hlth Org.* **37**, 773.

Monsur, K. A. (1961) *Trans. R. Soc. trop. Med. Hyg.* **55**, 440.

Moor, C. E. de (1938) *Bull Off. int. Hyg. publ.* **30**, 1510; (1949) *Bull. World Hlth Org.* **2**, 5.

Mosley, W. H. (1969) *Tex. Rep. Biol. Med.* **27**, Suppl. No. 1, p. 227; (1970) In: *Principles and Practice of Cholera Control*, Publ. Hlth Papers, World Hlth Org., No. 40, p. 23.

Mosley, W. H. *et al.* (1969) *Bull. World Hlth Org.* **40**, 177; (1970) *J. infect. Dis.* **121**, Suppl. p. 1; (1972) *Bull. World Hlth Org.* **47**, 229.

Mukerjee, S. (1962) *Ann. Biochem. exp. Med., Calcutta*, **22**, 9.

Munson, E. L. (1915) *Philipp. J. Sci. B*, **10**, 1.

Murata, N. (1904) *Zbl. Bakt.* **35,** 605.

Nalin, D. R. *et al.* (1979) *Trans. R. Soc. trop. Med. Hyg.* **73,** 10.

Neogy, K. N. and Chatterjee, B. D. (1971) *Lancet* **i,** 347.

Neogy, K. N., Mukherjee, M. K. and Manji, P. N. (1968) *Bull. Calcutta Sch. trop. Med.* **16,** 77.

Neoh, S. H. and Rowley, D. (1970) *J. infect, Dis.* **121,** 505.

Noble, J. E. (1964) *J. Hyg., Camb.* **62,** 11.

Oseasohn, R. O., Benenson, A. S. and Fahimuddin, M. (1965) *Lancet* **i,** 450.

Panse, M. V. and Dutta, N. K. (1961) *J. infect. Dis.* **109,** 81.

Panse, M. V., Jhala, H. I. and Dutta, N. K. (1964) *J. infect. Dis.* **114,** 26.

Peterson, J. S. (1946) *Chin. med. J.* **64,** 276.

Peterson, J. W., LoSpalluto, J. J. and Finkelstein, R. A. (1972) *J. infect. Dis.* **126,** 617.

Pierce, N. F., Banwell, J. G., Gorbach, S. L., Mitra, R. C., Mondal, A. and Manji, P. M. (1969) *Indian J. med. Res.* **57,** 706.

Pierce, N. F., Sack, R. B. and Mahalanabis, D. (1970) In: *Principles and Practice of Cholera Control.* Publ. Hlth Papers, World Hlth Org., No. 40, p. 61.

Pollitzer, R. (1959) *Cholera,* World Hlth Org., Geneva.

Raška, K. (1970) In: *Principles and Practice of Cholera Control,* Publ. Hlth Papers, World Hlth Org., No. 40, p. 111.

Read, W. D. B. and Pandit, S. R. (1941) *Indian J. med. Res.* **29,** 403.

Report (1968) *Bull. World Hlth Org.* **38,** 917; (1969*a*) *Tech. Rep. Ser. World Hlth Org.* No. 413; (1969*b*) *Ibid.* No. 414; (1972) *Commun. Dis. Rep., PHLS, Lond.* No. 6, p. 3; (1980) *Bull. World Hlth Org.* **58,** 353.

Rogers, L. (1911) *Cholera and its Treatment.* Oxford; (1928) *Indian med. Res. Mem.* No. 9; (1957) *Brit. med. J.* **ii,** 1193.

Russell, A. J. H. and Sudararajan, E. R. (1928) *Indian med. Res. Mem.* No. 12.

Sack, R. B. and Barua, D. (1964) *Indian J. med. Res.* **52,** 848.

Sack, R. B., Gorbach, S. L., Banwell, J. G., Jacobs, B., Chatterjee, B. D. and Mitra, R. C. (1971) *J. infect. Dis.* **123,** 378.

Sakazaki, R., Gomez, C. Z. and Sebald, M. (1967) *Jap. J. med. Sci. Biol.* **20,** 265.

Sanyal, A., Dastidar, S. G., Chakrabarty, A. N. and Manji, P. M. (1968) *Indian J. exp. Biol.* **6,** 112.

Savas, C. (1914) *Bull. Off. int. Hyg. publ.* **6,** 1653.

Sinha, R., Deb, B. C., Ghosh, A. K. and Shrivastava, D. L. (1969) *Indian J. med. Res.* **57,** 1636.

Smith, H. L., Freter, R. and Sweeney, F. J. (1961) *J. infect. Dis.* **109,** 31.

Snow, J. (1855) *On the Mode of Communication of Cholera,* 2nd ed. John Churchill, London.

Tanamal, S. (1959) *Amer. J. trop. Med. Hyg.* **8,** 72.

Taylor, J. (1941) *Cholera Research in India 1934–1940.* Job Press, Cawnpore.

Uylangco, C., Fabie, A. E., Primicias, P. and Geronimo, A. (1962) *J. Philipp. med. Ass.* **38,** 1.

Venkatraman, K. V. and Ramakrishnan, C. S. (1941) *Indian J. med. Res.* **29,** 681.

Verwey, W. F. *et al.* (1969) *Tex. Rep. Biol. Med.* **27,** Suppl. No. 1, p. 243.

Wallace, C. K. *et al.* (1966) *Brit. med. J.* **ii,** 447; (1967) *Lancet* **i,** 865.

West, P. A., Lee, J. V. and Bryant, T. N. (1983) *J. appl. Bact.* **55,** 263.

Wilson, R. *et al.* (1981) *Amer. J. Epidem.* **114,** 293.

Woodward, W. E. (1971) *J. infect. Dis.* **123,** 61.

Zafari, Y., Zarifi, A. H., Rahmanzadeh, S. and Shayan, H. (1968) *Bull. World Hlth Org.* **39,** 492.

71

Acute enteritis

Roger J. Gross

Introductory

Acute infectious enteritis is a major cause of morbidity throughout the world. The disease is often mild and self-limiting in healthy adults, but in the malnourished, in the aged and in young children the symptoms may be severe. The World Health Organization reported that diarrhoeal diseases were among the ten major causes of death among young children in eleven African, Asian and South American countries (Report 1974*a*). Rohde and Northrup (1976) calculated that in a single year about 500 million episodes of acute diarrhoea may have occurred among children under the age of 5 years in Africa, Asia and Latin America. As a result, between 5 and 18 million children may have died. Even in the developed countries for which figures were available enteritis was among the ten most common causes of death in infants and children under 5 years of age (Report 1974*b*). In Britain, enteritis in children is of less importance than it was 10 years ago. In the early 1970s the number of deaths due to diarrhoeal diseases among infants in England and Wales was 40 per 100 000, but by 1979 this figure had fallen to 9 per 100 000. In 1971 the number of children in the age group 0–4 years admitted to hospital in England and Wales for the treatment of gastro-enteritis was 855 per 100 000. This had been halved by 1977. Nevertheless, gastro-enteritis is still an important cause of mortality in the first year of life and the British mortality figures compare unfavourably with those for many other Common Market countries (Wharton 1981).

Increasingly large numbers of people travel from country to country each year. Many of them, especially those who visit the tropics, suffer from diarrhoea either during their visit or soon after returning home. The illness is usually, but by no means invariably, mild and self limiting. Nevertheless, *travellers' diarrhoea*

has important consequences in terms of human suffering and because of its economic effect on the tourist industry and its military implications.

Earlier in this volume we have considered enteritis caused by such well known bacterial pathogens as *Shigella* (Chapter 69) and *Vibrio cholerae* (Chapter 70), and in Chapter 72 we shall describe as 'food poisoning' the diarrhoea or vomiting, or both, due to the action of *Salmonella*, *Staphylococcus aureus*, *Clostridium perfringens*, *Bacillus cereus* and *Vibrio parahaemolyticus*. Our knowledge of other causes of diarrhoeal disease is incomplete but has increased greatly during the last decade. In particular, the role of *Escherichia coli* as a major cause of diarrhoea in all age groups has been established and the importance of rotaviruses as a cause of enteritis in infancy recognized. We shall include in this chapter an account of diarrhoea caused by *Campylobacter* (though other campylobacter infections were dealt with in Chapter 56), and a brief mention of other possible bacterial causes of diarrhoea. (For pathogenesis of diarrhoea, see Lambert 1979*a*)

Diarrhoea due to *Escherichia coli*

Enteropathogenic, enterotoxigenic and enteroinvasive strains

The organism now known as *Esch. coli* was first isolated by Escherich (1885) from the stools of infants with enteritis. It was soon realized that this organism could also be isolated from the faeces of healthy infants and adults, and the importance of distinguishing avirulent strains from those capable of causing diarrhoea was recognized. Lesage (1897) showed that serum from patients with infantile enteritis agglutinated *Esch. coli* isolated from other cases in the same epidemic but not those from healthy subjects. Goldschmidt (1933) used slide-agglutination techniques for the study of epidemic infantile enteritis in institutions and in doing so anticipated the serotyping methods now used in routine investigations. Bray (1945) investigated an outbreak of infantile enteritis in a London hospital and showed that the epidemic strain of *Esch. coli* belonged to a serogroup which was subsequently recognized as *Esch. coli* O111. Similar studies in Aberdeen during 1947 and 1948 implicated two serogroups of *Esch. coli* as the causative organisms of epidemics of infantile enteritis (Giles and Sangster 1948, J. Smith 1949). These strains were initially designated Aberdeen α and Aberdeen β but were subsequently recognized as *Esch. coli* O111 and O55 respectively. Further studies were greatly facilitated by Kauffmann's (1947) description of a comprehensive serotyping scheme for *Esch. coli*. Investigations of epidemics of infantile enteritis continued, and in 1961 Taylor considered

that members of 17 O serogroups of *Esch. coli* had been implicated in such epidemics.

During the 1950s a rabbit ileal-loop technique was used to demonstrate the production of an enterotoxin by *Vibrio cholerae* (De and Chatterje 1953) and the same method was subsequently used to show enterotoxin production by certain *Esch. coli* strains isolated from adults with diarrhoea in Calcutta (De *et al.* 1956). Various ileal-loop tests were used successfully for the study of *Esch. coli* diarrhoea in animals, and H.W. Smith and Gyles (1970*a*) used pig intestine to show that strains pathogenic for swine produced two different enterotoxins. One was antigenic and heat labile (LT) and the other non-antigenic and heat stable (ST). In 1972 a test was described in which intragastric challenge of infant mice was used to demonstrate ST produced by strains of *Esch. coli* that had previously been shown to be enterotoxigenic by means of the rabbit ileal-loop test (Dean *et al.* 1972). A little later, two tissue-culture methods for the detection both of LT and of the enterotoxin of *V. cholerae* were introduced, in which characteristic histological changes (p. 467) appeared in Y1 mouse-adrenal cells (Donta *et al.* 1974) and in Chinese-hamster ovarian cells (Guerrant *et al.* 1974). Strains of *Esch. coli* may form ST only, LT only, or both; these are referred to collectively as *enterotoxigenic strains*. The genetic determinants for the production of these enterotoxins are carried on transferable plasmids (Skerman *et al.* 1972, So *et al.* 1975, 1979). The use of the infant-mouse and tissue-culture tests in epidemiological studies soon established the importance of enterotoxigenic strains as causes of human diarrhoeal disease but, as shown by Gross and his colleagues (1976*a*), strains that had caused epidemics of infantile enteritis in Great Britain were not enterotoxigenic as judged by either of these tests. It thus became clear that *Esch. coli* could cause diarrhoea by more than one means.

The strains responsible for infantile enteritis—the so-called *enteropathogenic strains*—had originally been characterized by serological methods, and so belonged, by definition, to a limited range of serotypes. Ørskov and his colleagues (1976) pointed out that enterotoxigenic strains generally belonged to serotypes different from those of the enteropathogenic strains. This has since been confirmed by other workers (Merson *et al.* 1979).

A third group of strains of *Esch. coli* belonging to other serotypes cause dysentery-like symptoms and resemble *Shigella* in their ability to cause experimental keratoconjunctivitis in guinea-pigs and to penetrate HeLa cells in tissue culture (Sakazaki *et al.* 1967, Mehlman *et al.* 1977, Formal and Hornick 1978); these are called *enteroinvasive strains*. Rowe and his colleagues (Rowe *et al.* 1977, Rowe 1979) therefore suggested that *Esch. coli* strains that cause diarrhoea can be considered as forming three groups, as follows.

1 Enteropathogenic *Esch. coli* (EPEC)
    Common serogroups: O26,   O55,   O86,   O111,
                        O114, O119, O125, O126,
                        O127, O128, O142.

2 Enterotoxigenic *Esch. coli* (ETEC)
    Common serogroups: O6,   O8,   O15,   O25,   O27,
                        O63, O78, O115, O148,
                        O153, O159, O167.

3 Enteroinvasive *Esch. coli* (EIEC)
    Common serogroups: O28ac,   O112ac,   O124,
                        O136,   O143,   O144,   O152,
                        O164.

Enteropathogenic and enterotoxigenic *Esch. coli* are important causes of acute enteritis; we shall consider them in some detail in this chapter. Enteroinvasive strains of *Esch. coli* will not be discussed further here, as they are more appropriately considered as a cause of bacillary dysentery (Chapter 69).

Infantile diarrhoea due to enteropathogenic strains (EPEC)

Epidemiology

Until the 1930s, annual outbreaks of 'summer diarrhoea' occurred among babies in Europe and the United States and were associated with a high mortality. The aetiological agent was never determined, but it was considered likely that the babies were infected by contaminated cows' milk. The decline in the incidence of the disease during the 1920s and 1930s was

difficult to explain, though contributing factors were thought to include the introduction of dried and pasteurized milk and general improvements in hygiene and sanitation (*Lancet* 1936). During the 1940s, it was noticed that outbreaks of infantile enteritis were occurring with increased frequency in hospitals and nurseries and that these outbreaks were more likely to take place during the winter months. At this time the aetiological agent was not recognized nor was the manner of spread of infection determined. Nevertheless, it was observed that outbreaks could be controlled by the scrupulous sterilization of feeding bottles, the pasteurization or boiling of milk feeds shortly before use, and the strict application of measures to control the spread of infection from patient to patient. Hinden (1948*a*, 1948*b*) reported that although the highest incidence of infantile enteritis in hospitals occurred during the winter months, cases in the community were more evenly distributed throughout the year. During their investigations of outbreaks of infantile enteritis in Aberdeen, Giles and Sangster (1948) noticed two peaks in the seasonal incidence of the disease. The highest incidence—in March and April—was mainly due to cases arising in hospital and was accompanied by a high mortality. A lesser peak in July was mainly due to cases in the general community. It appeared that an outbreak in the general community had occurred before or at the same time as the hospital outbreaks and that cross-infection in the hospital had followed the introduction of the infection by babies admitted while excreting the causative organism.

Although there was much debate about the aetiology of infantile enteritis, the use of serotyping showed that epidemics were caused by strains of *Esch. coli* belonging to particular O serogroups, members of which became known as enteropathogenic *Esch. coli* (EPEC). During the 1950s many epidemics due to EPEC were reported among babies in hospitals and nurseries in Europe and North America (see 6th Edition for references); further serious outbreaks occurred in Great Britain and Ireland in the 1960s and early 1970s. In late 1967 there was an epidemic in several hospitals in Teesside in which the mortality rate was high (*Lancet* 1968); two *Esch. coli* strains were responsible: O119H6 and O128H2. A year or so later a similar outbreak due to *Esch. coli* O114H2 occurred in hospitals in the Manchester area (Jacobs *et al.* 1970). In late 1970 and early 1971 outbreaks due to *Esch. coli* O142H6 occurred in several hospitals in the Glasgow area (Love *et al.* 1972, Kennedy *et al.* 1973), and later in 1971 the same serotype was responsible for an outbreak in a Dublin hospital (Hone *et al.* 1973). In Dublin, it appeared that the infection spread by cross-infection in the hospital after the admission of infected infants from the general community, thus confirming the observations of Giles and Sangster in 1948. Since 1971 no serious epidemics of EPEC enteritis have been reported in Great Britain or the United

States, but a satisfactory explanation for this has not been put forward. In the absence of epidemics, the incidence of sporadic cases of diarrhoea in babies in Great Britain shows a peak in the summer months.

Although EPEC enteritis now appears to be of relatively little importance in temperate areas with good standards of hygiene, it is still common in tropical countries (Maiya *et al.* 1977) and in communities in which hygiene is poor (Gurwith and Williams 1977). The epidemiology of EPEC enteritis in tropical countries differs in some respects from that in Europe and North America (Rowe 1979). Thus, although outbreaks in institutions are often reported, sporadic cases and outbreaks occur very frequently in the general community. Many authors have stressed the protective effect of breast feeding; Scrimshaw and his colleagues (1968) showed that the peak incidence of enteritis in various countries was always in the few months after the beginning of the weaning period. Since weaning often takes place late in developing countries, the age distribution of EPEC enteritis differs from that observed in Europe and North America.

The importance of EPEC as a cause of enteritis in adults is difficult to evaluate since few laboratories look for these organisms in patients over 3 years of age. However, a few outbreaks have been described. Schroeder and his colleagues (1968) reported a water-borne outbreak due to *Esch. coli* O111 that affected adults attending a conference centre in the United States; and two food-borne outbreaks have been reported in Great Britain (Vernon 1969, Report 1974c). Neter and his colleagues (1955) showed that about 50 per cent of children had acquired haemagglutinating antibody to EPEC by the age of 1 year. It is therefore possible that most babies are infected with EPEC early in life and acquire some immunity.

Pathogenesis

The ability of EPEC strains to cause diarrhoea has been confirmed by oral administration of the organisms to babies and adults, but the mechanism by which these organisms exert their effect remains obscure (Ferguson and June 1952, Levine *et al.* 1978). Most strains do not produce ST or LT and are non-invasive, but a few strains of serogroups O114 (Burnham *et al.* 1976) and O128 (Reis *et al.* 1979, Ryder *et al.* 1979) have been shown by means of the infant-mouse and tissue-culture tests to be enterotoxigenic. Several workers have reported that other EPEC strains cause the accumulation of fluid in ligated loops of rabbit ileum (Taylor *et al.* 1958, H.W. Smith and Gyles 1970b). More recently, Klipstein and his colleagues (1978) examined strains of EPEC from outbreaks of infantile enteritis that did not produce ST or LT and found that extracts of them caused a net efflux of water in the perfused rat gut. This suggested that EPEC produced an enterotoxin that was not detected by the standard tests for ST and LT.

Konowalchuk and co-workers (1977) showed that culture filtrates of certain strains of *Esch. coli* had a cytotoxic effect on monolayers of Vero cells; this contrasted with the 'cytotonic' effect of LT on cells in tissue culture (p. 467). The cytotoxin (Vero toxin or VT) differs from LT and ST in that it has no action on the Y1 adrenal or Chinese-hamster ovarian cell lines commonly used to detect LT and causes no fluid accumulation in the infant-mouse test used to detect ST. Partly purified VT does, however, induce some fluid accumulation in the rabbit ileal-loop test. A relation between VT and the toxin demonstrated by Klipstein and co-workers (1978) has yet to be shown. Konowalchuk and his colleagues (1977) reported VT production by seven of 35 EPEC strains isolated from infants with diarrhoea. Three of the seven strains belonged to O group 26, and subsequently a particular association of VT production with strains of this O group has been reported. Scotland and co-workers (1980a) found 25 VT-producing strains among 253 EPEC strains of 11 O groups isolated from infants with diarrhoea in the United Kingdom; they included 20 members of serotype O26:H11, two of serotype O26:H − and two of O128:H2. Interest in VT has recently been increased by reports that strains of *Esch. coli* O157:H7 that produce it are associated with outbreaks of haemorrhagic colitis in the United States and Canada (Johnson *et al.* 1983, Riley *et al.* 1983). VT-forming strains have also been isolated from sporadic cases of haemolytic-uraemic syndrome in Canada (Karmali *et al.* 1983). Further investigations indicate that VT closely resembles 'Shiga toxin' (Chapter 36), a toxin formed by strains of *Shigella dysenteriae* serotype 1 (O'Brien *et al.* 1983). The role of VT in the classical form of infantile enteritis is uncertain; this disease affects the small bowel and so differs from haemorrhagic colitis.

It has been shown in necropsy studies (Thomson 1955) and by the intubation technique (Koya *et al.* 1954) that colonization of the upper part of the small intestine occurs in infantile enteritis associated with EPEC. Adhesion to the mucosa of the small intestine is an important factor in the pathogenesis of diarrhoea caused by ETEC (p. 462), but less is known about the mechanism of colonization by EPEC. Cravioto and his colleagues (1979a) reported adhesion to HEp-2 cells in tissue culture by a significantly higher proportion of strains belonging to EPEC serogroups and isolated from outbreaks of infantile enteritis than by strains from healthy human subjects. The ability to adhere to HEp-2 cells was uncommon in ETEC strains, even when they possessed the fimbrial colonization factors CFA/I and CFA/II. Adhesion to HEp-2 cells is mediated by non-fimbrial adhesins (Scotland *et al.* 1982). Recent clinical evidence emphasizes the importance of adhesion to intestinal epithelial cells in EPEC diarrhoea. Ulshen and Rollo (1980) described the isolation of *Esch. coli* O125 from the duodenal aspirate of a 7-day-old baby with protracted diarrhoea; the organism was non-enterotoxic and non-invasive, but the examination of biopsy material showed that it adhered closely to the epithelium of the small bowel. A further report described a strain of *Esch. coli* O111 isolated from the small bowel of two infants with chronic diarrhoea (Clausen and Christie, 1982). Again the strain was non-enterotoxigenic and non-invasive. Small-bowel biopsy revealed focal adhesion of the bacteria to the epithelium, accompanied by inflammation of the tissue; and in-vitro tests showed that the organism adhered to HEp-2 tissue-culture cells in densely packed aggregates. The authors suggested that the inflammatory changes might result solely from the close attachment of the organisms to the intestinal epithelium. In a study of 15 infants with severe diarrhoea (Rothbaum *et al.* 1982), *Esch. coli* O119 was present in cultures of fluid from the small intestine, and jejunal biopsy specimens from all of the patients showed that the bacteria were adherent to mucosal cells. The organisms did not produce ST, LT or VT. Certain strains of *Esch. coli* become attached to intestinal epithelial cells and cause diarrhoea in rabbits (Cantey and Blake, 1977). The strains responsible for rabbit diarrhoea—so-called RDEC strains—resemble human EPEC strains in not forming ST or LT and in being non-invasive. RDEC diarrhoea may therefore be a useful animal model for diarrhoea caused by EPEC in man.

It is not yet known how frequently the production of VT and the ability to adhere to enterocytes or to HEp-2 cells in tissue culture may be found together. One strain of *Esch. coli* O26, designated H19, adhered to HEp-2 cells (Cravioto *et al.* 1979*a*) and to human fetal small intestine (McNeish *et al.* 1975), and also produced VT (Scotland *et al.* 1980*b*), but the adhesive strain of *Esch. coli* O119 studied by Rothbaum and co-workers (1982) failed to form this toxin.

It is clear that there is considerable diversity among the strains previously designated EPEC. A few are enterotoxigenic and are best regarded as ETEC. Some produce Vero cytotoxin and some are strongly adhesive to intestinal epithelial cells. The distribution of these two properties among strains that cause infantile enteritis has yet to be determined, and other toxic substances may be discovered in the future.

Detection of enteropathogenic strains

It is usual in clinical laboratories to look for EPEC in stool specimens from children with diarrhoea who are less than 3 years old. Specimens are plated on non-selective media such as MacConkey's or eosin methylene blue agar and several colonies are examined by slide agglutination with polyvalent antisera for the EPEC serogroups. Those giving positive reactions are subcultured and again tested by slide agglutination with monovalent antisera; identification is finally confirmed by tube-agglutination tests with heated suspensions of the organism (Edwards and Ewing 1972). Strains provisionally serogrouped in this way should also be identified as *Esch. coli* by means of biochemical tests because there is widespread sharing of antigens among the various genera of the Enterobacteriaceae.

In the event of an outbreak of infantile enteritis in which EPEC are not found by routine methods the assistance of a reference laboratory should be sought. Complete serotyping by means of over 160 O antisera and over 50 H antisera may lead to the recognition of new EPEC serogroups (Rowe *et al.* 1974).

Diarrhoea due to enterotoxigenic strains (ETEC)

The circumstances under which ETEC cause diarrhoea are diverse and will be considered separately below. In general, these organisms are infrequently seen in geographical regions where hygiene and nutrition are good, but occasional outbreaks of diarrhoea, either among babies in hospital or subjects of any age in the general community, have been encountered. When persons from these regions visit countries in which hygiene is less good they often suffer from *travellers' diarrhoea* of which ETEC are the most common cause. In the developing countries, diarrhoeal diseases, usually with underlying malnutrition, are a major cause of death in children under 5 years of age; ETEC cause a proportion of these. The most severe manifestation of ETEC infection is a cholera-like disease that is difficult to distinguish clinically from that due to *V. cholerae*. This is seen occasionally in western countries but is common in areas in which true cholera is endemic.

These categories are, however, not entirely distinct. Thus, in groups of holidaymakers travelling together to the tropics, the incidence of travellers' diarrhoea may be amplified by person to person spread or by the simultaneous infection of a number of them from a common source. Also, some sub-populations in developed countries have a relatively poor standard of hygiene; they may suffer many ETEC infections, particularly when the climate is warm. For example, ETEC were found in the stools of 16 per cent of Apache Indian children from a reservation in Arizona who had been admitted to hospital with diarrhoea (Sack *et al.* 1975).

Infantile enteritis in developed countries

Strains identified as ETEC have so far been found in only a few hospital outbreaks. The first of these to be described (Gross *et al.* 1976*b*) occurred in a hospital nursery in Glasgow; 25 babies were affected and 5 required intravenous therapy, but there were no fatalities. The causative organism belonged to a previously undescribed serogroup, *Esch. coli* O159, and produced ST but not LT. In the same year an outbreak affected 55 of 205 infants admitted to the special-care nursery of a large hospital in the USA (Ryder *et al.* 1976a). The epidemic strain belonged to serogroup O78 and again produced ST but not LT. A further outbreak in Great Britain affected 10 of 18 babies in the special-care unit of a Gloucester hospital (Rowe *et al.* 1978). Five of the 10 babies with diarrhoea and 1 of the 8 who remained free of symptoms were shown to be excreting a strain of *Esch. coli* O6 which produced ST and LT. The source and route of transmission of infection in these outbreaks were uncertain. The pattern of the Gloucester outbreak suggested that the index case was a premature baby who developed diarrhoea 4 days after birth and that further cases resulted from cross-infection. The outbreak in the United States continued for 9 months and investigations revealed heavy contamination of the patients' environment with *Esch. coli* O78. The epidemic strain was also isolated from milk feeds but not from unopened containers of the same material.

Diarrhoea in the general community in developed countries

Surveys in Boston, USA (Echeverria *et al.* 1975), Canada (Gurwith and Williams 1977), Sweden (Bäck *et al.* 1977) and Britain (Gross *et al.* 1979) indicate that ETEC are probably not an important cause of sporadic diarrhoea in developed countries. There are, however, two reports from the USA of a high frequency of ETEC in babies and children with enteritis. Gorbach and Khurana (1972) recorded ETEC in 80 per cent of faecal specimens from children under 4 years of age in Chicago, and Rudoy and Nelson (1975) obtained a similar isolation rate in infants and children in Texas. These reports have not gained wide acceptance. In the first study an unorthodox infant-rabbit test was used rather than the well known tissue-culture tests or the infant-mouse test (p. 459). In the second, inadequate criteria for the interpretation of the infant-mouse test were used and the strains were not made available to other labortories for confirmation (Gangarosa 1978).

In the summer of 1975 more than 2000 staff and visitors at a national park in the United States suffered from diarrhoea (Rosenberg *et al.* 1977). The causative organism was a strain of *Esch. coli* O6 that produced ST and LT. The source of the infection was drinking water contaminated with sewage. Kudoh and his colleagues (1977) studied seven outbreaks of adult enteritis in Tokyo during the period 1969 to 1974. Six were caused by *Esch. coli* O159 and one was due to *Esch. coli* O11. In all cases the organisms produced ST but not LT. Two of these outbreaks were caused by contaminated water supplies and two others were probably food-borne.

Travellers' diarrhoea

This is a worldwide illness, usually of brief duration, often beginning with the rapid onset of loose stools and accompanied by variable symptoms, including nausea, vomiting and abdominal cramps. It occurs most frequently among those travelling from areas of good hygiene and temperate climate to areas with lower standards, particularly in the tropics.

Rowe and his colleagues (1970) demonstrated a relationship between *Esch. coli* and travellers' diarrhoea in a study of British troops in South Arabia. In this study, serotyping provided evidence that about 50 per cent of the diarrhoea cases among the new arrivals were due to *Esch. coli* O148, a previously undescribed serogroup. Later it was discovered that this strain produced ST (Rowe *et al.* 1976). Enterotoxigenic *Esch. coli* O148 was also found to be a cause of diarrhoea among United States soldiers in Vietnam (DuPont *et al.* 1971). ETEC belonging to a number of serotypes were the most common pathogens found among US military personnel with diarrhoea in the Philippines (Echeverria *et al.* 1979a), but among US troops in South Korea, an area with a relatively temperate climate, ETEC infections were uncommon even though 55 per cent of the soldiers suffered from diarrhoea (Echeverria *et al.* 1979b).

There have been several studies of travellers' diarrhoea in Mexico where 'turista' has an attack rate of 29–48 per cent; in one, ETEC were found in 45 per cent (Merson *et al.* 1976) and in another in 72 per cent of sufferers (Gorbach *et al.* 1975).

Civilian travellers from Europe to tropical or even to Mediterranean countries may develop diarrhoea due to ETEC. In studies in Sweden (Bäck *et al.* 1977) and Britain (Gross *et al.* 1979) ETEC were found in 11 per cent of persons who had contracted diarrhoea while abroad or soon after their return.

Epidemics of diarrhoea occur with particular frequency on cruise liners. These are often caused by *Salmonella* or *Shigella*, but outbreaks on two ships have been associated with ETEC. Hobbs and her colleagues (1976) reported that a strain of *Esch. coli* O27 that formed ST was isolated from 55, 61 and 20 per cent of patients with diarrhoea on three successive cruises by the same ship. The evidence suggested that the infections were food-borne. In a study on a different ship an *Esch. coli* O25 strain that produced ST was isolated from 83 per cent of diarrhoea cases in outbreaks occurring during two successive cruises. Numerous deficiencies in catering practices were discovered on this ship. After these had been dealt with no outbreaks of diarrhoea occurred on subsequent cruises (Report 1976).

Diarrhoea in developing countries

There is ample evidence that ETEC are an important

cause of diarrhoea in all age groups in areas of poor hygiene, particularly in the tropics. ETEC were found at the following frequencies in the faeces of babies or small children with diarrhoea: in Ethiopia, 14 per cent of 354 (Wadström *et al.* 1976); in black infants in South Africa, 26 per cent of 34 (Schoub *et al.* 1977); in Taiwan, 16 per cent of 57 (Echeverria *et al.* 1977); in Mexico, 16 per cent of 50 admitted to hospital (Donta *et al.* 1977), 40 per cent of 71 in a retrospective study (Evans *et al.* 1977*a*), and 47 per cent of 62 in outpatients (Evans *et al.* 1977*c*); in the Philippines, 11 per cent on admission to hospital (Echeverria (1978*a*).

Clinical cholera characterized by profuse 'rice-water' stools is traditionally associated with *V. cholerae* (serogroup O1) and may less often be caused by NAG or 'non-cholera' vibrios (Chapter 27). Before the discovery of ETEC the aetiology of cases of this type of disease frequently remained obscure. Sack and his colleagues (1971) used the rabbit ileal-loop test to demonstrate LT production by ETEC isolated from jejunal aspirates of four adult patients with cholera-like diarrhoea in Calcutta, thus confirming the earlier findings of De and his colleagues (1956). A survey in Bangladesh showed that LT-producing ETEC could be isolated from 19 per cent of all diarrhoea cases studied, from 55 per cent of hospital in-patients with diarrhoea, and from 70 per cent of those severely ill with 'non-vibrio' cholera (Nalin *et al.* 1975). In a further study in rural Bangladesh, Ryder and his colleagues (1976*b*) failed to find ETEC in 22 patients less than 2 years old, but found them in 56 per cent of 18 patients who were more than 10 years old. Sack and his colleagues (1977) isolated ETEC from 35 per cent of 65 patients with acute cholera-like disease in Dacca, Bangladesh. Diarrhoea associated with ETEC affected persons of all ages and could not be distinguished from cholera on clinical grounds.

The sources and modes of spread of ETEC infection in developing countries are not well understood. In one Bangladesh study, ETEC was found in one of 39 drinking-water sources (Ryder *et al.* 1976*b*), but in the Philippines such strains were not found in foods or surface waters, although they were isolated from a small number of pigs and water buffaloes (Echeverria *et al.* 1978*b*). A more recent Bangladesh study showed that in 9 per cent of households in which there were patients with ETEC diarrhoea, the drinking water supply was contaminated with ETEC of the same serotype as that causing the illness (Black *et al.* 1981). The rate of infection of household members was highest in houses with a contaminated water supply. It seems likely that water contaminated by human or animal sewage plays an important part in the spread of ETEC infection, but further studies are needed to confirm this.

Nature and mode of action of the enterotoxins

Heat-labile enterotoxin (LT)

Molecular weight estimations for purified *Esch. coli* LT range from 20 000 for toxin released by polymyxin B treatment to 200 000 for that from culture supernates. Dorner and his colleagues (1979) studied LT synthesized in a cell-free system directed by DNA of plasmid origin and showed that LT consisted of two polypeptide subunits. Subunit A had a molecular weight of 23 000–26 000 and stimulated adenylate-cyclase activity; subunit B had a molecular weight of 11 500. Under electrophoresis in sodium dodecyl sulphate polyacrylamide gel, polypeptide B migrated at a rate identical with that of subunit B of the cholera toxin. These results are in close agreement with those of Dallas and his colleagues (1979) who studied LT extracted from *Esch. coli* containing various LT plasmids. Two polypeptides of molecular weights 25 500 and 11 500 were found to be encoded by the LT region of the plasmid DNA. The larger molecule stimulated adenylate-cyclase activity and the smaller molecule was shown to form aggregates of five monomers which were adsorbable by Y1 adrenal cells.

Immunodiffusion and biological neutralization techniques show a close immunological relationship between LT and cholera toxin (Clements and Finkelstein 1978). The amino-acid sequences of the B subunits of *Esch. coli* LT and *V. cholerae* enterotoxin show significant homology (Dallas and Falkow 1980). It is clear that there is both structural and functional similarity between *Esch. coli* LT and cholera toxin. In both cases subunit B is responsible for binding to the Gm1 ganglioside of the epithelial cells of the small intestine; subunit A stimulates adenylate-cyclase activity, thereby increasing the concentration of cyclic adenosine monophosphate (cAMP) in the cells. The increased level of cAMP appears to act in two ways to cause fluid and electrolyte loss into the gut lumen. In the villus cells cAMP inhibits the absorption of Na^+ and hence of Cl^- and water. In the crypt cells, cAMP exerts a direct secretory effect by increasing Na^+ secretion and consequently causing loss of Cl^- and water (Field 1979). (See Figs. 71.1 and 71.2.)

Heat-stable enterotoxin (ST)

Estimates of the molecular weight of the ST polypeptide vary. Alderete and Robertson (1978) described a polypeptide of molecular weight 4000–5000 from an *Esch. coli* strain of porcine origin; Giannella and Staples (1979) used a different purification scheme to obtain ST of molecular weight 1900 from a strain of human origin. In an in-vitro system of ST synthesis directed by a cloned DNA template the ST produced by *Esch. coli* of porcine and bovine origin was a polypeptide of molecular weight 7000 (Lathe *et al.* 1980).

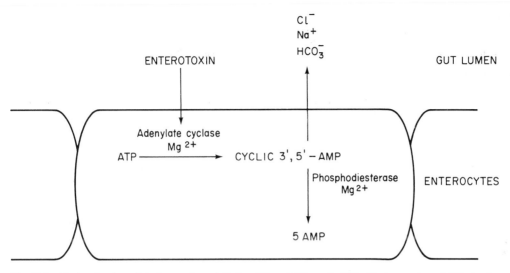

Fig. 71.1 Mode of action of cholera toxin and *Esch. coli* heat-labile toxin (LT). Cholera toxin and LT stimulate adenylate cyclase in the enterocyte, causing a rise in cAMP concentration which is responsible for the loss of water and electrolytes. Phosphodiesterase normally inactivates cAMP.

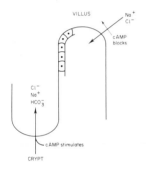

Fig. 71.2 Mode of action of cAMP in the small intestine. In the mature cells of the villi cAMP inhibits the influx of Na^+ and Cl^-; in the crypts it stimulates secretion.

This anomaly was somewhat clarified when it was shown that more than one form of ST exists: STa is methanol soluble and active in the infant-mouse test and in newborn piglets, but STb is methanol insoluble, inactive in infant mice and active in the ligated intestinal loop of older piglets and rabbits. *Esch. coli* strains may produce STa, STb or both (Burgess *et al.* 1978). The ST produced by strains isolated from different host species differs both in molecular weight and in amino-acid sequence (Rao *et al.* 1981). For example, a strain of porcine origin produced ST of molecular weight 3580 with 33 amino acid residues, but ST from a strain of human origin had a molecular weight of 1972 with 18 residues.

ST has no detectable effect on the concentration of cAMP but has been shown to stimulate the activity of guanylate cyclase and therefore to cause an increase in cGMP levels (Field *et al.* 1978, Newsome *et al.* 1978). This occurs only in intestinal epithelial cells and not in other tissues or in cell lines; it seems possible, therefore, that a unique toxin receptor is present in intestinal cells (Field 1979). The mechanism by which increased cGMP leads to a net secretion of water and electrolytes is not well understood, but ST appears to be mainly anti-absorptive in its effects, without the secretory activity of LT and cholera toxin.

Adhesive factors

It was first shown in studies of piglet enteritis that the ability to produce enterotoxin is not sufficient to enable an *Esch. coli* strain to cause diarrhoea (H.W. Smith and Linggood 1971). The organism must also be able to adhere to the mucosal surface of the epithelial cells of the small intestine. This adhesion is mediated by pilus-like filamentous protein structures which bind to specific receptors in the cell membrane. Adhesive factors are antigenic and can be recognized by means of agglutination or immunodiffusion tests with specific antisera prepared in rabbits. Their presence can also be demonstrated by haemagglutination, by experimental colonization of animal intestines or by tissue- or organ-culture methods. In contrast to the haemagglutination due to type 1 pili of *Esch. coli*, that due to adhesive factors associated with diarrhoeal disease is not inhibited in the presence of mannose. The genetic determinants controlling the production of *Esch. coli* adhesive factors are carried by plasmids which may be transferable. Plasmids that simultaneously carry genes for both an adhesive factor and enterotoxin production have been described (H.R. Smith *et al.* 1979).

In animal strains

The first adhesive factor found to be of importance in *Esch. coli* diarrhoea was recognized as an antigen (K88) which was shown to be controlled by a transferable plasmid (Ørskov and Ørskov 1966). K88 could also be detected by its ability to cause mannose-resistant haemagglutination (MRHA) of guinea-pig red-blood cells (Jones and Rutter 1974). Its importance in the pathogenesis of piglet enteritis was demonstrated by H.W. Smith and Linggood in 1971. These authors showed that the removal of the K88 plasmid from a strain of *Esch. coli* O141 was accompanied by the loss of its capacity to cause diarrhoea when given orally to piglets. This was restored by introducing a K88 plasmid from another strain of *Esch. coli*. Rutter and his colleagues (1975) later showed that there is a gene in pigs which is inherited in a simple Mendelian manner and determines the presence of receptors for K88 in the pig intestinal epithelium. Pigs lacking this receptor are resistant to colonization of the small intestine by *Esch. coli* strains with the K88 antigen.

A further antigen designated 987P has also been shown to be important in porcine infections (Isaacson *et al.* 1977); and the K99 antigen causes MRHA of sheep red-blood cells and is of importance in diarrhoea of sheep and cattle (Ørskov *et al.* 1975).

In human strains

In 1975 an enterotoxin-producing strain of *Esch. coli* O78 (H10407) isolated from a patient with cholera-like diarrhoea was found to possess an adhesive factor now known as colonization-factor antigen I (CFA/I) (Evans *et al.* 1975). When the organism was injected in small numbers into the duodenum of infant rabbits it multiplied greatly and diarrhoea developed. A laboratory-passaged derivative (H10407-P) failed to exhibit an increase in population and did not cause diarrhoea. Antiserum against strain H10407 absorbed with the derivative H10407-P agglutinated H10407 but not H10407-P. The agglutination depended on the presence of a pilus-like surface structure (Fig. 71.3), the loss of which was associated with the loss of a 60×10^6 dalton plasmid. Strains of *Esch. coli* which possess CFA/I cause MRHA of human group-A erythrocytes (Evans *et al.* 1977*b*), but the value of this method of testing for CFA/I is limited because strains of *Esch. coli* from extra-intestinal sources frequently cause group-A MRHA but are not enterotoxigenic and do not possess CFA/I (Cravioto *et al.* 1979*b*). The presence of CFA/I appears to be confined to certain serogroups of ETEC including O25, O63, O78 and O153 (Gross *et al.* 1978). The frequency with which CFA/I strains are found in surveys therefore depends

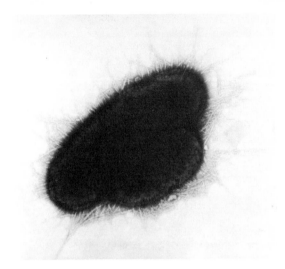

Fig. 71.3   A strain of *Esch. coli* O78 possessing colonization factor antigen I (CFA/I). ×20 000. (Electronmicrograph kindly provided by Dr A. Field.)

on the distribution of serotypes among the ETEC in the area studied. It has been confirmed by oral challenge experiments in human volunteers that strain H10407 causes diarrhoea, but the H10407-P derivative does not (Satterwhite *et al.* 1978). Nevertheless, strains of ETEC belonging to other serotypes and lacking CFA/I have also been shown to cause diarrhoea in volunteers (Levine *et al.* 1980).

A second colonization factor (CFA/II) differs from CFA/I antigenically and in its ability to cause MRHA of bovine but not human red-blood cells (Evans and Evans 1978). CFA/II was found in ETEC belonging to serogroups O6 and O8 and was common in strains of these serogroups isolated from patients with travellers' diarrhoea in Mexico. A third factor (CFA/III) has been described in a single ETEC strain belonging to serogroup O128. It resembles CFA/I in causing MRHA of human and bovine erythrocytes but is claimed to be antigenically distinct (Darfeuille *et al.* 1983). Another antigenically distinct colonization factor provisionally designated E8775 has been found in ETEC of serogroups O25, O115 and O167 (Thomas *et al.* 1982). It remains to be seen whether further antigenically distinct colonization factors are found in ETEC of other serogroups.

Laboratory techniques

Detection of LT

Tests in tissue cultures of Y1 mouse-adrenal cells (Donta *et al.* 1974) and Chinese-hamster ovary (CHO) cells (Guerrant *et al.* 1974) are accepted as standard methods for the detection of LT, but other cell lines, including Vero monkey kidney cells, are also sensitive to the toxin (Speirs *et al.* 1977). In these methods, exposure of the cells to culture supernates containing LT or cholera toxin results in an increase in the intracellular concentration of cAMP and leads to a morphological response which can be seen by microscopy. CHO cells respond by elongation, and Y1 cells become rounded. The response can be regarded as stimulatory or 'cytotonic' in contrast to the cytotoxic effect of supernates of some *Esch. coli* O26 and O126 cultures on Vero cells (Konowalchuk *et al.* 1977, Scotland *et al.* 1980*a*).

Intradermal injection of LT or cholera toxin in rabbits causes a local increase in vascular permeability which can be detected as an area of induration and blue colouration after the intravenous injection of Evans Blue dye. This reaction has formed the basis of the permeability factor (PF) test; this is of limited use for the detection of LT in surveys but has proved more valuable as a means of assaying antitoxin to LT or cholera toxin (Evans *et al.* 1973).

LT antisera can be prepared in animals, and a wide range of immunological techniques is available for the detection of the toxin. A passive immune haemolysis method has been described in which sheep erythrocytes sensitized with LT are lysed by exposure to antitoxin and complement (Evans and Evans 1977). The tests are performed in Microtiter trays or in tubes and the results can be read spectrophotometrically. Yolken and his colleagues (1977*a*) devised an enzyme-linked immunosorbent assay (ELISA) with cholera antitoxin prepared in guinea-pigs and a goat anti-guinea-pig serum conjugated with alkaline phosphatase. This method was used in a study of diarrhoea in adults in Bangladesh and Kenya. A solid-phase radioimmunoassay (RIA) for *Esch. coli* LT was described by Ceska and his colleagues (1978); an LT antitoxin prepared in goats was bound to polystyrene tubes and LT was assayed by competitive inhibition of the binding of radioiodinated LT. A technique described by Bramucci and Holmes (1978) enabled passive immune haemolysis tests to be performed in media solidified with agar. In this way large numbers of individual colonies of *Esch. coli* or *V. cholerae* could be tested for LT production and mutants could be selected for genetic study. The discovery that the Gm1 ganglioside binds LT and cholera toxin led to the developement of a ganglioside ELISA test for the detection of these toxins (Sack *et al.* 1980). Microtiter wells were coated with Gm1 ganglioside obtained from commercial sources and the test samples were then added. Bound toxin was detected by the addition of guinea-pig cholera antitoxin followed by enzyme-labelled goat anti-guinea-pig serum. This method can also be used to measure antibodies that block the binding of the toxins to Gm1. A precipitin test (the Biken test) performed directly on bacterial colonies growing on a special agar medium has been evaluated

and may become widely used in field laboratories (Honda *et al.* 1981). Finally, a staphylococcal co-agglutination test (Ronnberg and Wadström 1983) and a latex-particle agglutination test (Finkelstein and Yang 1983) have been described; these may also prove to be valuable as simple screening tests for use in routine laboratories.

Detection of ST

Though few strains produce LT but not ST, almost one-half of enterotoxigenic strains of *Esch. coli* produce only ST (Merson *et al.* 1980). Thus it is essential to include tests for ST production in any survey of enterotoxigenicity. Unfortunately the usual tests are cumbersome, time-consuming and extravagant in their use of experimental animals.

Injection of enterotoxin preparations, both LT and ST, into ligated ileal loops of rabbits, pigs, calves and dogs leads to the accumulation of fluid (De and Chatterje 1953, De *et al.* 1956, Smith and Halls 1967, Nalin *et al.* 1978). The action of ST can be distinguished from that of LT both by its relative stability to heat and by the rapidity of its action. Nevertheless, such tests have not been widely used in investigations of human disease. So far it has proved impossible to devise tissue-culture tests for ST and the most widely used method is the infant-mouse test described by Dean and his colleagues (1972). Culture supernates are injected directly into the milk-filled stomach; after 4 hr the mice are killed and the intestines examined for dilatation with fluid. The intestines are also removed and the ratio of gut weight to remaining body weight is used as an objective measure of fluid accumulation. Although the test appears to work well for *Esch. coli* of human and bovine origin, some strains of porcine origin produce a heat-stable enterotoxin that is detectable by means of a ligated intestinal-loop test in pigs but not by the infant-mouse test (Gyles 1979).

The non-antigenic nature of ST has until recently prevented the development of immunological tests. Frantz and Robertson (1981) have now prepared antiserum to purified ST by coupling the toxin to a bovine serum-albumin carrier and the antiserum has been used in a radioimmunoassay. Further studies should soon lead to the development of more convenient test methods, including ELISA.

Antisera for the identification of ETEC

Surveys show that many ETEC, especially those that produce both ST and LT, fall within a restricted range of serogroups. It has been suggested that serogrouping might therefore provide a simple test for the preliminary recognition of ETEC in clinical laboratories. In a study in Bangladesh, 12 antisera were used for the examination of *Esch. coli* from 618 patients with acute diarrhoea. The survey yielded 798 *Esch. coli* isolates, of which 130 produced ST and LT, 115 produced ST and 9 produced LT. The use of agglutination tests in the local laboratory identified 110 of the 130 ST/LT strains (85 per cent) but only 46 of the 115 ST strains (40 per cent). Further studies are required in other geographical areas to evaluate the diagnostic value of ETEC antisera and to determine whether any alterations to the constituent O groups might improve the sensitivity of the method (Merson *et al.* 1980).

Genetic probes for the detection of ST and LT

Radiolabelled specific DNA fragments from enterotoxin genes may be used as 'probes' to detect homologous DNA sequences in *Esch. coli* strains. In this way the genes coding for ST or LT may be detected in ETEC strains or in stool or water samples containing ETEC. The method is complicated by the existence of more than one kind of ST, namely STa and STb (see p. 465). Furthermore, cloning experiments have shown that two different STa genes exist and two different probes are therefore required for their detection (Moseley *et al.* 1980, 1982).

Prevention and treatment of *Esch. coli* enteritis

The reader is referred to books edited by Lambert (1979*b*) and DuPont and Pickering (1980) for information about the treatment of *Esch. coli* enteritis. It should be noted, however, that the early correction of fluid and electrolyte imbalance is the most important single factor in preventing the death of the patient in severe infections (Report 1983*b*).

General preventive measures

The most important of these is to prevent exposure to the infecting agent. Contaminated food and water are probably the chief means by which ETEC are transmitted in developing countries, and the provision of safe supplies of water and the inculcation of hygienic practice in the handling of food, particularly that given to recently weaned children, are of the utmost importance. The traveller to these countries should take local advice before selecting eating-places and should depend upon hot food and drinks, or water in firmly sealed bottles, preferably aerated. Self-peeled fruits are probably safe, but salads should be avoided. Unheated milk should always be considered unsafe.

The spread of infantile enteritis in hospitals and nurseries is mainly from patient to patient by the indirect route—generally on the hands of attendants—or from contaminated infant feeds. It can be prevented only by very strict hygienic precautions. Infected patients, and recently admitted patients suspected of being infected, must be isolated by strict barrier-nursing techniques. In some cases outbreaks can be halted only by closing the ward or nursery and cleaning thoroughly all accessible surfaces before it is reopened.

Vaccination

Relevant studies on the use of vaccines have so far been made mainly in the veterinary field. Dobrescu and Huygelen (1976) showed that LT preparations administered to pregnant sows protected newborn piglets against challenge with LT or ST. Pierce (1977) showed that immunization of rats with cholera toxoid protected against challenge with LT. A report in 1978 described the use of heated antigens derived from a number of enteropathogenic serotypes of *Esch. coli* as an additive to the diet of pigs at weaning. Although such preparations would not be expected to contain enterotoxin in an immunogenic form, it was found that the pigs became insensitive to ST and LT (Linggood and Ingram 1978). In the same year it was shown that vaccination of pregnant sows with purified K99 or 987P pili could protect suckling pigs against subsequent challenge with ETEC strains possessing these adhesive factors (Morgan *et al.* 1978). More recently, a potential vaccine for human use has been prepared by cross-linking a synthetically produced ST with the non-toxic B subunit of LT. Tests in the rat showed that the vaccine protected against subsequent challenge with ST or LT or with organisms that produced ST or LT (Klipstein *et al.* 1983). Future vaccines for human use may be based on toxoids or on preparations of colonization-factor antigens. Alternatively, genetic-engineering studies may soon lead to the development of non-pathogenic *Esch. coli* strains that produce colonization-factor antigens or non-toxic forms of ST and LT; these strains may prove valuable as live vaccines.

Inhibition of enterotoxin activity

A number of substances have been shown to inhibit or reverse the secretory effects of enterotoxins in experimental animals and may be of value in the prevention or treatment of diarrhoea. Ohgke and Wagner (1977) showed that phenylbutazone administered subcutaneously to the infant mouse 30 min before intragastric injection of ST resulted in a significant reduction in fluid accumulation. Drucker and his colleagues (1977) showed that activated attapulgite or charcoal given before or simultaneously with LT or cholera toxin prevented fluid accumulation in the rabbit ileal loop. Nicotinic acid is known to lower tissue levels of cAMP; Turjman and his colleagues (1978) showed that subcutaneous or intraluminal injection of nicotinic acid in rabbits blocked the rise in cAMP and fluid secretion caused by cholera toxin in the ileal loop. Madsen and Knoop (1978) examined the effect of indomethacin on the secretion of fluid caused by ST in the infant mouse and reported a significant decrease. Subsalicylate bismuth inhibits the effect of crude *V. cholerae* and *Esch. coli* enterotoxins in the rabbit intestinal loop. DuPont and his colleagues (1980) undertook a controlled trial of daily doses of this substance for the prophylaxis of travellers' diarrhoea in young adults visiting Mexico from the United States. Treated persons experienced fewer intestinal complaints and were less likely to pass loose stools than control subjects. Once diarrhoea developed, enteropathogens were less commonly found in stools of those receiving subsalicylate bismuth than in those receiving a placebo.

Further clinical trials of such agents appear to be warranted, especially in view of the contraindications to the widespread use of antimicrobial drugs for prophylaxis (see below).

(For the prevention and cure of diarrhoea due to *Esch. coli* in animals, see Smith and Huggins 1983.)

Antimicrobial prophylaxis

There is little doubt that a number of antimicrobial drugs reduce the incidence of diarrhoea in travellers to tropical areas. These include pthalylsulphathiazole, neomycin, streptotriad and doxycycline (Kean *et al.* 1962, Kean 1963, Turner 1967, Sack *et al.* 1978). Nevertheless, the widespread use of antibiotic prophylaxis has been criticized both on the grounds of drug toxicity and the possibility that the development and spread of drug resistance might be encouraged among a variety of enteropathogenic organisms. Plasmids have been discovered which carry the genetic determinants of both drug resistance and enterotoxin production and these plasmids can be transferred from one strain of *Esch. coli* to another (McConnell *et al.* 1979, Scotland *et al.* 1979). Antibiotics might therefore encourage the spread of enterotoxigenicity as well as drug resistance and thus increase the incidence of drug-resistant enterotoxigenic strains of *Esch. coli*.

Campylobacter enteritis

This disease—now the most frequently reported form of human bacterial diarrhoea in Britain—went largely unrecognized until, in the early 1970s, methods for isolating the causative organism from faeces were devised. As explained in Chapter 27, the thermophilic campylobacter responsible for enteritis in man is known under the American system of classification (Smibert 1974, 1978) as *C. fetus* subsp. *jejuni*, and under the French system (Véron and Chatelain 1973) as *C. jejuni* (and *C. coli*).

Butzler and his colleagues (1973) in Belgium and Skirrow (1977) in England found the organism in about 5–7 per cent of children and adults with diarrhoea. It has occasionally been found in as many as 14 per cent. In 1981 the number of reported isolations in England and Wales rose to 12 496. The disease is now known in many parts of the world, including the tropics. Infections occur in all age groups and at all times of the year but are commonest in older children and young adults and, in temperate climates, in the summer. They are a frequent cause of travellers' diarrhoea.

The incubation period is usually 2–5 days but can be as long as 10 days. Fever is accompanied by abdominal pain and acute diarrhoea lasting 2 or 3 days, and campylobacters can sometimes be cultured from the blood. The small intestine is the main site of infection, but the colon is sometimes affected. Blood is often present in the faeces. The organism persists in the faeces of untreated patients for periods ranging from a few days to 7 weeks after clinical recovery but usually for 2–3 weeks. Symptomless or mild infections are not uncommon in those closely associated with typical cases.

On rare occasions the abdominal pain is severe enough to lead to exploratory laparotomy. Other unusual complications include cholecystitis, grand-mal seizure, and lingering sterile arthritis. Bacteraemia, when it occurs, is probably confined to the early phase of the infection.

Agglutinin to the infecting strain appears on about the 5th day of the disease and quickly reaches its maximal titre, which can be as high as 10 240. Antibody subsides over a period of several months.

Sources of infection

Animals probably constitute the main reservoir. The organism has been isolated from the intestinal contents or faeces of normal calves, sheep, pigs, horses, dogs, cats, simian primates, poultry, wild birds, and other animals (see Chapter 56). The caecal contents of as many as 35 per cent of migratory waterfowl were found by Luechtefeld and her colleagues (1980) to be infected. Many animal isolates are indistinguishable by current typing methods from human strains, but whether they are all potentially pathogenic for man is still uncertain. The organism is present in the faeces of less than 1 per cent of normal human beings.

The origin of sporadic infections in man is seldom discovered, but infection is probably transferred from animals or raw animal products to man by means of food. Poultry carcasses are frequently contaminated. A small proportion of human cases, especially in children, may be attributable to direct infection from puppies with diarrhoea. The disease may occasionally occur as the result of transmission from one person—

usually a child—to another. Contaminated water supplies (Mentzing 1981, Vogt *et al.* 1982) and unpasteurized milk (P. H. Jones *et al.* 1981, Robinson and D. M. Jones 1981) are known to have caused serious outbreaks.

A comprehensive system for subdividing strains of the organism would be of great value in tracing the sources of human infections. For progress in serotyping methods see Report (1982, 1983a), Skirrow (1982), and Chapter 27.

Laboratory diagnosis

Darkground microscopical examination of faeces will often reveal the rapid darting movements, axial spinning, and spiral form of the causative campylobacter.

A method used earlier in veterinary laboratories (Shepler *et al.* 1963, Smibert 1965) was adopted for the isolation of campylobacters from human faeces (Dekeyser *et al.* 1972, Butzler *et al.* 1973). Filtration through a 0.65-μm Millipore filter removed most bacteria other than campylobacters. Cultures could then be made on solid media with or without selective substances.

A more convenient and sensitive method is the direct culture of faeces on Butzler's or Skirrow's selective medium at 42–43° in an atmosphere of reduced oxygen and increased carbon dioxide (see Butzler and Skirrow 1979, and Chapter 27). The diagnosis depends on the growth of flat, glossy, effuse colonies containing organisms that are oxidase positive and possess characteristic campylobacter morphology and motility.

Treatment

Most patients will recover satisfactorily without antimicrobial treatment. Erythromycin, tetracyclines, and furazolidone are among the more useful agents. Erythromycin resistance occurs in a small proportion of strains, and in septicaemic or seriously ill patients the use of gentamicin or chloramphenicol is advisable.

Other campylobacter infections in man

The organism referred to as *C. fetus* subsp. *intestinalis* by Smibert (1974, 1978) and *C. fetus* subsp. *fetus* by Véron and Chatelain (1973) occasionally causes systemic infections in patients whose resistance has been reduced, for example by cirrhosis, diabetes, cancer, or immunosuppressive therapy.

(For further information on campylobacter infections in man, especially enteritis, see Butzler and Skirrow 1979, Rettig 1979, Report 1982, 1983a, and Skirrow 1982.)

Other bacterial causes of acute enteritis

Organisms other than *Esch. coli* and *V. cholerae* have been found to produce enterotoxins and may cause acute diarrhoeal disease. Thus, in a study of Ethiopian children with diarrhoea, Wadström and his colleagues (1976) reported the isolation of enterotoxigenic strains of *Aeromonas*, *Citrobacter*, *Enterobacter*, *Klebsiella*, *Proteus* and *Serratia*. For the most part the significance of such organism awaits confirmation.

Aeromonas

The production of an enterotoxin by strains of *Aeromonas* has been confirmed (Annapurna and Sanyal 1977, Ljungh *et al.* 1977, Donta and Haddow 1978), but until recently the evidence that this organism was responsible for diarrhoea was scanty. In a prospective study of diarrhoea in children in Perth, Western Australia, Gracey and his colleagues (1982) isolated enterotoxigenic strains of *Aeromonas* from 10 per cent of over 1000 affected children but from only 0.6 per cent of age- and sex-matched controls. The disease was most frequent in children under 2 years of age and in the warmer months. It was usually characterized by watery diarrhoea, but one-fifth of the sufferers had bloody stools. The disease was often of short duration but lasted for more than 2 weeks in nearly 40 per cent of the patients.

Yersinia enterocolitica

Evidence accumulated in recent years suggests that *Y. enterocolitica* may be an important cause of diarrhoeal disease in many countries, especially in Scandinavia (Bottone 1977). The disease affects mainly infants and young children. In addition to sporadic cases, outbreaks may occur among persons in close contact, such as families, schools and hospitals. Infected patients may have dysentery-like symptoms or a profuse, watery diarrhoea. *Y. enterocolitica* is invasive in HeLa cells (Lee *et al.* 1977, Pedersen *et al.* 1979) and also produces a heat-stable enterotoxin (Pai *et al.* 1978, Velin *et al.* 1980). The dysenteric form of the disease may reflect the invasive nature of the organism and the profuse diarrhoea the small-intestinal secretion induced by enterotoxin activity. The symptoms may also include severe abdominal pain suggestive of acute appendicitis and this may lead to laparotomy, particularly in children and young adults. Enteritis associated with *Y. enterocolitica* is most commonly due to members of serogroups 3 and 9 and, less frequently, serogroup 8. However, in the United States, serogroup 8 appears to be the most common, and serogroups 3 and 9 are rarely reported. (See also Chapter 55).

Viral enteritis

Bacteriological studies fail to demonstrate an aetiological agent in a substantial proportion of cases of acute diarrhoeal disease. However, such studies rarely include a search for all the known bacterial enteropathogens and the role of bacteria is likely to be underestimated. Nevertheless, it has long been assumed that viruses are responsible for a high proportion of cases of acute diarrhoea. Only recently has the application of electron microscopy to the examination of negatively stained extracts of faecal specimens led to a confirmation of the role of viruses, particularly rotaviruses and parvovirus-like agents. These two groups of viruses each cause a clinically and epidemiologically distinct type of disease. (For further information on the role of viruses as a cause of enteritis see Chapter 100.)

Rotavirus infections

Rotaviruses were first recognized in duodenal-biopsy specimens from young children with acute enteritis in a Melbourne hospital (Bishop *et al.* 1973). Similar particles were soon detected in faecal extracts from children with acute enteritis in many parts of the world (Bishop *et al.* 1974, Kapikian *et al.* 1974, *Lancet*, 1975). The virus has a distinctive structure, with a smooth outer capsid surrounding spoke-like inner capsid subunits which radiate outwards from the core (Fig. 71.4). The genome consists of 11 segments of

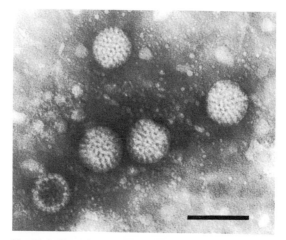

Fig. 71.4   Rotavirus, showing rough and smooth particles; bar represents 100 nm. (Electronmicrograph kindly provided by Dr E. O. Caul.)

double-stranded RNA. Because of this and other physico-chemical characters the rotaviruses are now included in the family *Reoviridae* (Kalica *et al.* 1976, Schnagl and Holmes 1976).

Epidemiology

Rotaviruses are the most common cause of acute non-bacterial enteritis in infants and children. In contrast to 'winter vomiting disease' (p. 472) the symptoms may be severe, necessitating admission to hospital and the parenteral administration of fluids, but in the newborn the disease is usually mild (Chrystie *et al.* 1978). In temperate climates, infection is most frequent among children aged between 6 months and 2 years; the great majority of cases occur during the winter. During this period up to 75 per cent of children admitted to hospital with acute enteritis may be excreting rotaviruses (Davidson *et al.* 1975). Rotaviruses are rarely detected in symptomless children. The virus is excreted in large numbers in the stools of infected patients and the consequent environmental contamination may lead to cross-infection (*Lancet*, 1975). Rotaviruses may also infect adults though symptoms are usually mild or absent; most older children and adults have antibody (Elias 1977).

Rotaviruses are an important cause of acute enteritis in tropical countries, especially in children under 2 years of age. For example, Ryder and his colleagues (1976*b*) found in a survey in Bangladesh that they were the most frequent cause of severe diarrhoea in children under 2 years of age causing 55 per cent of cases in this group. Rotaviruses were not found in older children or adults with diarrhoea. In contrast, enterotoxigenic strains of *Esch. coli* were not found in children under 2 years of age but were associated with 56 per cent of cases among older persons. The seasonal incidence of rotavirus infection is less obvious in the tropics; in many countries infections occur throughout the year, but the incidence may be higher in the cooler or rainy months (Cruikshank and Silberg 1976, Maiya *et al.* 1977, Schnagl *et al.* 1977).

Pathogenesis

The main feature of rotavirus infection is a loss of the absorptive cells lining the small intestine. The earliest lesion is detected at the proximal end of the small intestine, but its whole length may be affected in severe cases. The diarrhoea is due to a disordered electrolyte transport and defective absorption resulting from the migration of immature cuboidal cells from the crypts to the villi (Tallet *et al.* 1977). Experimental infections are less severe than natural ones. It has been suggested that the presence of *Esch. coli* in the intestine may potentiate the pathogenic effects of rotaviruses in some way (Banatvala 1979).

Laboratory diagnosis   Electron microscopy is the most widely used method for detecting rotaviruses in faeces, but it is inconvenient for the examination of large numbers of specimens. Serological methods are also available, including complement fixation (Zissis *et al*. 1978) and ELISA (Yolken *et al*. 1977*b*). Human rotaviruses have recently been cultured *in vitro* (Wyatt *et al*. 1980), but this is not done routinely. However, if faecal filtrates are centrifuged on to continuous cell monolayers, rotavirus antigen can be detected by immunofluorescence, but infectious virus is not produced (Banatvala *et al*. 1975, Totterdell *et al*. 1976).

Infections due to parvovirus-like agents

Zahorsky (1929) first described as '*winter vomiting disease*' epidemic nausea and vomiting affecting both children and adults and occurring most commonly during the winter. Symptoms may include fever, anorexia, nausea, vomiting, vertigo, myalgia, colic and diarrhoea. The incubation period is 24–48 hr and the symptoms persist for 2–3 days. Secondary cases are common in household contacts. The infectious nature of the disease was confirmed by its transmission to volunteers who ingested filtrates of faeces from affected patients (Reimann *et al*. 1945, Jordan *et al*. 1953). Kapikian and his colleagues (1972) first detected the virus by immune electron microscopy of stool specimens during the investigation of an outbreak of enteritis in a primary school in Norwalk, Ohio. Morphologically similar 25–27-nm virus particles were subsequently found in patients in small family outbreaks of enteritis in Montgomery County, Maryland, and in Hawaii (Wyatt *et al*. 1974). Human challenge experiments and immune electron microscopy showed that the Norwalk and Montgomery County agents were immunologically related to each other but were distinct from the Hawaii agent (Wyatt *et al*. 1974, Thornhill *et al*. 1977). Similar virus particles have also been reported in outbreaks of winter vomiting disease in England. The 'W' agent was found during the investigation of an outbreak in an English boarding school (Clarke *et al*. 1972) and the 'Ditchling' agent was identified during an outbreak of acute enteritis in children and staff in a primary school near Brighton (Appleton *et al*. 1977). The 'cockle' agent was associated with food-poisoning outbreaks affecting persons who had recently eaten shell-fish, but virus particles could not be found in this food (Appleton and Pereira 1977). Immune electron microscopy suggests that the 'W' and 'Ditchling' agents are immunologically related but differ from the 'cockle' agent, and that all three differ from the Norwalk, Montgomery County and Hawaii agents (Appleton and Pereira 1977, Appleton *et al*. 1977).

On the basis of their morphology, buoyant density, and heat, acid and ether stability, it has been suggested that these particles should be regarded as parvovirus-like agents. However, their definitive classification awaits further studies, including nucleic-acid analysis.

Radioimmunoassay and immune-adherence haem-agglutination techniques have been used to study the sero-epidemiology of the Norwalk agent. Kapikian and his colleagues (1978) showed that in the United States antibody is acquired gradually with increasing age, so that by the age of 40–50 years about 50 per cent of persons have antibody. A more complete understanding of the epidemiology of these agents awaits the application of similar techniques to other members of the group in different areas of the world.

Other viruse diarrhoeas

A number of other viruses have also been found more frequently in the faeces of patients with diarrhoea than in healthy persons, but further evidence is required before their significance can be properly assessed.

Adenoviruses   Outbreaks of acute viral enteritis in children have been attributed to adenoviruses (Flewett *et al*. 1975, Chiba *et al*. 1983).

Coronaviruses   These are an important cause of enteritis in pigs and calves, but their role in human diarrhoeal disease has yet to be established. Coronavirus-like particles have been observed in patients with both acute and chronic diarrhoea but are also found in the absence of symptoms. That the particles are true viruses has been demonstrated by their propagation in cell and organ cultures (Caul and Clarke 1975, Caul and Egglestone 1977).

'Small round' viruses   A number of morphologically distinct viruses 25–30 nm in diameter have been observed in the faeces of diarrhoea patients. Astroviruses, so called because of their star-shaped surface configuration (Fig. 71.5), have been implicated as a cause of diarrhoea in babies in Glasgow (Madeley 1979) and in an outbreak affecting children and staff in a hospital ward elsewhere (Kurtz *et al*. 1977). Volunteers who ingested faecal filtrates containing astroviruses showed mild symptoms, excreted the virus and showed an immune response (Reed *et al*. 1978). Caliciviruses are picornaviruses which are known to be animal pathogens. Calicivirus-like particles have been found in both ill and

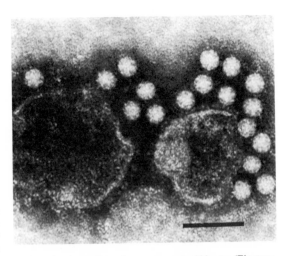

Fig. 71.5   Astrovirus; bar represents 100 nm. (Electronmicrograph by courtesy of Dr E. O. Caul.)

symptomless children (Madeley 1979), and McSwiggan and his colleagues (1978) reported their detection in children and staff in an outbreak of 'winter vomiting' in an English primary school. Finally, certain particles with an indistinct outline and termed 'fuzzy-wuzzies' have been detected in faecal filtrates, but their association with diarrhoeal disease is uncertain (Flewett 1978). (See Chapter 100.)

Enteritis in animals

Diarrhoeal disease of young domestic animals resembles human infantile enteritis in many respects. Indeed our understanding of the role of both enterotoxigenic *Esch. coli* and viruses in human enteritis owes a great deal to studies in the veterinary field. Frequent mention of animal enteritis has therefore been made in the preceding parts of this chapter and only a short summary will be included here.

Acute enteritis of newborn pigs, calves and lambs is sometimes known as '*enterotoxic colibacillosis*'. Experimental evidence for the role of enterotoxigenic *Esch. coli* in the disease was obtained by Smith and Halls (1967) who used the ligated intestinal-loop technique to demonstrate the enterotoxins produced by *Esch. coli* isolated from various animals with diarrhoea. Further studies established that in order to cause diarrhoea an enterotoxigenic strain of *Esch. coli* must also be able to colonize the small intestine. This ability depends on the possession of host-specific adhesive factors (p. 465) including the K88 antigen of strains causing piglet enteritis and the K99 antigen of strains affecting calves and lambs (Ørskov and Ørskov 1966, Smith and Linggood 1971, Jones and Rutter 1974, Ørskov *et al.* 1975).

Enterotoxaemic colibacillosis, which includes *oedema disease of swine*, is characterized by the proliferation of *Esch. coli* in the small intestine, where a toxin is produced which acts elsewhere. Oedema disease causes sudden death in pigs aged 8–12 weeks with oedema of the internal organs resulting from an increased vascular permeability. The *Esch. coli* strains responsible usually belong to serogroups O138, O139 and O141 (Rees 1959) and the disease has been reproduced by oral challenge (Smith and Halls 1968). Clugston and Nielsen (1974) undertook the preliminary characterization of the toxin prepared *in vitro*, and Clugston and colleagues (1974) emphasized its vasotoxicity, referring to it as the 'oedema disease principle'.

The *calf rotavirus* was the first rotavirus to be detected by electron microscopy, having been found in the stools of gnotobiotic calves inoculated by intraduodenal tube with filtrates of diarrhoeal calf faeces (Mebus *et al.* 1969). Rotaviruses have also been detected in many other mammalian species including infant mice, piglets, lambs, foals, rabbits, deer, goats and apes (Ashley *et al.* 1978, Flewett and Woode 1978, Scott *et al.* 1978). Infection probably occurs in animals of all ages but is most frequent and severe between birth and weaning. Rotaviruses are the commonest cause of calf diarrhoea (Woode 1976) and are of considerable economic importance among pigs and calves, since infection is often fatal (Flewett and Woode 1978). In addition, coronaviruses are an important cause of the severe diarrhoea known as *transmissible gastro-enteritis of pigs and calves* (Tajima 1970, Stair *et al.* 1972).

References

Alderete, J. F. and Robertson, D. C. (1978) *Infect. Immun.* **19**, 1021.
Annapurna, E. and Sanyal, S. C. (1977) *J. med. Microbiol.* **10**, 317.
Appleton, H., Buckley, M., Thom, B. T., Cotton, J. L. and Henderson, S. (1977) *Lancet* **i**, 409.
Appleton, H. and Pereira, M. S. (1977) *Lancet* **i**, 780.
Ashley, C. R., Caul, E. O., Clarke, S. K. R., Corner, B. D. and Dunn, S. (1978) *Lancet* **ii**, 477.
Bäck, E., Blomberg, S. and Wadström, T. (1977) *Infection* **5**, 2.
Banatvala, J. E. (1979) *Clin. Gastroenterol.* **8**, 569.
Banatvala, J. E., Totterdell, B., Chrystie, I. L. and Woode, G. N. (1975) *Lancet* **ii**, 821.
Bishop, R. F., Davidson, G. P., Holmes, I. H. and Ruck, B. J. (1973) *Lancet* **ii**, 1281; (1974) *Ibid.* **i**, 149.
Black, R. E., Merson, M. H., Rowe, B., Taylor, P. R., Abdul Alim, A. R. M., Gross, R. J. and Sack, D. A. (1981) *Bull. World Hlth Org.* **59**, 263.
Bottone, E. J. (1977) *CRC Crit. Rev. Microbiol.* **5**, 211.
Bramucci, M. G. and Holmes, R. K. (1978) *J. clin. Microbiol.* **8**, 252.
Bray, J. (1945) *J. Path. Bact.* **57**, 239.
Burgess, M. N., Bywater, R. J., Cowley, C. M., Mullan, N. A. and Newsome, P. M. (1978) *Infect. Immun.* **21**, 526.
Burnham, G. M., Scotland, S. M., Gross, R. J. and Rowe, B. (1976) *Brit. med. J.* **ii**, 1256.
Butzler, J. P., Dekeyser, P., Detrain, M. and Dehaen, F. (1973) *J. Pediat.* **82**, 493.
Butzler, J. P. and Skirrow, M. B. (1979) *Clin. Gastroenterol.* **8**, 737.
Cantey, J. R. and Blake, R. K. (1977) *J. infect. Dis.* **135**, 454.
Caul, E. O. and Clarke, S. K. R. (1975) *Lancet* **ii**, 953.
Caul, E. O. and Egglestone, S. I. (1977) *Arch. Virol.* **54**, 107.
Ceska, M., Grossmüller, F. and Effenberger, F. (1978) *Infect. Immun.* **19**, 347.
Chiba, S., Nakata, S., Nakamura, I., Taniguchi, K., Urasawa, S., Fujinaga, K. and Nakao, T. (1983) *Lancet* **ii**, 954.
Chrystie, I. L., Totterdell, B. M. and Banatvala, J. E. (1978) *Lancet* **i**, 1176.
Clansen, C. R. and Christie, D. L. (1982) *J. Pediat.* **100**, 358.
Clarke, S. K. R. *et al.* (1972) *Brit. med. J.* **ii**, 86.
Clements, J. D. and Finkelstein, R. A. (1978) *Infect. Immun.* **22**, 709.
Clugston, R. E. and Nielsen, N. O. (1974) *Canad. J. comp. Med.* **38**, 22.
Clugston, R. E., Nielsen, N. O. and Smith, D. L. T. (1974) *Canad. J. comp. Med.* **38**, 34.
Cravioto, A., Gross, R. J., Scotland, S. M. and Rowe, B. (1979a) *Curr. Microbiol.* **3**, 95; (1979b) *FEMS Microbiol. Lett.* **6**, 41.
Cruikshank, J. G. and Zilberg, B. (1976) *S. Afr. J. Med.* **50**, 1895.

Dallas, W. S. and Falkow, S. (1980) *Nature, Lond.* **288**, 499.

Dallas, W. S., Gill, D. M. and Falkow, S. (1979) *J. Bact.* **139**, 850.

Darfeuille, A., Lafeuille, B., Joly, B. and Cluzel, R. (1983) *Ann. Microbiol. Inst. Pasteur* **A134**, 53.

Davidson, G. P., Bishop, R. F., Townely, R. R. W., Holmes, I. H. and Ruck, B. J. (1975) *Lancet* **i**, 242.

De, S. N., Bhattacharya, K. and Sarkar, J. K. (1956) *J. Path. Bact.* **71**, 201.

De, S. N. and Chatterje, D. N. (1953) *J. Path. Bact.* **66**, 559.

Dean, A. G., Ching, Y., Williams, R. G. and Harden, L. B. (1972) *J. infect. Dis.* **125**, 407.

Dekeyser, P., Gossuin-Detrain, M., Butzler, J. P. and Sternon, J. (1972) *J. infect. Dis.* **125**, 390.

Dobrescu, L. and Huygelen, C. (1976) *Zentbl. Vet.* **B23**, 79.

Donta, S. T. and Haddow, A. D. (1978) *Infect. Immun.* **21**, 989.

Donta, S. T., Moon, H. W. and Whipp, S. C. (1974) *Science* **183**, 334.

Donta, S. T., Wallace, R. B., Whipp, S. C. and Olarte, J. (1977) *J. infect. Dis.* **135**, 482.

Dorner, F., Hughes, C., Nahler, G. and Högenauer, G. (1979) *Proc. nat. Acad. Sci., Wash.* **76**, 4832.

Drucker, M. M., Goldhar, J., Ogra, P. L. and Neter, E. (1977) *Infection* **5**, 211.

DuPont, H. L. and Pickering, L. K. (Eds) (1980) *Infections of the Gastrointestinal Tract*. Plenum Publishing Co., New York.

DuPont, H. L. *et al.* (1971) *New Engl. J. Med.* **285**, 1; (1980) *J. Amer. med. Ass.* **243**, 237.

Echeverria, P., Blacklow, N. R. and Smith, D. H. (1975) *Lancet* **ii**, 1113.

Echeverria, P., Blacklow, N. R., Zipkin, K., Vellet, J. J., Olson, J. A., DuPont, H. L. and Cross, J. H. (1979a) *Amer. J. Epidem.* **109**, 493.

Echeverria, P., Ramirez, G., Blacklow, N. R., Ksiazek, T., Cukor, G. and Cross, J. H. (1979b) *J. infect. Dis.* **139**, 215.

Echeverria, P., Verhaert, L., Basaca-Sevilla, V., Banson, T., Cross, J., Ørskov, F. and Ørskov, I. (1978b) *J. infect. Dis.* **138**, 87.

Echeverria, P. *et al.* (1977) *J. infect. Dis.* **136**, 383; (1978a) *Ibid.* **138**, 326.

Edwards, P. R. and Ewing, W. H. (1972) *Identification of Enterobacteriaceae*, 3rd edn. Burgess Publishing Co., Minneapolis.

Elias, M. M. (1977) *J. Hyg., Camb.* **79**, 373.

Escherich, T. (1885) *Fortschr. Med.* **3**, 515.

Evans, D. G. and Evans, D. J. (1978) *Infect. Immun.* **21**, 638.

Evans, D. G., Evans, D. J. and DuPont, H. L. (1977a) *J. infect. Dis.* **136**, S118.

Evans, D. G., Evans, D. J. and Tjoa, W. (1977b) *Infect. Immun.* **18**, 330.

Evans, D. G., Olarte, J., DuPont, H. L., Evans, D. J., Galindo, E., Portnoy, B. L. and Conklin, R. H. (1977c) *J. Pediat.* **91**, 65.

Evans, D. G., Silver, R. P., Evans, D. J., Chase, D. G. and Gorbach, S. L. (1975) *Infect. Immun.* **12**, 656.

Evans, D. J. and Evans, D. G. (1977) *Infect. Immun.* **16**, 604.

Evans, D. J., Evans, D. G. and Gorbach, S. L. (1973) *Infect. Immun.* **8**, 725.

Ferguson, W. W. and June, R. C. (1952) *Amer. J. Hyg.* **55**, 155.

Field, M. (1979) *Rev. infect. Dis.* **1**, 918.

Field, M., Graf, L. H., Laird, W. J. and Smith, P. L. (1978) *Proc. nat. Acad. Sci., Wash.* **75**, 2800.

Finkelstein, R. A. and Yang, Z. (1983) *J. clin. Microbiol.* **18**, 23.

Flewett, T. H. (1978) *J. Amer. vet. med. Ass.* **173**, 538.

Flewett, T. H., Bryden, A. S., Davies, H. and Morris, C. A. (1975) *Lancet* **i**, 4.

Flewett, T. H. and Woode, G. N. (1978) *Arch. Virol.* **57**, 1.

Formal, S. B. and Hornick, R. B. (1978) *J. infect. Dis.* **137**, 641.

Frantz, J. C. and Robertson, D. C. (1981) *Infect. Immun.* **33**, 193.

Gangarosa, E. J. (1978) *J. infect. Dis.* **137**, 634.

Giannella, R. A. and Staples, S. J. (1979) *Proc. 15th Jt Conf. Cholera*. US Department of Health, Education and Welfare, National Institute of Health, Bethesda, Maryland.

Giles, C. and Sangster, G. (1948) *J. Hyg., Camb.* **46**, 1.

Goldschmidt, R. (1933) *Jb. Kinderh.* **89**, 318.

Gorbach, S. L., Kean, B. H., Evans, D. G., Evans, D. J. and Bessudo, D. (1975) *New Engl. J. Med.* **292**, 933.

Gorbach, S. L. and Khurana, C. M. (1972) *New Engl. J. Med.* **287**, 791.

Gracey, M., Burke, V. and Robinson, J. (1982) *Lancet* **ii**, 1304.

Gross, R. J., Cravioto, A., Scotland, S. M., Cheasty, T. and Rowe, B. (1978) *FEMS Microbiol. Lett.* **3**, 231.

Gross, R. J., Rowe, B., Henderson, A., Byatt, M. E. and MacLaurin, J. C. (1976b) *Scand. J. infect. Dis.* **8**, 195.

Gross, R. J., Scotland, S. M. and Rowe, B. (1976a) *Lancet* **i**, 629; (1979) *Brit. med. J.* **i**, 1463.

Guerrant, R. L., Brunton, L. L., Schnaitman, T. C., Rebhun, L. I. and Gilman, A. G. (1974) *Infect. Immun.* **10**, 320.

Gurwith, M. J. and Williams, T. W. (1977) *J. infect. Dis.* **136**, 239.

Gyles, C. L. (1979) *Canad. J. comp. Med.* **43**, 371.

Hinden, E. (1948a) *Arch. Dis. Childh.* **23**, 27; (1948b) *Ibid.* **23**, 33.

Hobbs, B. C., Rowe, B., Kendall, M., Turnbull, P. C. B. and Ghosh, A. C. (1976) *J. Hyg., Camb.* **77**, 393.

Honda, T., Taga, S., Takeda, Y. and Miwatani, T. (1981) *J. clin. Microbiol.* **13**, 1.

Hone, R., Fitzpatrick, S., Keane, C., Gross, R. J. and Rowe, B. (1973) *J. med. Microbiol.* **6**, 505.

Isaacson, R. E., Nagy, B. and Moon, H. W. (1977) *J. infect. Dis.* **22**, 771.

Jacobs, S. I., Holzel, A., Wolman, B., Keen, J. H., Miller, V., Taylor, J. and Gross, R. J. (1970) *Arch. Dis. Childh.* **45**, 656.

Johnson, W. M., Lior, H. and Bezanson, E. S. (1983) *Lancet* **i**, 76.

Jones, G. W. and Rutter, J. M. (1974) *J. gen. Microbiol.* **84**, 135.

Jones, P. H., Willis, A. T., Robinson, D. A., Skirrow, M. B. and Josephs, D. S. (1981) *J. Hyg., Camb.* **87**, 155.

Jordan, W. S., Gordon, I. and Dorrance, W. R. (1953) *J. exp. Med.* **98**, 461.

Kalica, A. R., Garon, C. F., Wyatt, R. G., Mebus, C. A., Kirk, D. H. van, Chanock, R. M. and Kapikian, A. Z. (1976) *Virology* **74**, 86.

Kapikian, A. Z., Wyatt, R. G., Dolin, R., Thornhill, R. S., Kalica, A. R. and Chanock, R. M. (1972) *J. Virol.* **10**, 1075.

Kapikian, A. Z. *et al.* (1974) *Science* **185**, 1049; (1978) *J. med. Virol.* **2**, 281.

Karmali, M. A., Petric, M., Steele, B. T. and Lin, C. (1983) *Lancet* **i**, 619.

Kauffmann, F. (1947) *J. Immunol.* **57**, 71.

Kean, B. H. (1963) *Ann. intern. Med.* **59**, 605.

Kean, B. H., Schaffner, W., Brennan, R. W. and Waters, S. R. (1962) *J. Amer. med. Ass.* **162**, 367.

Kennedy, D. H., Walter, G. H., Fallon, R. J., Boyd, J. F., Gross, R. J. and Rowe, B. (1973) *J. clin. Path.* **26**, 731.

Klipstein, F. A., Engert, R. F., Clements, J. D. and Haughten, R. A. (1983) *J. infect. Dis.* **147**, 318.

Klipstein, F. A., Rowe, B., Engert, R. F., Short, H. B. and Gross, R. J. (1978) *Infect. Immun.* **21**, 171.

Konowalchuk, J., Speirs, J. I. and Stavric, S. (1977) *Infect. Immun.* **18**, 775.

Koya, G., Kosakai, N., Kono, M., Mori, M. and Fukasawa, Y. (1954) *Jap. J. med. Sci. Biol.* **7**, 197.

Kudoh, Y., Zen-Yoji, H., Matsushita, S., Sakai, S. and Maruyama, T. (1977) *Microbiol. Immunol.* **21**, 175.

Kurtz, J. B., Lee, T. W. and Pickering, D. (1977) *J. clin. Path.* **30**, 948.

Lambert, H. P. (1979*a*) In: *Recent Advances in Infection*, p. 109. Ed. by D. Reeves and A. Geddes. Churchill Livingstone, Edinburgh; (Ed.) (1979*b*) *Clin. Gastroenterol.* **8**, No. 3.

Lancet Editorial (1936) **ii**, 85; Annotation (1968) **i**, 32; Editorial **i**, (1975) 257.

Lathe, R., Hirth, P., Wilde, M. de, Harford, N. and Lecocq, J.-P. (1980) *Nature, Lond.* **284**, 473.

Lee, W. H., McGrath, P. P., Carter, P. H. and Eide, E. L. (1977) *Canad. J. Microbiol.* **23**, 1714.

Lesage, A. A. (1897) *C. r. Séanc. Soc. Biol.* **49**, 900.

Levine, M. M., Rennels, M. B., Daya, V. and Hughes, T. P. (1980) *J. infect. Dis.* **141**, 753.

Levine, M. M. *et al.* (1978) *Lancet* **i**, 119.

Linggood, M. A. and Ingram, P. L. (1978) *Res. vet. Sci.* **25**, 113.

Ljungh, A., Popoff, M. and Wadström, T. (1977) *J. clin. Microbiol.* **6**, 96.

Love, W. C., Gordon, A. M., Gross, R. J. and Rowe, B. (1972) *Lancet* **ii**, 355.

Luechtefeld, N. A. W., Blaser, M. J., Reller, L. B. and Wang, W.-L. L. (1980) *J. clin. Microbiol.* **12**, 406.

McConnell, M. M., Willshaw, G. A., Smith, H. R., Scotland, S. M. and Rowe, B. (1979) *J. Bact.* **139**, 346.

McNeish, A. S., Turner, P., Fleming, J. and Evans, N. (1975) *Lancet* **ii**, 946.

McSwiggan, D. A., Cubitt, D. and Moore, W. (1978) *Lancet* **i**, 1215.

Madeley, C. R. (1979) *J. clin. Path.* **32**, 1.

Madsen, G. L. and Knoop, F. C. (1978) *Infect. Immun.* **22**, 143.

Maiya, P. P., Pereira, S. M., Mathan, M., Bhat, P., Albert, M. J. and Baker, S. J. (1977) *Arch. Dis. Childh.* **52**, 482.

Mebus, C. A., Underdahl, N. R., Rhodes, M. B. and Twiehaus, M. J. (1969) *Univ. Nebraska Coll. Agric. Home Econ. Res. Bull.*, No. 233, p. 1.

Mehlman, I. J., Eide, E. L., Sanders, A. C., Fishbein, M. and Aulisio, C. C. G. (1977). *J. Ass. off. analyt. Chem.* **60**, 546.

Mentzing, L.-O. (1981) *Lancet* **ii**, 352.

Merson, M. H., Black, R. E., Gross, R. J., Rowe, B., Huq, I. and Eusof, A. (1980) *Lancet* **ii**, 224.

Merson, M. H., Ørskov, F., Ørskov, I., Sack, R. B., Huq, I. and Koster, F. T. (1979) *Infect. Immun.* **23**, 325.

Merson, M. H. *et al.* (1976) *New Engl. J. Med.* **294**, 1299.

Morgan, R. L., Isaacson, R. E., Moon, H. W., Brinton, C. C. and To, C.-C. (1978) *Infect. Immun.* **22**, 771.

Moseley, S. L., Echeverria, P., Seriwatana, J., Tirapat, C., Chaicumpe, W., Sakuldaipeara, T. and Falkow, S. (1982) *J. infect. Dis.* **145**, 863.

Moseley, S. L., Huq, I., Alim, A. R. M. A., So, M., Samadpour-Motalebi, M. and Falkow, S. (1980) *J. infect. Dis.* **142**, 892.

Nalin, D. R., Levine, M. M., Young, C. R., Berquist, E. J. and McLaughlin, J. C. (1978) *J. clin. Microbiol.* **8**, 700.

Nalin, D. R., Rahaman, M., McLaughlin, J. C., Yunus, M. and Curlin, G. (1975) *Lancet* **ii**, 1116.

Neter, E., Westphal, O., Lüderitz, O., Gino, R. M. and Gorzynski, E. A. (1955). *Pediatrics, Springfield* **16**, 801.

Newsome, P. M., Burgess, M. N. and Mullan, N. A. (1978) *Infect. Immun.* **22**, 290.

O'Brien, A., Lively, T. A., Chen, M. E., Rothman, S. W. and Formal, S. B. (1983) *Lancet* **i**, 702.

Ohgke, H. and Wagner, D. (1977). *Zbl. Bakt.*, I Abt. Orig. **A238**, 350.

Ørskov, F., Ørskov, I., Evans, D. J., Sack, R. B., Sack, D. A. and Wadström, T. (1976) *Med. Microbiol. Immunol.* **162**, 73.

Ørskov, I. and Ørskov, F. (1966) *J. Bact.* **91**, 69.

Ørskov, I., Ørskov, F., Smith, H. W. and Sojka, W. J. (1975) *Acta path. microbiol. scand.* **B83**, 31.

Pai, C. H., Mors, V. and Toma, S. (1978) *Infect. Immun.* **22**, 334.

Pedersen, K. B., Winblad, S. and Bitsch, V. (1979) *Acta path. microbiol. scand.* **B87**, 141.

Pierce, N. (1977) *Infect. Immun.* **18**, 338.

Rao, M. C., Orellana, S. A., Field, M., Robertson, D. C. and Giannella, R. A. (1981) *Infect. Immun.* **33**, 165.

Reed, S. E., Kurtz, J. B. and Lee, T. W. (1978) *Abstr. 4th int. Congr. Virol.*, p. 461. Centre for Agricultural Publishing and Documentation, Wageningen.

Rees, T. A. (1959) *J. comp. Path.* **69**, 334.

Reimann, H. A., Price, A. H. and Hodges, J. H. (1945) *Proc. Soc. exp. Biol. Med.* **58**, 8.

Reis, M. H. L., Castro, A. F. P., Toledo, M. R. F. and Trabulsi, L. R. (1979) *Infect. Immun.* **24**, 289.

Report (1974*a*) *World Hlth Stat. Rep.* **27**, 150; (1974*b*) *Ibid.* **27**, 563; (1974*c*) *Rep. for 1973, CMO, Dept Hlth Soc. Secur.* HMSO, London; (1976) *Morbid. Mortal. wkly Rep.* **25**, 229; (1982) *Campylobacter: Epidemiology, Pathogenesis and Biochemistry*. Ed. by D. G. Newell. (Proc. int. Workshop, University of Reading, 1981.) MTP Press, Lancaster, Boston, The Hague; (1983*a*) *Campylobacter II*. (Proc. 2nd int. Workshop, Brussels, Sept. 1983.) Ed. by A. D. Pearson, M. B. Skirrow, B. Rowe, J. Davies and D. M. Jones. PHLS, London; (1983*b*) *The Management of Diarrhoea and the Use of Oral Rehydration Therapy*. World Health Organization, Geneva.

Rettig, P. J. (1979) *J. Pediat.* **94**, 855.

Riley, L. W. *et al.* (1983) *New Engl. J. Med.* **308**, 681.

Robinson, D. A. and Jones, D. M. (1981). *Brit. med. J.* **282**, 1374.

Rodhe, J. E. and Northrup, R. S. (1976). In: *Acute Diarrhoea in Childhood*, p. 339. Ciba Foundation Symposium No. 42. Elsevier, Amsterdam.

Ronnberg, B. and Wadström, T. (1983) *J. clin. Microbiol.* **17**, 1021.

Rosenberg, M. L., Koplan, J. P., Wachsmuth, I. K., Wells, J. G., Gangarosa, E. J., Guerrant, R. L. and Sack, D. A. (1977) *Ann. intern, Med.* **86**, 714.

Rothbaum, R., McAdams, A. J., Giannella, R. and Partin, J. C., (1982) *Gastroenterology*, **83**, 441.

Rowe, B. (1979) *See* Lambert, H. P., p. 625.

Rowe, B., Gross, R. J., Lindop, R. and Baird, R. B. (1974) *J. clin. Path.* **27**, 832.

Rowe, B., Gross, R. J. and Scotland, S. M. (1976) *Lancet* **ii**, 37.

Rowe, B., Gross, R. J., Scotland, S. M., Wright, A. E., Shillom, G. N. and Hunter, N. J. (1978) *J. clin. Path.* **31**, 217.

Rowe, B., Scotland, S. M. and Gross, R. J. (1977) *Lancet* **i**, 90.

Rowe, B., Taylor, J. and Bettelheim, K. A. (1970) *Lancet* **i**, 1.

Rudoy, R. C. and Nelson, J. D. (1975) *Amer. J. Dis. Child.* **129**, 668.

Rutter, J. M., Burrows, M. R., Sellwood, R. and Gibbons, R. A. (1975) *Nature, Lond.* **257**, 135.

Ryder, R. W., Kaslow, R. A. and Wells, J. G. (1979) *J. infect. Dis.* **140**, 626.

Ryder, R. W., Wachsmuth, I. K., Buxton, A. E., Evans, D. G., DuPont, H. L., Mason, E. and Barrett, F. F. (1976*a*) *New Engl. J. Med.* **295**, 849.

Ryder, R. W. *et al.* (1976*b*) *Lancet* **i**, 659.

Sack, R. B., Gorbach, S. L., Banwell, J. G., Jacobs, B., Chatterje, B. D. and Mitra, R. C. (1971) *J. infect. Dis.* **123**, 378.

Sack, R. B., Hirschhorn, N., Brownlee, I., Cash, R. A., Woodward, W. E. and Sack, D. A. (1975) *New Engl. J. Med.* **292**, 1041.

Sack, D. A., Huda, S., Neogi, P. K. B., Daniel, R. R. and Spira, W. M. (1980) *J. clin. Microbiol.* **11**, 35.

Sack, D. A., McLaughlin, J. C., Sack, R. B., Ørskov, F. and Ørskov, I. (1977) *J. infect. Dis.* **135**, 275.

Sack, D. A. *et al.* (1978) *New Engl. J. Med.* **298**, 758.

Sakazaki, R., Tamura, K. and Saito, M. (1967) *Jap. J. med. Sci. Biol.* **20**, 387.

Satterwhite, T. K., DuPont, H. L., Evans, D. G. and Evans, D. J. (1978) *Lancet* **ii**, 181.

Schnagl, R. D. and Holmes, I. H. (1976) *J. Virol.* **19**, 267.

Schnagl, R. D., Holmes, I. H., Moore, B., Lee, P., Dickinson-Jones, F. and Gust, I. D. (1977) *Med. J. Aust.* **i**, 259.

Schoub, B. D., Greeff, A. S., Lecatsas, G., Prozesky, O. W., Hay, I. T., Prinsloo, J. G. and Ballard, R. C. (1977) *J. Hyg., Camb.* **78**, 377.

Schroeder, S., Caldwell, J. R., Vernon, T. M., White, P. C., Grainger, S. I. and Bennett, J. V. (1968) *Lancet* **i**, 737.

Scotland, S. M., Day, N. P. and Rowe, B. (1980*a*) *FEMS Microbiol. Lett.* **7**, 15.

Scotland, S. M., Day, N. P., Willshaw, G. A. and Rowe, B. (1980*b*) *Lancet* **i**, 90.

Scotland, S. M., Gross, R. J., Cheasty, T. and Rowe, B. (1979) *J. Hyg., Camb.* **83**, 531.

Scotland, S. M., Richmond, J. E. and Rowe, B. (1982) *Soc. gen. Microbiol. Quart.* **9**, M8.

Scott, A. C., Luddington, J., Lucus, M. and Gilbert, F. R. (1978) *Vet. Rec.* **103**, 145.

Scrimshaw, N. S., Taylor, C. E. and Gordon, S. E. (1968). *Interactions of Nutrition and Infection.* World Health Organization, Geneva.

Shepler, V. M., Plumer, G. J. and Faber, J. E. (1963) *Amer. J. vet. Res.* **24**, 749.

Skerman, F. J., Formal, S. B. and Falkow, S. (1972) *Infect. Immun.* **5**, 622.

Skirrow, M. B. (1977) *Brit. med. J.* **ii**, 9; (1982) *J. Hyg., Camb.* **89**, 175.

Smibert, R. M. (1965) *Amer. J. vet. Res.* **26**, 320; (1974) In: *Bergey's Manual of Determinative Bacteriology*, 8th ed.,

p. 207. Ed. by R. E. Buchanan and N. E. Gibbons. Williams and Wilkins, Baltimore; (1978) *Annu. Rev. Microbiol.* **32**, 673.

Smith, H. R., Willshaw, G. A., McConnell, M. M., Scotland, S. M., Gross, R. J. and Rowe, B. (1979) *FEMS Microbiol. Lett.* **6**, 255.

Smith, H. W. and Gyles, C. L. (1970*a*) *J. med. Microbiol.* **3**, 387; (1970*b*) *Ibid.* **3**, 403.

Smith, H. W. and Halls, S. (1967) *J. Path. Bact.* **93**, 499; (1968) *J. med. Microbiol.* **1**, 45.

Smith, H. W. and Huggins, M. B. (1983) *J. gen. Microbiol.* **129**, 2659.

Smith, H. W. and Linggood, M. A. (1971) *J. med. Microbiol.* **4**, 467.

Smith, J. (1949) *J. Hyg., Camb.* **47**, 221.

So, M., Crandall, J. F., Crosa, J. H. and Falkow, S. (1975) In: *Microbiology 1974*, p. 16. American Society for Microbiology, Washington, D.C.

So, M., Heffron, F. and McCarthy, B. J. (1979) *Nature, Lond.* **277**, 453.

Speirs, J. I., Stavric, S. and Konowalchuck, J. (1977) *Infect. Immun.* **16**, 617.

Stair, E. L., Rhodes, M. B., White, R. G. and Mebus, C. A. (1972) *Amer. J. vet. Res.* **33**, 1147.

Tajima, N. (1970) *Arch. Virol.* **29**, 105.

Tallet, S., Mackenzie, C., Middleton, P. J., Kerzner, B. and Hamilton, R. (1977) *Pediatrics, Springfield* **60**, 217.

Taylor, J. (1961) *J. appl. Bact.* **24**, 316.

Taylor, J., Maltby, M. P. and Payne, J. M. (1958) *J. Path. Bact.* **76**, 491.

Thomas, L. V., Cravioto, A., Scotland, S. M. and Rowe, B. (1982) *Infect. Immun.* **35**, 1119.

Thomson, S. (1955) *J. Hyg., Camb.* **53**, 357.

Thornhill, T. S., Wyatt, R. G., Kalica, A. R., Dolin, R., Chanock, R. M. and Kapikian, A. Z. (1977) *J. infect. Dis.* **135**, 20.

Totterdell, B. M., Chrystie, I. L. and Banatvala, J. E. (1976) *Arch. Dis. Childh.* **51**, 924.

Turjman, M., Gotterer, G. S. and Hendrix, T. R. (1978) *J. clin. Invest.* **61**, 1155.

Turner, A. C. (1967) *Brit. med. J.* **iv**, 653.

Ulshen, M. H. and Rollo, J. L. (1980) *New Engl. J. Med.* **302**, 99.

Velin, D., Emody, L., Pacsa, S. and Kontrohr, T. (1980) *Acta microbiol. hung.* **27**, 299.

Vernon, E. (1969) *Publ. Hlth, Lond.* **83**, 205.

Véron, M. and Chatelain, R. (1973) *Int. J. syst. Bact.* **23**, 122.

Vogt, R. L., Sours, H. E., Barrett, T., Feldman, R. A., Dickinson, R. J. and Witherell, L. (1982) *Ann. intern. Med.* **96**, 292.

Wadström, T., Aust-Kettis, A., Habte, D., Holmgren, J., Meeuwisse, G., Möllby, R. and Söderlind, O. (1976) *Arch. Dis. Childh.* **51**, 865.

Wharton, B. A. (1981) *Brit. med. J.* **283**, 1277.

Woode, G. N. (1976) *Vet. Ann.* **16**, 30.

Wyatt, R. G. *et al.* (1974) *J. infect. Dis.* **129**, 709; (1980) *Science* **207**, 189.

Yolken, R. H., Greenberg, H. B., Merson, M. H., Sack, R. B. and Kapikian, A. Z. (1977*a*) *J. clin. Microbiol.* **6**, 439.

Yolken, R. H., Kim, H. W., Clem, T., Wyatt, R. G., Kalica, A. R., Chanock, R. M. and Kapikian, A. Z. (1977*b*) *Lancet* **ii**, 263.

Zahorsky, J. (1929) *Arch. Pediat.* **46**, 391.

Zissis, G., Lambert, J. P. and Kegal, D. de (1978) *J. clin. Path.* **31**, 175.

72

Food-borne diseases and botulism

Richard J. Gilbert, Diane Roberts and Geoffrey Smith

Introductory

The consumption of unwholesome food may give rise on the one hand to bacterial, viral or helminthic infections or on the other to poisoning caused by toxins that are present in the food before ingestion. In this latter type of disease distinction must be made between toxins of bacterial and of chemical origin. Though this chapter deals with bacterial food poisoning, chemical poisoning must be mentioned if only for purposes of differential diagnosis.

Chemical food poisoning

Certain foods may be inherently poisonous, such as: toadstools; cereals infected with the fungus *Claviceps purpurea* which forms the poison responsible for ergotism; incorrectly stored scombroid fish; heavy metals, particularly lead, arsenic, zinc, cadmium and mercury; powders that may be mistaken for flour, such as nicotinic acid and sodium fluoride; mussels and some fish, mainly in warm latitudes, that have fed on marine dinoflagellates and as a consequence are capable of causing paralytic shell-fish and ciguatera poisoning; flour or oils that have been placed in barrels previously used for insecticides, herbicides or fungicides; and foods containing aflatoxin (see Jarvis 1982).

In Great Britain chemical food poisoning is responsible for probably not more than one per cent of all food poisoning episodes, but in eastern countries it is said to be commoner (for USA figures, see Hughes *et al.* 1977). Apart from one or two neurotoxins such as ortho-tricresyl phosphate, its main characteristic is the production of vomiting within a few minutes to half an hour of ingestion of the food. In this respect it differs strikingly from poisoning of bacterial origin, which rarely manifests itself in under two hours and usually not till much later. Recovery is usually more rapid, though with certain poisons the case-fatality rate may be quite high.

Bacterial food poisoning

The term 'bacterial food poisoning' is conventionally restricted to acute gastro-enteritis due to the presence of bacteria, usually in large numbers, or their products in food. It may follow (1) the in-vivo multiplication of bacteria, e.g. *Salmonella* spp. or *Vibrio parahaemolyticus*, taken in with the food—*infection type*; (2) ingestion of food in which a toxin, e.g. staphylococcal enterotoxin or *Bacillus cereus* emetic toxin, has already been formed—*toxin type*; or (3) release of toxin in the bowel by bacteria, e.g. *Clostridium perfringens*, that do not produce it readily in food—*intermediate type*. In Britain, though not in the United States, it is customary to exclude from food poisoning those outbreaks of the infection type due to *Shigella* or to certain strains of *Salm. paratyphi B*, though they are often clinically indistinguishable from outbreaks due to the 'food poisoning' salmonellae. (See Chapters 68 and 69.) *Botulism* is also a consequence of bacterial proliferation in food, but is caused by a preformed toxin that, when ingested, acts on the nervous system; it will therefore be discussed·in this chapter in a separate section. A number of organisms other than those already mentioned have also been implicated as causes of bacterial food poisoning, although their exact role and mechanism of action have not as yet been fully elucidated, e.g. *Yersinia enterocolitica*, *Campylobacter*, and the enterococci.

For general accounts of food poisoning and other food-borne infections and intoxications see Williams and Hobbs (1975), Hobbs and Gilbert (1978), Riemann and Bryan (1979) and Gilbert (1983); and for the microbiology of foods and food processing, see Nickerson and Sinskey (1972), Frazier and Westhoff (1978), Ayres and his colleagues (1980), and International Commission on Microbiological Specifications for Foods (ICMSF; 1980).

Symptomatology

It is possible to recognize clinically some types of food poisoning by their typical symptomatology. Table 72.1 summarizes the clinical and epidemiological features of the most common forms of bacterial food poisoning encountered in Great Britain. Forms that are due to organisms which multiply in the body, such as those caused by salmonellae, do not give rise to symptoms for several hours. They are characterized by fever, abdominal pain, diarrhoea, and to a lesser extent by nausea and vomiting. Their duration is usually 2 to 5 days. Forms, however, in which toxin is preformed in the food or is produced soon after ingestion have a shorter incubation period, of 2 to 6 hours, are non-febrile, and are characterized by nausea and vomiting and to a lesser extent diarrhoea. Recovery is usually complete within 24 hours.

Epidemiology

Incidence

It is difficult to assess the real incidence of food poisoning. Such figures as were available for England and Wales between 1931 and 1981 are given in Table 72.2 (Public Health Laboratory Service 1980*a*, *b*, 1981, 1982). Before 1939, only the larger outbreaks reached the notice of the authorities and much of the apparent increase observed between 1941 and 1955 may be attributed to more frequent reporting of outbreaks and sporadic infections by practitioners and to the growth of a comprehensive bacteriological service during these years. There is a general impression, however, that some increase did occur and was attributable to the growth of communal feeding, the use of pre-cooked foods and made-up dishes, and the bulk handling of foodstuffs. The upward trend of notifications

Table 72.1 Some clinical and epidemiological features of bacterial food poisoning

Organism	Incubation period (hours)	Duration and symptoms	Sources of organisms
Salmonella spp.	6–48 (usually 12–36)	1–7 days. Diarrhoea, abdominal pain, vomiting, fever nearly always present.	Human and animal excreta. Raw meat and poultry, feeding meals etc.
Clostridium perfringens	8–20	12–24 hours. Diarrhoea, abdominal pain, nausea but rarely vomiting, no fever.	Dust, soil, human and animal excreta. Raw meat and poultry, dried foods, herbs and spices.
Clostridium botulinum	Usually 18–36	Death in 24 hours to 8 days, or slow convalescence over 6–8 months. Symptoms variable, but usually include disturbances of vision and difficulties in speaking and swallowing. Mucous membranes of mouth, tongue and pharynx usually very dry. Progressive weakness and respiratory failure.	Soil, mud, fish, poorly preserved foods; uncooked, fermented or lightly smoked fish and other sea-foods.
Staphylococcus aureus	2–6	6–24 hours. Nausea, vomiting, diarrhoea and abdominal pain, but no fever. Collapse and dehydration in severe cases.	Anterior nares and skin of man and animals, septic lesions e.g. boils, carbuncles, whitlows. Raw milk of cows and goats, cream and cheese made from raw milk.
Vibrio parahaemolyticus	2–48 (usually 12–18)	2–5 days. Profuse diarrhoea often leading to dehydration, abdominal pain, vomiting and fever.	Raw and cooked sea-foods, e.g. fish, prawns, crabs, and other shellfish.
Bacillus cereus	(a) 8–16, diarrhoeal syndrome. (b) 1–5, vomiting syndrome.	(a) 12–24 hours. Abdominal pain, diarrhoea and sometimes nausea. (b) 6–24 hours. Nausea and vomiting, and sometimes diarrhoea.	Common in soil and vegetation. (a) Meat products, soups, vegetables, puddings, and sauces. (b) Rice.

reached a peak in 1955 and was followed by a modest improvement in the situation with the number of incidents reported falling from 8961 in 1955 to 3744 in 1966. Since that date there has been a gradual increase apart from one low point in 1972 (4095 incidents) until in 1981 there were reports of 7976 incidents. The increase in the 1970s was due to sporadic cases, and not to outbreaks in families or other groups. It can be attributed to the growth in intensive farming practices. This has made some meats, in particular poultry,

Table 72.2 Bacterial food poisoning in England and Wales (from the records of Ministry of Health and Public Health Laboratory Service)

Years	Average annual no. of incidents	Presumed causal agents				
		Salmonella spp.	*Staphylococcus aureus*	*Clostridium perfringens*	Other organisms	Unknown
1931–35	59	32	21	—		6
1936–40	81	38	8	—	7	33
1941–45	324	270	15	—	3	60
1946–50	1747	1119	45	5	34	545
1951–55	5422	3131	109	41	10	2131
1956–60	7271	4475	120	93	5	2578
1961–65	4565	3125	98	80	<1	1262
1966–68	4361	3184	49	68	1	1059
1969–72	4827	4769	20	36	2	Not recorded
1973–75	5686	5590	20	51	25	,, ,,
1976–78	6307	6196	17	71	22	,, ,,
1979–81	7800	7718	17	53	12	,, ,,

The term 'incident' includes outbreaks and sporadic cases, each presumably due to a different source of infection; it does not refer to the total number of persons affected, which is of course much larger, because in some outbreaks scores or hundreds of persons were affected. Outbreaks and cases caused by dysentery bacilli are excluded.

available to the consumer in increased quantity and at a cheaper price. Intensive methods of husbandry, however, can increase the spread of infectious disease among animals. Further dissemination of organisms occurs during detention of the animals in lairages, slaughter and the processing of carcasses. Thus meat from intensively reared animals is more likely than that from free-range stock to be contaminated with, for example, *Salmonella* spp.

Information on food poisoning statistics from other countries is available but is even less complete than that from England and Wales, and it would be unwise to attempt an estimate of the relative frequency of food poisoning in different parts of the world. Todd (1978) and Turnbull (1979) have provided some comparative information for various countries.

In Great Britain the incidence of food poisoning parallels the atmospheric temperature. Much of our food becomes contaminated with potentially pathogenic organisms but, if these are not permitted to grow, the risk of food poisoning is small. The three main classes of food poisoning due to specific organisms vary in their seasonal incidence, but staphylococcal food poisoning is more strictly confined to the warmer months of the year than are the others.

The real distribution of bacterial food poisoning cannot be determined with certainty. The methods of recording tend to exaggerate the frequency of salmonella food poisoning relative to that due to other agents because it can easily be recognized by bacteriological methods and is frequently classified as food poisoning even when it cannot be attributed to the consumption of a particular food; a single outbreak is therefore often recorded as a large number of separate sporadic 'incidents'. Other forms of bacterial food poisoning, such as those caused by *Staph. aureus*, *Cl. perfringens* and *Bac. cereus*, are more difficult to recognize in the absence of an obvious outbreak, and sporadic attacks are seldom recorded. In 1981 salmonella infection accounted for 99 per cent of all incidents but for only 89 per cent of cases; staphylococcal and clostridial poisoning—which each caused less than 1 per cent of incidents—resulted in 1 and 9 per cent of cases respectively. Of 184 outbreaks affecting members of more than one family, 68 per cent were due to salmonellae, 24 per cent to clostridia, 5 per cent to *Bac. cereus* and other *Bacillus* spp., and 3 per cent to staphylococci. A number of incidents occur in which none of the recognized bacterial agents can be found. Some of these, particularly those with a longer incubation period of up to 48 hours, are undoubtedly of viral origin (Appleton *et al.* 1981).

Morbidity and mortality

The attack rate varies greatly in different outbreaks, owing partly to the uneven distribution of the infecting organism or toxin in the food and partly to variation in individual susceptibility of those exposed to risk. On the whole it tends to be high. Geiger (1923) estimated it at 75–100 per cent, but more recent experience would tend to lower this figure. Savage and White (1925) state that in salmonella infections conveyed by milk the attack rate is only 40–50 per cent; this also holds true of outbreaks due to many other foods. Hobbs (1971) lists a number of outbreaks associated with the consumption of poultry, with attack rates of 8–86 (mean 20) per cent for *Salmonella*, 17–100 (mean 25) per cent for *Cl. perfringens*, and 6–92 (mean 23) per cent for *Staph. aureus*.

In outbreaks of unknown aetiology in which the agent is possibly a virus, the attack rates tend to be higher, with a range of 20–86 per cent; in 10 outbreaks cited by Appleton and co-workers (1981) the mean attack rate was 47 per cent.

The case-fatality rate is generally low. Estimates in the past have been 0.5 per cent in the USA (Geiger 1923), 1.5 per cent in Great Britain (Savage 1920) and 4.5 per cent in Germany (Lentz 1924). This last figure probably included cases of botulism and is therefore too high. In the period 1962–1971 the case-fatality rate for salmonellosis in the USA was 0.41 per cent (Brachman *et al.* 1973). In 1980 in England and Wales, 27 patients with salmonella food poisoning died out of a total of 9540, and 2 affected with other forms died out of a total of 1316, giving case-fatality rates of 0.3 and 0.15 per cent respectively, and for all forms of bacterial food poisoning 0.27 per cent (Public Health Laboratory Service 1981). Sex has, apparently, little effect on case-fatality rates, but age is of importance, most of the deaths occurring in the very young and the old.

Bacteriology and epidemiology of salmonella food poisoning

Very many different serotypes of *Salmonella* are concerned in the production of food poisoning. Their frequency distribution varies in different countries. Table 72.3 lists those most commonly encountered in England and Wales since 1966 as revealed by the reports on food poisoning and salmonellosis from the Public Health Laboratory Service (Vernon 1967, 1969, 1970, 1977, Vernon and Tillett 1974, Hepner 1980).

Until 1942 all but a few of the strains incriminated belonged to 14 (indigenous) types, the foremost of which were *Salm. typhimurium*, *Salm. enteritidis*, *Salm. newport* and *Salm. cholerae suis*. In 1942, a change became apparent; 10 new types were isolated, the commonest of which was *Salm. oranienburg*. In 1943 a further eight new types were identified. Nearly all of these types had been first described in other countries, many of them in the United States. There is little doubt that the majority were introduced into Great Britain by American spray-dried egg, which

Table 72.3 Rank order of frequency of salmonellae in incidents (outbreaks and sporadic cases) of food poisoning in England and Wales

Order of frequency	*Salmonella* spp. isolated in years			
	1966–68 (9550)*	1969–72 (29 192)	1973–75 (26 574)	1976–78 (27 894)
1	typhimurium (55%)	typhimurium (35%)	typhimurium (32%)	typhimurium (27%)
2	panama	enteritidis	agona	hadar
3	enteritidis	panama	enteritidis	enteritidis
4	stanley	agona	heidelberg	heidelberg
5	virchow	heidelberg	anatum	agona
6	brandenburg	virchow	indiana	virchow
7	dublin	saint paul	newport	senftenberg
8	indiana	indiana	infantis	indiana
9	anatum	bredeney	hadar	newport
10	heidelberg	stanley	bredeney	anatum

*Total number of incidents in parenthesis.

was first distributed on a large scale in the middle of 1942 (Report 1947). From then on the situation became more complex with increasing bacteriological investigation of cases of food poisoning. Six of the 14 types found in England and Wales before 1941, the 'indigenous' types, continued to be common in subsequent years; these were *Salm. typhimurium, Salm. enteritidis, Salm. thompson, Salm. newport, Salm. dublin*, and *Salm. bovismorbificans*. The proportion of infections with *Salm. typhimurium* rose from just under half in the years 1923–45 to over three-quarters in 1946–55. However, between 1962 and 1966 there was a decline in the isolations of *Salm. typhimurium* followed by a slight rise after 1966 (Lee 1973). Since 1977 there has been a continuing increase in *Salm. typhimurium* isolations (Public Health Laboratory Service 1982). Though several of the other 'indigenous' types were isolated with increased frequency, they accounted for a much smaller proportion of all infections than previously, owing to an upsurge in infections with 'new' serotypes (Table 72.3).

Most of the 'exotic' types first seen between 1941 and 1945 continued to be prevalent in later years, but only *Salm. anatum* has held its position continuously since. From 1951 onwards numerous types not previously encountered in this country, or only seen rarely, have become prevalent, and a few—notably *Salm. heidelberg, Salm. saint paul, Salm. brandenburg, Salm. bredeney, Salm. panama, Salm. virchow, Salm. indiana, Salm.* 4,12:d: −, *Salm. agona* and *Salm. hadar* have established themselves as widespread or regular causes of food poisoning. The remarkable rise to prominence of *Salm. heidelberg*, which was second only to *Salm. typhimurium* in frequency in Britain between 1956 and 1966, was experienced throughout Western Europe and in parts of North America. A similar rise was seen with *Salm. agona* in 1970–71 (Center for Disease Control 1973, Clark *et al.* 1973). Between 1977 and 1980 *Salm. hadar*, a serotype seldom identified before 1971, became the second most commonly isolated serotype in England and Wales. The main reservoir of the organism was turkeys. In 1981 the number of isolations of *Salm. hadar* fell dramatically, but an increase in *Salm. enteritidis* and *Salm. virchow* was noted. (For a discussion of the trends in salmonella food poisoning in England and Wales see Lee 1974 and McCoy 1975; and in the United States and Canada see Bryan 1981.)

The frequency of isolation of certain serotypes, as well as the number of isolations of all serotypes, varies between countries, but on the whole experience in Western Europe is similar to that in Britain. In Western Europe, America and Australasia *Salm. typhimurium* is nearly always the commonest isolate. In tropical Africa, where many sporadic cases of salmonella gastro-enteritis are caused by a great diversity of serotypes, of which none predominates, *Salm. typhimurium* is seldom seen. For further information about the situation in various parts of the world see van Oye (1964) and Williams and Hobbs (1975).

Animal reservoirs of salmonellae

Though human infection with *Salmonella* is widespread and frequent, the majority of food-poisoning outbreaks due to this group of organisms follow the consumption of food directly or indirectly associated with infection in some animal. Except for the host-specific types (*Salm. dublin* in cattle, *Salm. cholerae suis* in pigs, *Salm. abortus ovis* in sheep, *Salm. gallinarum* and *Salm. pullorum* in chickens, and *Salm. typhi* and *Salm. paratyphi B* in man) and *Salm. typhimurium*, the sequence of events for salmonella food poisoning can be described as: contaminated animal feed → animal → food → man. This sequence is perpetuated by the conversion of salmonella-infected animal remains into further contaminated foodstuff.

The role of animals, particularly food animals, in

the salmonella cycle has long been recognized (Buxton 1957, Hobbs 1961, van Oye 1964, Rowe 1973). It is well established that pigs and poultry are major reservoirs, and that cattle are the main source of *Salm. typhimurium* infection. Most salmonella infections in animals are symptomless. The prevalence of infection in animals is reflected by the frequency with which the organism can be detected in products of animal origin. Food animals are not, however, the only source; salmonella infection occurs quite frequently in wildlife and in a variety of pet animals including terrapins, frogs and fish, and can be transmitted either directly or by means of contaminated food. Such sources are probably of less importance than food animals, and the risk they present is more easily controlled. The incidence of salmonella infection in food and other animals has been reviewed by Williams and Hobbs (1975), Bryan and co-workers (1979), and Roberts (1982a).

Chickens Like wild birds, chickens and other poultry are often infected with salmonellae. Apart from *Salm. gallinarum* and its *pullorum* variety, which have a fairly high degree of specificity for chickens, numerous other serotypes give rise to acute fatal infections in chicks; adult birds are seldom ill but may suffer from chronic or latent infection (Smith and Buxton 1951, Blaxland *et al.* 1958). In Britain infection with *Salm. typhimurium* has for many years been an important cause of serious epidemics in chicks. Infection with some serotypes, for example *Salm. saint paul* (Galbraith *et al.* 1962a) and *Salm. virchow* (Pennington *et al.* 1968), may be endemic in hatcheries without causing many serious infections. The situation may remain unsuspected until an outbreak of human infection follows the consumption of adult birds that come from an infected hatchery. The low incidence of salmonellosis recorded in surveys of chickens—0.4 per cent (Smith and Buxton 1951) and 1.2 per cent (Brown *et al.* 1973)—is not a true reflection of the actual degree of salmonella infection in these birds, as the figures are based on post-mortem examination. Since birds may harbour without symptoms one or more serotypes for a period ranging from a few days to many months (Buxton 1957), the true incidence of salmonella infection in market chickens is likely to be much higher.

Ducks *Salm. typhimurium* is responsible not only for serious outbreaks among ducklings but also for a septicaemic infection in adult birds. It is the type that most commonly causes disease in the duck (Blaxland *et al.* 1958, Marthedal 1977) and in persons who consume infected duck meat. In ducks salmonellae are apt to gain access to the oviduct with the result that eggs may be infected before they are laid. Thus the rate of contamination of duck eggs with salmonellae is likely to be higher than that of hen eggs. The percentage of duck eggs containing salmonellae has been variously reported as 0.15 per cent (Report 1954) and 0–20 per cent (Hobbs and Gilbert 1978).

Turkeys Turkey poults suffer seriously from salmonella infection and as wide a variety of types has been isolated from them as from chickens, *Salm. typhimurium* and *Salm. enteritidis* being the most common. Rowe and his colleagues (1980) encountered *Salm. hadar*. Cases of food poisoning spanning an 8-year period and traced to a common source were reported by Payne and Scudamore (1977). Ten of the incidents described, mainly due to *Salm. enteritidis* phage-type 2, were traced to the consumption of turkey meat. All the birds originated from a single turkey breeding and rearing establishment. Although many farms were incriminated, all had obtained their poults from the same source.

Other birds may be infected with salmonellae, pigeons, gulls and sparrows having all been shown to excrete the organisms. Gulls feeding in polluted estuaries may excrete typhoid, paratyphoid and other salmonellae (Steiniger 1953, Williams *et al.* 1976, 1977, Johnston *et al.* 1979). Fenlon (1981) reported that 12.9 per cent of 1242 samples of seagull faeces contained salmonellae, the number of positive samples being significantly higher than that from the faeces of other small birds near sewage outfalls. Sparrows gaining access to a large hospital kitchen were the source of salmonellae that caused outbreaks of gastro-enteritis in a large mental hospital (Penfold *et al.* 1979).

Eggs These may be infected during their formation in the oviduct, as in ducks (above), or by passage of the organisms through pores, abrasions or cracks in the shell from the faecal and other debris on the exterior. Contamination of the egg contents may also occur in manufacturing plants at the time the shell is broken. Penetration of the shell depends on a number of factors. Undamaged dry eggs will remain edible for many months, but when a warm newly laid egg comes into contact with cool contaminated fluid bacteria are readily sucked through the shell (Haines and Moran 1940). The two main factors that prevent the spoilage of eggs are (1) the physical barrier of the shell and its membranes, and (2) the numerous factors in the egg albumen that make it a poor medium for growth. The microbiology of eggs and egg products has been reviewed by Elliott and Hobbs (1980). (See also p. 485.)

Pigs Fowls are rivalled by pigs in the frequency with which they are infected with salmonellae. Intensive rearing and the associated methods of feeding both serve to increase the rate of infection in a pig herd. Porcine products, such as sausages, which contain meat particles derived from many animals, may show a high rate of contamination (Roberts *et al.* 1975). Salmonellae may frequently be found in mesenteric lymph nodes and the types present are often those associated with infections in the human population. Excreter boars and sows can perpetuate infection on the breeding farm; and distribution of progeny to other farms for further breeding or for fattening will disseminate the organism widely (Ghosh 1972).

Cattle Until recently these animals were thought to be a less important source of human infection than poultry and pigs.

Before 1960, nearly all bovine salmonellosis in Britain was due to *Salm. dublin*, but more recently there has been a great increase in *Salm. typhimurium* infection in calves. The introduction of intensive methods for the rearing of calves provided opportunities for the wide dissemination of infection and was followed by an increase in the number of human infections that could be attributed directly to the consumption of veal (Anderson *et al.* 1961, Anderson 1968*a*). The ill-considered use of antibiotics for the treatment or prophylaxis of gastro-enteritis in calves led to the accumulation of transferable drug resistance in strains of *Salm. typhimurium* and this further accelerated the spread of the organism. Multi-resistant strains, notably of *Salm. typhimurium*, continue to appear and give rise to infections in animals and man (Anderson 1968*b*, Threlfall *et al.* 1978*a, b*, Rowe *et al.* 1979). Some of these strains, such as multi-resistant phage types 204 and 193 of *Salm. typhimurium*, have been isolated from minced meat and sausage; this indicates that they are well established in the food chain (Rowe *et al.* 1979).

Other animals Salmonellae cause disease in sheep and goats in many countries. *Salm. abortus ovis* causes abortion in ewes (Jack 1968); other serotypes, especially *Salm. typhimurium*, *Salm. dublin*, *Salm. oranienburg* and *Salm. java*, cause fatal infections in lambs (Prost and Riemann 1967, Gitter and Sojka 1970).

Dogs and other pets may suffer from salmonella infections or act as symptomless excreters. Direct transmission to man may occur, the greatest risk being associated with terrapins and dogs and the least with fish. Various carrier rates in pets have been reported, e.g. 4.5–20 per cent in dogs in Australia (Frost *et al.* 1969), and 1 per cent in dogs and 1.4 per cent in cats in Britain. Salmonellae were found in mesenteric lymph nodes from 5 of 200 healthy cats and 9 of 20 healthy dogs (Smith 1969). Dogs housed in kennels tend to have high faecal carrier rates. Galton and her colleagues (1952) found carrier rates ranging from 0 to 100 per cent in the USA. Many of the serotypes found in dog faeces were also found in dehydrated dog meal (Galton *et al.* 1955). Raw meat sold as dog food frequently contained salmonellae (Galbraith *et al.* 1962*b*). In North America, pet terrapins have been shown to constitute a hazard of some importance owing to the way in which they are handled, particularly by children (Williams and Helsdon 1965). A warning of the hazards of ownership was issued in the United Kingdom when the practice of keeping terrapins as pets began to increase (Anon. 1969, Jephcott *et al.* 1969).

Rats and mice can suffer naturally from infections with *Salm. typhimurium*, *Salm. enteritidis*, and other salmonellae. Surveys of carrier rates in rat populations have given results varying from 4 to 7 per cent, with *Salm. enteritidis*, usually of the *danysz* variety, as the commonest organism (Khalil 1938, Ludlam 1954, Brown and Parker 1957). Rodents that gain access to farms may spread salmonellae to and from animal feeds via their droppings (Edel *et al.* 1973). Ludlam (1954) observed that the proportion of infected rats caught in and around a meat factory rose at one time to 40 per cent and that many different serotypes were present. House mice were thought to be to blame for sporadic cases of salmonellosis in an old people's home in the Netherlands, and eradication of the mice resulted in the cessation of new cases (Beckers *et al.* 1982).

Salmonellae have been isolated from a variety of biting and non-biting arthropods, including fleas (Jadin 1951), ticks (Reitler and Mentzel 1946, Jadin 1951, Floyd and Hoogstraal 1956), human lice (Liu *et al.* 1937), animal lice (Messerlin and Courzi 1942, Milner *et al.* 1957), and cockroaches (Mackerras and Pope 1948). Flies and other insects can introduce salmonellae into the farm environment and play a part in their dissemination (Edel *et al.* 1976, 1978). Hobbs and Gilbert (1978) cite an outbreak of salmonellosis among children in a hospital in Australia, in which flies, cockroaches and mice were all found to be carrying the organism responsible for the infection. It was not suggested that they were the original source but that they had picked up the bacteria from the contaminated environment and conveyed them on their feet and bodies.

Epidemiology of salmonella infection among domestic animals

Many domestic animals suffer from salmonella infection, but the number of serotypes causing illness in a particular species at any one time and in a particular country is usually small. These indigenous salmonella infections, for example *Salm. dublin* in cattle and *Salm. typhimurium* in calves, appear to behave not unlike human enteric infection. They are probably acquired from a rather small dose; many animals become severely ill, others permanent carriers. Good animal husbandry is necessary to minimize the spread of infection, particularly on farms practising intensive rearing. Once infection has been introduced into a herd or flock, rapid spread can occur from the ingestion of food contaminated with faeces. Transmission can also occur when susceptible and infected animals are brought into contact with each other, for example, when they are mixed in pens, or at service; or by contact with the faeces of infected animals; or by placing stock in inadequately disinfected pens; or by the transfer of contamination from pen to pen by the movement of animals or other means (Linton 1979); or by cross-infection between animals being transported to market or kept in lairages awaiting slaughter.

Infection is perpetuated on poultry farms by polluted drinking water, and inadequate cleaning of equipment and premises (Snoeyenbos 1969). Transfer of salmonellae in both directions between litter and the intestinal tract is thought to be of significance in maintaining infection; this occurs more commonly in unchanged new litter (Fanelli *et al.* 1970).

Animals grazing on pasture recently sprayed with slurries of animal wastes can become infected from this source (Jack and Hepper 1969). The degree of infectivity of such pasture will depend on the persistence and concentration of the organisms and the interval between spraying and grazing (Taylor and Burrows 1971, Taylor 1973).

Some domestic animals, notably pigs and poultry, also harbour various salmonellae that rarely give rise to illness. Infection is usually short lived and confined mainly to the bowel. The origin of these infections is frequently certain ingredients of animal feeding stuffs imported into Europe and North America from tropical countries. Various exotic salmonellae have been demonstrated in these products and many appear in both the animal and human population. In a number of well documented instances a particular serotype causing an infection in human patients has been traced back to the meal given to food animals. Such instances include human disease caused by *Salm. typhimurium* in pork (Miller *et al.* 1969), *Salm. virchow* in spit-roasted poultry (Pennington *et al.* 1968), and *Salm. agona* in chickens and pigs originating from contaminated fish meal (Clark *et al.* 1973). Between 1976 and 1978 all of the 10 salmonella serotypes that occur most frequently in man were isolated from poultry, sausages and made-up meats, meat, and offal; and also from various animal feeds and feed ingredients (Hepner 1980). The events that lead from the use of contaminated animal feed to human infection have been described for many serotypes (Hobbs and Hugh-Jones 1969, Lee 1973, Rowe 1973). Many surveys have been carried out to determine the prevalence of salmonellae in feed and feed ingredients. In five surveys in the USA the rate of contamination varied from 1 to 86 per cent (Bryan *et al.* 1979). Similar rates have been reported in the United Kingdom and other parts of the world (Roberts 1982*a*).

The use of feeding meals containing low concentrations of antibiotic and chemotherapeutic agents has enabled pigs, poultry and calves to be fattened more quickly and economically. However, the indiscriminate use of such agents, for example penicillins, tetracyclines, and sulphonamides, for growth promotion or prophylactic purposes has resulted in the appearance of bacterial strains resistant to these agents. The emergence of multi-resistant strains of *Salm. typhimurium* has been reviewed by Anderson (1968*a* and *b*). The resistance spectrum as well as the proportion of drug-resistant strains increased dramatically in the early 1960s. The Swann Committee (Report 1969)

examined the use of antibiotics in veterinary medicine and animal husbandry. They and a working party of the World Health Organization (1974) recommended that only those antimicrobial agents that do not have a therapeutic value, e.g. bacitracin, flavomycin and virginiamycin, should be used as growth-promoting agents in animal feeds. Despite these recommendations multi-resistant strains, notably of *Salm. typhimurium*, continue to appear and to infect animals and man (Threlfall *et al.* 1978*a, b*, Rowe *et al.* 1979). Some of these strains have become established in the food chain, as evidenced by their isolation from minced meat and sausage (Rowe *et al.* 1979).

When animals and birds are crowded together during transport or while being held in markets or slaughterhouses, the rate of salmonella excretion is increased owing to the stresses produced by the overcrowding, shortage of water, and anxiety (Burns *et al.* 1965, Williams and Newell 1970, Williams and Spencer 1973). Galton and her colleagues (1954) showed that the excretion rate in pigs increased from 7 per cent on the farm to 25 per cent in the holding pens at the abattoirs, and to 51 per cent on the killing floor. Delays in the transport of poultry from farm to processing plant can result in increased excretion of salmonellae (Seligmann and Lapinsky 1970).

Source of infection in man

Type of food

Our information about the types of food most often responsible for salmonella food poisoning is deficient, as in only a small proportion of outbreaks is the food vehicle of infection ever identified, food remnants usually being disposed of before the onset of symptoms (18–36 hr after the meal). In food poisoning due to *Cl. perfringens*, *Staph. aureus* and *Bac. cereus* the incubation period is shorter and a specific food can usually be incriminated. In England and Wales from 1976–78 the food vehicle was traced in only 12 per cent of general and family salmonella outbreaks but in 85, 97 and 95 per cent respectively of those caused by *Cl. perfringens*, *Bac. cereus* and *Staph. aureus* (Hepner 1980). Of 400 outbreaks of salmonella food poisoning in England and Wales in the years 1969–1978 in which the probable vehicle was identified, 319 were traced to meat and poultry, 72 to milk and cream, four to eggs, four to sweets, desserts and cakes, and one to sea-food. In the 319 meat and poultry outbreaks the meat was derived from cattle in 14, pigs in 51 and poultry in 232. It is evident that poultry now play a major role in salmonella food poisoning. The striking rise in the number of outbreaks in the 1960s reflected the increased consumption of poultry meat— a food that had become cheaper and more readily available with the derationing of feedstuffs, and the adoption of intensive rearing techniques and highly efficient methods of slaughtering and carcass processing. Eggs and sweet dishes are no longer common causes of salmonella outbreaks; in the period 1951–1968 outbreaks associated with such foods numbered 180, compared with only 8 between 1969 and 1978. Raw milk continues to be an important and even increasing cause of salmonellosis, with 39 outbreaks in 1951–1968, and 72 in 1969–1978. However, meat, and in particular poultry, remains the major food incriminated in outbreaks of salmonella food poisoning. The risks associated with other foods have been reviewed by Gilbert and Roberts (1979). Canned foods are now only occasionally associated with food poisoning, and very rarely with salmonella food poisoning. Gilbert, Kolvin and Roberts (1982) cite 169 outbreaks of food poisoning and food-borne infections associated with freshly opened canned food between 1929 and 1980, of which only 16 were caused by salmonellae, 8 being outbreaks of typhoid fever and 8 of salmonella food poisoning. All occurred before 1970 and all but one were associated with meat products. Among foods that seldom give rise to poisoning in Great Britain are bread, vegetables, fish, freshly cooked meat and freshly opened cans, but in some countries vegetables may be contaminated with cysts of *Giardia* and *Entamoeba* and eggs of *Ankylostoma*, *Trichocephalus*, *Hymenolepis* and *Taenia* species (Marzochi 1977).

Mode of infection of the food

A food of animal origin may contain salmonellae because (1) one of the animals from which it was derived was infected, (2) the animal was slaughtered in close proximity to infected animals, or (3) it was contaminated with salmonellae from a source such as raw meat on the same work surface. It is often difficult to distinguish between intrinsically infected and extrinsically contaminated animal products.

Meat

The frequency with which salmonellae are found in meat varies widely from place to place, between different species of animal, and at different times. Many surveys have been carried out on the occurrence of salmonellae in meat animals both on the farm and at the slaughterhouse; and in meat at the abattoir, processing plant and retail shop. The results of surveys on red meat, offal and poultry have been summarized by Bryan (1979) and Roberts (1982a). The salmonella contamination rates reported range from less than 1.0 to 70 per cent, sausages, minced meat and offal having the highest rates. Of the various animal species, poultry and pigs have the worst record; meat from cattle and sheep on the whole has a lower contamination rate, probably because of differences in feeding practices and slaughter procedures. Recent surveys in retail

shops in England and Wales showed salmonella contamination rates of 5.1 and 1.7 per cent for minced beef and pork, respectively (Public Health Laboratory Service 1980c, Turnbull and Rose 1982), 12 per cent for sausages, and 79 per cent for chickens (Gilbert 1983). The bulking of meat from many carcasses tends to increase the rate of contamination. Thus boneless meat, which is imported in large frozen blocks, generally has a higher contamination rate than carcass meat (Hobbs 1965). 'Continuous line' procedures for the evisceration and handling of carcasses, and the use of large volumes of water in processing, help to increase the spread of contamination. Widespread contamination of a slaughterhouse during hot weather may lead to heavy contamination of the meat, such as was responsible for an outbreak in Sweden due to *Salm. typhimurium* in which there were 9000 cases and 90 deaths (Lundbeck *et al.* 1955). The effects of slaughter practices on the bacteriology of red-meat carcasses have been reviewed by Roberts (1980). The greatest source of surface contamination of meat and poultry is undoubtedly the gut of the animal itself. An experiment in which serratiae were injected into a chicken before it was eviscerated in a processing plant showed that the next 150 birds slaughtered were also contaminated, thus demonstrating the degree of bacterial spread from carcass to carcass (Stewart 1965). The practices of bulking poultry viscera and subsequently redistributing them at random among the carcasses, and immersing eviscerated birds in iced water, also serve to spread salmonellae from carcass to carcass (see also Barnes and Mead 1971, Patterson 1972, Bryan *et al.* 1979).

Milk

Salmonellae may be excreted in the milk by a cow with septicaemic infection or an udder lesion, or may gain entry to the milk by contamination from faeces or from a human case or carrier. Of 132 outbreaks of milk-borne salmonella food poisoning that occurred between 1951 and 1980, 22 were attributed to excretion of the organisms in the milk, 49 to faecal contamination, and 16 to infection in calves. In seven outbreaks in previously symptomless herds the source of infection was suggested as contaminated animal feed (4), water (1), and pasture flooded with effluent (1); seven outbreaks were thought to be due to human infection (Galbraith *et al.* 1982). In one outbreak in Scotland in which *Salm. dublin* was isolated from the milk at least 700 persons were affected (Small and Sharp 1979).

Eggs

We have noted (p. 482) that eggs may become contaminated with salmonellae either in the oviduct or after laying. In Great Britain 0.15 per cent of duck eggs and a much smaller proportion of hen eggs contain salmonellae. Not only duck eggs, but also hen and pigeon eggs have at times been responsible for food poisoning (Clarenburg and Dornickx 1932, Lovell 1932, Scott 1933, Hohn and Herrmann 1935, Seligmann 1935, Jansen 1936, Garrod and McIlroy 1949), but the frequency of infection from these sources is peculiarly difficult to ascertain.

Experience in Great Britain with spray-dried hen egg (Report 1947) drew attention to the fact that bulked egg products contained salmonellae more frequently than did individual eggs. Of 7000 samples of spray dried egg, mainly from the United States, 9.9 per cent contained salmonellae, representing 33 serotypes. Only a few organisms were present in each gramme of powder, and food poisoning was therefore unlikely to occur unless there was opportunity for bacterial multiplication after rehydration. Much home-produced liquid egg is used in the baking industry in Great Britain, and here also the rate of contamination is much higher than in shell eggs. Murdock (1954) reported that 48 per cent of liquid duck egg and 2 per cent of hen egg in a factory in Northern Ireland contained salmonellae, but corresponding figures found in a survey in England and Wales (Report 1955) were lower (3.1 per cent and 0.3 per cent respectively). Imported egg products from China were shown to be linked with outbreaks of paratyphoid fever and salmonellosis (Newell *et al.* 1955, Report 1957, 1958, Knowles 1971) and several surveys showed that these products contained salmonellae (Smith and Hobbs 1955, Rohde and Adam 1956, Albert 1957, Report 1958). In most countries bulked egg products are now pasteurized or otherwise heat treated and these measures have contributed to the decrease since 1960 in the frequency of salmonella infection from confectionery. Between 1961 and 1970 the average contamination rate for eggs (all types) imported into England and Wales was 3.7 per cent, but for the next seven-year period it fell to 1 per cent (Gilbert and Roberts 1979).

Other products

Certain animal products administered orally as medicaments may contain salmonellae and may result in outbreaks of infection. Products incriminated include dried thyroid powder (Kallings *et al.* 1966), pancreatin (Glencross 1972, Rowe and Hall 1975) and carmine dye (Lang *et al.* 1967).

Chocolate products, which normally give little cause for concern because of their low a_w (water activity), have recently caused outbreaks of salmonellosis. More than 170 cases of *Salm. eastbourne* infection were reported in Canada and the United States in 1973–74 after the consumption of chocolate products (Craven *et al.* 1975, D'Aoust *et al.* 1975,) and in 1982 imported Italian chocolate containing *Salm. napoli*

gave rise to many cases of disease in England and Wales (Gill *et al.* 1983). The interesting feature in both of these outbreaks was that the dose sufficient to cause illness was low, probably no more than 1000 organisms in the *Salm. eastbourne* outbreak. Numbers of *Salm. napoli* found in the Italian chocolate ranged from 2–23 or more per gramme. (For detailed information on the survival of micro-organisms in various foods, see Engley 1956.)

Contamination of food from extraneous sources

A food may be derived from a source free from salmonellae but become contaminated in the course of manufacture, transport or sale. The range of foods liable to contamination includes any item which will permit multiplication of salmonellae. Sometimes the source of salmonellae is another foodstuff; cooked meats may be contaminated from raw meat, or even from egg products (Graham *et al.* 1958), either by direct contact or via utensils, work surfaces, or hands. Fish and shell-fish from polluted rivers and estuaries may contain salmonellae (Floyd and Jones 1954, Gulasekharam *et al.* 1956). Several dried vegetable products, including herbs and spices, have been shown to contain salmonellae in small numbers (Roberts *et al.* 1982). Outbreaks of typhoid, paratyphoid and salmonellosis have been attributed to the consumption of a number of products of plant origin, including salad vegetables (celery, watercress, lettuce, endive), rhubarb, water melon, desiccated coconut, and pepper. The organisms contaminating these foods may come from sources such as soil, water and fertilizers, and from faeces, particularly when poor hygiene is practised (Geldreich and Bordner 1971, Bryan 1977, Roberts *et al.* 1982). These products do not, however, lead to many outbreaks of food poisoning. Before 1961 desiccated coconut was responsible for a number of outbreaks (Wilson and MacKenzie 1955, Kovács 1959, Galbraith *et al.* 1960, Winkle *et al.* 1960). In 1959–1960 a survey of this product imported into Britain from Ceylon showed that 9 per cent of samples were contaminated with salmonellae. Hygiene regulations were then introduced in Ceylon (Anon. 1961) to improve the production methods. The contamination rates of this product are now low (0.9 per cent; Gilbert and Roberts 1979).

Salmonellae may be introduced into food by infected animals, or by human excreters. Rats and mice have always been considered to be vectors (see p. 483), not because they are more susceptible to salmonella infection than other animals but because of their habits in feeding and defaecation. The part played by the human excreter is difficult to evaluate. In 121 outbreaks of food-borne salmonella infection investigated by Savage (1932) there were only five in which a human carrier seemed to have been responsible.

Similarly, a recent study by Roberts (1982*b*) showed that, although infected food handlers were discovered in 126 of 396 salmonella outbreaks in England and Wales between 1970 and 1979, in only 9 was there evidence to suggest that they were the original source of the contaminating organisms. The food handlers had either recently returned from holidays abroad or had continued to prepare food while suffering symptoms of gastro-enteritis. Other workers have been inclined to attribute a more important role to the human excreter (Rubenstein *et al.* 1944, Felsenfeld and Young 1949). The opinion is now widely held that food is less often infected from human than from animal sources. In most incidents of salmonella food poisoning the food handlers are victims not sources, and become infected through their frequent contact with contaminated raw food, from tasting during preparation, or from eating left-over contaminated food. Patients who have suffered from salmonella infection usually excrete the causative organism for several weeks. Half have ceased to do so by the end of the 4th week, and about 90 per cent by the 8th week (Perry and Tidy 1919, Mosher *et al.* 1941, Rubenstein *et al.* 1944, Kwantes 1952, Schäfer 1958). According to Lennox and co-workers (1954) the proportion of patients excreting the organism shows little change for the first three weeks, and then falls steeply until the middle of the 7th week. After this the rate of fall again becomes slow. The disappearance of salmonellae from the faeces during convalescence is thus described by a sigmoid curve. Excretion tends to last longer in infants than in older persons (Rubenstein *et al.* 1944, Szanton 1947). Many excreters of salmonellae give no history of recent gastro-intestinal illness; the infectivity of such *symptomless excreters* lasts as long as that of persons convalescing from food poisoning. In England and Wales workers whose jobs include the handling of food and who are shown to be carriers of infection are excluded from work until three negative stool samples have been obtained (Public Health [Infectious Diseases] Regulations 1968; Public Health Laboratory Service 1983). Pether and Scott (1982) suggest that clinically recovered food handlers, with formed stools, should be allowed to return to work, as they do not present a hazard provided that normal hygienic practices are observed.

Many workers dealing with contaminated raw food ingredients become symptomless excreters of salmonellae. Carriage of salmonellae may be intermittent and the serotype may change with different batches of raw food. Contamination of workers' hands by raw materials, especially of animal origin, is of much more importance than human faecal contamination. De Wit and Kampelmacher (1981) showed salmonellae on the hands of 5–36 per cent of workers in factories in which raw materials of animal origin were processed. In factories dealing with 'clean' food ingredients the same authors found low rates of contamination with *Esch.*

coli only; and in those not dealing with food *Esch. coli* was absent.

In spite of the widespread occurrence of infection in domestic animals, and in animal products used for food, salmonella food poisoning is still relatively infrequent in man. This is probably because, as experiments have shown, large doses of salmonellae are usually required for human infection. The dose necessary and the type of illness vary, however, with the invading species and serotype. This statement is supported by human volunteer experiments (Hormaeche *et al.* 1936, McCullough and Eisele 1951 *a–d*). Some volunteers became ill after ingesting 125 000 *Salm. bareilly* or 152 000 *Salm. newport*, others resisting doses of 1 700 000 and 1 350 000 organisms respectively. Infection with *Salm. derby* required an infecting dose of 15 000 000 organisms. One strain of *Salm. anatum* had an infecting dose of less than a million, whereas for another strain the dose required was about 50 000 000. Investigations by Blaser and Newman (1982) suggested that in 6 of 11 outbreaks of human salmonellosis the ingested doses were small—less than 1000 organisms; outbreaks in which the doses were larger showed high attack rates and short periods of incubation. Minimal infective doses vary with age and state of health; in the young they are very low.

Antibiotic treatment is seldom justified in salmonella gastro-enteritis. The administration of drugs to which the infecting organism is sensitive *in vitro* does not shorten the illness; and there is some evidence that it may prolong the period of faecal excretion of the organism (Dixon 1965, Aserkoff and Bennett 1969).

For general accounts of the epidemiology of salmonella infection see van Oye (1964), Bryan *et al.* (1979), and Turnbull (1979).

Clostridium perfringens type-A food poisoning

Cl. perfringens type A is widely distributed in soil. It is commonly found in the faeces of man and animals and in a variety of foods, particularly meat, poultry, and their products. Spores of *Cl. perfringens* can survive cooking; and during slow cooling and unrefrigerated storage they germinate, forming vegetative cells that multiply rapidly. Under optimal growth conditions at temperatures between 43° and 47° the organism has a generation time of only 10–12 min.

Although the first account of a suspected outbreak of *Cl. perfringens* food poisoning was reported by Klein (1895), it was not until Knox and Macdonald (1943) described outbreaks affecting children after the consumption of school meals that serious consideration was given to the organism as a cause of foodborne disease. McClung (1945) described three outbreaks in the USA associated with chicken dishes cooked the day before consumption, and Osterling (1952) reported 15 outbreaks in Sweden. Conclusive proof that

Cl. perfringens caused food poisoning came from Hobbs and her colleagues (1953) who investigated 18 outbreaks in great detail. Volunteer feeding studies and serological typing of isolates by means of eight antisera prepared against some of the outbreak strains provided the proof. The disease has been reported from many countries including Japan (Itoh 1972), Sweden (Fabiansson and Normark 1976), Finland (Raevuori 1976), Italy (Caroli *et al.* 1977), the USA (Bryan 1978), and Canada (Todd 1978). The subject has been extensively reviewed by Hobbs (1979).

Cl. perfringens food poisoning is characterized by the onset of diarrhoea and abdominal pain 8–20 hr after the ingestion of food containing large numbers of vegetative cells. Vomiting is uncommon, and pyrexia, shivering and headache are rare. The duration of the illness is short and symptoms usually disappear within 10–24 hr. Fatalities are rare but have occurred among debilitated persons, particularly the elderly.

Outbreaks are almost invariably caused by meat and poultry dishes that have been cooked hours in advance and then cooled slowly, or even allowed to stand at room temperature for several hours before being served. Large masses of cooked meat provide a favourable anaerobic medium for the proliferation of the organism from any spores that survive cooking. High viable counts (10^6 or more per g) are reached in a few hours. After the ingestion of large numbers of vegetative cells of *Cl. perfringens* in a food, multiplication occurs in the intestine for a brief period followed by sporulation and the production of an enterotoxin.

It was at first assumed that only the 'heat-resistant' strains, i.e. those whose spores survived heating at 100° for 60 min, were capable of causing food poisoning. Later it became clear that 'heat-sensitive' isolates were equally capable of causing outbreaks (Taylor and Coetzee 1966, Sutton and Hobbs 1968). The presence of large numbers (more than 10^6 per g) of *Cl. perfringens* in the faeces of those suffering from diarrhoea and suspected of being affected in an outbreak is an important diagnostic feature. In the faeces of normal healthy persons counts are usually of the order of 10^3–10^4 per g, but in some aged, healthy, institutionalized patients the numbers may be as high as 10^8–10^9 per g (Yamagishi *et al.* 1976, Stringer 1981).

Viable counts of *Cl. perfringens* (spores and vegetative cells) should be carried out on samples of suspected foods associated with outbreaks. Similar counts should also be made on faecal specimens collected from ill persons. In addition counts of *Cl. perfringens* spores alone should be determined on faecal suspensions diluted 1 in 10 and heated at 80° for 10 min. Methods for the isolation and enumeration of *Cl. perfringens* in food and faeces have been the subject of two comprehensive studies by the International Commission on Microbiological Specifications for Foods (Hauschild *et al.* 1977, 1979).

Isolates of *Cl. perfringens* from food and faeces can be serologically typed to determine whether they are the same. The type specific antigens reside in the capsular polysaccharide. Stringer and co-workers (1980) applied serological typing to 524 outbreaks in the United Kingdom and 37 in other countries. Of the 7245 strains concerned in these 561 outbreaks, 5554 (77 per cent) were typable with a set of 75 English antisera; in 354 (63 per cent) of the outbreaks a specific serotype was shown to be responsible. Serological typing has been extensively reviewed by Stringer and his colleagues (1980, 1982). Bacteriocine typing (Watson *et al.* 1982) can be a valuable complement to serotyping in the laboratory investigation of food poisoning outbreaks, especially when the causative strain is serologically untypable.

The enterotoxin produced by *Cl. perfringens* is a spore-related toxin which is formed in the intestine at the time of sporulation. In the laboratory, enterotoxin is produced in sporulation media but not in ordinary growth media. The toxin produces fluid accumulation in the ligated rabbit-ileal loop (Duncan *et al.* 1968), erythema in the skin of guinea-pigs when injected intradermally (Stark and Duncan 1971), increased capillary permeability in guinea-pig skin (Stark and Duncan 1972), and diarrhoea when fed orally to monkeys, rabbits and human volunteers (Duncan and Strong 1969, 1971, Skjelkvåle and Uemura 1977*b*). The biological responses in animals have been used to measure the levels of enterotoxin in culture filtrates or fractions.

Enterotoxin has been extracted from sporulated cells by ultrasonic disintegration and purified by gel and ion exchange chromatography (Hauschild and Hilsheimer 1971). The molecular weight of the toxin is $36\,000 \pm 4000$ and the isoelectric point is at pH 4.3. Sakaguchi and his colleagues (1973) described a simplified purification procedure in which enterotoxin was extracted from sonicated cells with 40 per cent saturated ammonium sulphate solution at pH 7.0, and then subjected to 'differential solubilization' and repeated gel filtration on Sephadex G200. Affinity chromatography has also been used to purify the toxin (Uemura and Skjelkvåle 1976). Purification has led to a wide range of serological techniques for the detection of *Cl. perfringens* enterotoxin. They include double gel diffusion, electroimmunodiffusion, counterimmunoelectrophoresis and reversed passive haemagglutination. The tests, their application, and the relative sensitivities of the serological and biological detection methods have been reviewed by Stringer and co-authors (1982).

Skjelkvåle and Uemura (1977*a*) investigated two episodes of food poisoning and were unable to detect *Cl. perfringens* enterotoxin in the sera of acutely ill patients; a rising antitoxin titre was detected, however, during the following two months. In a later study Uemura and Skjelkvåle (1976) showed that a dose of

8 mg of purified enterotoxin administered orally was necessary to cause diarrhoea in human volunteers. A dose of more than 10 mg resulted in a measurable concentration (2 μg per g) in faeces. Studies on serum before and after dosing indicated that the measurement of anti-enterotoxin titres was of little diagnostic value.

The administration of cell extracts of an enterotoxigenic strain of *Cl. perfringens* to various animals led Niilo (1971) to conclude that enterotoxin caused increases in capillary permeability, vasodilation, and intestinal mobility. McDonel and Duncan (1975) showed that, in the presence of enterotoxin at concentrations capable of causing fluid accumulation in the ligated rabbit-ileal loop, epithelium was denuded from the tips of the intestinal villi. The mode of action of *Cl. perfringens* enterotoxin has been reviewed by McDonel (1980).

In addition to *Cl. perfringens* type-A strains, enterotoxin production has been reported for type C (Skjelkvåle and Duncan 1975) and type D strains (Uemura and Skjelkvåle 1976). The chemical, physical and immunological properties of the enterotoxins produced by the type A, C and D organisms are identical.

(For a description of enteritis necroticans—or *pigbel* as it is called in New Guinea—caused by *Cl. perfringens* type C, see Chapter 63.)

Staphylococcus aureus food poisoning

The ability of some staphylococcal strains to produce a toxic substance capable of irritating the gastro-intestinal tract was discovered three times before 1920—by Denys (1894) in Belgium, by Owen (1907) in the United States, and by Barber (1914) in the Philippines—but attracted little attention until Dack and co-workers (1930) in Chicago described it for a fourth time. During the next 30 years little was added to the basic information obtained by Dack and his colleagues from their studies on human volunteers about the production of enterotoxin by staphylococci, mainly because a reliable and widely applicable method of assaying it was not available. Nevertheless, the study of epidemics of staphylococcal food poisoning by orthodox epidemiological and bacteriological methods, supplemented by a limited number of experiments—none of them strictly quantitative—on human volunteers and monkeys, had provided a general picture of the conditions under which staphylococci multiply and form enterotoxin in food. Since 1960, several immunological methods of detecting and measuring enterotoxin have been devised and are now yielding information that confirms and extends the conclusions reached earlier.

Six serologically distinct enterotoxins (A to F) have so far been recognized, of which enterotoxin A is by far the most often incriminated in food-poisoning outbreaks. Enterotoxin F has recently been identified (Bergdoll *et al.* 1981), as a result of its association with 'toxic shock syndrome' (see Chapter 60), and little is known about its role in food poisoning. The enterotoxins are simple proteins with a di-sulphide bridge, are resistant to most proteolytic enzymes and, to some extent at least, are resistant to heat.

Although heating at 100° lessens the toxicity of the enterotoxins, normal boiling of foods is insufficient to inactivate them completely. The temperatures and times used for processing canned food will, however, usually destroy any enterotoxin present. Some studies show that enterotoxin is inactivated to a greater extent at 70° or 80° than at 90°–100°. Enterotoxin B is somewhat more resistant than enterotoxin A to heat. Tatini (1976) reported that there might be an increase in biological activity of enterotoxin after heat treatment. Reichert and Fung (1976) described partial reactivation of heated enterotoxin B. The degree of inactivation by heat is also dependent on the medium and the purity of the toxin preparation used in the experiments. Lee and co-workers (1977) investigated the role of protein during heat inactivation and found that beef broth contained a dialysable factor that was associated with increased thermal stability of the enterotoxins. Notermans and his colleagues (1983) reported that some components of minced beef, such as gelatin, increased the heat sensitivity of enterotoxins B, C and E. Despite the relative stability of the enterotoxins, heated foods are seldom implicated in outbreaks unless recontamination and multiplication of the organism occur.

The presence of other organisms in the food affects the production of enterotoxin, apparently by limiting the multiplication of the staphylococcus. Besides being affected by temperature and time, the production of enterotoxin is influenced by the composition of the food, pH, moisture content, and the atmospheric conditions (Chordash and Potter 1976, Troller 1976, Lotter and Leistner 1978). The degree of initial contamination and the type of enterotoxin are additional factors likely to affect toxin production. Although enterotoxin is believed to be produced only by coagulase-positive staphylococci (*Staph. aureus*) there are a few reports of coagulase-negative enterotoxin-positive variants of the organism (Lotter and Genigeorgis 1975, 1977).

Staphylococcal food poisoning has been caused by a wide range of foods, whose only common property appears to be the ability to support vigorous bacterial growth. The optimum temperature for the production of enterotoxin is 35–40°, and the lower the temperature the smaller the amount of enterotoxin produced. This is why staphylococci, though often abundant in milk, seldom give rise in temperate climates to food poisoning; they are overgrown by other organisms unless the temperature is 24° or above.

The main clinical signs are nausea, vomiting, abdominal pain and diarrhoea, developing 2 to 6 hr after the ingestion of contaminated food. Clinically, the illness closely resembles severe sea sickness. The incubation time and the severity of symptoms depend on the amount of enterotoxin consumed and the sensitivity of the person; in severe cases dehydration and collapse may occur and intravenous therapy may be necessary. Recovery is rapid, usually within 24 hr.

The foods most often implicated in outbreaks depend on the food habits of the country. About 70 per cent of the outbreaks in England and Wales are associated with meat and poultry. The products, which are usually eaten cold, include ham, tongue, meat pies and chicken. Cold sweets, such as trifle and cream cakes, have also been incriminated on many occasions. In the United States baked ham and sweet dishes are often implicated and in Japan hand-prepared rice balls have caused many outbreaks.

Outbreaks associated with canned foods and meat pastes in jars also occur. Most of them are due to contamination of the contents after the can or jar has been opened, but staphylococcal food poisoning sometimes follows the consumption of food from freshly opened cans. Only a few cans in each batch cause illness. The evidence suggests that the staphylococci enter the can during the cooling process through minute defects in the seam. Large numbers of *Staph. aureus* can usually be isolated from the remainder of the contents of the can, but sometimes the organism is present in only small numbers, or may have disappeared entirely.

In some countries, especially those with a warm climate, raw milk and raw milk products such as cheese continue to be responsible for many outbreaks. Numerous outbreaks due to fresh cow or goat milk have been described. These outbreaks are the only ones in which the staphylococcus may come directly from a source other than the human carrier. It is usual to find that at least one cow in a dairy herd is excreting *Staph. aureus* from the udder, often without clinical evidence of mastitis (Barber 1914, Steede and Smith 1954). In a few outbreaks the organism may have reached the milk from a human source.

Ice-cream has been incriminated in many countries but no outbreaks have been recorded in Great Britain since the introduction of the Ice Cream (Heat Treatment) Regulations in 1947.

Man is the most important source of *Staph. aureus* implicated in food poisoning. It is usually the food handler who contaminates the food, and under favourable conditions the staphylococcus will multiply and produce enterotoxin in the product. Food handlers may be nasal carriers or carry the organism on their hands. Between 20 and 50 per cent of the population are carriers of *Staph. aureus* (Williams 1963) and about 15 per cent carry enterotoxigenic strains. Lesions such as boils, carbuncles and whitlows are also foci of staphylococcal infection. In only a small proportion of hand-carriers are septic lesions present on the hands. Many, however, have apparently non-septic cuts and abrasions from which the organism can be isolated in much greater numbers than from the rest of the hand (Hobbs and Thomas 1948).

In a typical outbreak of staphylococcal food poisoning the strains of *Staph. aureus* isolated from specimens of vomit and faeces are identical with those from the implicated food, and from the hands and often the nose of a food handler. The incriminated food often contains more than 10^6 *Staph. aureus* per g. Phage typing has been used for the identification of strains of *Staph. aureus* for more than 30 years and has proved extremely useful for tracing the sources of outbreaks of food poisoning (de Saxe *et al.* 1982). Strains implicated in food poisoning are usually lysed by phages of group III or groups I and III. Because many strains of *Staph. aureus* are enterotoxigenic, tests for the production of enterotoxin may not be sufficient alone to identify the strain responsible for an outbreak. However, enterotoxin tests are essential to confirm that a strain, considered from the results of phage typing to be the cause of an outbreak, is in fact enterotoxigenic. When sufficient food is available a sample of 20 to 100 g can be tested for the presence of enterotoxin. As little as $1.0 \, \mu g$ of enterotoxin can be sufficient to produce illness in man. Because the double diffusion slide test will not detect enterotoxin present at less than $0.25 \, \mu g$ per ml, it is necessary to separate the enterotoxin from the food constituents and to concentrate it. The homogenized food is therefore centrifuged and the supernatant fluid is extracted with chloroform; it is then adsorbed on to carboxymethyl cellulose or Amberlite CG-50 ion exchange resin, eluted, and concentrated (Casman 1967, Reiser *et al.* 1974). With the more sensitive radio-immunoassay or enzyme-linked immunosorbent assay techniques for the detection of enterotoxin, a less vigorous extraction method can be used.

Information from various countries indicates that strains of *Staph. aureus* producing enterotoxin A, or both enterotoxins A and D, are responsible for about 70 per cent of outbreaks of staphylococcal food poisoning (Casman 1967, Bergdoll 1979). For example, de Saxe and co-workers (1982), in their study of 255 strains of *Staph. aureus* from separate outbreaks in the UK, reported enterotoxin production as follows: type A, 50 per cent; A and D, 21 per cent; C and D, 10 per cent; D, 6 per cent; and other combinations, 8 per cent. The remaining 5 per cent of strains tested did not produce enterotoxins A, B, C, D or E.

Reports from many countries indicate that between 15 and 70 per cent of strains of *Staph. aureus* isolated from various sources, including a wide variety of foods, produce one or more enterotoxins (see Wieneke 1974). For example, about 15 per cent of strains from bovine mastitis produced enterotoxin C or D but none produced A or B (Olson *et al.* 1970). All studies agree, however, that enterotoxin A production predominates only among strains from food poisoning.

(For a review of staphylococcal food poisoning see Bergdoll 1979; and for detailed information on the characters of *Staph. aureus* and its behaviour in food see Minor and Marth 1971, 1972, 1976.)

Bacillus cereus food poisoning

Reports of food-borne disease attributed to anthracoid *Bacillus* species resembling *Bac. cereus* have appeared in the European literature since 1906 (Goepfert *et al.* 1972). However, *Bac. cereus* was first conclusively implicated as an aetiologic agent by Hauge (1950). It is now recognized that two types of illness result from the ingestion of foods contaminated with *Bac. cereus*: the 'diarrhoeal syndrome', characterized by an incubation period ranging from 8–16 hr (average 10–12 hr), abdominal pain, profuse watery diarrhoea, rectal tenesmus, and moderate nausea seldom resulting in vomiting, with clinical recovery generally within 12 hr of onset; and an 'emetic syndrome' characterized by a rapid onset (1–5 hr), nausea and vomiting; diarrhoea occasionally follows, lasting 6–24 hr. The symptoms of the first type are similar to those of *Cl. perfringens* food poisoning and those of the second to *Staph. aureus* food poisoning (Gilbert 1979).

The classic description of the diarrhoeal syndrome followed four outbreaks in Norway that affected some 600 persons and were attributed to a vanilla sauce prepared and stored for 24 hr at room temperature before being served (Hauge 1955). Samples of sauce taken after one episode contained between 25 and 110 million *Bac. cereus* per ml. The heating used during preparation had been insufficient to destroy all the spores in one of the ingredients, corn starch, leaving those surviving to germinate and multiply when conditions became favourable. Outbreaks of this type were subsequently reported throughout Europe and from Australia, Canada and the United States. A notably higher incidence exists in northern and eastern Europe, attributed to a combination of the widespread culinary use of spices—often heavily contaminated with *Bac. cereus*—in meat dishes, followed by storage under unsatisfactory conditions after cooking (Ormay and Novotny 1969). Although cooked meat and poultry products are the most frequent causes of diarrhoeal syndrome episodes, a variety of other foods has been incriminated, including pastas, soups, vegetable dishes, various desserts, sauces and milk.

The emetic syndrome of *Bac. cereus* gastro-enteritis was first described in detail by Mortimer and McCann (1974); by 1982 more than 120 episodes had been recorded in Great Britain, most of them associated with fried or boiled rice from Chinese restaurants or 'take-away' shops. The practice in these establishments of preparing large bulks of boiled rice that are often left to cool slowly and held for long periods at ambient temperature before being fried leads to the proliferation of heat-resistant strains of *Bac. cereus*. The spores of certain strains of *Bac. cereus* can survive the boiling and frying procedures (Gilbert *et al.* 1974). Vegetative cell growth is rapid in cooked rice stored at room temperature, and is enhanced by the addition of beef, chicken or egg (Morita and Woodburn 1977). In the majority of outbreaks in Britain large numbers of *Bac. cereus* (ca 10^6–10^9 per g) were found in remnants of cooked rice or faecal specimens, or both. Similar outbreaks have also been reported from Australia, Canada, Finland, India, Japan, the Netherlands, Singapore and the United States. In less than 10 per cent of outbreaks food vehicles other than cooked rice were concerned, including cooked pasta, vanilla slices, pasteurized cream, and infant feed.

The rapid onset of symptoms, the afebrile nature of the illness, and its short duration suggest that *Bac. cereus* food poisoning is an intoxication rather than an infection. Several studies, reviewed by Turnbull (1982), have been made on the nature and properties of the toxic factors concerned. The toxin responsible for diarrhoeal symptoms has been shown to be a true enterotoxin capable of causing fluid accumulation in rabbit ileal loops, altering vascular permeability in rabbit skin, and killing mice when injected intravenously. It is synthesized and released during the logarithmic growth phase of the organism (Spira and Goepfert 1975). Oral administration of crude preparations to rhesus monkeys caused diarrhoea (Goepfert 1974). Turnbull and colleagues (1979*b*) reported the enterotoxin to be a thermolabile antigenic protein of medium molecular weight, produced to some degree by the majority of strains. Its dermonecrotic property and necrotizing effect on the intestinal epithelium are believed to be of relevance also to the pathogenesis of non-gastrointestinal *Bac. cereus* infections (Turnbull *et al.* 1979*a*).

Melling and his colleagues (1976) carried out feeding trials with rhesus monkeys to determine whether a separate enterotoxigenic factor was responsible for the emetic syndrome. Only strains isolated from vomiting outbreaks were capable of causing vomiting; furthermore, this response could be elicited only if the test strains were grown in rice culture. The toxin has now been characterized as a highly stable molecule of low molecular weight (Melling and Capel 1978).

Bac. cereus belongs to the morphological group 1 of Smith and co-workers (1952) and grows within the temperature range of 10–48°, with an optimum between 28 and 35°. The spores and vegetative cells are ubiquitous and may be readily isolated from the air, soil, natural waters, vegetation, and many kinds of food, including milk, cereals, spices, meat and poultry (Goepfert *et al.* 1972, Norris *et al.* 1981). Ghosh (1978) isolated *Bac. cereus* in low numbers from 14 per cent of single faecal specimens obtained from a general population of healthy adults. Association of a particular food-borne illness with *Bac. cereus* requires the isolation of large numbers ($>10^5$ per g) of this organism from the implicated food together, if possible, with its detection in acute-phase specimens of faeces and vomit from affected persons. Enrichment procedures are generally of little value in the laboratory diagnosis of *Bac. cereus* food poisoning.

Methods for the detection and identification of *Bac. cereus* in foods and clinical specimens have been reviewed by Kramer and colleagues (1982). The use of selective media (see Donovan 1958, Mossel *et al.* 1967, Kim and Goepfert 1971, Holbrook and Anderson 1980) relies upon the suppression of gram-negative organisms by polymyxin and gives a presumptive identification by means of the lecithinase reaction and, in certain instances, failure to ferment mannitol.

A serological typing scheme has been devised to facilitate the epidemiological investigation of *Bac. cereus* food poisoning (Taylor and Gilbert 1975, Gilbert and Parry 1977). Based on the type-specificity of the flagellar (H) antigens the scheme comprises 23 agglutinating antisera prepared against prototype strains from selected foods and clinical specimens. In approximately 90 per cent of outbreaks the causative serotypes can be established. Of 135 episodes of the emetic syndrome in various parts of the world 70 per cent were caused by *Bac. cereus* type 1 (Kramer *et al.* 1982). Spores of type 1 strains differ from those of other serotypes in possessing a strikingly greater resistance to heat (Parry and Gilbert 1980). (For reviews of *Bacillus cereus* food poisoning, see Gilbert and Taylor 1976, Gilbert 1979.)

Vibrio parahaemolyticus food poisoning

The halophilic vibrio, *V. parahaemolyticus*, described in Chapter 27, is recognized as a cause of food poisoning throughout the world. In Japan it is responsible for over 70 per cent of all cases (Sakazaki 1979), incidents being mainly confined to the warmer months of the year (May to October). The illness is characterized by acute diarrhoea, with mild fever and much abdominal pain, and usually begins about 12 hr after the ingestion of contaminated food, although the interval may be as short as 2 or as long as 48 hr. In most outbreaks the food incriminated is marine in origin. Raw fish and shell-fish are the most important vehicles in Japan, but sea-foods recontaminated after cooking are the most likely sources of infection in countries in which fish are not usually eaten raw (Barker 1974, Barrow and Miller 1976).

V. parahaemolyticus is widely distributed in shallow coastal waters throughout the world (Ayres and Barrow 1978, Sakazaki 1979). Coastal fish and shell-fish are frequently contaminated and the organism has been isolated on numerous occasions from cooked Malaysian prawns, imported into the United Kingdom in the frozen state (Turnbull and Gilbert 1982). It is not found generally in deep sea fish, but contamination often occurs in the fish market.

Illness associated with *V. parahaemolyticus* was first recorded in the United Kingdom in 1972, when a group of air travellers, having eaten hors d'œuvres containing dressed crab prepared in Bangkok, became ill on arrival in London (Peffers *et al.* 1973). Between 1972 and 1980 some 200 cases were reported, including three outbreaks affecting 9, 24 and 40 persons who became ill after eating prawn cocktails made with imported cooked prawns that had been thawed and then served without further cooking. In another outbreak dressed crab from one fishmonger at a seaside resort caused illness in at least 12 holidaymakers (Hooper *et al.* 1974). In most of the sporadic cases reported the patients were infected abroad, usually in the Far East or Africa, and in some there was a definite history of consumption of seafood.

There is reason to believe that this organism is responsible for some cases of the diarrhoea said to be due to 'non-cholera' vibrios in Calcutta (Chatterjee *et al.* 1970).

Whether the enteropathogenic activity of *V. parahaemolyticus* is due to a single factor or not is unknown. Sakazaki (1979) suggests that the essential enteropathogenic factor is the ability of Kanagawa-positive strains to multiply more rapidly than Kanagawa-negative strains in the intestine; Kanagawa-positive and -negative strains are usually found in patients' faeces and food, respectively. Enterotoxic substances other than those associated with the Kanagawa reaction may contribute to the virulence of strains.

Other forms of bacterial food poisoning

Numerous outbreaks have been described in which *Salmonella*, *Cl. perfringens*, *Staph. aureus*, *Bac. cereus* and *V. parahaemolyticus* were not isolated, but a food eaten by the sufferers contained large numbers of another organism. Taken alone, evidence of this sort should not be accorded undue significance. Samples are often collected many hours after the suspected meal was eaten, and the food may have been left at room temperature in the meantime. The claims of each organism must be considered separately. The repeated isolation of a particular bacterium from foods thought to have caused illness, especially when the cultures are almost pure, coupled with the isolation of the same agent from the faeces or vomit of the patients, strengthens the suspicion that it is a pathogen. The final proof, however, must depend upon the production of the disease in human volunteers. Bryan (1979) has reviewed extensively the evidence incriminating a wide range of these other bacteria in episodes of gastroenteritis. Table 72.4 summarizes data relating to some, particularly *Campylobacter jejuni*, that are currently

Table 72.4   Examples of other bacterial agents incriminated in food poisoning

Organism	Incubation period	Food vehicle	Reference
Bacillus subtilis	1–6 hr	Cooked meat and pastry	Kramer *et al.* (1982)
Bacillus licheniformis	4–15 hr	Cooked meat dishes	Kramer *et al.* (1982)
Campylobacter spp.‡	2–10 days	Raw milk	Robinson and Jones (1981)
			Jones *et al.* (1981)
		Undercooked poultry	Skirrow (1977)
			Brouwer *et al.* (1979)
Escherichia coli‡	12–72 hr	Cheese, and various	Marier *et al.* (1973)
			Kornacki and Marth (1982)
Non-O I *Vibrio cholerae*†	5–48 hr	Various	Roberts and Gilbert (1979)
*Yersinia enterocolitica**	24–36 hr	Milk	Black *et al.* (1978)
			Shayegani *et al.* (1983)

See also Chapters *55, †70, and ‡71.

being recognized with increasing frequency as agents of food poisoning.

Non-bacterial agents clearly play a part in some types of food poisoning, such as viral gastro-enteritis, scombrotoxic fish poisoning, ciguatera poisoning, and poisoning by red kidney beans. These illnesses have recently been reviewed by Gilbert (1983).

Investigation of outbreaks of food poisoning

The immediate objects of the investigation of an outbreak of food poisoning are as follows. (1) To verify that there is an outbreak of illness and that the causative agent was food-borne. (2) To determine the nature of the agent and the foodstuff(s) by which it was transmitted. (3) To determine the way in which the food was contaminated. (4) To ensure that all cases or carriers of the agent are identified. (5) To stop the outbreak if it is continuing (Department of Health and Social Security 1982.) The procedures to be followed are: (1) to secure a complete list of sufferers, with clinical histories and a full list of the foods consumed in the previous 2–3 days; (2) to record details of the origin, and the mode of preparation and storage, of the suspected foods; (3) to collect specimens for laboratory examination (below); and subsequently (4) to endeavour to find out how the food or foods thought to have been responsible were contaminated and to trace the reservoir of the causative organism. Materials collected for examination should include (a) the actual food consumed, (b) the vomit and faeces of the patient, and (c) the blood, spleen, liver and intestine of fatal cases. As soon as sufficient evidence has been obtained, either from the clinical investigation or from the preliminary bacteriological examination, to indicate the probable vehicle and the causal agent, it will be possible to decide what further specimens are required. These may include, according to circumstances, specimens from contacts or food handlers, samples of food ingredients, and swabs and washings from the premises in which the food was prepared. The steps taken in the investigation of food outbreaks

in England and Wales have been summarized by Pinegar and Suffield (1982).

Salmonella outbreaks

In salmonella outbreaks the organisms can frequently be demonstrated in the faeces of patients; vomited matter is much less satisfactory. If the incriminated food is available it may be possible to demonstrate the presence of the organisms here also, but in a large proportion of outbreaks the food vehicle is not determined, often because symptoms did not appear before the food remnants were discarded. Cultural methods for the isolation of salmonellae were discussed in Chapter 68. The occurrence of uncommon serotypes of *Salmonella* often affords a strong indication of the probable vehicle or reservoir of infection; with more common serotypes, such as *Salm. typhimurium*, the use of a phage-typing system, the determination of drug resistance patterns, and the ability to transfer resistance, may provide valuable assistance (Chapter 37).

After an attack of salmonella food poisoning about 50 per cent of patients excrete the causative organism in the faeces for 4–5 weeks or longer. It is, therefore, always worth while attempting to make a retrospective diagnosis by examining the faeces of convalescents.

From foods that have caused illness, salmonellae may be isolated by the methods used for the culture of faeces. Both food and faeces are likely to contain large numbers of the organism, which may be detected by direct plating procedures. However, for the examination of food ingredients, material from abattoirs, and environmental samples, a more elaborate procedure is required, capable of detecting small numbers of pathogenic bacteria in the presence of other non-lactose-fermenting organisms. Many factors will influence the selectivity of enrichment broths for salmonellae. For example, selectivity is diminished by the presence of protein (Silliker and Taylor 1958); selenite broth may be rendered toxic by the ingredients of some foods; and it may be necessary to add an agent such as sodium heptadecyl sulphate to the medium to disperse and emulsify fat, so as to improve the isolation of salmonellae from fatty or oily foods (Morris and Dunn 1970).

A wide range of liquid enrichment and selective agar media have been devised for the isolation of salmonellae. Various additions to enrichment broths have been recommended, such as cystine (North and Bartram 1953) or brilliant green (Stokes and Osborn 1955) to selenite broth; and sodium lauryl sulphate (Jameson 1961) to selenite or tetrathionate broth. Incubation at an elevated temperature such as 43° has also been shown to improve salmonella isolation rates from various materials (Harvey and Thomson 1953, Edel and Kampelmacher 1968, 1969, Harvey and Price 1968).

Salmonellae in dried or frozen foods are often in a poor state of viability, and few in number, and can seldom be isolated by the direct inoculation of enrichment media. A preliminary resuscitation period in a mild non-selective medium is required to enable damaged organisms to recover and begin to multiply freely (van Schothorst and Kampelmacher 1968).

The ICMSF (1978) give a six-stage scheme for the enrichment, isolation and identification of salmonellae. They recommend a range of media for pre-enrichment (lactose broth, buffered peptone water, distilled water) according to the product under examination; two selective enrichment media (selenite cystine and tetrathionate brilliant green broth at 43°); and three selective agar media (bismuth sulphite and brilliant green agars in addition to a medium of the 'laboratory's choice').

A series of studies organized from the Netherlands has led to the formulation of a standard technique for isolating salmonellae from a variety of products in an attempt to eliminate many of the variables in salmonella isolation procedures (van Leusden *et al.* 1982). This method, which consists in pre-enrichment in buffered peptone water, selective enrichment in tetrathionate broth at 43°, and subculture on brilliant green agar, has been adopted by the International Organization for Standardization (ISO; 1975) and other bodies as a standard method. Recent studies show that a modified Rappaport's broth known as Rappaport Vassiliadis malachite green magnesium chloride enrichment medium (RV medium) is significantly more effective in the isolation of salmonellae than the Müller-Kauffmann tetrathionate broth recommended by the International Organization for Standardization (Vassiliadis *et al.* 1981).

A wide range of alternative procedures can be applied to the isolation of salmonellae from various products and to the selection of particular serotypes. These include the use of 'selective motility enrichment', and the incorporation of antisera into semisolid agar (see Harvey and Price 1982).

The disadvantage with most standard salmonella isolation methods is the length of time needed to complete the procedure before a result can be given; with a pre-enrichment procedure the minimum time in which a positive result can be reported is 4 days. A number of rapid methods, mainly designed to exclude the possibility that salmonellae are present in a food, reduce the delay in reporting results. Such methods include immunofluorescence (Georgala and Boothroyd 1964, Georgala *et al.* 1965, Fantasia 1969), enrichment serology (Sperber and Deibel 1969), radiometry (Stewart *et al.* 1980), enzyme immunoassay (Minnich *et al.* 1982), and the use of hydrophobic grid membrane filters (Entis *et al.* 1982). Cultural procedures may, however, still be necessary to eliminate false positives and to identify particular serotypes.

Clostridium perfringens outbreaks

When investigating these outbreaks it should be remembered that, after multiplying in meat, *Cl. perfringens* is mainly in the vegetative state. Unheated material should therefore be cultured on blood agar or neomycin blood agar. Qualitative tests for *Cl. perfringens* in faeces are of limited diagnostic significance, and counts of viable organisms should therefore be performed.

At least *one* of the following criteria should be satisfied for the laboratory confirmation of an outbreak.
a) The number of *Cl. perfringens* in the epidemiologically incriminated food is $> 10^6$/g.
b) The median faecal spore count of *Cl. perfringens* is $> 10^6$/g.
c) Strains isolated from the faeces of most of the patients belong to the same serotype.
d) Isolates from the incriminated food and faecal specimens belong to the same serotype.
For further information see page 488.

Staphylococcus aureus outbreaks

Staph. aureus should be sought in the vomit or faeces of sufferers within 24 hours of the onset of illness. Strains from the food, from the patients, and from the hand and nose swabs of food handlers should be examined by the bacteriophage-typing method (see Chapter 30). The mere presence of *Staph. aureus* in the faeces of patients, and in the nose and on the hands of a food handler is of no particular significance. The finding of the same straphylococcal strain in the faeces of several patients with typical symptoms is suggestive that an outbreak is staphylococcal but cannot be taken as definite proof unless the same organism is also present in large numbers in the food. If the suspected food is not available for examination, bacteriological information obtained from the patients alone is seldom conclusive. A particular food handler cannot be incriminated with any certainty unless a staphyloccal strain from his hands or nose is identical with one present in large numbers in a food eaten by patients suffering from a clinically typical staphylococcal illness. Strains of *Staph. aureus* should, where possible, be examined for their ability to produce enterotoxin, and if sufficient food is available this can be tested for the presence of enterotoxin (see page 490).

In most staphylococcal outbreaks the organism is present in large numbers (10^6 to 10^9 per g) in the food. The presence of less than 10^5 per g is of little significance unless a food—such as pasteurized or spray-dried milk—was heated before it was consumed; or—such as matured cheese—was kept for a long period after manufacture.

Bacillus cereus outbreaks

Because the spores of *Bac. cereus* are so widely distributed in the environment, the presence of the organism in small numbers in many types of food and in some faecal specimens is to be expected. It therefore follows that the mere detection of *Bac. cereus* in a food sample is insufficient to implicate either the food or the organism as the cause of an outbreak.

At least *one* of the following criteria should be satisfied for the laboratory confirmation of an outbreak.
a) *Bac. cereus* strains of the same serotype are present in the epidemiologically incriminated food and in the faeces or vomit of the affected persons.
b) Large numbers ($> 10^5/g$) of *Bac. cereus* of an *established food poisoning serotype* are isolated from the incriminated food, *or* faeces *or* vomit of the affected persons.
c) Large numbers ($> 10^5/g$) of *Bac. cereus* are isolated from the incriminated food, and organisms are detected in the faeces, vomit, or both, of the affected persons.
For further information see page 492.

Outbreaks due to halophilic vibrios

In outbreaks of food poisoning associated with fish or shell-fish, halophilic vibrios should be borne in mind. These organisms can easily be recognized on media used for the detection of *V. cholerae*, and one of these should be included as a routine when faecal specimens from outbreaks or sporadic cases of diarrhoea of uncertain cause are examined. On thiosulphate citrate bile-salt sucrose agar *V. parahaemolyticus* appears as colonies 2–3 mm in diameter with a green or blue centre. For the isolation of vibrios from sea-water and fish see Sakazaki (1979) and Furniss and co-workers. (1979). (For the highly pathogenic *V. vulnificus*, see Report 1984.)

Prophylaxis

Each of the main forms of bacterial food poisoning is due to the ingestion of large numbers of a specific organism produced by multiplication in a foodstuff.

Food poisoning would not occur if all foods in which bacteria can grow readily were cooked while fresh and eaten immediately; in practice this is not always possible, particularly when large numbers of people are being catered for at the same time. In an analysis of more than 1000 outbreaks of food poisoning that occurred between 1970 and 1979 in England and Wales, Roberts (1982*b*) showed that the most common contributory factors were storage of food at ambient temperature, inadequate cooling or reheating, warm holding and undercooking. As a result, foods were often held for considerable periods at temperatures that permitted rapid bacterial multiplication. Much can be done by an intelligent combination of refrigeration, cooking and hot storage, but the complete safety of all potentially dangerous foods is difficult to ensure by these means alone. For this reason, attempts should also be made to reduce the frequency with which pathogens gain access to food—particularly to food in bulk—and to find a means of freeing contaminated foods of bacteria without impairing their quality. Green foods that are to be eaten raw should be thoroughly washed; and shell-fish should preferably be taken from non-polluted areas and kept for two days in clean chlorinated water as described in Chapter 9. Shrimps and prawns should be boiled and shelled.

The prevention of salmonella infection is the most difficult and complicated problem in the control of food poisoning in western countries, owing to the widespread consumption of meat and poultry, often undercooked. The means by which food animals become infected with salmonellae and by which the organisms are spread at slaughter and during meat processing have already been discussed (see pp. 483–5). Despite the routine inspection of meat—which cannot, of course, detect carrier animals—and an understanding of the relevant cycles of infection, too much contaminated material still enters the kitchen in the form of raw ingredients. Gilbert (1983) reports that 79 per cent of frozen chickens, 7.5 per cent of minced beef and pork, and 12 per cent of pork sausages are contaminated with salmonellae. Though these organisms are killed by thorough cooking there is a possibility that other foods, which are to be eaten without further heating, may become cross-contaminated by hands, surfaces and utensils used for preparing both raw and cooked foods. Meat should be regarded as potentially dangerous and should never be eaten raw. Duck eggs should be boiled for 15 minutes (Clarenburg and Burger 1950); frying that does not ensure coagulation of the yolk is insufficient to destroy all salmonellae. All cooking utensils should, after cleaning, be thoroughly dried before being put away.

It is often difficult to make foods safe by heating without impairing their palatability. That is one of the dangers of made-up meat dishes. The penetration of heat into a large joint of meat or a large turkey is slow, and the interior may not reach, or may not be kept long enough at, a temperature sufficient to kill all organisms. Though most food poisoning organisms are not specially heat-resistant, apart from sporing forms such as *Cl. perfringens* and *Bac. cereus*, there is variation in this respect between strains. The degree of heat necessary to destroy them is also influenced by the nature of the medium in which they are contained (see Anellis *et al.* 1954, Genigeorgis and Riemann 1979). The safety of a food to the consumer will depend not only on the effectiveness of the cooking procedure but also on the handling and storage of food after cooking. Heat resistant spores of *Cl. perfringens* may be activated by the cooking process and begin to germinate rapidly during long slow cooling and warm holding periods.

Canning of food on a large scale needs careful supervision. Not only must the heat treatment be sufficient to destroy all pathogenic organisms, but the water in which the cans are cooled should be of drinking standard; otherwise pathogenic organisms may enter the can through the seams, which tend to gape during autoclaving and to allow organisms to be sucked in as cooling proceeds.

The regular bacteriological examination of food handlers would contribute little to the control of salmonella food poisoning (see Howie 1970, 1979). A strict code of personal hygiene should be enforced in all catering establishments, particular attention being paid to the washing of hands after each visit to the closet and after handling raw foods such as meat and poultry. Thorough cleaning and disinfection of food utensils and surfaces after use will reduce the risk of cross-contamination between foods. In a large catering establishment separate areas and personnel should be allocated for the preparation of raw and cooked foods.

Strains of *Staph. aureus* implicated in staphylococcal food poisoning are usually of human origin. Control of this type of food poisoning depends, therefore, on 'no-handling' techniques for the preparation or processing of the foods usually associated with the disease, together with adequate refrigeration.

Current methods of processing make it inevitable that spores of *Cl. perfringens* will often be present in meat and poultry; they will also be found on a wide range of other ingredients, such as herbs, spices, vegetables and soup mixes. However, the organism is dangerous only when ingested in large numbers; thus, prevention is concerned not only with its destruction, but also with the control of spore germination and of the subsequent multiplication of vegetative cells in cooked foods. Such control can be achieved either by serving food hot immediately after cooking, or by cooling rapidly and storing in a refrigerator until required, when it should be served cold. If required hot it should be thoroughly re-heated to boiling-point. Rapid cooling of large bulks of food is often difficult and special plant employing forced draughts of cold air, or cold rooms with extraction fans, may be necessary. Large quantities of food will cool more rapidly if broken down into smaller portions or, if liquid, into shallow layers. The same precautions are required with foods associated with outbreaks of *Bac. cereus* food poisoning, for example large quantities of cooked rice.

Botulism

Botulism (from Latin *botulus*, a sausage), sometimes known as allantiasis, or ichthyosismus, is a disease of man and animals. Its peculiarities were recognized by medical authorities in Europe towards the end of the eighteenth century, particularly in Germany (Dickson 1920). The causative bacterium was isolated and described by van Ermengem in 1896 (1896, 1897). After a festive gathering of a music club at the village of Ellezelles in Belgium, several of the members were taken ill, and three of them died within a week. The disease was confined to those who had partaken of a certain piece of raw ham; the other ham and the remainder of the animal had been previously consumed without causing trouble. The contaminated ham was paler and softer than normal and smelt rancid. When fed to cats it caused mydriasis, partial paresis, secretory disturbances, aphonia, and other symptoms. From it van Ermengem cultivated a strictly anaerobic organism, which secreted a powerful toxin, giving rise to the same symptoms in cats as those caused by the original ham.

Since that date numerous outbreaks of human botulism have been recorded in Europe, America, Japan, and elsewhere. In Great Britain only nine incidents have been described; of the total of 14 deaths, eight occurred in the Loch Maree outbreak, in which the affected persons all died within a week (Leighton 1923). The most recent British outbreak occurred in Birmingham in 1978; it affected four persons—two fatally—who had eaten canned Alaskan salmon (Ball *et al.* 1979).

Symptomatology and pathology

The incubation period is usually between 12 and 36 hours after the consumption of the affected food, but may be as short as 8 hours and as long as 8 days. Prominent among the symptoms are vomiting, constipation, ocular pareses, thirst, pharyngeal paralysis, the secretion of thick, viscid saliva, and sometimes aphonia. Less usual features are mentioned by Hughes and his colleagues (1981). General consciousness and sensibility remain intact till near the end, which is preceded by coma or delirium. The temperature is generally subnormal—35.5–36.7°. Later in the illness it may rise owing to the onset of broncho-pneumonia (Dickson 1918). Death may occur within 24 hours from the time of onset, or may be delayed for a week. The immediate cause of death is usually asphyxia, consequent on paralysis of the respiratory musculature. In patients who survive, complete recovery, particularly of the ocular movements, may not take place for 6 or 8 months. At autopsy the kidneys, liver, and meninges are congested. (See also Table 72.1.)

Bacteriology

Botulism is usually an intoxication, not an infection. The causative organism, *Clostridium botulinum*, multiplies in the food before it is consumed, and produces a powerful soluble toxin (see Sugiyama 1980), which on ingestion is absorbed mainly from the upper part of the intestine, and gives rise to the characteristic disease.

The toxins produced by the different types of *Cl. botulinum* differ in antigenic structure and in toxicity for a variety of experimental animals; but all are neural toxins, with a similar mode of action. These toxins are among the most powerful known. That intoxication by the oral route is peculiarly effective under natural conditions appears to depend on the readiness with which the organisms in the foodstuff synthesize toxin in concentrations such that, although much of the ingested toxin is destroyed or otherwise antagonized in the digestive tract, enough penetrates the mucosa to produce intoxication. Thus in the mouse, the lethal dose of type A toxin by the oral route is 50 000–250 000 times that by the intraperitoneal. The ratio of oral to intraperitoneal lethal doses for tetanus toxin, however, is of the same order, $200–1200 \times 10^3$:1, suggesting that tetanus exotoxin does not induce food poisoning because the conditions for effective oral dosage do not arise in nature (Lamanna 1959, 1960, Lamanna and Carr 1967).

The recorded cases of human botulism are mainly due to types A, B and E; human type C, D (see L. DS. Smith 1977) and F intoxications are rare. Type G intoxication is unknown except in experimental animals (Ciccarelli *et al.* 1977), but Sonnabend and co-workers (1981) demonstrated type G organisms and toxin in autopsy specimens from several persons who had died suddenly and unexpectedly.

Though the disease is usually a pure intoxication, a suspicion has existed for many years that multiplication of the bacillus occasionally takes place in the gut, gut wall, and even in the tissues. Several investigators (Coleman and Meyer 1922, Orr 1922, Starin and Dack

1925) observed that large doses of spores—of the order of 10^9—washed or heat-treated to remove toxin, may be lethal to guinea-pigs inoculated experimentally, and that *Cl. botulinum* is recoverable from the viscera *post mortem*. The results are compatible with a form of infection, but the deaths may have been due to the traces of toxin known to be present in the spores (Keppie 1951); and the post-mortem distribution of organisms to dissemination of the original spores rather than of proliferating bacilli. Moreover it is possible that the heat treatment activated the spores and caused the production of vegetative cells. Shvedov (1959) protected laboratory animals against a few MLD of *Cl. botulinum* type A culture more effectively with oral than with intramuscular antitoxin, and deduced that continuing production of toxin in the intestine is an important factor in the disease. The work of Shvedov (1960), Minervin (1967) and others (see Dolman 1961, Petty 1965) provided further evidence for the elaboration of toxin in the alimentary tract; it seemed possible that some degree of infection with consequent toxin production was responsible for the delayed deaths and relapses that sometimes occurred. Miyazaki and Sakaguchi (1978) showed that toxin production could occur in the caecum of the chicken.

Cl. botulinum is occasionally isolated from *wounds* in man. The few recorded cases of fatal botulism, appearing 4–14 days after injury, in persons with *Cl. botulinum* in the wound (see Merson and Dowell 1973), afford unequivocal evidence of an *infectious* botulism. It is noteworthy that the organism has been isolated from imperfectly sterilized catgut (Tardieux *et al.* 1952).

Infant botulism (see p. 499) also provides clear evidence of intra-intestinal toxin production (Midura and Arnon 1976, Arnon *et al.* 1977, Turner *et al.* 1978). The diets fed to infants less than six months old virtually preclude the possibility of ingestion of pre-formed toxin. *Cl. botulinum* organisms and toxin are sometimes present in the faeces of infants for many weeks after clinical recovery. Wilke and his colleagues (1980) found that in four patients the numbers of *Cl. botulinum* in the faeces ranged from 3.3×10^3 to 6.0×10^8 per g.

Mode of intoxication The detailed investigations of the neurotoxicity of the toxin are reviewed by Wright (1955) and by Stevenson (1958). (See also Hughes and Whaler 1962, Zacks and Sheff 1970.) Its action is peripheral, not central, and affects both the efferent autonomic nervous system supplying muscles and glands and the somatic nerves supplying skeletal muscles. Moreover, only cholinergic nerves, acting through the release of acetylcholine, are sensitive to the toxin, the adrenergic nerves acting through the release of adrenalin being resistant to it. Neither the muscles themselves nor the conduction in the nerves are affected. The site of poisoning appears to be at the terminal unmyelinated portion of the neurofibrils, and

Table 72.5 Epidemiological features of botulism in man and animals

Type	Designated by	Animal species chiefly affected	Vehicles	Main geographical occurrence
A	Burke (1919*b*)	Man, occasionally chickens and mink	Home-preserved vegetables and fruits; meats; fish	Western USA, Canada, USSR
B	Burke (1919*b*)	Man, occasionally horses and cattle	Home-preserved meat, especially pork, and vegetables; fish	Europe (Poland, Germany, France, Norway and elsewhere), USSR, Eastern USA, Canada
C	Bengtson (1922) and Seddon (1922)—Cα and Cβ; see Gunnison and Meyer (1929)	Waterfowl (Cα), cattle and horses (Cβ), chickens, pheasants, mink	Rotting vegetation and invertebrates in alkaline lakes; dipterous fly larvae; forage; carrion	In waterfowl—western USA, Canada, South America, South Africa, Australia, New Zealand, Japan, Europe. In cattle—Australia, South Africa, occasionally Europe and USA
D	Theiler *et al.* (1926–27*b*), Meyer and Gunnison (1928)	Cattle and horses	Carrion	South Africa
E	Gunnison *et al.* (1936)	Man, farmed fish	Uncooked products of fish and marine mammals	Northern Japan, Canada, USA (especially Alaska), USSR, occasionally Europe
F	Dolman and Murakami (1961)	Man (few episodes recorded)	Liver paste	Denmark, USA
G	Giménez and Ciccarelli (1970)	None recorded	...	Argentina (soil isolate)

the effect a block in conduction proximal to the site of acetylcholine release, though there is some evidence of central action resembling that of tetanus toxin. It appears that the toxin interferes with the acetylcholine-release process itself (Kao *et al.* 1976).

Both clinical and experimental evidence indicate that the brain and spinal cord are resistant to the toxin, though cholinergic structures are presumably present. The resistance is not due to any inability of the large toxin molecule to reach the central nervous system, or to the blood-brain barrier, since direct injection of the toxin into the brain stem has no effect.

In isolated nerve-muscle preparations, the toxin is rapidly fixed and induces paralysis, within minutes when strong toxin is used; after this fixation, added antitoxin has no protective action (see Burgen *et al.* 1949).

As regards absorption from the gut, we have already noted that in experimental animals it takes place chiefly in the upper part of the small intestine. In the rabbit type A toxin reaches the blood stream via the lymphatics and the thoracic duct (May and Whaler 1958). It appears to pass into the tissues from the lumen of the gut by the filtration mechanisms normally operative. There may, however, be some disaggregation of the large-molecular native toxin; when preparations with a sedimentation constant of 17.9 are put into the gut, toxin with a constant of 7.9 is found in the lymph (Heckley *et al.* 1960, Hildebrand *et al.* 1961).

A disease closely simulating botulism in man can be reproduced by the experimental inoculation or feeding of cats, monkeys, and certain other animals with toxic material (see Chapter 42).

Epidemiology

Botulism is due to the consumption of food in which *Cl. botulinum* has been growing and occurs in sporadic episodes limited to those who have partaken of the contaminated food. As the usual form of the disease is not an infection, secondary cases do not occur, but it is common for cats, dogs, and especially chickens that are fed on the remnants of the food to develop symptoms of poisoning.

Botulism is not a common disease. Though it is widely distributed, its real prevalence in the world is unknown, as nearly all our information comes from the more highly developed countries (Table 72.6). Most outbreaks affect only a few persons, but occasional large ones do occur when the same toxic foodstuff is consumed by many (see Gordon and Murrell 1967). Of the types of botulism encountered type A, associated particularly in the United States with the eating of home-canned or preserved vegetables, is the commonest; but of late years increasing numbers of outbreaks caused by type E have been reported from Japan and Canada, usually associated with marine products. Type B is most frequent in countries where pork products are consumed, such as France, Germany and Norway. Types C and D are rarely met with in man. Type C botulism has been reported from the USA (Meyer *et al.* 1955), France (Prévot *et al.* 1955), and the USSR (Matveev *et al.* 1967); and Type F from

Table 72.6 Occurrence of human botulism in certain countries (after Gordon and Murrell 1967, Matveev *et al.* 1967, Iida 1970)

Country	Period	Episodes	No. of cases	No. of deaths	Case fatality per cent *ca*
USA	1899–1964	651	1670	1008	60.3
France	1940–48	500	>1000	15	1.5
Germany	1898–1948	434	1294	179	13.8
Denmark	1901–64	12	34	14	41.2
Norway	1934–64	13	63	1	1.6
Sweden	1932–64	7	16	2	12.5
Canada	1919–64	36	110	62	56.4
Japan	1951–67	57	321	85	26.4
USSR	1958–64	95	328	95	30.0
British Isles*	1922–64	7	17	13	76.5

*Dolman (1964) quotes 11 outbreaks with 21 cases and 16 deaths.

Denmark (Møller and Scheibel 1960) and the USA (Condit and Renteln 1966).

Of the 240 USA outbreaks recorded in Table 72.6 in which the toxin was identified, 69 per cent were due to type A, 18 per cent to type B, and 13 per cent to type E. In Canada, on the other hand, 57 per cent were due to type E, and in Japan about 98 per cent.

Any seasonal distribution of the disease appears to be determined by the eating habits of a given community. In the USA the disease is commoner during the winter months, when more preserved foods are eaten. The case of type E botulism is peculiar (see Dolman and Chang 1953, Dolman 1957*b*, 1960). The incidence is heaviest in regions bordering the sea, especially on the shores of the northern Pacific Ocean, in Japan, Alaska and Canada. In most instances the disease is fish-borne; poisoning by other foods is rare. The contributory factors are the preference of the local inhabitants for raw and sometimes rotten marine foods; and the low temperature—down to 6°—at which toxigenesis may occur in the foods (see also Dolman and Iida 1963).

Infant botulism

This disease (see below and pp. 497, 500, 501, 503, 504) was first recognized by Pickett and his colleagues (1976). Up to the beginning of 1979 about 100 cases had been recorded, mainly in the USA (see Arnon *et al.* 1978), but also in England (Turner *et al.* 1978) and Australia. The ages of affected infants ranged from 3 to 26 weeks. These cases were caused by toxigenesis in the intestine, following ingestion of type A or B spores (Arnon *et al.* 1977). Environmental sources of spores were thought to include honey; about 10 per cent of honey samples contained type A or B spores (Sugiyama *et al.* 1978, Arnon *et al.* 1979). A case of type F botulism in an infant aged 14 days has been recorded (Report 1980).

Morbidity and case-fatality rate

In any episode the morbidity rate is high. As a rule all who partake of the contaminated food contract the disease. As will be seen from Table 72.6 the case-fatality rate varies in different outbreaks, depending partly on the type of toxin concerned and partly on the quantity in which it is present in the food. It is usually high in type A and B outbreaks associated with processed vegetables, but low in type B outbreaks associated with pork and ham products. In Canada the fatality rate from type E is double that in Japan, though in both countries the toxin is conveyed in marine dishes.

The comparative potencies of the different toxins may be gathered from the figures given by Dolman and Murakami (1961; see also Gunnison and Meyer 1930). The intragastric lethal doses for monkeys, in terms of approximate MLD for mice, were as follows: B 180; A 650; C_β and E 1500–2500; F 50 000–75 000; $C\alpha$ 100 000–250 000; and D 600 000.

The high fatality rates sometimes observed in type E botulism may be due to enhancement of toxicity of the toxin by proteolytic strains of certain other contaminating bacteria (Sakaguchi and Tohyama 1955); or by tryptic enzymes (Duff *et al.* 1956) operative *in vivo* (Dolman 1957*a*, Iida *et al.* 1958). The rarity of type E botulism in cooked foods is attributable to the relatively high heat-lability of the spores of *Cl. botulinum* type E.

It should be noted that not all raw fish poisoning is due to type E; for example, type B intoxication is also recorded (Skulberg 1958, Dolman *et al.* 1960).

The earlier the symptoms appear, the higher is the fatality rate. Thus it is recorded that, of those developing symptoms in 24 and 72 hours, 84 and 55 per cent respectively die; and of those alive after 8 days, only 20 per cent die (Burke *et al.* 1921).

Mortality from *infant botulism* (above) is low in cases admitted to hospital. Arnon and his colleagues (1978) produced evidence to suggest that the disease was the cause of a small proportion of sudden infant deaths. A further study (Arnon *et al.* 1982) suggested that it was more severe in artificially fed than in breast-fed patients.

Type of food

Most of the reported outbreaks of botulism have been caused by food that has been smoked, pickled or canned, allowed to stand for a time and eaten without cooking, or has been inadequately cooked (Jordan 1917). No cases have yet followed the consumption of fresh food, cooked or uncooked. In several instances those who consumed the affected food after cooking escaped, whereas those who consumed it before cooking were attacked (Geiger *et al.* 1922). The kind of food responsible for the poisoning varies with the eating habits of the communities concerned. In Europe it is sausages, ham, preserved meats, game pâtés and brawn. In America, on the other hand, the contaminated foods have generally been home-canned fruits and vegetables, such as olives (Armstrong *et al.* 1919, Edmondson *et al.* 1920), string beans (Geiger *et al.* 1922, Geiger 1924, Stricker and Geiger 1924), corn (Geiger *et al.* 1922, Geiger 1924), spinach (Geiger 1920), peas (Editorial 1926) and beets (see Sutherland 1960). A few have been due to cooked meat or fish (Editorial 1926), and some to cheese (Nevin 1921). Home-canned string beans alone accounted for 19 out of 55 outbreaks (Editorial 1926). In the United States, Meyer (1956) notes that, of 445 bacteriologically confirmed episodes, vegetable foods were implicated in 381; L.DS. Smith (1977) noted that vegetables were implicated in 76 per cent of 127 outbreaks. Sturgeon and salmon eggs, salted fish and preparations of raw fish are implicated in Russia, Japan, British Columbia and Alaska; since its identification in 1935 as a distinct type, the marine *Cl. botulinum* type E has usually proved to be the contaminating organism in these foods (Dolman 1957*b*). In Japan most of the outbreaks caused by type E have followed the consumption of *izushi*—a fermented food made of boiled rice, vegetables, and raw fish meat allowed to mature for up to two months (Nakano and Kodama 1970). The safety of farmed fish has been questioned after an outbreak in Germany associated with smoked trout fillets (Bach *et al.* 1971).

As a rule the preserved foods have been noticeably spoiled; though it must be emphasized that in many cases no spoilage is obvious. The cans are often blown and show numerous gas bubbles on opening; the solid part of the food has a mushy or disintegrated appearance, and smells like rancid butter or cheese (Burke 1919*c*). With ham the flesh is often paler and softer than normal, and has a rancid odour (van Ermengem 1897, Savage and White 1925).

Distribution of *Cl. botulinum* in nature

The first major survey of the distribution of *Cl. botulinum* in nature was reported in 1922 by Meyer and Dubovsky (1922*a*, *b*, *c*); the results of subsequent investigations were summarized by Meyer (1956) and L.DS. Smith (1977). (See also Report 1967, 1970.)

The earlier surveys were made largely in terms of *Cl. botulinum* types A and B. In North America type A spores predominate west of the Rocky Mountains, and type B in the east; types C, D and E also occur (L.DS. Smith 1978). Meyer and Dubovsky (1922*b*) found type A or B spores in 7.5 to 31.8 per cent of samples of American farm produce such as mouldy hay, string beans, decayed vegetation, ensilage, corn husks, beets, and tomato plants. In European soil type B spores are more prevalent than type A. The prevalence of both is low in British and Swedish soil. Types A–F occur in Russia (Kravchenko and Shishulina 1967, Bulatova *et al.* 1973), the prevalence being higher in the south than in the north. Type B occurs in South African soil (Knock 1952) and type D in mud from lakes in Zululand (Mason 1968). In Mauritius type C appears to be common in mud from inland waters, and in Nigeria types C and D (G. R. Smith 1982, 1984.) Takashi (1953) found type A in one of 717 Japanese soil samples. In India, Pasricha and Panja (1940) found type A in four of eight soil samples.

The low heat resistance of type E spores may explain why they have only recently been demonstrated in soils. Type E has been found in the soil of the USSR, the USA, Japan, and Canada. Larger numbers are found in sand and sludge of shallow coastal waters, especially in western North America, northern Japan, and Sweden. They also accumulate in sea-bottom sludge—not necessarily in regions near human habitation; and their presence in marine environments may be inferred from the occurrence of type E botulism among North American Eskimos, and of type E spores in sturgeon from the Sea of Azov (Dolman 1957*a*, Report 1970). Fish appear to acquire the spores by ingestion, and perhaps become superficially contaminated through damaged skin (Cann *et al.* 1968). Types other than E sometimes occur in marine environments. (For a review of *Cl. botulinum* in relation to fishery products see G. Hobbs 1976.) Haagsma (1974) found that 30 per cent of mud samples from Dutch inland waters contained type B, C or E, but that type E predominated. Types B–E occurred in the mud of British lakes and waterways (see G.R. Smith 1982), the prevalence being much higher than in British soil (G.R. Smith and Young 1980). Most London lakes were contaminated, more often with type B than with C, D or E (G. R. Smith and Moryson 1975); in the Norfolk Broads types B, C and E each occurred in more than 50 per cent of mud samples (Borland *et al.* 1977). Parts of the Mersey estuary were heavily contaminated with type C (G.R. Smith *et al.* 1982). Cann and his colleagues (1975) found that fish from 13 of 17 British trout farms were contaminated with type B, C, E or F. Certain muds are inhibitory to *Cl. botulinum* (Sugiyama *et al.* 1970, G.R. Smith *et al.* 1977, J.M. Graham 1978).

Cl. botulinum is found occasionally in the faeces and viscera of animals such as pigs, cattle, horses, and waterfowl (Burke 1919*a*, Easton and Meyer 1924, Meyer 1931, Gunderson 1933, Dolman and Chang 1953, Prévot and Brygoo 1953, Haagsma 1974); but there is no evidence that types A, B and E are common intestinal inhabitants. The organism is rarely present in the faeces of healthy human beings (Easton and Meyer 1924, Kahn 1924), including, according to all recent studies except one (Thompson *et al.* 1980), healthy infants (Arnon 1982). The occasional demonstration of *Cl. botulinum* in the intestinal tract and viscera of healthy animals (see Meyer 1956, Dolman 1964) suggests that meat foods may be potentially botulinogenic because of a 'carrier' state of this kind in the animals. Waterfowl may assist in spreading *Cl. botulinum* from one aquatic environment to another, but farm animals are likely to have only a slight influence on the distribution of the organism in soil; G.R. Smith and Milligan (1979) found, however, types B, C, D and E in soil on the site of the former Metropolitan Cattle Market, London.

Factors influencing the development of *Cl. botulinum* in foods

In spite of the wide distribution of *Cl. botulinum* in the soil and on fruits and vegetables, botulism is uncommon. As the organism does not usually grow in the animal body, fresh foods, cooked or uncooked, may be eaten with impunity. For botulism to result, the organism must multiply and form its toxin in the food before consumption. But even so, the incidence of the disease due to canned or otherwise stored foods is less than might be expected. Proper heat sterilization of the food is undoubtedly an important factor in the prevention of botulism. Nevertheless, the spores of *Cl. botulinum*, except those of type E, are relatively heat resistant. They are reported to withstand 100° for up to 22 hours, and 120° for up to 20 minutes (see Esty and Meyer 1922, Tanner and Twohey 1926). In the Loch Maree outbreak the duck paste had at one stage been exposed to 115° for 2 hours. Heat alone is therefore not sufficient to explain why the majority of canned foods are innocuous. Other factors must be concerned.

The hydrogen-ion concentration of the food has an important bearing on the efficacy of sterilization. Esty and Meyer (1922) found that at pH 7.0 the spores were destroyed in 330 minutes at 100°; at pH 5.05 in 45 minutes, and at pH 3.7 in 10 minutes. The higher the hydrogen-ion concentration, the shorter is the time required for destruction. The thermal resistance of the

spores decreases with increasing concentration of sodium chloride in the food (Weiss 1921). Spores from young cultures are said to be more resistant to heat than those from old (Weiss 1921, Esty and Meyer 1922). These are some of the factors influencing the sterilization of the food.

It has also been found that spores of *Cl. botulinum* may remain dormant for days, weeks or months before germinating (Dickson *et al.* 1925). The type of food influences germination; string beans, spinach, corn, peas and salmon appear to be specially favourable for growth; acid fruits and vegetables are less so (Koser *et al.* 1921, Geiger and Benson 1923, Schoenholz *et al.* 1923, Bachmann 1924). Below pH 4.5 germination is inhibited in most fruits. Such products can safely be processed in boiling water, whereas above pH 4.5 the processing must be sufficient to destroy even the most resistant spores (Esty 1935); and in a number of other foods, germination does not occur at pH values of 4.6 or less (see, *e.g.*, Townsend *et al.* 1954, Wagenaar and Dack 1954). Raatjes and Smelt (1979) described laboratory experiments in which toxigenesis occurred in protein-rich media at pH values below 4.6. In pickled foods at least 2 per cent of vinegar is necessary to destroy the spores, and in salted fish, salt concentrations of 10 per cent or more are needed to inhibit growth of the organisms. Other organisms, especially those which are frankly proteolytic—such as *Cl. sporogenes*—when present prevent the accumulation of toxin, perhaps by destroying it (Dack 1926; see also Crisley and Helz 1961) or, as with *B. licheniformis*, by the production of bacitracin which inhibits growth of the organisms (Wentz *et al.* 1967). *Streptococcus lactis*, and to a lesser extent lactobacilli, also prevent type A toxin formation in experimentally contaminated food (Saleh and Ordal 1955).

It should be noted that anaerobic conditions suitable for growth may obtain even in foods not in closed containers. Toxin can form, for example, in loose wet mincemeat exposed to the atmosphere (Aitken *et al.* 1936).

Given very favourable conditions, *Cl. botulinum* may produce detectable toxin within 12 hours, but usually it takes 2–14 days, depending on the temperature of storage of the food and other factors. B.C. Hobbs and Spooner (1966), for example, found that in experimentally cooked chickens contained in polythene bags type A organisms multiplied and formed toxin within one day at 35°, even in the presence of a competing flora (see also Pivnick and Bird 1965). Except with type E organisms little or no growth occurs below 10°, but food stored above this temperature may quickly become dangerous. Again with the exception of type E the optimum temperature for growth and toxin production appears to be 35–40°; no growth occurs at 45° (Tanner *et al.* 1940, Report 1953, Ohye and Scott 1953). Type E bacilli may grow slowly at as low a temperature as 5°, though toxin formation may not be detectable for 8 weeks or so (Ohye and Scott 1957). Vacuum-packed herring, however, inoculated with a total of 100 spores was found to be toxic after 15 days at this temperature (Cann *et al.* 1965); and ground beef inoculated with 10 000 spores per gram became toxic even at 3° (Warnecke *et al.* 1967; see also Ajmal 1968). The addition of nitrite in sufficient quantity is said to inhibit the formation of toxin (Christiansen *et al.* 1973).

Many factors determine the survival, germination, multiplication and toxin production of *Cl. botulinum* in foods preserved by heating; and unless they are fortuitously combined, the food remains innocuous. The most important appear to be a fairly heavy initial contamination by the spores, insufficient heating, anaerobic conditions, too slow cooking; and the use of the food without final cooking. Home canning, being as a rule an ill controlled process, is far more likely than industrial canning to result in poisoned food. (For further information see Report 1967 and 1970, L.DS. Smith 1977.)

Diagnosis

The symptoms of botulism in man are generally so characteristic that the disease may be diagnosed on clinical grounds. To confirm this diagnosis the following procedure should be adopted.

Demonstration of toxin in food

A rapid diagnosis may be made by means of a neutralization test in mice. The food is extracted with a minimum of normal saline or gelatin phosphate buffer. The extract is clarified by centrifugation; if desired, it may be sterilized by membrane filtration, or penicillin, 100 units per ml, may be added. If antitoxins A–F are available, each is mixed with a portion of the extract at the rate of 1 unit of antitoxin, usually in a volume of 0.1 ml, per dose of extract. Each mixture is allowed to stand for 30 minutes before intraperitoneal injection into two mice. A further four mice receive extract but no antitoxin; of these, two are given untreated extract, and two extract heated for 10 minutes at 100° to destroy any toxin. The mice are observed for four days. Survival of only one pair of antitoxin-treated mice, and of the pair given heated extract, indicates *Cl. botulinum* intoxication and identifies the type of *Cl. botulinum* concerned. The food may be extracted more thoroughly by macerating at 4° overnight. Because the toxicity of type B, E and F toxin is sometimes increased by the action of trypsin, the extract may be examined in the trypsinized (Duff *et al.* 1956) as well as the untrypsinized state. Various serological methods for the detection of toxin *in vitro* have been tried, but so far none has replaced the neutralization test in mice (Crowther and Holbrook 1976, Notermans *et al.* 1979).

Meanwhile the food should be seeded into 2 per cent glucose broth, Hitchens' medium (0.2 per cent dextrose infusion broth containing 0.1 per cent agar), beef heart peptic digest liver broth, peptonized bullock's heart broth (de Lavergne and Abel 1925), pork infusion thioglycollate semisolid agar medium containing 0.1 per cent of soluble starch (Wynne and Foster 1948), or cooked meat medium, and incubated anaerobically at 30°. A membrane filtrate, or the supernate after high-speed centrifugation, of the culture is tested for toxin as above after 5–10 days. Some strains produce their maximum toxin after 2–3 days. Type F toxin is labile, deteriorating rapidly even at 37°.

Isolation of the bacillus from the food   Pure cultures may be obtained by heating broth cultures to 80° for half an hour to destroy contaminating non-sporing bacilli, and seeding on to horse blood agar plates or into deep agar cultures, which are incubated anaerobically at 30°, preferably in the presence of added CO_2. The characteristic colonies should be picked off, grown in liquid medium, and tested for toxin production. Plating on lactose egg yolk agar, on which *Cl. botulinum* colonies produce a pearly layer and opalescence (Chapter 42), may help in identification. Part of the culture should be plated without prior heating, or with lesser degrees of heating than 80° for 30 minutes, for the isolation of *Cl. botulinum* type E, whose spores have a low heat resistance; alternatively an equal volume of absolute alcohol may be added to prevent overgrowth by contaminating organisms (Johnston *et al.* 1964). Cultures for types C and D are best incubated at 35°. Fluorescent staining of the bacilli in the food or in cultures is useful provided that the numerous pitfalls that beset it can be avoided (see Walker and Batty 1967, Kautter *et al.* 1970). For the isolation of type C, preliminary heating at 71° for 15 minutes followed by cultivation on an egg meat medium with the addition of yeast extract, ammonium sulphate, and glucose is recommended (Segner *et al.* 1971). It should be remembered that types A and B may throw non-toxic variants indistinguishable from the toxic forms; that type E throws proteolytic non-toxic variants; that types A, B, E and F throw heavily sporulating non-toxic variants; and that type C and D strains may lose the bacteriophages that govern their toxicity (Chapter 42).

Other methods   Toxin may sometimes be demonstrated in the blood during life and in the liver at autopsy (Schneider and Fisk 1939, Skulberg 1957, Ralovich and Barna 1966, Sebald and Saimot 1973). At least 10 ml of serum should be collected, before the administration of any antitoxin, for immediate examination by a neutralization test in mice. It is useful to attempt isolation of the bacillus from the patient's faeces or vomit (see e.g., Wheeler and Humphreys 1924). Toxin may sometimes be demonstrated in the faeces (Dowell *et al.* 1977). Demonstration of the bacillus and its toxin in faeces is of particular importance in *infant botulism* (see p. 499) as toxin will be absent from the food and almost certainly from the blood (Turner *et al.* 1978). A selective medium of egg-yolk agar containing cycloserine, sulphamethoxazole, and trimethoprim is described by Dezfulian and co-workers (1981). Cultures from liver and spleen at autopsy are occasionally positive (Dubovsky and Meyer 1922, Stricker and Geiger 1924). Stricker and Geiger (1924) found specific type-A antitoxin in the serum of two patients on the 14th day after the consumption of the affected food. Sub-lethal doses of toxin may often be too small to provide an appreciable antigenic stimulus.

For the bacteriological establishment of an authentic case of botulism, it is essential to prove the specific toxicity of the food, the patient's serum, or both; and if possible to isolate a demonstrably toxigenic strain of *Cl. botulinum*. The various methods for the isolation and identification of *Cl. botulinum* are reviewed by McClung (1967) and Kautter and his colleagues (1970).

Prophylaxis

As already pointed out, the foods responsible for botulism are usually those that have received inadequate preliminary treatment, such as heating, salting, smoking, drying, or pickling, been allowed to stand at a temperature permitting the growth of *Cl. botulinum*, and been eaten without cooking. Since the conditions necessary for the destruction of spores or for inhibition of growth of the bacilli were worked out and adopted by the canning industry in the United States, botulism caused by food from freshly opened cans has been virtually abolished, and practically all outbreaks have been due to home-canned produce (Geiger *et al.* 1922). The measures necessary for adequate processing are too detailed to be given here; for these the reader is referred to papers by Perkins (1964) and Thatcher and co-workers (1967). Food having a pH less than 4.6 or greater than 9.0 may be preserved without heating, since botulinum spores do not germinate in highly acid or alkaline media. The requisite acidity may be attained by the use of 2 per cent acetic or citric acid. Fractional sterilization is unreliable, because spores may not germinate in the periods between heating (Burke 1919c). Heat has been generally relied on for the destruction of spores, but chlorine is said to be even more effective (Ito *et al.* 1967). Radiation, where it is permitted, may be used for the same purpose, though it does not inactivate the toxin as heat does (see Report 1970).

In most episodes the food has been abnormal in appearance and odour. Bulging of the tin, the presence of gas bubbles on opening, a disintegrated appearance of the food, and a rancid smell, should be sufficient to condemn the food. On no account should such food be eaten, even after cooking (Tanner and Twohey 1926). But in some toxin-containing cans no evidence of spoiling is detectable (Schoenholz *et al.* 1923). Some type B and F strains, and type E strains are not proteolytic and may produce little obvious change. But whether spoilage is evident or not, home-canned produce should be neither tasted nor eaten before it has been cooked (see Meyer 1956).

Prevention of botulism due to the consumption of made-up meat foods is particularly difficult. van Ermengem (1897) lays emphasis on the necessity in pickling ham of using strong brine, which is unfavourable to the development of the organism.

Farmed fish such as trout may be heavily contaminated with *Cl. botulinum*; contamination is reduced by keeping the fish in constantly changing water without food for up to 3 days before killing (Huss *et al.* 1974).

In handling suspected food, contact with cuts should be avoided; the toxin may be absorbed from broken skin areas, mucous surfaces, and fresh wounds (Geiger 1924).

Human botulism is so rare and sporadic a disease that prophylactic immunization is impracticable. Those who may be exposed to the hazards of deliberate intoxication, as in biological warfare, may be protected by immunization with toxoids. As judged by the production of circulating antitoxin in concentrations similar to those that protect immunized experimental animals, alum-precipitated crude formol toxoids are effective (Hottle *et al.* 1947, Reames *et al.* 1947). More recent work has resulted in potent, highly purified types A, B, C, D and E toxoids, all effective in a polyvalent mixture. With mixed, alum-precipitated A and B toxoids, immunization at 0, 2 and 10 weeks, with a reinforcing dose after 1 year, confers a substantial immunity lasting for over 2 years (see Cardella *et al.* 1958, 1960, Fiock *et al.* 1962).

Treatment

The toxin still in the stomach should be removed immediately by lavage with 2–5 per cent bicarbonate solution. The stomach should be kept alkaline, as the toxins are labile at high pH, and an alkaline milieu is probably unfavourable for the intragastric activation reported for type E toxin (Dolman 1957a).

Antitoxins are protective in experimental animals when given simultaneously with the toxin. Large doses may be effective when given later; thus, 300 000 neutralizing doses of homologous antitoxin prevented death of a guinea-pig 24 hours after the injection of toxin (Kempner 1897). In man, the early administration of antitoxin lowers the case-fatality rate (Meyer and Eddie 1951).

Therapeutic antitoxin, polyvalent for types A, B and E, or monovalent when the type of intoxication is known, is given intravenously in large daily doses—50 ml for monovalent, and more for polyvalent—in 5-

10 per cent glucose solution. Antitoxin cannot reverse established intoxication of the myoneural junctions, so that its success depends on the earliest possible administration of the correct type (see Dolman 1960) to neutralize the maximum amount of unfixed toxin in the body. Shvedov (1959) and Minervin (1967) recommend oral antitoxin by duodenal tube. A prophylactic dose of 10 ml antitoxin should be given intramuscularly to all who have eaten the poisoned food.

Various proposals have been made for the specification of toxic potency and of the neutralizing potency of therapeutic antitoxins. Bowmer (1962) made a detailed investigation of horse antitoxins of the five main types and (1963) described the preparation and mode of assay of those adopted as international standards. These are recommended for issue with the following provisional units per ml: types A and B, 50; types C and D, 100; and type E, 10. Assay is best made with toxin–antitoxin mixtures given intraperitoneally in mice, death within 4 days being taken as a quantal response. As with other bacterial exotoxin–antitoxin systems, the dilution effect may be pronounced, especially with type D antitoxin.

As regards other modes of neutralizing toxin in the alimentary tract, the administration of high enemas of soap and olive oil has been recommended, on the grounds that soap neutralizes toxin and the oil prevents its absorption (Burke *et al.* 1921). Ethanol precipitates the toxin, and brandy in frequent small doses has been recommended to do this *in vivo*. Experiments in mice, however, suggest that only impracticably large doses of brandy would be effective in this way; olive oil enhanced the toxicity of oral toxin, and with brandy and olive oil, the toxicity was equal to that of toxin alone (Lamanna and Meyers 1960).

For combating the respiratory paralysis during treatment, mechanical respirators may be necessary and, as in tetanus, supportive treatment may be life-saving (Oberst *et al.* 1968).

(For reviews of botulism see Report 1964, 1967, 1970, and L. DS. Smith (1977); and for a bibliography up to 1964 see McClung 1964. Infant botulism is reviewed by Arnon 1980.)

Botulism in animals

Considerable attention has been devoted to a study of the disease in domestic animals, characterized as a rule by fairly sudden onset, without fever, paralytic symptoms without loss of sensibility, and often death. In many instances it has been possible to demonstrate a relationship between the type of fodder used and the occurrence of the disease. Organisms of the botulinum type have frequently been isolated from the animals and from the fodder, and the toxin produced in culture has been shown to be capable of reproducing the symptoms of the disease when given by the mouth or inoculated subcutaneously. Of the various kinds of

botulism, that of cattle is of major economic importance.

As in human botulism, the factors determining natural intoxication depend mainly on the distribution of spores in the environment, the availability of situations in which the toxins can form, and the topographical distribution and feeding habits of the animals concerned. To some extent also the susceptibility of the animal to a given type of toxin may be a determining, though not a major, factor. As Table 72.5 shows, horses are recorded as sufferers mainly from types C, D and B intoxication, cattle from types D, C_β and B, mink and chickens from types C and A, and waterfowl from type $C\alpha$.

There has in the past been confusion about the various kinds of *Cl. botulinum* isolated, especially those now known as types C and D (see Sterne and Wentzel 1953, Gunnison 1955, Dolman and Murakami 1961). Seddon (1922, 1927) called his organism *B. parabotulinus*, and its toxin was designated C_β by Gunnison and Meyer (1929). Theiler and Robinson's (1927) organism isolated during an outbreak of botulism in mules, named *B. parabotulinus equi*, also proved to be type C. The organism originally isolated from South African carrion giving rise to the cattle disease *lamsiekte* was named *Cl. parabotulinum bovis*, and later designated *Cl. botulinum* type D by Meyer and Gunnison (1928). The specific epithet *parabotulinus* was originally used to distinguish proteolytic and non-proteolytic types of *Cl. botulinum*, but with the recognition that non-proteolytic and proteolytic strains occur within a given type (*e.g.* type B), the distinction merely served to confuse. The type system of nomenclature of a single species of *Cl. botulinum*, based on that of Meyer and Gunnison, according to the antigenic specificity of the neurotoxin, is clearly preferable.

In the many attempts to estimate the toxicity of the different neurotoxins, the results have varied with the provenance of the given type of toxin used, and with the animals of a given species that are tested. In the absence of an extensive systematic study, little more than broad and incomplete generalizations are possible. Among laboratory animals, on a weight-for-weight basis, guinea-pigs and rabbits are about equally susceptible to parenteral intoxication by types A, B, C_α, C_β, D and E toxins, and mice slightly less so. The order of decreasing susceptibility is in general A, B, C, D (see Meyer 1928, Gunnison and Meyer 1930, Gunnison *et al.* 1936). Experimentally the oral dose is difficult to determine, but is always much higher than the subcutaneous dose. Mink are susceptible to C, but less so to A, B, and E (Karashimada *et al.* 1970). Type C toxin is absorbed more readily than the toxins of other types through the gut wall of chickens and pheasants (Gross and Smith 1971). The oral toxicity of type D for cattle is high, and for man and laboratory animals low (Sterne and Wentzel 1953). Pigs are fairly susceptible to type A, but resist types B, C and E (Schiebner 1955). (See also Dolman and Murakami 1961.)

Botulism in horses

Forage poisoning was formerly a common disease in America, affecting several thousand horses annually; numerous cases have been recorded in Europe and in Egypt. Buckley and Shippen in 1917 observed that experimentally induced botulism in horses closely simulated naturally occurring forage poisoning. From toxic oat-hay and ensilage they isolated a bacillus resembling *Cl. botulinum*; experimental intoxication could be prevented by the administration of anti-botulinum serum. R. Graham and Brueckner (1919) isolated an organism resembling *Cl. botulinum* from a corn ensilage that had been responsible for forage poisoning. Horses fed with 2 ml of a broth culture of this organism manifested clinical symptoms and post-mortem appearances of forage poisoning. Horses given a prophylactic dose of botulinum antiserum on the previous day recovered. The American organism was identified as a type B (Burke 1919*b*); the disease is more commonly caused by types C and D. Forage poisoning in South African mules, by hay which contained a contaminated rat carcass, was described by Theiler and Robinson (Theiler 1928), and the organism designated as type C_β (Gunnison and Meyer 1929). Type C_β toxin is also associated with horse botulism in Australia (Bennetts and Hall 1938), Europe (Prévot and Brygoo 1953, Pilet *et al.* 1959), and Denmark; in Denmark the cause was often hay containing a decomposing cat carcass (Müller 1963).

Forage poisoning is a broad term covering a number of different disorders, and it is not to be thought that it is uniformly due to poisoning with the toxin of *Cl. botulinum*. In view of the widespread distribution of *Cl. botulinum*, it is obviously insufficient for the purposes of diagnosis to isolate the organism from the fodder or the intestinal canal; toxin must be demonstrated in the food and the intestinal contents, and if possible in the blood. Among the diseases that may have been mistaken for it are grass sickness (Walker 1929), *Cl. perfringens* intoxication (Gordon 1934), and equine encephalomyelitis (see Meyer 1956).

Botulism in cattle and sheep

Forage poisoning in cattle is commonest between the ages of 6 months and 2 years; the case-fatality rate is 2–10 per cent (R. Graham and Schwarze 1921*b*).

Dubovsky and Meyer (1922) isolated *Cl. botulinum* from the liver and mesenteric glands of two cows suffering from ictero-haemoglobinuria. In Australia Seddon (1922, 1927) investigated an epizoötic disease of cattle known by the names of Midland cattle disease, impaction paralysis, or dry bible; this disease appears to be identical with the *lamsiekte* of South

Africa. From animals dying of the disease he isolated the organism now designated the C_β type of *Cl. botulinum*. Experimentally its toxin gave rise to bulbar paralysis in cattle. The same type is responsible for the contagious bulbar paralysis of cattle in Denmark (Müller 1963).

Our knowledge of botulism in cattle was extended by Theiler and his colleagues (1926–27*a*) working on *lamsiekte* in South Africa. This disease, which is characterized by paresis and paralysis, principally of the locomotor system, but sometimes of the muscles of mastication and deglutition, occurred widely but sporadically, with a yearly case fatality approaching 100 000 (Sterne and Wentzel 1953). Theiler related the incidence to a deficiency, in some regions seasonal, of phosphorus in the pasturage, to remedy which the animals fed on the debris of carcasses, especially the bones. Feeding bone-meal cured the animals of this 'osteophagia'. Australian experience, however, suggests that the bone-meal may be effective in such circumstances as a general nutrient; carcass feeding in Australian cattle was not associated with mineral deficiencies, and was reduced by generous food supplements (Underwood *et al.* 1939). Botulism occurs when the carcasses are contaminated with *Cl. botulinum* from the soil. In South Africa the organism is generally that first isolated by Theiler, type D, though 10–20 per cent appear to be due to type C_β (Sterne and Wentzel 1953). Besides mammalian carcasses, those of tortoises may be sources of abundant toxin (Fourie 1946). In Western Australia, where the yearly losses may also be high, rabbit carcasses contaminated from the soil used to be a major source of intoxication. As little as 3 g of toxic rabbit may be fatal to sheep (Bennetts and Hall 1938). Type C_β intoxication is the commonest, but type D also occurs in soil and forage (Bennetts and Hall 1938, Eales and Turner 1952). (For a description of lamsiekte in South Africa see Henning 1956.)

Recovery from the disease is not associated with increased immunity (Theiler and Robinson 1927). Injection of two doses of alum-precipitated toxoid, types C and D, 6–8 weeks apart, confers a good immunity on cattle and sheep (see Bennetts and Hall 1938, Mason *et al.* 1938, Oxer 1940, Sterne and Wentzel 1950, 1953).Vaccination is now widely practised in South Africa and Australia.

Cases of type B botulism have occurred in the Netherlands as a result of feeding contaminated brewers' grains and silage to cattle (see Haagsma and ter Laak 1978, Notermans *et al.* 1981).

Botulism in birds

'Limberneck' in **chickens** is characterized by weakness, muscular inco-ordination, a drooping neck, anorexia, prostration and coma. Death frequently follows. Geiger and his colleagues, writing in 1922, collected records in the USA of 103 outbreaks since 1910 with

3500 deaths. Evidence leaves no doubt that limberneck is an avian type of botulism. The toxin of *Cl. botulinum* was demonstrated in the crop and intestinal contents of fowls dying of the disease (R. Graham and Schwarze 1921*a*, Geiger *et al.* 1922, R. Graham and Boughton 1923*a*). It was found that the feeding of toxic material gave rise to symptoms of limberneck (Dickson 1918). At one time the disease frequently followed the consumption of home-canned string beans (Meyer and Dubovsky 1922*b*); most such outbreaks investigated were due to type A (Geiger *et al.* 1922). Wilkins and Dutcher (1920) attributed limberneck to toxic larvae of the fly *Lucilia cæsar*. Bengtson (1922, 1923) isolated from this source the organism later designated type C_α, whose toxin in large doses reproduced the symptoms of limberneck. It is probable that sporadic outbreaks of limberneck are due mainly to this type. In Britain three outbreaks due to type C were reported among broiler chickens by Roberts and co-workers (1973), and in one of these 3500 birds died; in a further outbreak (Smart and Roberts 1977) several thousand type C organisms per gram of litter and intestinal contents were found. Possible sources of toxin were usually inapparent. Miyazaki and Sakaguchi (1978) found that the oral administration of 100 spores of type A, C or D invariably killed chicks in 6–13 days; the site of toxin production and absorption was the caecum.

Annual outbreaks of botulism in **ducks** and other waterfowl (Kalmbach and Gunderson 1934) have been recognized since about 1910 in certain areas of lakes and mud flats in the western states of the USA. Outbreaks of this disease, formerly known as *western duck sickness* or *alkali disease*, may be severe. In 1932, for example, it was estimated that 250 000 birds died on the Great Salt Lake in Utah. The disease is also known in Canada, Argentina, Uruguay, Mexico, South Africa, Australia, New Zealand, Japan, and Europe. In the 1969 outbreak in St James's Park, London, over 400 birds died. At least 21 species, mainly ducks, were affected (Keymer *et al.* 1972; see also Roberts *et al.* 1972, G.R. Smith 1982). The disease is almost always caused by type C_α (Meyer 1956); occasionally types C_β (Reilly and Boroff 1967) and E (Fay 1966) may be responsible. In an affected aquatic environment the mud remains permanently contaminated with spores. Sudden multiplication of the organism and consequent toxin production may be brought about by warm weather, thermal pollution from power stations (Haagsma *et al.* 1972), shallow stagnant water, alkalinity, and oxygen depletion associated with pond weed, algae and rotting vegetation (Kalmbach and Gunderson 1934, Coburn and Quortrup 1938, Quortrup and Holt 1941, Quortrup and Sudheimer 1942, 1943); an abundance of aquatic invertebrates, living and dead, may play a part (Bell *et al.* 1955, Jensen and Allen 1960, Duncan and Jensen 1976). Dipterous fly larvae from rotting bird carcasses

may contain lethal amounts of toxin. Toxin sometimes persists for many months in the environment and produces occasional outbreaks of botulism in winter and spring (J.M. Graham *et al.* 1978). Refuse dumps may provide a source of toxin for gulls. Some workers, but not all, have found the organism in the liver of affected birds (see J.M. Graham and Smith 1977). Diagnosis is best made by killing several affected birds and demonstrating type C toxin in the serum of at least one. In the absence of such an examination avian botulism is easily confused with poisoning by industrial or agricultural chemicals (G.R. Smith and Oliphant 1983). Birds that recover are not immune (Rosen 1971, Haagsma 1974). Control is difficult; it is based mainly on water manipulation and removal of organic matter including vegetation and avian carcasses (Sperry 1947, Rosen 1971). (For a review of botulism in waterfowl see G. R. Smith 1982.)

Botulism also attacks pheasants, dipterous fly larvae often playing an important role (Lee *et al.* 1962). High degrees of resistance can be induced by immunization with type C toxoid and adjuvant (see Boroff and Reilly 1959). Carrion-eating species such as turkey vultures, crows and coyotes may possess antitoxin in their serum (see Holdeman 1970, Ohishi *et al.* 1979); the neutralizing activity of the serum may embrace several different types of toxin.

Other animals

Sporadic cases of botulism occur in other vertebrate animals (see Robinson 1929). Cats, dogs and goats are among the species affected (Geiger *et al.* 1922), but dogs are reported to be fairly resistant to oral type A and B toxins; type C botulism has been reported in foxhounds (Barsanti *et al.* 1978, Marlow and Smart 1982). Outbreaks have occurred in mink and other animals bred for fur in North America, Japan, the USSR, and Scandinavia. Types A, B and most commonly C are implicated (Quortrup and Gorham 1949, Moberg 1953, Avery *et al.* 1959). In both the USSR and Japan type C is responsible (Bulatova *et al.* 1967, Karashimada *et al.* 1970). Skulberg (1961) reports a type E outbreak. Mink may be protected by immunization with preparations of type C toxoid (see von Dintner and Kull 1950, Appleton and White 1959). Smart and co-workers (1980) describe an outbreak of type C botulism in captive monkeys. Type E botulism in farmed trout may result from the feeding of spoiled marine trash fish (Huss and Eskildsen 1974); and huge losses in juvenile salmon reared in earth-bottomed ponds have been reported from the USA (Eklund *et al.* 1982).

Acknowledgements

We are grateful to Antonnette Wieneke, Michael Stringer and John Kramer for their help in the preparation of this chapter.

References

Aitken, R.S., Barling, B. and Miles, A.A. (1936) *Lancet* **ii**, 780.

Ajmal, M. (1968) *J. appl. Bact.* **31**, 120, 124.

Albert, O.H. (1957) *Berl. Münch. tierärztl. Wschr.* **70**, 165.

Anderson, E.S. (1968a) *Brit. med. J.* **ii**, 333; (1968b) *Annu. Rev. Microbiol.* **22**, 131.

Anderson, E.S., Galbraith, N.S. and Taylor, C.E.D. (1961) *Lancet*, **i**, 854.

Anellis, A., Lubas, J. and Rayman, M.M. (1954) *Food Res.* **19**, 377.

Anon. (1961) *Desiccated Coconut (Manufacture and Export) Regulations 1961.* Coconut Products Ordinance; (1969) *Brit. med. J.* **iv**, 758.

Appleton, G.S. and White, P.G. (1959) *Amer. J. vet. Res.* **20**, 166.

Appleton, H., Palmer, S.R. and Gilbert, R.J. (1981) *Brit. med. J.* **i**, 1801.

Armstrong, C., Story, R.V. and Scott, E. (1919) *Publ. Hlth Rep., Wash.* **34**, 2877.

Arnon, S.S. (1980) *Annu. Rev. Med.* **31**, 541; (1982) *Pediatrics, Springfield* **69**, 830.

Arnon, S.S., Damus, K., Thompson, B. and Midura, T.F. (1982) *J. Pediat.* **100**, 568.

Arnon, S.S., Midura, T.F., Clay, S.A., Wood, R.M. and Chin, J. (1977) *J. Amer. med. Ass.* **237**, 1946.

Arnon, S.S., Midura, T.F., Damus, K., Thompson, B., Wood, R.M. and Chin, J. (1979) *J. Pediat.* **94**, 331.

Arnon, S.S., Midura, T.F., Damus, K., Wood, R.M. and Chin, J. (1978) *Lancet* **i**, 1273.

Aserkoff, B. and Bennett, J.V. (1969) *New Engl. J. Med.* **281**, 636.

Avery, R.J. *et al.* (1959) *Canad. J. comp. Med.* **23**, 203.

Ayres, J.C., Mundt, J.D. and Sandine, W.A. (1980) *Microbiology of Foods.* W.H. Freeman, San Francisco.

Ayres, P.A. and Barrow, G.I. (1978) *J. Hyg., Camb.* **80**, 281.

Bach, R., Wenzel, S., Müller-Prasuhn, G. and Gläsker, H. (1971) *Arch. LebensmittHyg.* **22**, 107.

Bachmann, F.M. (1924) *J. infect. Dis.* **34**, 129.

Ball, A.P. and 14 others (1979) *Quart. J. Med.* New Series **48**, 473.

Barber, M.A. (1914) *Philipp. J. Sci.* **B9**, 515.

Barker, W.H. (1974) In: *Vibrio parahaemolyticus*, p. 47. Ed. by T. Fujino, G. Sakaguchi, R. Sakazaki and Y. Takeda. Saikon Publ. Co., Tokyo.

Barnes, E.M. and Mead, G.C. (1971) In: *Poultry Disease and World Economy*, p. 47. Ed. by R.F. Gordon and B.M. Freeman. Longman Group Ltd, Edinburgh.

Barrow, G.I. and Miller, D.C. (1976) In: *Microbiology in Agriculture, Fisheries and Food*, p. 181. Ed. by F.A. Skinner and J.G. Carr. Academic Press, London.

Barsanti, J.A., Walser, M., Hatheway, C.L., Bowen, J.M. and Crowell, W. (1978) *J. Amer. vet. med. Ass.* **172**, 809.

Beckers, H.J., Leusden, F.M. van, Coutinho, R.A., Schrickx, A.M. and Meerburg Snarenberg, P.J. (1982) *J. appl. Bact.* **53**, xiv.

Bell, J.F., Sciple, G.W. and Hubert, A.A. (1955) *J. Wildlife Mgmt* **19**, 352.

Bengtson, I.A. (1922) *Publ. Hlth Rep., Wash.* **37**, 164; (1923) *Ibid.* **38**, 340.

Bennetts, H.W. and Hall, H.T.B. (1938) *Aust. vet. J.* **14**, 105.

Bergdoll, M.S., Crass, B.A., Reiser, R.F., Robbins, R.N. and Davis, J.P. (1981) *Lancet* **i**, 1017.

Bergdoll, M.S. (1979) In: *Food-borne Infections and Intoxications*, 2nd edn, p. 443. Ed. by H. Riemann and F.L. Bryan. Academic Press, New York and London.

Black, R.E., Jackson, R.J., Tsai, T., Medvesky, M., Shayegani, M., Feeley, J.C., Macleod, K.I.E. and Wakelee, A.M. (1978) *New Engl. J. Med.* **298**, 76.

Blaser, M.J. and Newman, L.S. (1982) *Rev. infect. Dis.* **4**, 1096.

Blaxland, J.D., Sojka, W.J. and Smither, A.M. (1958) *Vet. Rec.* **70**, 374.

Borland, E.D., Moryson, C.J. and Smith, G.R. (1977) *Vet. Rec.* **100**, 106.

Boroff, D.A. and Reilly, J.R. (1959) *J. Bact.* **77**, 142.

Bowmer, E.J. (1962) Thesis. University of Liverpool; (1963) *Bull. World Hlth Org.* **29**, 701.

Brachman, P.S., Taylor, H., Gangarosa, E.J., Merson, M.H. and Barker, W.H. (1973) In: *The Microbiological Safety of Food*, p. 73. Ed. by B.C. Hobbs and J.H.B. Christian. Academic Press, London.

Brouwer, R., Mertens, M.J.A., Siem, T.H. and Katchaki, J. (1979) *Antonie van Leeuwenhoek* **45**, 517.

Brown, C.M. and Parker, M.T. (1957) *Lancet* **ii**, 1277.

Brown, D.D., Duff, R.H., Wilson, J.E. and Ross, J.G. (1973) *Brit. vet. J.* **129**, 493.

Bryan, F.L. (1977) *J. Food Protect.* **40**, 45; (1978) *Ibid.* **41**, 816; (1979) In: *Food-borne Infections and Intoxications* 2nd edn, p. 212. Ed. by H. Riemann and F.L. Bryan. Academic Press, New York; (1981) *J. Food Protect.* **44**, 394.

Bryan, F.L., Fanelli, M.J. and Riemann, H. (1979) In: *Food-borne Infections and Intoxications*, 2nd edn, p. 74. Ed. by H. Riemann and F.L. Bryan. Academic Press, New York.

Buckley, J.S. and Shippen, L.P. (1917) *J. Amer. vet. med. Ass.* **50**, 809.

Bulatova, T.I., Chulkova, I.F., Anisimova, L.I. and Kazdobina, I.S. (1973) *J. Microbiol., Moscou* **50**, part 10, 105.

Bulatova, T.I., Matveev, K.I. and Samsonova, V.S. (1967) See *Report* 1967, p. 391.

Burgen, A.S.V., Dickens, F. and Zatman, L.J. (1949) *J. Physiol.* **109**, 10.

Burke, G.S. (1919a) *J. Bact.*, **4**, 541; (1919b) *Ibid.* **4**, 555; (1919c) *J. Amer. med. Ass.* **72**, 88.

Burke, V., Elder, J.C. and Pischel, D. (1921) *Arch. intern. Med.* **27**, 265.

Burns, C., Mair, N.S. and Hooper, W.L. (1965) *Mon. Bull. Minist. Hlth Lab. Serv.* **24**, 9.

Buxton, A. (1957) In: *Salmonellosis in Animals.* Review series No. 5 of the Commonwealth Bureau of Animal Health. Commonwealth Agricultural Bureaux, Farnham, Bucks.

Cann, D.C., Taylor, L.Y. and Hobbs, G. (1975) *J. appl. Bact.* **39**, 331.

Cann, D.C., Wilson, B.B. and Hobbs, G. (1968) *J. appl. Bact.* **31**, 511.

Cann, D.C., Wilson, B.B., Hobbs, G. and Shewan, J.M. (1965) *J. appl. Bact.* **28**, 431.

Cardella, M.A. *et al.* (1960) *J. Bact.* **79**, 372.

Cardella, M.A., Fiock, M.A. and Wright, G.G. (1958) *Bact. Proc.* **58**, 78.

Caroli, G., Armani, G., Sciacca, A., Bargagna, M. and Levré, E. (1977) *Ann. Sclavo* **19**, 494.

Casman, E.P. (1967) *Hlth Lab. Sci.* **4**, 199.

Center for Disease Control (1973) *Salmonella Surv.*, Rep. No. 115, 2.

Chatterjee, B.D., Neogy, K.N. and Gorbach, S.L. (1970) *Bull. World Hlth Org.* **42**, 460.

Chordash, R.A. and Potter, N.N. (1976) *J. Food Sci.* **41**, 906.

Christiansen, L.N., Johnston, R.W., Kautter, D.A., Howard, J.W. and Aunan, W.J. (1973) *Appl. Microbiol.* **25**, 357.

Ciccarelli, A.S., Whaley, D.N., McCroskey, L.M., Giménez, D.F., Dowell, V.R. and Hatheway, C.L. (1977) *Appl. environm. Microbiol.* **34**, 843.

Clarenburg, A. and Burger, H.C. (1950) *Food Res.* **15**, 340.

Clarenburg, A. and Dornickx, C.G.J. (1932) *Ned. Tijdschr. Geneesk.* **76**, 1579.

Clark, G. McC., Kauffman, H.F., Gangarosa, E.J. and Thompson, M.A. (1973) *Lancet* **ii**, 490.

Coburn, D.R. and Quortrup, E.R. (1938) *Trans. 3rd N. Amer. Wildlife Conf.* 869.

Coleman, G.E. and Meyer, K.F. (1922) *J. infect. Dis.* **31**, 622.

Condit, P.K. and Renteln, H.A. (1966) *Morbid. & Mortal. Wkly Rep. NCDC, Atlanta* **15**, 349.

Craven, P.C., Mackel, D.C., Baine, W.B., Barker, W.H., Gangarosa, E.J., Goldfield, M., Rosenfeld, H., Altman, R., Lachapelle, G., Davies, J.W. and Swanson, R.C. (1975) *Lancet* **ii**, 788.

Crisley, F.D. and Helz, G.E. (1961) *Canad. J. Microbiol.* **7**, 633.

Crowther, J.S. and Holbrook, R. (1976) In: *Microbiology in Agriculture, Fisheries and Food*, p. 215. Ed. by F.A. Skinner and J.G. Carr. Academic Press, London.

Dack, G.M. (1926) *J. infect. Dis.* **38**, 165; (1956) *Food Poisoning*, University of Chicago Press.

Dack, G.M., Cary, W.E., Woolpert, O. and Wiggers, H. (1930) *J. prev. Med., Baltimore* **4**, 167.

D'Aoust, J.Y.D., Aris, B.J., Thisdele, P., Durante, A., Brisson, N., Dragon, D., Lachapelle, G., Johnston, M. and Laidley, R. (1975) *Canad. Inst. Food Sci. Technol. J.* **8**, 181.

Denys, J. (1894) *Bull. Acad. roy. Méd. Belg.* **8**, 496.

Department of Health and Social Security (1982) *Memo. 188 Med.* HMSO, London.

Dezfulian, M., McCroskey, L.M., Hatheway, C.L. and Dowell, V.R. (1981) *J. clin. Microbiol.* **13**, 526.

Dickson, E.C. (1918) *Monogr. Rockefeller Inst. med. Res.* No. 8; (1920) *Amer. J. publ. Hlth* **10**, 865.

Dickson, E.C., Burke, G.S., Beck, D. and Johnston, J. (1925) *J. infect. Dis.* **36**, 472.

Dinter, Z. von and Kull, K.E. (1950) *Nord. VetMed.* **2**, 906.

Dixon, J.M.S. (1965) *Brit. med. J.* **ii**, 1343.

Dolman, C.E. (1957a) *Canad. J. publ. Hlth* **48**, 187; (1957b) *Jap. J. med. Sci. Biol.* **10**, 373; (1960) *Arctic* **13**, 230; (1961) *Canad. med. J.* **84**, 191; (1964). See *Report* 1964c, p. 5.

Dolman, C.E. and Chang, H. (1953) *Canad. J. publ. Hlth* **44**, 231.

Dolman, C.E. and Iida, H. (1963) *Canad. J. publ. Hlth* **54**, 293.

Dolman, C.E. and Murakami, L. (1961) *J. infect. Dis.* **109**, 107.

Dolman, C.E. et al. (1960) *J. infect. Dis.* **106**, 5.

Donovan, K.O. (1958) *J. appl. Bact.* **21**, 100.

Dowell, V.R., McCroskey, L.M., Hatheway, C.L., Lombard, G.L., Hughes, J.M. and Merson, M.H. (1977) *J. Amer. med. Ass.* **238**, 1829.

Dubovsky, B.J. and Meyer, K.F. (1922) *J. infect. Dis.* **31**, 501, 595.

Duff, J.T., Wright, G.C. and Yarinsky, A. (1956) *J. Bact.* **72**, 455.

Duncan, C.L. and Strong, D.H. (1969) *Canad. J. Microbiol.* **15**, 765; (1971) *Infect. Immun.* **3**, 167.

Duncan, C.L., Sugiyama, H. and Strong, D.H. (1968) *J. Bact.* **95**, 1560.

Duncan, R.M. and Jensen, W.I. (1976) *J. Wildlife Dis.* **12**, 116.

Eales, C.E. and Turner, A.W. (1952) *Aust. J. exp. Biol. med. Sci.* **30**, 295.

Easton, E.J. and Meyer, K.F. (1924) *J. infect. Dis.* **35**, 207.

Edel, W. and Kampelmacher, E.H. (1968) *Bull. World Hlth Org.* **39**, 489; (1969) *Ibid.* **41**, 297.

Edel, W., Schothorst, M. van, Guinée, P.A.M. and Kampelmacher, E.H. (1973) In: *Microbiological Safety of Food*, p. 247. Ed. by B.C. Hobbs and J.H.B. Christian. Academic Press, London.

Edel, W., Schothorst, M. van and Kampelmacher, E.H. (1976) *Zbl. Bakt., I. Abt. Orig.* **A 235**, 476.

Edel, W., Schothorst, M. van, Leusden, F.M. van and Kampelmacher, E.H. (1978) *Zbl. Bakt., I. Abt. Orig.* **A 242**, 468.

Editorial (1926) *J. Amer. med. Ass.* **86**, 482.

Edmonson, R.B., Bord, G.G. de and Thom, C. (1920) *Abstr. Bact.* **4**, 10.

Eklund, M.W., Peterson, M.E., Poysky, F.T., Peck, L.W. and Conrad, J.F. (1982) *Aquaculture* **27**, 1.

Elliott, R.P. and Hobbs, B.C. (1980) In: *Microbial Ecology of Foods*. Vol. 2, *Food Commodities*. Academic Press, New York.

Engley, F.B. (1956) *Texas Rep. Biol. Med.* **14**, 313.

Entis, P., Brodsky, M.H., Sharpe, A.N. and Jarvis, G.A. (1982) *Appl. environm. Microbiol.* **43**, 261.

Ermengem, E. van (1896) *Rev. Hyg.*, **18**, 761; (1897) *Z. Hyg. InfektKr.* **26**, 1.

Esty, J.R. (1935) *Amer. J. publ. Hlth* **25**, 165.

Esty, J.R. and Meyer, K.F. (1922) *J. infect. Dis.* **31**, 650.

Fabiansson, S. and Normark, C. (1976) *Svensk Veterinärtidskrift* **28**, 687.

Fanelli, M.J., Sadler, W.W. and Brownell, J.R. (1970) *Avian Dis.* **14**, 131.

Fantasia, L.D. (1969) *Appl. Microbiol.* **18**, 708.

Fay, L.D. (1966) *Trans. 31st N. Amer. Wildlife Conf.* 139.

Felsenfeld, O. and Young, V.M. (1949) *Amer. J. trop. Med.* **29**, 483.

Fenlon, D.R. (1981) *J. Hyg., Camb.* **86**, 195.

Fiock, M.A. et al. (1962) *J. Immunol.* **88**, 277.

Floyd, T.M. and Hoogstraal, H. (1956) *J. Egypt. publ. Hlth Ass.* **31**, 119.

Floyd, T.M. and Jones, G.B. (1954) *Amer. J. trop. Med.* **3**, 475.

Fourie, J.M. (1946) *J. S. Afr. vet. med. Ass.* **17**, 277.

Frazier, W.C. and Westhoff, D.C. (1978) *Food Microbiology*, 3rd edn. McGraw-Hill, New York.

Frost, A.J., Eaton, N.F., Gilchrist, D.J. and Moo, D. (1969) *Aust. vet. J.* **45**, 109.

Furniss, A.L., Lee, J.V. and Donovan, T.J. (1979) *Vibrios.* Publ. Hlth Lab. Serv. Monograph No. 11. HMSO, London.

Galbraith, N.S., Forbes, P. and Clifford, C. (1982) *Brit. med. J.* **i**, 1761.

Galbraith, N.S., Hobbs, B.C., Smith, M.E. and Tomlinson, A.J.H. (1960) *Mon. Bull. Minist. Hlth Lab. Serv.* **19**, 99.

Galbraith, N.S., Mawson, K.N., Maton, G.E. and Stone, D.M. (1962*a*) *Mon. Bull. Minist. Hlth Lab. Serv.* **21**, 209.

Galbraith, N.S., Taylor, C.E.D., Cavanagh, P., Hagan, J.G. and Patton, J.L. (1962*b*) *Lancet* **i**, 372.

Galton, M.M., Harless, M. and Hardy, A.V. (1955) *J. Amer. vet. med. Ass.* **126**, 57.

Galton, M.M., Scatterday, J.E. and Hardy, A.V. (1952). *J. infect. Dis.* **91**, 1.

Galton, M.M., Smith, W.V., McElrath, H.B. and Hardy, A.V. (1954) *J. infect. Dis.* **95**, 236.

Garrod, L.P. and McIlroy, M.B. (1949) *Brit. med. J.* **ii**, 1259.

Geiger, J.C. (1920) *Publ. Hlth Rep., Wash.* **35**, 2858 (1923) *J. Amer. med. Ass.* **81**, 1275; (1924) *Amer. J. publ. Hlth* **14**, 309.

Geiger, J.C. and Benson, H. (1923) *Publ. Hlth Rep., Wash.* **38**, 1611.

Geiger, J.C., Dickson, E.C. and Meyer, K.F. (1922) *Publ. Hlth Bull.* No. 127.

Geldreich, E.E. and Bordner, R.H. (1971) *J. Milk Food Technol.* **34**, 184.

Genigeorgis, C. and Riemann, H. (1979). In: *Food-borne Infections and Intoxications*, 2nd edn, p. 613. Ed. by H. Riemann and F.L. Bryan. Academic Press, New York.

Georgala, D.L. and Boothroyd, M. (1964) *J. Hyg., Camb.* **62**, 319.

Georgala, D.L., Boothroyd, M. and Hayes, P.R. (1965) *J. appl. Bact.* **28**, 421.

Ghosh, A.C. (1972) *J. Hyg., Camb.* **70**, 151; (1978) *Ibid.* **80**, 233.

Gilbert, R.J. (1979) In: *Food-borne Infections and Intoxications*, 2nd ed., p. 529. Ed. by H. Riemann and F.L. Bryan. Academic Press, New York; (1983) In: *Food Microbiology: Advances and Prospects*, p. 47. Ed. by T.A. Roberts and F.A. Skinner. Academic Press, London.

Gilbert, R.J., Kolvin, J.L. and Roberts, D. (1982) *Hlth Hyg.* **4**, 41.

Gilbert, R.J. and Parry, J.M. (1977) *J. Hyg., Camb.* **78**, 69.

Gilbert, R.J. and Roberts, D. (1979) *Hlth Hyg.* **3**, 33.

Gilbert, R.J., Stringer, M.F. and Peace, T.C. (1974) *J. Hyg., Camb.* **73**, 433.

Gilbert, R.J. and Taylor, A.J. (1976) In: *Microbiology in Agriculture, Fisheries and Food*, p. 197. Ed. by F.A. Skinner and J.G. Carr. Academic Press, London.

Gill, O.N., Sockett, P.N., Bartlett, C.L.R., Vaile, M.S.B., Rowe, B., Gilbert, R.J., Dulake, C., Murrell, H.O. and Salamaso, S. (1983) *Lancet* **i**, 574.

Giménez, D.F. and Ciccarelli, A.S. (1970) *Zbl. Bakt., I. Abt. Orig.* **215**, 221.

Gitter, M. and Sojka, W.J. (1970) *Vet. Rec.* **87**, 775.

Glencross, E.J.G. (1972) *Brit. med. J.* **ii**, 376.

Goepfert, J.M. (1974) *Proc. IV int. Congr. Food Sci. Technol.* **3**, 178.

Goepfert, J.M., Spira, W.M. and Kim, H.U. (1972) *J. Milk Food Technol.* **35**, 213.

Gordon, R.A. and Murrell, W.G. (1967) *CSIRO Food Preservation Quarterly* **27**, 6.

Gordon, W.S. (1934) *Vet. Rec.* **14**, 1016.

Graham, J.M. (1978) *J. appl. Bact.* **45**, 205.

Graham, J.M., Payne, D.J.H. and Taylor, C.E.D. (1958) *Mon. Bull. Minist. Hlth Lab. Serv.* **17**, 176.

Graham, J.M. and Smith, G.R. (1977) *Res. vet. Sci.* **22**, 343.

Graham, J.M., Smith, G.R., Borland, E.D. and Macdonald, J.W. (1978) *Vet. Rec.* **102**, 40.

Graham, R. and Boughton, I.B. (1923*a*) *Abstr. Bact.* **7**, 29.

Graham, R. and Brueckner, A.L. (1919) *J. Bact.* **4**, 1.

Graham, R. and Schwarze, H. (1921*a*) *J. infect. Dis.* **28**, 317; (1921*b*) *J. Bact.* **6**, 69.

Gross, W.B. and Smith, L. DS. (1971) *Avian Dis.* **15**, 716.

Gulasekharam, J., Veludapillai, T. and Niles, G.R. (1956) *J. Hyg., Camb.* **54**, 581.

Gunderson, M.F. (1933) *Proc. Soc. exp. Biol., N.Y.* **30**, 747.

Gunnison, J.B. (1955) *6th int. Congr. Microbiol., Rome* **4**, 173.

Gunnison, J.B., Cummings, J.R. and Meyer, K.F. (1936) *Proc. Soc. exp. Biol., N.Y.* **35**, 278.

Gunnison, J.B. and Meyer, K.F. (1929) *J. infect. Dis.* **45**, 119; (1930) *Ibid.* **46**, 335.

Haagsma, J. (1974) *Tijdschr. Diergeneesk.* **99**, 434.

Haagsma, J. and ter Laak, E.A. (1978) *Neth. J. vet. Sci.* **103**, 910.

Haagsma, J., Over, H.J., Smit, T. and Hoekstra, J. (1972) *Neth. J. vet. Sci.* **5**, 12.

Haines, R.B. and Moran, T. (1940) *J. Hyg., Camb.* **40**, 453.

Harvey, R.W.S. and Price, T.H. (1968) *J. Hyg., Camb.* **66**, 377; (1982) In: *Isolation and Identification Methods for Food Poisoning Organisms*, p. 51. Ed. by J.E.L. Corry, D. Roberts and F.A. Skinner. Academic Press, London.

Harvey, R.W.S. and Thomson, S. (1953) *Mon. Bull. Minst. Hlth Lab. Serv.* **12**, 49.

Hauge, S. (1950) *Nord. hyg. Tidskr.* **31**, 189; (1955) *J. appl. Bact.* **18**, 591.

Hauschild, A.H.W., Desmarchelier, P., Gilbert, R.J., Harmon, S.M. and Vahlefeld, R. (1979) *Canad. J. Microbiol.* **25**, 953.

Hauschild, A.H.W., Gilbert, R.J., Harmon, S.M., O'Keefe, M.F. and Vahlefeld, R. (1977) *Canad. J. Microbiol.* **23**, 884.

Hauschild, A.H.W. and Hilsheimer, R. (1971) *Canad. J. Microbiol.* **17**, 1425.

Heckley, R.J., Hildebrand, G.J. and Lamanna, C. (1960) *J. exp. Med.* **111**, 745.

Henning, M.W. (1956) *Animal Diseases in South Africa*, 3rd ed. Cent. News Agency Ltd, S. Africa.

Hepner, E. (1980) *Publ. Hlth, Lond.* **94**, 337.

Hildebrand, G.L., Lamanna, C. and Heckley, R.J. (1961) *Proc. Soc. exp. Biol., N.Y.* **107**, 284.

Hobbs, B.C. (1961) *J. appl. Bact.* **24**, 340; (1965) *Mon. Bull. Minist. Hlth Lab. Serv.* **24**, 123; (1971) In: *Poultry Disease and World Economy*, p. 65. Ed. by R.F. Gordon and B.M. Freeman. Longman Group Ltd, Edinburgh; (1979) In: *Food-borne Infections and Intoxications*, 2nd edn, p. 131. Ed. by H. Riemann and F.L. Bryan. Academic Press, New York.

Hobbs, B.C. and Gilbert, R.J. (1978) *Food Poisoning and Food Hygiene*, 4th edn. Edward Arnold, London.

Hobbs, B.C. and Hugh Jones, M.E. (1969) *J. Hyg., Camb.* **67**, 81.

Hobbs, B.C., Smith, M.E., Oakley, C.L., Warrack, G.H. and Cruickshank, J.C. (1953) *J. Hyg., Camb.* **57**, 75.

Hobbs, B.C. and Spooner, J.A. (1966) *Mon. Bull. Minist. Hlth Lab. Serv.* **25**, 132.

Hobbs, B.C. and Thomas, M.E.M. (1948) *Mon. Bull. Minist. Hlth Lab. Serv.* **7**, 261.

Hobbs, G. (1976) *Advanc. Food Res.* **22**, 135.

Hohn, J. and Herrmann, W. (1935) *Zbl. Bakt.* **133**, 183.

Holbrook, R. and Anderson, J.M. (1980) *Canad. J. Microbiol.* **26**, 753.

Holdeman, L.V. (1970) *J. Wildlife Dis.* **6**, 205.

Hooper, W.L., Barrow, G.I. and McNab, D.J.N. (1974) *Lancet* i, 1100.

Hormaeche, E., Peluffo, C.A. and Aleppo, P.L. (1936) *Arch. urug. Med.* **9**, 113.

Hottle, G.A., Nigg, C. and Lighty, J.A. (1947) *J. Immunol.* **55**, 255.

Howie, J.W. (1970) *Brit. med. J.* ii, 420; (1979) *J. appl. Bact.* **47**, 233.

Hughes, J.D., Blumenthal, J.R., Merson, M.H., Lombard, G.L., Dowell, V.R. and Gangarosa, E.J. (1981) *Ann. intern. Med.* **95**, 442.

Hughes, J.M., Herwitz, M.A., Merson, M.H., Barker, W.H. Jr and Gangarosa, E.J. (1977) *Amer. J. Epidem.* **105**, 233.

Hughes, R. and Whaler, B.C. (1962) *J. Physiol.* **160**, 221.

Huss, H.H. and Eskildsen, U. (1974) *Nord. VetMed.* **26**, 733.

Huss, H.H., Pedersen, A. and Cann, D.C. (1974) *J. Food Technol.* **9**, 451.

Iida, H. (1970). See *Report* 1970, p. 357.

Iida, H. *et al.* (1958) *Jap. J. med. Sci. Biol.* **11**, 115.

International Commission on Microbiological Specifications for Foods (1978) *Microorganisms in Foods I. Their Significance and Methods of Enumeration*, 2nd edn. University of Toronto Press, Canada; (1980) *Microbial Ecology of Foods.* Vol. 2, *Food Commodities.* Academic Press, London.

International Organization for Standardization (1975) ISO (International Standard) 3565, *Meat and Meat Products: Detection of Salmonellae* (Reference Method).

Ito, K.A., Seslar, D.J., Mercer, W.A. and Meyer, K.F. (1967) See *Report* 1967, p. 108.

Itoh, T. (1972) *Annual Report of Tokyo Metropolitan Research Laboratory of Public Health* **34**, 7.

Jack, E.J. (1968) *Vet. Rec.* **82**, 558.

Jack, E.J. and Hepper, P.J. (1969) *Vet. Rec.* **84**, 196.

Jadin, J. (1951) *Rev. belge Path.* **21**, 8.

Jameson, J.E. (1961) *J. Hyg., Camb.* **59**, 1.

Jansen, J. (1936) *Tijdschr. Diergeneesk.* **63**, 140.

Jarvis, B. (1982) *Food Technol. Aust.* **34**, 508.

Jensen, W.I. and Allen, J.P. (1960) *Trans. 25th N. Amer. Wildlife Conf.* 171.

Jephcott, A.E., Martin, D.R. and Stalker, R. (1969) *J. Hyg., Camb.* **67**, 505.

Johnston, R., Harmon, S. and Kautter, D. (1964) *J. Bact.* **88**, 1521.

Johnston, W.S., Maclachlan, G.K. and Hopkins, G.F. (1979) *Vet. Rec.* **105**, 526.

Jones, P.H., Willis, A.T., Robinson, D.A., Skirrow, M.B. and Joseph, D.S. (1981) *J. Hyg., Camb.* **87**, 156.

Jordan, E.O. (1917) *Food Poisoning.* Univ. Chicago Press, Chicago.

Kahn, M.C. (1924) *J. infect. Dis.* **35**, 423.

Kallings, L.O., Ringertz, O., Silverstolpe, L. and Ernerfeld, T.F. (1966) *Acta pharm. suec.* **3**, 219.

Kalmbach, E.R. and Gunderson, M.F. (1934) *USDA tech. Bull.* No. 411, 1.

Kao, I., Drachman, D.B. and Price, D.L. (1976) *Science* **193**, 1256.

Karashimada, T. *et al.* (1970) See *Report* 1970, p. 376.

Kautter, D.A., Lynt, R.K., Solomon, H.M., Lilly, T. and Harmon, S.M. (1970) See *Report* 1970, p. 236.

Kempner, W. (1897) *Z. Hyg. InfektKr.* **26**, 481.

Keppie, J. (1951) *J. Hyg., Camb.* **49**, 36.

Keymer, I.F., Smith, G.R., Roberts, T.A., Heaney, S.I. and Hibberd, D.J. (1972) *Vet. Rec.* **90**, 111.

Khalil, A.M. (1938) *J. Hyg., Camb.* **38**, 75.

Kim, H.U. and Goepfert, J.M. (1971) *J. Milk Food Technol.* **34**, 12.

Klein, E. (1895) *Zbl. Bakt., I. Abt. Orig.* **18**, 737.

Knock, G.G. (1952) *J. Sci. Food Agric.* **3**, 86.

Knowles, N.R. (1971) In: *Poultry Disease and World Economy*, p. 87. Ed. by R.F. Gordon and B.M. Freeman. Longman Group Ltd, Edinburgh.

Knox, R. and Macdonald, E.K. (1943) *Med. Offr* **69**, 21.

Kornacki, J.L. and Marth, E.H. (1982) *J. Milk Food Protect.* **45**, 1051.

Koser, S.A., Edmondson, R.B. and Giltner, L.T. (1921) *J. Amer. med. Ass.* **77**, 1250.

Kovács, N. (1959) *Med. J. Aust.* i, 557.

Kramer, J.M., Turnbull, P.C.B., Munshi, G. and Gilbert, R.J. (1982) In: *Isolation and Identification Methods for Food Poisoning Organisms*, p. 261. Ed. by J.E.L. Corry, D. Roberts and F.A. Skinner. Academic Press, London.

Kravchenko, A.T. and Shishulina, L.M. (1967) See *Report* 1967, p. 13.

Kwantes, W. (1952) *Mon. Bull. Minist. Hlth Lab. Serv.* **11**, 239.

Lamanna, C. (1959) *Science* **130**, 763; (1960) *Ibid.* **131**, 1100.

Lamanna, C. and Carr, C.J. (1967) *Clin. Pharmacol. Ther.* **8**, 286.

Lamanna, C. and Meyers, C.E. (1960) *J. Bact.* **79**, 406.

Lang, D.J., Kunz, L.S., Martin, A.R., Schroeder, S.A. and Thompson, A. (1967) *New Engl. J. Med.* **276**, 829.

Lavergne, V. de and Abel, E. (1925) *Rev. Hyg.* **47**, 950.

Lawrence, G., Shann, F., Freestone, D.S. and Walker, P.D. (1979) *Lancet* i, 227.

Lawrence, G. and Walker, P.D. (1976) *Lancet* i, 125.

Lee, I.C., Stevenson, K.E. and Harmon, L.G. (1977) *Appl. environm. Microbiol.* **33**, 341.

Lee, J.A. (1973) In: *The Microbiological Safety of Food*, p. 197. Ed. by B.C. Hobbs and J.H.B. Christian. Academic Press, London; (1974) *J. Hyg., Camb.* **72**, 185.

Lee, V.H., Vadlamudi, S. and Hanson, R.P. (1962) *J. Wildlife Mgmt* **26**, 411.

Leighton, G.R. (1923) *Rep. Scot. Bd Hlth.* HMSO, London.

Lennox, M., Harvey, R.W.S. and Thompson, S. (1954) *J. Hyg., Camb.* **52**, 311.

Lentz, O. (1924) *Z. Hyg. InfektKr.* **103**, 121.

Leusden, F.M. van, Schothorst, M. van and Beckers, H.J. (1982) In: *Isolation and Identification Methods for Food Poisoning Organisms*, p. 35. Ed. by J.E.L. Corry, D. Roberts and F.A. Skinner. Academic Press, London.

Linton, A.H. (1979) *Brit. vet. J.* **135**, 109.

Liu, P.Y., Zia, S.H. and Chung, H.L. (1937) *Proc. Soc. exp. Biol., N.Y.* **37**, 17.

Lotter, L.P. and Genigeorgis, C.A. (1975) *Appl. Microbiol.* **29**, 152; (1977) *Zbl. Bakt., I. Abt. Orig.* A **239**, 18.

Lotter, L.P. and Leistner, L. (1978) *Appl. environm. Microbiol.* **36**, 377.

Lovell, R. (1932) *Nat. vet. med. Ass. G.B. 50th ann. Congr.*

Ludlam, G.B. (1954) *Mon. Bull. Minist. Hlth Lab. Serv.* **13**, 196.

Lundbeck, H., Plazikowski, U. and Silverstolpe, L. (1955) *J. appl. Bact.* **18**, 535.

McClung, L.S. (1945) *J. Bact.* **50**, 229; (1964) See *Report* 1964*c*, p. 257; (1967) See *Report* 1967, p. 431.

McCoy, J.H. (1975) *J. Hyg., Camb.* **74**, 271.

McCullough, N.B., and Eisele, C.W. (1951*a*) *J. infect. Dis.* **88**, 278; (1951*b*) *J. Immunol.* **66**, 595; (1951*c*) *J. infect. Dis.* **89**, 209; (1951*d*) *Ibid.* **89**, 259.

McDonel, J.L. (1980) *Food Tech.,Champaign* **34**, 91.

McDonel, J.L. and Duncan, C.L. (1975) *Infect. Immun.* **12**, 1214.

Mackerras, I.M. and Pope, P. (1948) *Aust. J. exp. Biol. med. Sci.* **26**, 465.

Marier, R., Wells, J.C., Swanson, R.C., Callahan, W. and Mehlman, I.J. (1973) *Lancet* **ii**, 1376.

Marlow, G.R. and Smart, J.L. (1982) *Vet. Rec.* **111**, 242.

Marthedal, H.E. (1977) *Proceedings of the International Symposium on Salmonella and Prospects for Control*, p. 78. University of Guelph, Ontario.

Marzochi, M.C. de A. (1977) *Rev. Inst. Med. trop., São Paulo* **19**, 148.

Mason, J.H. (1968) *J.S. Afr. vet. med. Ass.* **39**, 37.

Mason, J.H., Steyn, H.P. and Bisschop, J.H.R. (1938) *J. S. Afr. vet. med. Ass.* **9**, 65.

Matveev, K.L., Nefedjeva, N.P., Bulatova, T.I. and Sokolov, I.S. (1967) See *Report* 1967, p. 1.

May, A.J. and Whaler, B.C. (1958) *Brit. J. exp. Path.* **39**, 307.

Melling, J. and Capel, B.J. (1978) *FEMS Microbiol. Lett.* **4**, 133.

Melling, J., Capel, B.J., Turnbull, P.C.B. and Gilbert, R.J. (1976) *J. clin. Path.* **29**, 938.

Merson, M.H. and Dowell, V.R. (1973) *New Engl. J. Med.* **289**, 1005.

Messerlin, A. and Courzi, G. (1942) *Bull. Inst. Hyg. Maroc.* **2** (N.S.), 15.

Meyer, K.F. (1928) In: *Kolle and Wassermann's Handbuch der Pathogenen Mikroorganismen*, 3te Aufl., iv. 1269; (1931) *J. prev. Med., Baltimore* **5**, 261; (1956) *Bull. World Hlth Org.* **15**, 281.

Meyer, K.F. and Dubovsky, B.J. (1922*a*) *J. infect. Dis.* **31**, 541; (1922*b*) *Ibid* **31**, 559; (1922*c*) *Ibid.* **31**, 600.

Meyer, K.F. and Eddie, B. (1951) *Z. Hyg. InfektKr.* **133**, 255.

Meyer, K.F. and Gunnison, J.B. (1928) *Proc. Soc. exp. Biol., N.Y.* **26**, 88.

Meyer, K.F. *et al.* (1955) *6th int. Congr. Microbiol., Rome* **4**, 123.

Midura, T.F. and Arnon, S.S. (1976) *Lancet* **ii**, 934.

Miller, A.R., Elias-Jones, T.F., Nicolson, N. and Wilson, T.S. (1969) *Med. Offr* **121**, 223.

Milner, K.C., Jellison, W.L. and Smith, B. (1957) *J. infect. Dis.* **101**, 181.

Minervin, S.M. (1967) See *Report* 1967, p. 336.

Minnich, S.A., Hartman, P.A. and Heimsch, R.C. (1982) *Appl. environm. Microbiol.* **43**, 877.

Minor, T.E. and Marth, E.H. (1971) *J. Milk Food Technol.* **34**, 557; (1972) *Ibid.* **35**, 21, 77, 228; (1976) *Staphylococci and their Significance in Foods.* Elsevier Scientific Publishing Company, Oxford, New York.

Miyazaki, S. and Sakaguchi, G. (1978) *Jap. J. med. Sci. Biol.* **31**, 1.

Moberg, K. (1953) *Proc. 15th int. vet. Congr.* **1**, 62.

Møller, V. and Scheibel, I. (1960) *Acta path. microbiol. scand.* **48**, 80.

Morita, T.N. and Woodburn, M.J. (1977) *J. Food. Sci.* **42**, 1232.

Morris, G.K. and Dunn, C.G. (1970) *Appl. Microbiol.* **20**, 192.

Mortimer, P.R. and McCann, G. (1974) *Lancet* **i**, 1043.

Mosher, W.E., Wheeler, S.M., Chant, H.L. and Hardy, A.V. (1941) *Publ. Hlth Rep., Wash.* **56**, 2415.

Mossel, D.A.A., Koopman, M.J. and Jongerius, E. (1967) *Appl. Microbiol.* **15**, 650.

Müller, J. (1963) *Bull. Off. int. Epizoot.* **59**, 1379.

Murdock, C.R. (1954) *Mon. Bull. Minist. Hlth Lab. Serv.* **13**, 43.

Nakano, W. and Kodama, E. (1970) See *Report* 1970, p. 388.

Nevin, M. (1921) *J. infect. Dis.* **28**, 226.

Newell, K.W., Hobbs, B.C. and Wallace, E.J.G. (1955) *Brit. med. J.* **ii**, 1296.

Nickerson, J.T. and Sinskey, A.J. (1972) *Microbiology of Foods and Food Processing.* Elsevier, New York.

Niilo, L. (1971) *Infect. Immun.* **3**, 100.

Norris, J.R., Berkeley, R.C.W., Logan, N.A. and O'Donnell, A.G. (1981) In: *The Prokaryotes—a Handbook on Habitats, Isolation and Identification of Bacteria*, Vol. II, p. 1711. Ed. by M.P. Starr, H. Stolp, H.G. Trüper, A. Balows and H.G. Schlegel. Springer-Verlag, Berlin.

North, W.R. and Bartram, M.T. (1953) *Appl. Microbiol.* **1**, 130.

Notermans, S., Boot, R., Tips, P.D. and De Nooij, M.P. (1983) *J. Food Protect.* **46**, 238.

Notermans, S., Dufrenne, J. and Kozaki, S. (1979) *Appl. environm. Microbiol.* **37**, 1173.

Notermans, S., Dufrenne, J. and Oosterom, J. (1981) *Appl. environm. Microbiol.* **41**, 179.

Oberst, F.W., Crook, J.W., Cresthull, P. and House, M.J. (1968) *Clin. Pharmacol. Ther.* **9**, 209.

Ohishi, I., Sakaguchi, G., Riemann, H., Behymer, D. and Hurvell, B. (1979) *J. Wildl. Dis.* **15**, 3.

Ohye, D.D. and Scott, W.J. (1953) *Aust. J. biol. Sci.* **6**, 178; (1957) *Ibid.* **10**, 85.

Olson, J.C., Casman, E.P., Baer, E.F. and Stone, J.E. (1970) *Appl. Microbiol.* **20**, 605.

Ormay, L. and Novotny, T. (1969) In: *The Microbiology of Dried Foods*, p. 279. Ed. by E.H. Kampelmacher, M. Ingram and D.A.A. Mossel. International Association of Microbiological Societies.

Orr, P.F. (1922) *J. infect. Dis.* **30**, 118.

Osterling, S. (1952) *Nord. hyg. Tidskr.* **33**, 173.

Owen, R.W.G. (1907) *Phys. & Surg., Ann. Arbor* **29**, 289 (quoted from Dack, G.M. 1956).

Oxer, D.T. (1940) *Aust. vet. J.* **16**, 4.

Oye, E. van (Ed.) (1964) *The World Problem of Salmonellosis.* W. Junk Publishers, The Hague.

Parry, J.M. and Gilbert, R.J. (1980) *J. Hyg., Camb.* **84**, 77.

Pasricha, C.L. and Panja, G. (1940) *Indian J. med. Res.* **28**, 49.

Patterson, J.T. (1972) *Rec. agric. Res.* **20**, 1.

Payne, D.J.H. and Scudamore, J.M. (1977) *Lancet* **i**, 1249.

Peffers, A.S.R., Bailey, J., Barrow, G.I. and Hobbs, B.C. (1973) *Lancet* **i**, 143.

Penfold, J.B., Amery, H.C.C. and Morley Peet, P.J. (1979) *Brit. med. J.* **iii**, 202.

Pennington, J.H., Brooksbank, N.H., Poole, P.M. and Seymour, F. (1968) *Brit. med. J.* **iv**, 804.

Perkins, W.E. (1964) See *Report* 1964*c*, p. 187.

Perry, H.M. and Tidy, H.L. (1919) *Spec. Rep. Ser. med. Res. Coun., Lond.*, No. 24.

Pether, J.V.S. and Scott, R.J.D. (1982) *J. Infect.* **5**, 81.

Petty, C.S. (1965) *Amer. J. med. Sci.* **249**, 345.

Pickett, J., Berg, B., Chaplin, E. and Brunstetter-Shafer, M. (1976) *New Engl. J. Med.* **295**, 770.

Pilet, C., Cazabat, H. and Ardonceau, R. (1959) *Bull. Acad. vét. Fr.* **32**, 297.

Pinegar, J.A. and Suffield, A. (1982) In: *Isolation and Identification Methods for Food Poisoning Organisms*, p. 1. Ed. by J.E.L. Corry, D. Roberts and F.A. Skinner. Academic Press, London.

Pivnick, H. and Bird, H. (1965) *Food Tech., Champaign* **19**, 132.

Prévot, A.R. and Brygoo, E.R. (1953) *Ann. Inst. Pasteur* **85**, 544.

Prévot, A.R. *et al.* (1955) *Bull. Acad. nat. Méd.* **139**, 355.

Prost, E. and Riemann, H. (1967) *Annu. Rev. Microbiol.* **21**, 495.

Public Health Laboratory Service (1980a) *Brit. med. J.* **ii**, 817; (1980b) *Ibid.*, **ii**, 1360; (1981) *Ibid.* **ii**, 924; (1982) *Ibid.* **ii**, 1127; (1980c) *Environm. Hlth* **88**, 123; (1983) *Commun. Med.* **5**, 152.

Quortrup, E.R. and Gorham, J.R. (1949) *Amer. J. vet. Res.* **10**, 268.

Quortrup, E.R. and Holt, A.L. (1941) *J. Bact.* **41**, 363.

Quortrup, E.R. and Sudheimer, R.L. (1942) *Trans. 7th N. Amer. Wildlife Conf.* 284; (1943) *J. Bact.* **45**, 551.

Raatjes, G.J.M. and Smelt, J.P.P.M. (1974) *Nature, Lond.* **281**, 398.

Raevuori, M. (1976) *Suom. Elainlaakaril.* **82**, 501.

Ralovich, B. and Barna, K. (1966) *Zbl. Bakt.* **200**, 509.

Reames, H.R. *et al.* (1947) *J. Immunol.* **55**, 309.

Reichert, C.A. and Fung, D.Y.C. (1976) *J. Milk Food Technol.* **39**, 516.

Reilly, J.R. and Boroff, D.A. (1967) *Wildlife Dis.* **3**, 26.

Reiser, R., Conaway, D. and Bergdoll, M.S. (1974) *Appl. Microbiol.* **27**, 83.

Reitler, R. and Mentzel, R. (1946) *Trans. R. Soc. trop. Med. Hyg.* **39**, 523.

Report (1947) *Spec. Rep. Ser. med. Res. Coun., Lond.*, No 360; (1953) Dehydrated Meat. *Dept sci. industr. Res. Food Investgn Spec. Rep.* No. 57; (1954) *Mon. Bull. Minist. Hlth Lab. Serv.* **13**, 38; (1957) *Ibid.* **16**, 233; (1958) *Ibid.* **17**, 252; (1964) *Botulism, Proceedings of a Symposium*. Ed. by K.H. Lewis and K. Cassel. *Publ. Hlth Serv. Publ.* No. 999–FP-1; (1967) *Botulism 1966*. Ed. by M. Ingram and T.A. Roberts. Chapman and Hall, London; (1969) Joint Committee on the use of Antibiotics in Animal Husbandry and Veterinary Medicine: Report., *Cmnd 4190*. HMSO, London. (1970) *Toxic Micro-organisms; Mycotoxins, Botulism*. Ed. by M. Herzberg. UJNR Joint Panels, Washington D.C.; (1980) *Morbid. Mortal.* **29**, 85 (1984) *J. Amer. med Ass.*, **251**, 323.

Riemann, H. and Bryan, F.L. (1979) *Food-borne Infections and Intoxications*, 2nd edn. Academic Press, New York.

Roberts, D. (1982a) In: *Meat Microbiology*, p. 319. Ed. by M.H. Brown. Applied Science Publishers, London; (1982b) *J. Hyg., Camb.* **89**, 491.

Roberts, D., Boag, K., Hall, M.L.M. and Shipp, C.R. (1975) *J. Hyg., Camb.* **75**, 173.

Roberts, D. and Gilbert, R.J. (1979) *J. Hyg., Camb.* **82**, 123.

Roberts, D., Watson, G.N. and Gilbert, R.J. (1982) In: *Bacteria and Plants*, p. 169. Ed. by M. Rhodes-Roberts and F.A. Skinner. Academic Press, London.

Roberts, T.A. (1980) *R. Soc. Hlth J.* **100**, 3.

Roberts, T.A., Keymer, I.F., Borland, E. and Smith, G.R. (1972) *Vet. Rec.* **91**, 11.

Roberts, T.A., Thomas, A.I. and Gilbert, R.J. (1973) *Vet. Rec.* **92**, 107.

Robinson, D.A. and Jones, D.M. (1981) *Brit. med. J.* **i**, 1374.

Robinson, E.M. (1929) *15th Rep. Director vet. Educ. Res., S Africa* Section iii, p. 111.

Rohde, R. and Adam, W. (1956) *Zbl. Bakt.* **166**, 329.

Rosen, M.N. (1971) In: *Infectious and Parasitic Diseases of Wild Birds*, p. 100. Ed. by J.W. Davis, R.C. Anderson, L. Karstad and D.O. Trainer. Iowa State University Press, Ames.

Rowe, B. (1973) In: *The Microbiological Safety of Food*, p. 165. Ed. by B.C. Hobbs and J.H.B. Christian. Academic Press, London.

Rowe, B. and Hall, M.L.M. (1975) *Brit. med. J.* **iv**, 51.

Rowe, B., Hall, M.L.M., Ward, L.R. and De Sa, J.D.H. (1980) *Brit. med. J.* **ii**, 1065.

Rowe, B., Threlfall, E.J., Ward, L.R. and Ashley, A.S. (1979) *Vet. Rec.* **105**, 468.

Rubenstein, A.D., Feemster, R.F. and Smith, H.M. (1944) *Amer. J. publ. Hlth* **34**, 841.

Sakaguchi, G. and Tohyama, Y. (1955) *Jap. J. med. Sci. Biol.* **8**, 247.

Sakaguchi, G., Uemura, T. and Riemann, H.P. (1973) *Appl. Microbiol.* **26**, 762.

Sakazaki, R. (1979) In: *Food-borne Infections and Intoxications*, 2nd edn, p. 173. Ed. by H. Riemann and F.L. Bryan. Academic Press, New York.

Saleh, M.A. and Ordal, Z.J. (1955) *Food Res.* **20**, 340.

Savage, W.G. (1920) *Food Poisoning and Food Infections*. Cambridge; (1932) *J. prev. Med., Baltimore* **6**, 425.

Savage, W.G. and White, P.B. (1925) *Spec. Rep. Ser. med. Res. Coun., Lond.* No. 92.

Saxe, M. de, Coe, A.W. and Wieneke, A.A. (1982) In: *Isolation and Identification Methods for Food Poisoning Organisms*, p. 173. Ed. by J.E.L. Corry, D. Roberts and F.A. Skinner. Academic Press, London.

Schäfer, W. (1958) *Zbl. Bakt.* **172**, 272.

Schiebner, G. (1955) *Dtsch. tierärztl. Wschr.* **62**, 355.

Schneider, H.J. and Fisk, R. (1939) *J. Amer. med. Ass.* **113**, 2299.

Schoenholz, P., Esty, J.R. and Meyer, K.F. (1923) *J. infect. Dis.* **33**, 289.

Schothorst, M. van and Kampelmacher, E.H. (1968) In: *The Microbiology of Dried Foods*, p. 193. Ed. by E.H. Kampelmacher, M. Ingram and D.A.A. Mossel. Grafische Industrie, Harlem, The Netherlands.

Scott, W.M. (1933) *Bull. Off. int. Hyg. publ.* **25**, 828.

Sebald, M. and Saimot, G. (1973) *Ann. Inst. Pasteur* **124A**, 61.

Seddon, H.R. (1922) *J. comp. Path.* **35**, 147; (1927) *Aust. vet. J.* **3**, 136.

Segner, W.P., Schmidt, C.F. and Boltz, J.K. (1971) *Appl. Microbiol.* **22**, 1017.

Seligmann, E. (1935) *Schweiz. med. Wschr.* **65**, 550.

Seligmann, R. and Lapinsky, Z. (1970) *Refuah. vet.* **27**, 7.

Shayegani, M., Morse, D., DeForge, I., Root, T., Malmberg Parsons, L. and Maupin, P.S. (1983) *J. clin. Microbiol.* **17**, 35.

Shvedov, L.M. (1959) *J. Microbiol. Epidemiol. Immunobiol.* **30**, 95; (1960) *Ibid.* **31**, 106.

Silliker, J.H. and Taylor, W.I. (1958) *Appl. Microbiol.* **6**, 228.

Skirrow, M.B. (1977) *Brit. med. J.* **iii**, 9.

Skjelkvåle, R. and Duncan, C.L. (1975) *Infect. Immun.* **11**, 563.

Skjelkvåle, R. and Uemura, T. (1977a) *J. appl. Bact.* **42**, 355; (1977b) *Ibid.* **43**, 281.

Skulberg, A. (1957) *Nord. hyg. Tidskr.* **38**, 263; (1958) *Ibid.* **39**, 133; (1961) *Nord. VetMed.* **13**, 87.

Small, R.G. and Sharp, J.C.M. (1979) *J. Hyg., Camb.* **82**, 95.

Smart, J.L. and Roberts, T.A. (1977) *Vet. Rec.* **100**, 378.

Smart, J.L., Roberts, T.A., McCullagh, K.G., Lucke, V.M. and Pearson, H. (1980) *Vet. Rec.* **107**, 445.

Smith, G.R. (1982) *Symp. zool. Soc. Lond.* No. 50, 97; (1984) To be published.

Smith, G.R. and Milligan, R.A. (1979) *J. Hyg., Camb.* **83**, 237.

Smith, G.R. and Moryson, C.J. (1975) *J. Hyg., Camb.* **75**, 371.

Smith, G.R., Moryson, C.J. and Walmsley, J.G. (1977) *J. Hyg., Camb.* **78**, 33.

Smith, G.R. and Oliphant, J.C. (1983) *Vet. Rec.* **112**, 457.

Smith, G.R., Oliphant, J.C. and White, W.R. (1982) *J. Hyg., Camb.* **89**, 507.

Smith, G.R. and Young, A.M. (1980) *J. Hyg., Camb.* **85**, 271.

Smith, H.W. (1969) *R. Soc. Hlth J.* **89**, 271.

Smith, H.W. and Buxton, A. (1951) *Brit. med. J.* **i**, 1478.

Smith, L. DS. (1977) *Botulism. The Organism, its Toxins, the Disease.* American Lecture Series, No. 997. Ed. by A. Balows. Charles C Thomas, Springfield; (1978) *Hlth Lab. Sci.* **15**, 74.

Smith, M.E. and Hobbs, B.C. (1955) *Mon. Bull. Minist. Hlth Lab. Serv.* **14**, 154.

Smith, N.R., Gordon, R.E. and Clark, F.E. (1952) *Aerobic Sporeforming Bacteria*, Monograph No. 16. United States Department of Agriculture, Washington DC.

Snoeyenbos, G.H. (1969) *Feedstuffs* **41**, 57.

Sonnabend, O.., Sonnabend, W., Heinzle, R., Sigrist, T., Dirnhofer, R. and Krech, U. (1981) *J. infect. Dis.* **143**, 22.

Sperber, W.H. and Deibel, R.H. (1969) *Appl. Microbiol.* **17**, 533.

Sperry, C.C. (1947) *Trans. 12th N. Amer. Wildlife Conf.* 228.

Spira, W.M. and Goepfert, J.M. (1975) *Canad. J. Microbiol.* **21**, 1236.

Starin, W.A. and Dack, G.M. (1925) *J. infect. Dis.* **36**, 383.

Stark, R.L. and Duncan, C.L. (1971) *Infect. Immun.* **4**, 89; (1972) *Ibid.*, **5**, 147.

Steede, F.D.F. and Smith, H.W. (1954) *Brit. med. J.* **ii**, 576.

Steiniger, F. (1953) *Zbl. Bakt.* **160**, 80.

Sterne, M. and Wentzel, L.M. (1950) *J. Immunol.* **65**, 175; (1953) *Rep. 14th int. vet. Congr.* **3**, 329.

Stevenson, J.W. (1958) *Amer. J. med. Sci.* **235**, 317.

Stewart, B.J., Eyles, M.J. and Murrell, W.G. (1980) *Appl. environm. Microbiol.* **40**, 223.

Stewart, D.J. (1965) *Proc. 1st int. Congr. Food Sci. Tech.* **2**, 485.

Stokes, J.L. and Osborne, W.W. (1955) *Appl. Microbiol.* **3**, 217.

Stricker, F.D. and Geiger, J.C. (1924) *Publ. Hlth Rep., Wash.* **39**, 655.

Stringer, M.F. (1981) PhD Thesis, University of London.

Stringer, M.F., Turnbull, P.C.B. and Gilbert, R.J. (1980) *J. Hyg., Camb.* **84**, 443.

Stringer, M.F., Watson, G.N. and Gilbert, R.J. (1982) In: *Isolation and Identification Methods for Food-poisoning Organisms*, p. 111. Ed. by J.E.L. Corry, D. Roberts and F.A. Skinner. Academic Press, London.

Sugiyama, H. (1980) *Microbiol. Rev.* **44**, 419.

Sugiyama, H., Bott, T.L. and Foster, E.M. (1970). See *Report* 1970, p. 287.

Sugiyama, H., Mills, D.C. and Cathy Kuo, L. (1978) *J. Food Protect.* **41**, 848.

Sutherland, H.P. (1960) *J. Amer. med. Ass.* **172**, 1266.

Sutton, R.G.A. and Hobbs, B.C. (1968) *J. Hyg., Camb.* **66**, 135.

Szanton, V.L. (1947) *Pediatrics, Springfield* **20**, 795.

Takashi, W. (1953) *Kitasato Arch. exp. Med.* **25**, 163.

Tatini, S.R. (1976) *J. Milk Food Technol.* **39**, 432.

Taylor, A.J. and Gilbert, R.J. (1975) *J. med. Microbiol.* **8**, 543.

Taylor, C.E.D. and Coetzee, E.F.C. (1966) *Mon. Bull. Minist. Hlth Lab. Serv.* **25**, 142.

Taylor, R.J. (1973) *Brit. vet. J.* **129**, 354.

Taylor, R.J. and Burrows, M.R. (1971) *Brit. vet. J.* **127**, 536.

Tanner, F.W., Beamer, P.R. and Rickher, C. (1940) *Food Res.* **5**, 323.

Tanner, F.W. and Twohey, H.B. (1926) *Zbl. Bakt.* **98**, 136.

Tardieux, P., Huet, M. and Björklund, B. (1952) *Ann. Inst. Pasteur* **82**, 763.

Thatcher, F.S., Erdman, I.E. and Pontefract, R.D. (1967) See *Report* 1967, p. 511.

Theiler, A. (1928) *13th and 14th Rep. Director vet. Educ. Res.*, S. Africa Part I, p. 47.

Theiler, A. and Robinson, E.M. (1927) *Z. InfektKr. Haustiere* **31**, 165.

Theiler, A. et al. (1926-27a) *11th and 12th Rep. Director vet. Educ. Res.*, S. Africa. Part II, p. 821; (1926-27b) *Ibid.* p. 1201.

Thompson, J.A., Glasgow, L.A., Warpinski, J.R. and Olson, C. (1980) *Pediatrics, Springfield* **66**, 936.

Threlfall, E.J., Ward, L.R. and Rowe, B. (1978a) *Brit. med. J.* **iv**, 997; (1978b) *Vet. Rec.* **103**, 438.

Todd, E.C.D. (1978) *J. Food Protect.* **41**, 559.

Townsend, C.T., Yee, L. and Mercer, W.A. (1954) *Food Res.* **19**, 536.

Troller, J.A. (1976) *J. Milk Food Technol.* **39**, 499.

Turnbull, P.C.B. (1979) *Clinics in Gastroenterology* **8**, 663; (1981) *Clin. Pharmacol. Ther.* **13**, 453.

Turnbull, P.C.B. and Gilbert, R.J. (1982) In: *Adverse Effects of Foods*, p. 297. Ed. by E.F.P. Jellife and D.B. Jelliffe. Plenum Press, New York.

Turnbull, P.C.B., Jørgensen, K., Kramer, J.M., Gilbert, R.J. and Parry, J.M. (1979a) *J. clin. Path.* **32**, 289.

Turnbull, P.C.B., Kramer, J.M., Jørgensen, K., Gilbert, R.J. and Melling, J. (1979b) *Amer. J. clin. Nutr.* **32**, 219.

Turnbull, P.C.B. and Rose, P. (1982) *J. Hyg., Camb.* **88**, 29.

Turner, H.D., Brett, E.M., Gilbert, R.J., Ghosh, A.C. and Liebeschuetz, H.J. (1978) *Lancet* **i**, 1277.

Uemura, T. and Skjelkvåle, R. (1976) *Acta. path. microbiol. scand. B* **84**, 414.

Underwood, E.J., Beck, A.B. and Shier, F.L. (1939) *Aust. J. exp. Biol. med. Sci.* **17**, 183.

Vassiliadis, P., Kalapothaki, V., Trichopoulos, D., Mavrommatti, C. and Serie, C. (1981). *Appl. environm. Microbiol.* **42**, 615.

Vernon, E. (1965) *Mon. Bull. Minist. Hlth Lab. Serv.* **24,** 321; (1966) *Ibid.* **25,** 194; (1967) *Ibid.* **26,** 235; (1969) *Publ. Hlth, Lond.* **83,** 205; (1970) *Ibid.*, **84,** 239; (1977) *Ibid.*, **91,** 225.

Vernon, E. and Tillett, H.E. (1974) *Publ. Hlth Lond.* **88,** 225.

Wagenaar, R.O. and Dack, G.M. (1954) *Food Res.* **19,** 521.

Walker, A.B. (1929) *Brit. J. exp. Path.* **10,** 352.

Walker, P.D. and Batty, I. (1967) See *Report* 1967, p. 482.

Warnecke, M.O., Carpenter, J.A. and Saffle, R.L. (1967) *Food Tech., Champaign.* **21,** 115A.

Watson, G.N., Stringer, M.F., Gilbert, R.J. and Mahony, D.E. (1982) *J. clin. Path.* **35,** 1361.

Weiss, H. (1921) *J. infect. Dis.* **28,** 70.

Wentz, M.W., Scott, R.A. and Vennes, J.W. (1967) *Science* **155,** 89.

Wheeler, M.W. and Humphreys, E. (1924) *J. infect. Dis.* **35,** 305.

Wieneke, A.A. (1974) *J. Hyg., Camb.* **73,** 255.

Wilcke, B.W., Midura, T.F. and Arnon, S.S. (1980) *J. infect. Dis.* **141,** 419.

Wilkins, S.D. and Dutcher, R.A. (1920) *J. Amer. vet. med. Ass.* **57,** 653.

Williams, B.M., Richards, D.W. and Lewis, J. (1976) *Vet. Rec.* **98,** 51.

Williams, B.M., Richards, D.W., Stephens, D.P. and Griffiths, T. (1977) *Vet. Rec.* **100,** 450.

Williams, E.F. and Spencer, R. (1973) In: *The Microbiological Safety of Food*, p. 41. Ed by B.C. Hobbs. and J.H.B. Christian. Academic Press, London.

Williams, L.P. Jr and Helsdon, H.L. (1965) *J. Amer. med. Ass.* **192,** 347.

Williams, L.P. and Hobbs, B.C. (1975) In: *Disease Transmitted from Animals to Man*, 6th edn, p. 33. Ed. by W.T. Hubbert, W.F. McCulloch and P.R. Schnurrenberger. Thomas, Springfield, Illinois.

Williams, L.P. and Newell, K.W. (1970) *Amer. J. publ. Hlth* **60,** 926.

Williams, R.E.O. (1963) *Bact. Rev.* **27,** 56.

Wilson, M.M. and Mackenzie, E.F. (1955) *J. appl. Bact.* **18,** 510.

Winkle, S., Rohde, R. and Adam, W. (1960) *Zbl. Bakt.* **179,** 583.

Wit, J.C. de, and Kampelmacher, E.H. (1981) *Zbl. Bakt. I. Abt. Orig.* **B 172,** 390

World Health Organization (1974) *The Public Health Aspects of Antibiotics in Feedstuffs.* WHO, Copenhagen.

Wright, G.P. (1955) *Pharmacol. Rev.* **7,** 413.

Wynne, E.S. and Foster, J.W. (1948) *J. Bact.* **55,** 61, 69.

Yamagishi, T., Serikawa, T., Morita, R., Nakamura, S. and Nishida, S. (1976) *Jap. J. Microbiol.* **20,** 397.

Zacks, S.I. and Sheff, M.F. (1970) See *Report* 1970, p. 348.

73

Miscellaneous diseases

Granuloma venereum, soft chancre, cat-scratch fever, Legionnaires' disease, bartonella infections, and Lyme disease

Graham Wilson

Granuloma venereum

SYNONYM: Granuloma inguinale

This disease, which is not to be confused with lympho-granuloma venereum (see Chapter 76), is characterized by a slowly progressive ulceration of the tissues in the genital region. It is widespread in the tropics. Whether it is of venereal origin is in doubt (Goldberg 1959). Both sexes are attacked. The incubation period varies from a few days to 2 or 3 months. Spontaneous cure is uncommon, but the infection is susceptible to treatment with antimony.

Many years ago Donovan (1905) described the constant presence in smears from the ulcerated lesions of characteristic intracellular bodies, which he regarded as parasites, and which now usually bear his name. They are often contained in giant mononuclear cells. They resemble bacilli of the Friedländer group, are gram negative and are surrounded by a well defined capsule, which can be demonstrated by Wright's stain. Non-capsulated forms, however, are also present. The precise agent responsible for this disease remained obscure till Anderson, DeMonbreun and Goodpasture (1945) succeeded in cultivating it in the yolk sac of the developing chick embryo.

Dulaney, Guo and Packer (1948) were successful in growing it on a medium consisting of equal parts of Locke's solution and of egg yolk taken from 5- to 8-day chick embryos, sloped, and heated at 80° for 15 minutes; and Rake and Oskay (1948) on a modified

Levinthal agar. On this medium translucent shiny colonies appeared in 48 hours increasing in size up to 1.5 mm in diameter and gradually becoming greyish-brown in colour. Subculture was necessary every 7–12 days. Goldberg (1959) grew the organism in a synthetic medium, and showed that it required a peptide-like substance present in lactalbumin hydrolysate and that anaerobic or microaerophilic conditions were required.

In cultures the organism is moderately pleomorphic, capsulated and uncapsulated forms, long bipolar-stained bacilli, safety-pin forms, and short chains of irregularly stained rods being met with. In sealed test-tubes, cultures may remain viable for 4 or 5 weeks at 25°. Injection of cultures into a variety of animals, including monkeys, fails to set up progressive disease. Anderson, DeMonbreun and Goodpasture (1945) therefore concluded that the causative organism of granuloma venereum was a specific human bacterial parasite and suggested for it the name *Donovania granulomatis* (Fig. 73.1).

In a further paper, Anderson, Goodpasture and DeMonbreun (1945) described the preparation from chick-embryo yolk cultures of a washed bacterial suspension which gave rise, on intradermal injection into patients suffering from the disease, to a red oedematous reaction reaching its height in 24 hours and disappearing within another 2 days. They were also able to extract with weak alkali a mucoid material, probably of capsular origin, which gave specific precipitin

515

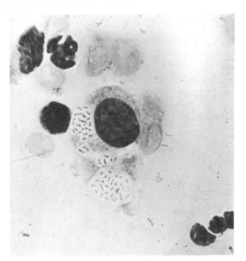

Fig. 73.1   Donovan bodies in a macrophage.

and complement-fixation reactions with the sera of granuloma patients. Rake (1948*a*, *b*) found that an antigen prepared from this organism fixed complement with sera from 87 per cent of clinical cases of granuloma inguinale and from 14 per cent of patients having varicose or decubitus ulcers. Much the same results were obtained with an antigen prepared from Friedländer's bacillus and from the scleroma bacillus. Rake is therefore of the opinion that *D. granulomatis* is merely a specially adapted form of Friedländer's bacillus. Goldberg (1962) isolated a bacterium resembling *D. granulomatis* from the faeces of a patient with granuloma venereum. Antigen prepared from this reacted with sera from 14 of 33 patients with the disease, and the organism was shown to possess common antigens with the prototype Anderson strain. He suggests (1964) that *Donovania* spp. are inhabitants of the gut. The organism is now regarded as belonging to the genus *Calymmatobacterium*, as suggested by Aragão and Vianna (1913), and is therefore designated *Cal. granulomatis*.

Streptomycin, aureomycin, and terramycin are all of value in treatment of the disease. (See Greenblatt 1953.)

Soft chancre or ulcus molle

Soft chancre is a non-syphilitic lesion of the external genitals and neighbouring regions due to infection with *Haemophilus ducreyi*, a small gram-negative bacillus first described by Ducrey (1889), and generally known by his name. For the differential diagnosis of this disease from syphilis by clinical methods, textbooks on venereal diseases must be consulted.

There appears to be little doubt that Ducrey's bacillus is responsible for soft chancre, and for the buboes which are sometimes associated with the pri-

mary lesion. Tomasczewski (1903) was successful in reproducing the disease in human subjects with pure cultures. Ulcerative lesions have likewise followed the inoculation of monkeys and rabbits with cultures several generations removed from primary isolation (Reenstierna 1921, Nicolle 1923). In the diagnosis of the disease, microscopical and cultural examinations should be made in the way already indicated in Chapter 39. Cultures are more often successful from the primary lesion than from associated buboes, and are best made from material obtained by biopsy (Heyman *et al.* 1945). Borchard and Hoke (1970) have described the use of the patient's own serum as a culture medium; they reported 21 isolations from 24 patients with lesions clinically suggestive of chancroid. Oberhofer and Bock (1982) recommend Mueller-Hinton sheep blood or chocolate agar supplemented by 1 per cent IsoVitalex, with incubation in 5–7 per cent CO_2 at 35°. Direct smears should be stained by Gram's and by Pappenheim's method, but the organisms are usually scanty. A skin test is often of value in diagnosis (Reenstierna 1923, Cole and Levin 1935). The antigen is prepared by washing off a blood agar culture with 0·5 per cent phenolized saline and keeping it in the ice-chest till it is sterile. For injection 0·2 ml is used. A positive reaction is characterized by the development of an area of induration 6 mm or more in diameter, reaching its maximum in about 48 hours. The reaction is positive in about two-thirds of the cases. An objection to it is that it may remain positive for years and cause confusion with the Frei test made for the diagnosis of lymphogranuloma inguinale. A positive reaction to one of these tests is of more diagnostic value than a reaction to both (Dulaney 1937, Greenblatt and Sanderson 1937).

The usual treatment is by sulphonamides or one of the broad-spectrum antibiotics. (See also Greenblatt 1953.) When treated early with streptomycin or sulphamethoxazole soft chancre may be expected to heal in 10 to 20 days (Morel *et al.* 1982).

Cat-scratch fever

Cat-scratch fever is a disease characterized by a suppurative subacute enlargement of one or more lymph nodes healing spontaneously in the end. The incubation period is about a week. A red papule may appear at the site of the scratch followed a day or two later by swelling and tenderness of the regional lymphatic nodes. Fever may or may not be present and is usually irregular. The lymphatic nodular inflammation may resolve, but more commonly it goes on to suppuration with discharge of pus through the skin and the formation of a fistula that may take weeks or months to heal up. Recovery is gradual. The disease is said to be widespread in the United States of America (Foshay 1952). Cases have been reported in Europe (Debré *et al.* 1950, Campbell and Wheaton 1952) and in Australia (Tonge *et al.* 1953). Other injuries besides those inflicted by a cat may be responsible. The disease has to be distinguished from tularaemia, rat-bite fever, infectious mononucleosis, pasteurella infection, tuberculous and mycotic lymphadenitis, Hodgkin's disease, and lymphosarcoma. The blood often shows a polymorphonuclear leucocytosis. A skin test made with pus from a bubo heated to 60° for an hour or two on 2 successive days gives rise to a strong reaction reaching its maximum in 24 hours and persisting sometimes for weeks or months. The disease appears to be caused by a pleomorphic gram-negative bacillus, 0.6–3.0 by 0.3–1.0 μm that can be seen, in sections stained by the Warthin-Starry silver impregnation method, in the walls of capillaries and in macrophages lining the sinuses in or near the germinal centres of lymph nodes (Wear *et al.* 1983). The organisms have not yet been cultivated. Mollaret and his colleagues (1950, 1951) described the successful transmission of the disease to *Cercopithecus* monkeys by intracutaneous inoculation. In both human and monkey material acidophilic granulo-corpuscles resembling those seen in psittacosis can be seen histologically in cells of the affected tissues. Patients affected with cat-scratch fever do not react to the Frei test, nor do lymphogranuloma patients react to the cat-scratch fever antigen. Aureomycin is said to be of value in treatment.

Epidemic epididymo-orchitis

Reference may be made here to this disease whose aetiology is completely obscure. An account is given by Tunbridge and Gavey (1946) of an outbreak occurring among members of the Services in Malta during the summer of 1943.

Erythema infectiosum (Fifth disease)

Epidemics of erythema infectiosum, sometimes known as *fifth disease*, have occurred in America and Central Europe, but seldom in Great Britain. An outbreak in Bristol in 1979 affected at least 100 persons aged 18 months to 39 years (Lovell *et al.* 1980). Clinically, the disease is characterized by an initial mild fever or sore throat followed by the 'slapped cheek' appearance. A reticulate migratory rash appears two to three days later starting on the upper arms or thighs, spreading to the trunk and lasting a mean of 11 days. There may be cervical lymphadenopathy. The disease lasts 1–2 weeks. As judged by the presence of specific IgM in the sera of patients, the disease appears to be due to a parvovirus (Anderson *et al.* 1983).

Kawasaki disease

This disease, called alternatively *mucocutaneous lymph node disease*, is very common in children in Japan where it is a major cause of heart disease, but is also met with in the United States and elsewhere. Clinically, it starts with acute fever, rash, conjunctival injection, strawberry tongue, and occasionally lymphadenopathy. Then comes a subacute stage marked by arthritic swelling of joints, and in some patients thrombocytosis and heart disease. During the convalescent stage in the fourth week there is peeling of the skin of the fingers and toes with the appearance of transverse grooves on the finger and toe nails. The disease does not appear to be contagious. Its cause is unknown; and there is no specific treatment (Editorial 1981). (For its original description, see Kawasaki *et al.* 1974.)

Legionnaires' disease

History

Interest in this disease was first aroused by an outbreak of a febrile respiratory illness that affected members attending an American Legion Convention at Philadelphia in the late summer of 1976 (see Fraser *et al.* 1977, Sanford 1979). Among 4500 persons present, 182 cases were recognized, of which 29 proved fatal. From four of these an unknown pleomorphic gram-negative bacillus was isolated; this was later characterized and named *Legionella pneumophila* (see Chapter 43). Retrospective clinical and serological scrutiny disclosed a number of sporadic cases and outbreaks that had occurred during the previous ten years or so. In 1965, for example, over 80 cases of the disease with 12 deaths occurred in a psychiatric hospital in Washington DC. Three years later there was an outbreak affecting the staff and persons visiting the County

Health Department at Pontiac, Michigan. This outbreak was unusual in being very mild and causing no deaths; patients suffered from myalgia but not from pneumonia. In 1973 a party of holiday makers returning to Glasgow from Spain suffered from the disease (Lawson *et al.* 1977). The following year witnessed an outbreak among a group of the Independent Order of Oddfellows attending a convention in Philadelphia (Terranova *et al.* 1978). It soon became apparent that Legionnaires' disease was widespread in the United States, Canada, the Mediterranean and other parts of Europe, affecting both healthy and immunologically deficient persons such as those with renal transplants or under treatment with cortico-steroids (Beaty *et al.* 1978, Bock *et al.* 1978). In England and Wales 202 cases, all of the pneumonic type, were reported in 1980.

Clinical picture

After an incubation period of 2–10 days the disease starts with malaise, fever, muscular pains, rigors, headache and mental confusion, followed by a non-productive cough and dyspnoea (Tsai *et al.* 1979). Pneumonic symptoms, accompanied sometimes by chest pains, dyspnoea, and bradycardia, increase, rendering the patient severely ill. Not infrequently renal insufficiency, diarrhoea, delirium and profound cerebral disturbances ensue. Sputum is not abundant but is extremely viscid and tenacious. Death, when it occurs, results after about seven days from respiratory failure. In patients admitted to hospital, the case-fatality rate is about 17 per cent. In non-fatal cases the fever falls after 7–11 days and recovery gradually occurs. The disease can range in severity from a mild respiratory illness without pneumonia (Pontiac fever) to one affecting not only the lungs but also the liver, intestine, kidney, and the central nervous system (see Blackmon *et al.* 1981). *X-ray examination* of the chest discloses patchy interstitial infiltration or areas of consolidation in one lobe, spreading to other lobes and the opposite lung, and persisting often for several weeks.

Pathological picture

At post-mortem examination pulmonary consolidation without any consistent segmental distribution is seen (Winn *et al.* 1979). As a rule both lungs are affected. The alveoli are filled with neutrophils, macrophages, fibrin, and protein material. The presence of coagulative necrosis of the lung parenchyma and oedema of the alveolar septa helps to distinguish the disease from lobar pneumonia; and the absence of interstitial inflammatory cellular infiltration from viral and rickettsial pneumonia. In sections of consolidated tissue stained by Dieterle's silver impregnation method gram-negative organisms can be seen. Immunofluorescent microscopy shows that antigenic material preponderates in the lumen of the terminal bronchioles, suggesting that infection has occurred by inhalation. Secondary retrograde invasion of the larger bronchioles follows; and the organisms are carried to other parts of the body by the lymphatic and blood vessels (Chandler *et al.* 1979, Hernandez *et al.* 1980, Hicklin *et al.* 1980).

Epidemiology

The disease is commonest in the summer. In the pneumonic form (Legionnaires' disease) men are affected more often than women, usually at 55 to 60 years of age. In the non-pneumonic form (Pontiac fever) the incubation period is only one to two days, the patients are younger—between 30 and 40 years of age—and in outbreaks the attack rate is much higher.

Where the organisms come from, and how infection is spread, are questions not yet fully answered. Evidence so far points to air-borne transmission and against case-to-case infection. An association with water has been demonstrated in more than one outbreak. The organisms can apparently survive in tap water for a year, and may be found in the water of cooling towers, in the evaporating pans of air conditioning plants (see Tobin *et al.* 1981), and in the various species of amoebae inhabiting the water of humidifiers (see Rowbotham 1980). It seems probable that dissemination of aerosols from such sources is responsible for human infection (Report 1979). The disease is not highly infectious. In those presumably exposed to infection the attack rate is only 2 per cent, and of these many are suffering from previous illness or are immunologically deficient in some way or other. Infection is probably far more widespread than is indicated by the prevalence of pneumonia. The occurrence of symptomless cases seems to be proved by the finding of antibodies in the serum of healthy persons, both young and old, including even children under four years of age (Foy *et al.* 1979). (For an illustrated description of cooling towers, see Kurtz *et al.* 1982.)

Diagnosis

The diagnosis of Legionnaires' disease can be made by demonstrating the causative organism in pathological specimens by either the Dieterle staining method or the direct fluorescent-antibody technique (see Beaty *et al.* 1978, Cherry *et al.* 1978). Alternatively *L. pneumophila* may be grown in culture from the blood (Edelstein *et al.* 1979, Greaves *et al.* 1979), from a transtracheal (Edelstein and Finegold 1979) or a bronchial aspirate (Fisher-Hoch *et al.* 1979), or from consolidated lung tissue at autopsy. The material should be inoculated on to Mueller-Hinton chocolate agar supplemented by 2 per cent IsoVitalex, or into the yolk sac of fertile eggs (see also Fallon 1982); alternatively it may be inoculated into guinea-pigs and cultured from the peritoneal fluid (Tobin *et al.* 1981). If the fluorescent-antibody test used for identifying the organism gives an equivocal result (see Edelstein *et al.* 1980), the answer may be provided by gas-liquid chromatography of the cellular fatty acids of the isolate (Fisher-Hoch *et al.* 1979). For routine purposes the indirect fluorescent-antibody technique is preferred on account of its simplicity, but is necessarily retrospective. A fourfold rise in titre is generally accepted as diagnostic. Antibody, however, is slow in developing, so that such a rise may not be reached for three weeks to two months. The specificity of the test depends on the antigen employed. The CDC etherized antigen was found to be less specific than a formolized yolk-sac antigen (Taylor *et al.* 1979). Preference should be given to a heat-killed antigen, so as to distinguish Legionnaires' disease from pneumonia due to

Mycoplasma pneumoniae (Taylor *et al.* 1980). Diagnosis of active disease cannot be made with certainty on a single specimen of serum because, as already pointed out, antibodies are not infrequently present in normal persons. However, a titre greater than 1/256 is strong presumptive evidence of activity. Various other serological methods are recommended by different workers, such as the microagglutination (Farshy *et al.* 1979), the microhaemagglutination (Edson *et al.* 1979), and the enzyme-linked immunosorbent assay for detecting antigen in the sputum or urine (Tilton 1979). See also Tang *et al.* (1984).

Treatment

Among the antibiotic agents erythromycin and rifampicin were the only common ones that were able to prevent death in infected guinea-pigs (Fraser *et al.* 1978). Observations on human patients likewise point to the value of erythromycin in treatment. This drug often has a remarkable therapeutic effect. In patients who do not respond to it rifampicin may be given in addition.

(For a review, see Broome and Fraser 1979, Sanford 1979, Blackmon *et al.* 1981, Lattimer and Ormsbee 1981; and for the proceedings of a Workshop, see Report 1983*a* and for 9 serogroups, see Report 1983*b*.)

Bartonella infections

These diseases are caused by minute micro-organisms intimately associated with the red blood corpuscles, and belonging either to the *Bartonella*, *Haemobartonella* or *Eperythrozoon* groups (see Chapter 43).

Oroya fever

Oroya fever and *Verruga peruana* are two stages of the same disease, caused by *Bartonella bacilliformis*. They are limited to the tropical zone of western South America, namely Peru, Colombia and Ecuador; and occur at an altitude of between 2500 and 9000 feet. Owing to the widespread epidemics that may occur and the high case fatality of the disease—somewhere about 40 per cent—Oroya fever is to be regarded as one of the most serious infections of South America. In the Colombian outbreak of 1938, 4000 deaths occurred among a total population of 100 000 (see Weinman 1944).

Oroya fever, sometimes known as Carrión's disease, is characterized by fever and a severe progressive anaemia, the red cell count sometimes falling within a few days to 1 million per mm^3. Examination of the blood shows the presence on the red corpuscles of small rod, dumb-bell, and coccoid bodies, varying considerably in number and staining red or reddish-purple with Giemsa. In severe cases 90 per cent of the red corpuscles may be affected. No haemolysin is detectable in cultures of the organism, there is no haemoglobinuria, and the method of haemolysis is unknown. The organisms multiply within the cytoplasm of the lining cells of the blood and lymph capillaries causing the cells to bulge into the lumen of the vessel. The incubation period of the disease is usually about 20 days. In fatal cases death occurs as a rule in 3 to 4 weeks. In patients that recover, convalescence is established after about 5 or 6 weeks, but is often succeeded in a month's time by an eruption of verruga. Second attacks of Oroya fever appear to be uncommon.

Verruga peruana

This is a disease characterized by the appearance on any part of the body surface of vivid red wart-like eruptions. The disease follows a short time after an attack of Oroya fever, and lasts as a rule for 4 to 6 months. Second attacks may occur, and sometimes a latent infection persists after the subsidence of the skin lesions. The case fatality is very low. There is generally a moderate degree of anaemia, but the parasites in the blood are too few to be seen microscopically, though they can often be revealed by culture. *Bartonella* can be readily cultivated from the local vascular granulomata, and in sections of excised tissue can be seen microscopically in the cytoplasm of the endothelial cells. Pure cultures, or juice from the nodules, inoculated intradermally above the eyebrow of monkeys give rise after an incubation period of 9–20 days to a local verruga papule. Experimental infection is followed by immunity. In the monkey the spleen appears to play no part in the defence mechanism of the host, such as it does in the rat infected with *Haemobartonella muris*.

The relation between Oroya fever and verruga was rendered evident by the unfortunate experience of Carrión in Peru who, wishing to define the symptoms in the pre-eruptive stage of the skin lesions, was inoculated with verruga material and died of Oroya fever 39 days later (Schultz 1968, Weinman 1968). In both diseases agglutinins may be found in the blood in low titre, 1/20–1/80 (Howe 1942).

Oroya fever and verruga are carried by the sandfly *Phlebotomus verrucarum* (Shannon 1929, Hertig 1937). The organisms are present in the mouth parts and intestinal tract of the insects but no cycle of development has been demonstrated. Prevention of the disease is readily accomplished by treating the inside and outside of the houses and the breeding places of the sandflies with DDT. In treatment of Oroya fever itself organic compounds and sulphonamides are useless, but penicillin, the tetracyclines and chloramphenicol have a dramatic effect. (For further description see Noguchi 1926, Kikuth 1931, 1934, Pittaluga 1938, Wigand 1958, and the comprehensive reviews of Weinman 1944, 1968.)

Infectious anaemia of rats

This is caused by *Haemobartonella muris* and is a disease that is precipitated by splenectomy. Except for certain breeds, such as the Wistar strain, a large proportion of adult rats, both wild and tame, appear to suffer from a latent infection with this organism. In such animals removal of the spleen is followed, usually in 4 or 5 days, by general illness, emaciation, and a severe progressive anaemia. Frequently haemoglobinuria develops and the animal dies, generally within 14 days. Recovery may, however, occur, but is liable to be followed by relapses at irregular intervals. Examination of the blood during the acute stage of the disease reveals a high proportion of red cells infected with *Haemobartonella* (Volume 2, Fig. 43.4). In animals that recover they disappear from the blood in 1–5 weeks. Sucking rats are comparatively resistant to infection, even after splenectomy; but if infected blood is injected into a normal adult rat at the time of splenectomy acute and fatal anaemia follows. Injection a fortnight before splenectomy leads to a much lower mortality, indicating the partial development of a latent immunity (Kessler 1943). Inoculation of an uninfected non-splenectomized animal produces only a mild disease. Anaemia can also be reproduced in young rabbits, guinea-pigs, and white mice by inoculation. Natural infection is transmitted by the rat louse, *Polyplax spinulosa*.

The part played by the spleen is decisive, though in what way it acts is still a matter of conjecture. A quarter of the spleen left *in situ* is sufficient to protect the animal against the disease (Perla and Marmorston-Gottesman 1930). Perla and Marmorston-Gottesman (1932) prepared an aqueous lipid extract of the spleen that neutralized the effect of splenectomy. The same workers found that, when rats were fed before splenectomy on a diet containing an adequate amount of iron and copper, anaemia failed to develop in a large proportion of them. Sandberg and Perla (1934) showed that splenectomy in non-infected albino rats was followed by an increased retention of iron but an increased elimination of copper. The suggestion is therefore that the spleen plays an important part in the utilization of copper, and that in animals on a normal diet the amount of this substance is insufficient to prevent the development of a severe anaemia when the spleen is removed. Faulkner and Habermann (1957), on the other hand, think that the controlling effect of the spleen is by virtue of its influence on the production of leucocytes in the bone marrow and their liberation into the blood stream. Certain trypanosome infections, poisons such as toluylenediamine, pyridine, and phenylhydrazine (Lauda and Marcus 1928), irradiation, and possibly some bacterial infections, may activate a latent haemobartonella infection even in the presence of an intact spleen.

In non-splenectomized bartonella-free rats complement-fixing antibodies appear in the blood about 5 days after infection, reach their maximum in 3–5 weeks, and persist for a considerable time. After treatment of the animals with neoarsphenamine the antibodies gradually decrease. There is also a temporary fall in infected animals after splenectomy and during relapses. They are of value for revealing the existence of latent infection (Wigand 1958).

Some degree of immunity develops under natural conditions, but it is of the nature of an infection immunity. Complete cure can however be brought about by treatment with organic compounds of arsenic and antimony. The sulphonamides and many of the antibiotics are without action, but the tetracyclines are effective. Administered before splenectomy, arsenic compounds prevent the development of anaemia. (References: Mayer 1921, Ford and Eliot 1928, Eliot and Ford 1929, Perla and Marmorston-Gottesman 1930, 1931, 1932, Marmorston-Gottesman and Perla 1930, 1931, 1932*a*, *b*, Lwoff and Vaucel 1931, Roth 1932, Kikuth 1934, McCluskie and Niven 1934, Wigand 1958).

Infectious anaemia of dogs

This is caused by *Haemobartonella canis*. The organisms were first observed by Kikuth (1929) in the blood of a splenectomized dog. The course of the disease resembles that in rats. The parasites multiply up to a certain point, then suddenly disappear from the peripheral blood, only to reappear after a few days and again work up to a maximum. Remissions of this type may occur for months. The length of the relapses and of the remissions varies from dog to dog (Regendanz and Reichenow 1932). The natural disease seldom occurs in healthy dogs; the anaemia produced is mild, death is infrequent, and spontaneous recovery leaves behind it a substantial degree of immunity. Infection appears to be transmitted from dog to dog by fleas. Neosalvarsan exercises a specific effect on the course of the disease. At the height of the illness *Haemob. canis* is present in large numbers. Its extracellular position may account for the fact that the parasites often disappear from the blood without any serious diminution occurring in the number of red blood corpuscles. The organisms are very pleomorphic, and may become agglutinated into characteristic masses or chains at the periphery of the cells.

Infectious anaemia of cats

This is caused by *Haemobartonella felis*. The disease is characterized by depression, weakness, anorexia, loss of weight, mucosal pallor, and cachexia. The organisms appear on the red blood corpuscles as ring and coccoid forms and sometimes rods. Splenectomy has little influence on the course of infection, and the effect of treatment is doubtful. By itself the organism causes no obvious illness, and latent infections are not uncommon (Flint *et al.* 1958, Seamer and Douglas 1959, Kreier and Ristic 1968).

Pigs are among numerous other animals that suffer from infectious anaemia; the organism that infects them is known as *Haemob. tyzzeri*.

Lyme disease

This disease was first recognized in Connecticut in 1975 (Steere *et al.* 1977), but is now known to occur in other parts of the States, in Europe and in Australia. It begins with a peculiar skin lesion—erythema chronicum migrans—accompanied by fever, chills, headache, stiff neck and myalgia, followed by arthritis of one or more joints. The skin lesion appears at the site of a tick bite (*Ixodes dammini*) and is caused by a spirochaete 4–30 μm long and 0.2 μm in diameter, having 4–8 flagella and a slime layer around the outer membrane and cell wall. Though not easy to cultivate, the spirochaete has been isolated from ticks and from the skin lesions, blood and cerebrospinal fluid of patients (Benach *et al.* 1983, Steere *et al.* 1983). IgM antibody reaches a peak in 3–6 weeks after onset of the disease; IgG antibody rises more slowly, not reaching its maximum concentration for some months. Attacks of arthritis generally recur over a period of some years. The disease responds to penicillin and tetracycline. Ryberg and his colleagues in Sweden (1983) reported on 11 patients suffering from lymphocytic meningoradiculitis. Six of 9 patients tested had antibodies to the Lyme disease spirochaete in their serum. They conclude, therefore, that this disease is probably a tick-borne infection of spirochaetal origin.

References

Anderson, K., DeMonbreun, W. A. and Goodpasture, E. W. (1945) *J. exp. Med.* **81**, 25.

Anderson, K., Goodpasture, E. W. and DeMonbreun, W. A. (1945) *J. exp. Med.* **81**, 41.

Anderson, M. J. *et al.* (1983) *Commun. Dis. Rep., Lond.* 83/23.

Aragão, H. de B. and Vianna, G. (1913) *Mem. Inst. Oswaldo Cruz* **5**, 211.

Beaty, H. N., Miller, A. A., Broome, C. V., Goings, S. and Phillips, C. A. (1978) *J. Amer. med. Ass.* **240**, 127.

Benach, J. L. *et al.* (1983) *New Engl. J. Med.* **308**, 740.

Blackmon, J. A. *et al.* (1981) *Amer. J. Path.* **103**, 427.

Bock, B. V. *et al.* (1978) *Lancet* **i**, 410.

Borchard, K. A. and Hoke, A. W. (1970) *Arch. Derm.* **102**, 188.

Broome, C. V. and Fraser, D. W. (1979) *Epidem. Reviews* **1**, 1.

Campbell, E. S. K. and Wheaton, C. E. W. (1952) *Lancet* **i**, 975.

Chandler, F. W. *et al.* (1979) *Amer. J. clin. Path.* **71**, 43.

Cherry, W. B. *et al.* (1978) *J. clin. Microbiol.* **8**, 329.

Cole, H. N. and Levin, E. A. (1935) *J. Amer. med. Ass.* **105**, 2040.

Debré, R., Lamy, M., Jammet, M-L., Costil, L. and Mozziconacci, P. (1950) *Bull. Soc. méd. Hôp. Paris* **66**, 76.

Donovan, C. (1905) *Indian med. Gaz.* **40**, 411.

Ducrey, A. (1889) *Mschr. prakt. Derm.* **9**, 387.

Dulaney, A. D. (1937) *Amer. J. Syph.* **21**, 667.

Dulaney, A. D., Guo, K. and Packer, H. (1948) *J. Immunol.* **59**, 335.

Edelstein, P. H. and Finegold, S. M. (1979) *J. clin. Microbiol.* **9**, 457.

Edelstein, P. H., Meyer, R. D. and Finegold, S. M. (1979) *Lancet* **i**, 750.

Edelstein, P. H. *et al.* (1980) *J. infect. Dis.* **141**, 652.

Editorial (1981) *J. Amer. med. Ass.* **246**, 819.

Edson, D. C. *et al.* (1979) See *Symposium* p. 691.

Eliot, C. P. and Ford, W. W. (1929) *Amer. J. Hyg.* **10**, 635.

Fallon, R. J. (1982) *Quart. Soc. gen. Microbiol.* **9**, Part 2, p. 31.

Farshy, C. E. *et al.* (1979) See *Symposium* p. 690.

Faulkner, R. R. and Habermann, R. T. (1957) *J. infect. Dis.* **101**, 62.

Fisher-Hoch, S., Hudson, M. J. and Thompson, M. H. (1979) *Lancet* **ii**, 323.

Flint, J. C., Roepke, M. H. and Jensen, R. (1958) *Amer. J. vet. Res.* **19**, 164.

Ford, W. W. and Eliot, C. P. (1928) *J. exp. Med.* **48**, 475.

Foshay, L. (1952) *Lancet* **i**, 673.

Foy, H. M. *et al.* (1979) *Lancet* **i**, 767

Fraser, D. W. *et al.* (1977) *New Engl. J. Med.* **299**, 1189; (1978) *Lancet* **i**, 175.

Goldberg, J. (1959) *Brit. J. vener. Dis.* **35**, 266; (1962) *Ibid.* **38**, 99; (1964) *Ibid.* **40**, 110.

Greaves, F. G., Sharp, G. and Macrae, A. D. (1979) *Lancet* **i**, 551.

Greenblatt, R. B. (1953) *Management of Chancroid, Granuloma Inguinale and Lymphogranuloma Venereum.* U.S. Public Health Service, Publication No. 255.

Greenblatt, R. B. and Sanderson, E. S. (1937) *Arch. Derm. Syph., N.Y.* **36**, 486.

Hernandez, F. J., Kirby, B. D. and Stanley, T. M. (1980) *Amer. J. clin. Path.* **73**, 488.

Hertig, M. (1937) *Proc. Soc. exp. Biol., N.Y.* **37**, 598.

Heyman, A., Beeson, P. B. and Sheldon, W. H. (1945) *J. Amer. med. Ass.* **129**, 935.

Hicklin, M. D., Thomason, B. M., Chandler, F. W. and Blackmon, J. A. (1980) *Amer. J. clin. Path.* **73**, 480.

Howe, C. (1942) *J. exp. Med.* **75**, 65.

Kawasaki, T., Kosaki, F. and Ogawa, S. *et al.* (1974) *Pediatrics*, Springfield **54**, 271.

Kessler, W. S. (1943) *J. infect. Dis.* **73**, 65, 77.

Kikuth, W. (1929) *Zbl. Bakt.* **113**, 1; (1931) *Z. ImmunForsch.* **73**, 1; (1934) *Proc. R. Soc. Med.* **27**, 1241.

Kreier, J. P. and Ristic, M. (1968) *Infectious Blood Diseases of Man and Animals*, Vol. ii, p. 387. Ed. by D. Weinman and M. Ristic. Academic Press, New York and London.

Kurtz, J. B., Bartlett, C. L. R., Newton, U. A., White, R. A. and Jones, N. L. (1982) *J. Hyg.. Camb.* **88,** 369.

Lattimer, G. L. and Ormsbee, R. A. (1981) Legionnaires' Disease, *Infect. Dis. antimicrob. Agents Series,* **1,** Oct. Marcel Dekker A.G.

Lauda, E. and Marcus, F. (1928) *Zbl. Bakt.* **107,** 104.

Lawson, J. H., Grist, N. R., Reid, D. and Wilson, T. S. (1977) *Lancet* **ii,** 1083.

Lovell, C. R. *et al.* (1980) *Brit. J. Dermatol.* **103,** Suppl. 18, p. 24.

Lwoff, A. and Vaucel, M. (1931) *Ann. Inst. Pasteur.* **46,** 258.

McCluskie, J. A. W. and Niven, J. S. F. (1934) *J. Path. Bact.* **39,** 185.

Marmorston-Gottesman, J. and Perla, D. (1930) *J. exp. Med.* **52,** 121; (1931) *Ibid.* **53,** 877; (1932*a*) *Ibid.* **56,** 763; (1932*b*) *Proc. Soc. exp. Biol., N.Y.* **29,** 989.

Mayer, M. (1921) *Arch. Schiffs- u. Tropenhyg.* **25,** 150.

Mollaret, P., Reilly, J., Bastin, R. and Tournier, P. (1950) *Bull. Soc. Méd., Paris* **66,** 424; (1951) *Presse méd.* **59,** 681, 701.

Morel, P., Casin, I., Gandiol, C., Vallet, C. and Civalte, J. (1982) *Nouv. Pr. méd.* **11,** 655.

Nicolle, C. (1923) *C. R. Soc. Biol.* **88,** 871.

Noguchi, H. (1926) *J. exp. Med.* **44,** 533, 697, 715, 729.

Oberhofer, T. R. and Bock, A. E. (1982) *J. clin. Microbiol.* **15,** 625.

Perla, D. and Marmorston-Gottesman, J. (1930) *J. exp. Med.* **52,** 131; (1931) *Ibid.* **53,** 869; (1932) *Ibid.* **56,** 777, 783.

Pittaluga, G. (1938) *Bull. Inst. Pasteur* **36,** 961.

Rake, G. (1948*a*) *Amer. J. Syph.* **32,** 150; (1948*b*) *J. Bact.* **55,** 865.

Rake, G. and Oskay, J. J. (1948) *J. Bact.* **55,** 667.

Reenstierna, J. (1921) *Acta derm.-venereol., Stockh.* **2,** 1; (1923) *Arch. Inst. Pasteur, Tunis* **12,** 273.

Regendanz, P. and Reichenow, E. (1932) *Arch. Schiffs- u. Tropenhyg.* **36,** 305.

Report (1979) *Lancet* **i,** 991; (1983*a*) *Zbl. Bakt.,* I. Abt. Orig. A **255,** 1–155; (1983*b*) *Commun. Dis. Rep., Lond.* 83/49.

Roth, H. (1932) *Z. ImmunForsch.* **74,** 483.

Rowbotham, T. J. (1980) *J. clin. Path.* **33,** 1179.

Ryberg, B., Nillson, B., Burgdorffer, W. and Barbour, A. A. G. (1983) *Lancet,* **ii,** 519.

Sandberg, M. and Perla, D. (1934) *J. exp. Med.* **60,** 395.

Sanford, J. P. (1979) *New Engl. J. Med.* **300,** 654.

Schultz, M. G. (1968) *New. Engl. J. Med.* **278,** 1323.

Seamer, J. and Douglas, S. W. (1959) *Vet. Rec.* **71,** 405.

Shannon, R. C. (1929) *Amer. J. Hyg.* **10,** 78.

Steere, A. C. *et al.* (1977) *Arthritis Rheum.* **20,** 7; (1983) *New Engl. J. Med.* **308,** 733.

Symposium (1979) In: *Ann. intern. Med.* **90,** 491–703.

Tang, P. W. *et al.* (1984) *J. clin. Microbiol.* **19,** 30.

Taylor, A. G., Harrison, T. G., Andrews, B. E. and Sillis, M. (1980) *Lancet* **i,** 764.

Taylor, A. G., Harrison, T. G., Dighero, M. W. and Bradstreet, C. M. P. (1979) See *Symposium* p. 686.

Terranova, W., Cohen, M. L. and Fraser, D. W. (1978) *Lancet* **ii,** 122.

Tilton, R. C. (1979) See *Symposium* p. 697.

Tobin, J. O'H., Swann, R. A. and Bartlett, C. L. R. (1981) *Brit. med. J.* **i,** 515.

Tomaszewski, E. (1903) *Z. Hyg. InfektKr.* **42,** 327.

Tonge, J. I., Inglis, J. A. and Derrick, E. H. (1953) *Med. J. Aust.* **ii,** 81.

Tsai, T. F. *et al.* (1979) *Ann. intern. Med.* **90,** 509.

Tunbridge, R. E. and Gavey, C. J. (1946) *Lancet* **i,** 775.

Wear, D. J. *et al.* (1983) *Science,* **221,** 1403.

Weinman, D. (1944) Infectious Anaemias due to Bartonella and Related Red Cell Parasites. *Trans. Amer. philosoph. Soc.* **33,** Pt. III, 243; (1968) *Infectious Blood Diseases of Man and Animals,* Vol. ii, p. 3. Ed. by D. Weinman and M. Ristic. Academic Press, New York and London.

Wigand, R. (1958) *Morphologische biologische und serologische Eigenschaften der Bartonellen.* Georg Thieme Verlag, Stuttgart.

Winn, W. C. *et al.* (1979) See *Symposium* p. 548.

Spirochaetal and leptospiral diseases

Graham Wilson and Joyce Coghlan

Spirochaetosis

Relapsing fever

SYNONYMS: Famine fever; tick fever.

Relapsing fever is the name given to a disease the chief characteristic of which is the occurrence of one or more relapses after the subsidence of the primary febrile paroxysm. There are two main forms differing in so many respects that we shall consider them separately. One, the epidemic form, is transmitted to man by lice, the other, the endemic form, by ticks.

Louse-borne relapsing fever

Clinically, after an incubation period of 2–12, usually 5–8, days, the disease sets in abruptly with shivering, headache, body pains, sometimes nausea or vomiting, and fever reaching a temperature of 38·9–40°. The spleen and liver are enlarged and tender. Jaundice occurs in 20–60 per cent and bronchitis in 40–60 per cent of those attacked. In favourable cases the temperature falls by crisis in 3–9 days. A relapse follows 11–15 days later, less severe and shorter than the primary attack. Not all cases relapse—in some epidemics not more than half. A second relapse occurs in 10–40 per cent of cases, a third or fourth in only 1–2 per cent (Megaw 1955). Pregnant women abort. Under good conditions the case-fatality rate is low—5 per cent or so—but in times of war, famine and other forms of distress it may reach 60–70 per cent. Death is due to hepatic damage, lobar pneumonia, subarachnoid haemorrhage, or splenic rupture. As seen at autopsy, the organs chiefly affected are the liver, brain, spleen, and lungs (Salih *et al.* 1977). Post-mortem examination reveals splenomegaly, hepatitis, jaundice, and petechial haemorrhages in the viscera.

Epidemiology Louse-borne relapsing fever is said to be the most epidemic of the epidemic diseases. Though at one time it was widespread in Europe, Asia, and Africa, it is becoming less and less common and now seldom spreads across national boundaries (Report 1969). Until the discovery of the causative organism, *Borrelia recurrentis*, by Obermeier in 1873, relapsing fever was often confused with typhus fever. During the past hundred years, however, the epidemic history of the disease is fairly well documented. Devastating outbreaks, along with typhus fever, occurred in Ireland during the famine years of 1846–50 (Mac-Arthur 1957). Infection was introduced into the United States by Irish immigrants, leading to outbreaks in Philadelphia in 1844 and 1869 and in New York in 1847 and 1871 (Moursund 1942). During the first world war great epidemics occurred in Poland and Rumania, and were followed during the years 1919–23 by widespread outbreaks in Russia and Central Europe. Relapsing fever of great violence broke out in North Equatorial Africa in 1921 and swept across the continent causing hundreds of thousands of deaths. It went on to attack the Sudan, where at least 200 000 deaths occurred. During the second world war a serious epidemic, beginning in North Africa in 1943, spread to the Eastern Mediterranean and Europe. Since 1964 Ethiopia is the only country that has continuously reported a large number of cases (see Sparrow 1958, Report 1969; and for further information on distribution see Simmons *et al.* 1954, Report 1956, Bryceson *et al.* 1970).

Credit for recognizing the louse to be the vector of relapsing fever must go to Mackie (1907) in India, though it was Charles Nicolle and his associates (1912, 1913) in Tunis who worked out the epidemiology of the disease (see also numerous further papers in *C.R. Acad. Sci.* and *Arch. Inst. Pasteur Tunis*) between 1913 and 1932; for exact references see Felsenfeld (1971). Only the body and the head louse are responsible for transmission, not the pubic louse. Mackie (1920) likewise pointed out that transmission from person to person might occur by access of freshly shed infected blood to abrasions on the skin or to the conjunctiva.

Like typhus fever, relapsing fever is associated with poor sanitation and personal hygiene, particularly overcrowding, under-nutrition, and dirty, infested clothing.

Causative organism The causative organism of the disease was discovered by Obermeier in the Berlin epidemic of 1867–68, though his observations were not published till 1873. He noticed the presence of thread-like bodies during the febrile stage, their absence during the intermission, and their reappearance during the relapse. Because of their motility and wave-like form he regarded them as belonging to the group of spirochaetes. In the blood they were seen as straight forms, S-forms, or even as circles; towards the end of the paroxysm he noticed their tendency to aggregation in rosettes and their granular disintegration. The organisms occurred only in relapsing fever patients; they were never found in the blood of normal persons nor of patients suffering from other fevers. These observations were among the first to establish the microbial theory of infection, and are therefore of special historical interest.

The organism described by Obermeier is now known as *Borrelia recurrentis* (see Chapter 44). As seen by dark-ground illumination in the blood of patients during a febrile paroxysm, it is a motile spiral organism with a series of five to ten fairly regular but loose primary waves. During rest its axis is generally straight, but when active it momentarily assumes various curved and bizarre forms. In a non-viscous medium it is extremely motile, darting rapidly across the field; but in wet blood films movement is slower, consisting of a to-and-fro passage over a distance not more than two or three times its own length (Novy and Knapp 1906). Rotation occurs around the long axis. During the first febrile paroxysm spirochaetes in the blood are numerous—several to a field—but as the fever declines their numbers diminish, they become less motile, and they may assume irregular shapes or accumulate in rosettes (see Fig. 44.8); these changes are interpreted as indicative of lysis or agglutination resulting from the appearance of circulating antibodies in the host. After the subsidence of the fever they can no longer be found microscopically, though inoculation of the blood into a susceptible animal may show that a few are still present. At the onset of the relapse they are again seen in the blood, though not in such large numbers as in the first attack. The fever and many of the symptoms are ascribed by Bryceson and his colleagues (for references see Sanford 1976) to release of a potent endotoxin. On the other hand, Butler and his colleagues (1979) bring evidence of its being caused by a heat-stable, non-endotoxic, particulate pyrogen.

The spirochaetes are best demonstrated by dark-ground illumination or by fluorescent microscopy; but they can be stained with methylene blue, or preferably Leishman or Giemsa, with both of which they are coloured blue. Films should be wet-fixed so as to prevent the distortion that occurs when the organisms are dried in air. The organisms are difficult to culture in either the test-tube or the fertile egg, and are best conserved by animal inoculation. Passage from animal to animal should be made during the spirochaetaemic phase.

Spirochaetes may be demonstrated in the stomach of the louse for about 24 hours after an infective meal. They can then no longer be found; but after 6 or 8 days they reappear in the fluids of the body cavity, where they multiply and whence they spread to other parts of the body including the legs and antennae but not the salivary glands, the gut, or the genital organs. Once the louse is infected, it remains so for the rest of its life—usually about 3 weeks. The disease is transmitted to man not by the bite of the louse, which appears to be harmless, but by contamination of the wounds, made by the bite itself or by scratching, with the body fluids of the louse. Occasionally, as in the laboratory, the disease results from contamination of the nasal or conjunctival mucosa, or of a skin abrasion, with the blood of an infected animal. In contrast to ticks, lice are not susceptible to hereditary transmission of infection (see Wolman and Wolman 1945).

B. recurrentis is less virulent than the tick-borne strains for rodents, though, as Baltazard, Bahmanyar and Mofidi (1948) have shown, it can readily infect newborn animals. (For reproduction of the disease in experimental animals, see p. 526.) In contrast to tick-borne relapsing fever, the louse-borne disease has only one known reservoir, namely man (Felsenfeld 1971).

Tick-borne relapsing fever

This form of relapsing fever differs from the louse-borne disease in a number of ways. Clinically, after an incubation period, varying with the locality, of 2–14 days, there are usually three or more spells of fever, each lasting not more than 1–4 days and occurring at shorter but irregular intervals, often 4–5 days; the temperature may be higher; nervous symptoms are commoner and include meningitis, spastic paraplegia, hemiparesis, facial palsy, aphasia, deafness, and optic neuritis; spirochaetes are far fewer in the blood; and the response to arsenic is not as good. Epidemiologically, the tick-borne disease is seen mainly in Asia, Africa, and Central and South America. It is essentially a 'place' disease, occurring in certain localities, and not causing the disastrous widespread epidemics that are so characteristic of the louse-borne form. Again, whereas in relapsing fever carried by the louse man is the exclusive mammalian host, the tick-borne disease, with the exception of East African fever, is maintained by a variety of rodents. Man may be infected not only from his fellows, but also from lower animals through the intermediation of ticks.

Credit for the discovery of the tick-borne origin of relapsing fever is due to Dutton and Todd (1905), who showed that the tick fever of the Congo Free State was spread by *Ornithodorus moubata*. Infected ticks were able to transmit the disease to monkeys; and in one experiment young ticks, newly hatched in the laboratory from eggs laid by infected parents, successfully conveyed the disease to monkeys, showing that infection could be spread hereditarily. (See also Davis 1943.) In Uganda, Ross and Milne (1904) brought evidence to show that the disease was spread by the bite of a tick, *O. savigny*. Besides *Borrelia duttoni*, the organism responsible for East African fever, several other tick-borne species or varieties of spirochaete have been described, such as *B. persica* of Asia, *B. hispanica* of Spain and Spanish Africa, *B. crocidurae* of Senegal, and *B. turicatae*, *B. parkeri*, and *B. hermsi* of North America. The organisms themselves, which are distinguished by differences in antigenic structure, in pathogenicity for experimental animals, in specificity for their intermediate host, and in cross-immunizing ability, are briefly referred to in Chapter 44. Some of them, such as *B. crocidurae* isolated from shrews and *B. microti* isolated from field mice, are very closely related and are carried by the same tick, *O. erraticus*. As a rule there is a specific relation between the organism and the vector tick, such as exists between *B. hermsi* and *O. hermsi*, *B. turicatae* and *O. turicatae*, *B. parkeri* and *O. parkeri*; but sometimes a given organism may be carried by more than one tick, as, for example, *B. microti* by *O. erraticus*, *O. lahorensis*, and *O. canestrini*; and sometimes a given tick may carry more than one species of organism, as, for example, *O. erraticus* which carries *B. crocidurae*, *B. microti* and *B. dipodilli*. *O. tholozani* is the common vector to man in Central Asia, *O. erraticus* in Spain and Africa, and *O. rudis* in Central and South America (Davis 1948). In spite of their host specificity, all species of tick-borne borrelias, with the exception of *B. persica*, have been shown experimentally to be capable of development in human lice (Mooser 1958). The severity of the disease in man varies with different species of *Borrelia*, and so does the frequency of transovarian passage of infection through the tick. In some ticks, such as *O. turicata*, ovarian transmission is the rule, whereas in others, such as *O. hermsi*, it is the exception. Working with *O. hermsi*, Wheeler (1942) found that, after ingestion of infected blood by the tick, spirochaetes invaded the coelomic cavity within 3 days, and by the tenth day were demonstrable in the muscles. They also invade the coxal and salivary glands and the genital organs, including the gonads, thus making possible transovarian transmission. Ticks remain infected throughout their life—sometimes as long as 5 years (Francis 1938).

The exact mode by which the tick conveys infection to man is in some doubt, and may perhaps vary with different species of tick. The three possible methods are by (*a*) the secretion of the salivary glands, (*b*) the coxal fluid, and (*c*) the material exuded from the anus, which may be regarded as faeces or, as with *O. moubata*, the secretion of the malpighian tubules (Davis 1948). It is generally assumed that the infected fluid exuded by the coxal glands contaminates the wound made by the bite of the tick. In some ticks, however, such as *O. hermsi* and *O. turicatae*, the coxal fluid is not infective (Herms and Wheeler 1936, Francis 1938), and infection probably results from the bite itself, particularly from bites of the larvae. The mammalian reservoir of infection is constituted by a variety of animals, especially rodents, but including pigs, porcupines, opossums, and armadillos (see Sautet 1937). *B. duttoni*, as already indicated, is the only tick-borne spirochaete whose exclusive mammalian host is man (see Mooser 1958).

In the following sections the louse-borne and the tick-borne forms of the disease will be considered together.

Reproduction of relapsing fever in animals

Infection can be transmitted to monkeys, rats, mice and, with some species of spirochaete, to guinea-pigs; adult rabbits are usually refractory. Newborn animals are much more susceptible than adults, and may die of infection (Baltazard *et al.* 1948). *B. recurrentis* is less virulent than the tick-borne strains for rodents, and can seldom infect them without previous passage through a monkey. The disease in **monkeys** runs much the same course as in man; 2 or 3 days after subcutaneous inoculation with the patient's blood a pyrexial attack occurs lasting for 3 or 4 days; two, three, or four relapses may follow at intervals of 2-8 days, each relapse lasting 1-4 days (Norris *et al.* 1906). After intraperitoneal inoculation of **mice** spirochaetes appear in the blood within 24 hours and persist for 3 or 4 days; they then disappear for several days, after which a relapse may occur; three or four relapses may follow each other, separated by an interval of about 7 days (Novy and Knapp 1906). As many as 10-50 organisms may be present per microscope field during the first infection, but not more than 1 or 2 as a rule in the relapses. Intraperitoneal inoculation of **white rats** is followed by the appearance of spirochaetes in the blood in about 40 hours; they disappear in 2 days or so. Infection is never fatal. Novy and Knapp (1906) state that rats do not have relapses, but Bohls and Irons (1942) deny this. The discrepancy in the two statements may well be due to differences in virulence of the spirochaetes used. **Guinea-pigs** are susceptible to some tick-borne strains, such as *B. persica* and *B. hispanica*, but not to louse-borne strains (Delpy and Rafyi 1939, Coghill and Gambles 1948). **Chick embryos**, but not hatched chickens, are susceptible to *B. duttoni* (Oag 1940). Animals infected with the tick-borne borrelias may remain infected for a long time,

the spirochaetes settling in the brain. Residual infection of the brain is not seen with *B. recurrentis* (Mooser 1958).

Immunity

The disappearance of the spirochaetes from the blood stream at the end of the first pyrexial attack is regarded by many workers as due to the development of antibodies. This view was to some extent substantiated by the work of Novy and Knapp (1906).

Working with the rat, they found that the organisms would live in defibrinated blood kept at room temperature for 30 or 40 days, provided the blood was removed from the animal during the early stage of infection; if it was taken during the later stages, they died in 24 hours. When the organisms were examined *in vivo* during the decline phase of the infection they were observed to become sluggish, to show end-to-end agglutination, and even to form small tangles. This was even more evident when blood from an immune animal was inoculated into an infected rat; half an hour later the spirochaetes were accumulated in tangled masses of 10 to 20 members, and showed end-to-end agglutination; 1 hour later they were agglutinated into perfect radiating rosettes, and 2 hours later they were very scarce and were mostly immobile. Blood serum from a rat, which had been hyperimmunized by a course of 26 injections of infective blood, had strong immobilizing and agglutinative properties, even when diluted to 1/100. In rats that had recovered naturally from infection Pfeiffer's phenomenon could be produced *in vivo*. Intraperitoneal injection of infective blood into such animals was followed by agglutination and granular degeneration of the spirochaetes; in 10 minutes no free spirochaetes could be found. The altered organisms were rapidly ingested by phagocytes. In hyperimmunized rats the spirochaetes completely disappeared from the peritoneal cavity in 2 minutes. In passively immunized rats the spirochaetes were agglutinated into rosettes, but later these broke up, and free organisms once more became numerous; the animals however, did not contract infection. The course of antibody production in human beings seems to resemble that in experimental animals (see Cunningham and Fraser 1935).

From these and other experiments, it would appear that during the course of the natural disease antibodies develop—chiefly agglutinins, spirochaeticidins and lysins—which are sufficiently powerful to overcome the blood infection and lead to the disappearance of the spirochaetes from the circulation. These organisms remain latent in the brain and other tissues, and when the circulating antibodies have decreased, they once more enter the blood and give rise to a relapse. This stimulates the production of fresh antibodies, which again lead to the disappearance of spirochaetes from the blood. After one or more relapses the active immunity produced by the host is sufficient to prevent further invasion of the blood by the spirochaetes, and an apparent cure results. Whether a true cure results, in the sense that the body is completely rid of spirochaetes, is doubtful. Animal experiments suggest rather that, even though the organisms give no token of their presence, they may yet remain alive in the tissues for weeks or months. Immunity in relapsing fever appears to be an infection-immunity; this corresponds to a state of the host in which, together with a humoral immunity, there is a latent infection of the tissues, which is capable under certain conditions of breaking down the existing immunity (Heronimus 1928). It is a state in which a working equilibrium is established between host and parasite, and like other equilibria is liable to disturbance. (For further discussion, see Felsenfeld 1976.)

There is a considerable amount of evidence to show that after the first attack the spirochaetes in the tissues undergo an antigenic change which renders them insusceptible to the antibodies produced by the host against the original strain. This enables them to invade the blood a second time and give rise to a relapse. The production of antibodies to the so-called 'relapse' or 'serum-fast' strain is followed by the disappearance of the organisms from the blood once more. A further antigenic change may occur in the tissues enabling the organisms to invade the blood for a third time. As Schuhardt (1942) points out, in-vitro serological proofs of antigenic variation are more reliable than those depending on cross-resistance tests *in vivo*, unless precautions are taken to prevent relapses, and therefore further possible variation in the immunized test animal. Up to nine distinguishable antigenic phases have been reported in a single tick-borne strain (Cunningham *et al.* 1934). Tick-borne strains appear to have greater potentialities for variation than louseborne strains. Successive episodes of variation with relapse occur after the inoculation of a single spirochaete (Schuhardt and Wilkerson 1951). Relapses in which no obvious antigenic variation occurs are also reported. In some instances it appears that the relapse strain reverts to the parent type and, when the circulating antibodies generated during the first attack have diminished sufficiently, another invasion of the blood may occur (Aristowsky and Wainstein 1929*a*).

The antigenic specificity of the relapsing fever spirochaetes is highly developed. This accounts not only for the fact that patients who have been infected with one type of relapsing fever, for example the European, can be infected with another type, such as the Indian or African, but that, as just noted, a patient who has recovered from invasion with a given strain may suffer from a relapse due to an antigenic variant of the same strain. There is nevertheless serological and immunological evidence of a broad antigenic similarity among strains of diverse origin and character (see Stein 1944, Chen *et al.* 1945).

Diagnosis

Blood should be taken during the pyrexial period and examined either by dark-ground illumination or after staining with Wright, Giemsa or gentian violet (see Beck 1936, Bohls and Irons 1942). If the staining technique is adopted, it is usually wise to make a thick film preparation by stirring 2 or 3 drops of blood on one slide with the corner of another slide till they cover an area about 1 cm in diameter. Unless they are stained at once, they should be laked in distilled water or 1 per cent acetic acid before staining. Thin films to be stained by Giemsa or gentian violet should be previously fixed for 3 minutes in methyl alcohol. Microscopically, the spirochaetes are usually seen in large numbers in blood taken during the febrile period, but later they may be scarce and are often coiled up or clumped together. Cultivation is not a reliable means of diagnosis.

For confirmation of the microscopical result, or for enrichment if this is negative, 1–2 ml of fresh whole blood, or of clot ground up in saline, should be injected intraperitoneally into a susceptible animal, either the newborn mouse or rat. Spirochaetes appear in the blood in a day or two, usually in large numbers, and remain for 5 or 6 days if the animal does not die. In adult animals the degree of spirochaetaemia is much less, and may not be detectable for several days.

Spirochaetes may sometimes be demonstrated by animal inoculation in the blood of patients for weeks after the primary attack. According to Chung (1938) the urine and prostatic fluid are sometimes infective. In making rodent surveys, the spleen, heart, or brain tissue should be ground up in saline and inoculated subcutaneously into white mice; the blood is then examined daily for spirochaetes in the usual way. It is worth noting that the Wassermann reaction may be positive in relapsing fever; and that, in louse-borne infection, the Weil-Felix test may be positive for *Proteus* OXK (Robinson 1942, Zarafonetis *et al.* 1946). A serological method of diagnosis, with spirochaetal suspensions made from saponin-lysed blood of heavily infected animals, is described by Stein (1944); and a complement-fixation test by El-Ramly (1946) carried out with an alcoholic extract of the liver and spleen of persons dead of relapsing fever, and by Wolstenholme and Gear (1948) with an antigen prepared from the allantoic fluid of infected chick embryos. Numerous other methods may be used, but serological diagnosis suffers from the disadvantage of being of little help in the early stage of the disease, and in the later stages of being confused by strain variation of the organisms (Felsenfeld 1971, Burgdorfer 1976). (For cultivation from the blood, see Dodge 1973*a, b*.)

Prophylaxis and treatment

Prophylaxis consists in measures designed to eliminate lice or ticks (see Chapter 77). Promising results were recorded in the eradication of *O. moubata* in Africa by the systematic spraying of all human dwellings with a preparation of gammexane (Lovett 1956; see also Mooser 1958). Inoculation of living cultures, or sometimes of cultures killed by heating at 60° for 30 minutes, is said to give rise to the production of lysins. Persons so vaccinated may apparently resist infection with small doses of living spirochaetes (Aristowsky and Wainstein 1929*a*, *b*). For treatment of the patient arsenic preparations have been widely used, but the results are not always satisfactory and resistant strains may emerge. Penicillin has been reported on favourably, but again cannot be relied upon completely. In experimental animals spirochaetes latent in the brain appear to be inaccessible to the drug (Schuhardt and O'Bryan 1944, 1945). The drug of choice for both forms of relapsing fever is tetracycline, preferably given intravenously to avoid vomiting. The injection of 300 000 units of procaine penicillin intramuscularly, followed the next day by 250 mg of tetracycline by mouth is recommended for the louse-borne form (Bryceson *et al.* 1970) or, alternatively, a single oral dose of 500 mg of either erythromycin or tetracycline (Butler *et al.* 1978). The serum of convalescent patients may be used therapeutically, but the little clinical and experimental evidence available suggests that this method of treatment is of no great value (Adler and Ashbel 1937, Wolman 1944). (For reviews of relapsing fever see Davis 1948, Mooser 1958, Felsenfeld 1965, 1971, Bryceson *et al.* 1970.) Attempts so far to prepare a satisfactory vaccine have proved a failure (Felsenfeld 1976). (For spirochaetal Lyme disease see Chapter 73.)

Avian spirochaetosis

In 1891 Sakharoff described a disease of **geese** that appeared every year in certain stations on the Transcaucasian railway, and resulted in a high mortality—80 per cent. Examination of the blood revealed the presence of spirochaetes closely resembling those of human relapsing fever. Clinically, the infected goose went off its feed, remained apathetic in a sitting-down posture, and died of exhaustion after a week or more; sometimes it suffered from diarrhoea, and its joints became affected. *Post mortem*, there was fatty degeneration of the heart and liver; the liver moreover showed miliary yellowish granules of caseous consistency; the spleen was soft and friable. During life actively motile spirochaetes were found in fairly large numbers in the blood at the beginning of the disease; later they collected into ray-forms, and finally into tangled ball-like masses. Before death they disappeared from the blood and at post-mortem examination they could be found neither in the blood nor in the organs. Subcutaneous inoculation of infected blood reproduced the disease in normal geese after an incubation period of 4 or 5 days.

Working with the spirochaete of goose septicaemia—*Borrelia anserina*—Gabritschewsky (1898) found that, if the blood serum of a goose which had recovered naturally from the disease was mixed with the blood of an infected goose, the spirochaetes were killed in a few minutes at 37°; if no antiserum was added, they lived for 18 hours. When the serum was heated to 60°, its bactericidal power was destroyed completely. Further experiments showed that, though blood serum from a peripheral vein or from the right side of the heart killed the spirochaetes in half an hour or less, juice sucked from the viscera—spleen, liver, kidney, and bone marrow—did not kill them for $2\frac{1}{2}$ to 9 hours. He also found that, in infected geese, spirochaetes could still be demonstrated in the internal organs after they had disappeared from the blood. He concluded therefore that the spirochaetes were destroyed in the blood by the spirochaeticidins, but remained latent in the viscera where the antibodies appeared to be less concentrated. This conception of immunity is similar to that reached in the case of human relapsing fever.

As well as the spirochaeticidins, there are also lysins present in the blood, and agglutinins, which are responsible for the clumping of the organisms. Phagocytosis occurs, apparently as a secondary phenomenon, after the death or immobilization of the spirochaetes, though according to Himmelweit (1933) it plays an active part in the destruction of the living organisms. Gabritschewsky further showed that naturally recovered geese remained immune to further infection, as long as bactericidins were present in the blood. Antispirochaetal serum, prepared by the inoculation of horses, had good protective properties when injected into geese simultaneously with, or 24 hours after, injection of spirochaetal blood; but once the organisms had appeared in the blood stream, it had no effect. That is to say, the serum had prophylactic, but not therapeutic, properties. By injection of normal geese with a single dose of antiserum, followed on the next day by injection of spirochaetal blood, he succeeded—by active immunization under cover of a passive immunity—in rendering the birds highly resistant to the disease. This method he recommended for combating natural outbreaks of the disease.

Al-Hilly (1969) found that a homologous antiserum yielded a precipitin line in the agar gel diffusion test with a suspension of minced liver from infected chickens. No other organ proved antigenic in this respect.

More widespread than the disease in geese is **spirochaetosis of fowls,** which was first described by Marchoux and Salimbeni (1903) in Brazil, and later by Balfour (1908) in the Sudan. The symptoms and general course of the disease appear to be much the same as in geese.

The disease can be transmitted by subcutaneous, intramuscular, or intraperitoneal injection of hens with infected blood; diarrhoea develops, the temperature rises to 43°, and spirochaetes appear in the blood in 24 hours. The temperature falls after 3 or 4 days to 40° or 41°, but the organisms continue to increase in the blood. Later, however, they become aggregated into groups, their movements become slower, and they form figures of 0 and 8. At this stage the bird dies, though sometimes a chronic disease develops, lasting for about a fortnight and followed by death. The disease can be transmitted to fowls, geese, ducks, guinea-fowls, and sparrows; pigeons are fairly refractory, monkeys and guinea-pigs completely so (Marchoux and Salimbeni 1903). Neither in the goose nor in the fowl do relapses occur; the bird either dies or recovers completely.

The natural disease is spread by ticks. In Brazil *Argas miniatus* was found to be responsible; in the old world *Argas persicus* is the tick that has been chiefly incriminated. Marchoux and Salimbeni (1903) found that ticks might remain infective for 5 months after biting a diseased fowl. According to Hindle (1912) ticks may transmit the infection to their progeny, and these again to the next generation, without having had an infective feed in the meantime. Loomis (1953), it may be mentioned, found no evidence, natural or experimental, to show that infection of turkeys in California was carried by *Argas persicus*; he concluded that the spirochaetes were ingested with infective faeces.

The disease occurs in **ducks** and in **turkeys** (Hoffman and Jackson 1946, McNeil *et al.* 1949) as well as in fowls and geese. It is probable that the causative organism is the same in each species; its proper name therefore is *Borrelia anserina* (Wenyon 1926). Those requiring further information on avian spirochaetosis are referred to the monograph of Knowles, Gupta, and Basu (1932) who, besides giving a bibliographical review of the subject, have made a number of observations themselves, particularly on the mechanism by which immunity develops; to the monograph by Lesbouyries (1941), and the paper by Loomis (1953).

Both penicillin and the organic arsenicals are reported to be effective therapeutically (see McNeil *et al.* 1949). Oxytetracycline, 2 mg per kilo body weight, may be given in a single dose intramuscularly.

Swine dysentery

Swine dysentery is caused by *Treponema hyodysenteriae*, possibly in association with other organisms. This spirochaete can be grown in liquid and on solid media under anaerobic conditions (see Lemcke *et al.* 1979). Cultivation is best suited for diagnosis, as it enables *Tr. hyodysenteriae* to be distinguished from another spirochaete that closely resembles it in both morphology and staining reactions (Hudson *et al.* 1976).

Clinically the disease is characterized by a mucohaemorrhagic diarrhoea, having a high morbidity and mortality in untreated animals. Lesions are confined to the colonic and caecal mucosa, which is covered with a copious secretion of mucus and blood. In severe cases, extensive ulceration, the formation of a necrotic pseudomembrane, and sloughing of the mucosa occur (Drasar and Hudson 1979). Carriers of infection may be detected by their reaction to the intradermal injection of killed spirochaetes. Experimentally, six intravenous injections of a formolized vaccine conferred a fairly good protection against intragastric challenge with the homologous strain of the organism (Glock *et al.* 1978). In Germany, where the disease is widespread, Meier and Amtsberg (1980) found that *Tr. hyodysenteriae* was excreted in the faeces not only of sick animals, but in those of normal and clinically suspicious animals. In both these groups, it was excreted for weeks on end. This organism cannot be distinguished from other intestinal spirochaetes by biochemical means or by agglutination with unabsorbed sera. Microtitre and growth-inhibition tests, however, are more specific (Lemcke and Burrows 1981). For its isolation Jenkinson and Wingar (1981) recommend a medium of tryptic soya blood agar containing spectinomycin 400 μg/ml.

Intestinal spirochaetosis Besides the campylobacters, a number of spiral organisms belonging to various genera, but so far imperfectly studied, are found in the colonic contents of both normal and dysenteric pigs. How far they are able to damage the intestine, either alone or in combination with other organisms, is doubtful In human beings there is some evidence that spirochaetes can cause ulceration of the colonic and rectal mucosa (see Drasar and Hudson 1979), but their role in this respect is still far from clear. (For further information on *Borrelia* infection of animals, see Felsenfeld 1979.)

Blood spirochaetoses in other animals

Cattle suffer from a spirochaetosis caused by *Borrelia theileri*; infection is conveyed, at least in South Africa, by the tick *Margaropus decoloratus*. A similar disease in horses and in sheep, probably due to the same organism, *B. theileri*, has also been reported by Theiler in South Africa. Spirochaetes of the relapsing fever type have been observed in the blood of elephants, camels, antelopes, monkeys, and some other mammals (Wenyon 1926).

Vincent's angina and certain related infections

There are certain necrotic and gangrenous infective processes in human beings such as ulcero-membranous gingivitis, hospital gangrene, trench mouth, noma, foetid bronchitis, gangrenous laryngitis, and tropical ulcer in which spirochaetes have frequently been demonstrated. Of these, one of the chief is the so-called spirillum described by Vincent (1896, 1899), now known as *Treponema vincenti* (see Chapter 44). It is not clear whether Vincent's spirillum is responsible for the necrotic lesions in which it is found, or whether it is a mere secondary invader. The fact that it is often present in scrapings of the gingivo-dental fold in apparently healthy mouths has led many observers to doubt its aetiological role in the inflammatory diseases just mentioned. Black (1938), for example, examined 33 healthy children under 12 years of age, none of whom had suffered from oral or dental disease, and found *Tr. vincenti* in 60 per cent and fusiform bacilli in 94 per cent. He regards these organisms as members of the normal flora of the mouth, which multiply when conditions are favourable, but which are without pathogenic action. Recent work (see Chapter 88) has shown that some of the gingivo-stomatitis of children is a primary herpetic infection; but there is nevertheless a group of diseases characterized by ulceration of the gums and throat, in which no virus has so far been demonstrated, but in which Vincent's spirilla are abundant. The organisms are often found associated with a characteristic fusiform bacillus, likewise described by Vincent (1896; see Chapters 26 and 62). Since we know that strict anaerobes are unable to grow in healthy tissue, the finding of large numbers of spirochaetes and fusiform bacilli in swabs taken from the ulcerated gums is an index of the presence of necrotic tissue or of non-living matter in the mouth. In fact, the disease is often associated with septic teeth, salivary calculi, or an erupting third molar. Why the disease becomes almost epidemic in war-time is still unknown. Shortage of ascorbic acid or of nicotinic acid in the diet has been suggested, but the evidence is not convincing. Infection from one person to another in camps or communal feeding centres probably plays some part. In histological preparations Klein (1950, 1952) found that the number of spirochaetes present varied in proportion to the severity of the disease; they were most abundant in the zone nearest to healthy tissue, suggesting that they were primarily responsible for the necrotic lesions. These results were confirmed by microscopical observations on specimens taken from the normal gums, in which spirochaetes were few, and from diseased gums in which they were present in large numbers. Klein concluded that these organisms undoubtedly were of significance in the causation of the various lesions of the gums with which they were associated. Arsenic has been used in treatment, but its value is doubtful. Measures directed to cleaning up the mouth, the removal of dead tissue, and of tartar on the teeth, and the use of hydrogen peroxide washes and zinc peroxide pastes, seem to be more effective. Klein (1952) found penicillin to be a specific poison for oral spirochaetes. It may be given parenterally, or in chewing gum or applied locally, in combi-

nation with surgical toilet of the mouth. To avoid the risk of sensitization or the emergence of resistant strains, penicillin may be replaced by metronidazole, given in 200 mg doses at 8-hour intervals for 3–4 days. (For papers on fusospirochaetal disease see Rosebury *et al.* 1950.)

Besides *Tr. vincenti*, spirochaetes belonging to the *Treponema* genus are common inhabitants of the mouth. Among the named species are *Tr. denticola*, *Tr. macrodentium*, and *Tr. oralis*. There are also unnamed species referred to as 'large spirochaetes'. These organisms are anaerobic and are confined to the gingival sulci about the teeth. Like *Tr. vincenti*, they increase in number in inflammatory lesions of the mouth; but whether this is cause or effect remains doubtful (see Loesche 1976).

Spirochaetes have also been isolated from the sputum of tuberculous patients (Benzançon and Etchegoin 1926), and from other bronchial and pulmonary lesions (Bacigalupo 1928); but we know nothing about the relation of these organisms to each other or to *Tr. vincenti*.

(For the role of treponemata in the causation of periodontal disease, see Loesche 1976.)

Rat-bite fever

As already pointed out (Chapter 49) there appear to be two distinct diseases known as rat-bite fever, one due to *Spirillum minus*, the other to an organism known as *Streptobacillus moniliformis*. We shall consider here only the spirillar form.

Rat-bite fever, or Sodoku as it is called in Japan, is a disease that occasionally supervenes in man on the bite of a rat, a cat, or some other animal (Mollaret and Bonnefoi 1938, Yamamoto 1939). The incubation period is generally 7 to 21 days, but may extend to weeks or even months. Illness is ushered in by a sharp febrile paroxysm accompanied by swelling of the lymph nodes and dark-red eruptions on the skin. Redness and swelling are noticeable at the site of the wound, which during the incubation period has generally healed satisfactorily. There are often pains in the limbs on the affected side. After 3 or 4 days the attack comes to an end, but is succeeded by another in a few days. These febrile paroxysms with intermittent afebrile periods may be repeated for months, or even years. The fatality of the disease varies from about 2 to 10 per cent.

The causative organism of the disease is a *Spirillum*, which was described in 1916 by Futaki and his coworkers in Japan (see Chapter 44). The organism has been found in cases of rat-bite fever in Great Britain (Robertson 1924), in Italy, Turkey, India, Indonesia, and the United States of America (Shattock and Theiler 1924, Brown and Nunemaker 1942, Beeson 1943). (For a review of the literature see Robertson 1924, Theiler 1926, McDermott 1927–28, Brown and

Nunemaker 1942.) The disease has been reproduced experimentally in human beings by inoculation with infected guinea-pig blood (see Hershfield *et al.* 1929). The clinical distinction of the disease caused by *Sp. minus* from that caused by *Streptobacillus moniliformis* was studied by Allbritten, Sheely and Jeffers (1940) and Brown and Nunemaker (1942), and is referred to in Chapter 44. In tsutsugamushi fever (see Chapter 77), with which rat-bite fever may be confused, the local ulcer is said to be larger, more irregular in outline, and painful, and the elements of the rash on the trunk are of greater size (Lewthwaite 1940).

In the human patient the organism is present in the swollen local lesion, the focal lymph glands, and the blood, but is very difficult to demonstrate except by animal inoculation; at autopsy it can be found also in the kidneys (Kaneko and Okuda 1917). Microscopical examination of the blood is usually negative. The Wassermann reaction is said to be positive in about half the cases (Brown and Nunemaker 1942); and agglutinins to *Proteus* OXK may be present (Lewthwaite 1940). The bacteriological diagnosis is best made by subcutaneous or intraperitoneal inoculation of blood taken at the height of a febrile paroxysm, or of serum expressed from the local lesion, into mice and guinea-pigs (Francis 1932). Aspirated peritoneal fluid or blood should be examined daily from the 5th to the 30th day by dark-ground illumination. Brown and Nunemaker (1942) recommend the inoculation of at least two guinea-pigs and four mice. The guinea-pigs should be inoculated intracutaneously, subcuta-

neously, and intraperitoneally. Intracutaneous inoculation is sometimes followed by a chancre, and subcutaneous inoculation by enlargement of the regional lymph nodes; in both of these sites spirilla may be demonstrated by dark-ground examination. It is important to remember that mice and less often guinea-pigs may be healthy carriers of *Sp. minus*, so that it is wise to take precautions against confusion resulting from this cause. Humphreys and his colleagues (1950) withdrew blood by cardiac puncture from the injected guinea-pigs into citrate, centrifuged the plasma, and examined a loopful just above the cellular layer by dark-ground microscopy. Examining 150 rats in Vancouver, they found 14 per cent of animals to be infected.

Numerous strains of *Spirillum minus* have been isolated from mice, rats, and human patients; according to Schockaert (1928) they are all similar, constituting a single species. In Japan about 3 per cent of house rats appear to carry the organism (Futaki *et al.* 1917). According to Manouélian (1940) the spirilla are present in considerable numbers in the muscles of the tongue, which may account for the infectivity of the bite. In the serum of patients who have recovered from rat-bite fever spirochaeticidal bodies have been found (Ido *et al.* 1917). Salvarsan or neoarsphenamine usually cuts the fever short, but complete cure of the disease is not always easy to obtain. More recent studies, however, suggest that it may respond to treatment with penicillin and streptomycin (Hudemann and Mücke 1951, Golstein 1972.)

Leptospirosis

Leptospirosis is a general term that denotes all infections of man and animals by spirochaetes of the genus *Leptospira*. At one time illnesses caused by such infections were given names such as seven-day fever, swamp fever or mud fever, autumnal fever, swineherds' disease, and cane-cutters' disease, which accorded either with their clinical syndrome or with some epidemiological feature. The term Weil's disease was first used by Goldsmidt (1887) to denote a severe febrile illness with jaundice and renal abnormalities that had been described by Adolph Weil, Professor of Medicine at Heidelberg (1886). It was not until many years later that Japanese workers (Inada *et al.* 1916) isolated a spirochaete from the blood of a patient suffering from Weil's disease and concluded that it was the causative organism. They named the organism *Spirochaeta icterohaemorrhagiae*, but the name was changed to *Leptospira icterohaemorrhagiae* when Noguchi (1918) created a new genus, *Leptospira*, to accommodate organisms that differed from other spirochaetes in their characteristic morphology and

motility. Weil's disease, whether in its complete or incomplete form, is one of the many manifestations of leptospiral infection in man. The emphasis on jaundice has tended to divert clinicians from a recognition of the fact that throughout the world other syndromes are caused by leptospiral infections, and that leptospirosis may mimic many other illnesses. In the British Isles jaundice occurs in less than half the diagnosed cases of leptospirosis; the remainder are characterized by aseptic meningitis, influenza-like illness, or PUO (pyrexia of unknown origin).

Investigations in many countries have shown that leptospires are widespread in their distribution, that a variety of wild and domestic animals may be infected, and that many immunologically distinct types (serotypes or serovars) of *Leptospira* are responsible for human infection. Each serotype tends to be associated with a particular vertebrate species that acts as the natural reservoir host. Consequently different countries and regions vary in respect of the types of leptospirosis that may occur, according to the local fauna.

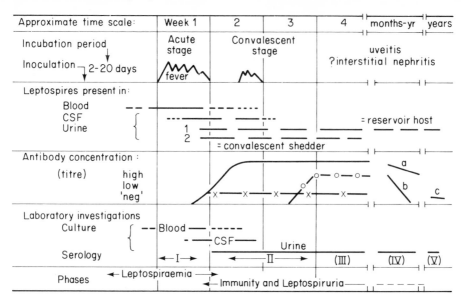

Fig. 74.1 Leptospirosis: phases and relevant diagnostic procedures. (Diagram prepared by Dr L H Turner.) (Reproduced from the *Transactions of the Royal Society of Tropical Medicine and Hygiene* by permission of the editor.)

The course of leptospiral infections

The clinical features of leptospirosis, the diagnostic procedures, treatment, and epidemiology are best understood by considering the course of infection. There are two overlapping phases: (1) leptospiraemia followed by rising antibody titres, and (2) leptospiruria caused by localization of the organisms within the kidney tubules (see Fig. 74.1).

Leptospiraemia

After penetrating the skin or mucous membranes of a fresh host, the leptospires invade the blood stream. They may be detected microscopically in drops of blood and cultured from it during the first week of infection. There is no obvious lesion at the site of entry. Very mild or even subclinical infection can occur, but usually after an incubation period of 7–13 days (occasionally 2 to 21 days or more) a febrile illness of varying severity develops. The leptospires may spread through the blood stream to any tissue or organ of the body, and give rise to minute focal lesions, the number and situation of which determine the different combinations of symptoms and signs that occur. The non-specific defence mechanisms of the body, aided by the increasing concentrations of specific antibodies that develop during the first three weeks, finally clear the leptospires from the blood and other tissues. The organism may, however, persist for a time in the convoluted tubules of the kidney—a site in which the effects of specific antibody and other natural defence mechanisms appear to be reduced.

Antibodies appear in detectable concentration about 4 to 5 days after the onset of illness and usually reach their peak during the third week. Thereafter they fall steadily; antibodies that are highly specific for the serotype of the infecting strain are sometimes detectable for many months or even years.

Leptospiruria

Leptospires that reach the kidney tissue may penetrate to the convoluted tubules in the renal cortex, where they multiply. Some of them escape in the urinary filtrate and are excreted. In convalescent patients and in incidental animal hosts, leptospiruria is transient, lasting for only a few weeks; but in some animals—maintenance or reservoir hosts—it may persist for life, constituting an important factor in the epidemiology of the disease.

Clinical features

The severity of the illness depends partly on the virulence of the infecting strain, and partly on the resistance of the patient. The strain may belong to any one of the 16 or more serogroups of *Leptospira interrogans*—the species that represents the pathogenic leptospires. The signs and symptoms can occur in a wide variety of combinations. Some of these represent common syndromes of which leptospirosis is an unusual cause; others, such as PUO, are less distinctive; and some may be quite unusual. Fort Bragg fever or pre-tibial fever, for example, which occurred among troops in North Carolina, USA, during the three summers 1942–1944, was characterized by headache, malaise, splenomegaly, and an erythematous rash limited to the pre-tibial area of both legs. The infective agent was identified as a leptospire of serotype *autumnalis* (Gochenour *et al.* 1952). A diagnosis of leptospiral infection can be made with certainty only by means of laboratory—mainly serological—tests.

Leptospirosis should be considered in the differential diagnosis of the following syndromes: aseptic meningitis; other febrile illnesses associated with affection of the CNS, e.g. encephalitis, non-paralytic poliomyelitis, transverse myelitis, peripheral neuritis; influenza-like illness; dengue-like illness; unconfirmed rickettsiosis (Q fever, tropical typhus); unconfirmed 'enteric' fevers (pseudo-typhoid); 'glandular fever' when infectious mononucleosis has been excluded; unconfirmed brucellosis and other relapsing fevers; jaundice with fever, neutrophil leucocytosis, renal abnormalities (albuminuria, casts, cells), haemorrhages, meningism, or any combination of these; pneumonitis; nephritis or nephrosis, especially when other organs or systems are affected; haemorrhagic states for which no obvious cause is found; PUO of less than three weeks' duration in untreated patients.

Survival of leptospires Pathogenic leptospires can survive outside the animal body for three weeks or longer in an environment that is moist, warm, and of roughly neutral pH. Such conditions occur during the summer and autumn in temperate climates. Adverse factors include desiccation, ultraviolet irradiation (e.g. exposure to direct sunlight), pH values outside the range 6.2–8.0, salinity, and chemical pollution.

Modes of transmission

Infection is acquired through contact with an environment contaminated by urine from carrier (reservoir) hosts or other infected animals. The leptospires enter the body through the skin or mucous membranes, especially if these are abraded. Infection may arise through bathing or accidental immersion in the fresh water of lakes, rivers, or canals polluted with the urine of infected livestock that use the water for drinking, or where the water may be contaminated from rodent nests in the banks, especially after heavy rainfall and flooding. Certain occupational groups are at risk, e.g. dairy farmers through inhalation of droplets of infected urine in cattle sheds during milking, and abattoir workers and butchers through direct contact with carcasses or organs of animals after slaughter. Venereal transmission, and transplacental infection of the fetus have been reported in man and some animals; otherwise man-to-man transmission is rare. On the basis of experimental work insects are not considered to be important epidemiologically, although they may transport leptospires passively and transmit infection by contaminating food or water, or by being accidentally rubbed into the skin of the host. Infections in man have followed the bites of animals such as dogs,

rats, and ferrets; and cases of so-called mud fever (Schlammfieber) were reported by Schüffner and Bohlander (quoted by Alston and Broom 1958) in children bitten by voles (*Microtus arvalis*) that were shown to be natural carriers of serotype *grippotyphosa* in the Netherlands. Leptospires are not excreted in the saliva of infected animals, and in such cases the source of infection was probably the fur of the animal contaminated with urine; or the actual urine of a frightened animal; or, when dogs were implicated, the teeth contaminated with the infected viscera of rats or other rodents. The bite wounds would undoubtedly allow the ready entry into the body of leptospires from such sources.

Reservoir hosts

Of prime importance in the epidemiology of leptospiral infections are the chronic persistent carriers and shedders of virulent leptospires. These are the maintenance or reservoir hosts. Babudieri (1958), following Pavlovsky's work in Soviet Russia (see Chapter 44), stressed the state of biological equilibrium which is established between some strains of *Leptospira* and their hosts. A particular species of host may be a transient carrier (incidental host) of some serotypes, but a maintenance host for others. The importance of an animal species as a host of *Leptospira* can be measured by the ratio between the 'culture rate' and the 'serology rate'. The culture rate is the proportion of animals examined from whose kidneys the particular serotype of *Leptospira* is isolated. The serology rate is the proportion of animals examined having the relevant antibodies. The higher the ratio of culture rate to serology rate, the more likely is that species of host to be an important reservoir of infection. (This subject has been discussed by Smith *et al.* 1961, Roth *et al.* 1963, and Emanuel *et al.* 1964.)

The role of rats and dogs in the spread of leptospiral infections is well known. However, mammals of numerous other kinds are more important reservoir hosts in some habitats, localities, or regions (see pages 534, 539).

Occupational and recreational risks

Infection may spread from wild animals to livestock, pets, and laboratory animals, thereby bringing the risk closer to man. The animals themselves may suffer from no obvious ill health, but human infections derived from such sources may be severe, even fatal. The inter-relations of the principal categories of animal hosts are shown in Fig. 74. 2.

The likely hazards, which are numerous, can be deduced from this Figure. Any person whose work or recreation brings him into the habitats of the animal hosts may contract infection. Because of improvements in personal and environmental hygiene based

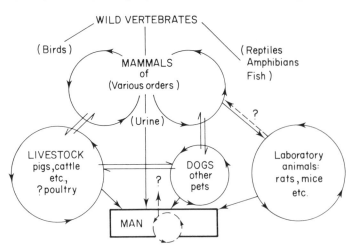

Fig. 74.2   Relationships of the principal reservoir hosts of *Leptospira*. (Diagram prepared by Dr L H Turner.) (Reproduced from the *Transactions of the Royal Society of Tropical Medicine and Hygiene* by permission of the editor.)

on greater awareness of the risks, such occupations as mining, underground sewer work, and fish-cleaning, which formerly gave rise to many rat-transmitted cases of icterohaemorrhagiae infection, are no longer particularly hazardous in Great Britain and some other countries. Farmers, especially dairy and cattle farmers, are now the occupational group with the highest case incidence in the British Isles (Coghlan 1979, Report 1981). Slaughterers, butchers, abattoir workers, those engaged in meat-processing, and veterinarians, are likewise exposed to risk. In some parts of the world, including Italy and Spain, where infection is carried by field mice, vaccines prepared from local strains afford protection to the rice-field workers (see Altava *et al.* 1955). In 1979 forty-three cases of leptospiral infection were reported in England, Wales, and Ireland, nearly all in males. Patients infected with serotypes *icterohaemorrhagiae* or *copenhageni* suffered from typical Weil's disease, whereas those infected with *hebdomadis* had malaise and pyrexia only. Three patients infected with *canicola* had meningitis (Report 1980).

Diagnosis

As already mentioned none of the syndromes caused by leptospiral infections can be regarded as pathognomonic and the diagnosis must therefore be made in the laboratory. There are three lines of investigation: demonstration of leptospires by direct microscopical observations of the specimen; isolation of the strain; and serological examination.

Microscopical demonstration of leptospires

Leptospires can be demonstrated in infected material by dark-field microscopy, silver-impregnation

methods, and fluorescent-antibody techniques. In practice these methods have certain limitations. The organisms must be present in large numbers; artefacts are commonly mistaken for leptospires; and some of the drugs used in treatment may distort the organism. Wolff (1954) found that, even with the double centrifugation method, he could demonstrate the organisms in the blood of only 32 of 100 patients in whom leptospiraemia was proved by blood culture. For examination blood is collected in an anticoagulant; the blood cells are deposited by low-speed centrifugation; the supernate is removed and centrifuged at high speed (10 000 rpm) to deposit the leptospires; and the material in the 'button' is examined by dark-field microscopy. The findings must be regarded as provisional, pending the results of cultural and serological tests.

Isolation of the strain

The infecting strain can often be isolated in culture provided that suitable material is obtained before antibiotics have been administered. Early in the course of illness—during the leptospiraemic phase—the inoculum of preference is blood or cerebrospinal fluid; later—during the phase of leptospiruria—it is urine. It is usually impracticable in the routine laboratory to attempt the isolation of strains from every patient with a febrile illness. However, a suitable inoculum may be conserved for this purpose by mixing 5.0 ml of the acute-phase blood specimen with 1.0 ml of 1.0 per cent 'liquoid' (polyanethol sulphonate), an anticoagulant that also counteracts the natural bactericidal action of the blood, and retaining it at a low temperature until serological findings indicate leptospiral infection. This method was used to transport blood from Central Africa to Antwerp in Belgium (van Riel *et al.* 1956).

In fatal cases leptospires may be isolated from the brain, renal cortex, and urine as well as from blood, liver and spleen. When investigating wild animals as reservoir hosts, the renal cortex and urine are sampled. Except in the early stages of infection, the blood of such animals is likely to be free of leptospires.

Isolation may be attempted by inoculating the samples directly into suitable media, or into laboratory animals which are later examined by cultural methods. From sick laboratory animals heart blood should be examined and from those that have died overnight suspensions of liver, spleen and kidney, i.e. vascular organs. Leptospires do not survive for long in dead tissue and therefore attempted isolation from post-mortem material should not be delayed.

Various types of media are used for culturing leptospires. Some, such as those of Fletcher (1928), Korthof (1932), and Stuart (1946a), have 'conventional' formulae supplemented with pooled normal-rabbit serum (7–10 per cent). The medium of Ellinghausen and McCullough (1965) on the other hand contains bovine albumin (fraction V) and Tween 80, and in a modified form (Johnson and Harris 1967) is available commercially (see Chapter 44). Inocula should be small in relation to the volume of the medium, so as to dilute any specific antibodies, lipids, or other inhibitory substances that may be present in the specimen. For example, 0.03 ml of blood may be inoculated into 5.0 ml of medium (Gochenour *et al.* 1953); or 2–3 drops are expelled into the medium from a gauge-18 needle. At least three tubes of medium should be used for each specimen. Attempts to isolate from urine directly into the medium are apt to be frustrated by contamination with other micro-organisms. Serial 10-fold dilutions of urine have been advocated to overcome this difficulty. To control the growth of contaminants, disks impregnated with neomycin sulphate are sometimes added to culture media; alternatively pyrimidine 5-fluorouracil at a final dilution of 100 μg per ml may be used (see Turner 1970). Cultures should be incubated at 28–30° if appropriate incubators are available; otherwise they should be kept at room temperature, despite the reduced rate of leptospiral multiplication that occurs at *ca* 20°. A temperature as high as 37° is not well tolerated, may be lethal for some strains, and should not be used.

Laboratory animals are used not only for isolating strains, but also for maintaining newly isolated strains until they can be adapted to laboratory medium. Young animals, preferably weanlings, should be used because older, more resistant, animals may survive the infection. Stocks must be free from endemic leptospiral infection. Guinea-pigs, hamsters, gerbils, young rabbits, Swiss white mice, the albino American deer mice (*Peromyscus maniculatus gambeli*) and 1–3 day-old chicks may be used. The material should be inoculated intraperitoneally through one of the lower quadrants of the abdominal wall. The animals must be examined carefully, preferably twice daily, for evidence of infection. A drop of peritoneal fluid should be withdrawn with a fine capillary pipette and examined microscopically with dark-field illumination for active leptospires, from the third to the seventh day (Wolff 1954). It must be remembered that the strain concerned may not be pathogenic for the species of animal used. (For further details and discussion see Galton *et al.* 1962, Alexander *et al.* 1970, and Turner 1970.)

Serological methods

Reliable serological diagnosis is now within the capacity of most general-duty laboratories. The wide use of tests on sera from febrile patients, whatever the clinical signs may be, is likely to reveal leptospiral infection in patients whose illness might otherwise remain undiagnosed.

Many tests have been reported in the past, but in current practice they fall into two categories—'genus-specific', and those that are more highly specific.

Genus-specific tests

Such tests distinguish leptospirosis from other microbial causes of febrile illness. They provide enough information for the clinician because treatment does not depend on the exact identification of the infecting strain of *Leptospira*. Genus-specific tests include a sensitized-erythrocyte lysis (SEL) or haemolytic (HL) test, a complement-fixation test, a biflexa agglutination test, and an indirect immunofluorescence test. Enzyme-linked immunosorbent assay may be used to detect specific anti-leptospiral IgM and IgG, and thus to indicate the stage of infection (Adler *et al.* 1980, Terpstra *et al.* 1980).

Sensitized-erythrocyte lysis test   Soluble substances that are adsorbed on to erythrocytes have been extracted from various micro-organisms and are known as erythrocyte-sensitizing substances (ESS). Two types of reaction can be demonstrated when ESS-treated cells are mixed with sera containing homologous antibodies. They are haemagglutination or sensitized-erythrocyte agglutination (HA or SEA), and haemolysis (HL or SEL). Chang and McComb (1954) first reported on leptospiral ESS. Various authors later introduced modifications, notably the use of Biflexa strains for preparing ESS (for a brief review see Turner 1968). The SEL test was used successfully in Malaysia for many years as the routine test for leptospiral infections in man.

Complement-fixation test   Numerous antigen preparations have been reported. Some were regarded as 'group-specific', others as 'type-specific', but none found general favour until Rumanian workers, who investigated various Biflexa strains for suitability as a 'genus-specific' antigen, selected a strain known as Patoc 1. They have used this strain to their satisfaction in complement-fixation tests since 1960 (Nicolescu and Borsai 1972). In the UK Turner and Mohun (unpublished) adopted the Patoc 1 complement-fixation test, but augmented the antigen with a few additional strains to produce a compound screening antigen (CSA), which they regarded as more sensitive than an antigen derived solely from Patoc 1. The qualitative results obtained with this antigen show a high correlation with the agglutination test in the examination of sera from patients in the acute and convalescent stages of infection. This antigen is now widely used in the UK, together with viral antigens, in the complement-fixation screening test on sera from febrile patients. The test reliably distinguishes leptospirosis from other more common causes of jaundice. It is not suitable for use in surveys to detect residual antibodies.

Biflexa agglutination test   Living or killed (formolized) suspensions of some strains of *Leptospira biflexa*, especially strain Patoc 1, are agglutinated readily by sera from patients in the acute and convalescent stages of infection. Patoc 1 may be used in a macroscopic (slide) test in which a drop of a dense suspension of the leptospires killed by boiling for 30 minutes is mixed with a drop of the patient's serum. A positive result is indicated by agglutination, clearly visible to the naked eye, regardless of the serotype of the infecting strain. The test is a modification of one described by Mailloux and co-workers (1974).

Immunofluorescence test   Torten and his colleagues (1966) reported on the use of strain Patoc 1 as an antigen for an indirect immunofluorescence test.

Serogroup- and serotype-specific tests

In this category fall the macroscopic and microscopic agglutination tests. The former rely on suspensions of formalin-killed organisms of a number of reference strains, concentrated by centrifugation so that the reaction can be read with the naked eye. The test was introduced by Galton and her colleagues (1958, 1962). The antigens are now available commercially. The microscopic agglutination tests rely on well-grown cultures, living or killed with formalin; and the reactions are read by low-power dark-field microscopy. The use of microscopic agglutination tests is usually confined to reference laboratories capable of maintaining strains representing all the serotypes likely to be encountered.

Leptospires possess numerous agglutinogenic factors (Kmety 1967). Each serogroup is associated with a set of 'major' (main) factors which account for the cross-reactions of some serotypes at high titres and are the basis of their grouping. These main factors are present, in various combinations, in each serotype within a particular serogroup; each serotype, therefore, is characterized by its own unique combination of main factors. Other, so-called minor, less specific factors probably account for the cross-reactions at lower titres between strains of different serogroups. At one time agglutination tests were regarded as serotype-specific. This view is no longer wholly tenable. The probable identity of a strain can be determined more accurately by a series of mirror-test agglutinin-absorption studies with all the recognized serotypes in a particular serogroup—a procedure which is impracticable as a routine. The agglutination reaction, however, can often indicate the serogroup to

which the infecting strain belongs, and will usually suggest the likely epidemiological factors concerned in the causation of any particular case. (For a discussion of serological tests, see Turner 1968.)

Robinson and his colleagues (1982) made use of bacterial restriction endonuclease DNA analysis—a technique not based on antigenic relationships—to distinguish serovars *hardjo*, *balcanica*, and *tarassovi*.

Prevention

The prevention of leptospiral infections is based on the control, if not the eradication, of reservoir hosts; on environmental and personal hygiene; and, where the risk is high, on vaccination. Environmental hygiene requires the thorough cleansing of workplaces and animal stalls, including treatment with disinfectants such as sodium hypochlorite solution (1 in 4000). Personal hygienic measures taken by those whose occupation places them at special risk should include the wearing of protective clothing, particularly boots and gloves, and dressing of all skin wounds. Vaccination has proved to be valuable when large numbers of workers are exposed to infection in circumstances that preclude satisfactory control of the reservoir hosts, e.g. in rice fields. The vaccine should be prepared from local strains of *Leptospira*.

Treatment

With the exception of chloramphenicol, most of the usual antibiotics are efficacious, especially when they are administered in the first few days of illness. Owing, however, to the usual delay in diagnosis this is seldom possible. Penicillin is generally preferred but tetracycline is said to have the merit of eliminating the leptospires from the kidney (Stoenner 1976). Munnich and Lakatos (1976) strongly recommend ampicillin and amoxicillin for the treatment of human infections. The dosage of crystalline penicillin should be at least 2 mega-units per day and, in severe illnesses, 8.0 or 10.0 mega-units daily; the treatment should be continued for at least one week (Turner 1967). Treatment within five days of infection may provoke a Jarisch and Herxheimer reaction, which lasts for 24 hours; this should not be allowed to interrupt the course of antibiotics. Early treatment reduces the duration of the fever and relieves the symptoms rapidly. However, even when administered later, treatment is beneficial in preventing complications and relapse (Mackay-Dick and Robinson 1957).

Particular attention must be given to the kidneys. If their function is impaired, peritoneal dialysis or the use of the artificial kidney should not be delayed, because renal failure is the commonest cause of death. For further details see Gsell (1966, 1968) and Turner (1967).

Leptospirosis in animals

From what has already been said it is clear that, although rats are the main reservoir hosts of serotypes of the Icterohaemorrhagiae serogroup, leptospirosis is a common although generally inapparent infection of many wild and domestic animal species; and it is caused by a very large number of different serotypes. Infected animals may act as direct or indirect sources of infection for man and other animals. The increasing awareness that leptospirosis is widespread in domestic animals has stimulated in recent years a considerable amount of investigational work to determine the extent of the infections, the serotypes responsible, and the disease manifestations that sometimes occur. The following paragraphs summarize the present knowledge of leptospirosis as it affects different domestic species.

Cattle

Bovine leptospirosis is common throughout the world. Strains of many different serogroups including Grippotyphosa, Pomona, Hebdomadis, Icterohaemorrhagiae and Canicola have been isolated from cattle suffering from acute, chronic or subclinical infection (Amatredjo and Campbell 1975). The cattle may become infected by drinking or wading in streams, ponds or canals contaminated with the urine of cattle or other farm animals or wild species. Infection may cause abortion, and calves infected with serotype *pomona* may suffer from red water fever, an acute disease with jaundice, haemorrhages, and a high mortality rate. *Pomona* infection in cattle is widespread in some parts of the world, where it can cause considerable economic loss. Control by vaccination is widely practised.

In many parts of the world including the British Isles members of the Hebdomadis serogroup, especially serotype *hardjo*, are being increasingly recognized as a cause of abortion, infertility, and mastitis. The organisms have been isolated from milk and from aborted fetuses (Ellis *et al.* 1976*b* and *c*) as well as from the kidneys of apparently normal cattle collected at an abattoir (Orr and Little 1979). The latter authors believe that cattle act as the normal reservoir host for serotype *hardjo*, since in endemically infected herds infection appears to be entirely subclinical. This is borne out by the work of Ellis and his colleagues (1981) who found no evidence of infection in badgers or mice trapped on farms where leptospires related to *hardjo* were isolated from a high percentage of cattle examined.

In New Zealand where the incidence of leptospirosis is high among dairy farmers, *hardjo* and *pomona* are the predominant serotypes and both are widespread among the cattle. The infections are probably transmitted from cattle to man through inhalation of droplets of infective urine in cow sheds (Christmas *et al.* 1974).

Pigs

Leptospiral infection of pigs occurs throughout the world and pigs are now recognized as the reservoir hosts for *pomona* and *tarassovi*, the two most common porcine serotypes in many European countries, North America, Australia, and New Zealand. Many infections are subclinical, but abortion is common and vaccination is widely practised (Kemenes and Suveges 1976, Weber and Fenske 1978). Sporadic cases and small outbreaks of infection in pigs due to leptospires of the Icterohaemorrhagiae serogroup have been reported in various countries including England (Field and Sellers 1951), Scotland (Nisbet 1951), and Ireland (McErlean 1973). The resulting disease is frequently severe, with jaundice and rapid death. A serological survey in southern England showed that 15 per cent of the animals had leptospiral antibodies. The organisms against which these were directed were *bratislava*, *copenhageni*, *zanoni*, *cynopteri*, and *autumnalis* (Hathaway *et al.* 1981*b*).

Subclinical infection of pigs with serotype *canicola* has been reported (Coghlan *et al.* 1957) and has been associated with many overt human cases of canicola fever (Lawson 1972, Shenberg *et al.* 1977). Once infection has been established in a piggery it may last for many years, thereby providing a source of infection for other animals and their attendants. Moist conditions resulting from the type of husbandry that relies on wet swill feeding undoubtedly contributed to the widespread and prolonged infections that occurred in Scottish pig farms during the 1950s.

Infection with strains belonging to the Australis serogroup has recently been found by serological methods in many English pig herds, and may be associated with infertility (Hathaway and Little 1981). Similar infections were reported from Holland (Hartman *et al.* 1975) and Germany (Weber and Fenske 1978), the Dutch pigs yielding serotype *lora* of the Australis serogroup. Leptospires from British pigs were identified as serotype *muenchen* (Hathaway, personal communication, 1981). The sporadic nature of these Australis serogroup infections may reflect contact with free-living wild animals that are the natural reservoir hosts, rather than endemic infection by pig-adapted strains.

Horses

There have been reports from many parts of the world of high antibody titres to various leptospiral serotypes in apparently normal horses. For example 35 per cent of 500 clinically healthy horses in England had leptospiral antibodies, most commonly against serotypes of the Icterohaemorrhagiae and Australis serogroups

(Hathaway *et al.* 1981*a*). Although disease manifestations are rare and when they occur are usually mild, leptospiral infections in horses have been associated with several clinical syndromes, the most common of which are pyrexia and icterus, and periodic ophthalmia (Hanson *et al.* 1969). Abortion has also been reported as a possible sequel to leptospiral infection, and Ellis and his colleagues (1976*a*) isolated both leptospires and equine herpesvirus type 1 from six fetuses during an outbreak of abortion in Northern Ireland which affected 10 out of 40 mares.

Endemic infection apparently does not occur in horses, and the spectrum of leptospiral antibodies present in equine sera seems to reflect exposure to serotypes maintained by other animals, domestic or wild. For example antibodies to *pomona* are common in horses in the USA where that serotype is present in domestic stock and in wildlife (Damude *et al.* 1979); whereas *grippotyphosa* antibodies are common in horses in Europe where the organism is probably maintained by the common vole (*Microtus arvalis*).

Sheep

Little is known about the prevalence or clinical significance of leptospiral infection of sheep. In New Zealand and the USA outbreaks of disease which appeared to be due to serotype *pomona* were associated with fatal haemolytic anaemia and haemoglobinuria in lambs, and sometimes with abortion in ewes (Hartley 1952, Davidson and Hirsh 1980). Acute illness has also been associated with serotypes *ballum* (Cacchione *et al.* 1963) and *grippotyphosa* (Amjadio and Abourai 1975). A serological survey suggested that infection with serotype *hardjo* was widely distributed in England and Wales (Hathaway *et al.* 1982). Strains of *hardjo* have been isolated from sheep in Australia and New Zealand (Bahaman *et al.* 1980, Gordon 1980) and infection has sometimes led to fatal acute hepatitis, chronic nephritis, and possibly abortion (Andreani *et al.* 1974). Strains of the Australis serogroup were isolated for the first time in England from the urine and kidney of apparently normal sheep by Little and co-workers (1981). It is believed that they originated from wildlife, since hedgehogs, voles, mice, and badgers are known to carry them (Salt and Little 1977).

Goats

Goats are susceptible to leptospiral infections of various kinds. van der Hoeden (1953) reported that *grippotyphosa* was responsible for many cases of jaundice, haemoglobinuria, and abortion among goats in Israel; and Torten (1979), also in Israel, listed strains of serogroups Australis, Grippotyphosa, Hebdomadis, and Pomona that were isolated from goats. Damude and co-workers (1979) reported that in Barbados antibodies to the Autumnalis serogroup predominated in goats and other livestock; this reflects the high rate of infection by strains of the same serogroup in rodents and mongooses on the island.

Dogs

The disease in dogs is predominantly due to infection with *icterohaemorrhagiae* or *canicola*, although other serotypes have been implicated. Okell and his colleagues (1925) first demonstrated that 'yellows' in dogs, a disease characterized by an acute and often fatal jaundice, was caused as a rule by serotype *icterohaemorrhagiae*. The disease primarily affects dogs up to three years of age, and epizootics are commonest among dogs housed together in rat-infested kennels. (Stuart 1946*b*). Infection usually results from contact with rats' urine, but can also be transmitted from dog to dog and from dog to man. Dogs also suffer from a form of leptospirosis that gives rise to uraemia. Klarenbeek and Schüffner (1933) isolated the causative organism from the urine of a dog in Utrecht and showed that it differed serologically from *icterohaemorrhagiae*; this organism was subsequently called *Leptospira canicola* (present nomenclature *L. interrogans*, serotype *canicola*). The infection is often more chronic than that produced by *icterohaemorrhagiae*, and is thought to be a common cause of subacute or chronic nephritis. It is seldom accompanied by jaundice but is nevertheless sometimes difficult to distinguish from the disease produced by *icterohaemorrhagiae*. *Canicola* is more frequently excreted in the urine than *icterohaemorrhagiae* and is more common in dogs than bitches, possibly owing to the common habit in dogs of sniffing the urine or external genitals of other dogs.

From the results of serological tests carried out on blood samples from over 12 000 dogs Ryu (1976) found that the prevalence of infection varied greatly in different parts of the world. In early surveys in Great Britain 25 to 40 per cent of dogs showed serological evidence of infection (Baber and Stuart 1947, Broom 1951), whereas Ryu found that only 4.5 per cent of 354 British dogs were serologically positive. A similar decrease has apparently occurred in other countries and may reflect the widespread use of vaccines. Serological and cultural tests indicate that dogs may become infected with serotypes other than the two already mentioned. For example, serotype *pomona* was isolated from the urine of a dog in the USA (Murphy *et al.* 1958) and Little (personal communication) isolated a leptospire of the Tarassovi serogroup from the urine of a hound in England in 1978.

Laboratory diagnosis follows the general lines already laid down for human disease. It is most easily made by demonstrating a rise in the concentration of agglutinating antibodies in the serum.

Penicillin is probably efficacious in the treatment of

infection, especially when administered in the early stages. It may not, however, eradicate the leptospires from the kidneys and prevent their being shed in the urine. Dihydro-streptomycin given intramuscularly may be more effective in eliminating leptospiruria (Hubbert and Shotts 1966). Vaccination is of value in protecting dogs from the clinical effects of infection. It is not certain that it always affords protection against invasion of the kidneys and the subsequent development of the carrier state. Kerr and Marshall (1974), however, showed experimentally that, if the vaccine was of a suitable standard, complete protection was obtained—not only against clinical disease, but also against the renal carrier state. Commercial vaccines usually consist of inactivated suspensions of serotypes *icterohaemorrhagiae* and *canicola* grown in a serum-free medium. Immunity is produced by two subcutaneous injections of 2 ml with an interval of 14 days between each dose. An annual reinforcing injection is also recommended.

Cats

Although cats are susceptible to experimental infection with a variety of leptospiral serotypes and may subsequently excrete leptospires in their urine, they seldom suffer from clinical leptospirosis (Fessler and Morter 1964). Serotypes *icterohaemorrhagiae* (Merten 1938), *bataviae* and *javanica* (Esseveld and Collier 1938), *grippotyphosa* (Vysotskii *et al.* 1960), *pomona* (Gordon Smith *et al.* 1961), and *bratislava* (Bryson and Ellis 1976) have been isolated from naturally infected cats; and serological evidence of infection, usually based on low titres, has been obtained in Britain and other countries. Infection is no doubt derived from the rodents that are hunted by cats. Shophet and Marshall (1980) succeeded in infecting cats with serotype *ballum* by feeding them with infected mice.

Other mammals

Bats of the genus *Cynopterus* were found to be carriers of serotype *schüffneri*, a pathogenic serotype related to *canicola*. Two other strains isolated from bats were found to constitute a new serotype named *cynopteri* (Collier and Mochtar 1939). Opossums, grey foxes, racoons, striped skunks, and wild cats have also been found to be infected (McKeever *et al.* 1958).

Poultry

On the whole, birds, including domestic poultry, are resistant to leptospiral infection. When experimentally infected by mouth, they react as a rule only by the formation of agglutinins (Chalquest 1957). Wading birds, however, may act as passive carriers of infection (Babudieri *et al.* 1960).

Amphibians

Strains of *Leptospira* subsequently identified as new serotypes were isolated from the kidneys of frogs (*Rana pipiens*) in the USA (Diesch *et al.* 1966) and of toads (*Bufo marinus*) in the Philippines (Babudieri *et al.* 1973). They were named serotype *ranarum* of serogroup Ranarum, and *bufonis* of the serogroup Carlos, respectively. Their role in the epidemiology of human leptospirosis is not clear. However, Everard and his colleagues (1980) have shown that the toad may be an important source of Autumnalis infections in Trinidad. In Barbados, strains of the same serogroup have been isolated from the kidneys of small whistling frogs (*Leptodactylus martiniquensis*); these animals occur commonly on the island and are especially active during heavy rainfall when the incidence of human cases tends to increase (Everard, personal communication).

The chief animal hosts are listed in Table 74.1. A more comprehensive list by country and serotype, based on the identification of isolates, is given in Report (1975).

Table 74.1   Chief animal hosts of leptospires

Rodentia: rats of many species (including *Rattus norvegicus*, the brown rat), voles, gerbils, coypus, field mice, domestic mice, and the harvest mouse (*Micromys minutus sorcinus*) found by Babudieri (1958) to be the reservoir host of the serotype *bataviae*, the cause of infection in rice-field workers in Italy.
Insectivora: hedgehogs, shrews
Carnivora: jackals, mongooses, skunks, racoons, civets—as well as dogs
Herbivora: domestic cattle
Marsupialia: bandicoots, opossums
Cheiroptera: bats
Artiodactyla: deer, pigs
Lagomorpha: hares, rabbits
Edentata: armadillos
Amphibia: frogs, toads

References

Adler, B., Murphy, A. M., Locarnini, S. A. and Faine, S. (1980) *J. clin. Microbiol.* **11**, 452.

Adler, S. and Ashbel, R. (1937) *Ann. trop. med. Parasit.* **31**, 89.

Alexander, A. D., Gochenour, W. S. Jr, Reinhard, K. R., Ward, M. K. and Yager, R. H. (1970) In: *Diagnostic Procedures and Reagents*, 5th edn, p. 382. American Public Health Association, New York.

Al-Hilly, J. N. A. (1969) *Amer. J. vet. Res.* **30**, 1877.

Allbritten, F. F., Sheely, R. F. and Jeffers, W. A. (1940) *J. Amer. med. Ass.* **114**, 2360.

Alston, J. M. and Broom, J. C. (1958) *Leptospirosis in Man and Animals*. E. & S. Livingstone Ltd, Edinburgh.

Altava, V., Barrera, M., Villalonga, I., Gil, P., Marin, C. and Babudieri, B. (1955) *Rev. Sanid. Hig. públ., Madr.* **29**, 167.

Amatredjo, A. and Campbell, R. S. F. (1975) *Vet. Bull.* **43**, 875.

Amjadio, A. R. and Abourai, P. (1975) *Arch. Razi Inst.* **27**, 71.

Andreani, E., Santarelli, E. and Diligenti, R. (1974) *Ann. Fac. Med. vet., Univ. Pisa* **27**, 33.

Aristowsky, W. M. and Wainstein, A. B. (1929a) *Z. Immun-Forsch*, **61**, 296; (1929b) *Ibid.* **63**, 240.

Baber, M. D. and Stuart, R. D. (1947) *Lancet* ii, 594.

Babudieri, B. (1958) *Ann. N.Y. Acad. Sci.* **70**, 393.

Babudieri, B., Addamiano, L., Buusinello, E., Giusti, E. and Salvi, A. (1960) *Rep. int. Symp., Microbiol. Comm. Polish Acad. Sci., Minist. Hlth, Wroclaw*, p. 58.

Babudieri, B., Carlos, E. R. and Carlos, E. T. (1973) *Trop. geogr. Med.* **25**, 297.

Bacigalupo, J. (1928) *C. R. Soc. Biol.* **99**, 1622.

Bahaman, A. R., Marshall, R. B., Blackmore, D. K. and Hathaway, S. C. (1980) *N.Z. vet. J.* **28**, 171.

Balfour, A. (1908) *Rep. Wellcome Res. Lab.* **3**, 38.

Baltazard, M., Bahmanyar, M. and Mofidi, C. (1948) *Ann. Inst. Pasteur* **73**, 1066.

Beck, M. D. (1936) *Dept publ. hlth, St. Calif., spec. Bull.* No. 61, p. 19.

Beeson, P. B. (1943) *J. Amer. med. Ass.* **123**, 332.

Bezançon, F. and Etchegoin, E. (1926) *C. R. Soc. Biol.* **94**, 1056.

Black, W. C. (1938) *Amer. J. Dis. Child.* **56**, 126.

Bohls, S. W. and Irons, J. V. (1942) *Publ. Amer. Ass. Advanc. Sci.* No. 18, p. 48.

Broom, J. C. (1951) *Mon. Bull. Minist. Hlth Lab. Serv.* **10**, 258.

Brown, T. M. and Nunemaker, J. C. (1942) *Johns Hopk. Hosp. Bull.* **70**, 201.

Bryceson, A. D. M., Parry, E. H. O., Perine, L. L., Vukotich, L. L., Warrel, D. A. and Leithead, C. S. (1970) *Quart. J. Med.* **39**, 129.

Bryson, D. G. and Ellis, W. A. (1976) *J. small Anim. Pract.* **17**, 450.

Burgdorfer, W. (1976) In: *The Biology of Parasitic Spirochetes*, p. 225. Ed. by R. C. Johnson. Academic Press, New York and London.

Butler, T., Hazen, P., Wallace, C. K., Awoka, S. and Habte-Michael, A. (1979) *J. infect. Dis.* **140**, 665.

Butler, T., Jones, P. K. and Wallace, C. K. (1978) *J. infect. Dis.* **137**, 573.

Cacchione, R. A., Cedro, V. C., Bulgini, M. J., Cascelli, E. S. and Martinex, E. S. (1963) *Proc. 17th World vet. Congr., Hanover* **1**, 427.

Carminati, G. M. (1957) *Boll. 1st. sieroter. Milano* **36**, 197.

Chalquest, R. R. (1957) *J. Poultry Sci.* **36**, 110.

Chang, R. S. and McComb, D. E. (1954) *Amer. J. trop. Med. Hyg.* **3**, 481.

Chen, Y. P., Zia, S. H. and Anderson, H. H. (1945) *Amer. J. trop. Med.* **25**, 115.

Christmas, B. W., Till, D. G. and Braggen, J. (1974) *N.Z. med. J.* **79**, 904.

Chung, H. L. (1938) *Proc. Soc. exp. Biol., N.Y.* **38**, 97.

Coghill, N. F. and Gambles, R. M. (1948) *Ann. trop. Med. Parasit.* **42**, 113.

Coghlan, J. D. (1979) *Brit. med. J.* ii, 872; (1981) *Commun. Dis. Rep., Colindale*.

Coghlan, J. D., Norval, J. and Seiler, H. E. (1957) *Brit. med. J.* i, 257.

Collier, W. A. and Mochtar, A. (1939) *Geneesk. Tijdschr. Ned. Ind.* **79**, 226.

Cunningham, J. and Fraser, A. G. L. (1935) *Indian J. med. Res.* **22**, 595.

Cunningham, J., Theodore, J. H. and Fraser, A. G. L. (1934) *Indian J. med. Res.* **22**, 105.

Damude, D. F., Jones, C. J. and Myers, D. M. (1979) *Trans. Roy. Soc. trop. Med. Hyg.* **73**, 161.

Davidson, J. N. and Hirsh, D. C. (1980) *J. Amer. vet. med. Ass.* **176**, 124.

Davis, G. E. (1942) *Publ. Amer. Ass. Advanc. Sci.* No. 18, pp. 41, 67; (1943) *Publ. Hlth Rep., Wash.* **68**, 839; (1948) *Annu. Rev. Microbiol.* **2**, 305.

Delpy, L. and Rafyi, A. (1939) *Ann. Parasit. hum. comp.* **17**, 45.

Diesch, S. L., McCulloch, W. F., Braun, J. L. and Ellinghausen, H. C. (1966). *Nature, Lond.* **209**, 939.

Dodge, R. W. (1973a) *Appl. Microbiol.* **25**, 935; (1973b) *Infect. Immun.* **8**, 891.

Drasar, B. S. and Hudson, M. J. (1979) In: *Recent Advances in Infection*, p. 41. Ed. by D. Reeves and A. Geddes. Churchill Livingstone, Edinburgh.

Dutton, J. E. and Todd, J. L. (1905) *Brit. med. J.* ii, 1259.

El-Ramly, A. H. (1946) *J. Egypt. publ. Hlth Ass.* **21**, 1, 150, 166.

Ellinghausen, H. C. and McCullough, W. G. (1965) *Amer. J. vet. Res.* **26**, 45.

Ellis, W. A., Bryson, D. G. and McFerran, J. B. (1976a) *Vet. Rec.* **98**, 218.

Ellis, W. A., O'Brien, J. J. and Cassells, J. (1981) *Vet. Rec.* **108**, 555.

Ellis, W. A., O'Brien, J. J., Neill, S., Hanna, J. and Bryson, D. G. (1976b) *Vet. Rec.* **99**, 458.

Ellis, W. A., O'Brien, J. J., Pearson, J. K. L. and Collins, D. S. (1976c) *Vet. Rec.* **99**, 368.

Emanuel, M. L., Mackerras, I. M. and Smith, D. J. W. (1964) *J. Hyg., Camb.* **62**, 451.

Esseveld, H. and Collier, W. A. (1938) *Z. ImmunForsch.* **93**, 512.

Everard, C. O. R., Sulzer, C. R., Bhagwandin, L. J., Fraser-Chanpong, G. M. and James, A. C. (1980) *Int. J. Zoon.* **7**, 90.

Felsenfeld, O. (1965) *Bact. Rev.* **29**, 46; (1971) *Borrelia. Strains, Vectors, Human and Animal Borreliosis*. Warren H. Green, Inc., St. Louis, Missouri; (1976) In: *The Biology of Parasitic Spirochetes*, p. 351. Ed. by R. C. Johnson. Academic Press, New York and London; (1979) In: *CRC Handbook Series in Zoonoses*, Vol. 1, p. 79. Ed. by H. Jones and J. H. Steele. CRC Press, Inc., Boca Raton, Florida.

Fessler, J. F. and Morter, R. L. (1964) *Cornell Vet.* **54**, 176.

Field, H. I. and Sellers, K. C. (1951) *Vet. Rec.* **63**, 78.

Fletcher, W. (1928) *Trans. R. Soc. trop. Med. Hyg.* **21**, 265.

Francis, E. (1932) *Trans. Ass. Amer. Phycns* **47**, 143; (1938) *Publ. Hlth Rep., Wash.* **53**, 2220.

Futaki, K., Takaki, I., Taniguchi, T. and Osumi, S. (1916) *J. exp. Med.* **23**, 249; (1917) *Ibid.* **25**, 33.

Gabritschewsky, G. (1898) *Zbl. Bakt.* **23**, 365, 439, 635, 721, 778.

Galton, M. M., Menges, R. W., Shotts, E. B., Nahmias, A. J. and Heath, C. W. (1962) *Leptospirosis: Epidemiology, Clinical Manifestations in Man and Animals, and Methods in Laboratory Diagnosis*. U.S. Publ. Hlth Serv., Pubn No. 951, Washington D.C.

Galton, M. M., Powers, D. K., Hale, A. D. and Cornell, R. G. (1958) *Amer. J. vet. Res.* **19**, 505.

Glock, R. D., Schwartz, K. J. and Harris, D. L. (1978) *Amer. J. vet. Res.* **39**, 639.

Gochenour, W. S., Smadel, J. E., Jackson, E. B., Evans, L. B. and Yager, R. H. (1952) *Publ. Hlth Rep., Wash.* **67**, 811.

Gochenour, W. S., Yager, R. H., Wetmore, P. W. and Hightower, J. A. (1953) *Amer. J. publ. Hlth* **43**, 405.

Goldsmidt, F. (1887) *Dtsch. Arch. klin. Med.* **40**, 238.

Golstein, E. (1972) In: *Infectious Diseases*, p. 1149. Ed. by P. D. Hoeprich. Harper and Row, New York and London.

Gordon, L. M. (1980) *Aust. vet. J.* **56**, 348.

Gordon Smith, C. E., Turner, L. H., Harrison, J. L. and Broom, J. C. (1961) *Bull. World Hlth Org.* **24**, 5, 23.

Gsell, O. (1966) *Ann. Soc. belge. Méd. trop.* **46**, 203; (1968) In: *Infektionskrankheiten*, Bd. ii, 826. Springer, Verlag, Berlin.

Hanson, L. E., Martin, R. J., Gibbons, R. W. and Schnurrenberger, P. R. (1969) *Proc. 73rd Ann. Meeting U.S. Anim. Hlth Ass.* 169.

Hartley, W. J. (1952) *Aust. vet. J.* **28**, 160.

Hartman, E. G., Brummelman, B. and Dikken, H. (1975) *Tijdschr. Diergeneesk.* **100**, 421.

Hathaway, S. C. and Little, T. W. A. (1981) *Vet. Rec.* **108**, 224.

Hathaway, S. C., Little, T. W. A., Finch, S. M. and Stevens, A. E. (1981*a*) *Vet. Rec.* **108**, 396.

Hathaway, S. C., Little, T. W. A. and Stevens, A. E. (1981*b*) *Res. vet. Sci.* **31**, 169; (1982) *Vet. Rec.* **110**, 99.

Herms, W. B. and Wheeler, C. M. (1936) *Dept publ. Hlth, St. Calif., spec. Bull.* No. 61, p. 24.

Heronimus, E. S. (1928) *Zbl. Bakt.* **105**, 394.

Hershfield, A. S. *et al.* (1929) *J. Amer. med. Ass.* **92**, 772.

Himmelweit, F. (1933) *Z. Hyg. InfektKr.* **115**, 710.

Hindle, E. (1912) *Proc. Camb. phil. Soc.* **16**, Part 6, p. 457.

Hoeden, J. van der (1953) *J. comp. Path.* **63**, 101.

Hoffman, H. A. and Jackson, T. W. (1946) *J. Amer. vet. med. Ass.* **109**, 481.

Hubbert, W. T. and Shotts, E. B. (1966) *J. Amer. vet. med. Ass.* **148**, 1152.

Hudemann, H. and Mücke, D. (1951) *Z. Hyg. InfektKr.* **132**, 292.

Hudson, M. J., Alexander, T. J. L. and Lysons, R. J. (1976) *Vet. Rec.*, **99**, 498.

Humphreys, F. A., Campbell, A. G., Driver, M. W. and Hatton, G. N. (1950) *Canad. J. publ. Hlth* **41**, 66.

Ido, Y., Ito, H., Wani, H. and Okuda, K. (1917) *J. exp. Med.* **26**, 377.

Inada, R., Ido, Y., Hoki, R., Kaneko, R. and Ito, H. (1916) *J. exp. Med.* **23**, 377.

Jenkinson, S. R. and Wingar, C. R. (1981) *Vet. Rec.* **109**, 384.

Johnson, R. C. and Harris, V. G. (1967) *J. Bact.* **94**, 27.

Kaneko, R. and Okuda, K. (1917) *J. exp. Med.* **26**, 363.

Kemenes, F. and Suveges, T. (1976) *Acta vet. hung.* **26**, 395.

Kerr, D. D. and Marshall, V. M. (1974) *Vet. Med/Small Anim. Clin.* **69**, 1157.

Klarenbeek, A. and Schüffner, W. A. P. (1933) *Ned. Tijdschr. Geneesk.* **77**, 4271.

Klein, H. S. (1950) *Österreich. Z. Stomatol.* **47**, 339; (1952) *Ibid.* **49**, 185.

Kmety, E. (1967) *Faktorenanalyse von Leptospiren der Icterohaemorrhagiae und einiger verwandter Serogruppen.* Slovakian Academy of Sciences, Bratislava.

Knowles, R, Gupta, B. M. D. and Basu, B. C. (1932) *Indian med. Res. Mem.* No. 22.

Korthof, G. (1932) *Zbl. Bakt., I. Abt. Orig.* **125**, 429.

Lawson, J. H. (1972) *Scot. med. J.* **17**, 220.

Lemcke, R. M., Bew, J., Burrows, M. R. and Lysons, R. J. (1979) *Res. vet. Sci.* **26**, 315.

Lemcke, R. M. and Burrows, M. R. (1981) *J. Hyg., Camb.* **86**, 173.

Lesbouyries, G. (1941) *La Pathologie des Oiseaux.* Vigot Frères, Paris.

Lewthwaite, R. (1940) *Lancet* **i**, 390.

Little, T. W. A., Parker, B. N. J., Stevens, A. E., Hathaway, S. C. and Markson, L. M. (1981) *Res. vet. Sci.* **31**, 386.

Loesche, W. J. (1976) In: *The Biology of Parasitic Spirochetes.* Ed. by R. C. Johnson. Academic Press, New York and London.

Loomis, E. C. (1953) *Amer. J. vet. Res.* **14**, 612.

Lovett, W. C. D. (1956) *Trans. R. Soc. trop. Med. Hyg.* **50**, 157.

MacArthur, W. (1957) *Brit. med. Bull.* **13**, 146.

McDermott, E. N. (1927-8) *Quart. J. Med.* **21**, 433.

McErlean, B. A. (1973) *Irish vet. J.* **27**, 157

Mackay-Dick, J. and Robinson, J. F. (1957) *J. R. Army med. Cps* **103**, 186.

McKeever, S., Gorman, G. W., Chapman, J. F., Galton, M. M. and Powers, D. K. (1958) *Amer. J. trop. Med. Hyg.* **7**, 646.

Mackie, F. P. (1907) *Brit. med. J.* **ii**, 1706; (1920) *Ibid.* **i**, 380.

McNeil, E., Hinshaw, W. R. and Kissling, R. E. (1949) *J. Bact.* **57**, 191.

Mailloux, M., Mazzonelli, J. and Dorta de Mazzonelli, G. T. (1974) *Zbl. Bakt., I. Abt. Orig.* **A229**, 238.

Manouélian, Y. (1940) *C.R. Soc. Biol.* **133**, 582.

Marchoux, E. and Salimbeni, A. (1903) *Ann. Inst. Pasteur* **17**, 569.

Megaw, J. (1955) *British Encyclopaedia of Medical Practice*, 2nd edn, p. 647. Butterworth & Co., London.

Meier, C. and Amtsberg, G. (1980) *Berl. Münch. tierärztl. Wschr.* **93**, 402.

Merten, W. K. (1938) *Ned. Ind. Bl. Diergeneesk.*, **50**, 78.

Mollaret, P. and Bonnefoi, A. (1938) *Bull. Soc. Path. exot.* **31**, 855.

Mooser, H. (1958) *Ergebn. Mikrobiol. ImmunForsch. exp. Ther.* **31**, 184.

Moursund, W. H. (1942) *Publ. Amer. Ass. Advanc. Sci.* No. 18, p. 1.

Munnich, D. and Lakatos, M. (1976) *Chemotherapy* **22**, 372.

Murphy, L. C., Cardeilhac, P. T., Alexander, A. D., Evans, L. B. and Marchwicki, R. H. (1958) *Amer. J. vet. Res.* **19**, 145.

Nicolescu, M. and Borsai, L. (1972) *Arch. roum. Path. exp. Microbiol.* **31**, 209.

Nicolle, C., Blaizot, L. and Conseil, E. (1912) *C. R. Acad. Sci.* **155**, 481; (1913) *Ann. Inst. Pasteur* **27**, 204.

Nisbet, D. I., (1951) *J. comp. Path.* **61**, 155.

Noguchi, H. (1918) *J. exp. Med.* **27**, 575.

Norris, C., Pappenheimer, A. M. and Flournoy, T. (1906) *J. infect. Dis.* **3**, 266.

Novy, F. G. and Knapp, R. E. (1906) *J. infect. Dis.* **3**, 291.

Oag, R. K. (1940) *J. Path. Bact.* **51**, 127.

Obermeier (1873) *Berl. klin. Wschr.* **10**, 152, 378, 391, 455.

Okell, C. C., Dalling, T. and Pugh, L. P. (1925) *Vet. J.* **81**, 3.

Orr, H. S. and Little, T. W. A. (1979) *Res. vet. Sci.* **27**, 343.

Report. (1956) *Welt-Seuchen-Atlas.* Teil II. Falk Verlag, Hamburg; (1969) *Wkly epid. Rec.* **44**, 425; (1975) *Leptospiral Serotype Distribution Lists.* U.S. Dept Hlth Educ. and Welfare, Wash.; (1980) *Commun. Dis. Rep.* No. 40, Colindale; (1981) *Brit. med. J.* **282**, 2066.

Riel, J. van, Szpajshendler, L. and Riel, M. van (1956) *Bull. Soc. Path. exot.* **49**, 118.

Robertson, A. (1924) *Ann. trop. Med. Parasit.* **18**, 157.

Robinson, A. J., Ramadass, P., Lee, A. and Marshall, R. B. (1982) *J. med. Microbiol.* **15**, 331.

Robinson, P. (1942) *Brit. med. J.* **ii**, 216.

Rosebury, T., Clark, A. R., Engel, S. G. and Tergis, F. (1950) *J. infect. Dis.* **87**, 217, 226, 234.

Ross, P. H. and Milne, A. D. (1904) *Brit. med. J.* **ii**, 1453.

Roth, E. E., Adams, W. V., Sanford, G. E. Jr, Moore, M., Newman, K. and Greer, B. (1963) *Publ. Hlth Rep., Wash.* **78**, 994.

Ryu, E. (1976) *Int. J. Zoonoses* **3**, 33.

Sakharoff, N. (1891) *Ann. Inst. Pasteur* **5**, 564.

Salih, S. Y., Mustafa, D., Wahab, S. M. A., Ahmed, M. A. M. and Omer, A. (1977) *Trans. R. Soc. trop. Med. Hyg.* **71**, 43.

Salt, G. F. H. and Little, T. W. A. (1977) *Res. vet. Sci.* **22**, 126.

Sanford, J. P. (1976) In: *The Biology of Parasitic Spirochetes*, p. 307. Ed. by R. C. Johnson. Academic Press, New York and London.

Sautet, J. (1937) *Marseille méd.* **74**, ii, 273.

Schockaert, J. (1928) *C. R. Soc. Biol.* **98**, 595.

Schuhardt, V. T. (1942) *Publ. Amer. Ass. Advanc. Sci.* No. 18, p. 58.

Schuhardt, V. T. and O'Bryan, B. E. (1944) *Science* **100**, 550; (1945) *J. Bact.* **49**, 312; *Ibid.* **50**, 127.

Schuhardt, V. T. and Wilkerson, M. (1951) *J. Bact.* **62**, 215.

Shattock, G. C. and Theiler, M. (1924) *Amer. J. trop. Med.* **4**, 453.

Shenberg, E., Birnbaum, S., Rodrig, E. and Torten, M. (1977) *Amer. J. Epidem.* **105**, 42.

Shophet, R. and Marshall, R. B. (1980) *Brit. vet. J.* **136**, 265.

Simmons, J. S. Whayne, T. F., Anderson, G. W., Horack, H. M., Thomas, R. A. *et al.* (1954) *A Geography of Disease and Sanitation*, Vol. 3. J. B. Lippincott Co., Philadelphia.

Sparrow, H. (1958) *Bull. Org. mond. Santé* **19**, 673.

Stein, G. J. (1944) *J. exp. Med.* **79**, 115.

Stoenner, H. G. (1976) In: *The Biology of Parasitic Infections*, p. 375. Ed. by R. C. Johnson. Academic Press, New York and London.

Stuart, R. D. (1946a) *J. Path. Bact.* **58**, 343; (1946b) *Vet. Rec.* **58**, 131.

Terpstra, W. J., Ligthart, G. S. and Schoone, G. J. (1980) *Zbl. Bakt., I. Abt. Orig. A* **247**, 400.

Theiler, M. (1926) *Amer. J. trop. Med.* **6**, 131.

Torten, M. (1979) In: *Handbook Series in Zoonoses*, Vol. 1, p. 363. Ed. by H. Jones and J. H. Steele. CRC Press Inc., Boca Raton, Florida.

Torten, M., Shenberg, E. and Hoeden, J. van der (1966) *J. infect. Dis.* **116**, 537.

Turner, L. H. (1967) *Trans. R. Soc. trop. Med. Hyg.* **61**, 842; (1968) *Ibid.* **62**, 880; (1970) *Ibid.* **64**, 623.

Vincent, H. (1896) *Ann. Inst. Pasteur* **10**, 488; (1899) *Ibid.* **13**, 609.

Vysotskii, B. V., Malykh, F. S. and Prokofiev, A. A. (1960) *Zh. Mikrobiol., Moscou* **31**, 140 (*Zh. Mikrobiol. Transl.* **31**, 364).

Weber, B. and Fenske, G. (1978) *Mh. Vet. Med.* **33**, 652.

Weil, A. (1886) *Dtsch. Arch. klin. Med.* **39**, 209.

Wenyon, C. M. (1926) *Protozoology, II.* London.

Wheeler, C. M. (1942) *Publ. Amer. Ass. Advanc. Sci.* No. 18, p. 89.

Wolff, J. W. (1954) *The Laboratory Diagnosis of Leptospirosis.* Charles C Thomas, Ill.

Wolman, B. and Wolman, M. (1945) *Ann. trop. Med. Parasit.* **39**, 82.

Wolman, M. (1944) *E. Afr. med. J.* **21**, 336.

Wolstenholme, B. and Gear, J. H. S. (1948) *Trans. R. Soc. trop. Med. Hyg.* **41**, 513.

Yamamoto, S. (1939) *Trop. Dis. Bull.* **36**, 774.

Zarafonetis, C. J. D., Ingraham, H. S. and Berry, J. F. (1946) *J. Immunol.* **52**, 189.

75

Syphilis, rabbit syphilis, yaws, and pinta

A. E. Wilkinson

Introductory

There is a group of diseases, sometimes referred to as the Treponematoses, which are caused by spirochaetes of the genus *Treponema* (see Chapter 44). Some are sporadic or mildly epidemic, some are endemic, some are spread by sexual contact, others are non-venereal. Though the endemic forms tend to be restricted to certain areas, the treponematoses as a whole are widely distributed throughout the world, giving rise to acute or chronic disease characterized by early and late stages separated by a latent period, by a charac-

teristic pathological response, and by their susceptibility to treatment by penicillin. In this chapter we shall concentrate on the chief member of the group, namely syphilis, and shall restrict ourselves principally to the venereal form of the disease. For information on the treponematoses in general, including endemic syphilis, the reader should consult T. B. Turner and Hollander (1957), L. H. Turner (1959), Cockburn (1961), Hudson (1961), and Willcox (1972).

Syphilis: history

The origin of syphilis is not definitely known. Its recognition in Europe dates from the end of the fifteenth century, and according to one view it was brought

back to Spain by Columbus's sailors returning from the New World and rapidly became disseminated throughout adjacent countries. Another view is that

543

the treponemal diseases syphilis, yaws, pinta, and endemic non-venereal syphilis form a continuous spectrum of an infection which has affected man for many thousands of years (Hackett 1963, Hudson 1965). Hudson groups these as various expressions of a single disease, treponematosis, the mode of transmission and clinical picture being governed by climate and the socio-economic conditions of the society in which it is present. From the early descriptions venereal syphilis (see Dennie 1962) appears to have been a severe, highly contagious disease which was often fatal. Now it is much milder, whether because of changes in the treponeme or the gradual development of some racial immunity is not known.

Edwin Klebs in 1875–77 was apparently the first to see spirochaetal bodies in syphilitic material and to transmit the disease to monkeys (see A. C. Klebs 1932). Haensell in 1881 produced a local lesion in rabbits by inoculation of the eye. Metchnikoff and Roux (1903, 1904*a, b*, 1905) transmitted the disease to apes, and found that not only primary but also secondary lesions developed in chimpanzees after inoculation with human syphilitic material. In 1905 Schaudinn and Hoffmann discovered the causative organism of the disease. In chancres and in the inguinal lymph nodes of syphilitic patients they demonstrated a spirochaete, now known as *Treponema pallidum*, which occurred both on the surface and in the depth of the tissue. They stained it by a modified Giemsa stain, and described its characteristic morphology and movements. In chancres it was frequently accompanied by another spirochaete—*Treponema refringens*—which was broader, less regular, and more refractile. But *Tr. refringens* was sometimes found in non-syphilitic lesions, such as gonorrhoeal papillomata, whereas *Tr. pallidum* was never found except in syphilis. Though Schaudinn and Hoffmann did not decide whether *pallidum* and *refringens* were different species, it has since been shown that it is *Tr. pallidum* alone that is responsible for the causation of syphilis. (For further history see Schuberg and Schlossberger 1930, Stokes 1931, Klebs 1932.)

The incidence of the disease is difficult to assess. Statistics are derived mainly from cases treated in clinics, but for various reasons these underestimate the extent of syphilitic infection in the community. In the United States, according to Brown (1971), there is an annual incidence of 75 000 fresh cases of the disease, and a reservoir of 510 000 cases of untreated syphilis. In England the rate per 100 000 population of new cases attending clinics with early syphilis was 4.92 in 1974 and 6.16 in 1979. In the latter year the total of new patients with syphilis at all stages was 4445 (Report 1981).

Bacteriology

Syphilis is transmitted chiefly by direct, but sometimes by indirect, contact. *Treponema pallidum* has the power of gaining entrance to the body through minute lesions of the skin or mucous membranes. It is, however, a very strict parasite, and its life outside the animal body is short—a fact which probably determines the predominantly venereal epidemiology of the disease. It is only in closed communities living in hygienically primitive conditions that endemic, non-venereal syphilis is found (see, e.g., Grin 1952, Hudson 1958). Infection may likewise be transmitted from mother to child via the placenta, resulting in the congenital type of syphilis. Except in this type, in which the disease is generalized from the start, infection is usually rendered evident by the development of a primary lesion or chancre. This appears within a month of infection, and is accompanied by enlargement of the local lymphatic nodes. From 6 to 12 weeks after the appearance of the primary chancre, the secondary stage of the disease sets in; this is marked by constitutional symptoms, cutaneous lesions, enlargement of the lymph nodes, and sometimes affections of the bones, joints, eyes, and other organs. Thereafter, the disease becomes latent and can be detected only by serological tests. This stage may persist for many years or for the patient's lifetime, but in a minority of patients late lesions appear. These may be relatively benign ulcerating lesions of the skin, mucous membranes or bones, or gummata of the internal organs. More serious are lesions of the heart and aorta producing aneurysms, or of the nervous system, of which tabes dorsalis and general paralysis are the most common; these last are sometimes called quaternary or parasyphilitic manifestations.

Up to the commencement of the secondary stage syphilis appears clinically as a localized infection, but dissemination of treponemes throughout the body occurs at a very early date. Bauer and his colleagues (1952) examined spinal fluid from patients with untreated early syphilis and found that 1.3 per cent of seronegative primary, 2.9 per cent of seropositive primary and 6.1 per cent of secondary cases showed abnormalities.

Treponemes have been isolated from the spinal fluid of patients with secondary syphilis (Chesney and Kemp 1924). In rabbits Collart and his colleagues (1974) demonstrated treponemes in the spinal fluid and aqueous humour 18–24 hours after infection. This experimental evidence of rapid dissemination shows that local prophylactic measures are unlikely to be of any value unless practised immediately after exposure to infection.

Invasion by only a few organisms may be sufficient to establish infection.

Magnuson, Eagle and Fleischman (1948) infected 50 per cent of rabbits receiving as few as four spirochaetes intracutaneously; intratesticularly an average of one spirochaete was regularly infective. During the primary stage of the disease in man, spirochaetes are found in the local chancre, and can sometimes be demonstrated in the blood. Uhlenhuth and Mulzer (1913) drew off the blood of patients with primary and secondary syphilis, defibrinated it, and injected it into the testicles and scrotum of rabbits; the whole operation was completed within 10 minutes. Syphilis developed in 67 per cent of the animals inoculated from patients with primary, and in 70 per cent of those inoculated from patients with secondary syphilis.

Spirochaetes are present in all the secondary lesions, and may be excreted in the semen (Uhlenhuth and Mulzer 1913). In the tertiary lesions such as gummata they are demonstrable, but only in small numbers; their virulence, however, appears to be maintained. Noguchi and Moore (1913) found them in the brain of patients dying of general paralysis; they were seen in all the layers of the cortex with the exception of the outer or neuroglial layer. In congenital syphilis spirochaetes are distributed in large numbers throughout the viscera, particularly the liver, lungs, spleen, and suprarenals. Both in congenital and acquired syphilis the organisms may remain latent for long periods of time without giving rise to any clinical manifestations of disease.

Immunity to syphilis

Even before the discovery of *Tr. pallidum* it was well known that second infections with syphilis were rare, suggesting the development of a substantial degree of immunity, although superinfection during the primary stage was recognized to be possible. Extensive studies on the development and nature of immunity were carried out, mainly with the rabbit as an experimental animal; this earlier work is reviewed by Chesney (1927) and Gastinel and Pulvenis (1934); for more recent reviews see Urbach and Beerman (1947), Magnuson (1948) and Cannefax (1965). Two somewhat different views emerged. The first, of which the main proponents were Neisser, Finger and Landsteiner, Kolle, Prigge and Truffi, was that immunity in syphilis was due to the persistence of the original infection—an 'infection' or 'chancre' immunity. The work of Chesney and his colleagues led to the view that a true immunity to re-infection does develop slowly in syphilis, and that this is not necessarily due to the persistence of the original infection. When the primary or immunizing infection was allowed to persist for more than three months before adequate treatment was given, subsequent challenge usually failed to produce an overt infection. The early workers used arsphenamines for treatment. Arnold and his colleagues (1947, 1950*a*, *b*) treated rabbits with early infections of only 6 to 8 weeks' duration with penicillin; node transfers from these animals gave negative results, suggesting that the original infection had been eradicated. The numerous re-infections seen in man in recent years after the treatment of primary syphilis would support

this. When rabbits similarly treated were re-inoculated 10 days later, 10 of 37 contracted darkground-positive lesions and the remainder subclinical infections as shown by subsequent positive node transfers. When the original infection was allowed to persist for 8 months before treatment, no lesions developed in 34 reinoculated animals, but 18 were infected subclinically. Immunity established after latent infections of 6 months' duration was found to persist for 6 to 8 months after adequate treatment with penicillin. Relatively large challenge inocula had been used in this and earlier work, and Magnuson and Rosenau (1948) introduced the use of graded doses of treponemes to challenge rabbits treated with penicillin after infections lasting 3 to 24 weeks. They concluded that some immunity was demonstrable as early as 3 weeks after infection and that it increased progressively during the 24 weeks the animals were studied, as shown by the need to give increasing numbers of organisms to produce a given response. Turner and Hollander (1957) think that resistance operates largely independently of the size of the challenge dose. The work of Gastinel and his colleagues (1938) suggests that immunity may wane; they found that in 3 of 9 rabbits challenged 10 to 20 months after treatment, symptomless infections developed compared with one only of 20 animals challenged within the first 10 months.

The number of infecting organisms may influence the degree of immunity that develops. Turner and Nelson (1950) infected rabbits with graded doses of treponemes and challenged them after treatment; infection with small numbers produced less resistance than that resulting from massive inocula. Prolongation of the incubation period by giving subcurative amounts of penicillin for 20 weeks did not affect the subsequent evolution of the disease (Hollander *et al.* 1952), despite the view of Magnuson and his colleagues (1950) that additional immunity might develop under such circumstances. The route of re-infection has an influence; animals refractory to a second intratesticular inoculation may be infected by applying treponemes to a granulating wound on the skin, and rabbits with systemic immunity may be successfully re-inoculated into the cornea or anterior chamber (Chesney *et al.* 1939). Turner and his colleagues (1947) showed that within 6 months of intratesticular inoculation of rabbits with *Tr. pallidum*, *Tr. pertenue* or *Tr. cuniculi* an appreciable degree of immunity had developed which protected against intracutaneous challenge with heterologous strains.

The observations of Magnuson and his colleagues (1956) on the experimental inoculation of human volunteers have given results confirming those from experimental animals. The 50 per cent infective dose by the intradermal route in normal persons was found to be about 57 treponemes. Syphilitic patients were inoculated with 100 000 treponemes, about 2000 times the ID50. None of 5 patients with untreated latent syphilis showed lesions or a rise in serological titre; in 11 patients previously treated for early syphilis, lesions developed at the site of inoculation, treponemes were present in 9, and all showed a rise in titre; in 10 of 26 patients previously treated for late latent syphilis, lesions developed but treponemes were found in only one; 4 of 5 patients with treated congenital syphilis showed local lesions, one of which contained treponemes. In one patient in each of the last two groups the lesion produced resembled a gumma, that is, a lesion contemporary with the stage of the disease at the time of inoculation.

Infectivity by sexual transmission to others in untreated human syphilis is greatest during the first two years of the disease, spanning the primary and secondary stages and early latency. After five years the risk of transmission is thought to be minimal. The risk of the female transmitting congenital syphilis also lessens with time. The lesions of late syphilis are not infectious to others. There is evidence that structures resembling *Tr. pallidum* may be present in the tissues of some patients with late or latent syphilis even after massive treatment with penicillin. These were first reported by Collart and his colleagues (1962*a, b*) and their observations have been confirmed by other workers (*inter al.* Yobs *et al.* 1964, 1968, Smith and Israel 1967, Rice *et al.* 1970; for review see Dunlop 1972). These forms have been proved to be infectious for experimental animals in only a very small proportion of the cases in which they have been observed; the nature and significance of those which have not given positive infectivity tests remains controversial (see Turner *et al.* 1969). They may provide an explanation for the long continued positivity of specific treponemal tests such as the immobilization and fluorescent treponemal antibody tests. For methods for detection of these forms see Miller (1971).

Humoral immunity in syphilis

Antibodies are present in syphilitic sera which are treponemicidal in the presence of complement (Turner 1939, Turner *et al.* 1948). Bishop and Miller (1976*b*) infected rabbits and challenged them intradermally at intervals with 10^3 *Tr. pallidum*; dilutions of sera from the infected animals were incubated with treponemes and their virulence was determined by intradermal inoculation of normal animals. Neutralization of virulence was detected a month after infection, and a close connection between neutralization and *Tr. pallidum* immobilization test titres was found. Attempts to confer passive immunity have been reported (Turner *et al.* 1973, Weiser *et al.* 1976, Bishop and Miller 1976*a*). Despite the administration of large amounts of 'immune' serum, complete protection was not conferred. The infection was, however, attenuated by comparison with that in control animals, as shown by an increased incubation period, and the production on

challenge of smaller lesions in which treponemes were difficult to find and which healed more rapidly. There is general agreement that antilipoidal antibody (reagin) is not related to immunity. In early syphilis, when immunity is low, this reaches high titres, whereas in late syphilis, when immunity is higher, reagin tests are often low in titre or even negative.

Normal human serum contains a heat-labile antibody of IgM class which immobilizes *Tr. pallidum* in the presence of complement and Ca^{++} and Mg^{++} ions (Hederstedt 1974, 1976*a–e*). Although the titre may rise during infection, it seems unlikely that this plays a major part in the immune response.

Cell-mediated immunity in syphilis

Electron-microscopical studies have suggested that *Tr. pallidum* can be phagocytosed by tissue cells. Lukehart and Miller (1978) have shown this to occur *in vitro* with rabbit macrophages. Metzger and Smogŏr (1975) transferred lymphocytes from the inguinal nodes of immune rabbits to normal rabbits; their results suggest that these animals acquired some protection against subsequent challenge. However, Baughn and his colleagues (1977) failed to confirm this by the transfer of splenic lymphocytes from immune rabbits, although resistance to *Listeria* and delayed hypersensitivity to tuberculin were produced in the recipients. Suppression of cell-mediated immunity by various methods in infected mice does not lead to an increase in the number of treponemes or to the suppression of anti-treponemal antibody (Wright *et al.* 1974). Studies on lymphoblastic transformation reactions and macrophage inhibition tests with treponemal antigens in infected patients and animals suggest that cell-mediated immunity is depressed during the early stages of syphilis, but later becomes effective. In this it differs from humoral immunity, which develops earlier. Both forms of response appear to be needed before the infection is controlled and the stage of latency is reached. (For reviews of work in this confusing field see Pavia *et al.* 1978, Metzger 1979, and Fitzgerald 1981*b*.)

Attempts to produce immunity in experimental animals by the injection of *Tr. pallidum* or commensal treponemes killed by heat or chemical agents have proved unsuccessful. Miller (1967) attenuated virulent *Tr. pallidum* by exposure to *γ*-radiation; the organisms became avirulent but retained their motility and their ability to stimulate the production of immobilizing antibody. Rabbits immunized for 24 weeks showed some resistance to challenge which was more pronounced when the immunization course was prolonged. Metzger and his colleagues (1969) produced evidence to show that a heat-labile protein antigen in *Tr. pallidum* played a part in the stimulation of immunity. Rabbits immunized intravenously over a 7-week period with suspensions of treponemes in which this antigen had been preserved showed a substantial protection; none of 16 animals manifested lesions on challenge although 10 had symptomless infections as shown by positive node transfers (Metzger and Smogôr 1969).

Diagnosis

Tr. pallidum can be demonstrated in the early lesions of syphilis. The lesion should be cleansed by swabbing with saline and squeezed gently until clear serum exudes; this is examined by dark-ground illumination. The morphology of *Tr. pallidum* has been described in Chapter 44. Features which help to distinguish it from the other treponemes which may be found on mucosal surfaces are the regularity of its spirals, about seven of which span the diameter of red cells, which are usually present in the preparation. It has sharply pointed ends and a characteristic silvery-white refractility. Although actively motile with a to-and-fro 'drilling' motion with occasional flexion of the body, it tends to remain in the same field of view. Other spirochaetes are usually thicker with rounded ends and more relaxed and easily deformed spirals; they may have a pinkish tinge and tend to move out of the field of view. The identification of *Tr. pallidum* needs considerable experience, especially in the examination of material from the mouth, as some oral spirochaetes closely resemble it. Confirmation of the diagnosis by serological tests should always be sought.

Tr. pallidum can also be found in material obtained by aspirating the lymph nodes draining the site of the primary lesion. They may be hard to find in the elements of the secondary rash on the dry skin, but are plentiful in the moist lesions at muco-cutaneous junctions, such as mucous patches or condylomata lata.

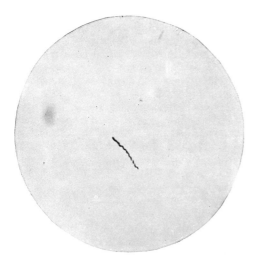

Fig. 75.1 *Treponema pallidum.* In material scraped from a hard chancre. Fontana. (× 1000.)

Fluorescent antibody methods, based on the use of a conjugated syphilitic serum made specific for *Tr. pallidum* by absorption with cultivable treponemes, have been described (Jue *et al.* 1967, Kellogg 1970, Wilkinson and Cowell 1971). The results compare favourably with those obtained by conventional dark-ground examination; and the method has the advantage that smears of exudate can be posted to a laboratory for examination.

Diagnostic serum reactions

Because of their clinical importance serological tests for syphilis have been intensively studied and an enormous body of literature has accumulated; this cannot be discussed in detail; attention will be mainly directed to the tests which are in current use. For earlier work in the field see the monographs by Eagle (1937), Kahn (1950) and Wilkinson (1975).

A number of different antibodies are produced in infection with *Tr. pallidum* and the other pathogenic treponemes. These, with the tests now in use for their detection, are summarized in Table 75.1 (see p. 548).

The Wassermann test

Wassermann, Neisser, and Bruck (1906) and Wassermann, Neisser, Bruck, and Schucht (1906) showed that sera from syphilitic monkeys and human patients, but not normal sera, were reactive in a complement-fixation test with a saline extract of fetal syphilitic liver as antigen. Later it was found that saline extracts of normal tissues provided equally good antigens (Marie and Levaditi 1907) and that alcoholic extracts were superior. These observations cast doubt on the specificity of the Wassermann reaction. The reactive substance in syphilitic serum was called 'reagin', as at the time it was not thought to be a true antibody. Browning and Mackenzie (1924) showed that the addition of cholesterol to the alcoholic tissue extract had a sensitizing effect, and many modifications of this type of antigen were devised for use in complement-fixation and flocculation tests. Pangborn (1942) isolated the active principle, cardiolipin, from ox heart muscle (the usual source of these antigens); this proved to be a phosphatidic acid (Macfarlane 1958). Although inactive by itself, it fixed complement and gave flocculation with syphilitic sera when mixed with suitable amounts of lecithin and cholesterol. The three components can be obtained in a pure state and thus provide reproducible antigens. By varying their proportions the sensitivity and, within limits, the specificity of an antigen can be adjusted. Cardiolipin antigens have now completely replaced the old crude tissue extracts because of their reproducibility and greater sensitivity and specificity. International standards of cardiolipin and lecithin have been established (Report 1954*a*, *b*).

Table 75.1   Antibodies produced in treponemal infection and tests used for their detection

Antibody	Tests
Anti-lipoidal (Reagin)	Complement-fixation: Wassermann reaction Flocculation: VDRL slide test Rapid plasma reagin (RPR) test Reagin screen (RS) test
Anti-treponemal (a) Group reactive (b) Specific	Reiter protein complement-fixation (RPCF) test *Tr. pallidum* immobilization (TPI) test Fluorescent treponemal antibody (FTA) test *Tr. pallidum* haemagglutination (TPHA) test

Whatever its immunological rationale, the Wassermann reaction has in practice proved a useful means of diagnosis when taken in conjunction with clinical findings. Methods for its performance are legion (for references see 5th Edition), and many modifications have been described to increase its reliability (Osler and Strauss 1952). In general, variations in technique include (a) prolonged incubation at 4° or for short periods at 37°, (b) testing a single dilution of serum against varying doses of complement or dilutions of serum against a single dose of complement, (c) complement doses based on 50 per cent lysis rather than complete lysis of the indicator cells. With any technique appropriate controls to detect anticomplementary properties of the serum under test and the antigen are essential. The use of buffered saline containing added Ca^{++} and Mg^{++} as diluent is preferable to normal saline as these ions, which are necessary for haemolysis, may not otherwise be present in optimal amounts. Fresh guinea-pig serum is labile, and the use of lyophilized or preserved complement (Richardson 1941) is advisable.

In this country the techniques described by Price (1950, 1953), Kolmer (Report 1959) and Bradstreet and Taylor (1962) or its modification by McSwiggan and Taylor (1972) have been widely used.

Flocculation tests

Flocculation of tissue extract antigens by syphilitic sera was first described by Michaelis (1907) and Jacobsthal (1910). Their observations led to the development of many tests of which the *Kahn test* was the most widely used. They are now of only historical interest and have been superseded by tests devised to use cardiolipin antigens. For information about these early tests see the monographs by Harrison (1931) and Vogelsang (1940).

Flocculation tests have obvious advantages over complement-fixation tests in that they provide a result within minutes and only two components are required, serum and antigen. They are more sensitive and specific than the Wassermann test; and the latter has been abandoned in most laboratories in favour of the simpler flocculation tests. Those which are in current use and are based on cardiolipin antigens will be described.

Venereal Disease Research Laboratory (VDRL) test (Harris *et al.* 1946)

In this test serum and antigen are mixed on a ringed slide and rotated for four minutes; the result is read at a magnification of × 100. Antigen for the test is prepared by adding buffered saline to the alcoholic solution of VDRL antigen; the resulting suspension remains usable for only a few hours. Careful attention to detail in its preparation is essential (see Report 1969). The test is reactive in about 75 per cent of sera from untreated patients with primary syphilis, in all during the secondary stage and in early latency, and in about 75 per cent of those with late latent or late overt infections. Prozones occasionally occur with strongly reactive sera and are said to be more common in patients with late syphilis (Pautrizel *et al.* 1961). Sera need inactivation at 56° for 30 minutes before testing in order to destroy a heat-labile inhibitory factor of IgM class which may be present in both normal and syphilitic sera (Lantz and Falcone 1968, Butler and Brenner 1978).

Rapid plasma reagin (RPR) test (Portnoy 1963)

It was found that previous inactivation of sera was not necessary if the VDRL antigen was suspended in a buffered solution of choline chloride and EDTA (Portnoy and Garson 1960). This *Unheated Serum Reagin test* was later modified by adding finely divided carbon particles to the antigen to make the flocculation more easily visible. The RPR test is performed on disposable cards and is read without magnification. The antigen suspension is stable and does not need preparation for each batch of tests. In evaluations by Tio (1970), Walker (1971) and Scrimgeour and Rodin (1973), the RPR test was found to be both more sensitive and more specific than the VDRL test. RPR test antigen has been used with AutoAnalyser equipment in the *Automated RPR test* (McGrew *et al.* 1968). This method was found more sensitive than the VDRL test in screening patients attending a venereal disease clinic (Wilkinson *et al.* 1972).

Reagin screen (RS) test   This is similar to the RPR test except that the lipid antigen is stained with a blue dye instead of having carbon particles added to it.

Minimally reactive sera are said to be more easily distinguished. The test appears to be highly specific (Dyckman *et al.* 1976). In a comparison of the RPR and RS tests on 2300 sera Stevens and his colleagues (1978) found the card tests to be more specific than the VDRL test and to offer a number of practical advantages.

The behaviour of reagin antibody in treponemal infection

Antibody becomes detectable about 7 to 10 days after the appearance of the primary chancre and the titre rises to a high level during the secondary stage. About 75 per cent of patients with darkground-positive primary lesions have reactive VDRL tests (Garner 1968, Wende *et al.* 1971). The order of appearance of reactivity in various tests has been described by Lassus and his colleagues (1967). Tests for reagin are reactive in most cases of untreated latent and late syphilis, but experience with the more sensitive treponemal tests has shown that they may be negative in an appreciable proportion. After treatment of early syphilis, tests for reagin become negative within 12 to 24 months in almost all cases; the rate of reversal depends on the initial titre and the duration of infection before treatment (Lassus *et al.* 1970, Fiumara 1977a, b). In patients who have been reinfected the rate of sero-reversal is slower (Fiumara 1977c). The course after treatment of latent or late syphilis is less predictable; most patients show a slow decline in titre and some eventually become sero-negative; in others the titre may decline and become fixed for many years (Wassermann fastness). For comparisons of the relative sensitivity of tests see Dyckman *et al.* (1976) and Stevens *et al.* (1978).

Reporting of results

Grading the strength of a reaction by an arbitrary number of plus signs or verbally as strongly or weakly positive is apt to confuse the clinician, who is adequately served by a simple notation of results as positive, doubtful or negative, with the implication that doubtful results should not be given diagnostic significance but call for a repetition of the test. It is preferable for tests to be performed quantitatively and the titre expressed as the highest dilution of serum giving a definitely positive result. The World Health Organization has established an international standard for use as a reference serum (Bentzon and Krag 1961).

The nature of reagin antibody

This is still controversial. Sachs and his colleagues (1925) held that it was produced in response to lipids set free by tissue breakdown which acted in con-

junction with spirochaetal protein; this brings it into the class of auto-antibodies. Kahn (1950) sees it as part of a spectrum of antilipoidal antibodies which result from antigens set free during tissue breakdown and which show different patterns of flocculation under the conditions of his universal serological reaction. Eagle and Fleischman (1948) consider reagin to be a true antibody to a treponemal antigen which cross-reacts with tissue lipids. *Tr. pallidum* has been shown to contain lipids, including cardiolipin (Matthews *et al.* 1979). Hardy and Nell (1955) have shown that experimentally produced reagin will agglutinate *Tr. pallidum*. Any explanation of the nature of reagin must take into account the presence of this type of antibody in some normal human and animal sera and its presence in conditions associated with the biological false positive reaction.

In early syphilis, reagin antibody may be present in both IgG and IgM fractions of serum (Aho 1967, Julian and others 1969, Shannon and Booth 1977). In latent and late syphilis the IgG antibody predominates, although Aho (1967) has found IgM reagin to be the main component in some cases of late syphilis.

False positive reactions

Apart from those due to errors in technique, false positive reactions may be due to the presence of antibody in the serum—the *biological false positive* (*BFP*) *reactions*. Attempts have been made to distinguish these from true reactions, due to treponemal infection, by means of verification tests in which lipoidal antigens are used (see Kahn 1950). These methods have not proved reliable, and differentiation with certainty has been possible only since the development of specific tests in which *Tr. pallidum* is used as antigen.

BFP reactions are of two types: (a) acute reactions which are transient and last for less than six months; (b) chronic reactions which persist for years. The acute type is often associated with infection, especially when this produces fever; bacterial or viral pneumonia or malaria are examples. It may be found after immunization procedures, particularly smallpox vaccination (Lynch *et al.* 1960, Grossman and Peery 1969). Quaife and Gostling (1971) described a spate of these reactions associated with serological evidence of enterovirus infection. Acute reactions have been said to be frequent in glandular fever, but a survey by Carbrera and Carlson (1968) does not support this. They may be found in normal persons, no precipitating cause being apparent; and are not thought to have any adverse significance for the patient.

The only infection known to be associated with the chronic type of BFP reaction is leprosy (see Ruge *et al.* 1960). Reactions are often found in auto-immune disease, particularly systemic lupus erythematosus (see Harvey 1962, Harvey and Shulman 1966). They are more common in women than in men; and there is some evidence that genetic factors may be concerned, as familial occurrence has been reported (Putkonen and Lassus 1965, Kostant 1972). They have been observed in a strain of inbred mice in which auto-immune haemolytic anaemia develops spontaneously (Norins *et al.* 1970). This type of BFP reaction may occur in apparently normal persons, in some of whom signs of auto-immune disease develop subsequently. Besides the BFP reaction, sera may show other abnormalities, such as the presence of antinuclear factor, rheumatoid factor or cryoglobulins (Mustakallio *et al.* 1967, Johansson *et al.* 1972). Chronic BFP reactions are fairly common in elderly persons (Carr *et al.* 1966, Johansson *et al.* 1970), and may perhaps be due to the process of ageing itself. They are common among drug addicts, but are not accompanied by other immunological abnormalities in these patients (Tuffanelli 1968). Studies on the immunoglobulins responsible for chronic BFP reactions have shown that these are usually IgM in type (Aho 1968, Tringali *et al.* 1969).

Treponemal tests

Group anti-treponemal antibody

The Reiter protein complement-fixation (RPCF) test   Failure to grow virulent *Tr. pallidum* in artificial media led to the study of the cultivable, commensal treponemes as antigens for diagnostic tests. Interest has largely been centred on the Reiter treponeme, originally isolated from a human genital lesion in the twenties (see Reiter 1960). D'Alessandro and his colleagues (1949) showed that it contained four antigens, a heat-labile protein, a heat-stable polysaccharide and two of lipoidal nature. The antigen in current use appears to be a lipopolysaccharide–protein complex (de Bruijn 1959), and may be derived from the axial filaments (Hardy *et al.* 1975); purified preparations of these gave very sensitive and specific results in a counterimmunoelectrophoresis technique (Nell and Hardy 1978). Requirements for the optimal performance of the test have been discussed by Bekker and his colleagues (1966). They stress the need for testing undiluted serum, and for the interpretation of results in conjunction with those of tests for reagin. Evaluations of the RPCF test have been reviewed by Förström (1967).

The antibody detected in the RPCF test differs from reagin; it appears at about the same time as the latter in primary syphilis but may persist longer after treatment. The RPCF test is rather more sensitive than tests for reagin in latent and late syphilis. The specificity of the RPCF test is higher than that for reagin but less than that of the TPI and FTA–ABS tests (see p. 551). False positive reactions are thought to be due to antibodies reacting with lipopolysaccharide impurities in the antigen (Dupouey 1972). The use of the RPCF test in conjunction with a sensitive test for reagin is a useful combination for screening purposes, as it permits the detection of more reactive sera than either test alone, although in many laboratories the RPCF test has been superseded by the more sensitive *Tr. pallidum* haemagglutination test.

Specific anti-treponemal antibody

***Treponema pallidum* immobilization (TPI) test**   Nelson and Mayer (1949) and Nelson and Diesendruck (1951) described the conditions for detecting a specific antibody acting directly on *Tr. pallidum*.

Treponemes are harvested from the sliced testes of rabbits infected 6 to 8 days previously and transferred to a survival medium in which the organisms remain actively motile and virulent for several days. The suspension is freed from tissue debris by centrifugation and mixed with inactivated patient's serum and fresh guinea-pig serum as a source of complement; the control is a mixture of treponeme suspension, patient's serum and inactivated guinea-pig serum. After incubation at 35° for 18 hours in an atmosphere of 19 parts N_2 and 1 part CO_2, the motility of the treponemes in the two tubes is compared. A positive result is shown by immobilization of at least 50 per cent of the treponemes in the test compared with those in the control tube. For details of the technique see Report (1959). The reaction is dependent on complement, and for a negative result to be valid, free complement must be demonstrable at the end of the test. In-vivo sensitization of the treponemes by antibody from the rabbit can largely be avoided by treating the animal with cortisone after infection; this also increases the yield of treponemes. The test can be made more sensitive by prolonging the incubation period or by the addition of lysozyme to the suspension (Metzger *et al.* 1961, Fribourg-Blanc 1962). The presence of penicillin, tetracycline or metronidazole in serum may interfere with the test. For a review of technical aspects of the test see Fribourg-Blanc (1957).

The TPI test is now accepted as the most specific test available for treponemal disease, but neither it nor any other test will distinguish syphilis from the other treponemal diseases. (See Nelson 1952, Nielsen and Metzger 1959, Fiumara 1960 and the review by Nielsen and Reyn 1956.) Immobilizing antibody, which is IgG in nature (Tringali *et al.* 1968), develops late in the primary stage, usually after tests for reagin have become positive, and it may not be unequivocally positive until the secondary stage is reached. It is positive during latency and in almost all cases of late syphilis. Once positive, the test is little influenced by treatment unless this is given in the primary or secondary stages, when a slow reversal to negativity may occur. Atwood and his colleagues (1968) have shown that reversal may also occur in about 10 per cent of cases of treated latent or late syphilis. The test is of greatest value in distinguishing latent syphilis from biological false positive reactions occurring in tests for reagin, and in the diagnosis of late syphilis when the latter tests are negative or equivocal. It is not used as a means of follow-up after treatment.

Fluorescent treponemal antibody (FTA) test Deacon, Falcone and Harris (1957) described an indirect test in which a 1 in 5 dilution of serum is applied to a fixed film of *Tr. pallidum* and the union of antibody with treponemes demonstrated by subsequent treatment with an anti-human globulin serum labelled with fluorescein. Although very sensitive, both group and specific anti-treponemal antibodies are detected under these conditions. Dilution of the serum 1 in 200 (FTA-200 test) makes the method more specific but at the expense of sensitivity (Deacon *et al.* 1960). In the absorbed FTA test (FTA-ABS test), group-reactive antibody, which usually predominates over specific antibody (Wilkinson and Rayner 1966), is removed from serum by absorption with ultrasonically disintegrated Reiter treponemes (Hunter *et al.* 1964) or with a heated culture filtrate of these organisms (Deacon *et al.* 1966). The predominant anti-treponemal antibody at most stages of syphilis is of the IgG class. For this reason a monospecific anti-IgG conjugate should preferably be used in the test; this is thought to increase specificity and to reduce the number of borderline reactions (Hunter *et al.* 1976). For a technical review of the reagents used in the test see Hunter (1975). This method has been widely used and has a high sensitivity and specificity; doubts have been expressed about the mechanism of action and efficiency of the culture filtrate used as sorbent (Király *et al.* 1967, Cannefax *et al.* 1968, Tringali and Cox 1970, Wilkinson and Wiseman 1971, Hardy and Nell 1972). In contrast to the TPI test, antibody is detectable very early in the primary stage, rendering the FTA-ABS test positive in about 85 per cent of cases. It is positive in all secondary and latent cases, and is more sensitive than the TPI test in late syphilis. Like the latter, it is little influenced by treatment unless this is given very early in the disease. (For evaluations of the test see Deacon *et al.* 1966, Garner *et al.* 1968, Hunter *et al.* 1968, and Johnston and Wilkinson 1968.) False positive reactions have been reported in sera containing abnormal macroglobulins; and some sera from patients with lupus erythematosus may give an atypical beaded pattern of staining of treponemes (Kraus *et al.* 1971). Goldman and Lantz (1971) found less than 1 per cent probable false positive reactions in tests on 250 nuns.

The FTA test has been used to study the immunoglobulin class of anti-treponemal antibodies. In primary syphilis IgM antibody appears first and is rapidly followed by IgG, which soon predominates. IgM antibody is present in untreated primary, secondary and early latent syphilis but its presence in late syphilis is disputed. Atwood and Miller (1970) and Müller and Loa (1974) found it in some sera. Shannon and his

colleagues (1980), who studied sera fractionated by ultracentrifugation, were unable to detect IgM anti-treponemal antibody although IgM anti-lipoidal antibody was consistently present. After treatment of early syphilis IgM anti-treponemal antibody usually disappears within 6 to 18 months, whereas IgG persists (O'Neill and Nicol 1972, Luger *et al.* 1977, Shannon and Booth 1977). The significance of the persistence of IgM antibody after treatment is still uncertain. O'Neill and Nicol (1972) suggest that it may be due to relapse or reinfection and so mean continued activity of the disease. Wilkinson and Rodin (1976) found IgM antibody persisting for more than two years, usually at only a low titre, in 17 of 202 patients who had had adequate treatment by current standards; none of these 17 showed clinical evidence of active disease, although one had been reinfected.

The FTA-IgM test has also been used to distinguish passive transfer of maternal antibody from active infection in neonatal syphilis. Since IgM antibody does not cross the intact placenta, its presence in the baby's blood is thought to indicate active infection (Scotti and Logan 1968, Johnston 1972*a*, Rosen and Richardson 1975). In some cases the test does not become positive until several weeks after birth; so that serial examinations are needed if infection is to be excluded. Kaufman and his colleagues (1974) have reviewed the earlier reports and think that the value of the test is not yet clearly established and that reliance should still be placed on serial quantitative tests for reagin on the blood of mother and baby.

Anti-immunoglobulins such as rheumatoid factor, which Lassus (1969) found in 5 per cent of primary and 18 per cent of secondary syphilitic sera, may produce false positive FTA-IgM tests. Competitive inhibition by excess IgG antibody has been found to diminish the sensitivity of the test for IgM (Reimer *et al.* 1975, Müller 1977). These factors may account for some of the differences between reports on the test. Difficulties arising from them can be overcome by testing the IgM fraction of sera or by removal of the interfering antibodies by absorption procedures (Chantler *et al.* 1976).

***Treponema pallidum* haemagglutination (TPHA) test** (Tomizawa and Kasamatsu 1966) This is based on work by Rathlev (1967); it provides a simple method of detecting specific anti-treponemal antibody which does not need specialized equipment and is within the compass of general laboratories. It is performed in microtitre plates and only small volumes of reagents are needed. The antigen is a suspension of formolized tanned sheep cells sensitized with a sonicate of *Tr. pallidum*. Sera are diluted in an absorbing diluent to remove group-reactive and other interfering antibodies. The TPHA test is rather under-sensitive in primary syphilis, but beyond this stage it is as sensitive as the FTA-ABS test and more so than the TPI test. As with these, the TPHA test remains positive after treatment unless this is given very early in the disease. It has a high specificity but occasional false positive reactions occur; these do not seem to be linked to any definite clinical ailment. A small minority of sera give inconclusive results owing to agglutination of both the sensitized (test) and unsensitized (control) cells; such results can usually be clarified by retesting the sera after absorption with sheep cells. (For reviews of the test see Johnston 1972*b*, Király and Prerau 1974, Shore 1974, and Rudolph 1976.)

Other tests for specific antibody based on agglutination, immune adherence, complement-fixation and FTA-inhibition have been described but have not been widely used; for references to these see the 6th Edition.

Tests on spinal fluid

These should be performed on all patients suspected of having neurosyphilis and on patients with latent syphilis; abnormalities may be found in some of the latter patients in the absence of clinical signs of neural involvement. Tests used on serum, such as the Wassermann and VDRL tests, can be applied to the spinal fluid. They may rarely be found positive when negative on the serum; this is probably not true of the more sensitive specific tests such as the TPI, TPHA, and FTA–ABS tests. The cell count and protein content of the fluid are the most sensitive indices of activity of infection in the neuraxis; these may be abnormal although the serological tests are negative. After treatment, the cell count is the first to revert to normality, followed by the protein content; reversion of the serological tests and colloidal gold curve is considerably slower.

Choice of serological tests and their interpretation

Tests are used as a means of diagnosis and also to assess the response to treatment. For the former purpose the combination of a quantitative test for reagin such as the VDRL or RPR test with a test for specific anti-treponemal antibody provides the most effective screening procedure. The combination of a VDRL test with the RPCF test has been widely used in the past, but the latter test has now been replaced by the more sensitive and specific TPHA test in most laboratories. Positive results with either of these tests should be confirmed by the FTA–ABS test unless there is definite clinical evidence of syphilis. It should be remembered that, whatever combination of tests is used, a small proportion of cases of early primary syphilis will escape detection; these can be identified only by dark-ground examination of suspected lesions and by repetition of the serological tests after an interval. Since the introduction of the FTA–ABS and TPHA tests recourse to the TPI test is now rarely needed; it may be of value when the results of the former tests disagree, or in patients with suspected late syphilis on whose sera these tests give inconclusive or negative results. There is at present no serological method that will distinguish syphilis from the other treponematoses—yaws, pinta and endemic non-venereal syphilis. Because of the persistence of the antibody detected by the specific tests, they do not give much help in deciding whether the disease process is still active in a given patient or whether the positive result is merely a 'serological scar' of no prognostic moment. The possible significance of persisting IgM anti-treponemal antibody has already been mentioned. To assess activity, the clinical evidence and the amount and duration of treatment must be considered. Tests for reagin may be more helpful in assessing activity; this should be suspected if they are positive to a high titre or if serial tests show a rise in titre, but it must be stressed that syphilis may be clinically progressive when tests for reagin are low in titre or even negative. For the follow-up of patients after treatment, quantitative tests for reagin are adequate; the specific tests, because of their persistent reactivity, do not contribute useful information in this connection.

Prophylaxis and treatment

We do not propose to discuss the various hygienic and socio-epidemiological measures that have been advocated for the prevention and control of syphilis.

At present there is no effective means of immunization against syphilis in man, although the animal experiments discussed on p. 545 suggest that this may eventually be possible. Should this be so, and should immunization be practised on a wide scale, the value of existing serological tests as diagnostic procedures might be diminished.

Incubating syphilis can be aborted by very small doses of penicillin. The fall in incidence of early syphilis since 1946 is probably due in part to the widespread use of penicillin for gonorrhoea and other illnesses, unsuspected early infections being aborted at the same time. The treatment of symptomless contacts of patients with infectious early syphilis is advocated on epidemiological grounds by many, but this view is controversial.

The arsenicals have been completely superseded by penicillin. The curative effect of this antibiotic in man was first demonstrated by Mahoney, Arnold and Harris (1943). In early infections a course of 10–12 daily injections of 0.6 mega unit of aqueous procaine penicillin will produce cure in more than 95 per cent of cases. In patients who are thought likely to default a single dose of 2.4 mega units of benzathine penicillin can be given. In late syphilis the aim of treatment is to prevent progression of disease. Longer courses, such as 0.6 mega unit of aqueous procaine penicillin daily for 20 days or three injections of three mega units of benzathine penicillin at weekly intervals, are recommended. Penicillin penetrates poorly into the cerebrospinal fluid after such courses of treatment (Dunlop *et al.* 1979, Polnikorn *et al.* 1980). In neurosyphilis the injection of 0.5 mega unit of crystalline penicillin with 0.5 g probenecid orally every six hours for 15–20 days has been advocated. If penicillin cannot be given because of sensitization, 500 mg tetracycline or erythromycin four times daily for 15 days in early syphilis and 30 days in later infections are alternatives. Experience of the long-term results of these last two methods is still inadequate; there have been reports of the failure of erythromycin to prevent fetal infection when it was given during pregnancy. For details of the treatment and management of syphilis see King, Nicol and Rodin (1980) and for syphilis in pregnancy, Thompson (1976).

Whatever form of treatment is given, it is essential that the patient is kept under clinical observation and progress monitored by regular serological tests. Fever therapy, including fever induced by infection with malaria, was formerly used to treat neurosyphilis, but has been superseded by penicillin.

The initial treatment of syphilis may be accompanied by a Jarisch–Herxheimer reaction. This is common in early syphilis in which it manifests itself in the form of fever, headache and an exacerbation of the early lesions or rash. It is transient and unimportant at this stage of the disease. In some forms of late syphilis with lesions in vital structures, such as gumma of the larynx or involvement of the coronary ostia, serious results have been attributed to it, and preliminary treatment with steroids has been advocated. The reaction may be a hypersensitivity due to the sudden release of treponemal products (Heyman *et al.* 1952); see also Bryceson (1970).

For further information on the different aspects of syphilis, the annual reviews of the literature presented by Beerman and his colleagues (1960) should be consulted, together with those by R. R. Willcox beginning in 1950.

Rabbit syphilis

Rabbit syphilis is a naturally occurring venereal disease of rabbits due to the spirochaete *Treponema cun-*

iculi. The disease was first described by Ross (1912) and the spirochaetes were first observed by Bayon (1913). The incidence of the disease in wild rabbits varies considerably; according to some authorities as many as 20–40 per cent in Great Britain are affected. In hutch rabbits the disease is less common; Adams, Cappell and McCluskie (1928) noticed it in 14 out of 228 rabbits. The lesions consist of small scaly patches, often slightly eroded and covered with a brownish crust, situated on the genitals or in the perineal region. Sometimes the nostrils or eyelids are affected. The spirochaetes are found in large numbers in scrapings from the lesions and in sections; they are confined apparently to the superficial layers. The disease can be reproduced in normal rabbits by inoculation of an infective tissue suspension on to the scarified skin of the genital region; transmission can also be effected by mating, but with less constancy (Graves *et al.* 1980). The incubation period is from 2 to 8 weeks; once established, infection persists for months. In males spontaneous cure usually occurs, but in females the disease is very chronic. Rabbits suffering from rabbit syphilis can be infected with human syphilitic material, indicating that the two diseases are not identical (Noguchi 1922). *Tr. cuniculi* is morphologically indistinguishable from *Tr. pallidum,* and infection is accompanied by the production of similar anti-lipoidal and anti-treponemal antibodies. Penicillin is curative. Infection of rabbits with *Tr. cuniculi* gives some protection against both syphilis and yaws (Turner and Hollander 1957). Graves (1981) found there was an initial depression of immunity followed after five months by limited protection against challenge with low doses of *Tr. pallidum.* Inoculation of a human volunteer intradermally with 2×10^7 virulent *Tr. cuniculi* produced a transient skin lesion and the Wassermann and TPHA tests became weakly positive for a short time (Graves and Downes 1981). It seems that *Tr. cuniculi* does not produce a strong immune response in man, although it does so in rabbits, and is therefore unlikely to provide a vaccine effective against human syphilis. (For further information on the disease and the properties of the causative organism see McLeod and Turner 1946, Turner and Hollander 1957, Smith and Pesetsky 1967.)

Yaws, pinta, and endemic syphilis

Yaws is a contagious inoculable disease characterized by the appearance of papules, which generally develop into a fungating, encrusted, granulomatous eruption (Manson 1914). It is widely diffused throughout the tropics, and occurs in its typical form only among the dark-skinned races. Alternative names for it are *pian* (French), *framboesia* (German and Dutch), *buba* (Spanish), and *bouba* (Portuguese). Infection occurs in

childhood. Its spread is favoured by a warm humid atmosphere, scanty clothing, poor personal cleanliness and a low standard of living. Latent cases are 2–5 times as common as clinically manifest disease, and the patients are apt to suffer from infectious relapses in the early years after infection (Report 1960). The disease is not of venereal origin. It is spread by contact, and possibly by insects. Congenital infection is unknown. The commonest site for the primary lesion is the lower extremities. The secondary eruption usually appears in another 2–4 weeks. In the late stages of yaws, bone lesions and ulcerative skin lesions develop. The causative organism, *Tr. pertenue*, was discovered by Castellani in 1905. It is indistinguishable from *Tr. pallidum*. Serologically, yaws cannot be distinguished from syphilis, even by the TPI test, thus indicating the extremely close relation of the two organisms. Treponemes have been isolated from monkeys captured in areas in which yaws is prevalent; the relation of these strains to *Tr. pertenue* is still under investigation (see Sepetjian *et al.* 1969). Extensive campaigns have been waged against yaws in many tropical countries under the general auspices of the World Health Organization (see Report 1960, Guthe 1969, Guthe *et al.* 1972). Infection with yaws provides considerable immunity against syphilis, although this is not absolute and may wane with time. In areas where treatment campaigns have been conducted young adults may no longer have the protection given by childhood yaws; there is evidence that venereally acquired syphilis is invading some of these regions. (For information on the nomenclature, pathology, epidemiology, diagnosis and treatment of yaws, see Symposium 1953, Hackett 1957, and Turner's Monograph 1959.)

Pinta, like yaws, is a contagious inoculable disease. It is restricted almost entirely to Mexico, Central America, parts of sub-tropical South America, and Cuba. Infection occurs by contact under conditions similar to those favouring the spread of yaws. Children 10–15 years of age are most commonly affected. The skin bears the brunt of the disease. The lesions are dry and scaly and variously coloured (hence the name from the Spanish word *pintar*, meaning to paint); they progress slowly and run a prolonged course. Other tissues of the body are seldom attacked. The causative organism, *Tr. carateum*, was first demonstrated by Saenz and his colleagues (1938). It is morphologically indistinguishable from *Tr. pallidum*. The disease has been transmitted to chimpanzees (Kuhn *et al.* 1970). Wassermann and flocculating antibodies appear in a low proportion of patients in the primary stage, in about 60 per cent of those in the secondary stage, and in most of those in the late stages (see Rein *et al.* 1952). Penicillin is curative (Rein *et al.* 1952).

Numerous forms of **endemic syphilis** are known, frequently called by different names in different countries, such as **Bejel** of the Middle East, and **Skerljevo** of Bosnia (see Grin 1953, Hudson 1961). Transmission is usually by contact. Venereal spread is uncommon. Primary chancres are rare. The outstanding features are the frequency of mucous patches in the mouth, and the occurrence of skin and bone lesions in both early and late stages of the disease.

The relation to each other of the various treponematoses and their causative organisms has still to be worked out. According to Turner and Hollander (1957), all the treponemal strains seem to cross-immunize to some extent, and no antigenic differences can be made out between them by the TPI test. In man too few observations have been made on cross-protection to justify any generalization.

References

Adams, D. K., Cappell, D. F. and McCluskie, J. A. (1928) *J. Path. Bact.* **31**, 157.

Aho, K. (1967) *Brit. J. vener. Dis.* **43**, 259; (1968) *Ibid.* **44**, 49.

Arnold, R. C., Mahoney, J. F. and Cutler, R. C. (1947) *Amer. J. Syph.* **31**, 264, 489.

Arnold, R. C., Wright, R. D. and Levitan, S. (1950a) *Amer. J. Syph.* **34**, 324.

Arnold, R. C., Wright, R. D. and MacLeod, C. (1950b) *Amer. J. Syph.* **34**, 327.

Atwood, W. G. and Miller, J. L. (1970) *Int. J. Derm.* **9**, 259.

Atwood, W. G., Miller, J. L., Stout, G. W. and Norins, L. C. (1968) *J. Amer. med. Ass.* **203**, 549.

Bauer, T. J., Price, E. V. and Cutler, J. C. (1952) *Amer. J. Syph.* **38**, 447.

Baughn, R. E., Musher, D. M. and Simmons, C. (1977) *Infect. Immun.* **17**, 535.

Bayon, H. (1913) *Brit. med. J.* ii, 1159.

Beerman, H., Nicholas, L., Schamberg, I. L. and Greenberg, M. S. (1960) *Arch. intern. Med.* **105**, 145, 324.

Bekker, J. H., Bruijn, J. H. de and Miller, J. N. (1966) *Brit. J. vener. Dis.* **42**, 42.

Bentzon, M. W. and Krag, P. (1961) *Bull. World Hlth Org.* **24**, 257.

Bishop, N. H. and Miller, J. N. (1976a) *J. Immunol.* **117**, 191; (1976b) *Ibid.* 197.

Bradstreet, C. M. P. and Taylor, C. E. D. (1962) *Mon. Bull. publ. Hlth Lab. Serv.* **21**, 96.

Brown, W. J. (1971) *J. infect. Dis.* **124**, 428.

Browning, C. H. and Mackenzie, I. (1924) *Recent Methods in the Diagnosis and Treatment of Syphilis.* 2nd Edn, London.

Bruijn, J. H. de (1959) *Leeuwenhoek ned. Tijdschr.* **24**, 41.

Bryceson, A. D. M. (1970) *Proceedings of the International Symposium on Immune Complex Diseases.* Milan.

Butler, G. H. and Brenner, L. J. (1978) *Z. ImmunForsch.* **154**, 442.

Cannefax, G. R. (1965) *Brit. J. vener. Dis.* **41**, 260.

Cannefax, G. R., Hanson, A. W. and Skaggs, R. (1968) *Publ. Hlth Rep., Wash.* **83**, 411.

Carbrera, H. A. and Carlson, J. (1968) *Amer. J. clin. Path.* **50**, 643.

Carr, R. D., Becker, S. W. and Carpenter, C. M. (1966) *Arch. Derm.* **93**, 393.

Castellani, A. (1905) *Brit. med. J.* ii, 1280.

Chantler, S., Devries, E., Allen, P. R. and Hurn, B. A. L. (1976) *J. immunol. Meth.* **13**, 367.

Chesney, A. M. (1927) *Immunity in Syphilis.* Williams & Wilkins, Baltimore.

Chesney, A. M. and Kemp, J. E. (1924) *J. Amer. med. Ass.* **83**, 1725.

Chesney, A. M., Woods, A. C. and Campbell, A. D. (1939) *J. exp. Med.* **69**, 103.

Cockburn, T. A. (1961) *Bull. World Hlth Org.* **24**, 22.

Collart, P., Borel, L.-J. and Durel, P. (1962a) *Ann. Inst. Pasteur* **102**, 596, 693; (1962b) *Ibid.* **103**, 953.

Collart, P., Franceschini, P., Poitevin, M., Dunoyer, F. and Dunoyer, M. (1974) *Brit. J. vener. Dis.* **50**, 251.

D'Alessandro, G., Oddo, F., Comes, R. and Dardanoni, L. (1949) *Riv. Ist. Sieroter. Ital.* **24**, 134.

Deacon, W. E., Falcone, V. H. and Harris, A. (1957) *Proc. Soc. exp. Biol., N.Y.* **96**, 477.

Deacon, W. E., Freeman, E. M. and Harris, A. (1960) *Proc. Soc. exp. Biol., N.Y.* **103**, 827.

Deacon, W. E., Lucas, J. B. and Price, E. V. (1966) *J. Amer. med. Ass.* **198**, 624.

Dennie, C. C. (1962) *A History of Syphilis.* Charles C Thomas, Ill.

Dunlop, E. M. C. (1972) *Brit. med. J.* ii, 577.

Dunlop, E. M. C., Al-Egaily, S. S. and Houang, E. T. (1979) *J. Amer. med. Ass.* **241**, 2538.

Dupouey, P. (1972) *Ann. Inst. Pasteur* **122**, 283.

Dyckman, J. D., Wende, R. D., Gantenbein, D. and Williams, R. P. (1976) *J. clin. Microbiol.* **4**, 145.

Eagle, H. (1937) *The Laboratory Diagnosis of Syphilis.* Kimpton, London.

Eagle, H. and Fleischman, R. (1948) *J. exp. Med.* **87**, 369.

Fitzgerald, T. (1981*a*) *Bull. World Hlth Org.* **59**, 787.

Fitzgerald, T. J. (1981*b*) *Annu. Rev. Microbiol.* **35**, 29.

Fiumara, N. J. (1960) *Publ. Hlth Rep., Wash.* **75**, 1011; (1977*a*) *Sex. trans. Dis.* **4**, 92; (1977*b*) *Ibid.* 96; (1977*c*) *Ibid.* 132.

Förström, L. (1967) *Acta derm-venereol., Stockh.* **47**, Suppl. 59.

Fribourg-Blanc, A. (1957) *Ann. Derm. Syph.* **84**, 286, 410; (1962) *Ann. Inst. Pasteur* **102**, 460.

Garner, M. F. (1968) *Med. J. Aust.* **i**, 672.

Garner, M. F., Grantham, N. M., Collins, C. A. and Roeder, P. J. (1968) *Med. J. Aust.* **i**, 404.

Gastinel, P. and Pulvenis, R. (1934) *La Syphilis Expérimentale.* Masson et Cie, Paris.

Gastinel, P., Pulvenis, R. and Collart, P. (1938) *C. R. Soc. Biol., Paris* **128**, 739.

Goldman, J. N. and Lantz, M. A. (1971) *J. Amer. med. Ass.* **217**, 53.

Graves, S. (1981) *Brit. J. vener. Dis.* **57**, 11.

Graves, S. and Downes, J. (1981) *Brit. J. vener. Dis.* **57**, 7.

Graves, S. R., Edmonds, J. W. and Shepherd, R. C. H. (1980) *Brit. J. vener. Dis.* **56**, 381.

Grin, E. I. (1952) *Bull. World Hlth Org.* **7**, 1; (1953) *World Hlth Org. Monograph* No. 11.

Grossman, L. J. and Peery, T. M. (1969) *Amer. J. clin. Path.* **51**, 375.

Guthe, T. (1969) *Acta derm-venereol., Stockh.* **49**, 343.

Guthe, T., Ridet, J., Vorst, F., D'Costa, J. and Grab, B. (1972) *Bull. World Hlth Org.* **46**, 1.

Hackett, C. J. (1957) *World Hlth Org., Monograph Ser.* No. 36; (1963) *Bull. World Hlth Org.* **29**, 7.

Haensell, P. (1881) *v. Graefes Arch. Ophthal.* **27**, 93.

Hardy, P. H., Fredericks, W. H. and Nell, E. E. (1975) *Infect. Immun.* **11**, 380.

Hardy, P. H. and Nell, E. E. (1955) *J. exp. Med.* **101**, 367; (1972) *Amer. J. Epidem.* **96**, 141.

Harris, A., Rosenberg, A. A. and Riedel, L. M. (1946) *J. vener. Dis. Inform.* **27**, 169.

Harrison, L. W. (1931) *A System of Bacteriology in Relation to Medicine* **8**, 251. H.M.S.O., London.

Harvey, A. M. (1962) *J. Amer. med. Ass.* **182**, 513.

Harvey, A. M. and Shulman, L. E. (1966) *Med. Clin. N. America* **50**, 1271.

Hederstedt, B. (1974) *Acta path. microbiol. scand.* **82B**, 185; (1976*a–e*) *Ibid.* **84C**, 135, 142, 148, 245, 250.

Heyman, A., Sheldon, W. H. and Evans, L. D. (1952) *Brit. J. vener. Dis.* **28**, 50.

Hollander, D. H., Turner, T. B. and Nell, E. E. (1952) *Bull. Johns Hopk. Hosp.* **90**, 105.

Hudson, E. H. (1958) *Non-venereal Syphilis.* E. and S. Livingstone Ltd, Edinburgh and London; (1961) *Arch Derm., Chicago* **84**, 545; (1965) *Bull. World Hlth Org.* **32**, 735.

Hunter, E. F. (1975) *CRC crit. Rev. clin. Lab. Sci.* 315.

Hunter, E. F., Deacon, W. E. and Meyer, P. E. (1964) *Publ. Hlth Rep., Wash.* **79**, 410.

Hunter, E. F., Maddison, S. E., Larsen, S. A., Felker, M. B. and Feeley, J. C. (1976) *J. clin. Microbiol.* **4**, 338.

Hunter, E. F., Norins, L. C., Falcone, V. H. and Stout, G. W. (1968) *Bull. World Hlth Org.* **39**, 873.

Jacobsthal, E. (1910) *Z. ImmunForsch.* **8**, 107.

Johansson, E. A., Lassus, A., Apajalahti, A. and Aho, K. (1970) *Ann. clin. Res.* **2**, 47.

Johansson, E. A., Lassus, A. and Salo, O. P. (1972) *Acta derm-venereol., Stockh.* **52**, 196.

Johnston, N. A. (1972*a*) *Brit. J. vener. Dis.* **48**, 464; (1972*b*) *Ibid.* **48**, 474.

Johnston, N. A. and Wilkinson, A. E. (1968) *Brit. J. vener. Dis.* **44**, 287.

Jue, R., Puffer, J., Wood, R. M., Schochet, G., Smartt, W. H. and Ketterer, W. A. (1967) *Amer. J. clin. Path.* **47**, 809.

Julian, A. J., Logan, L. C. and Norins, L. C. (1969) *J. Immunol.* **102**, 1250.

Kahn, R. L. (1950) *Serology with Lipid Antigen.* Ballière, Tindall & Cox, London.

Kaufman, R. E., Olansky, D. C. and Wiesner, P. J. (1974) *J. Amer. vener. Dis. Ass.* **1**, 79.

Kellogg, D. S. (1970) *Hlth Lab. Sci.* **7**, 34.

King, A., Nicol, C. and Rodin, P. (1980) *Venereal Diseases.* 4th Edn, Ballière Tindall, London.

Király, K., Jobbágy, A. and Kováts, L. (1967) *J. invest. Derm.* **48**, 98.

Király, K. and Prerau, H. (1974) *Acta derm-venereol., Stockh.* **54**, 303.

Klebs, A. C. (1932) *Science* **75**, 191.

Kostant, G. H. (1972) *J. Amer. med. Ass.* **219**, 45.

Kraus, S. R., Haserick, J. R., Logan, L. C. and Bullard, J. E. (1971) *J. Immunol.* **106**, 1665.

Kuhn, U. S. G., Medina, R., Cohen, P. G. and Vegas, M. (1970) *Brit. J. vener. Dis.* **46**, 311.

Lantz, M. A. and Falcone, V. H. (1968) *Amer. J. med. Tech.* **34**, 551.

Lassus, A. (1969) *Int. Arch. Allergy. appl. Immunol.* **36**, 515.

Lassus, A., Johansson, E. and Förström, L. (1970) *Acta derm-venereol., Stockh.* **50**, 148.

Lassus, A., Mustakallio, K. K., Aho, K. and Putkonen, T. (1967) *Acta path. microbiol. scand.* **69**, 612.

Luger, A., Schmidt, B. and Spendlingwimmer, I. (1977) *Brit. J. vener. Dis.* **53**, 287.

Lukehart, S. A. and Miller, J. N. (1978) *J. Immunol.* **121**, 2014.

Lynch, F. W., Kimball, A. C. and Kernan, P. D. (1960) *J. invest. Derm.* **34**, 219.

Macfarlane, M. G. (1958) *Nature, Lond.* **182**, 946.

McGrew, B. E., Du Cross, J. F., Stout, G. W. and Falcone, V. H. (1968) *Amer. J. clin. Path.* **50**, 52.

McLeod, C. and Turner, T. B. (1946) *Amer. J. Syph.* **30**, 442, 455.

McSwiggan, D. A. and Taylor, C. E. D. (1972) Laboratory Diagnosis of Venereal Disease. *Public Health Laboratory Service Monograph Series* No. 1. H.M.S.O., London.

Magnuson, H. J. (1948) *Amer. J. Med.* **5**, 641.

Magnuson, H. J., Eagle, H. and Fleischman, R. (1948) *Amer. J. Syph.* **32**, 1.

Magnuson, H. J., Halbert, S. P. and Rosenau, B. J. (1947) *J. vener. Dis. Inform.* **28**, 203.

Magnuson, H. J. and Rosenau, B. J. (1948) *Amer. J. Syph.* **32**, 203, 418.

Magnuson, H. J., Thomas, E. W., Olansky, S., Kaplan, B. I., Mellow, L. de and Cutler, J. C. (1956) *Medicine, Baltimore* **35**, 33.

Magnuson, H. J., Thompson, F. A. and Rosenau, B. J. (1950) *Amer. J. Syph.* **34**, 219.

Mahoney, J. F., Arnold, R. C. and Harris, A. (1943) *Amer. J. publ. Hlth* **33**, 1387.

Manson, P. (1914) *Tropical Diseases.* London.

Marie, A. and Levaditi, C. (1907) *Ann. Inst. Pasteur* **21**, 138.

Matthews, H. M., Yang, T-K. and Jenkin, H. M. (1979) *Infect. Immun.* **24**, 713.

Metchnikoff, E. and Roux, E. (1903) *Ann. Inst. Pasteur* **17**, 809; (1904*a*) *Ibid.* **18**, 1; (1904*b*) *Ibid.* **18**, 657; (1905) *Ibid.* **19**, 673.

Metzger, M. (1979) *Brit. J. vener. Dis.* **55**, 94.

Metzger, M., Hardy, P. H. and Nell, E. E. (1961) *Amer. J. Hyg.* **73**, 236.

Metzger, M., Michalska, E., Podwinska, J. and Smogor, W. (1969) *Brit. J. vener. Dis.* **45**, 299.

Metzger, M. and Smogôr, W. (1969) *Brit. J. vener. Dis.* **45**, 308; (1975) *Arch. Immunol. Ther. exp.* **23**, 625.

Michaelis, L. (1907) *Berl. klin. Wschr.* **44**, 1477.

Miller, J. N. (1967) *J. Immunol.* **99**, 1012; (1971) *Spirochaetes in Body Fluids and Tissues.* Charles C Thomas, Ill.

Müller, F. (1977) *Immun. Infekt.* **5**, 109.

Müller, F. and Loa, P. L. (1974) *Infection* **2**, 127.

Mustakallio, K. K., Lassus, A., Putkonen, T. and Wager, O. (1967) *Acta derm-venereol., Stockh.* **47**, 249.

Nell, E. H. and Hardy, P. H. (1978) *J. clin. Microbiol.* **8**, 148.

Nelson, R. A. (1952) *Brit. J. vener. Dis.* **28**, 160; (1953) *Science* **118**, 733.

Nelson, R. A. and Diesendruck, J. A. (1951) *J. Immunol.* **66**, 667.

Nelson, R. A. and Mayer, M. M. (1949) *J. exp. Med.* **89**, 369.

Nielsen, H. A. and Metzger, M. (1959) *Brit. J. vener. Dis.* **35**, 241.

Nielsen, H. A. and Reyn, A. (1956) *Bull. World Hlth Org.* **14**, 263.

Noguchi, H. (1922) *J. exp. Med.* **35**, 391.

Noguchi, H. and Moore, J. W. (1913) *J. exp. Med.* **17**, 232.

Norins, L. C., Logan, L. C. and Lantz, M. A. (1970) *J. Immunol.* **105**, 1108.

O'Neill, P. and Nicol, C. S. (1972) *Brit. J. vener. Dis.* **48**, 460.

Osler, A. G. and Strauss, J. H. (1952) *Amer. J. Syph.* **36**, 140.

Pangborn, M. C. (1942) *J. biol. Chem.* **143**, 247; (1945) *Ibid.* **161**, 71.

Pautrizel, R., Szersnovicz, F. and Toulza, M. (1961) *Ann. Biol. clin.* **19**, 707.

Pavia, C. S., Folds, J. D. and Baseman, J. B. (1978) *Brit. J. vener. Dis.* **54**, 144.

Polnikorn, N., Witoonpanich, R., Varachit, M., Vejjajiva, S. and Vejjajiva A. (1980) *Brit. J. vener. Dis.* **56**, 363.

Portnoy, J. (1963) *Amer. J. clin. Path.* **40**, 473.

Portnoy, J. and Garson, W. (1960) *Publ. Hlth Rep., Wash.* **75**, 985.

Price, I. N. O. (1950) *Brit. J. vener. Dis.* **26**, 172; (1953) *Ibid.* **29**, 12, 78, 175.

Putkonen, T. and Lassus, A. (1965) *Dermatologica* **130**, 332.

Quaife, R. A. and Gostling, J. V. T. (1971) *J. clin. Path.* **24**, 120.

Rathlev, T. (1967) *Brit. J. vener. Dis.* **43**, 181.

Reimer, C. B., Black, C. M., Phillips, C. J., Logan, L. C., Hunter, E. F., Pender, B. J. and McGrew, B. E. (1975) *Ann. N.Y. Acad. Sci.* **254**, 77.

Rein, C. R. *et al.* (1952) *J. invest. Derm.* **18**, 137.

Reiter, H. (1960) *Brit. J. vener. Dis.* **36**, 18.

Report (1954*a*) *World Hlth Org., Tech. Rep. Ser.* No. 86, p. 11; (1954*b*) *Ibid.* No. 79; (1959) *Manual of Serologic Tests for Syphilis.* U.S. pub. Hlth Serv.; (1960) *World Hlth Org., Tech. Rep. Ser.,* No. 190; (1969) *Manual of Serologic Tests for Syphilis.* U.S. Pub. Hlth Service; (1981) Annu. Rep. C.M.O. Dept Hlth & Social Security, Lond., *Brit. J. vener. Dis.* **57**, 402.

Rice, N. S. C. *et al.* (1970) *Brit. J. vener. Dis.* **46**, 1.

Richardson, G. M. (1941) *Lancet* **ii**, 696.

Rosen, E. U. and Richardson, N. J. (1975) *J. Pediat.* **87**, 38.

Ross, E. H. (1912) *Brit. med. J.* **ii**, 1653.

Rudolph, A. H. (1976) *J. Amer. vener. Dis. Ass.* **3**, 3.

Ruge, H. G. S., Fromm, G., Fühner, F. and Guinto, R. S. (1960) *Bull. World Hlth Org.* **23**, 793.

Sachs, H., Klopstock, A. and Weil, A. J. (1925) *Dtsch. med. Wschr.* **51**, 589.

Saenz, B., Grau Triana, J. and Alfonso Armenteros, J. (1938) *Arch. esp. Med. interna* **4**, 112.

Schaudinn, F. and Hoffmann, E. (1905) *Arb. Reichsge-sundhAmt.* **22**, 527.

Schuberg, A. and Schlossberger, H. (1930) *Klin. Wschr.* **9**, 499.

Scotti, A. T. and Logan, L. (1968) *J. Pediat.* **73**, 242.

Scrimgeour, G. and Rodin, P. (1973) *Brit. J. vener. Dis.* **49**, 342.

Sepetjian, M., Guerraz, F. T., Salussola, D., Thivolet, J. and Monier, J. C. (1969) *Bull. World Hlth Org.* **40**, 141.

Shannon, R. and Booth, S. D. (1977) *Brit. J. vener. Dis.* **53**, 281.

Shannon, R., Copley, C. G. and Morrison, G. D. (1980) *Brit. J. vener. Dis.* **56**, 372.

Shore, R. N. (1974) *Arch. Dermatol.* **109**, 854.

Smith, J. L. and Israel, C. W. (1967) *J. Amer. med. Ass.* **199**, 980.

Smith, J. L. and Pesetsky, B. R. (1967) *Brit. J. vener. Dis.* **43**, 117.

Stevens, R. W., Gombel, K. and Gaafar, H. A. (1978) *Hlth Lab. Sci.* **15**, 81.

Stokes, J. H. (1931) *Science* **74**, 502.

Symposium (1953) 1st int. Symp. Yaws Control. *Monogr. Ser., World Hlth Org.* No. 15.

Thompson, S. E. (1976) *J. Amer. vener. Dis. Ass.* **3**, 159.

Tio, B. S. (1970) *Brit. J. vener. Dis.* **46**, 287.

Tomizawa, T. and Kasamatsu, S. (1966) *Jap. J. med. Sci. Biol.* **19**, 305.

Tringali, G. R. and Cox, P. M. (1970) *Brit. J. vener. Dis.* **46**, 313.

Tringali, G., Del Carpio, C. and Zaffiro, P. (1968) *Riv. Ist. Sieroter. Ital.* **43**, 161.

Tringali, G. R., Julian, A. J. and Halbert, W. M. (1969) *Brit. J. vener. Dis.* **45,** 202.

Tuffanelli, D. L. (1968) *Acta derm-venereol., Stockh.* **48,** 542.

Turner, L. H. (1959) *Inst. med. Res., Malaya* Bull. No. 9.

Turner, T. B. (1939) *J. exp. Med.* **69,** 867.

Turner, T. B., Hardy, P. H. and Newman, B. (1969) *Brit. J. vener. Dis.* **45,** 183.

Turner, T. B., Hardy, P. H., Newman, B. and Nell, E. E. (1973) *Johns Hopkins med. J.* **133,** 241.

Turner, T. B. and Hollander, D. H. (1957) Biology of the Treponematoses. *World Hlth Org., Monograph Ser.,* No. 35.

Turner, T. B., Kluth, F. C., McLeod, C. and Winsor, C. P. (1948) *Amer. J. Hyg.* **48,** 173.

Turner, T. B., McLeod, C. and Updyke, E. L. (1947) *Amer. J. Hyg.* **46,** 287.

Turner, T. B. and Nelson, R. A. (1950) *Trans. Ass. Amer. Physcns* **63,** 112.

Uhlenhuth, P. and Mulzer, P. (1913) *Arb. ReichsgesundhAmt.* **44,** 307.

Urbach, E. and Beerman, H. (1947) *Amer. J. Syph.* **31,** 192.

Vogelsang, T. M. (1940) *Séro-diagnostic de la Syphilis.* J. W. Eides, Bergen.

Walker, A. N. (1971) *Brit. J. vener. Dis.* **47,** 259.

Wassermann, A., Neisser, A. and Bruck, C. (1906) *Dtsch. med. Wschr.* **32,** 745.

Wassermann, A., Neisser, A., Bruck, C. and Schucht, A. (1906) *Z. Hyg. InfektKr.* **55,** 451.

Weiser, R. S., Erichson, D., Perine, P. L. and Pearsall, N. N. (1976) *Infect. Immun.* **13,** 1402.

Wende, R. D., Mudd, R. L., Knox, J. M. and Holder, W. R. (1971) *Sthn med. J.* **64,** 633.

Wilkinson, A. E. (1975) *Recent Advances in Sexually Transmitted Diseases,* p. 127. Ed. by R. S. Morton and J. R. W. Harris. Churchill Livingstone, Edinburgh.

Wilkinson, A. E. and Cowell, L. P. (1971) *Brit. J. vener. Dis.* **47,** 252.

Wilkinson, A. E. and Rayner, C. F. A. (1966) *Brit. J. vener. Dis.* **42,** 8.

Wilkinson, A. E. and Rodin, P. (1976) *Brit. J. vener. Dis.* **52,** 219.

Wilkinson, A. E., Scrimgeour, G. and Rodin, P. (1972) *J. clin. Path.* **25,** 437.

Wilkinson, A. E. and Wiseman, C. C. (1971) *Proc. roy. Soc. Med.* **44,** 422.

Willcox, R. R. (1950) *Bull. Hyg.* **25,** 331; (1972) *Trans. St Johns Hosp. derm. Soc.* **58,** 21.

Wright, D. J. M., Gangas, J. M. and Rees, R. J. W. (1974) *Guy's Hosp. Rep.* **123,** 385.

Yobs, A. R., Clark, J. W., Mothershed, S. E., Bullard, J. C. and Artley, C. W. (1968) *Brit. J. vener. Dis.* **44,** 116.

Yobs, A. R., Rockwell, D. H. and Clark, J. W. (1964) *Brit. J. vener. Dis.* **40,** 248.

76

Chlamydial diseases

L. H. Collier and G. L. Ridgway

Introductory

The genus *Chlamydia* contains two species, *Chlam. trachomatis* and *Chlam. psittaci*, which resemble each other in their morphology and mode of replication and in their possession of a common complement-fixing antigen. They differ greatly, however, in their host range and organ specificity: *Chlam. trachomatis* infects almost exclusively the eye and urogenital tract of man, whereas *Chlam. psittaci* primarily affects birds and mammals, in whom it causes a wide range of syndromes including pneumonitis, polyarthritis, encephalomyelitis, and infections of the gut and placenta.

Diseases caused by *Chlam. trachomatis*

In man, the syndromes caused by *Chlam. trachomatis* fall into three groups, each of which is associated with a particular set of serotypes: (a) ophthalmic trachoma (mainly serotypes A, B, Ba and C); (b) oculogenital and, occasionally, more general infections (mainly serotypes D-K); and (c) lymphogranuloma venereum (serotypes L1, L2, and L3).

Much recent information on the clinical and laboratory aspects of chlamydial infections of man is contained in Symposium (1983*a*, *b*).

Trachoma

In its classical form, trachoma is defined as 'a specific communicable keratoconjunctivitis, usually of chronic evolution ... characterized by follicles, papillary hyperplasia, pannus and, in its later stages, cicatrization' (World Health Organization 1962). These lesions are more pronounced in, or confined to, the upper segment of the eye. Untreated, the disease runs a prolonged course and in severe cases there may be partial or total loss of vision.

History

Trachoma is one of the earliest recorded diseases. The Ebers papyrus (*ca* 1500 BC) refers to an affliction that was almost certainly trachoma and to its alleviation with copper salts, a form of treatment that persisted well into the twentieth century. Trachoma was familiar to the ancient Greeks and Romans. The name is Greek (τράχωμα = roughness) and refers to the characteristic conjunctival follicles.

Accounts of military campaigns from the Crusades to the Napoleonic wars refer to severe ophthalmic infections acquired in the Middle East and it is quite possible that trachoma was disseminated to Europe and elsewhere by returning soldiery; but some of these outbreaks may have been due, at least in part, to bacterial infections. Clinical descriptions became more exact nearer our own time and there is no doubt that trachoma was prevalent in immigrants from eastern Europe to the United Kingdom in the latter part of the nineteenth century; but the infection did not spread to the general population and eventually disappeared, even before the introduction of specific treatment.

Epidemiology

Geographical distribution   Trachoma is endemic primarily in tropical and subtropical countries; those worst affected are North Africa, the Middle East, and the northern part of the Indian subcontinent. The disease is also prevalent in sub-Saharan Africa, the Far East, Australasia, and Latin America; it is still to be found among North American Indians. Its prevalence and severity vary considerably from country to country and in different areas within the same country. Trachoma is associated with poor living standards and hygiene and thus tends to be more prevalent in rural than in urban settings.

The worldwide prevalence of trachoma is often quoted as 400–500 million cases, but because of the paucity of information from many areas, this figure is a very crude estimate. Only a minority suffer severe visual impairment, but even so the cases in this category must number several million. The epidemiology of trachoma is reviewed by Reinhards (1969). For detailed statistics of the prevalence of trachoma from 1955–69 in most countries of the world, see World Health Organization (1971).

Mode of spread   Classical trachoma is spread from eye to eye, but the mechanism is not certain and may vary in different circumstances; what evidence there is is largely circumstantial. Flies contaminated with infective discharges are often blamed, and Jones (1975) showed that fluorescein-stained ocular discharge can be transferred in this way between children in the same household. This mode of transmission is more likely to occur in areas such as North Africa and the Middle East, where ocular discharges are copious because of associated bacterial infections, and flies are numerous. On the other hand, trachoma is highly prevalent in The Gambia, where neither epidemic bacterial conjunctivitis nor fly infestation is serious; in such areas, poor hygiene, close physical contact and infected fomites may be the main factors in transmission (Sowa *et al.* 1965, pp. 71–76). Shared eye cosmetics such as mascara (Thygeson and Dawson 1966) and *kohl* have also been implicated in the spread of trachoma. The *incubation period* of naturally acquired trachoma is uncertain since the onset is insidious and it is difficult or impossible to establish the date of contact. In experimental infections of volunteers the incubation period is 2–7 days; on the premiss that under natural conditions the inoculum is smaller, it seems reasonable to assume a period of 7–14 days.

Age incidence   The higher the prevalence of trachoma in a given community, the younger is the age of onset. Thus in a Gambian village, the prevalence was highest (91 per cent) in the 5–9 year age group, and thereafter declined to about 50 per cent in those aged more than 20 years (Sowa *et al.* 1965, pp. 25–56). In a Tunisian community, all children had acquired the infection by the age of 2 years (Dawson *et al.* 1976).

Clinical and pathological features

For detailed descriptions of physical signs see Mac-Callan (1936), Nataf (1952), and World Health Organization (1962).

Subepithelial infiltration with lymphocytes and neutrophil leucocytes is the earliest change and is most pronounced in the upper tarsal and bulbar conjunctivae.
Conjunctival papillae result from hypertrophy of the normal papillary processes; when numerous they give the palpebral conjunctiva a reddened, velvety appearance. *Follicles* composed of lymphoid cells and with germinal centres occur in the palpebral and sometimes in the bulbar conjunctivae, where they are seen as pale round elevations—the so-called 'sago-grain' appearance (Fig. 76.1). Follicles at the limbus are pathog-

Fig. 76.1   Trachoma, Stage II, showing severe follicular hyperplasia. (Photograph by courtesy of J. Sowa).

nomic of trachoma, as are the depressed pigmented areas ('Herbert's pits') left on regression.
Corneal lesions mainly affecting the upper segment usually appear early on; their subsequent course is

variable. Superficial *punctate keratitis* is usually transient. *Pannus* (L. = a veil) is a cellular infiltration of the cornea accompanied by new blood vessels growing in from the limbus; gross pannus extending over the pupil may impair vision.
Cicatrization of the subepithelial tissues of the eyelid may depend in extent upon the severity and duration of the preceding inflammation and follicular hyperplasia; severe scarring may so distort the lids, particularly the upper, that the border turns inward (entropion) or more rarely outward (ectropion). Entropion results in continual trauma to the cornea by the inwardly directed eyelashes (trichiasis) and hence in ulceration, opacities, and visual impairment. Stenosis of the lacrimal duct may cause xerophthalmia and further corneal damage. On the other hand, the disease often resolves with little or no scarring.

Cytology

The pathological changes in the conjunctiva are reflected in the cytology of Giemsa-stained scrapings taken at intervals after infection (see Collier 1967). At first, polymorphonuclear leucocytes are numerous, but within a week or so start to be replaced by lymphocytes derived from the intensive subepithelial infiltration and later from the follicles; as the latter mature, lymphoblasts from their germinal centres appear, sometimes in large numbers. The epithelial cells undergo degenerative changes: their nuclei are enlarged and eosinophilic, and the cytoplasm becomes vacuolated and is shed, giving rise to much basophilic debris. Macrophages containing ingested debris ('Leber cells') are often present. These cytological changes are characteristic of trachoma and persist for long periods in the absence of treatment. The cause of the abnormalities in the epithelial cells is unknown; it may be related to the presence of soluble antigen, which, being a lipopolysaccharide, is probably cytotoxic.

Inclusions can be detected as early as 48 hr after infection and by the 7th day are usually fairly numerous. In baboons inoculated experimentally with inclusion conjunctivitis agent, their numbers are related to the degree of inflammation but not to that of follicular hyperplasia (Collier 1967). As the inflammatory process resolves, they diminish in number and eventually disappear.

Clinical classification

MacCallan (1936) distinguished four stages (TrI–IV) in the course of trachoma; in brief, they are those of onset, established trachoma, cicatrization, and final resolution. This classification was useful, but gave no information about the severity of infection or the incidence of sequelae, both of which are important in studies of epidemiology and control measures. A WHO Expert Committee on Trachoma (World Health Organization 1962) therefore made a distinction between the *relative intensity* and the *relative gravity* of trachoma; the first term refers to the degree of activity in an individual at a given time, and the second to the degree of disabling complications and sequelae, or to lesions that, untreated, are liable to lead to such sequelae. Intensity and gravity are not necessarily related, since highly active disease of short duration may have a better outcome than chronic trachoma of low intensity (Assaad and Maxwell-Lyons 1967). Assessments of relative gravity are much the more important for estimating the impact of trachoma on a community and the effects of treatment.

Reinfection and reactivation

The prolonged course of trachoma in endemic areas has been ascribed both to periodic reinfections and to reactivations; but because the number of chlamydial serotypes causing endemic trachoma is small, the value of change in serotype as a marker of reinfection is limited. Nevertheless, the balance of opinion is toward repeated reinfection rather than reactivation as a factor in chronicity.

Intercurrent eye infections

In some areas where trachoma is highly endemic, notably N. Africa and the Middle East, outbreaks of bacterial conjunctivitis, especially in the spring and summer, are common. The worst of these are caused by *Haemophilus aegyptius*; infections with *Haemophilus influenzae* and to a lesser extent *Neisseria* spp. (including gonococci), *Moraxella*, streptococci and staphylococci also occur. It is generally held that the increased volume of discharge occasioned by such infections assists the spread of trachoma, possibly through the medium of flies; and that repeated bacterial infections increase the severity of trachoma. These suppositions are probably valid in countries where bacterial conjunctivitis occurs in epidemic form, but the presence of potentially pathogenic bacteria does not necessarily exacerbate trachoma. Thus Sowa *et al.* (1965, pp. 53–57, 83–85) found that only 3 per cent of 384 Gambian villagers had bacteriologically sterile conjunctivae: streptococci, of which 36 per cent were *Str. pneumoniae*, staphylococci, of which 10 per cent were coagulase-positive, and *Haemophilus* spp. were present in 30–50 per cent of this population and were more often isolated from trachomatous than from non-trachomatous subjects. Epidemic conjunctivitis was not seen, however, in The Gambia and there was no evidence that bacterial infection affected the course of trachoma; in this population at least, the presence of trachoma might predispose to bacterial infection rather than *vice versa*.

Somewhat similar findings in American Indian school children were reported by Wood and Dawson (1967); of the bacterial isolates, only *Moraxella* spp. occurred more frequently when trachoma was present.

The possibility of intercurrent infection with viruses such as type 8 adenovirus, which causes epidemic keratoconjunctivitis, should be borne in mind.

Differential diagnosis

Trachoma is usually distinguishable from inclusion conjunctivitis by differences in epidemiological and clinical features, and in the infecting serotype of *Chlam. trachomatis* (see p. 559). For details of its differentiation from other forms of follicular conjunctivitis see World Health Organization (1962). On epidemiological, clinical and microbiological grounds it is not usually difficult to distinguish trachoma and inclusion conjunctivitis from infections with adenoviruses or molluscum contagiosum, from Axenfeld-type chronic folliculitis in which no micro-organisms have been demonstrated, or from Parinaud's oculoglandular syndrome.

Inclusion conjunctivitis

Inclusion conjunctivitis, or keratoconjunctivitis when corneal lesions are present, has features in common with trachoma, but differs in its epidemiology, propensity to spontaneous resolution without disabling sequelae, and in the infecting serotypes of *Chlam. trachomatis*.

Epidemiology

Whereas trachoma mainly occurs in the less advanced rural societies and is spread from eye to eye, inclusion conjunctivitis is primarily a sexually transmitted disease found in more advanced urban environments. In adults, infection is by direct or indirect contact with the genital area of another person or by autoinfection, presumably transmitted by the hands. Infants may acquire the infection from the maternal genital tract at birth. The infection may also be transmitted in inadequately chlorinated swimming pools, and gynaecologists have become infected from splashing of genital material during surgical operations (Thygeson and Stone 1942). By contrast with trachoma, which is almost always caused by serotypes A to C, the 'genital' serotypes D to K are responsible for inclusion conjunctivitis.

Clinical features

The incubation period is usually 1–2 weeks, but may be less in the newborn. The onset is more acute than that of trachoma, and by the third week there is intense hyperaemia, mucopurulent discharge and follicular hyperplasia, which, by contrast with trachoma, is more pronounced in the lower lid. Diffuse punctate keratitis is not uncommon; pannus of limited extent is occasionally observed. In the acute phase, preauricular lymphadenopathy and upper respiratory tract symptoms may occur; the middle ear may be affected and *Chlam. trachomatis* has been isolated from fluid drained at myringotomy (Dawson and Schachter 1967). Untreated, the disease runs a fluctuating course, but ultimately resolves without conjunctival cicatrization, although there may be some residual corneal scarring.

Immune responses in trachoma and inclusion conjunctivitis

The pathological and clinical differences between trachoma and inclusion conjunctivitis may be due in part to variations in immune responses that are as yet obscure, but the same general immunological considerations probably apply to all eye infections with trachoma-inclusion conjunctivitis (TRIC) agents; the following account of immunity in trachoma may be taken as applying to both syndromes.

Recovery from infection

Although the slow resolution of untreated trachoma must be mediated by immune mechanisms, the way in which they operate remains obscure. In the eye, the causal agent multiplies only in epithelial cells. The transient enlargement of the preauricular lymph nodes sometimes seen in the early stages may indicate replication in this tissue; and by analogy with neonatal inclusion conjunctivitis, Schachter and Dawson (1979) suggested that trachoma in young infants might affect the gut and respiratory tract, but this remains to be demonstrated. With these possible exceptions, trachoma is a highly localized infection that must be dealt with by immune mechanisms operating at or near epithelial surfaces.

Antibody responses

The appearance of antichlamydial antibodies in the conjunctival secretions (CS) and sera of patients with trachoma and of monkeys after experimental eye inoculation is well documented (see McComb and Nichols 1969, Hathaway and Peters 1971, Jawetz *et al.* 1971, Wang and Grayston 1971, Collier *et al.* 1972). In these studies, antibodies were usually detected by indirect immunofluorescence; some sera were tested by complement fixation. The findings vary somewhat according to circumstances but may be summarized as follows. Antibodies both in serum and CS are serotype-specific. IgM antibody is of inconstant appearance in the serum and is not detectable in CS; IgA of the secretory type is regularly present in CS,

and IgA has also been detected at low titre in serum (Jawetz *et al.* 1971). IgG antibody is present in serum and, usually at lower titre, in CS. IgE antibody was found in CS by Hathaway and Peters (1971). Several authors noted that the amount of antibody in CS is directly related to the severity of infection in terms of physical signs or numbers of inclusions.

The site of production of these antibodies is uncertain. IgA immunoglobulin is probably made locally by lymphoid cells in or around the follicles; the ability of conjunctival follicles to produce antibodies against conjunctival molluscum contagiosum, cat-scratch disease, and adenovirus infection was demonstrated by Jones (1971). Antichlamydial IgG antibody may also be made locally, but the finding that its concentration is usually higher in the blood than in CS suggests that some at least may be made elsewhere, perhaps in regional lymph nodes. Collier *et al.* (1972) found that IgG antibody was detectable in CS only when it was also present in serum, suggesting the possibility of transudation from serum to CS through the inflamed conjunctiva.

Although there is some evidence that the amount of antibody in serum and CS is related to protection against conjunctival challenge with live chlamydiae (McComb *et al.* 1971), its role in recovery from infection is not established.

Cellular responses

The function of the numerous neutrophil leucocytes present in the early stages of infection is unknown; their rapid diminution in number suggests that they play little part in the recovery process. Macrophages, which are constantly present in active trachoma, engulf cell debris; but neither their ability to ingest chlamydiae *in vivo* nor the role of antibody in such a process has been established. *In vitro*, mouse peritoneal macrophages ingest chlamydial elementary bodies, which then lose infectivity but retain toxicity; but uptake of even very few elementary bodies severely damages the gross appearance, metabolism, and phagocytic activity of the macrophages (Taverne and Blyth 1971). The ability of chlamydiae to impair macrophage function may account in part for the relative inefficiency of the immune response in trachoma; nevertheless, caution is needed in extrapolating the results of in-vitro experiments to events in the conjunctiva, particularly as treatment of the chlamydiae with antibody—which is usually present in the trachomatous eye—prevents their deleterious effects on macrophages (Taverne *et al.* 1974).

In experiments on Taiwan monkeys, Wang and Grayston (1967) and Gale and co-workers (1971) found that conjunctival inoculation induced pannus and cicatrization only in animals previously inoculated conjunctivally or parenterally with chlamydiae. These authors concluded that corneal lesions and scarring were due to a hypersensitivity reaction requiring previous sensitization; it should not be forgotten, however, that in man pannus may appear soon after primary infection, an observation confirmed by inoculation of a volunteer in whom cicatrization also occurred (Collier *et al.* 1960).

Interferon The ability of TRIC agents both to induce and to be inhibited by interferon was referred to in Chapter 45; the relevance of these findings to recovery from infection is not known.

Lysozyme Mull and Peters (1971) found no evidence that secretion of lysozyme in children in Saudi Arabia, an area of high endemicity, was deficient by comparison with that in American infants.

Immunization against trachoma

The wide prevalence of trachoma, its major role as a cause of visual disability, and the difficulties of mass treatment led to much research on immune responses to chlamydial antigens injected parenterally, and eventually to field trials of vaccines prepared by a number of laboratories. For accounts of animal experiments see Symposium (1967, 1971) and for reviews of field trials Collier (1966) and Schachter and Dawson (1978, pp. 83–88). In brief, both live and inactivated *Chlam. trachomatis* antigens induce antibodies detectable by, *inter alia*, neutralization of infectivity and toxicity, complement fixation, immunofluorescence, and gel diffusion; they can also protect monkeys and baboons against conjunctival inoculation with infective organisms. Chlamydial antigens can induce delayed hypersensitivity in guinea-pigs, as measured by skin tests; and some experimental vaccines, including those of low potency, enhance the response of monkeys or baboons to conjunctival challenge. These findings were reflected in the results of trials in man, in which different vaccines either conferred short-term protection, had no discernible effect, or were actually deleterious: some vaccines appeared to enhance the severity of trachoma in those who subsequently acquired the infection, and others increased rather than diminished the attack rate. Because of these disappointing results the idea of a trachoma vaccine has been generally abandoned, at least for the present.

Laboratory diagnosis of trachoma and inclusion conjunctivitis

For a manual on this subject, see World Health Organization (1975).

Infections with *Chlam. trachomatis* may be diagnosed by finding inclusion bodies in epithelial scrapings, isolating the organisms in cell cultures or chick embryos, and by serological methods. For infections with TRIC agents the latter are less reliable than direct demonstration of the causal agent; the methods used depend both on the purpose of the test—individual diagnosis or mass screening—and on the laboratory facilities available.

Staining of conjunctival inclusions

Staining by Giemsa's method is the classical technique and has the advantage of revealing the cytological appearances and the morphology of the inclusions in relation to that of the epithelial cells (Fig. 76.2). Its

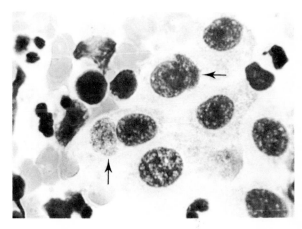

Fig. 76.2 Two conjunctival inclusions (arrowed). Giemsa's stain. (× 1000).

drawbacks are that, when inclusions are scanty, prolonged searching with a high-power objective may be needed; furthermore, inclusions may be hidden in masses of densely stained cells. Inexperienced observers may mistake nuclear extrusions, melanin granules, bacteria, or phagocytosed debris in macrophages for inclusions.

Iodine stains the glycogen matrix of mature inclusions a deep coppery brown. Its advantages are that the inclusions, even when buried in clumps of cells, are readily visible against a pale yellow background so that scrapings can be scanned with a low-power objective within a few minutes (Fig. 76.3); and only the

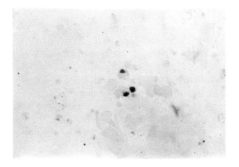

Fig. 76.3 Low power view of three conjunctival inclusions stained with iodine (× 350).

simplest equipment is needed. If desired, these preparations can be decolourized with alcohol and re-stained with Giemsa's stain for cytological studies. The disadvantage of iodine is its inability to stain immature inclusions. For optimal results, the stained scrapings must be blotted dry without washing and covered with a thin film of immersion oil before being examined under the microscope; a suitable method is that of Gilkes and co-workers (1958) as modified by Collier (1961). The wet-mounted preparations sometimes advocated, for example by the World Health Organization (1975, p. 18) and by Schachter and Dawson (1978, p. 191) are unsatisfactory.

Immunofluorescence The indirect method is usually employed (Fig. 76.4). Broadly reactive sera

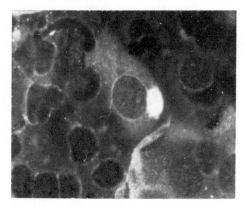

Fig. 76.4 Conjunctival inclusion stained by indirect immunofluorescence (× 1000). (From Sowa *et al.* 1971, with permission.)

can be obtained by repeated inoculation of rabbits with any strain of TRIC agent; Darougar and his colleagues (1971) used sera from patients with *Chlam. trachomatis* infections. This technique reveals inclusions at all stages of development; its disadvantages are that a higher degree of technical skill, better laboratory facilities, and more expensive reagents are required than for the other two methods.

Isolation of chlamydiae

Inoculation of chick embryos with conjunctival scrapings or material collected on swabs is an effective method for diagnosing trachoma in areas of high endemicity where shedding of organisms is heavy. It can be undertaken where facilities for cell culture are not available, but the time from inoculation to diagnosis is considerably longer than with cell cultures, which are generally more reliable (see Chapter 45).

Comparison of methods for direct demonstration of chlamydiae

Darougar and his colleagues (1971) found that the

proportion of inclusion-positive conjunctival scrapings from trachoma was higher with immunofluorescence than with Giemsa's stain, and higher with Giemsa's stain than with iodine, but the differences were not great. Sowa and co-workers (1971) tested conjunctival scrapings from trachomatous Gambian children: indirect immunofluorescence gave a somewhat higher proportion of positive results than iodine and revealed significantly higher numbers of inclusions in duplicate scrapings from the same individual. Thus, for staining, immunofluorescence is the method of choice when facilities are available, but for field use iodine has the advantages of speed and simplicity.

Of all methods for detecting TRIC agents directly, isolation in cell culture is probably the most efficient, and has the advantage of providing isolates for serotyping.

Serological methods Complement-fixation tests are of little or no value in diagnosing ocular infections with chlamydiae, but the indirect immunofluorescence technique detects antibodies of the IgG and IgA classes in conjunctival secretions and of the IgG and IgM classes in sera (Collier *et al.* 1972; Wang *et al.* 1977). There is a reasonably good correlation between the presence of antibody in blood and present or past ocular infection. The finding of IgM antibody suggests recent or current infection; but this technique demands more expertise than direct demonstration of chlamydiae and even in experienced hands a higher proportion of false negative results is to be expected.

Skin tests The results of skin tests in trachoma and inclusion conjunctivitis are so variable as to make them useless for diagnosis.

Prevention and treatment

Trachoma

The control of endemic trachoma depends on the improvement of living standards, which in itself can significantly reduce the prevalence; on antimicrobial treatment campaigns designed to eliminate ocular infections both with chlamydiae and with bacteria such as *H. aegyptius*; and on surgical correction of potentially blinding lid deformities. These aspects are well reviewed in handbooks published by the World Health Organization (1973, 1979). WHO has done much to assist the planning, execution and evaluation of treatment campaigns, which have resulted in the virtual or complete eradication of trachoma in countries where such efforts have received effective support from the national health authorities; in less favoured areas, where logistic, economic and organizational difficulties are more intractable, trachoma remains a major public health problem.

The planning of any treatment campaign must begin with an evaluation of the prevalence of trachoma and its complications in the various age groups. In areas of high endemicity, specific therapy is given by mass treatment with tetracycline eye ointment or oily suspension; intermittent schedules (e.g. twice daily for 5 consecutive days every 6 months) enable large numbers of persons to be treated by small mobile teams. Where the prevalence is lower, selective treatment, e.g. of school children only, may be employed. Oral therapy with antibiotic is much more rapidly effective than topical treatment, but problems of expense and supervision make it impracticable for mass use. For treatment of active disease on an individual basis a course of tetracycline by mouth, 250 mg four times a day for 3 weeks, may be given; erythromycin in equivalent dosage, e.g. erythromycin stearate 500 mg twelve-hourly, is a suitable alternative. Such treatment results in rapid subsidence of inflammation, but follicles may persist for weeks or months after the infection is eliminated. Unduly short courses of treatment may result in relapse.

Inclusion conjunctivitis

The association between inclusion conjunctivitis and genital infection both of patient and consort must be borne in mind when considering therapy, which should always be systemic rather than local; adults should be treated with oral tetracycline or erythromycin as for trachoma. The treatment of infants is described on p. 567.

Genital tract infections of men

In this section diseases due to TRIC agents and other non-gonococcal infections are conveniently considered together. For recent reviews see Schachter and Dawson (1978) and Oriel and Ridgway (1982).

Non-gonococcal urethritis (NGU)

Over 80 000 cases are reported annually in England and Wales; this syndrome is at least twice as common as gonococcal urethritis. Typically there is a history of urethral irritation and dysuria, starting 7–14 days after intercourse and followed by the appearance of a mucopurulent discharge; a gram-stained smear contains more than five polymorphonuclear leucocytes per × 100 field. Threads of polymorphs are present in an early morning urine specimen. These symptoms and signs may resolve without treatment in a matter of weeks.

Aetiology *Chlam. trachomatis* is the only agent firmly implicated in the causation of NGU. The use of cell cultures for isolation has greatly helped in the many studies made, the results of which are remarkably uniform, isolation rates from NGU being about 30–60 per cent (Oriel and Ridgway 1982). In matched controls without urethritis the carriage rate is 3–5 per cent. The interpretation of serological studies in NGU

is more difficult: 70–100 per cent of men with proved chlamydial infection have specific IgG antibody in their sera and 28–30 per cent have IgM antibody (Reeve *et al.* 1974, Bowie *et al.* 1977); but the corresponding figures for men with chlamydia-negative NGU are 18–44 and 3–8 per cent, probably because of a past or an existing but undiagnosed infection. Nevertheless, support for the aetiological role of chlamydiae in NGU is provided by the beneficial results of anti-chlamydial therapy and the finding that 'genital' strains can cause urethritis in primates (Digiacomo *et al.* 1975).

Ureaplasma urealyticum may account for a small proportion of cases of NGU (see Chapter 78). All in all, the results of various studies suggest its implication in 10–15 per cent of cases. *Herpesvirus hominis*, *Mycoplasma hominis*, and cytomegalovirus may account for at most another 10 per cent. *Trichomonas vaginalis* has been isolated from some patients with NGU but in a careful investigation Holmes and co-workers (1975) found no evidence for its aetiological role.

Diagnosis Although the incubation period is usually longer and the onset less acute than in gonorrhoea, these criteria are unreliable. Gonococcal infection is excluded by the examination of smears, with or without culture. Chlamydial infection is diagnosed by cell culture of a specimen obtained from within the urethra; searches for inclusions are usually unrewarding and serological tests are unreliable.

Post-gonococcal urethritis (PGU) The persistance or recurrence of symptoms and signs following effective treatment for gonorrhoea is mostly if not always due to concurrent infection with another agent; *Chlam. trachomatis* can be isolated from 11–30 per cent of men with gonococcal urethritis and probably accounts for 50–60 per cent of PGU.

Other infections of the male genital tract *Epididymitis* is an important cause of infertility and chlamydiae have been isolated from a substantial proportion of cases, particularly in younger men (Harnisch *et al.* 1977, Berger *et al.* 1978). There is little evidence that chlamydiae are implicated in acute or chronic *prostatitis*, which nowadays is mainly associated with coliforms and other urinary tract pathogens. Goldmeir and Darougar (1977) isolated *Chlam. trachomatis* from the rectums of two men with *proctitis*.

Treatment of genital infections in men

Since mild or symptomless infections may lead to complications, treatment is always indicated. Because of the possibility of multiple infection, the drug of choice should be effective both against *Chlam. trachomatis* and other likely genital pathogens. Two week courses of tetracycline, 250 mg four times a day, or erythromycin stearate, 500 mg twelve-hourly (Oriel *et al.* 1977), are effective against both chlamydia-positive and -negative NGU; even so, symptoms persist in 10–15 per cent of patients. Coufalik and co-workers (1979) found minocycline to be more effective than rifampicin in effecting clinical cure. The importance of prostatic penetration of the drug is unclear, but minocycline, erythromycin, and doxycycline are effective in this respect. Treatment may need to be prolonged beyond 2–3 weeks when there are complications such as prostatitis or epididymitis. As with any sexually transmitted disease the tracing and treatment of partners is important. Schachter and Dawson (1978, p. 132) emphasize the need for concurrent treatment of the consort in order to avoid reinfection.

Genital tract infections of women

Sexually acquired infections in women may involve not only the genital tract but also the urethra and rectum; perihepatitis may follow infection of the fallopian tubes.

Cervical infection

In the United Kingdom, *Chlam. trachomatis* can be isolated from the cervices of up to 30 per cent of women attending venereal disease clinics (Ridgway and Oriel 1977) and of more than 80 per cent of primary sexual contacts of culture-positive men; indeed, this organism can now be isolated more frequently than the gonococcus, with which it is often found in association. With the microimmunofluorescence test Oriel and co-workers (1978) found serum antibody in 78 per cent of culture-positive women attending a clinic for sexually transmitted diseases, and in 31 per cent of culture-negative women.

There are no distinctive symptoms and signs: the cervix may look normal, or at the other extreme be oedematous and inflamed with a mucopurulent discharge. In a careful study, Rees and her colleagues (1977) found that isolation rates were particularly high from women with 'hypertrophic' cervical erosions and that such erosions lost their hypertrophic appearance after tetracycline treatment. There is, however, no direct evidence that chlamydial infection causes erosions; it may be that the increased area of columnar epithelium available predisposes to infection.

Colposcopy sometimes reveals microfollicles (Dunlop *et al.* 1966) which are regarded by some as diagnostic of chlamydial infection. Cytological changes are non-specific and include degeneration of epithelial cells, an increase in parabasal cells, and the presence of polymorphonuclear leucocytes, lymphocytes, and large mononuclear cells.

The diagnosis of cervical infections with *Chlam. trachomatis* can be made with certainty only in the laboratory. Specimens for cell culture are taken with a swab from the squamo-columnar junction, after wiping away any exudate. Demonstration of inclusions is sometimes possible, but neither this nor serological techniques are recommended for diagnostic purposes. Because of the strong possibility of multiple infections, specimens must also be tested at least for *Neisseria gonorrhoeae, Treponema pallidum, Trichomonas vaginalis* and *Candida* spp.

Salpingitis

Dunlop and co-workers (1966) noted an association between salpingitis in some mothers and inclusion conjunctivitis in their babies. Salpingitis is not infrequently seen in women with confirmed chlamydial infection of the cervix. The demonstration of *Chlam. trachomatis* in tubal material obtained at laparoscopy (Mårdh *et al.* 1977, Møller *et al.* 1979) and experimental infection of the fallopian tubes of grivet monkeys (Møller and Mårdh 1980) provide additional evidence of its implication; but the prevalence of tubal infection is unknown and serological tests (Treharne *et al.* 1979) are of little help in assessing it. Unlike gonorrhoea, chlamydial infection does not cause extensive epithelial damage and loss of ciliary activity in organ culture (Hutchinson *et al.* 1979). There is good evidence that the acute perihepatitis that sometimes complicates salpingitis may result from chlamydial as well as gonococcal infection (Wang *et al.* 1980).

Other infections   *Chlam. trachomatis* has been isolated from the urethra both in the presence and absence of cervical infection, from inflamed Bartholin's ducts, and from the rectum, but the incidence of such infections is not known. The extent to which *Chlam. trachomatis* is implicated in infertility, abortion or fetal abnormality is also obscure.

Treatment of genital infections in women   Antimicrobial therapy for known or suspected chlamydial infections is similar to that used for men (see p. 565); some physicians prescribe metronidazole in addition to tetracycline or erythromycin.

Reiter's syndrome

The complex question of chlamydial infection in Reiter's syndrome and sexually acquired reactive arthritis (Keat *et al.* 1978) is reviewed and discussed by Schachter and Dawson (1978, pp. 141–152). Although *Chlam. trachomatis* can be demonstrated in both urethral and synovial material from a proportion of patients, a direct causative role for this micro-organism has not been established; it may be that the strong association of the histocompatibility antigen HLA-B27 with Reiter's syndrome predisposes to an unusual response to chlamydial and other infections.

Neonatal infections

Babies born through a maternal genital tract infected by *Chlam. trachomatis* may become colonized with the organism during the first few weeks of life. The sites affected are the conjunctiva, nasopharynx, middle ear, trachea, lung, rectum, and vagina. In some of these sites the association with clinical disease is uncertain, whereas in others, for example the eye, the physical signs are more distinctive. There are no precise estimates of the risk of infection, but 20–60 per cent of babies born to infected mothers acquire ocular infections, and in 25–50 per cent of these the respiratory tract is also colonized.

Ocular infection

Chlamydial infection of the eye in the newborn is more common than gonococcal ophthalmia although precise figures for the United Kingdom are not available. In the United States Schachter and his associates (1979) estimate the incidence as 14 per 1000 live births.

Clinical features   The onset occurs 3–13 days after birth. The physical signs range from florid conjunctivitis with pronounced conjunctival and periorbital oedema and purulent or mucopurulent discharge to little more than a 'sticky eye'. Pseudomembranes caused by adherence of exudate to the conjunctiva may be seen. If untreated, the condition usually resolves spontaneously within 2–3 months, but a benign outcome is by no means invariable; conjunctival scarring and pannus were described in 10 of 38 babies observed by Freedman and co-workers (1966). The possibility of co-existent gonococcal infection must always be borne in mind. Afebrile pneumonia was observed in 41 out of 56 infants by Tipple and her colleagues (1979).

Diagnosis   Inclusions are usually numerous and may be found in scrapings from the lower conjunctival fornix (see p. 563 *et seq.*). Scrapings may also be inoculated into cell cultures.

Treatment   Crédé's silver nitrate drops are of no value in the prophylaxis of chlamydial infection. The disease responds to topical therapy with chlortetracycline 1 per cent and oral erythromycin (30 mg/kg/day), which should be given concurrently for three weeks. When present, gonococcal infection should be treated before beginning antichlamydial therapy. Follow-up should include culture of conjunctival scrapings. The parents must be referred for investigation and treatment.

Pneumonia

This syndrome was described by Schachter and co-workers (1975) and by Beem and Saxon (1977).

Clinical features   The onset occurs characteristically 4–12 weeks after birth. The main features are dyspnoea, a staccato cough, and sometimes a mucoid nasal discharge. There may be evidence of otitis media. Chest radiography shows hyperexpansion with symmetrical diffuse interstitial and patchy alveolar infiltration. Untreated, the disease runs a benign but protracted course, leading to apparent recovery. About half such patients have already had an overt conjunctival infection, and those who have not frequently possess antibodies to *Chlam. trachomatis* in the conjunctival secretions. Nothing is known of the long term consequences of untreated disease.

Diagnosis   Chlamydiae may be isolated from nasopharyngeal or tracheal secretions. High titres of IgG antibody are demonstrable by microimmunofluorescence and IgM antibody is also sometimes present; serological tests are more likely to be useful than in most other infections with *Chlam. trachomatis*. The concentrations of non-specific IgA, IgM and IgG are also raised, and there may be eosinophilia.

Treatment   There are few reports on this subject. Beem and Saxon (1977) treated a number of infants either with sulfisoxazole 150 mg/kg/day, or with erythromycin ethyl succinate 40 mg/kg/day, for up to 14 days. All stopped shedding the organism within a few days and improved clinically.

Respiratory and cardiac infections of adults

Chlam. trachomatis may cause pneumonitis in adults. Tack and co-workers (1980) isolated it from the lower respiratory tract of six patients with chest infections, four of whom were receiving immunosuppressive drugs. Necropsy of a patient who died when 30 weeks pregnant (Bel-Khan *et al.* 1978) revealed endocarditis; chlamydiae were demonstrated in the vegetations by electron microscopy and there was good evidence of recently acquired infection with *Chlam. trachomatis*, probably of serotype F. This woman had had a cervical discharge earlier in her pregnancy.

Lymphogranuloma venereum

Lymphogranuloma venereum (LGV) differs considerably from other syndromes due to *Chlam. trachomatis*; it is caused by serotypes L1, L2 and L3, which attack lymphatic rather than epithelial tissues; there may be extensive fibrosis in the later stages. Although less prevalent than other sexually transmitted diseases in industrialized countries, it has a wide geographical distribution, the prevalence being highest in the tropics and sub-tropics.

Clinical features

Transmission occurs by sexual contact; the incubation period is usually 1–3 weeks. In men the primary lesion is a vesicle or ulcer on the penis; rectal infections occur in homosexuals. In women the commonest site is the fourchette. In both sexes the lesion may pass unnoticed. Primary lesions occasionally occur on extra-genital sites, for example fingers or tongue. Soon after the primary lesion has healed, the secondary stage, characterized by swelling of the inguinal lymph nodes, begins. Such swelling occurs more frequently in men than in women. Enlargement of the nodes both above and below the inguinal ligament sometimes results in the characteristic 'groove sign'. In women, the deeper nodes are affected unless the primary lesion was on the vulva. The nodes become matted, fluctuant and fixed to the skin; they may break down and discharge through multiple sinuses. In the acute phase there may be constitutional disturbances such as fever and joint pains. Complications include conjunctival infection and preauricular adenopathy, meningitis, synovitis, pneumonitis, and cardiac abnormalities. If untreated, the disease progresses after some years to the third stage, which is usually more serious in women than in men. Ulceration, proctitis, rectal strictures, and rectal or rectovaginal fistulae may occur. The vulva may become grossly affected by ulceration and granulomatous hypertrophy ('esthiomène'). Elephantiasis of the vulva or scrotum may also develop.

Diagnosis

Because of the variable clinical features, clinical diagnosis may be difficult. Syphilis, chancroid, genital herpes, granuloma inguinale, filariasis, and malignant disease enter into the differential diagnosis; it is particularly important to test for syphilis.

Skin tests   The Frei test depends on the production of a delayed hypersensitivity response, i.e. induration 48–72 hr after intracutaneous injection of partly purified LGV antigen prepared from chick-embryo yolk sac. A positive reaction indicates either present or past infection. False negative results are not uncommon. The test is used less than formerly, and the antigen is sometimes difficult to obtain.

Serological tests   Complement-fixation tests are usually done with genus-specific ('group') antigen (see Chapter 45), and hence may be positive as a result of other chlamydial infections, such as TRIC-agent infection or ornithosis. Nevertheless, a titre of 1/64 or more supports a diagnosis of LGV in patients with suggestive signs. In the microimmunofluorescence test LGV antigens react with antibodies to other chlamydiae; it therefore plays, like the complement-fixation test, a mainly confirmatory role in diagnosis.

Isolation of LGV agent   The organism may be isolated from pus aspirated from a suppurating lymph node; the older method of mouse-brain inoculation is now largely superseded by yolk-sac inoculation or, preferably, isolation in cell cultures.

Histology   In experienced hands, histological examination is a useful adjunct to diagnosis. Sections of lymph nodes contain multinucleate giant cells and masses of epithelioid cells, at the centres of which are degenerate polymorphonuclear leucocytes. Chromatophil 'Gamna-Favre bodies' may be seen within the cytoplasm of mononuclear cells.

Treatment

In general, early treatment has the best chance of success; fibrosis and other changes in the later stages make the disease less amenable to medical treatment. Published data on controlled clinical trials are scanty; both sulphonamides and tetracycline given for at least three weeks are effective (Greaves *et al.* 1957). Rifampicin has been successfully used in at least one study (Menke *et al.* 1979), and erythromycin may also be useful but has not been extensively evaluated.

Mouse pneumonitis

A number of workers have isolated *Chlam. trachomatis* from laboratory mice with signs of pneumonitis (see Meyer 1967). Such infections are important as a source of error when passaging other chlamydiae in mouse lung. Their presence in a colony should be excluded by macroscopic examination of the lungs and inoculation of material into cell cultures, if necessary after several 'blind' mouse-to-mouse passages by the intranasal route.

Diseases caused by *Chlam. psittaci*

By contrast with *Chlam. trachomatis, Chlam. psittaci* mainly affects birds and non-primate mammals (see review by Storz 1971); although infections of man are rare by comparison, the association between pneumonitis and contact with psittacine birds has long been recognized. The term 'psittacose' (psittacosis) was first used by Morange (1895). Chlamydial infections in other avian species are referred to as 'ornithosis'. Neither term is appropriate for infections of mammals and Storz (1971, p. 132) suggested 'chlamydiosis' to designate the generalized infections caused by *Chlam. psittaci*.

Infections of birds

Clinical and pathological features

Infection is usually acquired by the respiratory or oral route but in some species may be transmitted vertically *via* the egg. Ticks (Meyer 1967) and lice and mites infesting poultry (Eddie *et al.* 1962) may harbour chlamydiae, but there is no evidence that the organisms replicate within them, and their role in transmission is unknown. Inapparent infection is much more common than overt disease, which is often precipitated by stress, e.g. exposure to dampness or overcrowding.

The clinical signs may vary in nature and severity from outbreak to outbreak, between species, and by age, younger birds being the more severely affected; such signs include loss of condition, diarrhoea, respiratory distress, and conjunctivitis. If unchecked, the mortality rate in domestic flocks may reach 30 per cent.

Chlamydiosis of birds is a generalized infection that may affect all the major systems, especially the reticuloendothelial, alimentary, and respiratory. At necropsy, the lungs are consolidated; the liver and spleen are enlarged and may be haemorrhagic; serous surfaces are inflamed and covered with exudate; the gut lining is inflamed; and there may be signs of meningoencephalitis. The histological findings include oedema, haemorrhage, and extensive infiltration with lymphocytes and histiocytes. In poultry, there may be intercurrent infection with bacteria, e.g. salmonellae and pasteurellae.

Laboratory diagnosis

Demonstration of chlamydiae Strains of *Chlam. psittaci* that infect birds are likely to be highly infectious and pathogenic for man; birds with suspected ornithosis and diagnostic specimens from them must therefore be handled only in laboratories equipped to deal with Category B1 pathogens (Report 1978).

Elementary bodies stained by the Giemsa or Machiavello methods may be detected in impression preparations made from the viscera, particularly the spleen, or from exudates. The organism may also be isolated from organs, blood or faeces in the chick-embryo yolk sac or in cell cultures; since the inclusions of *Chlam. psittaci* do not contain glycogen, cell monolayers should be stained with Giemsa's stain and not with iodine.

Serological tests Because inapparent infections are comparatively common, serological tests are of little value in diagnosing acute disease in individual birds, but may be useful for determining the prevalence of chlamydial infection in flocks. The complement-fixation test is usually employed, but since the sera of chickens, ducks and turkeys do not fix guinea-pig complement, an indirect test (Karrer *et al.* 1950*a*, *b*) must be used for these species.

Control of avian chlamydiosis

The high prevalence of *Chlam. psittaci* in avian species makes its eradication an unrealistic proposition. Strict quarantine procedures should be applied to imported psittacine birds; but the high commercial value of some show birds sometimes leads to smuggling. Chlamydiosis in poultry can be minimized by keeping and transporting birds under good conditions. Particularly in the USA, attempts have been made with some success to control the infection in poultry and pet-bird breeding establishments by incorporating tetracycline

in the feed. The efficacy of such measures depends on their efficient application and the maintenance of adequate blood concentrations; but even under the best conditions it would be prudent to regard them as suppressive rather than therapeutic. Furthermore, the use of tetracycline in this way carries the risk of producing antibiotic-resistant bacteria. No effective immunization procedure for birds has as yet been devised.

Infections of animals

Infections with *Chlam. psittaci*, often inapparent, are widespread in mammalian species; they are of economic importance in farm animals and may also affect pets. (For reviews see Meyer 1967, Storz 1971, Wachendörfer and Lohrbach 1980.) As in birds, there is a wide range of syndromes: intrauterine and enteric infections appear to be caused primarily by serotype 1, and polyarthritis, encephalomyelitis and conjunctivitis by type 2 (see Chapter 45).

Enteric infections

Chlamydiae may be excreted in the faeces of apparently healthy sheep and cattle; overt enteritis is, however, uncommon except as part of a more generalized chlamydiosis. The main importance of such infections is that they disseminate organisms, which may then give rise to more serious disease in animals that ingest or inhale them.

Uterine infections

Chlam. psittaci causes ovine and bovine abortion, usually enzoötic in character but sometimes occurring in epizoötics. Up to 30 per cent of ewes in a flock may be affected. In animals that have aborted, the immune response usually prevents a repetition. In sheep, the most important factor in transmission is the shedding of large numbers of organisms in the products of conception at the time of abortion; the importance of faecal-oral transmission is not clear, and the roles of sexually acquired infection and of transmission by arthropods (Meyer 1967) remain uncertain. Abortions usually occur in the final stages of pregnancy, and result from placentitis consequent upon haematogenous spread of chlamydiae. The fetus may look more or less normal on external examination; lesions found at autopsy include generalized petechial haemorrhages and lymphadenopathy, focal necrosis of the liver, and ascites.

The diagnosis may be confirmed by microscopic or cultural demonstration of chlamydiae in the placenta and fetus. Paired serum samples from both ewes and cows taken at the time of abortion and 2-4 weeks later show a characteristic rise of 4-fold or more in the titre of complement-fixing antibody (Storz and McKercher

1970). The presence of other pathogens that cause abortion, notably *Brucella abortus* and *Campylobacter fetus* must be considered.

Control measures include segregation of aborting animals and careful disposal of the products of conception. Chemoprophylaxis with tetracycline has been attempted with varying success. Both live and inactivated vaccines, usually prepared from chick-embryo yolk sacs, have been tried and are available commercially. The results of a number of trials, reviewed by Storz (1971), were variable but suggest that some vaccines confer a moderate degree of protection.

Pulmonary infections

Pneumonitis due to *Chlam. psittaci* has been reported in sheep, goats, pigs, cats, mice, and other species. *Feline pneumonitis* (cat distemper) is a highly infectious disease, spread by the respiratory route and characterized by sneezing, coughing, anorexia, and mucopurulent discharge from the nose and eyes. Except in kittens and elderly animals it is not usually fatal, but may cause debility lasting for a month or so. *Mouse pneumonitis* may affect laboratory stocks, in which the infection is often inapparent. Chlamydial pneumonitis is characterized by clearly demarcated reddish-grey areas of consolidation, sometimes affecting a whole lobe or lung. Histologically, there is interstitial infiltration with lymphocytes and macrophages; the alveoli are filled with fluid exudate and inflammatory cells, and the lining cells are swollen.

Ophthalmic infections

Conjunctivitis or keratoconjunctivitis occurs in farm and small domestic animals, either alone or as part of a more generalized infection. Murray (1964) isolated a guinea-pig inclusion conjunctivitis (GP-IC) agent from animals with conjunctivitis, and this infection has since been detected in a number of laboratory colonies. In natural infections the clinical signs are inconspicuous, but artificial inoculation results in severe keratoconjunctivitis with numerous inclusions. Infections induced in the guinea-pig eye and genital tract with GP-IC have been used extensively as models for those caused by *Chlam. trachomatis* in man. Another *Chlam. psittaci* agent that may serve a similar purpose is that of feline keratoconjunctivitis which occurs sporadically in domestic cats and may be endemic in catteries. The organism has been isolated from the genital tract of a female cat; and infections with features in common with those caused in man by *Chlam. trachomatis* can be induced by inoculation of the feline eye and genital tract (Darougar *et al.* 1977).

Other infections

Polyarthritis in lambs and calves is a well-recognized manifestation of chlamydiosis and is of interest in relation to the pathogenesis of Reiter's syndrome associated with *Chlam. trachomatis* in man (see p. 566). Chlamydiosis in various avian and mammalian species may also take the form of *encephalomyelitis*. In bulls, *epididymitis* and *seminal vesiculitis* may impair fertility (Storz *et al.* 1968).

Infections of man: psittacosis and ornithosis

There is no evidence of any constant difference in the clinical course of infections acquired from psittacine birds and those transmitted from other avian species; the description that follows applies to both types of infection.

Epidemiology

Psittacosis first acquired prominence during the widespread outbreaks in 1929-30, which started in Latin America and thence spread to Europe, affecting about 600 people. This episode originated from parrots sold as pets. Because most if not all avian species may be infected, often subclinically, it follows that persons whose occupation or recreation brings them into continual contact with birds are those at greatest risk. They include poultry farmers and processors, owners of pet shops and of racing pigeons, and workers in zoos. In the Faroe Islands, ornithosis was reported in women who pluck and process fulmar petrels for human consumption (Bedson 1940). Serological surveys show that inapparent infections are not uncommon in persons at risk (Meyer and Eddie 1962). There have been a number of cases, some fatal, among laboratory workers handling avian strains of *Chlam. psittaci*. Palmer (1982) described an outbreak of psittacosis among workers in a duck-processing factory and another in veterinarians, both of which appeared to be due to liberation of aerosols during evisceration of the birds.

In man, infection is acquired by the respiratory route; some strains are highly infectious and the disease may be contracted by the most casual contact with a sick bird or its environment. Case-to-case infection is unusual, but may occur with particularly virulent strains. An outbreak of chlamydial pneumonia in the Bayou region of Louisiana was remarkable in that 18 persons became infected by secondary or tertiary spread from a single fatal case, the wife of a trapper. There were seven more deaths, all in contacts who had attended fatal cases during the final 48 hr of their illness (Olson and Treuting 1944). The detailed clinical (Treuting and Olson 1944) and pathological (Binford and Hauser 1944) observations on these patients are of considerable interest.

Clinical and pathological features

The incubation period is usually 1–2 weeks, but may be as long as a month. The onset is sometimes insidious with malaise and pains in the limbs, but is more often abrupt, with high fever, rigors, and headache. Patients usually have a cough, but sputum is scanty and respiratory distress is unusual except in severe cases. Characteristically, the pulse rate is slow in relation to the temperature; a high rate is held by some to indicate a poor prognosis. Physical signs in the chest are often limited to râles with little or no evidence of consolidation, and are thus at variance with the extensive signs of pneumonitis seen on radiological examination. Epistaxis is not infrequent and rose spots on the skin, somewhat like those in typhoid, are sometimes seen. The spleen may be enlarged and occasionally there is frank hepatitis with jaundice. Severe cases may be complicated by meningitis or meningoencephalitis, myocarditis, and, more rarely, endocarditis. In the acute phase the white cell count is often normal, but leucopaenia is evident in about 25 per cent of cases.

In patients who recover without treatment, the infection usually resolves slowly within 2 or 3 weeks, or sometimes longer; chlamydiae may be shed for quite long periods. Severely affected patients may become drowsy or stuporous if the central nervous system is affected. Death is usually the result of cardiovascular and respiratory insufficiency. The disease sometimes has the characteristics of a severe toxaemia.

Because of the incidence of subclinical and mild unrecognized infections, the true death rate is difficult to estimate; before the advent of chemotherapy it varied from 15 to 40 per cent of diagnosed cases, and was higher in middle-aged and elderly persons than in younger subjects. At post-mortem examination, extensive pneumonitis with areas of consolidation is a constant finding. The alveoli are packed with exudate containing erythrocytes, mononuclear cells, and polymorphonuclear leucocytes; characteristically, the alveolar cells are swollen. The enlarged spleen shows loss of the normal architecture and areas of focal necrosis may be seen in the liver. There may also be signs of inflammation in the meninges, and congestion of the cerebral parenchyma. Hyaline necrosis and haemorrhages in the rectus muscles may occur.

Diagnosis

Except during an outbreak, diagnosis on clinical grounds alone is difficult, especially in the absence of a history of contact with birds. The differential diagnosis must take account of other causes of atypical pneumonia such as *Mycoplasma pneumoniae* infection, Q fever, brucellosis, and influenza. The skin rash, when present, may at first suggest typhoid.

Chlamydiae may be isolated from the blood and respiratory secretions, but this should not be attempted in laboratories without adequate facilities for handling dangerous pathogens. The diagnosis is usually made by demonstrating a significant rise in antibody in paired sera taken on admission and 10–20 days later; either the complement-fixation or immunofluorescence test may be used. Low titres of antibody (around 1/16) are not infrequently found in normal persons and are of no diagnostic significance in single samples of serum. A titre of 1/64 or greater in a single sample is suggestive of active infection and the finding of specific IgM antibody in such a serum is good presumptive evidence of a current or recent infection; even so, it is preferable to obtain evidence of a rise in titre.

Treatment

The antibiotic of choice is tetracycline in a dose of at least 250 mg four times a day. Schachter and Dawson (1978, p. 25) recommend that treatment be continued for 21 days; shorter courses may be followed by relapse. In severely ill patients, measures must be taken as appropriate to maintain the fluid balance and to support the cardiac and respiratory systems.

Control measures

These are directed at preventing the importation of infection by quarantine measures, and at avoiding the activation of latent infections by rearing, keeping and transporting birds under satisfactory conditions. Chemoprophylaxis may be useful for controlling infections in commercially reared pet birds, but is less so for poultry. Persons in frequent contact with birds should be made aware of the signs and risks of ornithosis by publicity channelled through employers, societies and other organizations. In hospitals, person-to-person spread must be prevented by appropriate isolation measures and proper disposal of sputum, which is to be considered highly infective. Reference has already been made to the precautions needed for handling specimens in the laboratory.

Other infections

Overt infections of man with strains of *Chlam. psittaci* other than those of avian origin are rare; taking account of their wide distribution it is clear that strains derived from the lower mammals are of low virulence for man. There are records of two cases of conjunctivitis in man acquired from cats infected with feline pneumonitis (Ostler *et al.* 1969, Schachter *et al.* 1969). Barwell (1955) reported a laboratory infection, with clinical and radiological resemblances to ornithosis, by the agent causing enzoötic abortion of ewes; and there was good serological evidence for the implication of this agent in a human abortion (Roberts *et al.* 1967).

It is often asserted that cat-scratch disease (see Chapters 45 and 73) is a chlamydial infection, but the evidence is unconvincing. Schachter and Dawson (1978, pp. 154–156) failed to isolate chlamydiae from 20 patients with clinically diagnosed cat-scratch disease and found no evidence of rising antibody titres in paired sera from another series of 19 patients.

References

Assaad, F. A. and Maxwell-Lyons, F. (1967) *Amer. J. Ophthal.* **63**, 1327.

Barwell, C. F. (1955) *Lancet* **ii**, 1369.

Bedson, S. P. (1940) *Lancet* **ii**, 577.

Beem, M. O. and Saxon, E. M. (1977) *New Engl. J. Med.* **296**, 306.

Bel-Khan, J. M. van der, Watanakunakorn, C., Menefee, M. G., Long, H. D. and Dicter, R. (1978) *Amer. Heart J.* **95**, 627.

Berger, R. E., Alexander, E. R., Monda, G. D., Ansell, J., McCormick, G. and Holmes, K. K. (1978) *New Engl. J. Med.* **298**, 301.

Binford, C. H. and Hanser, G. H. (1944) *Publ. Hlth Rep., Wash.* **59**, 1363.

Bowie, W. R. *et al.* (1977) *J. clin. Invest.* **59**, 735.

Collier, L. H. (1961) *Lancet* **i**, 795; (1966) *Bull. World Hlth Org.* **34**, 233; (1967) *Arch. ges. Virusforsch.* **22**, 280.

Collier, L. H., Duke-Elder, S. and Jones, B. R. (1960) *Brit. J. Ophthal.* **44**, 65.

Collier, L. H., Sowa, J. and Sowa, S. (1972) *J. Hyg., Camb.* **70**, 727.

Coufalik, E. D., Taylor-Robinson, D. and Csonka, G. W. (1979) *Brit. J. vener. Dis.* **55**, 36.

Darougar, S., Dwyer, R. St C., Treharne, J. D., Harper, I. A., Garland, J. A. and Jones, B. R. (1971) In: *Trachoma and Related Disorders caused by Chlamydial Agents*, pp. 445–460. Ed. by R. L. Nichols. Excerpta Medica, London.

Darougar, S., Monnickendam, M. A., El-Sheikh, H., Treharne, J. D., Woodland, R. M. and Jones, B. R. (1977) In: *Non-gonococcal Urethritis and Related Infections*, pp. 186–198. Ed. by D. Hobson and K. K. Holmes. American Society for Microbiology, Washington.

Dawson, C. R., Daghfous, T., Messadi, M., Hoshiwara, I. and Schachter, J. (1976) *Brit. J. Ophthal.* **60**, 245.

Dawson, C. R. and Schachter, J. (1967) *Amer. J. Ophthal.* **63**, 1288.

Digiacomo, R. F., Gale, J. L., Wang, S. P. and Kiviat, M. D. (1975) *Brit. J. vener. Dis.* **51**, 310.

Dunlop, E. M. C. *et al.* (1966) *Brit. J. vener. Dis.* **42**, 77.

Eddie, B., Meyer, K. F., Lambrecht, F. L. and Furman, D. P. (1962) *J. infect. Dis.* **110**, 231.

Freedman, A. *et al.* (1966) *Trans. ophthal. Soc.* **86**, 313.

Gale, J. L., Wang, S. P. and Grayston, J. T. (1971) In: *Trachoma and Related Disorders caused by Chlamydial Agents*, pp. 489–493. Ed. by R. L. Nichols. Excerpta Medica, London.

Gilkes, M. J., Smith, C. H. and Sowa, J. (1958) *Brit. J. Ophthal.* **42**, 473.

Goldmeier, D. and Darougar, S. (1977) *Brit. J. vener. Dis.* **53**, 184.

Greaves, A. B., Hillman, M. R., Taggart, S. R. Bankhead, A. B. and Field, M. (1957) *Bull. World Hlth Org.* **16**, 277.

Harnisch, J. P., Berger, R. E., Alexander, E. R., Monda, G. and Holmes, K. K. (1977) *Lancet* **i**, 819.

Hathaway, A. and Peters, J. H. (1971) In: *Trachoma and Related Disorders caused by Chlamydial Agents*, pp. 260–268. Ed. by R. L. Nichols. Excerpta Medica, London.

Holmes, K. K. *et al.* (1975) *New Engl. J. Med.* **292**, 1199.

Hutchinson, G. R., Taylor-Robinson, D. and Dourmashkin, R. R. (1979) *Brit. J. vener. Dis.* **55**, 194.

Jawetz, E., Dawson, C. R., Schachter, J., Juchau, V., Nabli, B. and Hanna, L. (1971) In: *Trachoma and Related Disorders caused by Chlamydial Agents*, pp. 233–242. Ed. by R. L. Nichols. Excerpta Medica, London.

Jones, B. R. (1971) In: *Trachoma and Related Disorders caused by Chlamydial Agents*, pp. 243–253. Ed. by R. L. Nichols. Excerpta Medica, London; (1975) *Trans. ophthal. Soc., U.K.* **95**, 16.

Karrer, H., Meyer, K. F. and Eddie, B. (1950*a*) *J. infect. Dis.* **87**, 13; (1950*b*) *Ibid.* **87**, 24.

Keat, A. C., Maini, R. N., Nkwazi, G. C., Pegrum, G. D., Ridgway, G. L. and Scott, J. T. (1978) *Brit. med. J.* **i**, 605.

MacCallan, A. F. (1936) *Trachoma*, Butterworth, London.

McComb, D. E. and Nichols, R. L. (1969) *Amer. J. Epidem.* **90**, 278.

McComb, D. E., Peters, J. H., Fraser, C. E. O., Murray, E. S., Macdonald, A. B. and Nichols, R. L. (1971) In: *Trachoma and Related Disorders caused by Chlamydial Agents*, pp. 396–406. Ed. by R. L. Nichols. Excerpta Medica, London.

Mårdh, P.-A., Ripa, T., Svensson, L. and Weström, L. (1977). *New Engl. J. Med.* **296**, 1377.

Menke, H. E., Schuller, J. L. and Stolz, E. (1979) *Brit. J. vener. Dis.* **55**, 379.

Meyer, K. (1967) *Amer. J. Ophthal.* **63**, 1225.

Meyer, K. F. and Eddie, B. (1962) *Ann. N. Y. Acad. Sci.* **98**, 288.

Møller, B. R. *et al.* (1979) *Brit. J. vener. Dis.* **55**, 422.

Møller, B. R. and Mårdh, P.-A. (1980) *Acta path. scand., Section B* **88**, 107.

Morange, A. (1895) *De la Psittacose, ou Infection Spéciale Déterminée par des Perruches.* Thesis, Académie de Paris.

Mull, J. D. and Peters, J. H. (1971) In: *Trachoma and Related Disorders caused by Chlamydial Agents*, pp. 211–216. Ed. by R. L. Nichols. Excerpta Medica, London.

Murray, E. S. (1964) *J. infect. Dis.* **114**, 1.

Nataf, R. (1952) *Le Trachome.* Masson, Paris.

Olson, B. J. and Treuting, W. L. (1944) *Publ. Hlth Rep., Wash.* **59**, 1299.

Oriel, J. D., Johnson, A. L., Barlow, D., Thomas, B. J., Nayyar, K. and Reeve, P. (1978) *J. infect. Dis.* **137**, 443.

Oriel, J. D. and Ridgway, G. L. (1982) *Genital Infection by Chlamydia trachomatis.* Edward Arnold, London.

Oriel, J. D., Ridgway, G. L. and Tchamouroff, S. (1977) *Scot. med. J.* **22**, 375.

Ostler, H. B., Schachter, J. and Dawson, C. R. (1969) *Arch. Ophthal., N. Y.* **82**, 587.

Palmer, S. R. (1982) *J. roy. Soc. Med.* **75**, 262.

Rees, E., Tait, I. A., Hobson, D. and Johnson, R. W. A. (1977) In: *Non-gonococcal Urethritis and Related Infections*, pp. 67–76. Ed by K. K. Holmes and D. Hobson. Amer. Soc. Microbiol., Washington, D.C.

Reeve, P., Gerloff, R. K., Casper, E., Philip, R. N., Oriel, J. D. and Powis, P. A. (1974) *Brit. J. vener. Dis.* **50**, 136.

Reinhards, J. (1969) *Rev. int. Trachome* 1969/70, p. 211.

Report (1978) *Code of Practice for the Prevention of Infection in Clinical Laboratories and Post-mortem Rooms*. United Kingdom Health Departments. H.M.S.O. London.

Ridgway, G. L. and Oriel, J. D. (1977) *J. clin. Path.* **30**, 933.

Roberts, W., Grist, N. R. and Giroud, P. (1967) *Brit. med. J.* **iv**, 37.

Schachter, J. and Dawson, C. R. (1978) *Human Chlamydial Infections*. PSG Publishing Company Inc., Littlejohn, Massachusetts; (1979) *Lancet* **i**, 702.

Schachter, J., Grossman, M., Holt, J., Sweet, R., Goodner, E. and Mills, J. (1979) *Lancet* **ii**, 377.

Schachter, J., Ostler, H. B. and Meyer, K. F. (1969) *Lancet* **i**, 1063.

Schachter, J. *et al.* (1975) *J. Amer. med. Ass.* **231**, 1252.

Sowa, J., Collier, L. H. and Sowa, S. (1971) *J. Hyg., Camb.* **69**, 693.

Sowa, S., Sowa, J., Collier, L. H. and Blyth, W. A. (1965) *Trachoma and Allied Infections in a Gambian Village*. Spec. Rep. Ser. med. Res. Coun., Lond., no. 308. H.M.S.O., London.

Storz, J. (1971) *Chlamydia and Chlamydia-Induced Diseases*. Charles C Thomas, Springfield, Ill.

Storz, J., Carrol, E. J., Ball, L. and Faulkner, L. C. (1968) *Amer. J. vet. Res.* **29**, 549.

Storz, J. and McKercher, D. G. (1970) *Cornell Vet.* **60**, 192.

Symposium (1967) Conference on Trachoma and Allied Diseases. *Amer. J. Ophthal.* **63**, 1027; (1971) *Trachoma and Related Disorders Caused by Chlamydia Agents*. Ed. by R. L. Nichols. International Congress Series No. 223. Excerpta Medica, London; (1983*a*) *Chlamydial Infections*. Ed. by P. -A. Mårdh, K. K. Holmes, J. D. Oriel, P. Piot and J. Schachter. Elsevier Biomedical Press, Oxford; (1983*b*) *Brit. med. Bull.* **39**, 107–208.

Tack, K. J. *et al.* (1980) *Lancet* **i**, 116.

Taverne, J. and Blyth, W. A. (1971) In: *Trachoma and Related Disorders caused by Chlamydial Agents*, pp. 88–107. Ed. by R. L. Nichols. Excerpta Medica, London.

Taverne, J., Blyth, W. A. and Ballard, R. C. (1974) *J. Hyg., Camb.* **72**, 297.

Thygeson, P. and Dawson, C. R. (1966) *Arch. Ophthal.*, N.Y. **75**, 3.

Thygeson, P. and Stone, W. (1942) *Arch. Ophthal.*, *N.Y.* **27**, 91.

Tipple, M. A., Beem, M. O. and Saxon, E. M. (1979) *Pediatrics, Springfield* **63**, 192.

Treharne, J. D., Ripa, K. T., Mårdh, P.-A., Svensson, L., Weström, L. and Darougar, S. (1979) *Brit. J. vener. Dis.* **55**, 26.

Treuting, W. L. and Olson, B. J. (1944) *Publ. Hlth Rep., Wash.* **59**, 1331.

Wachendörfer, G. and Lohrbach, W. (1980) *Berl. Münch. tierärtzl. Wschr.* **93**, 248.

Wang, S. P., Eschenbach, D. A., Holmes, K. K., Wager, G. and Grayston, J. T. (1980) *Amer. J. Obstet. Gynec.* **138**, 1034.

Wang, S. P. and Grayston, J. T. (1967) *Amer. J. Ophthal.* **63**, 1133; (1971) In: *Trachoma and Related Disorders Caused by Chlamydial Agents*, pp. 217–232. Ed. by R. L. Nichols. Excerpta Medica, London.

Wang, S. P., Grayston, J. T., Kuo, C. C., Alexander, E. R. and Holmes, K. K. (1977) In: *Non-gonococcal Urethritis and Related Infections*, pp. 237–248. Ed. by D. Hobson and K. K. Holmes. American Society for Microbiology, Washington.

Wood, T. R. and Dawson, C. R. (1967) *Amer. J. Ophthal.* **63**, 1298.

World Health Organisation (1962) *Expert Committee on Trachoma: Third Report*. Tech. Rep. Ser. World Hlth Org. No. 234; (1971) *World Hlth Statist. Rep.* **24**, No. 4; (1973) *Field Methods for the Control of Trachoma*, ed. by M. L. Tarizzo. World Health Organization, Geneva; (1975) *Guide to the Laboratory Diagnosis of Trachoma*. World Health Organization, Geneva; (1979) *Guidelines for Programmes for the Prevention of Blindness*. World Health Organization, Geneva.

77

Rickettsial diseases of man and animals

Barrie P. Marmion

Introductory

Though epidemic typhus has been known for centuries and during this time has affected the course of history (Zinsser 1935, Drew 1965), not till 1909 did Nicolle and his colleagues (1911) show that infection was spread by the body louse. In 1916 da Rocha-Lima gave the name *Rickettsia prowazeki* to the aetiological agent in honour of Ricketts and Prowazek both of whom died of the disease.

Earlier, Ricketts (1907) in the United States had succeeded in transmitting the disease, Rocky Mountain spotted fever, to guinea-pigs, but its relation to typhus remained obscure until 1916 when Wolbach (1925) demonstrated that the causative agent was also a rickettsia. Antigenically distinct from *Rick. prowa-*

zeki this organism turned out to be the first identified member of a group of tick-borne rickettsiae associated with human disease in many parts of the world. Likewise during the 1914–18 war, and again in the second world war, a louse-borne disease emerged, known as trench fever, which was found to be caused by a rickettsia-like organism that differed from *Rick. prowazeki* (Swift 1919–1920, Mooser *et al.* 1948, 1949).

Subsequently the improvement in methods of cultivating rickettsiae, and in classifying them, led to the recognition of sporadic (murine) typhus, and the recrudescent forms of epidemic typhus (Brill–Zinsser disease). It also clarified the ecology and infective cycles of the organisms and their relation to arthropods such as fleas, ticks, mites and lice, as well as the part played by small wild mammals as reservoirs in nature.

Finally, in the 1930s Derrick, Burnet and their colleagues (Burnet and Freeman 1937, Derrick 1937, 1953) added another rickettsial disease, Q (for Query) fever, to the list, when they described it in meat and farm workers in Brisbane, Australia. Q fever was exceptional in that it not only had a cycle of maintenance in ticks and small bush animals but also spread among cattle, sheep and goats by contact and by the respiratory route. The epidemiology resembled that of brucellosis in a number of ways.

Brief descriptions of the various rickettsial diseases of man and animals are now given. The order followed is that in which the organisms are classified in Chapter 46 (Table 46.1), namely typhus fever, the spotted fever group, scrub typhus, trench fever, Q fever, and rickettsia-like infections of animals. Particular emphasis is placed on clinical features, laboratory diagnosis, epidemiology, treatment, and prevention.

Accounts of the rickettsial diseases with useful clinical illustrations may be found in the chapters by Snyder (1965) on typhus fever, Woodward and Jackson (1965) on spotted fever, and Smadel and Elisberg (1965) on scrub typhus. Woodward (1981) provides some interesting historical anecdotes on the pathogenesis of certain rickettsial diseases and on the recognition of murine typhus. Fiset (1978) provides a synoptic account of the clinical and laboratory diagnosis of rickettsial diseases of man, and Robertson (1976) of infections in animals in the tropics. Practical details of the choice and processing of specimens for laboratory diagnosis, isolation methods, and serological testing are described by Elisberg and Bozeman (1979). For a general account of rickettsial diseases, see Burgdorfer and Anacker (1981) and Report (1982).

Typhus fevers

Two main varieties of typhus fever are known. The first is often referred to as the classical or epidemic type; it is caused by *Rick. prowazeki* and appears to be invariably louse-borne between human beings; a further cycle of maintenance has recently been discovered in flying squirrels and their ectoparasites. The second variety of typhus fever is called the murine (endemic) type; it is caused by the related though not identical *Rick. typhi* (syn. *mooseri*), and is usually carried to man by the rat flea. Much of the early work on the separation of these two infections was done by Maxcy (1929), Mooser and his colleagues (Mooser 1928, 1932, Mooser *et al.* 1931) and Dyer and coworkers (1931).

Classical (epidemic) typhus fever *Synonyms:* Fleckfieber, Exanthematische typhus

Exanthematic typhus, as the Germans call it, was distinguished on clinical and pathological grounds from typhoid fever—the German abdominal typhus—by Gerhard (1837) during the 1836 outbreak in Philadelphia. Though epidemic typhus is now uncommon in most countries, it has been in the past one of the great scourges of the world. Like dysentery, it appeared whenever large numbers of people were herded together under insanitary conditions; for this reason it was common in gaols—hence its alternative name 'gaol fever'—in military campaigns, 'camp fever'— and during times of famine. It was responsible for the plague of Athens in 430–428 BC (see MacArthur 1959); and was a major factor in Napoleon's defeat in Russia in 1812 (Drew 1965) and in other campaigns. Along with relapsing fever and dysentery it spread ruin and misery during the Irish famine of 1845–50 (MacArthur

1956). It was prevalent in Russia and Central Europe during the 1914–18 war and again during the 1939–45 war, when it spread back into the concentration camps in Germany and also appeared in Italy. Fortunately during the second war vaccine and control measures were much improved and the use of DDT helped to check the spread of the disease. Those who wish to realize something of the havoc epidemic typhus has brought among the human population and of the significant part it has played in determining the outcome of great military campaigns should read the fascinating account given by Zinsser (1935).

Clinical features

After an incubation period of 10 to 14 days there is an abrupt onset with severe headache, chills, generalized muscular aching, high fever (39–41°), and vomiting. A severe and increasing frontal headache develops. Four to seven days later a rash appears and may be followed by respiratory and nervous manifestations. The rash appears first on the trunk and, in the course of a few days, covers the body except for face, palms and soles. The lesions start as macules and progress to a darker, sometimes purpuric rash in the second week. In the second and third weeks the patients may become comatose or delirious. A patchy pneumonia may be present. Gangrene of the toes, feet, tips of fingers, ear lobes, nose, penis, scrotum or vulva may occur. The case fatality varies with age. In severe epidemics it is negligible in children except infants, seldom more than 5 per cent below 20 years of age, about 10–15 per cent at 40 years, 50 per cent at 50 years, and generally fatal over 60 years of age (Megaw 1942).

Despite the severity of the clinical disease the survivors recover their mental and physical capacities rapidly and seldom show severe sequelae. There is a reduction in red cell count and haemoglobin during the acute disease; the white cell count is slightly lowered. There are red cells, albumin and granular casts in the urine, and there may be an increase in urea and non-protein nitrogen concentrations in the blood; high concentrations indicate a poor prognosis. For a detailed account of the clinical course of the disease and illustrations of the rash, the reader is referred to Snyder (1965).

The widespread nature of the pathological changes in typhus reflects the basic lesion of the disease—the multiplication of rickettsiae in endothelial cells lining small blood vessels. Thrombosis follows, with accumulation of polymorphonuclear leucocytes, macrophages, and lymphoid cells. Such lesions are numerous in the skin, central nervous system, and myocardium. Histologically, rickettsial pneumonia is characterized by an inflammatory interstitial infiltration of lung tissue, unlike bacterial pneumonia.

Epidemiology of epidemic typhus

Despite recent findings implicating a non-human reservoir for *Rick. prowazeki* (see below) human epidemic typhus spreads mainly by the 'man to louse to man' sequence of infection (Fig. 77.1).

Nicolle and his colleagues (1911) showed experimentally that the disease was spread by the body louse,

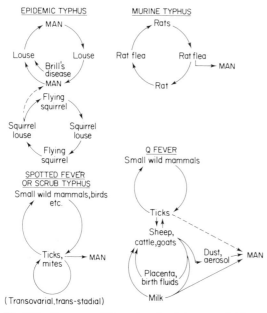

(Transovarial, trans-stadial)

Fig. 77.1 Summary of the cycles of maintenance of rickettsiae and *Coxiella burneti* in small wild animals, arthropods, and domestic animals; and routes of transmission to man.

Pediculus humanus corporis, which was infected by ingesting the blood of a patient. About a week after infection lice can transmit the disease to normal persons because the rickettsiae, having multiplied in the gut, are excreted in the faeces. The transmission usually takes place through the skin by scratching, though occasionally it may occur from inhalation of dust containing dried but still viable rickettsiae. da Rocha-Lima (1920) demonstrated the infectivity of lice for at least 24 days and believed that they remained infectious for the rest of their lives. However, most infected lice die within 14 days. Head lice as well as body lice, but not pubic, may act as natural vectors. The monkey louse *Pedicinus longiceps* can also transmit typhus.

Between epidemics the organism survives in latent human infections. Brill-Zinsser cases (see below) serve as a source of fresh outbreaks, given conditions of social disruption, louse infestation and so forth.

Reiss-Gutfreund (1956, 1966), Imam (1968) and Philip (1968) brought evidence to suggest that a reservoir of infection exists in domestic animals in certain countries, for example Ethiopia and Egypt, and is spread by ticks. Antibodies against *Rick. prowazeki* and agglutinins to *Proteus* OX19 have been found in sera from cattle, sheep, camels and goats, and strains of *Rick. prowazeki* isolated from such animals and from the ticks that infest them. These observations have not been confirmed, but Bozeman and coworkers (1975) clearly identified a non-human reservoir of *Rick. prowazeki* in the southern flying squirrel, *Glaucomys volans*, in the United States. Strains of rickettsiae isolated from these animals are serologically identical with classical strains of *Rick. prowazeki* and share with them a number of biological characters (Woodman *et al.* 1977, Dasch *et al.* 1978). Sporadic cases of human infection have been identified in persons living in houses visited by flying squirrels (Duma *et al.* 1981). Contact infection does not occur between squirrels, and the vector is probably the squirrel louse, *Neohaematopinus sciuropteri*, or possibly the squirrel flea, *Orchopeas howardi* (Bozeman *et al.* 1981).

Treatment

Apart from the general supportive measures well described by Snyder (1965), specific treatment rests on the use of tetracycline or chloramphenicol. Ley and Smadel (1954) showed that these compounds were effective in experimental typhus. Experience with some of the newer antibiotics in experimental systems is reported by Spicer and his colleagues (1981).

Brill–Zinsser disease

The evidence that this disease represented a recrudescence of a previous infection with epidemic typhus was marshalled by Zinsser (1934). Brill, a physician working with immigrant Russians and Poles in New York, observed a mild form of typhus (Brill 1910). The clinical phase was of shorter duration than in classical typhus, and the case-fatality rate lower; there was a severe headache and sometimes a rash. The Weil–Felix reaction was usually negative and the specific antibody response is now known to have been in immuno-globulin class IgG, rather than IgM where it is found in a primary attack of typhus. Zinsser found that infection was due to *Rick. prowazeki*, which must al-most certainly have been acquired before the patients had left Europe. The explanation appears to be that, although the immunity which follows typhus is in general long lasting, in a few instances it wanes suffi-ciently to allow the occurrence of a further mild attack. This occurs usually 10–20 years or more after the primary attack and is therefore noted in adults, parti-cularly in older persons. *Rick. prowazeki* has been isolated from the inguinal lymph nodes of persons with a past history of typhus (Price *et al.* 1958). Recru-descent typhus has been recognized in other parts of the world also; it is not restricted to those who have migrated from areas where epidemic typhus is preva-lent. Thus in 1950 Murray and his colleagues (1951) identified a substantial number of cases of Brill's dis-ease in Yugoslavia, in an area in which epidemic ty-phus occurred in 1944–5. In other studies they showed that lice fed on patients with Brill's disease became infected with *Rick. prowazeki* and that the organisms were virulent. Infection of lice from patients with re-crudescent typhus is clearly a mechanism whereby the organism can survive from one epidemic to another; observations in Yugoslavia disclosed several small outbreaks of epidemic typhus originating from cases of Brill–Zinsser disease.

Murine (endemic) typhus fever

This mild form of typhus is caused by *Rickettsia typhi* (syn. *mooseri*) (Lewthwaite 1952). Although it had a world-wide distribution, more recently it has been seen mainly in parts of Africa and South America. It resem-bles the louse-borne disease, with fever and a rash after an incubation period of 6–14 days, but it is insi-dious in onset, milder, and of shorter duration, and has fewer complications. It is non-contagious, occurs sporadically, and has a case-fatality rate of under 1 per cent. The rat acts as a reservoir and the rickettsiae are passed from rat to rat by the rat louse, *Polyplax spinulosa*, and by the rat flea, *Xenopsylla cheopis*. The latter is also the vector for man.

Epidemiology of murine typhus

Maxcy (1929), after investivating a form of typhus in the Southern United States, concluded that it was associated with a rodent reservoir. The causative rick-ettsia was isolated from rat brain during the course of

outbreaks in Mexico (Mooser *et al.* 1931) and from rat fleas in Baltimore (Dyer *et al.* 1931). It was eventu-ally named *Rickettsia typhi*. It differs from *Rick. prow-azeki* in the lesions (scrotal reaction) produced in the male guinea-pig after intraperitoneal inoculation, in the species specificity of its corpuscular antigen, and in the exhibition of only partial sequence homology with *Rick. prowazeki* (Chapter 46). At one time there was some confusion between various diseases that in-cluded Mexican typhus, known by its Spanish name of *Tabardillo*, Brill's disease, and murine (endemic) typhus. It was largely in Mexico, where both epidemic and endemic typhus have existed for centuries, that the distinction between them was made by epidemiol-ogical, cultural, and finally by serological investiga-tions (Zinsser and Castañeda 1932) which confirmed the antigenic differences between the two types of rick-ettsia. There is no satisfactory evidence of transfor-mation in the field from murine to louse-borne typhus (Mooser 1959); indeed, given the level of sequence homology between the genomes of the two rickettsiae, transformation would perhaps be unexpected (but see Price *et al.* 1958).

The disease is endemic in many parts of the world and occurs during the summer and autumn in persons living or working in rodent-infested areas. In rats, which are the main source of spread, the infection is mild but persistent, with rickettsiae circulating in the blood and being excreted in the urine for prolonged periods. Direct transfer to man by means of urine-contaminated food is therefore possible, but more usually the flea acts as intermediate host. The flea ingests infected blood and acquires a rickettsial infec-tion in its alimentary tract. The organisms are excreted in the flea faeces but are not present in the saliva. Once infected, fleas are not incapacitated and will go on excreting rickettsiae for long periods particularly if they have regular blood meals. Infection is not the direct result of a bite but occurs from implantation of infected flea faeces into abrasions through scratching. It may also occur through inhalation of dried flea faeces. The ecology of murine typhus is reviewed by Traub and his colleagues (1978).

Laboratory diagnosis of epidemic and murine typhus

Rickettsia prowazeki and *Rick. typhi* may be isolated from the blood of patients with epidemic and murine typhus respectively by inoculation of male guinea-pigs, or animals such as cotton-rats or white rats (Chapter 46). In practice, however, as with Q fever, this is a hazardous procedure except in specially equipped laboratories and with staff protected by vac-cination. The distinctive properties of the rickettsiae of epidemic and murine typhus in experimental ani-mals are described by Elisberg and Bozeman (1979) and in Chapter 46. As a means of diagnosis, however, animal inoculation is slower and much more expensive than detection of the patient's antibody response. Serological methods consist of a combination of the Weil–Felix test (see below) with a complement-fixation, microagglutination, or microimmunofluorescence test for specific antibody.

The Weil–Felix test

In 1916 Weil and Felix cultivated from the urine of a typhus patient a proteus-like organism that was agglutinated not only by the patient's serum but also by the sera of other patients with typhus. Similar organisms isolated from other typhus patients were called *Proteus* X strains, the most highly agglutinable and the most specific for typhus being the *Proteus* OX19 strain (Felix 1916). This was agglutinated to a titre varying from 1 in 50 to 1 in 5000 by serum from typhus patients, and only rarely to a titre as high as 1 in 25 by serum from others. Agglutinins appeared in the blood on about the 4th day of illness and reached a peak by about the end of the second week. During convalescence they declined rapidly and after 5 months had disappeared almost entirely. Usually a titre of 1 in 50 or higher, regarded as positive, was reached by the 8th day. The test is based on agglutination of the *Proteus* O antigen alone. The rapid rise and fall in agglutinin concentration is a valuable feature.

Though a titre of 1 in 100 or higher during the second week of an illness is suggestive of typhus, the general usefulness of the Weil–Felix test is subject to certain limitations, and results must be interpreted strictly in relation to the clinical findings. The most satisfactory diagnosis is always one based on a rising titre. Low titres may be found in some healthy persons, in pregnant women, and in sufferers from genito-urinary infections due to strains of *Proteus* with minor common antigens of the OX19 type. Raised titres may appear in other diseases such as brucellosis and the enteric fevers. The reaction is very weak or negative in the recrudescent typhus of Brill's disease.

Although devised originally as a means of confirming the diagnosis of typhus fever, the modification of the Weil–Felix test to include *Proteus* OXK and OX2 suspensions extended its application to include the scrub typhus and spotted fever groups (see below).

The Weil–Felix reaction is negative in Q fever and rickettsialpox. Since its introduction, several explanations of how the Weil–Felix reaction works have been given. The *Proteus* X strains cannot induce any of the rickettsial diseases, and are sometimes isolated from patients with other diseases. However, X strains possess some glycolipid surface antigens in common with the appropriate rickettsial species and these are sufficient to react with the rickettsial antibody.

Other serological methods of diagnosis

In the diagnosis of rickettsial diseases agglutination of rickettsial suspensions and, more recently, complement-fixation tests have been widely used. Complement fixation is probably the most effective single means of precise identification, particularly in sporadic cases. The rickettsiae are usually grown in the yolk sac of the developing chick embryo and antigens are prepared from extracts. A soluble antigen extracted from *Rick. prowazeki* and *Rick. typhi* with ether has group characteristics and reacts with the sera of patients with the epidemic or murine typhus. Complement-fixing antibodies appear 8–10 days after the onset of the disease and reach a peak by 20–30 days; they may remain at a significant titre for many years. The early antibodies are IgM and are sensitive to 6-mercaptoethanol or ethanethiol; in Brill-Zinsser disease the complement-fixing antibodies appear earlier, are IgG, and are resistant to ethanethiol. The immunoglobulin class can be identified by the use of the anti-IgG and -IgM conjugates in the microimmunofluorescence test (Murray and Gaon 1963, Ormsbee *et al.* 1977).

Distinction between epidemic and murine infections by serological tests with patients' sera is difficult and requires rickettsial suspensions washed free from the soluble, group-specific, surface antigen. Such antigens are not available commercially and are found only in a limited number of laboratories engaged in rickettsial research. With such washed suspensions microagglutination or microimmunofluorescence tests may be used to distinguish, on the basis of the height of antibody titres, the two forms of typhus (Philip *et al.* 1976, Ormsbee *et al.* 1977). This method may present difficulties (see for example data in Duma *et al.* 1981) and may be supplemented by toxin neutralization tests.

Control and prevention of epidemic and murine typhus

As in other insect-borne diseases the prevention of rickettsial infections is mainly a matter of the control or elimination of the insect vector. In epidemic typhus, for example, it is necessary to interrupt the spread of body lice from man to man. This is best done by reducing louse infestation to a minimum and by the prevention of overcrowding. Throughout its history this type of typhus has had a close relation to poverty, malnutrition, overcrowding, wretchedness, dirt, and war. During the 1939–45 war, lice and certain other insects were successfully controlled by long-acting insecticides such as DDT (dichloro-diphenyl-trichloroethane), not only in impregnated clothing but also sprayed on houses and over potentially infected areas generally, and by repellents such as dibutyl phthalate. The control of endemic typhus is a more difficult problem because of the wide distribution of the causative agent and its vectors and the sporadic nature of the disease. However, in addition to steps to control the rat population, insecticides may usefully assist in controlling the flea population.

Vaccination

Because long-lasting immunity usually follows rickettsial infections many attempts to produce effective vaccines against them have been made. Considerable success, particularly against typhus, was achieved during the 1939-45 war. Cox (1938*a*, 1941) found that the rickettsiae of typhus, spotted fever and Q fever would grow well in the yolk sac of the developing chick embryo, providing suspensions 100 to 1000 times more potent than those from mammalian tissues. The method was applicable to large-scale production of the various vaccines and is still widely used. The vaccines were considered to lower the incidence of the disease though they did not always prevent it; when infection did occur its course was modified. Since then, diminished incidence, effective methods for controlling vector populations, and the usefulness of antibiotic therapy have lessened the interest in vaccines. Their value in certain circumstances such as the protection of nurses, doctors and laboratory workers is, however, still considerable.

Both living attenuated and inactivated rickettsial vaccines have been tested but the latter have been used most. Renewed attempts to prepare live typhus vaccines have centred around the avirulent strain E of *Rick. prowazeki* (Clavero del Campo and Gallardo 1949). The strain E originated during serial yolk sac passage of a strain of *Rick. prowazeki*. At the 11th passage it lost its capacity to infect guinea-pigs and even after more than 300 passages did not regain this property. Inoculation into volunteers in suitable dosage resulted in inapparent but immunizing infections, there being a close correspondence between the presence of neutralizing antibody and resistance to infection (Fox *et al.* 1957). A field trial with this vaccine was carried out in Peru. Early and late reactions were fairly severe. The protective effect was difficult to determine because of the number of doubtful and inapparent infections (Fox *et al.* 1959).

Spotted fevers

There is a world-wide group of rickettsial diseases in which the aetiological agents, although not identical, possess a common soluble antigen. The names applied to these diseases include Rocky Mountain spotted fever, Indian tick typhus, East African tick fever, South African tick fever, Lone Star or Bullis fever, Queensland tick typhus, and rickettsialpox. The diseases are of varying severity, but are all characterized by fever and most of them by a rash. The infection is spread by a variety of ticks except in the case of rickettsialpox, in which a mite is the vector. In fièvre boutonneuse, rickettsialpox and the North Asian, African and Australian fevers there is a primary sore at the site of the initial bite accompanied by inflammation of the regional lymph nodes. Agglutinins to *Proteus* OX19 or to *Proteus* OX2 may be found in the spotted fever group of diseases, particularly in the more severe cases, but not in rickettsialpox.

Rocky Mountain spotted fever

This disease, caused by *Rickettsia rickettsi*, was first recognized in the Rocky Mountain region of the United States in 1896 and in Eastern USA in 1931 (Report 1965). Now cases are reported mainly from the latter area, particularly the vicinity of the Appalachian Mountains, with an annual incidence of between 500 and 1000. It affects those engaged in outdoor pursuits and is commonest during the summer months. Its severity varies greatly.

A history of tick bite is present in over 80 per cent of patients, but local lesions (eschars) representing the site of tick attachment are uncommon. The incubation period is about 7 days, with a range of 3 to 12 days. The onset of illness is abrupt, with severe general or frontal headache, rigors, muscular pains, particularly in the back and leg muscles, nausea, vomiting, and fever (39.4-40°). Fever may last for 15-20 days in untreated patients. The rash usually appears on the fourth day of fever, initially on the wrists, ankles, palms, and forearms. It spreads centripetally to the axilla, buttocks, trunk, neck, and face. At first macular, it becomes maculopapular, petechial and, in grave cases, haemorrhagic. As the disease progresses there may be circulatory failure, coma, muscular rigidity, hemiplegia and other neurological signs or symptoms. A specific rickettsial interstitial pneumonitis sometimes develops. The course of all these manifestations may be strikingly shortened by antibiotic treatment (Woodward 1959). Whereas antibiotic therapy may greatly improve the clinical state of patients with severe typhus even well on in the disease, treatment in Rocky Mountain spotted fever must be early, and should preferably be given on clinical suspicion. Convalescence may be prolonged. In the west, where older persons have been more often affected, a case-fatality rate as high as 70 per cent has been recorded, but in the east where most cases appear to occur in persons under 20 years of age the fatality rate has been much less, often *ca* 6 per cent. As with typhus infections, the small blood vessels are the primary site of pathological change. This results in endothelial swelling, proliferation and degeneration; thrombus formation; and fibrinoid degeneration of the muscular layers of the arteriolar wall and infiltration with polymorphonuclear leucocytes and mononuclears. Thrombosis and changes in the musculature of vessels are more common than in typhus (see Woodward and Jackson 1965 for further details).

Epidemiology

The basic cycle of maintenance of the spotted fever rickettsiae in nature is shown in Fig. 77.1. *Rickettsia rickettsi* is found in various natural vertebrate hosts—rodents, rabbits, and dogs. It also infects various species of tick: *Dermacentor andersoni* in North-West USA; *Dermacentor variabilis* and *Haemaphysalis leporis-palustris* in Eastern USA and Canada; the Lone Star Tick *Amblyomma americanum*, in South and South-West USA; and *Rhipicephalus sanguineus* and *Amblyomma cajennense* in Mexico and South America. There is persistent infection in these ticks, without obvious disease, and the rickettsia is transmitted via the ovary and between stages in the tick life cycle. New lines of ticks are infected by feeding on animals with rickettsaemia. The organism is excreted with saliva and faeces. Man is an incidental victim when he intrudes into the cycle of maintenance in small wild animals and their ticks.

Infections with other members of the spotted fever group

Rickettsialpox

This benign febrile disease with a papulo-vesicular rash somewhat resembling that of chickenpox was reported from New York, Boston and other cities in the United States (Huebner *et al.* 1946, Fuller 1954). A similar disease occurring at about the same time in Russia was named 'vesicular rickettsiosis' (see Zdrodovskii and Golinevich 1960). The causative organism, named *Rickettsia akari* by the American workers, is spread by a mite, *Allodermanyssus sanguineus*, that infests house mice (*Mus musculus*). It possesses the common soluble antigen of the spotted fever group of rickettsias, but although a member of the group it is unusual in being transmitted by mites rather than by ticks. It provokes cross-immunity in guinea-pigs to *Rickettsia australis*, but in complement-fixation tests with washed rickettsial suspensions the two organisms are distinguishable (Lackman *et al.* 1965). Clinically, a papulo-vesicular lesion occurs at the site of the bite, and the regional lymph nodes enlarge. This is followed in 3 to 10 days by fever lasting a few days during which a generalized papulo-vesicular rash develops. The initial lesion persists as a black eschar for several

weeks. During the fever the rickettsia can be isolated from the patient's blood by intraperitoneal inoculation of mice or guinea-pigs. In the course of the disease in man specific complement-fixing antibody develops but the Weil–Felix reaction remains negative. The rickettsia has been isolated from naturally infected house mice (Huebner *et al.* 1947).

Fièvre boutonneuse   *Synonyms:* Marseilles fever, Indian tick typhus, East and South African tick fever

Fièvre boutonneuse, first described by Conor and Bruch (1910) as occurring in Tunis, was later identified in several countries bordering the Mediterranean and Black Seas and its relation to the other named diseases gradually established. The name 'boutonneuse' refers to the raised character of the maculopapular rash that develops on the 3rd to 5th day of illness. A black spot or *tache noire* having a necrotic centre is normally present at the site of the tick bite in the early stage of the disease. The regional lymph nodes are enlarged. The clinical course is relatively brief with intermittent or remittent fever lasting up to 10–12 days, and the fatality is low. In some instances, as has been observed in East Africa, there is no obvious primary sore. For the most part infection is carried by the dog tick, *Rhipicephalus sanguineus*. Dogs, many of which are latently infected, serve as a reservoir for the causative organism, *Rickettsia conori* (Brumpt 1932). Once ticks are infected they remain so, and can pass on the agent transovarially to their progeny.

In South Africa infection has been transmitted by several ixodid ticks including the dog tick, *Haemaphysalis leachi leachi* (see Gear 1941). Man is infected by the nymphs, rats by the larvae, and dogs by the adult ticks. The main rodent reservoir is the striped mouse, *Rhabdomys pumilio*, and the vlei rat, *Otomys irroratus* (Gear 1954). *Rickettsia conori* shares the soluble antigen of spotted fever rickettsiae but can be separated into a distinct subgroup by the complement-fixation test when washed rickettsial suspension is used as antigen (Lackman *et al.* 1965).

North Asian tick-borne spotted fever

The tick typhus fever of Northern Asia is a febrile disease lasting about 8–10 days with an infiltrated sore at the site of the bite, regional adenitis, and a fairly abundant, sometimes haemorrhagic, roseolar papular rash. Endemic foci may still persist in rural areas in Siberia and the Far East. Infection is transmitted by a variety of ixodid ticks such as *Dermacentor nuttalli* and *Haemaphysalis concinna* that are parasitic on rodents. The causative organism, *Rickettsia sibirica*, again belongs to the spotted fever group but can be distinguished by the specific complement-fixation test (Lackman *et al.* 1965).

North Queensland tick typhus

This disease resembles a mild form of Rocky Mountain spotted fever (Andrew *et al.* 1946) with an eschar at the site of the tick bite, regional adenitis, fever, and rash. Tick transmission is presumed, with the scrub tick *Ixodes holocyclus* as the prime suspect. The causative organism is known as *Rick. australis*. It possesses the common soluble antigen of the spotted fever group. Guinea-pigs immunized with it are completely protected against *Rick. akari*, which causes the disease rickettsialpox, but not against *Rick. rickettsi*. In specific complement-fixation tests, however, with washed rickettsial suspensions as antigens, *Rick. australis* is distinct from *Rick. akari*, though more closely related to it than to other rickettsial species.

Laboratory diagnosis

The general approach is the same as with the typhus group. Isolation of the rickettsia is not an economical procedure and is hazardous; serological methods are rapid and effective. The Weil–Felix reaction is similar to that in typhus, sometimes with higher titres to *Proteus* OX2; it is negative in rickettsialpox. Complement-fixation tests with the group-specific antigen become positive in the second or third week. A microimmunofluorescence test with purified *Rick. rickettsi* antigen has proved to be highly sensitive for detecting and measuring antibodies. Other serological techniques, such as indirect haemagglutination and latex agglutination are also effective and have recently been compared with immunofluorescence and complement-fixation reactions (Kleeman *et al.* 1981). Woodward and his colleagues (1976) describe a rapid method for detecting *Rick. rickettsi* by immunofluorescence in biopsy specimens of skin from the area of the rash as early as four days after the onset of the disease.

Treatment and control

Antibiotic treatment is given as with typhus fever. Anti-tick measures and vaccination are described by Woodward and Jackson (1965).

Scrub typhus

This disease is now known to occur in many areas of Central Asia and the Far East. It appears in foci, sometimes referred to as ecological islands, which vary in the size and complexity of the terrain and extend from West Pakistan to Burma, Japan and the Philippines, and from Korea and the southern tip of Siberia to Malaysia, Indonesia and Northern Queensland in Australia (see Traub 1974). In the past it has been known by various names including *tsutsugamushi fever*, *Japanese river fever* or *kedani* in Japan; *scrub typhus* in Malaysia and Australia; and *mite typhus* in Indonesia. The evidence now is that all these illnesses are due to the same organism, *Rickettsia tsutsugamushi*. The foci in which it may be present include temperate climatic areas as in Japan, moist tropical areas which may be subject to flooding, rain forest, primary jungle, grassland, and semi-arid desert. Outbreaks have been recognized in new areas after alteration of the natural vegetation by human activities. (For a description of ecological niches in rickettsial diseases, see Biocenosis in Chapter 48.)

The disease is transmitted through the bite of larvae or chiggers of trombiculid mites, all being species of the genus *Leptotrombidium*. The classical vectors are *L. akamushi*, *L. deliense*, *L. pallidum*, and *L. scutellare*, but others may also be concerned (Traub and Wisseman 1968). These mites infest a range of mammals, particularly small rodents, including field voles, rats, tree-shrews, bandicoots, and mongooses; and also birds, including migratory birds, all of which can act as the host reservoirs. Only at the larval stage, when it usually feeds once, is the mite parasitic; the infection it spreads has therefore been transmitted transovarially. The disease is widespread because of the variety of situations in which the different species of mite may be found. Of the main vectors *L. akamushi* (= red mite) is associated with grassland, whereas *L. deliense* is associated with afforested areas. Both species need moisture and become abundant under warm wet conditions, but some mites still survive in a dry or even cold environment. Given warmth and dampness extensive outbreaks of scrub typhus can develop in a very short time in heavily infested areas; in dry terrain cases may still occur, but they are few and localized to sites in which the mite larvae still persist.

After an abrupt onset the disease, like classical typhus, is characterized by fever, headache, rash, conjunctival infection, generalized lymphadenopathy, and deafness. There is localized swelling at the site of the puncture wound, at first macular then papular, and finally, in some instances, ulceration. The adjacent lymph nodes enlarge, as does the spleen. There is no appreciable itching at the site of the bite and the lesion, unless severe, may not be noticed in dark-skinned patients. The severity of the lesion may depend on the degree of immunity resulting from previous infection. The incubation period ranges from 6–21 days, generally 10–12, somewhat longer than in typhus and spotted fever. This, together with the very variable mortality rate in different endemic areas, may reflect strain differences (Smadel and Elisberg 1965). In Japan the case-fatality rate has been as high as 60 per cent and in Burma and Malaysia has been reported as about 10 per cent (Lewthwaite 1952).

Laboratory diagnosis

The laboratory confirmation of a diagnosis of scrub typhus presents a number of difficulties and in practice rests heavily on the Weil–Felix reaction to detect *Proteus* OXK agglutinins. The rickettsia may be isolated by intraperitoneal inoculation of blood clot or tissue into white mice, which may become sick or die 10–24 days later. Rickettsiae may be seen in impression smears made from the surface of the spleen or liver. However, not all strains are highly pathogenic for mice. (For histological changes in mice, see Catanzaro *et al.* 1976.) Clearly animal inoculation has limitations as a routine diagnostic procedure. Specific serological tests are limited by the antigenic heterogeneity of strains in different geographical areas, or even in the same area (see Shirai *et al.* 1979). Complement-fixation tests with ether-extracted antigens from infected yolk sacs give reactions that are strain, rather than group, specific; in this *Rick. tsutsugamushi* differs from the causative rickettsiae of the typhus and spotted fever groups (see Chapter 46, Table 46.2). The use of purified rickettsial suspensions from the prototype Karp, Kato and Gilliam strains has proved satisfactory in complement-fixation tests for diagnostic and survey purposes (Shishido 1962); such suspensions may also be used in immunofluorescence tests (Elisberg and Bozeman 1966). The value of the test systems is, however, still limited by strain specificity and the very restricted availability of such purified rickettsial suspensions. Consequently most laboratories confine their tests to measurement of agglutinins to *Proteus mirabilis*, OXK strain. Such agglutinins appear at the end of the second week of illness, reach a maximum by the third week, and disappear rapidly. Agglutinin levels are lower than those seen with OX19 or OX2 in typhus and spotted fever, and reactions are confined to the OXK strain. Unfortunately, a positive Weil–Felix scrub typhus reaction can be found in only about 50 per cent of patients. A further complication is that some patients with louse-borne relapsing fever (Zarafonetis *et al.* 1946) or leptospirosis (Carley *et al.* 1955) may possess OXK agglutinins.

Treatment and control

Clinical cases of scrub typhus respond well to treatment with chloramphenicol (Smadel *et al.* 1948*b*) or tetracyclines. Killed vaccine has been found ineffective because it protects only against the homologous strain and not against natural infection with diverse field strains (Cord and Walker 1947). Protection by chemoprophylaxis and by vector control are described in detail by Woodward and Jackson (1965).

Trench fever *Synonyms:* Wolhynian fever, 5-Day fever

Trench fever first attracted attention in 1915 when it broke out in epidemic form among troops in the European theatre of war. It was a major cause of sickness, rivalling influenza in its impact. Later it was recognized in Egypt, Syria, and Mesopotamia. It reappeared in epidemic form on the eastern European front in the 1939–45 war. It could be transmitted experimentally to healthy volunteers by parenteral inoculation with blood, sputum or urine from patients (McNee *et al.* 1916), the blood being infectious from the first day of illness until at least the 51st day. Epidemiological evidence indicated that the disease was spread by body lice. These became infectious 5 to 9 days after feeding on a trench fever patient and remained so for life. The infectious agent was present in the louse excreta in considerable numbers, so that 0.1 mg sufficed to cause the disease when inoculated into man. The agent could be destroyed by dry heat at 100° in 20 minutes and by moist heat at 70° in 30 minutes. Töpfer (1916) described the finding of minute rods, often arranged in pairs and often showing bipolar staining, in lice from trench fever patients but not from healthy persons. These findings were confirmed by others. The organism was first named *Rickettsia quintana*, but when it was found to grow in cell-free media, and to be cell dependent rather than an obligate intracellular parasite, it was renamed *Rochalimaea quintana*. In the infected lice the organisms were found in large numbers crowding the region of the epithelial cells lining the gut; they were always extracellular in position. In the 1939–45 war Mooser and his associates (1948, 1949) isolated a strain of the organism from a trench fever sufferer in Yugoslavia by feeding lice on the patient. It was not possible to transmit the organism to mice, rats, guinea-pigs, hamsters, chick embryos, or tissue cultures. Finally, Vinson (1966) grew the organism on modified blood-agar medium (see Chapter 46).

It is very doubtful whether laboratory animals are susceptible to the disease, except perhaps the rhesus monkey (*Macaca mulatta*). Guinea-pigs may suffer from a low fever after inoculation with infectious material, but infection cannot be transmitted by further passage. Unlike animals, human beings are almost all susceptible to the disease. One attack seems to produce only a limited immunity. Some patients, having recovered from one attack, have been re-infected 4 or 5 months later.

Clinical features

After an incubation period of 14–30 days the disease has a sudden onset with chills, headache, dizziness, retro-orbital pain, nystagmus, engorgement of the conjunctival blood vessels, severe pains in the back and legs, and successive crops of erythematous macules or papules on the chest, abdomen or back. Fever sometimes lasts for several weeks with relapses. The infection is mild and rarely fatal. Biopsy examination of skin with rash shows inflammation around blood vessels, but in trench fever, unlike typhus and spotted fever, the endothelial cells are unaffected.

Laboratory diagnosis

The organism may be isolated from the blood early in the illness by inoculation of solid or liquid versions of the media developed by Vinson (Vinson 1966, Varela *et al.* 1969). However, as with the true rickettsial diseases, serological methods of diagnosis are safer and more economical. The Weil–Felix reaction is negative, but specific antibody can be measured by complement fixation, microagglutination or immunofluorescence with suspensions of the organism grown in liquid medium.

Treatment and control

The general approach is the same as that with rickettsial infections, namely treatment of the illness with tetracycline antibiotics and control of louse infestation.

Q fever

This rickettsial disease was first observed in Australia. Derrick (1937) described a disease, under the name Q (Query) fever, affecting abattoir workers and dairy farmers in Queensland. The outstanding features were severe headache, high fever, and slight cough; there was no primary sore, a rash was unusual, and the Weil–Felix reaction was negative. Derrick, perhaps influenced by methods for the diagnosis of leptospirosis, a disease also found in Brisbane meatworkers, inoculated guinea-pigs intraperitoneally with blood from febrile patients. The guinea-pigs in turn became febrile and an infectious agent was detected in suspensions of spleen or liver, or in blood taken during the fever, when these materials were passaged to fresh animals. Once they had recovered from the infection, the animals were resistant to challenge with suspensions of infected liver or spleen that produced fever in unexposed guinea-pigs. With this laborious but sensitive method of isolating and identifying the specific infective agent Derrick was able to distinguish Q fever from other abattoir fevers, and from typhus. However, he still did not know the nature of the infective agent. Burnet and Freeman (1937), working in collaboration with Derrick, adapted the agent Derrick had isolated in guinea-pigs to mice and saw microcolonies of rickettsia-like organisms in the macrophages of spleen sections stained by Giemsa or Mann's techniques. The rickettsiae were extracted from spleen tissue and used as the antigen preparation in an agglutination test; this simplified laboratory diagnosis and strain comparisons. Derrick then named the organism *Rickettsia burneti*.

At about the same time in the United States Davis and Cox (1938) reported the isolation of an agent from ticks, *Dermacentor andersoni,* collected near Nine Mile Creek, Montana. This had properties similar to those of the organism isolated by Derrick but appears to have been more virulent for guinea-pigs (Cox 1938*b*, Parker and Davis 1938). The Nine Mile isolate was not at first related to human disease until a laboratory infection occurred. Subsequent comparisons of *Rick. burneti* and the organism (*Rickettsia diaporica*) isolated in Montana showed that they were identical (Burnet and Freeman 1939). Later, Philip (1948) proposed the name *Coxiella burneti* (Derrick) in recognition of the differences between the Q fever organism and other rickettsiae (see Chapter 46).

Clinical features

Q fever, as a human disease, is now recognized to be present in most countries of the world (Kaplan and Bertagna 1955, Babudieri 1959). Derrick's original description of the clinical syndrome has been expanded in various ways (Derrick 1973). The incubation period usually ranges from 14–26 days under natural circumstances; it was as short as ten days in volunteers exposed to highly infective aerosols of the Q fever organism (Tigertt 1959). The onset of illness is often quite sudden, with fever, shivering and shaking, malaise, pains in the limbs or joints, and a frontal headache that may be so severe that it does not respond to ordinary analgesics. Retro-orbital pain and photophobia may be present. The patient may be pallid and grey rather than flushed from the fever and the sclera may have a yellow tinge that is not due to jaundice. Fever usually lasts for 9 to 14 days and tends to persist longer in older patients; there may be a weight loss of a stone (6.35 kg) or more. There is some evidence that the acute onset and generalized symptoms are associated with circulating immune complexes (Lumio *et al.* 1981). The fever follows one of a number of patterns. Commonly, the temperature is elevated for a number of days, then falls rapidly by lysis. In other patients there may be large swings in temperature or a relapse after the temperature has returned to normal. A substantial proportion of patients with Q fever have abnormal liver function tests, particularly that for alkaline phosphatase, and about 5 per cent have jaundice (Clark *et al.* 1951, Powell 1961). Liver biopsy shows characteristic small granulomas, sometimes with fibrinoid material, and epithelioid cells in annular arrangement (Pellegrin *et al.* 1980). The routine use and refinement of chest radiographs has revealed that 30 to 50 per cent of Q fever patients have pneumonia although this varies from one geographical area to another. The lesion usually consists of a minor area of pneumonitis—a so-called 'ground glass' change—at one or other lung base, but the consolidation may sometimes be substantial enough to mimic that of pneumococcal pneumonia. However, Q fever is more than a fever with hepatitis or pneumonitis. Any body system may be affected. Abnormalities include meningo-encephalitis, encephalitis with coma, cerebritis with cerebellar ataxia, myocarditis and pericarditis, peripheral thrombosis, granulomatous changes in the bone marrow, bone marrow necrosis, orchitis and placentitis, and transmission to the fetus (see references in Syrucek *et al.* 1958, Babudieri 1959, Okun *et al.* 1979, Brada and Bellingham 1980).

Experiments with cattle, sheep, and small laboratory animals have shown that *C. burneti* persists for long periods in tissues such as liver, spleen, lymph nodes, kidney, ovary, and brain. In animals that subsequently become pregnant the organism multiplies to produce large numbers in the placenta. Given such persistence in animals it was reasonable to expect a recrudescent illness of the Brill–Zinsser type as a late sequel to some human Q-fever infections. So far, however, such recrudescences have not been found although a few patients with placental infection after an earlier attack of Q fever have been described (Syrůček *et al.* 1958).

Chronic infection with *C. burneti* as *Q fever endocarditis* has been recognized for over 20 years; recently a second form of chronic infection affecting the liver but not the heart has been proposed (Peacock *et al.* 1983).

After some preliminary indications by Huebner cited by Marmion *et al.* (1953) and Marmion (1959), Robson and Shimmin (1959) and Andrews and Marmion (1959) published detailed accounts of a patient with Q fever endocarditis, including isolation of the organism from the blood during life, its demonstration in aortic vegetations, and a raised level of complement-fixing antibody to Phase 1 anti-

gen. Since then many case reports have been published; these have been collated and analysed from time to time (Marmion 1962, Turck *et al.* 1976). Q fever endocarditis may follow an apparent or inapparent infection and may affect the mitral or aortic valves, or the endocardium away from the heart valves (Willey *et al.* 1979). In some cases there may be predisposing factors such as previous rheumatic heart disease or aortic stenosis or calcification. Certain forms of artificial valve replacement have been colonized with *C. burneti*. In other patients there have been no obvious predisposing factors. The interval between infection and the clinical onset of endocarditis varies from 6 months to one or more years. There is a low grade pyrexia, night sweats, anaemia, joint pains, clubbing of fingers, liver dysfunction, and a developing or increasing heart murmur. Blood cultures are negative for bacteria, but coxiella and bacterial endocarditis may sometimes occur together on a valve replacement. Microcolonies of *C. burneti* are found in the vegetations from the valves or endocardium and may be stained by the Giemsa or Giménez (1964) method, or by immunofluorescence with serum containing Phase 1 antibody (see Fig. 77.2). Paradoxically, although *C. burneti* is thought to multiply in macrophages (Chapter 46) the microcolonies in heart valve vegetations rarely appear to be inside cells. Serum from Q fever endocarditis patients is characterized by high levels of complement-fixing antibody to both the Phase 1 and 2 antigens of *C. burneti*; by immunofluorescence the antibody is seen to be in the IgG and IgA fractions rather than in IgM as in acute Q fever (Kimborough *et al.* 1981). Turck and his colleagues (1976) draw particular attention to the frequent occurrence of chronic liver disease, sometimes even cirrhosis in patients with Q fever endocarditis. The erythrocyte sedimentation rate is raised, rheumatoid factor may be present, and thrombocytopaenia is also common.

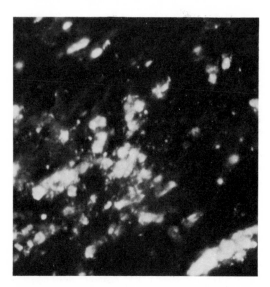

Fig. 77.2   Colonies of *Coxiella burneti* in vegetation on the heart valve in a fatal case of Q fever endocarditis. Indirect immunofluorescence with guinea-pig antiserum against Phase 1 and Phase 2 antigens.

Epidemiology

Q fever has a world-wide distribution but the cycles of maintenance in animal and arthropod hosts vary from country to country (Derrick 1953, Stoker and Marmion 1955*a*). As shown in Fig. 77.1 there are, in effect, two interacting cycles—one in wild animals with their ticks, the other in domestic animals (cattle, sheep and goats); the latter is not dependent on arthropod transmission. Human infections are associated with the second cycle and only very rarely with the first (McGuire and Williams 1948, Stoker and Marmion 1955*a*, Dwyer *et al*. 1960).

After describing the first human cases of Q fever, Derrick and his colleagues, doubtless influenced by the ecology of other rickettsiae, set about examining small bush animals and their parasites for evidence of infection. They shortly discovered evidence of infection in the bandicoot (*Isoodon torosus*) and its tick *Haemaphysalis humerosa*. Infection rates in bandicoots and ticks were high on Moreton Island off the coast near Brisbane. However, Q fever was not found among troops taking part in military exercises on the island. Links between the cycle in Australian bush animals and the disease in abattoir workers were difficult to envisage. Derrick suggested that the cattle tick, *Boophilus annulatus microplus*, might be infected from the natural cycle in bush animals and ticks, and that tick faeces on cattle hides might be the source of infection for the abattoir workers. However, although the tick could be infected experimentally from calves, many batches taken from cattle under natural conditions were uninfected.

The recognition of the second cycle of maintenance of Q fever—that in cattle, sheep and goats—gradually emerged in various ways. During the 1939–45 war there were numerous sporadic cases and outbreaks of respiratory illness (Balkan grippe) and atypical pneumonia in both Allied and German troops in Italy, Greece, and Corsica. Eventually some of the cases were identified as Q fever. There was frequently an association between outbreaks and farm environments. Robbins and co-workers (1946) and Caminopetros (1948) found that sheep and goats were sometimes infected and excreted the organism in their milk. The association of human infection with infected livestock again came into sharp focus in Los Angeles. The rapid growth of the city overran dairy farms, reducing them to small areas into which milking cattle were crowded and fed with imported feed. In these so-called 'dairies' straw and dung accumulated and was contaminated by the urine, placentas and birth fluids of calving animals. The dry conditions led to the wind-borne spread of dust. Cases of Q fever were observed in nearby residents having no immediate contact with animals (Bell *et al*. 1950). In the same investigation Luoto and Huebner (1950) showed that the placentas of infected dairy cattle sometimes contained large numbers of rickettsiae. Some of the milk from the dairies was sold raw and was considered to be the source of infection for persons outside the area of wind-borne infection. The high prevalence of Q fever was probably due to the proximity of non-immune city dwellers to cattle kept under conditions that maximized infection. Q fever was also common in northern California at this time but was related principally to infection in sheep and goats (Lennette *et al*. 1949); investigation revealed the infection of young sheep

from the environment, the persistence of infection, and its recrudescence in the later stages of pregnancy.

The relation between excretion of the organism at parturition and the generation of infective aerosols has been extensively investigated and is well summarized by Welsh and his colleagues (1958). Man becomes infected by inhaling infected droplets or dust contaminated by cattle, sheep or goats. The remarkable ability of *C. burneti* to survive in dust results in its carriage on working clothes, hair, vehicles, pelts, and other fomites. This in turn leads to the infection of persons at a distance from the infected animals (see Derrick 1953). The importance of the ingestion of infected raw milk as a source of clinical Q fever has been the subject of debate. There is little difficulty in demonstrating that the consumption of infected raw milk instead of pasteurized milk leads to the more frequent occurrence of *C. burneti* antibody in a population. Studies in South-East England suggested that milk was one source of clinical disease (Marmion and Stoker 1958). However it is notable that the attack rate among those exposed to infected raw milk is much lower than among those exposed to infective aerosols. The infectivity titres of milk are generally low and antibody may be present.

Q fever rarely spreads from one person to another, probably reflecting the limited proliferation of the organism in the lung or the limited production of sputum. Nevertheless Marmion and Stoker (1950) described infection of a nurse, two pathologists and a mortuary attendant, all of whom had had contact with a patient suffering from fatal rickettsial pneumonia (Whittick 1950). Further examples are summarized by Derrick (1953).

In Britain, as elsewhere, man is infected from cattle and sheep by the inhalation of droplets or dust, or by the ingestion of raw milk (Marmion and Stoker 1958). Infection of the tick, *Haemaphysalis punctata*, on sheep in Britain was observed by Stoker and Marmion (1955*b*), but a cycle in wild animals was not identified, although serological evidence of infection in poultry was noted. Syrůček and Raška (1956) obtained evidence of infection in hens in Czechoslovakia; it is uncertain whether this is associated with the cycle of maintenance in domestic animals or represents a separate cycle.

Laboratory diagnosis

C. burneti may be isolated in guinea-pigs or hamsters by intraperitoneal inoculation of heparinized blood samples or urine taken from patients during the fever, or from materials such as endocardial vegetations, lung, liver or spleen of fatal cases. The animals are bled before inoculation and 4–6 weeks afterwards; the paired sera are tested for antibody by complement-fixation or immunofluorescence tests with Q fever antigen. The formation of specific antibody is taken to indicate the presence of *C. burneti* in the inoculum. When material is to be obtained for passage, the rectal temperature of the guinea-pigs is determined each day, and when a temperature of 40° or more is reached blood or liver tissue is processed for the inoculation of fresh animals or the chick-embryo yolk sac. *It must be emphasized, however, that the serial passage of the organism, particularly in eggs, is associated with a high risk of laboratory infections.* It should not be undertaken in laboratories without proper means of containment and unless the staff have been vaccinated against the disease.

The serological examination of acute- and convalescent-phase serum specimens is safer, quicker and less laborious than animal inoculation. Complement-fixation tests with inactivated, purified suspensions of Phase 1 and 2 organisms from the chick-embryo yolk sac are widely used. A fourfold or greater increase in antibody to Phase 2 antigen suggests a current infection.

Recently, in line with serodiagnostic practice in other rickettsial infections, immunofluorescence tests on 'microdots' of purified suspensions of Phase 1 and 2 organisms, together with analysis of the immunoglobulin class of the antibody by the use of fluorescent conjugates against IgM, IgG and IgA, is supplementing or replacing the complement-fixation test. Antibody in an acute case of Q fever is predominantly in the IgM class and against Phase 2 antigen; IgG antibody to Phase 1 and 2 antigens appears later. The IgM response to Phase 2 antigen may be manifest well before the complement-fixation test becomes positive; further, the presence of IgM specific antibody may be of value in indicating a current or recent infection when the complement-fixation test shows only unchanging antibody titres. Finally the different kinetics of development of IgM or IgG specific antibody against Phase 1 or 2 antigens may also permit the observation of increasing concentrations of antibody when the older serological methods show, overall, only static titres.

Q fever antibody may be measured by other techniques such as microagglutination (Fiset *et al.* 1969), radio-immunoprecipitation (Tabert and Lackman 1965), neutralization tests in mice (Abinanti and Marmion 1957), or enzyme-linked immunosorbent assay. Of these methods microagglutination is the most commonly used alternative to complement fixation and immunofluorescence. Cellular immune responses such as lymphocyte transformation and skin hypersensitivity tests are not used for the diagnosis of acute cases of Q fever but may have some value in judging the need for and response to vaccination. The Weil–Felix reaction is negative.

Treatment and vaccination

Clark and Lennette (1952), in an early uncontrolled study of the treatment of Q fever, showed that chlortetracycline was superior to penicillin in that it shortened the duration of fever and effected an apparent improvement in the patients' condition. The favourable response of a smaller group of patients to chloramphenicol suggested that this antibiotic might be of value in patients not responding to tetracycline. Similar results were obtained by Tigertt and Benenson (1956) with oxytetracycline. Tetracycline was, however, found by Powell and his colleagues (1962) to be effective only if given early in the disease. Studies with experimental Q fever in volunteers showed that oxytetracycline was effective in treatment, and in preventing the development of clinical disease when given late in the incubation period (Tigertt 1959).

It is important to begin treatment as soon after the onset of illness as possible and to continue for 5 days after the disappearance of fever. Recent experimental studies (Spicer *et al.* 1981) in chick embryos show that doxycycline is inhibitory for *C. burneti*, and this long-acting tetracycline is now commonly used in treatment. The same workers found rifampicin and trimethoprim to be active against *C. burneti*; these compounds deserve further investigation for the treatment of chronic Q fever infections, such as endocarditis. Clindamycin, erythromycin, viomycin, cycloserine and cephalothin were, however, ineffective. The treatment of chronic Q fever endocarditis with tetracycline is generally ineffective although the symptoms and signs may be suppressed for a time. Combinations of tetracycline and lincomycin are reported (Turck *et al.* 1976) to have produced a long-term improvement. In the face of increasing damage, the infected valve is usually excised and replaced prosthetically; antibiotic treatment is given to prevent further colonization of the surrounding area.

Inactivated Q fever vaccine has been used to protect laboratory workers since the late 1940s (Smadel *et al.* 1948*a*, Meiklejohn and Lennette 1950). Severe local or systemic reactions sometimes occurred, leading on rare occasions to chronic abscess formation at the inoculation site. Experiments with this type of vaccine in volunteers challenged by inhalation of infective sprays of *C. burneti* showed clear evidence of protection (Benenson 1959). Since these early studies several advances have improved the vaccines available.

Abinanti and Marmion (1957), described a mouse protection test for the measurement of neutralizing antibody and showed that Phase 1 rather than Phase 2 antibody was protective. Ormsbee *et al.* (1964) showed that vaccines prepared from *C. burneti* in Phase 1 were 100–300 times as potent as Phase 2 vaccines in protecting guinea-pigs. The earlier 'Smadel vaccines' probably protected by virtue of the small content of Phase 1 organisms in a predominantly Phase 2 vaccine. Workers at the Rocky Mountain Laboratory (Lackman *et al.* 1962, Luoto *et al.* 1963) introduced the practice of intradermal sensitivity testing as a means of excluding those already immune, or with hypersensitivity to vaccine components. Finally, Ormsbee (1962) (see also Lackman *et al.* 1962) devised a continuous flow method of purifying *C. burneti* from infected yolk sacs in the presence of M/1 NaCl. This vaccine contained much less egg protein than previous ones. Q fever vaccines of this improved type, in Phase 1 and given in one dose of 30 μg, were shown to give significant protection to volunteers against respiratory challenge with living rickettsiae (Fiset 1967). Their use in laboratory workers exposed to *C. burneti* at the Rocky Mountain Laboratory, Hamilton, Montana, over a number of years also appeared to protect. Trials in two large South Australian abattoirs confirm the view that 'Ormsbee vaccine' is protective and show that it does not produce undesirable reactions (Marmion *et al.* 1983). A Phase 1 vaccine given to dairy cattle in California

has been shown to reduce the frequency of shedding of *C. burneti* in the milk (Biberstein *et al.* 1977).

Most experience with Q fever vaccine has been with formolized whole organisms, but Roumanian and Czech workers (Cracea *et al.* 1977, Kazar *et al.* 1982) have used trichloracetic acid extracts containing Phase 1 antigen. This appears to stimulate antibody formation, but information on protection against natural infection is not yet available. Finally, Russian workers (Genig 1965) have used a living attenuated strain (M44) of *C. burneti*, predominantly in Phase 2; seroconversion rates of more than 80 per cent, with few local or systemic reactions, have been observed.

Rickettsia-like infections in animals

Though animals are known to act as a reservoir for rickettsiae pathogenic for man, less is known of the extent to which other animal rickettsial infections occur.

Heartwater

This is an important disease of sheep, goats, cattle, and wild ruminants. It is widespread in many parts of Africa south of the Sahara. Infection, due to *Cowdria* (*Rickettsia*) *ruminantium* (Cowdry 1925, Haig 1955), is transmitted by several ticks of the genus *Amblyomma*, including *A. hebraeum*. In the tick transmission is trans-stadial but not transovarial. In the ruminant the organism occurs initially in cells of the reticulo-endothelial system and later in the vascular endothelial cells. Clinically the disease may vary from hyperacute to inapparent. The most severe cases often occur in animals recently introduced into heartwater areas. Adult animals in endemic areas are often immune. In sheep the incubation period is 7 to 14 days with, in the acute form, fever, gastro-enteritis, serous effusion into the pericardial, pleural and peritoneal cavities, and convulsions. The illness lasts 2 to 6 days and the case fatality rate is 50 to 100 per cent. Cases occur sporadically and in small groups, mainly in the wet season. The causative organism, 0.2 to 0.8 μm in diameter, is readily demonstrated by Giemsa's stain in the vascular endothelial cells, particularly of the jugular vein, kidney and brain. Laboratory animals seem relatively insusceptible. Prompt treatment with tetracyclines is effective. Sheep that recover remain carriers for a few weeks and are immune for at least a year, but reinfection may occur with a variable degree of reaction.

Tick-borne fever

In 1932 Gordon and his colleagues encountered a disease of sheep that was borne by the same tick, *Ixodes ricinus*, as was responsible for louping-ill. The

incubation period was 4–8 days; the fever lasted about 10 days and was followed by immunity, though this tended to be short lived (Hudson 1950). The causative agent was present in the blood and could be passaged in sheep but not in laboratory animals. The case-fatality rate was low and the only striking feature noted was splenic enlargement. Infection has been associated with abortion and with infertility in rams. It may predispose to staphylococcal infection in lambs, to louping-ill, and to pneumonia. Infected animals become carriers for some months and mild relapses can occur. Experimentally, haemorrhagic lesions in the large intestine of sheep associated with the occurrence of thrombocytopaenia have been observed (Foster *et al.* 1968). There is also a lymphocytopaenia followed by a profound neutropaenia. Cytoplasmic inclusions within granular leucocytes or monocytes of the blood and consisting of clusters of rounded or rod-shaped particles, ranging from 0.3 to more than 2.0 μm in diameter, led to the proposed name *Cytoecetes phagocytophilia* (Foggie 1951). Some would now classify the organism with *Ehrlichia*. It may remain alive in blood for 13 days at 4° and has been stored for 18 months at −79° in the presence of glycerol or dimethyl sulphoxide (Foggie *et al.* 1966). Cattle, goats, and wild ruminants such as deer are also susceptible to the disease. In Africa, *Hyalomma* spp. and *Rhipicephalus bursa* serve as tick hosts, and in Asia *Rhipicephalus haemaphysaloides*. Transmission in ticks is trans-stadial but not transovarial.

Bovine petechial fever *Synonym:* Ondiri disease

This sometimes fatal disease (Danskin and Burdin 1963) affects cattle and wild ruminants, especially bushbuck, in East Africa. The causative agent, *Cytoecetes* (*Ehrlichia*) *ondiri*, closely resembles that of tick-borne fever. The disease is characterized by sudden fever, reduced milk yield, abortion in pregnant animals, and widespread submucosal and subserosal haemorrhages. The organism, which can be found in circulating leucocytes, is present in large numbers in the spleen (Snodgrass 1975). A striking hyperplasia of the large cells of the lymphoid series occurs. Infection is followed by a carrier state lasting some months, and by an immunity of variable duration. A tick vector is suspected.

Canine rickettsiosis

This disease (Ewing 1969), which is caused by *Ehrlichia canis*, occurs in domestic dogs in many tropical and sub-tropical areas in Africa, Asia, and the USA. Wild dogs, jackals and other species may also be affected. The tick vector is *Rhipicephalus sanguineus*, in which transmission is trans-stadial but not trans-ovarial. The organism, whose diameter varies from 0.2 to 2.0 μm, occurs in clusters in leucocytes, particularly mononuclear cells. Pyrexia, purulent oculonasal discharge, enlarged spleen, transient leucopaenia, thrombocytopaenia, anaemia, and emaciation are common features, and the disease may be fatal. Treatment with tetracyclines may assist if given early. *Tropical canine pancytopaenia* (Buhles *et al.* 1974) is an often fatal immunological sequel whose onset is marked by epistaxis occurring several weeks after apparent recovery.

(For further information on rickettsial infections of animals see Robertson 1976; and for a broad ecological view of rickettsial infections Marchette 1982.)

The assistance of R.A. Ormsbee, late of the Rocky Mountain Laboratory, US Public Health Service, Hamilton, Montana, in reading Chapters 46 and 77 and in giving advice is gratefully acknowledged.

References

Abinanti, F.R. and Marmion, B.P. (1957) *Amer J. Hyg.* **66**, 173.

Andrew, R.R., Bonnin, J.M. and Williams, S. (1946) *Med. J. Aust.* **ii**, 253.

Andrews, P.S. and Marmion, B.P. (1959) *Brit. med. J.* **ii**, 983.

Babudieri, B. (1959) *Advanc. vet. Sci.* **5**, 82.

Bell, J.A., Beck, M.D. and Huebner, R.J. (1950) *J. Amer. med. Ass.* **142**, 868.

Benenson, A.S. (1959) In: *Symposium on Q fever*, Med. Sci. Pubn No. 6, Walter Reed Army Institute of Medical Research.

Biberstein, E.L., Riemann, H.P., Frank, C.E., Behymer, D.E., Ruppanner, E., Bushnell, R. and Crenshaw, G. (1977) *Amer J. vet. Res.* **38**, 189.

Bozeman, F.M., Masiello, S.A., Williams, M.S. and Elisberg, B.L. (1975) *Nature, Lond.* **255**, 545.

Bozeman, F.M., Sonenshine, D.E., Williams, M.S., Chadwick, D.P., Lauer, D.M. and Elisberg, B.L. (1981) *Amer. J. trop. Med. Hyg.* **30**, 253.

Brada, M. and Bellingham, A.J. (1980) *Brit. med. J.* **281**, 1108.

Brill, N.E. (1910) *Amer. J. med. Sci.* **139**, 484.

Brumpt, E. (1932) *C.R. Soc. Biol.* **110**, 1199.

Buhles, W.C., Huxsoll, D.L. and Ristic, M. (1974) *J. infect. Dis.* **130**, 357.

Burgdorfer, W. and Anacker, R.L. (1981) *Rickettsiae and Rickettsial Diseases*, Academic Press, New York.

Burnet, F.M. and Freeman, M. (1937) *Med. J. Aust.* **ii**, 299; (1939) *Ibid.* **ii**, 887.

Caminopetros, J. (1948) In: *Proc. 4th int. Congr. trop. Med. and Malaria*, Wash., p. 441.

Carley, J.G., Doherty, R.L., Derrick, E.H., Pope, J.H., Emanuel, M.L. and Ross, C.J. (1955) *Aust. Ann. Med.* **4**, 91.

Catanzaro, P.J., Shirai, A., Hildebrandt, P.K. and Osterman, J.V. (1976) *Infect. Immun.* **13**, 861.

Clark, W.H. and Lennette, E.H. (1952) *Ann. N.Y. Acad. Sci.* **55**, 1004.

Clark, W.H., Lennette, E.H., Railsback, O.C. and Romer, M.S. (1951) *Arch. intern. Med.* **88**, 155.

Clavero del Campo, G. and Gallardo, F.P. (1949) *Immunizacion contra el Tifus Exantematico con Vacuna Viva, Cepa E.* Minist. Hlth, Madrid.

Conor, A. and Bruch, A. (1910) *Bull. Soc. Path. exot.* **3**, 492.

Cord, W.I. and Walker, J.M. (1947) *Lancet* **i**, 481.

Cowdry, E.V. (1925) *J. exp. Med.* **42**, 231, 253.

Cox, H.R. (1938a) *Publ. Hlth Rep., Wash.* **23**, 2241; (1938b) *Publ. Hlth Rep., Wash.* **53**, 2270; (1941) *Science* **94**, 399.

Cracea, E., Dumitrescu-Constantinescu, S., Botez, D. and Ioanid, L. (1977) *Zbl. Bakt., 1 Abt. Orig. A* **238**, 413.

Danskin, D. and Burdin, M.L. (1963) *Vet. Rec.* **75**, 391.

Dasch, G.A., Samms, J.R. and Weiss, E. (1978) *Infect. Immun.* **19**, 676.

Davis, G.E. and Cox, H.R. (1938) *Publ. Hlth Rep., Wash.* **53**, 2259.

Derrick, E.H. (1937) *Med. J. Aust.* **ii**, 281; (1953) *Ibid.* **i**, 245; (1973) *Ibid.* **i**, 1051.

Drew, W.R.M. (1965) *J.R. Army med. Cps* **111**, 95.

Duma, R.J. *et al.* (1981) *J. Amer. med. Ass.* **245**, 2318.

Dwyer, R. St C., Tonge, J.I., Hoffman, T.R., Scott, W., Derrick, E.H. and Pope, J.H. (1960) *Med. J. Aust.* **ii**, 456.

Dyer, R.E., Rumreich, A. and Badger, L.F. (1931) *Publ. Hlth Rep., Wash.* **46**, 334.

Elisberg, B.L. and Bozeman, F.M. (1966) *Arch. Inst. Pasteur Tunis* **43**, 193; (1979) In: *Diagnostic Procedures for Viral, Rickettsial and Chlamydial Infections*, p. 1061. Ed. by E.H. Lennette and N.J. Schmidt. Amer. Publ. Hlth Ass., Washington, DC.

Ewing, S.A. (1969) *Advanc. vet. Sci.* **13**, 331.

Felix, A. (1916) *Wien. klin. Wschr.* **29**, 873.

Fiset, P. (1967) In: *Proc. 1st int. Conf. on Vaccines against Viral and Rickettsial Diseases of Man*, pp. 528–31. PAHO Science Publications 147. PAHO, Wash; (1978) In: *CRC Handbook Series in Clinical Laboratory Science, Virology and Rickettsiology*, Vol. 1, Part 2, 361–6. Ed. by G.D. Hsiung and R.H. Green. CRC Press, Boca Raton, Florida.

Fiset, P., Ormsbee, R.A., Silberman, R., Peacock, M. and Spielman, S.H. (1969) *Acta Virol.* **13**, 60.

Foggie, A. (1951) *J. Path. Bact.* **63**, 1.

Foggie, A., Lumsden, W.H.R. and McNeillage, G.J.C. (1966) *J. comp. Path.* **76**, 413.

Foster, W.N.M., Foggie, A. and Nisbet, D.I. (1968) *J. comp. Path.* **78**, 225.

Fox, J.P., Jordan, M.E. and Gelfand, H.M. (1957) *J. Immunol.* **79**, 348.

Fox, J.P., Montoya, J.A., Jordan, M.E., Cornejo, U. Jr, Garcia, J.L., Estrada, M.A. and Gelfand, H.M. (1959) *Arch. Inst. Pasteur Tunis* **36**, 449.

Fuller, H.S. (1954) *Amer. J. Hyg.* **59**, 236.

Gear, J. (1941) *Leech* **12**, No. 2; (1954) *S. Afr. J. clin. Sci.* **5**, 158.

Genig, V.A. (1965) *Vop. Virus.* **10**, 703.

Gerhard, W.W. (1837) *Amer. J. med. Sci.* **20**, 289.

Giménez, D.F. (1964) *Stain Tech.* **39**, 135.

Gordon, W.S., Brownlee, A., Wilson, D.R. and MacLeod, J. (1932) *J. comp. Path.* **45**, 301.

Haig, D.A. (1955) *Advanc. vet. Sci.* **2**, 307.

Hudson, J.R. (1950) *Brit. vet. J.* **106**, 3.

Huebner, R.J., Jellison, W.L. and Armstrong, C. (1947) *Publ. Hlth Rep., Wash.* **62**, 777.

Huebner, R.J., Stamps, P. and Armstrong, C. (1946) *Publ. Hlth Rep., Wash.* **61**, 1605.

Imam, I.Z.E. (1968) *J. Egypt. publ. Hlth Ass.* **43**, 165.

Kaplan, M.M. and Bertagna, P. (1955) *Bull. Wld Hlth Org.* **13**, 829.

Kazar, J., Brezina, R., Palanova, A., Tvrda, B. and Schramek, S. (1982) *Bull. World Hlth Org.* **60**, 389.

Kimborough, R.C., Ormsbee, R.A. and Peacock, M.G. (1981) In: *Rickettsia and Rickettsial Diseases*, p. 125. Ed. by W. Burgdorfer and R.L. Anacker. Academic Press, London and New York.

Kleeman, K.T., Hicks, J.L., Anacker, R.L., Philip, R.N., Casper, E.A., Heckemy, K.E., Wilfert, C.M. and MacCormack, J.N. (1981) In: *Rickettsiae and Rickettsial Diseases*, p. 171. Ed. by W. Burgdorfer and R.L. Anacker. Academic Press, London and New York.

Lackman, D.B., Bell, E.J., Bell, J.F. and Pickens, E.G. (1962) *Amer. J. Publ. Hlth* **52**, 87.

Lackman, D.B., Bell, E.J., Stoenner, H.G. and Pickens, E.G. (1965) *Hlth Lab. Sci.* **2**, 135.

Lennette, E.H., Clark, W.H. and Dean, B.H. (1949) *Amer. J. trop. Med.* **29**, 527.

Lewthwaite, R. (1952) *Brit. med. J.* **ii**, 826, 875.

Ley, H.R. and Smadel, J.E. (1954) *Antibiot. et Chemother.* **4**, 792.

Lumio, J., Penttinen, K. and Pettersson, T. (1981) *Scand. J. infect. Dis.* **13**, 17.

Luoto, L., Bell, J.F., Casey, M. and Lackman, D.B. (1963) *Amer. J. Hyg.* **78**, 1.

Luoto, L. and Huebner, R. (1950) *Publ. Hlth Rep., Wash.* **65**, 541.

MacArthur, W.P. (1956) *The Great Famine*, p. 263. Browne and Nolan, Dublin; (1959) *Trans. R. Soc. trop. Med. Hyg.* **53**, 423.

McGuire, T.J. and Williams, J.P. (1948) *Virginia med. Monthly* **75**, 225.

McNee, J.W., Renshaw, A. and Brunt, E.H. (1916) *Brit. med. J.* **i**, 225.

Marchette, N.J. (1982) *Ecological Relationships and Evolution of the Rickettsiae*, vol. 1 and 2. CRC Press, Boca Raton, Florida.

Marmion, B.P. (1959) *Proc. 6th int. Congr. trop. Med. and Malaria*, Lisbon, **5**, 711; (1962) *J. Hyg. Epidem. Microbiol. Immunol.* **6**, 79.

Marmion, B.P., Ormsbee, R.A., Kyrkou, M., Worswick, D., Cameron, S., Wright, J., Feary, B. and Collins, W. (1983) To be published.

Marmion, B.P. and Stoker, M.G.P. (1950) *Lancet* **ii**, 611; (1958) *Brit. med. J.* **ii**, 809.

Marmion, B.P., Stoker, M.G.P., McCoy, J.H., Malloch, R.A. and Moore, B. (1953) *Lancet* **i**, 503.

Maxcy, K.F. (1929) *Publ. Hlth Rep., Wash.* **44**, 1735.

Megaw, J.W.D. (1942) *Brit. med. J.* **ii**, 401, 433.

Meiklejohn, G. and Lennette, E.H. (1950) *Amer. J. Hyg.* **52**, 54.

Mooser, H. (1928) *J. infect. Dis.* **43**, 241; (1932) *Arch. Inst. Pasteur Tunis* **21**, 1; (1959) *Arch. Inst. Pasteur Tunis* **36**, 301.

Mooser, H., Castañeda, M.R. and Zinsser, H. (1931) *J. Amer. med. Ass.* **97**, 231.

Mooser, H., Leeman, A., Chao, S.H. and Gubler, H.V. (1948) *Schweiz. Z. Path. Bakt.* **11**, 513.

Mooser, H., Marti, H.R. and Leeman, A. (1949) *Schweiz. Z. Path. Bakt.* **12**, 476.

Murray, E.S. and Gaon, A.J. (1963) *Proc. VIIth Congr. trop. Med. and Malaria*, p. 322 (abstract).

Murray, E.S., Pšorn, T., Djaković, P., Sielski, S., Broz, V., Ljupša, F., Gaon, J., Pavlević, R. and Snyder, J.C. (1951) *Amer. J. Publ. Hlth* **41**, 1359.

Nicolle, C., Conor, A. and Conseil, E. (1911) *Ann. Inst. Pasteur* **25**, 97.

Okun, D.B., Sun, N.C. and Tanaka, K.R. (1979) *Amer. J. clin. Path.* **71**, 117.

Ormsbee, R.A. (1962) *J. Immunol.* **88**, 100.

Ormsbee, R.A., Bell, E.J., Lackman, D.B. and Tallent, G. (1964) *J. Immunol.* **92**, 404.

Ormsbee, R.A., Peacock, M., Philip, R., Casper, E., Plorde, J., Gabre-Kidan, T. and Wright, L. (1977) *Amer. J. Epidem.* **105**, 261.

Parker, R.R. and Davis, G.E. (1938) *Publ. Hlth Rep., Wash.* **53**, 2267.

Peacock, M.G., Philip, R.N., Williams, J.C. and Faulkner, R.S. (1983) *Infect. Immun.* **41**, 1089.

Pelligrin, M., Delsil, G., Auvergnat, J.C., Familiades, J., Faure, H., Guin, M. and Voigt, J.J. (1980) *Hum. Path.* **11**, 51.

Philip, C.B. (1948) *Publ. Hlth Rep., Wash.* **63**, 58; (1968) *Zbl. Bakt.* **206**, 343.

Philip, R.N., Casper, E.A., Ormsbee, R.A., Peacock, M.G. and Burgdorfer, W. (1976) *J. clin. Microbiol.* **3**, 51.

Powell, O.W. (1961) *Aust. Ann. Med.* **10**, 52.

Powell, O.W., Kennedy, K.P., McIver, M. and Silverstone, H. (1962) *Aust. Ann. Med.* **11**, 184.

Price, W.H., Emerson, H., Nagel, H., Blumberg, R. and Talmadge, S. (1958) *Amer. J. Hyg.* **67**, 154.

Reiss-Gutfreund, R.J. (1956) *Bull. Soc. Path. exot.* **49**, 946; (1966) *Amer. J. trop. Med. Hyg.* **15**, 943.

Report (1965) *Morbid. & Mortal.* **13**, 454; (1982) *Bull. World Hlth Org.* **60**, 157.

Ricketts, H.T. (1907) *J. infect. Dis.* **4**, 141.

Robbins, F.C., Gauld, R.L. and Warner, F.B. (1946) *Amer. J. Hyg.* **44**, 23.

Robertson, A. (Ed.) (1976) *Handbook on Animal Diseases in the Tropics*, 3rd ed., British Veterinary Association, London.

Robson, A.O. and Shimmin, G.D.L. (1959) *Brit. med. J.* **ii**, 980.

Rocha-Lima, H. da (1916) *Berl. klin. Wschr.* **53**, 567; (1920) *Prowazek's Hdb. path. Protozoen* **2**, 990.

Shirai, A. *et al.* (1979) *Jap. J. med. Sci. Biol.* **32**, 337.

Shishido, A. (1962) *Jap. J. med. Sci. Biol.* **15**, 308.

Smadel, J. E. and Elisberg, B.L. (1965) In: *Viral and Rickettsial Infections of Man*, p. 1059. Ed. by F.L. Horsfall and I. Tamm. Lippincott, Philadelphia.

Smadel, J.E., Snyder, M.J. and Robbins, F.C. (1948a) *Amer. J. Hyg.* **47**, 71.

Smadel, J.E., Woodward, T.E., Ley, H.L., Philip, C.B.,

Traub, R., Lewthwaite, R. and Savoor, S.R. (1948*b*) *Science* **108**, 160.

Snodgrass, D.R. (1975) *J. comp. Path.* **85,** 523.

Snyder, J.C. (1965) In: *Viral and Rickettsial Infections of Man*, p. 1059. Ed. by F.L. Horsfall and I. Tamm. Lippincott, Philadelphia.

Spicer, A.J., Peacock, M.G. and Williams, J.C. (1981) In: *Rickettsiae and Rickettsial Diseases*, p. 375. Ed. by W. Burgdorfer and R.L. Anacker. Academic Press, London and New York.

Stoker, M.G.P. and Marmion, B.P. (1955*a*) *Bull. World Hlth Org.* **13**, 781; (1955*b*) *J. Hyg., Camb.* **53**, 322.

Swift, H.W. (1919–1920) In: *The Harvey Lectures Series* 15, 58. Lippincott, Philadelphia.

Syrůček, L. and Raška, K. (1956) *Bull. World Hlth Org.* **15,** 329.

Syrůček, L., Sobeslavsky, O. and Gutvirth, I. (1958) *J. Hyg. Epidem. Microbiol. Immunol.* **2**, 29.

Tabert, G.G. and Lackman, D.B. (1965) *J. Immunol.* **94,** 959.

Tigertt, W.D. (1959) In: *Symposium on Q fever*, p. 39. Ed. by J.E. Smadel. Med. Sci. Pubn No. 6. Walter Reed Army Institute of Research, Washington DC.

Tigertt, W.D. and Benenson, A.S. (1956) *Trans. Ass. Amer. Phycns* **69**, 98.

Töpfer, H. (1916) *Münch. med. Wschr.* **63**, 1495.

Traub, R. (1974) *J. med. Entomol.* **11**, 237.

Traub, R. and Wisseman, C.L. (1968) *Bull. World Hlth Org.* **39**, 219.

Traub, R., Wisseman, C.L. and Farhaug-Azad, A. (1978) *Trop. Dis. Bull.* **75**, 237.

Turck, W.P.G., Howitt, G., Turnberg, L.A., Fox, H., Longson, M., Matthews, M.B. and Das Gupta, R. (1976) *Quart. J. Med.* **45**, 193.

Varela, G., Vinson, J.W. and Molina-Pasquel, C. (1969) *Amer. J. trop. Med. Hyg.* **18**, 708, 713.

Vinson, J.W. (1966) *Bull. World Hlth Org.* **35**, 155.

Weil, E. and Felix, A. (1916) *Wien. klin. Wschr.* **29**, 33, 974.

Welsh, H.H., Lennette, E.H., Abinanti, F.R. and Winn, J.F. (1958) *Ann. N.Y. Acad. Sci.* **70**, 528.

Whittick, J.W. (1950) *Brit. med. J.* **i**, 978.

Willey, R.F., Matthews, M.B., Plutherer, J.F. and Marmion, B.P. (1979) *Lancet* **ii**, 270.

Wolbach, S.B. (1925) *J. Amer. med. Ass.* **84**, 723.

Woodman, D.R., Weiss, E. and Dasch, G.A. (1977) *Infect. Immun.* **16**, 853.

Woodward, T.E. (1959) *Med. Clin. N. Amer.* **43**, 1507; (1981) In: *Rickettsiae and Rickettsial Diseases*, p. 17. Ed. by W. Burgdorfer and R.L. Anacker. Academic Press, London and New York.

Woodward, T.E. and Jackson, E.B. (1965) In: *Viral and Rickettsial Infections of Man*. Ed. by F.L. Horsfall and I. Tamm. Lippincott, Philadelphia.

Woodward, T.E., Pedersen, C.E., Oster, C.N., Bagley, L.R., Romberger, J. and Snyder, M.J. (1976) *J. infect. Dis.* **134**, 297.

Zarafonetis, C.J.D., Ingraham, H.S. and Berry, J.F. (1946) *J. Immunol.* **52**, 189.

Zdrodovskii, P.F. and Golinevich, H.M. (1960) *The Rickettsial Diseases.* Pergamon Press, Oxford.

Zinsser, H. (1934) *Amer. J. Hyg.* **20**, 513; (1935) *Rats, Lice, and History.* George Routledge & Sons, London.

Zinsser, H. and Castañeda, M.R. (1932) *J. exp. Med.* **56**, 455.

78

Mycoplasma diseases of animals and man

Geoffrey Smith

Introductory

The three genera, *Mycoplasma*, *Ureaplasma*, and *Acholeplasma*, already comprise more than 60 named species. Of these, most belong to the genus *Mycoplasma*. They are found mainly in the mouth, the upper respiratory tract, and the more distal parts of the genital tract of animals and man. The majority, though not all, are strikingly host-specific. A few cause severe diseases, others are associated with diseases of a less obvious nature, and many are non-pathogenic. Not surprisingly the pathogenic status of some species is uncertain. The isolation of a mycoplasma from healthy subjects does not prove that it is merely a commensal; conversely, isolation from diseased tissue, particularly when other micro-organisms are present, does not prove pathogenicity. Although mycoplasmas of veterinary importance can be subjected to experimentation in the natural host, a low degree of pathogenicity may be difficult to demonstrate, especially if its expression depends on a synergistic interaction between several micro-organisms, or on ill defined stress factors. The interpretation of experiments in conventionally reared animals may be complicated by the unknown microbiological status of the host; the use of gnotobiotic animals, on the other hand, raises difficulties by virtue of the host's artificiality. Mycoplasmas of medical importance have only rarely been studied by experiments in human volunteers.

(For a general description of the Mycoplasmas, see Chapter 47, and for a review of the factors concerned in pathogenicity, see Razin 1978.)

Mycoplasma infections in animals

Cattle

Contagious bovine pleuropneumonia (CBPP)

CBPP is a disease of cattle and water buffaloes caused by *Mycoplasma mycoides* subsp. *mycoides*, an organism characteristically giving rise to small colonies on artificial media. Large-colony strains (see p. 594 and Chapter 47), thought by some to merit inclusion in subspecies *mycoides* (Al-Aubaidi *et al.* 1972, Cottew and Yeats 1978) occur mainly in goats and are not isolated from CBPP. The disease became widespread in Europe from 1800 onwards (Turner 1959). In the middle of the nineteenth century it spread, by shipment of cattle, to England, South Africa, America, and Australia, and later to parts of Asia. By means of a slaughter policy it was eradicated from a number of countries before 1900. Australia was freed in 1973. The disease still occurs in more than 20 African countries south of the Sahara, and in a few parts of Asia such as Assam State (India), China, and Mongolia (Report 1979). Limited outbreaks occurred on the border between France and Spain in 1967, 1980 and 1982, and in Portugal in 1983.

The disease, which produces serious economic loss when left unchecked, is characterized by an exudative inflammation of the interlobular lymph vessels and of the alveolar and interstitial tissue of the lungs, with a serofibrinous pleurisy. According to Nocard and Roux (1898) the essential lesion is a distension of the connective tissue layers between the lobules of the lungs with a large quantity of yellowish, limpid, albuminous, highly infective fluid. In recovering or chronic cases, necrotic tissue may become walled off by fibrous tissue to form *sequestra*. The causative organism sometimes persists in such lesions for many months. In calves under six months of age infection may produce arthritis instead of pleuropneumonia.

Transmission of the disease normally requires close contact between infected and susceptible cattle, and is probably usually brought about by droplet infection. The risk of indirect transmission by organisms contaminating the immediate environment is slight. The incubation period is rarely less than three weeks but may be as long as three or more months. In an affected herd the course of the disease in individual animals may vary from hyperacute to subclinical; some cattle may remain uninfected (Campbell and Turner 1936). Mortality varies from less than 10 to more than 70 per cent. The insidious nature of CBPP is due to the occurrence of subclinical infections and infected sequestra; these may subsequently lead to overt disease and the spread of infection. A galactan, found both in artificial cultures and in the serum and urine of infected cattle, probably influences the organism's pathogenicity.

In the dead animal, laboratory diagnosis may be made by isolating the causative organism. The growth of post-mortem invaders can if necessary be controlled by the inclusion of such agents as penicillin and thallous acetate in the medium (Edward 1947). The presence of specific antigen in diseased lung tissue, pleural exudate, and sometimes serum, can be demonstrated by an agar-gel precipitation test, or by a tube precipitation test if a result is required quickly. In the living animal, diagnosis is usually based on the demonstration of serum antibody. In the late stages of the disease, antigen may be present in the blood in quantities sufficient to mask, or *eclipse*, the presence of antibody. Complement fixation is the most widely used serological method (Campbell and Turner 1953); Huddart's modification (see Hudson 1971) enables large numbers of cattle to be tested in the field. Complement fixation will detect acute cases, subclinical cases, and most chronic cases with infected sequestra. The agar-gel precipitation test may be adapted to detect either antibody or antigen in serum; alternatively a combined test for antigen and antibody may be used. Agar-gel precipitation is less sensitive than the complement-fixation test in detecting chronic cases (Gourlay 1965). Although subject to considerable inaccuracy, a slide test in which stained organisms are agglutinated by citrated blood is of some value as a rapid screening method for herds of cattle. Many other diagnostic tests have been tried with varying degrees of success (Cottew and Leach 1969, Hudson 1971).

Control measures include quarantine, slaughter, and vaccination. Vaccination is based on the method used effectively by Willems in Belgium from 1850 onwards (see Turner 1959). This consisted of injecting exudates from the thorax of naturally infected animals into the tail-tip of susceptible animals; the severity of the subcutaneous infection that resulted was limited by the unusual inoculation site. Currently, vaccines are prepared from certain well known strains that have been attenuated by repeated subculture in artificial media, or have been passaged through fertile eggs (Bennett 1932, Report 1971, Dyson and Smith 1976). It should be noted that the organism will also grow in infertile eggs. Strains that retain some degree of virulence are generally used; they must be injected into the tail-tip or muzzle in doses of not less than several million viable organisms. Vaccines, prepared locally in government laboratories, are often issued as broth cultures, with limited keeping qualities; some are lyophilized. Severe reactions that may even lead to loss of the tail are sometimes seen; their occurrence is unpredictable and depends in part on variations in susceptibility between different breeds and populations of cattle. In some vaccinated animals viable mycoplasmas persist for several months in the lymph nodes draining the inoculation site. Provided that care is exercised in the production and use of vaccine, immunity will last for a year or more, often after the disappearance of antibodies detectable by complement fixation and other serological methods. All strains of *M. mycoides* subsp. *mycoides* from CBPP belong to a single immunological type.

Antibody unrelated to complement-fixing or precipitating antibody can be demonstrated by a passive mouse-protection test in which serum from convalescent or vaccinated cattle is injected subcutaneously in doses of the order of 0.25 ml per mouse; the symptomless mycoplasmaemia that normally follows intraperitoneal infection of mice with a virulent strain of *M. mycoides* subsp. *mycoides* does not develop after pretreatment with immune serum (Smith 1967, 1971, Dyson and Smith 1975). The degree of attenuation of vaccine strains is reflected quantitatively in reduced ability to produce mycoplasmaemia in mice (Smith 1968, Dyson and Smith 1976). Masiga and his colleagues (1975) claim to have protected cattle from CBPP by the intravenous injection of large doses of serum (1 litre per 100 kg body weight) from immune animals. Delayed hypersensitivity and other cell-mediated immune reactions occur in CBPP (Gourlay 1964, Roberts *et al.* 1973), but there is no proof that they are concerned in increasing resistance to infection.

In experimental vaccine trials cattle are usually challenged by prolonged exposure in an enclosed building to donors previously infected by endobronchial intubation. A disadvantage of the method is the frequent failure to infect all non-immunized control animals. Assessment of the results is usually made on clinical, serological, and pathological grounds by a scoring method (Hudson and Turner 1963). (For reviews of the disease see Turner 1959, Seddon 1965, Cottew and Leach 1969, Hudson 1971; and for pleuropneumonia of calves due to *Pasteurella multocida* see Chapter 55.)

Other infections of cattle

In addition to *M. mycoides* subsp. *mycoides* 14 or more species of *Mycoplasma*, *Acholeplasma* and *Ureaplasma* have been found in healthy or diseased cattle. At least seven species have been isolated from mastitis, from cases of keratoconjunctivitis, and from each of the following sites in the presence or absence of disease: respiratory tract (upper and lower), and urogenital tract (male and female). Several species have been isolated from joints, and one from intestinal contents. Many, such as *M. bovirhinis* (Cottew and Leach 1969), have not been isolated from animals other than cattle, except possibly on rare occasions; a few such as *M. arginini*, *A. laidlawi* and ureaplasmas are not confined to cattle.

None of these organisms produces a disease as well defined as that caused by *M. mycoides* subsp. *my-*

coides, and some are probably completely non-pathogenic. Conclusive evidence of pathogenicity is lacking for all except a few such as *M. bovis*, an organism that undoubtedly causes mastitis (Boughton 1979, Jasper 1981). In many instances the only evidence that exists is indirect or circumstantial.

Mixtures of micro-organisms are frequently present in calf pneumonia; some strains of *M. dispar*, an organism commonly present in the upper respiratory tract of normal calves, and some ureaplasmas, are thought to play a pathogenic role (Gourlay and Howard 1979, Taylor-Robinson 1979, Stalheim 1983). *M. bovis* may sometimes contribute to the aetiology of calf pneumonia. *M. bovirhinis* is often present in the bovine respiratory tract; its pathogenic potential is uncertain. Organisms commonly found in the genital tract, particularly in the more distal parts in the female and in the prepuce and semen in the male, include *M. bovigenitalium* and ureaplasmas; their pathogenicity seems very low (Cottew and Leach 1969, Taylor-Robinson 1979). *M. bovigenitalium* may occasionally cause mastitis and be excreted in the milk (Stuart *et al.* 1963); the disease can be reproduced by the injection of organisms into the teat canal. Mastitis caused by *M. bovis* has been reported from many parts of the world; it affects lactating and dry cows, and can be reproduced experimentally with very small numbers of mycoplasmas. Organisms isolated from cases of keratoconjunctivitis include *M. bovoculi* (Langford and Leach 1973); there is no conclusive evidence that it is pathogenic.

Goats and sheep

Goats and sheep, like cattle, may be infected by a dozen or more species of *Mycoplasma*, *Acholeplasma* and *Ureaplasma* (Cottew 1979). Some seem of little or no pathogenic significance, but a few produce diseases that are well recognized and economically important.

Contagious caprine pleuropneumonia (CCPP)

Several mycoplasmas seem capable of producing severe pulmonary disease in goats. Descriptions have been given by Hutcheon (1881, 1889), Longley (1951), Turner (1959), and Cottew and Leach (1969). The disease occurs in Asia, Africa, and occasionally in European countries such as Greece and Spain. The incubation period ranges from several days to a month, and morbidity and mortality are often high. The classical post-mortem appearance is of lung tissue showing red and grey hepatization, interlobular septa distended with serous fluid, and severe fibrinous pleurisy.

Although *M. mycoides* subsp. *capri* (Edward 1953) has long been regarded as the classical causative agent, other mycoplasmas have more recently been isolated from caprine pleuropneumonia. Of these, some have been called *M. mycoides* subsp. *mycoides* (Al Aubaidi *et al.* 1972, Cottew and Yeats 1978) because of their close serological relation to the causative agent of contagious bovine pleuropneumonia (CBPP); similar organisms have been isolated from goats with other diseases such as polyarthritis and peritonitis. The so-called subspecies *mycoides* strains from goats fall into two main categories—large colony (LC) and small colony (SC)—both apparently indistinguishable from genuine CBPP strains by serological tests. The terms LC and SC, though not wholly satisfactory (Valdivieso-Garcia and Rosendal 1982), have come into common use. The LC strains, of which there exist many isolates from goats and a few from sheep and cattle, have been distinguished from SC (including CBPP) strains not only by their large colony size but also by their temperature resistance, action on casein and inspissated serum, fermentation of sorbitol, resistance to streptomycin, and behaviour in mycoplasmaemia and cross-protection tests in mice (Cottew and Yeats 1978, Hooker *et al.* 1979, Smith *et al.* 1980, Smith and Oliphant 1981*a*, 1981*b*, 1982, 1983). Apparently only a few caprine SC strains have ever been isolated; they remain indistinguishable from CBPP strains by all known methods. From cases of CCPP in Kenya MacOwan and Minette (1976) made numerous isolations of a mycoplasma, 'F38'; it has demanding growth requirements and may represent a new species (Ernø *et al.* 1979, McMartin *et al.* 1980). Caprine pleuropneumonia has been reproduced experimentally with *M. mycoides* subsp. *capri* (Watson *et al.* 1968) and with F38.

Contagious agalactia of sheep and goats

The causative organism, *M. agalactiae* (no longer called *M. agalactiae* var. *agalactiae*), was first cultivated by Bridré and Donatien (1923). The disease occurs in Europe, Asia and Africa but is commonest in Mediterranean countries. It is characterized by inflammatory lesions of the udder, eye, and joints. The milk yield in lactating animals is diminished, and later a dirty yellow serous fluid containing small clots is secreted (Galloway 1930). In male animals arthritis is the predominant lesion. Natural infection probably occurs by ingestion, and the blood becomes infected briefly. The case-fatality rate is said to average 15 per cent. The organism may be present for many months in secretions from the udder and the eye. Animals that have recovered from an attack of the disease have a high degree of immunity. Progress made in the study of vaccination is referred to by Cottew (1979). (For a review of the disease see Turner 1959.)

Organisms resembling *M. agalactiae* have been isolated from caprine pleuropneumonia in Australia (Cottew and Lloyd 1965) and from caprine granular vulvovaginitis in India (Singh *et al.* 1975).

Other infections of goats and sheep

Debonéra (1937) described oedema disease of goats, a fatal cellulitis affecting the limbs or head; the causative mycoplasma may have been *M. mycoides* subsp. *capri* (Edward 1953). *M. capricolum* (Tully *et al.* 1974) was almost certainly the causative agent of the fatal disease of goats, characterized by septicaemia and arthritis, described by Cordy and his colleagues (1955) in California. *M. conjunctivae* (Barile *et al.* 1972) has been isolated in Europe, America and Australia from keratoconjunctivitis in goats, sheep, and chamois; there is some evidence to suggest pathogenicity (Cottew 1979). *M. ovipneumoniae* is associated in Australia, Britain and elsewhere with certain forms of sheep pneumonia (Carmichael *et al.* 1972, Jones *et al.* 1979); it also occurs in the respiratory tract of normal sheep.

Pigs

Numerous mycoplasmas have been isolated from pigs, but only a few are known to be pathogenic.

Enzootic pneumonia

This disease, which was distinguished from swine influenza by Gulrajani and Beveridge (1951), is caused by *M. hyopneumoniae* (syn. *M. suipneumoniae*; see Rose *et al.* 1979) and is of great economic importance. It occurs in many countries and is usually chronic, though acute exacerbations may follow secondary infections with bacteria. In England about 50 per cent of bacon pigs are affected. The incubation period is usually 10–16 days. Morbidity is very high, but mortality generally low. The disease spreads by droplet infection and its onset tends to be insidious. Coughing is the most characteristic feature, and affected animals do not thrive as well as normal pigs. In slaughtered pigs the lesions are seen to be confined mainly to the ventral portions of the lungs. Early lesions appear as poorly defined areas of catarrhal bronchopneumonia; histological sections show mononuclear cells in large numbers and peribronchiolar lymphoreticular hyperplasia. The disease can be reproduced experimentally by introducing a high dilution of pneumonic lung-tissue suspension into the respiratory tract, or less easily by introducing culture. In infected lung tissue the organism survives storage at 4° for 14–17 days. It is susceptible to the tetracyclines but not to penicillin or the sulphonamides. In experienced hands the examination of impression smears stained by Giemsa's method is a useful diagnostic aid (Whittlestone 1973). Although difficult to grow, the organism can usually be isolated in a suitable selective broth medium (Farrington and Switzer 1977, Whittlestone 1979); it may then be identified by the usual methods, such as metabolism inhibition. Isolation is frequently complicated by the presence of *M. hyorhinis*, an organism that grows much more readily than *M. hyopneumoniae* in culture. Frozen sections of lung tissue may be examined by immunofluorescence. Various modifications of the complement-fixation test, and other serological tests such as the enzyme-linked immunosorbent assay (ELISA), have been used to measure the antibody response (Boulanger and L'Ecuyer 1968, Bruggmann *et al.* 1977); false-positive reactions may result from infection with *M. hyorhinis*. In spite of the susceptibility of the organism to the tetracyclines, treatment has so far proved unsatisfactory. All strains of *M. hyopneumoniae* are serologically closely related, and an immunity develops in animals that recover. Some hope lies in the development of protective vaccines (see Lam and Switzer 1971, Goodwin and Whittlestone 1973). The establishment of herds free from enzootic pneumonia is attempted in many countries, but breakdowns often occur, usually in the apparent absence of contact with infected pigs.

Other infections of pigs

M. hyorhinis, which appears to have a number of subtypes (Goiš *et al.* 1974), occurs commonly in the nose of young pigs, and as a secondary invader, especially in enzootic pneumonia (Switzer and Ross 1975). It may act as a primary pathogen in some piglet pneumonias, and in polyserositis and arthritis (Whittlestone 1979). *M. hyosynoviae* is commonly present in the nasopharynx; it is known to produce arthritis, especially in the USA (Switzer and Ross 1975). *M. flocculare* (Friis 1972) has been isolated from the nasal mucosa and pneumonic lung; it is difficult to grow in culture and its pathogenic significance is uncertain.

Birds

Mycoplasmas, originally called coccobacilliform bodies by Nelson (1936), occur commonly in birds of many species. A dozen or more *Mycoplasma* species or serotypes of avian origin are recognized; acholeplasmas and ureaplasmas have also been isolated.

Diseases of particular economic importance occur in chickens and turkeys. Both species are affected by the pathogens *M. gallisepticum* and *M. synoviae*, and in addition turkeys are susceptible to *M. meleagridis*. Different strains of these mycoplasmas vary in virulence and sometimes in their predilection for certain tissues. Young birds are often particularly susceptible. The course and severity of the diseases produced are greatly influenced by synergistic interactions of the causative mycoplasmas with viruses or bacteria, and by debilitating environmental factors. Transmission occurs by droplet infection and through the egg; *M. meleagridis* is in addition transmitted venereally. The organisms probably persist for no more than a few days outside the host. Infection-free flocks can be built up by the combined use of general hygienic measures and antibiotic- or heat-treatment of fertile eggs.

Respiratory disease

M. gallisepticum plays an important role in *chronic respiratory disease* of chickens, which occurs in many parts of the world and is associated particularly with intensive systems of husbandry. The organism acts either alone or synergistically with other agents such as *Escherichia coli* and wild or attenuated-vaccine strains of the viruses of Newcastle disease and infectious bronchitis. The incubation period is about 4-21 days. Turkeys, being more susceptible than chickens, are more frequently infected with *M. gallisepticum* alone, but other agents such as influenza A virus and *Esch. coli* sometimes play a part. Infection produces inflammation of the mucous membranes of the upper respiratory tract, and of the air-sacs; in turkeys the infraorbital sinuses are often affected, hence the name *infectious sinusitis*. Infection may persist for many months. Mortality is low and clinical signs are not always apparent, but serious economic loss results from a reduction in growth rate and in production and 'hatchability' of eggs. Laboratory diagnosis is made by isolating the mycoplasma and identifying it serologically, for example by plate or tube agglutination, haemagglutination inhibition, growth inhibition, or immunofluorescence (Rhoades 1977). In establishing infection-free flocks fertile eggs may be treated, either by inoculation or dipping, with a suitable antibiotic such as tylosine tartrate; alternatively they may be heated to 45° for 12-14 hours (Yoder 1970). Breeding stock may also be treated with antibiotics. Surveillance of such flocks is maintained by cultural and serological means. Plate agglutination, although subject to inaccuracy, is a useful screening test; haemagglutination inhibition is more specific.

M. synoviae produces a respiratory disease of chickens and turkeys analogous in most respects to that produced by *M. gallisepticum*. The organism may inhabit the respiratory tract of apparently normal birds for long periods. *M. meleagridis* is a common pathogen of young turkeys in many parts of the world. Infection results in inflammation of the air-sacs, abnormalities of the primary wing feathers, skeletal defects, reduced weight gain, and poor hatchability of eggs. *M. meleagridis* has a predilection for the cloaca and bursa of Fabricius in young birds, and for the oviduct and phallus in adults (Matzer and Yamamoto 1974). Infection may be present in apparently normal birds. Diagnostic methods resemble those used for *M. gallisepticum*.

Arthritis and synovitis

M. synoviae is a cause of *infectious synovitis* of chickens and turkeys in many countries. Osteoarthritis, synovitis, and tendovaginitis lead to lameness with painful swollen joints, especially in the hocks and feet. In turkeys the sternal bursae are frequently swollen. The disease often becomes chronic. Similar manifestations are occasionally produced by *M. gallisepticum*, and in turkeys by *M. meleagridis*. Infection with *M. synoviae* results in the formation of rheumatoid factors that can be demonstrated in the serum by latex γ-globulin agglutination (Cullen 1977). (A review of the literature on *M. synoviae* is given by Timms 1978; for reviews of avian mycoplasmosis see Hofstad *et al.* 1978, Jordan 1979, Stipkovits 1979.)

Rats and mice

Three species of *Mycoplasma* are of particular pathological importance; of these, *M. pulmonis* is by far the commonest cause of disease.

M. pulmonis infection

Chronic respiratory disease due to *M. pulmonis* (see Nelson 1967, Cassell and Hill 1979) occurs more commonly in rats than in mice. The organism, which is widespread in laboratory colonies, is capable of producing by itself all the lesions found in natural infections (Lindsey *et al.* 1971); it may on occasion be associated with other agents such as *Pasteurella pneumotropica* (see Chapter 38) or streptococci. Increased concentrations of ammonia in cages may predispose animals to infection (Broderson *et al.* 1976). Transmission takes place readily by droplet inhalation. Infection of young animals, occurring within a few weeks of birth, results in a slow, progressive, often lifelong disease. A prolonged period of subclinical infection may be followed by the occurrence of abnormal breathing sounds ('snuffling' in rats, and 'chattering' in mice), oculo-nasal discharges and encrustations, signs of inner-ear abnormalities, and gradual loss of condition. A septicaemic phase may occur. Histopathological changes, characterized by lymphoid hyperplasia, are often present in the absence of macroscopical lesions. Rhinitis, tracheitis, laryngitis, otitis media, purulent pneumonia with peribronchial infiltration of lymphoid cells, and bronchiectasis may all occur. *M. pulmonis* is sometimes present in the reproductive tract of female rats and mice. In rats the lesions produced include salpingitis and oophoritis, but abnormalities are often apparent only on histological examination. The organism sometimes produces arthritis. It is difficult to establish and maintain *M. pulmonis*-free breeding colonies of rats and mice. Rats can be protected against respiratory disease by vaccination (Cassell and Davis 1978).

M. arthritidis infection

M. arthritidis is the commonest cause of polyarthritis in rats. The animals affected are usually young. Natural infections of mice are unknown, but intravenous inoculation results in chronic arthritis. Intravenous inoculation of rats with large doses of organisms produces suppurative polyarthritis from which recovery usually takes place (see Cassell and Hill 1979, Cole and Ward 1979). *M. arthritidis* is sometimes present in the nasopharynx of apparently healthy rats. The factors responsible for initiating disease are poorly understood. The proportion of animals affected in an outbreak is usually small. *M. arthritidis* has on occasion been isolated from rats with ailments such as otitis media, abscesses, and respiratory infection.

M. neurolyticum infection

Latent infection is common in mice, particularly in the respiratory tract and brain. *M. neurolyticum* may sometimes be isolated from cases of epidemic conjunctivitis in mice (Nelson 1950); other organisms such as *M. pulmonis* and *Pasteurella pneumotropica* may or may not be isolated at the same time.

Rolling disease of mice due to *M. neurolyticum* was noted by Sabin (1938) and by Findlay and co-workers (1938) during the course of the intracerebral passage of mouse-brain homogenates in mice—a procedure that activated a latent infection. This rapidly fatal disease can be produced in mice and rats by the intravenous injection of *M. neurolyticum* culture or culture filtrate. It is characterized by cystic degeneration of the cerebrum and cerebellum, resulting from the action of a specifically neutralizable exotoxin that rapidly becomes fixed to the tissues (Thomas *et al.* 1966).

Other animal species

Many mycoplasmas, acholeplasmas and ureaplasmas have been isolated from the mouth, upper and lower respiratory tract, conjunctiva, and urogenital tract of other animal species, both in the presence and absence of disease. Included are *M. equirhinis* and *M. equigenitalium* from horses, human mycoplasmas such as *M. salivarium* from captive non-human primates, *M. canis* from dogs, and *M. felis* from cats. None of these organisms is known with certainty to be pathogenic; there is some evidence that *M. cynos* produces pneumonia in dogs, and that *M. felis* contributes to the aetiology of conjunctivitis in cats (Rosendal 1979).

Mycoplasma infections in man

Mycoplasmas, ureaplasmas and acholeplasmas are well known to occur in man, often in the absence of disease. At least 12 species have been isolated (Somerson and Cole 1979). In the oropharynx *M. orale* and *M. salivarium* are particularly common, the latter showing a predilection for the gingival sulci. *M. hominis* and *Ureaplasma urealyticum* are frequently present in the genito-urinary tract; although usually harmless, there is growing evidence that they sometimes produce disease. The evidence for the pathogenicity of *M. pneumoniae* is incontrovertible.

Atypical pneumonia

During the ten years or so before the second world war there came to be recognized cases of pneumonia, occurring sporadically or in outbreaks, that differed in certain respects from primary pneumococcal pneumonia. Experience of large groups of men during the war provided the opportunity for a careful study of such cases (see Dingle *et al.* 1943, 1944); and the distinction between the new disease and classical pneumonia was aided by the greater use of x-rays and the response to modern chemotherapy. The main features of the disease, as it occurs in man, are fever, a slow pulse, severe headache, oppressive pain behind the sternum, and a patchy pneumonic consolidation having a characteristic radiological appearance. The disease is seldom fatal, and though resistant to penicillin and sulphonamide, responds to streptomycin, aureomycin and the tetracyclines. The incubation period is probably 2–3 weeks, and the duration of the febrile disease about 10 days. The initial rigor, the pleural pain, and the rusty sputum of pneumonia are seldom present, nor is there a leucocytosis. The disease does not spread readily by contact, except in families and other semi-closed groups, it has no special seasonal incidence, and complications are uncommon. It came to be known as **primary atypical pneumonia.** In Seattle, Foy and his colleagues (1973) reported an incidence of 1 to 3 per 1000 of the population. It constituted about 10 per cent of all pneumonia cases, and occurred more frequently in children than in adults (see also Stevens *et al.* 1978). Its causation remained obscure till Eaton and his colleagues (1944) demonstrated the presence of a pneumotropic 'virus' by inoculation of sputum or lung tissue into the amnion of the developing chick embryo. The virus was not adaptable to growth in the yolk sac or extra-embryonic membranes, nor could it be grown in tissue culture. It proved to be mildly virulent to hamsters and cotton-rats, producing pulmonary consolidation. Antibodies neutralizing the virus were found in over half the patients in whom primary atypical pneumonia was diagnosed clinically.

Confirmation of Eaton's findings was hampered by the difficulty of demonstrating the virus, but this was largely overcome by the application of the fluorescent antibody technique by Liu, Eaton and Heyl (1956). These workers found that, after injection into the amnion of the chick embryo, the virus localized in the cytoplasm of the bronchial epithelial cells where it could be revealed by the fluorescent antibody technique. Using this method, Chanock, Mufson and their colleagues (1961) demonstrated the presence of the Eaton agent in 68 per cent of recruits suffering from primary atypical pneumonia, and obtained evidence that in a Service training centre infection was very common but usually remained latent. The nature of the virus or Eaton agent, as it was variously called, was doubtful. Its size, 180–250 nm, and the response of patients to antibiotics, suggested its classification in the lymphogranuloma-psittacosis group (see Chapter 45). Marmion and Goodburn (1961; see also Goodburn and Marmion 1962*a, b*), however, pointed out that it had many resemblances to the pleuropneumonia group of organisms. By an intensified Giemsa stain, they succeeded in showing the presence in the infected chick-embryo lung of small reddish-purple cocco-bacillary bodies, which were stained by the fluorescent antibody technique, and whose formation was suppressed by gold salts. The proof of their surmise was brought by Chanock, Hayflick and Barile (1962), who cultivated on a serum yeast-extract agar medium at 36° a mycoplasma, now known as *M. pneumoniae*, having all the properties attributable to the Eaton agent.

M. pneumoniae is now known to be one of the more important causes of lower respiratory-tract infection. The disease occurs in many countries. It has been reported to follow a cyclic pattern, with epidemic waves every 3–5 years (Freundt 1974, Noah 1976, Clyde 1979). Infection is often symptomless or accompanied only by mild signs of disease in the upper respiratory tract; clinical manifestations are uncommon in children less than four years old.

Diagnosis

Laboratory diagnosis entails the use of both cultural and serological methods. Other organisms capable of giving rise to pneumonic symptoms such as bacteria, rickettsiae, chlamydiae, and viruses must be excluded. Swabs taken to the laboratory in transport medium should be inoculated on to suitable media, preferably selective for *M. pneumoniae* (Smith *et al.* 1967, see Chapter 47). Cultures should be incubated at 37° and inspected at 5-day intervals. The organisms may be identified by their β-haemolysis of guinea-pig red cells, their fermentation of glucose, and by growth inhibition and immunofluorescence tests. For *serological diagnosis* the complement-fixation test is recommended. It is best carried out on two specimens of serum, one taken at the beginning of the illness and the other 3 weeks later. A fourfold or greater rise in the titre of antibodies may be considered diagnostic.

Besides these two specific methods, two non-specific reactions depending on a heterologous antibody response have been widely used. Several workers in the past pointed out that the serum of patients often contained *cold agglutinins* reacting with human group O red cells at 0° but not at 37° (Horstmann and Tatlock 1943, Peterson *et al.* 1943, Turner *et al.* 1943). These antibodies are absent in Q fever, but may be found in other diseases such as paroxysmal haemoglobinuria, tropical eosinophilia, trypanosomiasis, cirrhosis of the liver, and haemolytic anaemia, and after sulphonamide therapy. According to Savonen (1948) they may be present in all sorts of pneumonias, suggesting that they are dependent more on pulmonary inflammation *per se* than on any specific agent. Feizi and her colleagues (1969), on the other hand, ascribe their presence in atypical pneumonia to a reaction between *M. pneumoniae* and the I antigen of human red blood corpuscles with the production of an auto-antibody. Another heterologous agglutinin often found in this disease is that to *Streptococcus* MG (Thomas *et al.* 1945, Bedson 1950). Its formation probably depends on an antigen common to both the streptococcus and *M. pneumoniae*. Both types of agglutinin reach their maximum titre during early convalescence, and both increase in diagnostic significance when associated with a fourfold or greater rise in titre. (For cold agglutinins see also Garrow 1958.)

Treatment of the disease demands one of the broad-spectrum antibiotics. The tetracyclines and erythromycin are generally favoured.

Vaccination might be of value for certain groups, for example military recruits. Killed vaccines containing adjuvants, such as the vaccine studied by Mogabgab (1973), are known to give some protection. The immunity that follows the natural disease is not permanent.

(For further information on epidemiology, diagnosis, and vaccination see Couch 1969, Grayston *et al.* 1969, Whittlestone 1976, and Clyde 1979; and for other forms of non-bacterial pneumonia see Chapters 76, 77, 97, and 98.)

Genito-urinary infections

Colonization of the genital tract of both sexes with *M. hominis* and ureaplasmas (T mycoplasmas) sometimes occurs as a result of contamination at the time of birth. Such colonization is usually short-lived. In the prepubertal period these organisms are almost always absent from the genital tract of the male but occasion-

ally present in the female (Klieneberger-Nobel 1959, Lee *et al.* 1974, Foy *et al.* 1975). After puberty colonization of the male urethra and the vagina with *M. hominis* and ureaplasmas occurs not uncommonly, and is related to sexual activity, particularly in the female (McCormack *et al.* 1972, 1973a and b). Serum antibody often appears at about the time of puberty, probably as a result of colonization.

Non-gonococcal urethritis (NGU), a venereal disease of the male, has more than one cause. There is strong evidence that *Chlamydia trachomatis* (see Chapter 76) is sometimes responsible. The other causative agents are not known with certainty, but the evidence suggests that they include ureaplasmas.

Numerous workers (see Taylor-Robinson and McCormack 1979) have compared NGU patients with controls in respect of the isolation rates of large-colony mycoplasmas—most of which were probably *M. hominis*—and *Ureaplasma urealyticum*, of which there are at least 14 serotypes (Lin and Kass 1980). Many of the studies were invalid because the persons selected as controls were not genuinely comparable with the NGU patients, particularly as regards their sexual activity. There would seem little justification for concluding that either *M. hominis* or *U. urealyticum* occurs more frequently in NGU patients than in comparable controls (Holmes *et al.* 1975, Lee *et al.* 1976, Bowie *et al.* 1977b and c). It seems possible that, in patients free from both gonococci and chlamydiae, urethritis is associated with an increase in the ureaplasma infection rate and in the numbers of ureaplasmas present (Shepard 1974, Bowie *et al.* 1976, 1977a, b, and c); this could not be confirmed, however, by Taylor-Robinson, Evans and their colleagues (1979). NGU is only occasionally accompanied by a serum-antibody response to *U. urealyticum*. More convincing evidence that ureaplasmas are responsible for some cases of NGU is provided by studies in which patients infected with either ureaplasmas, chlamydiae, both, or neither, were treated with certain antibiotics; these included aminocyclitols, effective against ureaplasmas but not chlamydiae, sulphafurazole and rifampicin, effective against chlamydiae but not ureaplasmas, and minocycline (a tetracycline), effective against both groups of organisms (Bowie *et al.* 1976, Prentice *et al.* 1976, Coufalik *et al.* 1979). These studies showed that NGU patients infected with ureaplasmas but not chlamydiae often responded favourably to antibiotics active against ureaplasmas. The intraurethral infection of two human volunteers (Taylor-Robinson *et al.* 1977) provided further evidence of the pathogenicity of ureaplasmas. Urethritis developed in both volunteers, the urine contained 'threads' and polymorphonuclear leucocytes, and a transient serum-antibody response occurred; prostatitis developed in one volunteer. Taylor-Robinson and co-workers (1978) produced urethral infection with *U. urealyticum* in chimpanzees.

A new mycoplasma species, *M. genitalium*, has been isolated in 'SP-4 medium' from cases of NGU, but its pathogenic role, if any, remains to be proved (Tully *et al.* 1983).

There seems little doubt that *M. hominis* is responsible for certain cases of inflammatory disease of the upper reproductive tract of the female, particularly the fallopian tubes (Gotthardson and Melén 1953, Melén and Gotthardson 1955, Lemcke and Csonka 1962, Eschenbach *et al.* 1975, Mårdh *et al.* 1975, Weström and Mårdh 1975). Its isolation from the blood of patients with fever after abortion or childbirth also indicates pathogenicity (Tully *et al.* 1965, Harwick *et al.* 1970, 1971). There is some evidence that it may cause pyelonephritis. Shurin and co-workers (1975) showed a relation between chorioamnionitis and the presence of ureaplasmas. The relation between mycoplasmas, including ureaplasmas, and Reiter's disease is uncertain (Taylor-Robinson and McCormack 1979). There is a significant relationship between the presence of genital mycoplasmas in pregnancy and the occurrence of certain reproductive abnormalities, namely abortion and low birthweight (Braun *et al.* 1971), but its causal nature has not been established. The relation between genital mycoplasmas and infertility (Gnarpe and Friberg 1973, Cassell *et al.* 1983) is uncertain.

For *diagnosis* cultural methods are advisable. Urethral swabs should be taken into Stuart's (Stuart *et al.* 1954) transport medium, and plated on one of the special mycoplasma media and, for ureaplasmas, also inoculated into urea broth (see Chapter 47). The incubation of serial dilutions of the inoculum in liquid medium containing a suitable indicator system enables a rough estimate of the number of organisms to be made. *M. hominis* may be isolated on good-quality defibrinated blood agar, provided that incubation is continued for at least 4 days in a moist atmosphere containing extra CO_2, such as is provided by a candle jar. For isolation from the blood Stokes (1959) used Liquoid (sodium polyanethol sulphonate) in a concentration of 0.03–0.06 per cent; higher concentrations may be inhibitory. The organisms are best identified by the growth-inhibition test. Serological methods may be used for confirmation or in surveys. The complement-fixation test and the indirect haemagglutination test have both been used with success. (For a review of mycoplasma genito-urinary infections in man see Taylor-Robinson and McCormack 1979.)

No causal relationship between mycoplasmas and human rheumatoid arthritis or Reiter's disease has been proved (Taylor-Robinson and Taylor 1976, Cole and Ward 1979). (For reviews of mycoplasmal diseases in animals and man see Stanbridge 1976, Tully and Whitcomb 1979, and Cassell and Cole 1981.)

References

Al-Aubaidi, J. M., Dardiri, A. H. and Fabricant, J. (1972) *Int. J. syst. Bact.* **22,** 155.

Barile, M. F., Del Giudice, R. A. and Tully, J. G. (1972) *Infect. Immun.* **5,** 70.

Bedson, S. P. (1950) *Brit. med. J.* **ii,** 1461.

Bennett, S. C. J. (1932) *J. comp. Path.* **45,** 257.

Boughton, E. (1979) *Vet. Bull.* **49,** 377.

Boulanger, P. and L'Ecuyer, C. (1968) *Canad. J. comp. Med.* **32,** 547.

Bowie, W. R., Alexander, E. R., Floyd, J. F., Holmes, J., Miller, Y. and Holmes, K. K. (1976) *Lancet* **ii,** 1276.

Bowie, W. R., Pollock, H. M., Forsyth, P. S., Floyd, J. F., Alexander, E. R., Wang, S. P. and Holmes, K. K. (1977c) *J. clin. Microbiol.* **6,** 482.

Bowie, W. R., Wang, S. P., Alexander, E. R., Floyd, J., Forsyth, P. S., Pollock, H. M., Lin, J. S., Buchanan, T. M. and Holmes, K. K. (1977a) *J. clin. Invest.* **59,** 735.

Bowie, W. R., Wang, S. P., Alexander, E. R. and Holmes, K. K. (1977b) In: *Non-gonococcal Urethritis and Related Infections*, p. 19. Ed. by D. Hobson and K. K. Holmes. Amer. Soc. Microbiol., Washington, D.C.

Braun, P., Lee, Y. H., Klein, J. O., Marcy, S. M., Klein, T. A., Charles, D., Levy, P. and Kass, E. H. (1971) *New Engl. J. Med.* **284,** 167.

Bridré, J. and Donatien, A. (1923) *C. R. Acad. Sci., Paris* **177,** 841.

Broderson, J. R., Lindsey, J. R. and Crawford, J. E. (1976) *Amer. J. Path.* **85,** 115.

Bruggmann, S., Keller, H., Bertschinger, H. U. and Engberg, B. (1977) *Vet. Rec.* **101,** 109.

Campbell, A. D. and Turner, A. W. (1936) *Bull. Counc. sci. industr. Res. Aust.* No. 97, 11; (1953) *Aust. vet. J.* **29,** 154.

Carmichael, L. E., St George, T. D., Sullivan, N. D. and Horsfall, N. (1972) *Cornell Vet.* **62,** 654.

Cassell, G. H. and Cole, B. C. (1981) *New Engl. J. Med.* **304,** 80.

Cassell, G. H. and Davis, J. K. (1978) *Infect. Immun.* **21,** 69.

Cassell, G. H. and Hill, A. (1979) See Tully and Whitcomb (1979) p. 235.

Cassell, G. H., Younger, J. B., Brown, M. B., Blackwell, R. E., Davis, J. K., Marriott, P. and Stagno, S. (1983) *New Engl. J. Med.* **308,** 502.

Chanock, R. M., Hayflick, L. and Barile, M. F. (1962) *Proc. nat. Acad. Sci., Wash.* **48,** 41.

Chanock, R. M., Mufson, M. A., Bloom, H. H., James, W. D., Fox, H. H. and Kingston, J. R. (1961) *J. Amer. med. Ass.* **175,** 213.

Clyde, W. A. (1979) See Tully and Whitcomb (1979) p. 275.

Cole, B. C. and Ward, J. R. (1979) See Tully and Whitcomb (1979) p. 367.

Cordy, D. R., Adler, H. E. and Yamamoto, R. (1955) *Cornell Vet.* **45,** 50.

Cottew, G. S. (1979) See Tully and Whitcomb (1979) p. 103.

Cottew, G. S. and Leach, R. H. (1969) In Hayflick (1969) p. 527.

Cottew, G. S. and Lloyd, L. C. (1965) *J. comp. Path.* **75,** 363.

Cottew, G. S. and Yeats, F. R. (1978) *Aust. vet. J.* **54,** 293.

Couch, R. B. (1969) In Hayflick (1969) p. 683.

Coufalik, E. D., Taylor-Robinson, D. and Csonka, G. W. (1979) *Brit. J. vener. Dis.* **55,** 36.

Cullen, G. A. (1977) In: *Experimental Models of Chronic Inflammatory Diseases*, p. 240. Ed. by L. E. Glynn and H. D. Schlumberger, Bayer Symposium VI. Springer-Verlag, Berlin, Heidelberg and New York.

Debonéra, G. (1937) *Rec. Méd. vét.* **113**, 79.

Dingle, J. H. *et al.* (1943) *War Med.* **3**, 223; (1944) *Amer. J. Hyg.* **39**, 197, 269.

Dyson, D. A. and Smith, G. R. (1975) *Res. vet. Sci.* **19**, 8; (1976) *Ibid.* **20**, 185.

Eaton, M. D., Meiklejohn, G. and Herick, W. van (1944) *J. exp. Med.* **79**, 649.

Edward, D. G. ff. (1947) *J. gen. Microbiol.* **1**, 238; (1953) *Vet. Rec.* **65**, 873.

Ernø, H., Leach, R. H. and MacOwan, K. J. (1979) *Trop. Anim. Hlth Prod.* **11**, 84.

Eschenbach, D. A., Buchanan, T. M., Pollock, H. M., Forsyth, P. S., Alexander, E. R., Lin, J. S., Wang, S. P., Wentworth, B. B., McCormack, W. M. and Holmes, K. K. (1975) *New Engl. J. Med.* **293**, 166.

Farrington, D. O. and Switzer, W. P. (1977) See Report (1977) p. 72.

Feizi, T., Taylor-Robinson, D., Shields, M. D. and Carter, R. A. (1969) *Nature, Lond.* **222**, 1253.

Findlay, G. M., Klieneberger, E., MacCallum, F. O. and Mackenzie, R. D. (1938) *Lancet* **ii**, 1511.

Foy, H. M., Cooney, M. K., McMahan, R. and Grayston, J. T. (1973) *Amer. J. Epidem.* **97**, 93.

Foy, H., Kenny, G., Bor, E., Hammar, S. and Hickman, R. (1975) *J. clin. Microbiol.* **2**, 226.

Freundt, E. A. (1974)) *Path. et Microbiol., Basel* **40**, 155.

Friis, N. F. (1972) *Acta vet. scand.* **13**, 284.

Galloway, I. A. (1930) *A System of Bacteriology*, Med. Res. Counc. **7**, 335.

Garrow, D. H. (1958) *Brit. med. J.* **ii**, 206.

Gnarpe, H. and Friberg, J. (1973) *Nature, Lond.* **242**, 120.

Goiš, M., Kuksa, F., Franz, J. and Taylor-Robinson, D. (1974) *J. med. Microbiol.* **7**, 105.

Goodburn, G. M. and Marmion, B. P. (1962*a*) *J. gen. Microbiol.* **29**, 271; (1962*b*) *J. Hyg. Epidem., Prague* **6**, 176.

Goodwin, R. F. W. and Whittlestone, P. (1973) *Brit. vet. J.* **129**, 456.

Gotthardson, A. and Melén, B. (1953) *Acta path. microbiol. scand.* **33**, 291.

Gourlay, R. N. (1964) *J. comp. Path.* **74**, 286; (1965) *Ibid.* **75**, 97.

Gourlay, R. N. and Howard, C. J. (1979) See Tully and Whitcomb (1979) p. 49.

Grayston, J. T., Foy, H. M. and Kenny, G. E. (1969) In Hayflick (1969) p. 651.

Gulrajani, T. S. and Beveridge, W. I. B. (1951) *Nature, Lond.* **167**, 856.

Harwick, H. J., Purcell, R. H., Iuppa, J. B. and Fekety, F. R. (1970) *J. infect. Dis.* **121**, 260; (1971) *Obstet. & Gynec.* **37**, 765.

Hayflick, L. (Ed.) (1969) *The Mycoplasmatales and the L-Phase of Bacteria*. North-Holland Publ. Co., Amsterdam.

Hofstad, M. S., Calnek, B. W., Helmboldt, C. F., Reid, W. M. and Yoder, H. W. (Eds) (1978) *Diseases of Poultry*, 7th ed., p. 233. Iowa State University Press, Ames.

Holmes, K. K., Handsfield, H. H., Wang, S. P., Wentworth, B. B., Turck, M., Anderson, J. B. and Alexander, E. R. (1975) *New Engl. J. Med.* **292**, 1199.

Hooker, J. M., Smith, G. R. and Milligan, R. A. (1979) *J. Hyg., Camb.* **82**, 407.

Horstmann, D. M. and Tatlock, H. (1943) *J. Amer. med. Ass.* **122**, 369.

Hudson, J. R. (1971) *FAO Agricultural Studies* No. 86, FAO, Rome.

Hudson, J. R. and Turner, A. W. (1963) *Aust. vet. J.* **39**, 373.

Hutcheon, D. (1881) *Vet. J.* **13**, 171; (1889) *Vet. J.* **29**, 399.

Jasper, D. E. (1981) *Advanc. vet. Sci.* **25**, 121.

Jones, G. E., Buxton, D. and Harker, D. B. (1979) *Vet. Microbiol.* **4**, 47.

Jordan, F. T. W. (1979) See Tully and Whitcomb (1979) p. 1.

Klieneberger-Nobel, E. (1959) *Brit. med. J.* **i**, 19.

Lam, K. M. and Switzer, W. P. (1971) *Amer. J. vet. Res.* **32**, 1737.

Langford, E. V. and Leach, R. H. (1973) *Canad. J. Microbiol.* **19**, 1435.

Lee, Y. H., McCormack, W. M., Marcy, S. M. and Klein, J. O. (1974) *Pediat. Clin. N. Amer.* **21**, 457.

Lee, Y. H., Tarr, P. I., Schumacher, J. R., Rosner, B., Alpert, S. and McCormack, W. M. (1976) *J. Amer. vener. Dis. Ass.* **3**, 25.

Lemcke, R. and Csonka, G. W. (1962) *Brit. J. vener. Dis.* **38**, 212.

Lin, J.-S. L. and Kass, E. H. (1980) *Infection* **8**, 152.

Lindsey, J. R., Baker, H. J., Overcash, R. G., Cassell, G. H. and Hunt, C. E. (1971) *Amer. J. Path.* **64**, 675.

Liu, C., Eaton, M. D. and Heyl, J. T. (1956) *Bull. N.Y. Acad. Med.* **32**, 170.

Longley, E. O. (1951) *Colonial Res. Pubns* No. 7. H.M.S.O. London.

McCormack, W. M., Almeida, P. C., Bailey, P. E., Grady, E. M. and Lee, Y. H. (1972) *J. Amer. med. Ass.* **221**, 1375.

McCormack, W. M., Braun, P., Lee, Y. H., Klein, J. O. and Kass, E. H. (1973*a*) *New Engl. J. Med.* **288**, 78.

McCormack, W. M., Lee Y. H. and Zinner, S. H. (1973*b*) *Ann. intern. Med.* **78**, 696.

McMartin, D. A., MacOwan, K. J. and Swift, L. L. (1980) *Brit. vet. J.* **136**, 507.

MacOwan, K. J. and Minette, J. E. (1976) *Trop. Anim. Hlth Prod.* **8**, 91.

Mårdh, P.-A., Weström, L. and Colleen, S. (1975) In: *Genital Infections and their Complications*, p. 53. Ed. by D. Danielsson, L. Juhlin and P.-A. Mårdh. Almqvist and Wiksell, Stockholm.

Marmion, B. P. and Goodburn, G. M. (1961) *Nature, Lond.* **189**, 247.

Masiga, W. N., Roberts, D. H., Kakoma, I. and Rurangirwa, F. R. (1975) *Res. vet. Sci.* **19**, 330.

Matzer, N. and Yamamoto, R. (1974) *J. comp. Path.* **84**, 271.

Melén, B. and Gotthardson, A. (1955) *Acta path. microbiol. scand.* **37**, 196.

Mogabgab, W. J. (1973) *Ann. N.Y. Acad. Sci.* **225**, 453.

Nelson, J. B. (1936) *J. exp. Med.* **63**, 515; (1950) *Ibid.* **91**, 309; (1967) In: *Pathology of Laboratory Rats and Mice*, p. 259. Ed. by E. Cotchin and F. J. C. Roe. Blackwell Scientific Publications, Oxford and Edinburgh.

Noah, N. D. (1976) *Infection* **44**, Suppl. 1, 25.

Nocard, E. and Roux, E. (1898) *Ann. Inst. Pasteur* **12**, 240.

Peterson, O. L., Ham, T. H. and Finland, M. (1943) *Science* **97**, 167.

Prentice, M. J., Taylor-Robinson, D. and Csonka, G. W. (1976) *Brit. J. vener. Dis.* **52**, 269.

Razin, S. (1978) *Microbiol. Rev.* **42**, 415.

Report (1971) Fourth meeting of FAO/OIE/OAU *Expert Panel on Contagious Bovine Pleuropneumonia, Paris.* FAO, Rome; (1977) *Laboratory Diagnosis of Mycoplasmosis in Food Animals.* Ed. O. H. V. Stalheim. Amer. Ass. vet. Lab. Diagnosticians, Madison; (1979) *Animal Health Yearbook 1978,* FAO Animal Production and Health Series, No. 13. FAO, Rome.

Rhoades, K. R. (1977) See Report (1977) p. 89.

Roberts, D. H., Windsor, R. S., Masiga, W. N. and Kariavu, C. G. (1973) *Infect. Immun.* **8,** 349.

Rose, D. L., Tully, J. G. and Wittler, R. G. (1979) *Int. J. syst. Bact.* **29,** 83.

Rosendal, S. (1979) See Tully and Whitcomb (1979) p. 217.

Sabin, A. B. (1938) *Science* **88,** 189.

Savonen, K. (1948) Cold Haemagglutination Test and its Clinical Significance. *Ann. Med. exp. Biol. fenn* **26,** Suppl. No. 3.

Seddon, H. R. (1965) *Diseases of Domestic Animals in Australia,* 2nd ed., Commonw. Aust. Dept Hlth, Canberra.

Shepard, M. C. (1974) In: *Les Mycoplasmes de l'Homme, des Animaux, des Végétaux et des Insectes,* Colloq. No. 33, p. 375. Ed. by J. M. Bové and J. F. Duplan. INSERM, Paris.

Shurin, P. A., Alpert, S., Rosner, B., Driscoll, S. G., Lee, Y. H., McCormack, W. M., Santamarina, B. A. G. and Kass, E. H. (1975) *New Engl. J. Med.* **293,** 5.

Singh, N., Rajya, B. S. and Mohanty, G. C. (1975) *Cornell Vet.* **65,** 363.

Smith, C. B., Chanock, R. M., Friedewald, W. T. and Alford, R. H. (1967) *Ann. N.Y. Acad. Sci.* **143,** 471.

Smith, G. R. (1967) *J. comp. Path.* **77,** 203; (1968) *Ibid.* **78,** 267; (1971) *Ibid.* **81,** 267.

Smith, G. R., Hooker, J. M. and Milligan, R. A. (1980) *J. Hyg., Camb.* **85,** 247.

Smith, G. R. and Oliphant, J. C. (1981*a*) *J. Hyg., Camb.* **87,** 321; (1981*b*) *Ibid.* **87,** 437; (1982) *Ibid.* **89,** 521; (1983) *Ibid.* **90,** 441.

Somerson, N. L. and Cole, B. C. (1979) See Tully and Whitcomb (1979) p. 191.

Stalheim, O. H. V. (1983) *J. Amer. vet. med. Ass.* **182,** 403.

Stanbridge, E. J. (1976) *Annu. Rev. Microbiol.* **30,** 169.

Stevens, D. L., Swift, P. G. F., Johnston, P. G. B., Kearney, P. J., Corner, B. D. and Burman, D. (1978) *Arch. Dis. Childh.* **53,** 38.

Stipkovits, L. (1979) *Zbl. Bakt.,* I. Abt. Orig. A **245,** 171.

Stokes, E. J. (1959) *Brit. med. J.* **i,** 510.

Stuart, P., Davidson, I., Slavin, G., Edgson, F. A. and Howell, D. (1963) *Vet. Rec.* **75,** 59.

Stuart, R. D., Toshach, S. R. and Patsula, T. M. (1954) *Canad. J. publ. Hlth* **45,** 73.

Switzer, W. P. and Ross, R. F. (1975) In: *Diseases of Swine,* 4th edn, p. 741. Ed. by H. W. Dunne and A. D. Leman. Iowa State University Press, Ames.

Taylor-Robinson, D. (1979) *Zbl. Bakt.,* I. Abt. Orig. A **245,** 150.

Taylor-Robinson, D., Csonka, G. W. and Prentice, M. J. (1977) *Quart. J. Med.* **46,** 309.

Taylor-Robinson, D., Evans, R. T., Coufalik, E. D., Prentice, M. J., Munday, P. E., Csonka, G. W. and Oates, J. K. (1979) *Brit. J. vener. Dis.* **55,** 30.

Taylor-Robinson, D. and McCormack, W. M. (1979) See Tully and Whitcomb (1979) p. 307.

Taylor-Robinson, D., Purcell, R. H., London, W. T. and Sly, D. L. (1978) *J. med. Microbiol.* **11,** 197.

Taylor-Robinson, D. and Taylor, G. (1976) In: *Infection and Immunology in the Rheumatic Diseases,* p. 177. Ed. by D. C. Dumonde. Blackwell Scientific Publications Ltd, Oxford.

Thomas, L., Aleu, F., Bitensky, M. W., Davidson, M. and Gesner, M. (1966) *J. exp. Med.* **124,** 1067.

Thomas, L., Mirick, G. S., Curnen, E. C., Ziegler, J. E. and Horsfall, F. L. (1945) *J. clin. Invest.* **24,** 227.

Timms, L. M. (1978) *Vet. Bull.* **48,** 187.

Tully, J. G., Barile, M. F., Edward, D. G. ff., Theodore, T. S. and Ernø, H. (1974) *J. gen. Microbiol.* **85,** 102.

Tully, J. G., Brown, M. S., Sheagren, J. N., Young, V. M. and Wolfe, S. M. (1965) *New Engl. J. Med.* **273,** 648.

Tully, J. G., Taylor-Robinson, D., Rose, D. L., Cole, R. M. and Bove, J. M. (1983) *Int. J. syst. Bact.* **33,** 387.

Tully, J. G. and Whitcomb, R. F. (Eds) (1979) *The Mycoplasmas,* vol. 2. Academic Press, New York, San Francisco and London.

Turner, A. W. (1959) In: *Infectious Diseases of Animals: Diseases due to Bacteria,* vol. 2, p. 437. Ed. by A. W. Stableforth and I. A. Galloway. Butterworth, London.

Turner, J. C., Nisnewitz, S., Jackson, E. B. and Berney, R. (1943) *Lancet* **i,** 765.

Valdivieso-Garcia, A. and Rosendal, S. (1982) *Vet. Rec.* **110,** 470.

Watson, W. A., Cottew, G. S., Erdağ, O. and Arisoy, F. (1968) *J. comp. Path.* **78,** 283.

Weström, L. and Mårdh, P.-A. (1975) In: *Genital Infections and their Complications,* p. 157. Ed. by D. Danielsson, L. Juhlin and P.-A. Mårdh. Almqvist and Wiksell, Stockholm.

Whittlestone, P. (1973) *Advanc. vet. Sci.* **17,** 1; (1976) *Ibid.* **20,** 277; (1979) See Tully and Whitcomb (1979) p. 133.

Yoder, H. W. (1970) *Avian Dis.* **14,** 75.

Index

The index should be read in conjunction with the chapter contents lists.